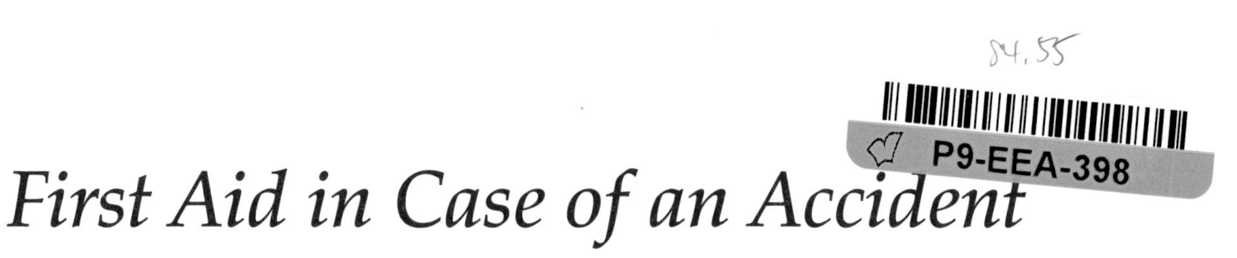

First Aid in Case of an Accident

The occurrence of an accident of any kind in the laboratory should be reported promptly to your instructor, even if it seems relatively minor.

FIRE

Your first consideration is to remove yourself from any danger, *not* to extinguish the fire. *If it is possible to do so without endangering yourself*, turn off any burners and remove containers of flammable solvents from the immediate area to prevent the fire from spreading. For the most effective use of a fire extinguisher, direct its nozzle toward the *base* of the flames. Burning oil may be put out with an extinguisher classified for use on "ABC" type fires.

If your clothing is on fire, DO NOT RUN; rapid movement will only fan the flames. Roll on the floor to smother the fire and to help keep the flames away from your head. Your neighbors can help to extinguish the flames by using fire blankets, laboratory coats, or other items that are immediately available. Do not hestitate to aid your neighbor if he or she is involved in such an emergency; a few seconds delay may result in serious injury. A laboratory shower, *if close by*, can be used to extinguish burning clothing, as can a carbon dioxide extinguisher, which must be used with care until the flames are extinguished *and only if the flames are not near the head.*

If burns are minor, apply a burn ointment. In the case of serious burns, do not apply any ointment; seek professional medical treatment at once.

CHEMICAL BURNS

Areas of the skin with which corrosive chemicals have come in contact should be immediately and thoroughly washed with soap and warm water. If the burns are minor, apply burn ointment; for treatment of more serious burns, see a physician.

Bromine burns can be particularly serious. These burns should first be washed with soap and warm water and then thoroughly soaked with 0.6 *M* sodium thiosulfate solution for three hours. Apply cod liver oil ointment and a dressing; see a physician.

If chemicals, in particular corrosive or hot reagents, come in contact with the eyes, immediately *flood* the eyes with water *from the nearest outlet*. A specially designed eyewash fountain is useful if available in the laboratory. *Do not touch the eye*. The eyelid as well as the eyeball should be washed with water for several minutes. In all instances where sensitive eye tissue is involved in such an accident, consult an ophthalmologist as soon as possible.

CUTS

Minor cuts may be treated by ordinary first-aid procedures; seek professional medical attention for serious cuts. If severe bleeding indicates that an artery has been severed, attempt to stop the bleeding with compresses and pressure; a tourniquet should be applied only by those who have received first-aid training. Arrange for emergency room treatment at once.

A person who is injured severely enough to require a physician's treatment *should be accompanied* to the doctor's office, or infirmary, even if he or she claims to be all right. Persons in shock, particularly after suffering burns, are often more seriously injured than they appear to be.

Discovery Experiments

Experimental Organic Chemistry

A Miniscale and Microscale Approach

FIFTH EDITION

John C. Gilbert

Santa Clara University

Stephen F. Martin

University of Texas at Austin

Australia • Brazil • Japan • Korea • Mexico • Singapore • Spain
United Kingdom • United States

***Experimental Organic Chemistry,*
Fifth Edition**
John C. Gilbert and Stephen F. Martin

Publisher/Executive Editor: Mary Finch
Acquisitions Editor: Lisa Lockwood
Developmental Editor: Rebecca Heider
Assistant Editor: Elizabeth Woods
Media Editor: Stephanie VanCamp
Marketing Manager: Amee Mosley
Marketing Assistant: Kevin Carroll
Marketing Communications Manager: Linda Yip
Content Project Management: Pre-Press PMG
Creative Director: Rob Hugel
Art Director: John Walker
Print Buyer: Linda Hsu
Rights Acquisitions Account Manager, Text: Tim Sisler
Rights Acquisitions Account Manager, Image: Scott Rosen
Production Service: Pre-Press PMG
Copy Editor: Pre-Press PMG
Cover Designer: Denise Davidson
Cover Image: ©Martyn F. Chillmaid/Photo Researchers, Inc.
Compositor: Pre-Pres PMG

For product information and technology assistance, contact us at
Cengage Learning Customer & Sales Support, 1-800-354-9706
For permission to use material from this text or product,
submit all requests online at **www.cengage.com/permissions**
Further permissions questions can be emailed to
permissionrequest@cengage.com

Library of Congress Control Number: 2009940488

ISBN-13: 978-1-4390-4914-3

ISBN-10: 1-4390-4914-9

Cengage Learning
20 Channel Center Street
Boston, MA 02210
USA

Cengage Learning is a leading provider of customized learning solutions with office locations around the globe, including Singapore, the United Kingdom, Australia, Mexico, Brazil, and Japan. Locate your local office at: **international.cengage.com/region**

Cengage Learning products are represented in Canada by Nelson Education, Ltd.

For your course and learning solutions, visit **academic.cengage.com**

Purchase any of our products at your local college store or at our preferred online store **www.CengageBrain.com.**

Printed in the United States of America
2 3 4 5 6 7 13 12 11

Contents in Brief

Table of Contents

Preface

The management and teaching of an introductory laboratory course in organic chemistry is ever-changing, even though the fundamental chemical principles remain the same. Some of the compelling reasons for innovation and change are linked to the increasing cost associated with purchase and disposal of the chemicals used. There is the added concern of their possible toxicological hazards, both to students and to the environment. These factors dictate that many experiments be performed on reduced scales according to procedures commonly termed as *miniscale* (sometimes called small-scale) and *microscale*. This edition of our textbook maintains our practice of providing both miniscale and microscale procedures for most experiments. This unusual feature gives instructors maximal flexibility in customizing the course for use of apparatus and glassware already on hand and to suit the specific needs of you, the student.

The experiments are thoughtfully selected to introduce you to the common laboratory practices and techniques of organic chemistry and to illustrate the chemistry of the wide range of functional groups that are present in organic molecules. Some experiments are designed to familiarize you with the kinetic and thermodynamic principles underlying chemical reactions. Others allow you to synthesize specific compounds—some of which are found in nature or are of commercial importance—using reactions that are fundamental to organic synthesis. Still others introduce you to *discovery-based* and *green-chemistry* approaches. The discovery-based procedures—there are over 40 of these in the new edition—allow you to develop your own protocols for addressing a particular question experimentally, as you might do in a research laboratory. Discovery experiments are listed inside the front cover and are indicated when they appear in the book with the magnifying glass icon shown in the margin. The four procedures involving green chemistry show you how some chemical transformations may be performed using more environmentally friendly procedures. Green chemistry experiments are indicated when they appear in the book with the leaf icon shown in the margin. Many of the chapters are accompanied by a Historical Highlight, an essay that focuses on interesting topics in organic chemistry and that we believe will broaden your interest in the subject. Overall, our hope is that your experiences in this course will inspire you to take additional laboratory and lecture courses in chemistry, to seize the opportunity to work in a research laboratory as an undergraduate student, and perhaps even to pursue a career in research.

Background Information

Our textbook is distinct from many other laboratory manuals because the focused discussions preceding each Experimental Procedure provide the essential theoretical and "how-to" background, so other sources need not be consulted in order to understand the mechanistic and practical aspects of the specific reactions and procedures being performed. These discussions offer the advantage of making the textbook self-contained, and because they focus on the experiments themselves, they also significantly augment the material found in your lecture textbook.

Experimental Procedures

The *miniscale approach* appeals to instructors who believe in the importance of performing experiments on a scale that allows isolation and characterization of products using conventional laboratory glassware. The quantities of starting materials used are usually in the range of 1–3 g, so the costs associated with purchasing and disposing of the chemicals are modest. The amounts of material may be easily handled, and it is possible to develop the techniques required to purify the products and characterize them by comparing their physical properties with those reported in the scientific literature. You will also be able to characterize the starting materials and products by spectroscopic techniques, so that you can see how their spectral properties differ. In short, you will be able to experience the real world of organic chemistry in which usable quantities of compounds are synthesized.

The *microscale approach* is especially attractive for minimizing the cost of purchasing and disposing of chemicals. The specialized glassware and other apparatus required for performing experiments on such small scales is now readily available. Indeed, many of the components found in a microscale kit are also found in the advanced organic laboratory, where trained researchers often work with minute amounts of material. The amounts of starting materials that are used in these procedures are often only 100–300 mg. Because of the small quantities of materials being handled, you must be meticulous in order to isolate products from microscale reactions. Purifying small quantities of materials by distillation or recrystallization is often tedious, so it will frequently be impractical to characterize pure products. Nevertheless, the experiments performed on the microscale should provide tangible quantities of material so that you can verify that the product was formed using chemical tests as well as some spectroscopic and analytical techniques.

Organization

The experiments we have included are intended to reinforce concepts given in the lecture course in organic chemistry and to familiarize you with the techniques that modern organic chemists routinely use. The basic types of apparatus you will need are described in Chapter 2. In addition, videos illustrating the steps required to assemble many of the set-ups are available at the optional Premium Companion Website at **www.cengage.com/login,** and we urge you to view these *prior* to going to the laboratory. In subsequent chapters, we provide figures in the margins of the pages to remind you how the assembled apparatus appears. The procedures in Chapters 3–6 are designed to introduce you to the different techniques for distillation, liquid-liquid and liquid-solid extraction, and thin-layer, column, and gas-liquid chromatography; the basic principles for these techniques are also described in their respective chapters. The spectroscopic methods that are fundamental to analyzing organic compounds are described in Chapter 8. Experiments that illustrate concepts such as selectivity of free-radical substitution (Chapter 9), kinetic and thermodynamic control of reactions (Chapter 13), kinetics of nucleophilic substitution reactions (Chapter 14) and electrophilic aromatic substitution reactions (Chapter 15), and the stereochemistry and regiochemistry of addition reactions (Chapters 10, 11, 12, and 17) are intended to provide a better understanding of these important subjects. Other experiments illustrate specific chemical transformations such as the generation, reactions, and rearrangements of carbocations (Chapters 10 and 15), electrophilic aromatic and nucleophilic substitution processes (Chapters 15 and 14, respectively), eliminations (Chapters 10 and 11), oxidations and reductions (Chapters 16 and 17, respectively), nucleophilic additions to carbonyl compounds and imines (Chapters 17 and 18, respectively), the generation and reactions of Grignard reagents (Chapter 19), and the formation of various carboxylic acid derivatives (Chapter 20). An experiment in the latter chapter allows you to observe the fascinating phenomenon of chemiluminescence. The value of enzymes for effecting enantioselective reactions is illustrated in Chapter 17. Because the current practice of organic chemistry in industry frequently involves multi-step transformations, several examples of multi-step synthesis are contained in Chapter 21. Experiments designed to introduce you to basic concepts of carbohydrate chemistry and polymer chemistry are provided in Chapters 22 and 23, respectively, and the experiments given in Chapter 24 give you an opportunity to explore one aspect of the world of bio-organic chemistry through synthesis of a dipeptide. A rational approach to solving the structures of unknown compounds with and without the aid of spectroscopic data is given in Chapter 25.

Textbook Website

This textbook is accompanied by an optional Premium Companion Website where students can access key material related to the experiments. This website provides the MSDSs and the ^{1}H NMR and IR spectra of the organic reactants and products for each experiment, as well as the Pre-Lab Exercises and technique videos. But there is more to be found there. For example, there are tutorials for analyzing ^{1}H and ^{13}C NMR, IR, and mass spectra, and tables of compounds and derivatives that are associated with qualitative organic chemistry (Chapter 25). Many laboratory manuals no longer include "qual organic" because of the availability of spectroscopic methods; however, we believe that this is a valuable component of the laboratory course because it will assist you in developing deductive skills so you can determine what functional groups are present in a compound whose identity is unknown to you. The website also includes links to additional information about experimental techniques, theoretical principles, and famous scientists related to each chapter. The icon for the website, shown in the margin here, alerts you to visit **www.cengage.com/login** to access this information. An access card for the website may be bundled with a new book, or students can purchase Instant Access at www.ichapters.com with ISBN 0538757140.

Spectroscopic Techniques

Spectroscopy may be the single most powerful tool for analyzing organic compounds. Consequently, thorough discussions of the theory and practical techniques for infrared, nuclear magnetic resonance (including ^{1}H and ^{13}C NMR), UV-Vis, and mass spectrometry are presented in Chapter 8. To reinforce the basic spectroscopic principles and to provide an opportunity for interpreting spectroscopic data, the infrared and nuclear magnetic spectra of all of the organic starting materials and products are provided in this textbook and at the website associated with it, on the optional Premium Companion Website at **www.cengage.com/login.** It is also possible for you to perform simple manipulations of the ^{1}H NMR and IR spectra that are available at the website. For example, you will be able to measure chemical shifts, integrals, and coupling constants directly on the ^{1}H NMR spectra. You will also be able to determine the position of an absorption in the IR spectrum that is associated with a specific functional group. This "hands-on" experience has proved an invaluable aid in teaching the basics of interpreting ^{1}H NMR and IR spectra and is unique to this laboratory textbook.

Safety and the Environment

Important sections entitled "Safety Alert" and "Wrapping It Up" are included with each experimental procedure. The information in the "Safety Alert" is designed to inform you and your instructor of possible hazards associated with the operations being performed. The abbreviated Material Safety Data Sheets (MSDSs) that are available at the optional Premium Companion Website at **www.cengage.com/login** provide additional information regarding flammability and toxicological properties of the chemicals being used and produced. Because of the flammable nature of the solvents and the chemicals that are handled in the laboratory, the use of *flameless* heating is emphasized and should be implemented in order to make the laboratory a safe workplace. The guidelines and methods in the "Wrapping It Up" section will familiarize you with the proper procedures for disposing of chemicals and other by-products after you have completed the experiment. Using these recommended methods will help protect the environment and lessen the costs associated with the ultimate disposal of these materials.

Essays

A feature of many of the chapters in the textbook is a Historical Highlight. These essays, some of which are biographical in nature, are designed to familiarize you with the lives of some of the chemical pioneers who have advanced science. These accounts will also provide you with a sense of the excitement and insights of individuals whose scientific observations form the basis for some of the experiments you will perform. Other essays are intended to relate organic chemistry to your everyday life. We hope they will whet your appetite for the subject of organic chemistry and enrich your experience as you further develop your scientific expertise.

Pre-Lab and Post-Lab Exercises

Each experiment is accompanied by two sets of questions. The Pre-Lab Exercises are provided at the optional Premium Companion Website at **www.cengage.com/login** and are designed to test your understanding of basic concepts, so you will be able to perform the experiments and techniques safely and successfully. Because these questions will assist you in preparing for work in the laboratory, we strongly recommend that you answer them *before* performing the experiment. The Post-Lab Exercises are found under the heading "Exercises" after each Experimental Procedure. These questions are written to reinforce the principles that are illustrated by the experiments and to determine whether you understand the observations you have made and the operations you have performed. Furthermore, questions on spectroscopy will help you develop the skills required to interpret IR and ^{1}H and ^{13}C NMR spectra.

Significant Changes from the Fourth Edition

This edition of the textbook includes 14 new discovery experiments and two new green chemistry procedures. Three additional Historical Highlight essays are provided, and many new Web-based references have been added to augment the vitality of the discussions. Furthermore, we have overseen the development and production of new videos that illustrate how to assemble the apparatus that is required to perform the various experimental procedures.

Feedback

As always, we seek your comments, criticisms, and suggestions for improving our textbook. Despite our best efforts, we are certain that there are typographical errors and the like that have escaped our notice, and we would appreciate your bringing them to our attention; our e-mail and snail-mail addresses are provided below. No matter how busy we might be, we shall respond to any messages you send.

Instructor's Manual

The Instructor's Manual is available to adopting instructors on the book's password-protected instructor companion website (accessible from www.cengage.com/gilbert) as downloadable Word and PDF files.

Acknowledgments

As with previous editions, a number of individuals contributed to making this one a reality. These include David Flaxbart (UT Austin), Donvan C. Haines (Sam Houston State University), Frederick J. Heldrich (College of Charleston), David Johnson (UT San Antonio), Chad Landrie (UI Chicago), Jason Serin (Glendale Community College), and G. Robert Shelton (University of North Texas). Prof. M. Robert Willcott kindly provided the MRI plots accompanying the Historical Highlight in Chapter 8. The capable staff at Cengage Learning provided invaluable support as we prepared this edition. We particularly acknowledge the efforts of Rebecca Heider for her assistance, wise counsel, and diligence, and Sara Arnold for her careful copyediting of the manuscript. We acknowledge the use of ^{13}C NMR spectral data from the Spectral Database for Organic Compounds of the Japanese National Institute of Advanced Industrial Science and Technology and also Aldrich Chemical Company, and of mass spectral data from the NIST database. We also thank Bill Vining of SUNY Oneonta and Bill Rohan for developing the Web-based information and John Colapret for conceptualizing and implementing the videos that accompany our textbook.

John C. Gilbert
jgilbert@scu.edu
Department of Chemistry & Biochemistry
Santa Clara University
Santa Clara, CA 95053

Stephen F. Martin
sfmartin@mail.utexas.edu
Department of Chemistry & Biochemistry
The University of Texas at Austin
Austin, TX 78712

CHAPTER 1

Introduction, Record Keeping, and Laboratory Safety

When you see this icon, sign in at this book's premium website at **www.cengage.com/login** to access videos, Pre-Lab Exercises, and other online resources.

This chapter sets the stage as you undertake the adventure of experimental organic chemistry. Although we may be biased, we think that this laboratory experience is one of the most valuable you will have as an undergraduate student. There is much to be learned as you progress from the relatively structured format of your first laboratory course in organic chemistry to the much less defined experimental protocols of a scientific research environment. The laboratory practices described in the following sections should serve you well in the journey.

1.1 INTRODUCTION

The laboratory component of a course in organic chemistry has an important role in developing and augmenting your understanding of the subject matter. The theoretical concepts, functional groups, and reactions presented in the lecture part of the course may seem abstract at times, but they are more understandable as a result of the experiments you perform. The successes, challenges, and, yes, frustrations associated with the "hands-on" experience gained in the laboratory, as you gather and interpret data from a variety of reactions, provide a sense of organic chemistry that is nearly impossible to communicate in formal lectures. For example, it is one thing to be told that the addition of bromine (Br_2) across the π-bond of most alkenes is a rapid process at room temperature. It is quite another to personally observe the *immediate* decoloration of a reddish solution of bromine in dichloromethane (Br_2/CH_2Cl_2) as a few drops of it are added to cyclohexene. The principles developed in the lectures will help you to predict what reaction(s) should occur when various reagents are combined in experimental procedures and to understand the mechanistic course of the process(es). Performing reactions allows you to test and verify the principles presented in lecture.

Of course, the laboratory experience in organic chemistry has another important function beyond reinforcing the concepts presented in lecture—to introduce you to the broad range of techniques and procedures that are important to the successful practice of experimental organic chemistry. You will learn how to handle a variety of chemicals safely and how to manipulate apparatus properly, talents that are critical to your success as a student of the chemical sciences. Along with becoming more skilled in the technical aspects of laboratory work, you should also develop a proper scientific approach to executing experiments and interpreting the results. By reading and, more importantly, *understanding* the concepts of this chapter, you will be better able to achieve these valuable goals.

1.2 PREPARING FOR THE LABORATORY

A common misconception students have about performing experiments is that it is much like cooking; that is, you merely follow the directions given—the "recipe"—and the desired product or data will result. Such students enter the laboratory expecting to follow the experimental procedure in a more or less rote manner. This unfortunate attitude can lead to inefficiencies, accidents, and minimal educational benefit and enjoyment from the laboratory experience.

To be sure, cooking is somewhat analogous to performing experiments. The successful scientist, just like a five-star chef, is a careful planner, a diligent worker, a keen observer, and is fully prepared for failures! Experiments may not work despite your best efforts, just as a cake may fall even in the hands of a premier pastry chef.

The correct approach to being successful in the laboratory is *never* to begin any experiment until you understand its overall purpose and the reasons for each operation that you are to do. This means that you must *study*, not *just read,* the entire experiment *prior* to arriving at the laboratory. Rarely, if ever, can you complete the necessary preparation in five or ten minutes, which means that you should not wait until just before the laboratory period begins to do the studying, thinking, and writing that are required. *Planning* how to spend your time in the laboratory is the key to efficient completion of the required experiments. Your performance in the laboratory will benefit enormously from proper advance work, and so will your grade!

The specific details of what you should do before coming to the laboratory will be provided by your instructor. However, to help you prepare in advance, we have developed a set of Pre-Lab Exercises for each of the experimental procedures we describe. These exercises are Web-based and are found at the URL given in the margin; you should bookmark this URL, as you will be visiting it frequently while preparing for each experimental procedure. In addition, the icon shown in the margin will appear whenever Web-based material is available.

Your instructor may require you to submit answers to the Pre-Lab Exercises for approval before authorizing you to proceed with the assigned experiments. Even if you are not required to submit the exercises, though, you will find that working them *prior* to the laboratory period will be a valuable educational tool to self-assess your understanding of the experiments to be performed.

You undoubtedly will be required to maintain a laboratory notebook, which will serve as a complete, accurate, and neat record of the experimental work that you do. Once more, your instructor will provide an outline of what specific information should appear in this notebook, but part of what is prescribed will probably necessitate advance preparation, which will further enhance your ability to complete the experiments successfully. The laboratory notebook is a *permanent record* of your accomplishments in the course, and you should take pride in the quality and completeness of its contents!

1.3 WORKING IN THE LABORATORY

You should be aware that experimental organic chemistry is *potentially* dangerous, because many of the chemicals used are toxic and/or highly flammable, and most of the procedures require the use of glassware that is easily broken. Careless handling of these chemicals and sloppy assembly of apparatus are sources of danger not only to you but also to those working near you. You should *not* be afraid of the chemicals and equipment that you will be using, but you *should* treat them with the respect and care associated with safe experimental practices. To facilitate this, there is an emphasis on the proper handling of chemicals and apparatus throughout the textbook, and the importance of paying particular attention to these subjects *cannot* be overemphasized. In a sense, laboratory safety is analogous to a chain, which is only as strong as its weakest link: the possibility that an accident will occur is only as great as the extent to which unsafe practices are followed. In other words, if you and your labmates adhere to proper laboratory procedures, the risk of an accident will be minimized.

It is important that you follow the experimental procedures in this textbook closely. There is a good reason why each operation should be performed as it is described, although that reason may not be immediately obvious to you. Just as it is risky for a novice chef to be overly innovative when following a recipe, it is *dangerous* for a beginning experimentalist to be "creative" when it comes to modifying the protocol that we've specified. As you gain experience in the organic laboratory, you may wish to develop alternative procedures for performing a reaction or purifying a desired product, but *always* check with your instructor *before* trying any modifications.

Note that rather detailed experimental procedures are given early in the textbook, whereas somewhat less detailed instructions are provided later on. This is because many of the basic laboratory operations will have become familiar to you in time and need not be spelled out. It is hoped that this approach to the design of procedures will decrease your tendency to think that you are essentially following a recipe in a cookbook. Moreover, many of the experimental procedures given in the literature of organic chemistry are relatively brief and require the chemist to "fill in the blanks," so it is valuable to gain some initial experience in figuring out some details on your own.

Most of your previous experience in a chemistry laboratory has probably required that you measure quantities precisely, using analytical balances, burets, pipets, and other precise measuring devices (Secs. 2.5 and 2.6). Indeed, if you have done quantitative inorganic analysis, you know that it is often necessary to measure weights to the third or fourth decimal place and volumes to at least the first. Experiments in organic chemistry that are performed at the **microscale** level, that is, experiments in which less than about 1 mL of the principal reagents is used and the amounts of solvents are less than 2 or 3 mL, also require relatively precise measuring of quantities. For example, if you are to use 0.1 g of a reagent and your measuring device only allows measuring to the nearest 0.1 g, you could easily have as much as about 0.15 g or as little as 0.05 g of the reagent. Such deviations from the desired quantity represent significant *percentage* errors in measurement and can result in serious errors in the proportions of reagents involved in the reaction. Consequently, weights should be accurate to within about 0.01 g and volumes to within about 0.1 mL. This requires the use of appropriate analytical balances and graduated pipets.

Experiments being performed at the **miniscale** level, which we define as involving 1–5 g of reagents and usually less than about 25 mL of solvent, normally do not require such precise measuring. Weighing reagents to the nearest tenth of a gram is usually satisfactory, as is measuring out liquids in graduated cylinders, which are accurate to ± 10%. For example, if you are directed to use 20 mL of diethyl ether as solvent for a reaction, the volume need *not* be 20.0 mL. In fact, it probably will make little difference to the success of the reaction whether anywhere from 15–25 mL of the solvent is added. This is not to say that care need not be exercised in measuring out the amounts of materials that you use. Rather, it means that valuable time need not be invested in making these measurements highly precise.

We've inserted markers in the form of **stars** (★) in many of the experimental procedures in this textbook. These indicate places where the procedure can be interrupted without affecting the final outcome of the experiment. These markers are designed to help you make the most efficient use of your time in the laboratory. For example, you may be able to start a procedure at a point in the period when there is insufficient time to complete it but enough time to be able to work through to the location of a star; you can then safely store the reaction mixture and finish the sequence during the next laboratory period. We've *not* inserted stars at every possible stopping point but only at those where it is not necessarily obvious that interruption of the procedure will have no effect on the experimental results. Consult your instructor if in doubt about whether a proper stopping point has been reached.

As noted above, a *carefully* written **notebook** and *proper* **safety procedures** are important components of an experimental laboratory course. These aspects are discussed further in the following two sections.

1.4 THE LABORATORY NOTEBOOK

One of the most important characteristics of successful scientists is the habit of keeping a complete and understandable record of the experimental work that has been done. Did a precipitate form? Was there a color change during the course of the reaction? At what temperature was the reaction performed, and for how long did the reaction proceed? Was the reaction mixture homogeneous or heterogeneous? On what date(s) was the work performed? These are observations and data that may seem insignificant at the time but may later prove critical to the interpretation of an experimental result or to the ability of another person to reproduce your work. All of them belong in a properly kept laboratory notebook. We make suggestions for such a document in the following two sections. Your instructor may specify other items to be included, but the list we give is representative of a good notebook.

1.5 GENERAL PROTOCOL FOR THE LABORATORY NOTEBOOK

1. Use a *bound* notebook for your permanent laboratory record to minimize the possibility that pages will be lost. If a number has not been printed on each page, do so manually. Some laboratory notebooks are designed with pairs of identically numbered pages so that a carbon copy of all entries can be made. The duplicate page can then be removed and submitted to your instructor or

put in a separate place for safekeeping. Many professional scientists use this type of notebook.

2. Reserve the first page of the notebook for use as a title page, and leave several additional pages blank for a Table of Contents.
3. Use as the main criterion for what should be entered in the notebook the rule that the record should be sufficiently complete so that anyone who reads it will know exactly what you did and will be able to repeat the work in precisely the way you originally did it.
4. Record all experimental observations and data in the notebook *as they are obtained.* Include the *date* and, if appropriate, the *time* when you did the work. In a legal sense, the information entered into the notebook *at the time of performance* constitutes the primary record of the work, and it is important for you to follow this principle. Many patent cases have been determined on the basis of dates and times recorded in a laboratory notebook. One such example is described in the Historical Highlight at the end of this chapter.
5. Make all entries in ink, and *do not delete anything you have written* in the notebook. If you make a mistake, cross it out and record the correct information. Using erasers or correction fluid to modify entries in your notebook is unacceptable scientific practice!

   Do not scribble notes on odd bits of paper with the intention of recording the information in your notebook later. Such bad habits only lead to problems, since the scraps of paper are easily lost or mixed up. They are also inefficient, since transcribing the information to your notebook means that you must write it a second time. This procedure can also result in errors if you miscopy the data.

   Finally, do not trust your memory with respect to observations that you have made. When the time comes to write down the information, you may have forgotten a key observation that is critical to the success of the experiment.
6. Unless instructed to do otherwise, do not copy *detailed* experimental procedures that you have already written elsewhere in your notebook; this consumes valuable time. Rather, provide a specific reference to the source of the detailed procedure and enter a *synopsis* of the written procedure that contains enough information that (1) you need not refer to the source while performing the procedure and (2) another chemist will be able to *duplicate* what you did. For example, when performing an experiment from this textbook, give a reference to the page number on which the procedure appears, and detail any *variations* made in the procedure along with the reason(s) for doing so.
7. Start the description of each experiment on a new page titled with the name of the experiment. The recording of data and observations from several different procedures on the same page can lead to confusion, both for yourself and for others who may read your notebook.

1.6 TYPES OF ORGANIC EXPERIMENTS AND NOTEBOOK FORMATS

There are two general classes of experiments, **investigative** and **preparative,** in this textbook. Investigative experiments normally involve making observations and learning techniques that are common to laboratory work in organic chemistry but do not entail conversion of one compound into another. Some examples are solubility tests, distillation, recrystallization, and qualitative organic analysis. In contrast, preparative experiments involve interconversion of different compounds. Most of the procedures described in this textbook fall into the latter category.

The format of the laboratory notebook is usually different for these two types of experiments. Once again, your instructor may have a particular style that is recommended, but we provide suggested formats below.

Notebook Format for Investigative Experiments

1. **Heading.** Use a new page of the notebook to start the entries for the experiment. Provide information that includes your name, the date, the title of the experiment, and a reference to the place in the laboratory textbook or other source where the procedure may be found.
2. **Introduction.** Give a brief introduction to the experiment in which you clearly state the purpose(s) of the procedure. This should require no more than one-fourth of a page.
3. **Summary of MSDS Data.** As directed by your instructor, either briefly summarize the Material Safety Data Sheet (MSDS) data (Sec. 1.10) for the solvents, reagents, and products encountered in the experiment or give a reference to where a printout of these data is located.
4. **Synopsis of and Notes on Experimental Procedure—Results.** Enter a one- or two-line statement for each part of an experiment. Reserve sufficient room to record results as they are obtained. As noted in Section 1.5 of "Notebook Format for Preparative Experiments," do *not* copy the experimental procedure from the textbook, but provide a synopsis of it.

   Much of this section of the write-up can be completed before coming to the laboratory, to ensure that you understand the experiment and that you will perform all parts of it.
5. **Interpretation of Instrumental Data.** If instructed to do so, discuss any instrumental data, such as gas-liquid chromatographic analyses and spectral data that you have obtained or are provided in the textbook.
6. **Conclusions.** Record the conclusions that can be reached, based on the results you have obtained in the experiment. If the procedure has involved identifying an unknown compound, summarize your findings in this section.
7. **Answers to Exercises.** Enter answers to any exercises for the experiment that have been assigned from the textbook.

A sample write-up of an investigative experiment is given in Figure 1.1.

Notebook Format for Preparative Experiments

1. **Heading.** Use a new page of the notebook to start the entries for the experiment. Provide information that includes your name, the date, the title of the experiment, and a reference to the place in the laboratory textbook or other source where the procedure may be found.
2. **Introduction.** Give a brief introduction to the experiment in which you clearly state the purpose(s) of the procedure. This should require no more than one-fourth of a page.
3. **Main Reaction(s) and Mechanism(s).** Write *balanced* equations giving the main reaction(s) for conversion of starting material(s) to product(s). The reason for balancing the equations is discussed in Part **4** below. Whenever possible, include the detailed mechanisms for the reactions that you have written.

1.

Your Name
Date

Separation of Green Leaf Pigments by TLC

Reference: *Experimental Organic Chemistry: A Miniscale and Microscale Approach,* 5th ed. by Gilbert and Martin, Section 6.2.

2. INTRODUCTION

The pigments in green leaves are to be extracted into an organic solvent, and the extract is to be analyzed by thin-layer chromatography (TLC). The presence of multiple spots on the developed TLC plate will indicate that more than a single pigment is contained in the leaves.

3. MSDS DATA

These data are available on the printouts inserted at the back of my lab book.

4. SYNOPSIS OF AND NOTES ON EXPERIMENTAL PROCEDURE—RESULTS

Procedure: Grind five stemless spinach leaves in mortar and pestle with 5 mL of 2:1 pet. ether and EtOH. Swirl soln. with 3 × 2-mL portions H_2O in sep. funnel; dry org. soln. for few min over anhyd. Na_2SO_4 in Erlenmeyer. Decant and concentrate soln. if not dark-colored. Spot 10-cm × 2-cm TLC plate about 1.5 mm from end with dried extract; spot should be less than 2 mm diam. Develop plate with $CHCl_3$. *Variances and observations:* Procedure followed exactly as described in reference. Org. soln. was dark green in color; aq. extracts were yellowish. Half of org. layer lost. TLC plate had five spots having colors and R_f-values shown on the drawing below.

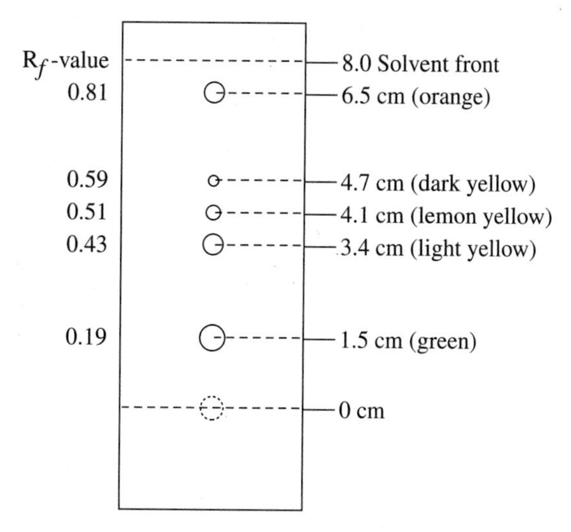

5. INTERPRETATION OF INSTRUMENTAL DATA

No data provided for this experiment.

6. CONCLUSIONS

Based on TLC analysis, the procedure used allows the extraction of at least five different pigments from the spinach leaves. Judging from colors, one of these is a carotene, three are xanthophylls, and the last is chlorophyll *b*.

7. ANSWERS TO EXERCISES

(Answers intentionally omitted.)

Figure 1.1
Sample notebook format for investigative experiments.

4. **Table of Reactants and Products.** Set up a Table of Reactants and Products as an aid in summarizing the amounts and properties of reagents and catalysts being used and the product(s) being formed. Only those reactants, catalysts, and products that appear in the main reaction(s) should be listed in the table; many other reagents may be used in the work-up and purification of the reaction mixture, but these should *not* be entered in the table.

   Your instructor will have specific recommendations about what should appear in the table, but the following items are illustrative.

   a. The name and/or structure of each reactant, catalyst, and product.

   b. The molar mass of each compound.

   c. The weight used, in grams, of each reactant and the volume of any liquid reactant. We recommend that the weight and/or volume of any catalysts used be entered for purposes of completeness.

   d. The molar amount of each reactant used; this can be calculated from the data in Parts **b** and **c.**

   e. The theoretical mole ratio, expressed in whole numbers, for the reactants and products; this ratio is determined by the *balanced* equation for the reaction, as given in Part **3.**

   f. Physical properties of the reactants and products. This entry might include data such as boiling and/or melting point, density, solubility, color, and odor.

   g. As directed by your instructor, either briefly summarize the MSDS data (Sec. 1.10) for the solvents, reagents, and products encountered in the experiment or give a reference to where a printout of these data is located.

5. **Yield Data.** Compute the maximum possible amount of product that can be formed; this is the **theoretical yield.** This can easily be calculated from the data in the Table of Reactants and Products as follows. First determine which of the reactants corresponds to the **limiting reagent.** This is the reagent that is used in the *least* molar amount relative to what is required theoretically. In other words, the reaction will stop once this reactant is consumed, so its molar quantity will define the maximum quantity of product that can be produced. From the number of moles of limiting reagent involved and the balanced equation for the reaction, determine the theoretical yield, in moles (written as "mol" when used as a unit, as in "g/mol"), of product. This value can then be converted into the theoretical yield in grams, based on the molar mass of the product.

   Once the isolation of the desired product(s) has been completed, you should also calculate the **percent yield,** which is a convenient way to express the overall efficiency of the reaction. This is done by obtaining the **actual yield** of product(s) in grams, and then applying the expression in Equation 1.1. Generally, the calculated value of percent yield is rounded to the nearest whole number. As points of reference, most organic chemists consider yields of 90% or greater as being "excellent," and those below 20% as "poor."

$$\text{Percent yield} = \frac{\text{Actual yield (g)}}{\text{Theoretical yield (g)}} \times 100 \qquad (1.1)$$

6. **Synopsis of and Notes on Experimental Procedure.** Provide an outline of the experimental procedure that contains enough detail so that you do not have to refer to the textbook repeatedly while performing the experiment.

Note any variations that you use, as compared to the referenced procedure, and observations that you make while carrying out the formation and isolation of the product(s).

7. **Observed Properties of Product.** Record the physical properties of the product that you have isolated in the experiment. Appropriate data under this heading might include boiling and/or melting point, odor, color, and crystalline form, if the product is a solid. Compare your observations with those available on the compound in various reference books (for example, the *CRC Handbook of Chemistry and Physics* or *Lange's Handbook of Chemistry*).
8. **Side Reactions.** If instructed to do so, list possible side reactions (those reactions leading to undesired products) that are likely to occur in the experiment. It is important to consider such processes because the by-products that are formed must be removed by the procedure used to purify the desired product. You may need to consult your lecture notes and textbook in order to predict what side reactions might be occurring.
9. **Other Methods of Preparation.** If instructed to do so, suggest alternative methods for preparing the desired compound. Such methods may involve using entirely different reagents and reaction conditions. Your lecture notes and textbook can serve as valuable resources for providing possible entries for this section.
10. **Method of Purification.** Develop a flow chart that summarizes the sequence of operations that will be used to purify the desired product. The chart will show at what stages of the work-up procedure unchanged starting materials and unwanted by-products are removed. By understanding the logic of the purification process, you will know why each of the various operations specified in the purification process is performed.

    Purifying the final product of a reaction can be the most challenging part of an experimental procedure. Professional organic chemists are constantly required to develop work-up sequences that allow isolation of a pure product, free from starting materials and other contaminants. They do this by considering the chemical and physical properties of both the desired and undesired substances, and it is important for you to gain experience in devising such schemes as well.
11. **Interpretation of Instrumental Data.** If instructed to do so, discuss any instrumental data, such as gas-liquid chromatographic analyses and spectral data you have obtained or that are provided in the textbook.
12. **Answers to Exercises.** Enter answers to any exercises for the experiment that have been assigned from the textbook.

A detailed example of the write-up for a preparative experiment involving the dehydration of cyclohexanol (Sec. 10.3) is given in Figure 1.2. You may not actually perform this reaction; nevertheless, you should carefully study the example in order to see how to prepare specific entries for the first eight items listed. The various entries in Figure 1.2 are labeled with circled, **boldface** numbers and are discussed further in the following paragraphs. It is assumed for illustrative purposes that an actual yield of 2.7 g is obtained.

1 Use a new page of the notebook to start the entries for the experiment. Provide information that includes your name, the date, the title of the experiment, and a reference to the place in the laboratory textbook or other source where the procedure can be found.

(1) **1.**

Your Name
Date

Dehydration of Cyclohexanol

Reference: *Experimental Organic Chemistry: A Miniscale and Microscale Approach,* 5th ed. by Gilbert and Martin, Section 10.3.

(2) **2.** INTRODUCTION

Cyclohexene is to be prepared by the acid-catalyzed dehydration of cyclohexanol.

(3) **3.** MAIN REACTION(S) AND MECHANISM(S)

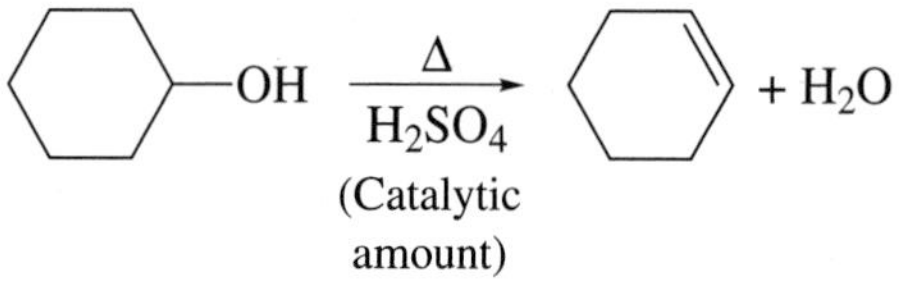

$$\text{Cyclohexanol-OH } (\mathbf{1}) \xrightarrow[\text{H}_2\text{SO}_4 \text{ (Catalytic amount)}]{\Delta} \text{Cyclohexene } (\mathbf{2}) + H_2O$$

(Mechanism intentionally omitted.)

(4) **4.** TABLE OF REACTANTS AND PRODUCTS

(5) Compound	(6) M.M.	(7a) Volume Used (mL)	(7b) Weight Used (g)	(8) Moles Used	(9) Moles Required	(10) Other Data
Cyclohexanol	100.2	5.2	5	0.05	1	bp 161 °C (760 torr), mp 25.1 °C, d 0.962 g/mL, colorless
Sulfuric acid (9 *M*)	98.1	2.5	#	#	0	d 1.84 (18 *M* H_2SO_4)
Cyclohexene	82.2	*	*	*	1	bp 83 °C (760 torr), mp 103.5 °C, d 0.810, colorless

#Entry left blank because this row is for the catalyst.

*Entry left blank because this row is for the product.

(11) LIMITING REAGENT: Cyclohexanol

(12) MSDS DATA

These data are available on the printouts inserted at the back of my lab book.

(13) **5.** YIELD DATA

Theoretical yield of cyclohexene = moles of limiting reagent (cyclohexanol) × M.W. of cyclohexene

= 0.05 mol × 82.2 g/mol

= 4.1 g

Actual yield = 2.7 g

(Continued)

Figure 1.2
Sample notebook format for investigative experiments.

Percent yield = [Actual yield (g)/theoretical yield (g)] × 100
= [2.7/4.1] × 100 = 66%

(14) **6.** SYNOPSIS OF AND NOTES ON EXPERIMENTAL PROCEDURE—RESULTS

Procedure: Put alcohol in 25-mL rb flask and add H_2SO_4. Mix, add stirbar, attach to fractional dist. apparatus. Heat with oil bath; heating rate such that head temp. stays below 90 °C. Stop when 2.5 mL remain in rxn. flask. Put distillate in 25-mL Erlenmeyer and add 1–2 g K_2CO_3.* Occasionally swirl mix. for 15 min and transfer liquid to 10-mL rb by decantation or pipet. Add stirbar and do simple distillation (no flames!); receiver must be close to drip tip of adapter to minimize losses by evaporation. Collect product at 80–85 °C (760 torr).
Variances and observations: Procedure followed exactly as described in reference. Distillate cloudy throughout dehydration step; formed two layers in receiver. Head temperature never exceeded 77 °C. Liquid in stillpot darkened as reaction proceeded. Addition of carbonate (1 g) to distillate caused evolution of a few bubbles of gas (CO_2?). Had to add about 0.5 g more of carbonate to get rid of cloudiness. Left solution over drying agent for one week (next lab period). Used pipet to transfer dried liquid to distillation flask. Collected cyclohexene in ice-cooled 10-mL rb flask attached to vacuum adapter protected with $CaCl_2$ tube. Stopped distillation when about 1 mL of yellowish liquid remained in stillpot.

(15) **7.** OBSERVED PROPERTIES OF PRODUCT

bp 80–84 °C (760 torr); colorless liquid; insoluble in water; decolorizes Br_2/CH_2Cl_2 solution and produces brown precipitate upon treatment with $KMnO_4/H_2O$.

(16) **8.** SIDE REACTIONS

2 (cyclohexyl)–OH $\xrightarrow[\text{H}_2\text{SO}_4 \text{ (Catalytic amount)}]{\Delta}$ (cyclohexyl)–O–(cyclohexyl)

3

(cyclohexyl)–OH + $HOSO_3H$ ⟶ (cyclohexyl)–OSO_3H

4

2 (cyclohexene) $\xrightarrow[\text{H}_2\text{SO}_4 \text{ (Catalytic amount)}]{\Delta}$ (cyclohexyl)–(cyclohexenyl) ⟶ Polymeric hydrocarbons

5 6

(Continued)

Figure 1.2 *(Continued)*

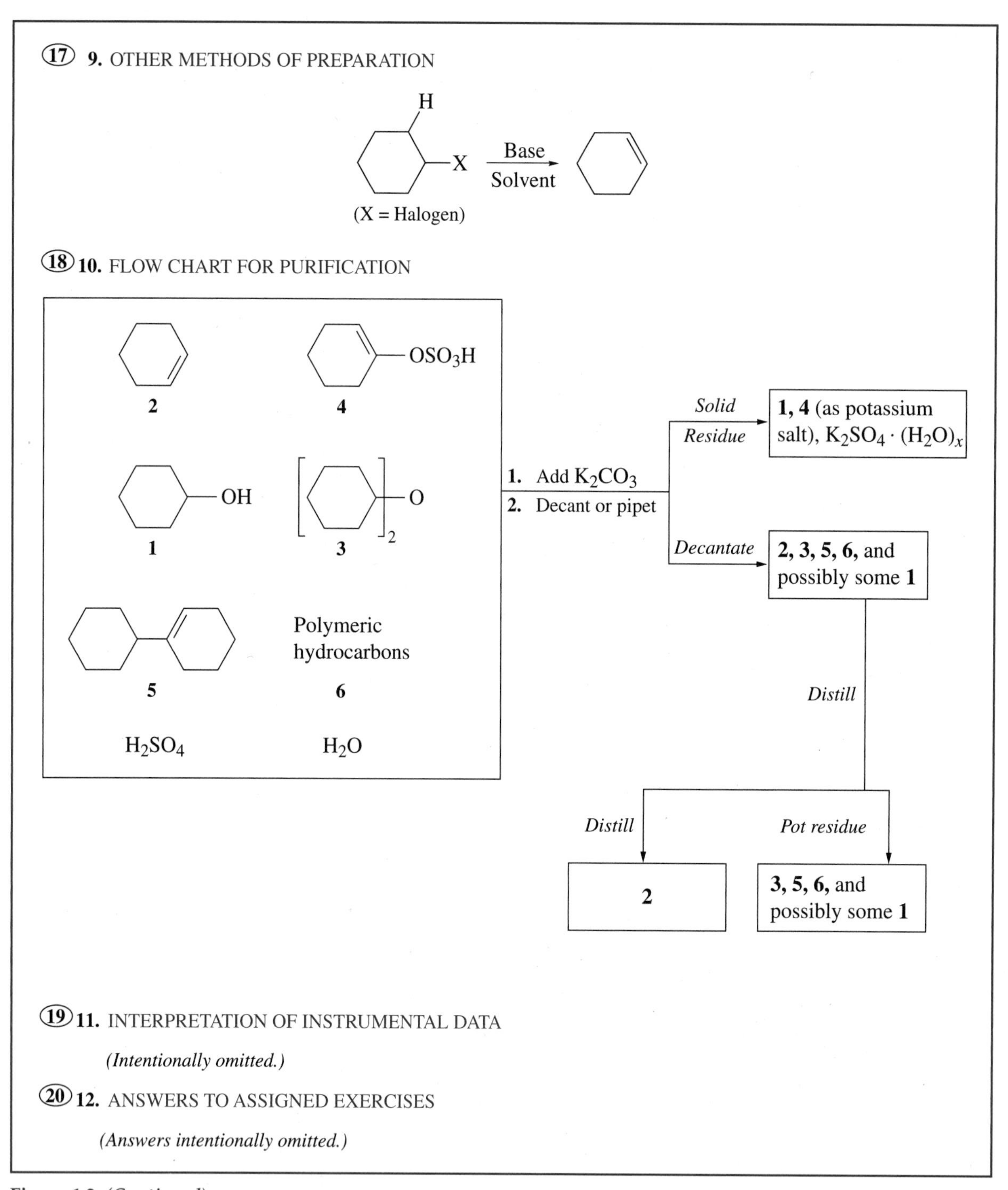

Figure 1.2 ***(Continued)***

2 Self-explanatory.

3 There is only a single reaction in our example, but in many cases more than one step is involved; write equations for *all* of the main reactions. A mechanism for the reaction is intentionally omitted in our example.

4 Use the illustrated format for the Table of Reactants and Products unless instructed to do otherwise.

5 Enter the name or structure of each reactant catalyst, if any, and desired product.

6 Record the molar mass (M.M.) of each reactant and desired product. For completeness, make an entry for any catalyst used, although this may be optional.

7a Give the volume, in milliliters (mL) or microliters (μL), of each *liquid* reactant and catalyst.

7b Record the weight, in grams (g) or milligrams (mg), of each reactant. This entry is optional for liquid catalysts but should be provided for reference purposes.

8 Calculate the moles used of each reactant. For completeness, a value for the catalyst is computed in our example.

9 Obtain the **theoretical ratio** for reactant(s) and product(s) by referring to the *balanced* main equation(s) for the reaction.

10 List selected physical properties of reactant(s) and product(s). The information needed is generally available in reference books.

11 Determine the **limiting reagent** in the following way. Compare the actual ratio of reactants used to that theoretically required. The reagent that is used in the least molar amount, relative to the theoretical amount, is the limiting reagent. In our example, there is only a single reagent, cyclohexanol, so it obviously must be the limiting reagent.

12 Self-explanatory. However, if you are instructed to provide MSDS data in the form of printouts from the website, be sure to read them. Ignorance is *not* bliss when it comes to handling chemicals!

13 Calculate the **theoretical yield** of the desired product both in moles and in grams. Knowing in our case that the limiting reagent is cyclohexanol and, from the main equation, that 1 mole of alcohol yields 1 mole of cyclohexene, it is clear that no more than 0.05 mole of the alkene can be formed.

Assuming that you were able to isolate 2.7 g of pure cyclohexene in the experiment, the **percent yield** would be calculated according to Equation 1.1.

14 Self-explanatory.

15 Self-explanatory.

16 Self-explanatory.

17 Self-explanatory.

18 Develop this diagram by considering what components, in addition to the desired product, may be present in the reaction mixture *after* the main reaction is complete. The chart shows how and where each of the inorganic and organic contaminants of the product is removed by the various steps of the work-up procedure. Ideally, pure cyclohexene results.

19 Self-explanatory.

20 Self-explanatory.

1.7 SAMPLE CALCULATIONS FOR NOTEBOOK RECORDS

Students frequently have difficulty in setting up Tables of Reactants and Products and calculating theoretical yields, so two hypothetical examples are provided for your reference.

Example 1

Problem Consider the reaction shown in Equation 1.2. Assume that you are to use 5 g (7.8 mL) of 1-pentene and 25 mL of concentrated HBr solution. Prepare a Table of Reactants and Products, determine the limiting reagent, and calculate the theoretical yield for the reaction.

$$CH_2{=}CHCH_2CH_2CH_3 + HBr \longrightarrow CH_3CH(Br)CH_2CH_2CH_3 \quad (1.2)$$

Answer First of all, note that the equation is balanced, because the "1" that signifies that 1 mole of each reactant will react to produce 1 mole of product is omitted by convention. Because an aqueous solution of HBr, rather than the pure acid, is being used, the amount of HBr present must be determined. Concentrated HBr is 47% by weight in the acid, and its density, d, is 1.49 g/mL, a value that would be recorded in the column headed "Other Data." Consequently, 25 mL of this solution contains 17.5 g of HBr (25 mL × 1.49 g/mL × 0.47). The needed data can then be entered into Table 1.1.

The limiting reagent is 1-pentene because theory requires that it and HBr react in a 1:1 molar ratio, yet they have been used in a ratio of 0.07:0.22. This means that no more than 0.07 mole of product can be formed, since theory dictates that the ratio between 2-bromopentane and 1-pentene also be 1:1. The calculation of the theoretical yield is then straightforward.

You may find it convenient to use units of milligrams (mg), microliters (μL), and millimoles (mmol) instead of grams, milliliters, and moles, respectively, in performing measurements and calculations when small quantities of reagents are used, as is the case for microscale reactions. For example, let's consider how Table 1.1 would be modified if 0.1 g of 1-pentene and 0.5 mL of concentrated HBr solution

Table 1.1 *Table of Reactants and Products for Preparation of 2-Bromopentane*

Compound	*M.M.*	*Volume Used (mL)*	*Weight Used (g)*	*Moles Used*	*Moles Required*	*Other Data*
1-Pentene	70.14	7.8	5	0.07	1	*
HBr	80.91	25	17.5	0.22	1	*
2-Bromopentane	151.05	†	†	†	1	*

*These entries have been intentionally omitted in this example.
†These entries are left blank because this line is for the product.
Limiting reagent: 1-pentene
Theoretical yield: 151.05 g/mol × 0.07 mol = 10.5 g

Table 1.2 *Table of Reactants and Products for Preparation of 2-Bromopentane*

Compound	*M.M.*	*Volume Used (mL)*	*Weight Used (g)*	*Mmols Used*	*Mmols Required*	*Other Data*
1-Pentene	70.14	0.16	100	1.4	1	*
HBr	80.91	0.5	358	4.4	1	*
2-Bromopentane	151.05	†	†	†	1	*

*These entries have been intentionally omitted in this example.
†These entries are left blank because this line is for the product.
Limiting reagent: 1-pentene
Theoretical yield: 151.05 g/mol × 1.4 mol × 211 mg

were used. If the calculations were done in grams and moles, the entries under "Moles Used" would be 0.0014 and 0.4, respectively. Errors can arise when making such entries, because a zero may inadvertently be added or dropped.

This potential problem is less likely if you enter the data as milligrams and millimoles. If you recognize that 0.1 g of alkene is 100 mg and 0.5 mL of HBr solution contains 358 mg of HBr (0.5 mL × 1490 mg/mL × 0.48), the entries would be those shown in Table 1.2. You may then determine the limiting reagent and calculate the theoretical yield as in Example 1. Note that the necessary cancellation of units occurs when the molar mass is expressed in mg/mmol.

Although the volume of 1-pentene to be used is expressed in milliliters, you may be measuring out this amount with a device that is calibrated in microliters (μL, 1 μL = 10^{-3} mL). Thus, in the present example, you would be using 160 μL of 1-pentene (0.16 mL × 10^3 μL/mL).

Example 2

Problem Now consider the transformation illustrated in Equation 1.3. Assume that you are to use 7 mL of ethanol and 0.1 mL of concentrated H_2SO_4 as the catalyst. Prepare a Table of Reactants and Products, determine the limiting reagent, and calculate the theoretical yield for the reaction.

$$2\,CH_3CH_2OH \xrightarrow[\text{(Catalytic amount)}]{H_2SO_4} CH_3CH_2OCH_2CH_3 + H_2O \quad (1.3)$$

Answer As in the previous example, a volumetric measurement must first be converted to a weight. The density of ethanol is 0.789 g/mL, information that would be entered under the column headed "Other Data," so that means 5.5 g is being used. Table 1.3 can then be completed. Note that the catalyst, although recorded in the table, is not used in any of the calculations because, by definition, it is not consumed during the reaction. Including it should help remind the experimentalist that it is indeed required to make the reaction occur!

Calculation of the theoretical yield is performed as in Example 1, with the important exception that a factor of 0.5 is incorporated to adjust for the fact that only one-half mole of diethyl ether would be produced for each mole of ethanol that is used.

Table 1.3 *Table of Reactants and Products for Preparation of Diethyl Ether*

Compound	*M.M.*	*Volume Used (mL)*	*Weight Used (g)*	*Moles Used*	*Moles Required*	*Other Data*
Ethanol	46.07	7	5.5	0.12	2	*
Sulfuric acid	†	0.1	†	†	†	*
Diethyl ether	74.12	‡	‡	‡	1	*

*These entries have been intentionally omitted in this example.
†These entries are left blank for reactants that serve only as catalysts.
‡These entries are left blank because this line is for the product.
Limiting reagent: ethanol
Theoretical yield: 74.12 g/mol × 0.12 mol · 0.5 = 4.4 g

1.8 SAFE LABORATORY PRACTICE: OVERVIEW

There is little question that one of the most important abilities that you, the aspiring organic chemist, can bring to the laboratory is a sound knowledge of how to perform experimental work safely. But just knowing *how* to work safely is insufficient! You must also make a *serious* commitment to follow standard safety protocols. In other words, having the knowledge about safety is useless if you do not put that knowledge into practice. What you actually do in the laboratory will determine whether you and your labmates are working in a safe environment.

Chemistry laboratories are potentially dangerous because they commonly house flammable liquids, fragile glassware, toxic chemicals, and equipment that may be under vacuum or at pressures above atmospheric. They may also contain gas cylinders that are under high pressure. The gases themselves may or may not be hazardous—for example, nitrogen is not, whereas hydrogen certainly is—but the fact that their containers are under pressure makes them so. Imagine what might happen if a cylinder of nitrogen fell and ruptured: you could have a veritable rocket on your hands, and, if the tank contained hydrogen, the "rocket" might even come equipped with a fiery tail! This is another way of saying *all* substances are hazardous under certain conditions.

Fortunately the laboratory need be no more dangerous than a kitchen or bathroom, but this depends on you *and* your labmates practicing safety as you work. Should you observe others doing anything that is unsafe, let them know about it in a friendly manner. Everyone will benefit from your action. We'll alert you repeatedly to the possible dangers associated with the chemicals and apparatus that you will use so that you can become well trained in safe laboratory practice. Mastery of the proper procedures is just as important in the course as obtaining high yields of pure products, and carefully reading our suggestions will assist you in achieving this goal. Some safety information will be contained in the text describing a particular experiment or in the experimental procedure itself. It will also appear in highlighted sections titled "**Safety Alert.**" These are designed to draw your special attention to aspects of safety that are of particular importance. We urge you to read these sections carefully and follow the guidelines in them carefully. You will then be fully prepared to have the fun and fulfillment of the laboratory experience.

1.9 SAFETY: GENERAL DISCUSSION

We highlight here, in the form of a **Safety Alert,** some general aspects regarding safe practices in the laboratory.

SAFETY ALERT

Personal Attire

1. ***Do not wear shorts or sandals in the laboratory;*** **the laboratory is *not* a beach! Proper clothing gives protection against chemicals that may be spilled accidentally. It is advisable to wear a laboratory coat, but in any case, the more skin that is protected by clothing the better.**
2. ***Always wear safety glasses or goggles in the laboratory.*** **This applies even when you are writing in your laboratory notebook or washing glassware, since nearby workers may have an accident. It is best *not* to wear contact lenses in the laboratory. Even if you are wearing eye protection, chemicals may get into your eyes, and you may not be able to get the contact lenses out before damage has occurred. Should you have to wear corrective glasses while working in the laboratory, make certain that the lenses are shatterproof. Wearing goggles over such glasses is recommended because the goggles give additional protection from chemicals entering your eyes from the sides of the lenses.**
3. ***Wear latex gloves when working with particularly hazardous chemicals.*** **Some reagents are especially hazardous if they come into contact with your skin. The ones you are most likely to encounter in the organic laboratory are concentrated acids and bases, and bromine and its solutions. Check with your instructor whenever you are uncertain whether you should be wearing gloves when handling reagents.**

General Considerations

1. ***Become familiar with the layout of the laboratory room.*** **Locate the exits from the room and the fire extinguishers, fire blankets, eyewash fountains, safety showers, and first-aid kits in and near your workspace. Consult with your instructor regarding the operation and purpose of each of the safety-related devices.**
2. ***Find the nearest exits from your laboratory room to the outside of the building.*** **Should evacuation of the building be necessary, use stairways rather than elevators to exit. Remain calm during the evacuation, and walk rather than run to the exit.**
3. ***Become knowledgeable about basic first-aid procedures.*** **The damage from accidents will be minimized if first aid is applied promptly. Read the section "First Aid in Case of an Accident" on the inside front cover of this book.**
4. ***Never work alone in the laboratory.*** **In the event of an accident, you may need the immediate help of a coworker. Should you have to work in the laboratory outside of the regularly scheduled periods, do so only with the express permission of your instructor and in the presence of at least one other person.**

Apparatus and Chemicals

1. ***Always check carefully for imperfections in the glassware that you will be using.*** **This should be done not only when checking into the laboratory for**

the first time but also when setting up the apparatus needed for each experiment. Look for cracks, chips, or other imperfections in the glass that weaken it. Use care in storing your glassware so that it is not damaged when you open or close the locker or drawer.

Pay particular attention to the condition of round-bottom flasks and condensers. The flasks often have "star" cracks (multiple cracks emanating from a central point) as a result of being banged against a hard surface. Heating or cooling a flask having this type of flaw may cause the flask to rupture with loss of its contents. This could result in a serious fire, not just loss of the desired product. To detect such cracks, hold the flask up to the light and look at all its surfaces closely. With respect to condensers, their most vulnerable points are the ring seals—the points where the inner tube and the water jacket of the condenser are joined. Special care must be given to examining these seals for defects, because if cracks are present water might leak into your apparatus and contaminate your product or, worse, cause violent reactions.

If you detect imperfections in your glassware, consult with your teacher immediately regarding replacement. Cracked or seriously chipped apparatus should always be replaced, but glassware with slight chips may still be safe to use.

2. *Dispose of glassware properly.* The laboratory should be equipped with a properly labeled special container for broken glassware and disposable glass items such as Pasteur pipets and melting-point capillaries. It is not appropriate to throw such items in the regular trash containers, because maintenance personnel may injure themselves while removing the trash. Broken thermometers are a special problem because they usually contain mercury, which is toxic and relatively volatile. There should be a separate, closed container for disposal of thermometers. If mercury has spilled as a result of the breakage, it should be cleaned up immediately. Consult with your instructor about appropriate procedures for doing so.

3. *Know the properties of the chemicals used in the experiments.* Understanding these properties helps you to take the proper precautions when handling them and to minimize danger in case of an accident. *Handle all chemicals with care.*

   Refer to MSDSs (Sec. 1.10) to learn about toxicity and other potential hazards associated with the chemicals you use. Most chemicals are at least slightly toxic, and many are *very* toxic and irritating if inhaled or allowed to come in contact with the skin. It is a good laboratory practice to wear plastic or latex gloves when handling chemicals, and there may be times when it is imperative to do so. Your instructor will advise you on the need for gloves.

   Should chemicals come in contact with your skin, they can usually be removed by a thorough and *immediate* washing of the affected area with soap and water. Do *not* use organic solvents like ethanol or acetone to rinse chemicals from your skin, as these solvents may actually assist the absorption of the substances into your skin.

4. *Avoid the use of flames as much as possible.* Most organic substances are flammable, and some are highly volatile as well, which increases their potential for being ignited accidentally. Examples of these are diethyl ether, commonly used as a solvent in the organic laboratory, and acetone. Occasionally, open flames

may be used for flame-drying an apparatus or distilling a high-boiling liquid. In such cases, a Safety Alert section will give special precautions for their use. Some general guidelines follow.

a. *Never use an open flame without the permission of your instructor.*

b. *Never use a flame to heat a flammable liquid in an open container.* Use a water or steam bath, hot plate, aluminum block, or similar electrical heat device instead. If a flammable liquid must be heated with an open flame, equip the container holding the liquid with a *tightly* fitting reflux condenser.

Information about the flammability of many commonly used organic solvents is provided in Table 3.1. Do *not* assume that a solvent is not flammable just because it is not listed in the table, however. In such cases, refer to the MSDSs (Sec. 1.10) or other sources to determine flammability.

c. *Do not pour flammable liquids when there are open flames within several feet.* The act of transferring the liquid from one container to another will release vapors into the laboratory, and these could be ignited by a flame some distance away.

d. *Do not pour flammable water-insoluble organic solvents into drains or sinks.* First of all, this is an environmentally unsound way to dispose of waste solvents, and second, the solvents may be carried to locations where there are open flames that could ignite them. Water-soluble solvents can be flushed down the drain if local regulations permit; consult with your instructor about this.

5. *Avoid inhaling vapors of organic and inorganic compounds.* Although most of the pleasant and unpleasant odors you encounter in everyday life are organic in nature, it is prudent not to expose yourself to such vapors in the laboratory. Work at a fume hood when handling particularly noxious chemicals, such as bromine or acetic anhydride, and, if possible, when performing reactions that produce toxic gases.

6. *Never taste anything in the laboratory unless specifically instructed to do so.* You should also never eat or drink in the laboratory, as your food may become contaminated by the chemicals that are being used.

7. *Minimize the amounts of chemicals you use and dispose of chemicals properly.* This aspect of laboratory practice is so important that we have devoted a portion of Section 1.10 to it. Read the relevant paragraphs *carefully* and consult with your instructor if there are any questions about the procedures.

1.10 SAFETY: MATERIAL SAFETY DATA SHEETS

The variety and potential danger of chemicals used in the organic chemistry laboratory probably exceed that of any laboratory course you have had. It is imperative to understand the nature of the substances with which you are working. Fortunately, the increased emphasis on the proper handling of chemicals has led to a number of publications containing information about the chemical, physical, and toxicological properties of the majority of organic and inorganic compounds used in the experiments in this textbook. A comprehensive source is *The Sigma-Aldrich Library of Chemical Safety Data* (Reference 8), and it or similar compilations should be available in your library or some other central location. The data provided by these sources are basically

Name	Ether	**Reviews and standards**	OSHA standard-air: TWA 400 ppm.
Other names	Diethyl ether	**Health hazards**	May be harmful by inhalation, ingestion, or skin absorption. Vapor or mist is irritating to the eyes, mucous membranes, and upper respiratory tract. Causes skin irritation. Exposure can cause coughing, chest pains, difficulty in breathing, and nausea, headache, and vomiting.
CAS Registry No.	60-29-7	**First aid**	In case of contact, immediately flush eyes or skin with copious amounts of water for at least 15 min while removing contaminated clothing and shoes. If inhaled, remove to fresh air. If not breathing, give artificial respiration; if breathing is difficult, give oxygen. If ingested, wash out mouth with water. Call a physician.
Structure	$(CH_3CH_2)_2O$	**Incompatibilities**	Oxidizing agents and heat.
MP	–116 °C	**Extinguishing media**	Carbon dioxide, dry chemical powder, alcohol, or polymer foam.
BP	34.6 °C (760 torr)	**Decomposition products**	Toxic fumes of carbon monoxide, carbon dioxide.
FP	–40 °C	**Handling and storage**	Wear appropriate respirator, chemical-resistant gloves, safety goggles, other protective clothing. Safety shower and eye bath. Do not breathe vapor. Avoid contact with eyes, skin, and clothing. Wash thoroughly after handling. Irritant. Keep tightly closed. Keep away from heat, sparks, and open flame. Forms explosive peroxides on prolonged storage. Refrigerate. Extremely flammable. Vapor may travel considerable distance to source of ignition. Container explosion may occur under fire conditions. *Danger:* Tends to form explosive peroxides, especially when anhydrous. Inhibited with 0.0001% BHT.
Appearance	Colorless liquid	**Spillage**	Shut off all sources of ignition. Cover with activated carbon adsorbent, place in closed containers, and take outdoors.
Irritation data	Human eye 100 ppm	**Disposal**	Store in clearly labeled containers until container is given to approved contractor for disposal in accordance with local regulations.
Toxicity data	Man, oral LDL_0 260 mg/kg		

Figure 1.3
Summary of MSDS for diethyl ether.

summaries of the information contained in the Material Safety Data Sheets (MSDSs) published by the supplier of the chemical of interest. Your instructor may be able to provide these sheets because by federal regulation an MSDS must be delivered to the buyer each time a chemical is purchased.

The information in an MSDS can be overwhelming. For example, the official MSDS for sodium bicarbonate is some six pages long. Even the summaries provided in most compilations are quite extensive, as illustrated in Figure 1.3, which contains specific data for diethyl ether. Entries regarding the structure and physical properties of the compound, including melting point (mp), boiling point (bp), and flash point (fp), are included, along with its CAS (Chemical Abstracts Service) Registry Number, which is unique for each different chemical substance (see Chap. 26 for a further discussion of CAS Registry Numbers), and RTECS (Registry of Toxic Effects of Chemical Substances) number. Further data are provided concerning its toxicity, the permissible levels set by OSHA for exposure to it in the air you breathe (time-weighted average of 400 ppm), and possible health consequences resulting from contact with the compound. For diethyl ether, the entry for "Toxicity Data" represents the *lowest* recorded *lethal* concentration for ingestion of the chemical. Valuable information is also given regarding first-aid procedures, classes of substances with which diethyl ether reacts and thus is "incompatible" with, products of its decomposition, and materials suitable for extinguishing fires involving ether. Finally, protocols for safe handling and storage are included, along with procedures for disposing and cleaning up spills of diethyl ether.

See MSDSs

Accessing MSDS information from commercial sources can be very time-consuming, although it is useful to refer to one or more of them if you need more complete MSDS information than is available at the URL associated with this textbook or if you are to use or produce a chemical that is not listed on it. We've developed the Web-based MSDSs to provide you with a rapid and convenient way to obtain important information on the chemicals you will be using or producing in the experimental procedures performed when using this textbook. The data we've provided on the website for this textbook are much more abbreviated than those in other sources, as seen in Figure 1.4. In developing our summaries of MSDS data, we've focused on just those data most relevant to your needs in the introductory organic laboratory.

We noted in the discussion of notebook formats (Sec. 1.6) that you may be required to summarize MSDS data in your laboratory book. This could be a daunting assignment, given the amount of information with which you might be faced, as illustrated in Figure 1.3. To assist you in doing this, we have provided one possible format for a summary in Figure 1.5. A summary for a particular chemical has to be provided only once and can be recorded at the end of your laboratory notebook on pages reserved for that purpose. Whenever the chemical is encountered in later experiments, you would only need to refer to the location of the summary of its MSDS information. *However,* you should reread the MSDS information so that you can continue to handle the chemical properly. This same recommendation applies if you have a file of MSDS-related printouts from the website for this textbook.

To summarize, you may think that reading about and recording data like those contained in Figures 1.3–1.5 is not a good investment of time. This is absolutely *wrong!* By knowing more about the chemicals that are used in the laboratory, you will be able to work safely and to deal with accidents, should they occur. The end result will be that you accomplish a greater amount of laboratory work and have a more valuable educational experience.

1.11 SAFETY: DISPOSAL OF CHEMICALS

The proper disposal of inorganic and organic chemicals is one of the biggest responsibilities that you have in the organic laboratory. Your actions, and those of your labmates, can minimize the environmental impact and even financial cost to your

Diethyl Ether
$C_4H_{10}O$

CAS No.	PS	Color	Odor	FP	BP	MP	d	VP	VD	Sol
60-29-7	Liquid	Colorless	Sweet	40	35	−116	0.7	442 @ 20	2.6	6.9 @ 20

Types of Hazards/Exposures	Acute Hazards/Symptoms	Prevention	First Aid/Fire
Fire	*Severe* fire hazard, *severe* explosion hazard; may form explosive peroxides; vapors or gases may ignite at distant ignition sources.	*No* flames, *no* sparks, *no* contact with hot surfaces.	Alcohol-resistant foam, carbon dioxide, regular dry chemical powder, water.
Inhalation	Central nervous system depression with drowsiness, dizziness, nausea, headache, and lowering of the pulse and body temperature.	Ventilation, local exhaust.	Remove from exposure immediately and seek medical advice.
Skin	Irritation, defatting, and drying of the skin.	Protective gloves and clothing.	Remove contaminated clothes/jewelry; thoroughly wash skin with soap and water; and seek medical advice.
Eyes	Painful inflammation.	Safety goggles.	Thoroughly flush eyes with water for several min, removing contact lenses if possible, and seek medical advice immediately.
Ingestion	Central nervous system depression with nausea, vomiting, drowsiness, dizziness; stomach may become promptly distended, which may hinder breathing.	Do *not* eat or drink in the laboratory.	Seek medical advice immediately.
Carcinogenicity	Not a known carcinogen.	**Mutagenicity**	Possible mutagen.

For more detailed information, consult the Material Safety Data Sheet for this compound.
Abbreviations: CAS No. = Chemical Abstracts Service Registry Number; PS = physical state; FP = flash point (°C); BP = boiling point (°C) @ 760 torr unless otherwise stated; MP = melting point (°C); d = density or specific gravity (g/mL); VP = vapor pressure (torr) at specified temperature (°C); VD = vapor density relative to air (1.0); Sol = solubility in water (g/100 mL) at specified temperature; N/A = not available or not applicable.

Figure 1.4
Example of MSDS data provided at the website for this textbook.

school for handling the waste chemicals that are necessarily produced in the experiments you do.

The experimental procedures in this textbook have been designed at a scale that should allow you to isolate an amount of product sufficient to see and manipulate, but

Compound	**Health Hazards, First Aid, Incompatibilities, Extinguishing Media, and Handling**
Diethyl Ether	May be harmful by inhalation, ingestion, or skin absorption. Avoid contact with eyes, skin, and clothing. In case of contact, immediately flush eyes or skin with copious amounts of water. Keep away from hot surfaces, sparks, and open flames. Extremely flammable. Vapor may travel considerable distance to source of ignition. If spilled, shut off all sources of ignition. Extinguish fire with carbon dioxide, dry chemical extinguisher, foam, or water.

Figure 1.5
Abstract of MSDS for diethyl ether.

they also involve the use of minimal quantities of reactants, solvents, and drying agents. Bear in mind, however, that minimizing the amounts of chemicals that are used is only the *first* part of an experimental design that results in the production of the least possible quantity of waste. The *second* part is to reduce the amounts of materials that you, the experimentalist, *define* as waste, thereby making the material subject to regulations for its disposal. From a legal standpoint, the laboratory worker is empowered to declare material as waste; that is, unneeded materials are not waste until you say they are! Consequently, a part of most of the experimental procedures in this textbook is reduction of the quantity of residual material that eventually must be consigned to waste. This means some additional time will be required for completion of the experiment, but the benefits—educational, environmental, and economic in nature—fully justify your efforts. The recommended procedures that should be followed are described under the heading **Wrapping It Up.**

How do you properly dispose of spent chemicals at the end of an experiment? In some cases this involves simply flushing chemicals down the drain with the aid of large volumes of water. As an example, solutions of sulfuric acid can be neutralized with a base such as sodium hydroxide, and the aqueous solution of sodium sulfate that results can safely be washed into the sanitary sewer system. However, the environmental regulations that apply in your particular community may require use of alternative procedures. *Be certain to check with your instructor before flushing any chemicals down the drain!*

For water-insoluble substances, and even for certain water-soluble ones, this option is not permissible under *any* circumstances, and other procedures must be followed. The laboratory should be equipped with various containers for disposal of liquid and solid chemicals; the latter should not be thrown in a trash can, because this exposes maintenance and cleaning personnel to potential danger, and it is environmentally unsound. The containers must be properly labeled as to what can be put in them, because it is very important for safety and environmental reasons that different categories of spent chemicals be segregated from one another. Thus, you are likely to find the following types of containers in the organic laboratory: hazardous solids, nonhazardous solids, halogenated organic liquids, hydrocarbons, and oxygenated organic liquids. Each student must assume the responsibility for seeing that her or his spent chemicals go into the appropriate container; otherwise dangerous combinations of chemicals might result and/or a much more expensive method of disposal be required.

REFERENCES

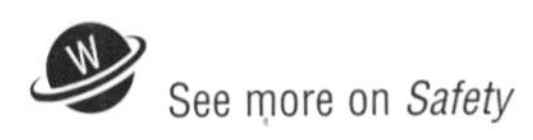
See more on *Safety*

1. Lunn, G.; Sansone, E. B. *Destruction of Hazardous Chemicals in the Laboratory,* 2nd ed., John Wiley & Sons, New York, 1994. A handbook providing procedures for decomposition of materials or classes of materials commonly used in the laboratory.
2. Committee on Hazardous Substances in the Laboratory. *Prudent Practices for Disposal of Chemicals from Laboratories,* National Academy Press, Washington, D.C., 1995. An excellent reference containing information for the minimization of waste generated in the laboratory and for the proper handling and disposal of waste chemicals, both organic and inorganic. Available at no cost online: www.nap.edu/catalog.php?record_id=4911#toc.
3. Young, J. A., ed. *Improving Safety in the Chemical Laboratory: A Practical Guide,* 2nd ed., John Wiley & Sons, New York, 1991. A book containing thorough discussions of the full range of safe practices in the laboratory.
4. Mahn, W. J. *Fundamentals of Laboratory Safety: Physical Hazards in the Academic Laboratory,* Van Nostrand Reinhold, New York, 1991.
5. Lide, D. A., ed. *CRC Handbook of Chemistry and Physics,* annual editions, CRC Press, Boca Raton, FL. Available online to subscribers: www.hbcpnetbase.com.
6. Speight, J. G., ed. *Lange's Handbook of Chemistry,* 16th ed., McGraw-Hill, New York, 2005. Available online to subscribers: knovel.com/web/portal/browse/display?_EXT_KNOVEL_DISPLAY_bookid=1347&VerticalID=0.
7. Luxon, S. G., ed. *Hazards in the Chemical Laboratory,* 16th ed., The Royal Society of Chemistry, London, 1992.
8. Lenga, R. E. and Votoupal, K. L., eds. *The Sigma-Aldrich® Library of Chemical Safety Data,* 2nd ed., Sigma-Aldrich, Milwaukee, WI, 1988.
9. O'Neil, M. J., ed. *Merck Index of Chemicals and Drugs,* 14th ed., Merck and Co., Rahway, NJ, 2006. Available online to subscribers: www.merckbooks.com/mindex/online.html.
10. Armour, M. A. *Hazardous Laboratory Chemical Disposal Guide,* 3rd ed., CRC Press, Boca Raton, FL, 2003.

HISTORICAL HIGHLIGHT

The Importance of Record Keeping

Alexander Graham Bell is a household name, whereas that of Elisha Gray, a formidable competitor of Bell's in the race to invent the telephone, is not. Therein lies a tale of the importance of accurate record keeping.

Both Bell and Gray were originally interested in inventions associated with transmitting information electrically, and Gray's accomplishments in this area prior to 1875 far exceeded those of Bell. Indeed, Gray's considerable inventive abilities had resulted in his being awarded a number of patents associated with telegraphy by 1870, and in 1872 he became a cofounder of what was to become the Western Electric Company. As early as the winter of 1866, Gray had been intrigued by the possibility of a "harmonic multiple telegraph system," which is the basis of telephonic communication because it permits transmitting tone or pitch in the form of electrical signals. However, he focused his attention on other telegraphic devices and managing the early growth of his company. It was not until 1874 that he could return to his research endeavors.

Meanwhile, Bell's lifelong interest in educating the deaf had led him toward studying the transmission of speech electrically and he, like Gray, became interested in the harmonic telegraph. His work, although

(Continued)

HISTORICAL HIGHLIGHT *The Importance of Record Keeping (Continued)*

not hindered by his having to manage an emerging company, was impeded by the demands made by his duties as a professor, tutor, and promoter of methods for teaching the deaf. Nonetheless, because of his research in telegraphy on November 23, 1874, he was able to write the following momentous words in a letter to his parents: "Please keep this paper as a *record of the conception of the idea* [emphasis our own] in case any one else should at a future time discover that the vibrations of a permanent magnet will induce a vibrating current of electricity in the coils of an electromagnet."

This concept of a "harp apparatus" to induce a variable current comprises a fundamental basis of telephonic speech, and Bell's foresight in instructing his parents to keep the *dated* letter as a record of his idea proved to be the crucial element in the legal wrangling that eventually developed between Bell and Gray as to who had precedent for this invention. Gray may have conceived of the principles for a harmonic multiple telegraph as early as the spring of 1874, but he failed to include his ideas in patent renewals he filed in January of 1875. However, he filed a patent application on February 23, 1876, that did outline his concept, predating Bell's own application by *two* days. Determining to whom the patent would be awarded, though, depended on the respective dates of conception by the two inventors. It was the written record, contained in Bell's letter to his parents, that led to his being awarded the first key patent for the harmonic multiple telegraph. Unfortunately for Gray, his claim of conceiving the same idea at an even earlier date had no corresponding documentation.

It was yet a second *dated* letter written by Bell that established his priority in conceiving the final fundamental principle for the modern telephone: variable resistance. This permits the *amplitude* as well as the pitch of a sound to be modulated electrically. Although he did not file a patent for this concept until February 14, 1876, an action again contested by Gray, Bell only three weeks later was awarded patent No. 174,465, which is generally considered the key patent for the invention of the telephone. The rest is history!

A fascinating description of the path that led Alexander Graham Bell to inventing the telephone is found online at www2.iath.virginia.edu/albell/homepage.html, and the definitive biography of his life is provided by the following reference: Robert V. Bruce, *Alexander Graham Bell and the Conquest of Solitude*, Ithaca, NY: Cornell University Press, 1973, 564 pp.

Relationship of Historical Highlight to Experiments

The importance of completely documenting the experiments you perform in the laboratory cannot be overemphasized. Writing down the observations you make and the date on which you make them may not be the key to winning litigation over patent rights, as was the case with Bell's inventions, but there is always the possibility that they will. A more likely outcome is that such documentation will demonstrate your competence in the laboratory and may result in winning the contest for the top grade in the class!

See more on *Alexander Graham Bell*
See more on *Elisha Gray*

Techniques and Apparatus

When you see this icon, sign in at this book's premium website at **www.cengage.com/login** to access videos, Pre-Lab Exercises, and other online resources.

In this chapter we introduce the basic experimental techniques and associated glassware and apparatus that are commonly used in the organic chemistry laboratory. In some instances, only the practical aspects of a particular technique are discussed in this chapter, and the theoretical principles underlying it are presented in later chapters.

It is important to skim the contents of this entire chapter, but we do not recommend that you read all parts of it in detail at this time. We make reference to specific techniques and/or apparatus as part of the procedures for many experiments; thus, when you are preparing for an experiment, *carefully* read the appropriate sections of this chapter. This should be done *prior* to entering the laboratory to perform the experimental procedure!

2.1 GLASSWARE: PRECAUTIONS AND CLEANING

Laboratory experiments in organic chemistry are commonly conducted in specialized glassware that is usually expensive. Since you are responsible for maintaining your glassware, you should follow proper procedures for safely handling and cleaning it. Failure to do so is likely to result in injury to yourself, breakage, or dirty glassware that is difficult to clean.

The cardinal rule in handling and using laboratory glassware is *Never apply undue pressure or strain to any piece of glassware.* Strained glassware may break at the moment the strain is induced, when it is heated, or even after standing for a period of time. When setting up a glassware apparatus for a particular experiment, be sure that the glassware is properly positioned and supported to avoid breakage (Sec. 2.4).

Sometimes it is necessary to insert thermometers or glass tubes into rubber stoppers or rubber tubing. *If you have to force it, do not do it!* Either make the hole slightly larger or use a smaller piece of glass. You may also lubricate the glass tube with a little glycerol prior to insertion into stoppers or tubing. When inserting a glass tube into a stopper or tubing, always grasp the glass piece as close as possible to the rubber part. It is also wise to wrap a towel around the glass tube and the rubber stopper while inserting the tube. This usually prevents a serious cut in the event the glass happens to break.

Glassware should be thoroughly cleaned *immediately* after use. Residues from chemical reactions may attack the surface of the glass, and cleaning becomes more difficult the longer you wait. Before washing glassware, it is good practice to wipe off any lubricant or grease from standard-taper ground-glass joints (Sec. 2.2) with a towel or tissue moistened with a solvent such as hexane or dichloromethane. This prevents the grease from being transferred during washing to inner surfaces of the glassware, where it may be difficult to remove. Most chemical residues can be removed by washing the glassware using a brush, special laboratory soap, and water. Acetone, which is miscible with water, dissolves most organic residues and thus is commonly used to clean glassware. In order to avoid unnecessary waste, use as little solvent as possible to do the job. Acetone should not be used to clean equipment that contains residual amounts of bromine, however, since bromoacetone, a powerful lachrymator, may form. A **lachrymator** is a chemical that adversely affects the eyes and causes crying, and may also affect the lungs and produce a burning sensation. Tear gas is an example of a lachrymator. After use, the spent wash acetone should be transferred to an appropriately marked container for disposal.

Stubborn residues sometimes remain in your glassware. These are often removable by carefully scraping the residue with a bent spatula in the presence of soap and water or acetone. If this technique fails, more powerful cleaning solutions may be required, but these must be used with great care as they are usually highly corrosive. *Do not allow these solutions to come into contact with your skin or clothing; they will cause severe burns and produce holes in your clothing.* Chromic acid, which is made from concentrated sulfuric acid and chromic anhydride or potassium dichromate, is often an effective cleaning agent, but because it is a strong oxidizing acid, it must be used with great care. When handling chromic acid, always wear latex gloves and pour it *carefully* into the glassware to be cleaned. After the glassware is clean, pour the chromic acid solution into a specially designated bottle, *not* into the sink.

Another powerful cleaning solution is alcoholic potassium hydroxide. This is prepared by putting some ethanol into the flask, adding a few pellets of solid potassium hydroxide, and warming the solution gently with swirling to facilitate dissolution. When the glassware is clean, pour the solution into a specially designated bottle, and rinse the glassware thoroughly with soap and water to complete the process. *Before using any cleaning solutions other than soap and water, consult your instructor for permission and directions concerning their safe handling and disposal.*

Brown stains of manganese dioxide that are left in the glassware can generally be removed by rinsing the apparatus with a 30% (4 *M*) aqueous solution of sodium bisulfite, $NaHSO_3$. If this fails, wash the equipment with water and then add a small amount of 6 *M* HCl. This must be done in a hood, since chlorine gas is evolved; try this technique only after obtaining permission from your instructor.

The fastest method for drying your glassware is to rinse the residual droplets of water from the flask with a small volume (<5 mL) of acetone; this acetone may be recovered as wash acetone to remove organic residues in the future. Flasks and beakers should be inverted to allow the last traces of solvent to drain. Final drying is also aided by directing a *gentle* stream of dry compressed air into the flask or beaker or by placing the glassware in a drying oven.

2.2 STANDARD-TAPER GLASSWARE FOR MINISCALE PROCEDURES

Your laboratory kit contains **standard-taper glassware** with **ground-glass joints.** Before the advent of standard-taper glassware, we had to bore corks or rubber stoppers to fit each piece of equipment, and the glass apparatus was assembled by connecting the individual pieces of equipment with glass tubing. This was not only time-consuming, but many chemicals reacted with and/or were absorbed by the

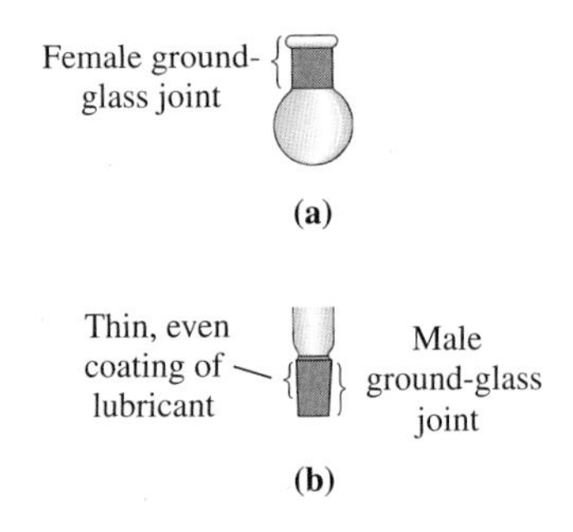

Figure 2.1
Standard-taper (₮) joints: (a) female; (b) male.

cork or rubber stoppers. Although standard-taper glassware has greatly simplified the task of assembling the apparatus required for numerous routine laboratory operations, it is expensive, so handle it carefully.

A pair of standard-taper joints is depicted in Figure 2.1. Regardless of the manufacturer, a given size of a male standard-taper joint will fit a female joint of the same size. The joints are tapered to ensure a snug fit and a tight seal. Standard-taper joints come in a number of sizes and are designated by the symbol ₮ followed by two sets of numbers separated by a slash, as in 14/20, 19/22, 24/40. The first number is the diameter of the joint in millimeters at its widest point, and the second is the length of the joint in millimeters. A standard-taper joint that is designated as ₮ 19/22 therefore has a widest diameter of 19 mm and a length of 22 mm.

When using glassware with standard-taper ground-glass joints, the joints should be properly lubricated so that they do not freeze and become difficult, if not impossible, to separate. To lubricate a joint properly, first spread a *thin* layer of joint grease around the outside of the upper half of the male joint. Then mate the two joints, and rotate them gently together to cover the surfaces of the joints with a thin coating of lubricant. *Applying the correct amount of grease to the joints is important.* If you use too much, the contents of the flask, including your product, may become contaminated with grease; if you use too little lubricant, the joints may freeze. As soon as you have completed the experiment, disassemble the glassware to lessen the likelihood that the ground-glass joints will stick. If the pieces do not separate easily, the best way to pull them apart is to grasp the two pieces as close to the joint as possible and try to loosen the joint with a *slight* twisting motion.

Sometimes the pieces of glass still cannot be separated. In these cases there are a few other tricks that can be tried. These include the following options: (1) Tap the joint *gently* with the wooden handle of a spatula, and then try pulling it apart as described earlier. (2) Heat the joint in hot water or a steam bath before attempting to separate the pieces. (3) As a last resort, heat the joint *gently* in the yellow portion of the flame of a Bunsen burner. Heat the joint slowly and carefully until the outer joint breaks away from the inner section. Wrap a cloth towel around the hot joint to avoid burning yourself, and pull the joint apart as described earlier. Consult your instructor before attempting this final option, as it is tricky.

The common pieces of standard-taper glassware with ground-glass joints for miniscale procedures are shown in Figure 2.2, and apparatus involving them are pictured throughout the textbook. The proper techniques for erecting and supporting apparatus are discussed in Section 2.4.

2.3 STANDARD-TAPER GLASSWARE FOR MICROSCALE PROCEDURES

As with miniscale procedures (Sec. 2.2), **standard-taper glassware** with **ground-glass joints** is commonly used for performing reactions at the microscale level. The apparatus is smaller, however, so the joints themselves are shorter, typically ₮ 14/10 (Fig. 2.3a).

When using glassware with standard-taper ground-glass joints, the joints should be properly lubricated so that they do not freeze and become difficult, if not impossible, to separate. To lubricate a joint properly, first spread a *thin* layer of joint grease around the outside of the upper half of the male joint. Then mate the two joints, and rotate them gently together to cover the surfaces of the joints with a thin coating of lubricant. *Applying the correct amount of grease to the joints is important.* If you use too

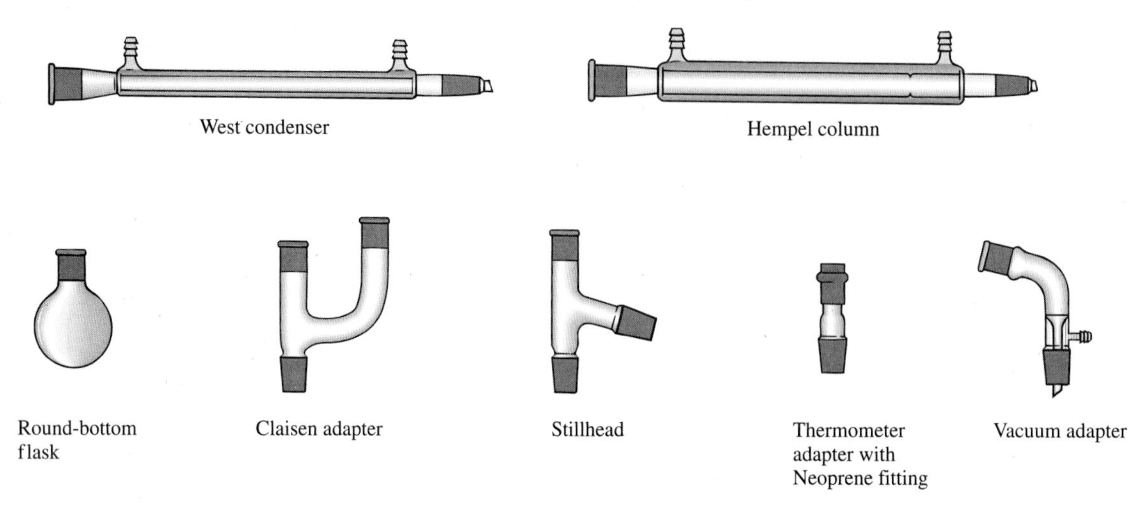

Figure 2.2
Miniscale standard-taper (₮) glassware.

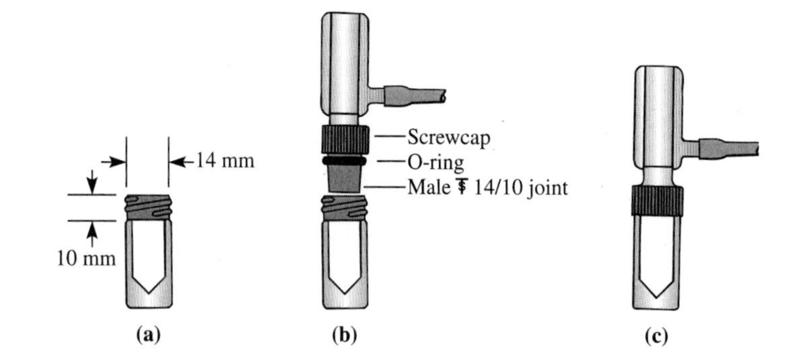

Figure 2.3
Standard-taper (₮) 14/10 joints and greaseless assembly.

much, the contents of the flask, including your product, may become contaminated with grease; if you use too little lubricant, the joints may freeze. As soon as you have completed the experiment, disassemble the glassware to lessen the likelihood that the ground-glass joints will stick. If the pieces do not separate easily, the best way to pull them apart is to grasp the two pieces as close to the joint as possible and try to loosen the joint with a *slight* twisting motion.

Sometimes the pieces of glass still cannot be separated. In these cases there are a few other tricks that can be tried. These include the following options: (1) Tap the joint *gently* with the wooden handle of a spatula, and then try pulling it apart as described earlier. (2) Heat the joint in hot water or a steam bath before attempting to separate the pieces. (3) As a last resort, heat the joint *gently* in the yellow portion of the flame of a Bunsen burner. Heat the joint slowly and carefully until the outer joint breaks away from the inner section. Wrap a cloth towel around the hot joint to avoid burning yourself, and pull the joint apart as described earlier. Consult your instructor before attempting this final option, as it is tricky.

The design of some microscale apparatus allows for *greaseless* mating of joints, and your kit of glassware may permit this option. As shown in Figure 2.3a, a female joint may have threads on the outside, which permits use of Teflon®-lined screw-caps

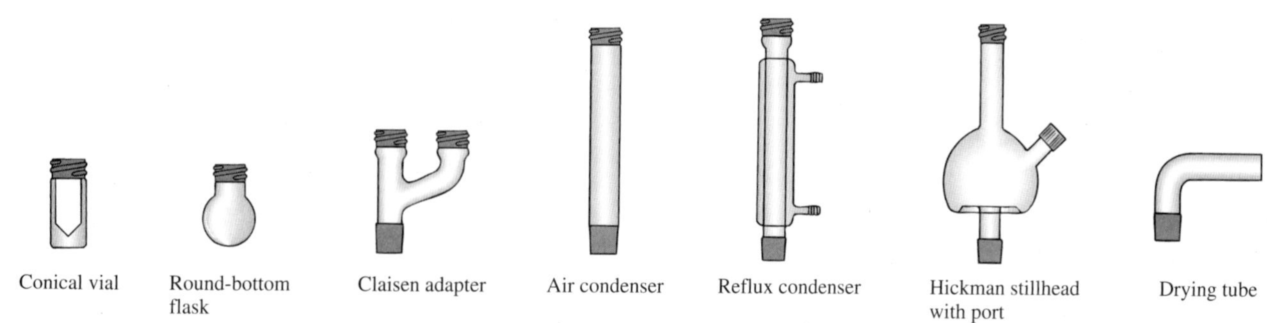

Figure 2.4
Microscale standard-taper (₹) glassware.

to join the pieces of glassware. The screw-cap contains a silicone rubber O-ring. The male joint is first inserted through the cap and O-ring (Fig. 2.3b) and then mated with the female joint. Tightening the cap compresses the O-ring, producing a gas-tight seal that holds the unit together (Fig. 2.3c). Not using grease has obvious advantages with respect to preventing contamination of reaction mixtures and simplifying clean-up of the glassware. You must disconnect greaseless joints *immediately* after completing use of the apparatus, however; otherwise, the joint will become frozen.

Various pieces of microscale glassware you may be using are depicted in Figure 2.4. A number of different assemblies using this type of glassware are portrayed in this chapter as well as others. The proper techniques for erecting and supporting apparatus are discussed in Section 2.4.

2.4 ASSEMBLING APPARATUS

To perform most of the experiments in the organic chemistry laboratory, you must first assemble the appropriate apparatus. In the sections that follow, you will be introduced to the various techniques. All experiments require certain items of glassware, various heating or cooling devices, and a means to mix the contents of a reaction vessel. The assembly of the various pieces of glassware and equipment according to the diagrams will seem awkward at first, but with practice you will soon be able to set up your own apparatus with ease.

Providing adequate support for your apparatus is paramount to preventing breakage. In general, you will accomplish this by clamping the apparatus to a ring stand or other vertical support rod using a three-finger clamp (Fig. 2.5a) or a jaw clamp (Fig. 2.5b). Plastic Keck clips (Fig. 2.5c) are then used as needed to secure the union of the glass joints. It is best to begin assembling the apparatus with the

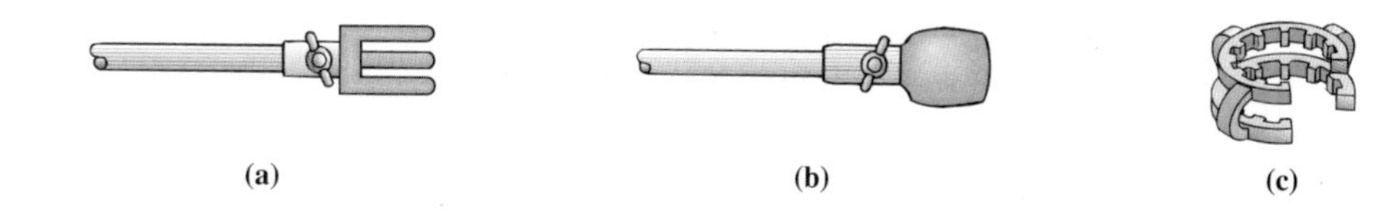

Figure 2.5
(a) Three-finger clamp. (b) Jaw clamp. (c) Keck clips.

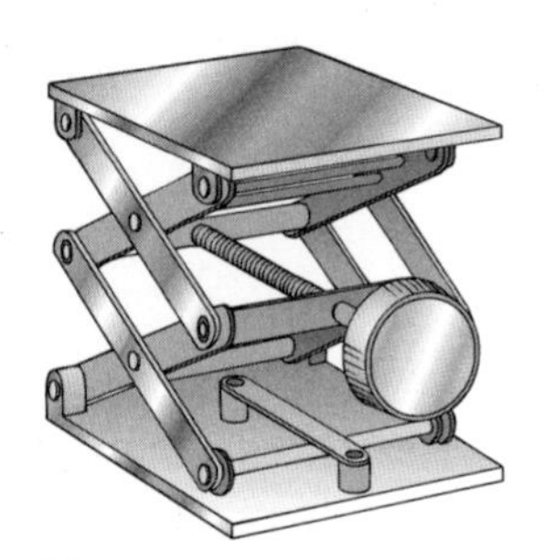

Figure 2.6
Laboratory jack.

reaction or distillation vessel and to use a single ring stand to minimize the manipulations that are required if the apparatus must be moved.

If the experimental technique requires heating or cooling the apparatus, then you should assemble the apparatus and support it about 15–25 cm above the base of the ring stand or bench top. You can then use a laboratory jack (Fig. 2.6) to raise and lower heating sources or cooling baths without moving the apparatus itself. In addition to providing this convenience, the laboratory jack allows you to lower a heating source quickly if an exothermic reaction begins to get out of control. If such lab jacks are not available, you should use an iron ring clamp to support the heating source or cooling bath; the ring clamp then may be lowered if necessary to remove the heat source or the cooling bath from the flask.

Miniscale Set-up

As a rule, when you assemble the apparatus for a miniscale experiment, you should *only* tighten the clamp to the round-bottom flask that contains your reaction or distillation mixture. The proper way to tighten a clamp is to place the neck of the flask or other piece of glassware being secured against the stationary side of the clamp, and then turn the wingnut or screw to bring the movable side of the clamp into contact with the glassware so that the piece of glassware fits snugly in the clamp. To avoid breaking the glassware, *do not overtighten the clamp.* Other clamps should be positioned to support the remainder of the apparatus and should be *loosely* tightened to avoid straining the glassware and risking breakage. Alternatively, plastic Keck clips (Fig. 2.5c) or rubber bands may be used in place of additional clamps. Proper locations of the clamps for the various laboratory set-ups are provided in the diagrams that accompany the description of the technique.

Microscale Set-up

Because the glassware required for *microscale* experiments is substantially smaller than that used in miniscale experiments, a single rigid clamp is usually sufficient to secure the apparatus. To avoid breaking the glassware, *do not overtighten the clamp.* The plastic screw-caps hold the different pieces of glassware together so additional clamping is unnecessary. Generally you will clamp a reflux condenser or a still-head by placing the glassware against the stationary side of the clamp and then turning the wingnut or screw to bring the movable side of the clamp into contact with the glassware so that the piece of glassware fits snugly in the clamp.

2.5 MEASURING AND TRANSFERRING LIQUIDS

Proper handling of liquid reagents and products is an important technique. It is particularly critical when no more than 1–2 mL of a liquid are involved. Indeed, when working at the microscale level, quantities of liquids being used often are less than 0.5 mL (500 μL). Even when larger amounts of liquids are being measured, knowing how to transfer them safely and accurately is important.

Graduated Cylinders, Beakers, Erlenmeyer Flasks, and Vials

Graduated cylinders (Fig. 2.7) are calibrated containers, which may be either glass or plastic, that are typically used to measure volumes in the range of 2–500 mL, depending on the size of the cylinder. The markings on the cylinder shaft represent the approximate volume contained when the bottom of the meniscus of the liquid is at the top of the line. The accuracy of calibration is only about ±10%, so graduated cylinders are not used for delivering precise quantities of liquids. As shown in Figure 2.7, a glass cylinder may be fitted with a plastic collar to minimize the chance of breakage if the cylinder is accidentally knocked over. You should

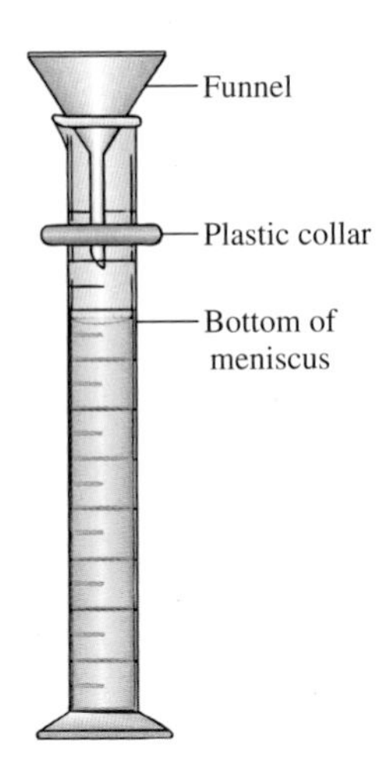

Figure 2.7
Graduated cylinder and funnel.

not fill graduated cylinders by pouring liquids directly from very large containers such as 3-L bottles. In such cases, you should first pour the liquid into a beaker, trying to estimate how much you will need so as to minimize waste, and then transfer the liquid from the beaker into the graduated cylinder. You may use a stemmed funnel to help avoid spillage.

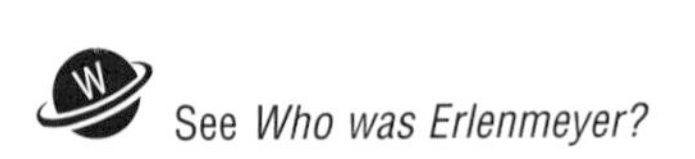

See *Who was Erlenmeyer?*

Beakers, Erlenmeyer flasks, and **conical vials** may also bear calibration marks. Keep in mind that, just as with graduated cylinders, volumetric measurements using these containers are only approximate, and other apparatus are used when more precise measurements are needed.

Pasteur Pipet

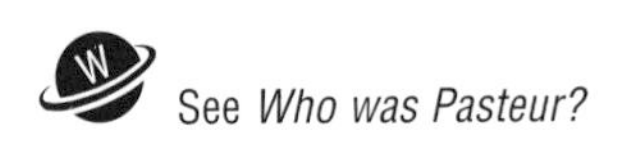

See *Who was Pasteur?*

The **Pasteur pipet** comprises a glass barrel, drawn out at one end to form a tip through which liquid is pulled with the aid of a latex suction bulb (Fig. 2.8). These pipets are commonly available in two lengths. Liquid is drawn into the pipet by first compressing the bulb and then inserting the tip of the pipet into the liquid to be transferred. To maximize the efficiency of the transfer, it is best not to compress the bulb more than is needed to draw the desired amount of liquid into the tip and barrel of the pipet. With practice, you will learn how much to compress the bulb to pull up a given volume of liquid.

Pasteur pipets are *not* used for quantitative measurements of volumes, but they can be used for qualitative measurements. This requires calibration of the pipet. Do this by first weighing, to the nearest 0.1 g, a specific amount of water (density 1.0 g/mL) into a test tube. You may also use a graduated cylinder to measure a given volume of liquid. Carefully draw the liquid into the pipet to be calibrated, and use a file to score (scratch) the pipet lightly at the level reached by the liquid (Fig. 2.8b). You may wish to place several calibration marks on the pipet, say at the 0.5-, 1-, and 1.5-mL levels. It is a good idea to calibrate several pipets at once since more than one may be needed in a given experimental procedure and because they are easily broken.

Although Pasteur pipets are often called "disposable pipets," they can be reused many times if given proper care. This involves cleaning (Sec. 2.1) and drying the pipet as soon as possible after use. When the pipet is no longer cleanable or the tip has become chipped or broken, it should be discarded in a special container labeled for broken glassware. Placing broken glassware in a wastebasket could lead to accidental injury to cleaning personnel.

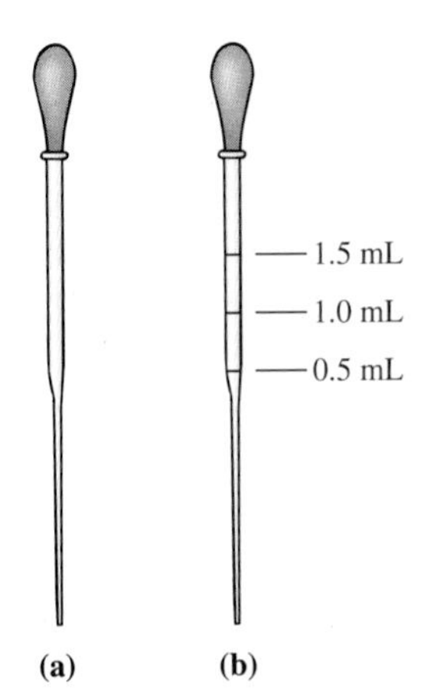

Figure 2.8
Pasteur pipets.

Modified Pasteur Pipets

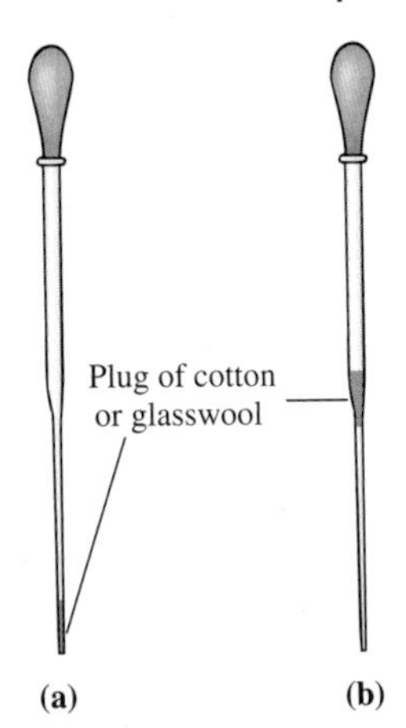

Figure 2.9
Modified Pasteur pipets:
(a) Pasteur filter-tip pipet;
(b) Pasteur filtering pipet.

The **Pasteur filter-tip pipet** (Fig. 2.9a) and **filtering pipet** (Fig. 2.9b) are two useful modifications of the Pasteur pipet. The filter-tip pipet is helpful for transferring liquids containing particularly volatile components like diethyl ether (bp 35 °C). The plug in the tip of the pipet resists the backpressure that develops when the liquid is drawn into the pipet and volatilizes. This backpressure may force the liquid out of the pipet and onto the bench. The filtering pipet has its plug at the base of the barrel rather than in the tip (Fig. 2.9b). This modification is useful for operations such as removing solids from solutions during microscale recrystallizations (Sec. 2.17).

Cotton is the material preferred for the plug because it packs more tightly than glasswool, making removal of finely divided solids more effective. However, glasswool is used when the particles to be filtered from a liquid are relatively large or when transferring or filtering strongly acidic solutions that may react with cotton.

You may prepare a filter-tip pipet by first inserting a small piece of the packing material in the top of a Pasteur pipet and using a length of heavy-gauge wire or a narrow rod to push it to the base of the barrel and then into the tip of the pipet. The second step is omitted when making a filtering pipet. For both types of pipets, the plug that results should neither be too large nor packed too tightly, because the flow of liquid into and out of the pipet may then be too slow.

Graduated Pipets

Quantitative volumetric measurements are possible with **graduated pipets,** of which there are a number of styles and sizes. Two examples are shown in Figure 2.10. The graduated "blow-out" pipet (Fig. 2.10a), sometimes labeled as a "to deliver" (TD) pipet, is calibrated so that the full volume of liquid in the pipet is dispensed only when the liquid remaining in the tip is forced out using a pipet bulb. Any drops remaining on the tip must also be added to the delivered liquid, and this is done by touching the drop to the inside of the container or to the surface of the liquid. Of course, you may measure volumes less than the maximum contained in the pipet by difference, in which case the entire contents of the pipet are *not* blown out.

The graduated pipet shown in Figure 2.10b is designed to deliver the full volume of liquid in the pipet without having to be blown out. Thus, a small amount of liquid will remain in the tip after dispensing. Proper use of this pipet requires that the bulb used to draw liquid into the pipet be removed as the contents are being delivered into a container. If there is a drop of liquid remaining on the tip, it too should be added to the dispensed liquid.

Syringes

Plastic or glass **syringes** (Fig. 2.11) are often used to deliver liquids into reaction mixtures. Syringes marked with gradations showing the amount of liquid contained in them normally have an accuracy of volumetric measurement on the order of ±5%, although the syringes originally designed for use in gas chromatography (Sec. 6.4) and those having volumes of 500 μL or less are more accurate. The

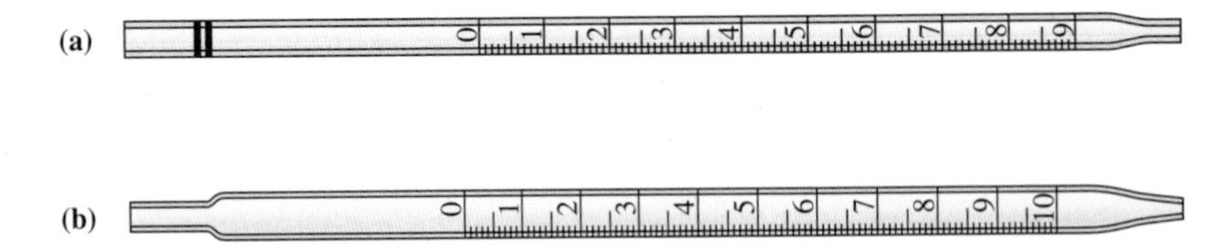

Figure 2.10
Graduated pipets:
(a) to deliver (TD);
(b) full-volume delivery.

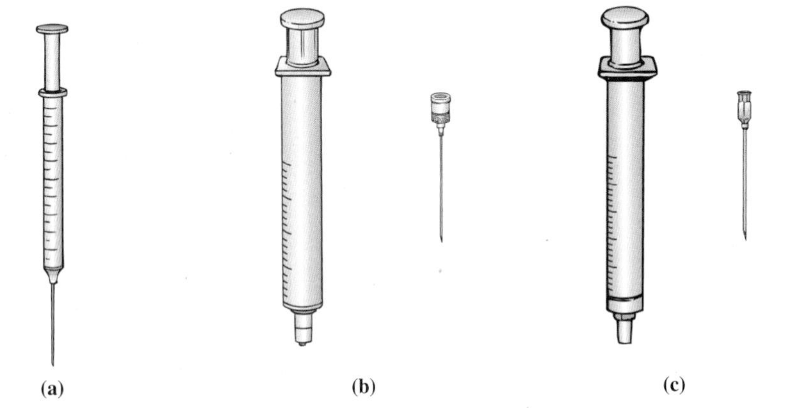

Figure 2.11
Fixed and demountable syringes: (a) fixed needle; (b) demountable, locking; (c) demountable, nonlocking.

needles on syringes are either fixed (Fig. 2.11a) or demountable (Fig. 2.11b, c). With the latter style of syringe, needles having various gauges can be affixed to the barrel. Removing and attaching the needles often require twisting the base of the needle to disengage it from or engage it to the barrel; this type of syringe is shown in Figure 2.11b. Otherwise, the needle simply slides on or off the plastic or ground-glass tip of the barrel (Fig. 2.11c).

Syringes are commonly used in the laboratory to transfer solutions or liquid reagents to reactions being performed under anhydrous conditions. The apparatus is usually fitted with a rubber septum, and the syringe needle is used to pierce the septum to deliver the liquid. You must be careful when inserting the needle into the septum to avoid bending it. The danger of bending can be minimized by supporting the sides of the needle with two fingers as you insert it into the septum.

A syringe is a sharp-pointed instrument, so you should be careful handling it. *Never* point an assembled syringe at yourself or anyone else, and do not leave needles pointed upward. You might get more than a simple needle prick as a result. You should clean a syringe immediately after use. This is done by pulling acetone or some other low-boiling solvent into the barrel and expelling it into an appropriate container. The rinsing should be repeated several times, whereupon the syringe should be dried by pumping the plunger to pull air through it or by attaching the barrel of the syringe to a water aspirator.

Dispensing Pumps and Automatic Pipets

Dispensing pumps and **automatic pipets** are devices used for accurately and quickly dispensing liquids. Because they are expensive, these devices are normally used to transfer liquids from containers, such as reagent bottles, that are being used by everyone in a laboratory.

Dispensing pumps (Fig. 2.12) are commonly used when quantities greater than about 0.5 mL must be dispensed. The plunger is made of Teflon, which allows you to use the unit with corrosive liquids and organic solvents, such as bromine in dichloromethane. The pump is loaded by pulling the plunger up as far as it will go, and the liquid that has been drawn into the glass reservoir is then expelled into the desired container by slowly depressing the plunger. Liquids of relatively low viscosity are usually expelled using only the weight of the plunger. More viscous liquids may require gently pushing the plunger down. If any liquid remains on the tip of the dispenser at the end of the transfer, the tip should be touched to the wall of the container to complete the operation.

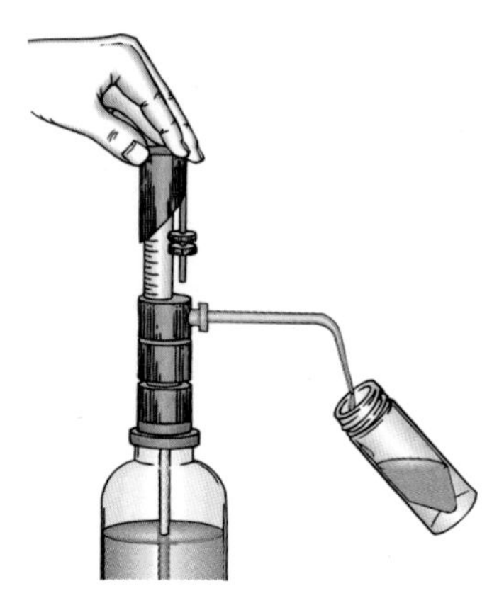

Figure 2.12
Dispensing pump.

The length the plunger travels, and thus the amount of liquid ultimately delivered, is adjustable. In cases in which this liquid is the limiting reagent in a reaction, it should be transferred into a tared container, which is then reweighed to determine the precise amount of the reagent. To ensure maximum accuracy in measuring the volume of liquid delivered by the pump, be certain that an air bubble has not developed as the liquid is drawn into the reservoir and that the spout has no air voids. If either of these conditions is not met, the proper volume of liquid will not be dispensed. You can correct these problems by raising and depressing the plunger until no air bubbles are visible; any liquid that is delivered by this operation should be collected and disposed of in the appropriate waste container.

Automatic pipets are used for accurately delivering 10–1000-μL (0.01–1-mL) quantities of liquid and are particularly valuable in microscale procedures. Although several styles of these pipets are available, their operation is basically the same. The pipet comprises a barrel, which contains a spring-loaded plunger whose length of travel may be variable, and a disposable plastic tip into which the liquid is drawn (Fig. 2.13a). There are two "detent" or stop positions on the plunger; these control the filling and dispensing stages of the liquid transfer. You *must* attach the plastic tip securely, and you should *not* totally immerse the tip in the liquid being transferred. These precautions are necessary because liquid should *never* come into contact with the barrel and plunger because this may damage the pipet. If liquid does accidentally touch these surfaces, notify your instructor immediately.

The operation of the pipet is straightforward. Holding the assembled pipet in the vertical position, depress the plunger to the *first* detent *before* inserting the tip into the liquid (Fig. 2.13a). Then insert the tip to a depth of 5–10 mm in the liquid and *slowly* release the plunger (Fig. 2.13b). If you allow the plunger to snap back to its original position, liquid may splash into the barrel of the pipet and damage the mechanism. Finally, dispense the liquid by placing the plastic tip, still held in a vertical position, against the inner wall of the tilted receiving vessel and slowly push

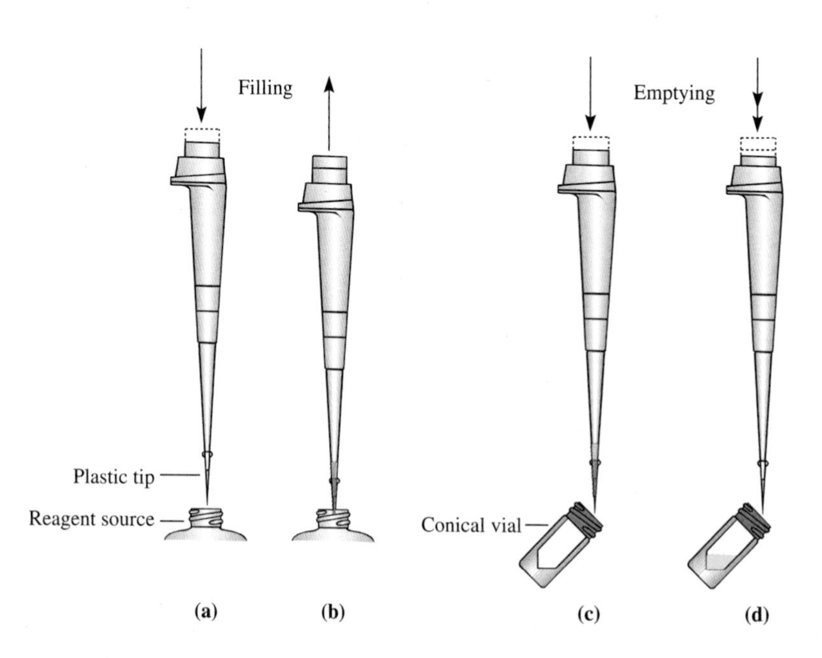

Figure 2.13
Automatic pipet: (a) depressing plunger to first detent; (b) releasing plunger to fill tip; (c) redepressing plunger to first detent to deliver liquid; (d) depressing plunger to second detent to complete delivery.

the plunger down to the *first* detent (Fig. 2.13c); a second or two later, further depress the plunger to the *second* detent to expel the final drop of liquid from the plastic tip (Fig. 2.13d). Remove the pipet from the receiver and follow the directions of your instructor with respect to any further steps to be taken with the pipet and its tip.

2.6 WEIGHING METHODS

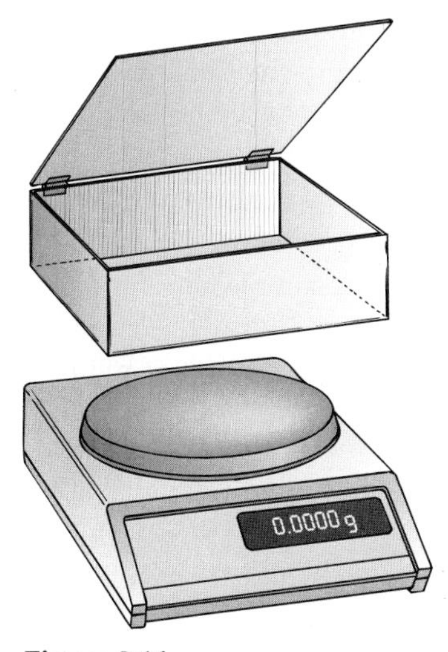

Figure 2.14
A top-loading balance with draft shield.

It will be necessary to measure quantities of reagents and reactants for the reactions you will perform. The techniques for measuring *volumes* of liquids are discussed in Section 2.5, and the *weight* of a liquid whose density is known is often most conveniently measured by transferring a known volume of the liquid to the reaction flask, using a graduated pipet or a syringe. This is especially true when the amount of a liquid to be weighed is less than 200–300 mg. However, for some liquids and all solids, weights are usually determined using a suitable **balance.** For quantities greater than 0.5 g (500 mg), a top-loading balance that reads accurately to the nearest 0.01 g (10 mg) is usually adequate. When performing reactions on the microscale, it is necessary to use a top-loading balance that has a draft shield and reads to the nearest 0.001 g (1 mg) (Fig. 2.14) or an analytical balance (Fig. 2.15) that reads to the nearest 0.0001 g (0.1 mg).

With modern electronic balances equipped with a taring device, it is possible to subtract the weight of a piece of paper or other container automatically from the combined weight to give the weight of the sample itself directly. For example, to weigh a liquid, tare a vial and remove it from the balance. Transfer the liquid from the reagent bottle to the vial with a pipet or a graduated pipet or syringe, and reweigh the vial. Be careful to avoid getting liquid on the outside of the vial or the balance pan. *Clean any spills promptly.*

To weigh a solid, place a piece of glazed weighing paper on the balance pan and press the tare device; the digital readout on the balance will indicate that the paper has "zero" weight. You should *never* weigh solids directly onto the balance pan. Using a microspatula, transfer the solid to be measured from its original container to the weighing paper until the reading on the balance indicates the desired weight. Do not

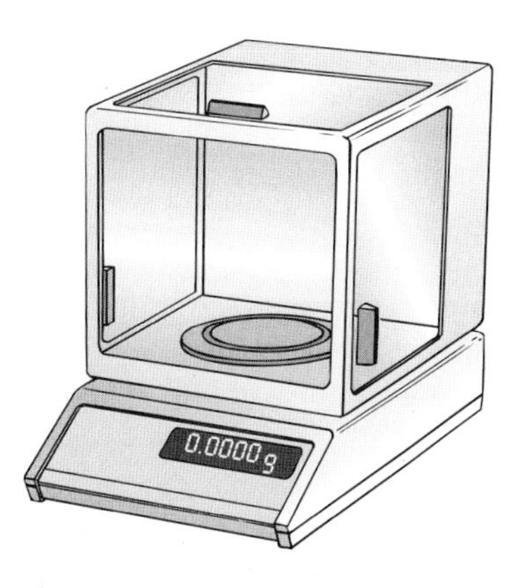

Figure 2.15
An analytical balance with draft shield.

pour the solid directly from the original bottle, because spills are more likely. Carefully transfer the solid to your reaction flask or other container. You may also weigh the solid directly into the reaction vessel by first taring the flask or conical vial. *Clean any spills promptly.*

2.7 MELTING-POINT METHODS AND APPARATUS

See video on *Melting Point Methods*

The laboratory practices and apparatus used to determine **melting points** of solids are discussed in this section, and the theory and use of melting points are described in Section 3.3. The task of determining the melting point of a compound simply involves heating a small amount of a solid and determining the temperature at which it melts. Many different types of heating devices can be used, but most utilize a capillary tube to contain the sample so that only a small amount of the sample is required.

Capillary Tubes and Sample Preparation

The first step in determining a melting point is transferring the sample into a melting-point capillary tube. Such tubes have one sealed end and are commercially available. The proper method for loading the sample into the capillary tube is as follows. Place a small amount of the solid on a clean watchglass and press the open end of the tube into the solid to force a small amount of solid (about 2–3 mm in height) into the tube (Fig. 2.16a); this operation should *not* be performed on filter paper, because fibers of paper as well as the solid may be forced into the tube. Then take a piece of 6–8-mm tubing about 1 m long, hold it vertically on a hard surface such as the bench top or floor, and drop the capillary tube down the larger tubing several times with the sealed end *down* (Fig. 2.16b). This packs the solid sample at the closed end of the capillary tube.

Melting-Point Determination

The melting point of the crystalline solid is determined by heating the packed capillary tube until the solid melts. Some representative devices for measuring melting points are presented in Figures 2.17–2.19. The most reproducible and accurate results are obtained by heating the sample at the rate of about 1–2 °C/min to ensure that heat is transferred to the sample at the same rate as the temperature increases and that the mercury in the thermometer and the sample in the capillary tube are in thermal equilibrium.

Many organic compounds undergo a change in crystalline structure just before melting, perhaps as a consequence of release of the solvent of crystallization. The solid takes on a softer, "wet" appearance, which may also be accompanied by shrinkage of the sample in the capillary tube. These changes in the sample should *not* be interpreted as the beginning of the melting process. Wait for the first tiny drop of liquid to appear.

Melting usually occurs over a range of a degree, perhaps slightly more. Accordingly, a **melting-point range** of a compound is typically reported with the lower temperature being that at which the first tiny drop of liquid appears and the higher temperature is that at which the solid has completely melted.

Melting-Point Apparatus

A simple type of melting-point apparatus is the **Thiele tube,** shown in Figure 2.17a. This tube is shaped such that the heat applied to a heating liquid in the sidearm by a burner is distributed evenly to all parts of the vessel by convection currents, so

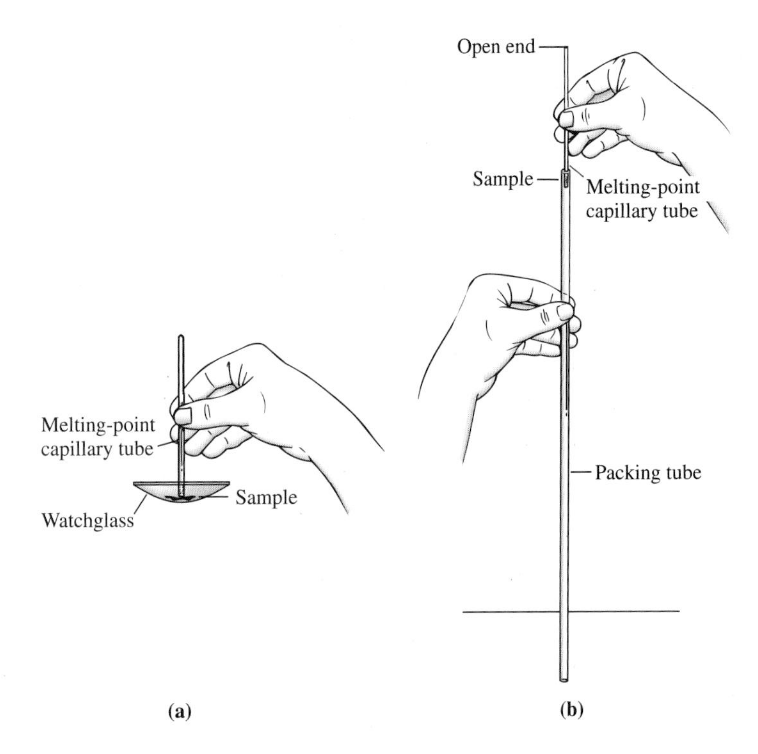

Figure 2.16
(a) Filling a melting-point capillary tube. (b) Packing the sample at the bottom of the tube.

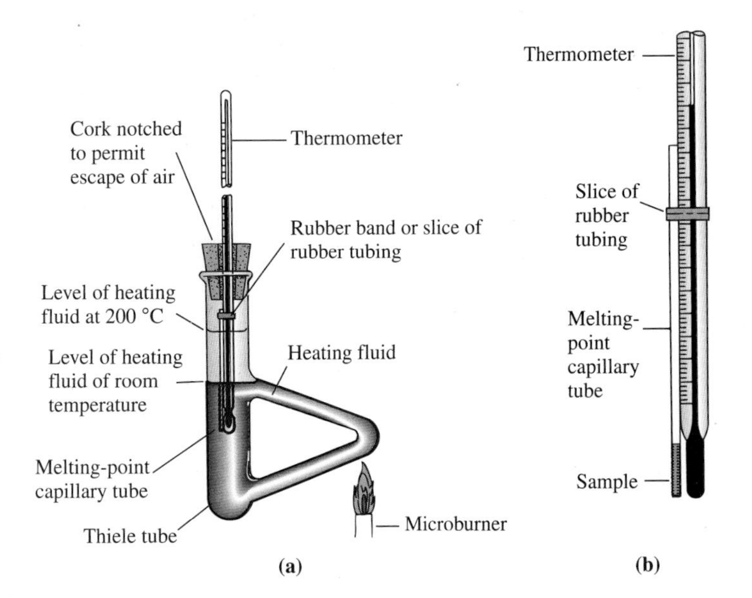

Figure 2.17
(a) Thiele melting-point apparatus. (b) Arrangement of sample and thermometer for determining melting point.

stirring is not required. Temperature control is accomplished by adjusting the flame produced by the microburner; this may seem difficult at first but can be mastered with practice.

Proper use of the Thiele tube is required to obtain reliable melting points. Secure the capillary tube to the thermometer at the position indicated in Figure 2.17b using either a rubber band or a small segment of rubber tubing. Be sure that the

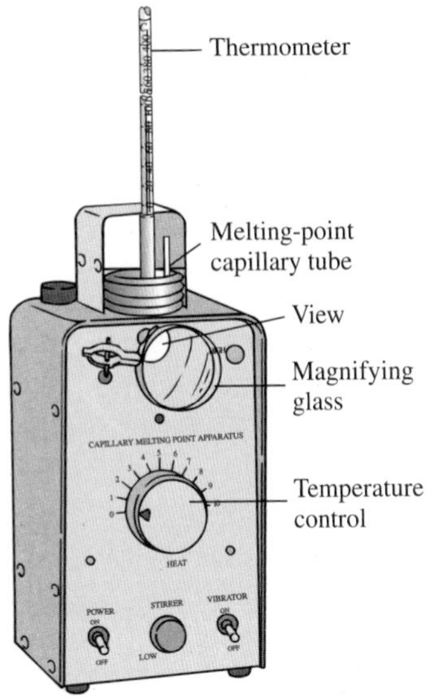

Figure 2.18
Thomas-Hoover© melting-point apparatus (courtesy of Arthur H. Thomas Company).

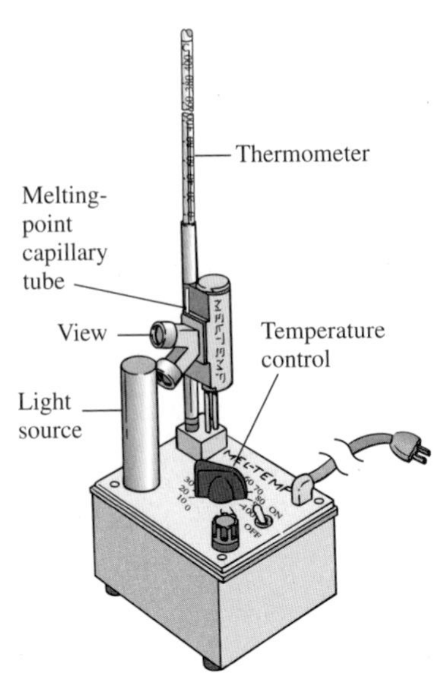

Figure 2.19
Mel-Temp© melting-point apparatus (courtesy of Laboratory Devices).

band holding the capillary tube on the thermometer is as close to the top of the tube as possible. Now support the thermometer and the attached capillary tube containing the sample in the apparatus either with a rubber stopper cork, as shown in Figure 2.17a, or by carefully clamping the thermometer so that it is immersed in the oil. The thermometer and capillary tube must *not* contact the glass of the Thiele tube. Since the oil will expand on heating, make sure that the height of the heating fluid is approximately at the level indicated in Figure 2.17a and that the rubber band is in the position indicated. Otherwise, the hot oil will come in contact with the rubber, causing the band to expand and loosen; the sample tube may then fall into the oil. Heat the Thiele tube at the rate of 1–2 °C/min in order to determine the melting point. The maximum temperature to which the apparatus can be heated is dictated by the nature of the heating fluid, a topic that is discussed in Section 2.9.

The Thiele tube has been replaced in modern laboratories by various **electric melting-point devices,** which are much more convenient to use. One common type of electric melting-point apparatus is the Thomas-Hoover melting-point unit shown in Figure 2.18. This particular unit has a built-in vibrating device to pack the sample in the capillary tube, and it also allows for the determination of the melting points of up to five samples simultaneously. The oil bath in this unit is electrically heated and stirred. An electrical resistance heater is immersed in a container of silicone oil. The voltage across the heating element is varied by turning the large knob in the front of the apparatus so that the oil is heated at a slow, controlled rate. A motor drives a stirrer in the oil bath to ensure even heating; the rate of stirring is controlled by a knob at the bottom of the unit. Some models are equipped with a movable magnifying lens system that gives the user a better view of the thermometer and the sample in the capillary tube. The capillary tube containing the sample is inserted into the apparatus as illustrated in Figure 2.18.

The Mel-Temp® apparatus shown in Figure 2.19 is another electrical unit that utilizes a heated metal block rather than a liquid for transferring the heat to the capillary tube. A thermometer is inserted into a hole bored into the block, and the thermometer gives the temperature of the block and the capillary tube. Heating is accomplished by controlling the voltage applied to the heating element contained within the block.

2.8 BOILING-POINT METHODS AND APPARATUS

There are several techniques that may be used to determine the **boiling point** of a liquid, depending upon the amount of material available. When multigram quantities are available, the boiling point is typically determined by reading the thermometer during a simple distillation, which is described in Section 2.13. However, for smaller amounts of liquid there is sometimes not enough sample to distill, so other techniques have been developed. Two of these are described here.

Miniscale Method

An accurate boiling point may be determined with as little as 0.5–1.0 mL of liquid using the method illustrated in Figure 2.20. Working at a hood, place the liquid in a long, narrow Pyrex® test tube, and add a small, black carborundum boiling stone; do not use a white marble chip, as bumping is more likely. Clamp the test tube and position the thermometer about 2 cm above the level of the liquid using a second clamp. Bring the liquid rapidly to a vigorous boil using a suitable heating device suggested by your instructor. You should see a reflux ring move up the test tube, and drops of the liquid should condense on the walls of the test tube. Control the amount of heating so the liquid does not boil out of the test tube. Be sure the bulb of the thermometer is fully immersed in the vapor of the boiling liquid long enough to allow the equilibration required for a good temperature reading to be obtained.

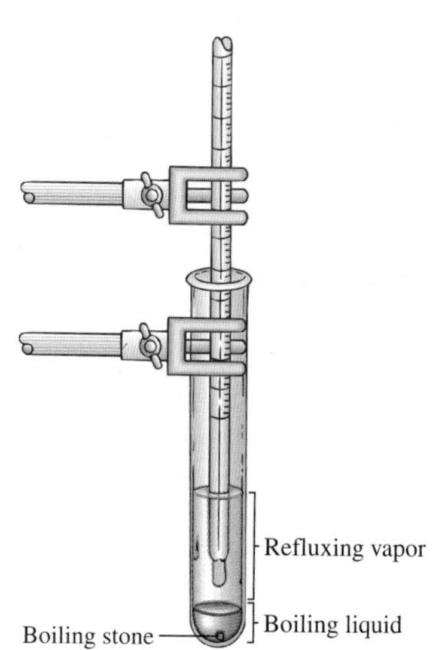

Figure 2.20
Miniscale technique to determine the boiling point of a liquid.

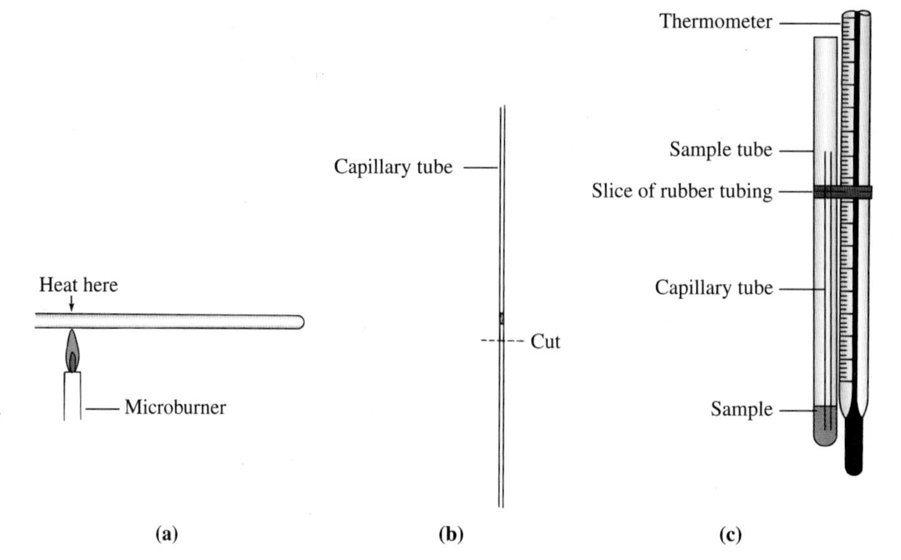

Figure 2.21
Micro boiling-point apparatus: (a) fusing a capillary tube; (b) joining capillary tubes and cutting off one end; (c) assembling micro boiling-point apparatus with correct placement of ebullition and sample tubes, and thermometer.

Microscale Method

The boiling point of a liquid may be determined on amounts less than 0.5 mL by constructing a simple **micro boiling-point** apparatus as follows. First, prepare a capillary ebullition tube by taking a standard melting-point capillary tube, which is already sealed at one end, and making a seal in it about 1 cm from the open end, using a hot flame (Fig. 2.21a). Alternatively, two melting-point capillary tubes can be joined by heating the closed ends in a hot flame; a clean cut is then made about 1 cm below the point where the tubes have been joined (Fig. 2.21b). Second, seal a piece of 4–6-mm glass tubing at one end using a hot flame and cut it to a length about 1 cm longer than the capillary ebullition tube. These tubes are prepared most easily by using a glassblowing torch, and they may be provided by your instructor.

Attach the 4–6-mm tube to a thermometer with a rubber ring near the top of the tube. The bottom of the tube should be even with the mercury bulb of the thermometer. Put the capillary ebullition tube into the larger glass tube, and with a Pasteur pipet add the liquid whose boiling point you wish to determine until the level of the liquid is about 2 mm above the seal of the capillary tube (Fig. 2.21c).

Immerse the thermometer and the attached tubes in a heating bath. A Thiele tube (Fig. 2.17) is convenient for this purpose, but other heating baths can be used. *Be sure that the rubber ring is well above the level of the oil in the heating bath.* Heat the oil bath at the rate of about 5 °C/min until a *rapid* and *continuous* stream of bubbles comes out of the capillary ebullition tube. Before this occurs, you may see some bubbles form in an erratic fashion. This is not caused by boiling; rather, these bubbles are due to the expansion of air trapped in the capillary tube. You should see a marked change from the slow evolution of air bubbles to the rapid evolution of bubbles resulting from the boiling action of the liquid as the boiling point is reached. *However, this is* ***not*** *the boiling point!* Remove the heating source and allow the bath to cool slowly. As the liquid starts to rise into the capillary tube, note the temperature measured by the thermometer; *this is the boiling point of the liquid.* If the liquid rises sufficiently slowly into the capillary tube, note the temperatures at which the liquid starts to rise and at which the capillary tube is full. This will be the **boiling-point**

range of the liquid. Remove the capillary ebullition tube and expel the liquid from the small end by gently shaking the tube. Replace it in the sample tube and repeat the determination of the boiling point by heating the oil bath at the rate of 1–2 °C/min when you are within 10–15 °C of the approximate boiling point as determined in the previous experiment. Observed boiling points may be reproduced to within 1 or 2 °C.

The physical basis of this technique is interesting. Before the liquid is heated, the capillary tube is filled with air. As the bath is heated, the air in the capillary tube is driven out and replaced with the vapor of the liquid. When the apparatus is heated until vigorous boiling of the liquid is observed, the actual boiling point of the liquid has been exceeded, and the air in the capillary tube has been completely replaced by the vapor of the liquid. On cooling, the vapor pressure of the liquid becomes equal to the external pressure, thus allowing the liquid to rise into the capillary tube. The temperature at which this occurs is, by definition, the boiling temperature of the liquid (Sec. 4.1).

2.9 HEATING METHODS

Heating is an important laboratory technique that is used in a variety of situations. For example, many chemical reactions require heating to proceed at a reasonable rate. Heating is also used to purify liquids by distillation, to remove volatile solvents during the work-up of a reaction, and to dissolve solids when purifying solid products by recrystallization.

Two general rules regarding heating are noted here. (1) Whatever device is being used to heat a liquid or solid, you must arrange the apparatus so that the heating source can be *rapidly* removed to prevent accidents that may occur by overheating. This normally means that the heating source is mounted on a ring clamp or a lab jack, either of which allows for quick removal of the device if necessary. (2) As a rule, the *safest* way to heat organic solvents is with a *flameless* heat source in a hood. This practice not only minimizes the chance of fire, but it also avoids filling the room with solvent vapors.

Heating with electrical devices is generally the method of choice in the organic laboratory, since it is much safer than using open flames. These devices are usually comprised of two essential components: (1) a resistance element that converts electrical into thermal energy, and (2) a variable transformer for controlling the voltage across the element. They differ, however, in the medium used to transfer heat from the element to the experimental apparatus. For example, air and oil are the heat transfer agents for a heat gun and an oil bath, respectively.

Although we describe below how burners are properly used for heating purposes, the experimental procedures in this textbook focus on *flameless* heating devices. Burners can be substituted for such devices in most cases, but not all. *Consult with your instructor if you have any questions regarding the suitability of substituting flames for other heating devices.*

Burners

Most chemistry laboratories are supplied with natural gas to fuel various types of burners. A burner provides the convenience of a rapid and reasonably inexpensive source of heat. However, many organic substances, especially solvents such as ether and hexane, are highly flammable, and you should always exercise

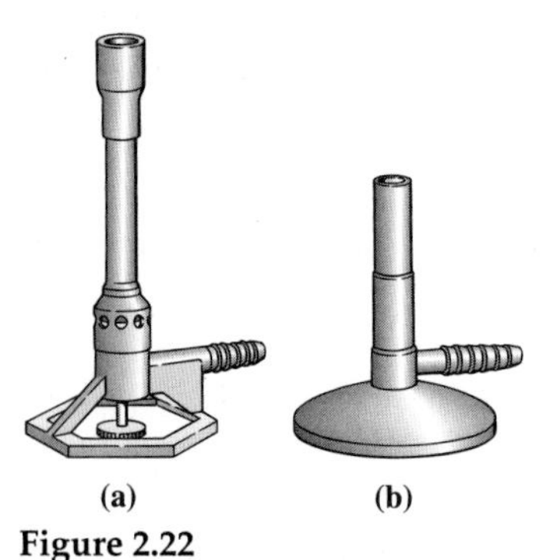

(a) (b)

Figure 2.22
Laboratory burners: (a) Bunsen burner; (b) microburner.

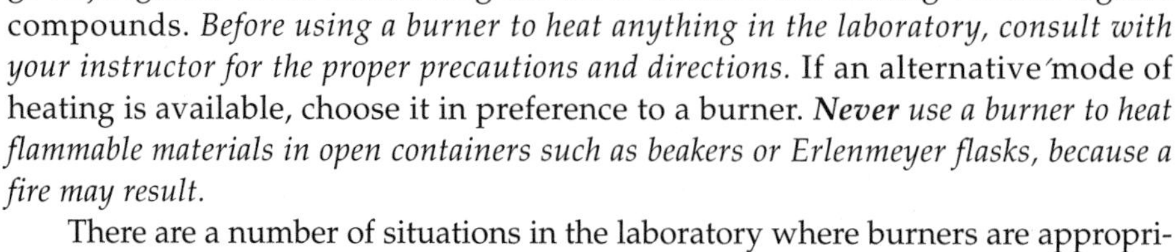

good judgment when considering the use of a burner for heating volatile organic compounds. *Before using a burner to heat anything in the laboratory, consult with your instructor for the proper precautions and directions.* If an alternative mode of heating is available, choose it in preference to a burner. ***Never** use a burner to heat flammable materials in open containers such as beakers or Erlenmeyer flasks, because a fire may result.*

There are a number of situations in the laboratory where burners are appropriate heating sources. They may be used to heat Thiele tubes in the determination of melting or boiling points of organic substances and to heat a water bath to obtain and maintain temperatures from ambient to about 80 °C. When volatile solvents are not being used in the laboratory, burners may be safely substituted for heat guns to dry apparatus so a reaction may be conducted under anhydrous conditions (Sec. 2.26). On occasion, you will need a burner to bend glass tubing or to fashion a piece of glass apparatus. Burners can also be used to heat aqueous solutions that do not contain flammable substances or to heat higher-boiling liquids that are completely contained in round-bottom flasks either fitted with a reflux condenser (Sec. 2.22) or equipped for distillation as discussed in Sections 2.13–2.16. In these instances, it is important to lubricate the joints of the apparatus with a hydrocarbon or silicone grease to minimize the danger of leaking vapors and reduce the likelihood of the joints freezing.

You must be aware of what others are doing in the laboratory. Although you might be using a burner to perform a completely safe operation, someone nearby may be working with a very volatile, flammable solvent, some of which can creep along the bench top for several feet! These vapors or others in the room may be ignited explosively by an open flame.

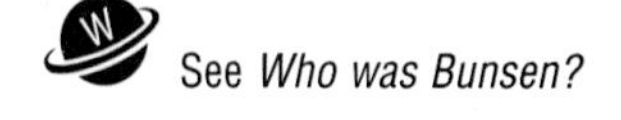

See *Who was Bunsen?*

Two common types of laboratory burners are pictured in Figure 2.22. The classic **Bunsen burner,** named after its inventor, is shown in Figure 2.22a. The needle valve at the bottom of the burner serves as a fine adjustment of the gas flow, and turning the barrel of the burner regulates the air flow; adjustment of gas and air flow provides control of the flame. In the **microburner** in Figure 2.22b, the air flow is adjusted at the baffle at the bottom of the burner, and the gas flow is adjusted at the gas valve on the laboratory bench.

Heating a flask with a burner may produce "hot spots" if most of the heat is applied to a small area on the bottom of the flask. Hot spots can lead to severe bumping, since the heat must be dispersed throughout the liquid by convection or by means of the turbulence caused by boiling. Hot spots can easily be avoided by holding the burner and slowly moving the flame over the bottom of the flask. Alternatively, a piece of wire gauze, which diffuses the heat reaching the flask, may be placed between the flame and the flask; the gauze is supported with an iron ring.

Heating Mantle

A common device for heating round-bottom flasks is a **heating mantle.** The mantle is typically fashioned either from a blanket of spun fiberglass partially covered with a flexible or rigid cover (Fig. 2.23a) or from a ceramic core contained in an aluminum housing (Fig. 2.23b). In either case, heat is provided by an electrical resistance coil embedded in the fiberglass or ceramic core, so when using these devices it is important not to spill liquid on them. The mantle may be equipped with a thermocouple so that its internal temperature can be monitored, but this is seldom done in undergraduate laboratories.

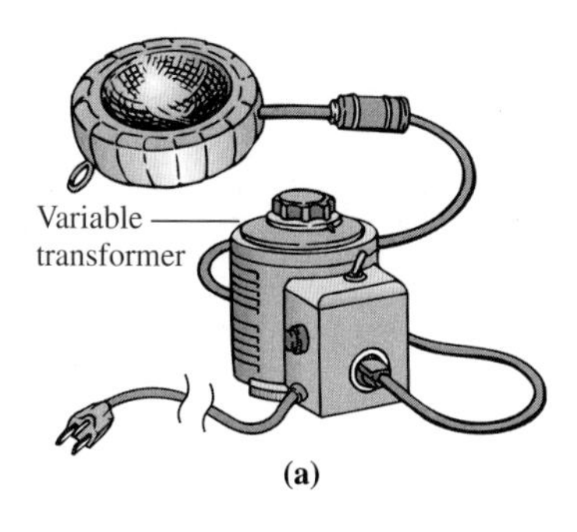

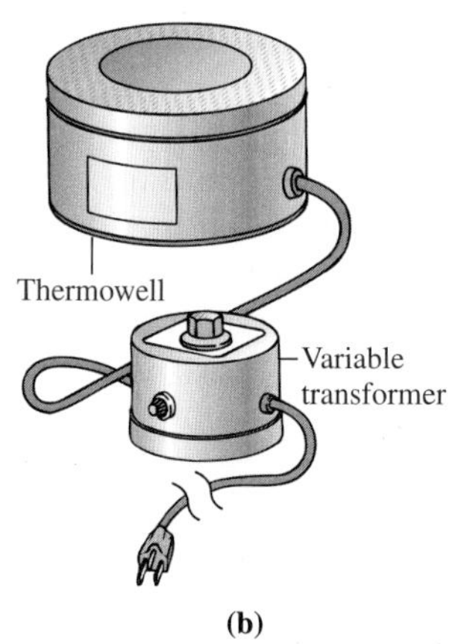

Figure 2.23
(a) Woven-glass heating mantle. (b) Heating mantle with ceramic core.

Because heating mantles are constructed of nonferrous materials, they can be used in conjunction with magnetic stirring (Sec. 2.11), so simultaneous heating and stirring of a reaction mixture is possible. A variable transformer connected to an electrical outlet provides the power for heating the mantle. The cord from the mantle itself must *always* be plugged into a transformer, *never* directly into a wall outlet!

Most heating mantles have a hemispherical cavity into which the round-bottom flask is fitted. With those mantles having a spun fiberglass interior, a different size is required for each size of flask. However, if the hemisphere is a rigid ceramic material, it is possible to put a granular heat-transfer agent such as sand in the cavity so that flasks of sizes smaller than the diameter of the cavity can be accommodated (see "**Sand Bath,**" p. 47).

There are some drawbacks to using heating mantles. They heat up rather slowly, and it is difficult to obtain a given temperature or maintain a constant temperature. Heating mantles have a high heat capacity, so if it becomes necessary to discontinue heating suddenly (for example, if a reaction begins to get out of control), *the heating mantle must be removed immediately from below the flask* to allow the flask to cool, either on its own or by means of a cooling bath; it is not sufficient simply to lower the voltage or turn off the electricity. After the mantle is removed, the electricity should be turned off at the transformer.

The amount of heat supplied by a heating mantle is moderated by the boiling liquid contained in the flask, because the hot vapors of the liquid transfer heat away from the mantle. If the flask becomes dry or nearly so during a distillation, the mantle can become sufficiently hot to melt the resistance wire inside, thus causing the mantle to "burn out." *Do not heat an empty flask with a heating mantle.* Moreover, most mantles are marked with a maximum voltage to be supplied, and this should not be exceeded.

Oil Bath

Electrically heated **oil baths,** which typically contain either mineral oil or silicone oil, are commonly employed in the laboratory. These baths are heated either by placing the bath container on a hot plate or by inserting a coil of Nichrome resistance wire in the bath. In the latter case, the resistance wire serves as the heating element and is attached to a transformer with an electrical cord and plug (Fig. 2.24). Thermal equilibration within the bath is maintained by placing a paper clip or stirbar in the bath and using a magnetic stirrer (Sec. 2.11).

Heating baths offer several important advantages over heating mantles. The temperature of the bath can be easily determined by inserting a thermometer in the liquid, and a given bath temperature may be obtained and accurately maintained by careful adjustment of the variable transformer. Although heat is transferred uniformly to the surface of the flask in the bath and there are no hot spots, there typically will be a temperature gradient of about 10 °C between the bath and the contents of the flask.

Some inconveniences are also encountered using oil baths. If the volume of heating liquid is fairly large, it may take a while to reach the desired bath temperature. The maximum temperature that may be safely attained in an oil bath is limited by the type of heating liquid being used. Silicone oils are more expensive but are generally preferable to mineral oils because they can be heated to 200–275 °C without reaching the **flash point,** the temperature at which a liquid can burst into flame, and without thickening through decomposition. Mineral oil should *not* be heated above about 200 °C because it will begin to smoke, and

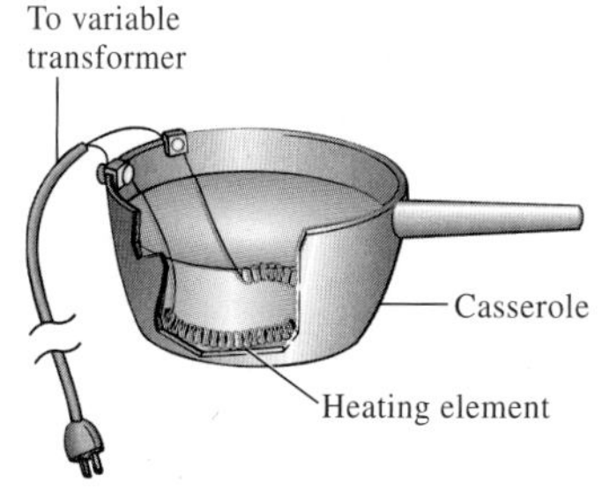

Figure 2.24
Electrically heated oil bath.

there is the potential danger of flash ignition of the vapors. Water must *not* be present in mineral and silicone oils, since at temperatures of about 100 °C, the water will boil, spattering hot oil. If water drops are present in the oil, change the heating fluid and clean and dry the container before refilling it.

A minor nuisance associated with oil baths is removing the film of mineral or silicone oils, both of which are water-insoluble, from the outer surface of the flask. This is best done by wiping the flask using a *small* amount of hexane or dichloromethane on a paper towel prior to washing the flask with soap and water.

Hot Plate

When flat-bottom containers such as beakers or Erlenmeyer flasks must be heated, convenient heat sources are **hot plates** (Fig. 2.25a) and **stirring hot plates** (Fig. 2.25b), which are hot plates with built-in magnetic stirrers; round-bottom flasks cannot be heated effectively with hot plates. The flat upper surface of the hot plate is heated by electrical resistance coils to a temperature that is controlled by a built-in voltage regulator, which is varied by turning a knob on the front of the unit. A hot plate generally should be limited to heating liquids such as water, mineral or silicone oil, and nonflammable organic solvents such as chloroform and dichloromethane. *Under no circumstances should a hot plate be used to boil highly flammable organic solvents,* because the vapors may ignite as they billow onto the surface of the hot plate or the electrical resistance coils inside the hot plate. Furthermore, many hot plates use a relay that turns the electricity on and off to maintain the desired temperature; because these relays are often not explosion-proof, they may produce sparks that can ignite fires.

Aluminum Block

An aluminum block is a convenient accessory used in conjunction with hot plates for reactions being performed in a microscale apparatus. The block has indentations of various sizes and shapes to accommodate conical vials and flasks (Fig. 2.26). There may also be a receptacle for inserting the bulb of a glass thermometer or the base of

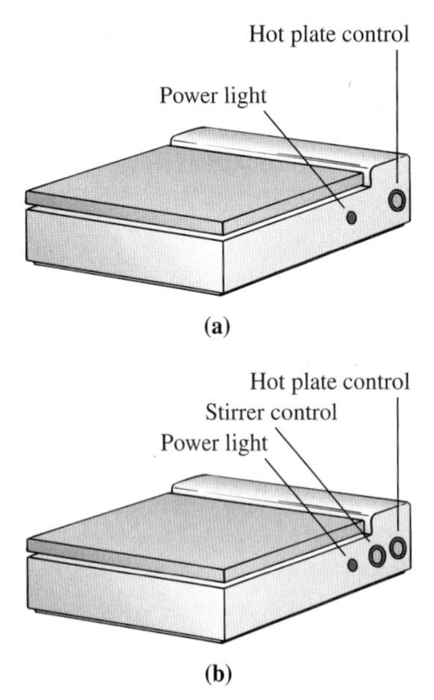

Figure 2.25
(a) Hot plate. (b) Stirring hot plate.

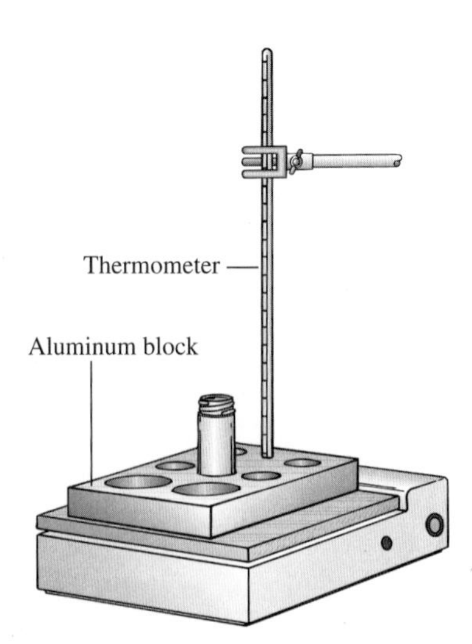

Figure 2.26
Aluminum block with conical vial in place.

a metal dial thermometer, so the temperature of the block can be monitored. Although a glass thermometer may remain vertical without any further support, it should be held in place by a clamp or a wire to prevent breakage. It is not necessary to clamp a metal dial thermometer.

The temperature of the block is controlled by varying the power to the hot plate, and it is helpful to have a correlation between the two. You may calibrate the block by selecting four or five evenly spaced settings on the control dial of the hot plate, determining the temperature of the aluminum block at each setting, and recording both the setting and the corresponding temperature in your notebook for future reference.

Heat Gun

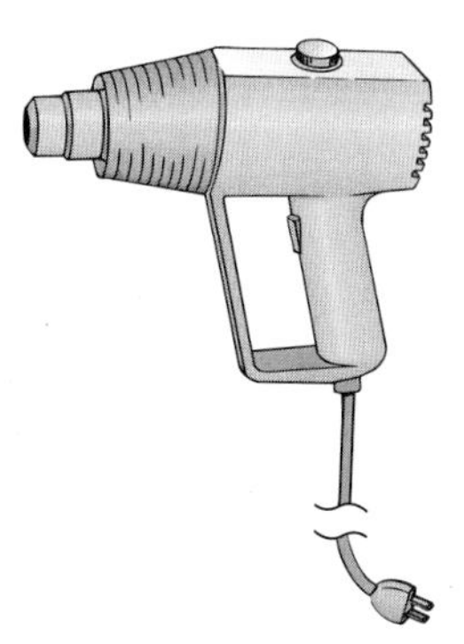

Figure 2.27
Heat gun.

A useful device for heating apparatus, but not reaction mixtures, is a heat gun (Fig. 2.27), which is essentially a high-powered hair dryer. It is particularly useful for drying assembled apparatus, as this eliminates the need to use a flame. The stream of hot air from the heat gun is directed at the apparatus to drive atmospheric moisture out through an open port, which, after the entire apparatus is heated, is then fitted with a drying tube (Sec. 2.27) so that air is dried as it enters the cooling apparatus. A slow stream of dry nitrogen gas may be introduced during the heating process to help expel moisture from the apparatus. A drying tube or some other device such as a bubbler containing mineral oil must then be attached to the apparatus to prevent intrusion of atmospheric moisture as the apparatus cools and during the course of the reaction itself. For best results, you should start heating of the apparatus at a point most distant from the open port and work toward the opening.

The temperature of the air stream emerging from a heat gun is commonly from 300 to 500 °C, far higher than that of the typical hand-held hair dryer. Consequently, you must use the heat gun prudently when drying apparatus held by plastic-containing devices such as Keck clips or clamps with rubber- or vinyl-coated jaws (Fig. 2.5), because the plastic can melt.

You should never point a heat gun in the direction of another person. Moreover, when using this device, you should be certain that there are no light-weight objects such as your starting material or reaction product in the vicinity that might become airborne on the powerful jet of air produced.

Sand Bath

Sand baths are convenient devices for heating small volumes of material in small flasks. These baths are easily prepared by putting about 1–3 cm of sand in a Pyrex crystallizing or petri dish and then placing the dish on a hot plate (Fig. 2.28). The temperature in the sand bath is controlled by varying the heat setting on the hot plate. Alternatively, the sand may be placed in the ceramic well of a heating mantle, such as that shown in Figure 2.23b. The temperature of the sand bath in this case is controlled by the variable transformer attached to the mantle. The temperature may be monitored with a thermometer that is inserted into the bath to the same depth as the flask or conical vial being heated. Because sand is not a good conductor of heat, there is a temperature gradient in the bath, with the highest temperature being closest to the hot plate or heating mantle. This gradient can be exploited: The flask or vial can be deeply immersed in the sand bath for rapid heating; once the mixture has begun to boil, the flask or vial can be raised to slow the rate of reflux or boiling.

Although sand is cleaner than mineral and silicone oil, there are some limitations to using sand baths. For example, sand baths in glass containers are normally

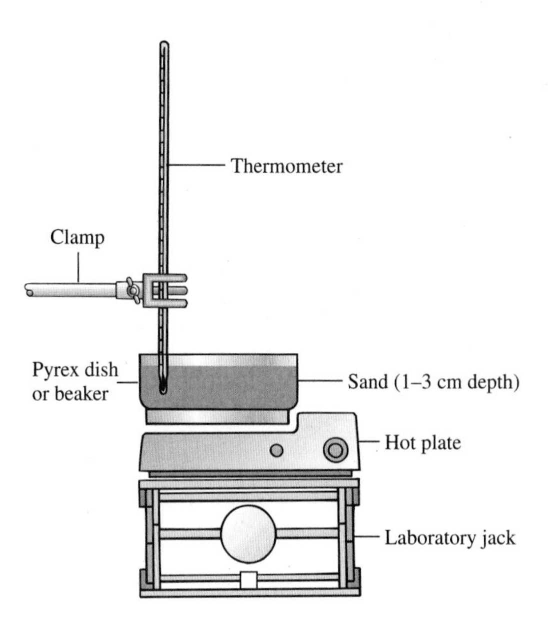

Figure 2.28
Sand bath heated by hot plate.

not used to heat flasks to temperatures higher than about 200 °C because the glass might break. Furthermore, temperature control is difficult because sand is a poor heat conductor, and it is not possible to stir the bath. Finally, different devices may heat the sand bath to a different temperature gradient, so you must calibrate each heating device/sand bath combination for reproducible results. A procedure for doing this is discussed under "**Aluminum Block.**"

Water Bath

A **water bath** may be used when temperatures no higher than about 80–90 °C are desired. The water is contained in a wide-mouth vessel such as a beaker or crystallizing dish. The bath is preferably heated with an electrical hot plate (Fig. 2.29a), but a Bunsen burner can also be used. The bath should be supported on a ring clamp bearing a wire mesh to diffuse the flame (Fig. 2.29b). A useful trick to minimize the loss of water through evaporation is to cover the open portions of the bath with aluminum foil.

Steam Bath

Steam provides a safe source of heat when temperatures up to 100 °C are required. The steam outlet is connected to either a **steam bath** or a **steam cone** (Fig. 2.30), both of which have an outlet at the bottom to drain the condensed water. The tops of steam baths and cones are typically fitted with a series of overlapping concentric rings that may be removed in succession to provide openings of variable size. For example, if a rapid rate of heat transfer to a flask is desired, the rings are removed until up to one-half of the surface of the flask is immersed in the steam (Fig. 2.30a); cloth towels may be wrapped around the upper portion of the flask to insulate it. If a slower rate of heating is desired, the opening may be adjusted so less of the flask is in direct contact with the steam. To heat beakers and Erlenmeyer flasks, the opening should be small enough so that the container sits directly on top of the steam bath with only its lower surface exposed to steam.

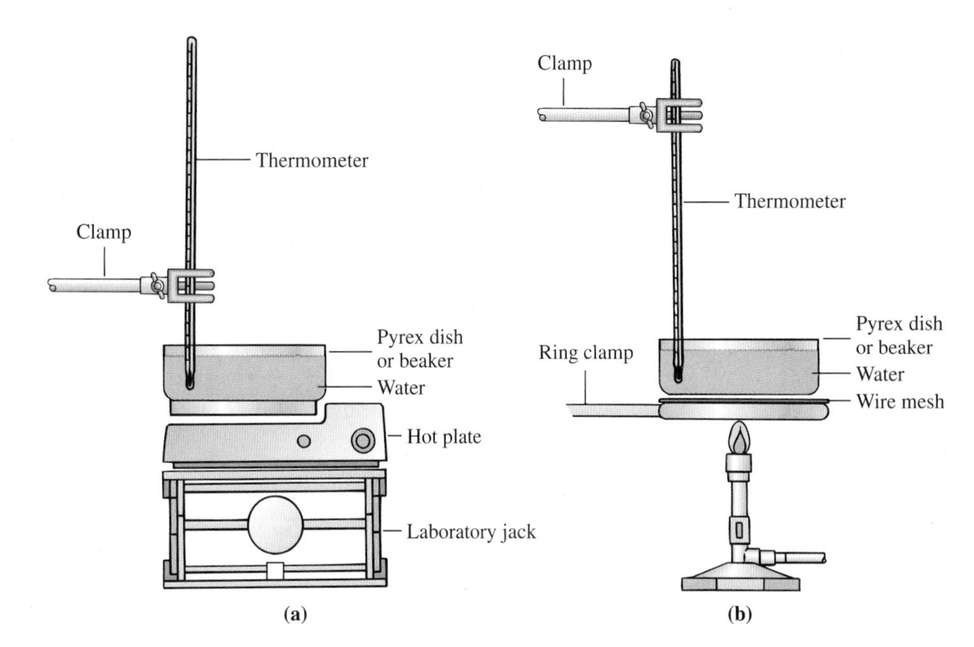

Figure 2.29
Water bath: (a) heated by hot plate; (b) heated by Bunsen burner.

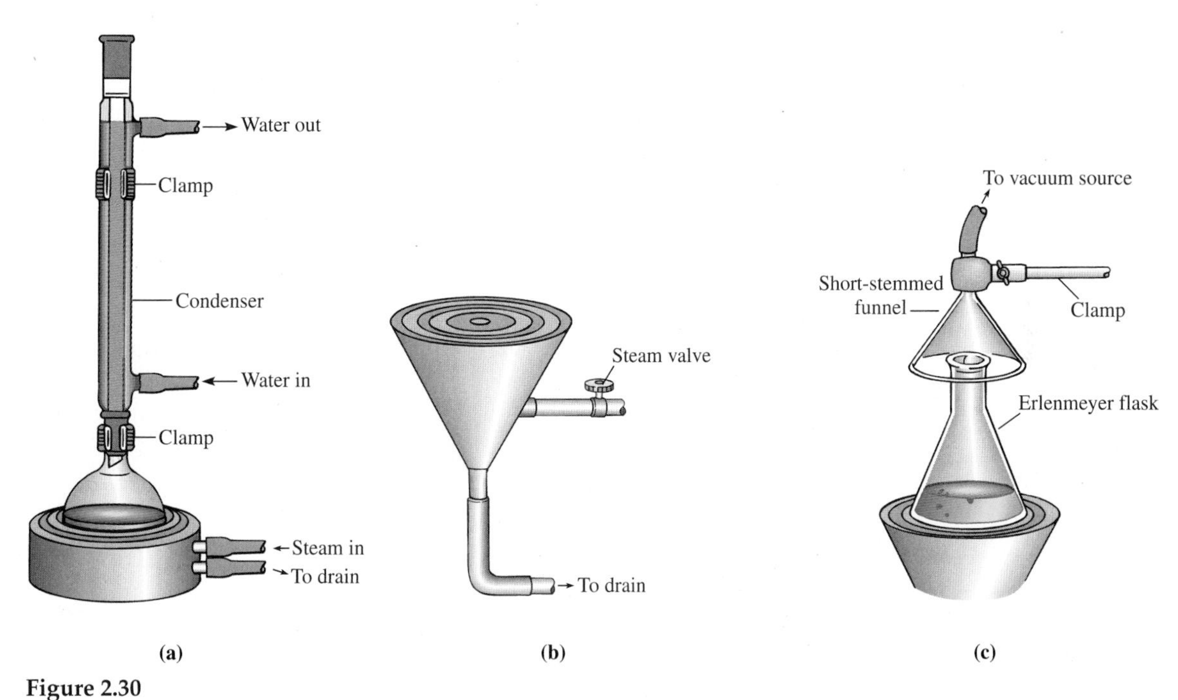

Figure 2.30
(a) Steam bath being used to heat a reaction mixture. (b) Steam cone. (c) Heating an Erlenmeyer flask on a steam cone, with a funnel attached to vacuum source for removing vapors.

When the steam valve is first opened, several minutes are usually required for the condensed water to drain out of the steam lines. Once the steam is issuing smoothly, the steam valve should be adjusted to provide a *slow,* steady flow of steam. There is no benefit to having a fast flow of steam, other than to fog your safety glasses and give everyone in the lab a sauna bath! Regardless of how fast the steam is flowing, the temperature will never exceed 100 °C. Another disadvantage of a fast flow of steam is that water will condense on your equipment and may even find its way into the reaction flask.

Steam baths are especially well suited as heat sources for the safe evaporation of volatile solvents used for a reaction or extraction (Sec. 2.29) or when purifying a solid product by recrystallization (Sec. 3.2). As an alternative to working at a hood, an inverted funnel attached to a vacuum source may be placed over the top of the container (Fig. 2.30c) to remove vapors and keep them from entering the room.

Microwave Ovens

Microwave ovens are attaining wider application as a heating source for organic reactions. Although these ovens rely upon the same technology as the microwave oven you may use at home to heat water or food, the ones used in the laboratory have been specially modified so they can be operated safely with organic solvents. Such ovens are explosion-proof and allow precise temperature control and uniformity of heating throughout the cavity of the oven. A number of commercial models that have a wide range of capacities and capabilities are available, and two different types are depicted in Figure 2.31. Some models feature temperature and pressure-feedback control, whereas others have modular accessories that allow for automated synthesis. Simple ovens are relatively inexpensive and are found in a number of academic laboratories, but more sophisticated models can be rather expensive and are normally found only in industrial laboratories.

The transfer of microwave energy to a molecule is extremely fast and results from the direct interaction of a molecule with the high-frequency, oscillating electric component of the microwave field. Because the absorption of microwave energy depends on the polarity of the molecule, the reactant molecules in an

(a)

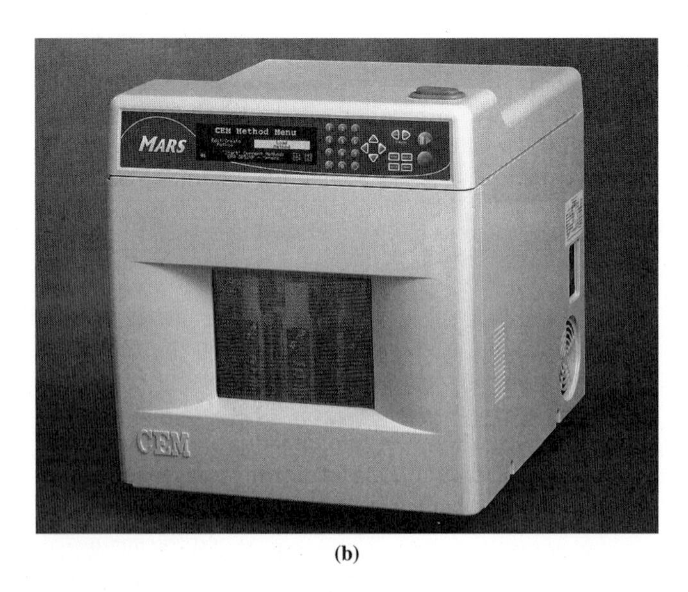

(b)

Figure 2.31
Commercial microwave reactors: (a) CEM Benchmate microwave synthesis instrument (courtesy of CEM Corporation); (b) CEM MARSXpress high throughput microwave system (courtesy of CEM Corporation).

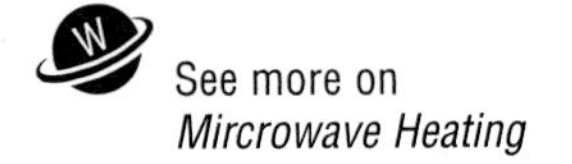
See more on *Mircrowave Heating*

organic reaction often absorb microwave energy better than the solvent. This preferential transfer of microwave energy to the reacting molecules results in an *instantaneous temperature increase*, which cannot be measured because of its short lifetime and molecular nature, and a rapid reaction then ensues. Reactions that may require hours of heating in an oil bath may be completed in a matter of minutes in a microwave oven.

The use of microwave ovens as heat sources not only drastically reduces the heating times of organic reactions, but the reactions often proceed more efficiently and selectively than when conduction heating methods are used. This is because microwave energy is transferred uniformly and almost simultaneously to the entire sample, thus eliminating any hot spots that may result in side reactions. If sealed reaction vessels are used, superheating can be easily achieved, so reactions that might be otherwise difficult to induce can be performed with relative ease. Another benefit to using microwave heating in synthesis is that less solvent is required than for traditional heating methods.

Reactions heated in a microwave oven do not necessarily require special apparatus. For example, they can be conducted in vessels that range from simple laboratory glassware such as an Erlenmeyer flask covered with a watchglass to flasks of various shapes that are fitted with standard-taper glass joints (Fig. 2.32a). Specially designed pressure vessels can also be used. A number of tubes can also be mounted on a rotor so that multiple reactions can be conducted simultaneously (Fig. 2.32b).

2.10 COOLING TECHNIQUES

Sometimes it is necessary to cool a reaction mixture either to avoid undesired side reactions or to moderate the temperature of exothermic reactions that could become uncontrollable. Cooling is also often used to maximize the recovery of solid

(a)

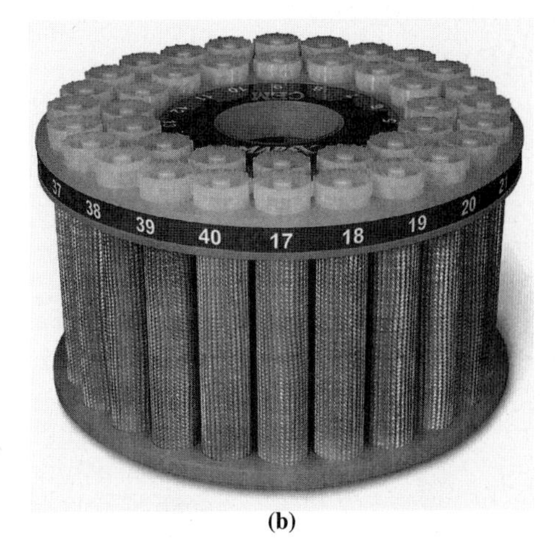

(b)

Figure 2.32
Reaction vessels: (a) assorted shapes of standard-taper ground glassware and sealable glass tubes (courtesy CEM Corporation); (b) array of special reaction tubes for multiple, simultaneous reactions (courtesy of CEM Corporation).

products during purification by recrystallization (Sec. 3.2) or to lower the temperature of hot reaction mixtures so that work-up procedures can be performed.

The most common cooling medium in the undergraduate laboratory is an ice-water bath. Liquid water is a more efficient heat transfer medium than ice, because it covers the entire surface area of the portion of the vessel that is immersed in the bath. Consequently, when preparing this type of bath, do not use ice alone.

An ice-water bath has an equilibrium temperature of 0 °C. For lower temperatures, an ice-salt bath can be prepared by mixing ice and sodium chloride in a proportion of about 3:1 to generate a temperature of approximately −20 °C. As the ice melts, the excess water should be removed and more ice and salt added to maintain this temperature. Still lower temperatures are possible using combinations of organic liquids and either dry ice (solid carbon dioxide) or liquid nitrogen; directions for preparing these types of baths can be found in Reference 5 of Chapter 1.

Immersing a vessel containing a warm reaction mixture in a container of tap water is useful for rapid cooling to room temperature so long as the glassware you are using is Pyrex. Consult with your instructor to ensure that this is the type of glass you have.

2.11 STIRRING METHODS

Heterogeneous reaction mixtures must be stirred to distribute the reactants uniformly and facilitate chemical reactions. Stirring also ensures thermal equilibration whenever the contents of a flask are being heated or cooled. If a mixture is boiling, the associated turbulence is usually sufficient to provide reasonable mixing; however, stirring a boiling mixture is an alternative to using boiling stones to maintain smooth boiling action and avoid bumping. Stirring is most effectively achieved using magnetic or mechanical stirring devices, but swirling is often sufficient.

Swirling

The simplest means of mixing the contents of a flask is swirling, which is accomplished by manually rocking the flask with a circular motion. If a reaction mixture must be swirled, carefully loosen the clamp(s) that support the flask and attached apparatus, and swirl the contents periodically during the course of the reaction. If the entire apparatus is supported by clamps attached to a single ring stand, the clamp(s) attached to the flask do not have to be loosened. Make sure all the clamps are tight, pick up the ring stand, and gently move the entire assembly in a circular motion to swirl the contents of the flask.

Magnetic Stirring

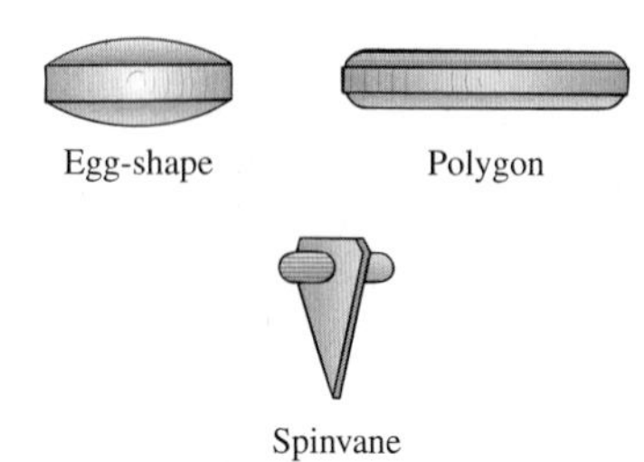

Figure 2.33
Magnetic stirbars and a spinvane.

Magnetic stirring is the most common technique for mixing the contents of a flask in the undergraduate laboratory. The equipment consists of a magnetic stirrer, which houses a large bar magnet that is spun by a variable-speed motor, and a stirbar (Fig. 2.33) that is contained in a round-bottom flask or conical vial. The metallic core of the stirbar is usually coated with a chemically inert substance such as Teflon, although glass is sometimes used for stirbars.

The stirbar is normally placed in the flask or vial *before* any other materials, such as solvents or reagents. In the case of introducing stirbars into a flask, you should *not* simply drop the stirbar into the flask, because you might crack or break it. Rather, tilt the flask and let the stirbar gently slide down the side.

A flat-bottom container such as a beaker or Erlenmeyer flask may be placed directly on top of the stirrer (Fig. 2.34a), whereas a round-bottom flask or a conical

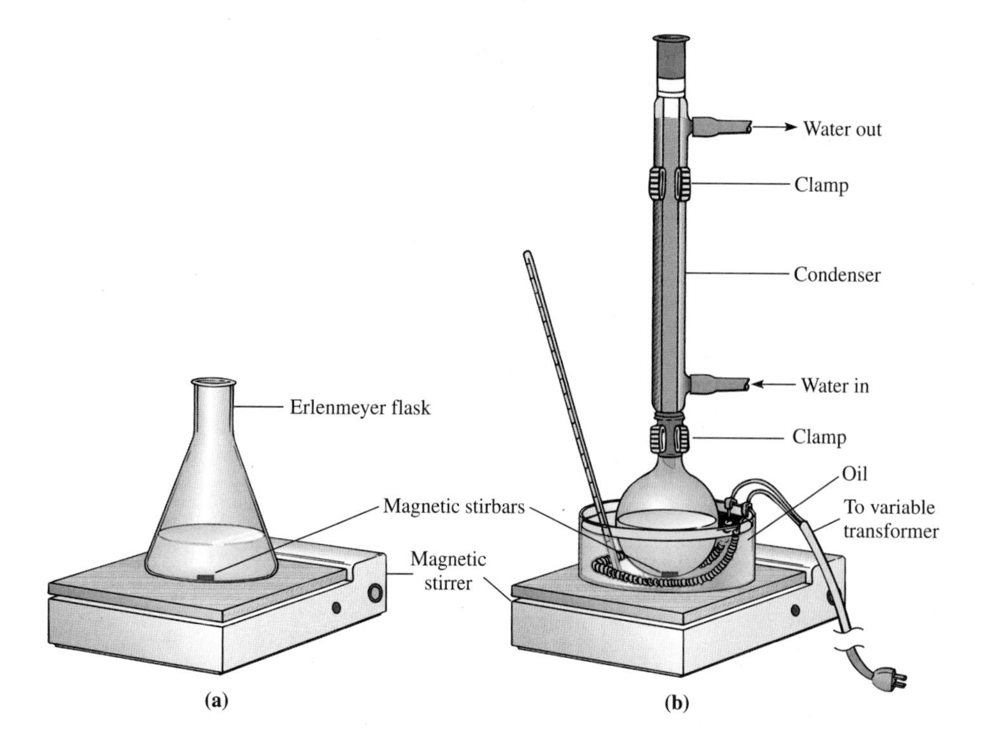

Figure 2.34
(a) Magnetic stirring of the contents of Erlenmeyer flask. (b) Using a heating source with magnetic stirring of reaction mixture.

vial must be clamped above the stirrer (Fig. 2.34b). A flask containing a stirbar should be *centered* on the magnetic stirrer so that the stirbar rotates smoothly and does not wobble. As the stirrer motor turns, the stirbar rotates in phase with the motor-driven magnet. The stirring rate may be adjusted using the control dial on the stirring motor, but excessive speed should be avoided because it often causes the stirbar to wobble or "jump" rather than to rotate smoothly. Because the shapes of a conical vial and a stirbar are matched, wobbling is not a problem. Nonetheless, the stirring rate must still be adjusted to minimize splashing.

The use of magnetic stirring in conjunction with heating is illustrated in Figure 2.34b. This figure depicts apparatus in which the contents of a reaction vessel are simultaneously being heated and magnetically stirred, a common operation in the organic laboratory. A large stirbar or paper clip is used to stir the bath itself; this maintains a homogeneous temperature throughout the heating fluid. Other heating devices such as a sand bath, aluminum block, or heating mantle can be used as well, although monitoring of the heating source is more difficult with a mantle.

Mechanical Stirring

Thick mixtures and large volumes of fluids are most efficiently mixed using a **mechanical stirrer;** a typical set-up is depicted in Figure 2.35. A variable-speed, explosion-proof, electric motor drives a stirring shaft and paddle that extend into the flask containing the mixture to be stirred. The motor should have high torque, so that it has sufficient power to turn the shaft and stir highly viscous mixtures. The stirrer shaft is usually constructed of glass, and the paddle, which agitates the contents of the flask, is constructed of an inert material such as stainless steel, Teflon, or glass. A glass paddle must be used to stir reaction mixtures containing active metals such as sodium or potassium. The paddle is easily removed from the

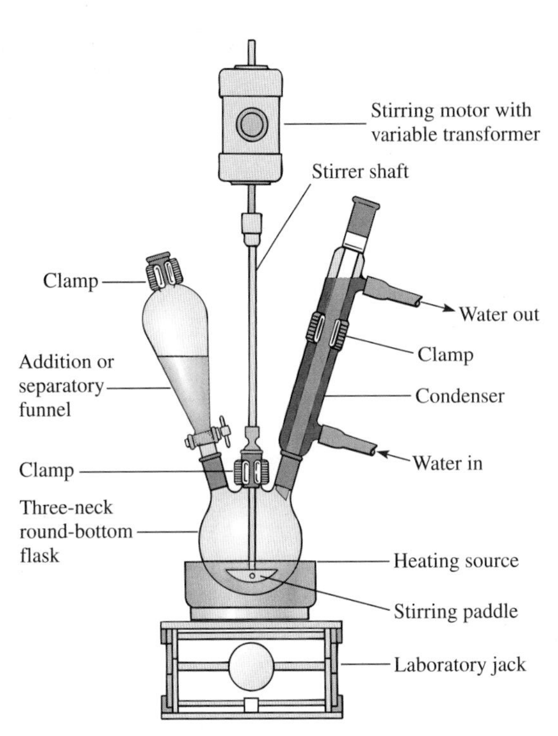

Figure 2.35
Flask equipped with mechanical stirring.

shaft to facilitate cleaning, and different-sized paddles can be used according to the size of the flask. The glass shaft and the inner bore of the standard-taper bearing are ground to fit each other precisely. A cup at the top of the bearing is used to hold a few drops of silicone or mineral oil, which lubricates the shaft and provides an effective seal.

The stirrer shaft is connected to the motor with a short length of heavy-walled rubber tubing that is secured with twisted copper wire or a hose clamp. The motor and shaft *must* be carefully aligned to avoid wear on the glass surfaces of the shaft and bearing and to minimize vibration of the apparatus that could result in breakage. The bearing is held in place in the flask with either a rubber band or a clamp so that it does not work loose while the motor is running. The rate of stirring is controlled by varying the speed of the motor with either a built-in or separate variable transformer.

Various operations can be performed while using mechanical stirring. For example, the flask in Figure 2.35 is a three-neck, standard-taper, round-bottom flask that is equipped with an addition funnel and a condenser. This apparatus could be used in cases where dropwise addition of a reagent to a stirred and heated reaction mixture is required.

2.12 CENTRIFUGATION

Centrifugation is a useful technique to facilitate the separation of two immiscible phases. The procedure simply involves spinning a sample in one or more tubes at high speed in a bench-top centrifuge (Fig. 2.36). It is important to balance the

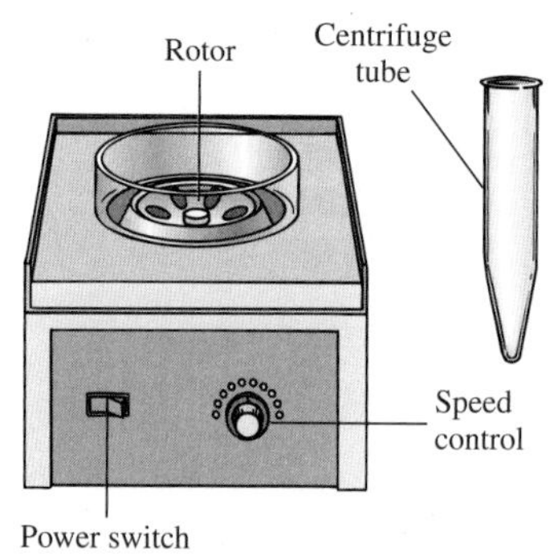

Figure 2.36
Centrifuge with centrifuge tube.

centrifuge before spinning to avoid vibrating the centrifuge during spinning. For example, if the sample to be centrifuged is placed in a single centrifuge tube, you should then fill a second tube with an equal volume of solvent and place this tube opposite the sample tube in the centrifuge. After the sample has been spun, the phases may be separated by decantation (Sec. 2.19) or by removing a liquid phase with a Pasteur or filter-tip pipet (Sec. 2.5).

Centrifugation is sometimes more effective than filtration for removing suspended solid impurities in a liquid, especially when the particles are so fine they would pass through a filter paper. It may also be used for Craig tube filtration during a microscale recrystallization (Sec. 2.17). Centrifugation will also aid in separating the organic and aqueous layers when performing an extraction (Sec. 2.21), especially when emulsions are formed.

2.13 SIMPLE DISTILLATION

See videos on
Simple Distillation

Simple distillation is a useful method for isolating a pure liquid from other substances that are not volatile. The experimental aspects of this technique are described in this section, whereas the theory and application of the method are discussed in detail in Section 4.3.

Miniscale Apparatus

Typical examples of laboratory apparatus for performing simple miniscale distillations are shown in Figure 2.37. The operation entails heating the liquid contained in the **stillpot,** also called the **distillation flask,** to its **boiling point,** which is defined as the temperature at which the total vapor pressure of the liquid is equal to the external pressure. The vapors then pass from the pot into a water- or air-cooled **condenser** where they condense to form the liquid phase that is collected in the **receiver.** The thermometer measures the temperature of the vapors. However, in order to obtain an accurate temperature reading, the thermometer must be carefully positioned so that the top of the mercury bulb is approximately even with the bottom of the sidearm outlet, as indicated in the inset in Figure 2.37a. To ensure an accurate temperature reading, a drop of condensate *must* adhere to the bottom of the mercury bulb. If a pure liquid is being distilled, the temperature read on the thermometer, which is termed the **head temperature,** will be identical to the temperature of the boiling liquid. Provided the pure liquid in the stillpot is not being superheated, the **pot temperature** will also be identical to the temperature of the boiling liquid, although the bath temperature will be higher. The head temperature corresponds to the boiling point of the liquid and will remain constant throughout the distillation. If a mixture of volatile compounds is being distilled, the head and pot temperatures will be different.

To avoid possible breakage and spillage, certain techniques must be followed when assembling the apparatus shown in Figure 2.37. The apparatus in Figure 2.37a is typically used for volumes of distillate in excess of about 10 mL. If a condenser is not necessary, the modified apparatus shown in Figure 2.37b, which comprises a distillation head and a bent vacuum adapter, is well suited for distilling quantities in the range of 1–10 mL; the shorter path length in this set-up minimizes material losses. *You should thoroughly familiarize yourself with the following general guidelines before attempting to set up your apparatus.*

1. Assemble the equipment so that the distillation flask is elevated 15 cm or so above the bench to allow placement of a suitable heat source. Read Section 2.4 for a general discussion.

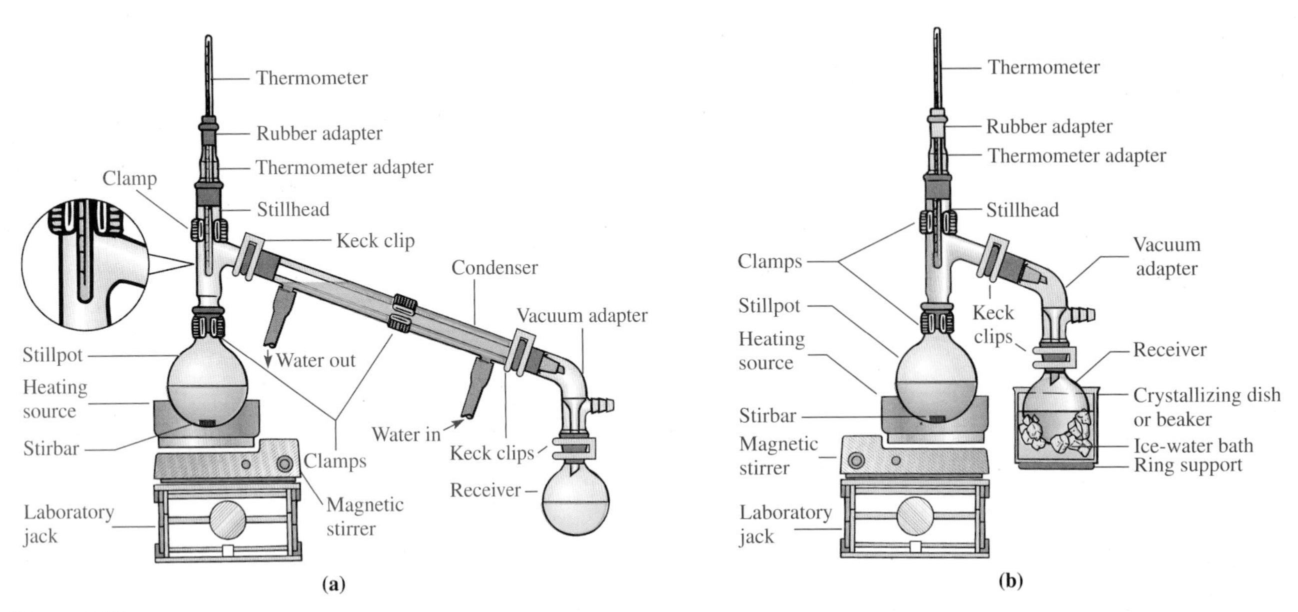

Figure 2.37
(a) Typical apparatus for miniscale simple distillation at atmospheric pressure or under vacuum. Inset shows correct placement of thermometer in stillhead. (b) Typical apparatus for miniscale shortpath distillation with icebath for cooling receiver.

2. Review the relevant heating methods discussed in Section 2.9. Heating mantles may be used to heat the stillpot, but an oil bath is generally the best way to heat the flask, since the temperature of the bath, and hence the stillpot, is easier to control. The level of the liquid in the distillation flask should be below the level of the oil in the heating bath to minimize the risk of bumping. *Do not use a flame unless directed to do so by your instructor.* The flask should *never* be heated *directly* with a flame, because this may produce hot spots and cause bumping; the flask may only be heated with a flame *if* wire gauze is placed under the flask to diffuse the heat.
3. When distilling materials that have boiling points in excess of 125 °C, wrap the stillhead with glasswool and then aluminum foil to prevent excessive heat loss; failure to do so usually means that it will be necessary to heat the distillation flask to a higher temperature than would otherwise be necessary to drive over the last portion of product.
4. Start assembling the apparatus by tilting the stillpot to one side and sliding a stirbar gently down the side. Then add the liquid to be distilled to the stillpot, making certain that the flask is no more than *half*-full. *Start assembling the equipment by clamping the stillpot in position,* and attach the rest of the glassware by next putting the stillhead in place, then the condenser, and finally the vacuum adapter at the end of the condenser. Place a thin film of lubricant (Sec. 2.2) on each of the standard-taper glass joints before mating them. Joints must fit snugly so that flammable vapors do not leak from the apparatus into the room.
5. Note the location of the rigid clamps in Figure 2.37; *do not "overclamp" the apparatus.* An apparatus that is too rigidly clamped may become stressed to the point that it breaks during its assembly or the distillation. Align the jaws of the clamp *parallel* to the piece of glass being clamped so that the clamp may be tightened without twisting or torquing the glass and either breaking it or pulling a joint

loose. Tighten the clamps only enough to hold each piece securely in place. *Do not tighten the clamp unless the piece of glassware is correctly positioned and the clamp is properly aligned.* Always practice the rule: *Do not apply undue pressure to the glassware.*

6. Use Keck clips to hold the condenser, vacuum adapter, and receiver in place. Provide additional support for flasks of 100-mL capacity or larger; smaller flasks may also require additional support if they become more than half-full during a distillation. Clamp the receiver. An iron ring holding a piece of wire gauze or a cork ring may be used to provide additional support underneath the receiving flask.
7. Note the location of the "water in" and "water out" hoses on the condenser. The tube carrying the incoming water is *always* attached to the lower point to ensure that the condenser is filled with water at all times.
8. Adjust the water flow through the condenser to a *modest* flow rate. There is no benefit to a fast flow, and the increased pressure in the apparatus may cause a piece of rubber tubing to pop off, spraying water everywhere. Showers in the laboratory should be restricted to the emergency shower! It is good practice to *wire* the hoses to the condenser and to the water faucet to minimize the danger that they will break loose.

Microscale Apparatus

The volume of distillate in microscale distillations is typically less than 500 μL (0.5 mL), so it is important to minimize the surface area of glass to avoid unnecessary loss of material. This is accomplished by combining the stillhead and collector in a single piece called a **Hickman stillhead,** which has a reservoir for containing distillate and may have a port to facilitate removal of distillate (Fig. 2.38).

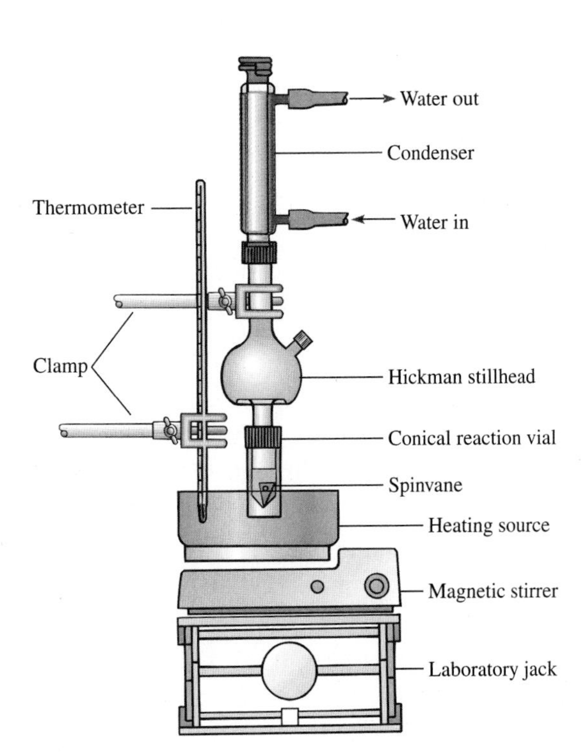

Figure 2.38
Microscale apparatus for distillation with external temperature monitoring.

Because the volume of distillate is small, it is somewhat difficult to measure the distillation temperature (boiling point) accurately at the microscale level. An *approximate* method is to determine the temperature of the heat source at the point when distillation occurs. This temperature is about 20 °C or so above the distillation temperature, so it is necessary to make the corresponding correction in the boiling point recorded for the distillate. Alternatively, a thermometer can be inserted through the condenser attached to the stillhead so the thermometer bulb is in the lower neck of the stillhead. A more accurate measurement of the boiling point of the distillate is possible using the technique of microscale boiling point determination (Sec. 2.8).

A Pasteur or filter-tip pipet (Sec. 2.5) is used to remove distillate from the Hickman stillhead. If the head has a port, the pipet can be inserted at this point. If not, the condenser must first be removed and the pipet inserted through the top of the stillhead. Depending on the geometry of the stillhead, it may be necessary to use a Pasteur pipet having a curved tip to allow the tip to reach the bottom of the reservoir. A Bunsen or microburner (Sec. 2.9) may be used to produce the necessary bend.

2.14 FRACTIONAL DISTILLATION

The technique of **fractional distillation** is a useful method for isolating the individual pure liquid components from a mixture containing two or more volatile substances. This technique is described in this section, whereas the theory and application of this method are discussed in detail in Section 4.4. An apparatus for fractional distillation at atmospheric pressure or vacuum is shown in Figure 2.39. The principal difference between an apparatus for fractional and simple distillation is the presence of a **fractional distillation column** (Fig. 2.40) between the stillpot and the stillhead. This column is similar to a condenser, the major difference being that a fractionating column has a large outside jacket and some indentations at the male end to hold the packing in place. Unpacked distillation columns such as the Hempel column shown in Figure 2.40a can also be used as condensers, but condensers lack the indentations and cannot serve as packed distillation columns.

Before assembling the apparatus, clean and dry the inner tube of the distillation column to avoid contaminating your product. It is not necessary to dry the water jacket, because traces of water in it will not affect the distillation. Pack the column by adding the desired column packing, a small quantity at a time, through the top of the column while holding it vertical. The column packing is an inert material such as glass, ceramic, or metal pieces in a variety of shapes (helices, saddles, woven mesh, and so on). Common packings include glass tubing sections, glass beads, glass helices, or a stainless steel sponge. Some column packings are sufficiently large that they will stop at the indentations. If the packing falls through the column, put a small piece of glasswool, wire sponge, or wire screen into the column just above the indentations by pushing it down the column with a wooden dowel or a piece of glass tubing. The column packing should extend to the top of the water jacket but should not be packed too tightly.

Assemble the apparatus using the general guidelines for simple distillation apparatus (Secs. 2.4 and 2.13). *Read those instructions carefully before proceeding with setting up the apparatus in Figure 2.39.* Start by putting the liquid to be distilled and a stirbar in the distillation flask, and then clamp the flask in place. Do not fill the flask more than *half*-full. Attach the distillation column and make sure that it is *vertical.* Lubricate and tighten all the joints after the stillhead and condenser are in place. Clamp only the distillation flask and condenser. *Do not run water through the jacket of the distillation column.*

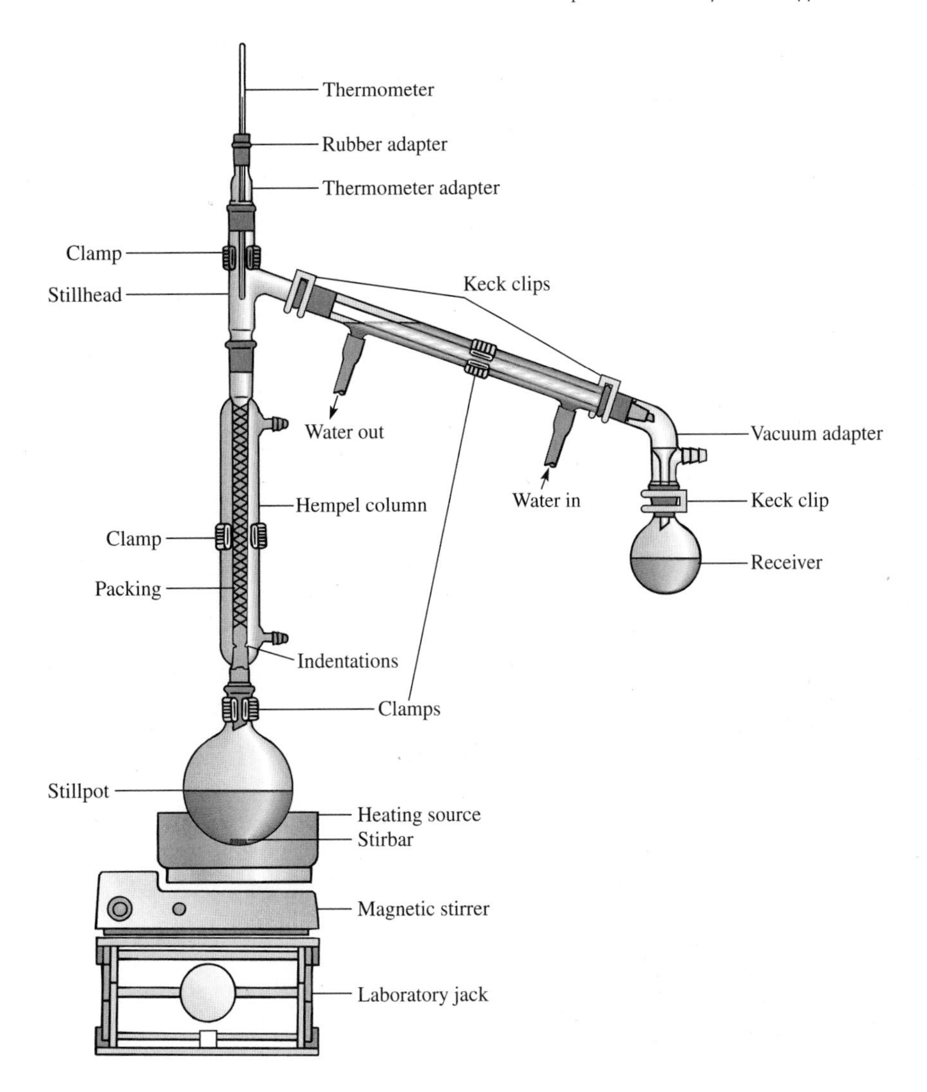

Figure 2.39
Miniscale apparatus for fractional distillation.

See *Who was Vigreux?*

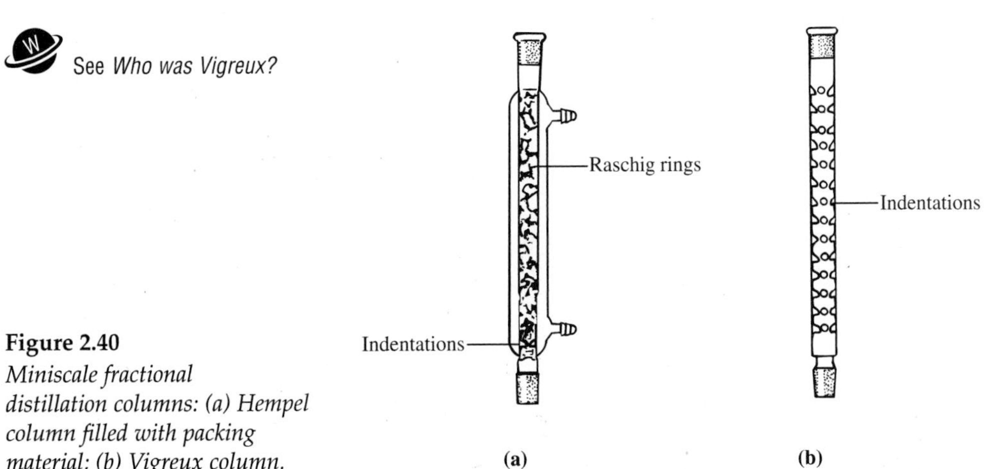

Figure 2.40
Miniscale fractional distillation columns: (a) Hempel column filled with packing material; (b) Vigreux column.

2.15 VACUUM DISTILLATION

See videos on *Vacuum Distillation*

It is most convenient to distill liquids at atmospheric pressure (760 torr), but compounds that have high molecular weights or numerous functional groups may decompose, oxidize, or undergo molecular rearrangement at temperatures below their atmospheric boiling points. These problems may frequently be circumvented if the distillation is conducted at a lower temperature. Since a liquid boils when its total vapor pressure is equal to the external pressure, it is possible to lower the boiling point of the liquid by performing the distillation at a pressure *less* than one atmosphere. The technique involved in such distillations is termed **vacuum distillation.**

Although accurate estimates of the effect of pressure upon the boiling point of a liquid may be made by use of charts or a nomograph (Fig. 2.41), two useful *approximations* of the effect of lowered pressure on boiling points are:

See more on *Using Nomographs*

1. Reduction from atmospheric pressure to 25 torr lowers the boiling point of a compound boiling at 250–300 °C at atmospheric pressure by about 100–125 °C.
2. Below 25 torr, the boiling point is lowered by about 10 °C each time the pressure is reduced by one-half.

Reduced pressures may be obtained by connecting a **water aspirator pump** or a **mechanical vacuum pump** to a vacuum adapter that is fitted between the condenser and the receiving flask. The vacuum produced by a water aspirator is limited by the vapor pressure of the water at the ambient temperature and the condition of the aspirator pump. Pressures as low as 8–10 torr may be obtained from a water aspirator with cold water, but pressures in the range of 15–25 torr are more common. A good mechanical vacuum pump can evacuate the apparatus to less than 0.01 torr; it is important to clean the oil periodically and maintain tight connections in the distillation apparatus to achieve the lowest possible pressures for a particular pump. Some

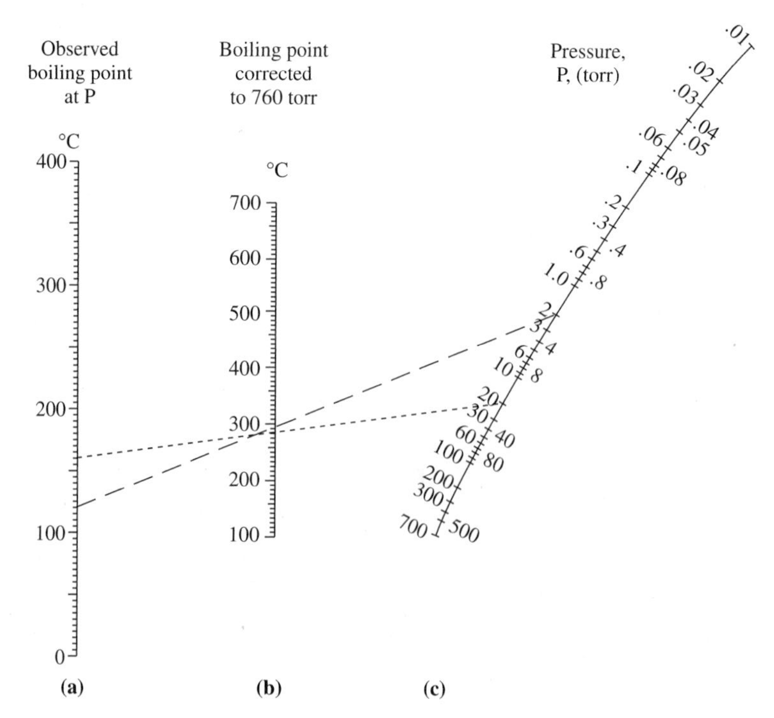

Figure 2.41

Pressure–temperature alignment nomograph (dashed lines added for illustrative purposes). How to use the nomograph: Assume a reported boiling point of 120 °C at 2 torr. To determine the boiling point at 20 torr, connect 120 °C (column **A**) *to 2 torr (column* **C**) *with a transparent plastic rule and observe where this line intersects column* **B** *(about 295 °C). This value corresponds to the normal boiling point. Next, connect 295 °C (column* **B**) *with 20 torr (column* **C**) *and observe where this intersects column* **A** *(160 °C). The approximate boiling point will thus be 160 °C at 20 torr.*

laboratories are equipped with "house" vacuum lines that are connected to a large central vacuum pump, but such vacuum systems seldom reduce the pressure to less than 50 torr. The pressure is measured by a device called a **manometer.**

The apparatus for vacuum distillations is a modification of that used for simple distillations (Fig. 2.37), and two options are shown in Figure 2.42. The set-up depicted in Figure 2.42a is typically used for volumes of liquid in excess of about 10 mL. If a condenser is not required, the modified apparatus of Figure 2.42b can be used for

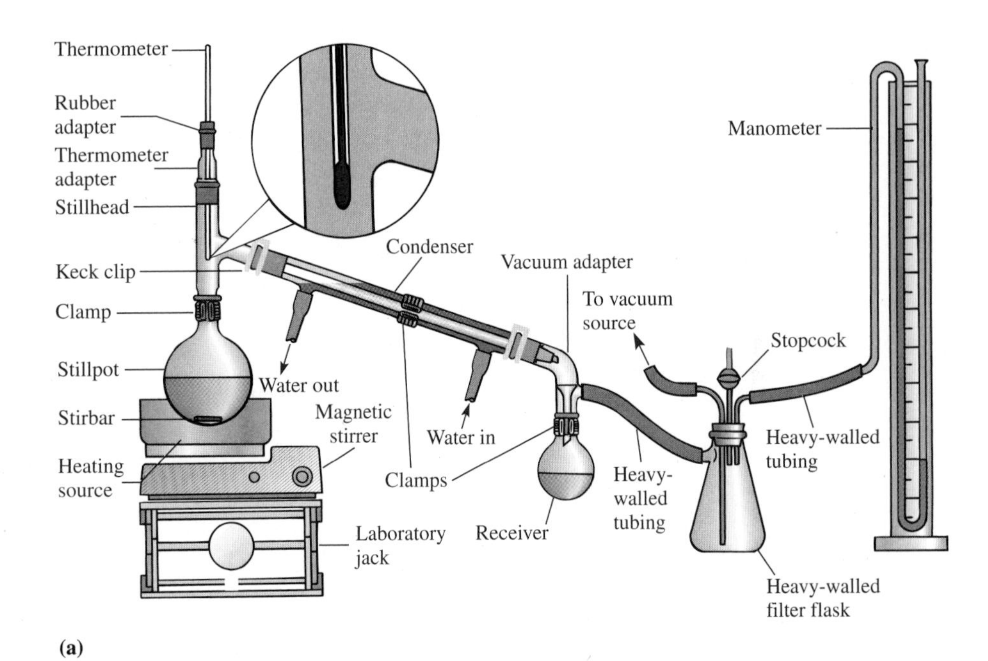

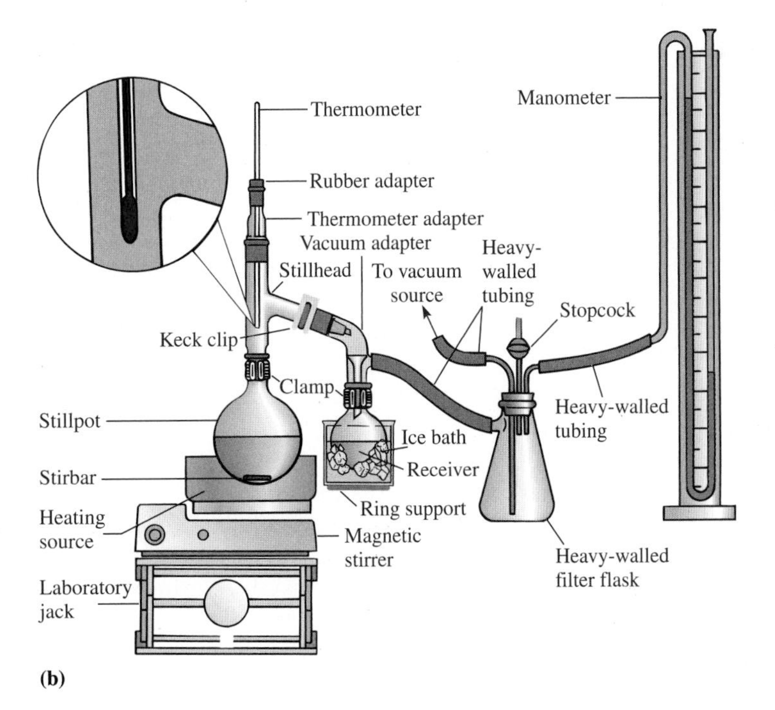

Figure 2.42
(a) Miniscale apparatus for vacuum distillation.
(b) Modified apparatus for miniscale shortpath vacuum distillation.

distilling volumes of liquid in the range of 1–10 mL; the shorter pathlength of this apparatus minimizes loss of material.

The primary difference between the apparatus for simple and vacuum distillations is the attachment of the vacuum adapter to a vacuum source. Rather than being directly connected to the vacuum source, the adapter is normally linked to a heavy-walled sidearm flask. This flask has several functions. For example, if a water aspirator is being used to produce a vacuum, the flask serves as a **trap** to prevent water from backing up into the apparatus, which may occur if there is a sudden decrease in the water pressure in the aspirator. If it is equipped with a stopcock release valve, the flask also provides a means for connecting the apparatus and the manometer to the source of the vacuum and releasing the vacuum. Further discussion of traps is provided under the topic "Vacuum Filtration" in Section 2.17.

The apparatus shown in Figure 2.42 is assembled (Sec. 2.4) according to the same general guidelines that are used for setting up the apparatus for simple distillation (Sec. 2.13). *Read those instructions carefully before proceeding with assembling the apparatus in Figure 2.42.* Because magnetic stirring (Sec. 2.11) is the most convenient means of obtaining an even rate of vaporization during the distillation, that is the method described here. However, if this option is not available, consult your instructor regarding alternative set-ups for vacuum distillation.

Start assembling the apparatus by tilting the stillpot to one side and sliding a stirbar gently down the side. Then add the liquid to the stillpot, making certain that the flask is no more than *half*-full. After lubricating all the ground-glass joints (Sec. 2.2), attach the stillhead and equip it with a thermometer adapter and thermometer. Continue the set-up as described in Section 2.13, clamping the condenser, if one is used, and the receiver. Make sure all of the joints are tightly mated. Finally, connect the apparatus to the sidearm flask, which in turn must be connected to the vacuum source and the manometer.

The volume and density of the vapor formed by volatilization of a given amount of liquid are pressure-dependent. For example, the volume of vapor formed from vaporization of a drop of liquid will be about 20 times as great at 38 torr as it would be at 760 torr. During a vacuum distillation, large quantities of vapor are produced in the distillation flask, and these vapors enter the condenser with high velocity prior to condensing. Since the density of the vapor is much lower at reduced pressure than at atmospheric pressure, it may be difficult to control the rate of vaporization and thereby minimize the difference in pressure between the distillation flask and the manometer where the pressure must be accurately measured.

The manometer measures the pressure at which the distillation is being conducted; this value is important and is reported with the boiling point. For example, benzaldehyde boils at 180 °C at atmospheric pressure and at 87 °C at 35 torr, and these two boiling points are reported in the format bp 180 °C (760 torr) and 87 °C (35 torr). To obtain an accurate measure of the pressure, however, you must carefully *control the rate of distillation.* The problem arises because a drop of condensate forms from a larger volume of vapor as the pressure is lowered, thereby causing higher vapor velocities to enter the condenser. This creates a back pressure so that the pressure in the distillation flask and stillhead is higher than that measured on the manometer, which is located beyond the condenser and receiver and is therefore less sensitive to back pressure. The difference between the actual and measured pressure can be minimized by distilling the liquid at a slow but steady rate. Superheating the vapor can be avoided by maintaining the oil bath at a temperature no more than about 25–35 °C higher than the head temperature.

The following paragraphs provide a general procedure for performing a vacuum distillation using the apparatus shown in Figure 2.42a. *Read these instructions carefully prior to executing a vacuum distillation.*

1. *Never use glassware with cracks, thin walls, or flat bottoms, such as Erlenmeyer flasks, in vacuum distillations.* Hundreds of pounds of pressure may be exerted on the *total* exterior surfaces of systems under reduced pressure, even if only a water aspirator pump is used. Weak or cracked glassware may *implode,* and the air rushing into the apparatus will shatter the glassware violently in a manner little different from that of an explosion. *Examine the glassware carefully.*
2. Lubricate and seal *all* glass joints carefully during assembly of the apparatus; this will help avoid air leaks and provide lower pressure. The rubber fittings holding the thermometer in place must be tight. If necessary, the neoprene fittings normally used with the thermometer adapters may be replaced with short pieces of heavy-walled tubing. Do not use rubber stoppers elsewhere in the apparatus, since direct contact between the rubber and the hot vapors during distillation may extract impurities from the rubber and contaminate your product. The three-holed rubber stopper in the safety trap should fit snugly to the flask and the pieces of glass tubing. The safety trap must be made of heavy-walled glass and wrapped with electrical tape as protection in case it implodes. Heavy-walled vacuum tubing must be used for all vacuum connections. *Check the completely assembled apparatus to make sure that all joints and connections are tight.*
3. Now begin stirring the liquid and turn on the vacuum; if you are using a water aspirator pump, you should *fully open the water faucet.* The release valve on the safety flask should be *open. Slowly* close the release valve, but be prepared to reopen it if necessary. The liquid often contains small quantities of low-boiling solvents, so foaming and bumping are likely to occur. If this occurs, adjust the release valve until the foaming abates. This may have to be done several times until the solvent has been removed completely. When the surface of the liquid in the flask is relatively quiet, fully evacuate the system to the desired pressure. The release valve may have to be opened slightly until the desired pressure is obtained; a needle valve from the base of a Bunsen burner may be used for fine control of the pressure. Check the manometer for constancy of pressure. Note and record the pressure at which the distillation is being performed. This pressure should be monitored throughout the course of the distillation, even when fractions are not being collected. Begin heating the flask, and maintain the temperature of the bath so that the distillate is produced at the rate of 3–4 drops per minute. *Do not begin heating the flask until the system is fully evacuated and the vacuum is stable.* The best way to heat the stillpot is with an oil bath, since it is easier to control the temperature in the bath and distillation flask; however, other techniques (Sec. 2.9) may be suggested by your instructor.
4. If it is necessary to use multiple receivers to collect fractions of different boiling ranges, the distillation must be interrupted to change flasks. Remove the heating source with caution and allow the stillpot to cool somewhat. *Slowly* open the vacuum release valve to re-admit air to the system. When atmospheric pressure is reached, change receivers, close the release valve, re-evacuate the system to the same pressure as previously, reapply heat, and continue distilling. The operation may result in a different pressure in the fully evacuated

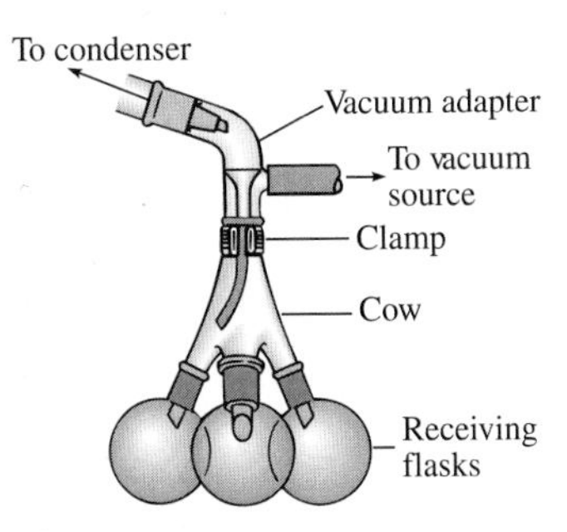

Figure 2.43
Multiflask adapter for miniscale vacuum distillation.

system. Periodically monitor and record the head temperature and the pressure, particularly just before and after changing receivers.

5. After the distillation is complete, discontinue heating, allow the pot to cool somewhat, slowly release the vacuum, and turn off the source of the vacuum.

One of the most inconvenient aspects of a vacuum distillation is the disruption of the distillation in order to change receivers. In order to eliminate this problem, you may wish to use a "cow" receiver that accommodates three or four round-bottom flasks (Fig. 2.43). These flasks are successively used as receivers by rotating them into the receiving position.

2.16 STEAM DISTILLATION

See video on *Steam Distillation*

Steam distillation is a mild method for separating and purifying volatile liquid or solid organic compounds that are immiscible or insoluble in water. This technique is not applicable to substances that react with water, decompose on prolonged contact with steam or hot water, or have a vapor pressure of less than about 5 torr at 100 °C. The practical features and the apparatus of this technique are described in this section; its theory and applications are discussed in detail in Sections 4.5 and 4.6.

The two basic techniques commonly used for a steam distillation in the laboratory are differentiated on the basis of whether the steam is introduced from an external source or generated internally. For larger-scale reactions, the most common and most efficient method for conducting a steam distillation involves placing the organic compound(s) to be distilled in a round-bottom flask equipped with a Claisen adapter, a stillhead, and a water-cooled condenser, as depicted in Figure 2.44. The combination of introducing steam from an external source into the distillation flask via the inlet tube and the turbulence associated with the boiling action tends to cause occasional violent splashing, and the Claisen adapter is necessary to prevent the mixture from splattering into the condenser. During the course of the distillation, water may condense in the distillation flask and fill it to undesirable levels. This problem can be avoided by gently heating the stillpot with an appropriate heat source.

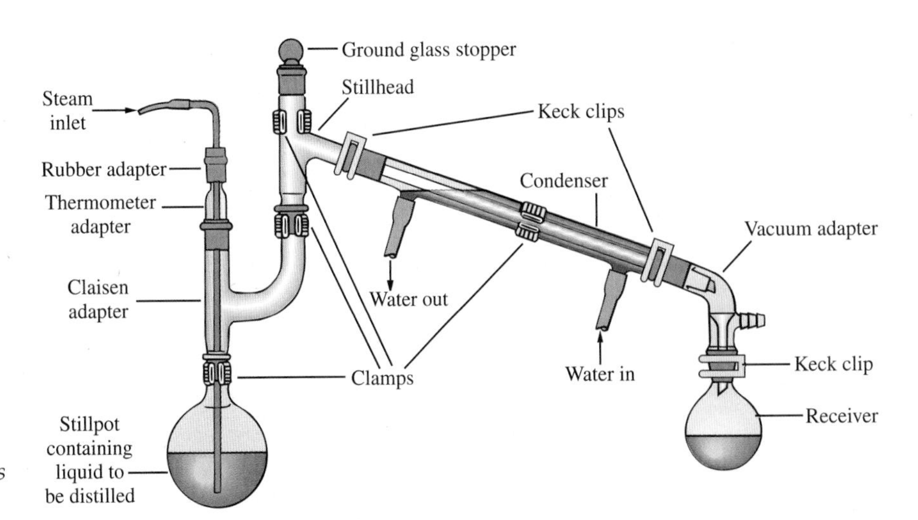

Figure 2.44
Apparatus for miniscale steam distillation. The steam tube is replaced with a stopper if steam is generated by direct heating.

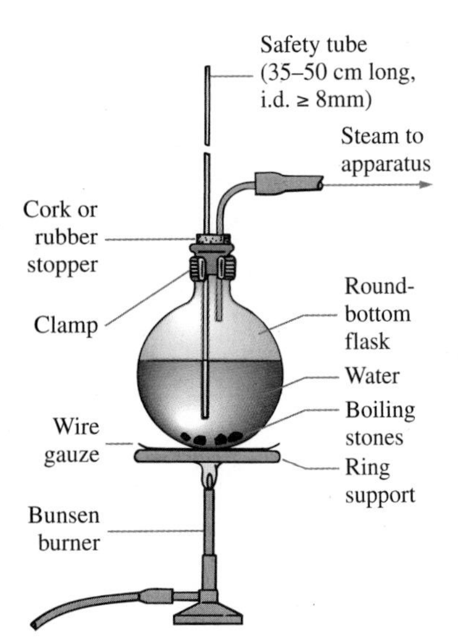

Figure 2.45
Steam generator.

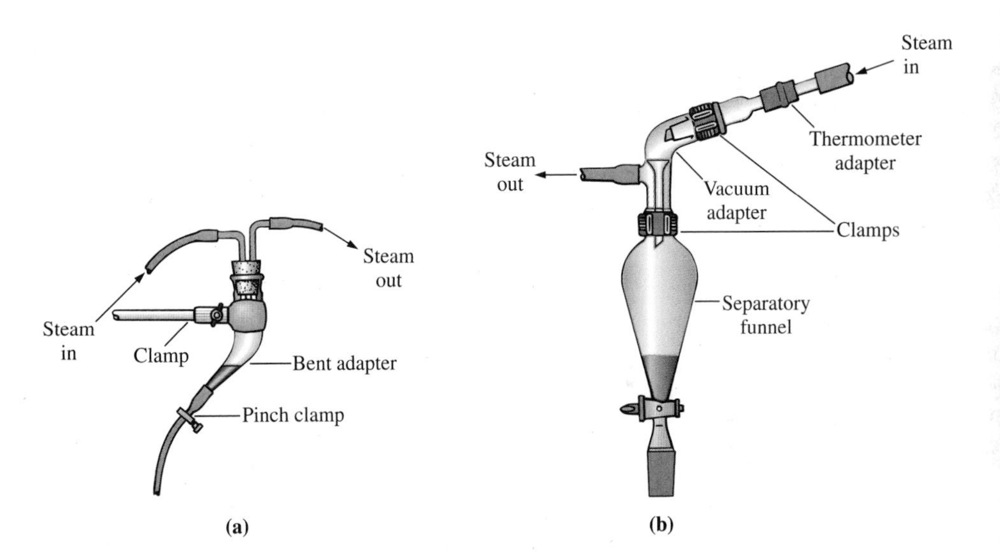

Figure 2.46
Water traps for use in steam distillations.

Steam may also be produced externally in a generator, as shown in Figure 2.45. The round-bottom flask is initially half-filled with water, and boiling stones are added *before* heating. The safety tube relieves internal pressure if steam is generated at too high a rate or the outlet tube to the apparatus becomes clogged. Steam is most conveniently obtained from a laboratory steam line. A trap (Fig. 2.46) is then placed between the steam line and the distillation flask to permit removal of any water and/or impurities present in the steam.

If only a small amount of steam is necessary to separate a mixture completely, a simplified method involving *internal steam generation* may be employed. Water is added directly to the distillation flask together with the organic compounds to be separated. The flask is equipped for steam distillation by setting up the apparatus as shown in Figure 2.44, *except* that the steam inlet tube is replaced with a stopper. The flask is then heated with a heating mantle or a Bunsen burner, and the distillate is collected. This technique is generally not applicable for distillations that require large amounts of steam, since it would be necessary to use an inappropriately large flask or to replenish the water in the flask frequently by using an addition funnel. However, this procedure is satisfactory and convenient for most miniscale reactions.

2.17 FILTRATION APPARATUS AND TECHNIQUES

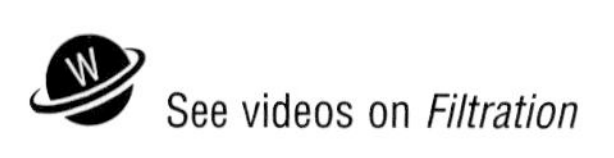

Filtration is the primary technique used to separate solids from liquids. It is important to perform filtrations properly to avoid loss or contamination of your product, regardless of whether it is a solid or liquid.

Conceptually, the process of filtration is simple and involves passing a liquid containing a solid material through a barrier that is permeable only to the liquid. The liquid is collected in an appropriate receiver, and the solid remains on the barrier. However, there are many nuances to how filtrations are performed, and some of them are discussed below.

Filtration Media

Various types of materials are used in conjunction with funnels to produce the barrier that is the key to separating solids from liquids. The most common of these is **filter paper,** which is available in different sizes and porosities. The more porous the paper, the faster the rate of filtration but also the greater the possibility that solid particles will pass through the paper. **Glasswool** and **cotton** may serve as filters, but they are considerably more porous than paper and thus are used only when relatively large particles are to be separated from a liquid.

Inert, finely divided solids such as silica gel and Celite®, both of which are forms of silicon dioxide, are sometimes used along with filter paper, glasswool, or cotton to assist filtration. These materials generally function by *adsorption* of solids and colloidal materials from the solution being filtered, thereby preventing their passing through the filter paper and/or clogging its pores. Applications of Celite as a "filteraid" are described in Section 2.18.

For microscale filtrations, there is a specialized technique that uses a Craig tube and depends on contact between the ground-glass strip on the inner shoulder of the glass tube and a glass or Teflon stopper (Fig. 2.47) to effect separation of liquid and solid. The ground-glass surface prevents complete contact with the glass or Teflon surface of the plug and thus allows for liquid to flow out of the tube upon centrifugation of the assembled unit. Very fine particles *are not* effectively removed by this technique.

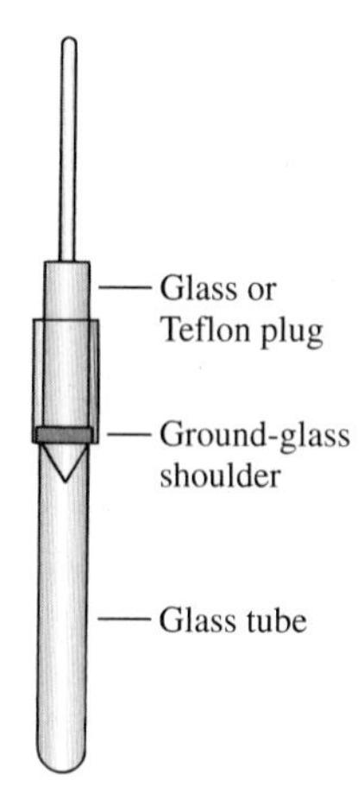

Figure 2.47
Craig tube assembly.

Gravity Filtration

Gravity filtration is the filtering technique most commonly used to remove solids such as impurities, decolorizing carbon (Sec. 2.18), or drying agents (Sec. 2.24) from liquids prior to crystallization, evaporation, or distillation. In addition to filter paper of the appropriate size, only a stemmed funnel, ring clamp, and a receiver (Fig. 2.48) are needed for this operation. The filter paper must be folded to fit into the funnel, and there are two ways of doing this. The simplest involves folding the circular piece of paper in half, then in half again so that corners 1 and 2 of Figure 2.49a are coincident, inserting the paper in the funnel, and opening the folded paper so that the solution can be poured through it. Before starting the filtration, it is helpful first to wet the filter paper with a small amount of the solvent being filtered. This serves to form a seal between the paper and the funnel and to keep the paper from collapsing on itself and closing the opening you have made in it.

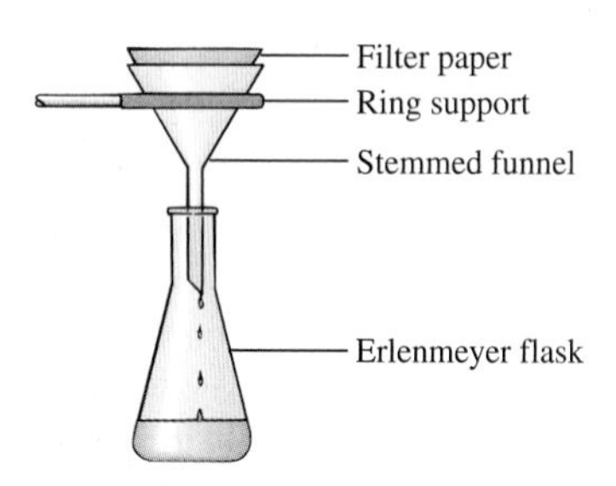

Figure 2.48
Apparatus for miniscale gravity filtration.

The rate and ease of filtration are increased if a piece of **fluted filter paper** (Fig. 2.49g) is used, since fluting increases the surface area of paper contacted by the

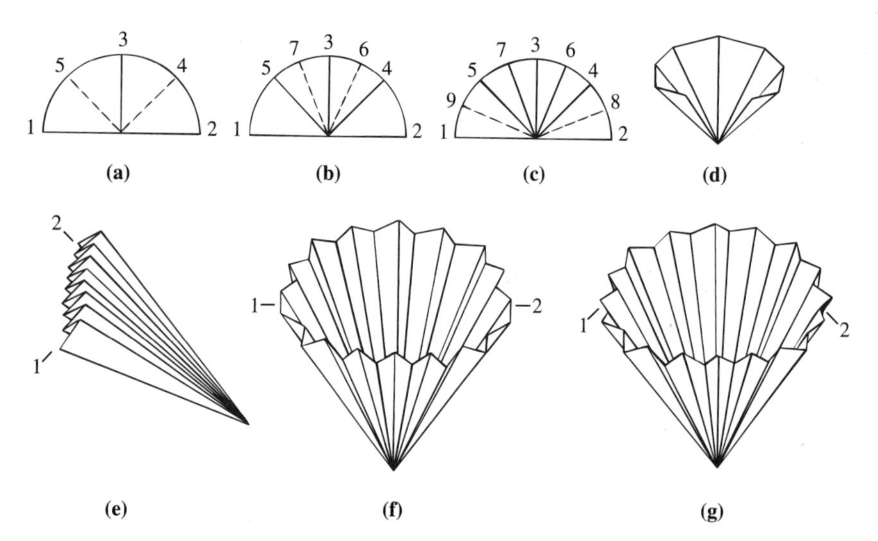

Figure 2.49
Folding filter paper.

liquid being filtered. Although such paper is commercially available, it may also be easily prepared from a flat piece of round filter paper. One method of fluting is shown in Figure 2.49 and involves the following sequence. Fold the paper in half, and then into quarters. Fold edge 2 onto 3 to form edge 4, and then 1 onto 3 to form 5 (Fig. 2.49a). Now fold edge 2 onto 5 to form 6, and 1 onto 4 to form 7 (Fig. 2.49b). Continue by folding edge 2 onto 4 to form 8, and 1 onto 5 to form 9 (Fig. 2.49c). The paper now appears as shown in Figure 2.49d. Do not crease the folds tightly at the center, because this might weaken the paper and cause it to tear during filtration. All folds thus far have been in the same direction. Now make folds in the *opposite* direction between edges 1 and 9, 9 and 5, 5 and 7, and so on, to produce the fanlike appearance shown in Figure 2.49e. Open the paper (Fig. 2.49f) and fold each of the sections 1 and 2 in half with reverse folds to form paper that is ready to use (Fig. 2.49g). Now that you have learned to fold filter paper, origami will seem simple!

To perform a filtration, the folded or fluted filter paper is first placed in a funnel and *wetted* with the appropriate solvent. The solution containing the solid is then poured onto the filter paper or fluted filter paper with the aid of a glass stirring rod to minimize dripping down the outside of the flask. As the solution passes through the paper, the solid remains on the filter paper. A typical experimental set-up is shown in Figure 2.50. If you do not use a ring stand to support the funnel, you should insert a paper clip or a bent piece of wire between the funnel and the lip of the flask to avoid forming a solvent seal between the flask and the funnel. Such a seal will interfere with the filtration, since there will be no vent for the air displaced by the filtrate.

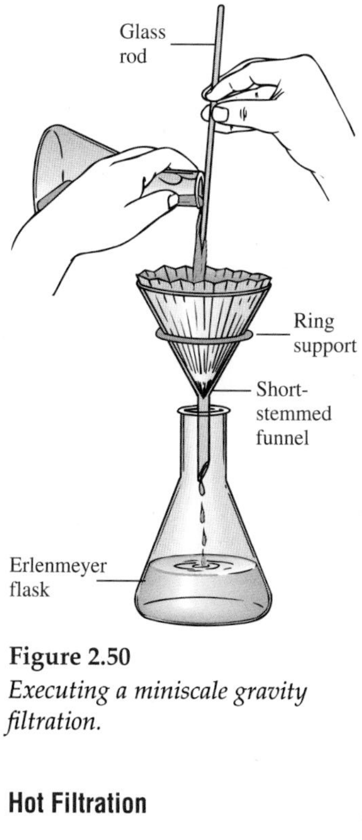

Figure 2.50
Executing a miniscale gravity filtration.

When volumes less than about 5 mL must be filtered, there is a danger of losing material on the filter paper, because the paper absorbs a significant volume of liquid. A Pasteur filtering pipet (Sec. 2.5) should be used in such cases. The liquid mixture to be filtered is transferred by pipet into the filtering pipet as shown in Figure 2.51. The plug should be rinsed with about 0.5 mL of solvent to maximize the recovery of the solute.

Hot Filtration

Sometimes it is necessary to remove solid impurities from an organic compound that is only sparingly soluble in the solvent at room temperature. In such cases, the mixture must be heated, and **hot filtration** must be performed to remove the insoluble solids without leaving the desired material in the funnel, too. Such a situation commonly arises during purification of an organic solid by recrystallization

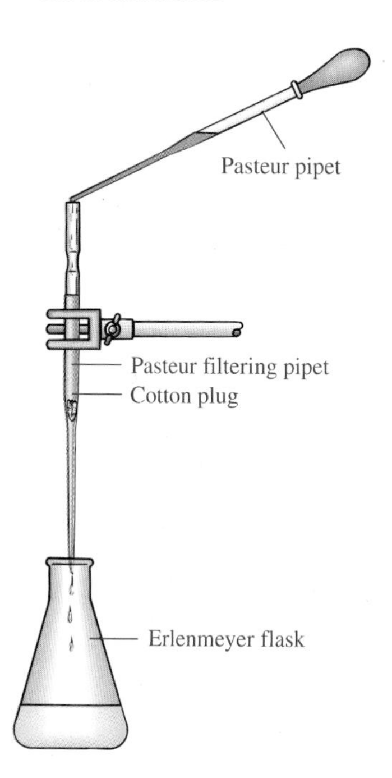

Figure 2.51
Gravity filtration with a Pasteur filtering pipet.

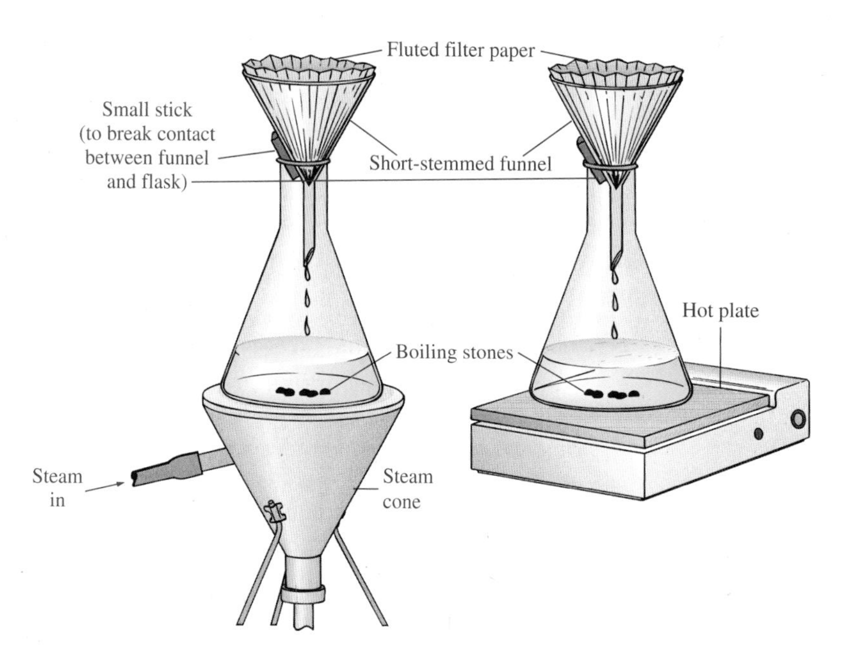

Figure 2.52
Miniscale hot filtration: (a) using steam cone for heating; (b) using hot plate for heating.

when decolorizing carbon (Sec. 2.18) or other insoluble residues must be removed; a filter-aid may be required to facilitate removal of finely divided solids.

Miniscale Apparatus

The apparatus for performing a miniscale hot filtration is shown in Figure 2.52. A small volume of the *pure* solvent is first placed in an Erlenmeyer flask containing a boiling stick or several boiling stones, and a short-stemmed or stemless filter funnel containing a piece of fluted filter paper is placed in the mouth of the flask. A small stick or piece of thick paper is inserted between the flask and the funnel to

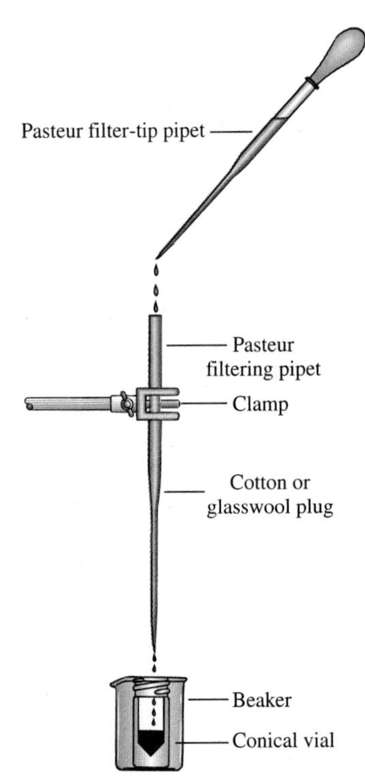

Figure 2.53
Microscale hot filtration.

allow the vapors of solvent to rise around the outside of the funnel. The flask is heated with either a steam bath or hot plate until the solvent boils and the vapors rise up the sides of the flask and around the funnel; the solvent vapors heat the filter funnel to minimize problems of crystallization of the desired material in the funnel. A steam bath is preferred for flammable solvents that boil below 90 °C, whereas a hot plate can be used for all nonflammable solvents and, with proper care and in a hood, for flammable solvents boiling above 90 °C.

When the flask and funnel are hot, the hot solution is poured onto the filter paper. Following this, several milliliters of hot solvent are added to the flask that contained the original recrystallization solute and solvent. The additional hot solvent is poured onto the filter paper to ensure the complete transfer of material to the flask containing the filtered solution.

Microscale Apparatus

A microscale hot filtration is performed using a combination of a Pasteur filter-tip pipet (Fig. 2.9a) and a filtering pipet (Fig. 2.9b). The goal is to filter the hot solution containing solid contaminants as rapidly as possible so that none of the solute crystallizes before the filtrate reaches the receiver. This requires preheating both pipets by pulling hot solvent up into the barrel several times and then *quickly* transferring the hot solution into the top of the filtering pipet, as shown in Figure 2.53. As with miniscale gravity filtration, the filter should be rinsed with the solvent to maximize the recovery of the solute, since some of it may have crystallized on the cotton plug. However, no more than 0.2–0.5 mL of additional solvent should be used.

Vacuum Filtration

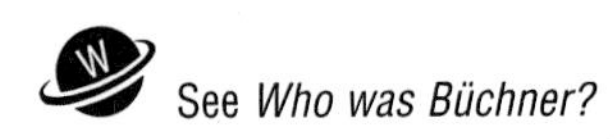

See *Who was Büchner?*

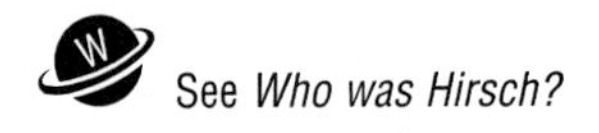

See *Who was Hirsch?*

Vacuum filtration is a technique for collecting crystalline solids from solvents after recrystallization or precipitation. A typical apparatus is shown in Figure 2.54. Either a **Büchner funnel** or a smaller **Hirsch funnel** is used, with the latter being better suited for isolating quantities of solid ranging from 100 to 500 mg. The funnel is fitted to a vacuum filter flask using a neoprene adapter or a rubber stopper. The sidearm of the filter flask is connected to an aspirator or house vacuum line through a **trap** using heavy-walled rubber tubing. The trap prevents water from the aspirator from backing up into the filter flask if there is a loss of water pressure. The trap should be a *heavy-walled* bottle or a second vacuum filter flask

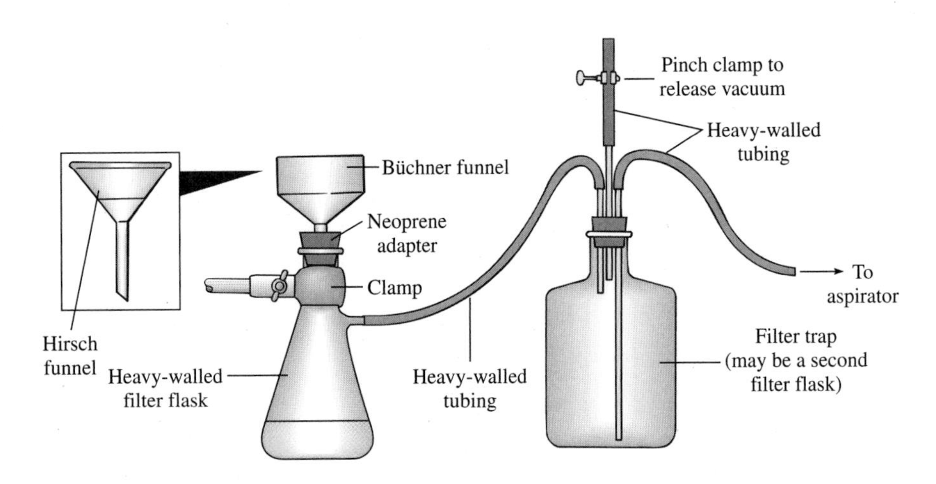

Figure 2.54
Apparatus for miniscale vacuum filtration; inset shows Hirsch funnel, used for small quantities of solids.

wrapped with electrical or duct tape. If a filter flask is used for the trap, it should be equipped with a two-holed stopper with its sidearm attached to the first filter flask with heavy-walled tubing. The glass tube that extends to within approximately 2–3 cm of the bottom of the filter trap is connected with heavy-walled tubing to the water aspirator pump so that any water that collects slowly in the trap is evacuated back through the aspirator pump and into the drain. If a sudden backup occurs, the vacuum should be released immediately so that the trap does not completely fill up with water. Any filtrate that may overflow the filter flask will also be collected in the trap. *The filter flask should be cleaned before doing the filtration,* since it may be necessary to save the filtrate.

A piece of filter paper is placed in the funnel so it lies flat on the funnel plate and covers all the small holes in the funnel. It should *not* extend up the sides of the funnel. A vacuum is applied to the system, and the filter paper is "wetted" with a small amount of pure solvent to form a seal with the funnel so that crystals do not pass around the edges of the filter paper and through the holes in the filter. The flask containing the crystals is swirled to suspend the crystals in the solvent, and the solution containing the crystals is poured slowly into the funnel. A stirring rod or spatula may be used to aid the transfer. The last of the crystals may be transferred to the funnel by washing them from the flask with some of the filtrate, which is called the **mother liquor.** When all the solution has passed through the filter, the vacuum is released slowly by opening the screw clamp or stopcock on the trap. The crystals are washed to remove the mother liquor, which contains impurities, by adding a small amount of *cold, pure solvent* to the funnel to just cover the crystals. Do not allow the filter paper to float off the surface of the funnel, as some of the crystals may be lost through the holes in the funnel. Vacuum is then reapplied to remove the wash solvent, and the crystals are pressed as dry as possible with a clean spatula or a cork while continuing to pull a vacuum on the funnel.

Most of the solvent may be evaporated from the crystals by allowing the vacuum to pull air through the crystals on the funnel for a few minutes. A clean spatula is used to scrape the crystals gently from the filter paper and then to transfer them to a clean watchglass. Be careful and do not scrape the filter paper too vigorously, because torn bits of filter paper or paper fuzz might contaminate your product. The crystals may be dried completely by allowing them to air-dry for a few hours or by leaving them loosely spread on a filter paper in a locker until the next class. The drying process may be accelerated by placing the crystals in an oven, but the temperature of the oven should be at least 20–30 °C below the melting point of the crystals. A vacuum desiccator can also be used to hasten the drying process. The vacuum desiccator should either be wrapped with electrical or duct tape or placed in a metal cage as protection should the desiccator implode. Specially designed desiccators permit heating samples under vacuum, but the temperature must again be kept about 20–30 °C below the melting point of the solid. Heat or vacuum desiccators should not be used to dry crystals of compounds that sublime readily (Sec. 2.20).

Vacuum filtration can also be used to remove undesired solids from a solution. If the solid is finely divided or colloidal, a pad of a filter-aid such as Celite is used to ensure complete removal of the solid. A pad of filter-aid can be formed on the filter funnel by first making a slurry of 0.5–1 g of filter-aid in a few milliliters of the solvent being used. The slurry is poured onto a filter paper in a Büchner or Hirsch funnel that is attached to a *clean* filter flask; vacuum is then slowly applied to draw the solvent through the filter paper, leaving a thin, even pad of the filter-aid; the pad should be about 2–3 mm thick. The solution containing the solid is then filtered as described above. Obviously this technique is useful for removing solid impurities *only* when the *solution* contains the substance of interest; a solid product is not collected in this manner, as it would be necessary to separate the desired material from the filter-aid. This would be a challenge even for Louis Pasteur (see Historical Highlight at the end of Chapter 7).

Craig Tube Filtration

Craig tube filtration is the microscale analog of vacuum filtration for isolating solid products. However, rather than using a vacuum to pull solvent through a filter, this technique depends on centrifugal forces to force the solvent through the tiny openings between a ground-glass surface and a plug seated against this surface.

The Craig tube itself consists of two components: a test tube-like glass vessel, the inner shoulder of which is ground-glass, and a plug made from glass or Teflon (Fig. 2.55a). In addition to the Craig tube, a centrifuge, centrifuge tubes, and a piece of wire are required.

The technique involves transferring a solution of the desired solid into the bottom portion of the Craig tube. After crystallization is complete, the top part of the Craig tube is attached, and a wire holder is affixed to the unit (Fig. 2.55b). The assembled unit is placed in an inverted centrifuge tube so that the top of the Teflon plug is at the bottom of the tube; the wire is looped over the lip of the tube (Fig. 2.55c). The entire assembly is then inverted carefully so that the two components of the Craig tube stay in close contact, thereby preventing solid from escaping past the plug into the centrifuge tube. After centrifugation, the solvent will be in the bottom of the centrifuge tube, whereas the desired solid remains in the Craig tube (Fig. 2.55d).

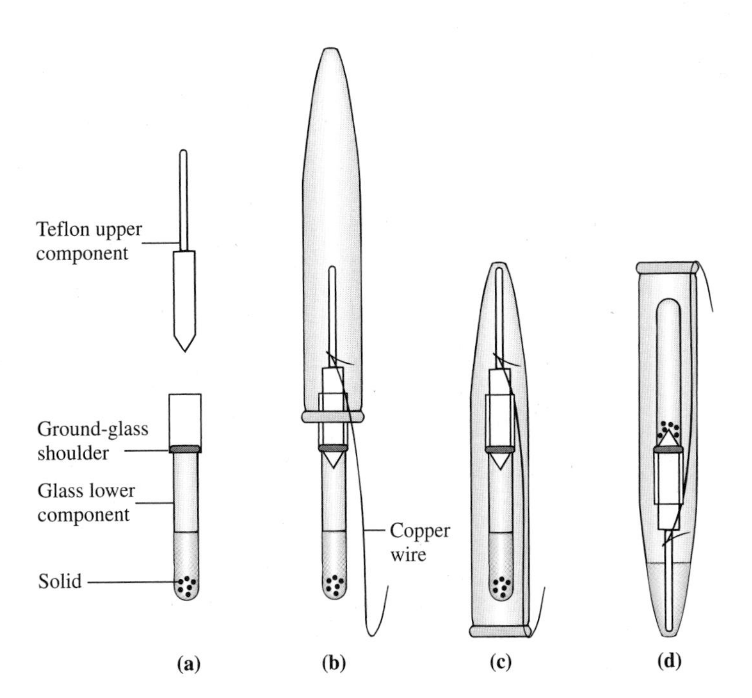

Figure 2.55
Filtration using the Craig tube.

2.18 DECOLORIZING CARBON

See videos on *Recrystallization*

A colored solution may result when a solid is dissolved in a suitable solvent for recrystallization. The color can arise from the compound itself or from the presence of colored impurities in the solution. Similarly, an organic liquid may be colored because of impurities or because of the nature of the compound itself. If you are in doubt about the cause of the color, you may try to remove it by adding activated **decolorizing carbon** to the solution or liquid and then removing the carbon by filtration; sometimes heating is required. Decolorizing carbon, which is commercially available under trade names such as Norite® or Darco®, is finely divided, activated charcoal with an extremely large surface area. Polar, colored impurities bind preferentially to the surface of decolorizing carbon; however, you should be aware that some of the desired compound may also be adsorbed, so excessive quantities of decolorizing carbon should *not* be used. After the decolorizing carbon is removed by filtration, the liquid or solution should be free of colored impurities; any remaining color is probably due to the compound itself.

Miniscale Technique

In a typical miniscale procedure, the solution, contained in an Erlenmeyer flask, is first brought to a temperature 10–20 °C *below* its boiling point. Decolorizing carbon is then added, and the hot solution is gently swirled. There are no firm rules concerning the amount of decolorizing carbon that should be used, but a good first approximation is to add about 0.5–1 g per 100 mL of solution. The decolorizing carbon is then removed by gravity filtration or hot gravity filtration, not vacuum filtration, using fluted filter paper; a filter-aid (Sec. 2.17) will facilitate complete removal of all of the carbon. Hot gravity filtration must be used whenever an organic solid is dissolved in a hot solvent, such as during a recrystallization. The solution or liquid must be refiltered if it contains any small black specks of carbon or any other solid material. Normally this extra step can be avoided if care is exercised when first filtering the mixture and a filter-aid is used. For example, the filter paper should not contain any tears or holes, and the mixture should be carefully poured onto the filter paper so that none of it runs down the side of the funnel. If the compound is known to be colorless and if the decoloration process does not remove the color, the procedure should be repeated.

Microscale Technique

The microscale procedure for decolorizing solutions is analogous to that used at the miniscale level. A small amount of powdered decolorizing carbon (10–20 mg) is added to the solution, which is contained in a conical vial or a test tube and is at a temperature 10–20 °C *below* its boiling point. After the mixture is gently swirled or stirred for several minutes, the decolorizing carbon is removed by hot filtration (Sec. 2.17). The particles of carbon are very fine, however, and some of them may pass through the cotton plug of the Pasteur filtering pipet, in which case the filtrations should be repeated, preferably with a clean filtering pipet. You may also make a small bed of a filter-aid such as Celite on top of the cotton plug to facilitate removal of the decolorizing carbon.

Pelletized carbon may be available in your laboratory, and you may be instructed to use it instead of the powdered form. Because of its larger size (pelletized carbon is about 10 mesh, or 2 mm in diameter, whereas powdered carbon is typically 100 mesh, or 0.14 mm in diameter), it is easier to separate from the solution. However, pelletized carbon is less efficient at removing colored contaminants because of its lower surface area.

2.19 DECANTING SOLUTIONS

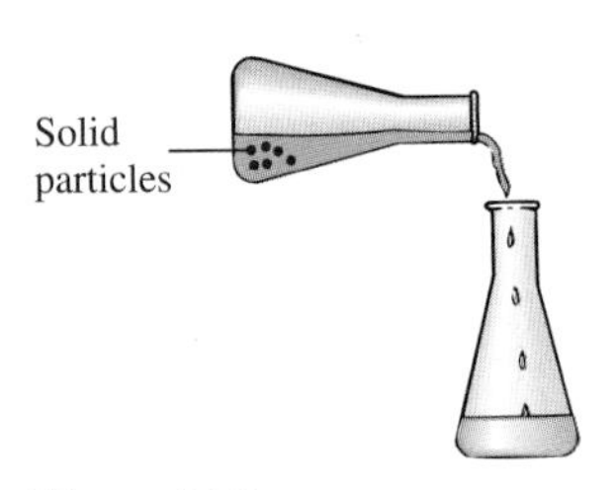

Figure 2.56
Decantation.

Small amounts of dense solids are sometimes removed by simple **decantation.** This technique is a viable alternative to filtration for removing solid drying agents (Sec. 2.24), but it cannot be used to remove finely dispersed solids such as decolorizing carbon. To decant a liquid from a solid, the solid should first be allowed to settle to the bottom of the container. The container is then *carefully* tilted, and the liquid is *slowly* poured into a clean container, possibly with the aid of a glass stirring rod to minimize dripping down the outside of the flask (Fig. 2.56); the solid should remain in the original container. Decantation is preferable to gravity filtration when working with very volatile organic liquids, since filtration is likely to result in considerable evaporation and loss of material. When decanting from an Erlenmeyer flask, a *loosely packed* ball of glasswool can be put in the neck of the flask to help keep the solid in the flask.

One major disadvantage of decantation is that some liquid will remain in the flask; this problem may be minimized by using a Pasteur pipet to transfer the last few milliliters and rinsing the residue with several milliliters of pure solvent that are also transferred with a pipet.

2.20 SUBLIMATION

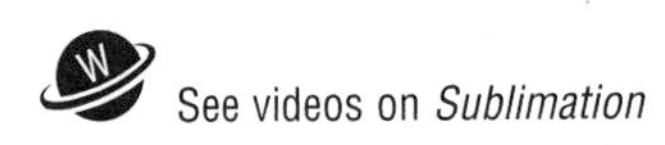

See videos on *Sublimation*

Like the vapor pressure of a liquid, the vapor pressure of a solid increases with temperature. If the vapor pressure of a solid is greater than the ambient pressure at its melting point, then the solid undergoes a direct-phase transition to the gas phase without first passing through the liquid state. This process is called **sublimation.**

The ease with which a molecule may escape from the solid to the vapor phase is determined by the strength of the intermolecular attractive forces between the molecules of the solid. Symmetrical structures have relatively uniform distributions of electron density, and they have small dipole moments compared to less symmetrical molecules. Since electrostatic interactions are the strongest intermolecular forces in the crystal lattice, molecules with smaller dipole moments will have higher vapor pressures. Sublimation thus is generally a property of relatively nonpolar substances having fairly symmetrical structures. Van der Waals forces, which are weaker than electrostatic ones, are also important and increase in magnitude with increasing molecular weight; hence, large molecules, even if symmetrical, tend not to sublime.

To be purified by sublimation, a compound must have a relatively high vapor pressure, and the impurities must have vapor pressures significantly lower than that of the compound being purified. If the impurities in a compound have similar vapor pressures, recrystallization (Sec. 3.2) or column chromatography (Sec. 6.3) may be used to purify the compound. Since few organic solids exhibit vapor pressures high enough to sublime at atmospheric pressure, most sublimations are performed at reduced pressure.

Two common types of sublimation apparatus are shown in Figure 2.57. In each case there is a chamber that may be evacuated using a water aspirator, house vacuum, or vacuum pump and a cold-finger in the center of the vacuum chamber to provide a surface upon which the sublimed crystals may form. The cold-finger is cooled by water (Fig. 2.57a) or another cooling medium such as ice/water or dry ice/acetone (Fig. 2.57b). A simple sublimation apparatus may be assembled

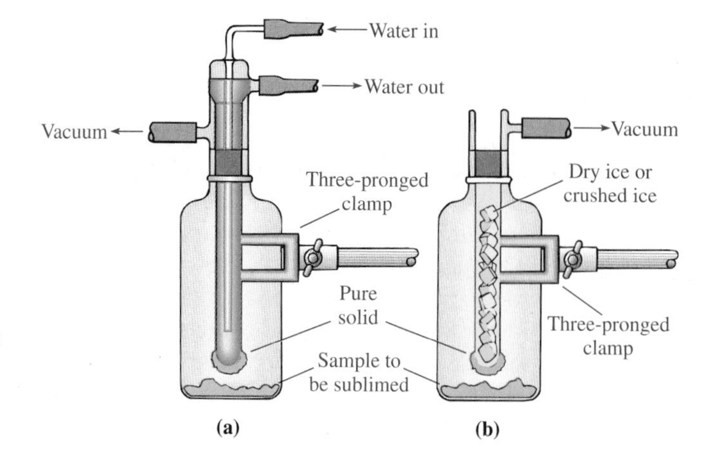

Figure 2.57
Miniscale sublimation apparatus: (a) using water as the coolant; (b) using low-temperature coolants.

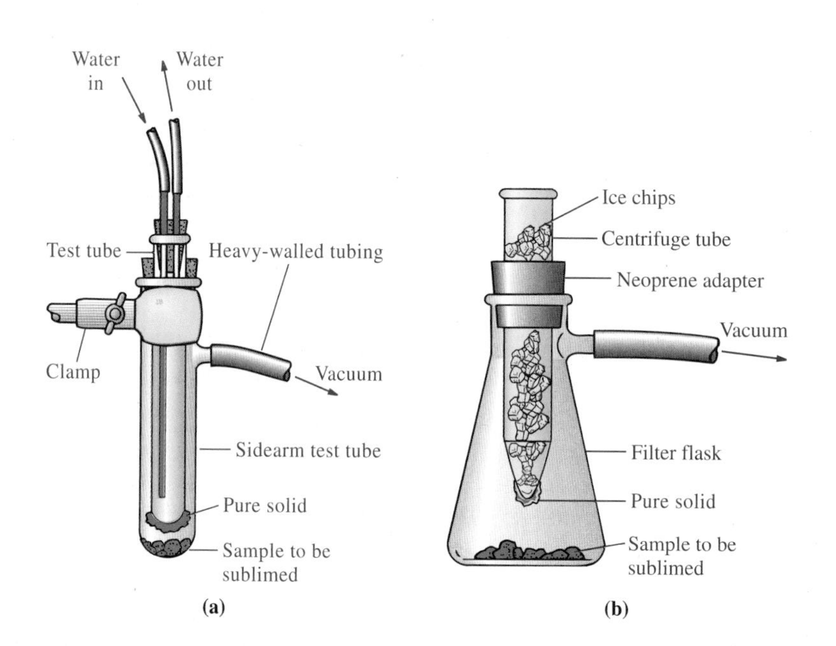

Figure 2.58
Simple sublimation apparatus: (a) test tube sublimator; (b) filter-flask sublimator.

inexpensively from two test tubes, one having a sidearm and rubber stoppers (Fig. 2.58a). In place of the test tube with a sidearm, a small vacuum-filtration flask may also be used (Fig. 2.58b). The cold-finger may be cooled either with running water or with ice chips. If the cold-finger is cooled with running water, the water hoses must be securely attached to the inlet and outlet of the condenser by means of a piece of twisted copper wire or a hose clamp.

Commercial sublimation adapters such as those shown in Figure 2.59 are available for microscale work. A Winston adapter is normally used with a sublimation chamber such as a 3- or 5-mL conical vial (Fig. 2.59a), whereas adjustable adapters may be used with 5- or 10-mL round-bottom flasks (Fig. 2.59b). In both cases a cold-finger that may be connected to a vacuum source is fitted to the chamber containing the substance to be sublimed, and the connection is secured with a standard-taper joint that is often accompanied by an O-ring and threaded collar (Fig. 2.59). The cold-finger can be filled with ice or dry ice. The advantage of using

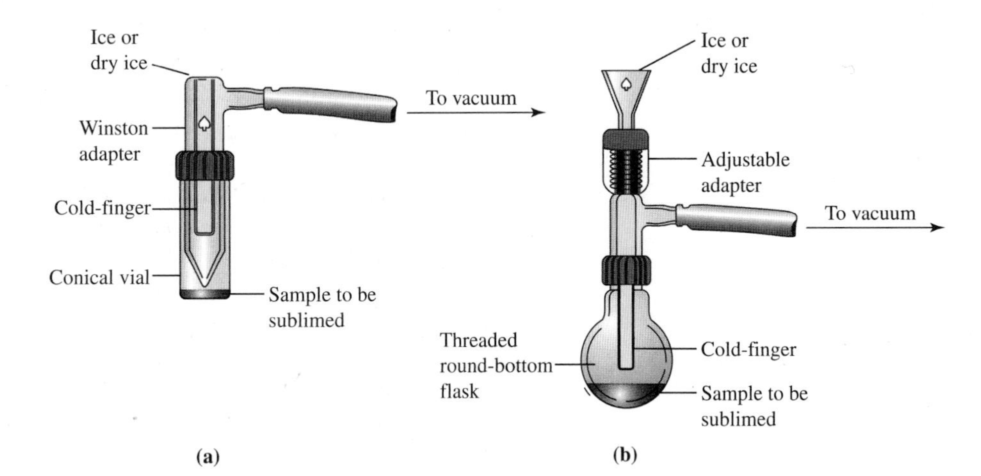

Figure 2.59
Commercial microscale sublimation apparatus: (a) conical vial with Winston adapter; (b) round-bottom flask with adjustable adapter.

these adapters is that air- and water-tight seals are easier to obtain because of the standard-taper connections, and the one-piece construction makes it easier to remove the crystals from the cold-finger. However, they suffer the disadvantage of being more expensive than the simple apparatus previously described.

In order to purify an impure substance by sublimation, it is first placed at the bottom of the sublimation chamber. The sample is then heated under reduced pressure, using an oil or sand bath or a small flame, to a temperature *below the melting point* of the solid. The solid will be vaporized and transferred via the vapor phase to the surface of the cold-finger, where it condenses to form a pure solid. After the sublimation is complete, the pressure must be released carefully to avoid dislodging the crystals from the cold-finger with a surge of air. For similar reasons, care must be exercised when removing the cold-finger from the sublimation apparatus. The pure crystals are scraped from the cold-finger with a spatula.

2.21 EXTRACTION

See videos on *Extraction*

A common technique for isolating and purifying the product of a chemical reaction involves liquid-liquid extraction, or simply **extraction;** the theory and applications are discussed in Chapter 5. This process involves transferring a solute from one solvent into another, because of its greater solubility in the second. The two solvents must be immiscible and form two distinct layers, and in general one layer is aqueous and the other is an organic solvent such as diethyl ether, hexane, or dichloromethane. Depending upon the amounts of material, the physical separation of the two immiscible phases will be performed in **separatory funnels** or **conical vials.**

Extractions Using Separatory Funnels: Miniscale Technique

Separatory funnels are available in many different shapes, ranging from pear-shaped to nearly conical (Fig. 2.60); the more elongated the funnel, the longer the time required for the two liquid phases to separate after the funnel is shaken. Although separatory funnels may be equipped with either a glass or Teflon stopcock, the latter does not require lubrication and thus is preferred because the solutions being separated do not get contaminated with stopcock grease. Separatory funnels are most commonly used during work-up procedures after completion of a chemical reaction. For example, they are used for extracting the desired product

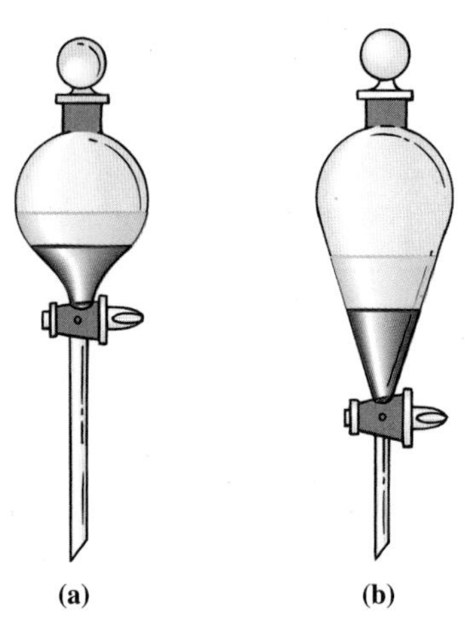

Figure 2.60
Separatory funnels: (a) conical; (b) pear-shaped.

from one immiscible liquid phase into another (Secs. 5.2 and 5.3) and for "washing" organic layers to remove undesired substances such as acids or bases from the desired organic compound.

There are a number of general guidelines for using separatory funnels that merit discussion:

1. **Filling Separatory Funnels.** The stopcock should be closed and a clean beaker placed under the funnel before any liquids are added to the funnel in case the stopcock leaks or is not completely closed. A separatory funnel should never be more than three-quarters full. The upper opening of the funnel is then stoppered with a ground-glass, plastic, or rubber stopper; most separatory funnels are now fitted with a ground-glass or plastic stopper.
2. **Holding and Using Separatory Funnels.** If the contents of the funnel are to be shaken, it is held in a specific manner. If the user is right-handed, the stopper should be placed against the base of the index finger of the left hand and the funnel grasped with the first two fingers and the thumb. The thumb and the first two fingers of the right hand can then be curled around the stopcock (Fig. 2.61a). Holding the funnel in this manner permits the stopper and the stopcock to be held tightly in place during shaking. A left-handed person might find it easier to use the opposite hand for each position. If you are lucky enough to be ambidextrous, take your choice!
3. **Shaking Separatory Funnels.** A separatory funnel and its contents should be shaken to mix the immiscible liquids as intimately as possible (Fig. 2.61b). The shaking process increases the surface area of contact between the immiscible liquids so that the equilibrium distribution of the solute between the two layers will be attained quickly; however, overly vigorous or lengthy shaking may produce **emulsions** (discussed below).

   The funnel must be vented every few seconds to avoid the buildup of pressure within the funnel. Venting is accomplished by *inverting the funnel with the stopcock pointing upward and away from you and your neighbors* and slowly opening it to release any pressure (Fig. 2.61c). If the funnel is not carefully vented, liquid may be violently expelled, covering you and your laboratory partners with the contents. Venting is particularly important when using volatile, low-boiling solvents such as diethyl ether or methylene chloride; it is also necessary whenever an acid is neutralized with either sodium carbonate or sodium bicarbonate, because CO_2 is produced. If the funnel is not vented frequently, the stopper may be blown out accidentally; under extreme circumstances the funnel might blow up. Any sudden release of pressure is likely to result in the contents being lost and spattered on you and your coworkers—not a good way to make friends. The funnel may be vented simply by holding it as described previously and opening the stopcock by twisting the fingers curled around it without readjusting your grip on the funnel.

   At the end of the period of shaking (1–2 min are usually sufficient if the shaking is vigorous), the funnel is vented a final time. It is then supported on an iron ring (Fig. 2.62), and the layers are allowed to separate. In order to prevent the iron ring from cracking or breaking the separatory funnel, the ring should be covered with a length of rubber tubing. This may be accomplished by slicing the tubing along its side and slipping it over the ring. Copper wire may be used to fix the tubing permanently in place. The lower layer in the separatory funnel is then carefully dispensed into a flask through the stopcock

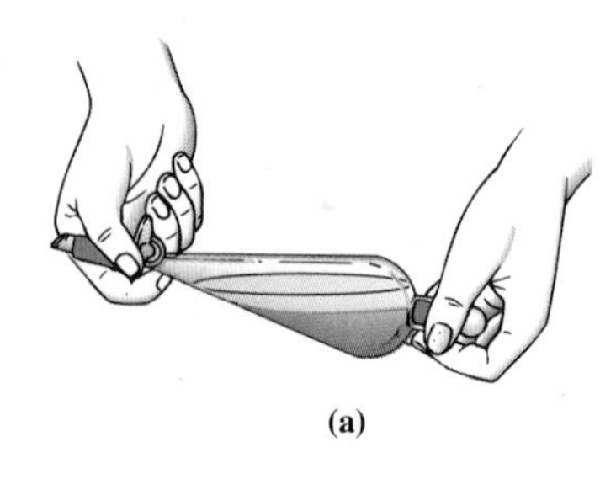

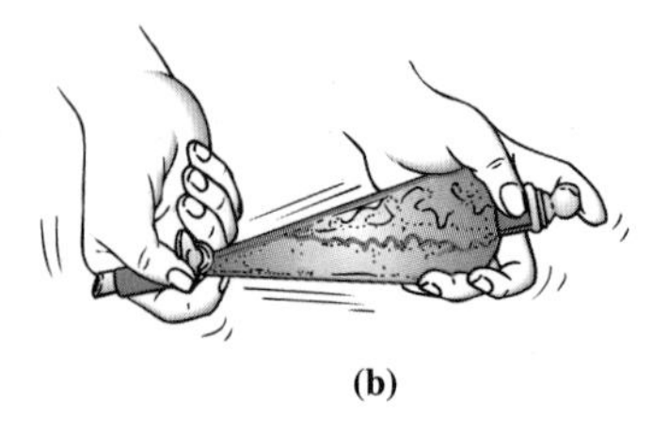

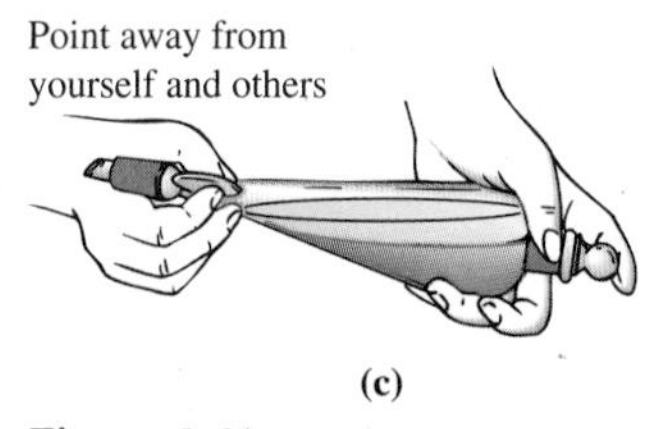

Figure 2.61
(a) Proper method for holding separatory funnel. (b) Shaking a separatory funnel. (c) Venting a separatory funnel.

while the interface between the two layers is watched. If small quantities of insoluble material collect at the boundary between the layers and make it difficult to see this interface, it is best to remove these solids with the *undesired* liquid layer; a small amount of the desired layer is inevitably lost by this procedure. An alternative procedure is to remove the solids by gravity or vacuum filtration before separating the layers.

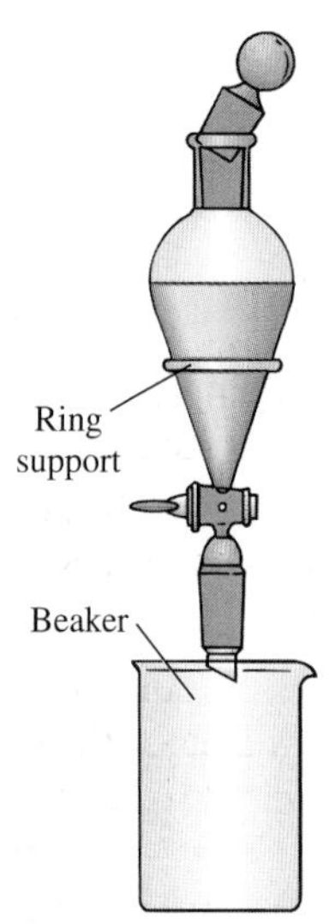

Figure 2.62
Separatory funnel positioned on iron ring with beaker located to catch liquid if funnel leaks.

4. **Layer Identification.** It is important to ascertain which of the two layers in a separatory funnel is the aqueous layer and which is the organic. Because the layers should separate so that the denser solvent is on the bottom, knowledge of the densities of the liquids being separated provides an important clue as to the identity of each layer. This generalization is not foolproof, however, because a high concentration of a solute in one layer may reverse the relative densities of the two liquids. You must not confuse the identity of the two layers in the funnel and then discard the layer containing your product. *Both layers should always be saved until there is no doubt about the identity of each and the desired product has been isolated.*

   Because one of the layers is usually aqueous and the other is organic, there is a simple and foolproof method to identify the two layers. Withdraw a few drops of the upper layer with a pipet and add these drops to about 0.5 mL of water in a test tube. If the upper layer is aqueous, these drops will be miscible with the water in the test tube and will dissolve, but if the upper layer is organic, the droplets will not dissolve.

5. **Emulsions.** Occasionally the two immiscible liquids will not separate cleanly into two distinct layers after shaking, because an **emulsion** may form that results from a colloidal mixture of the two layers. If prior experience leads you to believe that an emulsion might form, do not shake the funnel too vigorously; instead, swirl it gently to mix the layers. Encountering an emulsion during an experiment can be extraordinarily frustrating because there are no infallible, convenient procedures for breaking up emulsions. An emulsion left unattended for an extended period of time sometimes separates. However, it is usually more expedient to attempt one or more of the following remedies:

   a. Add a few milliliters of a *saturated* solution of aqueous sodium chloride, commonly called brine, to the funnel and gently reshake the contents. This increases the ionic strength of the water layer, which helps force the organic material into the organic layer. This process can be repeated, but if it does not work the second time, other measures must be taken.

   b. Filter the heterogeneous mixture by vacuum filtration through a thin pad of a filter-aid (Sec. 2.17), and return the filtrate to the separatory funnel. If a filter-aid is not used, the pores of the filter paper may become clogged, and the filtration will be slow. Sometimes emulsions are caused by small amounts of gummy organic materials whose removal will often remedy the problem.

   c. Add a *small* quantity of water-soluble detergent to the mixture and reshake the mixture. This method is not as desirable as the first two techniques, particularly if the desired compound is in the water layer, because the detergent adds an impurity that must be removed later.

   d. Intractable emulsions that appear to be stabilized by small trapped air bubbles are sometimes encountered during the work-up of phase-transfer reactions. If the separatory funnel is thick-walled, apply a gentle vacuum with a water aspirator to speed the separation of the phases.

e. If all these procedures fail, it may be necessary to select a different extraction solvent or seek the advice of your instructor.

Extractions Using Conical Vials and Centrifuge Tubes: Microscale Technique

The volumes of solvent are too small to use separatory funnels for extractions when reactions are conducted on a microscale level. In these instances, conical vials may be employed for volumes up to about 4 mL, and screw-cap centrifuge tubes may be used for volumes up to approximately 10 mL (Fig. 2.63). To avoid accidental spills, place the conical vial or centrifuge tube in a small beaker when you are not handling it.

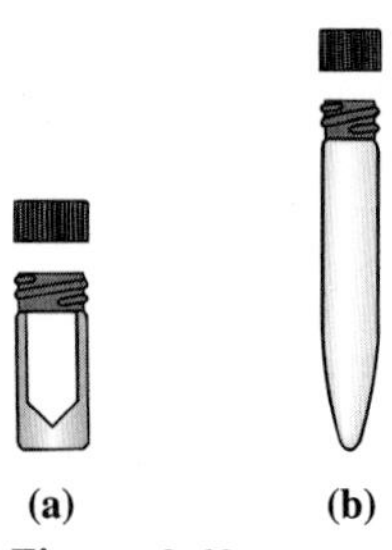

Figure 2.63
Apparatus for microscale extraction: (a) conical vial; (b) screw-cap centrifuge tube.

Before using either a conical vial or a centrifuge tube for an extraction, ensure that it does not leak when shaken. This may be done easily by placing about 1 mL of water in the vial or tube and screwing the cap on securely. Shake the vial or tube vigorously and check for leaks.

Most extractions involve an aqueous and an immiscible organic phase, and it is important to mix the two layers thoroughly when performing an extraction. This may be achieved in several ways, the simplest of which is to shake the vial or tube vigorously. Carefully vent the vial or tube by loosening the cap to release any pressure. A vortex mixer works well if one is available in your laboratory. Adequate mixing can sometimes be achieved by stirring the mixture with a magnetic stirbar or a microspatula for 5–10 min. Another technique involves drawing the mixture into a Pasteur pipet and rapidly squirting it back into the vial or tube; be careful not to squirt so hard that the mixture spills out of the vial and onto the bench. Repeat this operation several times to ensure adequate extraction. Multiple extractions with fresh solvent should be performed to obtain maximum recovery of your product.

Depending upon the solvent used, the organic phase may be either the upper or lower layer, and a slightly different procedure is used for separating each. The more common situation is one in which the organic solvent is less dense than water, and it is this layer that is removed. For example, if you extract an aqueous solution with diethyl ether (ether), the organic layer is the upper one because ether is less dense than water. Alternatively, when a solvent that is heavier than water, such as dichloromethane, is used for the extraction, the lower layer is the organic layer, and it is the one removed.

Microscale Technique for Removing the Upper Organic Layer

The procedure for removing the upper organic layer using conical vials is outlined in Figure 2.64; centrifuge tubes may be substituted for the conical vials for larger volumes.

1. Place the aqueous phase containing the dissolved product in a 5-mL conical vial (Fig. 2.64a).
2. Add about 1 mL of diethyl ether, cap the vial securely, and mix the contents by vigorous shaking or by one of the other techniques outlined previously. Unscrew the cap slightly to vent the pressure in the vial. Allow the phases to separate completely so you can see the two layers. If you are using centrifuge tubes, centrifugation may facilitate this separation. The diethyl ether phase is the upper layer (Fig. 2.64b).

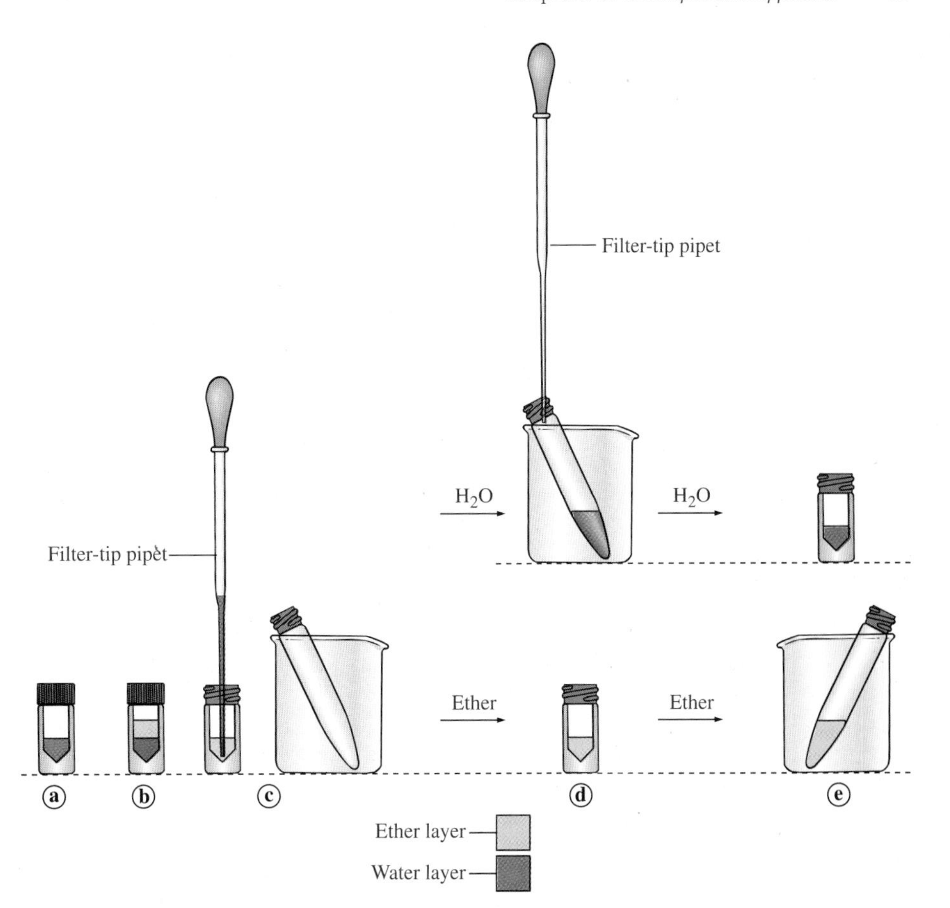

Figure 2.64
Extracting an aqueous solution using an organic solvent such as diethyl ether that is less dense than water: (a) the desired product is in the aqueous solution; (b) the aqueous phase is extracted with diethyl ether; (c) the lower (aqueous) phase is withdrawn; (d) the aqueous phase is transferred to a test tube; the ether layer remains in the conical vial; (e) the ether phase is transferred to a test tube, and the aqueous phase is returned to the extraction vial.

3. Place an empty test tube or conical vial adjacent to the extraction vial. You will need a beaker or test tube rack to hold a test tube. Attach a rubber bulb to a Pasteur filter-tip pipet (Fig. 2.9a), squeeze the bulb, and insert the tip of the pipet into the vial until it touches the bottom. Slowly withdraw the lower aqueous layer into the pipet while keeping the tip of the pipet on the bottom of the conical vial (Fig. 2.64c). Be sure to avoid transferring any of the organic layer or any emulsion that may remain at the interface of the two phases. Although a regular Pasteur pipet may also be used for this transfer, a filter-tip pipet is sometimes easier to control when removing small volumes of aqueous phase from the vial.
4. Transfer the aqueous phase you have drawn into the pipet into the empty test tube or conical vial, leaving the ether layer behind in the extraction vial (Fig. 2.64d). It is best to position the second container as close as possible to the extraction vial to avoid messy transfers and loss of your product.
5. Use a clean filter-tip pipet to transfer the ether layer from the extraction vial to a test tube for storage. Although a regular Pasteur pipet may also be used for this transfer, a filter-tip pipet is easier to control when removing small volumes of volatile organic solvents such as diethyl ether from the vial. Return the aqueous layer to the extraction vial using a Pasteur pipet (Fig. 2.64e).
6. Repeat steps **2–5** one or two times to achieve complete extraction of the desired material from the aqueous layer. Dry the organic layer with a suitable drying agent such as anhydrous sodium sulfate or magnesium sulfate (Sec. 2.25).

Microscale Technique for Removing the Lower Organic Layer

The procedure for removing the lower organic layer using conical vials is outlined in Figure 2.65; centrifuge tubes may be substituted for the conical vials for larger volumes.

1. Place the aqueous phase containing the dissolved product in a 5-mL conical vial (Fig. 2.65a).
2. Add about 1 mL of dichloromethane, cap the vial securely, and mix the contents by vigorous shaking or one of the other techniques outlined previously. Unscrew the cap slightly to vent the pressure in the vial. Allow the phases to separate completely so you can see the two layers. If you are using centrifuge tubes, centrifugation may facilitate this separation. The dichloromethane layer is the lower layer. Tapping the vial or gently stirring it with a microspatula may also help the layers separate (Fig. 2.65b).
3. Place an empty test tube or conical vial adjacent to the extraction vial. You will need a beaker or test tube rack to hold a test tube. Attach a rubber bulb to a Pasteur filter-tip pipet (Fig. 2.9a), squeeze the bulb, and insert the tip of the pipet into the vial until it touches the bottom. Slowly withdraw the lower organic layer into the pipet while keeping the tip of the pipet on the bottom of the conical vial (Fig. 2.65c). Be sure to avoid transferring any of the aqueous layer or any emulsion that may remain at the interface of the two phases.

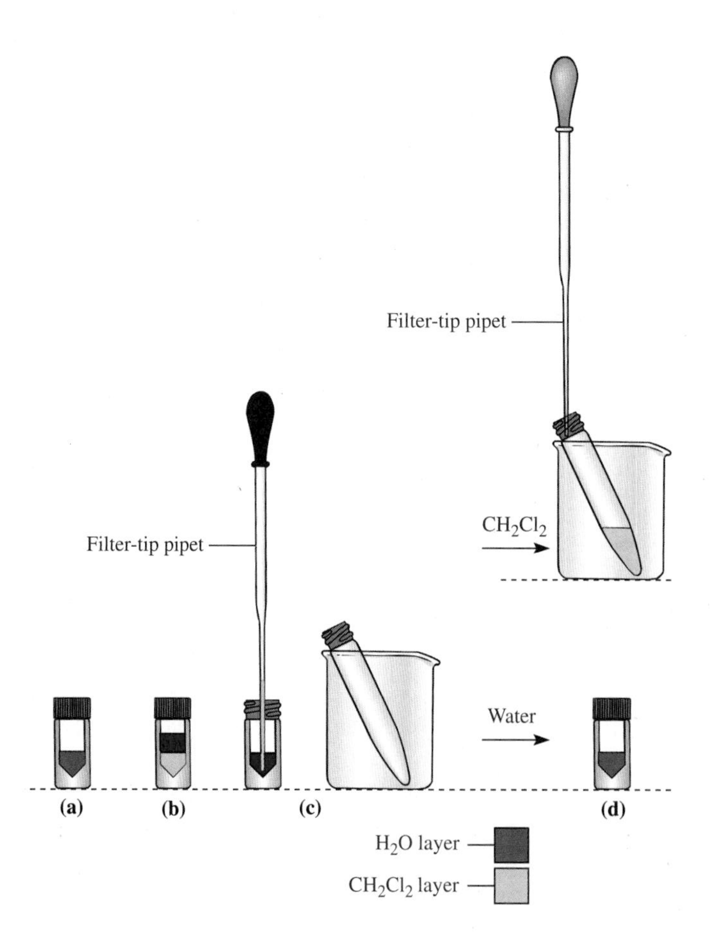

Figure 2.65
Extracting an aqueous solution using an organic solvent such as dichloromethane (CH_2Cl_2) that is denser than water: (a) the desired product is in the aqueous solution; (b) the aqueous phase is extracted with dichloromethane; (c) the lower (organic) phase is withdrawn; (d) the organic phase is transferred to a test tube; the aqueous phase remains in the conical vial.

Although a regular Pasteur pipet may also be used for this transfer, a filter-tip pipet is easier to control when removing small volumes of volatile organic solvents such as dichloromethane from the vial.

4. Transfer the organic phase you have drawn into the pipet into the *dry* empty test tube or *dry* conical vial, leaving the aqueous layer behind in the extraction vial (Fig. 2.65d). It is best to position the second container as close as possible to the extraction vial to avoid messy transfers and loss of your product.
5. Repeat steps **2–4** one or two times to achieve complete extraction of the desired material from the aqueous layer. Dry the organic layer with a suitable drying agent such as anhydrous sodium sulfate or magnesium sulfate (Sec. 2.25).

Layer Identification

It is important to ascertain which of the two layers in a conical vial is the aqueous layer and which is the organic. Because the layers will usually separate so that the denser solvent is on the bottom, knowledge of the densities of the liquids being separated provides an important clue as to the identity of each layer. This generalization is not foolproof, however, because a high concentration of a solute in one layer may reverse the relative densities of the two liquids. You must not confuse the identity of the two layers in the conical vial or test tube and then discard the layer containing your product. *Both layers should always be saved until there is no doubt about the identity of each and the desired product has been isolated.*

Since one of the layers is usually aqueous and the other is organic, there is a simple and foolproof method to identify the two layers: add a few drops of water to the layer you believe to be the aqueous one. Watch closely as you add the water drops to see whether they dissolve and increase the volume of the layer. If the drops form a new layer on either the top or bottom of the layer you are testing, then the test layer is organic. Similarly, you can test to see if a layer is organic by adding a few drops of the organic solvent you are using, for example diethyl ether or dichloromethane. If the volume of the layer increases, it is organic, but if a new layer forms, the original layer is aqueous.

2.22 HEATING UNDER REFLUX

See videos on *Heating under Reflux*

The term **heating under reflux** means that a reaction mixture is heated at its boiling point in a flask equipped with a **reflux condenser** to allow continuous return of the volatile materials to the flask; no solvent or reactant is removed or lost. Since the reaction can be conducted at higher temperatures using this technique, less time is required for its completion.

Miniscale Apparatus

A typical apparatus used for heating a miniscale reaction under reflux is shown in Figure 2.66a (see Sec. 2.4). The solvent, if any, and reactants are placed in a boiling flask, together with a stirbar; this flask should be set up about 15 cm or more from the bench top to allow for fast removal of the heating source if necessary. The flask is then fitted with a reflux condenser with water running slowly through it; the water hoses should be secured to the inlet and outlet of the condenser with copper wire or a hose clamp. *Do not stopper the top of the condenser, because a closed system should never be heated.* The heating source is then placed under the flask, and the contents are slowly heated to the boiling point of the mixture; slow heating makes

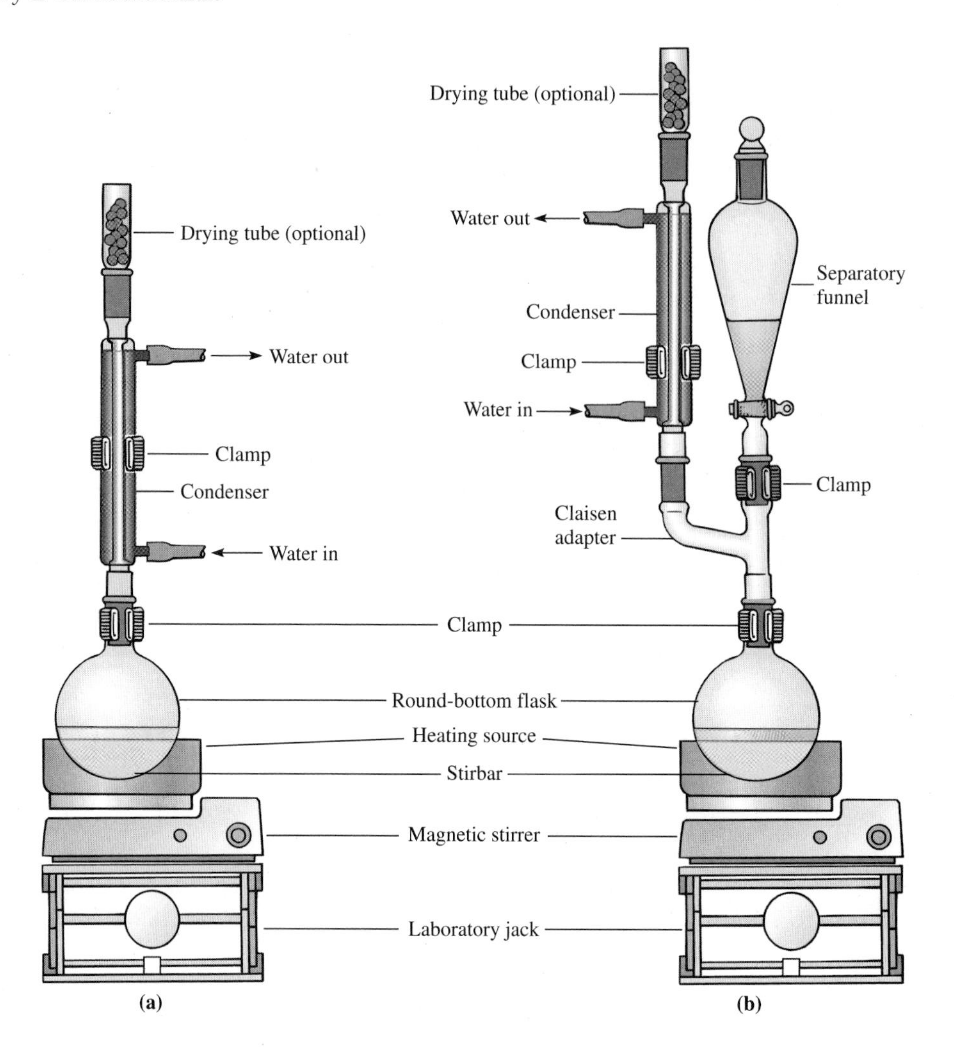

Figure 2.66
(a) Miniscale apparatus for heating a reaction mixture under reflux. (b) Miniscale apparatus for a reaction being heated under reflux and equipped for adding a liquid reagent or solution.

it possible to control any sudden exothermicity more readily. The volatile components of the mixture vaporize and reliquefy in the condenser, and the condensate returns to the boiling flask, where the process continues. A **reflux ring** appears in the condenser at the boundary of the two zones, and the condenser appears dry above this ring. The vapors should rise no more than about 3–6 cm above the bottom of the condenser. If they rise to the top of the condenser or are "seen" being emitted into the room, then you are probably overheating the flask. However, the flow of water through the condenser may also be insufficient, or the condenser may be too small to permit adequate cooling of the vapors. The problem may usually be corrected by lowering the temperature of the heat source or adjusting the flow of water through the condenser. Upon completion of the reaction, the heat is removed, and the flask is allowed to cool to room temperature. This ensures that all of the material in the condenser returns to the flask.

Many chemical reactions are executed by adding all the reagents, including solvent and catalyst, if any, to a reaction flask and heating the resulting mixture under

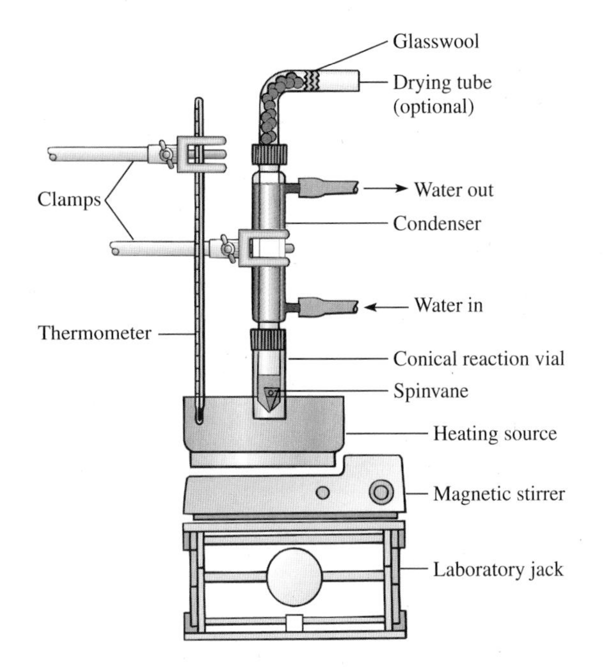

Figure 2.67
Microscale apparatus for heating a reaction mixture under reflux.

reflux; however, sometimes one of the reactants must be added to the reaction during reflux. For example, if a reaction is highly exothermic, its rate may be controlled by slowly adding one of the reagents. Other reactions may require that one of the reagents be present in high dilution to minimize the formation of side-products. Figure 2.66b shows a typical apparatus that allows a liquid reagent or solution to be added to the reaction flask using an addition funnel. The standard-taper separatory funnel available in many glassware kits can serve as an addition funnel, but there must be some provision for air to be admitted above the surface of the liquid to equalize the pressure inside; this can be accomplished by *not* placing the stopper on top of the funnel or by inserting a small piece of filter paper or copper wire between the stopper and the funnel. The flask can be heated to reflux during the addition, or the temperature in the flask can be maintained at room temperature or below with an ice-water bath and heated under reflux following completion of the addition.

Microscale Technique

The components of the apparatus for heating a reaction under reflux on the microscale are the same as for a preparative scale reaction. A common set-up is shown in Figure 2.67. It is generally best to use a water-jacketed condenser to prevent any vapors from escaping; a short-air condenser is frequently inadequate for the more volatile organic solvents such as diethyl ether and dichloromethane.

2.23 GAS TRAPS

Some organic reactions release noxious gases that *should not* escape into the laboratory atmosphere, as this would create an unpleasant and perhaps unsafe environment. Such reactions should be performed in the hood, if space is available. Alternatively, a **gas trap** can be used.

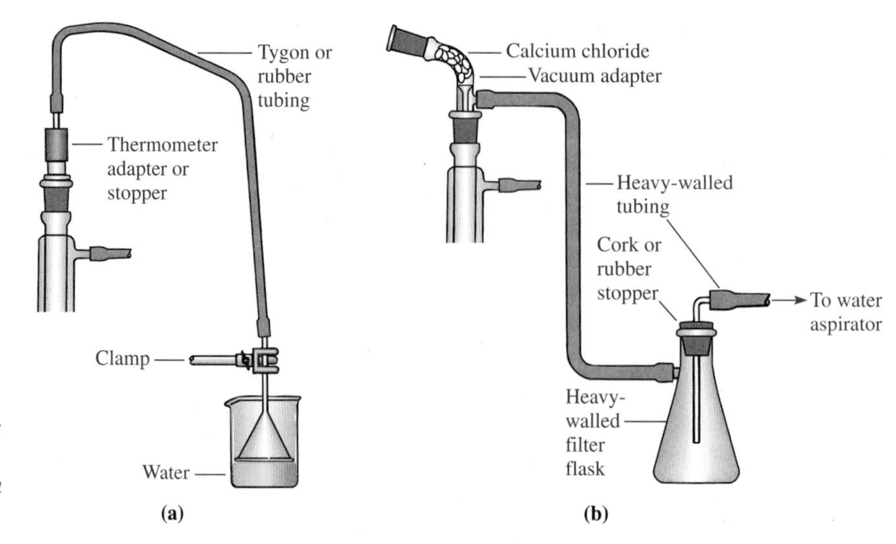

Figure 2.68
Gas traps for miniscale apparatus: (a) trapping water-soluble gas in water; (b) removing gases through water aspirator.

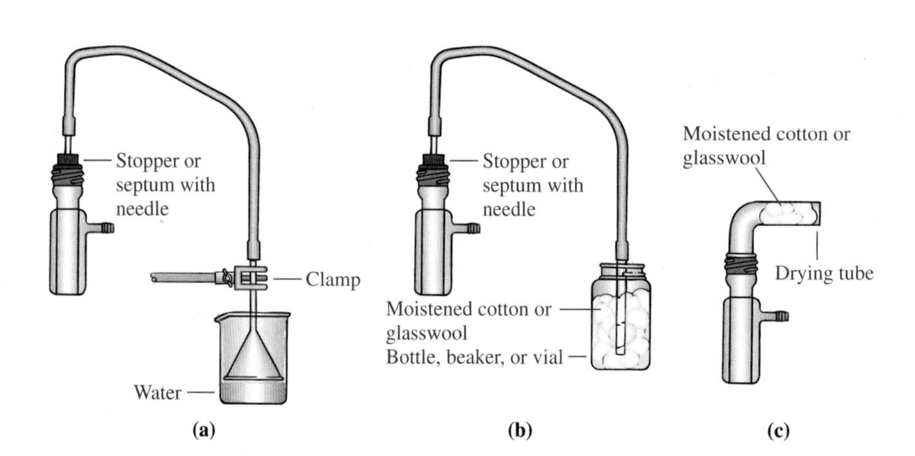

Figure 2.69
Gas traps for microscale apparatus: (a) trapping water-soluble gas in water; (b) trapping water-soluble gas in moistened cotton or glasswool; (c) simplified modification for trapping water-soluble gas in moistened cotton or glasswool.

Miniscale

If the gas being generated is water-soluble, the trap shown in Figure 2.68a may be suitable. In this example, the vapors are carried into a stemmed funnel whose rim is just above or resting on the surface of the water. A second modification evacuates the undesired water-soluble gas through a house vacuum or water aspirator protected by a trap (Fig. 2.68b). This style of gas trap uses a vacuum adapter attached at the top of a reflux condenser. If the reaction being performed is sensitive to atmospheric moisture, the adapter can be filled with anhydrous calcium chloride, which dries the air passing through it. Only a slight vacuum is needed to make this type of trap effective, so do *not* turn the house vacuum or water aspirator on full-force.

Microscale

The style of trap portrayed in Figure 2.68a can also be used with a microscale apparatus. A typical apparatus of this sort is shown in Figure 2.69a. A simplified analog is shown in Figure 2.69b, wherein the water-soluble gas is trapped in the water covering the surface of a cotton or glasswool packing. In some cases, simply putting moistened cotton or glasswool in a drying tube attached to the apparatus (Fig. 2.69c) will suffice to trap the gas being evolved.

2.24 DRYING AGENTS

Drying a reagent, solvent, or product is a task that must be performed at some stage of nearly every reaction conducted in the organic chemistry laboratory. The techniques of drying solids and liquids are described in this and the following sections.

Drying Agents and Desiccants

Most organic liquids are distilled at the end of the purification process, and any residual moisture that is present may react with the compound during the distillation; water may also co- or steam-distill with the liquid and contaminate the distillate. In order to remove these small traces of moisture before distillation, **drying agents,** sometimes called **desiccants,** are used. There are two general requirements for a drying agent: (1) neither it nor its hydrolysis product may react chemically with the organic liquid being dried, and (2) it must be *completely* and *easily* removed from the dry liquid. A drying agent should also be efficient so that the water is removed in a reasonably short period of time.

Some commonly used drying agents and their properties are listed in Table 2.1. These desiccants function in one of two ways: (1) the drying agent interacts *reversibly* with water by the process of adsorption or absorption (Eq. 2.1), or (2) it reacts irreversibly with water by serving as an acid or a base.

With drying agents that function by reversible hydration, a certain amount of water will remain in the organic liquid in equilibrium with the hydrated drying agent. The lesser the amount of water left at equilibrium, the greater the efficiency of the desiccant. A drying agent that forms a hydrate (Eq. 2.1) must be *completely* removed by gravity filtration (Sec. 2.17) or by decantation (Sec. 2.19) *before* the dried liquid is distilled, since many hydrates decompose with loss of water at temperatures above 30–40 °C.

$$\underset{\text{Anhydrous}}{\text{Drying agent (solid)}} + x\,H_2O\,\text{(liquid)} \rightleftharpoons \underset{\text{Hydrate}}{\text{Drying agent} \cdot x\,H_2O\,\text{(solid)}} \qquad (2.1)$$

Drying agents that remove water by an irreversible chemical reaction are very efficient, but they are generally more expensive than other types of drying agents. Such drying agents are sometimes more difficult to handle and are normally used to remove *small* quantities of water from reagents or solvents prior to a chemical reaction. For example, phosphorus pentoxide, P_2O_5, removes water by reacting vigorously with it to form phosphoric acid (Eq. 2.2). Desiccants such as calcium hydride (CaH_2) and sodium (Na) metal also react vigorously with water. When CaH_2 or Na metal is used as a drying agent, hydrogen gas is evolved, and appropriate precautions must be taken to vent the hydrogen and prevent buildup of this highly flammable gas.

$$P_2O_5\,\text{(solid)} + 3\,H_2O\,\text{(liquid)} \longrightarrow 2\,H_3PO_4 \qquad (2.2)$$

Of the drying agents listed in Table 2.1, *anhydrous* calcium chloride, sodium sulfate, and magnesium sulfate will generally suffice for the needs of this introductory laboratory course. Both sodium sulfate and magnesium sulfate have high capacities and absorb a large amount of water, but magnesium sulfate dries

Table 2.1 *Table of Common Drying Agents, Their Properties, and Their Uses*

Drying Agent	*Acid-Base Properties*	*Comments**
$CaCl_2$	Neutral	High capacity and fast action with reasonable efficiency; good preliminary drying agent; readily separated from dried solution because $CaCl_2$ is available as large granules; cannot be used to dry either alcohols and amines (because of compound formation) or phenols, esters, and acids (because drying agent contains some $Ca(OH)_2$).
Na_2SO_4	Neutral	Inexpensive, high capacity; relatively slow action and low efficiency; good general preliminary drying agent; preferred physical form is that of small granules, which may be easily separated from the dry solution by decantation or filtration.
$MgSO_4$	Weakly acidic	Inexpensive, high capacity, rapid drying agent with moderate efficiency; excellent preliminary drying agent; requires filtration to remove drying agent from solution.
H_2SO_4	Acidic	Good for alkyl halides and aliphatic hydrocarbons; cannot be used with even such weak bases as alkenes and ethers; high efficiency.
P_2O_5	Acidic	See comments under H_2SO_4; also good for ethers, aryl halides, and aromatic hydrocarbons; generally high efficiency; preliminary drying of solution recommended; dried solution can be distilled from drying agent.
CaH_2	Basic	High efficiency with both polar and nonpolar solvents, although inexplicably it fails with acetonitrile; somewhat slow action; good for basic, neutral, or *weakly* acidic compounds; cannot be used for base-sensitive substances; preliminary drying of solution is recommended; dried solution can be distilled from drying agent. *Caution:* Hydrogen gas is evolved with this drying agent.
Na or K	Basic	High efficiency; cannot be used on compounds sensitive to alkali metals or to base; special precautions and great care must be exercised in destroying excess drying agent in order to avoid fires; preliminary drying *required*; dried solution can be distilled from drying agent. *Caution:* Hydrogen gas is evolved with this drying agent.
BaO or CaO	Basic	Slow action but high efficiency; good for alcohols and amines; cannot be used with compounds sensitive to base; dried solution can be distilled from drying agent.
Molecular Sieve 3A or 4A†	Neutral	Rapid and highly efficient; preliminary drying recommended; dried solution can be distilled from drying agent if desired. *Molecular sieves* are aluminosilicates, whose crystal structure contains a network of pores of uniform diameter; the pore sizes of sieves 3A and 4A are such that only water and other small molecules such as ammonia can pass into the sieve; water is strongly adsorbed as water of hydration; hydrated sieves can be reactivated by heating at 300–320 °C under vacuum or at atmospheric pressure.

*Capacity, as used in this table, refers to the amount of water that can be removed by a given weight of drying agent; efficiency refers to the amount of water, if any, in equilibrium with the hydrated desiccant.
†The numbers refer to the nominal pore size, in Ångström units, of the sieve.

a solution more completely. Calcium chloride has a low capacity, but it is a more efficient drying agent than magnesium sulfate. Do *not* use an unnecessarily large quantity of drying agent when drying a liquid, since the desiccant may adsorb or absorb the desired organic product along with the water. Mechanical losses on filtration or decantation of the dried solution may also become significant. The amount of drying agent required depends upon the quantity of water present, the capacity of the drying agent, and the amount of liquid to be dried.

2.25 DRYING ORGANIC SOLUTIONS

When an organic solvent is shaken with water or an aqueous solution during an extraction procedure, it will contain some dissolved water. The amount of water varies with the organic solvent. Of the solvents commonly used for extraction, diethyl ether dissolves the most water. If this water is not removed prior to evaporation of the organic solvent, significant quantities of water will remain in the product and may complicate further purification. It is therefore necessary to dry the solution using a suitable drying agent, usually *anhydrous* sodium sulfate or magnesium sulfate.

Miniscale Technique

Place the organic liquid to be dried in an Erlenmeyer flask of suitable size so that it will be no more than half-filled. Start by adding a small spatula-tip full of drying agent such as *anhydrous* sodium sulfate or magnesium sulfate and swirl the flask gently. Swirling increases the surface area for contact between the solid and liquid phases and generally facilitates drying. If the drying agent "clumps" or if liquid still appears cloudy after the solid has settled to the bottom of the flask, add more drying agent and swirl again. Repeat this process until the liquid appears clear and some of the drying agent flows freely upon swirling the mixture. An amount of drying agent that covers the bottom of the flask should normally be sufficient. After drying is complete, remove the drying agent either by gravity filtration (Sec. 2.17) or by decantation (Sec. 2.19). Rinse the drying agent once or twice with a small volume of the organic solvent and transfer the rinse solvent to the organic solution.

Microscale Technique

Place the organic liquid to be dried in a conical vial or a test tube of suitable size so that it will be no more than half-filled. Start by adding a microspatula-tip full of drying agent such as *anhydrous* sodium sulfate or magnesium sulfate; granular anhydrous sodium sulfate is the most convenient for microscale manipulations. Stir the mixture with a microspatula. Stirring increases the surface area for contact between the solid and liquid phases and generally facilitates drying. If the drying agent "clumps" or if the liquid still appears cloudy after the solid has settled to the bottom of the flask, add more drying agent and stir again. Repeat this process until the liquid appears clear and some of the drying agent flows freely upon stirring the mixture. After drying is complete, use a *dry* Pasteur filter-tip pipet (Fig. 2.9a) or a *dry* Pasteur pipet to transfer the organic solution to a *dry* conical vial or test tube. Rinse the drying agent once or twice with a small volume of the organic solvent and transfer the rinse solvent to the organic solution.

2.26 DRYING SOLIDS

Solid organic compounds must be dried because the presence of water or organic solvents will affect their weight, melting point, quantitative elemental analysis, and spectra. Since sources of protons must be excluded from some reactions, it is also necessary to remove all traces of moisture or protic solvents from a solid prior to performing such a reaction.

A solid that has been recrystallized from a volatile organic solvent can usually be dried satisfactorily by allowing it to air-dry at room temperature, provided it is not **hygroscopic** and thus absorbs moisture from the air. After the solid is collected on a Büchner or Hirsch funnel by vacuum filtration (Sec. 2.17), it is first pressed as dry as possible with a clean spatula or cork while air is pulled through the funnel and solid. The solid is then spread on a piece of filter paper, which absorbs the excess solvent, or on a clean watchglass and allowed to stand overnight or longer at room temperature.

Water is sometimes removed from a solid by dissolving the solid in a suitable organic solvent such as chloroform or toluene, removing any water by azeotropic distillation, and then recovering the solid by removal of the solvent. However, water is more commonly removed from organic solids using desiccators containing desiccants such as silica gel, phosphorus pentoxide, calcium chloride, or calcium sulfate. The desiccator may be used at atmospheric pressure or under a vacuum; however, if a vacuum is applied, the desiccator must be enclosed in a metal safety cage or wrapped with electrical or duct tape. Desiccators or tightly stoppered bottles containing one of these desiccants may also be used to store dry solids contained in small vials.

If the sample is hygroscopic or if it has been recrystallized from water or a high-boiling solvent, it must be dried in an oven operating at a temperature 20–30 °C *below* the melting or decomposition point of the sample. The oven-drying process can be performed at atmospheric pressure or under vacuum. Air-sensitive solids must be dried either in an inert atmosphere, such as in nitrogen or helium, or under vacuum. Samples to be submitted for quantitative elemental analysis are normally dried to constant weight by heating them under vacuum.

2.27 DRYING TUBES

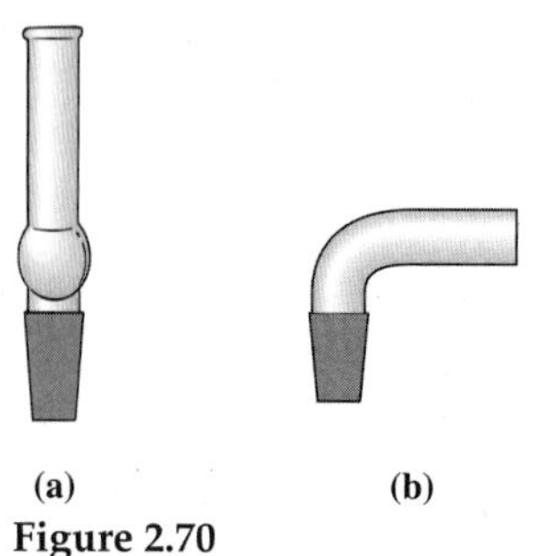

Figure 2.70
Drying tubes: (a) straight; (b) bent.

Frequently it is necessary to protect a reaction mixture from moisture. There are advanced techniques for doing this that involve performing the reaction under an inert, dry atmosphere of nitrogen or argon gas, but a simple and moderately effective procedure simply utilizes a drying tube containing a drying agent. Two types of drying tubes are shown in Figure 2.70. The straight tube (Fig. 2.70a) and the bent tube (Fig. 2.70b) are commonly found in glassware kits for miniscale and microscale experiments, respectively. A drying tube is prepared by placing a loose plug of glasswool at the bottom of the tube and filling the tube with a *granular* desiccant such as anhydrous calcium chloride or calcium sulfate. The drying agent often contains a blue indicator that turns pink when it is no longer effective. Although it is not necessary to cap the straight tube with glasswool, a loose plug of glasswool should be placed in the other end of the bent drying tube to prevent the drying agent from falling out. Neither the desiccant nor the glasswool should be packed too tightly.

2.28 DRYING APPARATUS

Reactions involving the use of water-sensitive reagents must be performed under anhydrous conditions. Thus, prior to conducting such a reaction and prior to introducing any reagents, as much moisture as possible must be removed from the apparatus itself.

The glass parts of an apparatus, such as round-bottom flasks, conical vials, and glass syringes, may be conveniently dried by placing them for 15–20 min in a drying oven heated at about 150 °C. Remove the glass components from the oven using a pair of tongs, *not* your bare hands, and if possible place them in a desiccator to keep them dry as they cool to room temperature. After the glassware is cool, remove it from the desiccator and assemble the apparatus as quickly as possible to avoid reintroduction of moisture on the glass surface. If a desiccator is not available, then allow the glassware to cool only to the point where you can handle it, and assemble the apparatus as quickly as possible to avoid reintroduction of moisture on the glass surface.

An alternative method that may be used is to "flame-dry" the apparatus. If you are instructed to use this technique, be sure that no one is working with flammable solvents in the room. The apparatus, which must have one opening to the atmosphere, is gently warmed with a Bunsen burner or microburner (Sec. 2.9), a process called "flaming." A heat gun (Sec. 2.9) may also be used to heat the apparatus. The apparatus is heated first at the point most remote from the opening to the atmosphere, and the flame is gradually worked toward the opening. The moisture in the apparatus will be driven through the opening. If a condenser is part of the apparatus, no water should be in or flowing through it. While the apparatus is still warm, a filled drying tube (Sec. 2.27) is inserted into the opening, and the system is allowed to cool to room temperature. As cooling occurs, air from the atmosphere is drawn into the apparatus through the drying agent, which removes the moisture, and the dried apparatus is ready for use.

2.29 EVAPORATING SOLVENTS

For many experiments, it is necessary to remove the excess solvents to recover the product. Although an Erlenmeyer flask or beaker containing the solution could simply be left unstoppered in the hood until the solvent has evaporated, this is rather slow and impractical unless you like spending excessive amounts of time in the laboratory. A variety of techniques may be used to speed the process. Several other methods for removing solvents in miniscale experiments are shown in Figure 2.71, and the related microscale techniques are illustrated in Figure 2.72.

Miniscale Methods

The solvents in solutions contained in Erlenmeyer flasks can be evaporated by adding several boiling stones or a wooden boiling stick to the flask and heating the flask with a suitable heat source such as a steam bath, a hot plate, or a sand or oil bath so that the solvent boils gently. Magnetic stirring may also be used to prevent bumping. The evaporation rate can be increased by directing a stream of air or nitrogen into the flask (Fig. 2.71a), or by applying a vacuum over the flask with an inverted funnel to help draw away solvent vapors (Fig. 2.71b). The solution may also be placed in a filter flask with a wooden boiling stick. After the flask is subjected to a gentle vacuum, it is swirled over the heat source to facilitate smooth evaporation and reduce the possibility of bumping (Fig. 2.71c).

Another common method for removing solvent is by simple distillation (Sec. 2.13). Discontinue heating when only a small amount of solvent remains, and *do not overheat the stillpot*, as your product may decompose if heated too strongly. The stillpot should be cooled to room temperature, and the receiving flask containing the distillate should be replaced with an empty round-bottom flask. The apparatus should then be set up for vacuum distillation (Fig. 2.42b) and attached to a vacuum source, such as a water aspirator, for a few minutes to remove the last traces of solvent.

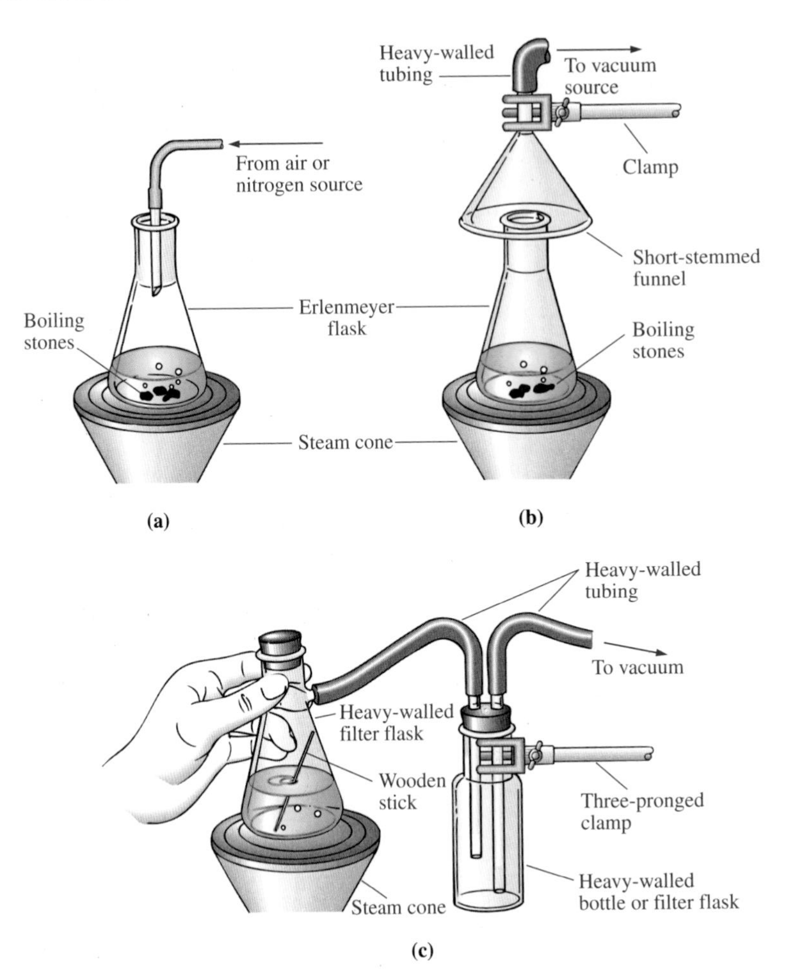

Figure 2.71
Miniscale techniques for evaporating solvents.

Microscale Methods

The solvents in solutions contained in a conical vial or test tube can be evaporated by heating the conical vial or test tube with a suitable heat source such as a sand bath or aluminum block. To prevent bumping, the solution should be agitated with magnetic stirring or with a boiling stone. The heat should also be adjusted so the solvent does not boil vigorously. The evaporation rate can be increased by directing a stream of air or nitrogen into the conical vial or test tube (Fig. 2.72a), or by applying a vacuum over the conical vial or test tube with an inverted funnel to help remove solvent vapors (Fig. 2.72b). The solution may also be placed in a test tube with a sidearm and a wooden boiling stick. After the test tube is subjected to a gentle vacuum, it is swirled over the heat source to facilitate smooth evaporation and reduce the possibility of bumping (Fig. 2.72c).

Another common method for removing solvent is by simple microscale distillation (Sec. 2.13). Discontinue heating when only a small amount of solvent remains, and *do not overheat the stillpot,* as your product may decompose if heated too strongly. The stillpot should be cooled to room temperature. Remove the Hickman stillhead and water condenser and insert an air condenser into the vial. Attach a thermometer adapter fitted with a glass tube to the top of the condenser and attach the tube to a vacuum source (Fig. 2.73). If you use a water aspirator, be sure to place

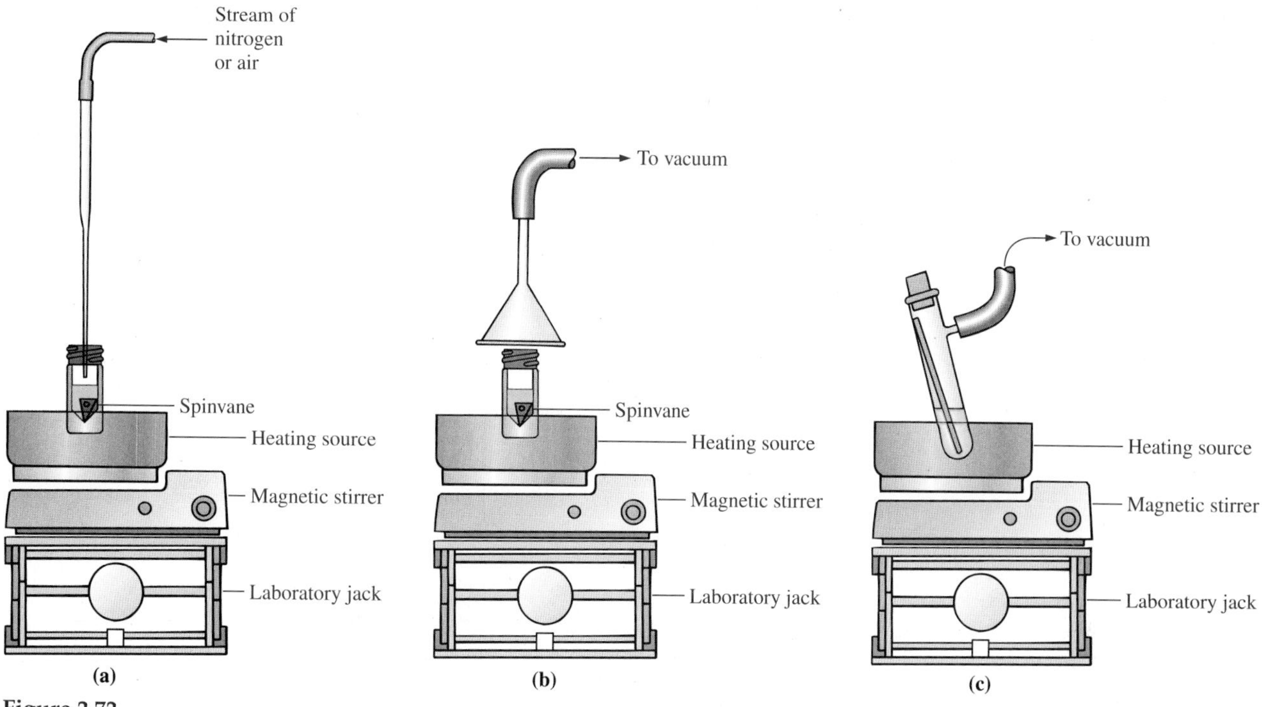

Figure 2.72
Microscale techniques for evaporating solvents.

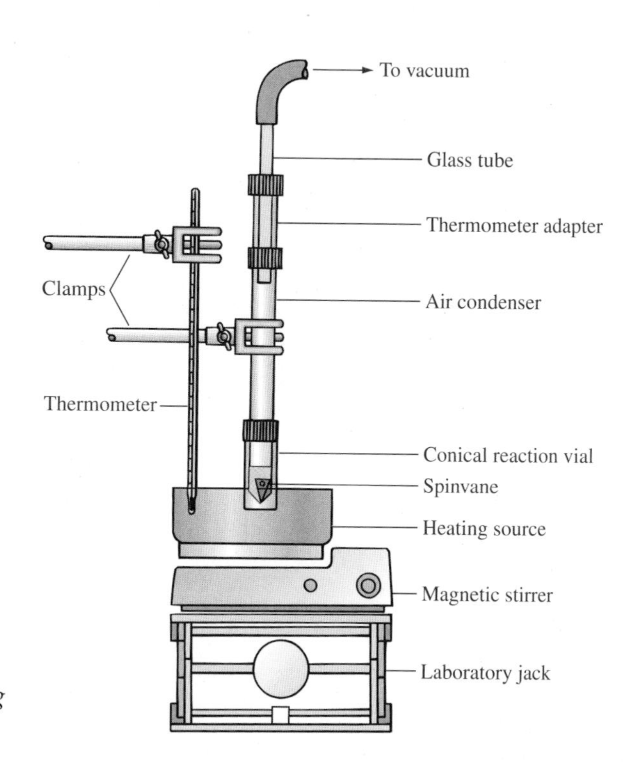

Figure 2.73
Microscale apparatus for removing solvent under reduced pressure.

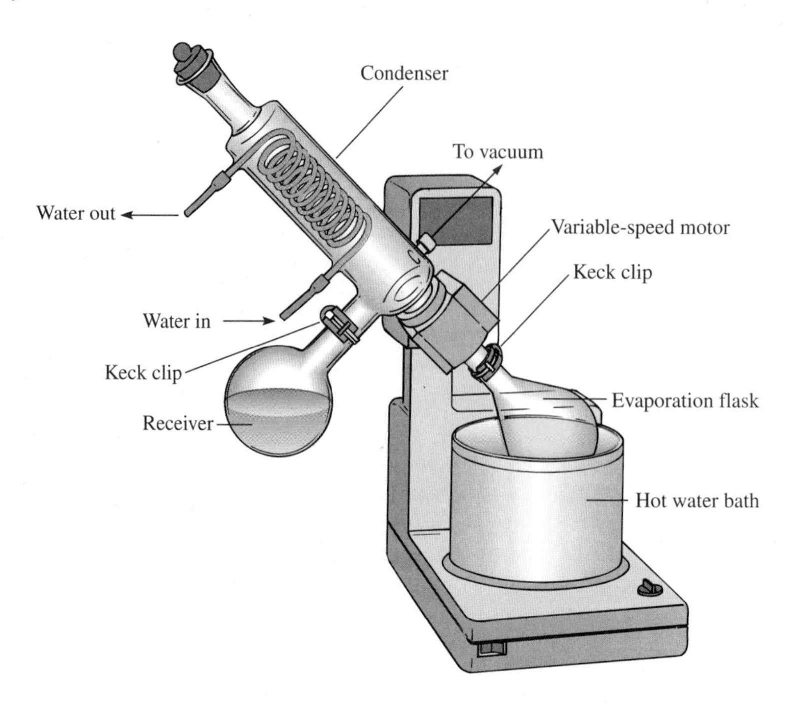

Figure 2.74
Rotary evaporator.

a trap between the condenser and the aspirator pump (Fig. 2.42b). While stirring the concentrated liquid, apply a gentle vacuum to the apparatus so the remainder of the organic solvent evaporates slowly under reduced pressure.

See videos on *Rotary Evaporation*

Rotary Evaporators

In a research laboratory and in many undergraduate laboratories, solvents are removed under reduced pressure, using an apparatus called a **rotary evaporator** (Fig. 2.74). This device is especially designed for the rapid evaporation of solvents without bumping. A variable-speed motor is used to rotate the flask containing the solvent being evaporated. While the flask is being rotated, a vacuum is applied and the flask may be heated. The rotation of the flask spreads a thin film of the solution on the inner surface of the flask to accelerate evaporation, and it also agitates the contents of the flask to reduce the problem of bumping. The rate of evaporation can be controlled by adjusting the vacuum, the temperature of the water bath, and the rate of rotation of the flask.

CHAPTER 3

Solids

Recrystallization and Melting Points

When you see this icon, sign in at this book's premium website at **www.cengage.com/login** to access videos, Pre-Lab Exercises, and other online resources.

One of the greatest visual pleasures for the practicing organic chemist is to observe the formation of a crystalline solid from solution. Obtaining beautifully formed crystals in myriad forms—prisms, cubes, platelets, rhomboids, and so on—gives a sense of accomplishment that is hard to match. This chapter gives the background to and experimental procedures for experiencing this thrill. Enjoy the experience!

3.1 INTRODUCTION

The organic chemist usually works with substances that are in the liquid or solid state. The purpose of this and the succeeding two chapters is to present an introduction to the theory and practice of the most important methods for separation and purification of mixtures of organic compounds in these two physical states. As you read these chapters, you should also refer to the corresponding sections of Chapter 2 that describe the apparatus used for the various techniques. References to the appropriate sections have been provided for your convenience.

A **pure compound** is a homogeneous sample consisting only of molecules having the same structure. However, substances believed to be pure on the basis of certain criteria may actually contain small amounts of contaminants. Indeed, the presence of impurities in what were believed to be pure compounds has sometimes led to incorrect structural assignments and scientific conclusions. The possibility of making such errors was of particular concern prior to the advent of the powerful chromatographic (Chap. 6) and spectral (Chap. 8) techniques that have been developed since the 1950s. It is now relatively easy for the chemist to purify solids and liquids and to demonstrate their purity.

A compound formed in a chemical reaction or extracted from some natural source is rarely pure when initially isolated. For example, a chemical transformation intended to produce a single product almost invariably yields a reaction mixture containing a number of contaminants. These may include the products of side reactions, unchanged starting materials, inorganic materials, and solvents. Unfortunately, even chemicals that have been purchased are not always pure, owing to the expense of the needed purification process or to decomposition that may occur during storage and shipment.

Organic chemists devote considerable effort to isolating pure products, the ultimate goal being to obtain a substance that cannot be further purified by any known experimental techniques. This chapter focuses on the purification of solids by recrystallization and their characterization by the physical property of melting points.

3.2 RECRYSTALLIZATION

See videos on *Recrystallization*

Recrystallization of solids is a valuable technique to master, because it is one of the most common methods used to purify solids. Other techniques for purifying solids include **sublimation** (Sec. 2.20), **extraction** (Chap. 5), and **chromatography** (Chap. 6). Nevertheless, even when one of these alternative methods of purification has been used, the solid material thus isolated may still be recrystallized to achieve a higher state of purity.

The process of recrystallization involves dissolving the solid in an appropriate solvent at an elevated temperature and allowing the crystals to re-form on cooling, so that any impurities remain in solution. This technique, called **solution recrystallization,** is discussed here. An alternative approach involves melting the solid in the absence of solvent and then allowing the crystals to re-form so that impurities are left in the melt. This method is not often used in the organic laboratory because the crystals often form from a viscous oil that contains the impurities and from which it is difficult to separate the desired pure solid. It is interesting to note, however, that this is the technique used to prepare the high-purity single crystals of silicon used in computer chips.

Almost all solids are *more* soluble in a *hot* than in a *cold* solvent, and solution crystallization takes advantage of this fact. Thus, if you first dissolve a solid in an amount of hot solvent insufficient to dissolve it when cold, crystals should form when the hot solution is allowed to cool. The extent to which the solid precipitates depends on the difference in its solubility in the particular solvent at temperatures between the extremes used. The upper extreme is determined by the boiling point of the solvent, whereas the lower limit is usually dictated by experimental convenience. For example, an ice-water bath is often used to cool the solution to 0 °C, whereas ice-salt and dry ice-acetone baths are commonly used to cool solutions to −20 °C and −78 °C, respectively (Sec. 2.10). The solid should be recovered with greater efficiency at these temperatures, provided the solvent itself does not freeze.

If the impurities present in the original solid mixture have dissolved and *remain* dissolved after the solution is cooled, isolation of the crystals that have formed should *ideally* provide pure material. Alternatively, the impurities may not dissolve at all in the hot solution and may be removed by filtration *before* the solution is cooled. The crystals that subsequently form should be purer than the original solid mixture. Solution recrystallization is seldom quite so simple in practice, but these two idealized generalizations do outline the basic principles of the technique.

Even after a solid has been recrystallized, it may still not be pure. Thus, it is important to determine the purity of the sample, and one of the easiest methods to do this is by determining the melting point of the solid. This technique is described in Section 3.3.

The technique of solution recrystallization involves the following steps:

1. **Selection** of an appropriate solvent.
2. **Dissolution** of the solid to be purified in the solvent near or at its boiling point.

3. **Decoloration** with an activated form of carbon, if necessary, to remove colored impurities and **filtration** of the hot solution to remove insoluble impurities and the decolorizing carbon.
4. **Formation** of crystalline solid from the solution as it cools.
5. **Isolation** of the purified solid by filtration.
6. **Drying** the crystals.

Each step of the sequence is discussed in the following subsections, and representative experimental procedures are presented at the end of the discussion.

Selection of Solvent

The choice of solvent is perhaps the most critical step in the process of recrystallization, since the correct solvent must be selected to form a product of high purity and in good recovery or yield. Consequently, a solvent should satisfy certain criteria for use in recrystallization: (**a**) The desired compound should be reasonably soluble in the *hot* solvent, about 1 g/20 mL (1 mg/20 μL) being satisfactory, and **insoluble** or *nearly* insoluble in the *cold* solvent. Often the reference temperature for determination of the solubility in "cold" solvent is taken to be room temperature. This combination of solute and solvent allows dissolution to occur in an amount of solvent that is not unduly large and also permits recovery of the purified product in high yield. A solvent having these solubility properties as a function of temperature would be said to have a favorable **temperature coefficient** for the desired solute. (**b**) Conversely, the impurities should either be insoluble in the solvent at all temperatures *or* must remain at least moderately soluble in the cold solvent. In other words, if the impurities are soluble, the temperature coefficient for them must be *unfavorable;* otherwise the desired product *and* the impurities would both crystallize simultaneously from solution. (**c**) The boiling point of the solvent should be low enough so that it can be readily removed from the crystals. (**d**) The boiling point of the solvent should generally be lower than the melting point of the solid being purified. (**e**) The solvent should not react chemically with the substance being purified.

The chemical literature is a valuable source of information about solvents that may be used for recrystallizing known compounds. If the compound has *not* been prepared before, it is necessary to resort to trial-and-error techniques to find an appropriate solvent for recrystallization. The process of selection can be aided by consideration of some generalizations about solubility characteristics for classes of solutes. Polar compounds are normally soluble in polar solvents and insoluble in nonpolar solvents, for example, whereas nonpolar compounds are more soluble in nonpolar solvents. Such characteristics are summarized by the adage "*like dissolves like.*" Of course, although a highly polar compound is unlikely to be soluble in a hot, nonpolar solvent, it *may* be very soluble in a cold, very polar solvent. In this case, a solvent of intermediate polarity *may* be the choice for a satisfactory recrystallization.

The solvents commonly used in recrystallizations range widely in polarity, a property measured by the **dielectric constants** (ϵ), listed in Table 3.1. Those solvents with dielectric constants in the range of 2–3 are considered **nonpolar** and those with constants above 10 as **polar.** Solvents in the range of 3–10 are of intermediate polarity. Of the various solvents listed, petroleum ether deserves special mention because of its confusing common name. This solvent does *not* contain the ether functional group at all; rather, it is a mixture of volatile aliphatic hydrocarbons obtained from refining petroleum. The composition and boiling point of the mixture depends on the particular distillation "cut" obtained. Thus, the boiling range of this type of solvent is usually given, as in the description "petroleum ether, bp 60 to 80 °C (760 torr)."

Table 3.1. *Common Solvents for Recrystallization**

Solvent	*Boiling Point, °C (760 torr)*	*Freezing Point, °C†*	*Water Soluble*	*Dielectric Constant (ε)*	*Flammable*
Petroleum ether‡	Variable		No	1.9	Yes
Diethyl ether	35		Slightly	4.3	Yes
Dichloromethane§	41		No	9.1	No
Acetone‡	56		Yes	20.7	Yes
Methanol‡	65		Yes	32.6	Yes
Tetrahydrofuran	65		Yes	7.6	Yes
Ethyl acetate‡	77		Yes	6.0	Yes
Ethanol (95%)‡	78		Yes	24.6	Yes
Cyclohexane	81	6	No	1.9	Yes
Water‡	100	0	N/A	78.5	No
1,4-Dioxane	101	11	Yes	2.2	Yes
Toluene‡	111		No	2.4	Yes

*Benzene has been omitted from this list, owing to its toxicity. Cyclohexane or toluene can usually be substituted for it.
†Freezing points not listed are below 0 °C.
‡Denotes solvents used most often.
§As a general rule, avoid the use of chlorocarbon solvents *if another equally good solvent is available.* Their toxicity is greater and their disposal is more difficult than for other types of solvents.

Occasionally a *mixture* of solvents is required for satisfactory recrystallization of a solute. The mixture comprises only two solvents; one of these dissolves the solute even when cold and the other one does not. The logic of this will become clear when you read about this technique under "Dissolution."

Chemical Structures of Common Recrystallization Solvents Found in Table 3.1

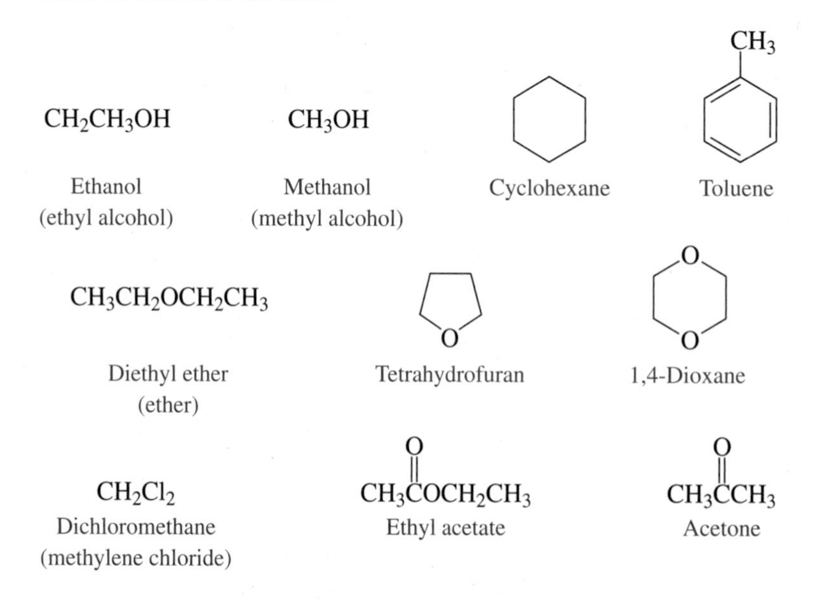

Dissolution

This step may involve the handling of relatively large volumes of volatile solvents. Although most solvents used in the organic laboratory are of relatively low toxicity, it is prudent to avoid inhaling their vapors. For this reason, the following operations are best performed in a hood. If this is not possible, clamping an inverted funnel over the recrystallization flask and connecting the stem of the funnel to a source of a vacuum (Fig. 2.71b, p. 90) helps to minimize exposure. Another precaution is to choose the source of heat for the recrystallization carefully. As noted in Table 3.1, many solvents are highly flammable and should *never* be heated with an open flame; rather, water, steam, and oil baths or hot plates (Sec. 2.9) should be used.

Single Solvents

The procedures for performing dissolutions in *single* solvents at the mini- and microscales differ slightly, as described below. Discussion of the strategies for using *mixed* solvents for recrystallizations is presented as well.

Miniscale Technique

The solid to be purified is weighed and placed in an Erlenmeyer flask of appropriate size. A beaker is *not* a suitable vessel for recrystallization, because it has a relatively large surface area to which recrystallized product may stick, lowering the efficiency of recovery. Some crystals of the impure solid should always be retained, for they may be needed as **"seeds"** to induce crystallization. The flask should be equipped for magnetic stirring (Sec. 2.11) or contain either boiling stones or a wooden stick to prevent bumping of the solution while boiling.

A few milliliters of solvent are added to the flask, and the mixture is then heated (Sec. 2.9) to the boiling point. More solvent is added to the mixture in *small* portions using a Pasteur pipet (Sec. 2.5) until the solid just dissolves. It is important to let boiling resume after each addition so that a minimum amount of solvent is used to effect dissolution; *using excessive amounts of solvent decreases the recovery of the solute.* If it is necessary to perform a hot filtration (Sec. 2.17), it is advisable to add an additional 2–5% of solvent to prevent premature crystallization during this operation.

If adding solvent fails to dissolve any more solid, it is likely that insoluble impurities are present. These can be removed by hot filtration. Thus, to avoid using too much solvent and risking poor recovery of the purified solute, you should observe the dissolution process *carefully.* This is particularly important when only a relatively small quantity of solid remains, as this may be the material that is insoluble.

Microscale Technique

The protocol for dissolution at the microscale level mimics that at larger scales. The solid is weighed and placed in a test tube or Craig tube, with a few crystals of impure material being retained as "seeds" to induce crystallization. A boiling stone should be added to prevent bumping of the solution upon boiling. Alternatively, smooth boiling can be promoted by twirling a microspatula in the mixture.

A few drops of solvent, normally totaling no more than a milliliter, are added, and the mixture is heated (Sec. 2.9) to the boiling point. Additional solvent, as needed, is added dropwise using a Pasteur pipet or filter-tip pipet (Sec. 2.5) to the boiling mixture until the solid just dissolves. It is important to let boiling resume after each addition so that a minimum amount of solvent is used to effect dissolution; *using excessive amounts of solvent decreases the recovery of the solute.* If it is necessary to perform a hot filtration (Sec. 2.17), it is prudent to add an additional 2–5% of solvent to prevent premature crystallization during this operation.

If adding solvent fails to dissolve any more solid, it is likely that insoluble impurities are present. These can be removed by hot filtration. Thus, to avoid using too much solvent and risking poor recovery of the purified solute, you should observe the dissolution process *carefully*. This is particularly important when only a relatively small quantity of solid remains, as this may be the material that is insoluble.

Mixed Solvents

The same general approach used for single-solvent dissolution is followed at the mini- and microscale levels when mixed solvents are employed. However, there are two options for effecting dissolution once the solvents have been selected. In one, the solid to be purified is first dissolved in a *minimum* volume of the hot solvent in which it is soluble; the second solvent is then added to the *boiling* solution until the mixture turns cloudy. The cloudiness signals initial formation of crystals, caused by the fact that addition of the second solvent results in a solvent mixture in which the solute is less soluble. Finally, more of the first solvent is added *dropwise* until the solution clears.

Two further aspects of this option should be noted. First, the solution must be cooled slightly below the lower boiling point before the second solvent is added *if* this solvent has a boiling point *lower* than the first; otherwise the addition of this solvent could cause sudden and vigorous boiling of the mixture, and hot solvent might spew from the apparatus. Second, hot filtration should be performed if needed *before* addition of the second solvent; this will prevent crystallization during the filtration step. A potential disadvantage of this method for mixed-solvent recrystallization is that unduly large volumes of the second solvent may be required if excessive amounts of the initial solvent have been used.

In the second option, the solute is added to the solvent in which it is insoluble, and the mixture is heated near the boiling point of the solvent; the second solvent is then added in small portions until the solid just dissolves. As with recrystallization from a single solvent, it is generally wise to add 2–5% of additional solvent to prevent premature crystallization of the solute during hot filtration, if this step is necessary. The use of this approach to mixed-solvent recrystallization also has the disadvantage that using too much of the first solvent will require the addition of undesirably large volumes of the second one.

Decoloration and Hot Filtration

After dissolution of the solid mixture, the solution may be colored. This signals the presence of impurities if the desired compound is known to be colorless. If the compound is colored, contaminants may alter the color of the solution. For example, impurities should be suspected if the substance is yellow but the solution is green. The decoloration step is unnecessary if the solution is colorless, of course.

Colored impurities may often be removed by adding a small amount of decolorizing carbon (Sec. 2.18) to the hot, *but not boiling,* solution. If it were added to a boiling solution, the liquid would likely froth over the top of the flask, resulting in loss of product and possible injury. After the decolorizing carbon is added, the solution is heated to boiling for a few minutes while being continuously stirred or swirled to prevent bumping.

Completely removing decolorizing carbon by hot filtration (Sec. 2.17) may be difficult if the powdered rather than the pelletized form is used, because the finely divided solid particles may pass through filter paper or the cotton of a Pasteur filtering pipet (Sec. 2.5). If this occurs, the dark particles should be visible in the filtrate.

The following steps are recommended to assist in removing the decolorizing carbon during filtration. The hot decolorized solution is allowed to cool slightly below its boiling point, a small amount of a filter-aid such as Celite (Sec. 2.17) is added to *adsorb* the carbon, the mixture is briefly reheated to boiling, and then it is subjected to hot filtration. It may be necessary to repeat this procedure if some of the decolorizing carbon still passes through the filter paper and remains in the filtrate.

This technique for decolorizing solutions works because colored impurities as well as the compound being purified are adsorbed on the surface of the carbon particles. For electronic reasons, the colored substances adsorb more strongly to the surface. You should avoid using too much decolorizing carbon because the desired product itself may be adsorbed by it, so less product will be recovered.

Insoluble impurities, including dust and decolorizing carbon, are removed by gravity filtration of the hot solution (Sec. 2.17); this step is not necessary if the hot solution is clear and homogeneous. Gravity filtration is normally preferred to vacuum filtration because the latter technique may cause cooling and concentration of the solution, owing to evaporation of the solvent, and this may result in premature crystallization. A short-stemmed or stemless glass funnel should be used to minimize crystallization in the funnel, and using fluted filter paper (Sec. 2.17) will minimize crystallization on the filter. To keep liquid from flowing over the top of the funnel, the top of the paper should not extend above the funnel by more than 1–2 mm.

Crystallization

The hot solution of solute is allowed to cool slowly to room temperature, and crystallization should occur. During cooling and crystallization, the solution should be protected from airborne contaminants by covering the opening with a piece of filter paper, an inverted beaker, or by loosely plugging it with a clean cork. Rapid cooling by immersing the flask in water or an ice-water bath tends to lead to the formation of very small crystals that may adsorb impurities from solution. Generally the solution should not be disturbed as it cools, since this also leads to production of small crystals. The formation of crystals larger than about 2 mm should also be avoided because some of the solution may become occluded or trapped *within* the crystals. The drying of such crystals is more difficult, and impurities may be left in them. Should overly large crystals begin to form, brief, gentle agitation of the solution normally induces production of smaller crystals.

Failure of crystallization to occur after the solution has cooled somewhat usually means that either too much solvent has been used or that the solution is **supersaturated.** A supersaturated solution can usually be made to produce crystals by *seeding*. A crystal of the original solid is added to the solution to induce crystallization, which may then be quite rapid. If no solid is available and a volatile solvent is being used, it is sometimes possible to produce a seed crystal by immersing the tip of a glass stirring rod or metal spatula in the solution, withdrawing it, and allowing the solvent to evaporate. The crystals that form on the end of the rod or spatula are then reinserted into the solution to initiate crystallization. Alternatively, crystallization can often be induced by using a glass rod to rub the inside surface of the crystallization vessel *at* or *just above* the air–solution interface. This should be done carefully, as excessive force may scratch or break the vessel or result in a broken rod.

Occasionally the solute will separate from solution as an oil rather than a solid. This type of separation, which is sometimes called **oiling out,** is undesirable for purification of solutes because the oils usually contain significant amounts of impurities. Two general approaches are helpful in solving this problem: (1) Oils may persist on cooling with no evidence of crystallization. These may often be induced to crystallize by scratching the oil against the side of the flask with a glass stirring rod at the interface of the oil and the solution. If this fails, several small seed crystals of the original solid may be added to the oil and the mixture allowed to stand for a period of time. Failure of these alternatives may necessitate separation of the oil from the solution and crystallization of it from another solvent. (2) Oils may form from the hot solution and then solidify to an amorphous mass at lower temperatures; in the meantime, crystals of the solute may precipitate from the mother liquor. Because the oil is not a pure liquid, the solid mass produced from it will be impure, as noted earlier. In a case such as this, the usual remedy is to reheat the entire mixture to effect dissolution, add a few milliliters of additional pure solvent, and allow the resulting solution to cool.

Filtration and Solvent Removal

The crystalline product is isolated by filtration. The technique for doing this varies depending on the scale on which the crystallization was performed.

Miniscale Technique

The solid product is isolated by vacuum filtration using a Büchner or Hirsch funnel and a clean, dry filter flask (Sec. 2.17). The crystals normally are washed with a small amount of *pure, cold* solvent, with the vacuum off; the vacuum is then reapplied to remove as much solvent as possible from the filter cake. Care must be taken in this step to ensure that the filter paper is not lifted off the bed of the filter while the vacuum is off, as this could result in loss of product when the vacuum is reapplied to remove the washes.

Microscale Technique

The crystalline product, contained in a Craig tube fitted with a plug, is isolated by centrifugation (Sec. 2.17). The solid is scraped off the plug back into the Craig tube and washed by stirring it briefly with a small amount of *pure, cold* solvent. The plug is then reinserted, and the solvent is once again removed by centrifugation.

Further cooling of the filtrate, sometimes called the **mother liquor,** in an ice-water or ice-salt bath (Sec. 2.10) may allow isolation of a second crop of crystals. The filtrate can also be concentrated by evaporating part of the solvent and cooling the residual solution. The crystals isolated as a second or even a third crop are likely to be less pure than those in the first. Consequently, the various crops should *not* be combined until their purity has been assessed by comparison of their melting points as described in Section 3.3.

Drying the Crystals

At the miniscale level, the final traces of solvent are removed by transferring the crystals from the filter paper of the Büchner or Hirsch funnel to a watchglass or vial. Alternatively, solids may also be transferred to fresh pieces of filter or weighing paper for drying. This is a less desirable option, however, because fibers of paper may contaminate the product when it is ultimately transferred to a container for submission to your instructor. For microscale operations, the solid may be left in the Craig tube used for the crystallization. It is good practice to protect the crystals from airborne contaminants by covering them with a piece of filter paper or using loosely inserted cotton or a cork to plug the opening of the vessel containing the solid.

Removing the last traces of solvent from the crystalline product may be accomplished by air- or oven-drying (Sec. 2.26). With the latter option, the temperature of the oven *must* be 20–30 °C below the melting point of the product; otherwise your crystals will turn into a puddle!

EXPERIMENTAL PROCEDURES

Recrystallization

Purpose To explore the techniques for recrystallizing solids.

SAFETY ALERT

1. **Do not use a burner in these procedures unless instructed to do so. Most solvents used for recrystallization are flammable (Table 3.1).**
2. **When using a hot plate, do not set it at its highest value. A moderate setting will prevent overheating and the resultant bumping and splashing of materials from the flask. Do not employ hot plates for heating volatile or flammable solvents; rather, use a steam bath.**
3. **Do not inhale solvent vapors. If a hood is not available to you, clamp an inverted funnel just above the Erlenmeyer flask in which you will be heating solvents. Attach this funnel to a source of vacuum by means of rubber tubing (Fig. 2.71b).**
4. **When pouring or transferring solutions, either wear latex gloves or avoid getting these solutions on the skin. Organic compounds are much more rapidly absorbed through the skin when they are in solution, particularly in water-soluble solvents such as acetone and ethanol. For this reason, do not rinse organic materials off your skin with solvents such as acetone; wash your hands thoroughly with soap and hot water instead.**
5. ***Never* add decolorizing carbon to a boiling solution; doing so may cause the solution to boil out of the flask. Add the carbon only when the temperature of the solvent is below the boiling temperature. This same precaution applies when using a filter-aid to assist in the removal of the carbon during the hot filtration step.**

A ■ *Solvent Selection*

Procedure

Preparation Sign in at **www.cengage.com/login** to answer Pre-Lab Exercises, access videos, and read the MSDSs for the chemicals used or produced in this procedure. Read or review Sections 2.5, 2.9, and 3.2, and review the solvent properties listed in Table 3.1. Although different criteria are used for defining solubility, plan to use the following definitions in this experiment: (a) soluble—20 mg of solute will dissolve in 0.5 mL of solvent; (b) slightly soluble—some but not all of the 20 mg of solute will dissolve in 0.5 mL of solvent; (c) insoluble—none of the solute appears to dissolve in 0.5 mL of solvent. *Be certain to record all your observations regarding solubilities in your notebook!*

Apparatus Test tubes (10-mm × 75-mm), hot-water (80–100 °C) or steam bath.

Protocol For *known* compounds, place about 20 mg (a microspatula-tip full) of the finely crushed solid in a test tube and add about 0.5 mL of water using a calibrated Pasteur pipet. Stir the mixture with a glass rod or microspatula to determine whether the solid is soluble in water at room temperature. If the solid is not completely soluble at room temperature, warm the test tube in the hot-water or steam bath, and stir or swirl its contents to determine whether the solid is soluble in hot water.

If any of your solutes is soluble in the hot solvent but only slightly soluble or insoluble at room temperature, allow the hot solution to cool slowly to room temperature and compare the quantity, size, color, and form of the resulting crystals with the original solid material.

Repeat the solubility test for the solutes using 95% ethanol and then petroleum ether (bp 60–80 °C, 760 torr). After completing these additional tests, record which of the three solvents you consider best suited for recrystallization of each of the solutes.

Compounds **1–4** contain a variety of functional groups that impart differing solubility properties to the molecules and are possible substrates on which to practice the technique of determining solubilities. Other known compounds may be assigned by your instructor.

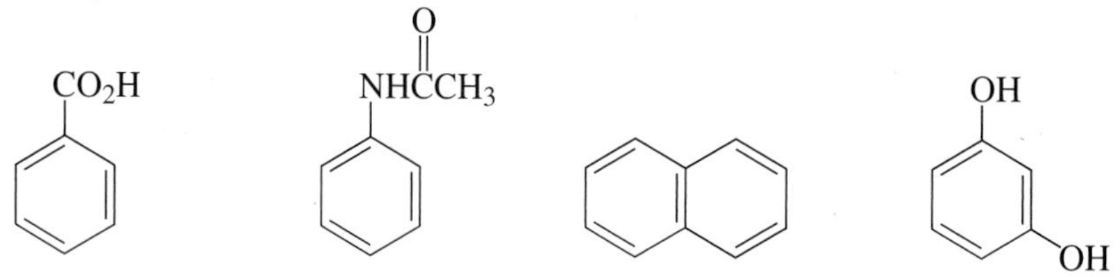

CO_2H **1** Benzoic Acid

$NHCCH_3$ (O) **2** Acetanilide

3 Naphthalene

OH, OH **4** Resorcinol

For *unknown* compounds, a *systematic* approach is important for determining their solubility. First, select the solvents from Table 3.1 to be used in the tests. It should not be necessary to test all the solvents, but you should consider trying those that are denoted with the symbol ‡ in the table. Your instructor may also suggest solvents to evaluate.

After selecting the solvents, obtain enough clean, dry test tubes so that there is one for each solvent to be tested. Place about 20 mg (a microspatula-tip full) of the finely crushed unknown in each test tube and add about 0.5 mL of a solvent to a tube containing the solid. Stir each mixture and determine the solubility of the unknown in each solvent at room temperature. Use the definitions of *soluble, slightly soluble,* or *insoluble* given earlier.

If the unknown is insoluble in a particular solvent, warm the test tube in the hot-water or steam bath. Stir or swirl the contents of the tube and note whether the unknown is soluble in hot solvent. If the solid is soluble in the hot solvent but only slightly soluble or insoluble at room temperature, allow the hot solution to cool to room temperature slowly. If crystals form in the cool solution, compare their quantity, size, color, and form with the original solid material and with those obtained from other solvents.

It is a good idea to test the solubility of a solute in a variety of solvents. Even though nice crystals may form in the first solvent you try, another one might prove better if it provides either better recovery or higher-quality crystals. To assist in determining the best solvent to use in recrystallizing an unknown, you should construct a table containing the solubility data you gather by the systematic approach described above.

If these solubility tests produce no clear choice for the solvent, mixed solvents might be considered. Review the discussion presented earlier in this section for the procedure for using a mixture of two solvents. Before trying any combinations of solvent pairs, take about 0.2 mL of each *pure* solvent being considered and mix them to ensure that they are miscible in one another. If they are not, that particular combination *cannot* be used.

B ■ *Recrystallizing Impure Solids*

MINISCALE PROCEDURES

Preparation Sign in at **www.cengage.com/login** to answer Pre-Lab Exercises, access videos, and read the MSDSs for the chemicals used or produced in this procedure. Read or review Sections 2.6, 2.7, 2.9, 2.11, 2.17, and 2.18. Plan to determine the melting points of both the purified and the crude solids as a way to show the benefits of recrystallization.

1. *Benzoic Acid*

Apparatus Two 50-mL Erlenmeyer flasks, graduated cylinder, apparatus for magnetic stirring, vacuum filtration, and *flameless* heating.

Dissolution Place 1.0 g of impure benzoic acid in an Erlenmeyer flask equipped for magnetic stirring or with boiling stones. Measure 25 mL of water into the graduated cylinder and add a 10-mL portion of it to the flask. Heat the mixture to a gentle boil and continue adding water in 0.5-mL portions until no more solid appears to dissolve in the boiling solution. Record the total volume of water used; no more than 10 mL should be required. *Caution:* Because the sample may be contaminated with insoluble material, pay close attention to whether additional solid is dissolving as you add more solvent; if it is not, *stop adding solvent.*

Decoloration Pure benzoic acid is colorless, so a colored solution indicates that treatment with decolorizing carbon (Sec. 2.18) is necessary. *Caution:* Do *not* add decolorizing carbon to a *boiling* solution! Cool the solution slightly, add a microspatula-tip full of carbon, and reheat to boiling for a few minutes. To aid in removing the finely divided carbon by filtration, allow the solution to cool slightly, add about 0.2 g of a filter-aid (Sec. 2.17), and reheat.

Hot Filtration and Crystallization If there are insoluble impurities or decolorizing carbon in the solution, perform a hot filtration (Sec. 2.17) using a 50-mL Erlenmeyer flask to receive the filtrate (Fig. 2.52). Rinse the empty flask with about 1 mL of *hot* water and filter this rinse into the original filtrate. If the filtrate remains colored, repeat the treatment with decolorizing carbon. Cover the opening of the flask with a piece of filter paper, an inverted beaker, or loose-fitting cork to exclude airborne impurities from the solution, and allow the filtrate to stand undisturbed until it has cooled to room temperature and no more crystals form.★ To complete the crystallization, place the flask in an ice-water bath for at least 15 min.

Isolation and Drying Collect the crystals on a Büchner or Hirsch funnel by vacuum filtration (Fig. 2.54) and wash the filter cake with two small portions of *cold* water. Press the crystals as dry as possible on the funnel with a clean cork or spatula. Spread the benzoic acid on a watchglass, protecting it from airborne contaminants with a piece of filter paper, and air-dry it at room temperature or in an oven. Be certain that the temperature of the oven is 20–30 °C below the melting point of the product!

Analysis Determine the melting points of the crude and recrystallized benzoic acid, the weight of the latter material, and calculate your percent recovery using Equation 3.1.

$$\text{Percent recovery} = \frac{\text{Weight of pure crystals recovered}}{\text{Weight of original sample}} \times 100 \qquad (3.1)$$

2. *Acetanilide*

Apparatus Two 50-mL Erlenmeyer flasks, graduated cylinder, apparatus for magnetic stirring, vacuum filtration, and *flameless* heating.

Dissolution Place 1.0 g of impure acetanilide in an Erlenmeyer flask equipped for magnetic stirring or with boiling stones. Measure 20 mL of water into the graduated cylinder and add a 10-mL portion of it to the flask. Heat the mixture to a gentle boil.

A layer of oil should form when the stated amount of water is added. (If you have not done so already, review the discussion of *Crystallization* in this section with emphasis on how to crystallize compounds that form oils.) This layer consists of a solution of water in acetanilide. More water must be added to effect complete solution of the acetanilide in water. However, even if a homogeneous solution is produced at the boiling point of the mixture, an oil may separate from it as cooling begins. The formation of this second liquid phase is known to occur only under specific conditions: the acetanilide-water mixture must have a composition that is between 5.2% and 87% in acetanilide and be at a temperature above 80 °C. Because the solubility of acetanilide in water at temperatures near 100 °C exceeds 5.2%, a homogeneous solution formed by using the minimum quantity of water meets these criteria. Such a solution will yield an oil on cooling to about 83 °C; solid begins to form below this temperature.

Continue adding 3–5 mL of water in 0.5-mL portions to the boiling solution until the oil has completely dissolved. *Caution:* Because the sample may be contaminated with insoluble material, pay close attention to whether additional solid is dissolving as you add more solvent; if it is not, *stop adding solvent.* Once the acetanilide has just dissolved, add an additional 1 mL of water to prevent formation of oil during the crystallization step. If oil forms at this time, reheat the solution and add a little more water. Record the total volume of water used.

Continue the procedure by following the directions for *Decoloration, Hot Filtration and Crystallization,* and *Isolation and Drying* given for benzoic acid in Part 1.

Analysis Determine the melting points of the crude and recrystallized acetanilide, the weight of the latter material, and calculate your percent recovery using Equation 3.1.

3. *Naphthalene*

Apparatus Two 50-mL Erlenmeyer flasks, graduated cylinder, apparatus for magnetic stirring, vacuum filtration, and *flameless* heating.

Dissolution Naphthalene may be conveniently recrystallized from methanol, 95% ethanol, or 2-propanol. Because these solvents are somewhat toxic and/or flammable, proper precautions should be taken. The sequence of steps up through the hot filtration should be performed in a hood if possible. Alternatively, if instructed to do so, position an inverted funnel connected to a vacuum source above the mouth of the flask being used for recrystallization (Fig. 2.71b).

Place 1.0 g of impure naphthalene in an Erlenmeyer flask equipped for magnetic stirring or with boiling stones and dissolve it in the minimum amount of boiling alcohol. *Caution:* Because the sample may be contaminated with insoluble material, pay close attention to whether additional solid is dissolving as you add more solvent; if it is not, *stop adding solvent.* Then add 0.5 mL of additional solvent to ensure that premature crystallization will not occur during subsequent transfers. Record the total volume of solvent used.

Continue the procedure by following the directions for *Decoloration, Hot Filtration and Crystallization,* and *Isolation and Drying* given for benzoic acid in Part 1; however, use the solvent in which you dissolved the naphthalene rather than water.

Analysis Determine the melting points of the crude and recrystallized naphthalene, the weight of the latter material, and calculate your percent recovery using Equation 3.1.

4. *Unknown Compound*

Apparatus Two 50-mL Erlenmeyer flasks, graduated cylinder, apparatus for magnetic stirring, vacuum filtration, and *flameless* heating.

Dissolution *Accurately* weigh about 1 g of the unknown compound and transfer it to an Erlenmeyer flask equipped for magnetic stirring or with boiling stones. Measure about 15 mL of the solvent you have selected on the basis of solubility tests into a graduated cylinder and add 10 mL of it to the flask. Bring the mixture to a gentle boil using *flameless* heating unless water is the solvent, add a 1-mL portion of the solvent, and again boil the solution. Continue adding 3-mL portions of solvent, one portion at a time, until the solid has completely dissolved. Bring the solution to boiling after adding each portion of solvent. *Caution:* Because the sample may be contaminated with insoluble material, pay close attention to whether additional solid is dissolving as you add more solvent; if it is not, *stop adding solvent.* Record the total volume of solvent that is added.

Continue the procedure by following the directions for *Decoloration, Hot Filtration and Crystallization,* and *Isolation and Drying* given for benzoic acid in Part 1.

Analysis Determine the melting points of the crude and recrystallized unknown, the weight of the latter material, and calculate your percent recovery using Equation 3.1.

5. Mixed-Solvent Crystallization

Apparatus Two 25-mL Erlenmeyer flasks, graduated cylinder or calibrated Pasteur pipet, apparatus for magnetic stirring, vacuum filtration, and *flameless* heating.

Dissolution Place 1.0 g of impure benzoic acid or acetanilide in an Erlenmeyer flask equipped for magnetic stirring or with boiling stones. Add 2 mL of 95% ethanol and heat the mixture to a gentle boil. If necessary, continue adding the solvent in 0.5-mL portions until no more solid appears to dissolve in the boiling solution. *Caution:* Because the sample may be contaminated with insoluble material, pay close attention to whether additional solid is dissolving as you add more solvent; if it is not, *stop adding solvent.* Record the total volume of 95% ethanol used.

Decoloration If the solution is colored, treat it with decolorizing carbon (Sec. 2.18) and a filter-aid according to the procedure given for benzoic acid in Part 1.

Hot Filtration and Crystallization If there are insoluble impurities or decolorizing carbon in the solution, perform a hot filtration (Sec. 2.17) using a 25-mL Erlenmeyer flask to receive the filtrate (Fig. 2.52). Rinse the empty flask with about 0.5 mL of *hot* 95% ethanol and filter this rinse into the original filtrate. If the filtrate remains colored, repeat the treatment with decolorizing carbon.

Reheat the decolorized solution to boiling and add water dropwise from a Pasteur pipet until the boiling solution remains cloudy or precipitate appears; this may require adding several milliliters of water. Then add a few drops of 95% ethanol to produce a clear solution at the boiling point. Remove the flask from the heating source, and follow the same directions as given for benzoic acid in Part 1 to complete both this stage of the procedure and *Isolation and Drying.*

Analysis Determine the melting points of the crude and recrystallized product, the weight of the latter material, and calculate your percent recovery using Equation 3.1.

MICROSCALE PROCEDURES

Preparation Sign in at **www.cengage.com/login** to answer Pre-Lab Exercises, access videos, and read the MSDSs for the chemicals used or produced in this procedure. Read or review Sections 2.5, 2.6, 2.7, 2.9, 2.17, and 3.3. Plan to determine the melting points of both the purified and the crude solids as a way to show the benefits of recrystallization.

1. Benzoic Acid

Apparatus A 10-mm × 75-mm test tube, a calibrated Pasteur pipet and Pasteur filtering and filter-tip pipets, apparatus for Craig tube filtration and *flameless* heating.

Dissolution Place 100 mg of benzoic acid in the test tube. Add 1 mL of water to the tube and heat the mixture to a gentle boil. To aid in dissolution, stir the mixture vigorously with a microspatula while heating; stirring also prevents bumping and possible loss of material from the test tube. If needed, add more solvent in 0.1–0.2 mL increments to dissolve any remaining solid. Bring the mixture to boiling and continue stirring after each addition. Once all the solid has dissolved, add an additional 0.1–0.2-mL portion of solvent to ensure that the solute remains in solution during transfer of the hot solution to a Craig tube. *Caution:* Because the sample may be contaminated with insoluble material, pay close attention to whether additional solid is dissolving as you add more solvent. If it is not, *stop adding solvent.* Record the approximate total volume of solvent used.

Decoloration If the solution is colored and the recrystallized product is known or suspected to be colorless, treatment with decolorizing carbon (Sec. 2.18) is necessary. *Caution:* Do *not* add decolorizing carbon to a *boiling* solution! Cool the solution slightly, add half of a microspatula-tip full of powdered carbon or, preferably, a pellet or two of decolorizing carbon, and reheat to boiling for a few minutes.

Hot Filtration and Crystallization Preheat a Pasteur filtering pipet by pulling hot solvent into the pipet. Then transfer the hot solution into this pipet with a Pasteur pipet or filter-tip pipet that has also been preheated with solvent, using the tared Craig tube as the receiver for the filtrate (Fig. 2.53). If decolorizing carbon or other insoluble matter appears in the Craig tube, pass the solution through the filtering pipet a second and, if necessary, a third time. Concentrate the clear solution in the Craig tube to the point of saturation by heating it to boiling. Rather than using a boiling stone to prevent superheating and possible bumping, continually stir the solution with a microspatula while heating. The saturation point will be signaled by the appearance of cloudiness in the solution and/or the formation of crystals on the microspatula at the air/liquid interface. When this occurs, add solvent dropwise at the boiling point until the cloudiness is discharged. Then remove the tube from the heating source, cap the tube with a loose plug of cotton to exclude airborne impurities, and allow the solution to cool to room temperature.★ If necessary, induce crystallization by gently scratching the surface of the tube at the air/liquid interface or by adding seed crystals, if these are available. To complete crystallization, cool the tube in an ice-water bath for at least 15 min.

Isolation and Drying Affix a wire holder to the Craig tube and, using the wire as a hanger, invert the apparatus in a centrifuge tube (Fig. 2.55). Remove the solvent by centrifugation, carefully disassemble the Craig tube, and scrape any crystalline product clinging to its upper section into the lower part. Protect the product from airborne contaminants by plugging the opening of the tube with cotton or by covering it with a piece of filter paper held in place with a rubber band. Air-dry the crystals to constant weight either at room temperature or in an oven. Be certain that the temperature of the oven is 20–30 °C below the melting point of the product!

Analysis Determine the melting points of the crude and recrystallized benzoic acid, the weight of the latter material, and calculate your percent recovery using Equation 3.1.

2. *Acetanilide*

Apparatus A 10-mm × 75-mm test tube, a calibrated Pasteur pipet and Pasteur filtering and filter-tip pipets, apparatus for Craig tube filtration and *flameless* heating.

Dissolution Place 100 mg of acetanilide in the test tube. Add 1 mL of water to the tube and heat the mixture to a gentle boil. To aid in dissolution, stir the mixture vigorously with a microspatula while heating; stirring also prevents bumping and possible loss of material from the test tube. A layer of oil should form when the stated amount of water is added. An explanation for this phenomenon is found under the *Dissolution* stage of Part 2 of the Miniscale Procedures. Continue adding water in 0.1-mL portions to the boiling solution until the oil has completely dissolved. *Caution:* Because the sample may be contaminated with insoluble material, pay close attention to whether additional solid is dissolving as you add more solvent; if it is not, *stop adding solvent.*

Once the acetanilide has just dissolved, add an additional 0.1 mL of water to prevent formation of oil during the crystallization step. If oil forms at this time, reheat the solution and add a little more water. Record the approximate total volume of water used.

Continue the procedure by following the directions for *Decoloration, Hot Filtration and Crystallization,* and *Isolation and Drying* given for benzoic acid in Part 1.

Analysis Determine the melting points of the crude and recrystallized acetanilide, the weight of the latter material, and calculate your percent recovery using Equation 3.1.

3. Naphthalene

Apparatus A 10-mm × 75-mm test tube, a calibrated Pasteur pipet and Pasteur filtering and filter-tip pipets, apparatus for Craig tube filtration and *flameless* heating.

Dissolution Naphthalene may be conveniently recrystallized from methanol, 95% ethanol, or 2-propanol. Because these solvents are somewhat toxic and/or flammable, proper precautions should be taken. If instructed to do so, position an inverted funnel connected to a vacuum source above the mouth of the test tube being used for recrystallization (Fig. 2.71b) to minimize release of vapors into the laboratory.

Place 100 mg of naphthalene in the test tube and dissolve it in a minimum amount of boiling solvent. *Caution:* Because the sample may be contaminated with insoluble material, pay close attention to whether additional solid is dissolving as you add more solvent; if it is not, *stop adding solvent.* Then add 0.1 mL of additional solvent to ensure that premature crystallization will not occur during subsequent transfers. Record the total volume of solvent used.

Continue the procedure by following the directions for *Decoloration, Hot Filtration and Crystallization,* and *Isolation and Drying* given for benzoic acid in Part 1; however, rather than water, use the solvent in which you dissolved the naphthalene.

Analysis Determine the melting points of the crude and recrystallized naphthalene, the weight of the latter material, and calculate your percent recovery using Equation 3.1.

4. Unknown Compound

Apparatus A 10-mm × 75-mm test tube, a calibrated Pasteur pipet and Pasteur filtering and filter-tip pipets, apparatus for Craig tube filtration and *flameless* heating.

Dissolution If instructed to do so, position an inverted funnel connected to a vacuum source above the mouth of the test tube being used for recrystallization (Fig. 2.71b) to minimize release of vapors into the laboratory. Place 100 mg of the unknown in the test tube and dissolve it in a minimum amount of boiling solvent you selected on the basis of solubility tests. *Caution:* Because the sample may be contaminated with insoluble material, pay close attention to whether additional solid is dissolving as you add more solvent; if it is not, *stop adding solvent.* Then add 0.1 mL of additional solvent to ensure that premature crystallization will not occur during subsequent transfers.Record the total volume of solvent used.

Continue the procedure by following the directions for *Decoloration, Hot Filtration and Crystallization,* and *Isolation and Drying* given for benzoic acid in Part 1; however, if you are not using water, use the solvent in which you dissolved the unknown.

Analysis Determine the melting points of the crude and recrystallized unknown, the weight of the latter material, and calculate your percent recovery using Equation 3.1.

5. *Mixed-Solvent Crystallization*

Apparatus A 10-mm × 75-mm test tube, a calibrated Pasteur pipet and Pasteur filtering and filter-tip pipets, apparatus for Craig tube filtration and *flameless* heating.

Dissolution Place 100 mg of benzoic acid or acetanilide in the test tube. Add 0.2 mL of 95% ethanol and heat the mixture to a gentle boil. If necessary, continue to add 95% ethanol dropwise by Pasteur pipet until a homogeneous solution is obtained. *Caution:* Because the sample may be contaminated with insoluble material, pay close attention to whether additional solid is dissolving as you add more solvent; if it is not, *stop adding solvent.* Once all the solid has dissolved, add an additional 0.1–0.2-mL portion of solvent to ensure that the solute remains in solution during transfer of the hot solution to a Craig tube and, if necessary, decolorize the solution according to the directions in Part 1 for benzoic acid.

Hot Filtration and Crystallization Preheat a Pasteur filtering pipet by pulling hot solvent into the pipet. Then transfer the hot solution into this pipet with a Pasteur pipet or filter-tip pipet that has also been preheated with solvent, using the tared Craig tube as the receiver for the filtrate (Fig. 2.53). If decolorizing carbon or other insoluble matter appears in the Craig tube, pass the solution through the filtering pipet a second and, if necessary, a third time.

Reheat the decolorized solution to boiling and add water dropwise from a Pasteur pipet until the boiling solution remains cloudy or precipitate forms. Rather than using a boiling stone to prevent superheating and possible bumping, continually stir the solution with a microspatula while heating. Then add a drop or two of 95% ethanol to restore homogeneity. Remove the tube from the heating source, cap the tube with a loose plug of cotton to exclude airborne impurities, and allow the solution to cool to room temperature.★ Follow the same directions as given for benzoic acid in Part 1 to complete both this stage of the procedure and *Isolation and Drying.*

Analysis Determine the melting points of the crude and recrystallized product, the weight of the latter material, and calculate your percent recovery using Equation 3.1.

Discovery Experiment

Formation of Polymorphs

Develop an experimental procedure for forming polymorphs (see Historical Highlight) of *trans*-cinnamic acid or derivatives thereof by referring to the following publications: Bernstein, H. I.; Quimby, W. C., *J. Am. Chem. Soc.* **1943**, *65*, 1845–1848; Cohen, M. D.; Schmidt, G. M. J.; Sonntag, F. I. *J. Chem. Soc.* **1964**, 2000–2013. Consult with your instructor before undertaking your proposed procedure.

WRAPPING IT UP

Flush any *aqueous filtrates* or *solutions* down the drain. With the advice of your instructor, do the same with the *filtrates* derived from use of *alcohols, acetone,* or other *water-soluble solvents.* Use the appropriate containers for the *filtrates* containing *halogenated solvents* or *hydrocarbon solvents.* Put *filter papers* in the container for nontoxic waste, unless instructed to do otherwise.

EXERCISES

1. Define or describe each of the following terms as applied to recrystallization:
   - **a.** solution recrystallization
   - **b.** temperature coefficient of a solvent
   - **c.** the relationship between dielectric constant and polarity of a solvent
   - **d.** petroleum ether
   - **e.** mixed solvents
   - **f.** solvent selection
   - **g.** filter-aid
   - **h.** hot filtration
   - **i.** seeding
   - **j.** vacuum filtration
   - **k.** mother liquor
   - **l.** filtrate
   - **m.** solute
   - **n.** solvent
   - **o.** occlusion
2. Provide the IUPAC name for ethyl acetate and acetone.
3. For which of the following situations is it appropriate to perform a recrystallization? More than one answer may be circled.
   - **a.** To purify an impure liquid.
   - **b.** To purify an impure solid.
   - **c.** The melting point of a compound is depressed and melts over a wide range.
   - **d.** The melting point of a compound is sharp and agrees with the literature value.
   - **e.** An impure solid is soluble in all possible recrystallization solvents that are available.
4. List the steps in the systematic procedure for miniscale recrystallization, briefly explaining the purpose of each step.
5. In performing a hot filtration at the miniscale level, what might happen if the filter funnel is not preheated before the solution is poured through it?
6. Describe the use of the following pieces of equipment during recrystallizations at the miniscale level.
   - **a.** filter flask
   - **b.** filter trap
   - **c.** Büchner funnel
7. Why should the filter flask not be connected directly to a water aspirator pump when performing a vacuum filtration?

8. List the steps in the systematic procedure for microscale recrystallization, briefly explaining the purpose of each step.
9. Why is a Pasteur filter-tip pipet preferred to a regular Pasteur pipet for transferring hot volatile solutions?
10. In performing a hot filtration at the microscale level, what might happen if the Pasteur filtering and filter-tip pipets are not preheated before filtration?
11. Why is a Craig tube filtration ineffective for removing finely divided particles from a solution?
12. Briefly explain how a colored solution may be decolorized.
13. Briefly explain how insoluble particles can be removed from a hot solution.
14. When might Celite, a filter-aid, be used in recrystallization, and why is it used?
15. List five criteria that should be used in selecting a solvent for a recrystallization.
16. How does the principle of *like dissolves like* explain the differing solubilities of solutes in various solvents?
17. The following solvent selection data were collected for two different impure solids:

Solid A

Solvent	*Solubility at Room Temperature*	*Solubility When Heated*	*Crystals Formed When Cooled*
Methanol	Insoluble	Insoluble	—
Ethyl acetate	Insoluble	Soluble	Very few
Cyclohexane	Insoluble	Soluble	Many
Toluene	Insoluble	Soluble	Very few

Solid B

Solvent	*Solubility at Room Temperature*	*Solubility When Heated*	*Crystals Formed When Cooled*
Water	Soluble	—	—
Ethanol	Soluble	—	—
Dichloromethane	Insoluble	Insoluble	—
Petroleum ether	Insoluble	Insoluble	—
Toluene	Insoluble	Insoluble	—

Based on these results, what solvents or mixture of solvents might you consider using to recrystallize solids A and B? Explain.

18. Briefly describe how a mixture of sand and benzoic acid, which is soluble in hot water, might be separated to provide pure benzoic acid.
19. Look up the solubility of benzoic acid in hot water. According to the published solubility, what is the minimum amount of water in which 1 g of benzoic acid can be dissolved?

20. The solubility of benzoic acid at 0 °C is 0.02 g per 100 mL of water, and that of acetanilide is 0.53 g per 100 mL of water. If you performed either of these recrystallizations, calculate, with reference to the total volume of water you used in preparing the hot solution, the amount of material in your experiment that was unrecoverable by virtue of its solubility at 0 °C.

21. Assuming that either solvent is otherwise acceptable in a given instance, what advantages does ethanol have over 1-octanol as a crystallization solvent? hexane over pentane? water over methanol?

22. In the course of a synthesis of an important antibiotic, an impure solid was obtained as one of the intermediates. The solubility of this material in various solvents is shown below.

	Water	*Ethanol*	*Toluene*	*Petroleum Ether*	*2-Butanone*	*Acetic Acid*
Cold	Insoluble	Soluble	Insoluble	Insoluble	Soluble	Slightly soluble
Hot	Slightly soluble	Soluble	Soluble	Insoluble	Soluble	Soluble

    a. Which of the solvents above would be the most suitable for recrystallization of the impure solid? Explain your reasoning.
    b. Provide a reason why each of the following solvents would *not* be a suitable solvent for recrystallization of this material: ethanol, petroleum ether.
    c. Write the chemical structure for each of the solvents in the table above.
    d. Which solvent is the most polar? least polar?

23. The goal of the recrystallization procedure is to obtain purified material with a maximized recovery. For each of the items listed, explain why this goal would be adversely affected.
    a. In the solution step, an unnecessarily large volume of solvent is used.
    b. The crystals obtained after filtration are not washed with fresh cold solvent before drying.
    c. The crystals referred to in (**b**) are washed with fresh hot solvent.
    d. A large quantity of decolorizing carbon is used.
    e. Crystals are obtained by breaking up the solidified mass of an oil that originally separated from the hot solution.
    f. Crystallization is accelerated by immediately placing the flask of hot solution in an ice-water bath.

24. A second crop of crystals may be obtained by concentrating the vacuum filtrate and cooling. Why is this crop of crystals probably less pure than the first crop?

25. Explain why the rate of dissolution of a crystalline substance may depend on the size of its crystals.

26. In the technique of recrystallization, one step involves heating the solvent containing the solute to its boiling point and, after the solution is cooled to room temperature, cooling it further in an ice-water bath. Why is it important to operate at these temperature extremes during a recrystallization?

27. In the process of a recrystallization, if crystals do not form upon cooling the solution, it is often recommended that the inside of the flask be scratched at the air–liquid interface with a glass stirring rod. What purpose does this serve, and how does it work? What else might be done to induce crystallization?
28. Should some loss of sample mass be expected even after the most carefully executed recrystallization? Explain.
29. In general, what solvent should be used to rinse the filter cake during the vacuum filtration step of a recrystallization? Should this solvent be cooled prior to use?
30. Why do you seldom see high-boiling solvents used as recrystallization solvents?
31. At the end of a recrystallization, where should the *impurities* be located?
32. A student has been asked to recrystallize 1.0 g of impure stilbene from ethanol. Provide a set of standard step-by-step instructions for recrystallization of this sample so as to maximize the purity and yield obtained.
33. An important product from a multistep synthesis must be recrystallized to remove a small amount of an impurity. However, all the available solvents each individually fail to be suitable recrystallization solvents. Offer a solution to this problem using only the available solvents. (*Hint:* Consider binary solvents.)
34. A suspension of decolorizing carbon (charcoal) is often administered to poison victims.
    a. Speculate on the purpose decolorizing carbon serves in this particular application. (*Hint:* It is similar to the way in which decolorizing carbon is used in a recrystallization.)
    b. How is the charcoal ultimately removed from the victim?

3.3 PHYSICAL CONSTANTS: MELTING POINTS

Physical Constants

See more on *Melting Point*

Physical constants of compounds are numerical values associated with measurable properties of these substances. These properties are *invariant* and are useful in the identification and characterization of substances encountered in the laboratory so long as accurate measurements are made under specified conditions such as temperature and pressure. Physical constants are useful only in the identification of *previously known* compounds, however, because it is not possible to predict the values of such properties accurately. Among the more frequently measured physical properties of organic compounds are **melting point (mp), boiling point (bp), index of refraction (n), density (d), specific rotation ([α]),** and **solubility.** Melting points, discussed below, boiling points, described in Section 4.2, and solubilities, outlined in Section 3.2, are the properties most commonly encountered. Index of refraction and density are mentioned in Chapter 25. Specific rotation is discussed in Chapters 7 and 23 but applies only to molecules that are **optically active.** Whether the substance is known or unknown, such values, along with other properties like color, odor, and crystal form, should be recorded in the laboratory notebook.

The values of one or two of the common physical properties *may* be identical for more than one compound, but it is most unlikely that values of several such properties will be the same for two different compounds. Consequently, a list of physical constants is a highly useful way to characterize a substance. Extensive compilations of the physical constants are available (Chap. 26). One of the most convenient is the *CRC Handbook of Chemistry and Physics,* which contains a tabulation of the physical constants and properties of a large number of inorganic and organic compounds. *The Handbook of Tables for Organic Compounds* is especially useful for organic compounds. Neither of these books is comprehensive; rather, they contain entries for only the more common organic and inorganic substances. So many compounds are known that multi-volume sets of books are required to list their physical properties (Chap. 26).

Melting Point of a Pure Substance

The melting point of a substance is defined as the temperature at which the liquid and solid phases exist in equilibrium with one another without change of temperature. Ideally, addition of heat to a mixture of the solid and liquid phases of a pure substance at the melting point will cause no rise in temperature until all the solid has melted. Conversely, removal of heat from the equilibrium mixture will produce no decrease in temperature until all the liquid solidifies. This means that the melting and freezing points of a pure substance are identical.

The melting point is expressed as the temperature *range* over which the solid starts to melt and then is completely converted to liquid. Consequently, rather than a melting *point,* what is actually measured is a **melting range,** although the two terms are used interchangeably. If a crystalline substance is pure, it should melt over a narrow or sharp range, which will normally be no more than 1 °C if the melting point is determined carefully. The melting ranges reported for many "pure" compounds may be greater than 1 °C because the particular compound was not quite pure or the melting point was not measured properly. The process of melting may actually begin by "softening," as evidenced by an apparent shrinking of the solid, but such softening is difficult to observe. Thus, for our purposes, the start of melting is defined as the temperature at which the first tiny droplet of liquid can be detected. Note that it is improper and inexact to report a single temperature, such as 118 °C, for a melting point; rather, a range of 117–119 °C or 117.5–118.0 °C, for example, should be recorded.

Effect of Impurities on Melting Points

Many solid substances prepared in the organic laboratory are initially impure, so the effect of impurities on melting-point ranges deserves further discussion. Although this topic is discussed in freshman chemistry textbooks, a brief review of its basic principles is given here.

The presence of an impurity generally *decreases* the melting point of a pure solid. This is shown graphically by the melting-point–composition diagram of Figure 3.1, in which points *a* and *b* represent the melting points of pure *A* and *B,* respectively. Point *E* is called the **eutectic point** and is determined by the equilibrium composition at which *A* and *B* melt in constant ratio. In Figure 3.1, this ratio is 60 mol % *A* and 40 mol % *B;* an impure solid composed of *A* and *B* in this ratio would be called a **eutectic mixture.** The temperature at the eutectic point is designated by *e.*

Now consider the result of heating a solid mixture composed of 80 mol % *A* and 20 mol % *B,* a sample that might be considered as "impure *A.*" As heat is applied to the solid, its temperature will rise. When the temperature reaches *e, A* and *B* will both begin to melt in the constant ratio defined by the composition at the eutectic point. Once all of the "impurity" *B* has melted, only solid *A* will be left in equilibrium with the melt. The remaining solid *A* will continue to melt as additional heat is supplied, and the percentage of *A* in the melt will increase, changing the composition of the

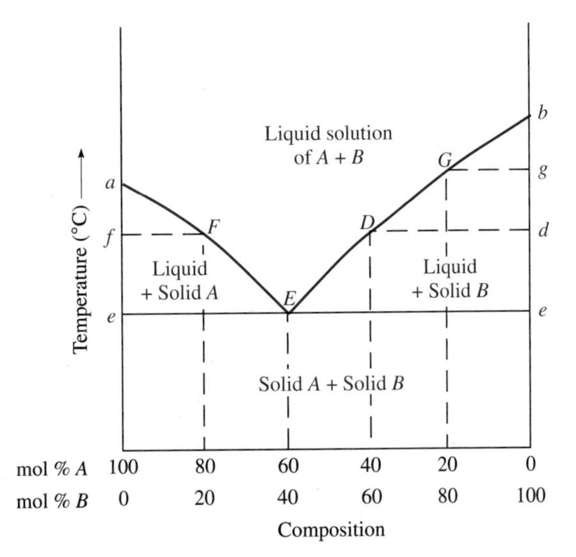

Figure 3.1
Melting-point–composition diagram for two hypothetical solids, A and B.

melt from that of the eutectic mixture. This increases the vapor pressure of *A* in the solution according to Raoult's law (Eq. 4.2) and raises the temperature at which solid *A* is in equilibrium with the molten solution. The relationship between the equilibrium temperature and the composition of the molten solution is then represented by curve *EF* in Figure 3.1. When the temperature reaches *f*, no solid *A* will remain and melting of the sample will be complete. The impure sample *A* exhibits a melting "point" that extends over the relatively broad temperature range *e–f*. Because melting both begins and ends below the melting point of pure *A*, the melting point of *A* is said to be *depressed.*

The foregoing analysis is easily extended to the case in which substance *B* contains *A* as an impurity. In Figure 3.1, this simply means that the composition of the solid mixture is to the right of point *E*. The temperature during the melting process would follow curve *ED* or *EG*, and the melting range would now be *e–d* or *e–g*.

A sample whose composition is exactly that of the eutectic mixture (point *E*, Fig. 3.1) will exhibit a sharp melting point at the eutectic temperature. This means a eutectic mixture can be mistaken for a pure compound, because both have a sharp melting point.

From a practical standpoint, it may be very difficult to observe the initial melting point of solid mixtures, particularly with the capillary-tube melting-point technique used in the Experimental Procedure that follows. This is because the presence of only a minor amount of impurity means that only a tiny amount of liquid is formed in the stage of melting that occurs at the eutectic temperature. In contrast, the temperature at which the last of the solid melts (points *d* and *g*, Fig. 3.1) can be determined accurately. Consequently, a mixture containing smaller amounts of impurities will generally have both a higher final melting point and a narrower observed melting-point range than one that is less pure.

The broadening of the melting-point range that results from introducing an impurity into a pure compound may be used to advantage for identifying a pure substance. The technique is commonly known as a **mixed melting-point** and is illustrated by the following example. Assume that an unknown compound *X* melts at 134–135 °C, and you suspect it is either urea, H_2NCONH_2, or *trans*-cinnamic acid, $C_6H_5CH{=}CHCO_2H$, both of which melt in this range. If *X* is mixed intimately with urea and the melting point of this mixture is found to be lower than that of the pure compound and pure urea, then urea is acting as an impurity, and the compound cannot be urea. If the mixture melting point is identical to that of the pure compound and of urea, the compound is identified as urea. Obviously, this procedure is useful in identifying compounds only when authentic samples of the likely possibilities are available.

A convenient and rapid method for ascertaining the purity of a solid is measuring its melting point. A narrow melting-point range ordinarily signals that the sample is *pure,* although there is a *low* probability that the solid is a eutectic mixture. If recrystallizing a sample changes an originally broad melting range to a narrow one, the reasonable conclusion is that the recrystallization was successful in purifying the solid. Should the melting-point range remain broad after recrystallization, the sample may be contaminated with solvent and additional drying is required. It is also possible that the recrystallization was not completely successful in removing impurities, in which case the solid should be recrystallized using the same solvent. If this fails to narrow the melting range satisfactorily, recrystallization should be performed with a different solvent.

Micro Melting-Point Methods

The determination of accurate melting points of organic compounds can be time-consuming. Fortunately, micro methods are available that are convenient, require negligible amounts of sample, and give melting-point data that are satisfactory for most purposes. The technique using the capillary-tube melting-point procedure is the one used most commonly in the organic laboratory.

There are practical considerations in determining melting points, and some of them are briefly noted here. First, the observed melting-point range depends on several factors, including the quantity of sample, its state of subdivision, the rate of heating during the determination, and the purity and chemical characteristics of the sample. The first three factors can cause the observed melting-point range to differ from the true value because of the time lag for transfer of heat from the heating medium to the sample and for conduction of heat within the sample. For example, if the sample is too large, the distribution of heat may not be uniform, and inaccurate melting ranges will result. A similar problem of nonuniform heat distribution is associated with using large crystals. It will be difficult to pack the sample tightly in the capillary melting tube, and the airspace that results causes poor conduction of heat. If the rate of heating is too fast, the thermometer reading will lag behind the actual temperature of the heating medium and produce measurements that are low. The chemical characteristics of the sample may be important if the compound tends to decompose on melting. When this occurs, discoloration of the sample is usually evident, and it may be accompanied by gas evolution. The decomposition products constitute impurities in the sample, and the true melting point is lowered as a result. The reporting of melting points for compounds that melt with decomposition should reflect this, as in "mp 195 °C (dec)."

In determining the melting point of a compound, valuable time can be wasted waiting for melting to occur if the proper slow rate of heating is being used on a sample whose melting point is unknown. It is considerably more efficient to prepare two capillary tubes containing the compound being studied and to determine the approximate melting point by rapidly heating one of them, and then allowing the heating source to cool 10–15 °C below this approximate melting point before obtaining an accurate melting point with the second sample.

The accuracy of any type of temperature measurement ultimately depends on the quality and calibration of the thermometer. A particular thermometer may provide accurate readings in some temperature ranges but may be off by a degree or two in others. Melting points that have been determined using a calibrated thermometer may be reported in the form "mp 101–102 °C (corr.)," where "corr." is the abbreviation for "corrected"; the corresponding abbreviation for values obtained with an uncalibrated thermometer is "uncorr." for "uncorrected."

Calibration involves the use of standard substances for the measurement of the temperature at a series of known points within the range of the thermometer and the comparison of the observed readings with the true temperatures. The difference between the observed and the true temperature measurement provides a correction that must be applied to the observed reading. Calibration over a range of temperatures is necessary because the error is likely to vary at different temperatures.

EXPERIMENTAL PROCEDURES

Melting Points

Purpose To determine melting points using the capillary-tube method.

SAFETY ALERT

1. **If a burner is used in this experiment, be sure that no flammable solvents are nearby. Keep the rubber tubing leading to the burner away from the flame. Turn off the burner when it is not being used.**
2. **Some kinds of melting-point apparatus, such as the Thiele tube, use mineral or silicone oils as the heat transfer medium. These oils may not be heated safely if they are contaminated with even a few drops of water. Heating these oils above 100 °C may produce splattering of hot oil as the water turns to steam. Fire can also result if spattered oil comes in contact with open flames. Examine your Thiele tube for evidence of water droplets in the oil. If there are any, either change the oil or exchange tubes. Give the contaminated tube to your instructor.**
3. **Mineral oil is a mixture of high-boiling hydrocarbons and should not be heated above 200 °C because of the possibility of spontaneous ignition, particularly when a burner is used for heating. Some silicone oils may be heated to about 300 °C without danger (Sec. 2.9).**
4. **Be careful to avoid contact of chemicals with your skin. Clean up any spilled chemicals immediately with a brush or paper towel.**
5. **If you use a Thiele tube, handle it carefully when you are finished, because the tube cools slowly. To avoid burns, take care when removing it from its support.**

A ■ Calibration of Thermometer

Procedure

Preparation Sign in at **www.cengage.com/login** to answer Pre-Lab Exercises, access videos, and read the MSDSs for the chemicals used or produced in this procedure. Read or review Sections 2.7 and 2.9.

Apparatus Capillary melting-point tubes, packing tube, and melting-point apparatus.

Protocol Carefully determine the capillary melting points of a series of standard substances. A list of suitable standards is provided in Table 3.2. The temperatures given in Table 3.2 correspond to the upper limit of the melting-point range for pure samples of these standards.

Plot the corrections in your notebook as deviations from zero versus the temperature over the range encompassed by the thermometer. This allows you to tell, for example, that at about 130 °C the thermometer gives readings that are 2 °C too low, or that at 190 °C the readings are about 1.5 °C too high. These values should then be applied to correct all temperature measurements taken. The corrections you obtain are valid only for the thermometer used in the calibration. If you break it, not only are you likely to be charged for a replacement, but you must repeat the calibration.

B ■ Determining Capillary-Tube Melting Points

Procedure

Preparation Sign in at **www.cengage.com/login** to answer Pre-Lab Exercises, access videos, and read the MSDSs for the chemicals used or produced in this procedure. Read or review Sections 2.7 and 2.9. Plan to prepare two melting-point tubes if you are determining the melting points of unknown compounds.

Apparatus Capillary melting-point tubes, packing tube, and melting-point apparatus.

1. Pure Compound

Protocol Select one or more compounds from a list of available compounds of known melting point and determine the melting-point ranges. Repeat as necessary until you obtain accurate results and are confident with the technique.

2. Unknown Compound

Protocol Accurately determine the melting range of an unknown pure compound supplied by your instructor.

3. Mixed Melting Points

Sample Preparation Prepare a sample for mixed melting-point determination by introducing 5–10% of a second substance as an impurity into a solid whose

Table 3.2 *Standards for Thermometer Calibration*

Compound	*Melting Point (°C)*
Ice water	0
3-Phenylpropanoic acid	48.6
Acetamide	82.3
Acetanilide	114
Benzamide	133
Salicylic acid	159
4-Chloroacetanilide	179
3,5-Dinitrobenzoic acid	205

melting point was determined in Part 1. Intimately mix the two components by grinding them together with a small mortar and pestle. Alternatively, use a small, clean watch glass and glass stirring rod or metal spatula to mix the components. Be careful not to apply too much pressure to the glass rod, however, because it is more fragile than a pestle and may break. Do not perform this operation on a piece of filter paper, because fibers from the paper may contaminate the sample.

Protocol Accurately determine the melting range of the sample and compare it to that of the major component of the mixture in order to study the effect of impurities on the melting range of a previously pure compound.

Discovery Experiment *Melting-Point Depression*

Obtain a pair of solids whose melting points are within 1 °C of one another and determine whether they indeed depress one another's melting points. Some possible examples are 3-furoic acid and benzoic acid, 1-naphthylacetic acid and 2,5-dimethylbenzoic acid, *trans*-cinnamic acid and 2-furoic acid, 2-methoxybenzoic acid and malic acid, or dimethyl fumarate, 9-fluorenone, and methyl 4-bromobenzoate. Consult with your instructor regarding these and other possibilities.

WRAPPING IT UP

Return any *unused samples* to your instructor or dispose of them in the appropriate container for nonhazardous organic solids. Discard the used *capillary tubes* in a container for broken glass; do *not* leave them in the area of the melting-point apparatus or throw them in wastepaper baskets.

EXERCISES

1. Describe errors in procedure that may cause an observed capillary melting point of a pure compound
   a. to be *lower* than the correct melting point.
   b. to be *higher* than the correct melting point.
   c. to be *broad* in range (over several degrees).
2. Briefly define the following terms:
   a. vapor pressure as applied to melting
   b. melting point or melting-point range
   c. mixture or mixed melting point
   d. eutectic point
   e. eutectic mixture
3. Describe on a molecular level the difference between the two physical changes "melting" and "dissolving."
4. Answer the following questions about melting points.
   a. Why is the melting "point" of a substance actually a melting "range" and therefore should never be recorded as a single temperature?
   b. In theory, does a melting "point" exist? Explain your answer.

5. Explain how a eutectic mixture could be mistaken for a pure substance, and comment on whether encountering a eutectic mixture would be a frequent or infrequent occurrence.
6. Compound A and compound B have approximately the same melting point. State two ways in which a mixed melting point of these two compounds would be different from the melting point of either pure A or pure B.
7. Filter paper is usually a poor material on which to powder a solid sample before introducing it into a capillary melting-point tube because small particles of paper may end up in the tube along with the sample. Why is this undesirable, and how might the presence of paper in the sample make the melting-point determination difficult?
8. Some solids sublime before they melt, making a determination of a melting point impossible using a standard melting-point capillary tube. How could you modify your capillary tube to obtain a melting point for such a compound?
9. Some solids, in particular many amino acids, decompose upon melting. These compounds are often reported in the literature with the term "dec" following their melting-point range.
   a. How might the melting process appear different for this type of compound?
   b. Look up and record the melting point and structure for a compound that decomposes upon melting. Use a chemical handbook or a chemical catalog as the source of this information.
10. Criticize the following statements by indicating whether each is *true* or *false*, and if false, explain why.
    a. An impurity always lowers the melting point of an organic compound.
    b. A sharp melting point for a crystalline organic substance always indicates a pure single compound.
    c. If the addition of a sample of compound *A* to compound *X* does not lower the melting point of *X*, *X* must be identical to *A*.
    d. If the addition of a sample of compound *A* lowers the melting point of compound *X*, *X* and *A* cannot be identical.
11. The melting points of pure benzoic acid and pure 2-naphthol are 122.5 °C and 123 °C, respectively. Given a pure sample that is known to be either pure benzoic acid or 2-naphthol, describe a procedure you might use to determine the identity of the sample.
12. A student used the Thiele micro melting-point technique to determine the melting point of an unknown and reported it to be 182 °C. Is this value believable? Explain why or why not.
13. The melting-point–composition diagram for two substances, *Q* and *R*, is provided in Figure 3.2, which should be used to answer the following questions.
    a. What are the melting points of pure *Q* and *R*?
    b. What are the melting point and the composition of the eutectic mixture?
    c. Would a mixture of 20 mol % *Q* and 80 mol % *R* melt if heated to 120 °C? to 160 °C? to 75 °C?
    d. A mixture of *Q* and *R* was observed to melt at 105–110 °C. What can be said about the composition of this mixture? Explain briefly.

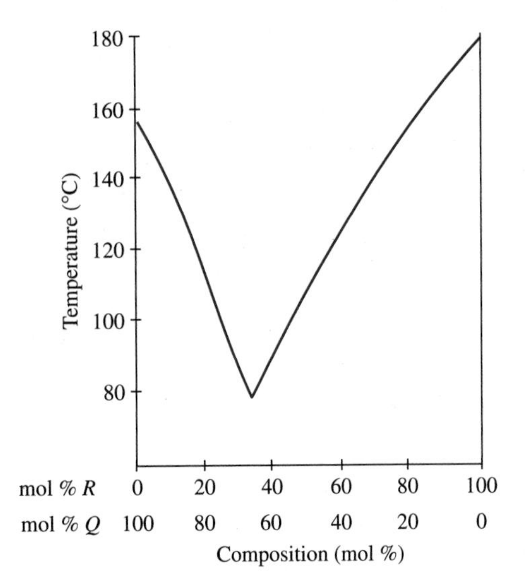

Figure 3.2
Melting-point–composition diagram for Exercise 13.

14. For the following melting points, indicate what might be concluded regarding the purity of the sample:

a. 120–122 °C

b. 46–60 °C

c. 147 °C (dec)

d. 162.5–163.5 °C

15. Suppose that the observed melting-point range of a solid was originally 150–160 °C, but this became 145–145.5 °C after recrystallization. How might you account for this observation?

16. An unknown solid was found to have a melting point of 160–161.5 °C. Given other information that was available for the unknown, it was thought to be one of the following compounds:

Salicyclic acid	158–160 °C
Benzanilide	161–163 °C
Triphenylmethanol	159.5–160.5 °C

If you had access to pure samples of the above compounds and could only use a melting-point apparatus, devise an experiment to determine which of these three compounds corresponds to the identity of the unknown.

HISTORICAL HIGHLIGHT

Polymorphism

The recrystallizations that you perform in the organic chemistry laboratory will often produce beautiful crystals that may be described as needles, plates, cubes, etc. These shapes characterize the ways that the individual molecules fit together within the crystal structure. A familiar albeit inorganic example is sodium chloride, which has the sodium and chloride ions arranged such that its physical form is a cube.

Although the recrystallizations you do in the laboratory may consistently yield solids having the same geometric form, this may not always be the case, and therein lies an interesting phenomenon. We'll focus on organic compounds in our discussion, but it applies to inorganic compounds as well.

In 1832, Wöhler and Liebig found that slowly cooling a "boiling hot" aqueous solution of benzamide yielded a mass of white silky needles. Over the course of days, the solid in the unfiltered mixture of needles and water was mysteriously transformed into crystals having a rhombic shape. One can imagine how perplexed these early chemists must have been upon seeing this transformation. We now understand that this change signals that the needles represent a metastable, thermodynamically less stable form of crystalline benzamide. Moreover, it is now known that this rhombic form is produced exclusively if a hot aqueous solution of benzamide is cooled at a rate of less than about 1 °C/minute, a rate presumably slower than that used by Wöhler and Liebig. A third, even less stable crystalline form of benzamide that transforms to the stable rhombic in hours rather than days has recently been reported and results from very rapid cooling of a solution of the compound (*Cryst. Growth Des.* 2005, *5,* 2218–2224). Thus, benzamide exists in at least three different crystalline forms, which are called polymorphs.

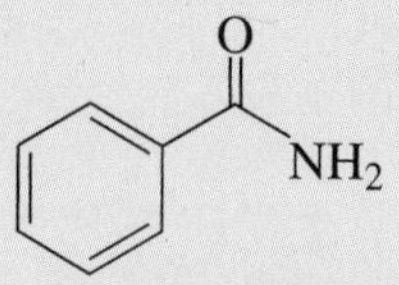

O

NH_2

Benzamide

The thermodynamic relationship among the polymorphs has recently been described: *Angew. Chem., Int. Ed.*, 2007, *46,* 6729–6731. As discussed in this article, the polymorphs have the benzamide molecules organized in the three orientations portrayed below, with Form III believed to be that originally isolated by Wöhler and Liebig and Form I being the structure to which both Forms III and II transform at differing rates. Thus, Form I is the most thermodynamically stable of the three, and Form III is the next most stable. Although it is believed that Form III has the most stabilization resulting from "π–π stacking," as symbolized by the boldfaced hydrogen atoms in the structure, Form I has more stabilization from hydrogen bonding between the carbonyl oxygen atom and a hydrogen atom on the amido nitrogen atom of another molecule. Form II also benefits from hydrogen bonding but may be destabilized by repulsions between the stacked aromatic π-systems.

(Continued)

HISTORICAL HIGHLIGHT *Polymorphism (Continued)*

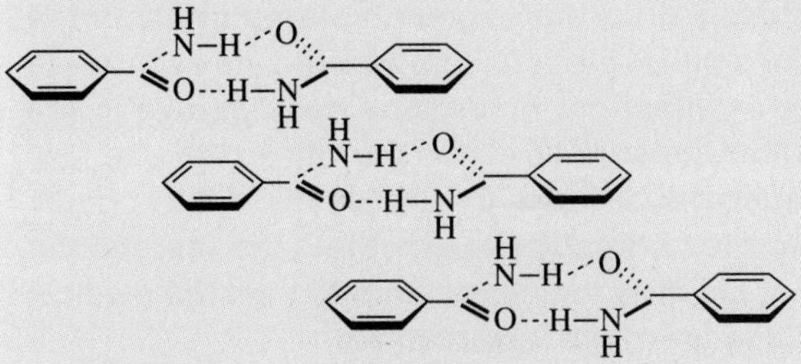

Form I

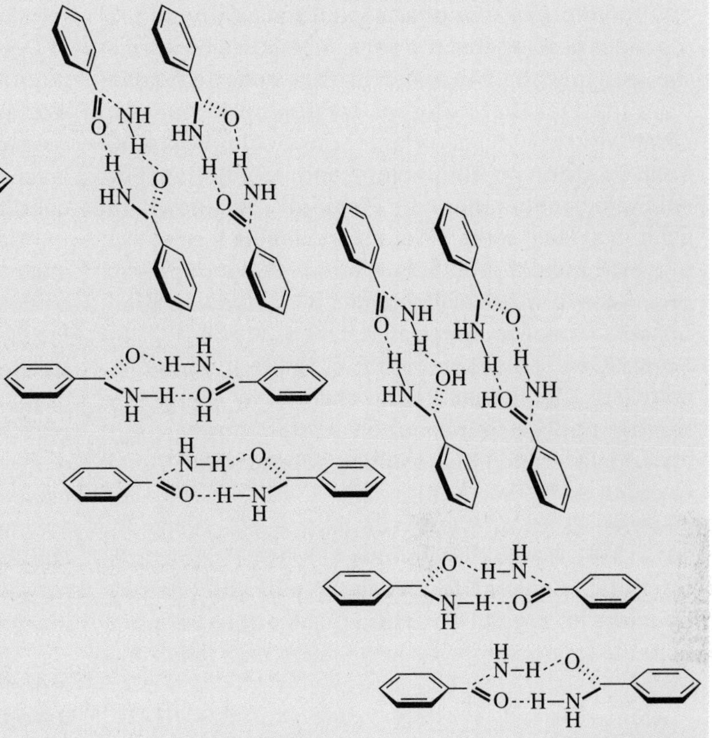

Form II

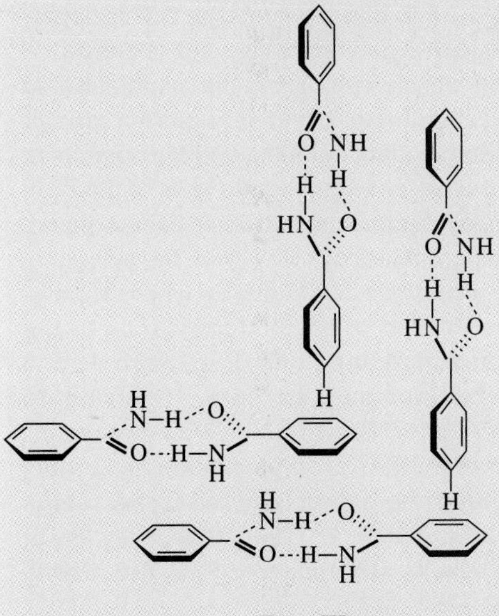

Form III

The existence of polymorphs can be of great significance. For example, in the case of pharmaceutical compounds, approval of a drug by the Food and Drug Administration is predicated on a *specific* crystalline form of the drug being subjected to the extensive testing required before the compound is deemed suitable for commercial use. What can and has happened in the course of the development and testing of a drug is that it may exist as two or even more polymorphs, and the original form that may already have been subjected to very expensive clinical testing may not be the most thermodyamically stable. If this is the case, a more stable polymorph of the compound may suddenly appear in the course of the manufacturing process, and it may be extremely difficult, if not nearly impossible, to reproduce the form that has been subjected to testing. For example, it may be necessary to construct the equivalent of a "clean room," essentially a new research and/or manufacturing facility, as part of attempt to reproduce the desired form, the thought being that seed crystals of the undesired polymorph are contaminating the facilities where the original form was produced.

(Continued)

HISTORICAL HIGHLIGHT *Polymorphism (Continued)*

You might wonder why the existence of different polymorphs of a pharmaceutical compound is of concern, given that the same molecules comprise the polymorphs. Answers to this question arise from the fact that different crystalline forms of a compound may have different physical properties, such as color, melting points, and solubilities. For example, acetaminophen (Tylenol®) is known to exist in at least three different crystalline forms, two of which are pictured below. Form I is a monoclinic prism, the form usually obtained by recrystallization of acetaminophen from water; it melts at 169–172 °C. Form II, an orthorhombic form, is thermodynamically less stable than Form I and can be formed by rapidly cooling pure molten acetaminophen followed by warming the resulting super-cooled liquid to about 80 °C (see Kauffman, J. F.; Batykefer, L. M.; Tuschel, D. D. *J. Pharm. Biomed. Anal.* 2008, *48*, 131–1315). Its melting point is 158–160 °C. It is this form that is desired for use in the pills you can buy because its crystal structure turns out to be more suitable for compression into tablets than is Form I.

HO—C_6H_4—NHC(=O)CH_3

Acetaminophen

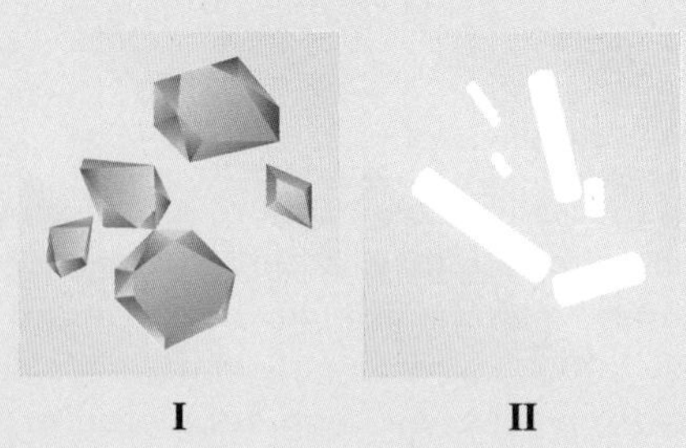

I II

Another factor of concern regarding polymorphs is that their therapeutic activity may differ, a fact you may find surprising because the molecules comprising the polymorphs are identical. In this case, one reason for differing activities is associated with another physical property, namely, solubility. Consider a drug that is administered orally in the form of a pill. A particular polymorph less soluble in saliva or intestinal fluids may not dissolve in the mouth or intestinal tract as rapidly as the one that was approved for use. Consequently, it may simply be excreted, thereby not providing the therapeutic level of the drug that is required to treat the medical condition for which it is intended.

Solubility can even be an issue when the substance is administered in solution because it may not be possible to obtain the necessary concentration of the drug in the amount of solvent that can be administered. Medications delivered as eye drops are a good example. It is possible that a polymorph that is appropriately soluble in the medium being used is converted to a less-soluble polymorph during the manufacturing process. The less-soluble form will require a greater volume of solvent to effect solution, and the quantity of liquid needed to deliver a therapeutic level of the drug may exceed the eye's capacity to hold it. Excess liquid, the drug included, will just overflow from the eye.

There are a number of examples wherein polymorphism has proved to be of significance in the development and marketing of a drug, but only one of them will be summarized here. Referring to the URLs found under "See more on *Polymorphism*" will lead to several other examples, such as Zantac, as will the following references: Dunitz, J. D.; Bernstein, J. *Acc. Chem. Res.* 1995, *28*, 193–200; Davey, R. J.; Blagden, N.; Potts, G. D.; Docherty, R. *J. Am. Chem. Soc.* 1997, *199*, 1767–1772.

Ritonavir (Norvir®) is a peptidomimetic drug that was introduced by Abbott in 1996 for use in treating HIV-1 infections. Initially, it was produced and marketed exclusively in a monoclinic crystalline form that had to be delivered as a capsule containing the drug in a water-alcohol ("hydroalcoholic") mixture because of low bioavailability of the compound when administered in tablet form. In 1998 it was noticed that some lots of the drug were failing to meet the specifications initially appropriate for dissolution and that a precipitate was forming in the capsules. The precipitate was ultimately identified as a polymorph having less than 50% of the solubility in the solvent originally used. This forced withdrawal of the

(Continued)

HISTORICAL HIGHLIGHT *Polymorphism (Continued)*

drug from the marketplace for a year, which not only cost Abbott an estimated $250 million from loss of sales but also denied patients access to the life-saving drug. Extensive investigations were pursued for understanding the basis for the shift from one polymorph to another, reformulating the solvent for the new polymorph, and identifying strategies for regenerating the original form of the compound. Ritonavir was again marketed in 1999. It is of interest to note that five polymorphs of this compound are now known (Morissette, S. L. et al., *Proc. Natl. Acad. Sci. U. S. A.* 2003, *100*, 2180–2184).

Ritonavir

Relationship of Historical Highlight to Experiments

As seen in this discussion, the crystalline form of a solid can vary and may do so as a result of the procedure followed for recrystallizing a compound. You may observe this phenomenon as you perform recrystallizations in the introductory organic laboratory, or you may find that the form of your crystals is different from that of other students in your laboratory. Even if you don't observe the formation of polymorphs, it is important to know of their existence.

See more on *Wöhler*

See more on *Liebig*

See more on *Acetaminophen*

See more on *Polymorphism*

See more on *Bioavailability*

See more on *Ritonavir*

CHAPTER 4

Liquids

Distillation and Boiling Points

When you see this icon, sign in at this book's premium website at **www.cengage.com/login** to access videos, Pre-Lab Exercises, and other online resources.

The odors of both natural and synthetic organic chemicals surround us in our everyday life. Some, like the smells of fruits and many spices, are pleasant, whereas others, like those of decaying meat and rancid butter, can be rather unpleasant. We detect these odors because many organic chemicals are volatile, and in this chapter we explore the theory and practice of distillation, a technique that depends on our ability to coax these chemicals into the gas phase.

4.1 INTRODUCTION

The purification of solids by recrystallization and the use of melting points as a criterion of their purity are discussed in Chapter 3. The techniques used for the purification of liquids involve **simple, fractional, steam,** and **vacuum distillation,** and these are topics of this chapter. Boiling points are also discussed as a physical property that can be used as one means of determining the purity and identity of liquids.

Two of the laws, Dalton's and Raoult's, that you have encountered in previous courses in chemistry and/or physics are critical to understanding the phenomenon of distillation. We'll review them in this chapter and show how they are applied to all forms of distillation that you will find in this textbook. As you will see, the apparatus for accomplishing distillations is considerably more complex than that needed for recrystallizations (Chap. 3), but with a little experience you will be able to complete a distillation in as little time as it takes to perform a recrystallization.

4.2 BOILING POINTS OF PURE LIQUIDS

The molecules of a liquid are constantly in motion, and those at the surface are able to escape into the vapor phase. The consequences of vaporization of a liquid contained in a closed, *evacuated* system are considered first, and then the situation in which the system is open to the atmosphere is discussed.

In a closed evacuated system, the number of molecules in the gas phase will initially increase until the rate at which they reenter the liquid becomes equal to the rate at which they escape from it. At this point, no further *net* change is observed, and the system is said to be in a state of **dynamic equilibrium.** The molecules in the gas phase are in rapid motion and continually collide with the walls of the vessel, which results in the exertion of pressure against the walls. The magnitude of this vapor pressure at a given temperature is called the **equilibrium vapor pressure** of the liquid at that temperature. Vapor pressure is temperature-dependent, as shown in Figure 4.1. As the temperature of the system rises, the average kinetic energy of the liquid molecules increases, thus facilitating their escape into the gas phase. The rate of reentry of gaseous molecules into the liquid phase also increases, and equilibrium is reestablished at the higher temperature. Because there are now more molecules in the gas phase than there were at the lower temperature, the vapor pressure of the system is greater.

There is a very important safety rule based on the fact that the pressure in a closed system *increases* as the temperature of the system rises. Such a system should *not* be heated unless an apparatus designed to withstand the pressure is used; otherwise, an *explosion* will result. For purposes of the first laboratory course in organic chemistry, the rule is quite simply *"Never heat a closed system!"*

See *Who was Dalton?*

Now suppose that a liquid sample at a particular temperature is placed in an *open* container so that the molecules of the vapor over the liquid are mixed with air. The **total pressure** above the liquid is defined by **Dalton's law of partial pressures** (Eq. 4.1) and is equal to the sum of the partial pressures of the sample and of air. The partial pressure of the sample is equal to its **equilibrium vapor pressure** at the given temperature. When the temperature of the liquid is raised, the equilibrium vapor pressure of the sample will rise, and the number of gas molecules that have escaped from the liquid by the process of evaporation will increase in the space above the liquid. This will have the net effect of displacing some of the air. At the higher temperature, the partial pressure of the sample will be a *larger* percentage of the *total* pressure. This trend will continue as the temperature of the liquid is further increased, until the equilibrium vapor pressure of the sample equals the total pressure. At this point, all of the air will have been displaced from the vessel containing the liquid. Entry of additional molecules from the liquid into the gas phase will only have the effect of further displacing those already in that phase; the partial pressure of the molecules of the sample will no longer increase.

$$P_{total} = P_{sample} + P_{air} \tag{4.1}$$

When the temperature of the liquid is such that the equilibrium vapor pressure of the sample equals the total pressure, the rate of evaporation increases dramatically, and bubbles form in the liquid. This is referred to as the boiling process, and the temperature associated with it is the **boiling point** of the liquid. Since the boiling

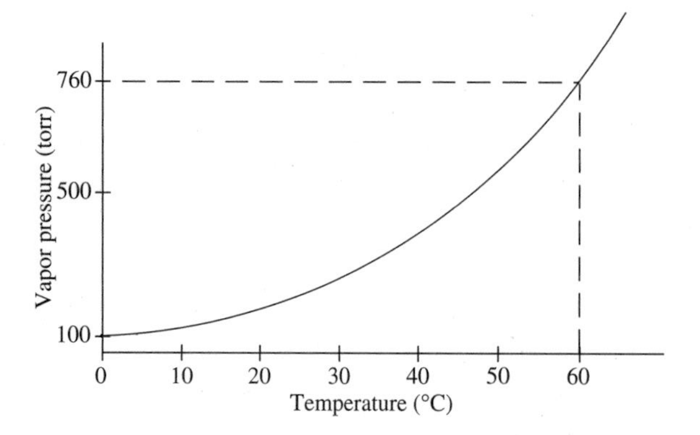

Figure 4.1
Graph of the dependence of vapor pressure on temperature for a liquid.

point is dependent upon the total pressure, that pressure must be specified when boiling points are reported, for example, "bp 132 °C (748 torr)." The boiling point of water is 100 °C *only* when the external pressure is 760 torr, or 1 atm. The **normal** boiling point of a liquid is measured at 760 torr (1 atm) and is shown by the dashed lines in Figure 4.1 to be 60 °C. This figure also allows determination of the boiling point of the liquid at various total pressures. For example, at 500 torr the boiling point will be about 50 °C; the decrease in temperature results from the decrease in the total pressure in the system.

The dependence of vapor pressure upon total pressure, as reflected by the boiling point of a liquid, can be used to advantage in the following way. Suppose that a liquid has a normal boiling point such that it decomposes appreciably when heated to this temperature. Reduction in the total pressure of the system reduces the boiling point of the sample to a temperature at which it no longer decomposes. This technique, called **vacuum distillation,** is discussed in Section 2.15.

A **pure** liquid generally boils at a constant temperature or over a narrow temperature range, provided the total pressure in the system remains constant. However, the boiling point of a liquid is affected by the presence of both nonvolatile and volatile impurities, and most **mixtures** of liquids boil over a fairly wide range. These effects are discussed in Section 4.3.

EXPERIMENTAL PROCEDURES

Boiling Points of Pure Liquids

Purpose To determine the boiling points of pure liquids at atmospheric pressure.

SAFETY ALERT

1. **Volatile organic liquids are flammable, so use open flames, if needed, with *great* care. Consult with your instructor *before* lighting a burner.**
2. **Use paper towels to clean up spilled liquids; discard the towels as directed by your instructor. Avoid contact of organic liquids with your skin; if this happens, wash the affected area thoroughly with soap and water.**

MINISCALE PROCEDURE

Preparation Sign in at **www.cengage.com/login** to answer Pre-Lab Exercises, access videos, and read the MSDSs for the chemicals used or produced in this procedure. Read or review Sections 2.8 and 2.9.

Apparatus Thermometer, 10-mm × 75-mm test tube, two clamps, and apparatus for *flameless* heating.

Setting Up Obtain a liquid from your instructor and determine its boiling point following the technique described in Section 2.8, using the apparatus shown in Figure 2.20 and *flameless* heating.

MICROSCALE PROCEDURE

Preparation Sign in at **www.cengage.com/login** to answer Pre-Lab Exercises, access videos, and read the MSDSs for the chemicals used or produced in this procedure. Read or review Sections 2.8 and 2.9.

Apparatus Thiele melting-point apparatus, micro boiling-point apparatus, and a Bunsen burner or microburner.

Setting Up Determine the boiling point of the liquid(s) assigned by your instructor. Follow the technique presented in Section 2.8 for using the micro boiling-point apparatus (Fig. 2.21). Use a Bunsen burner or a microburner for heating. In the event you do not know the boiling point of the liquid, *first* determine an approximate boiling-point range by heating the Thiele tube (Fig. 2.17a) *fairly rapidly.* Repeat the measurement by heating the tube until the temperature is 20–30 °C below the approximate boiling point, and then heat the sample at a rate of 4–5 °C/min to obtain an accurate value. It may be desirable to repeat this procedure to obtain a more reliable boiling point.

WRAPPING IT UP

Unless instructed otherwise, return the *organic liquids* to the appropriate bottle for either nonhalogenated liquids or halogenated liquids.

EXERCISES

1. Refer to Figure 4.1 and answer the following:
   a. What total pressure would be required in the system in order for the liquid to boil at 45 °C?
   b. At about what temperature would the liquid boil when the total pressure in the system is 300 torr?
2. Describe the relationship between escaping tendency of liquid molecules and vapor pressure.
3. Define the following terms:
   a. boiling point
   b. normal boiling point
   c. Dalton's law of partial pressures
   d. equilibrium vapor pressure
4. Using Dalton's law, explain why a fresh cup of tea made with boiling water is not as hot at higher altitudes as it is at sea level.
5. At a given temperature, liquid A has a higher vapor pressure than liquid B. Which liquid has the higher boiling point?
6. Explain why the boiling point at 760 torr of a solution of water, bp 100 °C (760 torr), and ethylene glycol ($HOCH_2CH_2OH$), bp 196–198 °C (760 torr), exceeds 100 °C. For purposes of your answer, consider ethylene glycol as a *nonvolatile* liquid.

7. Why should there be no droplets of water in the oil of a heating bath?
8. Why is the micro boiling-point technique not applicable for boiling points in excess of 200 °C if *mineral* oil rather than *silicone* oil is the heating fluid in the Thiele tube?
9. A *rotary evaporator* (Sec. 2.29) is a device frequently used in the laboratory to remove solvent quickly under vacuum.
   a. Why is it possible to effect the removal of solvent at temperatures below their normal boiling points using this device?
   b. Other than being faster than simple distillation for removing a given volume of solvent, why might this type of distillation be preferred for isolating a desired product?

4.3 SIMPLE DISTILLATION

Simple distillation allows separation of distillates from less-volatile substances that remain as pot residue at the completion of the distillation. In the ideal case, only a single component of the mixture will be volatile, so the distillate will be a pure compound. Real life is rarely ideal, however, and it is more common that several volatile components comprise the mixture. Simple distillation allows isolation of the various components of the mixture in acceptable purity if the *difference* between the boiling points of each pure substance is greater than 40–50 °C. For example, a mixture of diethyl ether, bp 35 °C (760 torr), and toluene, bp 111 °C (760 torr), could easily be separated by simple distillation, with the ether distilling first. Organic chemists frequently use this technique to separate a desired reaction product from the solvents used for the reaction or its work-up. The solvents are usually more volatile than the product and are readily removed from it by simple distillation.

To understand the principles of distillation, a review of the effect of impurities on the vapor pressure of a pure liquid is necessary. The discussion starts with consideration of the consequences of having *nonvolatile* impurities present and then turns to the more common case of contamination of the liquid with other *volatile* substances.

Consider a homogeneous solution composed of a nonvolatile impurity and a pure liquid; for the present purpose, these are taken as sugar and water, respectively. As a nonvolatile impurity, the sugar reduces the vapor pressure of the water because it lowers the concentration of the volatile constituent in the liquid phase. The consequence of this is shown graphically in Figure 4.2. In this figure, Curve 1 corresponds to the dependence of the temperature upon the

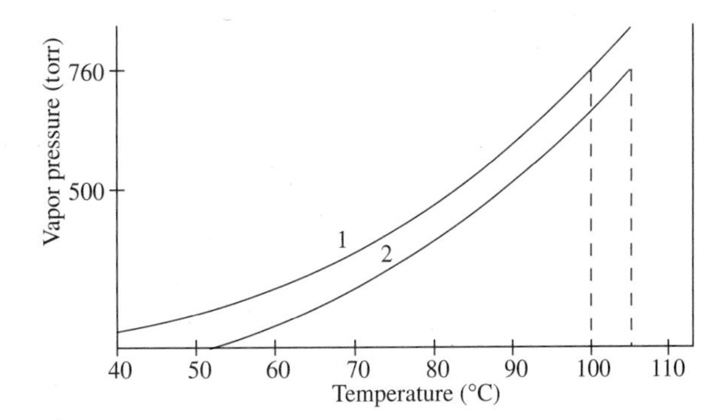

Figure 4.2
Diagram of dependence of vapor pressure on temperature. Curve 1 is for pure water, and Curve 2 is for a solution of water and sugar.

vapor pressure of *pure* water and intersects the 760-torr line at 100 °C. Curve 2 is for a *solution* having a particular concentration of sugar in water. Note that the presence of the nonvolatile impurity reduces the vapor pressure at any temperature by a constant amount, in accord with Raoult's law as discussed later. The temperature at which this curve intersects the 760-torr line is higher because of the lower vapor pressure, and consequently the temperature of the boiling solution, 105 °C, is higher.

Despite the presence of sugar in the solution, the **head temperature** (Sec. 2.13) in the distillation will be the same as for pure water, namely, 100 °C (760 torr), since the water condensing on the thermometer bulb is now uncontaminated by the nonvolatile impurity. The **pot temperature** will be *elevated,* however, owing to the decreased vapor pressure of the solution (Fig. 4.2). As water distills, the pot temperature will progressively rise because the concentration of the sugar in the stillpot increases, further lowering the vapor pressure of the water. Nevertheless, the head temperature will remain constant, just as though pure water were being distilled.

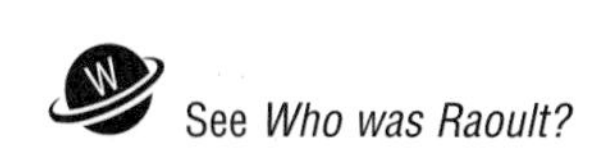

See *Who was Raoult?*

The quantitative relationship between vapor pressure and composition of homogeneous liquid mixtures is known as **Raoult's law** (Eq. 4.2). The factor P_X represents the partial pressure of component X, and it is equal to the vapor pressure, P°_X, of pure X at a given temperature times the mole fraction N_X of X in the mixture. The **mole fraction** of X is defined as the fraction of *all* molecules present in the liquid mixture that are molecules of X. It is obtained by dividing the number of moles of X in a mixture by the sum of the number of moles of all components (Eq. 4.3). Raoult's law is strictly applicable only to **ideal solutions,** which are defined as those in which the interactions between *like* molecules are the same as those between *unlike* molecules. Fortunately, many organic solutions approximate the behavior of ideal solutions, so the following mathematical treatment applies to them as well.

$$P_X = P^{\circ}_X N_X \tag{4.2}$$

$$N_X = \frac{nX}{nX + nY + nZ + \cdots} \tag{4.3}$$

Note that the partial vapor pressure of X above an ideal solution depends *only* on its mole fraction in solution and it is completely independent of the vapor pressures of the other volatile components of the solution. If all components other than X are nonvolatile, the total vapor pressure of the mixture will be equal to the partial pressure of X, since the vapor pressure of nonvolatile compounds may be taken as zero. Accordingly, the distillate from such a mixture will always be pure X. This is the case for the distillation of a solution of sugar and water, as discussed earlier.

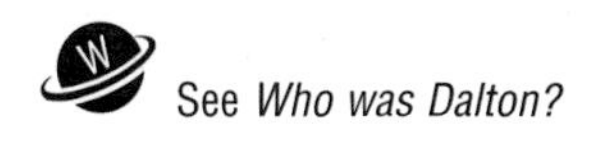

See *Who was Dalton?*

When a mixture contains two or more volatile components, the *total* vapor pressure is equal to the *sum* of the partial vapor pressures of each such component. This is known as **Dalton's law** (Eq. 4.4), where P_X, P_Y, and P_Z refer to the vapor pressures of the volatile components. The process of distilling such a liquid mixture may be significantly different from that of simple distillation, because the vapors above the liquid phase will now contain some of each of the volatile components. Separation of the liquids in this case may require the use of fractional distillation, which is discussed in Section 4.4.

$$P_{total} = P_X + P_Y + P_Z + \cdots \tag{4.4}$$

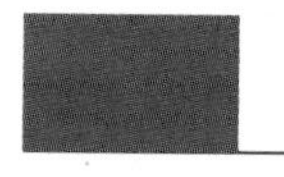

EXPERIMENTAL PROCEDURES

Simple Distillation

Purpose To demonstrate the technique for purification of a volatile organic liquid containing a nonvolatile impurity.

SAFETY ALERT

1. **Cyclohexane is highly flammable, so be sure that burners are not being used in the laboratory. Use *flameless* heating (Sec. 2.9).**
2. **Examine your glassware for cracks and other weaknesses before assembling the distillation apparatus. Look for "star" cracks in round-bottom flasks, because these can cause a flask to break upon heating.**
3. **Proper assembly of glassware is important in order to avoid possible breakage and spillage or the release of distillate vapors into the room. Be certain that all connections in the apparatus are tight before beginning the distillation. Have your instructor examine your set-up after it is assembled.**
4. **The apparatus used in these experiments *must* be open to the atmosphere at the receiving end of the condenser. *Never heat a closed system,* because the pressure buildup may cause the apparatus to explode!**
5. **Be certain that the water hoses are *securely* fastened to your condensers so that they will not pop off and cause a flood. If heating mantles or oil baths are used for heating in this experiment, water hoses that come loose may cause water to spray onto electrical connections or into the heating sources, either of which is potentially dangerous.**
6. **Avoid excessive inhalation of organic vapors at all times.**

MINISCALE PROCEDURE

Preparation When you see this icon, sign in at this book's premium website at **www.cengage.com/login** to access videos, Pre-Lab Exercises, and other online resources. Read or review Sections 2.2, 2.4, 2.9, 2.11, and 2.13.

Apparatus A 25-mL round-bottom flask and apparatus for simple distillation, magnetic stirring, and *flameless* heating.

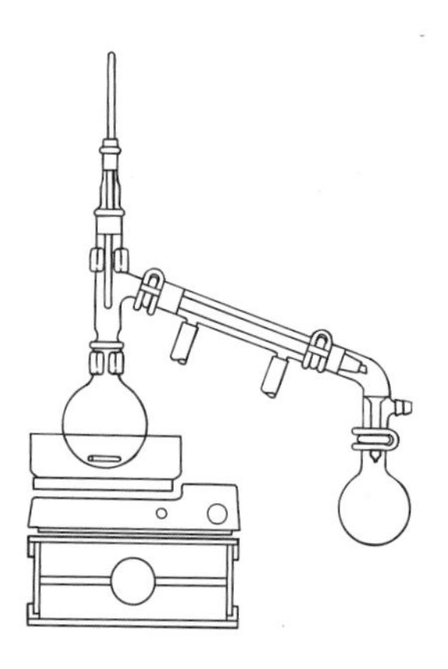

Setting Up Place 10 mL of cyclohexane containing a nonvolatile dye in the round-bottom flask. Add a stirbar to the flask to ensure smooth boiling, and assemble the simple distillation apparatus shown in Figure 2.37a. Be sure to position the thermometer in the stillhead so the *top* of the mercury thermometer bulb is level with the *bottom* of the sidearm of the distillation head. Have your instructor check your apparatus *before* you start heating the stillpot.

Distillation Start the magnetic stirrer and begin heating the stillpot. As soon as the liquid begins to boil *and the condensing vapors have reached the thermometer bulb,* regulate the heat supply so that distillation continues steadily at a rate of

2–4 *drops per second;* if a drop of liquid cannot be seen suspended from the end of the thermometer, the rate of distillation is too *fast.* As soon as the distillation rate is adjusted and the head temperature is constant, note and record the temperature. Continue the distillation and periodically record the head temperature. Discontinue heating when only 2–3 mL of impure cyclohexane remains in the distillation flask. Record the volume of distilled cyclohexane that you obtain.

Optional Procedure

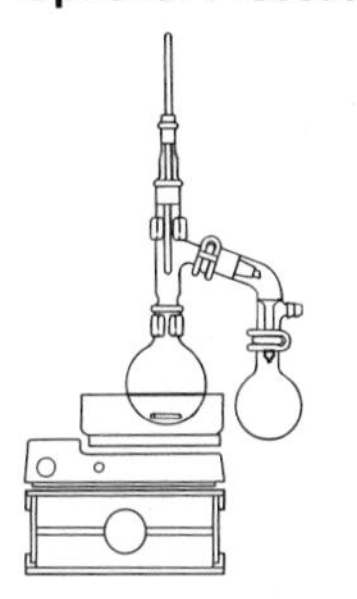

You may be required to perform this distillation using the shortpath apparatus discussed in Section 2.13 and illustrated in Figure 2.37b. After assembling the equipment, make certain that the *top* of the mercury in the thermometer bulb is level with the *bottom* of the sidearm of the distillation head. The preferred way to collect the distillate in this distillation is to attach a dry round-bottom flask to the vacuum adapter and put a drying tube containing calcium chloride on the sidearm of the adapter to protect the distillate from moisture. The receiver should be cooled in an ice-water bath to prevent loss of product by evaporation and to ensure complete condensing of the distillate.

MICROSCALE PROCEDURE

Preparation Sign in at **www.cengage.com/login** to answer Pre-Lab Exercises, access videos, and read the MSDSs for the chemicals used or produced in this procedure. Read or review Sections 2.3, 2.4, 2.9, 2.11, and 2.13.

Apparatus A Pasteur pipet and apparatus for simple distillation, magnetic stirring, and *flameless* heating.

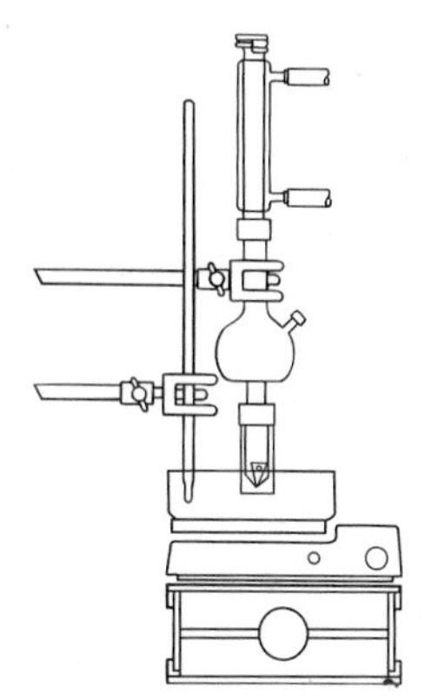

Setting Up Transfer 2 mL of impure cyclohexane containing a nonvolatile dye to a 5-mL conical vial. Equip the vial with a spinvane, the Hickman stillhead, and a condenser as shown in Figure 2.38. Place the apparatus in the heating source and have your instructor check your apparatus before you start heating the vial.

Distillation Start the magnetic stirrer and begin heating the stillpot. Increase the temperature of the heating source until vapors of distillate begin to rise into the Hickman stillhead and condense into the flared portion of the head. Control the rate of heating so that the vapor line rises no more than halfway up the upper portion of the head; otherwise, distillate may be lost to the atmosphere. Be certain to record the bath temperature at which the cyclohexane distills. You may also measure the approximate temperature of the vapors by inserting a thermometer through the condenser and stillhead to a point in the top third of the conical vial just above the boiling liquid. This will give you an approximate boiling point of the cyclohexane. Terminate heating when about 0.3 mL of the original solution remains. Disconnect the vial from the Hickman still-head and, using a Pasteur pipet, transfer the distillate to a properly labeled screw-cap vial.

WRAPPING IT UP

Unless directed otherwise, return the *distilled* and *undistilled cyclohexane* to a bottle marked "Recovered Cyclohexane."

EXERCISES

1. Define the following terms:
   a. simple distillation
   b. head temperature
   c. pot temperature
   d. Raoult's law
   e. ideal solution
   f. Dalton's law
2. Sketch and completely label the apparatus required for a simple distillation.
3. Why should you never heat a closed system, and how does this rule apply to a distillation?
4. Explain the role of the stirbar that is normally added to a liquid that is to be heated to boiling.
5. In a miniscale distillation, the top of the mercury bulb of the thermometer placed at the head of a distillation apparatus should be adjacent to the exit opening to the condenser. Explain the effect on the observed temperature reading if the bulb is placed (a) below the opening to the condenser or (b) above the opening.
6. Distillation is frequently used to isolate the nonvolatile organic solute from a solution containing an organic solvent. Explain how this would be accomplished using a simple distillation.
7. Using Raoult's and Dalton's laws, explain why an aqueous NaCl solution will have a higher boiling point than pure water.
8. At 100 °C, the vapor pressures for water, methanol, and ethanol are 760, 2625, and 1694 torr, respectively. Which compound has the highest normal boiling point and which the lowest?

4.4 FRACTIONAL DISTILLATION

It is easy to separate a volatile compound from a nonvolatile one by simple distillation (Sec. 4.3). The same technique may also be used to separate volatile compounds from one another if their boiling points differ by at least 40–50 °C. If this is not the case, the technique of fractional distillation is normally used to obtain each volatile component of a mixture in pure form. The theoretical basis of this technique is the subject of the following discussion.

Theory

For simplicity, we'll only consider the theory for separating *ideal solutions* (Sec. 4.3) consisting of two volatile components, designated X and Y. Solutions containing more than two such components are often encountered, and their behavior on distillation may be understood by extension of the principles developed here for a binary system.

The vapor pressure of a compound is a measure of the ease with which its molecules escape the surface of a liquid. When the liquid is composed of two volatile components, in this case X and Y, the number of molecules of X and of Y in a given volume of the vapor above the mixture will be proportional to their respective partial vapor pressures. This relationship is expressed mathematically by Equation 4.5, where N'_X/N'_Y is the ratio of the mole fractions of X and Y in the *vapor* phase. The mole fraction of each component may be calculated from the equations $N'_X = P_X/(P_X + P_Y)$

and $N'_Y = P_Y/(P_X + P_Y)$. The partial vapor pressures, P_X and P_Y, are determined by the composition of the liquid solution according to Raoult's law (Eq. 4.2). Since the solution boils when the sum of the partial vapor pressures of X and Y is equal to the external pressure, as expressed by Dalton's law (Eq. 4.4), the boiling temperature of the solution is determined by its composition.

$$\frac{N'_X}{N'_Y} = \frac{P'_X}{P'_X} = \frac{P^\circ_X N_X}{P^\circ_Y N_Y} \tag{4.5}$$

The relationship between temperature and the composition of the liquid and vapor phases of ideal binary solutions is illustrated in Figure 4.3 for mixtures of benzene, bp 80 °C (760 torr), and toluene, bp 111 °C (760 torr). The lower curve, the **liquid line,** gives the boiling points of all mixtures of these two compounds. The upper curve, the **vapor line,** is calculated using Raoult's law and defines the composition of the vapor phase in equilibrium with the boiling liquid phase at the same temperature. For example, a mixture whose composition is 58 mol % benzene and 42 mol % toluene will boil at 90 °C (760 torr), as shown by point *A* in Figure 4.3. The composition of the vapor in equilibrium with the solution when it *first* starts to boil can be determined by drawing a horizontal line from the *liquid line* to the *vapor line;* in this case, the vapor has the composition 78 mol % benzene and 22 mol % toluene, as shown by point *B* in Figure 4.3. This is a key point, for it means that at any given temperature the *vapor phase is richer in the more volatile component than is the boiling liquid with which the vapor is in equilibrium.* This phenomenon provides the basis of **fractional distillation.**

When the liquid mixture containing 58 mol % benzene and 42 mol % toluene is heated to 90 °C (760 torr), its boiling point, the vapor formed initially contains 78 mol % benzene and 22 mol % toluene. If this first vapor is condensed, the condensate would also have this composition and thus would be much richer in

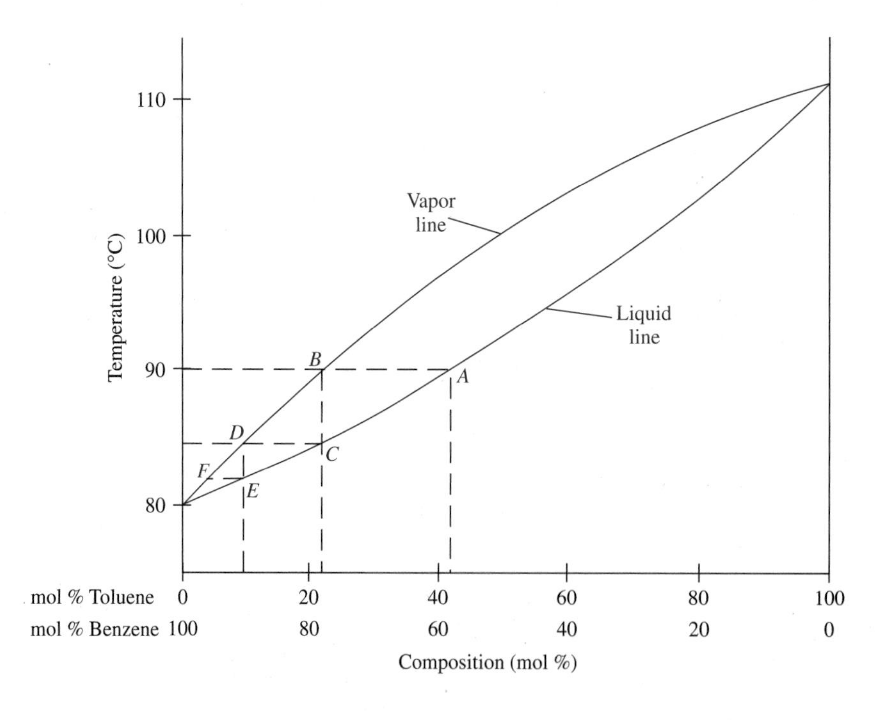

Figure 4.3
Temperature–composition diagram for binary mixture of benzene and toluene.

benzene than the original liquid mixture from which it was distilled. After this vapor is removed from the original mixture, the liquid remaining in the stillpot will contain a smaller mol % of benzene and a greater mol % of toluene because more benzene than toluene was removed by vaporization. The boiling point of the liquid remaining in the distilling flask will rise as a result. As the distillation continues, the boiling point of the mixture will steadily increase until it approaches or reaches the boiling point of pure toluene. The composition of the distillate will change as well and will ultimately consist of "pure" toluene.

Now let's return to the first few drops of distillate that are obtained by condensing the vapor initially formed from the original mixture. This condensate, as noted earlier, has a composition identical to that of the vapor from which it is produced. Were this liquid to be collected and then redistilled, its boiling point would be the temperature at point *C*, namely 85 °C; this boiling temperature is easily determined by drawing a vertical line from the vapor line at point *B* to the liquid line at point *C*, which corresponds to the composition of the distillate initially produced. The first distillate obtained at this temperature would have the composition *D*, 90 mol % benzene and 10 mol % toluene; this composition is determined from the intersection with the vapor line of the horizontal line from point *C* on the liquid line.

In theory, this process could be repeated again and again to give a very small amount of pure benzene. Similarly, collecting the *last* small fraction of each distillation and redistilling it in the same stepwise manner would yield a very small amount of pure toluene. If larger amounts of the initial and final distillates were collected, reasonable quantities of materials could be obtained, but a large number of individual simple distillations would be required. This process would be extremely tedious and time-consuming. Fortunately, the repeated distillation can be accomplished almost automatically in a single operation by using a **fractional distillation column,** the theory and use of which are described later in this section.

Most homogeneous solutions of volatile organic compounds behave as ideal solutions, but some of them exhibit *nonideal* behavior. This occurs because unlike molecules are affected by the presence of one another, thereby causing deviations from Raoult's law for *ideal* solutions (Eq. 4.2). When nonideal solutions have vapor pressures *higher* than those predicted by Raoult's law, the solutions are said to exhibit *positive* deviations from it; solutions having vapor pressures lower than predicted are thus considered to represent *negative* deviations from the law. In the present discussion, we'll consider only positive deviations associated with binary solutions, as such deviations are generally most important to organic chemists.

To produce positive deviations in a solution containing two volatile liquids, the forces of attraction between the molecules of the two components are *weaker* than those between the molecules of each individual component. The combined vapor pressure of the solution is thus *greater* than the vapor pressure of the pure, more volatile component for a particular range of compositions of the two liquids. This situation is illustrated in Figure 4.4, in which it may be seen that mixtures in the composition range between *X* and *Y* have boiling temperatures *lower* than the boiling temperature of either pure component. The *minimum-boiling* mixture, composition *Z* in Figure 4.4, may be considered as though it is a third component of the binary mixture. It has a *constant* boiling point because the vapor in equilibrium with the liquid has a composition *identical* to that of the liquid itself. The mixture is called a **minimum-boiling azeotrope.** Fractional distillation of such mixtures will *not* yield both of the components in pure form; rather, only the azeotropic mixture *and* the component present in *excess* of the azeotropic composition will be produced from the fractionation. For example, pure ethanol cannot be obtained by fractional

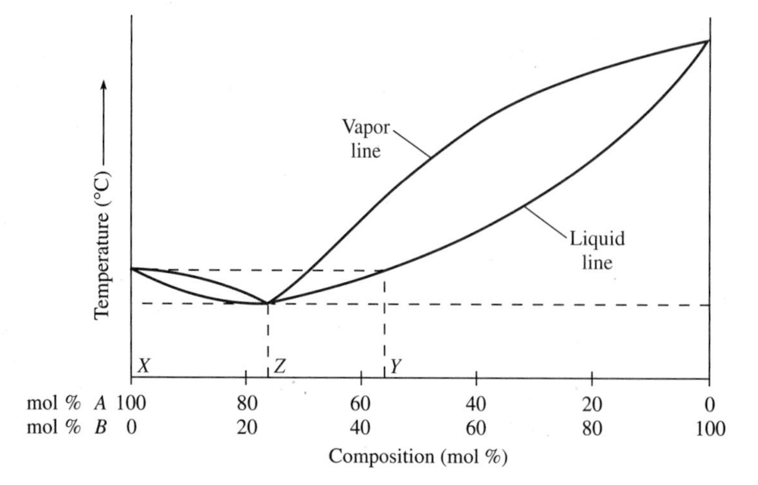

Figure 4.4
Temperature–composition diagram for minimum-boiling azeotrope.

distillation of aqueous solutions containing less than 95.57% ethanol, the azeotropic composition, even though the boiling point of this azeotrope is only 0.15 °C below that of pure ethanol. Since optimal fractional distillations of aqueous solutions containing less than 95.57% ethanol yield this azeotropic mixture, "95% ethyl alcohol" is readily available. Pure or "absolute" ethanol is more difficult to obtain from aqueous solutions. However, it can be prepared by removing the water chemically, through the use of a drying agent such as molecular sieves (Sec. 2.24), or by distillation of a ternary mixture of ethanol-water-benzene.

Azeotropic distillation is a useful technique for removing water from organic solutions. For example, toluene and water form an azeotrope having a composition of 86.5 wt % of toluene and 13.5 wt % water, and so distillation of a mixture of these two effectively removes water from a mixture. This technique is used in the Experimental Procedure of Section 18.4 for driving an equilibrium in which water is being formed to completion. Azeotropic distillation may also be used to dry an organic liquid that is to be used with reagents that are sensitive to the presence of water. This application is found in the Experimental Procedure of Section 15.2, in which anhydrous *p*-xylene is required for a Friedel-Crafts alkylation reaction.

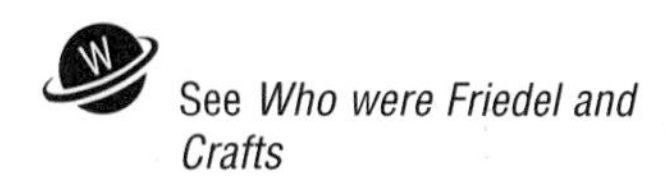

See *Who were Friedel and Crafts*

Fractional Distillation Columns and Their Operation

There are many types of fractional distillation columns, but all can be discussed in terms of a few fundamental characteristics. The column provides a vertical path through which the vapor must pass from the stillpot to the condenser before being collected in the receiver (Fig. 2.39). This path is significantly longer than in a simple distillation apparatus. As the vapor from the stillpot rises up the column, some of it condenses *in the column* and returns to the stillpot. *If the lower part of the distilling column is maintained at a higher temperature than the upper part of the column,* the condensate will be partially revaporized as it flows down the column. The uncondensed vapor, together with that produced by revaporization of the condensate in the column, rises higher and higher in the column and undergoes a repeated series of condensations and revaporizations. This repetitive process is equivalent to performing a number of simple distillations *within* the column, with the vapor phase produced in each step becoming increasingly richer in the *more* volatile component; the condensate that flows down the column correspondingly becomes richer in the less volatile component.

Each step along the path *A-B-C-D-E-F* of Figure 4.3 represents a *single* ideal distillation. One type of fractional distillation column, the bubble-plate column, was designed to effect one such step for each **plate** it contained. This led to the description of the efficiency of any fractional distillation column in terms of its equivalency to such a column in **theoretical plates.** Another index of the separating efficiency of a fractional distillation column is the **HETP,** which stands for ***h*eight *e*quivalent to a *t*heoretical *p*late** and is the vertical length of a column that is necessary to obtain a separation efficiency of one theoretical plate. For example, a column 60 cm long with an efficiency of 30 plates has an HETP value of 2 cm. Such a column would usually be better for research purposes than a 60-plate column that is 300 cm long (HETP = 5 cm) because of the small liquid capacity and **hold-up** of the shorter column. "Hold-up" refers to the condensate that remains in a column during and after distillation. When small amounts of material are to be distilled, a column must be chosen that has an efficiency, HETP, adequate for the desired separation and also a low to moderate hold-up.

As stated earlier, equilibrium between liquid and vapor phases must be established in a fractional distillation column so that the more volatile component is selectively carried to the top of the column and into the condenser, where the vapor condenses into the distillate. After all of the more volatile component is distilled, the less volatile one remains in the column and the stillpot; the heat supplied to the stillpot is then further increased in order to distill the second component. The most important requirements for performing a successful fractional distillation are (a) intimate and extensive contact between the liquid and the vapor phases in the column, (b) maintenance of the proper temperature gradient along the column, (c) sufficient length of the column, and (d) sufficient difference in the boiling points of the components of the liquid mixture. Each of these factors is considered here.

a. The desired contact between the liquid and vapor phases can be achieved by filling the column with an inert material having a large surface area. Examples of suitable packing materials include glass, ceramic, or metal pieces. Figure 2.40a shows a Hempel column packed with Raschig rings, which are pieces of glass tubing approximately 6 mm long. This type of column will have from two to four theoretical plates per 30 cm of length, if the distillation is carried out sufficiently slowly to maintain equilibrium conditions. Another type of fractional distillation column is the Vigreux column (Fig. 2.40b), which is useful for small-scale distillations of liquid where low hold-up is of paramount importance. A 30-cm Vigreux column will only have 1–2 theoretical plates and consequently will be less efficient than the corresponding Hempel column. The Vigreux column has the advantage of a hold-up of less than 1 mL as compared with 2–3 mL for a Hempel column filled with Raschig rings.

b. **Temperature gradient** refers to the difference in temperature between the top and bottom of the column. The maintenance of the proper temperature gradient within the column is particularly important for an effective fractional distillation. Ideally, the temperature at the bottom of the column should be approximately equal to the boiling temperature of the solution in the stillpot, and it should decrease continually in the column until it reaches the boiling point of the more volatile component at the head of the column. The significance of the temperature gradient is seen in Figure 4.3, where the boiling temperature of the distillate decreases with each succeeding step, for example, *A* (90 °C) to *C* (85 °C) to *E* (82 °C).

The necessary temperature gradient from stillpot to stillhead in most distillations will be established *automatically* by the condensing vapors *if* the rate of distillation is properly adjusted. Frequently, this gradient can be maintained only by insulating the column with a material such as glasswool around the outside of the column. Insulation helps reduce heat losses from the column to the atmosphere. Even when the column is insulated, an insufficient amount of vapor may be produced to heat the column if the stillpot is heated too slowly, so that little or no condensate reaches the head. This rate must then be increased, but it must be kept below the point where the column is flooded. A **flooded column** is characterized by a column or "plug" of liquid that may be observed within the distillation column, often at the joint between it and the stillpot.

Factors directly affecting the temperature gradient in the column are the rate of heating of the stillpot and the rate at which vapor is removed at the stillhead. If the heating is too vigorous or the vapor is removed too rapidly, the entire column will heat up almost uniformly, and there will be no fractionation and thus no separation of the volatile components. On the other hand, if the stillpot is not heated strongly enough and if the vapor is removed too slowly at the top, the column will flood with returning condensate. Proper operation of a fractional distillation column thus requires *careful* control of the heat supplied to the stillpot and of the rate at which the distillate is removed at the stillhead. This rate should be *no more than one drop* every 2–3 sec.

The ratio of the amount of condensate returning to the stillpot and the amount of vapor removed as distillate per unit time is defined as the **reflux ratio.** A ratio of 10:1, for example, means that 10 drops of condensate return to the stillpot for each drop of distillate that is obtained. In general, the higher the reflux ratio the more efficient the fractional distillation.

c. Correct column length is difficult to determine in advance of performing a fractional distillation. The trial-and-error technique must normally be used, and if a particular column does not efficiently separate a certain mixture, a longer column or a different type of column or column packing must be selected.

d. The difference in boiling points between the two pure components of a mixture should be no less than 20–30 °C in order for a fractional distillation to be successful when a Hempel column packed with Raschig rings or a similar type of packing is used. As mentioned previously, modifications in column length and type may result in the successful separation of mixtures having smaller boiling point differences.

In summary, the most important variables that can be controlled experimentally in a fractional distillation are proper selection of the column and column packing, adequate insulation of the column, and careful control of the rate of heating so as to provide the proper reflux ratio and a favorable temperature gradient within the column. Under such conditions, two different temperature **plateaus** will be observed in the fractional distillation of a typical binary mixture (Fig. 4.5). The head temperature should first rise to the normal boiling point of the more volatile component and remain there until that component is mostly removed (Fig. 4.5, Fraction 1). The head temperature may then drop somewhat, indicating that the more volatile component has largely been removed. As additional heat is provided to the stillpot, the less volatile component will begin to distill, and the head temperature will rise to the boiling point of the second component (Fig. 4.5, Fraction 2). If the separation is efficient, the volume of this fraction, which contains a mixture of the two components, will be small. The head temperature should then remain constant at the normal boiling point of the less volatile component until most of it has distilled (Fig. 4.5, Fraction 3).

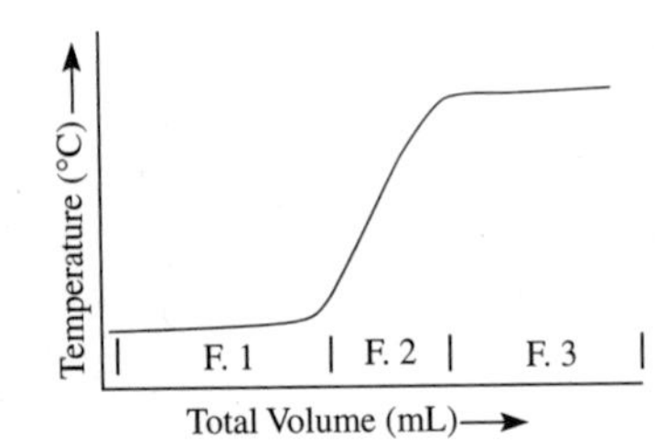

Figure 4.5
Progress curve for typical fractional distillation.

EXPERIMENTAL PROCEDURES

Fractional Distillation of a Binary Mixture

Purpose To demonstrate the technique for the separation of two volatile organic liquids.

SAFETY ALERT

1. **Cyclohexane and toluene are highly flammable, so be sure that burners are not being used in the laboratory. Use *flameless* heating (Sec. 2.9).**
2. **Examine your glassware for cracks and other weaknesses before assembling the distillation apparatus. Look with particular care for "star" cracks in round-bottom flasks, because these can cause a flask to break upon heating.**
3. **Proper assembly of glassware is important in order to avoid possible breakage and spillage and to avoid the release of distillate vapors into the room. Be certain that all connections in the apparatus are tight *before* beginning the distillation. Have your instructor examine your set-up after it is assembled.**
4. **The apparatus used in these experiments *must* be open to the atmosphere at the receiving end of the condenser. *Never heat a closed system,* because the pressure buildup may cause the apparatus to explode!**
5. **Be certain that the water hoses are securely fastened to your condensers so that they will not pop off and cause a flood. If heating mantles or oil baths are to be used, water hoses that come loose may cause water to spray onto electrical connections or into the heating sources, either of which is potentially dangerous to you and to those who work around you.**
6. **Avoid excessive inhalation of organic vapors at all times.**

MINISCALE PROCEDURE

Preparation Sign in at **www.cengage.com/login** to answer Pre-Lab Exercises, access videos, and read the MSDSs for the chemicals used or produced in this procedure. Read or review Sections 2.2, 2.4, 2.9, 2.11, and 2.14.

Apparatus A 50-mL round-bottom flask, apparatus for fractional distillation, magnetic stirring, and *flameless* heating.

Setting Up Place 10 mL of cyclohexane and 20 mL of toluene in the round-bottom flask, and add a stirbar to ensure smooth boiling. Equip this flask for fractional distillation as shown in Figure 2.39. Pack a Hempel or similar distillation column, using the type of packing specified by your instructor. When packing the column, *be careful not to break off the glass indentations at the base of the column.* Do not pack the column too tightly, because heating a fractional distillation apparatus equipped with a column that is too tightly packed is analogous to heating a closed system. Insulate the fractionating column by wrapping it with glasswool. The position of the thermometer in the stillhead is particularly

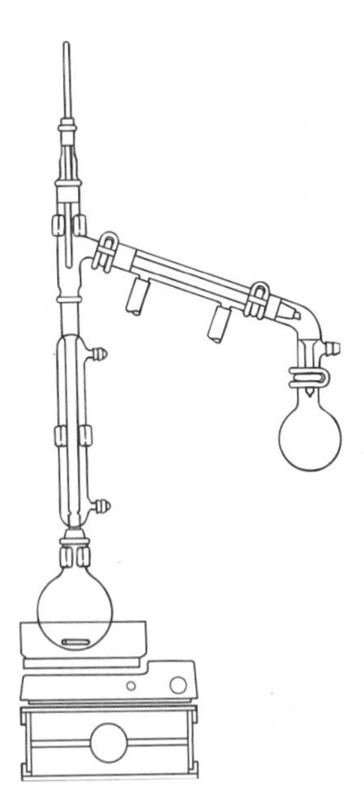

important; the *top* of the mercury thermometer bulb should be level with the *bottom* of the sidearm of the distillation head. Clean and dry three 25-mL containers, which may be bottles or Erlenmeyer flasks, for use as receiving flasks, and label them *A, B,* and *C.* Place receiver *A* so that the tip of the vacuum adapter extends inside the neck of the container to minimize evaporation of the distillate. Have your instructor check your assembled apparatus before heating the stillpot.

Distillation Start the magnetic stirrer and begin heating the stillpot using the heating method specified by your instructor. As the mixture is heated, the head temperature will rise to 81 °C (760 torr), which is the normal boiling point of cyclohexane, and distillation will begin. Regulate the heat so that distillation continues steadily at a rate *no faster* than *one drop of distillate every 1–2 sec;* if a drop of liquid cannot be seen suspended from the end of the thermometer, the rate of distillation is *too fast.*

The head temperature will remain at 81 °C for a period of time, but eventually it will either rise or drop slightly. Receiver *A* should be left in place until this increase or decrease in temperature is observed. As soon as the temperature deviates from 81 °C by more than ±3 °C, change to receiver *B* and increase the amount of heat supplied to the stillpot. The temperature will again start to rise, and more liquid will distill. Leave receiver *B* in place until the temperature reaches 110 °C (760 torr), which is the normal boiling point of toluene, and change to receiver *C.* Continue the distillation until 1–2 mL of liquid remains in the stillpot, and then discontinue heating.

Analysis Record the volumes of the distillate collected in each receiver by means of a graduated cylinder. Allow the liquid in the column to drain into the stillpot, then record the volume of this pot residue. If instructed to do so, submit or save 0.2-mL samples of each fraction *A, B,* and *C* for GLC or GC-MS analysis (Sec. 6.4).

Discovery Experiment

Comparative Fractional Distillations

Compare the results of a fractional distillation using a packed column, as described in the above experimental procedure, with those obtained using a simple distillation apparatus, an unpacked column, and one packed with a material different from that used originally.

Discovery Experiment

Fractionation of Alternative Binary Mixtures

Subject other binary mixtures to fractional distillation using a variety of apparatus. These might include 50:50 mixtures of acetone and 1,4-dioxane, hexane and heptane, hexane and toluene, heptane and ethyl benzene, acetone and toluene, and tetrahydrofuran and 1-butanol. You might also wish to explore the efficacy of fractional distillation for separating a binary mixture having components whose boiling points are separated by less than 30 °C. Examples might be tetrahydrofuran and acetone, ethyl acetate and 1,4-dioxane, and ethanol and methanol.

Discovery Experiment

Fractional Distillation of Unknowns

Obtain a 50:50 mixture of two unknown solvents from your instructor. These solvents will differ in boiling point by more than 20 °C. Possible solvents include hexane, cyclohexane, heptane, octane, toluene, ethyl benzene, acetone, methanol, 1-butanol, tetrahydrofuran, 1,4-dioxane, ethyl acetate, and others listed by your instructor. Look up the boiling points for each of these solvents. Perform a fractional distillation using a packed column as described in the above experimental procedure. Prepare a distillation curve with the boiling point on the vertical axis and the volume on the horizontal axis. Based upon your experimentally determined boiling points for the two liquids, identify the components of your mixture.

WRAPPING IT UP

Unless directed to do otherwise, pour the *pot residue* into the container for nonhalogenated organic liquids and return the *distillation fractions* to a bottle marked "Recovered Cyclohexane and Toluene."

EXERCISES

1. Define the following terms:

- **a.** fractional distillation
- **b.** head temperature
- **c.** pot temperature
- **d.** Raoult's law
- **e.** ideal solution
- **f.** mole fraction
- **g.** height equivalent to a theoretical plate (HETP)
- **h.** temperature gradient
- **i.** Dalton's law
- **j.** reflux ratio

2. Specify whether a simple distillation or a fractional distillation would be more suitable for each of the following purifications, and briefly justify your choice.

- **a.** Preparing drinking water from sea water.
- **b.** Separating benzene, bp 80 °C (760 torr), from toluene, bp 111 °C (760 torr).
- **c.** Obtaining gasoline from crude oil.
- **d.** Removing diethyl ether, bp 35 °C (760 torr), from *p*-dichlorobenzene (s), mp 174–175 °C.

3. Sketch and completely label the apparatus required for fractional distillation.

4. Explain why a packed fractional distillation column is more efficient than an unpacked column for separating two closely boiling liquids.

5. If heat is supplied to the stillpot too rapidly, the ability to separate two liquids by fractional distillation may be drastically reduced. In terms of the theory of distillation presented in the discussion, explain why this is so.

6. Explain why the column of a fractional distillation apparatus should be aligned as near to the vertical as possible.

7. Explain the role of the stirbar normally added to a liquid that is to be heated to boiling.

8. The top of the mercury bulb of the thermometer placed at the head of a distillation apparatus should be adjacent to the exit opening to the condenser. Explain the effect on the observed temperature reading if the bulb is placed (a) below the opening to the condenser or (b) above the opening.

9. Calculate the mole fraction of each compound in a mixture containing 15.0 g of cyclohexane and 5.0 g of toluene.

10. In the fractional distillation of your mixture of cyclohexane and toluene, what can be learned about the efficiency of the separation on the basis of the relative volumes of fractions *A*, *B*, and *C*?

11. **a.** A mixture of 60 mol % *n*-propylcyclohexane and 40 mol % *n*-propylbenzene is distilled through a simple distillation apparatus; assume that no fractionation occurs during the distillation. The boiling temperature is found to be 157 °C (760 torr) as the first small amount of distillate is collected. The standard vapor pressures of *n*-propylcyclohexane and *n*-propylbenzene are known to be 769 torr and 725 torr, respectively, at 157.3 °C. Calculate the percentage of each of the two components in the first few drops of distillate.

    b. A mixture of 50 mol % benzene and 50 mol % toluene is distilled under exactly the same conditions as in Part **a.** Using Figure 4.3, determine the distillation temperature and the percentage composition of the first few drops of distillate.

    c. The normal boiling points of *n*-propylcyclohexane and *n*-propylbenzene are 156 °C and 159 °C, respectively. Compare the distillation results in Parts **a** and **b.** Which of the two mixtures would require the more efficient fractional distillation column for separation of the components? Why?

12. Examine the boiling-point–composition diagram for mixtures of toluene and benzene given in Figure 4.3.

    a. Assume you are given a mixture of these two liquids of composition 70 mol % toluene and 30 mol % benzene and that it is necessary to effect a fractional distillation that will afford at least some benzene of greater than 99% purity. What would be the *minimum* number of theoretical plates required in the fractional distillation column chosen to accomplish this separation?

    b. Assume that you are provided with a 20-cm Vigreux column having an HETP of 10 cm in order to distill a mixture of 48 mol % benzene and 52 mol % toluene. What would be the composition of the first small amount of distillate that you obtained?

13. At 50 °C, the vapor pressures for methanol and ethanol are 406 and 222 torr, respectively. Given a mixture at 50 °C that contains 0.2 mol of methanol and 0.1 mol of ethanol, compute the partial pressures of each liquid and the total pressure.

14. Figure 4.6 shows a temperature (°C) vs. composition diagram for a mixture of acetone and ethyl acetate at 760 torr. You can calculate the mole fraction of ethyl acetate at any given point of the plot by subtracting the mole fraction of acetone from one because of the relationship mole fraction of acetone + mole fraction of ethyl acetate = 1. Answer the following questions using Figure 4.6.

    a. Specify the normal boiling point of acetone and of ethyl acetate.

    b. A mixture of acetone and ethyl acetate of unknown ratio begins to boil at 65 °C (760 torr). What is the composition of this binary mixture in terms of the mole fraction of acetone and that of ethyl acetate?

    c. For a mixture comprised of 50 mol % acetone and 50 mol % ethyl acetate:

    i. At what temperature will the mixture begin to boil?

    ii. When the mixture begins to boil, what is the mol % acetone in the vapor?

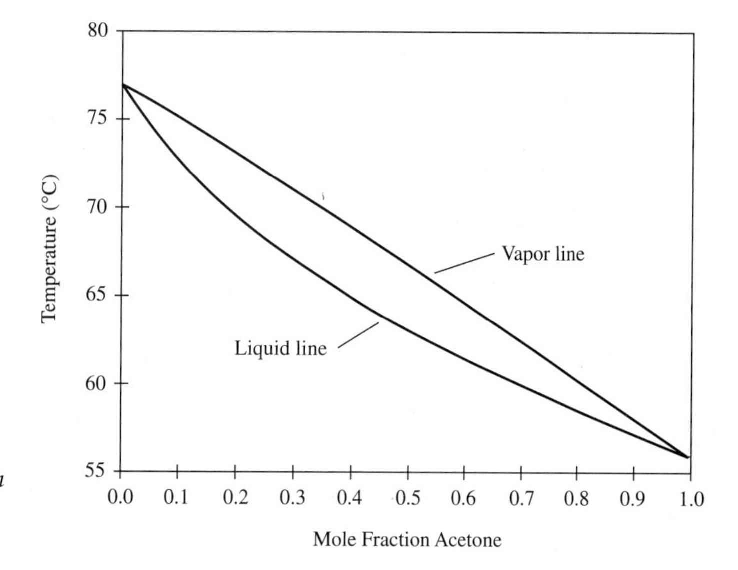

Figure 4.6
Temperature–composition diagram for acetone and ethyl acetate (Exercise 14).

iii. What is the boiling point of the liquid formed by condensation of the vapor from (**ii**) above?

iv. How many theoretical plates (distillation stages) will be necessary to isolate acetone of at least 90% purity?

d. A mixture of acetone and ethyl acetate of unknown molar ratio is heated to boiling. Immediately after boiling commences, a sample of the vapor is found to contain 0.8 mol fraction of ethyl acetate. What is the composition in mol % of the original binary mixture and what is its boiling point?

e. Would you expect a solution comprised of 30 mol % acetone and 70 mol % ethyl acetate to boil above or below *ca.* 67 °C in Denver, CO, which is at an elevation of one mile above sea level? Explain your answer.

4.5 STEAM DISTILLATION

The separation and purification of *volatile* organic compounds that are **immiscible** or nearly immiscible with water can often be accomplished by steam distillation. The technique normally involves the codistillation of a mixture of organic liquids and water, although some organic solids can also be separated and purified by this means. Of the various distillation methods, steam distillation is utilized least frequently, owing to the rather stringent limitations on the types of substances for which it can be used. These limitations as well as the virtues of this technique are revealed by considering the principles underlying steam distillation.

Theory and Discussion

The partial pressure P_i of each component i of a mixture of *immiscible,* volatile substances at a given temperature is equal to the vapor pressure P_i° of the pure compound at the same temperature (Eq. 4.6) and does not depend on the mole fraction of the compound in the mixture. In other words, each component of the mixture vaporizes independently of the others. This behavior contrasts sharply with that exhibited by solutions of miscible liquids, for which the partial pressure of each constituent of the mixture depends on its mole fraction in the solution (Raoult's law, Eq. 4.2).

$$P_i = P_i^\circ \tag{4.6}$$

You may recall that the total pressure, P_T, of a mixture of gases is equal to the sum of the partial pressures of the constituent gases, according to Dalton's law (Eq. 4.4). This means that the total vapor pressure of a mixture of immiscible, volatile compounds is given by Equation 4.7. This expression shows that the total vapor pressure of the mixture at any temperature is always *higher* than the vapor pressure of even the most volatile component at that temperature, owing to the contributions of the vapor pressures of the other constituents in the mixture. The boiling temperature of a mixture of immiscible compounds must then be *lower* than that of the lowest-boiling component.

$$P_T = P_a^\circ + P_b^\circ + \cdots + P_i^\circ \tag{4.7}$$

Application of the principles just outlined is seen in an analysis of the steam distillation of an immiscible mixture of water, bp 100 °C (760 torr), and bromobenzene, bp 156 °C (760 torr). Figure 4.7 is a plot of the vapor pressure versus temperature for each pure substance and for a mixture of these compounds. Analysis of this graph shows that the mixture should boil at about 95 °C (760 torr), the temperature at which the total vapor pressure equals standard atmospheric pressure. As theory predicts, this temperature is below the boiling point of water, which is the lowest-boiling component in this example.

The ability to distill a compound at the relatively low temperature of 100 °C or less by means of a steam distillation is often of tremendous value, particularly in the purification of substances that are heat-sensitive and would therefore decompose at higher temperatures. It is useful also in the separation of compounds from reaction mixtures that contain large amounts of nonvolatile residues such as inorganic salts.

The composition of the condensate from a steam distillation depends upon the molar masses of the compounds being distilled and upon their respective vapor pressures at the temperature at which the mixture steam-distils. To illustrate this, consider a mixture of two immiscible components, *A* and *B*. If the vapors of *A* and *B* behave as ideal gases, the ideal gas law can be applied and Equations 4.8a and 4.8b are obtained. In these two expressions, P° is the vapor pressure of the pure liquid, *V* is the volume

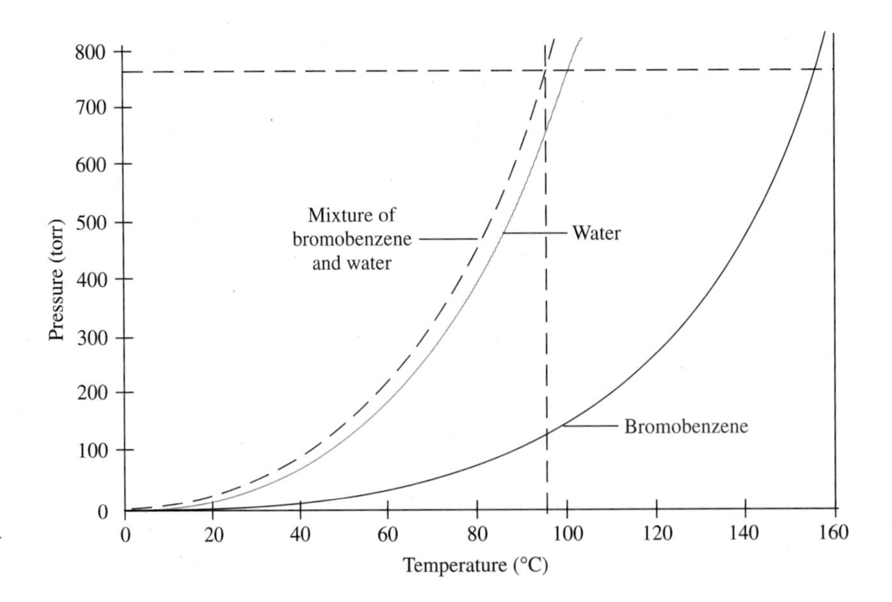

Figure 4.7
Vapor pressure–temperature graph for pure bromobenzene, pure water, and a mixture of bromobenzene and water.

in which the gas is contained, g is the weight in grams of the component in the gas phase, M is its molar mass, R is the gas constant, and T is the absolute temperature in kelvins (K). Dividing Equation 4.8a by Equation 4.8b gives Equation 4.9.

$$P^{\circ}_A V_A = (g_A/M_A)(RT) \tag{4.8a}$$

$$P^{\circ}_B V_B = (g_B/M_B)(RT) \tag{4.8b}$$

$$\frac{P^{\circ}_A V_A}{P^{\circ}_B V_B} = \frac{g_A/M_A(RT)}{g_B/M_B(RT)} \tag{4.9}$$

Because the RT factors in the numerator and denominator are identical and because the volume in which the gases are contained is the same for both ($V_A = V_B$), these terms in Equation 4.9 cancel to yield Equation 4.10.

$$\frac{\text{Grams of } A}{\text{Grams of } B} = \frac{(P^{\circ}_A)(\text{molar mass of } A)}{(P^{\circ}_B)(\text{molar mass of } B)} \tag{4.10}$$

Now let the immiscible mixture of A and B consist of bromobenzene and water, whose molar masses are 157 g/mol and 18 g/mol, respectively, and whose vapor pressures at 95 °C, as determined from Figure 4.6, are 120 torr and 640 torr, respectively. The composition of the distillate at this temperature can be calculated from Equation 4.10 as shown in Equation 4.11. This calculation indicates that on the basis of weight, *more* bromobenzene than water is contained in the steam distillate, even though the vapor pressure of the bromobenzene is *much* lower at the temperature of the distillation.

$$\frac{g_{\text{bromobenzene}}}{g_{\text{water}}} = \frac{(120)(157)}{(640)(18)} = \frac{1.64}{1} \tag{4.11}$$

Organic compounds generally have molar masses much higher than that of water, so it is possible to steam-distill compounds having vapor pressures of only about 5 torr at 100 °C with a fair efficiency on a weight-to-weight basis. Thus, *solids* that have vapor pressures of at least this magnitude can be purified by steam distillation. Examples are camphor, used in perfumes, and naphthalene, present in some brands of mothballs. The rather high vapor pressures of these solids is evidenced by the fact that their odors are detectable at room temperature.

In summary, steam distillation provides a method for separating and purifying moderately volatile liquid and solid organic compounds that are insoluble or nearly insoluble in water from nonvolatile compounds. Although relatively mild conditions are used in steam distillation, it cannot be used for substances that decompose on prolonged contact with steam or hot water, that react with water, or that have vapor pressures of 5 torr or less at 100 °C, all of which are significant limitations to the method.

4.6 STEAM DISTILLATION: ISOLATION OF CITRAL FROM LEMON GRASS OIL

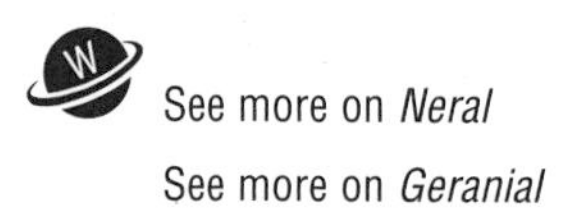

See more on *Neral*
See more on *Geranial*

Citral is a naturally occurring oil mainly comprised of two isomeric unsaturated aldehydes, geranial (**1**) and neral (**2**), that are extremely difficult to separate. These isomers differ only in the spatial orientation of the substituents about the carbon-carbon double bond that bears the aldehyde moiety (–CHO). This qualifies **1** and **2** as **stereoisomers** and more specifically as **diastereomers,** the definitions of which are found in

Section 7.1. In addition, they are sometimes also called **geometric isomers,** an older term that is used to describe stereoisomers that differ because the three-dimensional distribution of their substituents is the result of a double bond.

Citral possesses a lemonlike odor and taste. Although this odor and taste is pleasant to humans, it is less attractive to other species. For example, ants secrete citral to ward off potential predators, so citral is functioning as a **defense pheromone.** There are several commercial applications of citral. For example, it

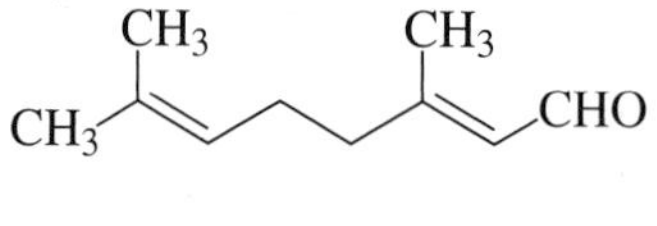

1
Geranial

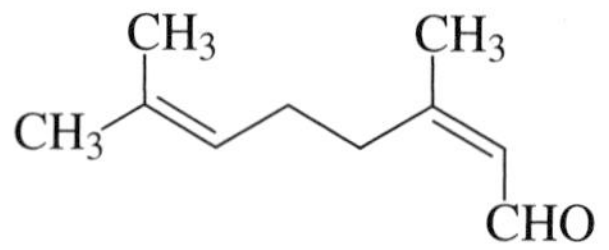

2
Neral

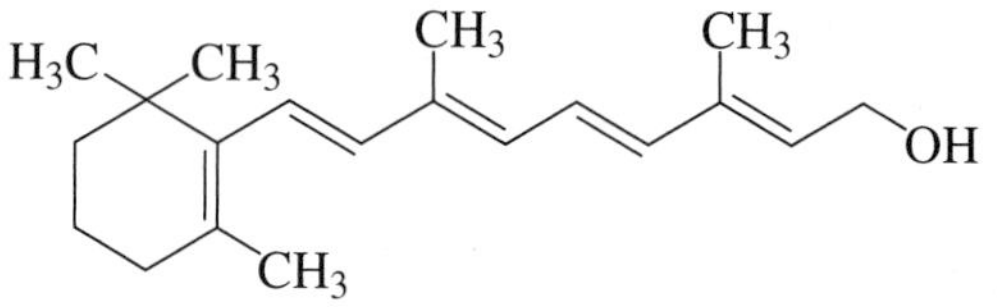

3
Vitamin A

See more on *Vitamin A*

may be added to perfumes whenever a lemonlike essence is desired, and it is used as an intermediate for the synthesis of vitamin A (**3**).

The commercial importance of citral has stimulated an extensive search for its presence in nature. One source is the oil from the skins of lemons and oranges, although it is only a minor component of this oil. However, citral is the major constituent of the oil obtained from lemon grass, and in fact 75–85% of the crude oil derived from pressing lemon grass is this natural product.

Citral is a chemically labile substance, and its isolation therefore presents a challenge. This task is simplified by the fact that citral, bp 229 °C (760 torr), is relatively volatile and has low solubility in water. These properties make it a suitable candidate for steam distillation, a technique that allows distillation of citral from crude lemon grass oil at a temperature below 100 °C, which is far below its normal boiling point. Neutral conditions are maintained in steam distillation, as is the partial exclusion of atmospheric oxygen, so the possibility of oxidation and/or polymerization of citral is minimized.

Steam distillation can be performed by two different methods, as described in Section 2.16. The first involves using an external steam source, which may be either a laboratory steam line or a steam generator, and passing the steam through a mixture of lemon grass oil and water. This method has some advantages, but it is experimentally more difficult than an alternative method, which involves heating a mixture of lemon grass oil and water and collecting the distillate. Although this latter procedure is simpler from the experimental standpoint, there are some limitations to its application. For example, in the steam distillation of only slightly volatile substances, a large initial volume of water will be required, or water must be added as the distillation proceeds, perhaps by means of an addition funnel. Nonetheless, the technique for internal generation of steam is well suited for this experiment.

EXPERIMENTAL PROCEDURE

Steam Distillation of Citral from Lemon Grass Oil

Purpose To isolate the natural product citral using the technique of steam distillation.

SAFETY ALERT

1. **Steam distillation involves the use of glassware that becomes very hot. Exercise care when handling hot glassware.**
2. **Be certain that the steam distillate is cooled below 30 °C before extracting it with diethyl ether; otherwise excessive pressure may develop in the separatory funnel and may blow out the stopper.**
3. **Diethyl ether is *extremely* flammable. Be certain that there are no flames in the vicinity during its use and its removal from the citral.**

MINISCALE PROCEDURE

Preparation Sign in at **www.cengage.com/login** to answer Pre-Lab Exercises, access videos, and read the MSDSs for the chemicals used or produced in this procedure. Read or review Sections 2.2, 2.4, 2.9, 2.13, 2.16, 2.19, 2.21, 2.29, and 4.4.

Apparatus A 125-mL separatory funnel, a 125-mL round-bottom flask, and apparatus for steam distillation using an internal steam source, simple distillation, and *flameless* heating.

Setting Up Determine the weight of 2.5 mL of lemon grass oil, and then place it and 60 mL of water in the round-bottom flask. Equip this flask for steam distillation as shown in Fig. 2.44, except replace the steam inlet tube with a stopper; use the heating source suggested by your instructor. Have your instructor check your assembled apparatus before continuing the experiment.

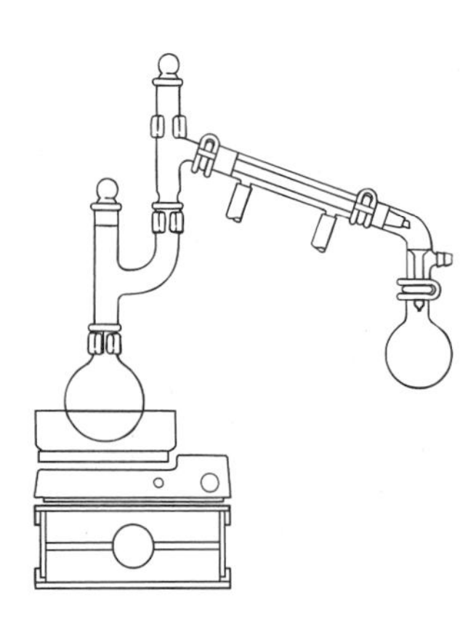

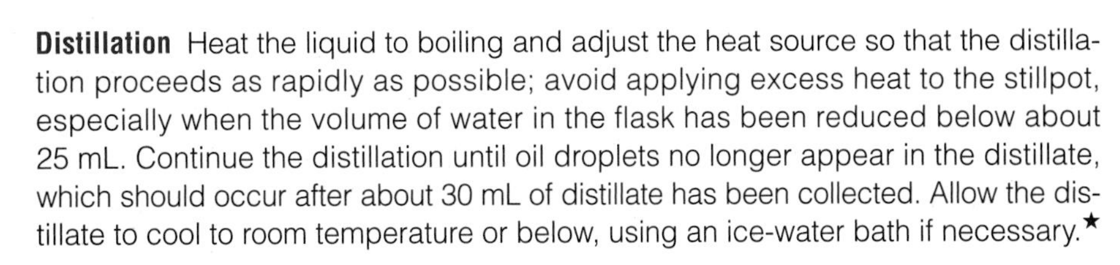

Distillation Heat the liquid to boiling and adjust the heat source so that the distillation proceeds as rapidly as possible; avoid applying excess heat to the stillpot, especially when the volume of water in the flask has been reduced below about 25 mL. Continue the distillation until oil droplets no longer appear in the distillate, which should occur after about 30 mL of distillate has been collected. Allow the distillate to cool to room temperature or below, using an ice-water bath if necessary.★

Isolation Transfer the cooled distillate to a separatory funnel and extract it sequentially with two 15-mL portions of diethyl ether. The funnel should be vented *frequently* to avoid the buildup of pressure. Transfer the organic layer from the separatory funnel to a 50-mL Erlenmeyer flask, and add about 0.5–1.0 g of anhydrous calcium chloride. Allow the ethereal solution to remain in contact with the drying agent until the organic layer is dry, as evidenced by its being completely clear. If the experiment is stopped at this point, loosely stopper the flask and store it in a hood; *never leave flasks containing diethyl ether in your locker drawer.*★

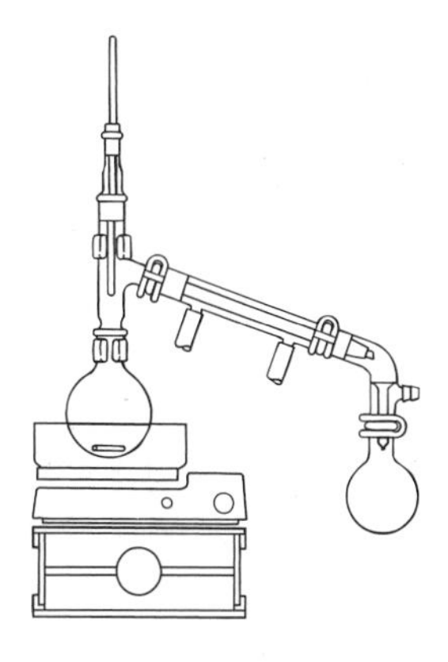

Decant the dried organic solution into a *tared* 125-mL round-bottom flask. Equip the flask for simple distillation and evaporate the solvent using simple distillation. Alternatively, one of the techniques in Section 2.29 as directed by your instructor may be used to concentrate the solution. After the ether is completely removed, the pot residue is crude citral.

Analysis Determine the weight of citral isolated and calculate the percentage recovery of citral based on the weight of the original sample of lemon grass oil. If instructed to do so, save or submit a sample of citral for GLC or GC-MS analysis (Sec. 6.4). You may be asked to perform one or more of the following tests on citral.

1. Test for a carbon-carbon double bond. Perform the tests for unsaturation described in Section 25.8.
2. Test for an aldehyde. Perform the chromic acid test outlined in Section 25.7D.
3. Analyze the citral using gas chromatography to assess the purity of the product.

WRAPPING IT UP

Flush the *aqueous solution* remaining in the distillation flask down the drain. Do the same with the *aqueous steam distillate* once you have completed the extraction. Pour the *solution from the test for unsaturation* into the container for halogen-containing liquids. Neutralize the *solution for the chromic acid test* and then pour it into the container for hazardous heavy metals.

EXERCISES

1. Why was a steam distillation rather than a simple distillation performed in the isolation of citral from lemon grass oil?
2. What type of product is expected from the reaction of citral with Br_2/CH_2Cl_2? With chromic acid?
3. Provide structures for the semicarbazone and the 2,4-dinitrophenylhydrazone of citral.
4. Why does the citral float on the surface of the aqueous distillate rather than sinking to the bottom?
5. Explain why the substitution of 1-propanol, bp 97 °C (760 torr), for water in a steam distillation would not work.
6. Suppose that you are to steam-distill a sample of a natural product whose vapor pressure at 100 °C is known to be half that of citral. What consequence would this have on the amount of distillate required per mole of the natural product present?
7. Both geranial (**1**) and vitamin A (**3**) are members of the class of natural products called **terpenes.** This group of compounds has the common characteristic of being biosynthesized by linkage of the appropriate number of five-carbon units having the skeletal structure shown below. Determine the number of such units present in each of these terpenes and indicate the bonds linking the various individual units.

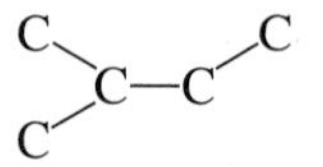

C C
C—C
C

8. In the reduction of nitrobenzene to aniline (Eq. 4.12), the product is readily isolated from the reaction mixture by steam distillation.

a. From this information, what do you know about the *miscibility* of aniline in water?

b. At approximately what temperature will aniline and water codistil?

c. What is the normal boiling point of aniline?

$$C_6H_5-NO_2 \xrightarrow[\text{(reduction)}]{[H]} C_6H_5-NH_2 \qquad (4.12)$$

9. Using Raoult's and Dalton's laws, explain how the boiling point of a binary mixture depends on the miscibility of the two liquids.

HISTORICAL HIGHLIGHT

Reducing Automobile Emissions

From a smog alert in Los Angeles, CA, to an "Ozone Action Day" in Austin, TX, it is clear that the air we breathe is under attack. The "aggressors" in this attack include volcanoes, plants, and trees. Volcanoes spew tons of noxious gases and particulate matter into the atmosphere during an eruption, and trees and plants create the haze that makes the Blue Ridge Mountains appear "blue." This haze results from photochemical reactions between atmospheric oxygen and the volatile hydrocarbons that are produced by and emitted from the flora that cover these beautiful mountains. Humans are also important contributors to air pollution through activities such as power generation and manufacturing processes. Primary sources of the volatile chemicals that contribute to degradation of air quality are the automobiles we drive, the buses we ride, and the trucks we depend on in many ways. These chemicals include carbon monoxide and carbon dioxide, oxides of nitrogen and sulfur, abbreviated as NO_x and SO_x, respectively, and hydrocarbons, HC.

Unfortunately, we do not presently have a complete picture of how these emitted substances interact with one another in the atmosphere to cause environmental problems, but some broad generalizations are possible. For example, carbon dioxide is judged to be the most significant "greenhouse gas" contributing to global warming. Carbon monoxide depletes hydroxyl radicals, HO, in the atmosphere, and this may lead to formation of ozone under certain atmospheric conditions. Although stratospheric ozone is beneficial because it blocks harmful radiation from penetrating the atmosphere—the existence of an ozone "hole" over the Earth's poles has received wide publicity and is a problem of great environmental concern—the presence of ozone at lower altitudes is not. Ozone reacts with carbon-carbon double bonds present in materials like rubber tires, thereby shortening their useful life. It also reacts with plant and animal tissues, including your skin, to cause health-related problems. The nitrogen oxides are formed by oxidation of atmospheric nitrogen at the high temperatures and pressures attending the combustion of gasoline and diesel fuel in air; they are primarily comprised of nitric oxide, NO. This oxide is converted to nitrogen dioxide, NO_2, by atmospheric oxygen and sunlight, which, in turn, reacts with hydrocarbons to form ozone, O_3, or with water to form nitrate, NO_3^-. Both O_3 and NO_2 are very environmentally damaging. Nitrate in the form of nitric acid contributes to acid rain, which has a major negative impact on natural habitats such as lakes and forests. Similarly, sulfur oxides, formed from oxidation of sulfur-containing components in fuels, may ultimately be converted to sulfuric acid, which is another contributor to acid rain.

In many urban areas, one-third to one-half or even more of the NO_x and HC pollutants are produced by motor vehicles having internal combustion engines. Indeed, on a national basis, some 30% of all smog-forming emissions are produced by automobiles, buses, and trucks, so they are obvious targets for addressing the matter of air pollution. It is

(Continued)

HISTORICAL HIGHLIGHT *Reducing Automobile Emissions (Continued)*

known that most pollutants are produced during the period when the engine is warming to its normal operating temperatures. This is because only the lower-molecular-weight hydrocarbons in gasoline are efficiently burned (oxidized) at lower engine temperatures; their higher-molar-mass relatives may not burn at all or are only partially oxidized to carbon dioxide and water. So a key to decreasing the emission of pollutants is to make the burning of gasoline more efficient in a cold engine.

The desired increased efficiency in burning can be effected through the technology illustrated on page 148. The gasoline in the regular tank is pumped into a chamber, in which the heat of the engine distills the gasoline. The lower-boiling hydrocarbons are selectively vaporized, condensed, and then stored in a special tank. With appropriate control devices, this more volatile "light" gasoline can be fed into the engine when the engine is cold; once it warms up, the "regular" gas is used, and the "light" gasoline in the special tank is replenished, ready for use the next time a "cold-start" is needed.

This simple system is remarkably effective in reducing emissions of pollutants: Emissions of hydrocarbon are decreased by some 50% and that of partially oxidized hydrocarbons by over 80%. Although the system has not yet been commercialized, Ford Motor Company is in the process of attempting to bring this technology to the marketplace, and other car, truck, and bus manufacturers are sure to follow if it works.

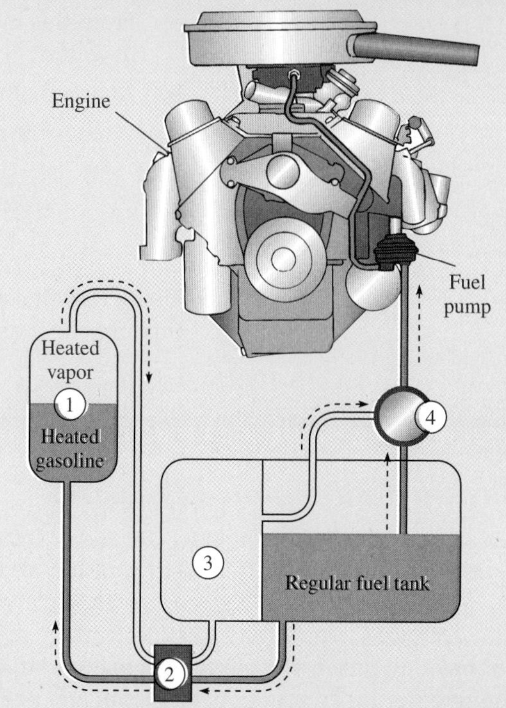

(1) Vapor separator that uses the heat from the running engine to vaporize "regular" gasoline; (2) condenser in which vapors are condensed; (3) alternate fuel tank where the distillate, or "light" gasoline, is stored; (4) valve controlled by vehicle's computer, which, based on the temperature of the engine, delivers either "light" or "regular" gasoline to the fuel pump and then to the engine.

Relationship of Historical Highlight to Experiments

The device illustrated in the figure depends on distillation to effect separation of the lower-molecular-weight components of gasoline. This is a fractional distillation, in effect, one that potentially has important implications for lessening air pollution. As you may know, chemists have long used this technique to separate and purify volatile liquids having differing boiling points.

CHAPTER 5

Extraction

When you see this icon, sign in at this book's premium website at **www.cengage.com/login** to access videos, Pre-Lab Exercises, and other online resources.

Extraction is a technique commonly used in organic chemistry to separate a material you want from those you do not. While the term *extraction* may be unfamiliar to you, the process is actually something you commonly perform. For example, many of you probably start the day, especially after a long night of studying, with an extraction when you brew a pot of coffee or tea. By heating coffee grounds or tea leaves with hot water, you extract the caffeine, together with other water-soluble compounds such as dark-colored tannins, from the solid material. You can then drink the liquid, which is certainly more enjoyable than eating coffee grounds or tea leaves, to ingest the caffeine and benefit from its stimulating effect. Similarly, when you make a soup, the largely aqueous liquid portion contains numerous organic and inorganic compounds that have been extracted from spices, vegetables, fish, or meat, and these give your culinary creation its distinctive flavor. In the procedures found in this chapter, you will have an opportunity to develop your existing experimental skills further by isolating organic compounds using different types of extractions.

5.1 INTRODUCTION

The desired compound from a reaction is frequently part of a mixture, and its isolation in pure form can be a significant experimental challenge. Two of the more common methods for separating and purifying organic liquids and solids are **recrystallization** and **distillation;** these procedures are discussed in Chapters 3 and 4, respectively. Two other important techniques available for these purposes are **extraction** and **chromatography.** As you will see here and in Chapter 6, both of these methods involve partitioning of compounds between two *immiscible* phases. This process is termed **phase distribution** and can result in separation of compounds if they distribute differently between the two phases.

Distribution of solutes between phases is the result of **partitioning** or **adsorption** phenomena. Partitioning involves the difference in solubilities of a substance in two immiscible solvents—in other words, *selective dissolution.* Adsorption, on the other hand, is based on the *selective attraction* of a substance in a liquid or gaseous mixture to the surface of a solid phase. The various chromatographic techniques depend on both of these processes, whereas the process of extraction relies only on partitioning.

Extraction involves *selectively* removing one or more components of a solid, liquid, or gaseous mixture into a separate phase. The substance being extracted will partition between the two immiscible phases that are in contact, and the ratio of its distribution between the phases will depend on the relative solubility of the solute in each phase.

5.2 THEORY OF EXTRACTION

Liquid-liquid extraction is one of the most common methods for removing an organic compound from a mixture. This process is used by chemists not only in the isolation of natural products but also in the isolation and purification of products from most chemical reactions.

The technique involves distributing a solute, A, between two immiscible liquids, S_x, the **extracting phase,** and S_o, the **original phase.** The immiscible liquids normally encountered in the organic laboratory are water and some organic solvent, such as diethyl ether, $(C_2H_5)_2O$, or dichloromethane, CH_2Cl_2. At a given temperature, the amount of A, in g/mL, in each phase is expressed *quantitatively* in terms of a constant, K, commonly called the **partition coefficient** (Eq. 5.1). Strictly speaking, the volumes of *solution* should be used in the definition of $[A]$, but if the solutions are dilute, only slight errors result if volumes of *solvent* are used. Furthermore, a close *approximation* of the partition coefficient K may be obtained by simply dividing the solubility of A in the extracting solvent S_x by the solubility of A in the original solvent S_o.

$$K = \frac{[A] \text{ in } S_x}{[A] \text{ in } S_o} \tag{5.1}$$

The process of liquid-liquid extraction can be considered a competition between two immiscible liquids for solute A, with solute A distributing or partitioning between these two liquids when it is in contact with both of them. The mathematical expression of Equation 5.1 shows that *at equilibrium,* the ratio of concentrations of A in the two phases will always be constant.

Equation 5.1 leads to several important predictions. For example, if $K > 1$, solute A will be mainly in the *extracting* solvent, S_x, so long as the *volume,* V_x, of this solvent is at least equal to the volume, V_o, of the original solvent S_o; the amount of solute remaining in S_o will depend on the value of K. By recasting Equation 5.1 into Equation 5.2, it is easily seen that if the volumes of S_x and S_o are equal, the value of K is simply the ratio of the grams of A in each solvent. Because the product of the ratios A_x/A_o and V_o/V_x must be constant for a given solvent pair, increasing the volume of extracting solvent S_x will result in a net *increase* in the amount of solute A in S_x.

$$K = \frac{\text{grams of } A \text{ in } S_x}{\text{grams of } A \text{ in } S_o} \times \frac{\text{mL of } S_o}{\text{mL of } S_x} = \frac{A_x}{A_o} \times \frac{V_o}{V_x} \tag{5.2}$$

There are practical, economic, and environmental reasons, however, that limit the quantities of organic solvents that can realistically be used in extractions. The question thus arises as to whether it is better to perform a single extraction with all of the solvent, or to perform several extractions with smaller volumes. For example, if an organic compound is dissolved in 10 mL of water, and only 30 mL of diethyl ether is used, will *three* extractions with 10-mL portions of ether provide better recovery of solute than a *single* one with 30 mL?

Applying Equation 5.2 would provide the answer, but the process is tedious for multiple extractions. This equation can be generalized, however, to accommodate multiple extractions in terms of the fraction, F_A, of solute A remaining in the original solvent S_o after n extractions, using a constant volume of an immiscible solvent S_x for each extraction. Thus, the amount of solute remaining in S_o after one extraction with S_x is obtained by rearranging Equation 5.3 to Equation 5.4. The fraction, F_A, of A still in the original solvent is obtained through Equation 5.5, in which C_i and C_f are the *initial* and *final* concentrations, respectively. For a second extraction,

Equations 5.6–5.8 result. Equation 5.8 can be generalized to Equation 5.9 when n extractions are performed.

$$K = \left(\frac{A_o - A_1}{V_x}\right)\left(\frac{V_o}{A_1}\right) \tag{5.3}$$

where A_o = amount (grams) of solute in S_o before extraction
$A_1 = (A_o - A_x)$ = amount (grams) of solute in S_o after extraction
V_o and V_x = volume (mL) of original and extracting solvents, respectively

$$A_1 = \frac{(A_oV_o)}{KV_x + V_o} \tag{5.4}$$

$$F_A = \frac{C_f}{C_i} = \frac{V_o}{KV_x + V_o} \tag{5.5}$$

where $C_i = A_o/S_o$ and $C_f = A_1/S_o$, respectively

$$K = \left(\frac{A_1 - A_2}{V_x}\right)\left(\frac{V_o}{A_2}\right) \tag{5.6}$$

$$K = A_o\left(\frac{V_o}{KV_x + V_o}\right)^2 \tag{5.7}$$

$$F_{A'} = \frac{C_f}{C_i} = \left(\frac{V_o}{KV_x + V_o}\right)^2 \tag{5.8}$$

$$F_A = \frac{C_f}{C_i} = \left(\frac{V_o}{KV_x + V_o}\right)^n \tag{5.9}$$

Performing calculations according to Equation 5.9 for the case of $K = 5$, we obtain $F_A = 1/16$ for a single extraction with 30 mL of diethyl ether, and 1/216 for three extractions with 10-mL portions. This means that 6.3% of A remains in the aqueous phase when *one* extraction is performed, whereas the value drops to only 0.5% in the case of *three* successive extractions with the *same* total volume of solvent. The amount of A that could be isolated is increased by some 6% with multiple extractions. Setting K at 2 gives corresponding values for F_A of 1/7 and 1/27, respectively, which translates to a 10% increase in the quantity of A that is removed from the aqueous layer with multiple extractions.

From these calculations, it is clear that multiple extractions become increasingly important as the value of K decreases. The improved recovery of solute from multiple extractions, though, must be balanced with the practical consideration that the relatively small increase in recovery may not justify the additional time and solvent required to perform multiple extractions unless the product is of great value. Importantly, the partition coefficients K are generally large for the extractions required in the procedures provided in this textbook, so only one or two extractions are usually required.

Selection of the appropriate extracting solvent is obviously a key for successfully using this technique to isolate and purify compounds, and important guidelines for making the correct choice are summarized here.

1. The extracting solvent *must not react* in a chemically irreversible way with the components of the mixture.

2. The extracting solvent *must be immiscible,* or nearly so, with the original solution.
3. The extracting solvent *must selectively remove* the desired component from the solution being extracted. That is, the partition coefficient K of the component being removed must be *high,* while the partition coefficients of all other components should be *low.*
4. The extracting solvent *should be readily* separable from the solute. Use of a volatile solvent facilitates its removal by simple distillation or one of the other techniques outlined in Section 2.29.

5.3 BASE AND ACID EXTRACTIONS

The discussion of Section 5.2 focuses on partitioning *one* substance between two *immiscible* solvents. Now consider what happens if a mixture of *two or more* compounds is present in a given volume of solvent S_0 and an extraction using a solvent S_x is performed. If the partition coefficient K of one component, A, is *significantly* greater than 1.0 and if those of other components are *significantly* less than 1.0, the majority of A will be in S_x, whereas most of the other compounds will remain in S_0. Physical separation of the two solvents will give a partial separation, and thus purification, of the solute A from the other components of the mixture.

Solutes differing significantly in polarity should have very different coefficients K for partitioning between nonpolar and polar solvents. For example, consider the distribution of two organic compounds, the first neutral and nonpolar, and the second ionic and polar, between a nonpolar solvent and a polar solvent. If a solution of these compounds in the nonpolar solvent is shaken with the polar solvent, the neutral compound will preferentially partition into the *nonpolar* phase, whereas the polar constituent will preferentially partition into the *polar* phase. Separating the two phases effects a separation of the two solutes.

Carboxylic acids and phenols (**1** and **3,** Eqs. 5.10 and 5.11, respectively) are two classes of organic compounds containing **functional groups** that are **polar** and **hydrophilic** (water-loving). Unless they contain fewer than about six carbon atoms, such compounds are generally either *insoluble* or only *slightly soluble* in water because of the **hydrophobic** (water-avoiding) properties of the carbon-containing portion, R or Ar, of the molecule. They are *soluble* in common organic solvents like dichloromethane or diethyl ether that have at least modest polarity (Table 3.1, p. 96). Consequently, carboxylic acids or phenols dissolved in diethyl ether, for example, will largely remain in that phase when the solution is extracted with water.

$$\underset{\mathbf{1}}{R{-}C({=}O){-}O{-}H} + B^- \,Na^+ (aq) \rightleftharpoons \underset{\mathbf{2}}{R{-}C({=}O){-}O^-} \;Na^+ + B{-}H \qquad (5.10)$$

1		**2**
A carboxylic acid Water *insoluble* $K_{water/org} < 1$	A base (as sodium salt)	A carboxylate (as sodium salt) Water *soluble* $K_{water/org} > 1$

$$\text{Ar—O—H} + \text{B}^- \text{Na}^+ \text{(aq)} \rightleftharpoons \text{Ar—O}^- \text{Na}^+ + \text{B—H} \quad (5.11)$$

3
A phenol (Ar = aryl)
Water *insoluble*
$K_{water/org} < 1$

A base
(as sodium salt)

4
A phenoxide
(as sodium salt)
Water *soluble*
$K_{water/org} > 1$

Now consider what happens if the organic solution is extracted with a *basic* aqueous solution. If the base, B^-, is strong enough, the organic acid **1** or **3** will be converted into the corresponding **conjugate base 2** or **4** (Eqs. 5.10 and 5.11). Because it is a salt, the conjugate base is highly polar, and $K_{water/org} > 1$. Thus the conjugate bases of organic acids may be selectively extracted from an organic phase into an aqueous phase. If the basic extract is then neutralized by an acid such as hydrochloric acid, the conjugate base **2** or **4** will be protonated to regenerate the organic acid **1** or **3** (Eqs. 5.12 and 5.13). Because the acid is water-insoluble, it will appear as either a precipitate or a second layer, if it is a liquid. The desired organic acid may then be recovered by filtration or separation of the layers.

$$\text{R—C(=O)O}^- \text{Na}^+ + \text{H—Cl (aq)} \rightleftharpoons \text{R—C(=O)O—H} + \text{Na}^+ \text{Cl}^- \quad (5.12)$$

2 Water *soluble*; **1** Water *insoluble*

$$\text{Ar—O}^- \text{Na}^+ + \text{H—Cl} \rightleftharpoons \text{Ar—O—H} + \text{Na}^+ \text{Cl}^- \quad (5.13)$$

4 Water *soluble*; **3** Water *insoluble*

Thus, two water-insoluble organic compounds, **HA,** which is acidic, and **N,** which is neutral, that are dissolved in an organic solvent may be separated by selectively extracting the acidic compound into a basic aqueous phase. After the aqueous and organic phases are separated, **HA** is recovered from the aqueous phase upon neutralization, and **N** is obtained by removing the organic solvent (Fig. 5.1).

The choice of the base is determined by the acidity of the organic acid **HA.** This is defined in aqueous solution by $\boldsymbol{K_a}$ in Equation 5.14 and is often expressed as $\boldsymbol{pK_a}$ (Eq. 5.15).

$$K_{a(HA)} = \frac{[A^-][H_3O^+]}{[HA]} \quad (5.14)$$

$$pK_a = -\log_{10} K_a \quad (5.15)$$

The next step in determining what base is needed is to consider the equilibrium shown in Equation 5.16, in which **HA** is the organic acid, $\mathbf{B^-}$ is the base being used, and **HB** is the **conjugate acid** of this base. The **equilibrium constant,** K_{eq}, for this process is given by Equation 5.17. Given the expressions for K_a for **HA** and **HB** (Eqs. 5.14 and 5.18, respectively), Equation 5.17 transforms to Equation 5.19. From this equation we see that $\log_{10}$ of the equilibrium constant, K_{eq}, equals $pK_{a(\mathbf{HB})} - pK_{a(\mathbf{HA})}$. We can predict how effective a particular base $\mathbf{B^-}$ will be in converting an organic

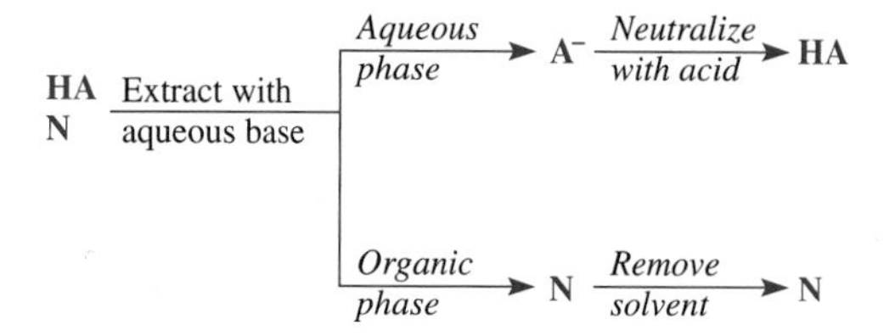

Figure 5.1
Separating an acidic compound and a neutral compound.

acid **HA** to its anion simply by knowing the relative acidities of **HA** and **HB.** The equilibrium for Equation 5.14 will lie to the right; namely, $\log_{10} K_{eq}$ will be > 1 whenever the acidity of **HA** is *greater* than that of **HB**.

$$HA + B^- \overset{K_{eq}}{\rightleftharpoons} A^- + HB \qquad (5.16)$$

$$K_{eq} = \frac{[A^-][HB]}{[HA][B^-]} \qquad (5.17)$$

$$K_{a(HB)} = \frac{[B^-][H_3O^+]}{[HB]} \qquad (5.18)$$

$$K_{eq} = \frac{K_{a(HA)}}{K_{a(HB)}} \text{ and } \log_{10} K_{eq} = \log_{10} K_{a(HA)} - \log_{10} K_{a(HB)} \qquad (5.19)$$

Let's now consider two specific examples of mixtures of acidic and neutral compounds. Assume that you have a solution of benzoic acid (**5**) and naphthalene (**7**), a neutral compound, in diethyl ether and that you wish to separate the two. Both compounds and the solvent are insoluble in water at room temperature. The strategy is to convert **5** to its water-soluble conjugate base **6** by using an aqueous base to extract the organic solution. The pK_a of benzoic acid is 4.2, so any base whose *conjugate acid* has a pK_a *greater* than this will yield a value of $\log_{10} K_{eq} > 1$ (Eq. 5.19). Aqueous hydroxide, whose conjugate acid is water (pK_a 15.7), would give $\log_{10} K_{eq} = 10.5$, which means that this base will be extremely effective for the extraction of **5** into the aqueous phase. Indeed, aqueous hydroxide is commonly used for extracting organic acids into an aqueous phase.

5 Benzoic acid　　**6** Benzoate　　**7** Naphthalene

Other aqueous bases could also be used. For example, carbonate (**8**) and bicarbonate (**9**) would also work because their conjugate acids, **9** and **10** (Eq. 5.20), have pK_as of 10.3 and 6.4, respectively. These bases are not as strong as hydroxide, so on the purely *thermodynamic* basis described by Equation 5.19, they would provide a lower value of $\log_{10} K_{eq}$. However, both **8** and **9** may be converted to carbonic acid (**10**) upon di- and monoprotonation, respectively (Eq. 5.20). Carbonic acid is *kinetically* unstable and decomposes to carbon dioxide and water (Eq. 5.21), and this disrupts the equilibrium of Equation 5.16, driving it to the *right* as **HB,** in the form of **10,** undergoes decomposition. The net result would be efficient deprotonation of **5** by these bases.

$$^{-}O-C(=O)-O^{-} \xrightarrow{H_3O^+} HO-C(=O)-O^{-} \xrightarrow{H_3O^+} HO-C(=O)-OH \quad (5.20)$$

8 Carbonate — **9** Bicarbonate — **10** Carbonic acid

$$H_2CO_3 \longrightarrow CO_2 + H_2O \quad (5.21)$$

It is important to recognize that the conversion of **10** into carbon dioxide and water presents a practical problem when **8** and **9** are used to deprotonate acids for purposes of extraction. If a separatory funnel is being used for the extraction, the gas pressure that results from the formation of carbon dioxide can blow out the stopcock and stopper, spraying the contents on you, your neighbors, or the floor. Alternatively, if a screw-cap vessel is being used, product may be ejected when the cap is loosened to relieve the gas pressure. Accordingly, it is preferable to use hydroxide for basic extractions as a general rule.

Turning to a second example, assume you have a solution of 2-naphthol (**11**), which is a phenol, and naphthalene (**7**) in diethyl ether. The pK_a of **11** is 9.5, so in this case either aqueous hydroxide or carbonate would give $K_{eq} > 1$; bicarbonate (**8**) is too weak a base to effect the deprotonation. Because aqueous hydroxide would provide $K_{eq} \gg 1$, it is the base of choice for this extraction.

11 2-Naphthol — **12** 2-Naphthoxide

The examples discussed so far have involved separation of compounds having different acidities from neutral compounds. However, organic bases, usually amines, must sometimes be separated from mixtures containing organic acids and neutral compounds. Before considering how to perform such separations, it is useful to regard amines as being derivatives of ammonia where the hydrogen atoms on the nitrogen atom are replaced with substituted carbon atoms that are designated as R or Ar. Like the functional groups of carboxylic acids and phenols, the amino functional group is polar and hydrophilic, and amines having more than six carbon atoms are either insoluble or only slightly soluble in water because of the hydrophobic properties of the R or Ar groups.

Amines are often soluble in aqueous acid at a pH < 4 because they are converted into their respective ammonium salts, the conjugate acids of the amines. This process is illustrated in Equation 5.22, wherein aqueous HCl serves as the acid. The enhanced ionic character of the ammonium salt, as we saw before with the salts of carboxylic acids and phenols, makes it water soluble ($K_{water/org} > 1$), whereas its parent base is not ($K_{water/org} < 1$). When the acidic solution is neutralized by adding aqueous base, the ammonium ion is deprotonated and converted into the original water-insoluble organic base (Eq. 5.23) that will now either precipitate from solution or form a separate layer. The amine may then be recovered by filtration or by separation of the layers.

$$R-\ddot{N}H_2 + H-Cl\,(aq) \longrightarrow R\overset{+}{N}H_2-H\ Cl^- \quad (5.22)$$

(Water *insoluble*) $K_{water/org} < 1$ — (Water *soluble*) $K_{water/org} > 1$

$$\underset{\substack{\text{(Water } soluble\text{)} \\ K_{water/org} > 1}}{R\overset{+}{N}H_2\text{—}H\ Cl^-} + HO^-\ Na^+(aq) \longrightarrow \underset{\substack{\text{(Water } insoluble\text{)} \\ K_{water/org} < 1}}{R\text{—}\ddot{N}H_2} + H_2O + NaCl \qquad (5.23)$$

The question now is: How can the two water-insoluble organic compounds, **B,** which is basic, and **N,** which is neutral, that are dissolved in an organic solvent be separated? This problem is analogous to the earlier one of separating two water-insoluble organic compounds, **HA,** which is acidic, and **N,** which is neutral (see Fig. 5.1). Namely, the basic compound can be selectively extracted into an acidic aqueous phase as the soluble salt HB^+. After the aqueous and organic phases are separated, **B** is recovered from the aqueous phase upon neutralization, and **N** is obtained by removing the organic solvent (Fig. 5.2).

One example of an amine is 4-nitroaniline (**13**), which is converted into the corresponding anilinium ion **14** by aqueous hydrochloric acid. Although 4-nitroaniline is a weak base with a pK_b of 13.0, its conjugate acid **14** has a pK_a of 1.0 and is a relatively strong organic acid. Hence, it is necessary to use a strong aqueous acid such as 6 *M* HCl, in which the acid is actually the hydronium ion, H_3O^+, having a pK_a of about −1.7, to convert **13** efficiently into its salt **14.**

NH_2 / NO_2 — **13** — 4-Nitroaniline

$\overset{\oplus}{N}H_3Cl$ / NO_2 — **14** — 4-Nitroanilinium hydrochloride

In the experimental procedures that follow, three types of separations of neutral, acidic, and basic compounds are described. The first involves separating a mixture of benzoic acid (**5**) and naphthalene (**7**), using aqueous hydroxide for the extraction. The flow chart for this separation corresponds to that of Figure 5.1. The second procedure involves separating a mixture of **5, 7,** and 2-naphthol (**11**). In this case, **5** is first *selectively* removed from the solution by extraction with aqueous bicarbonate, which *does not* deprotonate **11.** A second basic extraction using aqueous hydroxide removes **11** from the organic solution. The flow chart for this sequence is depicted in Figure 5.3.

The last procedure involves separating a three-component mixture containing benzoic acid (**5**), naphthalene (**7**), and 4-nitroaniline (**13**). First **13** is *selectively* removed from the organic solution by extraction with aqueous hydrochloric acid, and then **5** is removed by extraction with aqueous base, leaving **7** in the organic solution. The flow chart for this sequence of operations is shown in Figure 5.4. All procedures demonstrate the power of using the *pH*-dependence of water solubility of organic compounds as a basis for separation. Such procedures for the preparative separation of organic compounds are much easier than the chromatographic separations you will learn about in Chapter 6.

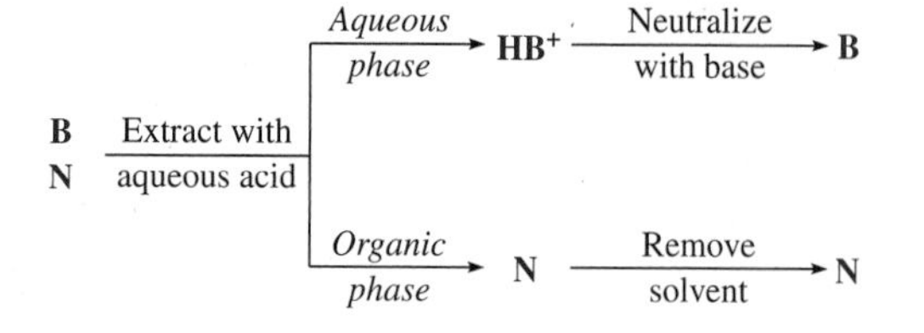

Figure 5.2
Separating a basic compound and a neutral compound.

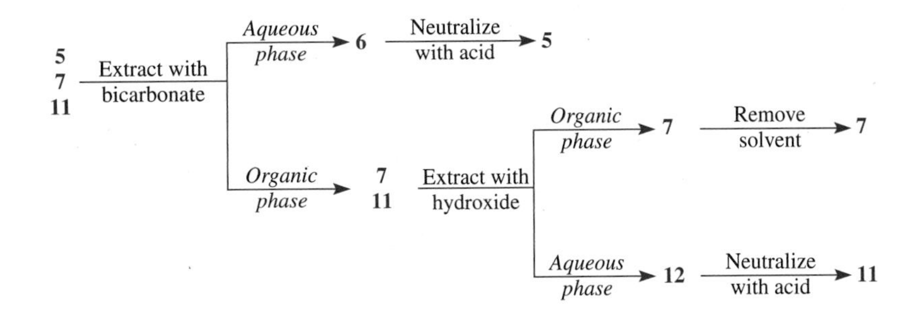

Figure 5.3
Separating a neutral compound and two compounds having differing acidities.

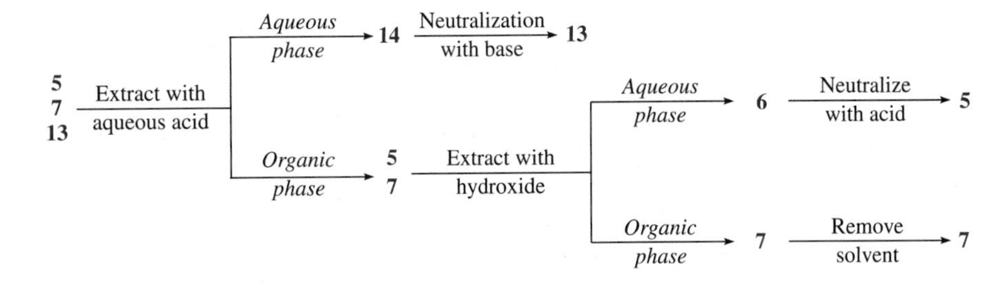

Figure 5.4
Separating an acidic, a basic, and a neutral compound.

EXPERIMENTAL PROCEDURES

Base and Acid Extractions

Purpose To separate multicomponent mixtures as a function of pH of solution.

SAFETY ALERT

1. **Do not allow any of the chemicals used in this experiment to come in contact with your skin. If they do, *immediately* wash the affected areas with soap and water.**
2. **Diethyl ether is *extremely* volatile and flammable. Be certain there are *no flames* in your vicinity when using it for extraction and when removing it from the "Neutral Compound."**
3. **Wear latex gloves for the extractions, and use caution in handling wet glassware, because it can easily slip out of your grasp.**
4. **When extracting a solution, vent the separatory funnel or vial *often* to avoid a buildup of pressure.**

MINISCALE PROCEDURE

Preparation Sign in at **www.cengage.com/login** to answer Pre-Lab Exercises, access videos, and read the MSDSs for the chemicals used or produced in this procedure. Read or review Sections 2.7, 2.10, 2.11, 2.13, 2.17, 2.21, 2.24, 2.25, 2.29, 3.2, and 3.3.

A. One-Base Extraction

Apparatus Separatory funnel, ice-water bath, apparatus for vacuum filtration, simple distillation, and *flameless* heating.

Dissolution Obtain 2 g of a mixture of benzoic acid and naphthalene. Dissolve the mixture by swirling it with 30 mL of diethyl ether in an Erlenmeyer flask. If any solids remain, add more diethyl ether to effect complete dissolution. Transfer the solution to the separatory funnel.

Extraction Extract the solution with a 15-mL portion of 2.5 *M* (10%) aqueous sodium hydroxide. *Be sure to hold both the stopcock and the stopper of the funnel tightly and frequently vent the funnel by opening its stopcock* (Fig. 2.61). Identify the aqueous layer by the method described in Section 2.21, transfer it to an Erlenmeyer flask labeled "Hydroxide Extract," and transfer the organic solution into a clean Erlenmeyer flask containing two spatula-tips full of anhydrous sodium sulfate and labeled "Neutral Compound." Let this solution stand for about 15 min, occasionally swirling it to hasten the drying process. If the solution remains cloudy, add additional portions of sodium sulfate to complete the drying process.★

Precipitating and Drying Cool the "Hydroxide Extract" in an ice-water bath. Carefully acidify this solution with 3 *M* hydrochloric acid, so that the solution is distinctly acidic to pHydrion™ paper.★ Upon acidification, a precipitate should form; cool the mixture for 10–15 min to complete the crystallization.

Collect the precipitate by vacuum filtration (Fig. 2.54) using a Büchner or Hirsch funnel. Wash the solid on the filter paper with a small portion of *cold* water. Transfer the solid to a labeled watchglass, cover it with a piece of filter or weighing paper, and allow the product to air-dry until the next laboratory period. Alternatively, dry the solid in an oven having a temperature of 90–100 °C for about 1 h. After drying, transfer the benzoic acid to a dry, *tared* vial.

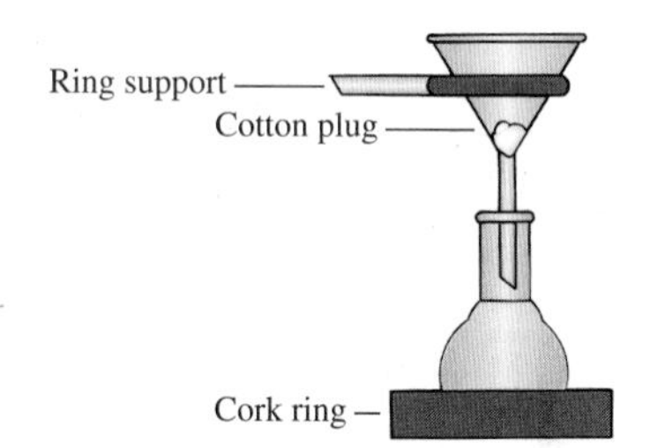

Figure 5.5
Miniscale gravity filtration through a cotton plug.

Separate the "Neutral Compound" from the drying agent by decantation (Fig. 2.56) or gravity filtration through a cotton plug (Fig. 5.5) into a 100-mL round-bottom flask. Remove the solvent by simple distillation (Fig. 2.37). Alternatively, use rotary evaporation or one of the other techniques described in Section 2.29. Allow the residue to cool to room temperature to solidify, scrape the contents of the flask onto a piece of weighing paper to air-dry, and then transfer it to a dry, *tared* vial.

Analysis and Recrystallization Determine the weight and melting point of each crude product. If you know the relative amounts of benzoic acid (**5**) and naphthalene (**7**) in the original mixture, calculate the percent recovery of each compound. Recrystallize them according to the general procedures provided in Section 3.2 and determine the melting points of the purified materials.

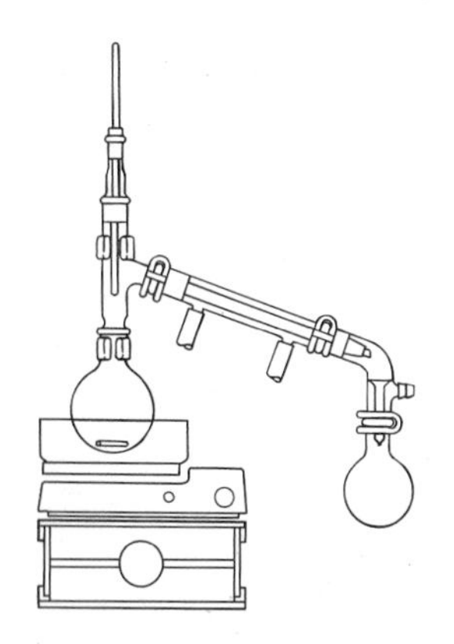

B. Two-Base Extraction

Apparatus Separatory funnel, ice-water bath, apparatus for vacuum filtration, simple distillation, and *flameless* heating.

Dissolution Obtain 2 g of a mixture of benzoic acid, 2-naphthol, and naphthalene. Dissolve the mixture by swirling it with 30 mL of diethyl ether in a 125-mL Erlenmeyer flask. If any solids remain, add more diethyl ether to effect complete dissolution.

Extraction Extract the solution with a 20-mL portion of 1.25 *M* (10%) aqueous sodium bicarbonate (Fig. 2.61). *Caution:* Gaseous carbon dioxide is generated! To prevent accidental loss of material, do the following: After adding the aqueous bicarbonate to the funnel, swirl the *unstoppered* funnel until all foaming has subsided. Then stopper the funnel and, *holding both the stopcock and the stopper of the separatory funnel tightly,* invert the funnel and *immediately* vent it by opening the *stopcock.* Finally, shake the funnel with *frequent* venting until gas is no longer evolved. Separate the layers, transferring the aqueous layer to a 125-mL Erlenmeyer flask labeled "Bicarbonate Extract."★ To ensure properly identifying the aqueous layer, consult Section 2.21. Return the organic solution to the separatory funnel.

Now extract the organic solution with a 20-mL portion of *cold* 2.5 *M* (10%) aqueous sodium hydroxide, venting the funnel frequently during the process. Transfer the aqueous layer to a 125-mL Erlenmeyer flask labeled "Hydroxide Extract." Transfer the organic solution to a 50-mL Erlenmeyer flask labeled "Neutral Compound" and containing two spatula-tips full of anhydrous sodium sulfate. Loosely cork the flask containing the organic solution, and let it stand for at least 15 min with occasional swirling to promote drying. If the solution remains cloudy, add additional portions of sodium sulfate to complete the drying process.★

Precipitating and Drying For the "Bicarbonate" and "Hydroxide Extracts," follow the directions given in Part A for the "Hydroxide Extract." Also apply the protocol provided in Part A for the "Neutral Compound" to the corresponding solution obtained in this procedure.

Analysis and Recrystallization Determine the weight and melting point of each crude product. If you know the relative amounts of benzoic acid (**5**), naphthalene (**7**), and 2-naphthol (**11**) in the original mixture, calculate the percent recovery of each compound. Recrystallize them according to the general procedures provided in Section 3.2 and determine the melting points of the purified materials.

C. Acid-Base Extraction

Apparatus Separatory funnel, ice-water bath, apparatus for vacuum filtration, simple distillation, and *flameless* heating.

Dissolution Obtain 1.5 g of a mixture of benzoic acid, 4-nitroaniline, and naphthalene. Dissolve the mixture by swirling it with about 40 mL of dichloromethane in an Erlenmeyer flask. Transfer the solution to the separatory funnel.

Extraction Extract the organic solution three times using 15-mL portions of 3 *M* hydrochloric acid. *Be sure to hold both the stopcock and the stopper of the funnel tightly and frequently vent the funnel by opening its stopcock* (Fig. 2.61). Identify the aqueous layer by the method described in Section 2.21. Combine the three acidic aqueous layers from the extractions in an Erlenmeyer flask labeled "HCl Extract." Return the organic layer to the separatory funnel and extract it two times using 15-mL portions of 3 *M* sodium hydroxide solution. Combine these two aqueous layers in a second flask labeled "Hydroxide Extract." Transfer the organic solution into a clean Erlenmeyer flask containing two spatula-tips full of *anhydrous* sodium sulfate and labeled "Neutral Compound." Let this solution stand for about 15 min, occasionally swirling it to hasten the drying process. If the solution remains cloudy, add additional portions of sodium sulfate to complete the drying process.★

Precipitating and Drying Cool the "HCl Extract" and the "Hydroxide Extract" in an ice-water bath. Neutralize the "HCl Extract" with 6 *M* sodium hydroxide and add a little excess base to make the solution distinctly basic to pHydrion paper. Neutralize the "Hydroxide Extract" with 6 *M* hydrochloric acid and add a little excess acid to make the solution distinctly acidic to pHydrion paper.★ Upon neutralization, a precipitate should form in each flask.

Collect the precipitate in the flasks labeled "HCl Extract" and "Hydroxide Extract" by vacuum filtration (Fig. 2.54) using a Büchner or Hirsch funnel. Wash each solid on the filter paper with a small portion of *cold* water. Transfer each solid to a labeled watchglass, cover it with a piece of filter or weighing paper, and allow the product to air-dry until the next laboratory period. Alternatively, dry the solid in an oven having a temperature of 90–100 °C for about 1 h. After drying, transfer the 4-nitroaniline and the benzoic acid each to a different dry, *tared* vial.

Separate the "Neutral Compound" from the drying agent by decantation (Fig. 2.56) or gravity filtration through a cotton plug (Fig. 5.5) into a 100-mL round-bottom flask. Remove the solvent by simple distillation (Fig. 2.37). Alternatively, use rotary evaporation or one of the other techniques described in Section 2.29. Allow the residue to cool to room temperature to solidify, scrape the contents of the flask onto a piece of weighing paper to air-dry, and then transfer it to a dry, *tared* vial.

Analysis and Recrystallization Determine the weight and melting point of each crude product. If you know the relative amounts of benzoic acid (**5**), naphthalene (**7**), and 4-nitroaniline (**13**) in the original mixture, calculate the percent recovery of each compound. Recrystallize them according to the general procedures provided in Section 3.2 and determine the melting points of the purified materials.

MICROSCALE PROCEDURE

Preparation Sign in at **www.cengage.com/login** to answer Pre-Lab Exercises, access videos, and read the MSDSs for the chemicals used or produced in this procedure. Sections 2.7, 2.9, 2.10, 2.17, 2.21, 2.25, 2.29, 3.2, and 3.3.

A. One-Base Extraction

Apparatus A 5-mL conical vial, Pasteur filter-tip pipets, ice-water bath, 3-mL Craig tube, apparatus for Craig tube filtration and *flameless* heating.

Dissolution Obtain 0.2 g of a mixture of benzoic acid and naphthalene. Dissolve the mixture by swirling it with 2 mL of diethyl ether in the conical vial. If any solids remain, use a filter-tip pipet to add enough diethyl ether to effect complete dissolution.

Extraction Add 0.5 mL of *cold* 2.5 *M* (10%) aqueous sodium hydroxide to the vial and cap it. Shake the vial vigorously for a minute or two, occasionally unscrewing the cap slowly to relieve any pressure. Allow the layers to separate and, using a Pasteur pipet, transfer the *aqueous* layer to a Craig tube labeled "Hydroxide Extract." To ensure properly identifying the aqueous layer, consult Section 2.21.

Precipitating and Drying Add a microspatula-tip full of sodium sulfate to the organic solution in the conical vial and dry the solution according to the directions provided in Section 2.25. Once the organic solution is dry, use a filter-tip pipet to transfer it to a test tube. Evaporate the diethyl ether using *flameless* heating; stir the solution with a microspatula or use a boiling stone to prevent bumping during evaporation. Perform this step at a hood or use a funnel attached to a vacuum source for removing vapors (Fig. 2.72b). Scrape the isolated naphthalene onto a piece of weighing paper to air-dry and then transfer it to a dry, *tared* vial.

Cool the "Hydroxide Extract" in an ice-water bath and, with stirring, add concentrated hydrochloric acid dropwise with a Pasteur pipet until the solution is distinctly acidic to pHydrion paper.★ Allow the acidified solution to remain in the bath to complete the crystallization. Then isolate the benzoic acid by Craig tube filtration (Fig. 2.55). Rinse the solid back into the Craig tube, with 0.5 mL of cold water, stir the mixture briefly, and repeat the filtration. Transfer the benzoic acid to a watchglass, cover it with a piece of filter paper, and allow the product to air-dry until the next laboratory period. After drying, put the isolated solid in a dry, *tared* vial.

Analysis and Recrystallization Determine the weight and melting point of each crude product. If you know the relative amounts of benzoic acid (**5**) and naphthalene (**7**) in the original mixture, calculate the percent recovery of each compound. Recrystallize them according to the general microscale procedures provided in Section 3.2 and determine the melting points of the purified materials.

B. Two-Base Extraction

Apparatus A 5-mL conical vial, Pasteur filter-tip pipets, ice-water bath, 3- and 5-mL Craig tubes, apparatus for Craig tube filtration and *flameless* heating.

Dissolution Obtain 0.3 g of a mixture of benzoic acid, 2-naphthol, and naphthalene. Dissolve the mixture by swirling it with 2 mL of diethyl ether in the conical vial. If any solids remain, use a filter-tip pipet to add enough diethyl ether to effect complete dissolution.

Extraction Extract the solution with a 1-mL portion of 1.25 *M* (10%) aqueous sodium bicarbonate. *Caution:* Gaseous carbon dioxide is generated! To prevent accidental loss of material, do the following: After adding the first portion of aqueous bicarbonate to the vial, stir the two-phase mixture vigorously with a microspatula or stirring rod until all foaming has subsided. Then cap the vial, shake it gently for 2–3 sec, and loosen the cap *slowly* to relieve any pressure. Retighten the cap and shake the vial vigorously for about 1 min, occasionally venting the vial by loosening its cap. Allow the layers to separate and, using a Pasteur pipet, transfer the *aqueous* layer to a 5-mL Craig tube labeled "Bicarbonate Extract." To ensure properly identifying the aqueous layer, consult Section 2.21.★

Now extract the organic solution with a 0.5-mL portion of *cold* 2.5 *M* (10%) aqueous sodium hydroxide, following the same protocol as that given in part **A,** placing the aqueous layer in a 3-mL Craig tube labeled "Hydroxide Extract."

Precipitating and Drying Use the same procedure as that given in part **A** for *separately* acidifying the "Bicarbonate" and "Hydroxide" extracts and isolating and drying the two solids that result.

Analysis and Recrystallization Determine the weight and melting point of each crude product. If you know the relative amounts of benzoic acid (**5**), naphthalene (**7**), and

2-naphthol (**11**) in the original mixture, calculate the percent recovery of each compound. Recrystallize the products according to the general microscale procedures provided in Section 3.2 and determine the melting points of the purified materials.

C. Acid-Base Extraction

Apparatus A 5-mL conical vial, Pasteur and filter-tip pipets, ice-water bath, four 13-mm × 10-mm test tubes, apparatus for vacuum filtration and *flameless* heating.

Dissolution Obtain 0.25 g of a mixture of benzoic acid, 4-nitroaniline, and naphthalene. Dissolve the mixture by swirling it with about 3 mL of dichloromethane in the conical vial.

Extraction Add 1.0 mL of 3 *M* hydrochloric acid to the vial and cap it. Shake the vial vigorously for a minute or two, occasionally unscrewing the cap to relieve any pressure. Allow the layers to separate and, using a Pasteur pipet, transfer the *aqueous* layer to a test tube labeled "HCl Extract." To ensure properly identifying the aqueous layer, consult Section 2.21. Repeat this extraction with 1.0 mL of 3 *M* hydrochloric acid, and again transfer the aqueous layer to the test tube labeled "HCl Extract." Now extract the organic solution once with 1.0 mL of *3 M* sodium hydroxide solution, and transfer the *aqueous* layer to a *second* test tube labeled "Hydroxide Extract." Transfer the organic solution into a clean test tube labeled "Neutral Compound."★

Precipitating and Drying Add two microspatula-tips full of anhydrous sodium sulfate to the organic solution in the test tube "Neutral Compound" and dry the solution according to the directions provided in Section 2.25. Once this organic solution is dry, use a Pasteur or a filter-tip pipet to transfer it to another test tube. Evaporate the dichloromethane using *flameless* heating; stir the solution with a microspatula or use a boiling stone to prevent bumping during evaporation. Perform this step at a hood or use a funnel attached to a vacuum source for removing vapors (Fig. 2.72b). Scrape the isolated naphthalene onto a piece of weighing paper to air-dry and then transfer it to a dry, *tared* vial.

Cool the test tubes containing the "HCl Extract" and the "Hydroxide Extract" in an ice-water bath. Neutralize the "HCl Extract" with 6 *M* sodium hydroxide and add a little excess base to make the solution distinctly basic to pHydrion paper. Neutralize the "Hydroxide Extract" with 6 *M* hydrochloric acid and add a little excess acid to make the solution distinctly acidic to pHydrion paper. Upon neutralization, a precipitate should form in each test tube; cool each tube for 10–15 min to complete the crystallization.★ Isolate the 4-nitroaniline and the benzoic acid by vacuum filtration using a Hirsch funnel (Fig. 2.54). Rinse each solid back into its *respective* test tube, with 0.5 mL of *cold* water, stir the mixture briefly, and repeat the filtration. Transfer the 4-nitroaniline and the benzoic acid each to a separate labeled watchglass, cover both with a piece of filter paper, and allow the products to air-dry until the next laboratory period. Alternatively, dry the solid in an oven having a temperature of 90–100 °C for about 1 h. After drying is complete, transfer the 4-nitroaniline and the benzoic acid each to a different dry, *tared* vial.

Analysis and Recrystallization Determine the weight and melting point of each crude product. If you know the relative amounts of benzoic acid (**5**), naphthalene (**7**), and 4-nitroaniline (**13**) in the original mixture, calculate the percent recovery of each compound. Recrystallize them according to the general procedures provided in Section 3.2 and determine the melting points of the purified materials.

Discovery Experiment

Separation of Unknown Mixture by Extraction

Obtain an unknown mixture comprising up to four different compounds to be separated using one-base extraction, two-base extraction, acid-base extraction, or a combination of these separation techniques. The mixture might contain, for example, one compound from each of the following four groups: (1) *N*-(4-nitrophenyl) benzamide, 9-fluorenone, fluorene, anthracene, phenanthrene, or naphthalene; (2) *p*-cresol, 2-naphthol, or 4-methoxyphenol; (3) benzoic acid, 2-methylbenzoic acid, 2-chlorobenzoic acid, 4-chlorobenzoic acid, or salicyclic acid; (4) 3-nitroaniline, 4-nitroaniline, or 2-methyl-4-nitroaniline. Devise a scheme that may be used for separating this unknown mixture into its individual components; the procedure you devise should work irrespective of the number of components in the mixture. After recrystallizing each of the components, determine their melting points so that you can identify each compound.

Discovery Experiment

Isolation of Ibuprofen

See more on *Ibuprofen*

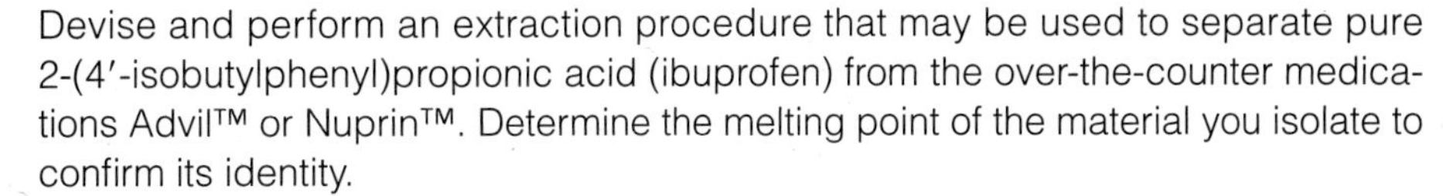

Devise and perform an extraction procedure that may be used to separate pure 2-(4′-isobutylphenyl)propionic acid (ibuprofen) from the over-the-counter medications Advil™ or Nuprin™. Determine the melting point of the material you isolate to confirm its identity.

WRAPPING IT UP

Place the used *filter papers* in a container for nontoxic solid waste. Flush the *acidic* and *basic* filtrates down the drain. Pour any *diethyl ether* that has been isolated into a container for nonhalogenated organic liquids. Pour any *dichloromethane* that has been isolated into a container for halogenated organic liquids.

EXERCISES

General Questions

1. Using Equation 5.9, show that three extractions with 5-mL portions of a solvent give better recovery than a single extraction with 15 mL of solvent when $K = 0.5$.
2. Show mathematically how Equation 5.17 can be converted to Equation 5.19.
3. Define the following terms:
   - **a.** immiscible liquid phases
   - **b.** distribution or partition coefficient
   - **c.** adsorption
   - **d.** absorption
   - **e.** conjugate base of an acid, HA
   - **f.** conjugate acid of a base, B^-
   - **g.** liquid-liquid extraction
   - **h.** hydrophilic
4. Explain why swirling or shaking a solution and its drying agent hastens the drying process.
5. Assume that the partition coefficient, *K*, for partitioning of compound *A* between diethyl ether and water is 3; that is, *A* preferentially partitions into ether.

a. Given 400 mL of an aqueous solution containing 12 g of compound *A*, how many grams of *A* could be removed from the solution by a *single* extraction with 200 mL of diethyl ether?

b. How many total grams of *A* can be *removed* from the aqueous solution with three successive extractions of 67 mL each?

6. Based on the principle of "like dissolves like," indicate by placing an "x" in the space those compounds listed below that are likely to be soluble in an organic solvent like diethyl ether or dichloromethane and those that are soluble in water.

Compound	*Organic Solvent*	*Water*
a. $C_6H_5CO_2Na$	______	______
b. Naphthalene (**7**)	______	______
c. Anthracene (**15**)	______	______
d. Phenol (**16**)	______	______
e. C_6H_5ONa	______	______
f. Aniline (**17**)	______	______
g. NaCl	______	______
h. CH_3CO_2H	______	______

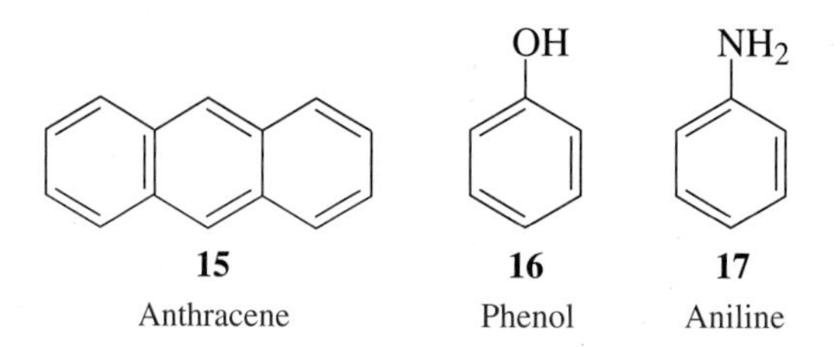

7. Benzoic acid (**5**) is soluble in diethyl ether but not water; however, benzoic acid is extracted from diethyl ether with aqueous sodium hydroxide.

a. Complete the acid-base reaction below by writing the products of the reaction.

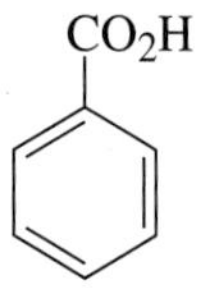

CO_2H

+ NaOH (aq) ⟶ ______ + ______

5

b. In the reaction of Part **a,** label the acid, the base, conjugate acid, and conjugate base.

c. Indicate the solubility of benzoic acid and its conjugate base in diethyl ether and in water.

8. Aniline (**17**), an amine, is soluble in diethyl ether but not water; however, aniline is extracted from diethyl ether with aqueous hydrochloric acid.

a. Complete the acid-base reaction below by writing the products of the reaction.

NH_2 (aniline, **17**) + HCl (aq) ⟶ ________ + ________

b. In the reaction of Part **a,** label the acid, the base, conjugate acid, and conjugate base.

c. Indicate the solubility of aniline and its conjugate acid in diethyl ether and in water.

9. Naphthalene (**7**) is soluble in diethyl ether, but it is insoluble in water regardless of the solution pH. Explain why this compound cannot be readily ionized in aqueous solution.

10. There are three common functional groups in organic chemistry that are readily ionized by adjusting the pH of the aqueous solution during an extraction. Name and write the chemical structure of these three functional groups, and show each of them in both their neutral and ionized forms.

11. a. The pK_a of benzoic acid (**5**) is 4.2. Show mathematically that this acid is 50% ionized at pH 4.2.

b. Use the result of Part **a** to explain why precipitation of **5** is incomplete if the pH of an aqueous solution of benzoate ion (**6**) is lowered only to pH 7 by adding acid.

12. Consider the base monosodium phosphate, $Na^+ \ (HO)_2P(=O)O^-$. (a) Write the structure of the conjugate acid of this base. (b) Given that the pK_a of this conjugate acid is 2.1, explain why an aqueous solution of monosodium phosphate would be ineffective for extracting benzoic acid (**5**) from a diethyl ether solution.

13. Provide a flowchart analogous to those in Figures 5.1–5.4 for separating a diethyl ether solution containing anthracene (**15**), benzoic acid (**5**), 2-naphthol (**11**), and 4-nitroaniline (**13**). All compounds are solids when pure.

14. The equilibrium for phenol (**16**), sodium phenoxide (**18**), sodium bicarbonate, and carbonic acid is shown below:

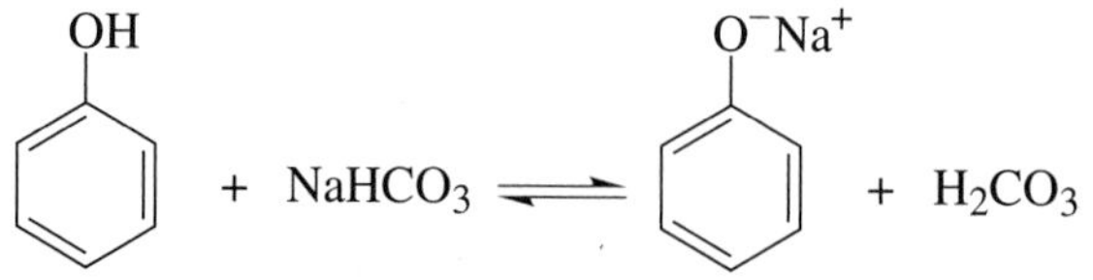

$$\text{Phenol (16)} + NaHCO_3 \rightleftharpoons \text{Sodium phenoxide (18)} + H_2CO_3$$

OH — **16** Phenol; Sodium bicarbonate; O^-Na^+ — **18** Sodium phenoxide; Carbonic acid

a. The pK_as for phenol and carbonic acid are 10.0 and 6.4, respectively. Determine the K_{eq} for this reaction.

b. Based upon your answer to Part **a**, would sodium bicarbonate be a suitable base for separating phenol from a neutral organic compound via an aqueous extraction?

One-Base Extraction

15. What practical consideration makes aqueous hydroxide rather than aqueous bicarbonate (**9**) the preferred base for extracting benzoic acid (**5**) from diethyl ether?

16. Naphthalene (**7**) has a relatively high vapor pressure for a solid (because of its volatility, naphthalene is the active ingredient in some brands of mothballs). In view of this, what might happen if you placed naphthalene under vacuum or in an oven for several hours to dry it?

17. When benzoic acid (**5**) is partitioned between diethyl ether and aqueous sodium hydroxide solution and the aqueous layer is separated, acidification of the aqueous solution yields a precipitate.

a. Using arrows to symbolize the flow of electrons, show the reaction of benzoic acid (**5**) with hydroxide and draw the structure of the product of this reaction.

b. Using arrows to symbolize the flow of electrons, show the reaction of the product of the reaction in Part **a** with aqueous acid and draw the structure of the product of this reaction.

c. Why does the organic product of the reaction in Part **b** precipitate from aqueous solution?

Two-Base Extraction

18. Why does the sequence for extracting the diethyl ether solution of benzoic acid (**5**), naphthalene (**7**), and 2-naphthol (**11**) start with aqueous bicarbonate and follow with aqueous hydroxide rather than the reverse order?

19. What would be the consequence of acidifying the basic extract containing **12** only to pH 10?

20. Why is caution advised when acidifying the bicarbonate solution of sodium benzoate (**6**)?

Acid-Base Extraction

21. Why is anhydrous sodium sulfate added to the organic solution remaining after the extractions with 6 *M* HCl and 6 *M* NaOH?
22. The pK_as of benzoic acid (**5**) and 4-nitroanilinium hydrochloride (**14**) are 4.2 and 1.0, respectively.
    - **a.** Determine the K_{eq} for the reaction of benzoic acid (**5**) with 4-nitroaniline (**13**).
    - **b.** Based upon your answer to Part **a,** do you think a significant amount of salt would form from mixing equimolar amounts of benzoic acid (**5**) and 4-nitroaniline (**13**)? Explain your reasoning.

5.4 EXTRACTION OF A NATURAL PRODUCT: TRIMYRISTIN

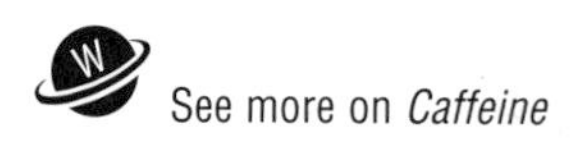

See more on *Caffeine*

Solvent extraction is a valuable method for obtaining a desired substance from its natural source. A familiar example is the hot-water extraction of caffeine (**19**) and the various oils that constitute the flavors of freshly brewed coffee and tea from coffee beans and tea leaves. As contrasted to the **liquid-liquid** extractions described in Section 5.2, this process is an example of **solid-liquid** extraction. The theory underlying it is the same, however. Because most organic compounds we wish to isolate are *insoluble* in water, organic solvents such as diethyl ether, dichloromethane, ethanol, and acetone are used for extracting **natural products,** that is, compounds found in nature.

19

Caffeine

Extraction of natural materials often produces complex mixtures of products, as described in the Historical Highlight at the end of this chapter, so additional operations are required to separate and purify individual compounds. These procedures may involve techniques in which the compounds are separated on the basis of their acidic or basic properties (Sec. 5.3) or by chromatographic methods (Chap. 6). Thus, the isolation of pure natural products normally involves a series of complex and time-consuming operations. Fortunately there are some exceptions to this general rule. The isolation of trimyristin (**20**) from nutmeg is one of them.

Trimyristin, which occurs in many vegetable **fats** and **oils,** is a **triester** of glycerol (**21**) and the **fatty acid** myristic acid (**22**). Most naturally occurring fats and oils are esters of **21** and straight-chain carboxylic acids, the most common of which contain 14 to 20 carbon atoms. In fact, the only difference between a fat and an oil is whether it is a solid or liquid at room temperature. The lack of any double bonds in the carbon chain of myristic acid makes **20** a member of the family of **saturated fats,** which allegedly increase the risk of heart disease if present in the diet in excessive amounts.

See more on *Saturated and Unsaturated Fats*

Nutmeg is a hard, aromatic seed of an East Indian tree (*Myristica fragrans*) that is the source of trimyristin in this experiment. It has been a valued spice ever since its discovery in the Spice Islands of Indonesia by Portuguese sea captains over four centuries ago. The chemical makeup of nutmeg is somewhat unusual in that extraction of the ground seeds with diethyl ether yields **20** in high purity without contamination by other structurally related esters of glycerol and fatty acids. Although the simple procedure described here is *not* typical of that usually required for isolating natural products, it nicely demonstrates the general technique of extracting biological materials from natural sources.

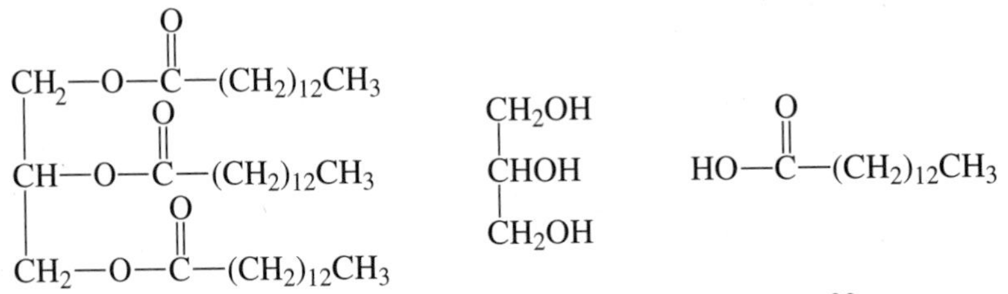

$CH_2{-}O{-}C(=O){-}(CH_2)_{12}CH_3$
$CH{-}O{-}C(=O){-}(CH_2)_{12}CH_3$
$CH_2{-}O{-}C(=O){-}(CH_2)_{12}CH_3$

20
Trimyristin

$CH_2OH{-}CHOH{-}CH_2OH$

21
Glycerol

$HO{-}C(=O){-}(CH_2)_{12}CH_3$

22
Myristic acid

EXPERIMENTAL PROCEDURES

Isolation of Trimyristin from Nutmeg

Purpose To apply the technique of extraction for isolating a natural product.

SAFETY ALERT

1. **Diethyl ether and acetone are both highly volatile and flammable solvents. *Use no flames or electrical dryers in the laboratory during this experiment.***
2. **Monitor the heating of diethyl ether under reflux so that the ring of condensate *remains in the lower third of the condenser.* This will ensure that no vapors escape from the condenser into the room. Very little heat will be required regardless of the heating source used to maintain reflux.**

MINISCALE PROCEDURE

Preparation Sign in at **www.cengage.com/login** to answer Pre-Lab Exercises, access videos, and read the MSDSs for the chemicals used or produced in this procedure. Read or review Sections 2.7, 2.9, 2.10, 2.11, 2.13, 2.17, 2.21, 2.22, 2.29, 3.2, and 3.3.

Apparatus A 25- and a 50-mL round-bottom flask, ice-water bath, apparatus for heating under reflux, magnetic stirring, simple distillation, vacuum filtration, and *flameless* heating.

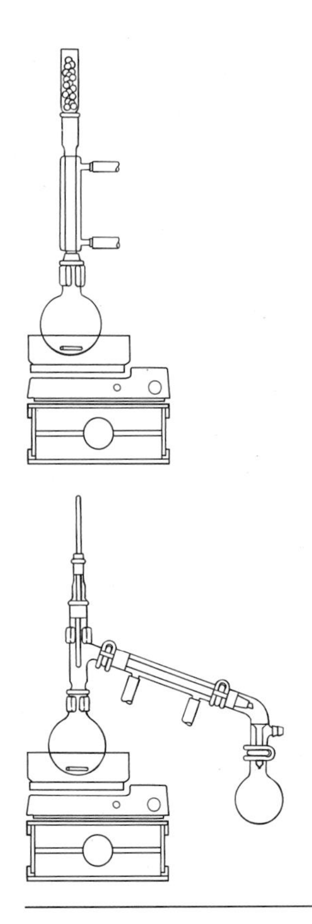

Setting Up Place about 4 g of ground nutmeg in the 50-mL round-bottom flask containing a spinvane and add 10 mL of diethyl ether. If only seeds of nutmeg are available, grind the seeds in a mortar and pestle or a blender. Attach a water-cooled condenser to the flask, and set up the apparatus for heating under reflux (Fig. 2.66a).

Extraction and Isolation Heat the mixture under gentle reflux for approximately 0.5 h. Allow the contents of the flask to cool to room temperature.★ Filter the mixture by gravity filtration through a fluted filter paper into the 25-mL round-bottom flask (Fig. 2.50), rinsing any residue remaining in the 50-mL flask onto the filter paper with an additional 2–4 mL of diethyl ether.

Equip the flask containing the filtrate for simple distillation (Fig. 2.37a), and remove the diethyl ether by distillation. Alternatively, use rotary evaporation or one of the other techniques described in Section 2.29. Dissolve the resulting yellow oil in 3–4 mL of acetone while warming the mixture (*no flames!*). Immediately pour the hot solution into a 25-mL Erlenmeyer flask. Allow the solution to stand at room temperature for about 0.5 h. If crystallization has not occurred during this time, gently scratch the flask at the air/liquid interface.★ Complete crystallization by cooling the flask in an ice-water bath for an additional 15 min. Collect the trimyristin on a Büchner or Hirsch funnel by vacuum filtration (Fig. 2.54). Transfer it to a watchglass or piece of clean filter or weighing paper for air-drying.

Analysis Weigh the trimyristin and determine its melting point. The reported melting point is 55–56 °C. Based on the original weight of nutmeg used, calculate the percent recovery.

WRAPPING IT UP

Once any residual diethyl ether has evaporated from the filter cake, place the used filter paper and its contents in a container for nonhazardous solids. Any diethyl ether that has been isolated should be poured into a container for nonhalogenated organic liquids. Either flush the acetone-containing filtrate down the drain or pour it into a container for nonhalogenated organic liquids.

MICROSCALE PROCEDURE

Preparation Sign in at **www.cengage.com/login** to answer Pre-Lab Exercises, access videos, and read the MSDSs for the chemicals used or produced in this procedure. Read or review Sections 2.5, 2.7, 2.9, 2.10, 2.13, 2.17, 2.21, 2.22, 2.29, 3.2, and 3.3.

Apparatus A 5-mL conical vial or 10-mL round-bottom flask, ice-water bath, Pasteur filter-tip and filtering pipets, tared 5-mL Craig tube, apparatus for heating under reflux, magnetic stirring, and *flameless* heating.

Setting Up Place 1 g of ground nutmeg in the vial or flask equipped for magnetic stirring. Add 2–3 mL of diethyl ether, attach the reflux condenser, and set up the apparatus for heating under reflux (Fig. 2.67).

Extraction and Isolation Heat the heterogeneous mixture for 0.5 h under *gentle* reflux with stirring, being certain not to lose diethyl ether through the top of the condenser.

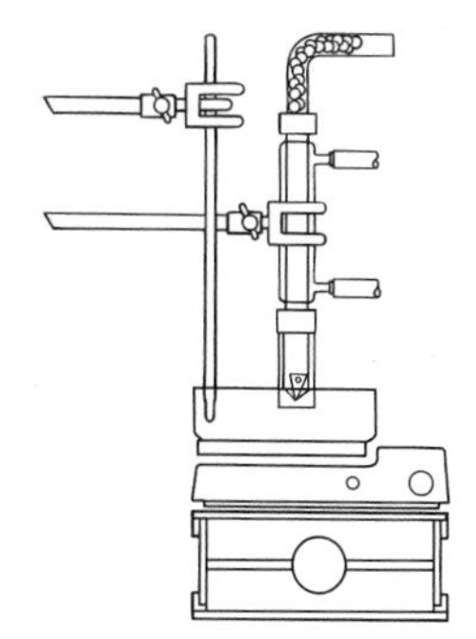

Allow the mixture to cool to room temperature★ and then gravity-filter the ethereal solution into a Craig tube, using Pasteur filter-tip and filtering pipets (Fig. 2.51). Use a filter-tip pipet to rinse the reaction vessel with 0.5–1 mL of technical diethyl ether, and use the pipet to transfer the rinsing to the Craig tube.★

Add a small boiling stone or a boiling stick to the Craig tube and, working at the hood, gently evaporate the diethyl ether using *flameless* heating. If hood space is not available, perform the evaporation under an inverted funnel attached to a water aspirator (Fig. 2.69b). After the diethyl ether has evaporated, add 1 mL of acetone to the residual yellow oil and warm the mixture briefly (*no flames!*) to effect solution. Allow the solution to cool to room temperature and remain at that temperature for about 0.5 h.★ If crystallization has not occurred during this time, gently scratch the tube at the air/liquid interface.

Complete crystallization by cooling the Craig tube in an ice-water bath for an additional 15 min, and then isolate the product by Craig tube filtration (Fig. 2.55). Carefully disassemble the Craig tube, scrape any crystalline material clinging to the glass or Teflon plug into the tube, and air-dry the product.

Analysis Weigh the trimyristin and determine its melting point. The reported melting point is 55–56 °C. Based on the original weight of nutmeg used, calculate the percent recovery.

WRAPPING IT UP

Either flush the *acetone-containing filtrate* down the drain or pour it into a container for nonhalogenated organic liquids.

EXERCISES

1. Why is diethyl ether rather than acetone chosen as the extraction solvent?
2. Why is finely divided rather than whole nutmeg specified in this extraction?
3. What are the structural features, functional groups, and physical properties of trimyristin that place it in the class of natural products known as saturated fats?
4. Trimyristin is only one component of nutmeg. How are the other components of nutmeg removed in this experiment?
5. Considering the solubility properties of trimyristin (**20**), would water be a good choice as an extracting solvent for obtaining **20** from nutmeg?
6. Judging from your examination of trimyristin (**20**), is this substance responsible for the odor of nutmeg?
7. Why is ether added to the nutmeg and then removed at the end of the experiment?
8. What is the purpose of adding acetone to the crude product?
9. Indicate which of the following are reasons for performing a simple distillation in this procedure (more than one answer may be correct):
   a. To obtain pure diethyl ether.
   b. To isolate the nonvolatile solute.
   c. To isolate a volatile impurity.

d. To make extra work for the student.

e. To remove the solvent from the solute.

10. A certain plant material is known to contain mainly trimyristin and tripalmitin in approximately equal amounts. Tripalmitin, the structure of which you should look up in a reference source, has mp 66–67 °C. Extraction of these two compounds from the plant material with diethyl ether gave an oil after removal of the solvent, and this oil was difficult to crystallize. Explain this result. (*Hint:* see Sec. 3.3.)

11. In the Miniscale Procedure,

a. why is gravity rather than vacuum filtration used to separate the ethereal extract from the residual nutmeg?

b. why is fluted rather than plain filter paper used for the filtration procedure?

12. In the Microscale Procedure,

a. why is it important that a gentle rather than vigorous reflux rate be used?

b. why is it preferred to transfer the ethereal extract with a Pasteur filter-tip pipet rather than a regular Pasteur pipet?

HISTORICAL HIGHLIGHT

Natural Products

The naturally occurring organic compounds produced by living organisms have fascinated human beings for centuries. Indeed, traditional folk medicines from many cultures are effective because individual organic compounds that occur in the native plants exhibit biological activities that may be used to treat various maladies. Compounds of nature have been widely used in modern medicine as analgesics to relieve pain and as antibacterial, antiviral, anticancer, and cholesterol-lowering agents to treat or cure numerous ailments and diseases. On the other hand, some plants produce compounds that are toxic and may even cause death. For example, Socrates drank a fatal potion prepared from an extract of hemlock. Other plants have attracted attention because of the flavors they possessed or the odors they released. Organic compounds found in these plants are often incorporated as additives in the food and perfume industries. Some natural products, such as those found in flowering plants, are highly colored, and these may be used to prepare the dyes that color paints and fabrics.

However, natural products have not only drawn attention because of their practical applications in our daily lives; they are also important for the various scientific challenges they present. Such challenges include understanding the basis of their sensory and biological properties and developing means for their chemical synthesis. Indeed, natural products provided chemists with many interesting experimental problems during the period when chemistry was emerging from alchemy into a more exact science. Today, the isolation, identification, and synthesis of natural products, especially those having interesting and useful biological activity, remain important areas of research for modern organic chemists.

Historically, most natural products were extracted from plants rather than from animals, but more recently, microorganisms are assuming increased importance as sources of natural products, particularly since the advent of molecular biology. The isolation of a natural product in pure form normally represents a significant challenge, mainly because even the simplest plants and microorganisms contain a multitude of organic compounds, often in only minute amounts. For example, a kilogram of plant material may only yield milligram quantities, or even less, of a particular natural

(Continued)

HISTORICAL HIGHLIGHT *Natural Products (Continued)*

product. The general approach used to isolate and purify a natural product is labor-intensive and typically starts with grinding a plant or other organism into fine particles. The resulting material is then extracted with a solvent or a mixture of solvents. Volatile natural products in the extract can be detected and isolated by gas chromatography. However, many natural products are relatively nonvolatile, and removing the solvent used for the extraction yields an oil or gum that requires further separation by chromatographic methods to obtain the various components in pure form. Often a series of chromatographic separations are required before the individual compounds are pure.

The next step is to determine the structure of the isolated products. Traditional approaches to structural elucidation involved performing standard qualitative tests for the various functional groups to identify those present in the molecule (Chap. 24). The compound was then subjected to a series of simple chemical degradations to obtain simpler compounds. If a known compound was formed as a result of these transformations, it was then possible to work backward and deduce the structure of the original natural substance. More recently, instrumental techniques including infrared and nuclear magnetic resonance spectroscopy (Chap. 8) and mass spectrometry have greatly facilitated the determination of structures. However, the most reliable means of determining the structure of an unknown substance is by X-ray diffraction. In this technique, a crystal of the sample is irradiated with X-rays to provide a three-dimensional map of the positions of each atom within the molecule. Applying this method requires that the unknown itself, or a suitable derivative of it, be a crystalline solid, and this is not always possible.

The final goal of organic chemists working in this exciting field is often the laboratory synthesis of the natural product. While the total synthesis of a natural product may represent mainly an intellectual challenge, more often the undertaking is a unique opportunity to develop and demonstrate the utility of new synthetic techniques in organic chemistry. In cases where the natural product has medicinal applications, the development of an efficient synthetic route may be of great importance, especially when the supply of the material from the natural source is severely limited or its isolation is difficult. Through synthesis, it is often possible to prepare derivatives of the natural product that possess improved biological properties.

There are a number of classes of natural products, and a detailed discussion of these compounds may be found in your lecture textbook. Representative compounds from the different classes range from the simple, as in the case of formic acid, HCO_2H, the irritating ingredient in ant venom, to the extraordinarily complex, as engendered in DNA, for example. The examples of reserpine (1), Taxol® (2), and calicheamicin γ_1^I (3) are illustrative of the diversity and complexity of the molecules in nature that have inspired the research efforts of contemporary organic chemists and tested their creativity.

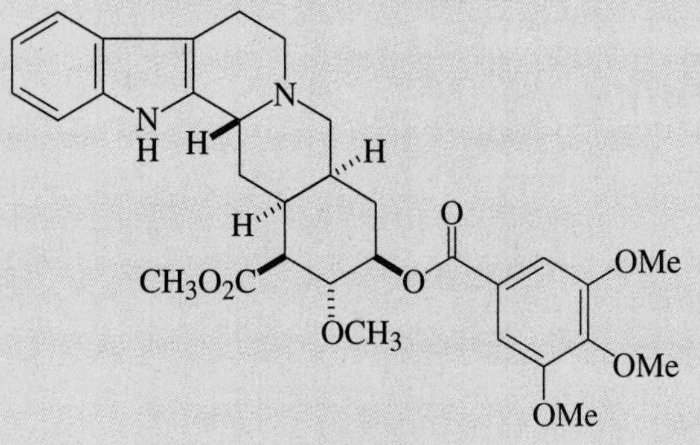

1
Reserpine

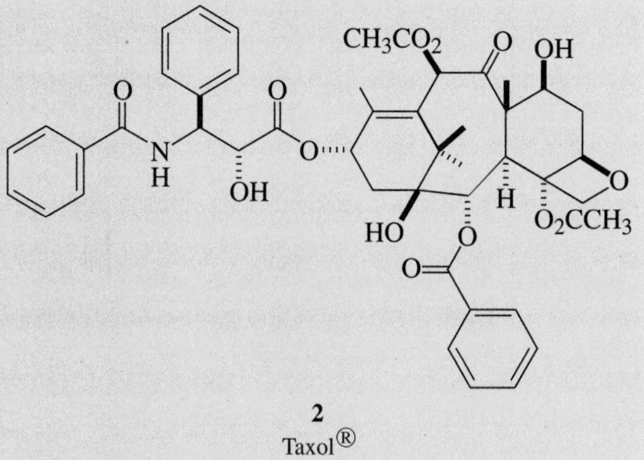

2
Taxol®

(Continued)

HISTORICAL HIGHLIGHT *Natural Products (Continued)*

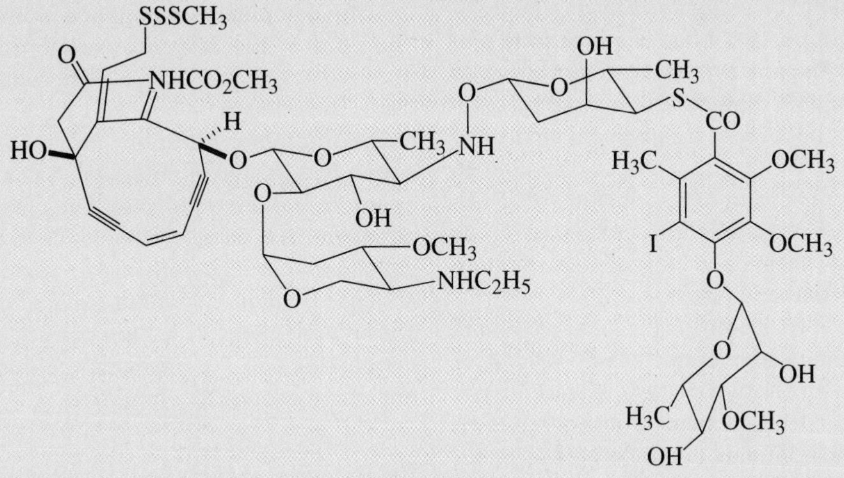

3

Calicheamicin γ_1^I

Reserpine (1) is an alkaloid that was isolated from the Indian snakeroot *Rauwolfia serpentina* Benth in 1952. This alkaloid has a potent effect upon the central nervous system, where it depletes the postganglionic adrenergic neurons of the neurotransmitter norepinephrine. This novel mode of action was responsible for its clinical use as an agent for treating high blood pressure. Recently, however, antihypertensive drugs with improved pharmacological profiles have replaced reserpine as a drug. It is normally more economical to isolate complex natural products from their source than it is to prepare them by *total synthesis*. Thus, it is noteworthy that the reserpine that was used for medical purposes was prepared by total synthesis using slight modifications of the original route that was developed by Professor Robert B. Woodward of Harvard University in 1956. A number of research groups subsequently prepared reserpine by innovative routes.

Taxol (2), also now known as paclitaxel, was isolated from the bark of the Pacific yew tree, *Taxus brevifolia*, in 1962 as part of an extensive program directed toward the discovery of novel anticancer agents that was conducted jointly by the Department of Agriculture and the National Cancer Institute. Because extracts isolated from the bark of this tree were observed to exhibit potent cytotoxic activity, there was an intense effort to identify the active principle, which was isolated in only minute quantities. Finally, in 1971, the structure of taxol, a complex diterpene, was elucidated by X-ray crystallography.

The clinical use of taxol under the registered trademark Taxol® was initially hampered by its low solubility, but its scarcity rendered its use even more problematic. For example, it was estimated that the bark of a 100-year-old yew tree would yield only enough taxol for a *single dose* for a cancer patient. Stripping the bark from a tree is a death sentence for the tree, and environmentalists became extremely concerned that the quest for sufficient amounts of taxol to treat needy cancer patients would destroy the ancient forests of the Pacific northwest, which are home to the endangered spotted owl. Fortunately for humans and spotted owls alike, a related diterpene was found in the needles and twigs, which are more readily renewable than the tree itself, of the European yew tree, *Taxus baccata*. This compound may be converted into taxol using organic reactions related to those you learn in the lecture portion of this course. Consequently, the drug taxol is now available in sufficient quantities for treating a number of cancers including ovarian, breast, and lung. The importance of this natural product as an anticancer drug is reflected in its annual sales, which presently exceed one billion dollars!

The biological importance of taxol coupled with its unusual and complex structure served as the

(Continued)

HISTORICAL HIGHLIGHT *Natural Products (Continued)*

stimulus for synthetic investigations by a large number of research groups, and it has been recently prepared by several elegant syntheses that start with commercially available materials unrelated to taxol itself. These efforts uncovered a significant amount of novel and useful chemistry that may be applied to other important problems in organic chemistry. However, they did not provide a *practical* synthesis of taxol; this goal remains a challenge to the ingenuity of future organic chemists—perhaps you.

Calicheamicin γ_1^I (3) is perhaps the most prominent member of a recently discovered class of anticancer agents that contain an enediyne moiety, which consists of a double and two triple bonds. Its discovery is an interesting story that typifies the ongoing search for medicinally active compounds in nature. Employees of many pharmaceutical companies make it a practice to collect soil samples during their travels around the world. When these scientists return, these samples are placed in different growth cultures to see whether the microorganisms living in the soil produce compounds with interesting biological activity. One such sample was collected by a scientist from some chalky rocks, which are commonly called caliche, along a highway in Central Texas. Bacteria (*Micromonospora echinospora* ssp. calichenis) lived within these rocks and produced calicheamicin and other related compounds upon cultivation in laboratory cultures.

The complexity of calicheamicin γ_1^I is awesome. The core of the bicyclic skeleton contains a constrained enediyne array, whose reactivity and DNA-cleaving ability is unleashed by the cellular cleavage of the trisulfide. Although the exact mechanism of action of 3 in vivo is unknown, it is a highly efficient agent for inducing the selective cleavage of double-stranded DNA, an event that leads to cell death. While calicheamicin γ_1^I itself is too toxic for use as an anticancer agent in humans, related compounds possessing enediyne arrays are currently attractive drug candidates and are involved in clinical trials at several pharmaceutical companies.

Relationship of Historical Highlight to Experiments

In one of the experiments in this chapter, you isolate the naturally occurring fat trimyristin from nutmeg, a valuable spice. In other experiments in this textbook, you have the opportunity to isolate citral, the odoriferous component of lemongrass oil (Sec. 4.6), and to separate different pigments found in green leaves (Sec. 6.2). You may also examine the two enantiomers of carvone, one of which is found in spearmint and the other in caraway seed oil (Sec. 7.4). The techniques you will use in these experiments serve as an introduction to the methods that are used in research laboratories to isolate natural products having potential value to society.

See more on *Paclitaxel*

See more on *Reserpine*

See more on *Calicheamicin*

See *Who was Socrates?*

See *Who was Woodward?*

CHAPTER 6

Chromatography

When you see this icon, sign in at this book's premium website at **www.cengage.com/login** to access videos, Pre-Lab Exercises, and other online resources.

Oh, no, you forgot to put the cap on your ballpoint pen, so you now have a big inkspot on the pocket of your best shirt as a "reward" for your negligence. You dab it with a wet cloth, only to see the spot grow even bigger. New colors seem to be appearing as the spot widens. Congratulations! You've just performed a form of chromatography in which you have crudely separated some of the dyes that comprise the ink. In this chapter, we'll be learning some of the basic chromatographic techniques that are used to separate mixtures by taking advantage of the differential distribution of the individual compounds between two immiscible phases—in the case of your shirt, these were its fibers and the water you used in an effort to remove the stain. These techniques are now so powerful and sophisticated that even enantiomers (Sec. 7.1) may be separated efficiently by such means.

6.1 INTRODUCTION

We described the common laboratory techniques of recrystallization, distillation, and extraction for purifying organic compounds in Chapters 3 through 5. In many cases, however, the mixtures of products obtained from chemical reactions do not lend themselves to ready separation by any of these techniques because the physical properties of the individual components are too similar. Fortunately, there are a number of chromatographic procedures available that we can use to effect the desired purification, and some of them are described in this chapter.

The word **chromatography** was first used to describe the colored bands observed when a solution containing plant pigments is passed through a glass column containing an adsorbent packing material. From that origin, the term now encompasses a variety of separation techniques that are widely used for analytical and preparative purposes.

See more on *Chromatography*

See more on *Chromatography/Nobel Prize*

All methods of chromatography operate on the principle that the components of a mixture will distribute unequally between two immiscible phases, which is also the basis for separations by extraction (Chap. 5). The **mobile phase** is generally a liquid or a gas that flows continuously over the fixed **stationary phase,** which may be a solid or a liquid. The individual components of the mixture have different affinities for the mobile and stationary phases, so a dynamic equilibrium is established in which each component is selectively, but temporarily, removed from the mobile phase by binding to the stationary phase. When the equilibrium concentration of that substance in the mobile phase decreases, it is released from the stationary phase and the process continues. Since each component partitions between the two phases with a different equilibrium constant or **partition coefficient,** the components divide

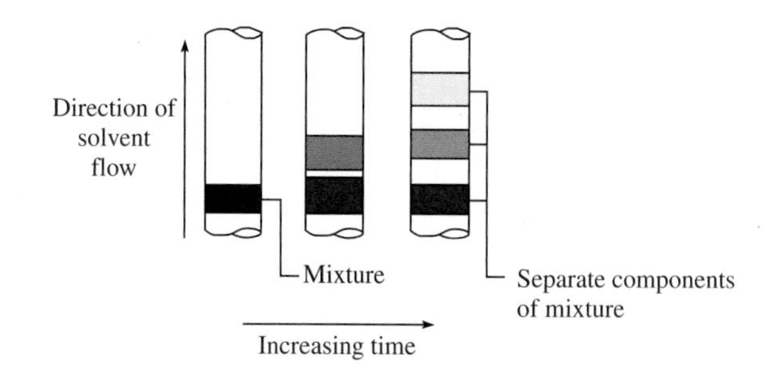

Figure 6.1
Separation of mixture by chromatography.

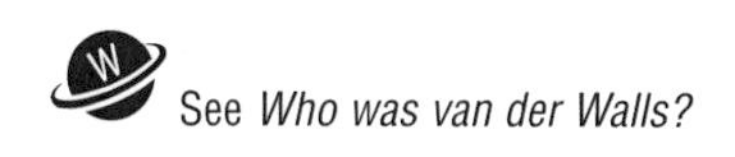
See *Who was van der Walls?*

See more on *HPLC*

into separate regions termed **migratory bands** (Fig. 6.1). *The component that interacts with or binds more strongly to the stationary phase moves more slowly in the direction of the flow of the mobile phase.* The attractive forces that are involved in this selective adsorption are the same forces that cause attractive interactions between any two molecules: electrostatic and dipole-dipole interactions, hydrogen bonding, complexation, and van der Waals forces.

The chromatographic methods used by modern chemists to identify and/or purify components of a mixture are characterized by the nature of the mobile and stationary phases. For example, the techniques of **thin-layer** (TLC), **column,** and **high-performance** (or **high-pressure**) **liquid chromatography** (HPLC) each involve *liquid-solid* phase interactions. **Gas-liquid partition chromatography** (GLC), also known as **gas chromatography** (GC), involves distributions between a mobile *gas* phase and a stationary *liquid* phase coated on a solid support. These important techniques can be used as tools to analyze and identify the components in a mixture as well as to separate the mixture into its pure components for preparative purposes. Although there are other chromatographic techniques, such as ion exchange and paper chromatography, a review of those methods is beyond the scope of this discussion.

6.2 THIN-LAYER CHROMATOGRAPHY

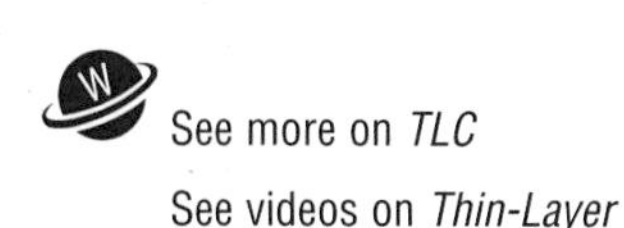
See more on *TLC*

See videos on *Thin-Layer Chromatography*

Thin-layer chromatography (TLC) is a form of **solid-liquid adsorption** chromatography and is an important technique in organic chemistry for rapid analysis of small quantities of samples, sometimes as little as 10^{-9} g. Thus, TLC is frequently used to monitor the progress of reactions and of preparative column chromatographic separations as well as to determine the optimal combinations of solvent and adsorbent for such separations (Sec. 6.3). An important limitation to this technique, and that of column chromatography as well, is that it cannot be used on volatile compounds having boiling points below about 150 °C (760 torr).

To execute a TLC analysis, a *small* amount of the sample to be analyzed, or a solution of it, is first applied to a solid **adsorbent** bound to a rectangular glass or plastic plate (Fig. 6.2a). The adsorbent serves as the stationary phase. Next, the plate, with its spotted end down, is placed in a closed jar, called a **developing chamber** (Fig. 6.3). The chamber contains a *saturated atmosphere* of a suitable **eluant** or **eluting solvent,** which is the mobile phase and may be comprised of either a single solvent or mixture of two or more. A folded filter paper is often used to help maintain solvent equilibration in the chamber. It is important that the level of

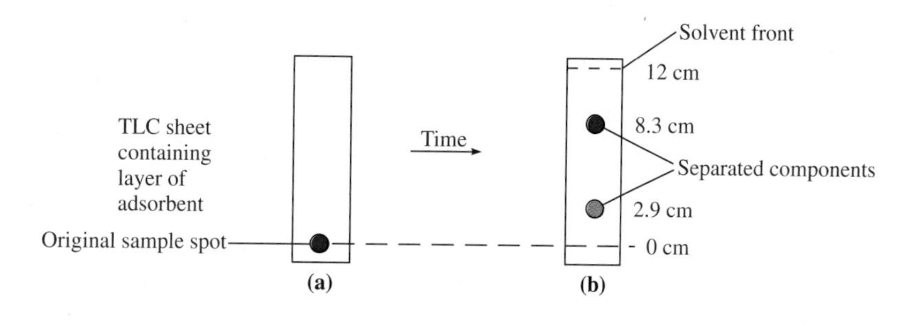

Figure 6.2
Thin-layer chromatography. (a) Original plate loaded with sample. (b) Developed chromatogram.

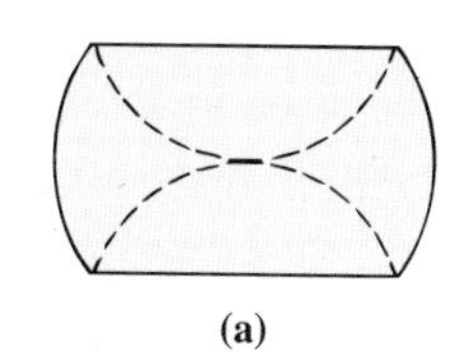

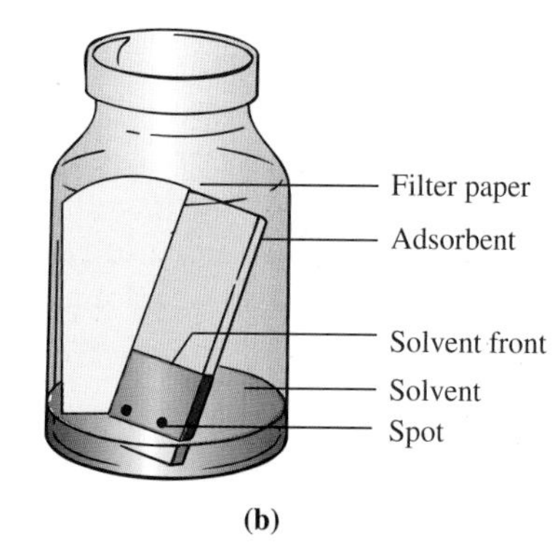

Figure 6.3
TLC chamber. (a) Folded filter paper to be placed in developing chamber for solvent equilibration. (b) Developing chamber with filter paper and TLC plate.

solvent in the chamber be *below* the location of the spot on the TLC plate. Otherwise, the sample would be removed from the plate by dissolution in the bulk solvent, thereby ruining the analysis.

When the plate is placed in the chamber, the solvent begins to ascend the plate. The individual components of the sample, which initially were at the bottom, or "origin," are carried up the plate along with the eluting solvent at a rate that is dependent on their relative affinities for the solid adsorbent: more weakly adsorbed compounds move up the plate faster than do those that are more strongly adsorbed. Ideally, when the solvent front has nearly reached the top of the TLC plate, which normally takes only a few minutes, each component of the original sample will appear as a separate spot on the plate (Fig. 6.2b). This completes the development of the chromatogram, and the plate is then removed from the developing chamber for analysis.

Adsorbents

The solid adsorbent in TLC is usually alumina (Al_2O_3) or silica gel (silicic acid, $SiO_2 \times H_2O$), both of which are polar. Alumina is the more polar of the two and is commercially available in three forms: neutral, acidic, and basic. Acidic and basic alumina are sometimes used to separate basic and acidic compounds, respectively, but neutral alumina is the most common form of this adsorbent for TLC. Silica gel, which is slightly acidic, is the adsorbent used in the experimental procedures described in this section.

As you might imagine, consideration of the acidic or basic character of the solid adsorbent used for a TLC experiment can be particularly important if the substances to be analyzed contain functional groups that are sensitive to acids or bases. In a worst-case scenario, the adsorbent may function as a catalyst to destroy the functionality by chemical reaction during the course of the analysis; this greatly complicates the interpretation of the TLC results.

Although TLC plates may be prepared in the laboratory, their ready commercial availability generally makes this unnecessary. The plates are produced by mixing the adsorbent with a small quantity of a **binder** such as starch or calcium sulfate and spreading the mixture as a layer approximately 250 μ thick on the supporting plate. The binder is needed for proper adhesion of the thin layer of adsorbent to the plate. TLC plates should be dried in an oven for an hour or more at 110 °C prior to use, to remove any atmospheric moisture adsorbed on them. This is necessary because the activity of the adsorbent and its effectiveness in separating the components of a mixture are decreased because water occupies binding sites on the surface of the solid.

The strength of adsorption of organic compounds to the stationary phase of the TLC plate depends on the polarity and nature of the adsorbent and the type of functional groups present in the compounds. Substances containing carboxyl groups and other polar functional groups are more strongly adsorbed than are those containing less polar moieties, such as those present in alkenes and alkyl halides, as the **elutropic series** shown in Figure 6.4 indicates. When a sample contains compounds having highly polar functionalities, more polar eluants may be required to effect a TLC analysis, as discussed next.

Eluant

Selecting the eluant or the mobile liquid phase is an important decision that must be made in planning a TLC analysis. The best solvent or combination of solvents is determined by trial and error, so several TLC plates may need to be prepared and developed using different eluants to determine the optimal conditions.

Certain criteria guide the selection of the eluant. For example, an effective eluting solvent must readily dissolve the solute but not compete with it for binding sites on the stationary phase. If the mixture to be separated is not soluble in the solvent, the individual components may remain adsorbed at the origin. Another criterion is that a solvent should not be too polar because it may bind strongly to the adsorbent and force the solute to remain in the mobile phase. In such circumstances, the components will move rapidly up the TLC plate, offering little opportunity to establish the solid-liquid equilibria required for separation. Consequently, the eluting solvent must be significantly *less* polar than various components of the mixture to obtain an effective separation. As a rule, the relative ability of different solvents to move a given substance up a TLC plate is termed **eluting power** and is generally found to follow the order shown in Figure 6.5.

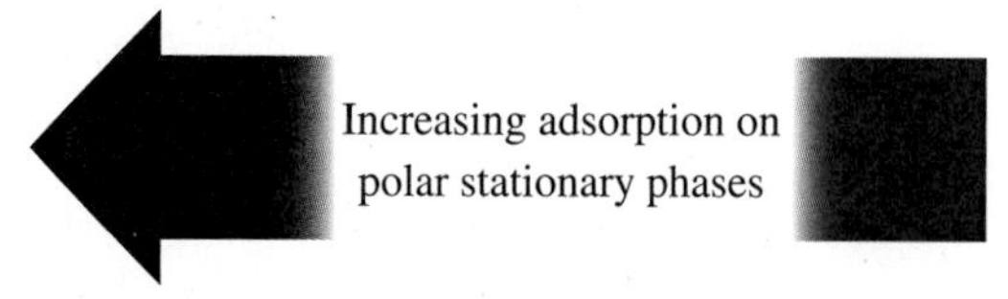

$$RCO_2H > ROH > RNH_2 > R^1R^2C{=}O > R^1CO_2R_2 > OCH_3 > RR^1R^2C{=}CR^3R^4 > RHal$$

Figure 6.4
Elutropic series for polar stationary phases.

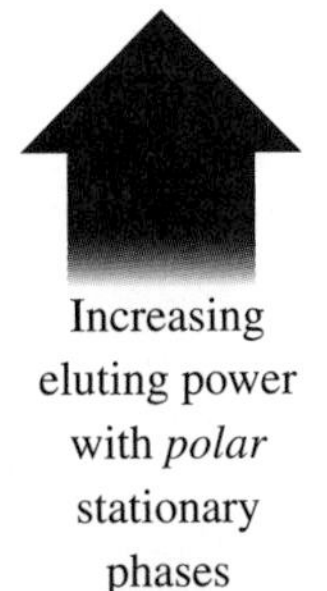

Water
Methanol
Ethanol
1-Propanol
Acetone
Ethyl acetate
Diethyl ether
Chloroform
Dichloromethane
Toluene
Hexane
Petroleum ether

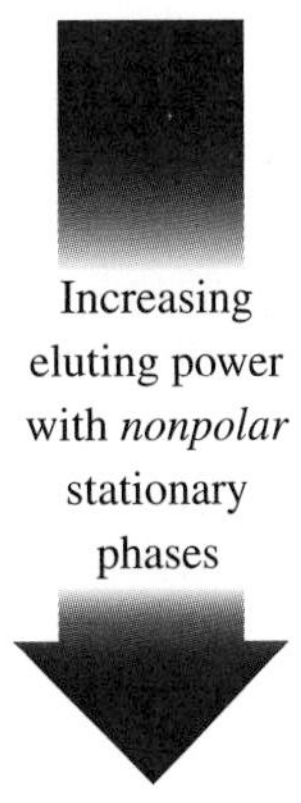

Figure 6.5
Eluting power of solvents as function of polarity of stationary phases.

Experimental Technique

The general protocol for preparing and developing a TLC plate was described earlier. We have not yet discussed the means for detecting or "visualizing" the separated components of a mixture in a TLC analysis. There are several ways of doing this. The easiest situation is that in which the compounds being separated are colored, so visual detection is easy, a "no-brainer," so to speak. Many organic compounds are colorless, however, and a variety of methods have been developed to detect their presence on the plate:

See more on *Fluorescence*

1. Compounds that **fluoresce** may be located by placing the plate under an ultraviolet light. Since the spots disappear when the light is removed, it is necessary to circle the spots with a pencil in order to have a permanent record of the chromatogram. There are also commercially available plates that contain a fluorescent material as part of their coating; compounds that do not fluoresce but do absorb ultraviolet light then appear as dark spots under ultraviolet light.
2. The chromatographic plate may be sprayed with a variety of reagents such as sulfuric acid, potassium permanganate, phosphomolybdic acid, and ninhydrin; these reagents react with the individual components to produce colored or dark spots.
3. The chromatographic plate may be exposed to iodine vapor by placing it in a closed chamber containing several crystals of iodine. As the iodine forms complexes with the various organic compounds, the spots become brown. Since the process is reversible and the spots fade, it is necessary to circle the spots with a pencil in order to have a permanent record of the chromatogram.

Once the separation of the components of the mixture is complete and the individual spots have been detected, the **retention factor** (R_f) of each compound may be calculated as shown below for the chromatogram pictured in Figure 6.2b:

$$R_f = \frac{\text{distance traveled by substance}}{\text{distance traveled by solvent}}$$

$$R_f(\text{compound 1}) = \frac{2.9\ \text{cm}}{12\ \text{cm}} = 0.24$$

$$R_f(\text{compound 2}) = \frac{8.3\ \text{cm}}{12\ \text{cm}} = 0.69$$

The R_f-value for a compound is a physical constant for a given set of chromatographic conditions, so the adsorbent and the eluting solvent should be recorded along with the experimentally determined R_f-values.

There are many important applications of TLC in modern organic chemistry. For example, TLC is commonly used to identify components of an unknown mixture by running chromatograms of the unknown sample side by side with known standards. Multiple aliquots of samples collected from chromatographic columns (Sec. 6.3) may be analyzed by TLC to follow the chromatographic separation. Alternatively, it is possible to follow the progress of a reaction by TLC by monitoring the disappearance of starting material or the appearance of product. Samples are simply withdrawn from a reaction mixture and subjected to TLC analysis.

Applications

Two experiments are presented here to demonstrate the TLC technique. The first involves the separation of the pigments present in spinach leaves. A variety of other sources, including crushed tomato pulp or carrot scrapings, as well as leaves from grasses, shrubs, and trees, may be substituted; however, *waxy* leaves are not acceptable.

In the second experiment, the **diastereomers** (geometric isomers) (Sec. 7.1) *syn*-azobenzene (**1**) and *anti*-azobenzene (**2**) are separated. Commercially available azobenzene consists predominantly of the more-stable *anti-* form, but this isomer may be isomerized to the less-stable *syn-* isomer by irradiation with ultraviolet light or sunlight. Since the colors of the two isomers differ, they may be detected visually. The course of the reaction and the effectiveness of the irradiation is followed by placing spots of irradiated and nonirradiated samples side by side on the TLC strip.

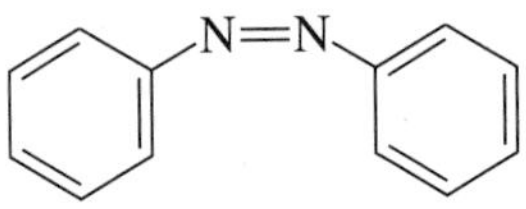

1
Syn-azobenzene

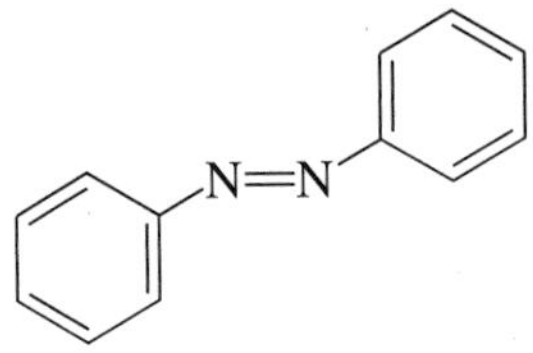

2
Anti-azobenzene

EXPERIMENTAL PROCEDURES

Separation of Spinach Pigments by TLC

Purpose To identify solvent mixtures that will separate the colored components in spinach leaves using thin-layer chromatography.

SAFETY ALERT

Petroleum ether, ethanol, and acetone are highly volatile and flammable solvents. Be certain there are *no flames* in the vicinity during this experiment.

Procedure

Preparation Sign in at **www.cengage.com/login** to answer Pre-Lab Exercises, access videos, and read the MSDSs for the chemicals used or produced in this procedure. Read or review Section 2.21.

Apparatus A wide-mouth bottle with a screw-top cap, small mortar and pestle, 2-cm × 10-cm silica gel TLC plates, and capillary pipets.

Setting Up Prepare a developing chamber by placing a folded filter paper lengthwise in a wide-mouth bottle (Fig. 6.3). As directed by your instructor or by working in teams, prepare several mixtures of eluants that contain different ratios of varying pairs of the following solvents: petroleum ether, bp 60–80 °C (760 torr), chloroform, acetone, and ethanol. For example, prepare 10 mL of a 70:30 mixture of petroleum ether and acetone to use as one eluant. Add an amount of the eluant to the developing chamber so that it forms a 1-cm layer on the bottom of the container. Screw the cap onto the bottle *tightly,* and *shake* the container *well* to saturate the atmosphere of the chamber with vapors of the solvent.

Preparing the Sample Using a small mortar and pestle, thoroughly grind a spinach leaf in a mixture of 4 mL of petroleum ether and 2 mL of ethanol. Transfer the liquid extract to a test tube using a Pasteur pipet and *swirl* the extract gently with an equal volume of water. Do not shake the test tube, because emulsions (Sec. 2.21) are easily formed, and these can be a source of great frustration to you. Remove and discard the aqueous layer; if you do not know which layer is the aqueous one, perform the necessary test (Sec. 2.21). Wash the organic layer with water two more times to remove the ethanol and any water-soluble materials that are present in the leaves. Transfer the petroleum ether extract to the Erlenmeyer flask and add several spatula-tips full of anhydrous sodium sulfate. After 5–10 min, decant the solution from the drying agent. If the solution is not deeply colored, concentrate it using a gentle stream of air or nitrogen to remove some of the solvent.

Preparing and Developing a Plate Obtain a 2-cm × 10-cm strip of silica gel chromatogram sheet *without* a fluorescent indicator. Handle the strip *only* by the sides to avoid contaminating the plate with oils from your hands. Place a pencil dot in the middle of the sheet about 1 cm from one end. Using a capillary pipet, apply a spot of pigment solution over the pencil dot by *lightly* and *briefly* applying the tip of the pipet to the surface of the plate; you may blow gently on the plate as the sample is applied. Do not allow the spot to diffuse to a diameter of more than 1–2 mm during application of the sample. Perform the spotting process an additional four or five times, allowing the solvent of each drop to evaporate before adding the next. When the spot has thoroughly dried, place the strip in the developing chamber containing the first eluant; be careful not to splash solvent onto the plate. The spot *must be above* the solvent level. Allow the solvent front to move to within 2–3 mm of the top of the strip and then remove the strip. Mark the position of the solvent front with a pencil, and allow the plate to air-dry. Repeat this process with other eluants as directed by your instructor.

Analysis A good separation will reveal as many as eight colored spots. These are the carotenes (orange), chlorophyll *a* (blue-green), the xanthophylls (yellow), and chlorophyll *b* (green). Calculate the R_f-values of all spots on the plate(s) you developed. Compile these data, together with those of others if you worked as a team, and determine which solvent mixture provided the best and which the worst separations. Draw a picture to scale of the plates you developed that showed the best and worst separations and include them in your notebook as a permanent record.

Discovery Experiment

Effect of Solvent Polarity on Efficiency of Separation

Explore the possible consequences of using only petroleum ether, a nonpolar solvent, for extracting the pigments from spinach leaves. For comparative purposes, use the same solvent mixture for developing the TLC plate as you did in the original procedure.

Discovery Experiment

Analysis of Plant Pigments from Various Sources

Investigate the distribution of pigments obtained from other sources of green leaves, such as beets, chard, trees, and shrubs. Avoid the use of waxy leaves, as their coating makes extraction and isolation of the pigments more difficult.

WRAPPING IT UP

Put the unused eluants containing only *petroleum ether, acetone,* and *ethanol* in the container for nonhalogenated organic liquids and any unused eluants containing *chloroform* in the container for halogenated organic liquids. Discard the dry chromatographic plates in the nonhazardous solid waste container.

EXPERIMENTAL PROCEDURES

Separation of Syn- *and* Anti-*Azobenzenes by TLC*

Purpose To identify solvent mixtures that will separate *syn-* and *anti*-azobenzenes using thin-layer chromatography.

SAFETY ALERT

1. **Petroleum ether, ethanol, and acetone are highly volatile and flammable solvents. Be certain there are *no flames* in the vicinity during this experiment.**
2. **Since azobenzene is a suspected carcinogen, avoid contacting it with your skin or ingesting it.**

Procedure

Preparation Sign in at **www.cengage.com/login** to answer Pre-Lab Exercises, access videos, and read the MSDSs for the chemicals used or produced in this procedure.

Apparatus A wide-mouth bottle with a screw-top cap, 3-cm × 10-cm silica gel TLC plates, and a capillary pipet.

Setting Up Prepare a developing chamber by placing a folded filter paper lengthwise in a wide-mouth bottle (Fig. 6.3). As directed by your instructor, prepare three or four mixtures of eluants that contain different ratios of varying pairs of the following solvents: petroleum ether, bp 60–80 °C (760 torr), chloroform, acetone, and ethanol. For example, prepare 10 mL of a 90:10 mixture of petroleum ether and chloroform to use as one eluant. Add an amount of the eluant to the developing chamber so that it forms a 1-cm layer on the bottom of the container. Screw the cap onto the bottle *tightly,* and *shake* the container *well* to saturate the atmosphere of the chamber with vapors of the solvent.

Preparing and Developing a Plate Obtain three or four 3-cm × 10-cm strips of silica gel chromatogram sheets *without* a fluorescent indicator. Handle the strip *only* by the sides in order to avoid contaminating the plate with oils from your hands. Place one pencil dot about 1 cm from the left side and about 1 cm from one end of one sheet and another about 1 cm from the right side the same distance from the bottom as the first. Using a capillary pipet, carefully apply a *small* spot of a 10% solution of commercial azobenzene in toluene, which you should obtain from your instructor, over one of the pencil dots. Do not allow the spot to diffuse to a diameter of more than 1–2 mm during application of the sample. Repeat this process for each strip. Allow the spots to dry and then expose the plates to sunlight for one to two hours (or a sunlamp for about 20 min).

When the irradiation is complete, apply another spot of the *original* solution on the plate over the second pencil dot in the same manner as just described and allow each strip to dry. Place a strip in the developing chamber, being careful not to splash solvent onto the plate. Both spots *must be above* the solvent level. Allow the solvent to move to within approximately 2–3 mm of the top of the strip and then remove the strip. Repeat this process for each additional strip using a different eluting solvent as directed by your instructor. Mark the position of the solvent front with a pencil, and allow the plate to air-dry.

Analysis Note the number of spots arising from each of the two original spots. Pay particular attention to the relative intensities of the two spots nearest the starting point in each of the samples; these are *syn*-azobenzenes. Calculate the R_f-values of each of the spots on your developed plate. In your notebook, include a picture of the developed plate drawn to scale as a permanent record. Identify the solvent mixture that gave the best separation of *syn*- and *anti*-azobenzene.

Discovery Experiment

Analysis of Analgesics by TLC

See more on *Acetaminophen*

See more on *Caffeine*

See more on *Excedrin*

See more on *Tylenol*

Design and execute an experimental procedure for testing over-the-counter analgesics such as Excedrin™, Tylenol™, and Advil™ for the presence of caffeine (**3**) and/or acetaminophen (**4**). A 50:50 (v:v) mixture of ethanol and dichloromethane can be used to extract the active ingredients.

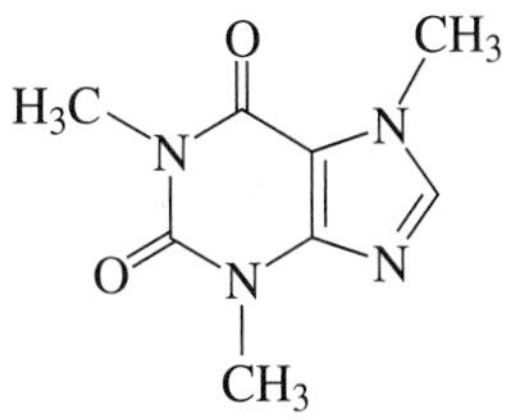

H_3C, O, CH_3, N, N, O, N, N, CH_3

3

Caffeine

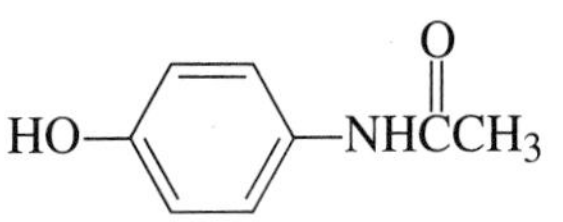

HO—C_6H_4—NHC(O)CH_3

4

Acetaminophen

WRAPPING IT UP

Put the *unused eluants* containing mixtures of only *petroleum ether, acetone,* and *ethanol* in the container for nonhalogenated organic liquids and any unused mixtures containing *chloroform* in the container for halogenated organic liquids. Put the dry chromatographic plates in the hazardous solid waste container, since they contain small amounts of azobenzene.

EXERCISES

1. Explain why TLC is not suitable for use with compounds that have boiling points below about 150 °C (760 torr).
2. What may occur if a mixture containing a component that is very sensitive to acidic conditions is subjected to a TLC analysis in which silica gel serves as the stationary phase?
3. In a TLC experiment, why should the spot not be immersed in the solvent in the developing chamber?
4. Explain why the solvent must not be allowed to evaporate from the plate during development.
5. Explain why the diameter of the spot should be as small as possible.
6. Which of the two diastereomers of azobenzene would you expect to be more thermodynamically stable? Why?
7. From the results of the TLC experiment with the azobenzenes, describe the role of sunlight.
8. A student obtained the silica gel TLC plate shown in Figure 6.6 by spotting samples of Midol™, caffeine, and acetaminophen on the plate and eluting with petroleum ether:chloroform (9:1 v:v).
   a. What are the R_f-values of acetaminophen and of caffeine, respectively?
   b. Based on this TLC analysis, what are the ingredients in a tablet of Midol™?
   c. What are the mobile and stationary phases, respectively, in this TLC experiment?
   d. No spots were observed visually when the TLC plate was removed from the developing chamber. How might the student effect visualization of the spots?
   e. Another student accidentally used Midol PM™ in her experiment and observed only one spot. Speculate as to which spot was absent and offer a possible explanation for the difference in this student's result.

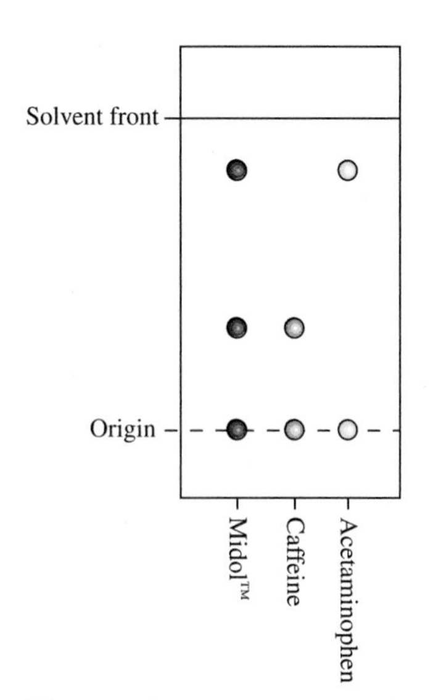

Figure 6.6
TLC analysis of mixture.

6.3 COLUMN CHROMATOGRAPHY

See more on *Column Chromatography*

See video on *Column Chromatography*

Column chromatography is another form of **solid-liquid adsorption** chromatography and depends on the same fundamental principles as does thin-layer chromatography (TLC, Sec. 6.2), as you will see from the discussion that follows. It has an advantage over TLC in that multigram amounts of mixtures can be separated but has the disadvantage that this technique requires considerably more time

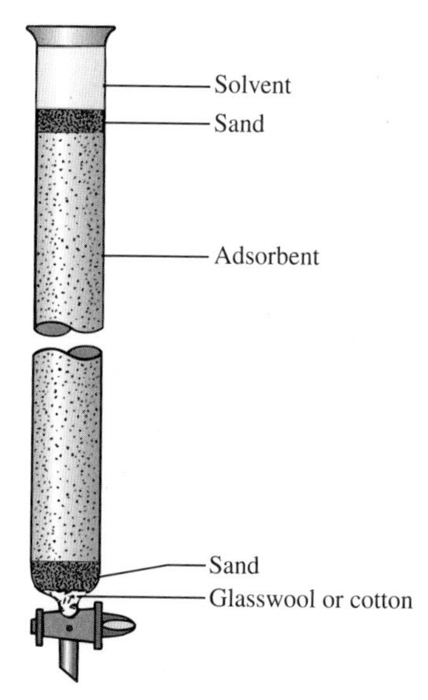

Figure 6.7
Chromatography column.

to perform. The larger scale of separations using this technique makes it valuable for purifying components of a reaction mixture so that one of them may be purified and used in a subsequent chemical reaction.

In column chromatography, a multicomponent mixture is typically dissolved in a small amount of an appropriate solvent and applied to the top of a packed column containing a finely divided, active solid **adsorbent** that serves as the stationary phase (Fig. 6.7). Next an **eluant** or **eluting solvent,** which is the mobile phase, is passed down the column. The individual components of the mixture, which were initially adsorbed on the stationary phase at the top of the column, begin to move downward with the eluting solvent (Fig. 6.8). These components travel at different rates depending on their relative affinities for the packing material; a more weakly adsorbed compound is eluted faster from the column than is a more strongly adsorbed compound. The individual components are collected in separate containers as they exit from the bottom of the column in bands. The solvent is then removed from each fraction by evaporation to provide the pure components, which are characterized and identified by determining their physical constants (Chaps. 3 and 4) and spectral properties (Chap. 8).

As with TLC (Sec. 6.2), the chromatographically separated bands are easily observed when all of the components of a mixture are colored. Many organic compounds are colorless, however, and other methods are required for detecting the banks as they elute from the column. For those organic compounds that absorb ultraviolet or visible light (Sec. 8.4), **electronic detectors** that measure differences in the absorption of light as the solvent exits the column are used to locate the bands of the individual components. Detectors that measure differences in the **refractive index** of the **eluate** are also used to identify the different bands; such detectors do not rely on absorption of light by the organic components.

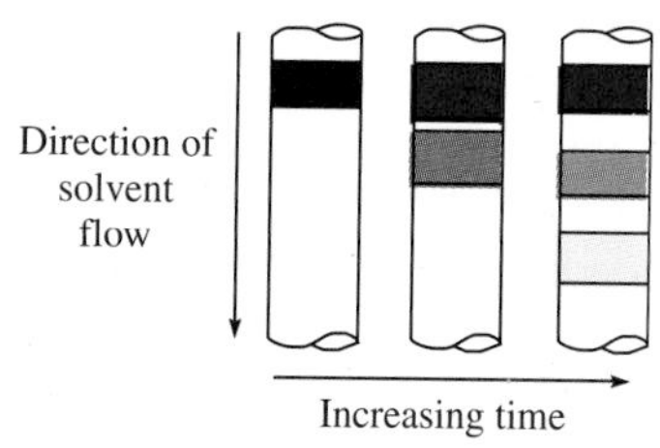

Figure 6.8
Separation of mixture by column chromatography.

If a detector is not available, the progress of the chromatographic separation can be conveniently followed using TLC to analyze the eluate at regular intervals. Another method, albeit more laborious, involves collecting small, equal fractions of the eluate from the column in a series of flasks. The solvent in each flask is then evaporated, and the presence or absence of a solute in the residue then provides a means of locating the bands of each component. If there is adequate separation of the different bands, a given flask will often contain only one constituent of the original mixture. That component will normally appear in a number of consecutive fractions, however.

Adsorbents

Alumina and silica gel are the most commonly used adsorbents for column chromatography, just as they are for TLC. The quality of these adsorbents is high in that they have *uniform particle size* and *high specific area.* The higher the specific area, the faster the equilibrium of the solute between the mobile and solid phases is established and the narrower the bands. High specific areas on the order of several hundred m^2/g are common for good grades of alumina and silica gel.

As noted in the discussion of TLC, the strength of the adsorption of an organic compound to the solid support depends on the polarity and nature of the adsorbent as well as on the nature of the functional groups present in the molecule. When **normal-phase column chromatography** is performed, a polar stationary phase such as alumina or silica gel is used in combination with organic solvents as the mobile phase or eluant. Under these conditions, the **elutropic series** described for TLC in Section 6.2 applies.

In **reverse-phase column chromatography,** the packing material for the stationary phase consists of glass beads coated with a nonpolar hydrocarbon film,

$RCO_2H < ROH < RNH_2 < R^1R^2C{=}O < R^1CO_2R^2 < ROCH_3 < R^1R^2C{=}CR^3R^4 < RHal$

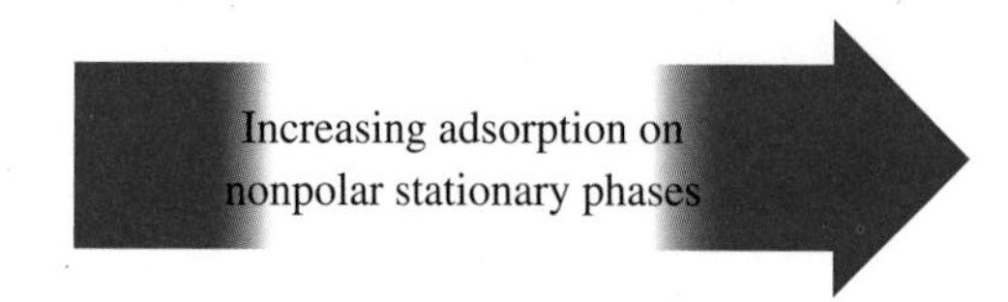

Figure 6.9
Elutropic series for nonpolar stationary phases.

and mixtures of water and organic solvents are generally used as the eluting solvents. Under these conditions, nonpolar organic molecules are more strongly attracted to the nonpolar stationary phase, whereas polar solutes are more strongly attracted to the mobile phase. The order of elution is then the *reverse* of that shown earlier, with the more polar components of a mixture eluting more rapidly than the less polar ones. The elutropic series then becomes the reverse (Figure 6.9). Reverse-phase chromatography may sometimes be used to separate mixtures that are inseparable by normal-phase chromatography.

Eluant

The most efficient method for determining the optimal solvent system(s) for a specific column chromatographic separation often is to perform a series of trial separations using TLC. These trials can be run quickly, and the amount of material needed is small.

The same criteria for selecting an eluant for TLC apply to normal-phase column chromatography. For example, if the mixture to be separated is not soluble or only slightly soluble in the eluant being used, the components will remain permanently adsorbed to the stationary phase near the top of the column. Conversely, if the components are too readily displaced from the adsorbent by the eluant, they will move down the column too rapidly to allow the needed equilibration between the stationary and mobile phases and will not be separated from one another. To allow equilibration in normal-phase column chromatography, the eluting solvent must be *less* polar than the components of the mixture. Although the relative eluting power of solvents for normal-phase column chromatography is the same as that for TLC when alumina or silica gel is used as an adsorbent, that for reverse-phase chromatography is the reverse (Fig. 6.10).

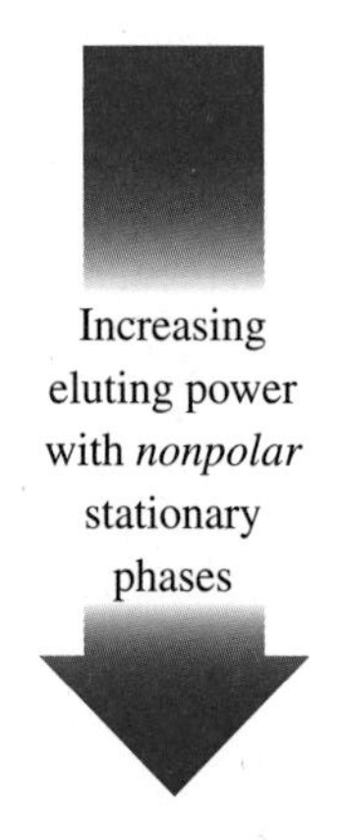

Water
Methanol
Ethanol
1-Propanol
Acetone
Ethyl acetate
Diethyl ether
Chloroform
Dichloromethane
Toluene
Hexane
Petroleum ether

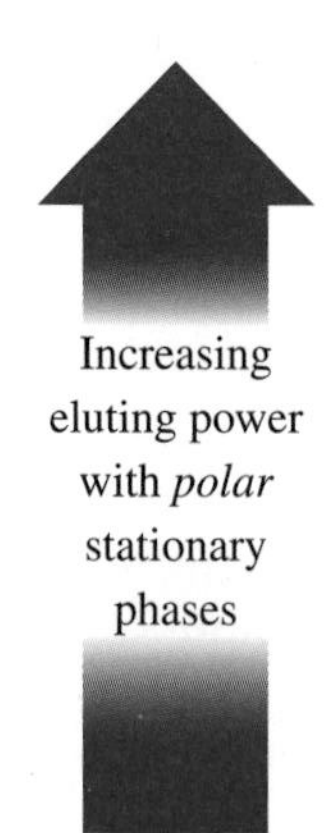

Figure 6.10
Eluting power of solvents as function of polarity of stationary phases.

Experimental Technique

With this general discussion as background, we can now discuss some of the experimental aspects of performing column chromatography. The optimal adsorbent and eluting solvent(s) typically are first determined using TLC, and then the column is packed with the adsorbent. The amount of adsorbent used to prepare the column varies according to the differences in partition coefficients and polarities of the individual components in the chromatographic system. For simple separations, it is possible to use as little as 10 g of adsorbent for each 1 g of the mixture, but when the components of the mixture have similar polarities, this ratio must be increased to as much as 100–200:1; a ratio of about 25:1 is a convenient starting point. As a general rule of thumb, the ratio of height-to-diameter for the packed column should be about 8:1.

The column is made of glass and is fitted with a stopcock or a segment of polyethylene tubing with a screw clamp to control the flow of solvent through the column. The column is prepared by first inserting a plug of cotton or glasswool into the small end of the column using a long glass rod or a piece of wire; this prevents the adsorbent from washing out of the bottom. A layer of white sand approximately 1 cm deep is then added to provide an even bed for the adsorbent (Fig. 6.7).

Proper packing of the column is vital to the success of column chromatography because this step ultimately determines the efficiency of separation. Two general protocols are followed for this operation. The first of these, the **dry-pack** method, involves pouring the dry adsorbent slowly into a vertical glass column half-filled with the solvent that will serve as the eluant. The other technique is the **wet-pack** method, in which a slurry of the adsorbent in the eluting solvent is added to the column; this is the preferred procedure when silica gel is the adsorbent. With both methods, the column must be constantly tapped as the solid settles through the liquid to ensure even and firm packing of the adsorbent and to remove any entrapped air bubbles. Some solvent may be drained from the column during this operation, but the liquid level in the column should *never* be allowed to fall below the top of the adsorbent. If this occurs, the air bubbles that form in the column will allow *channeling,* which results in poor separations because the components are eluted as ragged rather than sharp bands. *Uniform packing of the adsorbent is essential* so the solvent will move down the column with a horizontal front.

To complete packing the column, a layer of sand is normally placed on top of adsorbent, as shown in Figure 6.7. The purpose of the sand is twofold: (1) It allows the sample to flow evenly onto the surface of the adsorbent, and (2) it prevents disruption of the packing material as eluting solvent is added.

Mixtures of solids are typically dissolved in a *minimal* volume of a solvent before being transferred to the top of the column; a liquid mixture may be transferred directly to the column. It is important to distribute the sample evenly on the surface of the adsorbent and to use as little solvent as possible in loading the column. This procedure ensures that the bands that form during development of the chromatogram will be narrow, thereby providing the best possible separation. If too much solvent is used to dissolve the sample, the initial band will be broad, and poor resolution of the mixture may result.

Once the sample has been loaded onto the column, there are several different techniques that may be used to elute its components. In a **simple elution** experiment, a *single* solvent is passed through the column during the entire course of the separation. This procedure works well for the separation of mixtures containing only two or three compounds having similar polarities. However, the more common chromatographic procedure is **stepwise** or **fractional elution.** In this technique, a series of increasingly more polar solvents is used to elute the mixture from the column. A nonpolar solvent such as petroleum ether or hexane is first used to move the least polar component of the mixture down the column, while the others remain at or near the top of the column. After elution of the first band, the polarity of the eluant is increased using either pure solvents or combinations of mixed solvents so that the bands are individually eluted from the column. Systematic and gradual increases in solvent polarity are sometimes essential so that individual bands remaining on the column separate and do not co-elute. As the polarity of the solvent system is increased, those components of the mixture that are more tightly adsorbed on the column will begin to move. As a rule of thumb, a volume of solvent approximately equal to three times the column volume is passed through the column prior to switching to a solvent of higher polarity.

The separation is monitored using one of a variety of methods. Unfortunately, many organic compounds are not highly colored, as we noted above, and sophisticated devices for their detection are rarely available in the undergraduate laboratory. The most effective technique for following the separation is to collect fractions of equal volume in tared flasks, to concentrate the solvent, and to reweigh the flasks. The fractions containing the different bands may then be easily identified by the relative amounts of solute in each flask. One may also use TLC to monitor the separation.

In the experiment that follows, column chromatography will be used to separate fluorene (**5**) from an oxidation product, 9-fluorenone (**6**). One of these compounds is white, and the other is yellow. Consequently, the progress of the chromatography may be followed by evaporation of the solvent at periodic intervals as well as by observing the slower-moving yellow band.

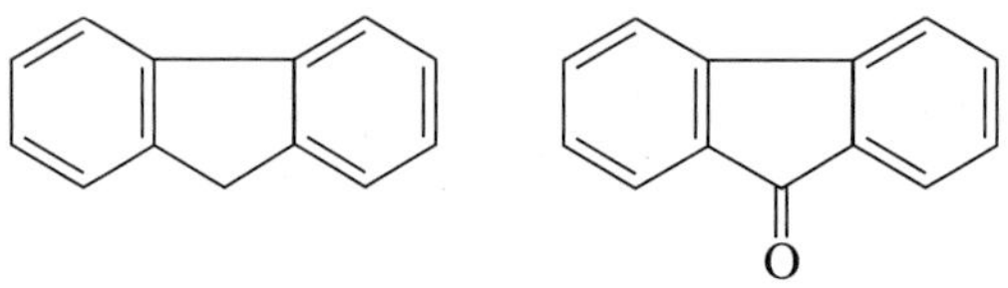

O

5
Fluorene
White

6
9-Fluorenone
Yellow

EXPERIMENTAL PROCEDURES

Column Chromatography

Purpose To separate fluorene and 9-fluorenone by column chromatography.

SAFETY ALERT

Petroleum ether is a highly volatile and flammable mixture of low-boiling hydrocarbons. During the preparation and development of the chromatographic column, be certain that there are *no flames* in the vicinity.

Procedure

Preparation Sign in at **www.cengage.com/login** to answer Pre-Lab Exercises, access videos, and read the MSDSs for the chemicals used or produced in this procedure. Read or review Sections 2.13 and 2.29.

Apparatus A *dry* 50-mL glass buret or chromatography column about 1 cm in diameter and 25 cm long and fitted with either a Teflon stopcock or a short piece of polyethylene tubing and a screw clamp, three 25- or 50-mL Erlenmeyer flasks, watchglass, Pasteur pipet, and apparatus for simple distillation. Consult with your instructor regarding whether the glassware, sand, and alumina require oven-drying prior to use.

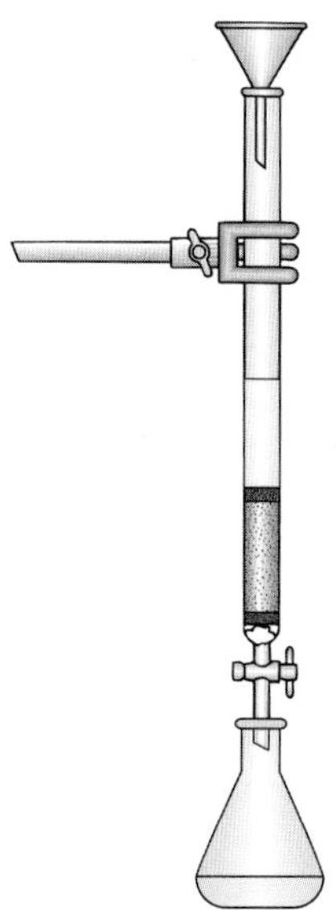

Figure 6.11
Set-up for column chromatography.

Setting Up Clamp the glass buret or chromatography column in a vertical position with its *ungreased* stopcock, preferably Teflon, *closed.* Using a piece of glass tubing, insert a small plug of cotton or glasswool loosely into the bottom of the column. Cover the cotton or glasswool plug with enough clean, *dry* sand to form a layer about 1 cm thick, and add approximately 25 mL of petroleum ether, bp 30–60 °C (760 torr). Place a funnel on top of the column, and *slowly* add 5 g of *dry* neutral alumina to the column while constantly tapping the buret. A rubber "tapping mallet" may be made by placing a pencil in a one-hole rubber stopper. When this process has been completed, wash the inner walls of the column with additional petroleum ether to remove any alumina that may adhere to the sides. Cover the alumina with a 1-cm layer of clean sand, and open the stopcock to allow the solvent to drain into an Erlenmeyer flask until the solvent level reaches just to the *top of the alumina.* The column is now ready for the addition of the sample mixture (Fig. 6.11).

Separation Obtain a sample of an approximately 1:1 mixture of fluorene (**5**) and 9-fluorenone (**6**) and accurately determine its melting-point range. In a small test tube, suspend about 0.1 g of this mixture in 0.5 mL of petroleum ether and slowly add just enough dichloromethane with a Pasteur pipet to effect solution. Using a Pasteur or filter-tip pipet, carefully transfer this solution directly to the top of the column. Open the stopcock until the liquid level is at the top of the alumina. *Do not allow the solvent to drain below the level of the alumina,* as air bubbles and channels might develop in the solid support. Add approximately 1–2 mL of fresh petroleum ether to the top of the column, and again allow the liquid to drain to the top of the alumina.

Fill the buret with approximately 20 mL of fresh petroleum ether, open the stopcock, and collect the eluant in an Erlenmeyer flask. Follow the progress of the chromatography by collecting a drop or two of eluant on a watchglass with every 5 mL that elutes from the column. When the solvent evaporates, any white solid present will be visible on the watchglass. You can determine when all of the white solid has been eluted using this visualization technique. Your instructor might also direct you to follow the chromatography by TLC using 15% dichloromethane in petroleum ether as the developing solvent (Sec. 6.2). Most of the white solid should elute in a volume of 15–20 mL of petroleum ether, and slow movement of a yellow band down the column should occur. Wash any of the white solid from the tip of the column into your collection flask with fresh petroleum ether.

When all of the white solid has eluted from the column, change the collection flask to another clean Erlenmeyer flask. Elute the column with about 5 mL of petroleum ether, and then change the eluant to dichloromethane, a more polar solvent. Watch the progress of the yellow band as it now proceeds rapidly down the column. When this yellow band just reaches the bottom of the column, change to a third clean Erlenmeyer flask. The intermediate fraction should not contain significant amounts of solid; verify this by evaporating a few drops on a watchglass.

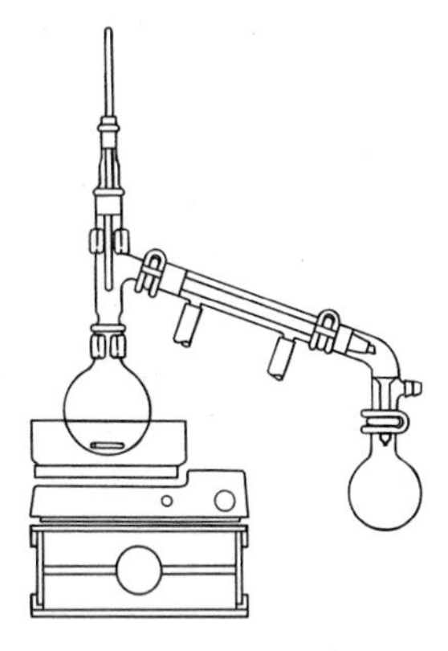

Continue eluting the column with dichloromethane until the eluant is colorless; approximately 10 mL will be required.

Isolation The first and third fractions should contain pure samples of fluorene and 9-fluorenone. Most of the solvent in these fractions may be removed by simple distillation (Sec. 2.13) or by one of the techniques outlined in Section 2.29. Attach the flask to a water aspirator or house vacuum to remove the last traces of solvent under reduced pressure.

Analysis When the crystals of each of the purified compounds are completely dry, determine their melting points and weights. Use your experimentally determined melting points to identify the order of elution of fluorene and fluorenone from the column and record your observations and conclusions in your notebook. Based upon the weight of the mixture used, calculate the percent recovery of fluorene and 9-fluorenone.

Discovery Experiment

Column Chromatographic Separation of Benzyl Alcohol and Methyl Benzoate

Design and execute an experimental procedure for separating benzyl alcohol (**7**) from methyl benzoate (**8**) by normal-phase column chromatography.

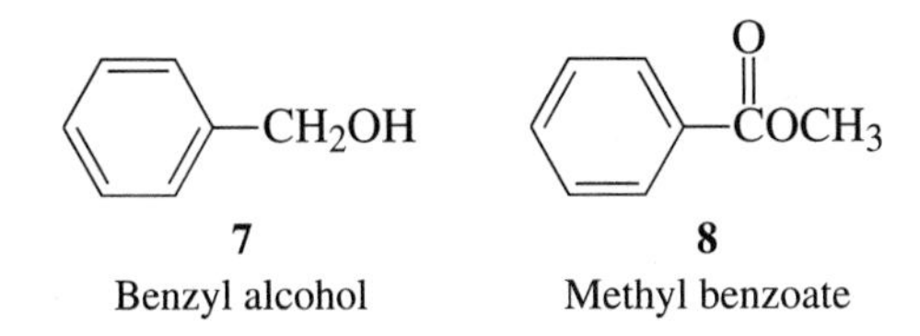

WRAPPING IT UP

Place recovered *petroleum ether* in the container for nonhalogenated organic liquids, but pour the recovered *dichloromethane* into a container for halogenated organic liquids. Spread out the *alumina adsorbent,* which is wet with organic solvent, in a hood to dry and then place it in the nonhazardous solid waste container.

EXERCISES

1. What are two important characteristics of the silica gel and alumina that are used for column chromatography?
2. Explain why it is unwise to use column chromatography on compounds having boiling points below about 150 °C (760 torr).
3. Define the following terms.
   a. eluate
   b. eluant
   c. adsorbent
   d. reverse-phase chromatography
4. In normal-phase column chromatography, which solvent has more eluting power: petroleum ether or dichloromethane? In what way is the eluting power of a solvent related to its polarity?

5. In reverse-phase column chromatography, which phase is the more polar: the stationary phase or the mobile phase?
6. How does the principle of "like dissolves like" explain the affinity of a compound for the mobile phase relative to the stationary phase?
7. When separating a mixture by normal-phase column chromatography, why is it better to change from a less-polar solvent to a more-polar solvent rather than the opposite?
8. If the polarity of the eluant is to be increased during column chromatography, the increase is normally made gradually. However, in this procedure, *pure* dichloromethane rather than intermediate mixtures of petroleum ether and dichloromethane was added to the column. Why is this variation from the usual technique acceptable in this case?
9. State two major differences between TLC and column chromatography.
10. If you had 5.0 g of material that needed to be purified, would you opt for using TLC or column chromatography to purify your material? Explain your answer.
11. Why is it preferable to use a Teflon or an ungreased stopcock rather than a greased stopcock on a column used for column chromatography?
12. Why should care be exercised in the preparation of the column to prevent air bubbles from being trapped in the adsorbent?
13. Why is a layer of sand placed above the cotton plug prior to the addition of the column packing material?
14. Does fluorene or 9-fluorenone move faster down the column when petroleum ether is used as the eluant? Why?
15. Consider the structures of fluorene (**5**) and 9-fluorenone (**6**).
    - **a.** Other than the aromatic rings, specify what functional group, if any, is present in **6**.
    - **b.** Predict which compound is more polar and explain why.
    - **c.** Which compound would you expect to elute from a normal-phase column first and why?
16. In the separation of fluorene and fluorenone by column chromatography, what is the stationary phase? The mobile phase?
17. The observed melting point of the 1:1 mixture of fluorene and 9-fluorenone should be relatively sharp, although lower than the melting point of either of the pure compounds. On the other hand, a 3:1 mixture has a broad melting-point range of about 60–90 °C. Explain these observations. (*Hint:* See Sec. 3.3.)
18. A mixture containing compounds **9–11** was separated by normal-phase column chromatography, using neutral alumina as the stationary phase and petroleum ether as the eluant. Predict the order in which **9–11** will elute from the column.

CH_3–C_6H_4–$CH(CH_3)OH$ (**9**) CH_3–C_6H_4–$COOH$ (**10**) CH_3–C_6H_4–$CH(CH_3)_2$ (**11**)

19. Most peptides (polymers of amides constructed from amino acids) are very polar molecules; therefore, they are most successfully purified by reverse-phase chromatography.
    a. In this type of column chromatography, is the mobile or stationary phase more polar?
    b. Provide an explanation for why reverse-phase chromatography is more appropriate in this application than normal-phase chromatography.
20. A typical eluting solvent for reverse-phase column chromatography is acetonitrile (CH_3CN), water, and a buffer. In what way is this mobile phase different from the one you used to separate fluorene (**5**) from 9-fluorenone (**6**)?

6.4 GAS-LIQUID CHROMATOGRAPHY

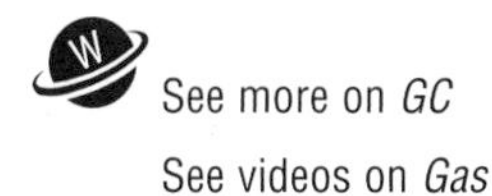

See more on *GC*

See videos on *Gas Chromatography*

Gas-liquid chromatography (GLC), which is also called **gas chromatography** (GC), is a technique that may be used to separate mixtures of volatile compounds whose boiling points may differ by as little as 0.5 °C. It can also be applied as an analytical tool to identify the components of a mixture or in preparative applications when quantities of the pure components are desired.

Basic Principles of Gas-Liquid Chromatography

Gas-liquid chromatography operates on the principle of partitioning the components of a mixture between a mobile gaseous phase and a stationary liquid phase. In practice, a sample is injected into a heated chamber where it is immediately vaporized and carried through a column by a flowing inert gas such as helium or nitrogen, which is called the **carrier gas.** This gaseous mixture is the **mobile phase.** The column is packed with a finely divided solid support that has been coated with a viscous, high-boiling liquid, which serves as the **stationary phase.** As the mobile phase moves through the column, its components are continuously partitioned between the two phases. Those components that show a higher affinity for the mobile phase move through the column more quickly, whereas those with a stronger attraction to the stationary phase migrate more slowly, and separation occurs. As with a fractional distillation column (Sec. 4.4), a GLC column may be characterized by the number of theoretical plates it provides. GLC columns typically have many more theoretical plates, however, so they can effect separations that would be impossible using fractional distillation.

The **retention time** of a component is the elapsed time required for the compound to pass from the point of injection to the detector, and it may be used for purposes of identification. The retention time of a component is *independent* of the presence or absence of other components in the sample mixture. There are four experimental factors that influence retention time of a compound: (1) the *nature* of the stationary phase, (2) the *length* of the column, (3) the *temperature* of the column, and (4) the *flowrate* of the inert carrier gas. Thus, for a particular column, temperature, and flowrate, the retention time will be the same for a specific compound.

Although a large number of **stationary liquid phases** are available, only a few are widely used (Table 6.1). Each liquid phase has a maximum temperature limit above which it cannot be used. This temperature depends upon the stability and volatility of the liquid phase; at higher temperatures, the liquid phase will vaporize and "bleed" from the column with the mobile phase.

The differences in the *partition coefficients* (Secs. 5.2 and 5.3) of the individual components of a mixture in GLC depend primarily upon the differences in solubility

Table 6.1 *Common Stationary Phases for Gas-Liquid Chromatography*

Liquid Phase	*Type*	*Property*	*Maximum Temperature Limit, °C*
Carbowax 20M	Hydrocarbon wax	Polar	250
OV-17	Methylphenyl silicone	Intermediate polarity	300
QF-1	Fluorosilicone	Intermediate polarity	250
SE-30	Silicone gum rubber	Nonpolar	375

of each of the components in the liquid phase. Two important factors determining the solubility of a gas in a liquid are its vapor pressure at the **ambient temperature,** which is the temperature of the column, and the magnitude of its interactions with the liquid phase. Regarding the first factor, the solubility of a gas in a liquid decreases as its vapor pressure increases. This means the more volatile components of a gaseous mixture tend to pass through the column more rapidly and elute before the less volatile ones. The impact of the second factor is understandable from the principle that *like dissolves like* (Sec. 3.2). Polar samples are most effectively separated by using a polar liquid phase, whereas nonpolar compounds are best separated using a non-polar liquid phase.

The stationary liquid phase is normally coated as a thin film on an inert solid support. The support is composed of small, uniformly meshed granules, so that a large surface area of the liquid phase is available for contact with the vapor phase to ensure efficient separation. Some common types of solid supports include Chromosorb P and Chromosorb W, which are composed of diatomaceous earth. The surface areas of these supports vary from 1 to 6 m^2/g. Columns are now commercially available with a wide variety of liquid phases on different solid supports. An alternative method of supporting the liquid phase is used in capillary columns. In these columns, the liquid is coated directly onto the inner walls of the tubing. These columns are highly efficient but relatively expensive.

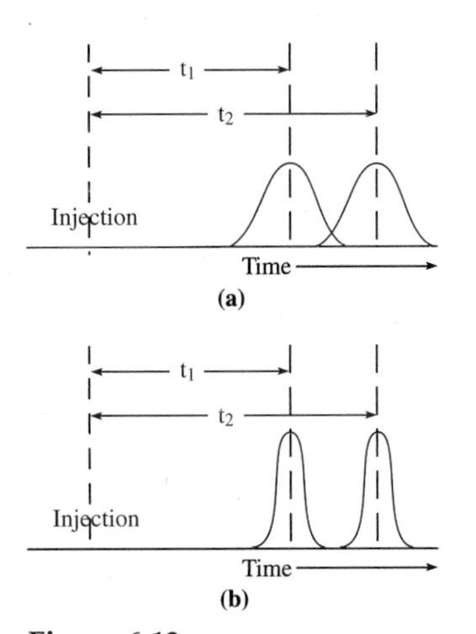

Figure 6.12
Effect of bandwidth on resolution. (a) Wide bands. (b) Narrow bands.

In general, the efficiency, or **resolution,** of a column increases with increasing length and decreasing diameter. Increasing the pathlength increases the difference in retention times between bands, whereas decreasing the diameter of the column gives rise to narrower bands. With a small band separation, as measured from the band centers, wide bands are more likely to overlap (Fig. 6.12a) than narrow bands (Fig. 6.12b).

The two other experimental factors that may be varied to alter the degree of separation of the bands are the *temperature* at which the column is maintained and the *flowrate* of the carrier gas. Increasing the temperature results in shorter retention times, because the solubility of gases in liquids decreases with increasing temperature. The partition coefficients are affected, and the bands move through the column at a faster rate. Higher flowrates also cause retention times to decrease. In spite of the decreased resolution and band separation obtained at higher temperatures and flowrates, these conditions are sometimes necessary for substances that would otherwise have very long retention times.

Instrumentation

All commercially available gas-liquid chromatographs (GLCs) have a number of basic features in common. These are illustrated schematically in Figure 6.13.

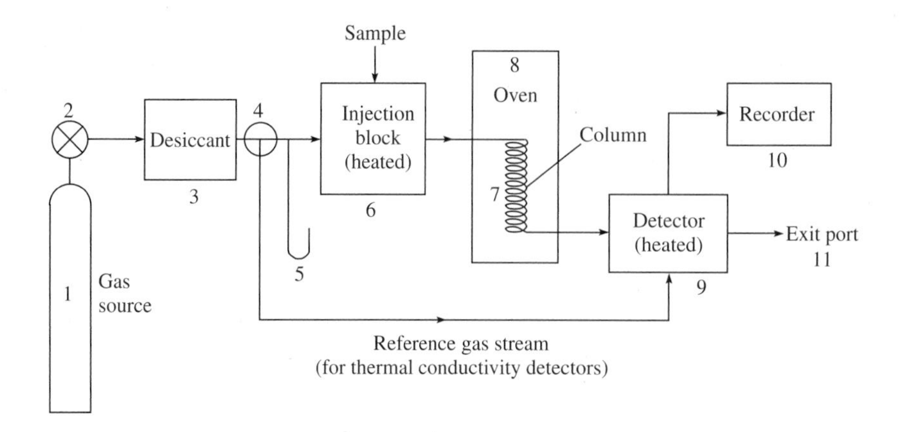

Figure 6.13
Schematic diagram of apparatus for GLC.

See more on *GC Sequence*

Parts **1–5** are associated with supplying the dry **carrier gas,** usually helium or nitrogen, and allowing an operator to control its flow. The mixture to be separated is injected using a gas-tight syringe through a septum into the **injection block** (**6**), an individually heated chamber in which the sample is immediately vaporized. The sample then enters the flowing stream of carrier gas and is swept into and through the **column** (**7**), which is located in an **oven** (**8**) and consists of coiled aluminum, stainless steel, or glass tubing containing an appropriate stationary phase. In the column, the individual components separate into bands that ultimately pass through a **detector** (**9**), producing a signal whose voltage is proportional to the amount of material other than carrier gas present in the mobile phase. One type of detector that is commonly used is the **thermal conductivity detector** (TCD), which operates on the basis of differences in the thermal conductivity of the mobile phase as a function of its composition. A **flame ionization detector** (FID) is much more sensitive and functions by detecting the number of ions produced by passing the mobile phase through a hydrogen flame. The **recorder** (**10**) plots the changes in voltage measured by the detector as a function of time to give the **gas chromatogram.** The vapors then pass from the detector into either the atmosphere or a collection device at the **exit port** (**11**).

The GLCs available in the introductory organic laboratory often can only be operated at a constant oven temperature, the so-called **isothermal** mode of operation. Some GLCs, however, have a temperature-programming option, which allows the temperature of the oven, and consequently that of the column, to be varied over a range of temperatures. This is particularly useful when the sample being analyzed contains components having widely varying boiling points. Thus, the oven temperature may be held at a constant temperature for a specified period of time at the beginning of an analysis but then may be increased to higher temperatures as a function of time. For example, the temperature program illustrated in Figure 6.14 involved an initial temperature of 100 °C. After 5 minutes, the temperature was ramped up to 125 °C at a rate of 5 °C/min, held at that temperature for 5 additional minutes, and then further ramped up at a rate of 10 °C/min to a final temperature of 175 °C for completing the analysis. The programming option allows the higher-boiling components of the mixture to elute in a reasonable period of time because rates of elution increase with increasing temperatures. If the analysis had been performed at the higher temperature, the lower-boiling components might have eluted too quickly to be separated from each other.

See more on *GC-MS*

Another modification of a GLC involves its direct connection to a mass spectrometer (MS) to produce a hybrid instrument commonly called a **GC-MS.** This

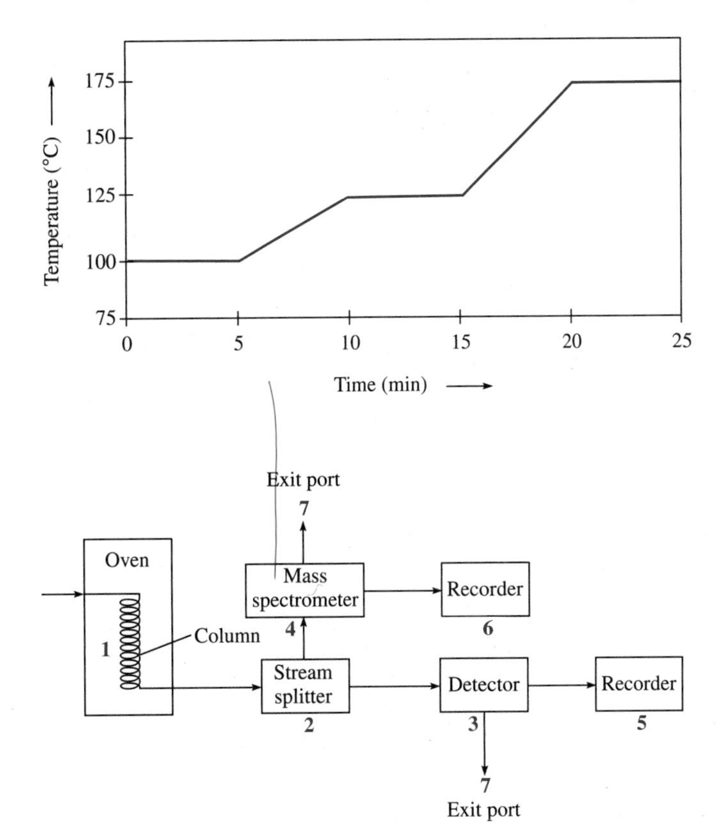

Figure 6.14
Example of temperature vs. time program for a GLC analysis.

Figure 6.15
Partial schematic diagram of a GC-MS.

combination provides a powerful analytical technique because the GC-MS combines the separating power of GLC with the ability of mass spectrometry (Sec. 8.5) to determine molar masses using very small amounts of material (see the Historical Highlight at the end of this chapter). Thus, a GC-MS allows the mass spectrum of each component of a mixture of volatile compounds to be obtained as each individual component emerges from the GLC. If the components of a mixture are known, as they would be if you were analyzing a fraction obtained by normal-phase column chromatography of a mixture of benzyl alcohol (**7**) and methyl benzoate (**8**) (Optional Discovery Experiment, Sec. 6.3), you could use the information provided by the mass spectrum, as discussed below, to confirm the identity and order of elution from the chromatographic column of each of these colorless compounds. If the components are unknown, on the other hand, knowledge of the molar mass of each compound, coupled with information about the origin of the mixture being analyzed, may allow assignment of structures to the unknown substances. Other spectral data (Chap. 8) are generally required before structural assignments can be made, however.

A partial schematic of a GC-MS is shown in Figure 6.15; the elements of the GLC that precede the column have been omitted but are the same as those shown in Figure 6.13. By comparing Figures 6.13 and 6.15, you see that a GC-MS has the outlet of the column (**1**) leading to a **stream splitter** (**2**), which sends part of the eluant to a **detector** (**3**), usually an FID, and part of it to a **mass spectrometer** (**4**); **recorders** (**5, 6**) provide the necessary records of when components are eluting from the GLC and what their mass spectra are. Both the GLC and the MS units are vented to the atmosphere (**7**).

Experimental Techniques

Qualitative Analysis. The *retention time* of a pure compound is constant under a specified set of experimental conditions, including the column, temperature, and flowrate. Consequently, this property may be used as a *first* step to identify an unknown compound or the individual components in a mixture. In a typical experiment, an unknown compound or mixture is injected into the injection port of a GLC, and the retention time(s) of the component(s) is (are) measured. A series of known samples are then injected under the same conditions. Comparison of the retention times of the standard samples with those of the unknown allows a preliminary identification of the component(s) of the unknown. A convenient way of confirming that the retention times of a standard and the unknown are the same involves injecting a sample prepared by combining equal amounts of the two. If a *single* peak is observed in the chromatogram, the *retention times* of the standard and the unknown are identical. However, observation of the same retention time for a known and an unknown substance is a *necessary but not sufficient* condition to establish identity, because it *is* possible for two different compounds to have the same retention time. Independent confirmation of the identity of the unknown by spectral (Chap. 8) or other means is imperative.

An example of the use of GLC as a qualitative, analytical tool is illustrated in Figure 6.16. These sets of peaks represent a gas chromatographic separation of the distillation fractions of a mixture of cyclohexane and toluene similar to those obtained in the distillation experiment described in Section 4.3. The notations *A*, *B*, and *C* refer to the three fractions taken in that experiment. The individual peaks

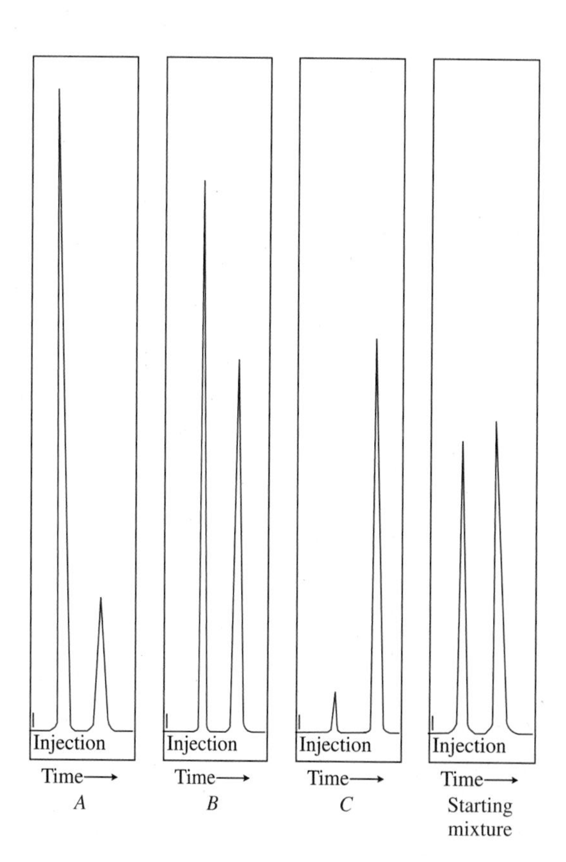

Figure 6.16
GLC analysis of the distillation fractions from the distillation experiment in Section 4.3.

in the mixture may be identified by comparing their retention times with those of pure cyclohexane and pure toluene; the peak with the shorter retention time in the mixture is cyclohexane, whereas the peak with the longer retention time is toluene.

The GC-MS analysis of a mixture of benzyl alcohol (**7**, bp 215.3 °C, 760 torr) and methyl benzoate (**8**, bp 199.6 °C, 760 torr) is provided in Figure 6.17. Figure 6.17a is the GLC trace of the mixture. Because the GLC column being used separates mixtures on the basis of the relative boiling points of its components, Peak A, which emerges first, should be **8**, whereas Peak B should be **7**. This could be confirmed with a GLC if the retention times of **7** and **8** were determined separately. Such additional analyses would not be necessary with GC-MS, however, because the masses of the two compounds are different; that of **7** is 108.14 atomic mass units (amu), and that of **8** is 136.15 amu. Figures 6.17b and 6.17c are the mass spectra corresponding to Peaks A and B, respectively. In Figure 6.17b we see that the molecular ion is at m/z 108, whereas it is at m/z 136 in Figure 6.17c. This proves that Peaks A and B are benzyl alcohol and methyl benzoate, respectively; our expectation of the reverse order of elution, based on considering boiling points, has been proven wrong. The stationary phase of the column must interact more strongly than expected with **8**. In any case, the GC-MS analysis has provided unambiguous identification of the two peaks through a *single* analysis of the mixture. Moreover, integration of the peaks in the GLC tract would provide a quantitative measure of their relative proportions in the mixture being analyzed.

A feature that may complicate the interpretation of the mass spectrum from a GC-MS analysis may be seen in Figure 6.17c. This feature is a small peak at m/z 146, a value that is not associated with either of the components contained in our known mixture of **7** and **8**. Although the substance that is the source of the peak could be an impurity in our sample, it is not. Rather, it is a peak produced from the "stationary" phase of the GLC column being used; the temperature at which the GLC was operating caused a small amount of the high-boiling material comprising this phase to "bleed" from the column into the MS, which results in minor peaks appearing in the spectrum. Such peaks can be confusing if you are not prepared for their possible presence.

A second example illustrating the power of an analysis by GC-MS rather than GLC alone is that of a mixture of methyl benzoate (**8**) and diethyl malonate (**12**), as seen in Figure 6.18. The boiling point of **12** (199.3 °C, 760 torr) is essentially identical to that of **8** (see above), and considering the differing polarities expected for these two esters, predicting their order of elution from a GLC column (Fig. 6.18a) is problematic. However, the mass spectra associated with the two peaks eluting at 150 and 164 seconds in the GLC trace show that Peak A must be diethyl malonate (Fig. 6.18b), whereas Peak B is methyl benzoate (Fig. 6.18c).

You may be perplexed by the fact that the molecular ion from **12** appears to have m/z 161 rather than the expected value of 160.17. This could be the result of poor calibration of the instrument, so that the m/z values are inaccurate; proper calibration of the instrument is critical for proper interpretation of GC-MS data. The actual explanation in this case, though, is that diethyl malonate apparently has a propensity to be protonated under the conditions of the mass spectrometric analysis, so it is this ion rather than the molecular ion that is being detected. This is yet another phenomenon that can make interpretation of GC-MS data difficult for a beginner. In any event, a single GC-MS analysis of the mixture of **8** and **12** allows

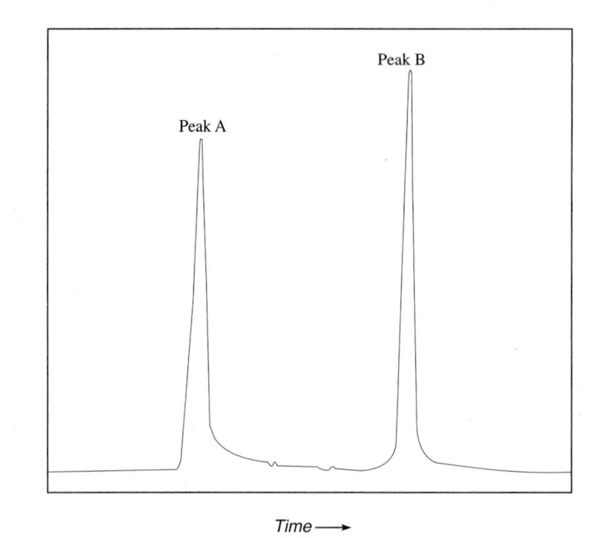

(a)

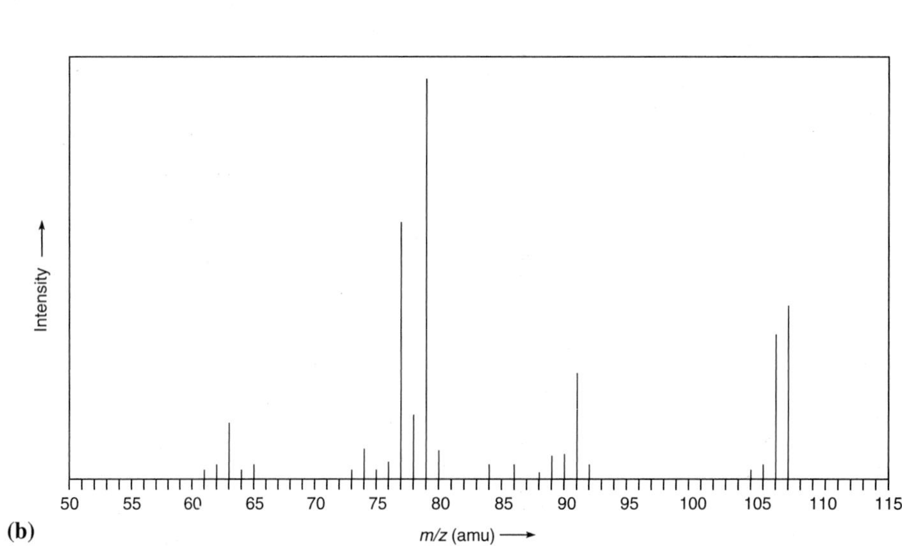

(b)

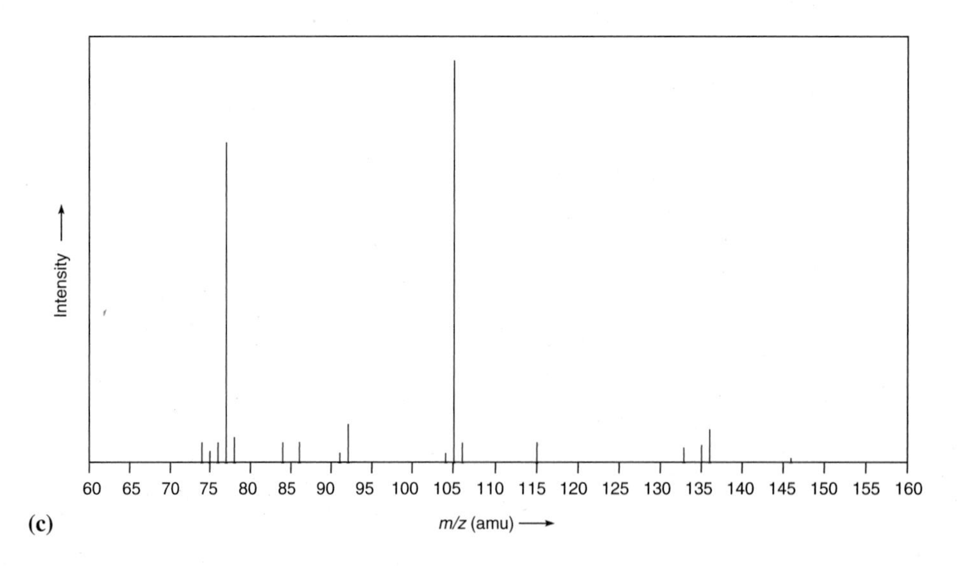

(c)

Figure 6.17
GC-MS analysis of a mixture of benzyl alcohol and methyl benzoate. (a) GLC trace. Column and conditions: 0.5-mm × 15-m, 0.25-μ film of BPX5 (DB-5); initial column temperature: 50 °C (2.0 min); ramp rate: 15 °C/min; final column temperature: 260 °C (30 min); flowrate: 1.5 mL/min. (b) MS spectrum of Peak A. (c) MS spectrum of Peak B.

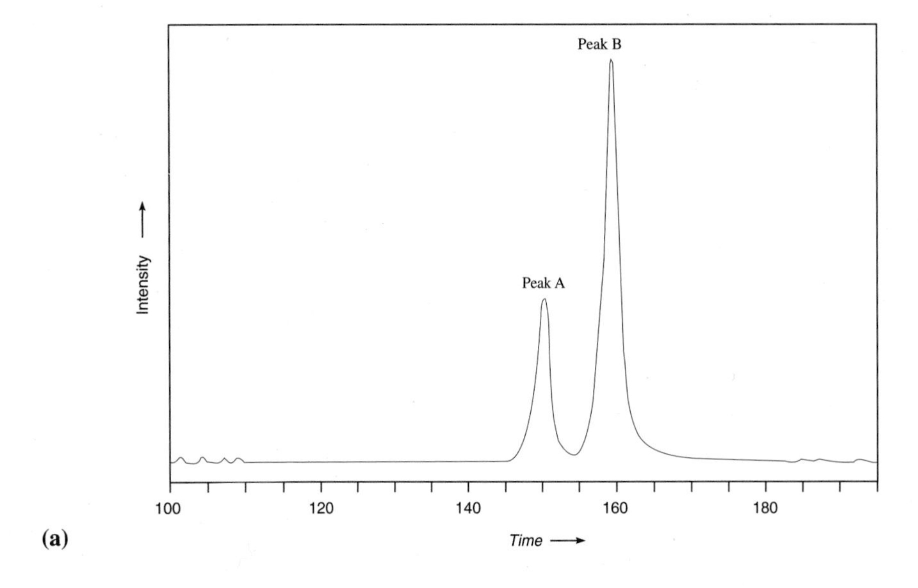

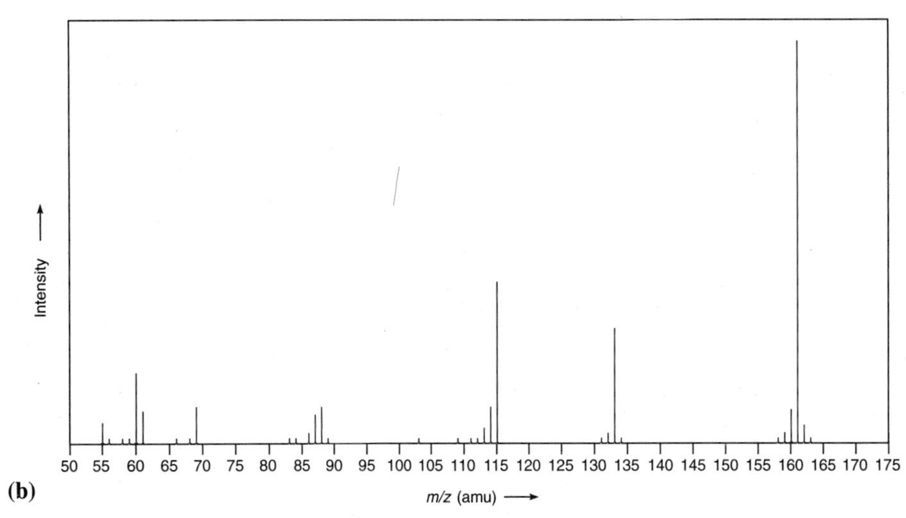

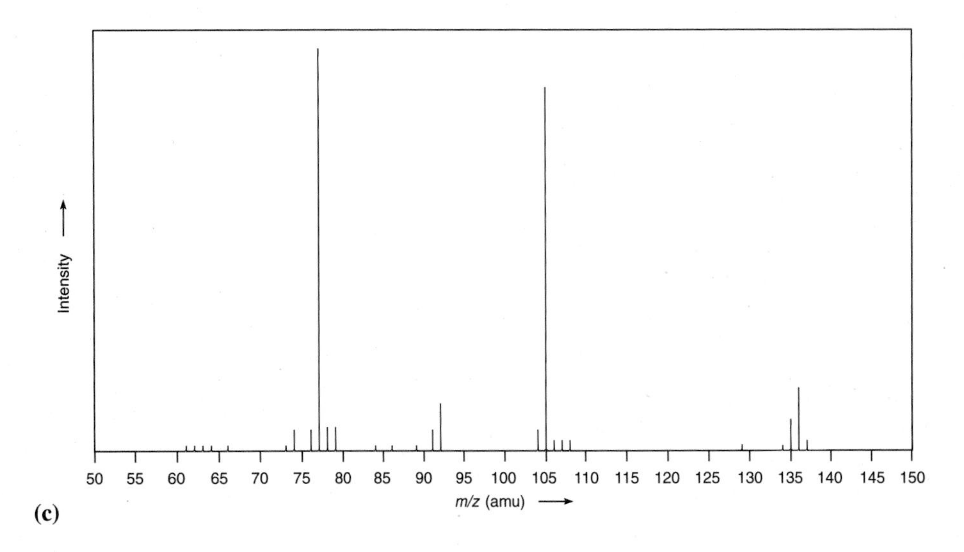

Figure 6.18
GC-MS analysis of a mixture of methyl benzoate and diethyl malonate. (a) GLC trace. Column and conditions: 0.5-mm × 15-m, 0.25-μ film of BPX5 (DB-5); initial column temperature: 50 °C (2.0 min); ramp rate: 15 °C/min; final column temperature: 260 °C (30 min); flowrate: 1.5 mL/min. (b) MS spectrum of Peak A. (c) MS spectrum of Peak B.

unambiguous determination of the order of elution from the GLC column of these esters, just as we saw in the analysis of the mixture of **7** and **8.**

$$CH_3CH_2O-C(=O)-CH_2-C(=O)-OCH_2CH_3$$

12

Diethyl malonate

Quantitative Analysis. The voltage output of the detector is related to the mole fraction of the material being detected in the vapor, so there is a correlation between the *relative areas* under the peaks in the chromatogram and the *relative amounts* of each of the components in the mixture. The quantitative evaluation of the chromatogram thus requires reliable methods for determining these peak areas.

An **electronic integrator,** which measures the intensity of detector output as a function of time, is the most accurate method for determining peak areas. However, these devices are expensive, so they are usually found only in research laboratories. Since the thickness and density of chart paper are reasonably uniform, another means of determining the relative areas involves carefully cutting the peaks out with scissors; the peak areas are then assumed to be proportional to their weight, as measured on an analytical balance. The original chromatogram should be saved as a permanent record, so the peaks should be cut from a photocopy of the chromatogram.

If the peaks are symmetrical, as are those shown in Figure 6.16, the areas may be approximated by assuming them to be equilateral triangles. The area of a symmetrical peak is then determined by multiplying the width of the peak at its half-height times its height. The percentage of each component in the mixture may be computed as the area of the peak corresponding to that component, expressed as a percentage of the sum of the areas of all peaks in the chromatogram. A sample calculation of this type is shown in Figure 6.19.

Although the peak areas are related to the mole fraction of the component in the mobile phase, they are not *quantitatively* related because the response of the detector varies with the class of the compound. Not all compounds have the same thermal conductivity (TCD), nor do they ionize in a hydrogen flame to form the same types or number of ions (FID). Thus, it is necessary to *correct* the measured areas in the chromatogram using the appropriate **response factor** to obtain an accurate quantitative analysis of the mixture. Although response factors for different compounds may be determined experimentally, approximate values are published in monographs on gas chromatography. The response factors for thermal conductivity and flame ionization detectors for compounds that you may encounter in this experiment are given in Table 6.2. Notice that the correction factors vary more widely for flame ionization detectors than for thermal conductivity detectors. In the experimental section, you have the opportunity to calculate the response factor for an unknown compound.

To analyze a mixture of substances quantitatively when **weight factors** are known, the peak area for each component is simply *multiplied* by the weight factor for that particular compound. The resulting *corrected* areas are used to calculate the percentage composition of the mixture according to the procedure outlined in Figure 6.19. Note that using these factors provides the composition on a *weight percentage* basis.

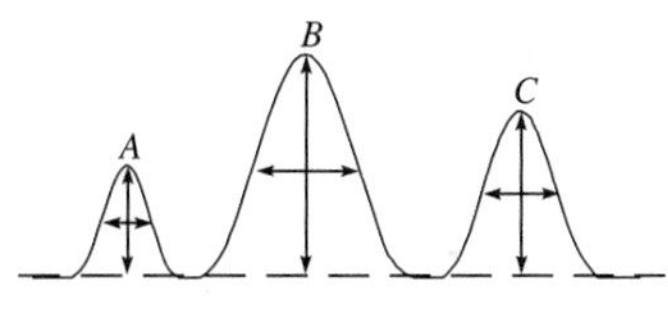

Areas: $A = 17 \times 8 = 136\ \text{mm}^2$

$B = 34 \times 17 = 578$

$C = 25 \times 12 = 300$

Total area = $1014\ \text{mm}^2$

$\%A = \frac{136}{1014} \times 100 = 13.4\%$

$\%B = \frac{578}{1014} \times 100 = 57.0\%$

$\%C = \frac{300}{1014} \times 100 = 29.6\%$

100%

Figure 6.19
Determination of percentage composition of a mixture by GLC.

Table 6.2 *Weight (W_f) and Mol (M_f) Correction Factors for Some Representative Substances**

	Thermal Conductivity		*Flame Ionization*	
Substance	W_f	M_f	W_f	M_f
Benzene	1.00	1.00	1.00	1.00
Toluene	1.02	0.86	1.01	0.86
Ethylbenzene	1.05	0.77	1.02	0.75
Isopropylbenzene	1.09	0.71	1.03	0.67
Ethyl acetate	1.01	0.89	1.69	1.50
n-Butyl acetate	1.10	0.74	1.48	0.99
Heptane	0.90	0.70	1.10	0.86
o-Xylene	1.08	0.79	1.02	0.75
m-Xylene	1.04	0.76	1.02	0.75
p-Xylene	1.04	0.76	1.02	0.75
Ethanol	0.82	1.39	1.77	3.00
Water	0.71	3.08	—	—

*McNair, H.M.; Bonelli, E.J. *Basic Gas Chromatography*, 5th ed., Varian Aerograph, Walnut Creek, CA, 1969.

Calculation of the composition of a mixture on a mole percentage basis requires the use of **mole factors,** M_f. These are obtained by dividing the weight factors by the molar masses of each component of the standard solution and normalizing the resulting numbers. A sample calculation utilizing mole correction factors is provided below in the analysis of a mixture of ethanol, heptane, benzene, and ethyl acetate with a GLC equipped with a thermal conductivity detector. The last column shows the percentage composition calculated directly from the measured peak areas, without correction for detector response. The dramatic differences in the calculated composition with and without this correction, as noted in the last two columns, underscore the importance of this correction for quantitative analysis.

Compound	*Area (A)* *(mm^2)*	*Uncorrected %* *(A/207.1 × 100)*	M_f	$A \times M_f$	*Mol %* *($A \times M_f$/194 × 100)*
Ethanol	44.0	21.2	1.39	61.2	31.5
Heptane	78.0	37.7	0.70	54.6	28.1
Benzene	23.2	11.2	1.00	23.2	11.9
Ethyl acetate	61.9	29.9	0.89	55.4	28.5
Total	207.1	100		194.4	100

Use of Syringes. Liquid samples are injected into the heated injection port of a gas chromatograph using a gas-tight syringe with a capacity of 1–10 μL. The sample is either injected neat or dissolved in a volatile liquid such as diethyl ether or pentane. The sample should not contain nonvolatile substances that may eventually clog the injection port or contaminate the stationary phase of the column.

Gas-tight syringes are precision-made and expensive. You should handle them carefully and adhere to the following procedure when using them. To fill the syringe, draw slightly more than the desired volume of the sample into the barrel by withdrawing the plunger as needed, point the syringe needle-up, push out the excess liquid, and then wipe the tip of the needle with a tissue. To inject the sample, insert the needle *straight* into the septum as far as it will go, *push the plunger all the way in with one quick motion,* and remove the syringe from the septum while holding the plunger in place. If the sample is injected slowly, it will produce a wider band on the column and the peaks will be broadened. Be careful to *avoid bending the needle or the plunger* during the injection. It is important to *clean the syringe immediately after use.* Rinse it with a volatile solvent such as acetone and dry it by drawing a stream of air through it.

EXPERIMENTAL PROCEDURES

A ■ *Qualitative and Quantitative Analyses of a Mixture of Compounds by GLC*

Purpose To identify individual components of a mixture and to quantify the relative amounts of each by gas-liquid chromatography.

SAFETY ALERT

The solvents and other liquids used in sample mixtures for GLC analysis are flammable and volatile. There should be *no open flames* used in the vicinity of these liquids.

Procedure

Preparation Sign in at **www.cengage.com/login** to answer Pre-Lab Exercises, access videos, and read the MSDSs for the chemicals used or produced in this procedure. Read or review Section 2.5.

Apparatus Gas chromatograph and a gas-tight syringe.

Setting Up Since there is considerable variation in the operating procedures of different commercial gas chromatographs, consult your instructor for specific directions for using the instrument. Obtain a mixture of liquid compounds from your instructor. A suitable mixture would be one containing varying amounts of some or all of the following compounds: ethanol, ethyl acetate, toluene, *n*-butyl acetate, ethylbenzene, and isopropylbenzene.

Injection Following the directions of your instructor, use the syringe to inject a 1–5-μL sample of your mixture into the gas-liquid chromatograph to obtain a gas chromatogram. In the same way, obtain a chromatogram of pure samples of ethanol, ethyl acetate, toluene, *n*-butyl acetate, ethylbenzene, and isopropylbenzene under the same instrumental conditions used for the mixture.

Qualitative Analysis Identify each of the components in your mixture as follows: (1) Compare the retention time of each of the components of the mixture with the retention times of the authentic samples of pure compounds. (2) Verify your assignments for one or more components of the mixture by preparing a series of new sample mixtures, each of which contains one volume of the pure known compound with two volumes of the original mixture. Prepare a gas chromatogram of this new mixture. The peak of the mixture that has been amplified in the new chromatogram corresponds to the pure compound that has been added.

Quantitative Analysis Use the quantitative method recommended by your instructor to determine the relative areas under the peaks in your chromatogram of the mixture. Correct these values by multiplying the measured area by the appropriate weight or molar correction factors for the type of detector in the GLC. These factors may be found in Table 6.2 or will be provided by your instructor. Determine the relative amounts, in weight and mole percentages, of each of the components of the mixture.

Discovery Experiment

Analysis of Factors Affecting Retention Times

Design and execute experiments that explore how factors such as flowrate and temperature influence the GLC retention times of a standard mixture of compounds.

Discovery Experiment

Effect of Stationary Phase on Separation of a Mixture

Design and execute experiments that explore how the nature of the stationary phase of a GLC column influences the separation of a mixture of volatile compounds having a range of polarities. Consult with your instructor regarding the variety of GLC columns that would be available for your investigation.

WRAPPING IT UP

Pour any unused *volatile organic compounds* into the appropriate container for nonhalogenated liquids or halogenated hydrocarbons; put any mixture that contains a halogenated hydrocarbon in the container for halogenated hydrocarbons.

B ■ *Determining GLC Response Factors*

Discovery Experiment

Purpose To determine the weight and mole correction factors for compounds using flame ionization or thermal conductivity detectors.

The solvents and other liquids that are used in sample mixtures for GLC analysis are flammable and volatile. *No flames* should be used in the vicinity of these liquids.

Procedure

Preparation Sign in at **www.cengage.com/login** to answer Pre-Lab Exercises, access videos, and read the MSDSs for the chemicals used or produced in this procedure. Read or review Section 2.5.

Apparatus Gas chromatograph, 1-mL syringe, and a gas-tight syringe.

Setting Up Since there is considerable variation in the operating procedures of different commercial gas chromatographs, consult your instructor for specific directions to use the instrument in your laboratory. Look up the densities of the compound whose weight and mole response factors are to be determined and the standard substance whose response factors are known. For example, the standard sample may be any of the compounds in Table 6.2. Prepare a test solution by combining 0.5 mL of the compound whose response factors are to be determined and 0.5 mL of the standard in a small vial or test tube, and swirl the solution to mix the two components *completely.* Assuming the volumes of the two liquids are additive, calculate the weight fraction and mole fraction composition of the mixture.

Injection Following the directions of your instructor, use a gas-tight syringe to inject a 1–5-μL sample of the test mixture into the GLC to obtain a chromatogram. Repeat this process for each of the compounds for which you are to obtain weight and mole response factors.

Analysis Use the quantitative method recommended by your instructor to determine the areas under the two peaks in your gas chromatogram of the test mixture. Divide the weight fraction and mole fraction of each of the two components by the area under the peak for that component. Normalize the value for the weight and mole factors for the unknown by the corresponding weight and mole factors for the standard, and record these values in your notebook.

Discovery Experiment

Molar Response Factors of Isomers

Using the protocol of this experimental procedure, determine whether the molar response factors, M_f, for either a pair of acyclic or a pair of cyclic isomers are identical.

Discovery Experiment

Molar Response Factors of Non-isomeric Compounds

Using the protocol of this experimental procedure, determine whether the molar response factors, M_f, are identical for a pair of non-isomeric compounds, one of which is a hydrocarbon and the other of which contains a heteroatom, that have the same molecular weight. Possible examples might be (a) cyclohexane, (b) hexane and 2- or 3-pentanone, and (c) ethylbenzene and benzaldehyde.

WRAPPING IT UP

Pour any unused *volatile organic compounds* into the appropriate container for nonhalogenated liquids or halogenated hydrocarbons; put any mixture that contains a halogenated hydrocarbon in the container for halogenated hydrocarbons.

EXERCISES

1. Define the following terms.

a. stationary phase
b. mobile phase
c. carrier gas
d. retention time
e. solid support
f. thermal conductivity

2. Refer to the GLC traces given in Figure 6.16. These are analyses of the various fractions collected during the fractional distillation of the mixture of cyclohexane and toluene. The weight and mole correction factors (flame ionization detector) for cyclohexane are 0.84 and 0.78, respectively, and those for toluene are given in Table 6.2. Accurately determine both the weight percent and the mole percent compositions of the distillation fractions *A*, *B*, and *C*.

3. Benzene (1 g, 12.5 mmol) is allowed to react with 1-chloropropane (1 g, 12.5 mmol) and $AlCl_3$. The product (1.2 g) is subjected to analysis on a GLC equipped with a thermal conductivity detector. The chromatogram shows two product peaks identified as *n*-propylbenzene (area = 65 mm^2; W_f = 1.06) and isopropylbenzene (area = 113 mm^2; W_f = 1.09). Calculate the percent yield of each of the two isomeric products obtained in this reaction. Note that since each of the products has the same molar mass of 120, the use of weight factors gives both weight and mole percent composition.

Benzene + $CH_3CH_2CH_2Cl$ $\xrightarrow{AlCl_3}$

Benzene 1-Chloropropane

$CH_2CH_2CH_3$ (*n*-Propylbenzene) + $CH(CH_3)_2$ (Isopropylbenzene)

n-Propylbenzene Isopropylbenzene

4. A gas chromatogram of the organic components of a sample of beer using a column that separates compounds on the basis of their relative boiling points provides a GLC trace with several peaks. Two of the smaller peaks, with retention times of 9.56 and 16.23 minutes, are believed to be ethyl acetate and ethyl butyrate, respectively.

   a. From the above information, which component of the sample, ethyl acetate or ethyl butyrate, elutes faster? What are the reported boiling points of these two substances?

   b. What GLC experiment(s) could you perform to confirm the identity of the peaks at 9.56 and 16.23 minutes?

   c. The major component of the sample elutes first. What organic compound present in beer do you suspect is responsible for this peak? Is your speculation consistent with the reported boiling point of the compound you are proposing? Explain.

   d. Suggest two ways in which you could adjust the conditions of the experiment so as to reduce the retention time for all components in the sample.

5. In the Olympics and other major athletic competitions, a GC-MS is used to screen athletes for their use of banned substances such as steroids (see Historical Highlight). If an analytical method has been developed and the retention time for a particular anabolic steroid has been determined, how might you use this method to determine the possibility of the use of anabolic steroids by an athlete?

HISTORICAL HIGHLIGHT

Who's Taking What? Analysis of Biological Fluids for Illegal Substances

A star athlete on your favorite team has just been suspended for the remainder of the season. A champion has been stripped of the medal won in an Olympic competition. You bet on a loser at the racetrack, but the horse won and you stood to win enough to pay for your college education—yet you did not collect a dime. Your friend's boss was just fired without warning and for no obvious reason.

What happened in all these cases? You probably can guess the answer: the athletes, including the horse, and the boss all failed a drug test because their urine or blood contained traces of illegal substances. In the case of the athletes, the drugs likely were performance-enhancing, whereas your friend's boss may have been using one or more "recreational drugs" like marijuana or cocaine. Using performance-enhancing or mood-affecting drugs, legal or illegal, is certainly not limited to athletes and executives, of course, as evidenced by the males who have made Viagra and its analogs "cash cows" for their producers and natives in the high Andes who chew coca leaves as a means to allow them to continue working in thin air and under brutally cold conditions (see the Historical Highlight section at the end of Chapter 20).

It is not only in modern times that humans have attempted to improve their performances through the use of various medicinal substances. Indeed, records from the second century BC show that participants in the first Olympiads consumed exotic mushrooms and seeds in an effort to be the "best of the best." Interestingly, public awareness of drug testing for athletes is mainly associated with the Olympic games, the Tour de France, and, most recently, Major League Baseball, but it is only since the 1960s that the International Olympic Committee (IOC) and other supervisory organizations have instituted random testing for substances that these groups have deemed as prohibited. The banned substances are determined by the World Anti-Doping Agency, and the number of prohibited substances is constantly increasing. At the time of the 2008 Summer Olympic Games in Beijing, there were some 200 compounds listed, whereas the figure was 125 six years earlier.

The number of Olympians subjected to random testing of their urine or blood has increased dramatically since the testing program was started in 1968, when a total of 753 tests were performed in the Winter and Summer Games. The corresponding combined figure for the 2006 Winter Games in Turin, Italy, and the 2008 Summer Games in Beijing was some 4800, an increase of over 600%. Unfortunately, the increased testing has been accompanied by corresponding growth in the number of participants—particularly those in the Summer Games—found to be using banned substances. For example, only one competitor was caught using a prohibited drug during the Mexico City Summer Games (1968), and this was a pentathlete who had consumed an alcoholic beverage to steady his hand in a shooting competition. Although you may not consider ethanol to be a sedative, it actually is when consumed in small amounts. The number of athletes violating the rules on banned substances during the summer games increased to seven at Munich (1972) and to 12 at Los Angeles (1984). Only two competitors were found to be using banned drugs in Atlanta (1996), but the games in Sydney (2000) were marred by the identification of seven participants who used prohibited substances; six of them lost their medals as a result. In the Athens games (2004) and those in Beijing (2008), a dozen athletes in each were disqualified because of failed drug tests, although in the latter case, four of the disqualifications were because the horses that the competitors were riding in equestrian events had been doped.

Many fewer violations have been discovered for athletes participating in the Winter Olympic Games, even considering that only about 20% as many individuals compete in them as compared to the Summer Games. Since 1968, a total of only seven athletes have been disqualified for violating the IOC's drug policy.

Classes of substances currently banned by the IOC include stimulants (amphetamines, cocaine, caffeine, pseudoephedrine), narcotics (heroin, methadone), anabolic agents (testosterone, nandrolone), diuretics (acetazolamide, spironolactone), and peptide hormones and their mimetics and analogs (erythropoietin, human growth hormone). Alcohol and marijuana at specified levels are also prohibited. For the latter, a level of as little as 15 ng/mL of 11-nor-Δ^8-tetrahydrocannabinol-9-carboxylic acid is a violation of the rules. A complete

(Continued)

HISTORICAL HIGHLIGHT *Who's Taking What? Analysis of Biological Fluids (Continued)*

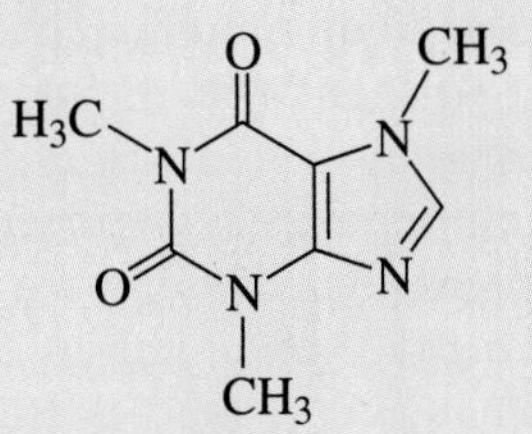

Caffeine

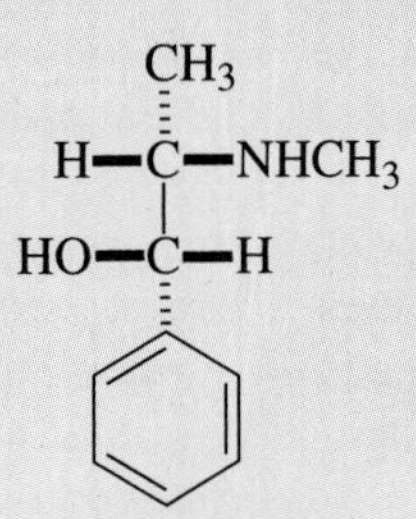

Pseudoephedrine (threo)

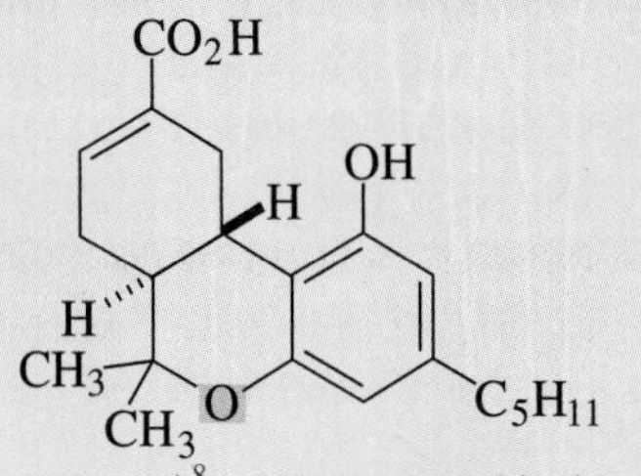

11-Nor-Δ^8-tetrahydrocannabinol-9-carboxylic acid (THCA)

list of banned substances as of 2009 is available at http:// www.wada-ama.org/en/prohibitedlist.ch2.

The presence of compounds like caffeine and pseudoephedrine on the list of banned drugs may surprise you. Caffeine is present in coffee, tea, and many types of soft drinks, among other liquids that an athlete might drink. However, the specified limit for caffeine makes it unlikely that a consumer would unwittingly exceed the limit: To violate the rules, about eight cups of coffee would have to be drunk within approximately two hours prior to collection of the urine sample. Pseudoephedrine is found in a number of over-the-counter nasal decongestants such as Dimetapp™ and Sudafed™, but once again, the levels that trigger a violation are such that it would be difficult to do so unwittingly. A violation associated with pseudoephedrine, incidentally, was the reason Andreea Raducan, a female Romanian gymnast, was stripped of her gold medal in the 2000 Summer Olympics in Sydney in a highly publicized and controversial ruling. Her team doctor prescribed the use of a product containing this substance because Raducan allegedly had a cold.

One of the more interesting recent controversies surrounding drug testing is associated with nandrolone (17β-hydroxy-19-nor-4-androsten-3-one). This is an anabolic steroid—a steroid that contributes to the building of muscle tissue—that naturally occurs in the body, albeit in miniscule amounts. The allowed threshold level set by the IOC for nandrolone in urine, in the form of its metabolite, 19-norandrosterone, is 2 ng/mL. Some athletes have been found to have levels some 100 times higher than this, a result that suggests illegal use of the steroid. Some creative alibis have been provided by implicated individuals: One claimed that his high reading was the result of having consumed a serving of spaghetti Bolognese in which the meat sauce contained beef from cattle that had been fed anabolic drugs; another cited his having had sex with his pregnant wife as the cause of his elevated value! However, most of those who have been accused cite their use of legal dietary supplements like "weight-gain" protein milkshakes and the α-amino acid creatine as the cause of their apparent violation, and recent studies suggest that this might be possible. In one of them, three volunteers gave urine specimens prior to taking the suspect dietary supplements; none of the specimens showed a level of 19-norandrostrone that exceeded 2 ng/mL. They then took the supplements and, 24 hours later, provided a second urine sample. Two of the volunteers, neither of

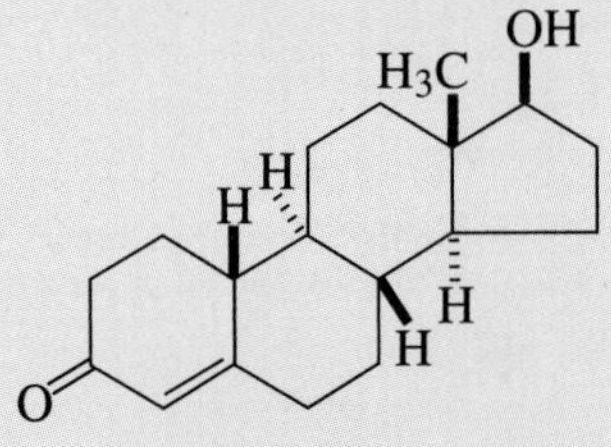

Nandrolone

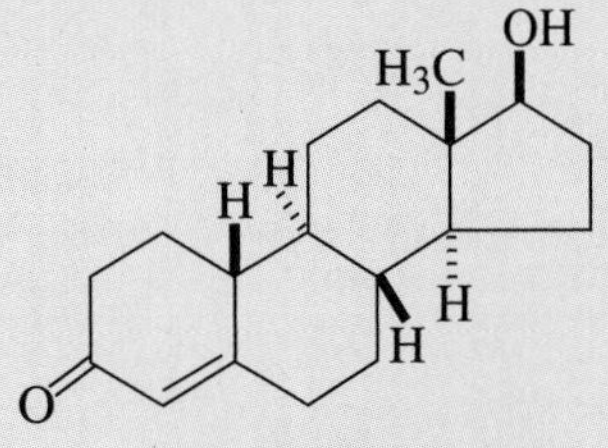

19-Norandrosterone

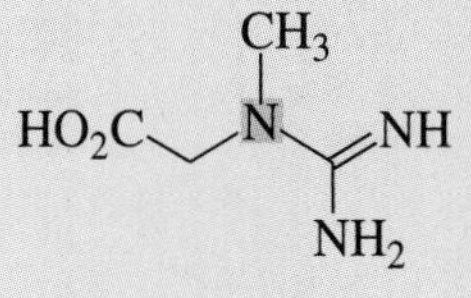

Creatine

(Continued)

HISTORICAL HIGHLIGHT *Who's Taking What? Analysis of Biological Fluids (Continued)*

whom had exercised during this period, again had only a low level of the metabolite in their urine. However, the third individual, who had exercised, had a level of 10 ng/mL, which exceeds the allowed threshold level. Thus there may be a relationship between consumption of the dietary supplements often used by athletes, their training regimens, and the production of nandrolone and its metabolite. Further studies are needed to clarify this confusing picture.

Just how are these drugs or their metabolites detected in urine and blood, fluids that contain myriad compounds? The answer to this is found in a powerful combination of the techniques of extraction (Chap. 5), gas-liquid chromatography (Sec. 6.4), and mass spectrometry (Sec. 8.5). The aqueous solution (urine or blood) is extracted with an organic solvent such as dichloromethane, and the extracts are subjected to analysis by GC-MS (Sec. 6.4). Illegal substances may then be identified through meticulous comparison of the retention times and fragmentation patterns of the analyte with those of known standards.

A major challenge to the analytical chemist in testing for the presence of performance-enhancing drugs is the fact that new substances are constantly becoming available, and experimental protocols must be developed for detecting them. Moreover, athletes are becoming more adept at masking their drug use by consuming substances that interfere with the analytical procedure, rendering it inconclusive. In other words, detecting violations of rules governing the use of performance-enhancing drugs is a moving target, one requiring continuing development of new methods to ensure that violators are identified unambiguously and that non-violators are not wrongly prosecuted because of a "false-positive" test.

Relationship of Historical Highlight to Chapter

The analysis of the components of complex mixtures remains one of the greatest challenges to the experimentalist. In this context, the development of new chromatographic techniques has contributed significantly to solving many of the most formidable problems in achieving analyses of such mixtures. The combination of gas chromatography with mass spectrometry in the form of GS-MS instruments represents an effective strategy for taking advantage of the separating ability of chromatography with the analytical power of a mass spectrometer. This strategy has been extended to liquid chromatography as well, in the form of instruments that link a liquid chromatograph with a mass spectrometer, a technique termed LC-MS.

See more on *Drug Testing*

See more on *Pseudoephedrine*

See more on *Nandrolone*

See more on *Norandrosterone*

See more on *Creatine*

See more on *Caffeine*

See more on *Erythropoietin*

See more on *Acetazolamide*

See more on *Spironolactone*

See more on *Tetrahydrocannabinol*

CHAPTER 7

Stereoisomers

When you see this icon, sign in at this book's premium website at **www.cengage.com/login** to access videos, Pre-Lab Exercises, and other online resources.

See more on *Chirality/ Nobel Prize*

See more on *Odor/ Nobel Prize*

Most organic molecules are three-dimensional, and the physical and chemical properties of such molecules are determined by the relative orientations of the various atoms and functional groups that comprise the molecule. Information about the shapes of organic molecules is important because it may help us understand how two molecules interact with each other; this knowledge may have practical applications. For example, an enzyme binds its substrate because of favorable bimolecular interactions that result from shape complementarity of the two three-dimensional molecules. In such cases, knowledge of the shape of the enzyme and the substrate can lead to the design and development of novel molecules that bind to the active site of an enzyme, thereby blocking its mode of action. These small organic molecules may be useful drugs for treating diseases. Similarly, our senses of smell and taste are dependent upon specific interactions between small organic molecules and three-dimensional receptors in our nose and on our tongue. Hence, our ability to enjoy a healthy life and to appreciate the odors and flavors we experience every day is a consequence of the shapes of molecules. So while you are studying stereoisomers, don't forget that it is the complementarity of molecular shapes that enables you to smell the roses and to enjoy a delicious meal!

7.1 INTRODUCTION

Understanding the three-dimensional properties of organic molecules is an essential part of organic chemistry. The experiments in this chapter are designed to provide an introduction to **stereoisomers.** Such isomers have molecular skeletons that are identical, but they differ in the three-dimensional orientation of their atoms in space. The two broad subclasses of stereoisomers that are of importance in organic chemistry are **conformational** isomers and **configurational** isomers.

Conformational isomers, as illustrated by the two **Newman projections 1** and **2** for 1,2-dibromoethane, are stereoisomers that are interconverted by rotation about single bonds. The interconversion of such isomers is usually a low-energy process and occurs rapidly at room temperature and below. Consequently, isolation of conformational isomers is *rarely* possible.

Configurational isomers, in contrast, are stereoisomers whose interconversion normally requires the breaking and remaking of a chemical bond, usually a high-energy process. An example of this type of stereoisomerism is seen with carvone, a natural product for which the two representations **3** and **4** can be written. These are related as **nonsuperimposable mirror images,** so **3** and **4** are defined as **enantiomers.** Molecules having a nonsuperimposable mirror image are said to be **chiral,** a term that is derived from the Greek word *cheir,* meaning "hand." Because hands are perhaps the best-known example of nonsuperimposable mirror images, we use the word "chiral" to characterize molecules that have a "handedness." A companion term, **achiral,** describes all molecules having **superimposable mirror images.**

See more on *Chirality*
See more on *Chirality Game*

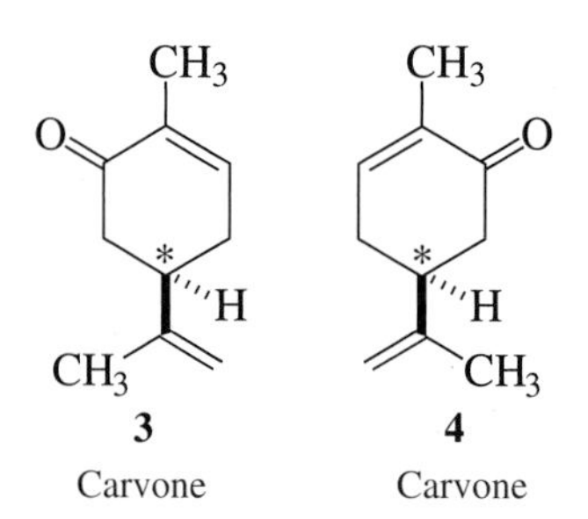

Enantiomers have identical physical and chemical properties in an achiral environment. Their differential effect on plane-polarized light, however, is an important exception to this general rule: Enantiomers rotate the plane of such light an equal number of degrees but in *opposite* directions. For this reason they are sometimes called **optical isomers** and are said to possess **optical activity.** Under most circumstances, the chiral compounds that you will prepare or use in the laboratory will be 50:50 mixtures of the two enantiomers, a composition referred to as the **racemate** or the **racemic mixture.** An equimolar mixture of **3** and **4** would thus be called a racemate and would produce *no* net rotation of plane-polarized light.

Frequently the source of chirality in an organic molecule is the presence of a **center of chirality,** also called a **stereocenter,** which is usually a carbon atom bearing four different substituents. This is the case for the atom marked with an asterisk in **3** and **4.** The substituents are a hydrogen atom, an isopropenyl group, and two methylene groups. The methylene groups are distinct from each other because one of them is bound to a carbon atom of a carbon-carbon double bond, and the other to a carbon atom doubly bound to oxygen.

A chiral molecule containing two or more stereocenters may exist as **diastereomers,** which are configurational isomers that are *not* enantiomers. Consider 1,2-cyclohexanediol, **5,** for example, which contains the stereocenters marked by asterisks. Each center consists of a carbon atom bound to a hydroxyl group, a hydrogen atom, a methylene group, and the carbon atom of the other center of chirality. Three configurational isomers, **6–8,** can then be written for **5.** The *cis* isomer **6** is clearly different from the *trans* isomers **7** and **8,** which are enantiomers of one another and diastereomers of **6.** Moreover, even though it contains two chirality centers, **6** has a *superimposable mirror image.* We could therefore variously describe this isomer as being achiral, optically inactive, or **meso.** The last term describes those configurational isomers that have two or more chirality centers *and* a superimposable mirror image. Although the identity of the physical and chemical properties of the enantiomers **7** and **8** might make their separation from one another difficult, the isolation of **6** free from contamination by **7** and **8** would be comparatively easy because *diastereomers have different physical and chemical properties as a general rule.*

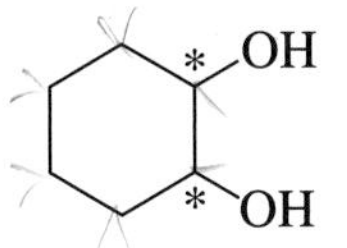

5 1,2-Cyclohexanediol

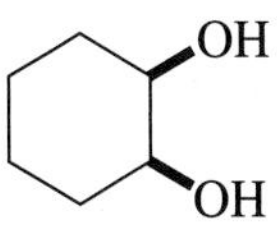

6 *cis*-1,2-Cyclohexanediol

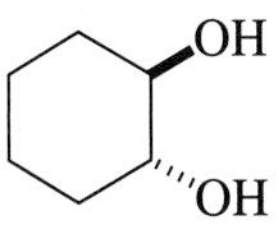

7 *trans*-1,2-Cyclohexanediol

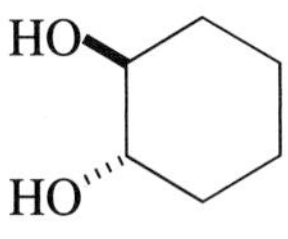

8 *trans*-1,2-Cyclohexanediol

Even isomeric pairs of *achiral* molecules may exist as separable stereoisomers. Examples include *cis-* and *trans-*, or (*Z*)- and (*E*)-, alkenes such as **9** and **10,** and cyclic compounds such as **11** and **12.** The molecules are nonsuperimposable, yet they are *not* mirror images; that is, they are *not* enantiomers. Rather, they are diastereomers and have different chemical and physical properties.

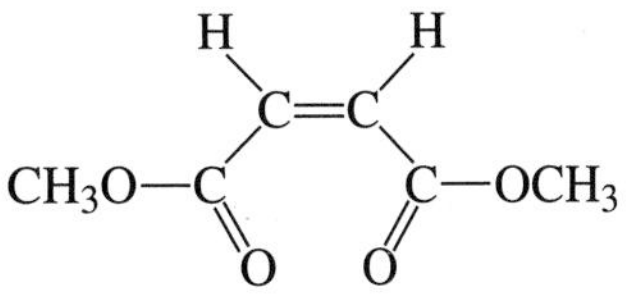

9 Dimethyl maleate

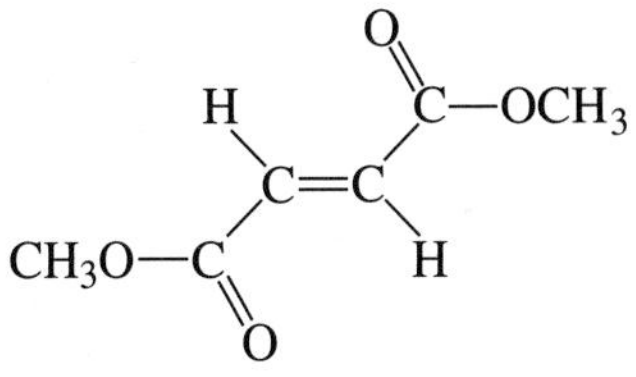

10 Dimethyl fumarate

11 *cis*-1,4-Dimethylcyclohexane

12 *trans*-1,4-Dimethylcyclohexane

In contrast to the case with diastereomers, separation or **resolution** of enantiomers is generally difficult because of the identity of their chemical properties; their differing effect on plane-polarized light is not helpful for separation. Nonetheless, various experimental techniques have been developed to accomplish resolutions, and one of them is presented in Section 7.6.

The procedures in this chapter are chosen to study (a) separating diastereomers by chromatography (Sec. 7.2), (b) converting one diastereomer into another (Sec. 7.3), (c) evaluating some chemical and physical properties of enantiomers (Sec. 7.4), and (d) resolving a racemate (Sec. 7.6). A description of the technique of polarimetry is given in Section 7.5 to support this last study.

7.2 SEPARATION OF DIASTEREOMERIC 1,2-CYCLOHEXANEDIOLS

Cyclic 1,2-diols can be produced stereoselectively from cycloalkenes. For example, a racemic mixture of the enantiomeric *trans*-1,2-cyclohexanediols **7** and **8** is produced by reaction of a peracid with cyclohexene and hydrolysis of the intermediate epoxide **13** (Eq. 7.1). Oxidation of this alkene by permanganate, on the other hand, gives *cis*-1,2-cyclohexanediol (**6**), as shown in Equation 7.2. The intermediate in this process is presumably the cyclic manganese-containing species **14**, which is not isolated.

Cyclohexene $\xrightarrow[\text{A peracid}]{RCO_3H}$ **13** Cyclohexene oxide $\xrightarrow{H_3O^+}$ **7** + **8** (7.1)

Cyclohexene + MnO_4^- (Permanganate ion) $\longrightarrow$ [**14**] $\xrightarrow{H_3O^+}$ **6** (7.2)

In this experiment, a commercial mixture containing the *meso*-isomer **6** and the *racemate* **7/8** will be separated into its components. This mixture is produced by the catalytic hydrogenation of catechol (**15**) (Eq. 7.3). Because the enantiomeric *trans*-diols **7** and **8** have identical physical properties, except for their effect on plane-polarized light, they will not be resolved from one another in the procedure of this section. The *meso*-isomer **6**, on the other hand, being diastereomeric to **7** and **8**, has physical and chemical properties that differ sufficiently to allow its separation from the other two stereoisomers.

15 Catechol + $3\ H_2$ $\xrightarrow[\text{Heat, pressure}]{\text{Catalyst}}$ **6** + **7** + **8** (7.3)

The separation of diastereomers from one another may often be done solely on the basis of differing solubilities, but this is not the case with the 1,2-cyclohexanediols because they have similar solubilities. The desired separation is possible by chromatographic means, however, and the technique of TLC (Sec. 6.2) is used here, although column (Sec. 6.3) or gas chromatography (Sec. 6.4) would also work.

A challenge to the experimentalist is determining which spot on the TLC plate corresponds to the racemic mixture of *trans* isomers and which to the *cis* diastereomer. In other words, it is one thing to separate different compounds and quite another to assign specific structures to them. This common problem can be addressed in a variety of ways. For example, you might extract the portions of the TLC plate containing each of the isomers and determine the melting point or IR or NMR spectral characteristics (Chap. 8) of the solid remaining after removal of solvent. Spectral characterization is difficult to do in this case because only extremely small amounts of material would be obtained from the TLC plate. In addition, the melting points of the diastereomers are similar, 100–101 °C for **6** and 103–104 °C for **7** and **8**; thus, traces of impurities in the *trans* isomer could easily lower its melting range (Sec. 3.3) to that of the *cis*, making unambiguous assignment of structure impossible.

The method used to make the needed structural assignment in this experiment is to compare the "unknown" compounds with an authentic specimen of "known" material. Specifically, by spotting the TLC plate with authentic samples of either *cis*- or *trans*-1,2-cyclohexanediol as well as the commercial mixture, you will be able to make a direct comparison of R_f-values and determine which component of the mixture is which.

EXPERIMENTAL PROCEDURES

Separation of Diastereomeric 1,2-Cyclohexanediols

Purpose To demonstrate that diastereomers may be readily separated by chromatography.

SAFETY ALERT

1. **Acetone and petroleum ether are *highly* flammable; *do not have any flames in the vicinity* when they are being used.**
2. **Do not allow iodine to come in contact with your skin and do not inhale its vapor; as it is corrosive and *toxic*.**

Procedure

Preparation Sign in at **www.cengage.com/login** to answer Pre-Lab Exercises, access videos, and read the MSDSs for the chemicals used or produced in this procedure. Read or review Section 6.2.

Apparatus Capillary pipets, 2-cm × 10-cm silica gel TLC plate, TLC chamber.

Setting Up Obtain or prepare approximately 5% solutions in acetone of a commercial mixture of *cis*- and *trans*-1,2-cyclohexanediol and of pure *trans*-1,2-cyclohexanediol. Place small spots of each solution side by side on the TLC plate.

Elution and Analysis Develop the chromatogram (Fig. 6.3b) using a mixture of 75% petroleum ether, bp 60–80 °C (760 torr), and 25% 2-propanol by volume. Remove the plate from the solvent mixture when the solvent front approaches the top of the plate, mark the position reached by the solvent front, air-dry the plate for a few minutes, and place it in a closed container with a few iodine crystals to make the spots visible. Remove the plate and circle the location of any spots. Record the R_f-value of each spot.

Discovery Experiment

Solvent Effects on R_f-Values

Investigate the effect of changing the ratios of the two solvents used for the separation of the 1,2-cyclohexanediols on the R_f-values for the diastereomers. One way to report the relative effectiveness of various solvent mixtures in effecting the separation would be to report values of ΔR_f for each mixture used. Consult with your instructor before carrying out any experiments.

WRAPPING IT UP

Transfer any excess of the *eluting solvents* to a container for nonhalogenated organic liquids. Dispose of the used *TLC plates* in a container for nonhazardous materials.

EXERCISES

1. The maximum number of stereoisomers that are possible for a given molecule can be determined from the formula 2^n, where n is the number of stereocenters in the molecule.

a. Using this formula, what is the maximum number of stereoisomers that could exist for 1,2-cyclohexanediol (**5**)?

b. Why are there only three stereoisomers of 1,2-cyclohexanediol (**5**) rather than the number you calculated in part **a?**

2. A *meso* compound is achiral by virtue of the fact that it contains an internal mirror plane of symmetry, so the molecule is superimposable on its mirror image and hence does not have an enantiomer. Use a dashed line to show the internal plane of symmetry present in the structure of *cis*-1,2-cyclohexanediol (**6**).

3. Indicate the stereochemical relationship (i.e., enantiomers or diastereomers) between each of the following pairs of stereoisomers of 1,2-dibromocyclohexane, and indicate whether each pair will have identical or different physical properties.

a. 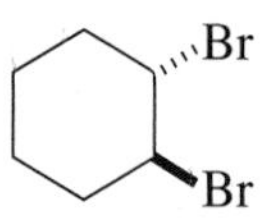 Br / Br and Br / Br

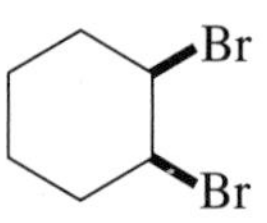

b. 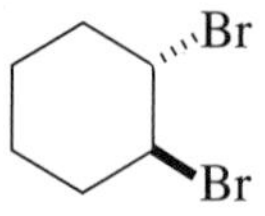 Br / Br and Br / Br

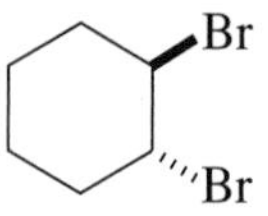

c. 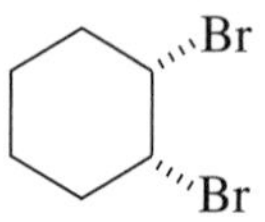 Br / Br and 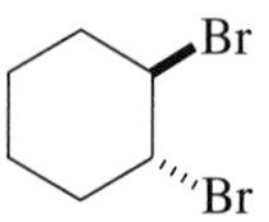 Br / Br

4. Which, if any, of the molecules in Exercise 3 are *meso* compounds?
5. By comparing the R_f-values of the separated spots from the mixture of the *cis* and *trans* diols with the R_f-value of the *trans* isomer, decide which isomer is adsorbed more strongly by the silica gel. Give a reason for the difference in adsorptivity of the two isomers.
6. It may happen that the separation of the isomers by TLC is not complete—one of the spots may be found to be smeared out to some extent. Considering this, which one of the isomers could more easily be obtained pure by column chromatography?
7. What would be the consequence of using pure petroleum ether as the eluting solvent?
8. What would be the consequence of using pure 2-propanol as the eluting solvent?
9. In Section 7.1, it was stated that interconversion of configurational isomers requires the cleavage and reformation of a chemical bond. Indicate any such bonds in *cis*-1,2-cyclohexanediol (**6**) that upon being broken and then reformed would afford the enantiomers **7** and **8.** Be sure to describe what must occur stereochemically during the process of cleavage and remaking of a bond to effect the interconversion of **6** to the other two diastereomers.
10. **a.** Consider the bond dissociation energies of the various bonds in **6,** and then predict which bond would be easiest to break for the conversion of **6** to **7** and **8.**

    b. Estimate the $\Delta H^{\ddagger}$ for this process.

7.3 ISOMERIZATION OF DIMETHYL MALEATE TO DIMETHYL FUMARATE

Dimethyl maleate (**9**) and dimethyl fumarate (**10**) are examples of *cis* and *trans* isomers, called **geometric isomers.** Their names do not reflect this isomeric relationship, however, because the parent diacids from which they can be prepared were discovered and named before this stereochemical fact was known. Being diastereomers, these diesters would be expected to have different physical and chemical properties, and indeed they do. The difference in their physical states at room temperature is the most dramatic: The maleate **9** is a *liquid* with a freezing point of about −19 °C, whereas the fumarate **10** is a *solid* with a melting point of 103–104 °C.

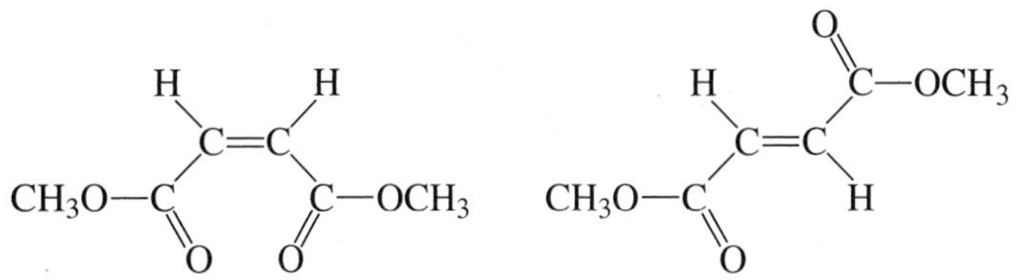

H H; C=C; CH_3O—C(=O) C(=O)—OCH_3

9

Dimethyl maleate

H C(=O)—OCH_3; C=C; CH_3O—C(=O) H

10

Dimethyl fumarate

As discussed in Section 7.1, the interconversion of configurational isomers involves the cleavage and subsequent remaking of a chemical bond. The diastereomers **9** and **10** could be equilibrated by breaking the π component of the carbon-carbon double bond, rotating about the single bond that remains, and remaking the π-bond. The required bond cleavage can be promoted (a) thermally, (b) photochemically, or (c) chemically. The latter possibility involves the reversible addition of a reagent to one of the carbon atoms involved in the carbon-carbon double bond, resulting in breaking of the π-bond.

The goal of this experiment is to formulate a mechanism by which dimethyl maleate (**9**) is isomerized to dimethyl fumarate (**10**) by the action of heat, light, or molecular bromine. Three different experiments will be performed to study the reaction. This involves exposing two samples of **9**, one of which contains a catalytic amount of bromine, to light and treating a third sample with bromine but keeping the mixture in the dark at room temperature. By thoughtful analysis of your results, you should be able to discover which of the three modes of breaking the π-bond, which were designated (a)–(c) in the previous paragraph, is responsible for the isomerization. The analysis will also allow you to determine whether polar intermediates or radical species are involved in the isomerization.

There will be residual bromine at the end of the procedure, which must be destroyed prior to isolating the crystalline **10**. A convenient way to do this is by adding cyclohexene, which reacts with bromine to produce a dibromide (Eq. 7.4), to the reaction mixture.

$$\text{Cyclohexene} + Br_2 \longrightarrow \text{Trans-1, 2-Dibromocyclohexane} \quad (7.4)$$

EXPERIMENTAL PROCEDURES

Isomerization of Dimethyl Maleate to Dimethyl Fumarate

Purpose To study the mechanism of a geometric isomerization.

SAFETY ALERT

1. **Bromine is a hazardous chemical that may cause serious chemical burns. Do not breathe its vapors or allow it to come into contact with the skin. Perform all operations involving the transfer of solutions of bromine at a hood and wear latex gloves when handling bromine or solutions of it. If you get bromine on your skin, *immediately* wash the affected area with soap and warm water and soak the skin in 0.6 *M* sodium thiosulfate solution, for up to 3 hr if the burn is particularly serious.**
2. **Bromine reacts with acetone to form α-bromoacetone, $BrCH_2COCH_3$, a powerful *lachrymator.* Do not rinse glassware that might contain residual bromine with acetone! Rather, follow the procedure described in "Wrapping It Up."**

MINISCALE PROCEDURE

Preparation Sign in at **www.cengage.com/login** to answer Pre-Lab Exercises, access videos, and read the MSDSs for the chemicals used or produced in this procedure. Read or review Sections 2.9, 2.17, and 3.2.

Apparatus Three 13-mm × 100-mm test tubes, incandescent light source, apparatus for vacuum filtration, and *flameless* heating.

Setting Up Measure 0.5 mL of dimethyl maleate into each of the test tubes. Working at the hood and wearing latex gloves, add enough of a 0.6 *M* solution of bromine in dichloromethane dropwise with the filter-tip pipet to *two* of the tubes to give an orange solution. Add an equal volume of dichloromethane to the third test tube. Loosely cork all the test tubes.

Isomerization and Isolation Place one of the tubes containing bromine in the dark, and expose the other two tubes to strong light. Place the tubes about 20 cm from the light source to prevent boiling the solvent and popping the cork off the tubes. If decoloration of a solution occurs, add an additional portion of the bromine solution. After 30 min, record in which test tube(s) crystals have appeared and add 1–2 drops of cyclohexene to the tubes containing bromine to destroy it. Add 2–3 mL of 95% ethanol to any tubes containing solid, and heat the mixture to effect dissolution. Then allow the solution to cool slowly to room temperature. Isolate the precipitate by vacuum filtration using a Hirsch funnel (Fig. 2.54). Press the precipitate as dry as possible on the filter disk with the aid of a clean cork or spatula, transfer it to a watchglass or piece of weighing paper, and air-dry it.

Analysis Weigh the dimethyl fumarate (**10**) and calculate its yield, based on the amount of dimethyl maleate used in the solution(s) that produced **10.** Determine its melting point and compare the result with the reported melting point. If the observed melting point is low, recrystallize the product from ethanol.

MICROSCALE PROCEDURE

Preparation When you see this icon, sign in at this book's premium website at **www.cengage.com/login** to access videos, Pre-Lab Exercises, and other online resources. Read or review Sections 2.9, 2.17, and 3.2.

Apparatus Two Craig tubes, 10-mm × 75-mm test tube, incandescent light source, apparatus for Craig tube filtration, and *flameless* heating.

Setting Up Using a Pasteur pipet, dispense two drops of dimethyl maleate into each of the Craig tubes and the test tube. Working at the hood, add enough of a 0.1 *M* solution of bromine in dichloromethane dropwise to the *Craig tubes* to produce an orange-colored solution. Add sufficient dichloromethane to the *test tube* so that the volume of solution in the tube approximately equals that in the Craig tubes. Loosely cork each of the tubes.

Isomerization and Isolation Place one of the Craig tubes in a dark location, and expose the other two solutions to strong light. Place the tubes about 20 cm from the light source to prevent boiling the solvent and popping the cork off the tubes. Should decoloration of a solution occur, add an additional portion of the bromine solution. After 20 min, record in which test tube(s) crystals have appeared and add 1 drop of cyclohexene to the tubes containing bromine to destroy it. Add 0.5–1 mL of 95% ethanol to any tubes containing solid, and heat the mixture to effect dissolution. Then allow the solution to cool slowly to room temperature.

Filter any solutions containing solids by Craig tube filtration (Fig. 2.55). Transfer the product to a watchglass or piece of weighing paper, and air-dry it.

Analysis Weigh the dimethyl fumarate (**10**). Determine its melting point and compare the result with the reported melting point. If the observed melting point is low, recrystallize the product from ethanol using a Craig tube.

Discovery Experiment

Iodine as a Catalyst for Isomerization

Devise a protocol for determining whether iodine will promote the isomerization of dimethyl maleate to dimethyl fumarate, and use the same three conditions as used with bromine. You may also wish to explore the relative rates at which isomerization occurs with the two different halogens. Consult with your instructor before carrying out any experiments.

Discovery Experiment

Assessing Purities of Dimethyl Maleate and Fumarate

Devise analyses that will allow you to assess the purity of the dimethyl maleate used and of the dimethyl fumarate produced in the isomerization experiment.

WRAPPING IT UP

Decolorize any *solutions or containers in which the color of bromine* is visible by dropwise addition of cyclohexene, and then discard the resulting solutions in a container for halogenated organic liquids. Flush the *ethanolic filtrate* from the recrystallization process down the drain with water.

EXERCISES

1. What observation allows exclusion of thermal cleavage of the π-bond as the initiation step for isomerization of **9** to **10**?
2. What observation allows exclusion of direct photoexcitation of **9** as the means of initiating its conversion to **10**?
3. Depending on the type of chemical reaction, molecular bromine (Br_2) can be a source of Br^+, Br^-, or $Br^•$.
   a. Provide a Lewis dot structure for each of the three species of mono atomic bromine.
   b. Using curved arrows to symbolize the flow of electrons, show how Br^+, Br^-, and $Br^•$ may be formed from Br_2.
   c. Based upon the results of this experiment, indicate whether Br^+, Br^-, or $Br^•$ initiates the isomerization of dimethyl maleate to dimethyl fumarate. Explain your reasoning.

4. Using curved arrows to symbolize the flow of electrons, write a stepwise mechanism for the isomerization of **9** to **10.** Be certain to account specifically for the roles, if any, that light and bromine play in promoting this process.
5. Explain why decoloration of a solution of bromine and **9** is slow.
6. State which one of the two diesters **9** and **10** is predicted to be more stable thermodynamically, and support your prediction.
7. What type of configurational isomers are **9** and **10,** enantiomers or diastereomers? How is your answer consistent with the fact that these two compounds have different physical properties?
8. The double bonds of **9** and **10** contain both a σ- and a π-component. Explain why rupture of the π-bond rather than the σ-bond is the more probable event in the isomerization process.

7.4 PROPERTIES OF THE ENANTIOMERS OF CARVONE

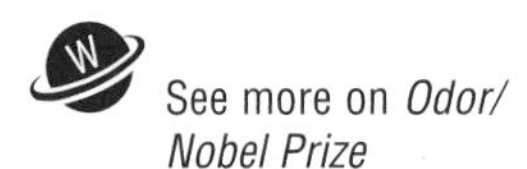
See more on *Odor/Nobel Prize*

Enantiomers have *identical* physical properties except with respect to their effect on plane-polarized light. Their chemical properties are also identical, with one *important* exception: One enantiomer undergoes a chemical change at a rate *different* from that of its mirror image *if* the reaction occurs in a stereochemically asymmetric environment, such as the active site of an enzyme or a biological receptor. This is because the active sites of enzymes and receptors are themselves chiral and, as a consequence, complex preferentially with one member of a pair of enantiomers. The complexed enantiomer will then undergo the enzyme-promoted reaction faster than will its mirror image.

A dramatic example of such differential complexation is the olfactory response to the two enantiomers of carvone. (*R*)-(−)-Carvone (**3**) smells like spearmint and is the principal component of spearmint oil, which also contains minor amounts of limonene (**16**) and α- and β-phellandrene (**17** and **18,** respectively). On the other hand, (*S*)-(+)-carvone (**4**), along with limonene, is found in caraway seed and dill seed oils and has been shown to be the compound largely responsible for the characteristic odor of these oils. This remarkable difference in how we sense these two enantiomers is because the odor receptors in our noses are chiral environments that are linked to the nervous system. The type of receptor site that complexes with a particular enantiomer will determine what odor is detected by the brain.

3 (*R*)-(−)-Carvone **4** (*S*)-(+)-Carvone **16** Limonene **17** α-Phellandrene **18** β-Phellandrene

It is unusual for both enantiomers of a compound to occur naturally, and it is extremely rare for each of them to be available from readily accessible sources such as spearmint and caraway oils. This fortunate situation enables the separate isolation of **3** and **4** from their respective sources by vacuum distillation (Sec. 2.15). Because of their ready isolation, both of these enantiomers are commercially available.

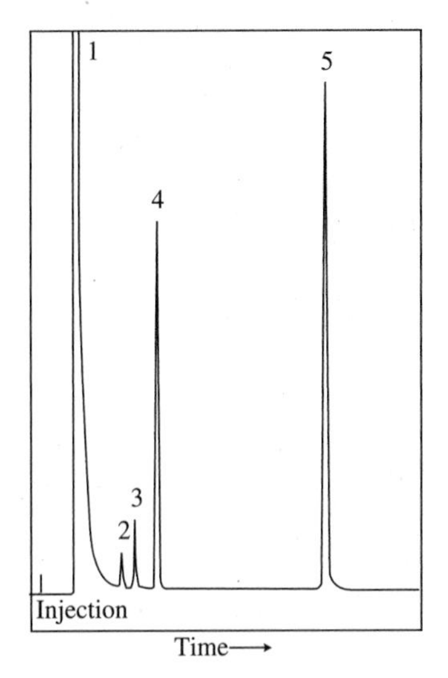

Figure 7.1
GLC analysis of spearmint oil (numbers in parentheses are retention times in minutes and peak areas, respectively). Peak 1: solvent; peak 2: unknown; peak 3: phellandrenes (2.16, 17819); peak 4: limonene (2.46, 84889); peak 5: (R)-(−)-carvone (5.13, 248316).

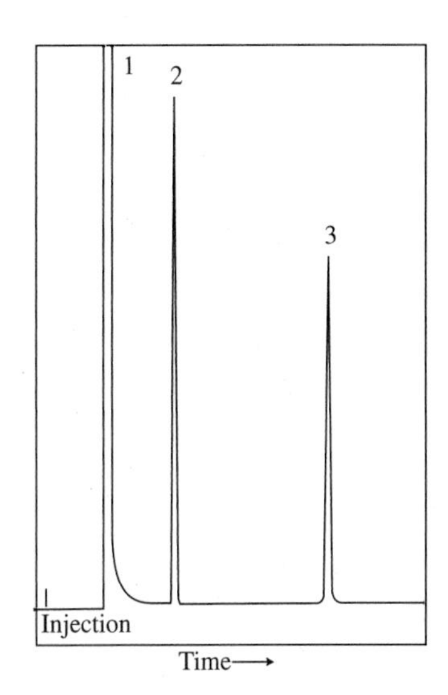

Figure 7.2
GLC analysis of caraway seed oil (numbers in parentheses are retention times in minutes and peak areas, respectively). Peak 1: solvent; peak 2: limonene (2.46, 125719); peak 3: (S)-(+)-carvone (5.13, 385950).

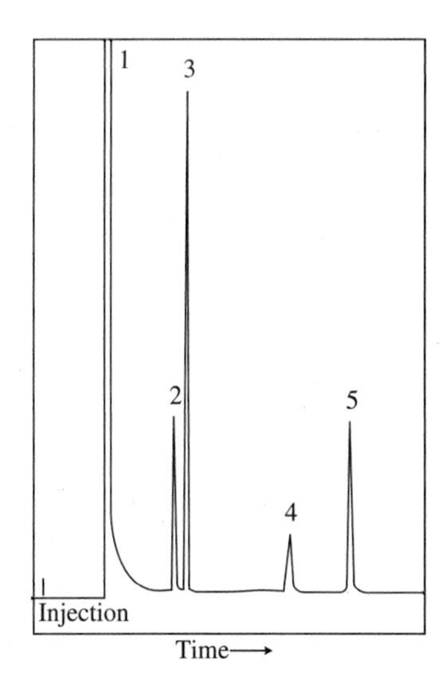

Figure 7.3
GLC analysis of dill seed oil (numbers in parentheses are retention times in minutes and peak areas, respectively). Peak 1: solvent; peak 2: unknown (2.27, 49418); peak 3: limonene (2.44, 139972); peak 4: unknown (4.15, 17091); peak 5: (S)-(+)-carvone (5.07, 78223).

The investigative experiments that you perform on the enantiomeric carvones will depend upon the time and facilities that are available. If samples of both enantiomers are available, you can verify their different odors. Bear in mind, however, that 8–10% of the population cannot detect the difference, and you may be in this group. Should a gas-liquid chromatograph (Sec. 6.4) be available, analyses may be performed on samples of spearmint, caraway oils, and dill seed as well as solutions containing one or both of the enantiomers **3** and **4.** Typical chromatograms are given in Figures 7.1–7.3.

Note that under the conditions of these analyses, limonene (**16**) and the phellandrenes **17** and **18** elute *before* **3** and **4** because their boiling points, 170–177 °C (760 torr), are lower than that of the enantiomers of carvone (bp 232–233 °C, 760 torr).

Carvone $\xrightarrow[\textit{Red}\text{ solution}]{2\,Br_2/CH_2Cl_2}$ tetrabromide (*Colorless*) (7.5a)

Carvone $\xrightarrow[\textit{Purple}\text{ solution}]{KMnO_4/H_2O}$ tetraol (*Colorless*) + $MnO_2\downarrow$ (*Brown* precipitate) (7.5b)

$$H_3C\text{-(carvone)} + H_2N\text{-}NH\text{-}C_6H_3(NO_2)_2 \longrightarrow \text{carvone 2,4-dinitrophenylhydrazone} \qquad (7.6)$$

2,4-Dinitrophenylhydrazine Carvone 2,4-dinitrophenylhydrazone

The presence of a carbon-carbon double bond in carvone can be shown by performing a test for unsaturation using bromine in dichloromethane (Eq. 7.5a) and aqueous potassium permanganate (Eq. 7.5b). Confirmation of the carbonyl function in **3** and/or **4** and of the structural assignment may be obtained by making the 2,4-dinitrophenylhydrazone (2,4-DNP) of carvone (Eq. 7.6) and comparing its melting point to that reported (193–194 °C) for authentic material. By performing a mixture melting point (Sec. 3.3) of the 2,4-DNPs prepared from both **3** and **4**, you can determine whether one enantiomeric derivative depresses the melting point of the other.

EXPERIMENTAL PROCEDURES

Properties of the Enantiomeric Carvones

Purpose To explore the physical and chemical properties of enantiomers.

SAFETY ALERT

1. **Bromine is a hazardous chemical that may cause serious chemical burns. Do not breathe its vapors or allow it to come into contact with the skin. Perform that all operations involving the transfer of solutions of bromine at a hood and wear latex gloves when handling bromine or solutions of it. If you get bromine on your skin, *immediately* wash the affected area with soap and warm water and soak the skin in 0.6 *M* sodium thiosulfate solution, for up to 3 h if the burn is particularly serious.**
2. **Bromine reacts with acetone to form α-bromoacetone, $BrCH_2COCH_3$, a powerful *lachrymator*. Do not rinse glassware that might contain residual bromine with acetone! Rather, follow the procedure described in Wrapping It Up.**
3. **Wear latex gloves to prevent solutions of potassium permanganate from contacting your skin. These solutions cause unsightly stains on your skin and last several days. If contact occurs, wash the affected area thoroughly with warm water.**

Procedure

Preparation Sign in at **www.cengage.com/login** to answer Pre-Lab Exercises, access videos, and read the MSDSs for the chemicals used or produced in this procedure. Read or review Sections 6.3, 25.7, and 25.8.

1. **Odor.** Compare the odor of the carvone obtained from spearmint oil with that of the carvone from caraway seed oil.
2. **Tests for Unsaturation.** Obtain samples of **3** and/or **4,** and test for the presence of a carbon-carbon double bond using procedures A and/or B in Section 25.8. Record your observations.
3. **Gas-Liquid Chromatography.** Obtain gas chromatograms of spearmint and caraway oils and of the enantiomeric carvones. Co-inject an authentic sample of **3** or **4** with one of the natural oils to determine which peak in the oil corresponds to carvone. Use the data to determine (a) the retention time of the carvones and the relative percentage of carvone **3** or **4** compared to the total amount of the three compounds **16–18** that may be present in the essential oils. A capillary column with dimensions of 0.25 mm × 25 m with 100% dimethylpolysiloxane as the stationary phase using a flowrate of 1.2 mL/min at an oven temperature of 110 °C is suitable for the analysis.
4. **Carvone 2,4-Dinitrophenylhydrazone.** Using a 0.5-g sample of **3** or **4,** or of the oils from which they are obtained, prepare a 2,4-dinitrophenylhydrazone derivative according to procedure A in Section 25.7 and determine its melting point. If recrystallization is necessary, use a mixture of ethanol and ethyl acetate as the solvent. Obtain a sample of the 2,4-DNP prepared from the enantiomer opposite from yours. Make a 1:1 mixture of the enantiomeric 2,4-DNPs by intimately mixing the two, and determine the melting point of the mixture. Compare the result with the melting point you originally obtained.

WRAPPING IT UP

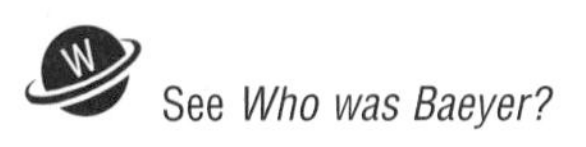

See *Who was Baeyer?*

Transfer the *pot residues* from the distillation of spearmint and/or caraway oil to the container for nonchlorinated organic wastes. After performing the prescribed tests on your distillate, put the isolated *natural products* into appropriately labeled containers. Place the *dichloromethane solution* from the bromine test for unsaturation in a container for halogenated organic liquids; put the *manganese dioxide* from the Baeyer test for unsaturation in a container for heavy metals. Neutralize and then filter any excess 2,4-dinitrophenylhydrazine solution. Put the *filter cake* in the container for nonhazardous solids. Flush the *filtrate* as well as *filtrates* obtained from recrystallization down the drain.

EXERCISES

1. Why would a positive test for unsaturation on spearmint, caraway seed, or dill seed oil *not* prove that the enantiomeric carvones contain unsaturation?
2. Bromine is known to add to carbon-carbon double bonds in an *anti*-fashion; namely, one bromine atom adds from one face of the carbon-carbon double bond and the other bromine adds from the opposite face. Draw all of the possible stereoisomeric forms of the tetrabromo products obtained upon the addition of bromine to (*R*)-(−)-carvone (**3**) (see Eq. 7.5a).
3. What is it about the properties of enantiomers that makes it easy, at least for most people, to distinguish the enantiomers of carvone by smell but difficult to distinguish them experimentally in the laboratory? What common laboratory technique could be used to distinguish the two enantiomers of carvone?

4. If you developed a new type of GLC column that *could* distinguish between *R*- and *S*-carvone, in what fundamental way would this column be different from the one you used for the GLC analysis in this experiment?
5. There is no *meso* form for carvone. What symmetry element is absent in carvone that would allow for the existence of a *meso* form?
6. Based on the results of the GLC analysis of the enantiomeric carvones, would you expect the *boiling point* of a *mixture* of the enantiomers to be the same as that of a *single* enantiomer? Explain.
7. Assume that the peak areas given in Figures 7.1 and 7.2 for limonene (**16**), the phellandrenes **17** and **18**, and the carvones **3** and **4** reflect the relative *molar* amounts of these compounds. Compute the percentage of the carvone present in spearmint and caraway seed oils.
8. Figure 7.3 is a gas chromatogram of dill seed oil.
   a. Using the same assumption as that in Exercise 7, compute the percentage of the carvone present in this oil.
   b. Explain why the retention time of the carvone in this oil does *not* allow assignment of which enantiomer is present.
9. When a student attempted to prepare the 2,4-DNP of (*S*)-(+)-carvone directly from dill seed oil, a red-orange solid was obtained that melted at 110–115 °C even after recrystallization. This result suggests that a mixture of DNPs has been produced and that they are not easily separated by recrystallization. How does the GLC analysis of this oil (Fig. 7.3) support the proposition that there could be a significant amount of a carbonyl-containing contaminant of carvone in dill seed oil? What might you do experimentally to support your proposal?
10. What melting point should you expect from a mixture of the 2,4-DNPs of (*R*)-(−)-carvone (**3**) and (*S*)-(+)-carvone (**4**) if both derivatives are pure?
11. Identify the center of chirality in limonene (**16**) and draw the (*R*)-enantiomer of this molecule.
12. Show that the phellandrenes **17** and **18** are chiral.
13. In Section 7.1 it was stated that interconversion of configurational isomers involves the cleavage and remaking of a chemical bond. Indicate the four different bonds in **3** that could be broken and then remade to produce **4.** For each bond, be sure to describe what must occur stereochemically during the process of bond cleavage/remaking that effects the interconversion of one enantiomer to the other.

7.5 POLARIMETRY

A solution of chiral molecules exhibits optical activity (Sec. 7.1) because there is a net rotation of the plane of polarized light as the light passes through the sample. An explanation of the physical basis for this phenomenon is beyond the scope of this discussion, but it is important to understand that achiral molecules do *not* exhibit this same property.

Of primary interest to organic chemists is the fact that each member of a pair of enantiomers rotates plane-polarized light by exactly the same amount but in *opposite* directions. This means that a solution containing equal amounts of two enantiomers will cause *no* net rotation of the light, because the rotation in one direction by molecules of one enantiomer will be exactly offset with an opposite rotation by those of the other. Thus, a solution containing enantiomers can produce a rotation only if the amount of one enantiomer exceeds that of its mirror image.

The **observed rotation** expressed by the value **α** depends on a number of factors: (a) the nature of the chiral compound; (b) its *concentration*, if in solution, or *density*, if a neat liquid; (c) the *pathlength* of the cell containing the sample; (d) the *temperature* at which the measurement is taken; (e) the nature of the *solvent*, if one is used; and (f) the *wavelength* of plane-polarized light used in the measurement. Factors (b) and (c) determine the average number of chiral molecules in the path of the light and, in combination with the wavelength, have the greatest effect on the magnitude of the rotation of a sample.

The sign of α is defined by convention. When you view through the sample *toward* the source of the plane-polarized light, a clockwise rotation of the plane of the light is taken as positive or **dextrorotatory,** whereas a counterclockwise rotation is negative or **levorotatory.**

Optical rotation represents a **physical constant** of a chiral compound *if* the variables listed above are considered. By specifying the temperature and wavelength at which the measurement is taken, and dividing the observed rotation by the factors that define the average number of molecules in the light path, a constant called the **specific rotation, [α]**, is obtained. This is expressed mathematically by Equation 7.7.

$$[\alpha]_{\lambda}^{T} = \frac{\alpha}{l \times c} \text{ or } \frac{\alpha}{l \times d} \qquad (7.7)$$

where

$[\alpha]$ = specific rotation (degrees)
λ = wavelength (nanometers, nm)
T = temperature (°C)
α = observed rotation (degrees)
l = pathlength of cell (decimeters, dm)
c = concentration (g/mL of solution)
d = density (g/mL, neat sample)

A specific rotation might be reported as $[\alpha]_{490}^{25} -19.6°$ (c = 1.8, $CHCl_3$). Alternatively, the specific rotation might be written as $[\alpha]_{D}^{23} + 39.2°$ (c = 12.1, CH_3OH). The "D" refers to the fact that a sodium lamp emitting radiation at 589 nm, the D line of sodium, was used as the light source. The chemical formulas given with c specify the solvent used for the measurement.

The basic components of all **polarimeters,** the instruments used for measuring optical rotations, are shown in Figure 7.4. Ordinary light first passes through a polarizer, such as a Nicol® prism or a sheet of Polaroid® film, and the plane-polarized light that emerges passes through the sample tube. If the tube contains an optically active substance, the plane of the light will be rotated either clockwise or counterclockwise. The degree of rotation can be determined by turning the analyzer, another Nicol prism or piece of Polaroid film, until the intensity of the light reaching the detector matches that observed in the absence of any of the

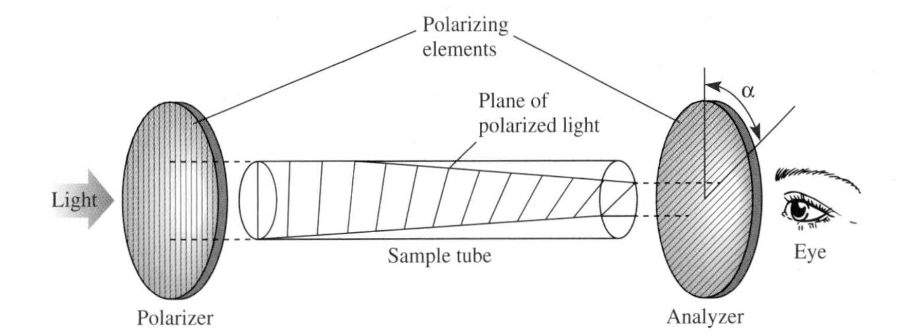

Figure 7.4
Schematic illustration of optical rotation ($\alpha = \sim-45°$).

optically active compounds in the sample tube. The human eye or a photoelectric cell constitutes the detector. Finally, using Equation 7.7 allows the conversion of the observed rotation α to the specific rotation $[\alpha]$.

7.6 RESOLUTION OF RACEMIC 1-PHENYLETHANAMINE

Enantiomers cannot normally be physically separated (see Historical Highlight at the end of this chapter) because they have identical properties, except toward plane-polarized light (Sec. 7.1). However, *diastereomers* do have different physical properties, such as solubility, melting and boiling points, and adsorptive characteristics in chromatography. Recognition of this characteristic of diastereomers has led to a general strategy for separating enantiomers that involves three steps: (1) conversion of the enantiomers to diastereomers by way of a chemical reaction; (2) separation of these diastereomers by any of a number of standard techniques, such as recrystallization and chromatography; and (3) regeneration of one or both of the pure enantiomers from the corresponding diastereomer. This is illustrated schematically in Equation 7.8, where ***R*** and ***S*** represent the enantiomers to be separated and R' stands for a *single* enantiomer of a compound, termed a **resolving agent,** with which both ***R*** and ***S*** can react. To convince yourself that ***RR′*** and ***SR′*** are actually *diastereomers,* you need only show that the *enantiomer* of RR', for instance, would be written as SS'.

$$\underset{\text{Racemic mixture}}{R + S} \xrightarrow{R'} \underset{\text{Diastereomeric mixture}}{RR' + SR'} \xrightarrow[\text{technique}]{\text{Separation}} \underset{\text{Separated diastereomers}}{\begin{cases} RR' \xrightarrow{\text{Regeneration}} R + R' \\ SR' \xrightarrow{\text{Regeneration}} S + R' \end{cases}} \quad \text{Pure enantiomers} \tag{7.8}$$

Nature is the source of a number of different resolving agents that are used to prepare separable diastereomeric pairs from enantiomers. In this experiment, we use (+)-tartaric acid (**19**), which is produced from grapes during the production of wine, to resolve racemic 1-phenylethanamine (**20**), as shown in the resolution scheme outlined in Figure 7.5. Conversion of the enantiomers of **20** into separable diastereomers involves an acid–base reaction with **19.** The diastereomeric salts that result have different solubilities in methanol and are separable by fractional

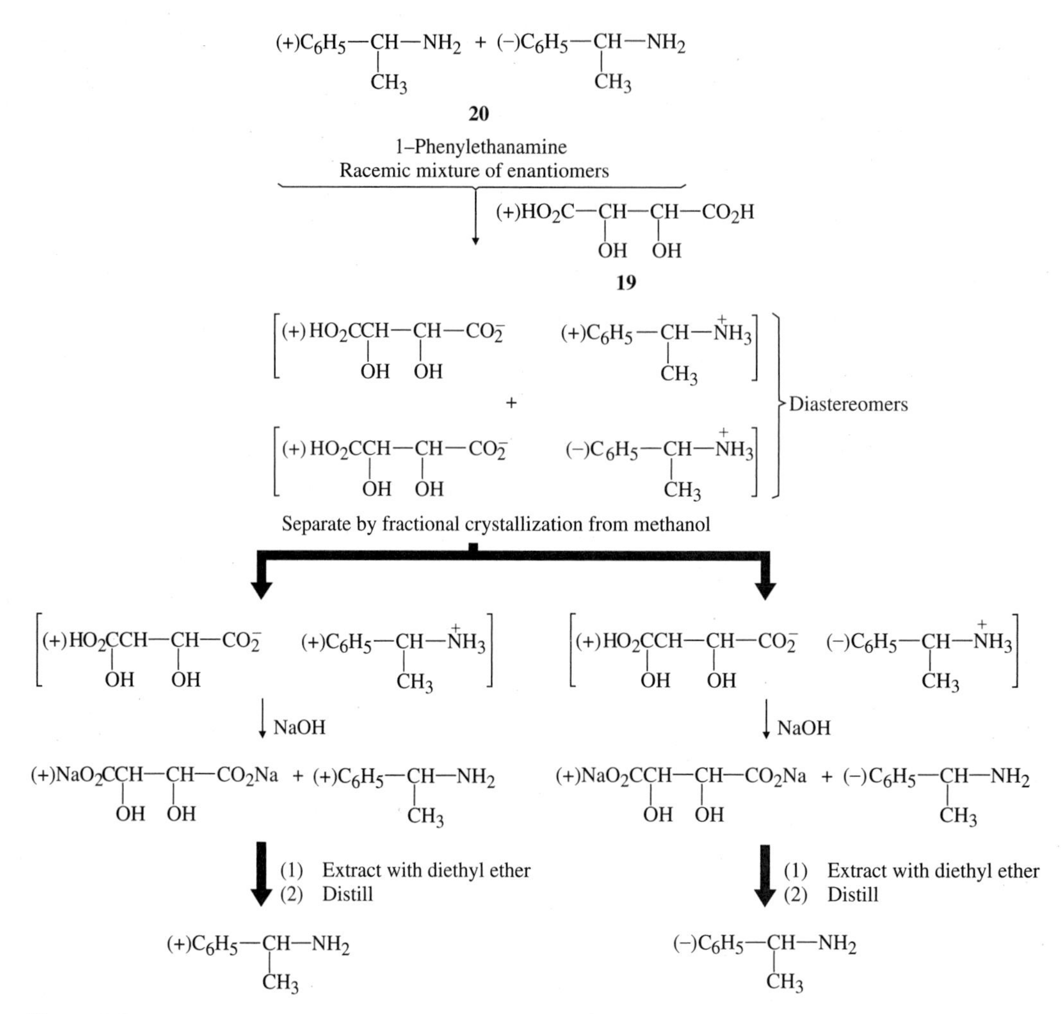

Figure 7.5
Resolution of (±)-1-phenylethanamine.

crystallization. The individual enantiomers of **20** are then regenerated by neutralization of the corresponding salts.

In principle, it would be possible to obtain both enantiomers of a chiral compound in high optical purity by this technique, but this generally would require an undesirably large number of crystallizations. Fortunately, one or two crystallizations usually suffice to allow isolation, in reasonable optical purity, of the enantiomer derived from the less-soluble salt.

Measuring the optical rotation (Sec. 7.5) of the 1-phenylethanamine recovered in this experiment allows you to determine whether it is the (+)- or (−)-enantiomer and the extent to which optical purification is achieved. If you are to do this part of the experiment, consult with your instructor for special experimental directions.

EXPERIMENTAL PROCEDURE

Resolution of Racemic 1-Phenylethanamine

Purpose To demonstrate a method for resolving enantiomers.

SAFETY ALERT

1. **Methanol is flammable; when heating it, use *flameless* heating.**
2. **Avoid contact of 14 *M* NaOH with skin, as this solution is corrosive. Wear latex gloves when handling this solution. If skin contact occurs, wash the area with copious amounts of water.**
3. **Diethyl ether is extremely volatile and flammable. When using it for extractions or distilling it, be certain that there are *no flames* in the vicinity.**

MINISCALE PROCEDURE

Preparation Sign in at **www.cengage.com/login** to answer Pre-Lab Exercises, access videos, and read the MSDSs for the chemicals used or produced in this procedure. Read or review Sections 2.6, 2.9, 2.11, 2.13, 2.15, 2.17, 2.25, and 2.29.

Apparatus A 500-mL Erlenmeyer flask, apparatus for magnetic stirring, vacuum filtration, and *flameless* heating.

Setting Up Dissolve an *accurately* weighed sample of approximately 5 g of racemic 1-phenylethanamine in 35 mL of methanol and determine its specific rotation [α] using a polarimeter and Equation 7.7. The amounts of 1-phenylethanamine and methanol specified here give satisfactory results when a polarimeter sample tube with a light path of 2 dm or more and a volume of 35 mL or less is used. If a tube of shorter light path and/or larger volume is used, either increase the amount of racemic 1-phenylethanamine or combine your resolved product with that of other students to give a solution having sufficient concentration to allow accurate measurement of the optical rotation. Consult with your instructor about the type of polarimeter sample tube that is available.

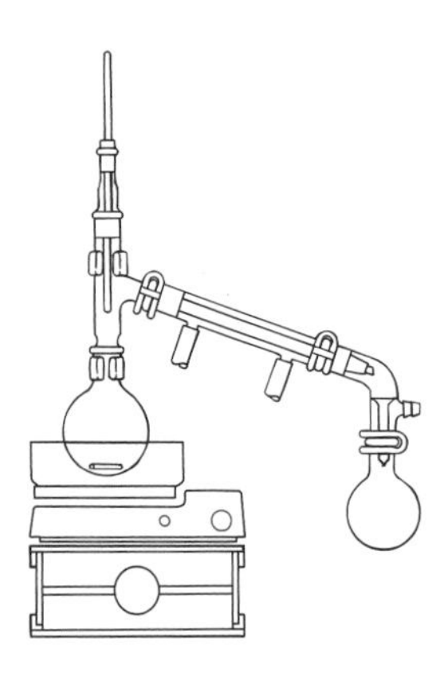

Diastereomer Formation and Isolation Place 15.6 g of (+)-tartaric acid and 210 mL of methanol in the Erlenmeyer flask and heat the mixture to boiling using *flame-less* heating. To the stirred hot solution, *cautiously* add the solution recovered from the polarimeter and 7.5 g of racemic 1-phenylethanamine. Allow the solution to cool slowly to room temperature and to stand *undisturbed* for 24 h or until the next laboratory period.★ The amine hydrogen tartrate should separate in the form of white prismatic crystals. If the salt separates in the form of needlelike crystals, the mixture should be reheated *until all* the crystals have dissolved and then allowed to cool slowly. If any prismatic crystals of the salt are available, use them to seed the solution.

Collect the crystals of the amine hydrogen tartrate by vacuum filtration using a Büchner funnel (Fig. 2.54); wash them with a small volume of *cold* methanol.

Do *not* discard the filtrate (see Wrapping It Up). Dissolve the crystals in four times their weight of water, and add 8 mL of 14 *M* sodium hydroxide solution. Extract the resulting mixture with four 35-mL portions of diethyl ether. Do *not* discard the aqueous solution (see Wrapping It Up). Wash the combined ethereal extracts with 25 mL of saturated aqueous sodium chloride, and then dry the ethereal solution over several spatula-tips full of anhydrous sodium sulfate for 10–15 min; swirl the flask; if the solution remains cloudy after this period of time, add additional portions of sodium sulfate. Decant or filter the dried solution from the desiccant.★

Remove the diethyl ether from the solution by simple distillation. (Alternatively, use one of the techniques discussed in Section 2.29.) Then distill the residual 1-phenylethanamine under vacuum with a shortpath apparatus (Fig. 2.37b). It will be necessary to use electrical heating for the distillation because the amine has a boiling point of 94–95 °C (28 torr).

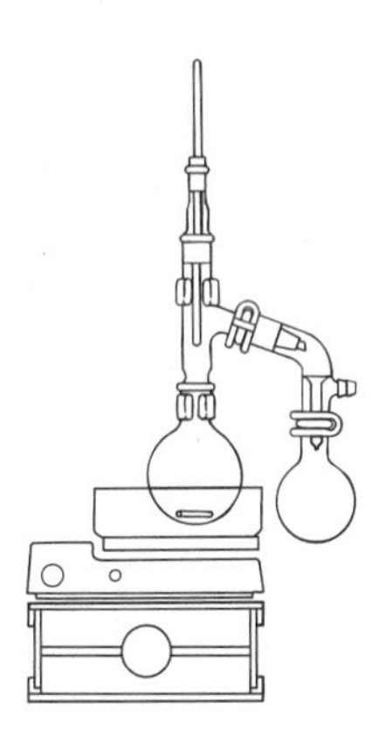

Analysis Determine the yield of the resolved 1-phenylethanamine (**20**). Combine your product with that of another student to give a total of 3 g of **20.** Weigh the combined product accurately, and then transfer it *quantitatively* into about 30 mL of methanol. Measure the volume of the methanolic *solution* accurately. It is the *weight,* in grams, of the 1-phenylethanamine divided by the *volume* of the *solution,* in mL, that provides the concentration *c* for Equation 7.7. Transfer the solution to the polarimeter sample tube, measure the observed rotation, and determine the specific rotation of the sample by using Equation 7.7. Optically pure (*R*)-(+)-1-phenylethanamine has $[\alpha]_D^{20}$ +30° (c = 10, CH_3OH).

WRAPPING IT UP

Transfer the recovered *diethyl ether* to a container for nonhalogenated organic liquids, and put the *methanolic filtrate* from isolation of the less-soluble amine tartrate either in this same container or in a special container from which the more-soluble amine tartrate could be recovered; consult your instructor for specific directions. Place the *methanolic solution of 1-phenylethanamine* either in a container so labeled or in the one labeled for nonhalogenated organic liquids. Transfer the *sodium hydroxide solution* into a container so labeled; the tartaric acid salt it contains can be recovered.

EXERCISES

1. The agent used in this experiment to resolve racemic 1-phenylethylamine is (+)-tartaric acid (**19**).

HO_2C–CH(OH)–CH(OH)–CO_2H

19

(+)-Tartaric acid

a. How many stereocenters does **19** contain?

b. Draw *all* possible stereoisomers for **19** and indicate whether each will be optically active or not. Indicate whether the possible pairs of stereoisomers of tartaric acid are enantiomers, diastereomers, or *meso*.

c. Which, if any, of the other stereoisomers of **19** might be used to resolve racemic 1-phenylethylamine?

d. Would there be any possible advantage to using a different stereoisomer of **19** to resolve racemic 1-phenylethylamine? Explain your answer.

e. Is the optical purity of the resolving agent important? Explain your answer.

2. Consider the reaction of racemic 1-phenylethylamine with racemic tartaric acid.

a. How many stereochemically distinct salts will be formed?

b. What is the stereochemical relationship (i.e., enantiomeric, diastereomeric, or identical) between the different possible pairs of products formed in this reaction?

c. What pairs of salts should be separable by crystallization and what pairs would be inseparable by crystallization?

3. When a label is prepared for the product that has been isolated, it should include the magnitude of the specific rotation as well as its sign. Suppose the observed rotation was found to be 180°. How could you determine whether the rotation was (+) or (−)?

4. Given the value for the specific rotation of optically pure 1-phenylethanamine, calculate the optical purity of the sample of resolved amine for which you obtained a rotation.

5. How might the enantiomeric purity of the product be increased?

6. Describe clearly the point in the experimental procedure at which the major part of the other enantiomer was removed from your product.

7. How could the (+)-tartaric acid be recovered so that it can be reused to resolve additional racemic amine?

8. The absolute configuration of one of the stereocenters in (+)-tartaric acid is *R*. Explain why the absolute configuration at the other stereocenter must also be *R* and provide a perspective drawing of the (*R, R*)-enantiomer of tartaric acid.

9. The absolute configuration of (+)-1-phenylethanamine is *R*. Make a perspective drawing of this configuration.

10. Suppose you had prepared a racemic organic *acid* and it was necessary to resolve it. Outline a procedure to illustrate how this could be done.

HISTORICAL HIGHLIGHT

Discovery of Stereoisomers

Louis Pasteur (1822–1895) is better known for his contributions to microbiology than to chemistry, although he began his professional career as a chemist. In 1848, when Pasteur was only 24 years old and had just received his doctorate from the Sorbonne in Paris, he undertook the study of a salt of racemic acid, a substance deposited on wine casks during the fermentation of grape juice. (The word *racemic* is derived from the Latin word *racemus,* meaning "a bunch of grapes.") Another chemist, E. Mitscherlich, had earlier reported that the sodium ammonium salt of racemic acid was identical in all ways to a salt of tartaric acid that is also found on wine casks, *except* that the salt of tartaric acid was optically active, whereas the salt of racemic acid was not.

Pasteur was puzzled by the fact that although the salts of tartaric acid and racemic acid were said to be identical in many ways, including chemical composition and even crystalline form, they had different effects on plane-polarized light. Specifically, the salt of racemic acid had *no* effect, whereas the salt of tartaric acid was dextrorotatory. When Pasteur examined the crystals of the salt of racemic acid produced according to Mitscherlich's description, he observed something that Mitscherlich had not. There were, in fact, *two* kinds of crystals present that were related to one another as a left hand is to a right hand. Pasteur carefully separated the "left-handed" crystals from the "right-handed" crystals, using tweezers under a microscope. When he had a sufficient quantity of each type of crystal, he did something that arose from either a hunch or a flash of genius. He separately dissolved some of each kind of crystal in water and placed the solutions in turn in a polarimeter. He found that *both* solutions were optically active, the solution of the "left-handed" crystals rotating the polarized light counterclockwise, and the solution of the "right-handed" crystals rotating the polarized light clockwise. The author Vallery-Radot, in *The Life of Pasteur* (1902), reported that the young scientist was so excited by his discovery that he, "not unlike Archimedes, rushed out of the laboratory exclaiming 'I have it!' "

When Pasteur carefully measured the amounts of each kind of crystal for making the solutions, he found that equal amounts produced exactly the same degree of rotation, *but in opposite directions.* Moreover, the magnitude of rotation by the "right-handed" crystals was the same as that given by a similar solution of the salt of tartaric acid. Thus, Pasteur demonstrated that his "right-handed" racemic acid salt was actually identical to the dextrorotatory tartaric acid salt, and his "left-handed" racemic acid salt was a previously unknown mirror-image form of tartaric acid salt, as shown below. Finally, Pasteur made a mixture of equal amounts of the two kinds of crystals and found, as he expected, that a solution of this mixture was optically inactive.

(+)-Tartrate	Mirror plane	(–)-Tartrate
CO_2^-		CO_2^-
H—C—OH		HO—C—H
HO—C—H		H—C—OH
CO_2^-		CO_2^-

By separating the two kinds of crystals of the salt of racemic acid, Pasteur accomplished the first and most famous example of what chemists now call resolution of a racemic mixture. The name of the specific acid, *racemic,* that Pasteur studied is now used generally to describe any equimolar mixture of enantiomers such as the "left-handed" and "right-handed" salts of Pasteur.

One is likely to think that the separation of the crystals of different shapes and the demonstration that their solutions rotate the plane of polarized light in opposite directions was a clever but rather trivial experiment. In a limited sense it was, but from a broader perspective the experiments had tremendous importance. This was the first time it had been shown that organic compounds exist in enantiomeric forms at the molecular level. Before Pasteur's experiments, the effect of quartz crystals on polarized light could be explained on the basis of the shape of the crystals themselves, because when the quartz crystals were melted, the optical activity disappeared. Pasteur's work showed that the difference in the crystalline form of the salts of racemic acid reflected a difference

(Continued)

HISTORICAL HIGHLIGHT *Discovery of Stereoisomers (Continued)*

in the three-dimensional shape of the molecules themselves. Thus, despite the fact that dissolving the salts of racemic acid in water destroyed their crystal structures just as that of quartz is destroyed by melting it, the resulting *solutions* of the two kinds of crystals of Pasteur's salt still exhibited optical activity. Pasteur later converted the separated salts of racemic acid into the corresponding acid forms and showed that there were indeed two isomeric forms, "left-handed" or (−)-tartaric acid and "right-handed" or (+)-tartaric acid.

Pasteur's observation of the different shapes of the crystals, his separation of them, and his deductions about the meaning of their opposite effects on polarized light were indeed acts of genius, but accidents played a large part in his discovery. There were two remarkable elements of chance that entered into the findings. The first was that the sodium ammonium salt of racemic acid, which was the one Pasteur examined, has been shown to be one of the only salts of this acid that crystallizes in mirror-image forms that are visually different and can be separated mechanically. Second, the crystallization in these two forms occurs only at temperatures below 26 °C (79 °F); above 26 °C, the crystals that form are identical and have no optical activity. To obtain his crystals, Pasteur had placed the flask containing the solution of racemic acid on a cool window ledge in his laboratory in Paris and left it there until the next day for the crystallization to occur. Except for the combination of a fortuitous choice of the proper salt of racemic acid and the cool Parisian climate, Pasteur would not have made his important observations.

By his discovery, he recognized that a direct relationship exists between molecular geometry and optical activity. This led him to propose that molecules that rotate plane-polarized light in equal but opposite directions are related as an object and its mirror image. However, it remained for two other chemists, Van't Hoff and Le Bel, to explain 25 years later exactly how the atoms could be assembled into such molecular structures. In the interim, Pasteur turned his attention with great success to the biological problems mentioned earlier, but it was his pioneering work on the resolution of racemic acid that led the way for other chemists to explain the relationship of chirality or "handedness" in molecular structure to biological activity, and this is the real significance of Pasteur's work.

Many commercially available drugs have one or more stereocenters. Those that are derived from natural sources are available as single enantiomers. However, those that are synthesized in the laboratory are usually obtained and marketed as racemic mixture because of the costs associated with preparing them in enantiomerically pure form. The physiological targets of many drugs include enzymes, nucleic acids, or carbohydrates, and because each of these biomolecules is chiral, the interactions between enantiomeric drugs and their macromolecular targets will be different. Enantiomers of a drug may also be metabolized to different products. Hence, depending on the situation, enantiomers may have physiological properties and biological activities that range from similar to very different. One may have toxic side effects, for example, whereas the other does not.

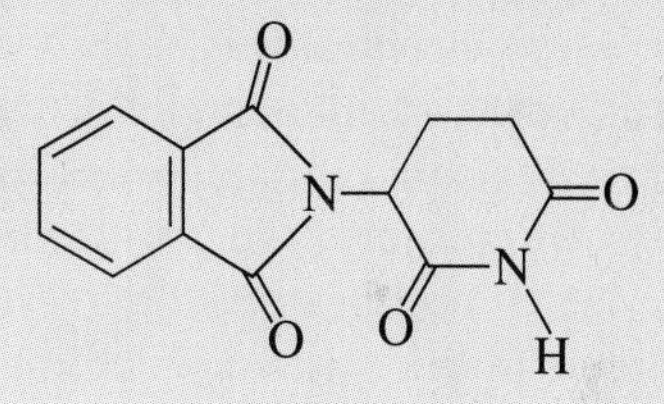

O, N, O, N, O, O, H

Thalidomide

CH_3, $(CH_3)_2CHCH_2$, $CHCO_2H$

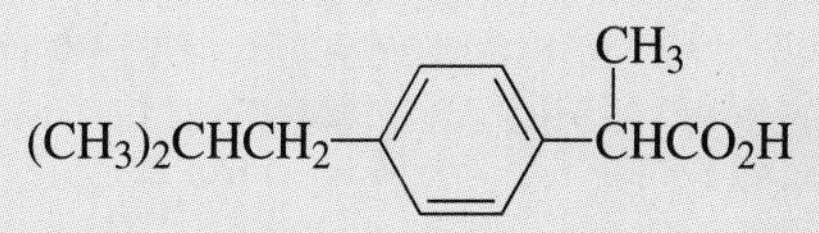

Ibuprofen

The importance of the chirality of molecules is tragically illustrated by the case of **thalidomide**, whose structure is shown above. In the 1950s and 1960s, this drug was prescribed to pregnant women in Europe as a sedative and as a treatment for morning sickness. The (+)-enantiomer had stronger sedative properties than the (−)-enantiomer, but the commercial drug was the racemic mixture. Unfortunately, it was found that the (−)-enantiomer was a potent **teratogen**, an agent that causes fetal deformities, when it was discovered that babies born to women who were administered the drug during the first trimester had numerous serious birth defects. It was later determined that the (+)-enantiomer also

(Continued)

HISTORICAL HIGHLIGHT *Discovery of Stereoisomers (Continued)*

exhibited mild teratogenic activity and that each of the two enantiomers underwent rapid racemization *in vivo.* Thalidomide was then banned from the marketplace, although it has recently been approved by the FDA for use in treating Hansen's disease (leprosy) and is currently used in combination with other drugs for greating multiple myeloma.

A less dramatic example of the importance of chiral drugs is presented by ibuprofen, a common analgesic (pain-suppressing) and nonsteroidal anti-inflammatory agent that is the pharmacologically active component in over-the-counter products such as Nuprin® and Advil®. Like thalidomide, ibuprofen has a single center of chirality, but only the *S*-enantiomer is effective for relief of pain. For example, the *S*-enantiomer provides relief in only 12 minutes, whereas the racemic drug requires 30 minutes. However, the *R*-isomer is slowly converted to the *S*-isomer in the body, so it too can be therapeutically useful. The commercially available drug is generally sold as the racemate, but the pure *S*-enantiomer is now also available.

That the different enantiomers of drugs may have vastly different pharmacological properties has led to a rapidly growing interest in developing efficient methods for the synthesis of single enantiomers. Another incentive in the context of "chiral synthesis" is the fact that the Food and Drug Administration requires that the pharmacological and toxicological properties of the enantiomers of a chiral compound must be evaluated *individually* even if the substance is to be sold as the racemate, and the costs for doing so are extremely high. A recent article describing the importance of enantiopure compounds in the development of new drugs and other materials is provided in *Chemical and Engineering News,* August 4, 2008, pages 12–20.

See more on *Pasteur*

See *Who was Van't Hoff?*

See *Who was Le Bel?*

See more on *Tartaric Acid*

See more on *Thalidomide*

See more on *Ibuprofen and Analogs*

See more on *Chiral Chemistry*

Relationship of Historical Highlight to Experiments

Enantiomers of chiral compounds can sometimes be distinguished not only by their equal, but opposite, effects on plane-polarized light, as first demonstrated by Pasteur, but also by their odors, if they are volatile enough. The enantiomers of carvone (Sec. 7.4) fit this description, and their differing odors can be distinguished by most persons because the nasal receptors responsible for the detection of smells are chiral. The manner in which chiral odor receptors differentiate between enantiomers can be illustrated in the following way. Assume that your receptors are right-handed, symbolized as (+). Combining these with the (+)- and (−)-enantiomers of carvone gives complexes that can be represented as (+)/(−) for spearmint oil and (+)/(+) for caraway seed oil and dill seed oil, respectively. These complexes are diastereomeric, and they trigger unique nervous responses that we sense as different aromas. Thus, your nose is able to distinguish between diastereomers just as a chromatographic plate or column is able to differentiate between the diastereomers of 1,2-cyclohexanediol (Sec. 7.2).

Pasteur's observation of the fortuitous resolution of the tartrates by crystallization of the racemate is an exception to the general methods required for separating enantiomers. The more usual case involves converting the racemic mixture into a mixture of diastereomers, which may then be separated on the basis of differing physical properties, such as solubility. The resolution of 1-phenylethanamine (Sec. 7.6) is representative of this approach.

CHAPTER 8

Spectral Methods

When you see this icon, sign in at this book's premium website at **www.cengage.com/login** to access videos, Pre-Lab Exercises, and other online resources.

See more on the *Ozone Hole/Nobel Prize*

See more on *MRI/Nobel Prize*

See more on the *Ozone Hole*

We live in a world of electromagnetic radiation, which have energies ranging from those carried by γ- and X-rays to those associated with radio waves and microwaves. We're worried about damage to ourselves and our environment that might be caused by undesired ultraviolet radiation, which now penetrates the atmosphere because of the "ozone hole" caused by the presence of man-made halocarbons in the stratosphere. And what about the concerns that the radiation from cell phones, which seem to be glued to the ears of many of our friends, may cause cancer? Yes, radiation may be harmful, but it obviously has myriad benefits as well. Where would we be without solar radiation, for example? You are surely aware of many practical applications of radiation of various wavelengths. For instance, *infrared* and *microwave* radiation are used for heating food, and radio waves are the key to *magnetic resonance imaging* (MRI), a noninvasive technique for assessing soft tissues in the body. A variety of spectroscopic techniques that depend on radiation of specific wavelengths are described in this chapter, and you will soon see that they comprise an important component in the arsenal of methods that we have for analyzing organic compounds.

8.1 INTRODUCTION

As a beginning organic chemist, you will first master the skills and techniques for executing organic reactions and isolating the products in pure form. The structures of the starting materials and products that you encounter will usually be known, so you will not be faced with the necessity of determining the structures of your products. However, you may be required to determine the structure of a compound whose structure is known by your instructor but not by you. Indeed, solving the structures of compounds is a challenge commonly faced by organic chemists, who must identify compounds that are isolated from new chemical reactions or from natural sources.

Historically, the method for determining the exact structures of organic compounds was a difficult and time-consuming undertaking. A variety of solubility and classification tests were first performed to identify the functional groups present (Chap. 25). Typically, the structure of the unknown substance was then elucidated by performing a series of chemical reactions to convert it into another compound whose structure was already known. Fortunately, powerful spectroscopic methods are now available that have greatly simplified the task of structural elucidation. Perhaps the techniques of greatest importance to the practicing organic chemist are **infrared** (IR) and **nuclear magnetic resonance** (NMR) spectroscopy,

See *Spectral Database for Organic Compounds*

although **ultraviolet-visible** (UV-Vis) spectroscopy, **mass spectrometry** (MS), and **X-ray diffractometry** are also widely used for structure determination. With the notable exception of X-ray diffractometry, unequivocal definition of a structure is seldom possible by performing only a single type of spectroscopic analysis; rather, a combination of different analyses is generally required. We emphasize IR and NMR spectroscopy (Sec. 8.2 and 8.3, respectively) in this textbook because the necessary instrumentation is more commonly available to students in the introductory organic laboratory course. However, in addition to these spectroscopic techniques, UV-Vis spectroscopy (Sec. 8.4) and MS (Sec. 8.5) will be discussed because of their importance to the practice of organic chemistry. You can find a number of invaluable resources and **spectral databases** on the Web.

Most of the spectral methods described in this chapter depend on the absorption of radiation to produce an excitation that transforms the original molecule to a state of higher energy. The particular type of excitation that occurs is determined by the amount of energy associated with the radiation. The relationship between the energy of radiation (in kcal/mol), E, and its frequency, v, or wavelength, λ, is expressed in Equation 8.1, where N is Avogadro's number (6.023×10^{23}), h is Planck's constant (1.5825×10^{-37} kcal/sec), c is the velocity of light (2.998×10^{14} μm/sec), and v and λ are in Hertz (1.0 Hz = 1.0 cycle/sec) and micrometers (μm = 10^{-6} m), respectively. Because of the reciprocal relationship between v and λ, the *higher* the frequency or the *lower* the wavelength, the *higher* is the energy associated with the radiation.

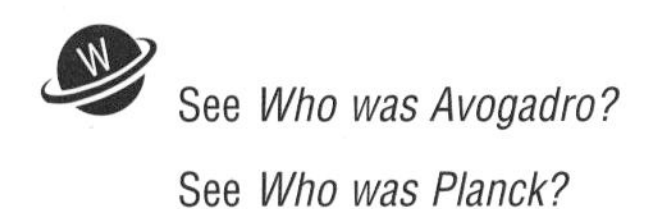
See *Who was Avogadro?*

See *Who was Planck?*

See *Who was Hertz?*

$$E = Nh\nu = Nhc/\lambda \tag{8.1}$$

In IR spectroscopy, the absorption of radiation corresponds to transitions among different molecular **vibrational-rotational levels** within the same electronic state, normally the ground state, of the molecule; these excitations require energies of 1–36 kcal/mol. The data from IR spectroscopy are most useful for determining the presence or absence of **functional groups** in a molecule. For instance, examining the appropriate regions of an IR spectrum will show whether or not carbon-carbon multiple bonds, aromatic rings, carbonyl groups, or hydroxyl groups are present. This technique does not give quantitative information regarding the elemental composition of a compound, nor does it allow assignment of an exact structure to an unknown compound unless the IR spectrum of the unknown is shown to be *identical* to that of a known compound.

The phenomenon associated with NMR spectroscopy involves transitions among **nuclear spin states** in a magnetic field; these excitations are of very low energy, less than 1 cal/mol. NMR spectroscopy provides information regarding the number and arrangement of various atoms in a molecule, although not all types of nuclei found in organic molecules are detectable by this method. Application of this technique to the analysis of hydrogen atoms in molecules, which is called ^{1}H NMR spectroscopy, allows determination of the number of **nearest neighbors** to a hydrogen atom as well as the presence of certain functional groups, such as carbon-carbon multiple bonds and carbonyl groups. A companion method involving carbon-13 atoms, and abbreviated as ^{13}C NMR spectroscopy, permits analysis of the different types of carbon atoms present in a molecule and may be useful for detecting certain functional groups whose presence is not indicated in the ^{1}H NMR spectrum.

To give you valuable experience in interpreting IR and NMR spectroscopic data, reproductions of these spectra for most of the reagents and products encountered in this book are provided, as are exercises associated with them. Moreover,

the spectra are available at the website for the book, and tools are available there for analyzing them in more detail than is possible for the reproductions. After you understand the basis for these two techniques, as discussed in the next two sections, you should use the spectra on the website to help you become an expert in interpreting IR and NMR data.

Ultraviolet-visible (UV-Vis) spectroscopy is based on electronic transitions, processes that require higher energies than either IR or NMR spectroscopy. For example, visible spectroscopy involves electronic excitations having energies of 38–72 kcal/mol, whereas those associated with ultraviolet spectroscopy are still greater, being 72–143 kcal/mol. UV-Vis spectroscopy is most useful for detecting the presence of systems of π-bonds that are **conjugated,** that is, systems in which two or more multiple bonds are directly linked by single bonds, as in 1,3-dienes, aromatic rings, and 1,3-enones. Because of this structural limitation, this technique is of less value to modern organic chemists than either IR or NMR spectroscopy. It finds use in the study of the kinetics of chemical reactions, however, because of the accuracy and rapidity with which the spectra can be obtained.

In Figure 8.1 we summarize the relationships between the energies associated with IR, NMR, and UV-Vis spectroscopies, the corresponding wavelengths of light providing these energies according to Equation 8.1, and the molecular effect associated with each technique. The microwave region is included for the sake of completeness, although organic chemists do not find this type of spectroscopy to be of much use. We do use the energy associated with radiation in the microwave region to heat reaction mixtures, using commercially available microwave ovens to do so (Sec. 2.9).

Mass spectrometry is a useful technique for determining the molar masses and elemental compositions of organic molecules. This technique involves ionizing the sample in a high vacuum and then detecting the gaseous ions after they have passed through a magnetic field, which "sorts" the ions according to their relative masses and charges. Such information is of great value in identifying the component being analyzed. Because MS requires that the sample be ionized, the energies required are much

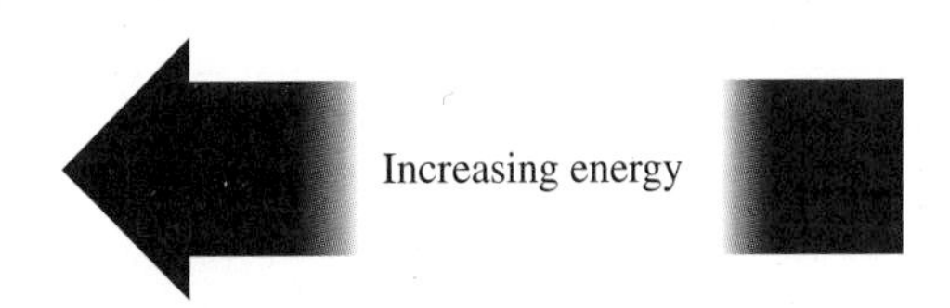

Spectral region	Ultraviolet	Visible	Infrared	Microwave	Radio
Wavelength (μm)	0.20–0.40 (200–400 nm)	0.40–0.75 (400–750 nm)	0.75–50	$50–5 \times 10^4$	$5 \times 10^4 - 2 \times 10^9$
Energy (kcal/mol)	143–71.5	71.5–38.1	38.1–0.57	$0.57–5.7 \times 10^{-4}$	$5.7 \times 10^{-4} - 1.4 \times 10^{-8}$
Molecular effect.	Promotion of electron from filled to unfilled orbital, e.g., $\pi \rightarrow \pi^*$.	Same as UV except that absorption of light is visible (colored) to the human eye.	Stretching and bending of interatomic bonds.	Molecular rotations.	Realignment of nuclear spins in a magnetic field.

Figure 8.1
A portion of the electromagnetic spectrum.

higher than those used in IR, NMR, and UV-Vis, none of which involves ionization. For example, producing ionization by bombarding the sample with an electron beam (Eq. 8.2) requires electrons having energies of some 70 electron volts (1600 kcal/mol).

$$\underset{\text{Molecule}}{P} + e^- \rightarrow \underset{\text{M+ (Parent ion)}}{P^{\dot{+}}} + 2\,e^- \tag{8.2}$$

8.2 INFRARED (IR) SPECTROSCOPY

Introduction

The span of the infrared portion of the electromagnetic spectrum is from about 0.8 μm (3.8×10^{14} Hz) to 50 μm (6×10^{12} Hz). Within this range, the region from 2.5–20 μm is of most interest to organic chemists because most of the common functional groups absorb IR radiation in this range. Although wavelength, in μm, was once used to express the location of an IR absorption, current practice is to do so with another unit, **wavenumber** or **reciprocal centimeter,** written as $\tilde{\nu}$ or cm^{-1}. The relationship between wavelength and wavenumber is expressed by Equation 8.3. Thus, the normal IR range of 4000–500 cm^{-1} corresponds to 2.5–20 μm. As with frequency, ν, the energy associated with radiation *increases* as $\tilde{\nu}$ increases. It is imprecise to call wavenumbers "frequencies," despite the mathematical relationship between the two, although we do so for convenience.

$$\text{Wavenumber } (cm^{-1}) = 10{,}000/\lambda\ (\mu m) \tag{8.3}$$

Principles

See more on *Molecular Vibrations*

See *Who was Hooke?*

The effect of IR radiation on an organic molecule may be qualitatively understood by imagining that the covalent bonds between atoms are analogous to groups of molecular springs that are constantly undergoing **stretching, twisting,** and **bending** (Fig. 8.2). The characteristic frequencies, and thus the energies, of these vibrations depend on the masses of the atoms involved and the type of chemical bond joining the atoms. Indeed, a good approximation of the *stretching* frequency of bonds can be obtained by applying **Hooke's law** for simple **harmonic oscillators** (Eq. 8.4). The value of the **force constant** k, which essentially reflects the strength of the bond between the atoms A and B, is about 5×10^5 dyne cm^{-1} for single bonds and twice and three times this figure for double and triple bonds, respectively.

$$\tilde{\nu} = \frac{1}{2\pi c}\sqrt{\frac{k}{m^*}} \tag{8.4}$$

where $\tilde{\nu}$ = frequency of absorption in cm^{-1}

c = speed of light

k = force constant of bond

m^* = reduced mass of atoms joined by bond = $\dfrac{m_A m_B}{m_A + m_B}$

The validity of Equation 8.4 for predicting stretching frequencies is illustrated by its use to give $\tilde{\nu}_{C\text{-}H}$ as 3040 cm^{-1} (see Exercise 2); the experimental value for this vibration is in the range of 2960–2850 cm^{-1}. Other consequences of Equation 8.4 are noteworthy. For instance, experimentally the wavenumber for stretching of a C–D bond is about 2150 cm^{-1}, a lower value than that for a C—H bond; this difference reflects the effect of atomic mass on the absorption frequency as accounted

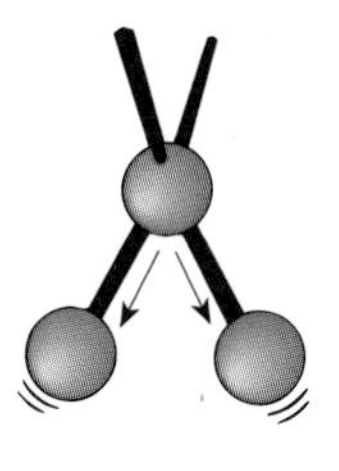

Symmetric stretch

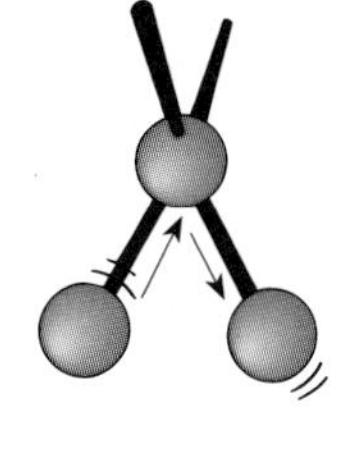

Asymmetric stretch

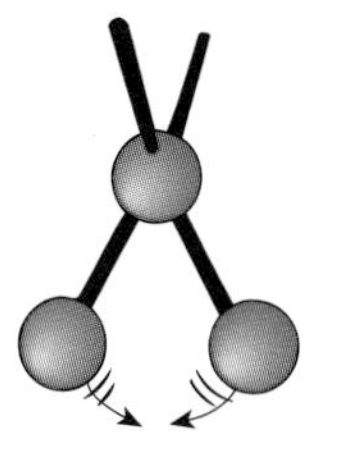

Symmetric in-plane bend

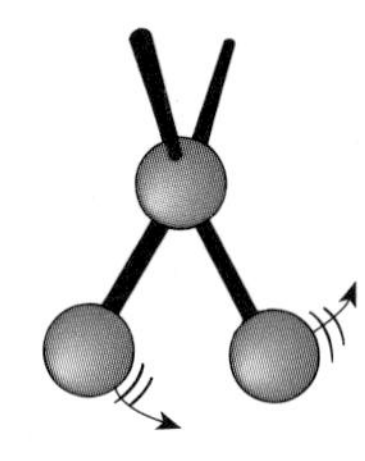

Asymmetric in-plane bend

Figure 8.2
Some vibrational modes of covalently bonded atoms.

for by m^* in Equation 8.3 (see Exercise 3). Similarly, the appearance of the stretching vibration of a carbon-carbon double bond at about 1620 cm^{-1} and that of a carbon-carbon triple bond at approximately 2200 cm^{-1} is because k (Eq. 8.4) approximately doubles in going from a double to a triple bond; this increases the energy required to stretch the bond between the atoms.

Because different energies are required to cause the molecular vibrations associated with various types of bonds, each functional group has a characteristic vibrational frequency. When the energy of the radiation corresponds to that associated with a particular stretching or bending vibration, energy from the incident light will be absorbed. The efficiency of this absorption depends on several factors that are beyond the scope of the present discussion. Suffice it to say that the greater the change in the **dipole moment** associated with a particular vibration, the more efficient the transfer of energy to the molecule and the stronger the observed absorption.

See more on *FT-IR Instrumentation*

Most modern IR instruments are designed to measure the amount of light transmitted *through* the sample, the **transmittance,** rather than the **absorbance** of light *by* the sample. Consequently, the **infrared spectra** portrayed in this textbook are plots of transmittance, expressed in *percent* as defined in Equation 8.5, versus the wavenumber of the **incident radiation** (I_0), as seen in Figure 8.3. The lower the %T, the greater the amount of incident radiation being absorbed by the sample. For example, in the IR spectrum shown in Figure 8.3, the strongest absorption occurs at about 1675 cm^{-1}. Although you might consider this absorption to be a minimum, it is referred to as a

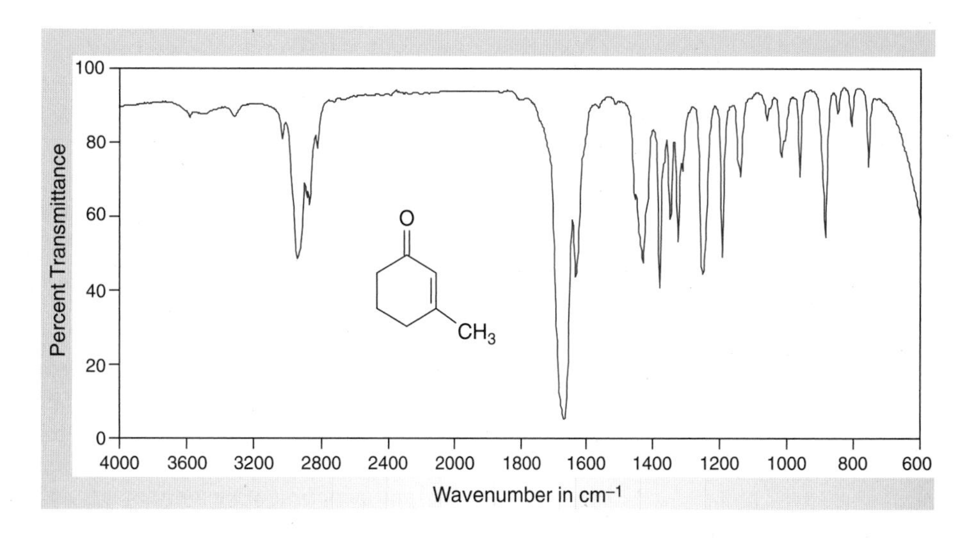

Figure 8.3
IR spectrum of 3-methyl-2-cyclohexenone (neat).

"peak" in the parlance of IR spectroscopy. The energy required to produce the observed molecular excitations *decreases* in going from left to right in the spectrum.

$$\%T = I/I_o \times 100 \tag{8.5}$$

I = intensity of radiation transmitted through sample at a particular wavelength
I_o = intensity of radiation incident on sample at same wavelength

What happens to a molecule once it has absorbed infrared radiation? The absorption causes an increase in the total energy of the molecule, thereby increasing the **amplitude** of the particular molecular vibration that was responsible for absorption of the radiation. However, no irreversible change occurs in the molecule, as the excess energy is quickly dissipated in the form of heat. Obtaining an IR spectrum thus does *not* permanently change the molecular structure.

Practical Considerations

Transmittance

As a general rule, the IR spectrum of a liquid or solid sample should show detectable peaks for relatively weak absorptions, but there should be no absorptions so strong that they go off-scale. Experimentally, a good operating principle is to modify the sample being analyzed so the *most* intense peak in its spectrum is close to 0%*T*; this maximizes the possibility that even weak peaks will be observable. This practice is exemplified in Figure 8.3, where the strongest absorption has a %*T* of nearly zero. The two experimental variables determining the absolute intensities of the peaks in an IR spectrum are the concentration of the sample and the length of the cell containing it. These factors define the number of molecules that are in the path of the radiation as it passes through the sample; the more molecules there are, the greater the intensity of an absorption. With modern IR spectrometers, the intensities of the peaks on the printed spectrum may also be amplified electronically.

Cells

There are two general types of sample cells used for IR spectroscopy: cells in which the pathlength or thickness of the sample is fixed, and those in which the pathlength is variable. The latter type of cell is expensive, but a simple version of it is two transparent salt plates, or windows, between which the sample is sandwiched (Figs. 8.4 and 8.5). The cell is easily prepared by putting a drop of sample on one plate and placing the second plate on top of the film that results. The assembled plates are then put into a holder that fits into the spectrometer. If a sample is too thick, some of it may be removed or the plates may be squeezed together—*carefully* to avoid breaking them—to produce a thinner film. However, if the sample is too thin, the plates can be separated and more material can be added. The plates are then rejoined so as to leave the film of the sample as thick as possible.

A **fixed-thickness cell** is constructed of two transparent salt plates separated by a plastic or metal gasket, which defines the thickness of the sample contained in the cell. One such cell is shown in Figure 8.6. A syringe is normally used to load the cell, which should be laid at an angle on top of a pen or pencil to keep an air bubble from forming within the cell as the sample is introduced. If a bubble is observed, more sample must be added until the bubble disappears.

Proper care and handling of the windows used in IR cells are *very* important. The plates are generally a clear fused salt such as sodium chloride or potassium bromide.

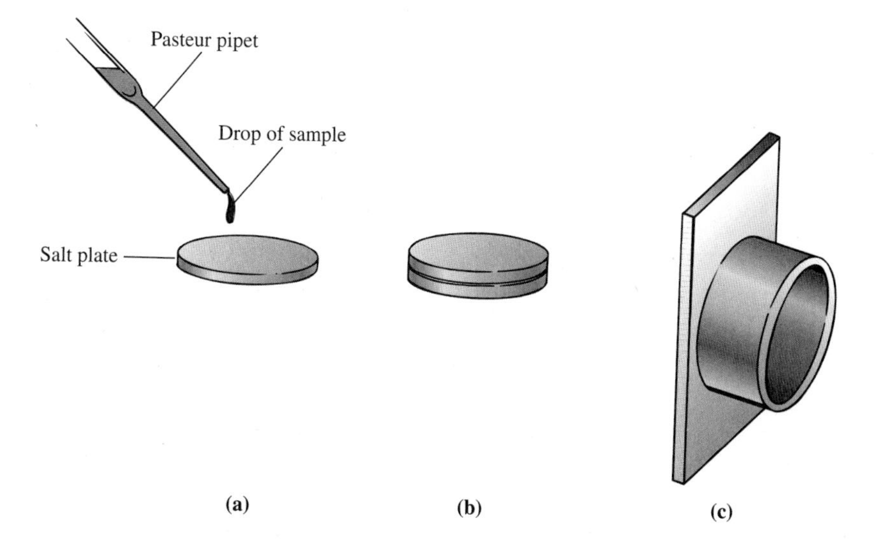

Figure 8.4
Simple demountable IR cell.
(a) Salt plate for sample.
(b) Assembled pair of salt plates.
(c) Cell holder.

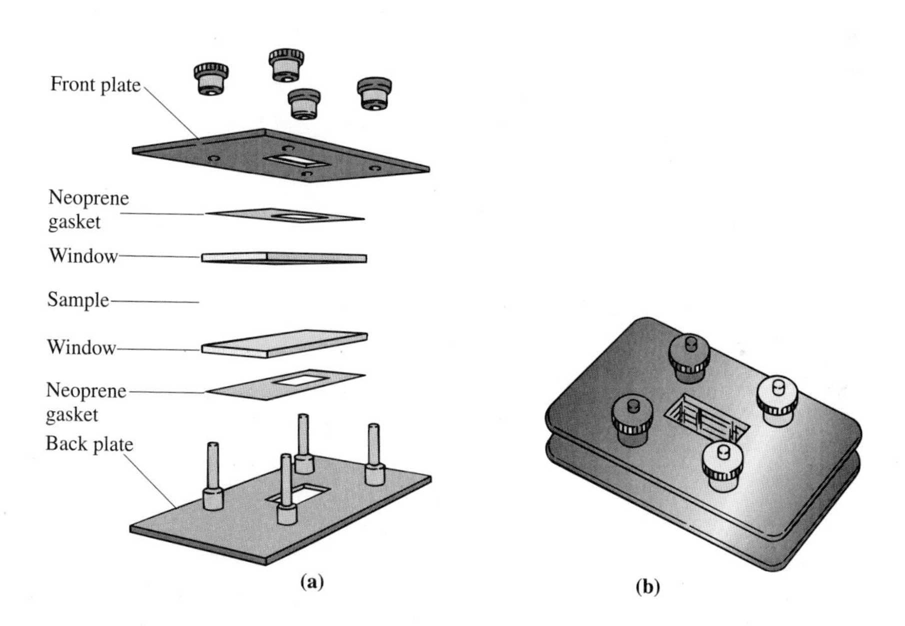

Figure 8.5
A demountable IR cell. (a) Detail showing correct assembly of cell. (b) Completely assembled cell.

Contact of these plates with moisture must be minimized to keep the windows from becoming cloudy, because cloudy plates decrease the amount of light transmitted through the cell. Consequently, avoid breathing on the windows or touching their faces when handling assembled IR cells or the salt plates themselves. Moreover, any substances, including samples, that come in contact with the cells must be *dry*. This means that any materials used to clean an IR cell or salt plates must also be *dry*. Solvents that are suitable for cleaning cells include absolute ethanol, dichloromethane, and chloroform. It goes without saying that woe be to the student who cleans an IR cell with water; this is a very expensive way to make saline solution!

Permanent fixed-thickness cells are dried by passing a slow stream of *dry* nitrogen or air through them, whereas demountable cells and salt plates may also be dried by *gently* wiping or blotting them with soft laboratory tissue to avoid scratching the

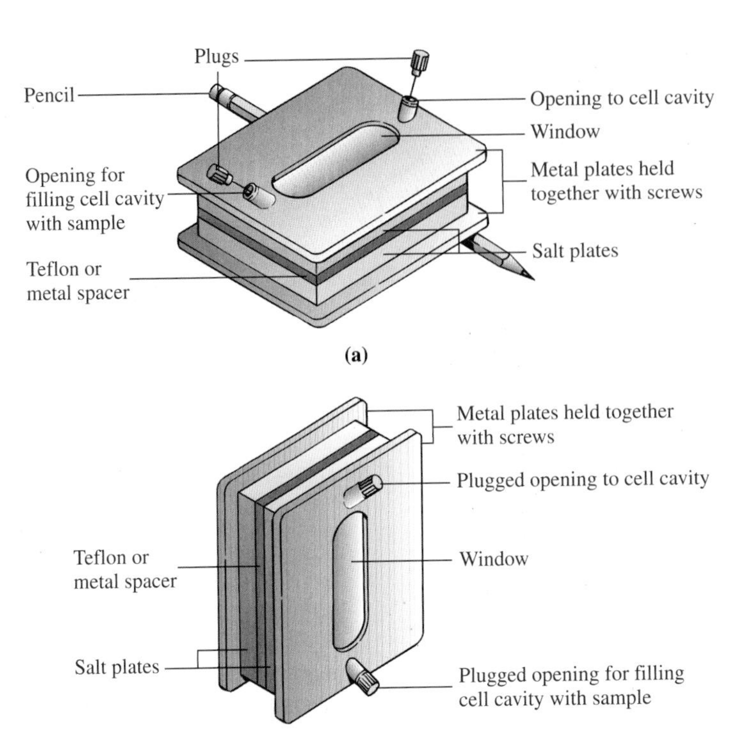

Figure 8.6
A fixed-thickness IR cell. (a) Cell with plugs removed for cleaning or filling. (b) Completely assembled cell.

faces of the cells or plates. The clean, dry cells or plates are stored in a desiccator for protection from atmospheric moisture.

Liquid Samples

The IR spectrum of a liquid sample is most conveniently obtained on the pure liquid, called **a neat sample.** Since concentration is not a variable in such samples, peak intensities may be modified only by changing the pathlength. This is done in one of two ways. One approach is to use a fixed-thickness cell with a pathlength of 0.020–0.030 mm. This short pathlength generally provides a spectrum in which strong absorptions, like those associated with carbonyl groups, give peaks that remain on the chart paper, rather than going to 0%*T*. If the spectrum is not suitable, however, different fixed thicknesses may be used. The other option is to use salt plates and to adjust the thickness of the film between them by gently pressing the plates together. Salt plates are *not* appropriate for use with volatile samples because some or all of the sample will evaporate as the spectrum is run, and inaccurate relative peak intensities will result. Alternatively, a liquid sample may be analyzed as a solution or absorbed into a microporous film, as described in the next section.

Solid Samples

There are several options for preparing a solid sample for IR spectroscopy. These include the preparation of **solutions, castings, KBr pellets,** and **microporous polyethylene films,** or the use of an accessory for obtaining **attenuated total reflectance** IR spectra. Each of these techniques is described in turn.

1. ***Solution.*** A common technique for obtaining an IR spectrum of a solid is to analyze it in solution. The preferred solvents are chloroform and carbon disulfide, in that order. The solubility of the sample in the selected solvent must be in the range of 5–10 weight percent because solutions of such concentration generally provide suitable spectra in a fixed-thickness cell having a pathlength of 0.1 mm.

   Two fixed-thickness cells having identical pathlengths, so-called **matched cells,** are used if a *double-beam* IR instrument is available. One cell contains the solution and is referred to as the **sample cell,** whereas the other contains pure solvent and is the **reference cell.** The purpose of the reference cell is to allow cancellation of the absorption bands due to the solvent itself so that they do not appear in the spectrum.

See more on *Fourier Transforms*

   If a **Fourier transform (FT) IR** instrument is used, a reference cell is not required because the IR spectrum of the pure solvent is previously stored in the memory of the spectrometer. The software that controls the instrument then *substracts* those absorptions that are due to the solvent from the spectrum. However, as is evident from the IR spectrum in Figure 8.7a, this subtraction may be imperfect; this is most notable at about 750 and 1220 cm^{-1}, where the solvent chloroform absorbs strongly.

2. ***Casting.*** A solution of a solid may be "cast" onto a sodium chloride disc to provide a thin film of solid upon evaporation of the solvent. Appropriate solvents include diethyl ether, dichloromethane, and chloroform.

   To prepare a casting, about two drops of a solution are applied evenly to the disc, and the solvent is allowed to evaporate. The disc is then placed in a holder in the spectrometer, and the spectrum is obtained (Figure 8.7b). If the film is too thin to provide a useful spectrum, additional solution may be added to the disc or a more concentrated solution of the solid may be used. On the other hand, if the sample is so thick that the absorption bands are too intense, the disc may need to be cleaned with a solvent and the casting procedure repeated with a more dilute solution. It is important to remember that sodium chloride is **hygroscopic,** so water or 95% ethanol should *not* be used for preparing solutions or cleaning the discs.

3. ***Potassium Bromide (KBr) Pellet.*** Potassium bromide may be fused to produce a nearly transparent salt plate. This is done by subjecting the KBr to high pressures, in the range of 5–10 tons/in^2, to generate the heat required for the salt to become plastic and flow; releasing the pressure then provides a clear plate. A KBr pellet containing the compound may thus be prepared by intimately mixing about 1 mg of sample with 100 mg of *dry,* powdered KBr and subjecting the resulting mixture to high pressure in a die. Mixing is done with a mortar and pestle made of agate rather than ceramic materials, because ceramics may be abraded and contaminate the pellet with nontransparent particles. Once a pellet having satisfactory transparency has been produced, the IR spectrum is obtained. If the intensity of the bands in the spectrum is unsatisfactory, another pellet must be prepared in which the concentration of the solid sample has been modified accordingly.

   Potassium bromide is *extremely* hygroscopic, so it should be stored in a tightly capped bottle in a desiccator except when needed for preparation of a pellet. Even with the greatest precautions, it is difficult to prepare a pellet that is completely free of water. Possible contamination of a KBr pellet with water may cause problems in interpreting IR absorptions in the range 3600–3200 cm^{-1}

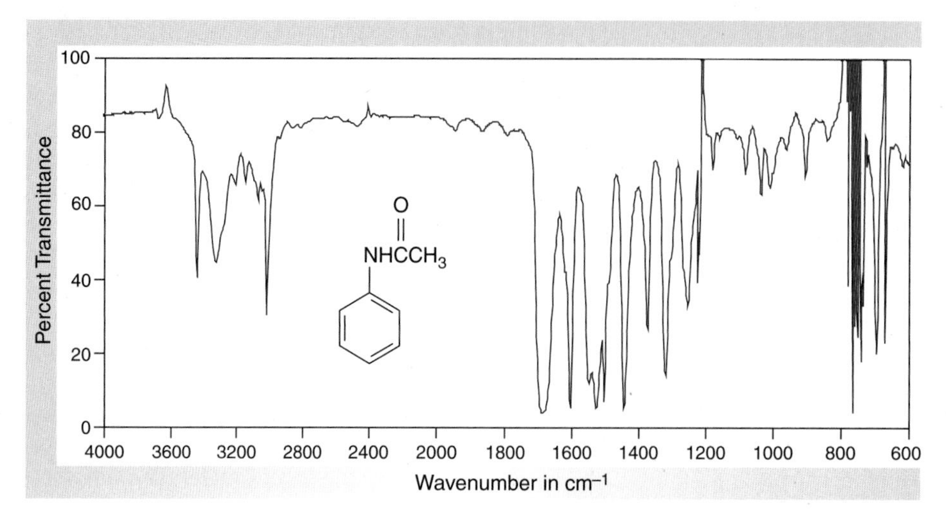

(a) In chloroform.

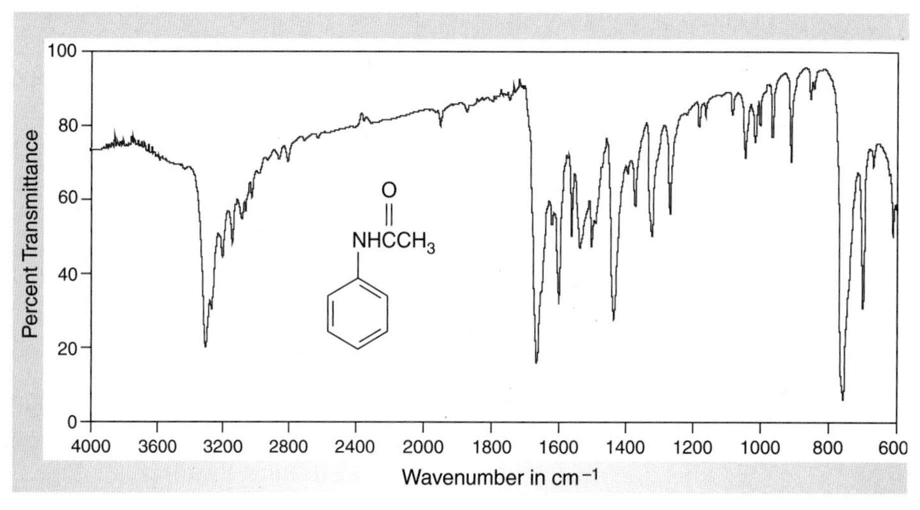

(b) As casting on NaCl disc.

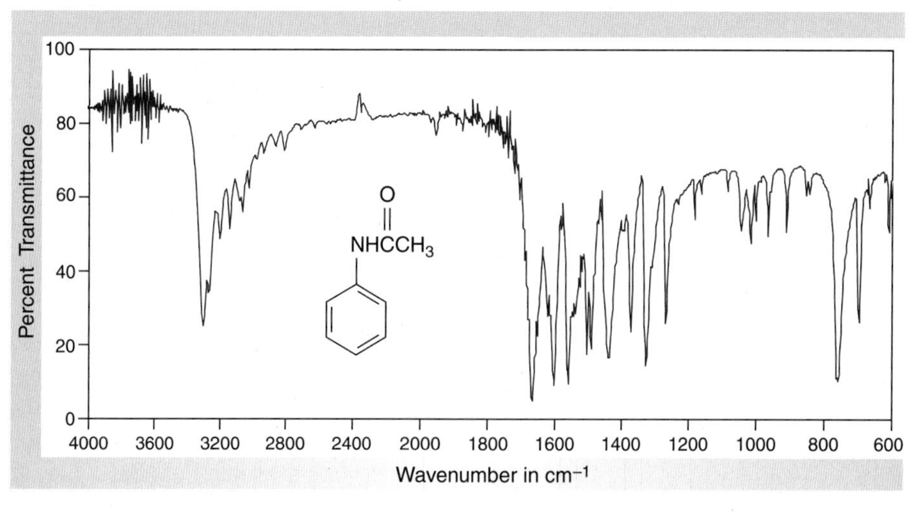

(c) As KBr pellet.

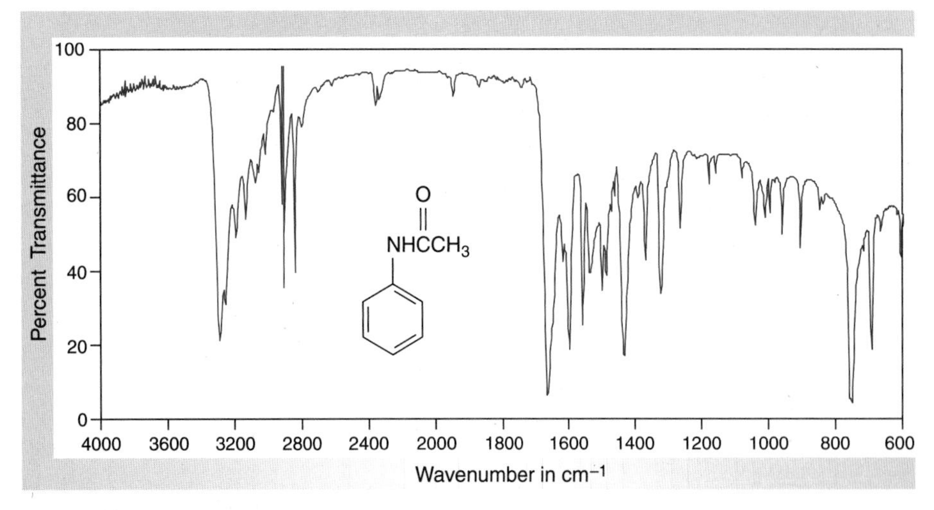

(d) On microporous polyethylene film (IR card).

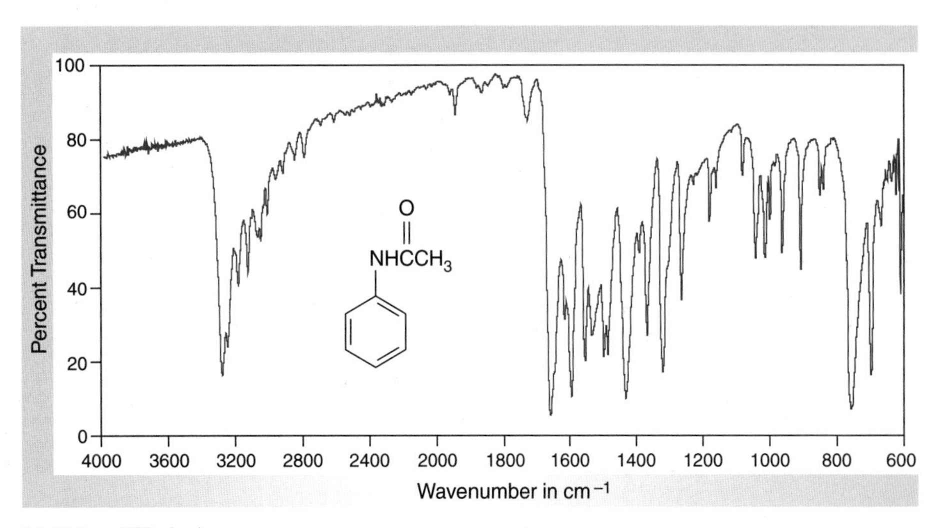

(e) Using ATR device.

Figure 8.7
IR spectra of acetanilide.

because the O–H stretching vibration of water occurs in this region. Consequently, moisture in the pellet may partially or completely obscure the absorptions due to the O–H and N–H stretching vibrations of alcohols and amines, which occur in this same region (Fig. 8.7c). Further, the appearance of an absorption in the aforementioned range may lead to an erroneous conclusion about the possible presence of O–H and N–H groups in the sample.

4. ***Microporous Polyethylene Films.*** Disposable IR cards are available to which samples can be directly applied for infrared analysis (Fig. 8.8). These convenient IR cards have two 19-mm circular apertures containing a thin microporous film of chemically resistant polyethylene. Cards having microporous polytetrafluoroethylene are also available for special applications. The cards with polyethylene films may be used for infrared analysis from 4000–400 cm^{-1} except for the region of aliphatic C–H stretching that occurs between 3000–2800 cm^{-1}. In this region of an FT-IR spectrum, there are several sharp

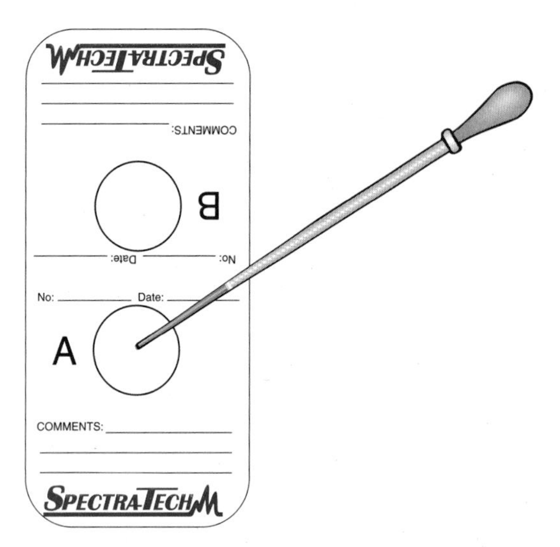

Figure 8.8
Applying sample to disposable IR card.

spikes at about 2840 cm^{-1} and 2910 cm^{-1} that result from the imperfect subtraction of the stored spectrum of the polyethylene film, as shown in Figure 8.7d.

To prepare a sample for IR analysis, two drops of a solution of the compound in a volatile organic solvent are simply applied to the film using a pipet or syringe. Because of the unique nature of the microporous films, the solution is absorbed into the film. The solvent evaporates within minutes at room temperature, leaving the solute evenly dispersed in the polyethylene matrix. Nonvolatile liquid samples may also be applied to the film. The cards are then mounted directly into the slot on the spectrometer that is normally used for holding reusable cells. Because the cards are expensive, you should use them twice before either taping them on your laboratory notebook or discarding them.

5. ***Reflectance Spectroscopy.*** The technique of **reflectance spectroscopy** is another means for obtaining IR spectra of solid and liquid samples, and we'll discuss one form of it, **attenuated total reflectance** (ATR). This method is based on monitoring radiation as it is *reflected from* a sample. The sample is placed on or between plates that focus the incident light on it and allow passage of the reflected light to the detector. The reflected light, called the *evanescent wave,* behaves as if it has penetrated the sample to depths varying from a fraction of a wavelength to several wavelengths. If the sample absorbs some of the evanescent wave, the intensity of the wave is decreased (attenuated) at the specific wavelengths that are characteristic of the absorption bands of the sample, much the same as when transmission of light *passing through* a sample is being measured (Figs. 8.7a–d), as discussed previously.

   Specialized adapters (Fig. 8.9) for this technique fit into the sample compartment of the FT-IR instrument, and the resulting spectra are called ATR spectra. As with transmission spectra, they are plotted as %T vs. cm^{-1}. The appearance of ATR spectra is similar but not identical to transmission spectra, as seen by comparing the ATR spectrum of acetanilide (Fig. 8.7e) with those of Figures 8.7a–d. Although the position and shapes of the peaks are the same,

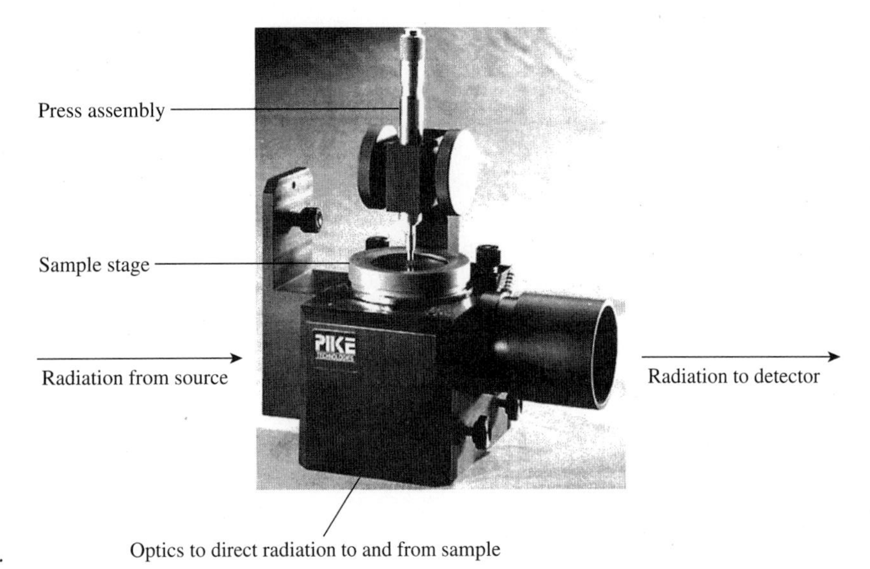

Figure 8.9
Device for obtaining ATR spectra.

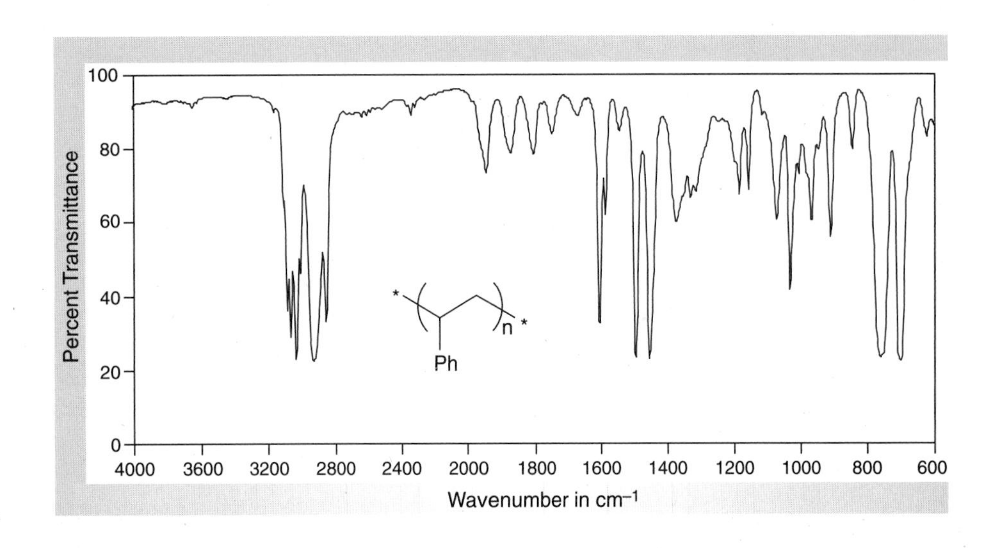

Figure 8.10
IR spectrum of polystyrene film.

the intensities of the peaks are not, and the overall range of $\%T$ may be less with ATR spectra. The peak intensities are different in the two techniques, because ATR spectra depend on such parameters as index of refraction, the angle of incident light, and the depth of wave penetration as a function of wavelength, none of which is important in transmission spectra. The range of $\%T$ for an ATR spectrum is less because of the limited depth of penetration of the evanescent wave into the sample and on other instrumental parameters. Since the shapes and positions of the peaks are unchanged, the protocol for interpreting an ATR spectrum is the same as that for a transmission spectrum.

To obtain an ATR spectrum using an adapter such as that shown in Figure 8.9, the sample is placed on top of the sample stage and the press is lowered to produce intimate contact between the sample and the crystal plate. The apparatus is now set up to obtain the desired spectrum.

Spectral Calibration

If you are not using an FT-IR spectrometer, it is good laboratory practice to calibrate the experimentally obtained IR spectra against a standard that has absorption bands at known positions. Calibration is necessary because problems such as misalignment of the chart paper in the instrument or mechanical slippage may result in absorptions appearing at wavenumbers that are incorrect. Polystyrene (Fig. 8.10) is a commonly used standard and is typically available as a thin, transparent film in a cardboard holder that fits into the spectrometer. After the spectrum of the sample is obtained, the standard is inserted in place of the sample cell; the reference cell, if used, is removed; and at least one band of the standard is recorded on the spectrum of the sample. In the case of polystyrene, the absorption at 1601 cm^{-1} is typically recorded, although other bands may be used.

Once the reference peak of the standard is measured, the reported positions of the absorptions of the sample are adjusted accordingly. For example, if the 1601 cm^{-1} peak of polystyrene appears at 1595 cm^{-1} on the spectrum, the wavenumbers for the bands in the sample are corrected by *adding* 6 cm^{-1} to the values recorded on the chart paper. Because the source of the error may vary over the range, it is not strictly correct to use the same correction factor throughout the spectrum, but for most purposes this is a satisfactory protocol. For more precise work, calibration peaks should be recorded at several different locations in the IR spectrum.

Use and Interpretation of IR Spectra

An IR spectrum of a compound may be used to *identify* the *functional groups* present in a molecule, *evaluate* the *purity* of a known compound, and *determine* the *structure* of an unknown substance. These valuable applications are discussed in the following paragraphs.

Analysis of the IR spectrum of an unknown compound provides useful information regarding the presence or absence of different functional groups. As a general rule, absorptions in the approximate range of 4000–1250 cm^{-1}, the so-called **functional group region,** involve vibrational excitations that are associated with particular functional groups. Thus, the stretching mode of the carbonyl group of a ketone appears in the region of 1760–1675 cm^{-1}, and that of a carbon-carbon double bond occurs in the vicinity of 1680–1610 cm^{-1}. A compilation of various functional groups and the region of the IR spectrum in which they absorb is provided in Table 8.1. Table 8.2 is a more detailed version of Table 8.1. Note that the *absence* of an absorption may have as much significance as the *presence* of one. For example, if there is no strong band in the carbonyl region of the spectrum, functional groups such as aldehyde, ketone, amide, and ester are probably absent.

Assessing purity involves comparing the spectrum of the sample with that of the pure material; the observation of extra bands in the sample is a sign that contaminants are present. On the whole, the technique is *not* sensitive to low levels of impurities, however, because levels of contamination in the range of 1–5% may easily go undetected.

Determining the structure of an unknown depends on the generally accepted premise that *if the IR spectra of two samples are superimposable, then the samples are identical.* The criterion of **superimposability** is a *very* demanding requirement in the strictest sense because it means that the intensity, shape, and location of *every* absorption in the two spectra *must be identical.* This is difficult to achieve unless

Table 8.1 *Abbreviated Table of Infrared Absorption Ranges of Functional Groups*

Bond	*Type of Compound*	*Frequency Range, cm^{-1}*	*Intensity*
C–H	Alkanes (>C=C<H)	2850–2970	Strong
C–H	Alkenes	1340–1470	Strong
		3010–3095	Medium strong
		675–995	
C–H	Alkynes (–C=C–H)	3300	Strong
C–H	Aromatic rings	3010–3100	Medium
		690–900	Strong
O–H	Monomeric alcohols, phenols	3590–3650	Variable
	Hydrogen-bonded alcohols, phenols	3200–3600	Variable, sometimes broad
	Monomeric carboxylic acids	3500–3650	Medium
	Hydrogen-bonded carboxylic acids	2500–3200	Broad
N–H	Amines, amides	3300–3500	Medium
C=C	Alkenes	1610–1680	Variable
C=C	Aromatic rings	1500–1600	Variable
C≡C	Alkynes	2100–2260	Variable
C–N	Amines, amides	1180–1360	Strong
C=N	Nitriles	2210–2280	Strong
C–O	Alcohols, ethers, carboxylic acids, esters	1050–1300	Strong
C=O	Aldehydes, ketones, carboxylic acids, esters	1675–1760	Strong
NO_2	Nitro compounds	1500–1570	Strong
		1300–1370	Strong

the *same* cells and instrument are used to obtain the IR spectra of *both* samples. Nevertheless, finding that the IR spectrum of an unknown material is very similar to the spectrum of a known material, even if the spectra are not obtained with the same instrument and cells, makes it probable that the structures of two samples are similar, if not identical. Alternative methods, such as NMR spectroscopy (Sec. 8.3), preparing derivatives (Secs. 25.6–18), and mixed melting points (Sec. 3.3), are then used to determine whether or not the samples are indeed identical.

Absorptions that occur in the region from 1250–500 cm^{-1} usually arise from a complex *vibrational-rotational* excitation of the *entire* molecule. This portion of the spectrum is typically *unique* for a particular compound and is aptly described as the **fingerprint region.** Although the IR spectra of similar molecules may be

Table 8.2 *Detailed Table of Characteristic Infrared Absorption Frequencies*

The hydrogen stretch region (3600–2500 cm^{-1}). Absorption in this region is associated with the stretching vibration of hydrogen atoms bonded to carbon, oxygen, and nitrogen. Care should be exercised in interpreting very weak bands because these may be overtones of strong bands occurring at frequencies from 1800–1250 cm^{-1}, which is one-half the value of the weak absorption. Overtones of bands near 1650 cm^{-1} are particularly common.

$\tilde{\nu}$(cm^{-1})	*Functional Group*	*Comments*
(1) 3600–3400	O–H stretching Intensity: variable	3600 cm^{-1} (sharp) unassociated O–H, 3400 cm^{-1} (broad) associated O–H; both bands frequently present in alcohol spectra; with strongly associated O–H (CO_2H or enolized β-dicarbonyl compound), band is very broad (about 500 cm^{-1} with its center at 2900–3000 cm^{-1}).
(2) 3400–3200	N–H stretching Intensity: medium	3400 cm^{-1} (sharp) unassociated N–H, 3200 cm^{-1} (broad) associated N–H; an NH_2 group usually appears as a doublet (separation about 50 cm^{-1}); the N–H of a secondary amine is often very weak.
(3) 3300	C–H stretching of an alkyne Intensity: strong	The *complete* absence of absorption in the region from 3300–3000 cm^{-1} indicates the absence of hydrogen atoms bonded to C=C or C≡C and *usually* signals the lack of unsaturation in the molecule. Because this absorption may be very weak in large molecules, care should be exercised in this interpretation. In addition to the absorption at about 3050 cm^{-1}, aromatic compounds frequently show *sharp* bands of medium intensity at about 1500 *and* 1600 cm^{-1}.
(4) 3080–3010	C–H stretching of an alkene Intensity: strong to medium	
(5) 3050	C–H stretching of an aromatic compound Intensity: variable; usually medium to weak	
(6) 3000–2600	OH strongly hydrogen-bonded Intensity: medium	A very broad band in this region superimposed on the C–H stretching frequencies is characteristic of carboxylic acids.
(7) 2980–2900	C–H stretching of an aliphatic compound Intensity: strong	As in previous C–H entries (3–5), *complete* absence of absorption in this region indicates absence of hydrogen atoms bound to tetravalent carbon atoms. The tertiary C–H absorption is weak.
(8) 2850–2760	C–H stretching of an aldehyde Intensity: weak	Either one or two bands *may* be found in this region for a single aldehyde function in the molecule.

The triple-bond region (2300–2000 cm^{-1}). Absorption in this region is associated with the stretching vibration of triple bonds.

$\tilde{\nu}$(cm^{-1})	*Functional Group*	*Comments*
(1) 2260–2215	C≡N Intensity: strong	Nitriles conjugated with double bonds absorb at *lower* end of frequency range; nonconjugated nitriles appear at *upper* end of range.
(2) 2150–2100	C≡C Intensity: strong in *terminal* alkynes, variable in others	This band is absent if the alkyne is symmetrical, and will be very weak or absent if the alkyne is nearly symmetrical.

Table 8.2 *Detailed Table of Characteristic Infrared Absorption Frequencies (Continued)*

The double-bond region (1900–1550 cm^{-1}). Absorption in this region is *usually* associated with the stretching vibration of carbon-carbon, carbon-oxygen, and carbon-nitrogen double bonds.

$\tilde{\nu}(cm^{-1})$	*Functional Group*	*Comments*
(1) 1815–1770	C=O stretching of an acid chloride Intensity: strong	Carbonyls conjugated with double bonds absorb at *lower* end of range; nonconjugated carbonyls appear at *upper* end of range.
(2) 1870–1800 and 1790–1740	C=O stretching of an acid anhydride Intensity: strong	*Both bands* are present; *each band* is altered by ring size and conjugation to approximately the same extent noted for ketones.
(3) 1750–1735	C=O stretching of an ester or lactone Intensity: very strong	This band is subject to all the structural effects discussed for ketones; thus, a conjugated ester absorbs at about 1710 cm^{-1} and a γ-lactone absorbs at about 1780 cm^{-1}.
(4) 1725–1705	C=O stretching of an aldehyde or ketone Intensity: very strong	This value refers to carbonyl absorption frequency of acyclic, nonconjugated aldehyde or ketone having no electronegative groups, for example, halogens, near the carbonyl group; because this frequency is altered in a predictable way by structural alterations, the following generalizations are valid: (a) *Effect of conjugation:* Conjugation of carbonyl group with an aryl ring or carbon-carbon double or triple bond *lowers* the frequency by about 30 cm^{-1}. If the carbonyl group is part of cross-conjugated system (unsaturation on each side of the carbonyl group), the frequency is lowered by about 50 cm^{-1}. (b) *Effect of ring size:* Carbonyl groups in six-membered and larger rings exhibit approximately the same absorption as acyclic ketones; carbonyl groups in rings smaller than six absorb at *higher* frequencies: for example, a cyclopentanone absorbs at about 1745 cm^{-1} and a cyclobutanone at about 1780 cm^{-1}. The effects of conjugation and ring size are additive: A 2-cyclopentenone absorbs at about 1710 cm^{-1}, for example. (c) *Effect of electronegative atoms:* An electronegative atom, especially oxygen or halogen, bound to the α-carbon atom of an aldehyde or ketone usually raises the position of the carbonyl absorption frequency by about 20 cm^{-1}.
(5) 1700	C=O stretching of an acid Intensity: strong	Conjugation *lowers* this absorption frequency by about 20 cm^{-1}.
(6) 1690–1650	C=O stretching of an amide or lactam Intensity: strong	Conjugation *lowers* the frequency of this band by about 20 cm^{-1}. The frequency of the band is *raised* about 35 cm^{-1} in γ-lactams and 70 cm^{-1} in β-lactams.
(7) 1660–1600	C=C stretching of an alkene Intensity: variable	Conjugated alkenes appear at *lower* end of range, and absorptions are medium to strong; nonconjugated alkenes appear at *upper* end of range, and absorptions are usually weak. The absorption frequencies of these bands are raised by ring strain but to a lesser extent than noted with carbonyl functions.
(8) 1680–1640	C=N stretching Intensity: variable	This band is usually weak and difficult to assign.

Table 8.2 *Detailed Table of Characteristic Infrared Absorption Frequencies (Continued)*

The hydrogen bending region (1600–1250 cm^{-1}). Absorption in this region is commonly due to bending vibrations of hydrogen atoms attached to carbon and to nitrogen. These bands generally do not provide much useful structural information. In the listing, the bands that are most useful for structural assignment are marked with an **asterisk**.

$\tilde{\nu}$(cm^{-1})	*Functional Group*	*Comments*
(1) 1600	NH_2 bending Intensity: strong to medium	In conjunction with bands in the 3300 cm^{-1} region, this band is often used to characterize primary amines and amides.
(2) 1540	NH bending Intensity: generally weak	In conjunction with bands in the 3300 cm^{-1} region, this band is used to characterize secondary amines and amides. This band, like the N–H stretching band in the 3300 cm^{-1} region, may be very weak in secondary amines.
(3)* 1520 and 1350	NH_2 coupled stretching bands Intensity: strong	This pair of bands is usually very intense.
(4) 1465	CH_2 bending Intensity: variable	Intensity of this band varies according to the number of methylene groups present; the more such groups, the more intense the absorption.
(5) 1410	CH_2 bending of carbonyl containing component Intensity: variable	This absorption is characteristic of methylene groups adjacent to carbonyl functions; its intensity depends on the number of such groups present in the molecule.
(6)* 1450 and 1375	CH_3 bending Intensity: strong	The band of lower frequency (1375 cm^{-1}) is usually used to characterize a methyl group. If two methyl groups are bound to one carbon atom, a characteristic doublet (1385 and 1365 cm^{-1}) is observed.
(7) 1325	CH bending Intensity: weak	This band is weak and often unreliable.

The fingerprint region (1250–600 cm^{-1}). The fingerprint region of the spectrum is generally rich in detail and contains many bands. This region is particularly diagnostic for determining whether an unknown substance is identical to a known substance, the IR spectrum of which is available. It is not practical to make assignments to all these bands because many of them represent combinations of vibrational modes and therefore are very sensitive to the overall molecular structure; moreover, many single-bond stretching vibrations and a variety of bending vibrations also appear in this region. Suggested structural assignments in this region must be regarded as tentative and are generally taken as corroborative evidence in conjunction with assignments of bands at higher frequencies.

$\tilde{\nu}$(cm^{-1})	*Functional Group*	*Comments*
(1) 1200	(cyclohexene ring)–O Intensity: strong	It is not certain whether these strong bands arise from C–O bending or C–O stretching vibrations. One or more strong bands are found in this region in the spectra of alcohols, ethers, and esters. The relationship indicated between structure and band location is only approximate, and any structural assignment based on this relationship must be regarded as tentative. Esters often exhibit one or two strong bands between 1170 and 1270 cm^{-1}.
(2) 1150	C–O Intensity: strong	
(3) 1100	CH–O Intensity: strong	

Table 8.2 *Detailed Table of Characteristic Infrared Absorption Frequencies (Continued)*

$\tilde{\nu}(cm^{-1})$	*Functional Group*	*Comments*
(4) 1050	CH_2–O Intensity: strong	
(5) 985 and 910	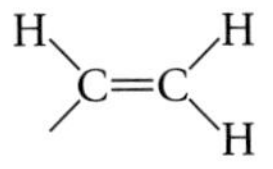 H(H)C=CH C–H bending Intensity: strong	This pair of strong bands characterizes a terminal vinyl group.
(6) 965	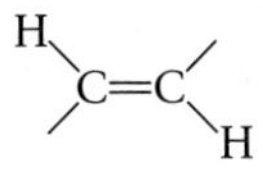 HC=CH (trans) C–H bending Intensity: strong	This strong band is present in the spectra of *trans*-1,2-disubstituted ethylenes.
(7) 890	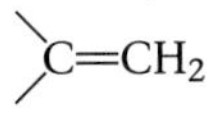 C=CH_2 C–H bending Intensity: strong	This strong band characterizes a 1,1-disubstituted ethylene group and its frequency may be *increased* by 20–80 cm^{-1} if the methylene group is bound to an electronegative group or atom.
(8) 810–840	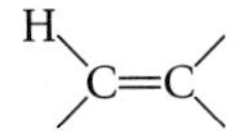 HC=C Intensity: strong	Very unreliable; this band is not always present and frequently seems to be outside this range because substituents are varied.
(9) 700	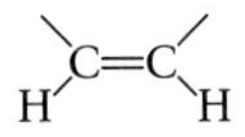 HC=CH (cis) Intensity: variable	This band, attributable to a *cis*-1,2-disubstituted ethylene, is unreliable because it is frequently obscured by solvent absorption or other bands.
(10) 750 and 690	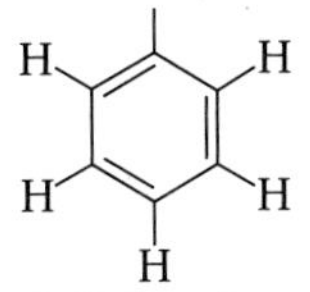 monosubstituted benzene (5 H) C–H bending Intensity: strong	These bands are of limited value because they are frequently obscured by solvent absorption or other bands. They are most useful when independent evidence leads to a structural assignment that is complete except for positioning of aromatic substituents.
(11) 750	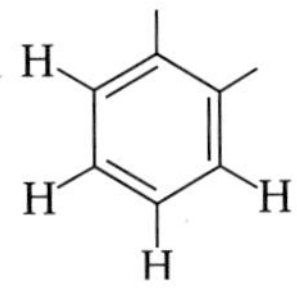 1,2-disubstituted benzene (4 H) C–H bending Intensity: very strong	
(12) 780 and 700	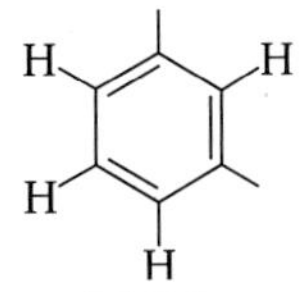 1,3-disubstituted benzene (4 H) and 1,2,3- Intensity: very strong	

Table 8.2 *Detailed Table of Characteristic Infrared Absorption Frequencies (Continued)*

$\tilde{\nu}(cm^{-1})$	*Functional Group*	*Comments*
(13) 825	H H H H and 1, 2, 4- Intensity: very strong	
(14) 1400–1000	C–F Intensity: strong	The position of these bands is quite sensitive to structure. As a result, they are not particularly useful because the presence of halogen is more easily detected by chemical methods. The bands are usually strong.
(15) 800–600	C–Cl Intensity: strong	
(16) 700–500	C–Br Intensity: strong	
(17) 600–400	C–I Intensity: strong	

comparable in the functional group region from 4000 to 1250 cm^{-1}, it is highly unlikely that this similarity will extend throughout the fingerprint region, as is illustrated in Figure 8.11.

It is important to learn to identify functional groups on the basis of the IR spectra of organic compounds. A useful way to do this is to examine the spectra of known compounds and correlate the key absorption bands with the functional groups responsible for them with the aid of Tables 8.1 and 8.2. Analysis of the IR spectra of the starting materials and products associated with the various experimental procedures in this textbook is an excellent way to sharpen your skills at interpreting such spectra.

EXERCISES

1. Derive the mathematical relationship between v and $\tilde{\nu}$ by using Equations 8.1 and 8.3 and point out why it is imprecise to use the terms "frequency" and "wavenumber" interchangeably.
2. Compute the reduced mass, m^*, for the C–H bond and use Equation 8.4 to confirm that $\tilde{\nu}_{C\text{-}H}$ is predicted to be 3040 cm^{-1}.
3. Compute the reduced mass, m^*, of a C–D bond and then use Equation 8.4 to determine the wavenumber at which the C–D stretching vibration occurs, assuming the corresponding vibration for the C–H group is 3000 cm^{-1}.
4. Assume that the force constant k for a carbon-carbon double bond is twice that of a carbon-carbon single bond. Use Equation 8.4 to determine the wavenumber at which the C–C stretching vibration occurs, assuming the corresponding vibration for the C=C group is 1640 cm^{-1}. Repeat this calculation for the case of a carbon-carbon triple bond, for which k is the appropriate multiple of that for the carbon-carbon double bond.

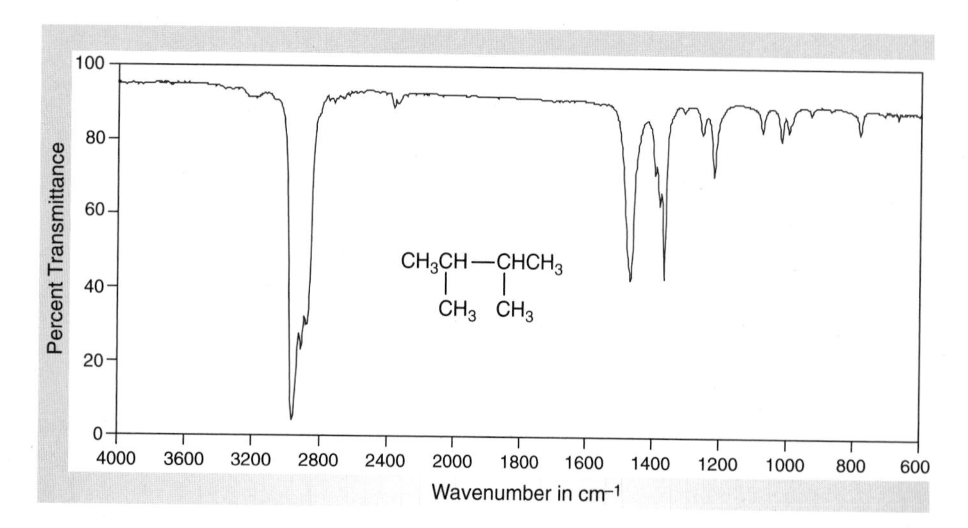

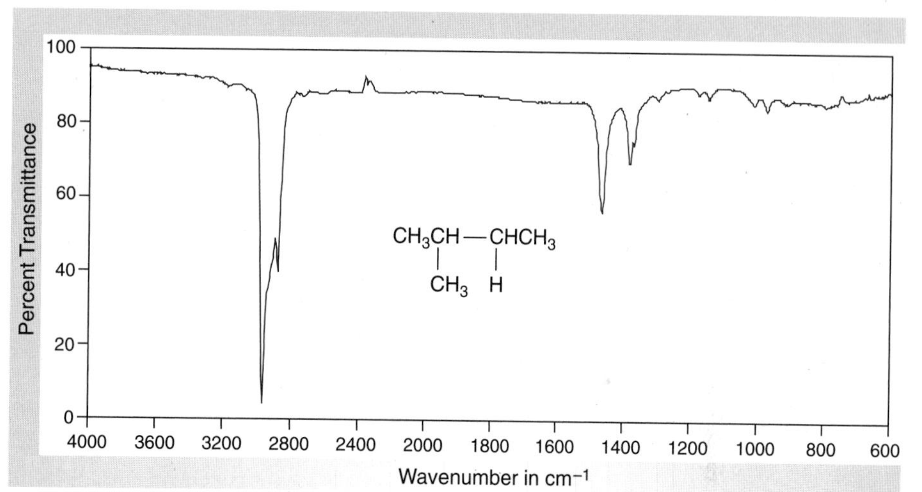

Figure 8.11
IR spectra of (a) 2, 3- dimethylbutane and (b) 2-methylbutane (neat).

5. Water has its O–H stretching mode at about 3600 cm^{-1}. Use Equation 8.5 to calculate at approximately what wavenumber "heavy" water, D_2O, would have the corresponding O–D stretch. Show your work.

6. **a.** If the strengths of a carbon-carbon triple bond and a carbon-nitrogen triple bond are assumed to be the same, as reflected in the force constant k (Eq. 8.4), predict which of the two functionalities would appear at the higher wavenumber in the IR spectrum and support your prediction.

   b. Is this in agreement with experimental fact? (See Table 8.2.)

7. Define the following terms:

   a. percent transmittance
   b. neat liquid
   c. hygroscopic
   d. fixed-thickness cell
   e. functional group
   f. stretching vibrational mode
   g. bending vibrational mode
   h. superimposability (as applied to IR spectra)

8. Explain why the windows of IR cells and KBr pellets should not be exposed to moisture.
9. Explain why the windows of IR cells should not be subjected to excessive pressure.
10. Why and when is an IR spectrum calibrated?
11. Name two solvents that are appropriate and two that are inappropriate for cleaning the faces of IR cells.
12. What are the two main drawbacks to obtaining IR spectra of solids as KBr pellets?
13. When obtaining the IR spectra of liquids or solids using IR cards with microporous polyethylene films, there are two sharp spikes at 2840 cm^{-1} and 2910 cm^{-1}. What causes these spikes?
14. What portions of an IR spectrum are referred to as the functional group region and the fingerprint region?
15. What difficulties would be associated with obtaining a solution IR spectrum on a double-beam spectrometer without using a reference cell containing the pure solvent?
16. What problems might be encountered in attempting to make a KBr pellet of an organic solid that has crystallized as a hydrate?
17. From one of the sources of IR spectra listed in the references at the end of this section, obtain a spectrum of acetanilide contained in a Nujol mull. (A Nujol mull is a slurry prepared by intimately mixing a solid sample with mineral oil.) Compare this spectrum to the ATR spectrum of Figure 8.7e and explain the source(s) of any differences.

Sign in at **www.cengage.com/login** and use the spectra viewer and Tables 8.1–8.8 as needed to answer the blue-numbered questions on spectroscopy.

18. Figures 8.12–8.18 are the IR spectra of the compounds shown on the individual spectra. Assign as many of the major absorptions as possible in the spectrum of each compound, using Tables 8.1 and 8.2.

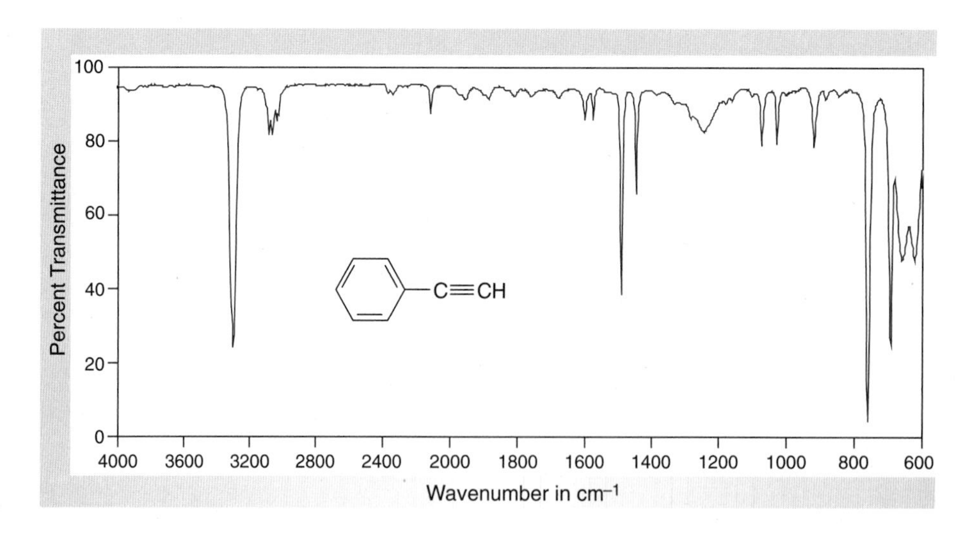

Figure 8.12
IR spectra of phenylethyne (neat).

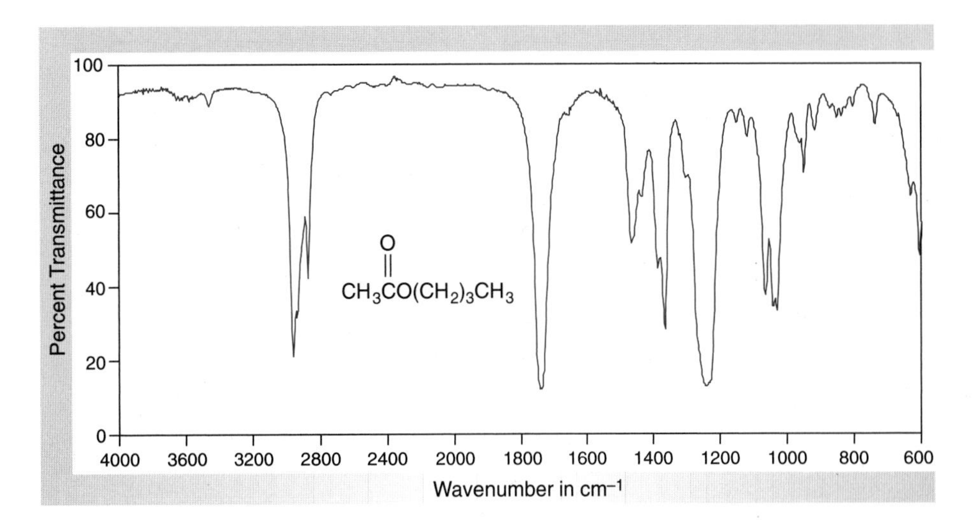

Figure 8.13
IR spectrum of n-*butyl acetate.*

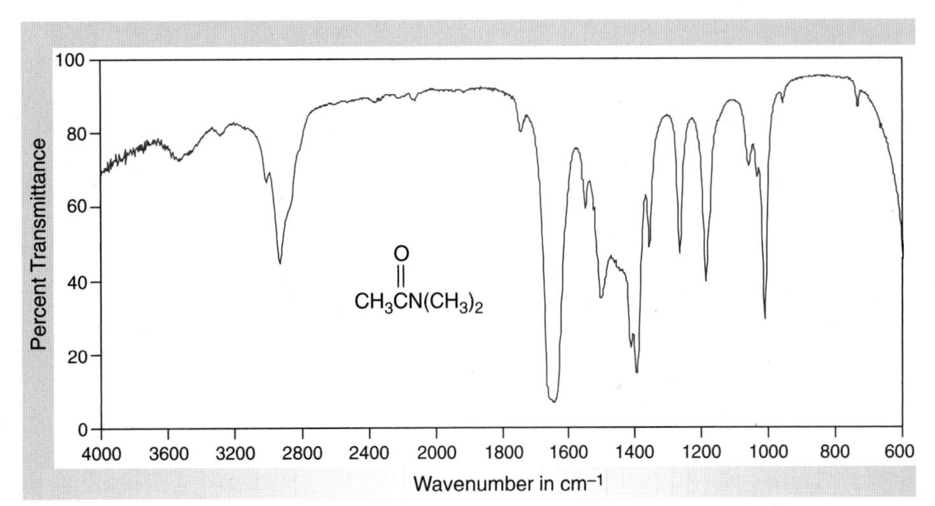

Figure 8.14
IR spectrum of N,N-*dimethylacetamide (neat).*

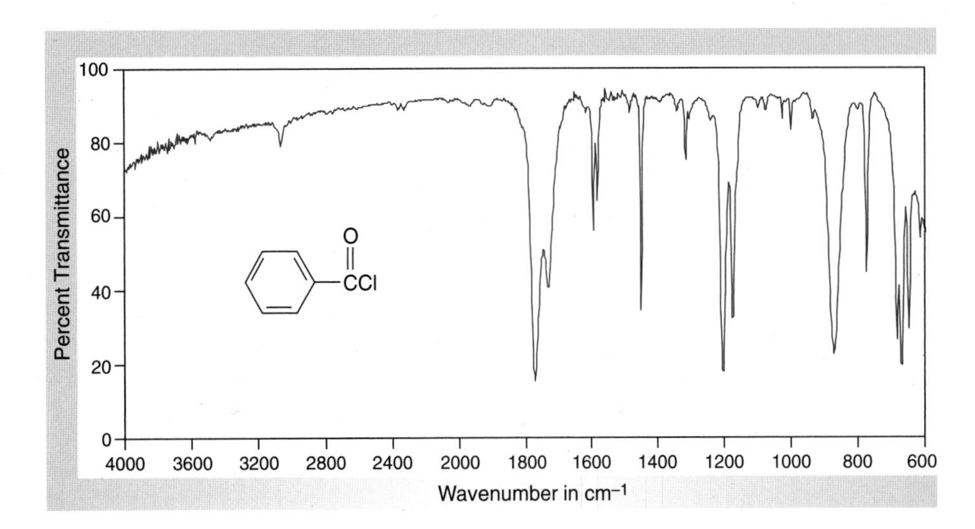

Figure 8.15
IR spectrum of benzoyl chloride (neat).

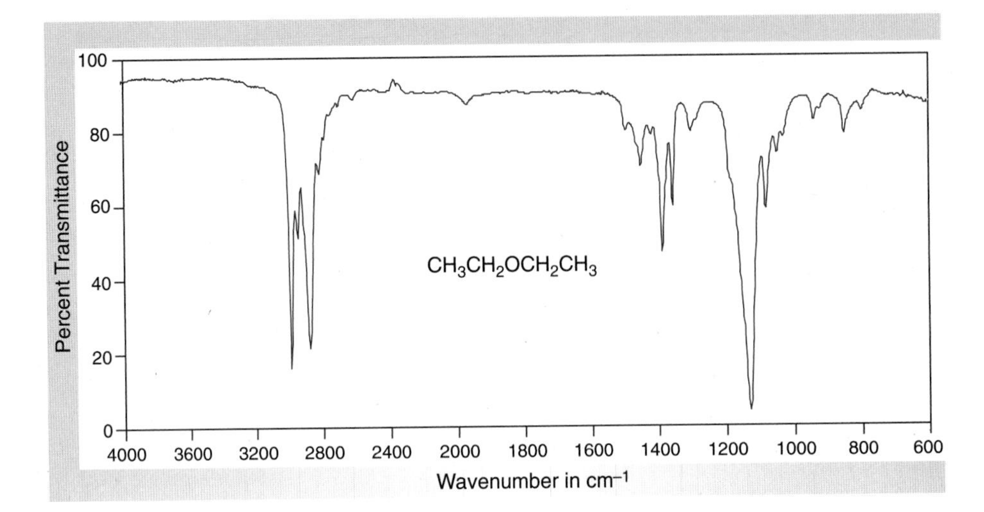

Figure 8.16
IR spectrum of diethyl ether (neat).

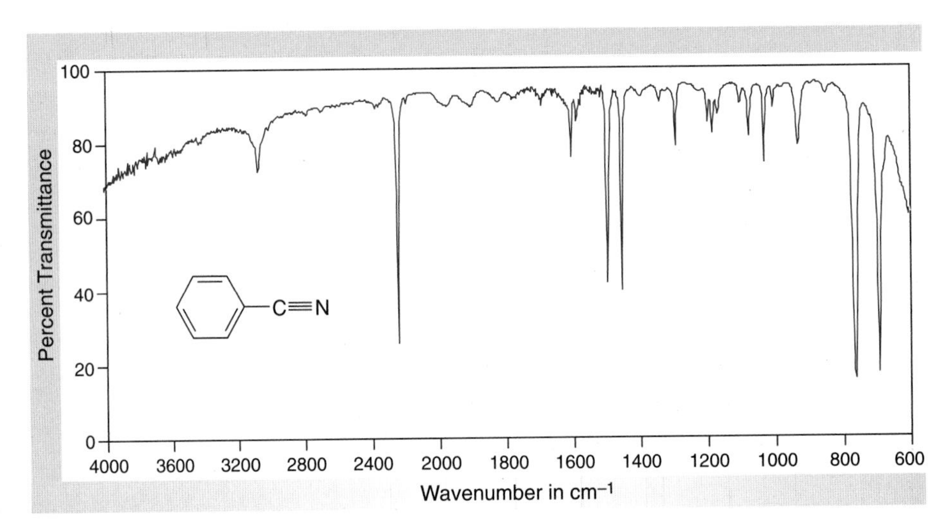

Figure 8.17
IR spectrum of benzonitrile (neat).

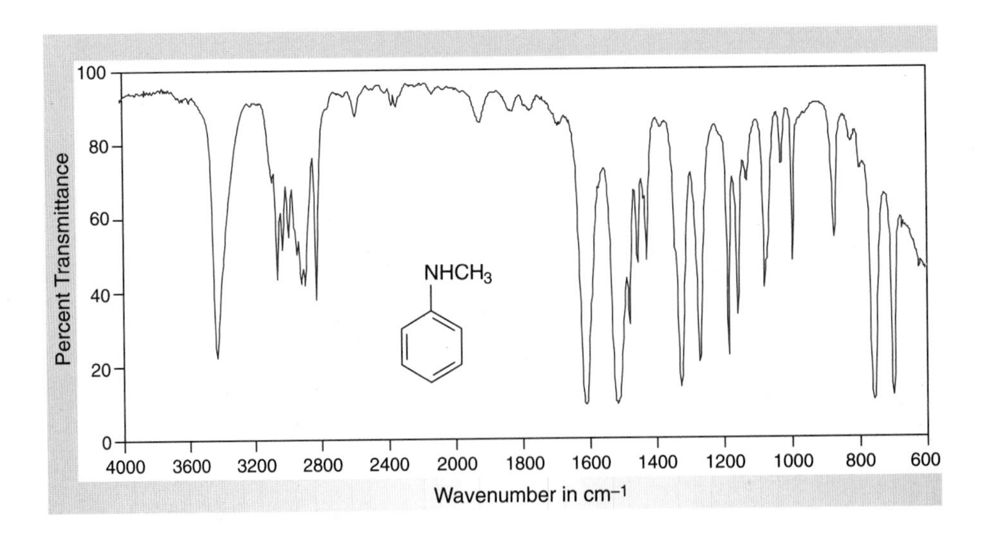

Figure 8.18
IR spectrum of N*-methylaniline (neat).*

REFERENCES

1. Colthup, N. B.; Daly, L. H.; Wiberley, S. E. *Introduction to Infrared and Raman Spectroscopy,* 3rd ed., Academic Press, New York, 1990.
2. Roelges, N. P. G. *A Guide to the Complete Interpretation of Infrared Spectra of Organic Compounds,* John Wiley & Sons, Chichester and New York, 1994.
3. Silverstein, R. M.; Webster, F. X.; Kiemle, D. *Spectrometric Identification of Organic Compounds,* 7th ed., John Wiley & Sons, New York, 2005.
4. Crews, P.; Rodriquez, J.; Jaspers, M. *Organic Structure Analysis,* Oxford University Press, New York, 1998.
5. Keller, R. J., ed. *Sigma Library of FT–IR Spectra,* Sigma Chemical Company, St. Louis, MO, 1986.
6. Pouchert, C. J., ed. *Aldrich Library of FTIR Infrared Spectra,* 2nd ed., vol. I–III, Aldrich Chemical Co., Milwaukee, WI, 1997. Compilation of over 18,000 spectra.
7. *Sadtler Standard Infrared Prism Spectra,* vol. 1–123, Sadtler Research Laboratories, Philadelphia, PA. Compilation of 91,000 spectra.
8. *Sadtler Standard Infrared Grating Spectra,* vol. 1–123, Sadtler Research Laboratories, Philadelphia, PA. Compilation of 91,000 spectra.

8.3 NUCLEAR MAGNETIC RESONANCE (NMR) SPECTROSCOPY

Introduction

Nuclear magnetic resonance (NMR) spectroscopy is probably the single most powerful spectroscopic method available to the modern organic chemist for analyzing compounds. This technique depends on the property of **nuclear spin** that is exhibited by certain nuclei when they are placed in a magnetic field. Some such nuclei commonly found in organic compounds are ^{1}H, ^{2}H, ^{19}F, ^{13}C, ^{15}N, and ^{31}P. Nuclei such as ^{12}C, ^{16}O, and ^{32}S do *not* have nuclear spin and thus cannot be studied by NMR spectroscopy.

Principles

The theory underlying the technique of NMR spectroscopy is presented in any comprehensive lecture textbook of organic chemistry and is not discussed in detail here. It is important to recall, however, that the observation of absorptions in the NMR spectrum of a particular nucleus is due to *realignment of nuclear spins* in an **applied magnetic field,** H_o, as shown in Parts (b) and (c) of Figure 8.19. The energy associated with the realignment of spin or spin flip, as it is often called, is expressed in Equation 8.6 and depends on the strength of the applied field H_0 and the **magnetogyric ratio** γ characteristic of the particular nucleus being examined. In a modern NMR spectrometer operating at 90–500 MHz, the value of H_o is in the range of 21,000 to 117,000 gauss (2.1–11.7 Tesla); this means that the energy required for the transition is less than 0.1 cal/mol for a nucleus such as ^{1}H.

$$\Delta E = h\gamma H_o/2\pi \tag{8.6}$$

Another way of illustrating the relationship between the strength of the magnetic field H_0 and the energy difference between spin states is given in Figure 8.20. Here it is seen that for a nucleus having two spin states, labeled $\pm 1/2$, where the plus and minus signs refer to nuclear spins aligned with and against the applied magnetic field, H_o, respectively; ΔE increases as H_0 increases. Energies of this sort are associated with radiation in the **radio-frequency (rf) range** of the electromagnetic spectrum; consequently, a radio-frequency oscillator is incorporated into the NMR

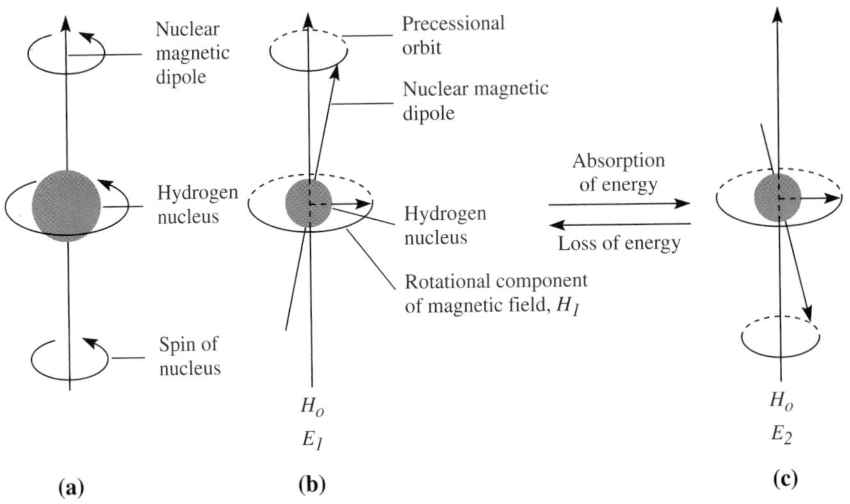

Figure 8.19
Spin properties of the hydrogen nucleus. (a) Rotation of hydrogen nucleus and its magnetic moment. (b) Magnetic moment of nucleus aligned with applied external magnetic field. (c) Magnetic moment of nucleus aligned against applied magnetic field.

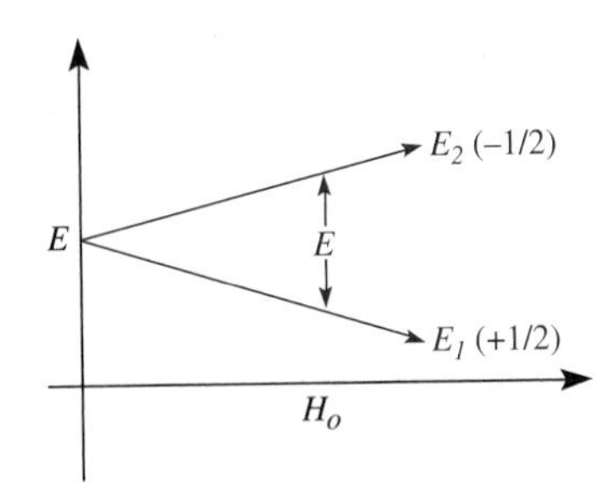

Figure 8.20
Energy difference between spin states as a function of strength of magnetic field, H_o.

spectrometer to provide the energy for the nuclear spin excitation. When the relationship between the magnetic field and the frequency of the oscillator is that defined by Equation 8.7, the **resonance condition** is achieved, and the transition from one nuclear spin state to the other can occur.

$$hv = h\gamma H_o/2\pi \quad \text{or} \quad v = \gamma H_o/2\pi \tag{8.7}$$

where ΔE = energy difference between two spin states
h = Planck's constant
γ = magnetogyric ratio (a constant characteristic of a particular type of nucleus)
H_o = strength of applied external magnetic field
v = frequency of oscillator

To meet the resonance condition of Equation 8.7, H_o could be held constant and v varied, or v could be kept constant and H_o changed. In **continuous-wave** (CW) spectrometers, the earliest form of commercial NMR instruments, the latter option is more common, and the spectrum is obtained by slowly sweeping through the range of field strengths required to produce resonance at a particular oscillator frequency. Modern instruments, called **Fourier transform** (FT) spectrometers, operate with a pulse technique in which all resonance frequencies are produced simultaneously while H_o is held constant; this technique allows collection of spectral data in much less time than with a CW instrument.

See more on *NMR Instrumentation*

Proton Magnetic Resonance (^{1}H NMR) Spectroscopy

The hydrogen nucleus, ^{1}H, has a nuclear spin, I_z, of 1/2 and the NMR spectral technique involving it is called **proton magnetic resonance** spectroscopy, abbreviated as ^{1}H NMR. Before discussing this type of NMR spectroscopy further, it is important to note that in the terminology of NMR spectroscopy, a covalently bound hydrogen atom is often imprecisely called a proton. Rather than being purists, we will frequently use this term for the sake of convenience.

Consideration of Equations 8.6 and 8.7 reveals that the *same* energy would be required to produce the resonance condition for *all* hydrogen nuclei in a molecule *if* all of these nuclei were *magnetically identical*. In other words, if the magnetic environment as defined by H_o were identical for all of the hydrogen atoms, the

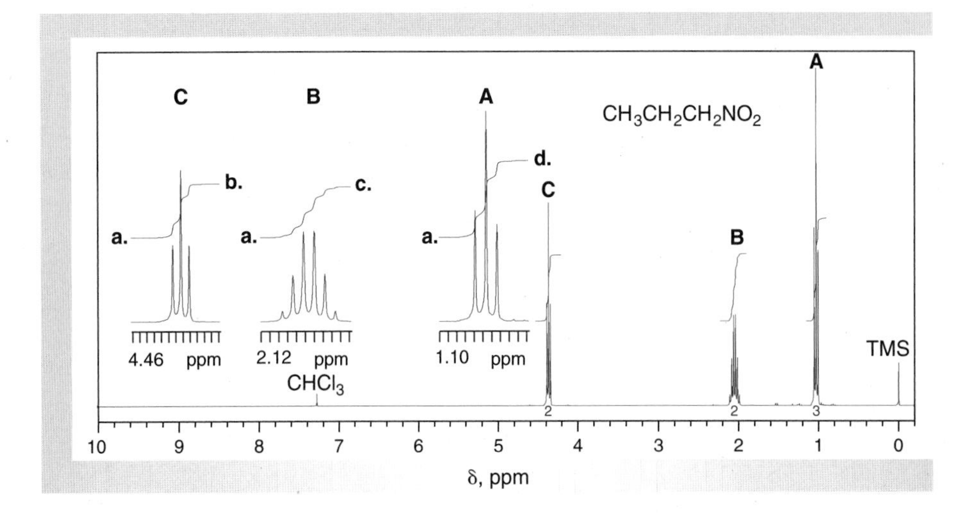

Figure 8.21
^{1}H NMR spectrum of 1-nitropropane (300 MHz, $CDCl_3$).

same frequency v would be needed for the spin flip. The consequence of this would be the appearance of a *single* absorption band in the ^{1}H NMR spectrum, a most uninformative result. Fortunately, the nuclei are *not* all magnetically equivalent because the three-dimensional electronic structure of a molecule produces variations in the magnetic environments of the hydrogen atoms in it (see "**Chemical Shift**" later in this section). This means that achieving the resonance condition requires *varying* amounts of energy and results in a spectrum containing valuable information for determining molecular structure.

Let's now analyze the ^{1}H NMR spectrum of 1-nitropropane (**1**) shown in Figure 8.21. The significant features illustrated in this spectrum are **chemical shift, spin-spin splitting,** and **peak integration,** each of which is discussed in detail. Before doing so, however, let's first take a brief overview of the general format for presenting NMR data.

$$\begin{array}{c} \quad H_a \; H_b \; H_c \\ H_a\!-\!C\!-\!C\!-\!C\!-\!NO_2 \\ \quad H_a \; H_b \; H_c \end{array}$$

1

In Figure 8.21, the strength of the applied magnetic field H_o is plotted along the horizontal axis and *increases* from left to right in the spectrum. This direction is called **upfield;** the opposite one is termed **downfield.** As an example, the group of peaks labeled B could be described as being upfield of the resonances labeled C. The *energy* needed to achieve resonance also *increases* in going from left to right. The intensity of the absorption band is plotted on the vertical axis and *increases* from the bottom, or baseline, of the plot.

A scale is printed horizontally across the bottom of the spectrum to define the location of peaks in the spectrum relative to those of a reference compound. The **delta- or δ-scale,** which is in units of **parts per million,** ppm, is used for the ^{1}H NMR spectra in this textbook. The reference compound most commonly used is **tetramethylsilane,** $(CH_3)_4Si$, which is abbreviated as TMS. TMS is an inert volatile liquid that is added directly to solutions of the sample and thus serves as an **internal standard.** The essential aspects of ^{1}H NMR spectra are presented under the headings "Chemical Shift," "Spin-Spin Splitting," and "Integration," respectively. The ^1H NMR spectra of most of the organic compounds you will be encountering in this and subsequent chapters are provided at the website associated with this textbook. These spectra may be integrated and expanded to facilitate their analysis.

Chemical Shift

Three groups of peaks, centered at about 1.0, 2.0, and 4.4 on the δ-scale, are seen in the spectrum of Figure 8.21. These values are the chemical shifts, in ppm, of the three chemically distinct types of hydrogen nuclei in 1-nitropropane (**1**). More precise values for chemical shifts may be determined by analyzing the spectrum using the power of the software available at the website. Predicting whether the various protons in a molecule are chemically distinct, and thus will *probably* appear at different chemical shifts, is an important part of understanding ^1H NMR spectra. Fortunately, determining how many chemically different protons are present in a molecule is straightforward.

The first step in the analysis is to group the protons according to their **connective equivalency,** placing in the same set all protons having identical molecular connectivities. The next step is to assess whether *connectively* equivalent nuclei are **chemically equivalent** and thus likely to have identical chemical shifts. This procedure is illustrated by further consideration of **1.** Examining **1** reveals that it contains three connectively unique sets of protons, labeled H_a, H_b, and H_c, respectively. *Hydrogen nuclei that are connectively distinct are chemically nonequivalent* and are defined as being **heterotopic.** As a general rule, such nuclei have *different* chemical shifts from one another, although some of the peaks may overlap.

The second step in the analysis involves evaluating connectively *identical* nuclei for their possible chemical nonequivalency. This is done by a **substitution test** in which it is supposed that a hydrogen nucleus is replaced with a **probe nucleus** that can be distinguished from other nuclei in the same set. The stereoisomeric relationship between such substituted species is then assessed.

The substitution procedure is presented in Figure 8.22 for protons of types H_a and H_c in **1.** As shown in Figure 8.22a, the hypothetical replacement of H_{a1}, H_{a2}, and H_{a3} with deuterium (**D**) as the probe nucleus provides three new representations that are *identical* to one another. The identity is established by rotating about the bond joining the methyl group adjacent to the methylene group and seeing that the resulting molecules are *superimposable* upon one another. Such hydrogens are defined as being **homotopic** and are *chemically equivalent.* Consequently, they will all have the *same* chemical shift in the ^1H NMR spectrum.

Applying the same substitution process to H_{c1} and H_{c2} gives *enantiomeric* representations, as seen in Figure 8.22b, and the two hydrogen atoms are described as being **enantiotopic.** Enantiotopic nuclei are also *chemically equivalent* and will have the same chemical shift, provided they are in an *achiral* environment, such as that provided by the solvents commonly used in NMR studies. It is left as an exercise to demonstrate that the hydrogen atoms of type H_b are also enantiotopic.

Another type of topicity is possible when a molecule contains a **center of chirality** (Sec. 7.1), or is subject to **restricted rotation,** a phenomenon usually associated with the presence of a π-bond or a ring. The first case is illustrated by considering 2-chlorobutane (**2**) as seen in Figure 8.23; although only a single enantiomer is shown, applying the same analysis to its mirror image would give the same result. The molecule has a chiral center at C2, the effect of which makes the connectively equivalent hydrogen atoms at C3 chemically *nonequivalent.* The nonequivalency is revealed by performing the substitution test on these protons, as

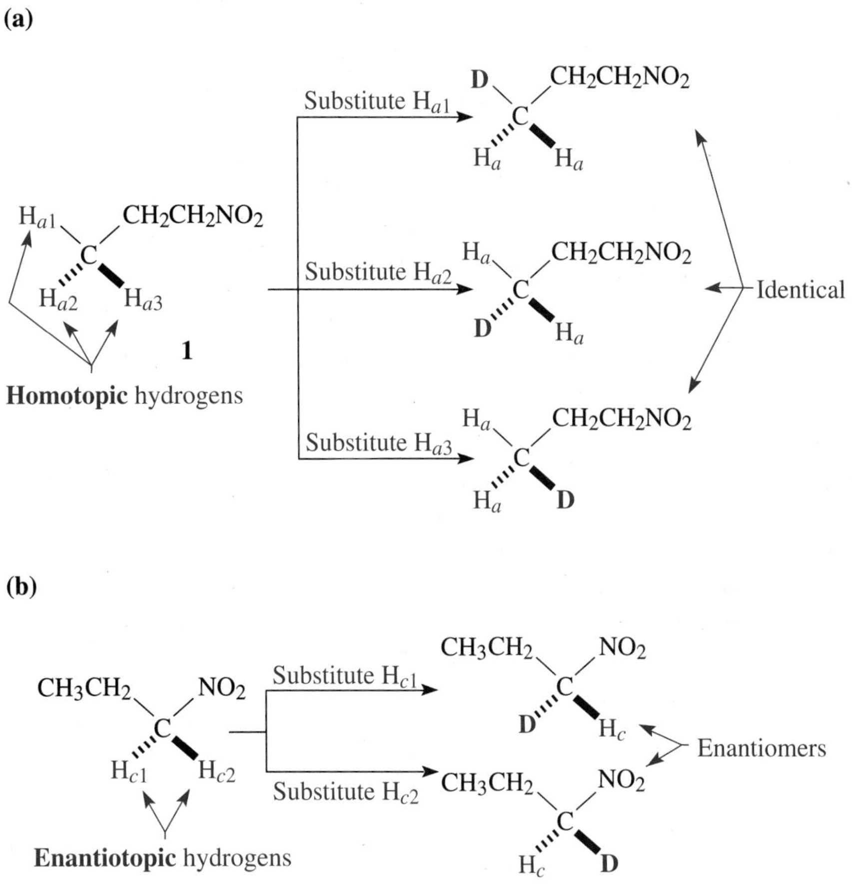

(a)

Substitute H_{a1}
D, $CH_2CH_2NO_2$, H_a, H_a

H_{a1}, $CH_2CH_2NO_2$, H_{a2}, H_{a3}

1

Homotopic hydrogens

Substitute H_{a2}
H_a, $CH_2CH_2NO_2$, D, H_a

Identical

Substitute H_{a3}
H_a, $CH_2CH_2NO_2$, H_a, D

(b)

CH_3CH_2, NO_2, H_{c1}, H_{c2}

Enantiotopic hydrogens

Substitute H_{c1}
CH_3CH_2, NO_2, D, H_c

Enantiomers

Substitute H_{c2}
CH_3CH_2, NO_2, H_c, D

Figure 8.22
Topicity analysis of 1-nitropropane by substitution.

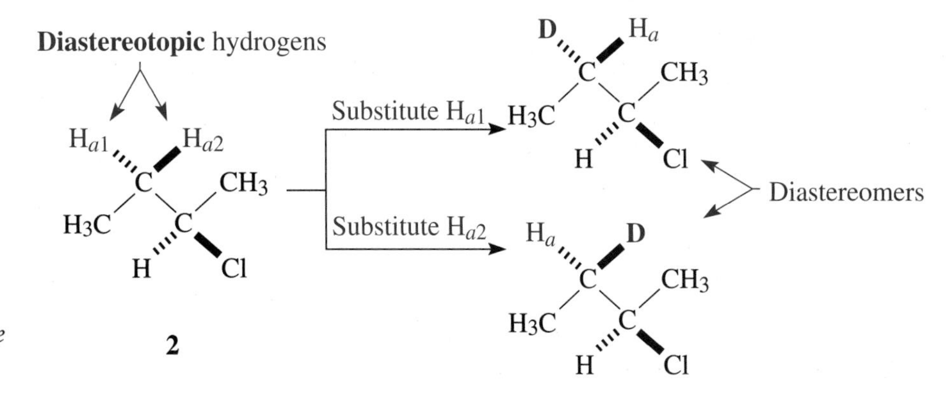

Figure 8.23
Topicity analysis of 2-chlorobutane by substitution.

illustrated in Figure 8.23. This generates **diastereomers,** thereby defining the two protons at C3 as **diastereotopic.** Because diastereotopic nuclei *may* have chemical shifts that are different from one another, there are *five* rather than four sets of chemically nonequivalent protons in **2.** These are the methine proton at C2, the two *diastereotopic* protons at C3, and the two different methyl groups, each of which has three *homotopic* protons.

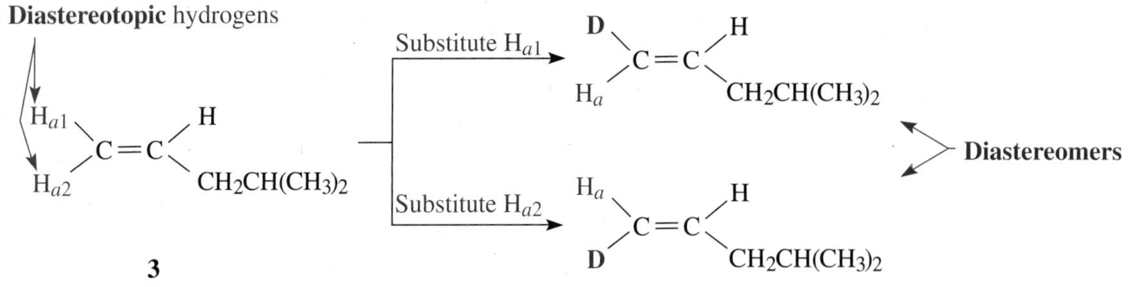

Diastereotopic hydrogens

H_{a1}, H, H_{a2}, $C{=}C$, $CH_2CH(CH_3)_2$

3

Substitute H_{a1} → D, H, H_a, $C{=}C$, $CH_2CH(CH_3)_2$

Substitute H_{a2} → H_a, H, D, $C{=}C$, $CH_2CH(CH_3)_2$

Diastereomers

Figure 8.24
Topicity analysis of 4-methyl-1-pentene by substitution.

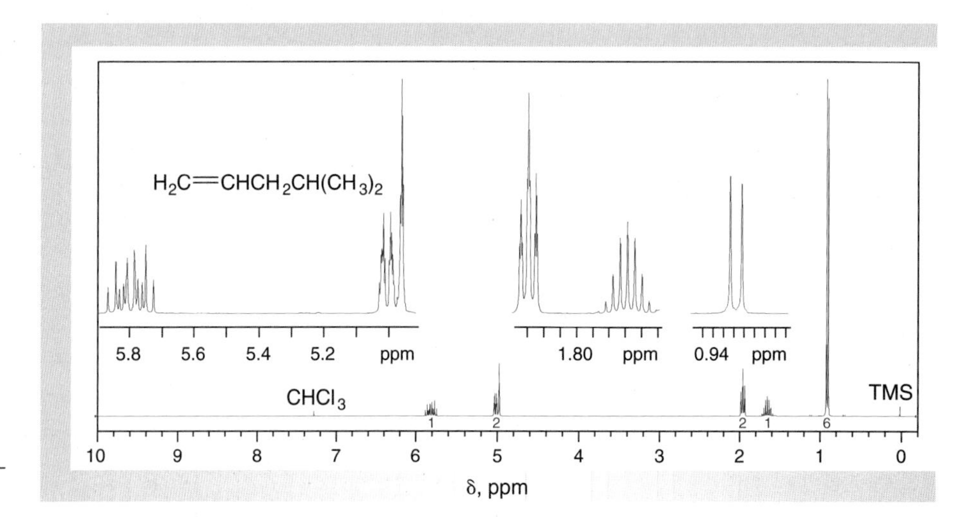

Figure 8.25
^1H NMR spectrum of 4-methyl-1-pentene (300 MHz, $CDCl_3$).

Chemical nonequivalency resulting from restricted rotation is seen in 4-methyl-1-pentene (**3**) (Fig. 8.24). Applying the substitution test shows the two vinylic hydrogen atoms at C1 are *diastereotopic* and thus chemically nonequivalent. This means that **3** contains *three* distinct sets of vinylic protons, each of which appears at a different chemical shift, as seen in Figure 8.25. Applying the substitution test in this instance is simply a formal way of proving that the geminal vinylic protons of an *unsymmetrically* substituted alkene, $\mathbf{H_2C{=}CR^1R^2}$, are *not* equivalent to one another. For example, relative to R^1, one of these hydrogen atoms will be *cis* and the other *trans;* their nonequivalency is determined rapidly and simply from this stereochemical relationship.

The presence of diastereotopic hydrogens in a molecule may complicate the appearance of a ^1H NMR spectrum. However, the simple procedure of looking for chiral centers and for the features that may enforce restricted rotation in the molecule minimizes erroneous analyses caused by the increased number of resonances associated with diastereotopicity.

Let's now return to our analysis of 1-nitropropane (**1**), in which we previously identified three chemically nonequivalent sets of protons. As a result, we predict that three different resonances should appear in its ^1H NMR spectrum, just as is observed (Fig. 8.21). The chemical shifts of the three groups of resonances vary according to the magnetic environment of each hydrogen nucleus in the molecule. This primarily depends on two factors: (1) the externally applied magnetic field, and (2) circulation of the electrons within the molecule, which provides an **electronic shield** about the atoms.

The field H_o is of *uniform* strength throughout the molecule, so it *cannot* be responsible for the magnetic nonequivalence of the different nuclei. On the other hand, the shielding effect produced by circulation of electrons is *not* uniform because the electron density varies within the molecule. This is a critical point because the **internal electric field** of the molecule induces an **internal magnetic field,** H_i, about the various nuclei in the molecule. The induced magnetic field acts to *oppose* the applied magnetic field, so the **effective magnetic field,** H_e, at a given nucleus is *less* than H_o (Eq. 8.8); this phenomenon is the source of **diamagnetic shielding** in the molecule.

$$H_e = H_o - H_i \tag{8.8}$$

The strength of H_i varies in the same way as that of the electric field and thus makes H_e different for the various types of protons in the sample. For example, the greater H_i is, the less H_e is, so a larger external field must be applied to achieve resonance. In other words, the more magnetic shielding experienced by a nucleus, the higher is the external field H_o required to produce the resonance condition at a particular oscillator frequency.

The consequences of the magnitude of H_i on the location of resonances in ^1H NMR spectra are seen in Figure 8.21. The nitro group, NO_2, serves as an electron sink, causing the electric field to be lower for those nuclei near it. Correspondingly, the value of H_i is less for these nuclei than for others more distant from the nitro function. The protons labeled H_c in **1** experience the *least* diamagnetic shielding and thus resonate at the *lowest* value of H_o. The methyl protons H_a are the *most* **shielded** because they are furthest from the nitro group and appear at the highest field (Fig. 8.21). Put another way, the nearer a proton is to the electronegative nitro group, the more **deshielded** it is relative to other hydrogen atoms in the molecule, and the further **downfield** it appears. On this basis, the various groups of peaks in the spectrum are assigned as shown below.

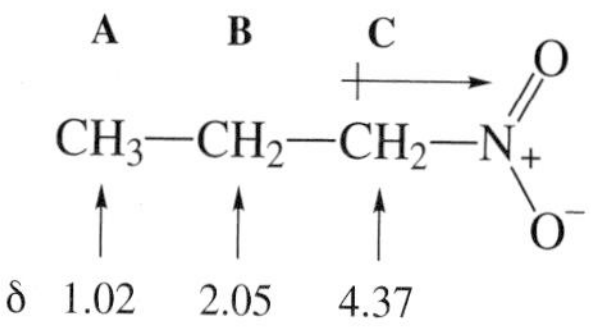

A B C

$CH_3—CH_2—CH_2—N^+(=O)O^-$

δ 1.02 2.05 4.37

The locations of the three groups of peaks in 1-nitropropane are expressed in units of ppm on the δ-*scale*, as was noted earlier. This scale, calculated according to Equation 8.9, is defined such that measurement of chemical shifts is *independent* of the frequency of the rf oscillator of the NMR instrument. The numerator is the shift, in Hz, of a resonance of the sample relative to that of the reference compound, which is usually TMS for ^1H NMR spectra. The sign of the shift by convention is taken to be *positive* if the resonance appears *downfield* of TMS and *negative* if it is *upfield*. The shift is divided by the corresponding oscillator frequency, and for convenience in reporting chemical shifts, the result is multiplied by 10^6 to provide units of ppm. Alternatively, the relative positions of peaks in the spectrum could simply be expressed in Hz relative to TMS. This is unsatisfactory because the chemical shift would then depend on the frequency of the oscillator being used in the spectrometer. Consequently, reporting shifts in Hz would require specifying the frequency at which the data were obtained.

$$\delta = \frac{\text{chemical shift (in Hz)} \times 10^6}{\text{oscillator frequency (in Hz)}} \tag{8.9}$$

To illustrate the difference between measuring chemical shifts in terms of δ versus Hz, consider the case in which a resonance is observed *100 Hz* downfield of TMS when a spectrometer operating at 100 MHz is used. On the δ-scale, this corresponds to a shift of *1.0 ppm.* Had the spectrum been measured on a 300-MHz instrument, the resonance would appear *300 Hz,* rather than *100 Hz,* downfield of TMS; however, the shift would remain at *1.0 ppm* on the δ-scale.

Almost all types of protons appear *downfield* of TMS and thus have chemical shifts that are *positive* values because of the sign convention noted above. However, some organic compounds contain hydrogen nuclei that are *more* shielded than those in TMS and therefore appear *upfield* of it. The chemical shifts are negative in such cases, as noted above.

The consistent effect that particular functional groups have on chemical shifts has led to tabulations correlating these shifts with the structural features responsible for them. Table 8.3 is one such compilation, and Table 8.4 is an expanded listing of typical chemical shifts for a variety of specific types of hydrogen atoms. Because the shifts are given in units of ppm, they apply to all NMR spectrometers regardless of the frequency at which the instrument is operating.

Table 8.3 *Chemical Shifts of Hydrogen Atoms Attached to Various Functional Groups*

Functional Group; Hydrogen Type Shown as H	*δ, ppm*	*Functional Group; Hydrogen Type Shown as H*	*δ, ppm*
TMS, $(CH_3)_4Si$	0	Alcohols, ethers	
Cyclopropane	0–1.0	HO–C–**H**	3.4–4
Alkanes			
RCH_3	0.9		
R_2CH_2	1.3	RO–C–**H**	3.3–4
R_3CH	1.5		
Alkynes		Acetals (–O)(–O)C–**H**	5.3
–C=C–**H** (vinyl)	4.6–5.9		
–C=C–C**H**$_3$ (allyl)	1.7	Esters	
Alkynes			
–C≡C–H	2–3	R–C(=O)–O–C–**H**	3.7–4.1
–C≡C–CH_3	1.8		
Aromatic			
Ar–**H**	6–8.5	RO–C(=O)–C–**H**	2–2.6
Ar–C–**H** (benzyl)	2.2–3	Carboxylic acids	
Fluorides, F–C–**H**	4–4.45	HO–C(=O)–C–**H**	2–2.6

Table 8.3 *Chemical Shifts of Hydrogen Atoms Attached to Various Functional Groups (Continued)*

Functional Group; Hydrogen Type Shown as H	*δ, pm*	*Functional Group; Hydrogen Type Shown as H*	*δ, pm*
Chlorides			
Cl–C–**H**	3–4	R–C(=O)–O–**H**	10.5–12
		Ketones	
Cl_2C–**H** (Cl–C(Cl)–**H**)	5.8	R–C(=O)–C–**H**	2–2.7
		Aldehydes	
Bromides, Br–C–**H**	2.5–4	R–C(=O)–**H**	9–10
Iodides, I–C–**H**	2–4	Amides	
		R–C(=O)–N–**H**	5–8
Nitroalkanes, O_2N–C–**H**	4.2–4.6	Alcohols, R–O–**H**	4.5–9
		Phenols, Ar–O–**H**	4–12
		Amines, R–NH_2	1–5

Table 8.4 *Compilation of* ^{1}*H NMR Absorptions for Various Molecules*

Listed below are the ^{1}H NMR chemical shifts observed for the protons of a number of organic compounds. The shifts are classified according to whether they are methyl, methylene, or methine types of hydrogen atoms. The atom shown in **bold** is responsible for the absorptions listed.

Methyl Absorptions

Compound	*δ, ppm*	*Compound*	*δ, ppm*
CH_3NO_2	4.3	CH_3CHO	2.2
CH_3F	4.3	CH_3I	2.2
$(CH_3)_2SO_4$	3.9	$(CH_3)_3N$	2.1
$C_6H_5COOCH_3$	3.9	$CH_3CON(CH_3)_2$	2.1
C_6H_5–O–CH_3	3.7	$(CH_3)_2S$	2.1
CH_3COOCH_3	3.6	CH_2=C(CN)CH_3	2.0
CH_3OH	3.4	CH_3COOCH_3	2.0
$(CH_3)_2O$	3.2	CH_3CN	2.0
CH_3Cl	3.0	CH_3CH_2I	1.9
$C_6H_5N(CH_3)_2$	2.9	CH_2=CH–C(CH_3)=CH_2	1.8

Table 8.4 *Compilation of* ^{1}H *NMR Absorptions for Various Molecules (Continued)*

Methyl Absorptions

Compound	*δ, ppm*	*Compound*	*δ, ppm*
$(\mathbf{CH_3})_2NCHO$	2.8	$(\mathbf{CH_3})_2C{=}CH_2$	1.7
$\mathbf{CH_3}Br$	2.7	$\mathbf{CH_3}CH_2Br$	1.7
$\mathbf{CH_3}COCl$	2.7	$C_6H_5C(\mathbf{CH_3})_3$	1.3
$\mathbf{CH_3}SCN$	2.6	$C_6H_5CH(\mathbf{CH_3})_2$	1.2
$C_6H_5CO\mathbf{CH_3}$	2.6	$(\mathbf{CH_3})_3COH$	1.2
$(\mathbf{CH_3})_2SO$	2.5	$C_6H_5CH_2\mathbf{CH_3}$	1.2
$C_6H_5CH{=}CHCO\mathbf{CH_3}$	2.3	$\mathbf{CH_3}CH_2OH$	1.2
$C_6H_5\mathbf{CH_3}$	2.3	$(\mathbf{CH_3}CH_2)_2O$	1.2
$(\mathbf{CH_3}CO)_2O$	2.2	$\mathbf{CH_3}(CH_2)_3$ X (X=Cl, Br, I)	1.0
$C_6H_5OCO\mathbf{CH_3}$	2.2	$\mathbf{CH_3}(CH_2)_4\mathbf{CH_3}$	0.9
$C_6H_5CH_2N(\mathbf{CH_3})_2$	2.2	$(\mathbf{CH_3})_3CH$	0.9

Methylene Absorptions

Compound	*δ, ppm*	*Compound*	*δ, ppm*
$EtOCOC(CH_3){=}\mathbf{CH_2}$	5.5	$C_6H_5\mathbf{CH_2}N(CH_3)_2$	10.0
$\mathbf{CH_2}Cl_2$	5.3	$CH_3\mathbf{CH_2}SO_2F$	9.9
$\mathbf{CH_2}Br_2$	4.9	$CH_3\mathbf{CH_2}I$	9.8
$(CH_3)_2C{=}\mathbf{CH_2}$	4.6	$C_6H_5\mathbf{CH_2}CH_3$	3.3
$CH_3COO(CH_3)C{=}\mathbf{CH_2}$	4.6	$CH_3\mathbf{CH_2}SH$	3.3
$C_6H_5\mathbf{CH_2}Cl$	4.5	$(CH_3\mathbf{CH_2})_3N$	3.1
$(CH_3O)_2\mathbf{CH_2}$	4.5	$(CH_3\mathbf{CH_2})_2CO$	2.6
$C_6H_5\mathbf{CH_2}OH$	4.4	$BrCH_2\mathbf{CH_2}CH_2Br$	2.4
$CF_3CO\mathbf{CH_2}C_3H_7$	4.3	Cyclopentanone (α-$\mathbf{CH_2}$)	2.4
$Et_2C(COO\mathbf{CH_2}CH_3)_2$	4.1	Cyclohexene (α-$\mathbf{CH_2}$)	2.4
$HC{\equiv}C{-}\mathbf{CH_2}Cl$	4.1	Cyclopentane	2.4
$CH_3COO\mathbf{CH_2}CH_3$	4.0	Cyclohexane	2.0
$CH_2{=}CH\mathbf{CH_2}Br$	3.8	$CH_3(\mathbf{CH_2})_4CH_3$	2.0
$HC{\equiv}C\mathbf{CH_2}Br$	3.8	Cyclopropane	1.5
$Br\mathbf{CH_2}COOCH_3$	3.7	$C_6H_5\mathbf{CHO}$	1.4
$CH_3\mathbf{CH_2}NCS$	3.6	$4\text{-}ClC_6H_4\mathbf{CHO}$	1.4
$CH_3\mathbf{CH_2}OH$	3.6	$4\text{-}CH_3OC_6H_4\mathbf{CHO}$	0.2
$CH_3CH_2\mathbf{CH_2}Cl$	3.5	$CH_3\mathbf{CHO}$	9.7
$(CH_3\mathbf{CH_2})_4N^+I^-$	3.4	Pyridine (α-**H**)	8.5
$CH_3\mathbf{CH_2}Br$	3.4	$1,4\text{-}C_6\mathbf{H_4}(NO_2)_2$	8.4

Table 8.4 *Compilation of ^1H NMR Absorptions for Various Molecules (Continued)*

Methylene Absorptions

Compound	*δ, ppm*	*Compound*	*δ, ppm*
$C_6H_5CH{=}CHCOCH_3$	7.9	*p*-Benzoquinone	6.8
C_6H_5CHO	7.6	$C_6H_5NH_2$	6.6
Furan (α-**H**)	7.4	Furan (β-**H**)	6.3
Naphthalene (β-**H**)	7.4	$CH_3CH{=}CHCOCH_3$	5.8
1,4-$C_6H_4I_2$	7.4	Cyclohexene (vinylic **H**)	5.6
1,4-$C_6H_4Br_2$	7.3	$(CH_3)_2C{=}CHCH_3$	5.2
1,4-$C_6H_4Cl_2$	7.2	$(CH_3)_2CHNO_2$	4.4
C_6H_6	7.3	Cyclopentyl bromide (**H** at C1)	4.4
C_6H_5Br	7.3	$(CH_3)_2CHBr$	4.2
C_6H_5Cl	7.2	$(CH_3)_2CHCl$	4.1
$CHCl_3$	7.2	C_6H_5CCH	2.9
$CHBr_3$	6.8	$(CH_3)_3CH$	1.6

The *magnitude* of chemical shifts is primarily related to two factors, namely, the *electronegativity* of any functional groups that are near the proton being observed and the nature of *induced magnetic fields, H_i,* in molecules having π-bonds. Both effects are associated with circulation of electrons in the molecule and the interaction of the resulting induced magnetic field with the applied field H_o (Eq. 8.8).

You can see the influence of electronegativity by comparing the chemical shifts of the methyl halides, CH_3–X, in which the δ-values are 4.3, 3.0, 2.7, and 2.2 for X = F, Cl, Br, and I, respectively (Table 8.4). That of π-electrons is seen in the relative chemical shifts of the aliphatic protons of alkanes versus the vinylic protons of alkenes, as in ethane (δ 0.9) and ethylene (δ 5.25). The hydrogen nuclei of ethylene are *deshielded* partly because an sp^2-hybridized carbon atom is more electronegative than one that is sp^3-hybridized. Moreover, the vinylic protons are further deshielded because they lie in a region of the induced field H_i where the lines of force *add* to H_o, as shown in Figure 8.26a. A similar analysis of the induced field accounts for the fact that an alkynic proton, as in acetylene (δ 1.80), resonates at *higher* field than does a vinylic hydrogen, even though an *sp*-hybridized carbon atom is the *most* electronegative of all carbon atoms found in organic compounds. In alkynes, the acetylenic proton is in a portion of the magnetic field induced by the circulating π-electrons where the lines of force *oppose* the applied field (Fig. 8.26b), thereby increasing the shielding of this type of nucleus. The protons of aromatic compounds, as in benzene (δ 7.27), resonate at lower fields than the vinylic protons of alkenes because the induced magnetic field responsible for deshielding is greater owing to the *cyclic* nature of the circulation of π-electrons (Fig. 8.26c); this phenomenon is called the **ring-current effect.**

Spin-Spin Splitting

Information beyond that regarding the magnetic environment of a nucleus is available from analysis of the NMR spectrum. For example, the **spin-spin splitting** pattern shown by a particular type of hydrogen atom provides information about

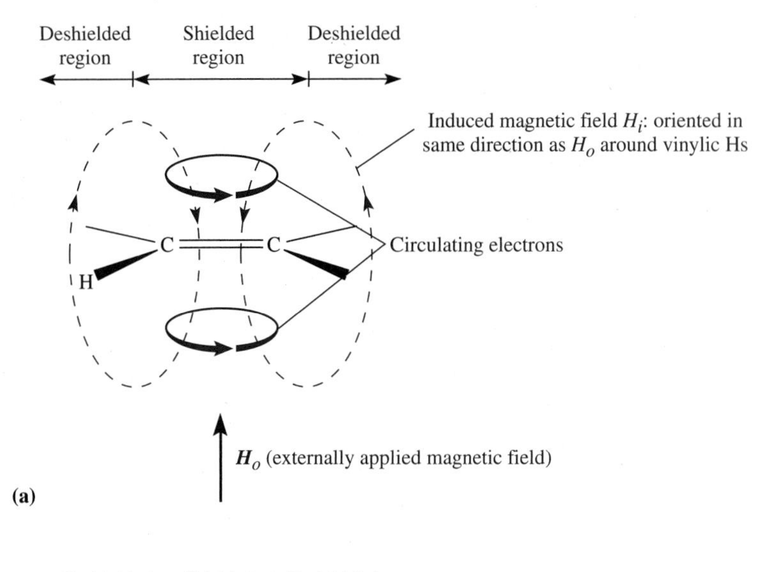

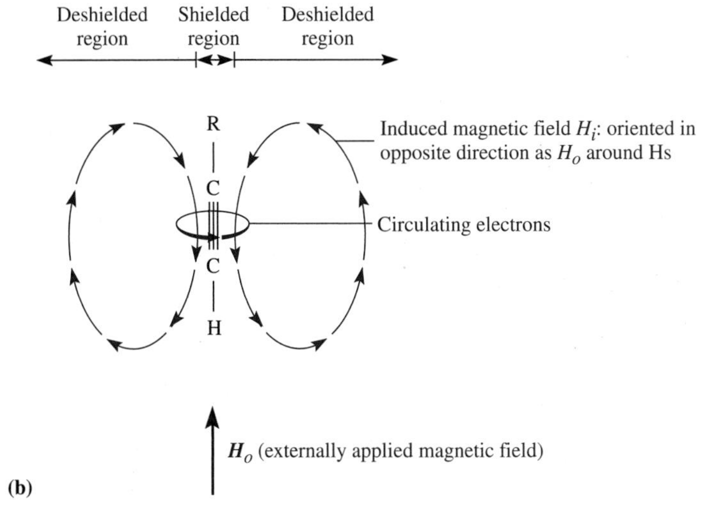

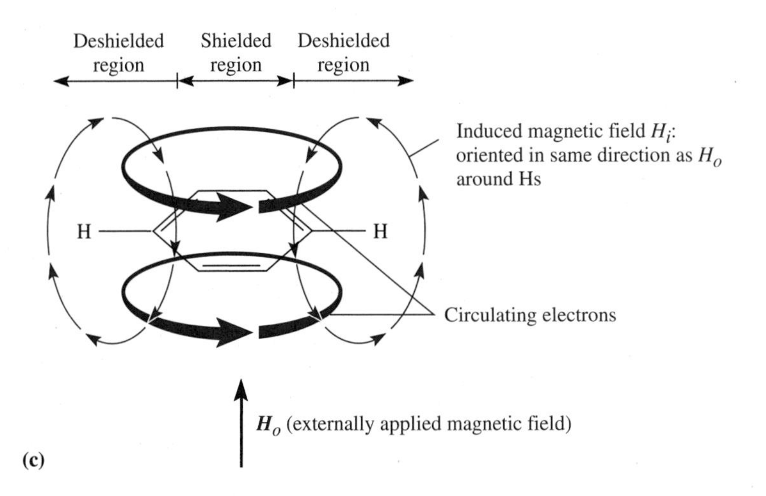

Figure 8.26
Effect on chemical shifts of induced field from circulation of π-electrons. (a) Deshielding of vinylic protons. (b) Shielding of acetylenic protons. (c) Deshielding of aromatic protons.

the *number* of its **nearest neighbors.** Nearest neighbors are elements that have nuclear spin and, in most cases, are *no more* than *three* bonds, or *two* atoms, away from the atom of interest. For our purposes, hydrogen is the primary element of interest for analyzing splitting patterns because compounds containing other magnetically active nuclei are only rarely encountered in the experiments in this textbook. The presence of ^{13}C in organic compounds is *not* a factor in the analysis because this isotope, which also has a nuclear spin of 1/2, is present at the level of only about 1%. Thus, peaks due to spin-spin splitting with it are generally too weak to be observed.

Observation of coupling between atoms having nuclear spins requires that the nuclei must *not* be magnetically equivalent. By definition, **magnetic equivalency** means that *every* nucleus in a particular set of chemical shift-equivalent nuclei is coupled equally to *every* other nucleus of the spin system. Chemical shift-nonequivalent nuclei are always magnetically nonequivalent and couple with one another. As discussed earlier, this type of nonequivalence is generally seen for protons that are chemically distinct by virtue of being *heterotopic* or *diastereotopic.* That protons may be chemically equivalent but magnetically nonequivalent is illustrated by a discussion later in this section.

Analysis of 1-nitropropane (**1**) in this context reveals that the type *a* hydrogen atoms have *two* nearest neighbors in the form of the two protons of type *b* that are magnetically active and chemically nonequivalent. Nuclei of type *c* also have two such nearest neighbors, whereas those of type *b* nuclei have *five.*

The **splitting** or **coupling pattern** of each chemically distinct type of nucleus may be predicted in the following way. In the general case, an atom *A* having nuclear spin I_z, where $I_z = 1/2, 1, 3/2$, and so on, and coupling with another atom *B* will split the resonance of *B* into the number of peaks given by Equation 8.10. When *A* is ^{1}H, ^{13}C, ^{19}F, or ^{31}P, this expression simplifies to the $n + 1$ rule expressed in Equation 8.11 because $I_z = 1/2$ for all these nuclei. The expected splitting patterns for 1-nitropropane (**1**), according to Equation 8.11, are *three* peaks each for H_a and H_c and six peaks for H_b. This is precisely the splitting pattern seen in Figure 8.21.

$$N = 2nI_z + 1 \qquad (8.10)$$

where N = number of peaks observed for absorbing atom B
n = number of magnetically equivalent nearest neighbor atoms A
I_z = nuclear spin of A

$$N = n + 1 \qquad (8.11)$$

Strictly speaking, the $n + 1$ rule for predicting splitting patterns of coupled hydrogen nuclei applies only under the following circumstances: (1) The values of the **coupling constants, *J*,** of all nearest neighbor hydrogen nuclei coupling with the nucleus of interest must be identical, and (2) the ratio of the difference in chemical shift $\Delta\nu$, in Hz, of the coupled nuclei to their coupling constant J must be greater than about 10 (Eq. 8.12). When these criteria are met, a **first-order analysis** of the multiplicity of each resonance is possible. Importantly, Equation 8.11 may be used to determine the number, n, of nearest neighbors to a hydrogen nucleus when ^{1}H NMR spectra are first-order. For example, if a *sextet* is observed in the ^{1}H NMR spectrum, there are *five* nearest neighbors to the proton(s) responsible for that sextet because N of Equation 8.11 is *six*.

$$n + 1 \text{ rule valid if: } \Delta v \text{ (Hz)}/J\text{(Hz)} \geqslant {\sim}10 \qquad (8.12)$$

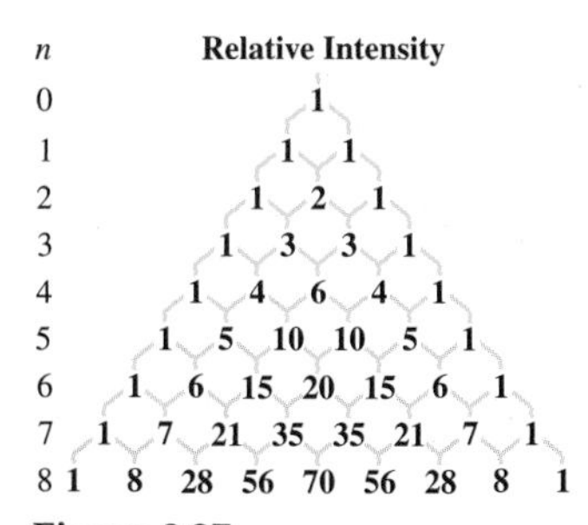

Figure 8.27
Pascal's triangle showing relative intensities of first-order multiplets; coupling constants of all nearest neighbors, n, are assumed to be equal.

The *relative intensities* of the peaks in a *first-order* multiplet that obeys Equation 8.11 may be approximated by applying Pascal's triangle (Fig. 8.27). For example, if a hydrogen atom has two nearest neighbors, then the triplet that results is predicted to have intensities in the ratio of about 1:2:1. If there are five nearest neighbors, the corresponding ratio for the sextet is 1:6:15:20:15:6:1. For reasons we'll not discuss, the prediction of relative peak intensities by this method becomes more accurate as the ratio $\Delta v/J$ increases.

Because the ^1H NMR spectra in this book were obtained at 300 MHz, many may be interpreted as first-order spectra, and the peak intensities of the multiplets are predictable by applying Pascal's triangle. For example, reconsidering the spectrum of Figure 8.21, we find that $J_{ab} = J_{bc} \cong 7$ Hz. The difference in chemical shifts for the sets of coupled nuclei are $\Delta\nu_{AB} = 315$ Hz, and $\Delta\nu_{BC} = 700$ Hz, respectively. Application of Equation 8.12 gives values of $315/7 = 45$ and $700/7 = 100$, respectively, for the ratios of interest. The fact that both ratios exceed the criterion of Equation 8.12 means that the splitting patterns are predicted by the $n + 1$ rule, as evidenced in Figure 8.21. Moreover, the relative intensities of each of the peaks in the multiplets in Figure 8.21 have the approximate values predicted from Pascal's triangle.

As the term $\Delta v/J$ decreases, the observed spectra become **second-order** and are more complicated. The splitting patterns may no longer be predictable by the $n + 1$ rule, and the intensities of the peaks in the multiplets cannot be approximated by using Pascal's triangle. This trend may be illustrated by examining the ^1H NMR spectra for 1-butanol (**4**) in Figures 8.28 and 8.29 and pentane (**5**) in Figure 8.30.

$H_e-C(H_e)(H_e)-C(H_d)(H_d)-C(H_c)(H_c)-C(H_b)(H_b)-OH_a$

4

1-Butanol

$H_a-C(H_a)(H_a)-C(H_b)(H_b)-C(H_c)(H_c)-C(H_b)(H_b)-C(H_a)(H_a)-H_a$

5

Pentane

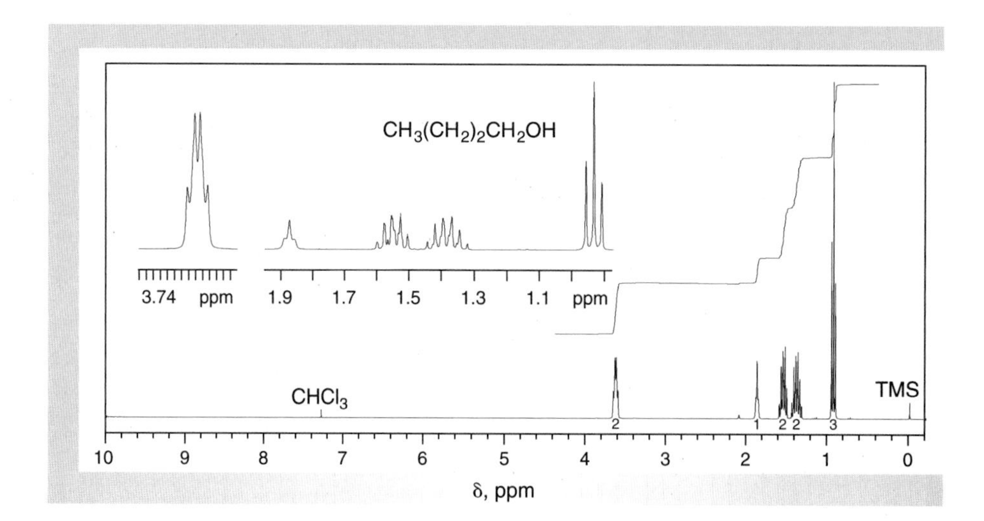

Figure 8.28
^1H NMR spectrum of 1-butanol in $CDCl_3$ (300 MHz).

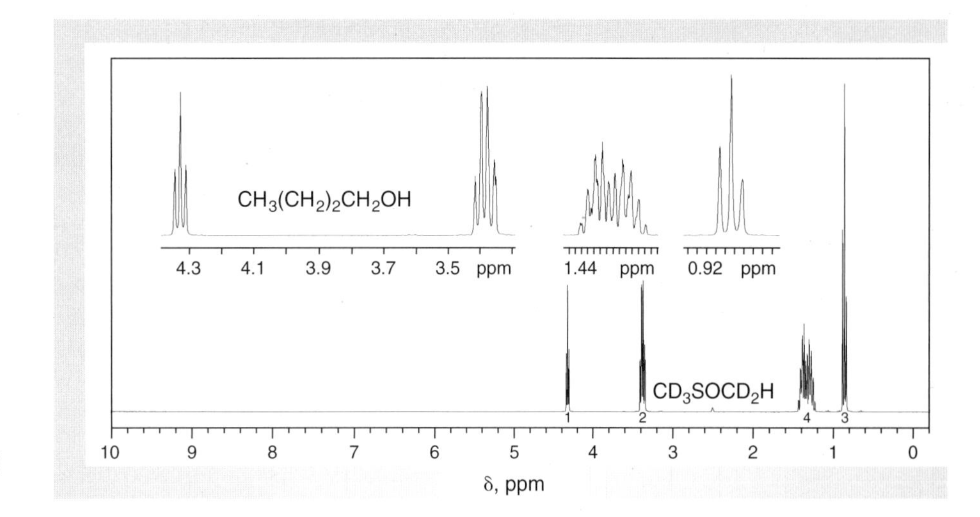

Figure 8.29
^1H NMR spectrum of 1-butanol in DMSO-d_6 (300 MHz).

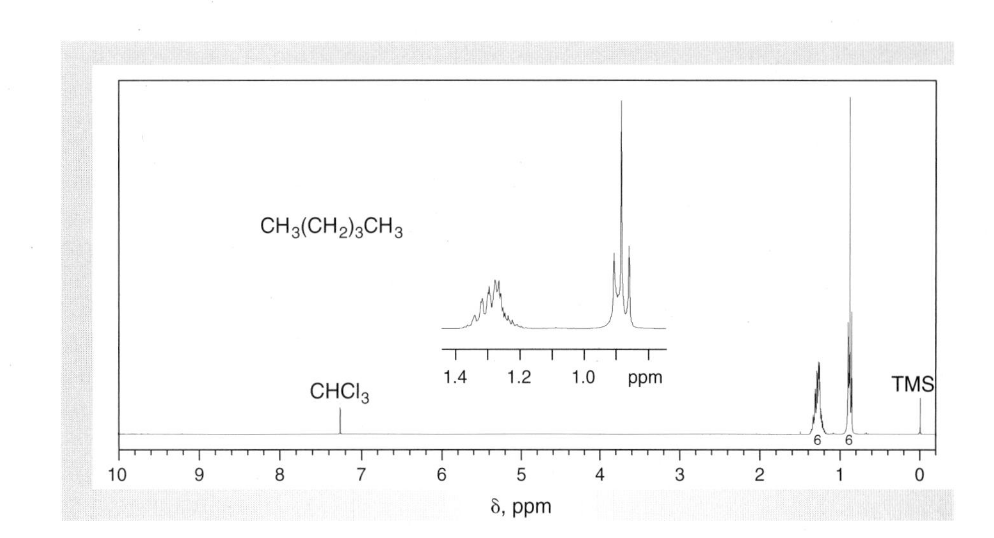

Figure 8.30
^1H NMR spectrum of pentane (300 MHz, $CDCl_3$).

In 1-butanol (**4**), nuclei of types H_e and H_d produce multiplets centered at δ 0.94 ppm and at about δ 1.4 ppm, respectively; J_{de} is about 7 Hz (Fig. 8.28). Using a 300-MHz spectrometer, the ratio $\Delta\nu/J_{de}$ is greater than 10, so the criterion of Equation 8.12 is satisfied. The nuclei of type H_e appear as a triplet, and the relative intensities of the three peaks approximate that expected from Pascal's triangle, namely, 1:2:1. In contrast, although the numbers of peaks in the multiplets for protons of types H_d and H_c correspond to those predicted by the $n + 1$ rule, the relative intensities of the peaks are no longer approximated by using Pascal's triangle. This is because the criterion of Equation 8.12 is *not* satisfied. The difference in chemical shifts of the two types of protons is only 48 Hz (300 × (1.56 − 1.40)) and their coupling constant, J_{cd}, is about 7, so the ratio $\Delta\nu/J_{cd}$ is about 7 (48/7), and the ratios of the peaks in the two multiplets cannot be predicted by applying simple first-order rules.

Assuming that $J_{bc} = J_{ab}$, applying the $n + 1$ rule to nuclei of type H_b and H_a leads to a predicted splitting pattern of a quartet for H_b and a triplet for H_a. However, what we actually observe are a somewhat broadened and distorted quartet and a triplet centered at δ 3.65 and 1.63 ppm, as seen in Figure 8.28. The broadening of these peaks is not due to any second-order effects. Rather, the spectrum was obtained under conditions in which the hydroxylic proton H_a is undergoing rapid exchange with other molecules of the alcohol or with adventitious water present in the sample. As a result of this rapid exchange, the spin information carried by the hydroxylic proton is partially lost, and the peaks arising from protons that are spin-coupled with the hydroxyl proton are broadened. Spectra obtained under conditions that suppress this exchange, such as using CD_3SOCD_3 (DMSO-d_6) rather than $CDCl_3$ as the solvent, allow us to observe the coupling, and the quartet at δ 3.38 for H_a and the triplet at δ 4.33 for H_b are sharp, as illustrated in Figure 8.28. Comparison of the spectra of **4** in Figures 8.28 and 8.29 also reveals that the chemical shifts for protons in a molecule *can* vary with the solvent, the most significant change being for $H_{a'}$, so it is always important to indicate the solvent used to prepare the sample when reporting NMR spectral data.

In the spectrum of pentane (**5**) (Fig. 8.30), we see that nuclei of type H_a resonate at about δ 0.90 ppm, whereas those of types H_b and H_c appear at about δ 1.30 ppm. Because the ratio $\Delta v/J_{ab}$ is greater than 10, the condition of Equation 8.12 is satisfied, and the peak for the H_a is a triplet according to the $n + 1$ rule. However, the ratio $\Delta v/J_{bc}$ is much less than 10, and the splitting pattern for the nuclei of types H_b and H_c is complex and cannot be interpreted by first-order analysis and the $n + 1$ rule.

Thus far, we have predicted splitting patterns by grouping hydrogens according to their chemical shift equivalence and then determining the number of nearest neighbors *n*. There are cases, however, in which nuclei may have equivalent chemical shifts but are *not* magnetically equivalent; this may complicate the appearance and interpretation of the spectrum. This phenomenon is often seen in the ^1H NMR spectra of aromatic compounds.

For example, analyzing 4-bromonitrobenzene (**6**) by the substitution test leads to the conclusion that there are two sets of homotopic hydrogen atoms, H_aH_a' and H_bH_b'. The two protons in each set have equivalent chemical shifts. However, the coupling constants between all members of the spin set are *not* equivalent. Namely, $J_{H_aH_b} \equiv J_{H_{a'}H_{b'}}$ and $J_{H_{ab'}} \equiv J_{H_{a'}H_{b}}$, but $J_{H_{a'}H_{b'}} \neq J_{H_{a'}H_b}$ and $J_{H_aH_b} \neq J_{H_aH_{b'}}$. This makes H_a and $H_{a'}$ **magnetically nonequivalent,** as are H_b and $H_{b'}$. We note that the pairs of nuclei $H_aH_{b'}$ and $H_{a'}$ H_b are not nearest neighbors, yet coupling may occur between them. Nuclear spin interactions that occur over more than three bonds are termed **long-range coupling,** a phenomenon that is most frequently observed in systems having **conjugated π-bonds.**

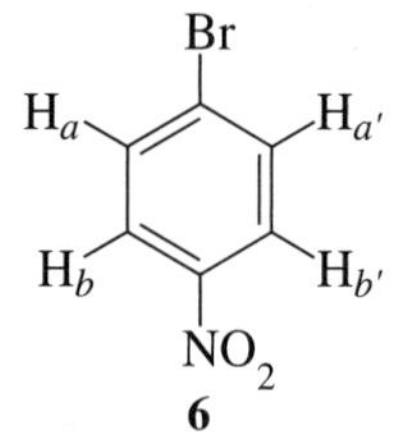

Br
H_a $H_{a'}$
H_b $H_{b'}$
NO_2

6

4-Bromonitrobenzene

The effect of magnetic nonequivalency is evident in the ^1H NMR spectrum of **6** (Fig. 8.31). Carefully examining the resonances for each of the doublets reveals additional splitting that is the result of magnetic nonequivalency within the set.

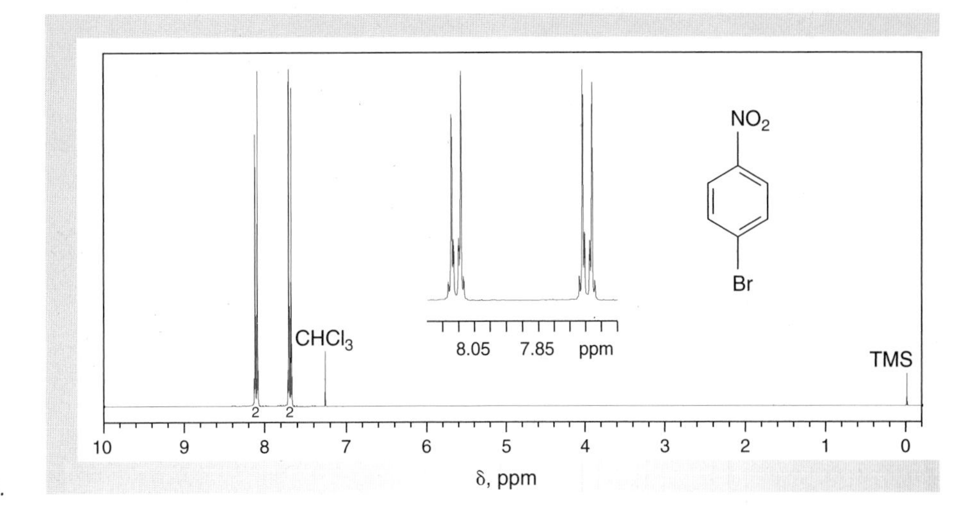

Figure 8.31
^1H NMR spectrum of 4-bromo-nitrobenzene (300 MHz, $CDCl_3$).

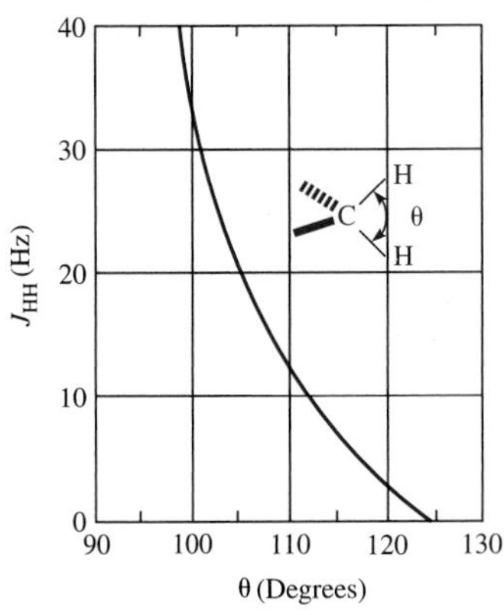

Figure 8.32
The geminal Karplus correlation. J_{HH} for CH_2 groups as a function of H–C–H angle, θ.

Nevertheless, the aromatic multiplets still have the overall appearance of the doublets expected from the simple first-order analysis using the $n + 1$ rule.

The *magnitude* of coupling constants, J_{HH}, is *independent* of the oscillator frequency, so units of Hz rather than δ are used. The measured values depend on two main factors: the number of bonds intervening between the coupled nuclei, and the bond angle or dihedral angle between the nuclei. The angular dependence of **geminal coupling,** which is coupling between magnetically different protons on the *same* atom, is shown graphically in Figure 8.32, in which θ is the bond angle between the coupled nuclei. The situation for **vicinal coupling,** in which the coupled nuclei are on *adjacent* atoms and have a **dihedral angle,** ϕ, is presented in Figure 8.33. Some typical ranges for coupling constants are provided in Figure 8.34.

Coupling constants provide valuable information about the structure of organic compounds. If two different protons are on adjacent carbon atoms, the coupling

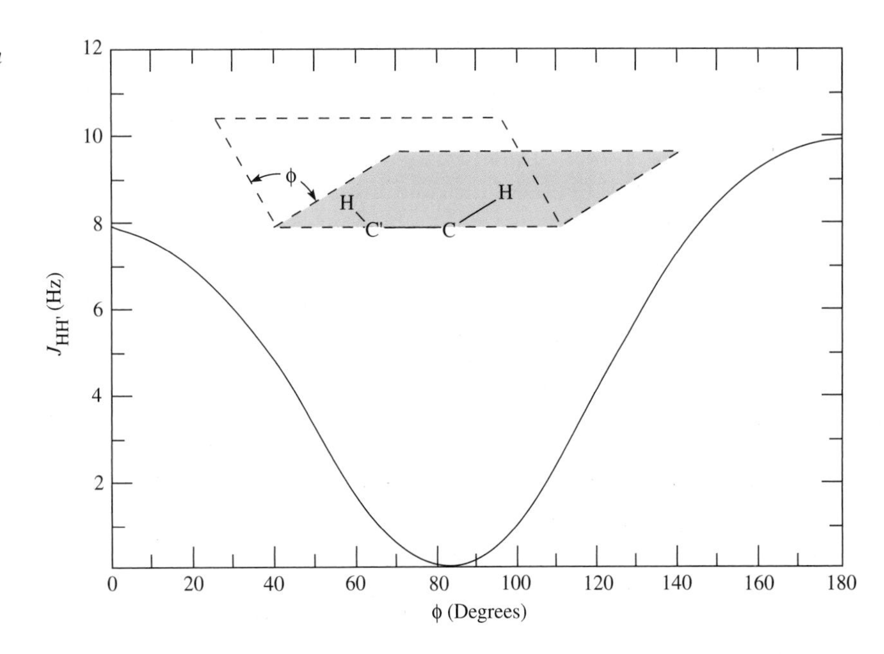

Figure 8.33
The vicinal Karplus correlation. Relationship between dihedral angle, ϕ, and coupling constant for vicinal protons.

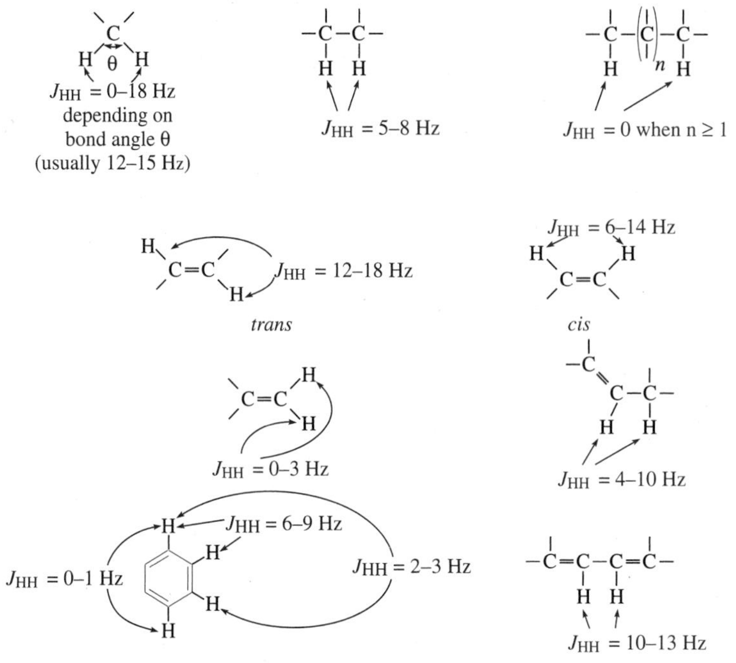

J_{HH} = 0–18 Hz depending on bond angle θ (usually 12–15 Hz)

J_{HH} = 5–8 Hz

J_{HH} = 0 when n ≥ 1

J_{HH} = 12–18 Hz

trans

J_{HH} = 6–14 Hz

cis

J_{HH} = 0–3 Hz

J_{HH} = 4–10 Hz

J_{HH} = 6–9 Hz

J_{HH} = 0–1 Hz

J_{HH} = 2–3 Hz

J_{HH} = 10–13 Hz

Figure 8.34
Examples of the values of coupling constants J_{HH} for selected functional groups.

constant between the two has only one value, that is, $J_{ab} \equiv J_{ba}$, because the protons are mutually coupled with each other. Thus, if the coupling constants of two multiplets suspected to be the result of two particular nuclei coupling with one another are *not* the same, the two nuclei are *not* coupled. This fact helps in making structural assignments.

It is difficult to measure coupling constants using the 300-MHz spectra reproduced in this textbook because the spectra have been reduced, making measurement of the distance between peaks subject to large error. However, using modern high-field NMR spectrometers, the values of coupling constants for simple splitting patterns are easily determined from digital printouts of chemical shifts of the various peaks. As noted previously, to enable you to gain experience in interpreting ^{1}H NMR data, the ^{1}H NMR spectra of all the organic reactants and products encountered in subsequent chapters are found at the website for this book. By expanding the spectra, you will be able to measure the coupling constants accurately.

Data provided in the form of $\Delta\delta$ must first be converted to Hz before computation of the coupling constant, *J*. For example, if the peaks of a quartet are found to differ by 0.04 ppm in a spectrum obtained on a 300-MHz instrument, the coupling constant would be 12 Hz by application of Equation 8.13, where $\Delta\delta$ is 0.04 and the *instrument frequency* is 300×10^6 Hz.

$$J(\text{Hz}) = (\Delta\delta \times 10^{-6}) \times \text{instrument frequency} \tag{8.13}$$

Integration

The areas of the peaks in a ^{1}H NMR spectrum are important measures of the *relative* numbers of the different types of hydrogen nuclei in a molecule because these peak areas are a linear function of the number of nuclei producing the peaks. The areas are determined by an electronic integration that is typically plotted as a stepped trace that appears on the spectrum, as seen in Figure 8.21.

In practice, you would first record the ^{1}H NMR spectrum and then plot the integration on the same sheet of paper. The vertical distance that the integration rises over a peak or group of peaks is a direct measure of the area under the peak(s) and thus of the relative number of protons producing the resonances. You would usually reset the integrator to zero after each set of resonances has been measured, as in Figure 8.21, but this need not be done, as seen in Figure 8.28. Resetting the integrator allows the height of each step in the integration curve to be higher, so the experimental error in measuring the step is less. In this textbook, the numerical values of the integrations rather than the plotted integrals will normally be provided below sets of peaks. You may verify these values by examining the ^{1}H NMR spectra at the website for this text.

You would then measure the height of each step and compute the *relative* number of hydrogen atoms producing the steps by dividing the integration heights by the smallest height measured; this gives a ratio of the various types of protons. For example, in the spectrum of 1-nitropropane (**1**), the heights of the steps in Figure 8.21 are represented by the distances *ab, ac,* and *ad.* The relative heights are measured to obtain the ratio of 1.55:1.0:1.0 for groups *A, B,* and *C* in the spectrum. Because the number of hydrogen atoms present in a molecule must be integral, this ratio is converted to whole numbers by multiplying *each* value by an *integer* that gives the **relative ratio** of the types of protons in whole numbers. *If* you know the molecular formula of the molecule, the integer is chosen to give a *sum* of the values in the ratio corresponding to the *absolute* number of hydrogen nuclei present. There are seven hydrogen atoms in 1-nitropropane, so *two* is used as the multiplier to provide the **absolute ratio** 3.10:2.0:2.0, the sum of which is 7.1. Because the accuracy of electronic integration is generally no better than 5–10%, this is within experimental error of the total number of protons in the molecule.

Analysis of ^{1}H NMR Spectra

General Procedure

The proper analysis of the ^{1}H NMR spectrum of a compound may provide a wealth of information regarding its structure. Determining the *chemical shift* of each peak or group of peaks allows speculation about the *type* of functional group to which the hydrogen nucleus producing the resonance is attached. The *spin-spin splitting pattern* and the $n + 1$ rule give information concerning the *number of nearest neighbor hydrogen atoms* for the proton(s) producing a particular absorption. *Integration* allows evaluation of the peak areas of each set of peaks and defines the *relative numbers of each type* of proton present in the molecule; if the molecular formula of the compound is known, the *absolute* number of each type may also be computed.

Appropriate steps for complete analysis of a ^{1}H NMR spectrum are as follows:

1. Determine the relative numbers of the different types of hydrogen nuclei by measuring the integrated step height for each peak or group of peaks. Convert the resulting *relative* ratio to an *absolute* ratio if the molecular formula is known. The *absolute* values of the integrations are provided for most of the spectra in this textbook by the numbers in parentheses located below the respective sets of peaks in the spectra.
2. Measure the *chemical shifts* of the different sets of resonances and use the data to try to deduce the functional groups that might be present in the molecule. This process is greatly aided if the molecular formula and/or other spectral data, such as IR and ^{13}C NMR spectra, are available.

3. Analyze the *spin-spin splitting* patterns of each group of resonances to determine the number of *nearest neighbors* associated with each type of hydrogen nucleus.

Applying these steps *may* allow assignment of the structure of an unknown compound. However, without additional information about a substance, such as its molecular formula or the chemical reaction(s) by which it was formed, only a partial structural determination may be possible. Other chemical or spectral data may be required to make a complete assignment. In any case, structural assignments based *exclusively* on interpretation of spectra *must* be confirmed by comparing the spectra of the unknown with those of an authentic specimen, if the compound is known, or by chemically converting the unknown to a known compound.

Analysis of an Unknown

An example of the use of this procedure for making a structural assignment follows. Other examples may be found in the exercises at the end of this section.

Problem

Provide the structure of the compound having the molecular formula C_4H_9Br and whose ^{1}H NMR spectrum is given in Figure 8.35.

Solution

For purposes of this analysis, the groups of peaks are referred to as representing nuclei of types *A–C* in going from left to right, or upfield, in the spectrum.

Integration. The *relative* ratio of the integrals for H_a, H_b, and H_c is 2.1:1.0:5.7, respectively. Given the molecular formula, the *absolute* ratio must be 2:1:6, for a total of nine hydrogen atoms. Thus, there are two hydrogen atoms of type *A*, *one* of type *B*, and *six* of type *C*.

Chemical Shifts. The δ-values at the center of the individual multiplets are approximately 3.3 (H_a), 2.0 (H_b), and 1.0 (H_c). The low-field doublet for H_a presumably reflects

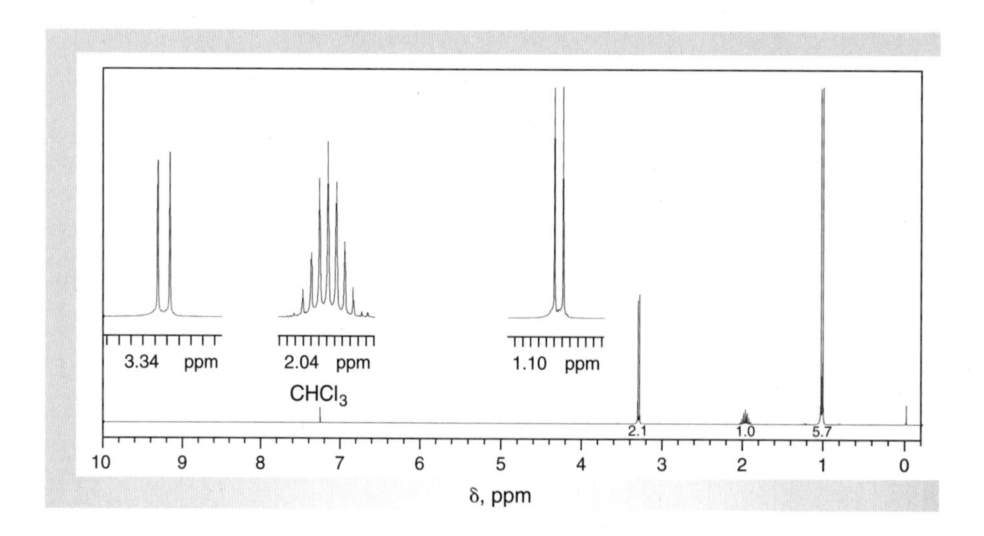

Figure 8.35
^{1}H NMR spectrum for unknown compound (300 MHz, $CDCl_3$).

the nearby presence of the electronegative bromine atom. Since there are *two* such nuclei, the partial structure, $-CH_2Br$, may be written. The appearance of H_c at about δ 1.0 is consistent with these nuclei being associated with a methyl group. Because there are six such hydrogen atoms, *two* methyl groups must be present. Another partial structure, $(CH_3)_2C-$, may then be written. Because these two partial structures account for *all* of the carbon atoms required by the molecular formula and all but *one* of the hydrogen atoms, the missing atom must be a *methine* hydrogen atom located on the carbon atom bearing the two methyl groups and the bromomethyl substituent. Although the chemical shift of this methine hydrogen atom is outside the range for such nuclei, as given in Table 8.3, reference to Table 8.4 shows that the methyl resonances are shifted *downfield* when there is a halogen atom β to the carbon atom bearing the nuclei of interest; this is evidenced in the δ-value of 1.7 for the methyl group of ethyl bromide, CH_3CH_2Br. The same effect applies to methine hydrogen atoms. Enough information is now available to enable assignment of the structure of the unknown as 1-bromo-2-methylpropane (isobutyl bromide), $(CH_3)_2CHCH_2Br$.

In this case, the determination of the structure may be made simply on the basis of molecular formula, chemical shifts, and integration; analysis of the spin-spin splitting patterns, as in Part (c), is required only to confirm the conclusion regarding the structure of the unknown and not for initially generating the structure. More typically, the information contained in the splitting patterns is crucial to a first formulation of the structure of an unknown.

Spin-Spin Splitting Patterns. The fact that the nuclei of types *A* and *B* appear as a doublet means that they have *one* nearest neighbor, which is consistent with the fact that both H_a and H_c in 1-bromo-2-methylpropane should be coupled with the methine hydrogen atom H_b but *no* others. Conversely, H_b must be coupled with both the other types of protons, which total eight. Application of the $n + 1$ rule dictates nine peaks in the splitting pattern of the methine hydrogen atom. The two "missing" members of this multiplet are not apparent because they are the weak, outermost peaks, and the amplification at which the spectrum was plotted was insufficient for them to be seen. If the multiplet for H_b is viewed at significantly higher amplification, all nine peaks may be observed.

There is an important lesson to be learned from the fact that fewer peaks are observed than would be expected from the $n + 1$ rule: When multiplets of six or more are predicted, one or more of the outermost absorptions of the multiplet is frequently too weak (see Pascal's triangle, Figure 8.27) to appear under the normal operating conditions for the spectrometer. This fact must be taken into account when using spin-spin splitting patterns to interpret the spectrum of an unknown compound.

Determination of Structure. Although the present problem was solved by using only two of the three basic types of information available from an integrated ^1H NMR spectrum, the usual case would require careful interpretation of all three. The strategy is to generate partial structures that are consistent with the available data and then to attach these pieces in various ways that are consistent with the observed splitting patterns and the valences of the bonded atoms. In a majority of cases, particularly with molecules having a total of only twenty atoms or so and one or more functional groups, the actual structure of the compound may be correctly assigned. Once proposed, a possible structure must be checked carefully to ascertain that it is consistent with all of the available data. If not, the structure is incorrect and alternative possibilities must be explored and tested against the data until one is found that is consistent with *all* of the spectral information.

EXERCISES

1. Define, explain, or give an example of the following:
   a. tetramethylsilane (TMS) as an internal standard
   b. (δ) delta-scale
   c. upfield shift
   d. $n + 1$ rule
   e. chemical shift
   f. integration curve
   g. *relative* ratio of integrated peaks
   h. *absolute* ratio of integrated peaks
   i. homotopic hydrogen atoms
   j. enantiotopic hydrogen atoms
   k. diastereotopic hydrogen atoms
   l. coupling constant
   m. chemically equivalent protons
   n. diamagnetic shielding
   o. heterotopic hydrogen atoms
2. Why is it important to specify the solvent that is used when reporting chemical shifts?
3. Demonstrate that the H_b atoms in 1-nitropropane (**1**) are enantiotopic.
4. Label the sets of connectively equivalent sets of protons in the molecules (**a**)–(**j**) according to whether they are homotopic, enantiotopic, or diastereotopic. Using Tables 8.3 and 8.4, predict the approximate chemical shifts, in δ, and the splitting patterns expected for each chemically distinct type of hydrogen atom that you identify. Assume in this analysis that the coupling constants of all nearest neighbors are identical.

a. 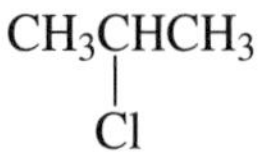 $CH_3CH(Cl)CH_3$

b. 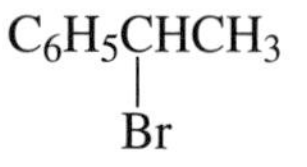 $C_6H_5CH(Br)CH_3$

c. $(CH_3)_2CHCH(CH_3)_2$

d. 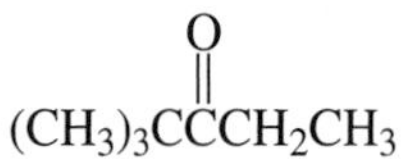 $(CH_3)_3CC(=O)CH_2CH_3$

e. 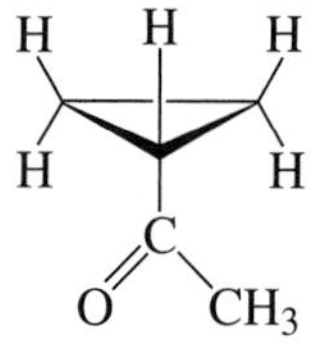 cyclopropane ring bearing five H atoms and a $C(=O)CH_3$ group

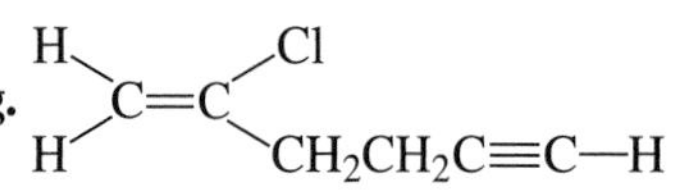

f. $CLCH_2CH_2CH_2CL$

g. $H_2C{=}C(Cl)CH_2CH_2C{\equiv}C{-}H$

h. $(CH_3)_2CHC(=O){-}OH$

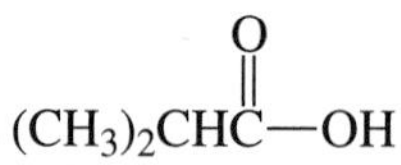

i. 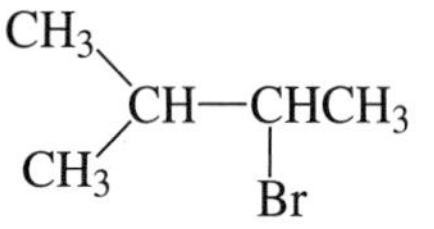 $(CH_3)_2CH{-}CH(Br)CH_3$

j. 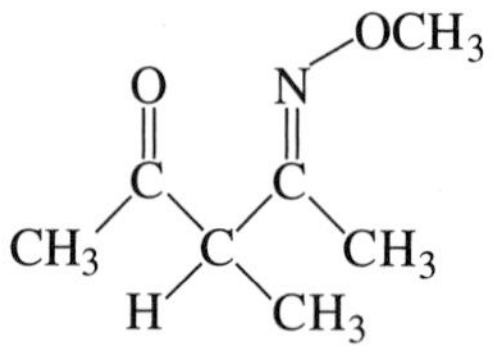 $CH_3C(=O){-}CH(CH_3){-}C(=N{-}OCH_3)CH_3$ (with H on the central C)

5. By applying the criterion of Equation 8.12, predict whether the multiplicities of the following sets of coupled nuclei, H_a and H_b, could be predicted by a first-order analysis. Note the units used to report the chemical shifts of the nuclei and the frequency of the *rf* oscillator for the instrument used to obtain the shifts.

a. δ_{H_a} 1.3 ppm, δ_{H_b} 3.9 ppm, J_{ab} = 8.5 Hz, 200-MHz instrument
b. δ_{H_a} 0.8 ppm, δ_{H_b} 1.0 ppm, J_{ab} = 2.5 Hz, 200-MHz instrument
c. δ_{H_a} 0.8 ppm, δ_{H_b} 1.0 ppm, J_{ab} = 8.5 Hz, 200-MHz instrument
d. δ_{H_a} 0.8 ppm, δ_{H_b} 1.0 ppm, J_{ab} = 8.5 Hz, 500-MHz instrument
e. δ_{H_a} 0.8 ppm, δ_{H_b} 1.0 ppm, J_{ab} = 2.5 Hz, 200-MHz instrument

6. The ^{1}H NMR chemical shifts, splitting patterns, and *relative* numbers of hydrogen atoms for three compounds are provided in **a–c.** Deduce one or more structures consistent with these data. You may find it helpful to sketch these spectra on a sheet of paper.
   a. C_4H_9Br: δ 1.04 (6 H) doublet, δ 1.95 (1 H) multiplet, δ 3.33 (2 H) doublet
   b. $C_3H_6Cl_2$: δ 2.2 (2 H) quintet, δ 3.75 (1 H) triplet
   c. $C_5H_{11}Br$: δ 0.9 (3 H) doublet, δ 1.8 (1.5 H) complex multiplet, δ 3.4 (1 H) triplet

7. Compute the *relative* and *absolute* ratios for the different types of hydrogens in each pair of molecules shown below.

   a. CH_3—CH(CH_2Br)—C(=O)H (1,1-dimethylcyclopentane: CH_3, CH_3) b. (cyclohex-2-enone) Cl(H)C=C(Cl)CH_2—NH_2

Sign in at **www.cengage.com/login** and use the spectra viewer and Tables 8.1–8.8 as needed to answer the blue-numbered questions on spectroscopy.

8. Figures 8.36–8.42 are of the compounds whose structures are shown on the spectra. Interpret these spectra as completely as possible in terms of the observed chemical shifts, integrations, and splitting patterns.

9. Provide structures of the compounds **7–9,** whose ^{1}H NMR spectra are given in Figures 8.43–8.45, respectively.
   a. compound **7,** $C_{10}H_{14}$
   b. compound **8,** C_3H_8O
   c. compound **9,** C_8H_9Cl

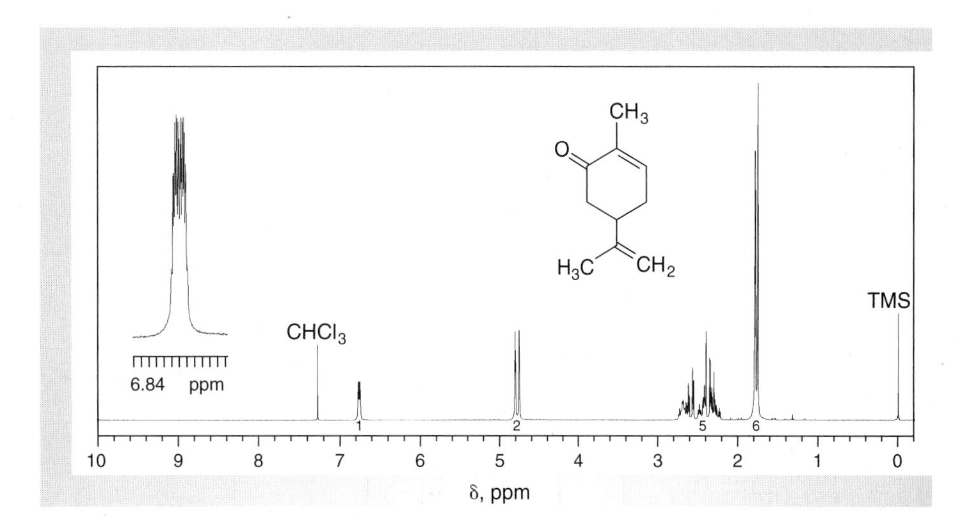

Figure 8.36
^{1}H NMR spectrum of carvone (300 MHz, $CDCl_3$).

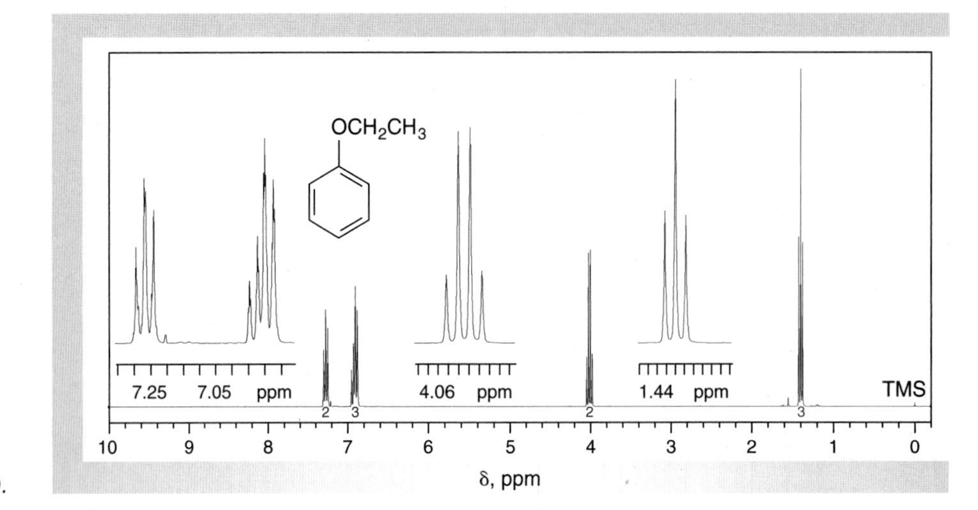

Figure 8.37
^1H NMR spectrum of ethyl phenyl ether (300 MHz, $CDCl_3$).

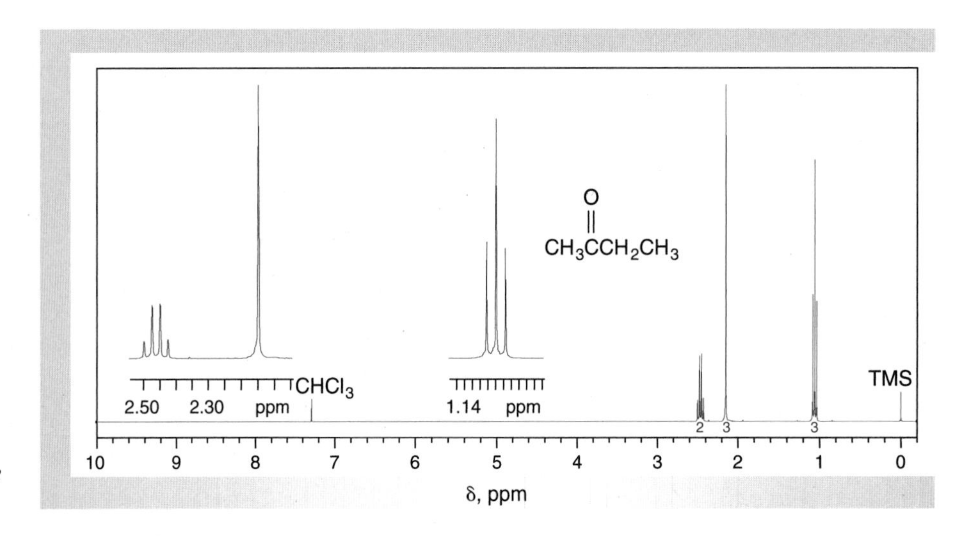

Figure 8.38
^1H NMR spectrum of 2-butanone (300 MHz, $CDCl_3$).

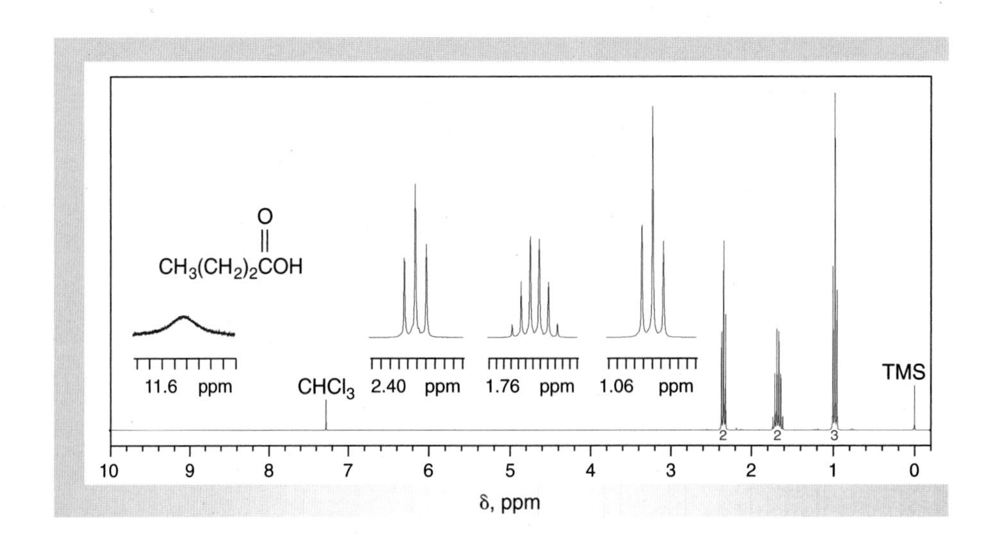

Figure 8.39
^1H NMR spectrum of butanoic acid (300 MHz, $CDCl_3$).

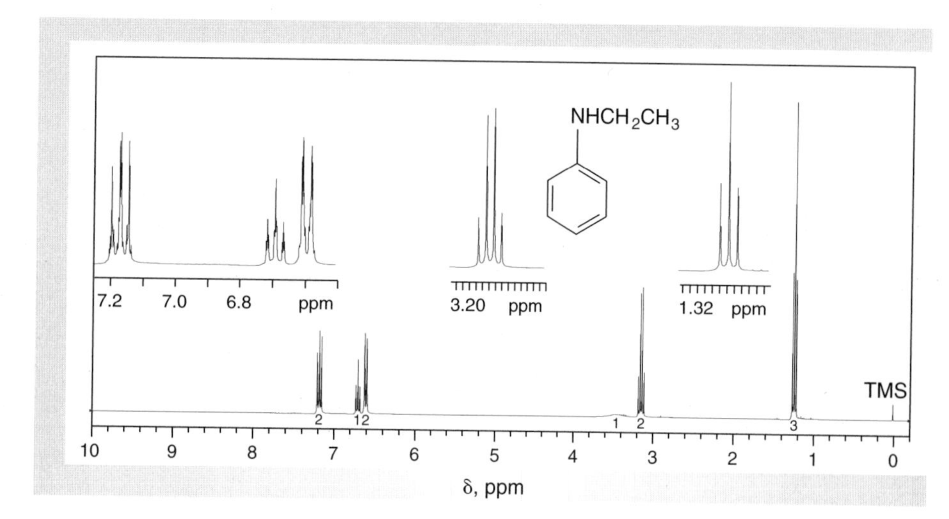

Figure 8.40
^1H NMR spectrum of N-ethylaniline (300 MHz, $CDCl_3$).

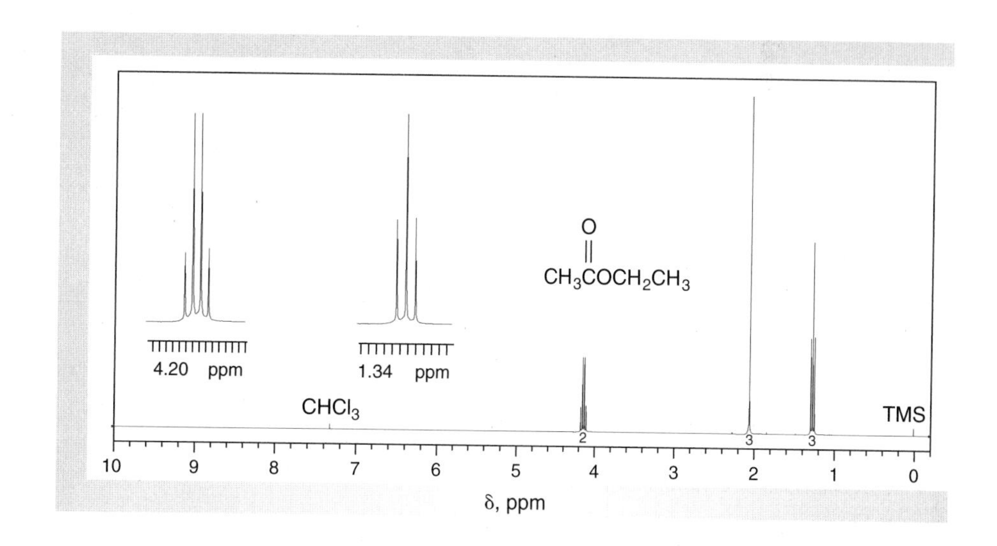

Figure 8.41
^1H NMR spectrum of ethyl acetate (300 MHz, $CDCl_3$).

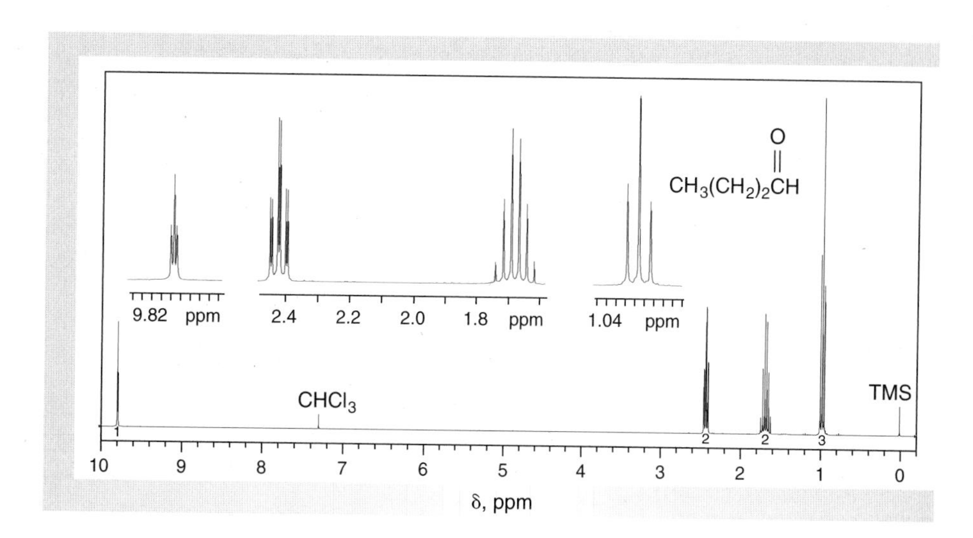

Figure 8.42
^1H NMR spectrum of butanal (300 MHz, $CDCl_3$).

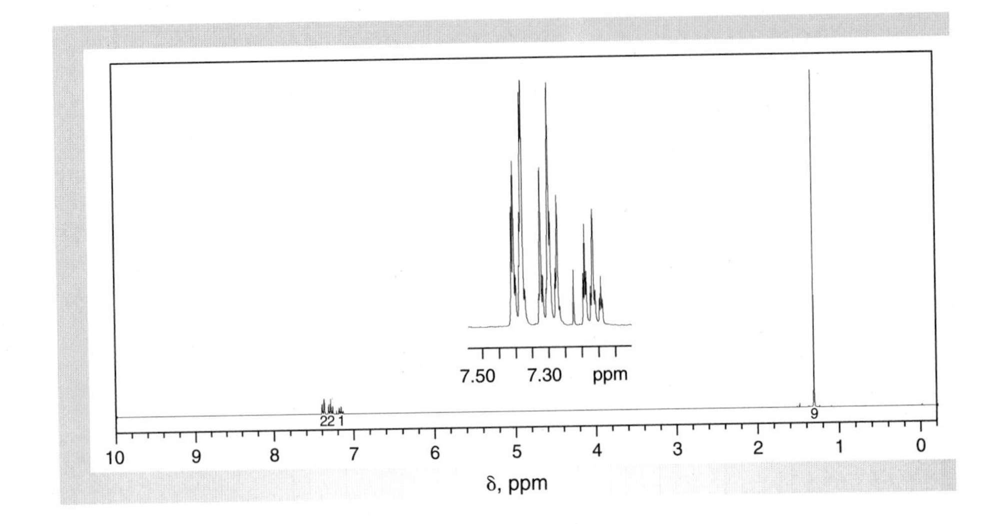

Figure 8.43
*^1H NMR spectrum (300 MHz, $CDCl_3$) for compound **7**.*

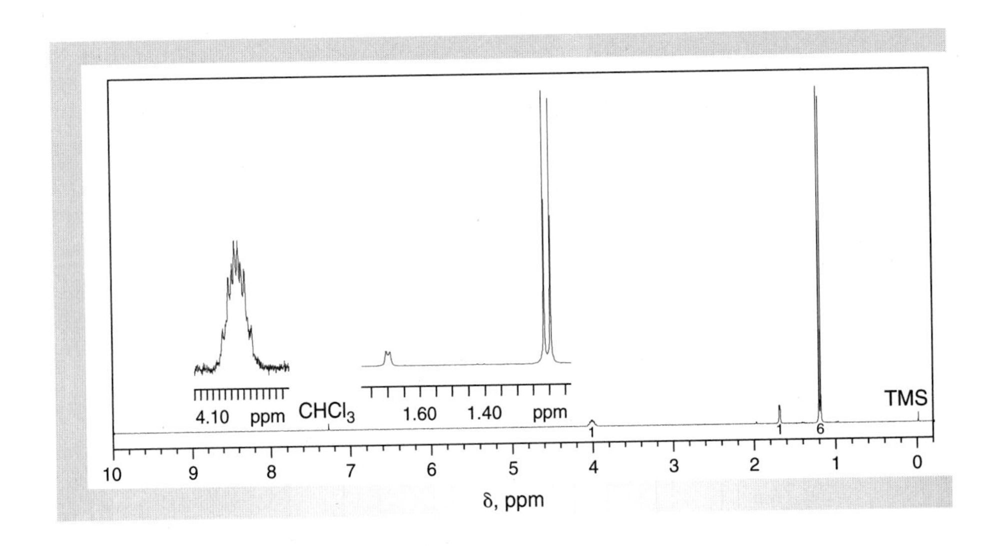

Figure 8.44
*^1H NMR spectrum (300 MHz, $CDCl_3$) for compound **8**.*

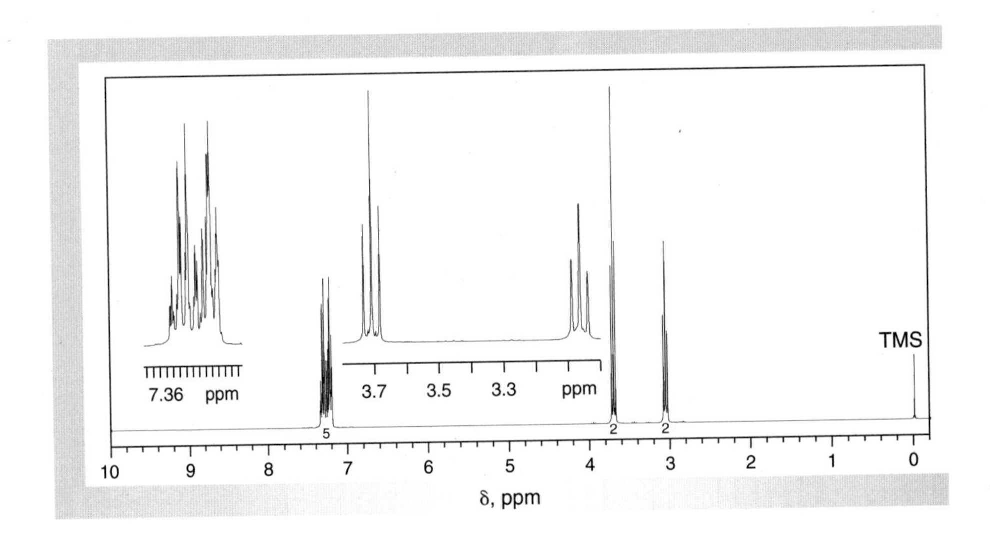

Figure 8.45
*^1H NMR spectrum (300 MHz, $CDCl_3$) for compound **9**.*

REFERENCES

1. Silverstein, R. M.; Webster, F. X.; Kiemle, D. *Spectrometric Identification of Organic Compounds,* 7th ed., John Wiley & Sons, New York, 2005.
2. Crews, P.; Rodriquez, J.; Jaspars, M. *Organic Structure Analysis,* Oxford University Press, New York, 1998.
3. King, R. W.; Williams, K. R. "The Fourier Transform in Chemistry-NMR," *J. Chem. Ed.* 1989, *66,* A213–A219, A243–A248.
4. King, R. W.; Williams, K. R. "The Fourier Transform in Chemistry-NMR," *J. Chem. Ed.* 1990, *67,* A93–A99, A100–A105, A125–A137.
5. Breitmaier, E. *Structure Elucidation by NMR in Organic Chemistry: A Practical Guide,* 3rd rev. ed., John Wiley & Sons, Chichester and New York, 2002.
6. Pouchert, C. J., ed. *Aldrich Library of NMR Spectra,* vol. I–II, Aldrich Chemical Co., Milwaukee, WI, 1983.
7. Pouchert, C. J.; Behnke, J., eds. *Aldrich Library of ^{13}C and ^{1}H NMR Spectra,* vol. I–III, Aldrich Chemical Co., Milwaukee, WI, 1993.
8. *Sadtler Nuclear Magnetic Resonance Spectra,* vol. 1–119, Sadtler Research Laboratories, Philadelphia, PA. Compilation of 64,000 ^{1}H NMR spectra.
9. *Sadtler 300 MHz Proton NMR Standards,* vol. 1–24, Sadtler Research Laboratories, Philadelphia, PA. Compilation of 12,000 300 MHz ^{1}H NMR spectra.

Carbon-13 Nuclear Magnetic Resonance (^{13}C NMR) Spectroscopy

Carbon-12, ^{12}C, has no **nuclear spin,** but its carbon-13 isotope, ^{13}C, has a nuclear spin of 1/2, as does hydrogen, ^{1}H. The **natural abundance** of ^{13}C is about 1.1%, so unless substances are prepared from precursors having artificially high levels of this isotope, only 1.1% of the carbon atoms in a compound will undergo the *spin flip* or *resonance* that is characteristic of the NMR experiment. In other words, only about one in a hundred carbon atoms of a sample having ^{13}C present at natural abundance will be the proper isotope for producing an absorption in the NMR spectrum.

This low level of the NMR-active isotope of carbon makes it more difficult to obtain a suitable **carbon nuclear magnetic resonance** (^{13}C NMR) spectrum. For example, whereas it is usually possible to measure a ^{1}H NMR spectrum in a few minutes, it may take tens of minutes or even hours to accumulate enough data to produce a ^{13}C NMR spectrum in which the signal-to-noise ratio is high enough for the resonances due to the carbon atoms to be seen. Nonetheless, modern spectrometers and the sophisticated computers associated with them allow acquisition of the data necessary for a ^{13}C NMR spectrum on samples of 1–5 mg, which is about an order of magnitude greater than the amount needed for a ^{1}H NMR spectrum.

The principles of ^{13}C NMR spectroscopy are the same as those for ^{1}H NMR spectroscopy. When placed in an external magnetic field, ^{13}C nuclei adopt one of two spin states, whose energy difference is determined by the strength of the field (Fig. 8.19). Exposing such nuclei to electromagnetic radiation that has the appropriate energy produces the resonance condition (Eq. 8.7). This means that the spin flip associated with transition from one energy level to another (Fig. 8.20) may occur.

Like hydrogen nuclei, the ^{13}C nuclei exist in different electronic environments in the molecule, and this gives rise to *chemical shifts* that are characteristic of the magnetic environments of the various types of carbon atoms. As in ^{1}H NMR spectroscopy, tetramethylsilane, $(CH_3)_4Si$, is commonly used as the reference

compound for measuring chemical shifts in ^{13}C NMR spectra. Most ^{13}C resonances are *downfield* from that of TMS and are given *positive* values by convention. These chemical shifts are computed according to Equation 8.8 and are reported as parts per million (ppm) on a δ scale (Fig. 8.46), again in analogy to the procedure used with ^{1}H NMR shifts.

Modern instruments are capable of measuring both ^{1}H and ^{13}C NMR spectra. When used for the latter purpose, the spectrometer is usually operated in a mode that permits decoupling of all of the proton(s) that are *nearest neighbors* to the ^{13}C atom, a process termed **broadband proton decoupling.** Applying this technique provides ^{13}C NMR spectra that do *not* show the effects of the spin-spin splitting observed in ^{1}H NMR spectra. Moreover, possible coupling between adjacent ^{13}C atoms is of no concern: Because of the low natural abundance of this isotope in organic compounds, the probability of having two such atoms bound to one another is only *one* in *ten thousand.* Proton-decoupled ^{13}C NMR spectra are thus extremely simple, as they consist of a single sharp resonance line for each magnetically distinct carbon atom.

This is seen in the spectrum of 2-butanone (Fig. 8.46) where the four different types of carbon atoms produce four separate resonances. Note that three peaks near δ 79.5 ppm in the spectrum are associated with the solvent, deuterochloroform, $CDCl_3$. The single carbon atom of this solvent produces more than one peak because of *deuterium*-carbon splitting, which is *not* eliminated by the broadband decoupling technique. Deuterium has a nuclear spin, I_z, of 1, so applying Equation 8.10 predicts a *triplet* for the carbon atom, as is observed experimentally. Rather than provide the broadband-decoupled spectra themselves, we will simply give the chemical shifts for the various carbon atoms of the starting materials and products in this textbook.

A specialized technique associated with ^{13}C NMR spectroscopy bears mention here because it provides information regarding the number of hydrogen atoms attached to a given carbon atom. This technique is referred to as **distortionless enhancement of polarization**, commonly abbreviated as DEPT. Although discussion of the basis for this technique is beyond the scope of this textbook, DEPT spectra are easy to interpret. Specifically, if there is an *odd* number of hydrogen atoms attached to a carbon atom, as is the case for a methyl (CH_3) or methane (CH) carbon atom, the DEPT spectrum will have a *positive* peak at the same chemical shift as observed in a normal ^{13}C NMR spectrum. If an *even* number of hydrogen atoms is attached, as with a methylene (CH_2) carbon atom, a *negative* peak is seen in the DEPT spectrum, and *no* peak is observed when *no* hydrogen atom is attached to a carbon atom. The DEPT spectrum of 2-butanone is provided in Figure 8.46. The lower trace is the usual ^{13}C spectrum, whereas the upper one is the DEPT spectrum. As expected, there are two positive peaks representing the resonances for the two magnetically distinct methyl groups and one negative peak for the methylene group of the molecule. Note that the peak at about δ 210 in Figure 8.45 is absent in the DEPT spectrum, proof that this resonance is due to the carbonyl carbon atom.

In contrast to decoupled spectra, *coupled* ^{13}C NMR spectra may be extremely complex and difficult to interpret because the magnitudes of ^{1}H–^{13}C coupling constants are large, on the order of 120–230 Hz, so overlapping of peaks may become a serious problem. However, special instrumental techniques provide the type of information that would normally be derived by analysis of splitting patterns, namely the number of hydrogen atoms bound to the carbon atom of interest. For example, **off-resonance decoupling** simplifies the spectrum as is illustrated in Figure 8.47, where the multiplets for each carbon atom correspond to the number of peaks predicted by the same $n + 1$ rule (Eq. 8.11) that is used to predict spin-spin

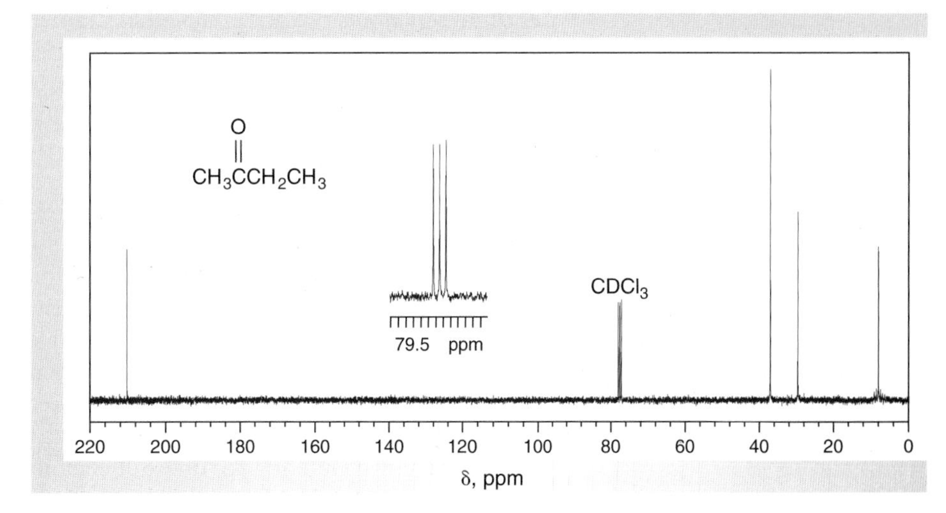

Figure 8.46
Broadband proton-decoupled ^{13}C NMR spectrum of 2-butanone (75 MHz, $CDCl_3$).

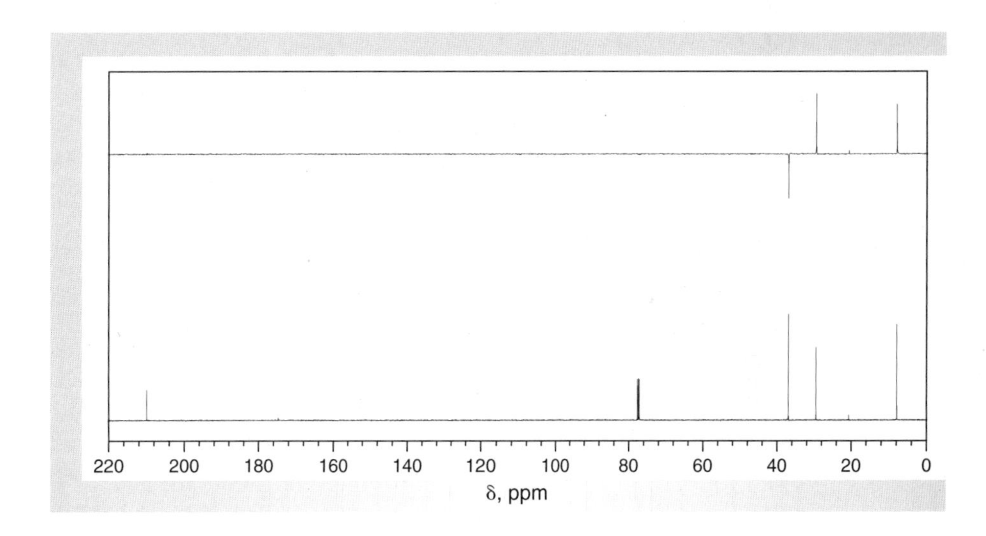

Figure 8.47
DEPT ^{13}C NMR spectrum of 2-butanone (75 MHz, $CDCl_3$).

splitting patterns in ^1H NMR spectra. The difference is that *n* corresponds to the number of hydrogen nuclei that are *directly bound* to the carbon atom rather than the number of *nearest neighbors* as defined for predicting the splitting patterns for proton spectra. Even *off-resonance decoupled* spectra may be complicated, so only the *chemical shifts* derived from *broadband decoupled* spectra are given for the starting materials and products in this textbook.

The chemical shifts of carbons atoms are *much more sensitive* to their molecular environments than are those of protons, so the range of chemical shifts for ^{13}C resonances is much greater than for ^1H. Specifically, whereas most types of hydrogen atoms resonate within 10 ppm downfield of TMS, the chemical shifts of carbon atoms occur over a range of some 220 ppm downfield of TMS. This means that it is unlikely that two different types of carbon atoms will resonate at exactly the same chemical shift. Consequently, the number of peaks in a proton-decoupled ^{13}C NMR spectrum may *tentatively* be interpreted as being equal to the number of connectively different types of carbon atoms present in the molecule. This principle is illustrated both by the observation of four peaks in the ^{13}C NMR spectrum of 2-butanone (Fig. 8.47) and of six peaks in that of methyl benzoate (Fig. 8.48). In the latter case, the carbon atoms *ortho* and *meta* to the ester function are magnetically equivalent, so there are only four resonances for the carbon atoms of the aromatic ring.

As with ^{1}H NMR spectroscopy, tables of chemical shifts have been developed for carbon atoms in different environments; one example is provided by Table 8.5. Examination of the data in the table shows that some of the structural features in a molecule that produce downfield chemical shifts in ^{1}H NMR spectra do the same in ^{13}C NMR spectra. For example, an electronegative substituent such as a carbonyl group or hetero-atom on a carbon atom causes a downfield shift in the resonance, relative to a saturated hydrocarbon, just as it would with hydrogen nuclei. This effect is caused by the *deshielding* effect of the electronegative moiety. A second factor in defining chemical shifts in ^{13}C NMR spectra is the hybridization of the carbon atom, as is seen by comparing the range of chemical shifts for the sp^3- (10–65 ppm), sp^2- (115–210 ppm), and sp-hybridized (65–85 ppm) carbon atoms of the general structures shown in Table 8.5.

Finally, the nature of a substituent G that is zero (α-effect), one (β-effect), or two (γ-effect) atoms away from the carbon atom of interest, as shown in **10**, may also affect the chemical shift. Such substituent effects are remarkably additive in nature, so it becomes possible to predict with reasonable accuracy the expected position of the chemical shifts for carbon atoms in a molecule whose spectrum has not been reported. Table 8.6 is a compilation of some of these additivity effects for acyclic alkanes. These may be used in conjunction with the data in Table 8.7 to predict chemical shifts for a compound. Similar tables of additivity effects for other classes of organic compounds are published in the references at the end of this section.

γ α β *G*

10

The following example illustrates the method for making these predictions. Consider 1-chlorobutane (**11**), $CH_3CH_2CH_2CH_2Cl$. The computed chemical shifts, δ, are

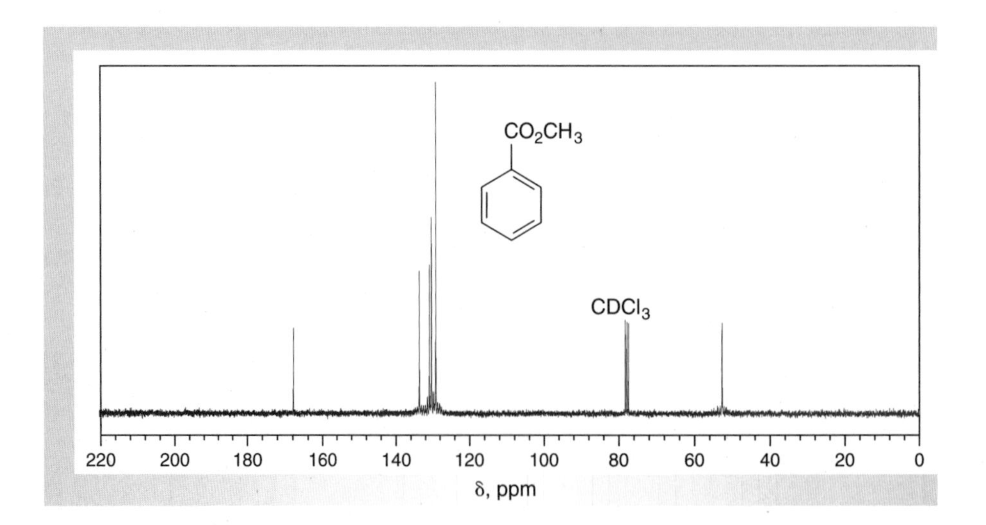

Figure 8.48
^{13}C NMR spectrum of methyl benzoate (75 MHz, $CDCl_3$).

Table 8.5 ^{13}C *Chemical Shifts in Carbon-13 Magnetic Resonance* (^{13}C *NMR*) *Spectroscopy*

Absorbing Carbon Atom (shown as **C** *or* **Ar***)*	*Approximate δ, ppm*
$RCH_2\mathbf{C}H_3$	13–16
$R\mathbf{C}H_2CH_3$	16–25
$R_3\mathbf{C}H$	25–38
$\mathbf{C}H_3C(=O)—R$	30–32
$\mathbf{C}H_3C(=O)—OR$	20–22
$R\mathbf{C}H_2$–Cl	40–45
$R\mathbf{C}H_2$–Br	28–35
$R\mathbf{C}H_2$–NH_2	37–45
$R\mathbf{C}H_2$–OH	50–65
$RC{\equiv}\mathbf{C}H$	67–70
$R\mathbf{C}{\equiv}CH$	74–85
$RCH{=}\mathbf{C}H_2$	115–120
$R\mathbf{C}H{=}CH_2$	125–140
$R\mathbf{C}{\equiv}N$	118–125
$\mathbf{Ar}H$	125–150
$R\mathbf{C}(=O)—NR'R''$	160–175
$R\mathbf{C}(=O)—OR'$	170–175
$R\mathbf{C}(=O)—OH$	175–185
$R\mathbf{C}(=O)—H$	190–200
$R\mathbf{C}(=O)—CH_3$	205–210

determined from the observed chemical shifts (Table 8.7) of 13.4 and 25.2 ppm for C1 (C4) and C2 (C3), respectively, for butane as follows:

C1: $13.4 + 31 = 44.4$ ppm C2: $25.2 + 11 = 36.2$ ppm

C3: $25.2 - 4 = 21.2$ ppm C4: $13.4 + 0 = 13.4$ ppm

The observed chemical shifts for **11** are provided below the structure. The agreement is remarkable and shows the impressive predictive power of this approach.

$$\underset{13.4}{CH_3}—\underset{20.4}{CH_2}—\underset{35.0}{CH_2}—\underset{44.6}{CH_2}—Cl$$

11

1-Chlorobutane

Table 8.6 *Incremental Substituent Effects (ppm) on Replacement of H by Y in Alkanes; Y is Terminal or Internal* (+ downfield, − upfield)*

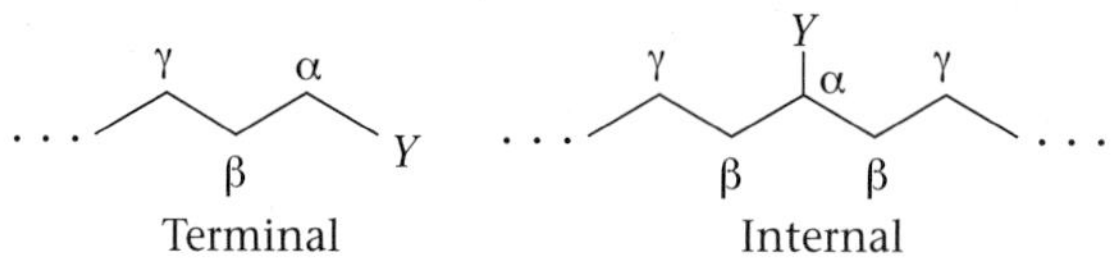

γ α β Y — Terminal

γ β Y α β γ — Internal

Y	α Terminal	α Internal	β Terminal	β Internal	γ
CH_3	+9	+6	+10	+8	−2
$CH{=}CH_2$	+20		+6		−0.5
$C{\equiv}CH$	+4.5		+5.5		−3.5
COOH	+21	+16	+3	+2	−2
COOR	+20	+17	+3	+2	−2
COR	+30	+24	+1	+1	−2
CHO	+31		0		−2
Phenyl	+23	+17	+9	+7	−2
OH	+48	+41	+10	+8	−5
OR	+58	+51	+8	+5	−4
OCOR	+51	+45	+6	+5	−3
NH_2	+29	+24	+11	+10	−5
NHR	+37	+31	+8	+6	−4
NR_2	+42		+6		−3
CN	+4	+1	+3	+3	−3
F	+68	+63	+9	+6	−4
Cl	+31	+32	+11	+10	−4
Br	+20	+25	+11	+10	−3
I	−6	+4	+11	+12	−1

*Add these increments to the shift values of the appropriate carbon atom in Table 8.7.
Source: Reference 1.

Returning now to the broadband-decoupled spectrum of 2-butanone (Fig. 8.46), the resonance that is far downfield near δ 210 is assignable to the carbonyl carbon atom (Table 8.5). The remaining three resonances near δ 37, 30, and 8 ppm may be assigned either (a) by recognizing that the carbonyl group causes the carbon atoms to which it is directly attached to be shifted farther downfield than those atoms more distant from it (α-effect) or (b) by using the data of Tables 8.6 and 8.7 to compute the expected chemical shifts. These computations are provided below. Again, comparison with the experimentally observed shifts provided in Figure 8.46 shows the value of the method of substituent additivity in assigning resonances to the carbon atoms associated with them. When data from DEPT spectra are available, the task of making ^{13}C assignments is greatly simplified. In the case of 2-butanone, the DEPT spectrum

Table 8.7 *13C Shifts for Some Linear and Branched-Chain Alkanes (ppm from TMS)*

Compound	*C1*	*C2*	*C3*	*C4*	*C5*
Methane	−2.3				
Ethane	5.7				
Propane	15.8	16.3	15.8		
Butane	13.4	25.2	25.2		
Pentane	13.9	22.8	34.7	22.8	13.9
Hexane	14.1	23.1	32.2	32.2	23.1
Heptane	14.1	23.2	32.6	29.7	32.6
Octane	14.2	23.2	32.6	29.9	29.9
Nonane	14.2	23.3	32.6	30.0	30.3
Decane	14.2	23.2	32.6	31.1	30.5
2-Methylpropane	24.5	25.4			
2-Methylbutane	22.2	31.1	32.0	11.7	
3-Methylpentane	11.5	29.5	36.9	18.8 (3-methyl group)	

Source: Reference 1.

(Fig. 8.47) allows immediate identification of the chemical shifts for the carbonyl and methylene carbon atoms; differentiation between the two methyl groups of the molecule, however, requires use of the strategies (a) and/or (b) described above.

C1: $-2.3 + 30 = 27.7$ ppm   C2: cannot be computed from data in Tables 8.6 and 8.7

C3: $5.7 + 31 = 36.7$ ppm   C4: $5.7 + 1 = 6.7$ ppm

The relative intensities of the various peaks are *not* identical in the ^{13}C NMR spectrum of 2-butanone, even though a single carbon atom is responsible for each resonance. The physical basis for this is beyond the scope of this discussion. Although there is *not* a 1:1 correlation between the number of carbon atoms producing a particular resonance and the intensity of that resonance, there is a *rough* correlation between intensities as reflected in peak heights and whether or not any hydrogen atoms are on the carbon atom resonating at a particular chemical shift. As a general rule, if there are *no* hydrogen atoms attached to the carbon atom, the intensity of the resonance associated with that atom will be relatively low. This relationship is seen in Figure 8.46, although you should note that the signal at δ 208.3, which is the resonance for the carbonyl carbon atom, has been electronically amplified by a factor of 2 relative to all other resonances in the spectrum. In contrast, note that the intensity of the methylene carbon atom is greater than that of either of the carbon atoms of the methyl groups in the molecule, a fact showing that there is *no general* relationship between the number of hydrogen nuclei on a carbon atom and the intensity of the absorption for that atom. In short, although integrated ^{1}H NMR spectra provide invaluable information about the number of the various types of hydrogen nuclei present, this type of information is not available for ^{13}C NMR spectra unless special techniques are used.

Consideration of the ^{13}C NMR spectrum (Fig. 8.48) of methyl benzoate (**12**) shows how assigning peaks may be done on the basis of information in Table 8.5 and use of intensities as a measure of whether or not a carbon atom bears any hydrogens. First of all, there are a total of *six* magnetically distinct carbon atoms in the molecule because of its symmetry. According to the data in Table 8.5, the resonance at δ 166.8 must be due to the carbonyl carbon atom, C1′, and, as expected, is of relatively low intensity. Similarly, the table supports assigning the absorption at δ 51.8 to the methoxy carbon atom, C2′.

The remaining four resonances are in the aromatic region of the ^{13}C NMR spectrum and may be assigned as follows. Based on its low intensity, the peak at δ 130.5 is probably due to C1 of the aromatic ring. The carbon atoms *ortho* and *para* to an ester group should be *downfield* of those for the *meta* carbon atoms because the ester moiety deshields the *ortho* and *para* positions by delocalizing π-electrons, as illustrated by resonance structures **12a–12c** and by inductively withdrawing electrons from the ring via σ-bonds. This means that the resonance at δ 128.5 is for the two meta carbon atoms. A *tentative* assignment of the remaining two resonances at δ 129.7 and 132.9 in the aromatic region is made on the basis that there are twice as many *ortho* as *para* carbon atoms. Thus the more intense resonance at δ 129.7 is for C2 and that at δ 132.9 for C4. These assignments are consistent with those calculated using the substituent additivity effects provided in Table 8.8.

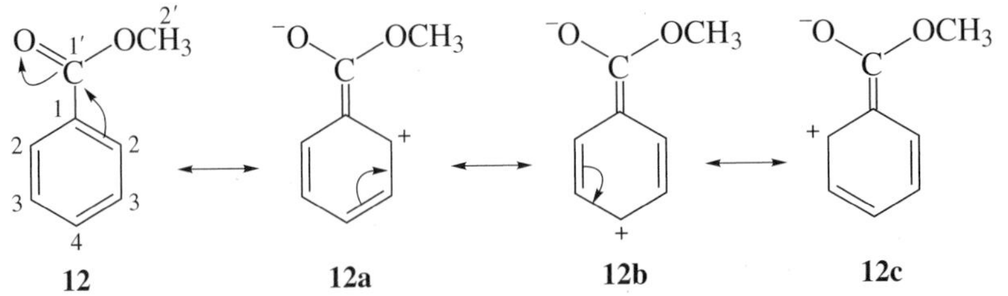

$O{=}C(OCH_3)$–Ph (atoms labelled 1′, 2′, 1, 2, 3, 4) ⟷ ^{-}O–C(OCH_3)= ring^{+} ⟷ ^{-}O–C(OCH_3)= ring^{+} ⟷ ^{-}O–C(OCH_3)= ring^{+}

12 **12a** **12b** **12c**

Methyl benzoate

Table 8.8 *Incremental Shifts of the Aromatic Carbon Atoms of Monosubstituted Benzenes (ppm from Benzene at 128.5 ppm, + downfield, − upfield); Carbon Atom of Substituents (ppm from TMS)*

Substituent	*C1 (Attachment)*	*C2*	*C3*	*C4*	*C of Substituent (ppm from TMS)*
H	0.0	0.0	0.0	0.0	
CH_3	9.3	+0.7	−0.1	−2.9	21.3
CH_2CH_3	+15.6	−0.5	0.0	−2.6	29.2 (CH_2), 15.8 (CH_3)
$CH(CH_3)_2$	+20.1	−2.0	0.0	−2.5	34.4 (CH), 24.1 (CH_3)
$C(CH_3)_3$	+22.2	−3.4	−0.4	−3.1	34.5 (C), 31.4 (CH_3)
$CH{=}CH_2$	+9.1	−2.4	+0.2	−0.5	137.1 (CH), 113.3 (CH_2)
$CH{\equiv}CH$	−5.8	+6.9	+0.1	+0.4	84.0 (C), 77.8 (CH)
C_6H_5	+12.1	−1.8	−0.1	−1.6	
CH_2OH	+13.3	−0.8	−0.6	−0.4	64.5
OH	+26.6	−12.7	+1.6	−7.3	
OCH_3	+31.4	−14.4	+1.0	−7.7	54.1
OC_6H_5	+29.0	−9.4	+1.6	−5.3	

Table 8.8 *Incremental Shifts of the Aromatic Carbon Atoms of Monosubstituted Benzenes (ppm from Benzene at 128.5 ppm, + downfield, −upfield); Carbon Atom of Substituents (ppm from TMS) (Continued)*

Substituent	*C1 (Attachment)*	*C2*	*C3*	*C4*	*C of Substituent (ppm from TMS)*
$OC(=O)CH_3$	+22.4	−7.1	−0.4	−3.2	23.9 (CH_3), 169.7 (C=O)
$C(=O)H$	+8.2	+1.2	+0.6	+5.8	192.0
$C(=O)CH_3$	+7.8	−0.4	−0.4	+2.8	24.6 (CH_3), 195.7 (C=O)
$C(=O)C_6H_5$	+9.1	+1.5	−0.2	+3.8	196.4 (C=O)
$C(=O)OH$	+2.9	+1.3	+0.4	+4.3	168.0
$C(=O)OCH_3$	+2.0	+1.2	−0.1	+4.8	51.0 (CH_3), 166.8 (C=O)
$C(=O)Cl$	+4.6	+2.9	+0.6	+7.0	168.5
C≡N	−16.0	+3.6	+0.6	+4.3	119.5
NH_2	+19.2	−12.4	+1.3	−9.5	
$N(CH_3)_2$	+22.4	−15.7	+0.8	−11.8	40.3
$NHC(=O)CH_3$	+11.1	−9.9	+0.2	−5.6	
NO_2	+19.6	−5.3	+0.9	+6.0	
F	+35.1	−14.3	+0.9	−4.5	
Cl	+6.4	+0.2	+1.0	−2.0	
Br	−5.4	+3.4	+2.2	−1.0	
I	−32.2	+9.9	+2.6	+7.3	
SO_2NH_2	+15.3	−2.9	+0.4	+3.3	

Source: Reference 1.

EXERCISES

1. Based on the substituent effects shown in Table 8.8, calculate the chemical shifts expected for the various carbon atoms of the aromatic ring of methyl benzoate (**12**).
2. The ^{13}C NMR spectrum of 3-methyl-2-butanone, $CH_3C(=O)CH(CH_3)_2$, exhibits peaks at δ 18.2, 27.2, 41.6, and 211.2. Provide assignments of the various resonances to the carbon atoms responsible for them, using Tables 8.5–8.7.
3. Compute the ^{13}C NMR chemical shifts expected for 2-methyl-1-propanol, $(CH_3)_2CHCH_2OH$.

REFERENCES

1. Whitesell, J. K.; Minton, M. A. *Stereochemical Analysis of Alicyclic Compounds by C-13 NMR Spectroscopy,* Chapman & Hall, London, 1987.
2. Silverstein, R. M.; Webster, F. X.; Kiemle, D. *Spectrometric Identification of Organic Compounds,* 7th ed., John Wiley & Sons, New York, 2005.
3. Crews, P.; Rodriquez, J.; Jaspers, M. *Organic Structure Analysis,* Oxford University Press, New York, 1998.
4. Breitmaier, E. *Structure Elucidation by NMR in Organic Chemistry: A Practical Guide,* 3rd rev. ed., John Wiley & Sons, Chichester and New York, 2002.
5. Pouchert, C. J.; Behnke, J., eds., *Aldrich Library of ^{13}C and ^{1}H NMR Spectra,* vol. I–III, Aldrich Chemical Co., Milwaukee, WI, 1993.
6. *^{13}C NMR Spectra,* vol. 1–210, Sadtler Research Laboratories, Philadelphia, PA. Compilation of 42,000 ^{13}C NMR spectra.

Preparing Samples for NMR Spectroscopy

Organic chemists normally obtain NMR spectra of liquid samples contained in special glass tubes of high-precision bore. Although spectra may be obtained of pure liquids having low viscosities, the substrate of interest, regardless of its normal physical state, is generally dissolved in an appropriate solvent. The spectra of viscous samples are unsatisfactory because broad absorptions may be observed, causing a loss of resolution between the peaks.

All modern NMR spectrometers operate in the pulsed Fourier transform (FT) mode; for technical reasons, this means that deuterated solvents must be used. Deuterochloroform, $CDCl_3$, is the most common such solvent, although a number of other, albeit generally more expensive, solvents such as acetone-d_6, $(CD_3)_2CO$, dimethyl sulfoxide-d_6, CD_3SOCD_3, and benzene-d_6, C_6D_6 are also available. Deuterium oxide, D_2O, may be used if the sample is water-soluble.

Proton-containing solvents are generally not appropriate because the intense resonance due to the hydrogen atoms of the solvent may obscure absorptions of the sample itself. Although deuterium, ^{2}H, also possesses the property of nuclear spin, its resonance appears in a region of the NMR spectrum different from that of ^{1}H. Replacing all of the hydrogen atoms of solvent with deuterium removes such resonances from the ^{1}H NMR spectrum, but for practical reasons, weak absorptions due to residual protium may still be observed. This is because common organic solvents containing 100 atom % D are too expensive for routine spectral work. For example, the approximate cost of $CDCl_3$ having 100 atom % D is \$1.97/g, whereas that having 99.8 atom % D is \$0.24/g.

Minor resonances due to solvent must therefore be taken into account when interpreting ^{1}H NMR spectra. Moreover, in solvents in which there is more than one deuterium atom on a carbon atom, the residual *protium* on that atom will have a splitting pattern that reflects the presence of the deuterium atoms, which have a nuclear spin, I_z, of 1. The resonance of the residual hydrogen present in deuteroacetone appears as a *quintet* centered at δ 2.17, for example. This pattern arises from the coupling of a single residual hydrogen with the two deuterium atoms that would be present on the same carbon atom if the acetone-d_6 has 99 atom % or greater deuterium content. You may confirm that the multiplet for the residual hydrogen should be a quintet by using Equation 8.10.

Of course, the *carbon* atom(s) of an NMR solvent appear in the ^{13}C NMR spectrum whether or not the solvent is deuterated. As with residual protium, splitting of the carbon resonance(s) of the solvent by attached deuterium atoms is observed.

The appearance of *three* peaks for the carbon atom of $CDCl_3$ illustrates this (Figs. 8.46–8.48). Care should therefore be exercised when interpreting ^{13}C NMR spectra to avoid mistaking peaks due to solvent for those of the sample itself. If overlap of sample and solvent peaks is suspected, it may be necessary to select a different solvent for a ^{13}C NMR spectrum.

The concentration of the solutions appropriate for obtaining NMR spectra is in the range of 5–15% by weight; about 1 mL of solution is needed to fill an NMR tube to the proper level, although as little as 0.6 mL can be used. The tube must be scrupulously clean and dry, and the solution must be free of undissolved solids arising from the sample itself or even dust. Furthermore, as a consequence of contact of the solute or the solvent with metals such as the iron in "tin" cans, trace amounts of ferromagnetic impurities may contaminate the solution; the result is a spectrum that has only broad, weak absorptions.

A solution that contains any solid material must be filtered *prior* to measuring its NMR spectrum. This is done most easily by inserting a small plug of glasswool into a Pasteur pipet and then filtering the solution through the plug into the NMR tube. Following their use, NMR tubes should be cleaned and thoroughly dried. They are best stored in a closed container or in an inverted position to minimize the possibility that particles of dust will contaminate the inside of the tube.

8.4 ULTRAVIOLET AND VISIBLE SPECTROSCOPY

Introduction

Absorption of light of the proper energy produces electronic transitions within molecules, and these transitions are the basis for UV-Vis spectroscopy. The absorptions occur in the ultraviolet and/or visible regions, which are adjacent to one another in the electromagnetic spectrum (Fig. 8.1). We can visually detect absorption of light by organic compounds in the visible region because the compounds appear colored as a result; in contrast, colorless compounds absorb energy in the UV region, and the phenomenon is not detectable by the human eye.

Principles

The phenomenon underlying UV-Vis spectroscopy is the excitation of an electron from a lower- to a higher-energy electronic state. The two kinds of electronic excitations of

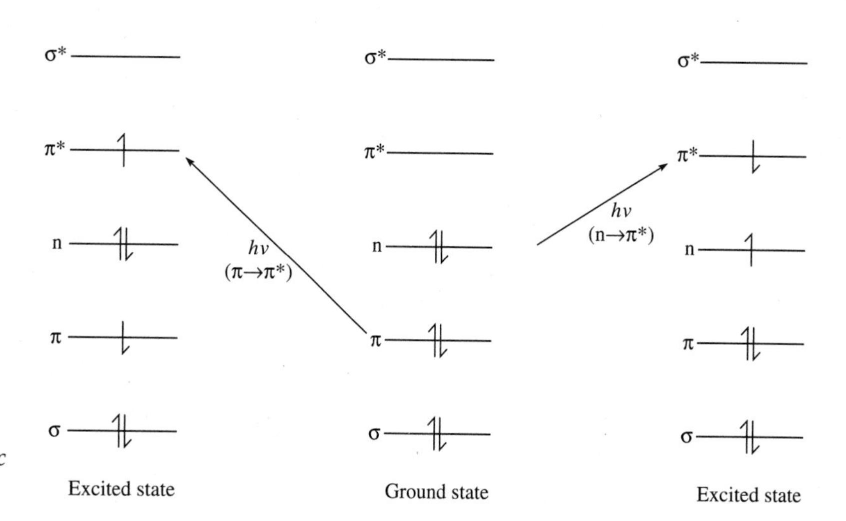

Figure 8.49
Energy diagram showing electronic transitions.

greatest value to organic chemists involve promotion of an electron originally in either a nonbonding molecular orbital (an *n*-electron) or a σ-bonding molecular orbital (a π-electron) into an antibonding molecular orbital (Fig. 8.49). By referring to Figure 8.1, you can see that the energies needed for such excitations range from 38 to more than 140 kcal/mol, values corresponding to light in the wavelength range of 750–200 nm (Eq. 8.1). Shorter wavelengths of light are required for excitation of σ-type electrons, and the resulting spectra, although important for understanding some fundamental molecular properties, are not of much value for assigning structures to organic molecules. As seen in Figure 8.49, π-type antibonding molecular orbitals, designated π*, are lower in energy than the corresponding σ*-orbitals. Consequently, the transitions caused by absorption of light in the UV-Vis region involve populating the π*-state by excitation of an electron from an *n*-orbital ($n \rightarrow \pi^*$) or a π-orbital ($\pi \rightarrow \pi^*$).

An excited state produced by promotion of an electron to an antibonding molecular orbital can return to the ground state in a process variously called *relaxation* or *decay*. Although the relaxation may involve radiationless loss of energy in the form of heat, it may also occur with emission of light. An interesting example of the latter is associated with the chemiluminescence of luminol (Sec. 20.4).

Practical Considerations

A UV-Vis spectrum is subject to a number of variables, among which are solvent, concentration of the solution being examined, and the pathlength of the cell through which the light passes. The amount of light absorbed by a particular solution is quantitatively defined by the Beer-Lambert law (Eq. 8.14).

See more on *Beer-Lambert Law*

$$A = \log \frac{I_o}{I} = \epsilon c l \qquad (8.14)$$

where A = absorbance or optical density
I_o = intensity of radiation incident on sample at specified wavelength
I = intensity of radiation transmitted through sample at same wavelength
ϵ = molar extinction coefficient
c = concentration of sample in mol/liter
l = length of cell in cm

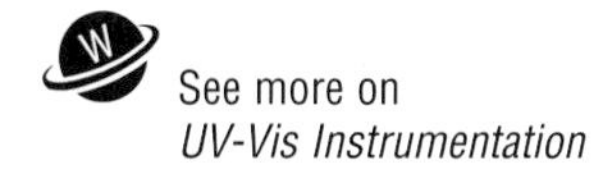

See more on *UV-Vis Instrumentation*

In a UV-Vis spectrum, absorbance, A, is plotted as a function of the wavelength, λ, of the incident radiation. The concentration (c) and cell length (l) are known, and the absorbance (A) can be determined from the spectrum at each wavelength, so the molar extinction coefficient (ϵ) may be calculated from Equation 8.14. These coefficients typically range from 10 to 100,000. Citral (**13**), a natural product that can be obtained by steam-distilling lemon grass oil (Sec. 4.5), has the UV spectrum provided in Figure 8.50. Because **13** is colorless, there is no absorption in the visible region of the spectrum. Two separate traces are shown on the spectrum because solutions having different concentrations were needed so that maxima for both strong and weak absorptions could be observed.

For reporting a UV-Vis spectrum, only the wavelength, λ_{max}, and intensity, ϵ or $\log \epsilon$, of any maxima are typically given, along with the solvent in which the measurement was made. The solvent should be stated because the values of both ϵ and λ_{max} are solvent-dependent. The critical information contained in Figure 8.50 could thus be expressed in the following way:

$$\lambda_{max}^{\text{cyclohexane}}\ 230\text{ nm}, \log \epsilon\ 4.06;\ 320\text{ nm}, \log \epsilon\ 1.74$$

The absorption maxima in **13** are due to the α,β-saturated carbonyl **chromophore,** where chromophore is defined as the functional group(s) responsible for the

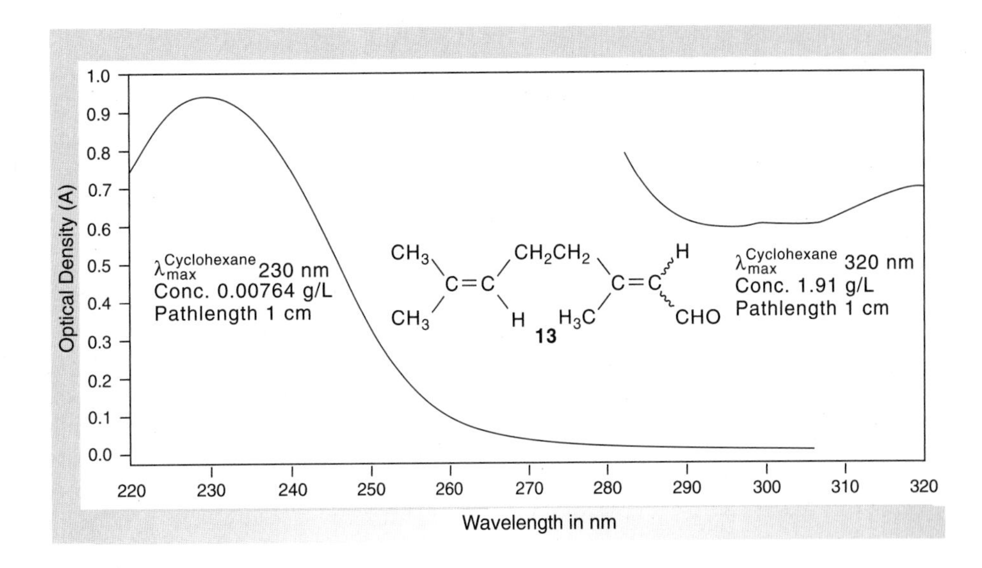

Figure 8.50
UV spectrum of citral.

absorption. The more intense absorption at 230 nm arises from a $\pi \rightarrow \pi^*$ electronic excitation, whereas that at 320 nm results from an $n \rightarrow \pi^*$ excitation.

Preparing Samples for Ultraviolet-Visible Spectroscopy

Because almost all UV-Vis spectra are obtained on solutions, preparing samples for analysis is straightforward, provided an appropriate solvent can be found. Fortunately, a wide variety of organic solvents as well as water may be used for UV-Vis spectroscopy. They share the common property of not absorbing significantly at wavelengths greater than about 200 nm, as indicated in Table 8.9. The wavelength provided for each solvent is the so-called "cutoff point," below which the solvent has appreciable absorption, making it unusable for this type of spectroscopy. "Technical" and "reagent-grade" solvents often contain light-absorbing impurities and should be purified before use, but more expensive "spectral-grade" solvents need no purification. The solvent used must not react with the solute, of course.

Table 8.9 *Solvents for Ultraviolet-Visible Spectroscopy*

Solvent	*Useful Spectra Range (Lower Limit in nm)*
Acetonitrile	<200
Chloroform	245
Cyclohexane	205
95% Ethanol	205
Hexane	200
Methanol	205
Water	200

The concentration of the solution should be such that the observed value of A is in the range of 0.3–1.5 in order to provide the greatest accuracy in the measurement. The approximate concentration should be estimated from the value of ϵ for any chromophores present, from which c can be determined using Equation 8.14. As a rough rule of thumb, 0.01–0.001 M solutions will give absorbances of appropriate magnitude for excitations of low intensity (log ϵ of about 1.0). This solution may then be diluted to permit more intense absorptions to be observed in the desired range for A.

The cells for containing the solutions must be quartz for ultraviolet spectra because other varieties of glass absorb light in the UV region of the spectrum; those for visible spectroscopy are made of less expensive borosilicate glass. This latter type of glass is opaque to light in the UV region and is therefore not suitable for UV cells.

The concentration of solute must be accurately known so that you may achieve precise measurements of A and hence of ϵ. This requires the solute to be accurately weighed, *quantitatively* transferred to a volumetric flask, and precisely diluted. Furthermore, introducing even a minute amount of an intensively absorbing impurity into the solution may have a dramatic effect on the observed UV spectrum, so it is critical to use clean glassware and other equipment needed for preparing the solution. To minimize contamination, the cells should be thoroughly rinsed both before and after use with the solvent being used for the measurement, and the outside optical surfaces of the cells must be clean and free of fingerprints. Moreover, you should not rinse quartz cells with acetone because trace residues of acidic or basic catalysts on the quartz may foster aldol condensation to form trace quantities of 4-methyl-3-penten-2-one (**14**); this enone has a high value of ϵ, and its presence may then compromise your goal of obtaining precise measurements of ϵ. The cells are best cleaned using the same pure solvent that is being used for the solute; they must also be rinsed thoroughly when changing from one concentration to another.

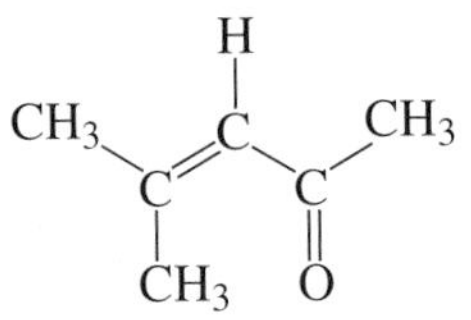

$$(CH_3)_2C{=}CH{-}C({=}O){-}CH_3$$

14

4-Methyl-3-penten-2-one

Uses of Ultraviolet-Visible Spectroscopy

As mentioned earlier, UV-Vis spectroscopy may be used to help determine the structure of an unknown compound provided that the compound contains a chromophore that absorbs in the UV-Vis region. A more common use of this spectroscopic technique, however, is for making quantitative measurements of the rate reactions. Equation 8.14 dictates that if the quantities λ, l, and A are known for a solution, it is possible to determine c, the concentration of the absorbing species in solution. You can determine the magnitude of ϵ experimentally by preparing solutions of the sample of known concentration and then determining A for each of these solutions in a cell of known pathlength. Because the extinction coefficient and the pathlength, l, are known and have constant values, any change in A for a particular solution over time must result from a variation in c. By monitoring A for a particular solution as a function of time, you may determine the time-dependence of the concentration of the absorbing species, thus allowing you to measure the rate at which it is formed or consumed.

Quantitative determination of the relative rates of electrophilic bromination of substituted benzenes (Eq. 8.15), as described in Section 15.5, relies on this application of UV-Vis spectroscopy. The rates of reaction are found by following the disappearance of molecular bromine, whose maximum is in the visible region at 400 nm.

$$G\text{-}C_6H_5 + Br_2 \longrightarrow G\text{-}C_6H_4\text{-}Br + HBr \quad (8.15)$$

EXERCISES

1. Define the following terms:

   a. $\pi \rightarrow \pi^*$ transition
   b. I_0
   c. conjugated π-system
   d. chromophore
   e. visible spectrum
   f. "cut-off" point of a solvent
   g. electronic transition
   h. absorbance

2. Give three properties of a good solvent for UV-Vis applications.
3. Why should acetone not be used to rinse quartz UV cells?
4. Suppose you were not careful in preparing a sample for UV-Vis analysis, so that its actual concentration, c, in methanol was 0.00576 g/L rather than the 0.00600 g/L that you thought it was.

   a. What is the error, in mg/L, in the mass of the sample present in solution?
   b. Suppose that a maximum was observed at a wavelength of 278 nm for this solution and that the value of A at this maximum was 0.623 when a 1-cm cell was used.

      i. Using the value of c that you thought you had for the solution, calculate ϵ for the sample, assuming that its molar mass is 100.
      ii. Show how the information regarding this maximum could be expressed.
      iii. What is the % error in ϵ associated with the inaccuracy in your knowing the actual concentration of the solution?

5. For each of the maxima given in the UV spectra of Figures 8.51–8.56, determine the value of A and of log ϵ.

REFERENCES

1. Silverstein, R. M.; Webster, F. X.; Kiemle, D. *Spectrometric Identification of Organic Compounds*, 7th ed., John Wiley & Sons, New York, 2005.
2. Crews, P.; Rodriquez, J.; Jasper, M. *Organic Structure Analysis*, Oxford University Press, New York, 1998.
3. *Sadtler Standard Ultraviolet Spectra*, vol. 1–170, Sadtler Research Laboratories, Philadelphia, PA. Compilation of 148,140 UV Spectra.
4. Phillips, J. P.; Bates, D.; Feuer, H.; Thyagarajan, B. S., eds., *Organic Electronic Spectra Data*, vol. I–XXXI, John Wiley & Sons, New York, 1996.

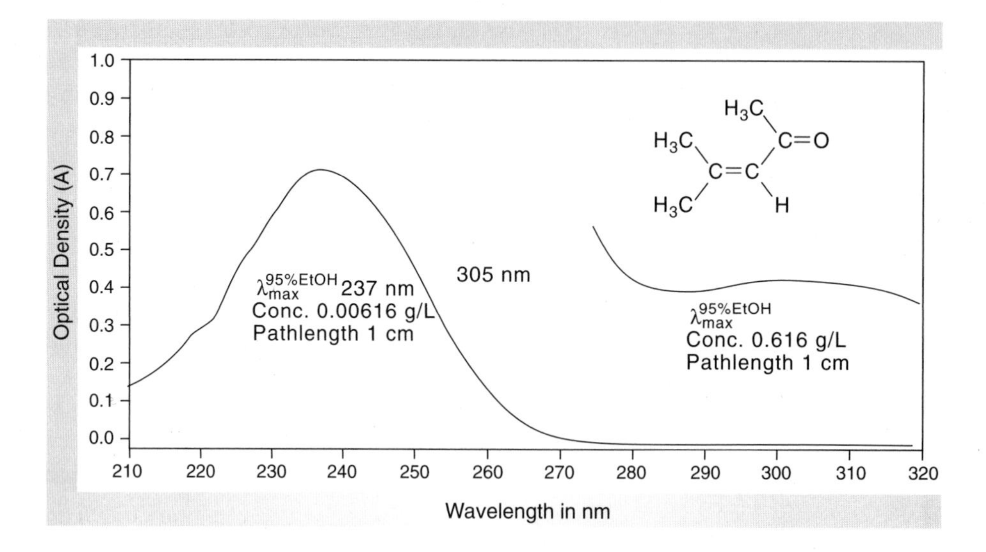

Figure 8.51
UV spectrum of 4-methyl-3-penten-2-one.

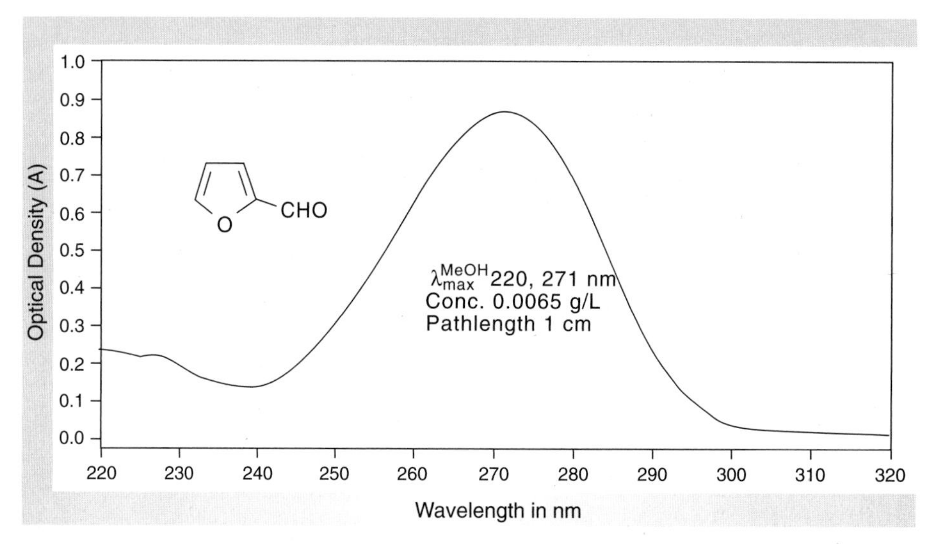

Figure 8.52
UV spectrum of 2-furaldehyde.

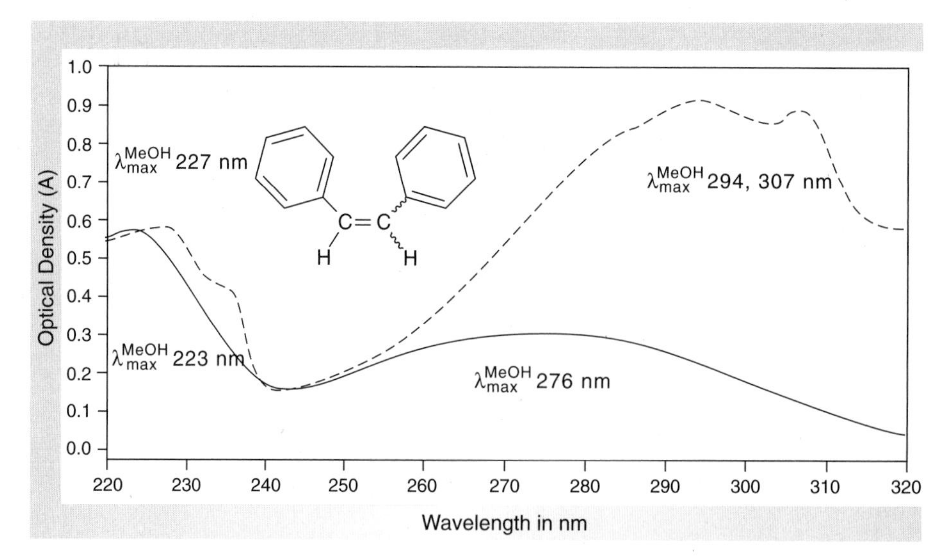

Figure 8.53
UV spectra of cis- *(solid line) and* trans- *(dashed line) stilbene. Concentration 0.0500 g/L; pathlength 1.0 cm.*

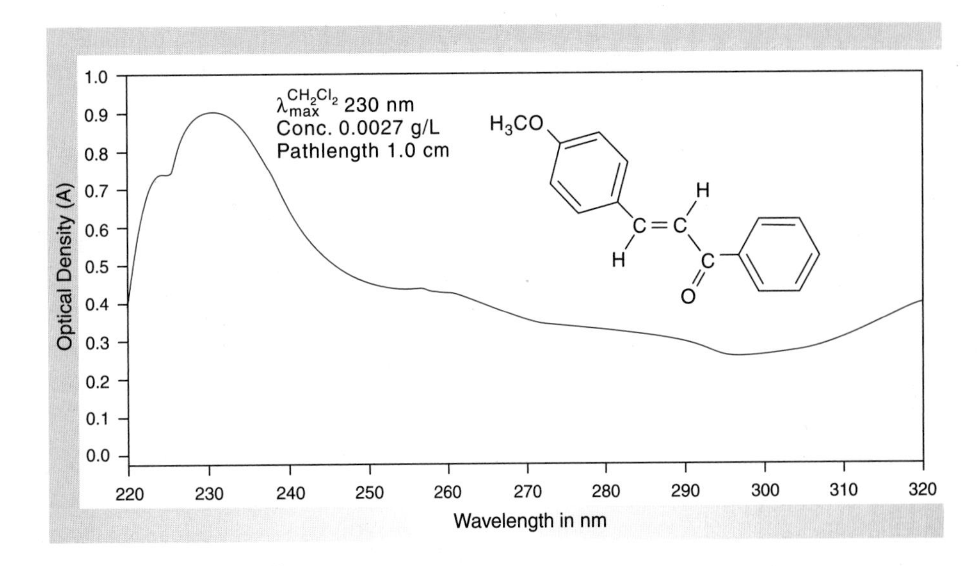

Figure 8.54
UV spectrum of trans-*p-anisalacetophenone.*

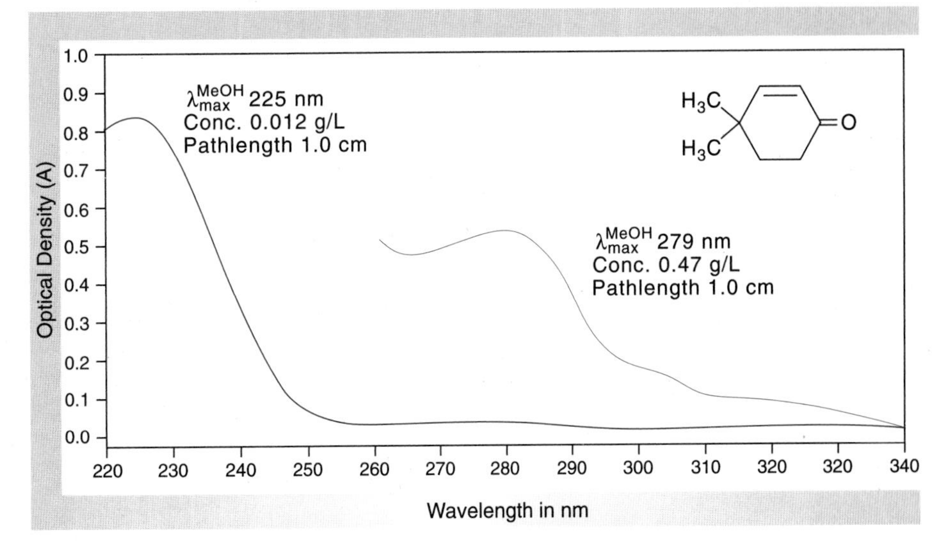

Figure 8.55
UV spectrum of 4,4-dimethyl-2-cyclohexen-1-one.

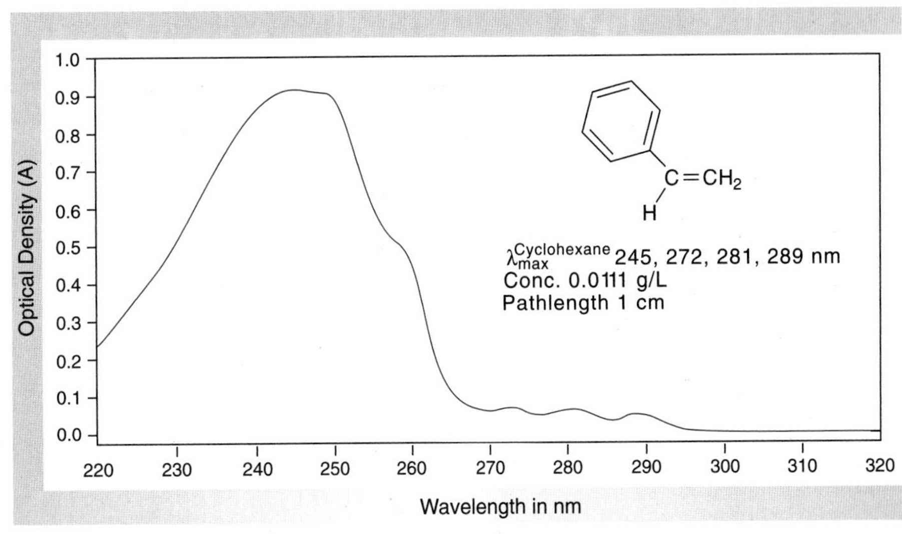

Figure 8.56
UV spectrum of styrene.

8.5 MASS SPECTROMETRY

Introduction

The characterization and identification of the structures of organic molecules are greatly aided by the application of various spectroscopic techniques such as IR (Sec. 8.2), NMR (Sec. 8.3), and UV-Vis (Sec. 8.4), all of which have the virtue of *not* destroying the sample being studied. Mass spectrometry (MS) is another important spectroscopic method used by organic chemists, but the small amount (~10^{-6} g) of sample needed for a measurement is destroyed in the process. Although this instrumental technique has a variety of applications, two of its most important are the determination of molar masses (molecular weights) and elemental compositions. It is *faster,* more *accurate,* and requires a much smaller amount of sample than the classic methods for obtaining these two types of data. As a consequence, it has almost completely replaced the classic approach in the contemporary era of organic chemistry. Mass spectrometers, because of their expense, are generally unavailable in the undergraduate organic laboratory. However, less expensive mass spectrometers that are combined with GLCs (GC-MS instruments, Sec. 6.5) are becoming more prevalent in teaching laboratories, so we'll give a brief introduction to mass spectrometry and the interpretation of mass spectra.

Principles

Instrumentation

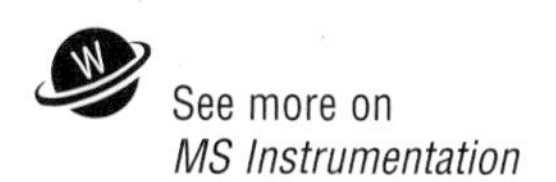
See more on *MS Instrumentation*

Before considering the specific structural information that a mass spectrometer can provide, let's develop a general understanding of how a mass spectrometer operates. Most mass spectrometers consist of a few basic components, the design, operation, and sophistication of which distinguish one type of mass spectrometer from another. These components are represented in Figure 8.57. Other instruments that can separate mixtures, such as a gas or a liquid chromatograph, may also be used in tandem with a mass spectrometer. As seen in the figure, many of the components are enclosed in an evacuated chamber where pressures as low as 10^{-7} torr are maintained. This is necessary because the analysis requires generation of highly reactive ionic species in the gas phase, and it is critical that their concentration remain low to minimize bimolecular reactions.

The inlet system allows introduction of a very small amount (μg or μL, for example) of a sample, which may be a solid, pure liquid, solution, or gas, into the ionization chamber. Ions are produced by one of several means, including bombardment with high-energy electrons, chemical methods, and irradiation with lasers. The ions are transported in the vapor phase into the mass analyzer. By means of a magnetic field, the analyzer sorts the ions in a manner so that they reach the detector according to their *mass-to-charge ratio,* designated *m/z*; only ions of a single mass reach the detector at any given time. This principle is illustrated in Figure 8.58.

The detector of a mass spectrometer converts the intensity of the ion beam into an electronic signal that can be transformed by the processor into useful formats for storage or display. The intensity of the ion beam and thus of the signal is determined by the number of the ions of a given *m/z* that reach the detector. A common way to display the processed information is as a **mass spectrum,** in which the intensity of the signal is displayed as a function of the *m/z* of the ions arriving at the detector (Fig. 8.59). From such a display, molar mass and, in some cases, structural features can be determined.

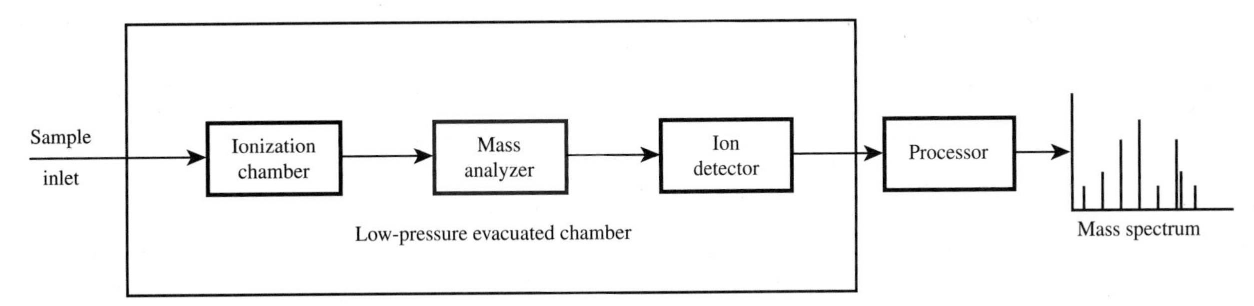

Figure 8.57
Block diagram of mass spectrometer.

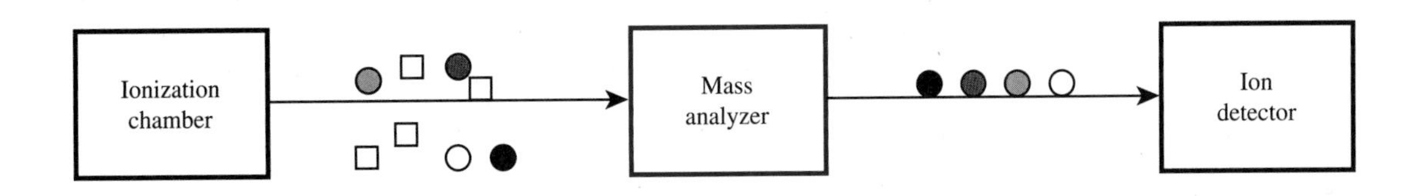

□ Ions/fragments other than +1.
○ +1 ions, larger m/z represented by darker fill color.

Figure 8.58
Function of mass analyzer.

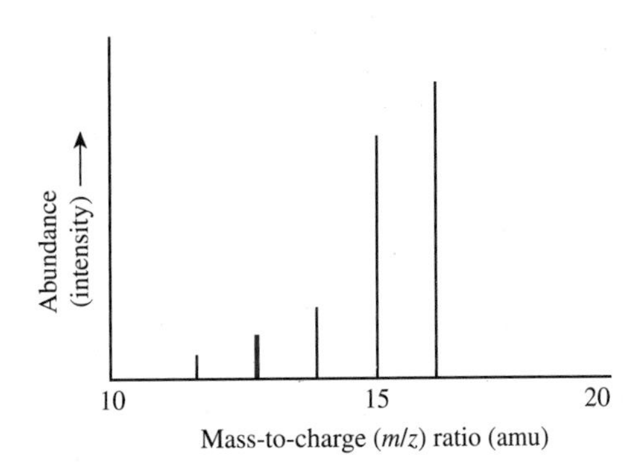

Figure 8.59
Mass spectrum of methane.

Mass Spectra

To illustrate the value of mass spectrometry, let's consider how the peaks in the mass spectrum of methane (Fig. 8.59) can be related to its structure. This simple example is presented to assist you in making the connection between the spectrum and the molecular structure. It also serves to introduce you to some of the basic principles of the technique.

When bombarded with high-energy electrons, for example, methane can be ionized to produce a **radical cation**, $CH_4^{\dot{+}}$, a positively charged species having an unpaired electron. This process is illustrated in general in Equation 8.2 and specifically in Equation 8.16.

$$\underset{\text{Methane}}{CH_4 + 1\,e^-} \longrightarrow \underset{\substack{\text{Mass} = 16 \\ (m/z = 16)}}{CH_4^{\dot{+}} + 1\,e^-} \qquad (8.16)$$

The $CH_4^{\dot{+}}$ ion has the same mass as the original molecule and is called the **molecular** or **parent ion**. This ion, with $z = +1$, reaches the detector and produces a signal at m/z 16, the molar mass of methane. The molecular ion is unstable, however, and can fragment in the ionization chamber into other cations, referred to as **daughter ions**, that are sorted according to their m/z before reaching the detector. In the case of methane, fragmentation occurs simply via loss of hydrogen atoms from the molecular ion as seen in Equation 8.17. The resulting mass spectrum (Figure 8.59) shows the parent and daughter ions predicted in Equation 8.17. The relative intensities of the various peaks is related to the probability of a particular ion being formed and reaching the detector.

$$\underset{m/z\ 16}{CH_4^{\dot{+}}} \xrightarrow{-H^{\cdot}} \underset{m/z\ 15}{CH_3^{+}} \xrightarrow{-H^{\cdot}} \underset{m/z\ 14}{CH_2^{\dot{+}}} \xrightarrow{-H^{\cdot}} \underset{m/z\ 14}{CH^{+}} \xrightarrow{-H^{\cdot}} \underset{m/z\ 12}{C^{\dot{+}}} \qquad (8.17)$$

As you might surmise, molecules more complex than methane produce more complicated fragmentation patterns, as illustrated by the mass spectrum of 2-methylbutane (Fig. 8.60). A generalized representation of the fragmentation of the parent ion and daughter ions into combinations of smaller charged and neutral species for such molecules is shown in Equations 8.18 and 8.19. In these equations and the subsequent discussion, the radical cations are represented in a simplified form in which we omit the symbol · for the unpaired electron.

$$P^{+} : A^{+} + B^{\cdot} \text{ or } P^{\cdot} + B^{+} \qquad (8.18)$$

$$A^{+} \longrightarrow C^{+} + D^{\cdot} \text{ or } C^{\cdot} + D^{+} \qquad (8.19)$$

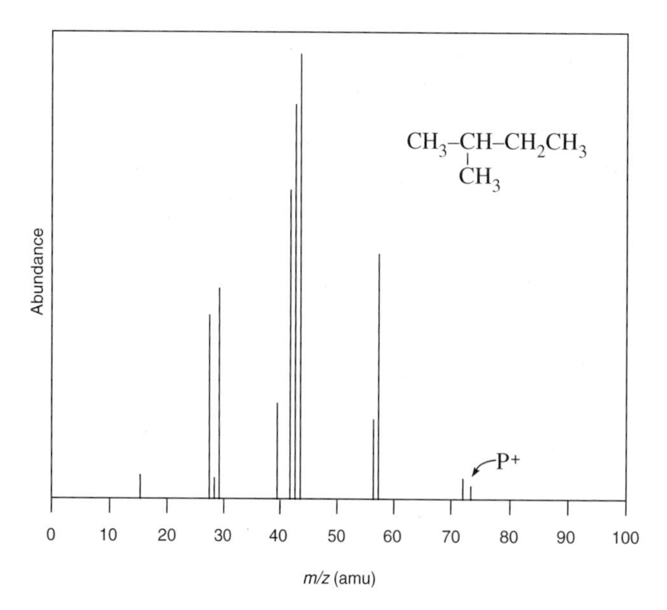

Figure 8.60
Mass spectrum of 2-methylbutane.

To assist your understanding of mass spectra, you need to be aware of an important phenomenon in MS that arises from the fact that many elements are comprised of more than one stable or nonradioactive isotope. For example, ^{13}C NMR spectroscopy (Sec. 8.3) depends on the presence of this isotope, along with its more common ^{12}C isotope, in organic molecules. Because mass spectrometers are readily able to distinguish between ions having a difference of m/z of one atomic mass unit (amu), a molecular ion and the various fragments derived from it may appear as multiple peaks because of the presence of stable isotopes. Thus, you may observe peaks in a spectrum having values of m/z that are one, two, or more atomic mass units greater than that of the molecular ion, P^+; such peaks are labeled $P^+ + 1$, $P^+ + 2$, and so on. The mass spectrum of carbon monoxide, for example, has a peak at m/z 28 that corresponds to the molar mass of P^+ for $^{12}C^{16}O$. However, the spectrum also has another peak at m/z 29; this is the $P^+ + 1$ peak and reflects the presence in ^{13}C and ^{17}O in some molecules of the sample. The intensity of this peak is small, approximately 1% of that for the molecular ion, and reflects the low **natural abundance** of these two isotopes.

See *MS Tutorial*

Table 8.10 lists the relative abundances of the stable isotopes of carbon and hydrogen, as well as those of some of the heteroatoms commonly found in organic molecules. The percentages are derived by defining the abundance of the isotope of lowest atomic mass as 100% and relating the abundances of its heavier isotopes to this figure. Referring to this table will be valuable when you are interpreting mass spectra, particularly when the sample contains bromine and chlorine, both of which are comprised of high percentages of two isotopes.

The fragmentation of the molecular ion normally affords a large number of peaks in the mass spectrum, making its detailed interpretation difficult. Consequently, your use of the spectral data may be limited to determining the molar mass of the sample. You can extract more information from the spectrum, however, by applying principles that you may have encountered in learning about organic reaction mechanisms. For example, there is a preference for a molecular ion to undergo bond breaking so as to provide an allyl, benzyl, or tertiary carbocation, each of which is a relatively stable ion. There is also a tendency for thermodynamically stable neutral molecules, such as water, ammonia, dinitrogen, carbon monoxide, carbon dioxide, ethylene, and acetylene, to be formed by fragmentation. These are not observed, because only ions are detectable in the mass spectrum, but their loss from a molecular ion to give the corresponding fragments gives clues to the nature of

Table 8.10 *Natural Abundance of Various Isotopes*

Element	*Relative Abundance (%) of Lowest-Mass Isotope*	*Relative Abundance (%) of Other Isotopes*	
Carbon	100 ^{12}C	1.08 ^{13}C	
Hydrogen	100 ^{1}H	0.016 ^{2}H	
Nitrogen	100 ^{14}N	0.38 ^{15}N	
Oxygen	100 ^{16}O	0.04 ^{17}O	0.20 ^{18}O
Sulfur	100 ^{32}S	0.78 ^{33}S	4.40 ^{34}S
Chlorine	100 ^{35}Cl	32.5 ^{37}Cl	
Bromine	100 ^{79}Br	98.0 ^{81}Br	

functional groups that may be present in the sample. Consequently, determining the m/z values of fragments may provide valuable information about the structural features present in an unknown compound.

Application of these principles is important for interpreting the mass spectrum of 2-methylbutane (**15**) (Fig. 8.60). Several peaks reveal information in support of its structure. For example, the molecular ion, P^+, is seen at m/z 72, which is the molar mass you would calculate for **15** by using atomic masses of 12.000 for carbon and 1.000 for hydrogen. The small $P^+ + 1$ peak at m/z 73 results from the presence of ^2H and ^{13}C in the compound. A $P^+ + 2$ peak is not observed because it is statistically unlikely that a molecule of 2-methylbutane would contain two atoms of either ^2H or ^{13}C or one of each of these isotopes. As you can see, the most intense peak in the spectrum is not the molecular ion, but rather is a fragmentation peak having m/z 43. This peak is defined as the **base peak;** its relative intensity is set as 100%, and the intensities of all other peaks in the spectrum are measured relative to it.

As seen in Scheme 8.1, the origin of several of the more intense daughter ions observed in Figure 8.60 may be explained by considering the possible ways in which the 2-methylbutane cation radical (**16**) may fragment. The ion **16** may lose a methyl group from C2 in two ways to form either 2-butyl cation radical (**17**) and methyl radical (**18**) or 2-butyl radical (**19**) and methyl cation radical (**20**). Based on the known relative stabilities of carbocations, we can reason that forming **17,** a secondary carbocation, is preferred to forming **20,** a primary carbocation. Indeed, the low-intensity peak at m/z 15 is due to **20,** with the much more intense peak at m/z 57 being **17.** The molecular ion **16** may also fragment through cleavage of the C2 and C3 bond to give either 2-propyl cation radical (**21**) and ethyl radical (**22**) or the 2-propyl radical (**23**) and the ethyl cation radical (**24**). The formation of **21,** m/z 43, is expected to be favored over that of **24,** m/z 29, based on the relative stabilities of these two carbocations. The other peaks in the spectrum result from phenomena that we will not discuss, but some of them are the $P^+ + 1$ and $P^+ + 2$ peaks of the various cation radicals.

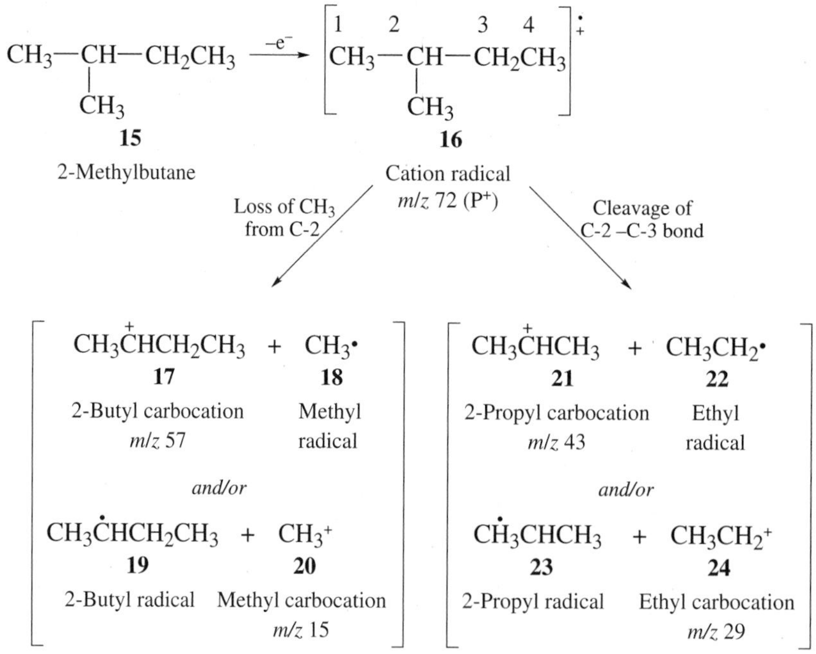

$CH_3-CH(CH_3)-CH_2CH_3$ **15** 2-Methylbutane $\xrightarrow{-e^-}$ $[CH_3-CH(CH_3)-CH_2CH_3]^{+\cdot}$ (carbons numbered 1 2 3 4) **16** Cation radical m/z 72 (P^+)

Loss of CH_3 from C-2:

$CH_3\overset{+}{C}HCH_2CH_3$ **17** 2-Butyl carbocation m/z 57 + $CH_3\cdot$ **18** Methyl radical

and/or

$CH_3\dot{C}HCH_2CH_3$ **19** 2-Butyl radical + CH_3^+ **20** Methyl carbocation m/z 15

Cleavage of C-2–C-3 bond:

$CH_3\overset{+}{C}HCH_3$ **21** 2-Propyl carbocation m/z 43 + $CH_3CH_2\cdot$ **22** Ethyl radical

and/or

$CH_3\dot{C}HCH_3$ **23** 2-Propyl radical + $CH_3CH_2^+$ **24** Ethyl carbocation m/z 29

Scheme 1
Fragmentation of 2-methylbutane cation radical.

Our discussion thus far has assumed that a mass spectrometer can distinguish between ions that differ in mass by one amu, where the m/z values are measured to the nearest whole number. In the jargon of MS, these are called *low-resolution* mass spectrometers. Much more expensive instruments, called *high-resolution* mass spectrometers, can measure the m/z ratios to ±0.001 or ±0.0001 amu and provide very accurate measurements of molar masses. These data may then be used to determine elemental composition, which is extremely valuable information when you are trying to assign a structure to an unknown compound.

EXERCISES

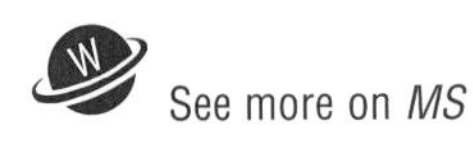

See more on *MS*

1. Define the following terms:
   a. base peak
   b. daughter ion
   c. stable isotope
   d. molecular ion
   e. fragmentation, as applied to MS
   f. natural abundance, as applied to isotopes
   g. cation radical
   h. molar mass
2. In the mass spectrum of 2,2-dimethylpropane, the base peak occurs at m/z 57, peaks of moderate intensity are found at m/z 29 and 43, and a weak peak occurs at m/z 15.
   a. What is m/z for P^{+}?
   b. Using curved arrows to symbolize the flow of electrons, account for the specified fragmentation peaks.
3. The mass spectrum of 3,3-dimethylheptane consists of peaks at the following values of m/z; relative peak intensities are in parentheses: 43 (100%), 57 (100%), 71 (90%), 29 (40%), 99 (15%), and 113 (very low). The parent peak is not observed.
   a. Indicate what fragments are responsible for each value given of m/z.
   b. Explain why the peak at 113 is of very low intensity.
   c. Calculate the value of m/z for the parent peak and suggest why it is not observed.
4. A student performed a reaction in which bromobenzene was converted to the corresponding Grignard reagent, which was then reacted with methyl benzoate (Eq. 8.20). Mass spectra (Figs. 8.61–8.63) were obtained of three components of the reaction mixture, which were expected to be triphenylmethanol (**25**), benzophenone (**26**), and biphenyl (**27**). Confirm that these products were indeed formed by assigning the appropriate mass spectrum to each compound. For each spectrum, (a) specify the parent and base peaks; (b) identify the compound represented by the mass spectrum; and (c) predict the structure of the ion for each of the peaks for which a value of m/z is provided.

$$C_6H_5Br \xrightarrow[Et_2O]{Mg} C_6H_5MgBr \xrightarrow{C_6H_5CO_2CH_3} (C_6H_5)_3COH + C_6H_5{-}C_6H_5 + C_6H_5COC_6H_5 \quad (8.20)$$

25 **26** **27**

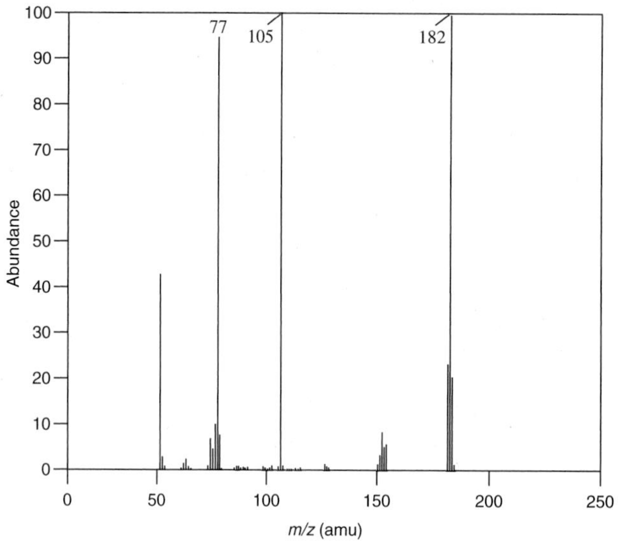

Figure 8.61
Mass spectrum of component of reaction mixture (Exercise 4).

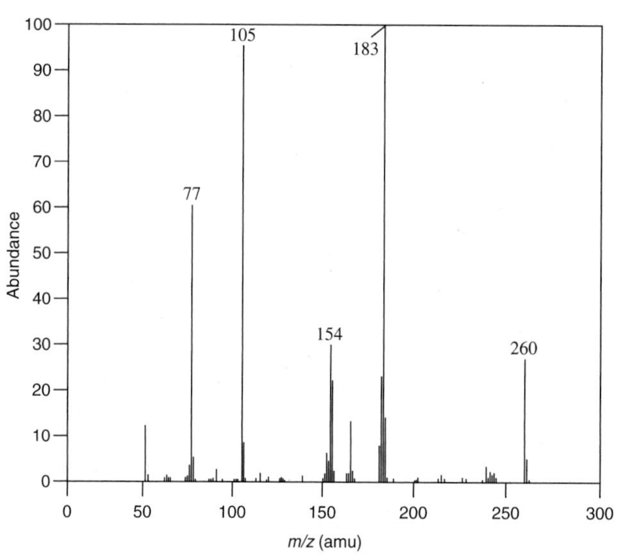

Figure 8.62
Mass spectrum of component of reaction mixture (Exercise 4).

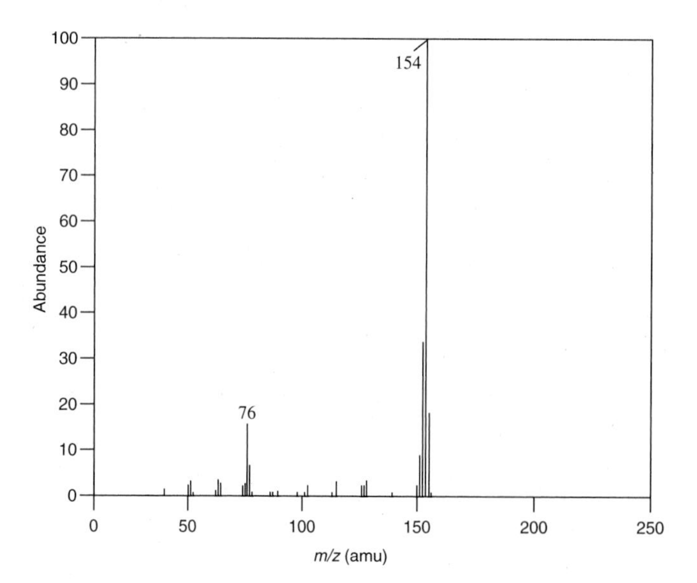

Figure 8.63
Mass spectrum of component of reaction mixture (Exercise 4).

REFERENCES

1. Silverstein, R. M.; Webster, F. X.; Kiemle, D. *Spectrometric Identification of Organic Compounds,* 7th ed., John Wiley & Sons, New York, 2005.
2. Crews, P.; Rodriquez, J.; Jasper, M. *Organic Structure Analysis,* Oxford University Press, New York, 1998.
3. Johnstone, R. A. W.; Rose, M. E. *Mass Spectrometry for Chemists and Biochemists,* 2nd ed., Cambridge University Press, Cambridge, UK, 1996.
4. McLafferty, F. W.; Stauffer, D. B. *The Wiley/NBS Registry of Mass Spectral Data,* vol. I–VII, John Wiley & Sons, New York, 1989.

HISTORICAL HIGHLIGHT

Medical Diagnostics via Nuclear Magnetic Resonance Spectroscopy

Sometimes discoveries made as part of doing "pure" or basic scientific research rather than "applied" or goal-oriented studies ultimately evolve into valuable practical applications. Such is the case with nuclear magnetic resonance (NMR) spectroscopy.

In the early 1940s, physicists observed that certain nuclei undergo transitions between two or more energy states, so-called "spin flips," when placed in an applied magnetic field; this is the so-called "NMR effect." Discovery of this fundamental property for nuclei prompted considerable experimental and theoretical research. By the late 1950s, chemists were hard at work making the phenomenon of nuclear spin flip the basis for the extremely powerful analytical tool that nuclear magnetic resonance spectroscopy is today. In fact, many present-day chemists believe that it is the single most powerful technique for elucidating the structures of unknowns containing nuclei susceptible to the NMR effect.

Throughout the 1960s, the focus of NMR spectroscopy was on analysis of solutions of samples contained in small glass tubes. The tubes were then inserted into a probe within the magnet of a spectrometer so that the NMR phenomenon could be observed. By the 1970s, the theory of NMR, the availability of powerful computers that could process the data, and the quality of the magnets had all progressed to the point that an exciting new application of the technique was envisioned. This new dimension of NMR involved the direct observation of NMR-active nuclei in *living* organisms.

All organisms are comprised of myriad organic compounds, and the human body is no exception. Among other elements, our bodies contain carbon, nitrogen, hydrogen, and phosphorus in molecules such as nucleic acids, proteins, fats, and carbohydrates, and certain isotopes of each of these elements have nuclear spin. Furthermore, water is a universal component of fluids throughout the body, and its protons are susceptible to the NMR effect. Our bodies thus serve as "containers" for a plethora of compounds having nuclei that respond to an externally applied magnetic field. This sets the stage for the medical diagnostic procedure termed "magnetic resonance imaging," or MRI, a name that is a surrogate for nuclear magnetic resonance spectroscopy but avoids the negative connotation the average citizen associates with the word "nuclear."

From a practical standpoint, the only NMR-active isotopes in the body present in concentrations high enough for detection in a typical MRI scan are ^{31}P and ^{1}H. As you know, the usual *in vitro* ^{1}H NMR experiment provides chemists with information regarding chemical shifts, relative numbers of protons in different magnetic environments, and the number of nearest neighbors to a particular type of proton. In contrast, the ^{1}H MRI experiment yields information on hydrogen density and spin-lattice (T_1) and spin-spin relaxation times (T_2), which define the rate and mode by which a nucleus relaxes to a lower-energy spin state.

The value of MRI for medical diagnostic purposes is seen in a variety of applications. For example, tumors may be detected because the relaxation times T_1 and T_2 for water molecules contained in certain types of tumors are different from those for water in healthy cells. In another application, anomalies in the brain are detectable because white and gray matter respond differently to the MRI experiment; a typical image is shown below. Moreover, certain metabolic irregularities may be diagnosed by monitoring the conversion of adenosine triphosphate (ATP) to adenosine monophosphate (AMP) and of organic

(Continued)

HISTORICAL HIGHLIGHT *Medical Diagnostics via Nuclear Magnetic Resonance (Continued)*

phosphate to inorganic phosphate. Yet another application is producing images of the heart, showing that the technique can be used for an organ that is constantly in motion and undergoing changes in shape.

Many, but not all, of the images of internal structures that are provided by the MRI technique are also available by using X-ray technology. However, the latter method, albeit currently less expensive, is less applicable to soft tissue structures such as the brain and internal organs. X-rays are also a high-energy form of radiation that can damage tissue; MRI does not suffer this disadvantage. Thus, MRI has become an exceedingly valuable tool in the arsenal of diagnostic methods available to medical researchers and practitioners, and there are thousands of MRI devices installed worldwide.

The quality of imaging that is possible is illustrated in the two figures on this page. They are both horizontal "slices" of the head of a living person and clearly show the soft tissues of the brain. One of the slices portrays tissue associated with the optic system as well. The images result from portraying a particular NMR property as a grayscale color.

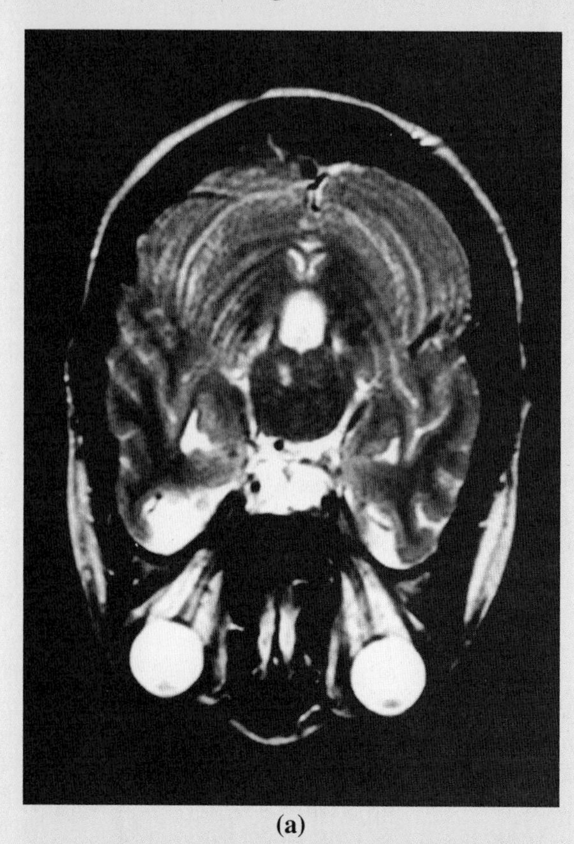

(a)

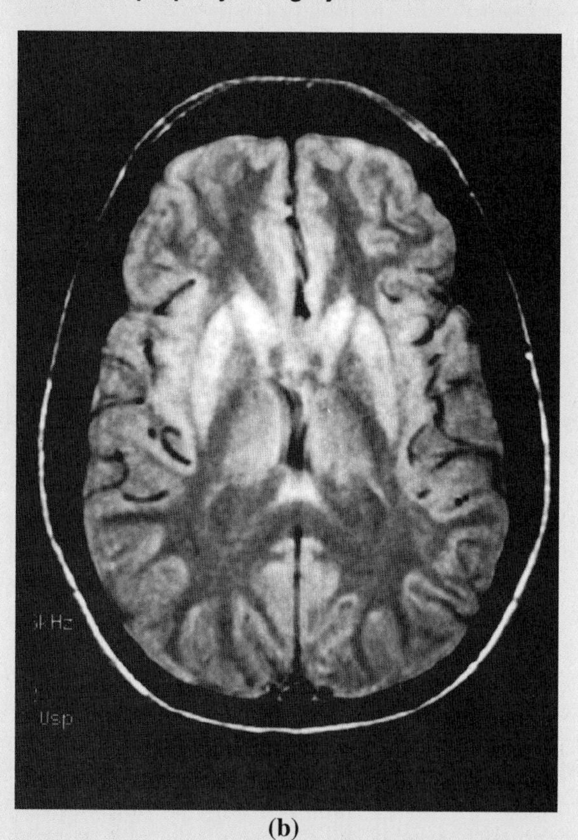

(b)

(a) MRI of the human brain, axial view, showing optic nerves. (b) MRI of the human brain, axial view. *(Courtesy of Professor M. R. Willcott, University of Texas, Galveston.)*

Relationship of Historical Highlight to Chapter

The theory and interpretation of ^1H and ^{13}C NMR spectra are presented in this chapter. The importance of this technique for confirming the structures of both known and unknown organic compounds is established, and NMR spectra of reactants and products are provided throughout the textbook and on the website that accompanies the text. This Historical Highlight describes the evolution of a phenomenon that was once primarily of interest to physicists to one that is an increasingly important tool for medical applications.

See more on *MRI*

Alkanes

When you see this icon, sign in at this book's premium website at **www.cengage.com/login** to access videos, Pre-Lab Exercises, and other online resources.

In this chapter you will be investigating the reactivity of the carbon-hydrogen bond in **acyclic (noncyclic)** and **cyclic alkanes.** This class of compounds (saturated hydrocarbons) contains only carbon-carbon single bonds and carbon-hydrogen bonds. To the extent there is a **functional group** in such molecules, it would be the C–H bond. But in reality, this type of bond is rather inert chemically unless it is made more reactive, or "activated," by being part of a molecule that has a functional group such as a carbon-carbon multiple bond (Chap. 10), halogen atom (Chap. 14), hydroxyl function (Chap. 16), or carbonyl group (Chap. 18) present. This lack of reactivity, even toward strong acids and bases, severely limits the chemical transformations that can be accomplished. However, we'll see that there are ways by which we can effect certain types of reactions of unactivated C–H bonds and produce more chemically interesting products, that is, "functionalized" compounds, as a result.

9.1 INTRODUCTION

See more on *Free Radicals*

See more on *Bond Dissociation Energy*

The fact that alkanes are chemically unreactive toward most reagents makes them of limited use as practical starting materials for most laboratory syntheses. However, chemical reactions that occur by **free-radical chain** mechanistic pathways do allow the introduction of certain functional groups into an alkane. For example, alkanes are converted into alkyl chlorides or bromides, R–X (X = Cl or Br, respectively), by a free-radical process in which a mixture of the alkane and halogen is heated at 200–400 °C or is irradiated with ultraviolet light (Eq. 9.1). Under these conditions, the σ-bond of molecular chlorine or bromine undergoes **homolytic cleavage** (Eq. 9.2) to generate chlorine and bromine atoms, which are **free radicals.** These conditions provide sufficient energy to promote homolytic cleavage of the σ-bond of molecular chlorine or bromine to generate chlorine and bromine atoms, respectively, which are free radicals (Eq. 9.2); the amount of energy required to effect this reaction is called the bond dissociation energy. Generating chlorine and bromine atoms is essential to initiating the reaction between an alkane and molecular chlorine or bromine to form alkyl halides. These may then be transformed into a variety of other functional groups.

$$\underset{X\,=\,\text{Cl or Br}}{\text{R—H} + \text{X}_2} \xrightarrow{\text{heat or hv}} \underset{\substack{\text{Alkyl halides}\\\text{(mixture)}}}{\text{R—X}} + \text{H—X} \tag{9.1}$$

$$\text{X—X} \xrightarrow{\text{heat or hv}} \text{X}^{\bullet} + \text{X}^{\bullet} \tag{9.2}$$

The procedures described in Sections 9.2 and 9.3 illustrate methods for transforming alkanes to alkyl chlorides and bromides by **free-radical substitution** reactions. These experiments give you an opportunity to analyze the mixtures of products that are obtained and to gain insights about the relative reactivities of different types of hydrogen atoms toward chlorine and bromine radicals.

9.2 CHLORINATION USING SULFURYL CHLORIDE

Chlorine atoms may be generated from molecular chlorine under mild conditions using a *catalytic* amount of an **initiator,** In–In. Thus, homolysis of a molecule of initiator occurs upon irradiation or gentle heat to give free radicals, In• (Eq. 9.3). These free radicals may then react with molecular chlorine to produce In–Cl and a chlorine atom (Eq. 9.4) to initiate the **free-radical chain reaction.** For safety and convenience, sulfuryl chloride, SO_2Cl_2, rather than molecular chlorine is used in this experiment as the source of chlorine radicals.

$$\text{In—In} \xrightarrow{\text{heat or h}\nu} \text{In}^\bullet + \text{In}^\bullet \quad (9.3)$$

$$\text{In}^\bullet + \text{Cl—Cl} \longrightarrow \text{In—Cl} + \text{Cl}^\bullet \quad (9.4)$$

The first step in our procedure for initiating the free-radical chain reaction is the homolysis of 1,1′-azobis[cyclohexanenitrile] (**1**), abbreviated as ABCN, to form nitrogen and the free radical **2** (Eq. 9.5). The rate of this reaction is sufficiently fast at 80–100 °C to generate enough chlorine atoms to initiate the chain process. The radical **2** then attacks sulfuryl chloride to generate chlorine atoms and SO_2 according to Equations 9.6 and 9.7. The series of reactions depicted in Equations 9.5–9.7 comprise the **initiation** steps of the reaction.

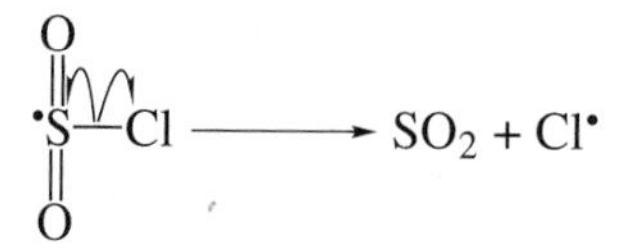

Initiation

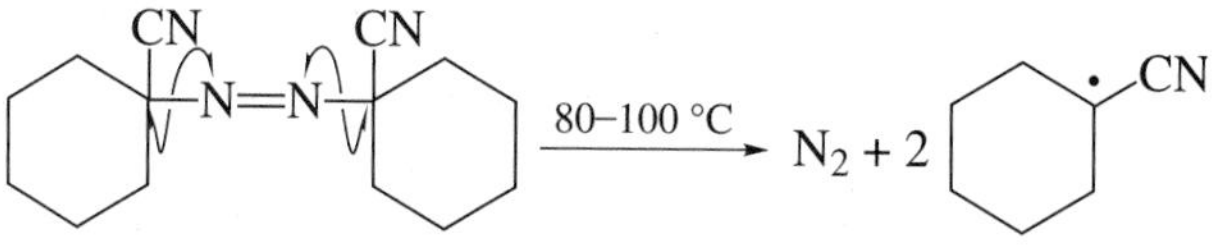

$$\mathbf{1}\ (\text{ABCN}) \xrightarrow{80\text{–}100\ ^\circ\text{C}} N_2 + 2\ \mathbf{2} \quad (9.5)$$

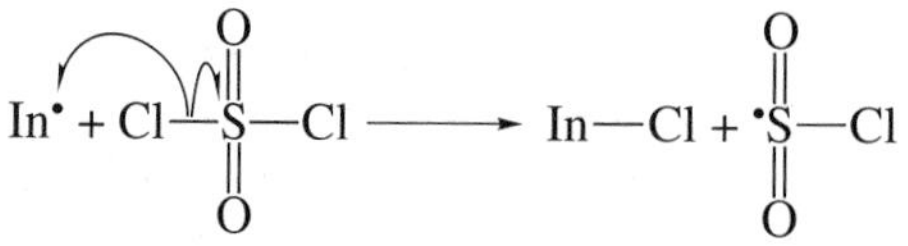

$$\text{In}^\bullet + \text{Cl—SO}_2\text{—Cl} \longrightarrow \text{In—Cl} + {}^\bullet\text{SO}_2\text{—Cl} \quad (9.6)$$

Sulfuryl chloride

$${}^\bullet\text{SO}_2\text{—Cl} \longrightarrow SO_2 + \text{Cl}^\bullet \quad (9.7)$$

Propagation steps are the next to occur in the overall process. These include abstraction of a hydrogen atom from the hydrocarbon by a chlorine atom to produce a new free radical, R• (Eq. 9.8), which attacks sulfuryl chloride to yield the alkyl chloride and the radical •SO_2Cl (Eq. 9.9). In the final propagation step, this radical fragments into SO_2 and another chlorine atom (Eq. 9.10), thereby producing a radical chain reaction.

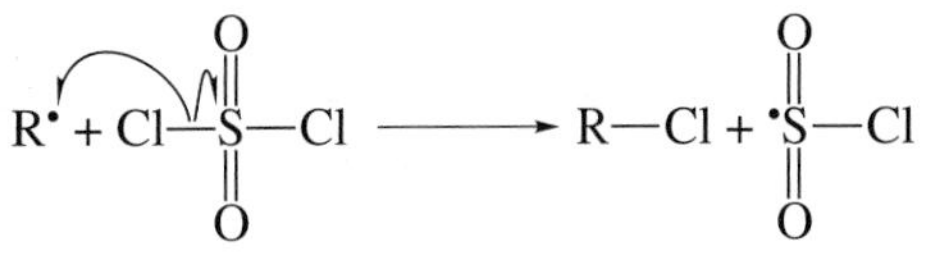

Propagation

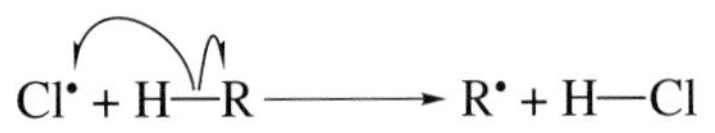

$$\mathrm{Cl^{\bullet} + H{-}R \longrightarrow R^{\bullet} + H{-}Cl} \tag{9.8}$$

$$\mathrm{R^{\bullet} + Cl{-}SO_2{-}Cl \longrightarrow R{-}Cl + {}^{\bullet}SO_2{-}Cl} \tag{9.9}$$

$$\mathrm{{}^{\bullet}SO_2{-}Cl \longrightarrow SO_2 + Cl^{\bullet}} \tag{9.10}$$

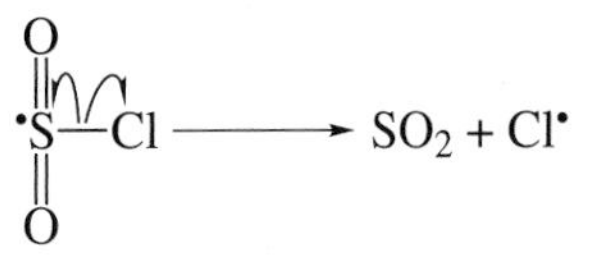

You might assume that this reaction, once initiated, continues until either sulfuryl chloride or the alkane, whichever is the **limiting reagent** (Sec. 1.6), is entirely consumed. In practice, the chain reaction is interrupted by a series of side reactions known as **termination** reactions (Eqs. 9.11–9.13), and the initiation process must be continued throughout the course of the reaction.

Termination

$$\mathrm{Cl^{\bullet} + {}^{\bullet}Cl \longrightarrow Cl{-}Cl} \tag{9.11}$$

$$\mathrm{R^{\bullet} + {}^{\bullet}Cl \longrightarrow R{-}Cl} \tag{9.12}$$

$$\mathrm{R^{\bullet} + {}^{\bullet}R \longrightarrow R{-}R} \tag{9.13}$$

In summary, the distinct stages of a **free-radical chain mechanism** are:

1. *Initiation:* Radicals are formed in *low* concentration from neutral molecules, resulting in a *net increase* in the concentration of free radicals within the system.
2. *Propagation:* Radicals produced in the first step react with molecules to yield new molecules and new radicals; there is *no net change* in the concentration of radicals.
3. *Termination:* Various radicals combine to give molecules, resulting in a *net decrease* in radical concentration and a decrease in the rate of the reaction.

Free-radical halogenation of saturated compounds generally yields products of both monosubstitution and of polysubstitution. For example, chlorination of

1-chlorobutane (**3**), the substrate you will use in the experimental procedure of this section, gives the *di*chloro compounds **4–7**, which are products of monosubstitution (Eq. 9.14), as well as *tri*chloro compounds. Polysubstitution may be minimized by using an excess of the substrate undergoing halogenation as compared to the amount of halogenating agent present. Thus the molar ratio of substrate to chlorinating agent is more than **2** in our procedure. This strategy has a logical basis: statistically speaking, a chlorine atom is more likely to collide with a molecule of **3** than with a dichlorobutane.

$$\underset{\substack{\mathbf{3}\\ \text{1-Chlorobutane}}}{CH_3CH_2CH_2CH_2Cl} \xrightarrow[SO_2Cl_2/\text{heat}]{\text{ABCN}} \underset{\substack{\mathbf{4}\\ \text{1,4-Dichlorobutane}}}{ClCH_2(CH_2)_2CH_2Cl} + \underset{\substack{\mathbf{5}\\ \text{1,3-Dichlorobutane}}}{CH_3\overset{\overset{Cl}{|}}{C}HCH_2CH_2Cl}$$

$$+ \underset{\substack{\mathbf{6}\\ \text{1,2-Dichlorobutane}}}{CH_3CH_2\overset{\overset{Cl}{|}}{C}HCH_2Cl} + \underset{\substack{\mathbf{7}\\ \text{1,1-Dichlorobutane}}}{CH_3(CH_2)_2CHCl_2} \quad (9.14)$$

See more on *Transition State*

The mechanistic step determining which product will ultimately be formed in a free-radical halogenation reaction is the one in which a hydrogen atom is abstracted (Eq. 9.8). In the case of 1-chlorobutane, the ratio of the different isomers **4–7** produced is dictated by the relative energies of the **transition states** for the various C–H abstractions. These energies, in turn, are controlled by a **statistical factor** and an **energy factor.** The statistical factor is determined by the number of hydrogen atoms whose replacement will give a specific *constitutional isomer,* whereas the energy factor is related to the strength of the type of C–H bond being broken and is called the **relative reactivity.** In hydrocarbons such as butane, $CH_3(CH_2)_2CH_3$, the relative reactivity of **secondary** (2°) and **primary** (1°) hydrogen atoms (Fig. 9.5, Sec. 9.3) toward chlorine atoms is 3.3:1.0 and reflects the fact that a secondary C–H bond is *weaker* than a primary one. However, the presence of the chloro substituent means that these values do *not* apply to the corresponding types of hydrogens in **3.** The electron-withdrawing effect of this substituent is known to *increase* the strength of the C–H bonds near to it, thereby *lowering* their relative reactivity. By analyzing the data from this experiment, you should be able to determine if this is true and to derive corrected values for the relative reactivities if it is.

Using sulfuryl chloride for chlorination results in formation of equimolar amounts of sulfur dioxide and hydrochloric acid (Eq. 9.15), both of which are lost as gases from the reaction mixture. The resulting loss of mass provides a convenient means for monitoring the progress of the chlorination because the reaction is complete when the theoretical loss of mass has occurred. This can be calculated on the basis that sulfuryl chloride is the *limiting reagent* in the reaction.

$$R{-}H + SO_2Cl_2 \xrightarrow{\text{Initiator}} R{-}Cl + SO_2 + HCl \quad (9.15)$$

EXPERIMENTAL PROCEDURES

Free-Radical Chain Chlorination of 1-Chlorobutane

Purpose To perform a free-radical chain reaction and determine relative reactivities of C–H bonds toward chlorine atoms.

SAFETY ALERT

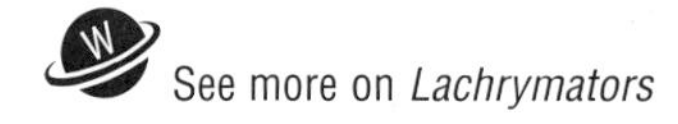

See more on *Lachrymators*

1. **Sulfuryl chloride is a potent lachrymator and reacts violently with water to generate sulfuric acid and hydrochloric acid. *Exercise great care* when handling this chemical: *Make sure that your apparatus is dry.* Do *not* allow it to contact your skin or inhale its vapors; measure out this chemical in a hood, and wear latex gloves when transferring it. Vessels containing sulfuryl chloride should be kept stoppered whenever possible!**
2. **Be certain that all connections in your apparatus are tight prior to heating the reaction mixture and that your gas trap is functioning to prevent venting of noxious gases into the laboratory.**
3. **Carbon dioxide is generated in the separatory funnel when the reaction mixture is washed with aqueous sodium carbonate. To relieve any gas pressure that develops, be sure to vent the funnel frequently when shaking it.**

MINISCALE PROCEDURE

Preparation Sign in at **www.cengage.com/login** to answer Pre-Lab Exercises, access videos, and read the MSDSs for the chemicals used or produced in this procedure. Review Sections 2.11, 2.14, 2.21, 2.22, 2.23, 2.25, 2.28, and 6.4.

Apparatus A *dry* 25-mL round-bottom flask, gas trap, separatory funnel, apparatus for magnetic stirring, heating under reflux, and *flameless* heating.

Setting Up Assemble the apparatus for heating under reflux, and attach a gas trap such as that in Figure 2.65 to the top of the condenser. Equip the round-bottom flask with a stirbar and add 0.1 g of 1,1′-azobis(cyclohexanenitrile) (ABCN). Working *at the hood,* sequentially measure 5 mL of 1-chlorobutane and 2 mL of sulfuryl chloride into the flask. Stopper and weigh the flask, and then attach it to the reflux condenser, working quickly to minimize introducing vapors of sulfuryl chloride into the atmosphere.

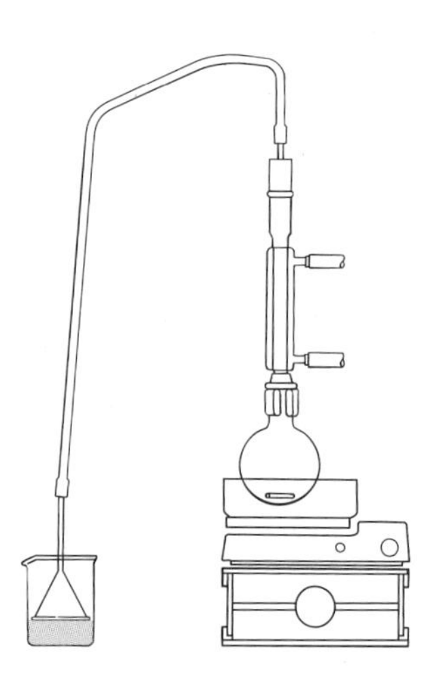

Reaction Heat the solution under gentle reflux for 20 min with stirring and then allow the reaction mixture to cool below the reflux temperature. Add a second 0.1-g portion of ABCN and continue heating the stirred mixture under reflux for another 10 min. Weigh the flask and consult with your instructor if the resulting mixture has not lost at least 90% of the expected mass after this second period of reflux.

Work-Up and Isolation Cool the reaction mixture in an ice-water bath and, working at the hood, *cautiously* pour it with stirring into 15 mL of ice-cold saturated aqueous sodium chloride (brine) contained in an Erlenmeyer flask. Transfer the resulting two-phase solution to a separatory funnel and separate the layers.★ Wash the organic layer in the funnel with 10 mL of 0.5 *M* sodium carbonate solution, venting the funnel *frequently.* Using pHydrion™ paper, determine whether the aqueous layer (Sec. 2.21) is basic. If not, repeat washing the organic layer with 10-mL portions of sodium carbonate solution until the aqueous washes are *basic.* Wash the organic layer with 15 mL of brine and transfer it to an Erlenmeyer flask containing several spatula-tips full of anhydrous sodium sulfate.★ Swirl the flask occasionally during 10–15 min to hasten drying, which is indicated by the liquid becoming clear. If the liquid remains cloudy, add additional portions of sodium sulfate with swirling. Carefully decant the organic layer into a dry, *tared* container.

Analysis Weigh the reaction mixture and use the result to calculate the percentage yield of material you recovered. This calculation requires that you first determine the theoretical mass of product dichlorides *and* unreacted 1-chlorobutane expected from the reaction.

Analyze the reaction mixture by gas chromatography or GC-MS or submit samples to your instructor for such analysis. The peak corresponding to 1-chlorobutane should be "off-scale" or nearly so, so that the peaks for the isomeric dichlorobutanes are sufficiently large for an accurate analysis of their areas. Assign the four product peaks to the isomers responsible for them and calculate the relative amounts of each isomer present in the mixture. A typical GLC trace of the reaction mixture is depicted in Figure 9.1.

Obtain IR and ^{1}H NMR spectra of your starting material and an NMR spectrum of your product mixture; compare them with those of authentic samples (Figs. 9.2–9.4).

MICROSCALE PROCEDURE

Preparation Sign in at **www.cengage.com/login** to answer Pre-Lab Exercises, access videos, and read the MSDSs for the chemicals used or produced in this procedure. Review Sections 2.4, 2.11, 2.14, 2.21, 2.22, 2.23, 2.25, 2.28, and 6.4.

Apparatus *Dry* 3-mL screw-cap conical vial, gas trap, 5-mL conical vial or screw-cap centrifuge tube, apparatus for magnetic stirring, heating under reflux, and *flameless* heating.

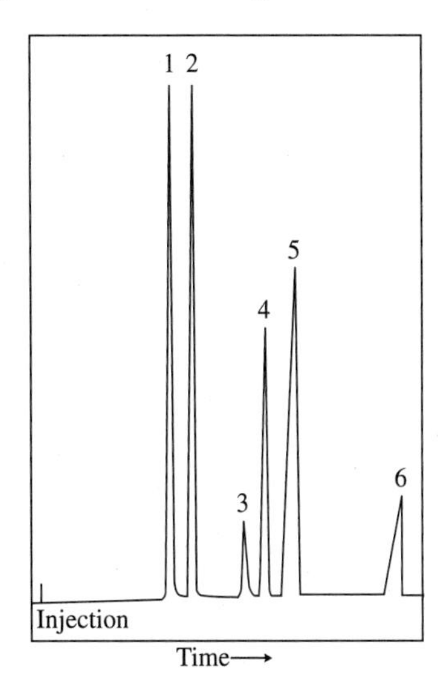

Figure 9.1
GLC analysis of mixture from chlorination of 1-chlorobutane (numbers in parentheses are boiling points at 760 torr and peak areas, respectively). Peak 1: solvent; peak 2: 1-chlorobutane (77–78 °C, 2719); peak 3: 1,1-dichlorobutane (114–115 °C, 487); peak 4: 1,2-dichlorobutane (121–123 °C, 1771); peak 5: 1,3-dichlorobutane (131–133 °C, 3367); peak 6: 1,4-dichlorobutane (161–163 °C, 1137); column and conditions: 0.5-mm × 30-m, 0.25-μ film of SE-54; 80 °C, 40 mL/min.

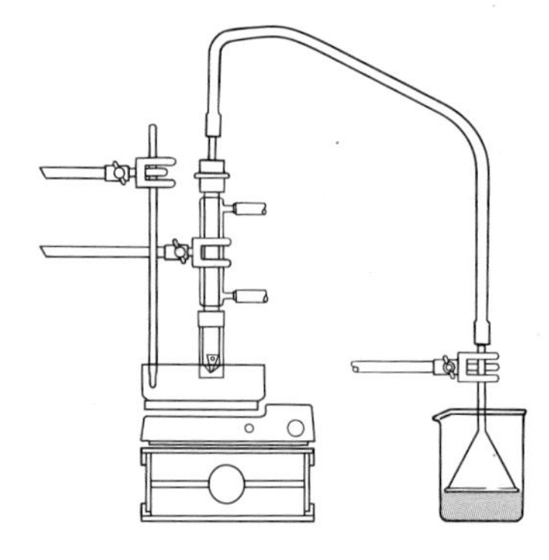

Assembly Assemble the apparatus for heating under reflux, and attach a gas trap, such as that of Figure 2.69, to the top of the condenser. Place a spinvane in the vial and add 10 mg (0.010 g) of 1,1′-azobis(cyclohexanecarbonitrile) (ABCN). Working *at the hood,* sequentially add 0.5 mL of 1-chlorobutane and 0.2 mL (200 μL) of sulfuryl chloride to the vial. Cap and weigh the vial, and then attach it to the reflux condenser, working quickly to minimize introducing vapors of sulfuryl chloride into the atmosphere.

Reaction Heat the solution under gentle reflux with stirring for about 20 min and then allow the reaction mixture to cool below the reflux temperature. Add a second 0.010-g portion of ABCN and continue heating the stirred mixture under reflux for another 10 min. Weigh the flask, and consult with your instructor if the resulting mixture has not lost at least 90% of the expected mass after this second period of reflux.

Work-Up and Isolation Remove the vial from the apparatus and cap it.★ Working at the hood, *cautiously* pour the reaction mixture with stirring into 1 mL of ice-cold saturated aqueous sodium chloride (brine) contained in the 5-mL conical vial or centrifuge tube. Allow the layers to separate, and stir or shake the mixture briefly. Remove the lower layer with a Pasteur pipet and determine whether it is the organic or aqueous layer (Sec. 2.21).★ *Do not* discard *either* layer until you know which one is the organic layer.

Carefully add 1 mL of aqueous 0.5 *M* sodium carbonate to the organic layer (*caution:* foaming), shake the mixture with venting as necessary, and remove the lower layer. Using pHydrion paper, determine whether the aqueous layer is basic. If not, repeat washing the organic layer with 1-mL portions of sodium carbonate solution until the aqueous washes are *basic.* Shake the organic layer with 1 mL of brine, and carefully transfer it to a dry vial containing one or two microspatula-tips full of anhydrous sodium sulfate. Stir the resulting mixture occasionally during 10–15 min to hasten drying, which is indicated by the liquid becoming clear. If the liquid remains cloudy, add additional portions of sodium sulfate with stirring. Carefully transfer the dried mixture to a dry, *tared* sample vial using a Pasteur filter-tip pipet.

Analysis Weigh the reaction mixture and use the result to calculate the percentage yield of material you recovered. This calculation requires that you first determine the theoretical weight of product dichlorides *and* unreacted 1-chlorobutane expected from the reaction.

Analyze the reaction mixture by gas chromatography or GC-MS or submit samples to your instructor for such analysis. The peak corresponding to 1-chlorobutane should be "off-scale" or nearly so, so that the peaks for the isomeric dichlorobutanes are sufficiently large for an accurate analysis of their areas. Assign the four product peaks to the isomers responsible for them and calculate the relative amounts of each isomer present in the mixture. A typical GLC trace of the reaction mixture is depicted in Figure 9.1.

Obtain IR and NMR spectra of your starting material and an NMR spectrum of your product mixture; compare them with those of authentic samples (Figs. 9.2–9.4).

Discovery Experiment

Chlorination of Heptane

Using the procedure described above as your model, design and execute an experimental protocol for monochlorinating heptane and determining the ratio of chloroheptanes produced. Also describe how you would analyze the mixture for the presence of polychloro isomers. Check with your instructor before undertaking any experimental work.

Discovery Experiment ***Chlorination of 2,3-Dimethylbutane***

Using the procedure described above as your model, design and execute an experimental protocol for monochlorinating 2,3-dimethylbutane and determining the ratio of chlorodimethylbutanes produced. Also describe how you would analyze the mixture for the presence of polychloro isomers. Check with your instructor before undertaking any experimental work.

WRAPPING IT UP

Combine *all of the aqueous solutions.* If the resulting solution is acidic, neutralize it carefully with solid sodium carbonate and flush it down the drain with a large excess of water. Rinse the *glasswool* from the gas trap with aqueous 0.5 *M* sodium carbonate and discard it in the trash. Put the *sodium sulfate* in a container for solids contaminated with alkyl halides.

EXERCISES

1. Define and provide a specific example of each of the following terms:
   a. propagation step
   b. initiation step
   c. bond homolysis
   d. limiting reagent
   e. secondary hydrogen atom
2. Why is 1-chlorobutane not classified as a hydrocarbon?
3. Which, if any, of the dichlorobutanes contain a center of chirality?
4. What is the fundamental difference between a reaction mechanism that uses single-headed arrows vs. one that uses half-headed (fish-hook) arrows?

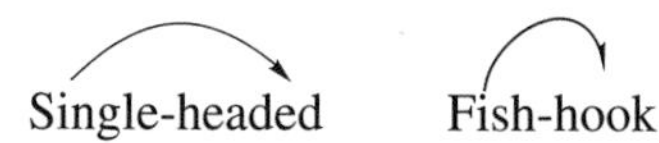

5. Using curved arrows to symbolize the flow of electrons, write the complete mechanism for the formation of 1,3-dichlorobutane from 1-chlorobutane using ABCN (**1**) and sulfuryl chloride. Clearly label each step as to whether it is part of the initiation, propagation, or termination steps.
6. Why is only a catalytic amount of ABCN (**1**) needed in this reaction?
7. Write at least one product that might be derived from ABCN (**1**) and explain why the potential contamination of the reaction mixture by such a product is of little concern.
8. Why is the amount of sulfuryl chloride used *less* than the amount theoretically required to convert all the starting material to monosubstituted products?

9. Why is it prudent to use caution when pouring the crude reaction mixture into brine? Why should the reaction mixture and brine be *ice-cold* for this step?

10. In the work-up procedure, the organic layer was washed with a solution of sodium carbonate.

a. Why was this wash necessary?

b. Write the reaction that occurs during this wash and specify the gas that is evolved.

c. Why is it especially important to vent a separatory funnel when performing this wash?

11. How might a loss of mass greater than theoretically calculated occur in a chlorination reaction performed according to the procedure of this section?

12. What physical property of 1,4-dichlorobutane accounts for it being the isomer with the longest retention time in the GLC analysis?

13. Using a relative reactivity for 1° to 2° hydrogen atoms of 1:3.3,

a. calculate the percentage of each chlorobutane expected from monochlorination of 1-chlorobutane, and

b. determine the ratio of 1° to 2° dichlorobutanes to be expected.

14. a. Using your own data or those provided in Figure 9.1, calculate the ratio of 1° to 2° dichlorobutanes obtained from monochlorination of 1-chlorobutane.

b. Why does this ratio *not* agree with that found in Exercise 13?

15. Based on your calculation of the percentage of the various dichlorobutanes formed in the reaction, determine the values for the relative reactivity for the various types of hydrogen atoms in 1-chlorobutane. Assign the relative reactivity of the methyl hydrogen atoms as 1.0.

16. Specify the factors determining the proportion of monochlorinated products derived from 1-chlorobutane.

17. Chlorine and ABCN (**1**) are good initiators for free-radical chain reactions.

a. What characterizes a molecule that is a good initiator?

b. Why is it unnecessary to use a stoichiometric amount of the initiator?

c. Two other common radical initiators are shown below. Using curved arrows to symbolize the flow of electrons, show the bond homolysis that occurs when these initiators are heated or irradiated.

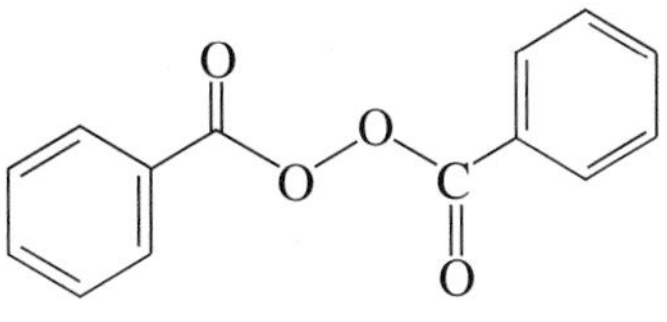

$$C_6H_5{-}C(=O){-}O{-}O{-}C(=O){-}C_6H_5$$

Benzoyl peroxide

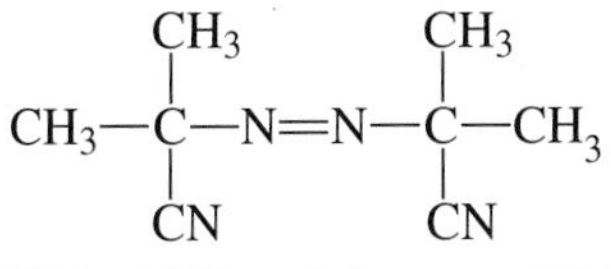

$$CH_3{-}C(CH_3)(CN){-}N{=}N{-}C(CH_3)(CN){-}CH_3$$

2,2′-Azobis[2-methylpropanenitrile]

18. A linear geometry is portrayed in the structures of both **1** and, as shown in Exercise 17, 2,2′-azobis [2-methylpropanenitrile]. Using "R" to represent the alkyl substituents on the nitrogen atoms, provide a representation of the thermodynamically more favored geometry of these azo compounds. Justify your answer.

19. Indicate from which of the three compounds shown below it should be easiest to prepare a *single* pure monochlorinated product, and explain why.

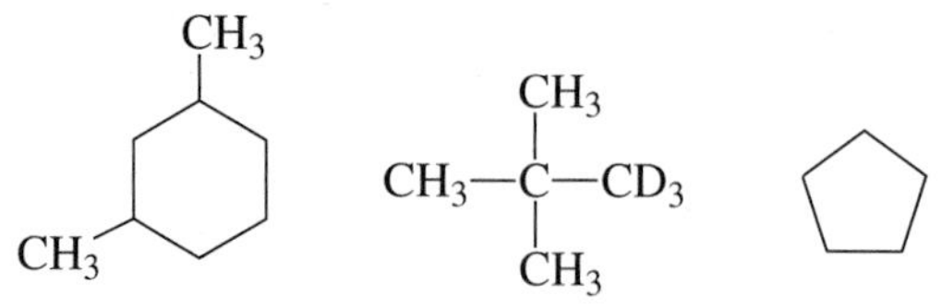

CH_3 CH_3 (cyclohexane) CH_3–C(CH_3)(CH_3)–CD_3 (cyclopentane)

20. Consider the monochlorination of heptane using ABCN (**1**) and sulfuryl chloride.

a. Write the structures for and the names of the possible monochlorination products that could be produced.

b. Circle the center(s) of chirality, if any, that are present in the isomers that you wrote in Part **a**.

c. Assume that you have been provided with a pure sample of each monochloroheptane but that the samples were unlabeled. Discuss how you could use a combination of ^1H and ^{13}C NMR spectroscopic data to prove the structure of each sample.

21. Chlorofluorocarbons (CFCs) have been found to contribute to the problem of ozone depletion. In the stratosphere, CFCs initiate a radical chain reaction that can consume 100,000 ozone molecules per CFC molecule.

a. Freon 12, CF_2Cl_2, is a typical CFC. Using curved arrows to symbolize the flow of electrons, show the radical initiation step that occurs when light interacts with Freon 12 in the stratosphere. (*Hint:* Identify the weakest bond in the molecule.)

b. Write the products formed when the radical produced in the reaction in Part **a** reacts with a molecule of ozone, O_3, in the first propagation step. (*Hint:* Oxygen, O_2, and another radical are produced.)

Sign in at **www.cengage.com/login** and use the spectra viewer and Tables 8.1–8.8 as needed to answer the blue-numbered questions on spectroscopy.

22. Consider the NMR data for 1-chlorobutane (Fig. 9.3).

a. In the ^1H NMR spectrum, assign the various resonances to the hydrogen nuclei responsible for them.

b. For the ^{13}C NMR data, assign the various resonances to the carbon nuclei responsible for them.

23. Figure 9.4 is the ^1H NMR spectrum of a mixture of 1,*x*-dichlorobutanes obtained by free-radical chlorination of 1-chlorobutane.

a. Assign the triplets centered near δ 1.0 and 1.1 and the doublet centered near δ 1.6 to the hydrogen nuclei of the three different isomers from which they arise. (*Hint:* Write the structures of the isomeric 1,*x*-dichlorobutanes and predict the multiplicity and approximate chemical shift of each group of hydrogens.)

b. The integrated areas of the selected multiplets in the spectrum are as follows: δ 1.0, 2 units, δ 1.1, 9 units, δ 1.6, 32 units, δ 3.6, 26.8 units, δ 3.7, 26.8 units, δ 4.0, 3.3 units, δ 5.8 (a triplet *not* visible in the spectrum), 0.7 units. Based on your assignments in Part **a** and on the integration data, calculate the approximate percentage of each isomer present in the mixture. Explain your determination as completely as possible.

24. Compare the ^{1}H NMR spectrum of the mixture of 1,*x*-dichlorobutanes (Fig. 9.4) with that of 1-chlorobutane (Fig. 9.3). What features in Figure 9.4 are consistent with the incorporation of a second chlorine atom in this experiment and with the proposal that a *mixture* of dichlorobutanes, rather than a single isomer, is formed?

SPECTRA

Starting Materials and Products

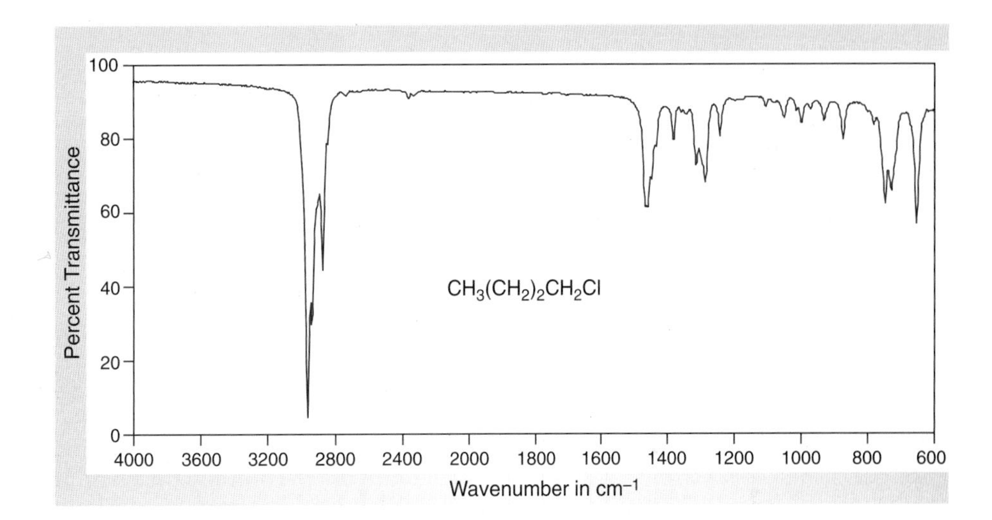

Figure 9.2
IR spectrum of 1-chlorobutane (neat).

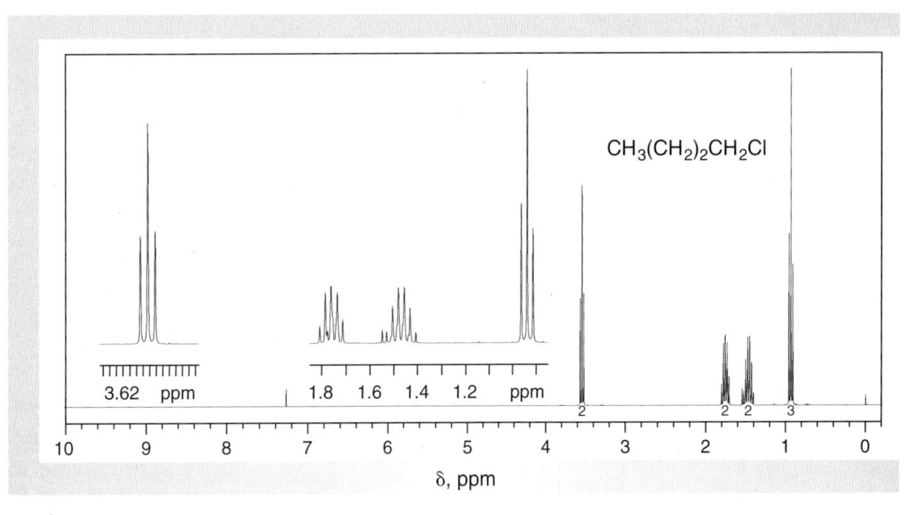

Figure 9.3
NMR data for 1-chlorobutane ($CDCl_3$).

(a) ^{1}H NMR spectrum (300 MHz).
(b) ^{13}C NMR data: δ 13.4, 20.3, 35.0, 44.6.

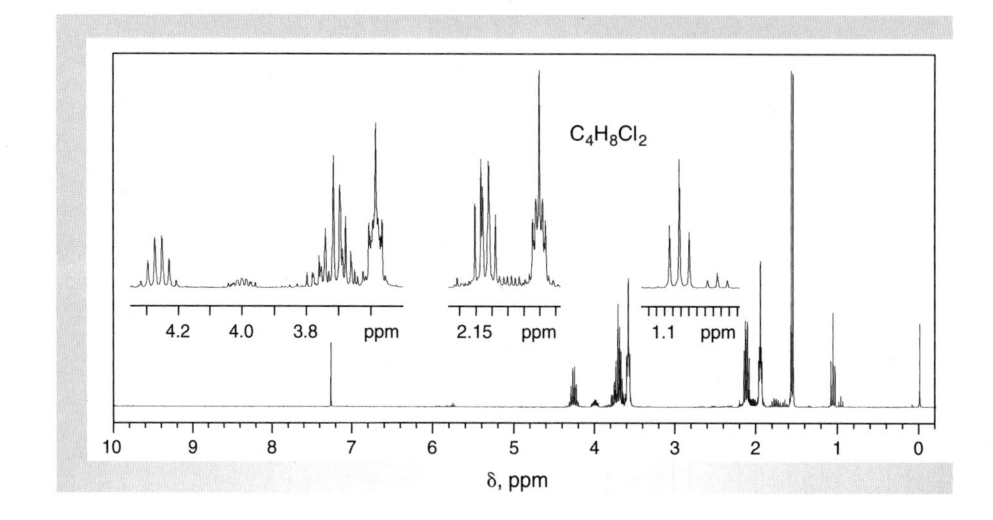

Figure 9.4
^1H NMR spectrum of mixture of 1,x-dichlorobutanes ($CDCl_3$, 300 MHz).

9.3 BROMINATION: SELECTIVITY OF HYDROGEN ATOM ABSTRACTION

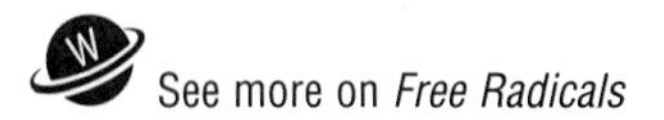

Hydrocarbons undergo free-radical chain bromination (Eq. 9.16) through a mechanism analogous to that for free-radical chlorination (Sec. 9.2). The **initiation** step involves the thermal or photochemical **homolysis** of molecular bromine to form bromine atoms, Br• (Eq. 9.17); the energy required for this is referred to as the **bond dissociation energy**. **Propagation** steps then produce the **substitution** product, R–Br, and regenerate Br• to continue the chain reaction (Eqs. 9.18 and 9.19). The **termination** steps are analogous to those of Equations 9.11–9.13 (Sec. 9.2).

$$\underset{\text{(colorless)}}{R\text{—}H} + \underset{\text{(reddish orange)}}{Br_2} \xrightarrow{h\nu} \underset{\text{(colorless)}}{R\text{—}Br} + \underset{\text{(colorless)}}{H\text{—}Br} \quad (9.16)$$

$$Br\text{—}Br \xrightarrow{\text{heat or } h\nu} Br^\bullet + Br^\bullet \quad (9.17)$$

$$Br^\bullet + H\text{—}R \longrightarrow R^\bullet + HBr \quad (9.18)$$

$$R^\bullet + Br\text{—}Br \longrightarrow R\text{—}Br + Br^\bullet \quad (9.19)$$

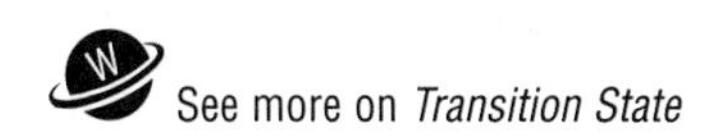

As with chlorination, the mechanistic step involving abstraction of a hydrogen atom by Br (Eq. 9.18) determines which product will be formed. The likelihood that a particular type of hydrogen atom will be successfully abstracted is defined by the relative energies of the **transition states** for the various C–H abstractions. These energies, in turn, are controlled by a **statistical factor** and an **energy factor.** The statistical factor is determined by the number of hydrogen atoms whose replacement will give a specific *constitutional isomer,* whereas the energy factor is related to the strength of

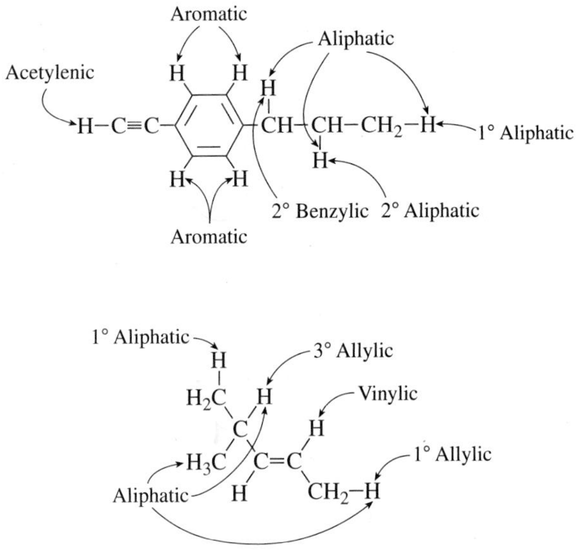

Figure 9.5
Classification of hydrogen atoms in hydrocarbons.

the type of C–H bond being broken and is called the **relative reactivity.** In the experiment of this section, you will determine the relative rates of the free-radical chain reaction of molecular bromine with several hydrocarbons containing different types of hydrogen atoms. You should be able to deduce the order of relative reactivity of *seven* different types of hydrogens toward Br· by carefully considering the results of your observations.

As summarized by the structures in Figure 9.5, hydrogen atoms in hydrocarbons are classified according to the type of carbon atom to which they are attached. Specifically, hydrogens bound to sp^3-hybridized carbons are categorized as **aliphatic;** all hydrogen atoms in the hydrocarbon family of alkanes belong to this group. Aliphatic hydrogens may be further classified as being **primary (1°), secondary (2°),** or **tertiary (3°)** depending upon the number of other carbon atoms attached to the reference carbon atom; primary carbon atoms are bound to one other carbon atom, secondary to two, and tertiary to three. An aliphatic hydrogen atom attached to an sp^3-hybridized carbon atom that in turn is bound to a vinylic or aromatic carbon atom is called **allylic** or **benzylic,** respectively. Combining these classifications gives the terms primary, secondary, or tertiary aliphatic, allylic, or benzylic hydrogen atoms. Hydrogen atoms attached to sp^2-hybridized carbon atoms are classified as either **vinylic** or **aromatic,** depending upon whether the carbon atom is part of a normal double bond or an aromatic ring. Hydrogen atoms are called **acetylenic** when bound to an sp-hybridized carbon atom.

The hydrocarbons **8–13** contain various types of hydrogen atoms and are the substrates for our experiment. Measuring their relative reactivities toward Br· in free-radical chain bromination is straightforward for two reasons. First, molecular bromine is colored, so the progress of bromination is easily monitored by observing the disappearance of color as the bromine is consumed (Eq. 9.16). Second, because the relative rates of hydrogen abstraction by Br· are highly sensitive to the strength of the C–H bond being broken, the substrates have rates of free-radical chain bromination that differ enough to allow ready determination of the qualitative order of their reactivity.

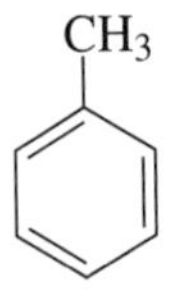

CH_3 — **8** Toluene

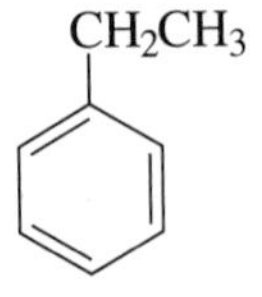

CH_2CH_3 — **9** Ethylbenzene

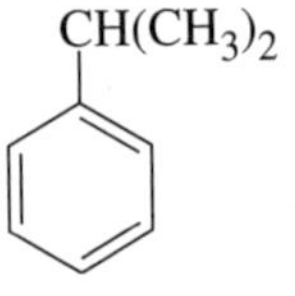

$CH(CH_3)_2$ — **10** Isopropylbenzene

$C(CH_3)_3$ — **11** *tert*-Butylbenzene

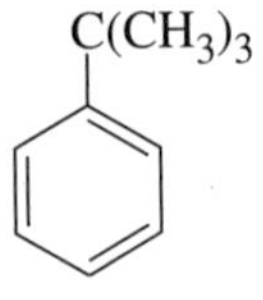

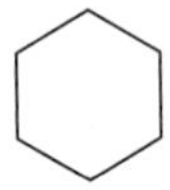

12 Cyclohexane

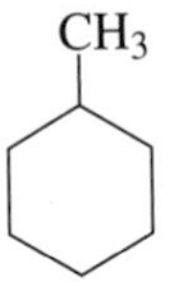

CH_3 — **13** Methylcyclohexane

Experimentally, you will allow an *excess* of a hydrocarbon to react with molecular bromine and will measure the time required for disappearance of the color of the bromine. The reactions are performed in dichloromethane solution using two different conditions: (a) room temperature with no illumination, and (b) room temperature with illumination. The period of time required for the decoloration depends upon the reaction conditions and the relative reactivities of the various hydrocarbons; some of the substrates may not react at all under the specified conditions. The ratios of the different reaction times thus represent the relative rates of the free-radical chain bromination reaction.

EXPERIMENTAL PROCEDURE

Relative Rates of Free-Radical Chain Bromination

Purpose To determine the relative reactivities of different types of hydrogen atoms toward bromine atoms.

SAFETY ALERT

1. ***Bromine is a hazardous chemical that may cause serious chemical burns.*** **Do *not* breathe its vapors or allow it to come into contact with skin. Perform all operations involving the transfer of the pure liquid or its solutions at a hood and wear latex gloves. If you get bromine on your skin, wash the area immediately with soap and warm water and soak the affected area in 0.6 *M* sodium thiosulfate solution for up to 3 h if the burn is particularly serious.**
2. **Keep the 1 *M* bromine in dichloromethane solution in the hood. Dispense 0.5-mL portions of it, using a pipet pump, calibrated Pasteur filter-tip pipet, or a buret fitted with a Teflon stopcock.**

3. **Bromine reacts with acetone to produce the powerful lachrymator α-bromoacetone, $BrCH_2COCH_3$. Do *not* rinse glassware containing residual bromine with acetone!**
4. **Avoid inhalation of the vapors of any of the materials being used in this experiment.**

Procedure

Preparation Sign in at **www.cengage.com/login** to answer Pre-Lab Exercises, access videos, and read the MSDSs for the chemicals used or produced in this procedure. Review Sections 2.5 and 2.10.

Construct a table in your notebook with the following four main headings: (a) "Hydrocarbon"; (b) "Types of Hydrogen Atoms," entries under which will include the terms 1° aliphatic, 2° benzylic, and so on; (c) "Conditions," with the subheadings RT, RT/hν; and (d) "Elapsed Time," entries under which will be the time required for reaction as measured by decoloration.

Carefully plan your execution of this experiment, because proper labeling of test tubes and managing of time once the reactions have been started are critical. You will perform two separate experimental trials, with one trial starting 5–10 min after the other.

Apparatus Fourteen 13-mm × 100-mm test tubes and corks, six 10-mm × 75-mm test tubes, six Pasteur pipets, filter-tip pipet, ice-water bath, light source.

Setting Up Organize *twelve* 13-mm × 100-mm test tubes into pairs and label each tube in the pair with the name or number of the hydrocarbon **8–13** that you will add to it. Measure 2.5 mL of dichloromethane into each of these tubes and then add 0.5 mL (10 drops) of the hydrocarbon for which it is labeled. Label the *two* remaining 13-mm × 100-mm test tubes "Control" and add 3 mL of dichloromethane to them. Place the *seven* pairs of labeled tubes in a beaker. Working at the hood, dispense 0.5 mL of a 1 *M* solution of bromine in dichloromethane into each of the *six* small test tubes and into one of the larger tubes labeled "Control."

Bromination In rapid succession, add the solution of bromine to each of the hydrocarbon-containing tubes. Do this with agitation to ensure good mixing and record the time of the additions. After all the additions are made, loosely cork each tube to prevent loss of solvent and bromine. Monitor all the tubes, including that labeled "Control," and record the elapsed time required for the discharge, if any, of the color of each solution.

After about 5 min, again dispense 0.5 mL of 1 *M* bromine solution into the six small test tubes used originally for this reagent *and* also into the remaining tube labeled "Control." Working as before, add the solution of bromine to each of the hydrocarbon-containing tubes. Place the six tubes containing the resulting solutions and the tube labeled "Control" in a beaker located 14–16 cm above an unfrosted 100- or 150-watt light bulb so that all the solutions are exposed to the same amount of light. Monitor these seven tubes and record the elapsed time required for the discharge, if any, of the color of each solution.

Terminate your observations of both sets of samples after about 1 h.

Analysis Evaluate the intensity of color remaining in any of the tubes relative to that of the "Control" samples and record the results in your notebook. Obtain IR and ^{1}H NMR spectra of your starting materials and compare them with those of authentic samples (Figs. 9.6–9.17).

WRAPPING IT UP

Decolorize any *solutions* in which the color of bromine is visible by the dropwise addition of cyclohexene; then discard the resulting solutions together with *all other solutions* in a container for halogenated organic liquids.

EXERCISES

1. Draw the structure of the major monobrominated product expected from each of the hydrocarbons used in this experiment.

2. Comment on the need for light in order for bromination to occur with some of the hydrocarbons.

3. Why is a "Control" sample needed in this experiment?

4. Perform the calculations necessary to demonstrate that bromine is indeed the limiting reagent (Sec. 1.6). The densities and molar mass needed to complete the calculations can be found in the abbreviated MSDSs on the website or various handbooks of chemistry.

5. How is polyhalogenation minimized in this experiment?

6. Using curved arrows to symbolize the flow of electrons, show the mechanism for the propagation steps that convert **10** into the corresponding monobromide, and provide two possible termination steps for the process.

7. Answer the questions below for substrates **8–13.**

a. Which is statistically the most likely to form a monobromide? Briefly explain your answer.

b. Which is the most likely to form a single isomer of a monobromide? Briefly explain your answer.

c. Which is expected to produce the *most* stable radical? Briefly explain your answer.

d. Which is expected to produce the *least* stable radical? Briefly explain your answer.

8. Arrange the six hydrocarbons **8–13** in *increasing* order of reactivity in free-radical chain bromination.

9. a. On the basis of the observed order of reactivity of the hydrocarbons, deduce the order of reactivity of the seven different types of hydrogens found in these compounds, that is, (1) primary aliphatic, (2) secondary aliphatic, (3) tertiary aliphatic, (4) primary benzylic, (5) secondary benzylic, (6) tertiary benzylic, and (7) aromatic.

b. Outline the logic you used to arrive at your sequence in Part **a.**

10. Consider the free-radical halogenation of an alkane. You may wish to refer to your lecture textbook and to tables of bond dissociation energies when answering these questions.

a. Why is the bromination much more regioselective than chlorination?

b. Why are fluorinations extremely dangerous? (*Hint:* Consider heats of reaction of the propagation steps.)

c. Why is it difficult to generate an alkyl iodide by free-radical chain halogenation?

See more on *Hammond Postulate*

Sign in at **www.cengage.com/login** and use the spectra viewer and Tables 8.1–8.8 as needed to answer the blue-numbered questions on spectroscopy.

11. How does the Hammond postulate explain
    a. the greater selectivity for bromination vs. chlorination of an alkane?
    b. the fact that the relative reactivity of C–H bonds of an alkane for free-radical chain halogenation is 3° > 2° > 1°?
12. Why is a 1° benzylic radical more stable than a 1° aliphatic radical?
13. Consider the IR spectra of compounds **8–11** (Figs. 9.6, 9.8, 9.10, and 9.12), all of which contain a phenyl ring. What absorption(s) in the spectra denote the presence of this ring?
14. Consider the NMR data for compounds **8–13** (Figs. 9.7, 9.9, 9.11, 9.13, 9.15, and 9.17).
    a. In the ^{1}H NMR spectra, assign the various resonances to the hydrogen nuclei responsible for them.
    b. In the ^{13}C NMR spectra of these compounds, assign the various resonances to the carbon nuclei responsible for them.

SPECTRA

Starting Materials

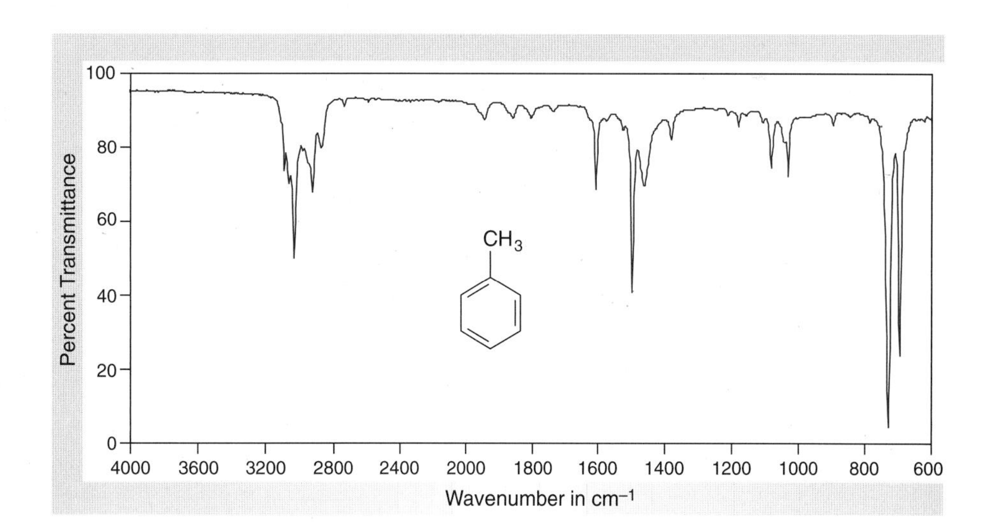

Figure 9.6
IR spectrum of toluene (neat).

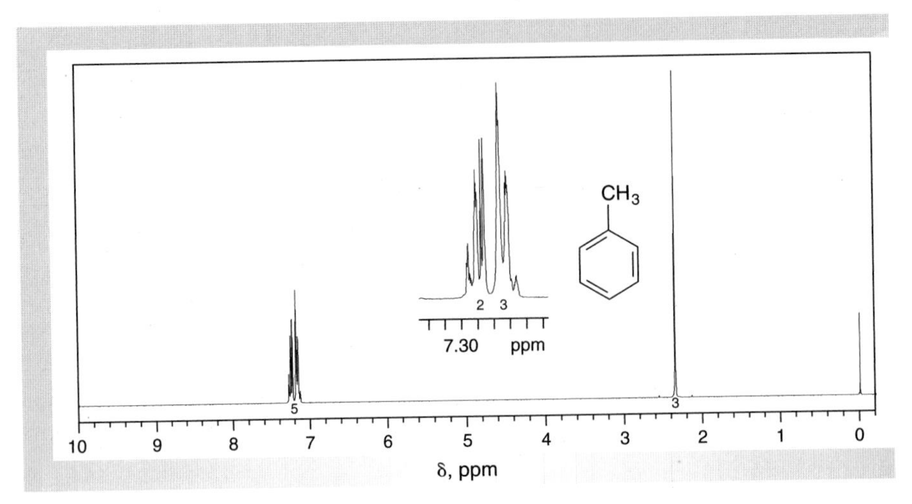

Figure 9.7
NMR data for toluene ($CDCl_3$).

(a) ^{1}H NMR spectrum (300 MHz).
(b) ^{13}C NMR data: δ 21.4, 125.3, 128.2, 129.0, 137.8.

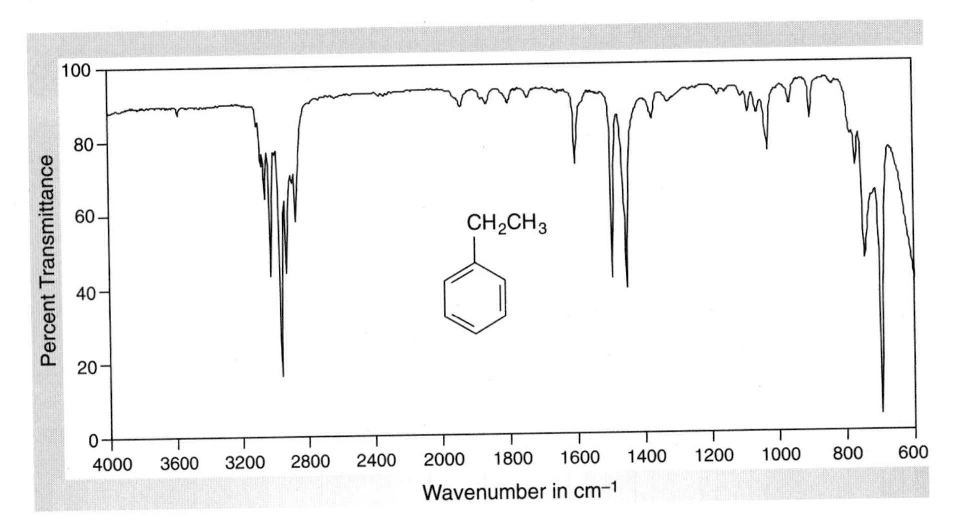

Figure 9.8
IR spectrum of ethylbenzene (neat).

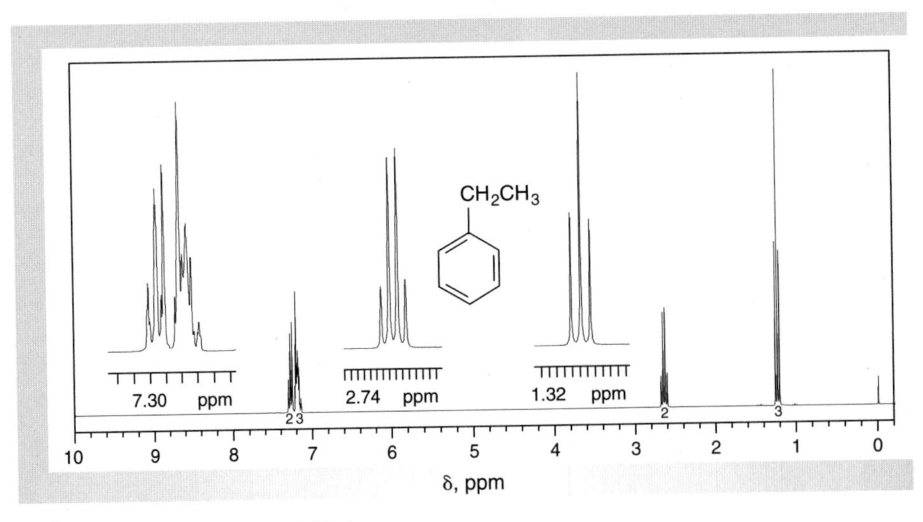

Figure 9.9
NMR data for ethylbenzene ($CDCl_3$).

(a) ^{1}H NMR spectrum (300 MHz).
(b) ^{13}C NMR data: δ 15.6, 29.1, 125.7, 127.9, 128.4, 144.2.

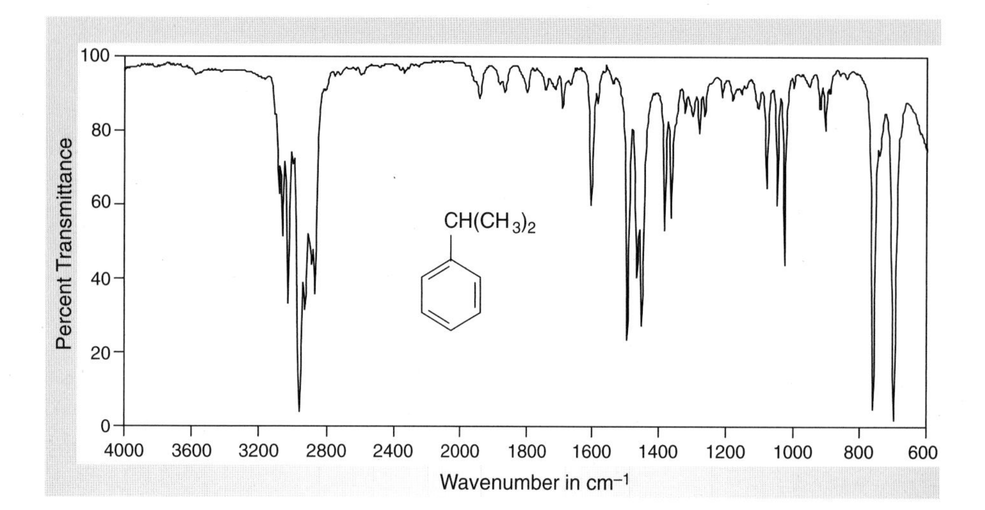

Figure 9.10
IR spectrum of isopropylbenzene (neat).

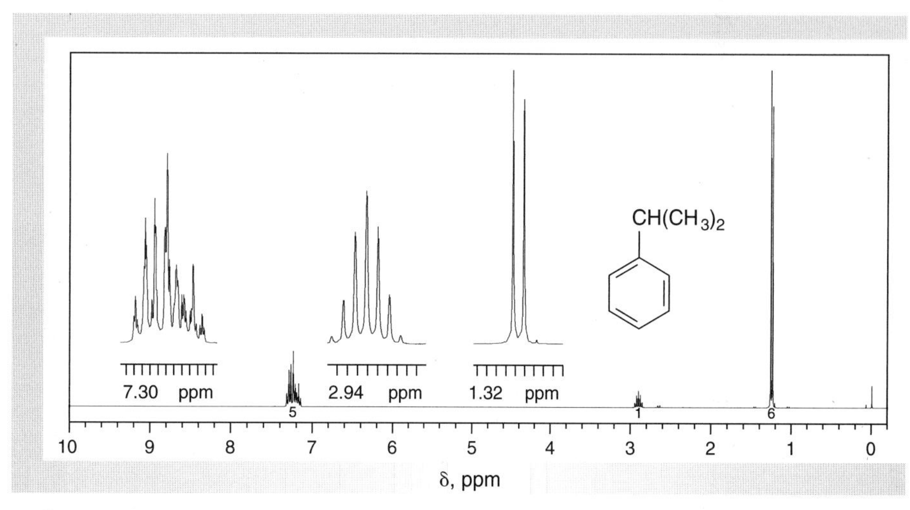

Figure 9.11
NMR data for isopropylbenzene ($CDCl_3$).

(a) ^{1}H NMR spectrum (300 MHz).
(b) ^{13}C NMR data: δ 24.1, 34.2, 125.8, 126.4, 128.4, 148.8.

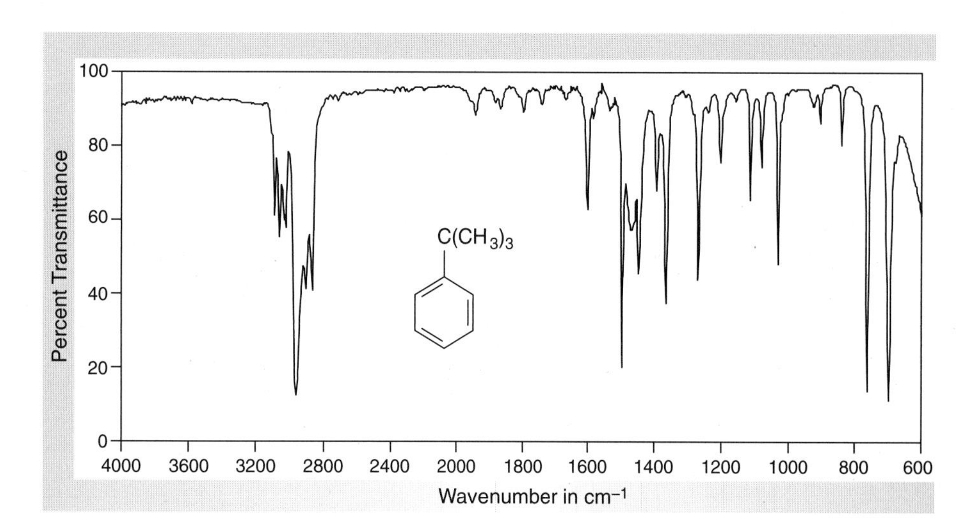

Figure 9.12
IR spectrum of tert-*butylbenzene (neat).*

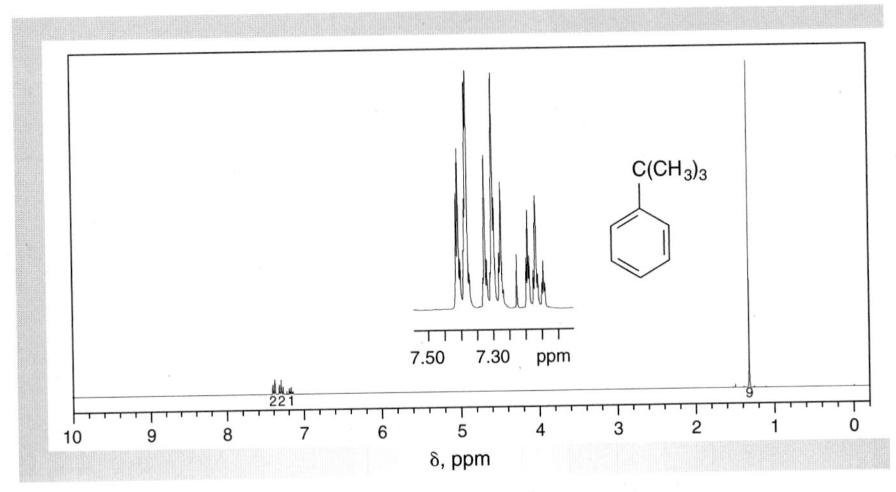

Figure 9.13
NMR data for tert-butylbenzene ($CDCl_3$).

(a) ^{1}H NMR spectrum (300 MHz).
(b) ^{13}C NMR data: δ 31.3, 34.5, 125.1, 125.3, 128.0, 150.8.

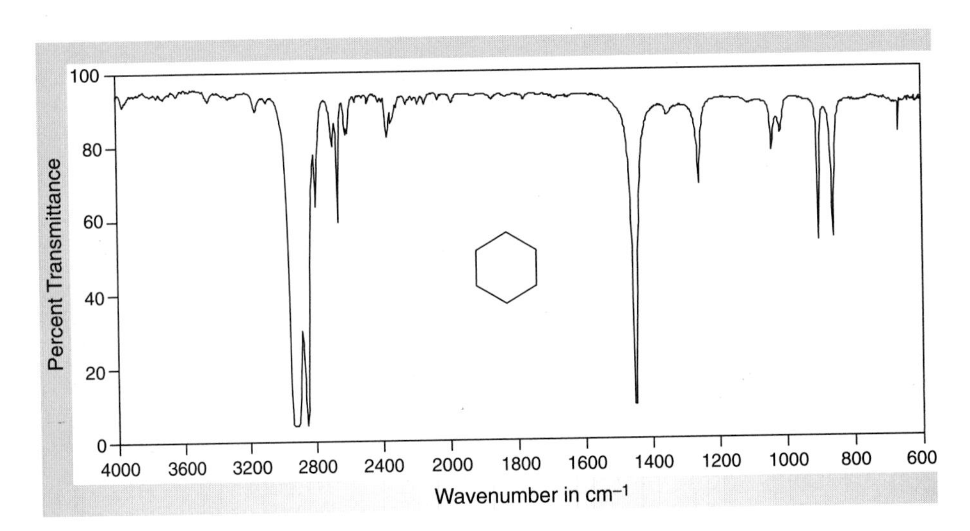

Figure 9.14
IR spectrum of cyclohexane (neat).

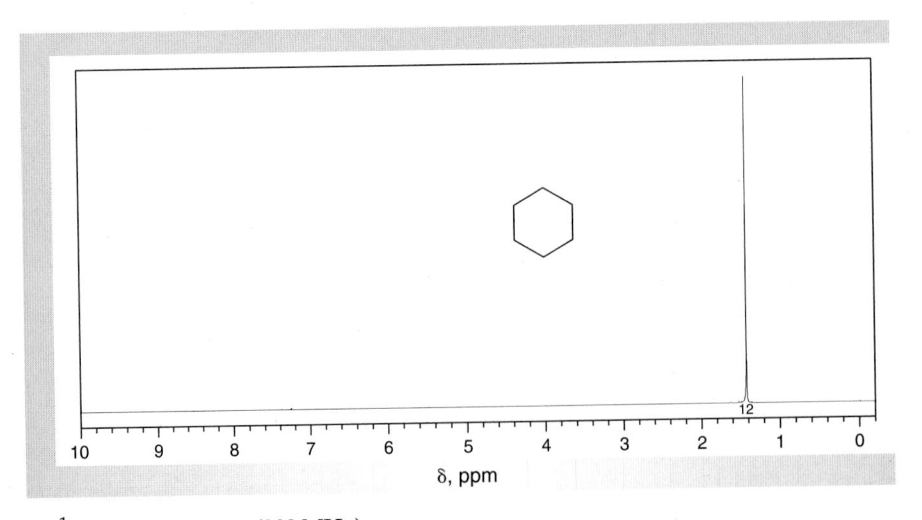

Figure 9.15
NMR data for cyclohexane ($CDCl_3$).

(a) ^{1}H NMR spectrum (300 MHz).
(b) ^{13}C NMR datum: δ 27.3.

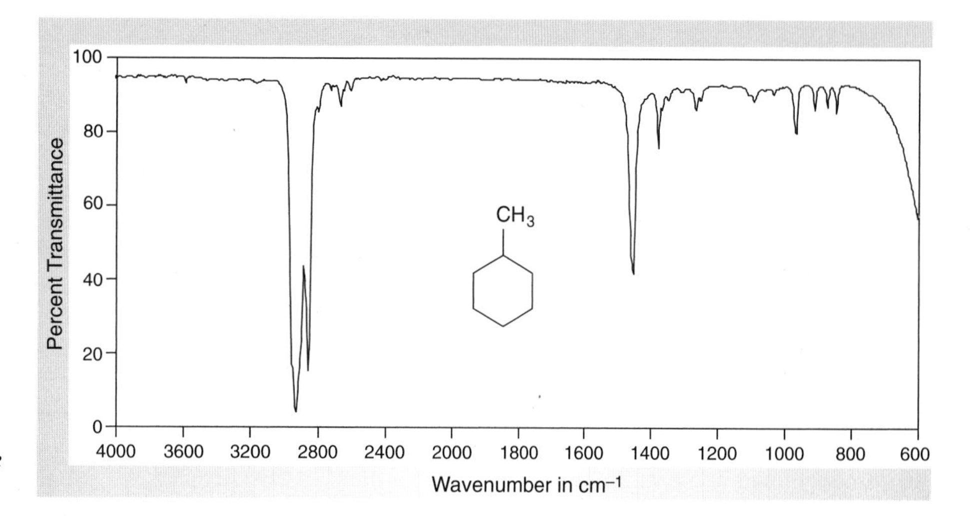

Figure 9.16
IR spectrum of methylcyclohexane (neat).

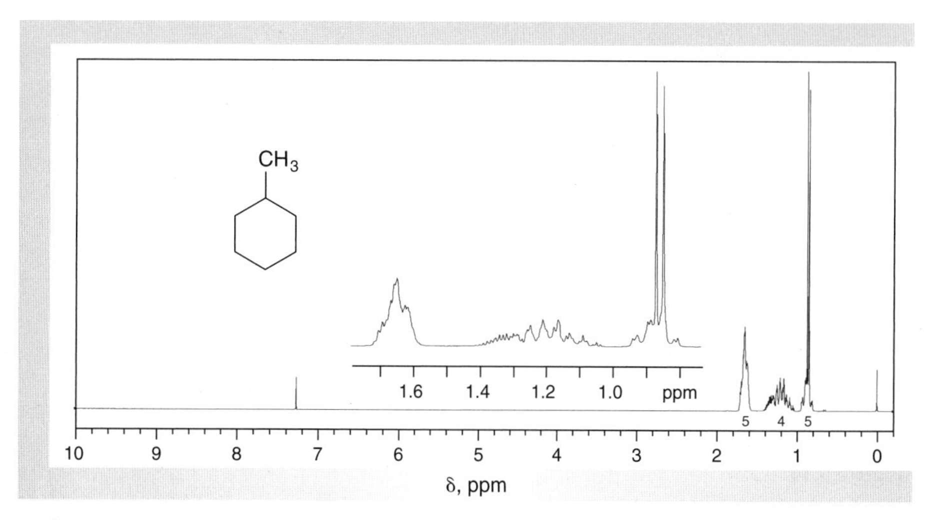

Figure 9.17
NMR data for methylcyclohexane ($CDCl_3$).

(a) ^{1}H NMR spectrum (300 MHz).
(b) ^{13}C NMR data: δ 22.9, 26.6, 26.7, 33.0, 35.6.

HISTORICAL HIGHLIGHT

Keeping It Cool

Humans have historically been interested in controlling the temperature of their living environments to enhance their comfort. Cave-dwellers used fires to maintain warmth, although we do not know whether they cooled their "residences" artificially. In more recent times, coal- and gas-fired fireplaces and furnaces provided heating, but cooling of living areas remained more difficult. Some early options for maintaining cooled environments involved constructing cellars for storing perishable foods and running water over walls of porous materials, which resulted in evaporative cooling. Evaporative cooling works well provided the relative humidity is low, and it is the basis for the operation of "swamp coolers," devices that are still in use today in many areas of the United States, such as the desert Southwest. In the northern United States, ice was harvested during the winter, stored, and used during the summer. Indeed, refrigerators were originally called iceboxes because a large block of ice was placed in an insulated unit to provide the desired cooling.

(Continued)

HISTORICAL HIGHLIGHT *Keeping It Cool (Continued)*

Mechanical refrigeration, which first became available in the mid-19th century, is based on the principle that converting a liquid to a gas requires heat and thereby leads to cooling of the surrounding environment. Thus, a low-boiling refrigerant is first condensed to the liquid state by a compressor and then passed into the coils of the cold section of the refrigerator, where the refrigerant vaporizes, resulting in cooling to a temperature that is determined by a thermostat. This same principle of liquefaction and vaporization is used in modern air conditioners.

In the early days of mechanical refrigeration, the commonly used coolants were ammonia, sulfur dioxide, and methyl chloride, all of which are toxic gases at room temperature and therefore dangerous. Indeed, people died as a result of leaks in the coolant systems of refrigerators in their residences. The hazards associated with these coolants prompted a search for safer alternatives that led to the discovery of the Freons, a trade name for chlorofluorocarbons (CFCs), in 1928. The inventor was Thomas Midgley, a mechanical engineer/chemist, who also discovered the value of tetraethyl lead as a gasoline additive for overcoming "knocking" in internal combustion engines. Although CFCs have the beneficial property of low toxicity, they, like tetraethyl lead, have been found to have negative environmental consequences.

Because of the chemical inertness of the CFCs, a number of applications were soon found for them, including their use as propellants in aerosol cans and inhalants, as refrigerants, and as "blowing agents" for producing polymeric foams such as Styrofoam. Modifying the chemical composition of a CFC provides a way to tailor the boiling point of the compound to its desired use. For example, the boiling points of dichlorodifluoromethane, CCL_2F_2, and trichlorofluoromethane, CCL_3F (historically the two most widely used CFCs), are −29.9 °C and +23.8 °C (760 torr), respectively.

So what's the problem with the CFCs? In laboratory tests, they were found to be chemically inert and nontoxic, so there was little concern that tons of these gases were being released into the environment. Ah, that's the rub, because these gases eventually reach the stratosphere (15–25 miles high), where they are anything but chemically benign. As reported in the mid-1970s, Rowland and Molina, of the University of California Irvine, and Crutzen, currently at the Max Planck Institute (in Mainz, Germany), showed that the CFCs absorb ultraviolet light to produce an electronically excited molecule, designated by the asterisk in Scheme 1. This species then decomposes to generate a chlorine atom, which subsequently reacts with ozone to yield the radical $^{\bullet}ClO_3$. Loss of O_2 provides a second radical, $ClO^{\bullet}$, that combines with another molecule of ozone, creating $^{\bullet}ClO_4$, yet another radical. Decomposition of this unstable species affords two molecules of oxygen and regenerates the chlorine atom. A free-radical chain reaction is thus underway, one that destroys two molecules of ozone for every cycle. Studies have indicated that generation of one chlorine atom, which may have a "lifetime" of one to two years in the upper atmosphere, can result in the destruction of 100,000 molecules of ozone! The three aforementioned chemists shared the 1995 Nobel Prize in Chemistry for their pioneering studies of the mechanism by which halogen atoms deplete the ozone layer.

Why is ozone an important component of the atmosphere? On the surface of our planet, it is a highly reactive chemical that is responsible for destruction of π-bonds, as in the degradation of the rubber in tires. In the atmosphere, however, it absorbs "UV-B" ultraviolet

$$F_2C(Cl)-Cl \xrightarrow{h\nu} [F_2C(Cl)-Cl]^* \longrightarrow F_2\dot{C}Cl + Cl^{\bullet} \xrightarrow{O_3} {}^{-}O-O^{+}(-O-Cl) \xrightarrow{-O_2} {}^{\bullet}O-Cl \xrightarrow{O_3} {}^{-}O-O^{+}(-O-O-Cl) \xrightarrow{-2\,O_2} Cl^{\bullet}$$

(Continued)

HISTORICAL HIGHLIGHT *Keeping It Cool (Continued)*

light, which corresponds to wavelengths of 280–320 nm that are harmful to both flora and fauna. For example, these wavelengths can be a cause of melanoma skin cancer in humans.

The discovery of a "hole" in the ozone layer over Antarctica in the 1980s led to an international treaty, formally termed the Montreal Protocol on Substances that Deplete the Ozone Layer, calling for a reduction of 50% in the use of CFCs by 2000 and a total ban on the manufacture of CFCs after 1995. The concentration of CFCs in the stratosphere has declined significantly since the treaty went into effect, but by some estimates the ozone layer will not be restored to normal for another century.

Alternatives to CFCs are available. Currently, hydrochlorofluorocarbons (HCFCs), which are more benign in the stratosphere than CFCs, are being used, but because they still contain chlorine, the HCFCs are considered only a transitional solution. By 2020, it is anticipated that hydrofluorocarbons (HFCs) such as 1,1,1,2-tetrafluoroethane (CF_3CH_2F), which are less susceptible to photochemical decomposition in the stratosphere, will have replaced HCFCs as propellants, refrigerants, and so on.

There are other halogen-containing compounds that are destructive to the ozone layer. These include methyl bromide and the halons, which are compounds composed of carbon, fluorine, bromine, and sometimes chlorine. Methyl bromide is a gas that has been widely used for fumigating soils and granaries, but it photochemically dissociates in the stratosphere to produce bromine atoms, which are even more efficient at destroying ozone than are chlorine atoms. This is because of a difference in the nature of the compounds formed by the chain-terminating step that leads to removal of the halogen radical from the stratosphere. In the case of a chlorine atom, the products of termination include HCl and $ClONO_2$, both of which are relatively resistant to photochemical dissociation so they diffuse into the troposphere, where they react with water vapor and descend to the earth in acidic raindrops. The corresponding compounds, HBr and $BrONO_2$, that are derived from bromine atoms are more susceptible to photochemical decomposition to regenerate bromine atoms, so removal of bromine atoms from the stratosphere by diffusion of the products of chain termination is slower, resulting in greater destruction of ozone. The use of methyl bromide is currently banned in some countries and is being phased out in others, such as the United States. According to the guidelines of the Montreal Protocol, the target date internationally for the total ban on using this substance is 2015.

Halons, which, like CFCS, are nonflammable, were primarily used as propellants in fire extinguishers. The fact that they contain bromine, however, makes these substances, like methyl bromide, highly destructive of the ozone layer. An amendment to the Montreal Protocol banned their use by 1994.

Relationship of Historical Highlight to Experiments

Free-radical chain halogenation is a valuable technique for functionalizing hydrocarbons, resulting in replacement of a C–H bond by a C–Hal bond in a process involving formation of chlorine and bromine atoms in a key step. Even though such reactions may be initiated photochemically, the wavelength of light used to effect the reaction does not promote cleavage of the C–Hal bond once it has been formed. This is not the case when shorter wavelengths of light are used, and the chlorine and bromine atoms that are produced initiate a free-radical chain reaction that results in destruction of ozone in the stratosphere, leading to the creation of the ozone hole.

See more on *CFCs*

See more on *Ozone Hole/Nobel Prize*

See more on *Ozone Hole*

See *Who was Thomas Midgley, Jr.?*

See more on *Ultraviolet Light*

See more on *Montreal Protocol*

See more on *Methyl Bromide*

CHAPTER 10

Alkenes

When you see this icon, sign in at this book's premium website at **www.cengage.com/login** to access videos, Pre-Lab Exercises, and other online resources.

As discussed in the preceding chapter, alkanes are saturated hydrocarbons. You encounter such compounds in your daily life, especially in the form of fuels to power automobiles, trains, planes, and many power plants that generate electricity. These fuels also contain compounds that are known as alkenes, which are unsaturated hydrocarbons. In this chapter, you will discover that alkenes are much more interesting organic compounds because they undergo a variety of different chemical reactions, not just the radical reactions that are typical of alkanes. Thus, although you may have considered the chemistry of alkanes "boring," you will find that the chemistry of alkenes is rich and characterized by a number of fascinating transformations of the carbon-carbon double bond, the functional group found in alkenes.

10.1 INTRODUCTION

A **functional group** is an atom or group of atoms that governs the chemical and physical properties of a family of compounds. The introduction and manipulation of these functional groups are major objectives in modern organic chemistry. In this chapter, we will explore the chemistry of **alkenes, 1,** which are organic compounds possessing a polarizable carbon-carbon double bond, a π-bond, as the functional group. Methods for introducing a carbon-carbon double bond into a molecule from alkyl halides and alcohols are presented first, and then some of the reactions characteristic of this functional group are examined.

Elimination reactions are among the most common ways to produce a carbon-carbon π-bond. For example, the elements of hydrogen halide, H–X, may be eliminated from an **alkyl halide, 2.** The functional group of an alkyl halide is a carbon-halogen, single bond, **C–X,** and the process by which the carbon-halogen bond and an adjacent carbon-hydrogen bond are converted into a carbon-carbon π-bond via **dehydrohalogenation** is an example of a **functional group transformation.** A carbon-carbon π-bond may also be formed by removing the elements of water from an **alcohol, 3,** in which a **C–OH** single bond is the functional group; this reaction is called **dehydration.** Although other aspects of the chemistry of alkyl halides and alcohols will be presented in Sections 14.1 to 14.5 and 16.2, a brief introduction to these families is essential to understanding how they may be used as starting materials for the synthesis of alkenes.

$$R^2R^1C{=}CHR^3 \qquad R^1{-}CH(R^2){-}\overset{\delta^+}{CH}(R^3){-}\overset{\delta^-}{X} \qquad R^1{-}CH(R^2){-}\overset{\delta^+}{CH}(R^3){-}\overset{\delta^-}{OH}$$

1 An alkene

2 X = Cl, Br, I An alkyl halide

3 An alcohol

10.2 DEHYDROHALOGENATION OF ALKYL HALIDES

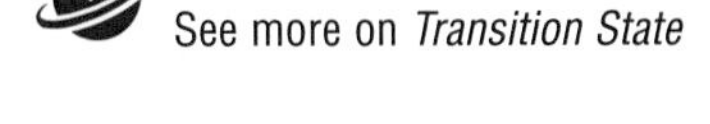
See more on *Transition State*

The electronegative halogen atom of an alkyl halide polarizes the carbon-halogen bond so the carbon atom bears a partial positive charge, δ^+, and the halogen atom a partial negative charge, δ^-. This polarization may be transmitted through the σ-bond network, a phenomenon referred to as an **inductive effect,** to enhance the acidity of hydrogen atoms on the β-carbon atom. Removing a proton from this atom, and simultaneous departure of the halide ion, which is a **leaving group,** from the α-carbon atom forms the carbon-carbon π-bond of the alkene in a **concerted reaction** (Eq. 10.1). An important characteristic of a good leaving group is that it should be a *weak* base; in contrast, the base B:⁻ should be *strong*. The **transition state 4** for the reaction is shown in Equation 10.2, in which the curved arrows symbolize the flow of electrons. As shown in **4,** the preferred **dihedral angle** between the β-hydrogen being removed and the leaving group X is 180°, an angular relationship called **anti-periplanar.** This transformation of alkyl halides to form alkenes is called **dehydrohalogenation.**

$$\underset{\text{An alkyl halide}}{R\overset{H}{\underset{\beta}{C}}H\underset{\alpha}{\overset{\delta^+}{C}H_2}{-}\overset{\delta^-}{X}} + B{:}^- \longrightarrow \underset{\text{An alkene}}{RCH{=}CH_2} + B{-}H + X{:}^- \qquad (10.1)$$

$$[\text{anti-periplanar alkyl halide} + B{:}^-] \longrightarrow [\text{transition state } \mathbf{4}]^{\ddagger} \longrightarrow RCH{=}CH_2 + B{-}H + X{:}^- \qquad (10.2)$$

4

Transition state

The concerted reaction depicted in Equation 10.2 is classified as an **E2** process, where **E** stands for **elimination** and **2** refers to the **molecularity** of the rate-determining step of the reaction. For E2 processes, the rate of the reaction depends upon the concentrations of the organic substrate and the base (Eq. 10.3), so *both* reactants are involved in the transition state of the rate-determining step. This **bimolecularity** is illustrated in **4** (Eq. 10.2).

$$\text{Rate} = k_2[\text{alkyl halide}][\text{B:}^-] \qquad (10.3)$$

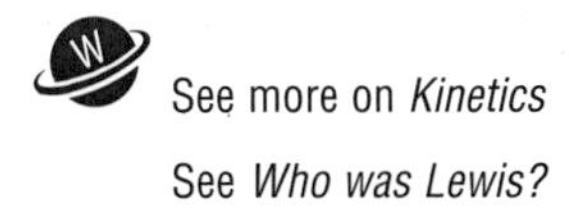
See more on *Kinetics*

See *Who was Lewis?*

Because it bears a partial positive charge, the α-carbon atom of an alkyl halide is **electrophilic** and thus also subject to attack by **nucleophiles,** which, as **Lewis bases,** are electron-rich and frequently anionic species. This process produces **substitution** rather than elimination products (Eq. 10.4), so a possible competition between the two types of reactions must be considered. Examining transition states **4** and **5** for the two reactions (Eqs. 10.2 and 10.4) suggests that steric factors may play a critical role in the competition. In support of this hypothesis, it is observed experimentally that elimination is favored as the degree of substitution on the α-carbon atom increases. The resulting steric hindrance inhibits direct attack of the base, or nucleophile, on the α-carbon atom, and removal of the β-proton and formation of elimination products becomes favored. Thus, when the α-carbon atom is *tertiary*, elimination to form alkenes is generally the only observable reaction.

$$B:^- + RCH_2(H)(H)C^{\delta+}-X^{\delta-} \longrightarrow \left[B^{\delta-}\cdots C(CH_2R)(H)(H)\cdots X^{\delta-}\right]^{\ddagger} \longrightarrow B-C(CH_2R)(H)(H) + X:^- \quad (10.4)$$

5
Transition state

Dehydrohalogenation may give a mixture of products if the halogen is unsymmetrically located on the carbon skeleton. For example, 2-bromo-2-methylbutane (**6**), the substrate you will use in this experiment, yields both 2-methyl-2-butene (**7**) and 2-methyl-1-butene (**8**) on reaction with strong base (Eq. 10.5). Because such elimination reactions are normally *irreversible* under these experimental conditions, the alkenes **7** and **8** do *not* undergo equilibration subsequent to their production. Consequently, the ratio of **7** and **8** obtained is defined by the relative rates of their formation. These rates, in turn, are determined by the relative free energies of the two transition states, **9** and **10**, respectively, rather than by the relative free energies of the alkenes **7** and **8** themselves.

$$CH_2(H)-C(CH_3)(Br)-CH(H)-CH_3 \xrightarrow{\text{base}} CH_2(H)-C(CH_3)=CH-CH_3 + CH_2=C(CH_3)-CH(H)-CH_3 \quad (10.5)$$

6 2-Bromo-2-methylbutane; **7** 2-Methyl-2-butene; **8** 2-Methyl-1-butene

$$\left[H_3C(H_3C)C_{\alpha}(Br^{\delta-})(CH_3)\cdots C_{\beta}(H)\cdots H\cdots B:^{\delta-}\right]^{\ddagger} \qquad \left[H_3C(H)(H)C_{\alpha}(Br^{\delta-})(CH_2CH_3)\cdots C_{\beta}\cdots H\cdots B:^{\delta-}\right]^{\ddagger}$$

9 **10**

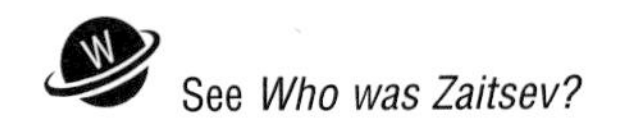
See *Who was Zaitsev?*

In the absence of complicating factors, the predominant product in an E2 elimination is the more highly substituted alkene, which is **7** in the present example; this observation is the source of Zaitsev's rule. The trend is observed because an increase in the number of alkyl substituents on the double bond almost always increases the stability—that is, lowers the free energy—of an alkene. Those factors that stabilize the product alkenes also play a role in stabilizing the respective transition states in which *partial* double-bond character is developing between the two carbon atoms. Thus, the enthalpy of activation, [$\Delta H^{\ddagger}$], for forming the more stable alkene **7** is less than that for the less stable alkene **8**.

The relative free energies of **transition states** of competing elimination reactions may be influenced by **steric factors** that increase the energies of some transition states relative to others. For instance, the less stable alkene may become the major product if steric factors raise the energy of the transition state leading to the more highly substituted alkene more than that for forming the less highly substituted. This may occur when sterically demanding substituents on the carbon atom from which the proton is being removed offer hindrance to the approaching base. Removal of a proton from a less substituted carbon to give a less substituted alkene then becomes more favorable. As more sterically demanding, bulky bases are used, abstraction of the more accessible proton will also be favored, thereby leading to the formation of even greater amounts of the less substituted alkene.

Transition states **9** and **10** serve as an illustrative example of these principles. In **9**, which is the transition state leading to the formation of **7**, there can be an unfavorable steric interaction between the methyl group on the β-carbon atom and the base. There is no comparable steric interaction in **10**, which produces the thermodynamically less stable **8**. Consequently, as the steric bulk of the base is increased, **8** will be formed in increased amounts.

In the two experiments that follow, you will explore these principles by performing base-promoted elimination reactions of 2-bromo-2-methylbutane (**6**). Using this tertiary alkyl bromide eliminates problems that might be associated with forming by-products from substitution reactions and allows a study of effects on product ratios when bases having different steric requirements are used. The dehydrohalogenation of **6** may be effected by using various base/solvent combinations. For example, some members of your class may use potassium hydroxide as the base and 1-propanol as the solvent for the dehydrohalogenation of **6**, whereas others may use potassium *tert*-butoxide as the base and *tert*-butyl alcohol as the solvent. By comparing the ratios of the alkenes produced by the two different methods, you can explore how varying the reagents and solvents can influence the course of a chemical reaction.

EXPERIMENTAL PROCEDURES

Base-Promoted Elimination of an Alkyl Halide

Purpose To demonstrate the formation of alkenes by dehydrohalogenation and assess product distributions using different bases.

SAFETY ALERT

1. **Most of the chemicals used in this experiment are highly flammable, so do not handle these liquids near open flames.**
2. **The solutions used in this experiment are *highly* caustic. *Do not allow them to come in contact with your skin. If this should happen, flood the affected area with water and then thoroughly rinse the area with a solution of dilute acetic acid. Wear latex gloves when preparing and transferring all solutions.***

A ■ Elimination with Alcoholic Potassium Hydroxide

MINISCALE PROCEDURE

Preparation Sign in at **www.cengage.com/login** to answer Pre-Lab Exercises, access videos, and read the MSDSs for the chemicals used or produced in this procedure. Review Sections 2.2, 2.4, 2.9, 2.10, 2.11, 2.14, 2.22, 2.27, and 6.4.

Apparatus A 10-mL and a 50-mL round-bottom flask, drying tube, apparatus for fractional distillation, magnetic stirring, and *flameless* heating.

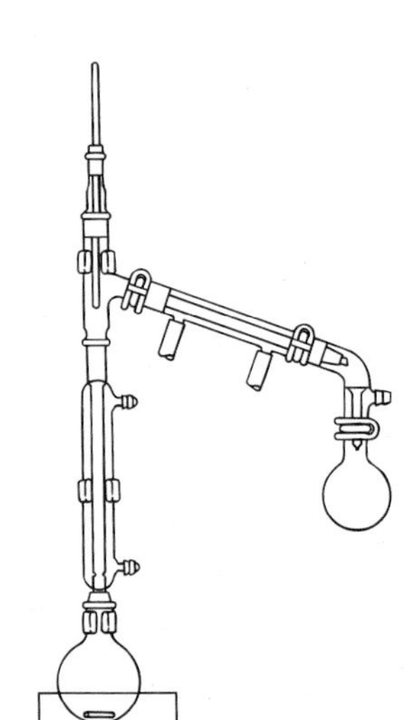

Setting Up Place 25 mL of a 4 *M* solution of potassium hydroxide in 1-propanol in the 50-mL flask containing a stirbar. Attach a drying tube to the flask. Add 2.5 mL of 2-bromo-2-methylbutane and then equip the flask for fractional distillation. Be sure to lubricate the ground-glass joint connecting the Hempel column to the flask with a thin layer of stopcock grease. Lubricating this joint is *particularly* important in this experiment because the strong base being used may cause the joint to freeze. To increase the cooling efficiency of the Hempel column, fill it with Raschig rings, coarsely broken glass tubing, coarse steel wool, or another packing material as directed by your instructor. Using a short piece of tubing, fit the vacuum adapter holding a 10-mL receiving flask with a drying tube and immerse the receiving flask in an ice-water bath. Attach water hoses to the *Hempel column* and circulate water through the jacket of the column during the period of reflux for this reaction.

Elimination and Isolation Heat the reaction mixture under gentle reflux with stirring for a period of 1–1.5 h.★ Allow the mixture to cool below its boiling point, and drain the water from the jacket of the Hempel column. Connect water hoses to the *condenser* so that the apparatus is now set for fractional distillation. Leave in the receiving flask any low-boiling distillate that has condensed and continue to cool this flask in an ice-water bath. Distill the product mixture, collecting all distillate boiling *below* 45 °C, and transfer the product to a tared sample bottle with a tight-fitting stopper or cap.

See *Who was Baeyer?*

Analysis Weigh the distillate and calculate the yield of products. Then put the container in an ice-water bath until all tests on the distillate have been completed. Test the distillate for unsaturation using the bromine and Baeyer tests (Secs. 25.8A and B, respectively). Analyze your distillate by GLC or submit a sample of it for analysis. After obtaining the results, calculate the relative percentages of the two isomeric alkenes formed; assume that the response factors are identical for the two alkenes. A typical GLC analysis of the products from this elimination is shown in Figure 10.1a. Obtain IR and NMR spectra of your starting material and product. Compare your spectra with those of authentic samples of 2-bromo-2-methylbutane (Figs. 10.2 and 10.3) and of the two possible alkenes (Figs. 10.4–10.7). The ^{1}H NMR spectrum of the product mixture obtained from a representative experiment is given in Figure 10.8.

MICROSCALE PROCEDURE

Preparation Sign in at **www.cengage.com/login** to answer Pre-Lab Exercises, access videos, and read the MSDSs for the chemicals used or produced in this procedure. Review Sections 2.3, 2.4, 2.9, 2.11, 2.13, 2.22, 2.27, and 6.4.

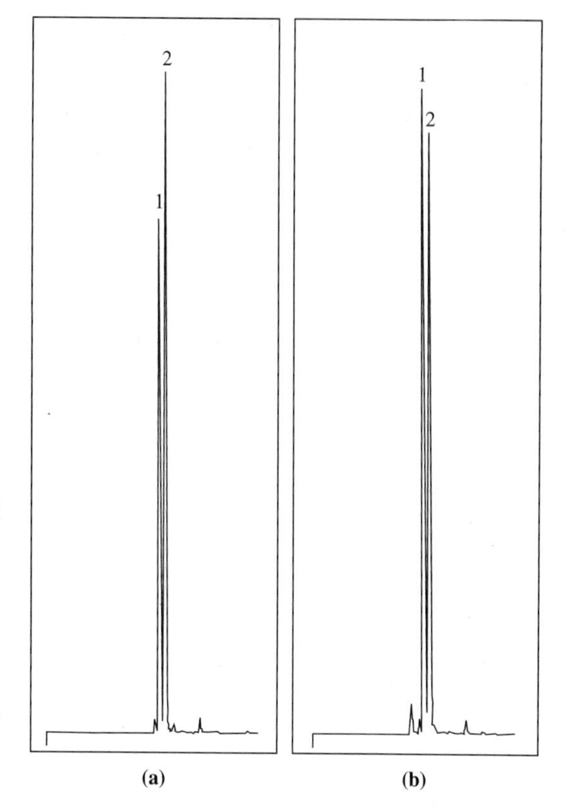

Figure 10.1
Typical GLC traces of the products of elimination of 2-bromo-2-methylbutane. Assignments and peak areas (in parentheses): peak 1: 2-methyl-1-butene; peak 2: 2-methyl-2-butene. (a) Elimination with KOH; peak 1 (82210), peak 2 (120706). (b) Elimination with $KOC(CH_3)_3$; peak 1 (94260), peak 2 (85140); column and conditions: 50-mm × 30-m, 0.25-μ film of SE-54; 80 °C, 40 mL/min.

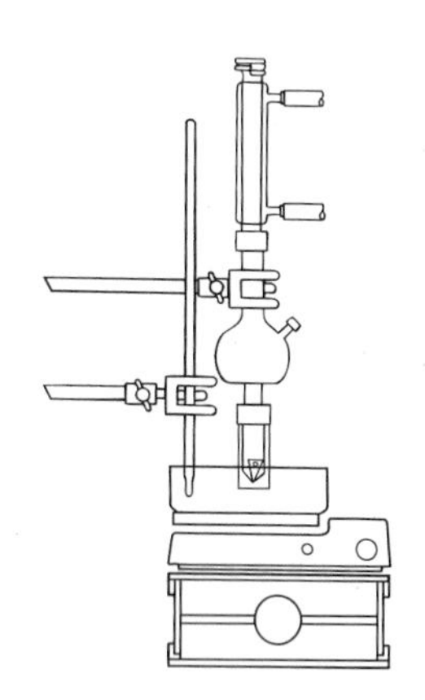

Apparatus A 5-mL conical vial, apparatus for simple distillation, magnetic stirring, and *flameless* heating.

Setting Up Place a spinvane in the conical vial. Using a calibrated pipet, add 2.5 mL of a 4 *M* solution of potassium hydroxide in 1-propanol. Avoid getting any of this solution on the ground-glass joint of the vial. With a clean syringe or calibrated pipet, measure 1 mL of 2-bromo-2-methylbutane into the vial. Thoroughly mix the contents by stirring for a few seconds. Equip the vial for microscale distillation and circulate cold water through the condenser. Be sure to lubricate the ground-glass joint connecting the Hickman distillation head to the vial with a thin layer of stop-cock grease. Lubricating this joint is *particularly* important because otherwise the strong base being used may cause the joint to freeze.

Elimination and Isolation Heat the reaction vial with a bath that has been preheated to about 75 °C (bath temperature). The vial should be positioned so that the levels of the top of the bath and of the liquid in the reaction vial are equal. Stir the mixture at 75 °C for 1 h and then slowly raise the temperature of the bath to about 90 °C over a period of about 20 min, collecting all of the distillate until about 2.5 mL of liquid remains in the vial. It should not be necessary to heat the bath higher than 90–95 °C. Using a Pasteur or a filter-tip pipet, transfer the distillate to a dry sample vial or a test tube with a tight-fitting cap or stopper.

See *Who was Baeyer?*

Analysis Weigh the distillate and calculate the yield of products. Then put the container in an ice-water bath until all tests on the distillate have been completed. Test the distillate for unsaturation using the bromine and Baeyer tests (Secs. 25.8A and B, respectively). Analyze your distillate by GLC or submit a sample of it for analysis. After obtaining the results, calculate the relative percentages of the two isomeric alkenes formed; assume that the response factors are identical for the two alkenes. A typical GLC trace of the products from this elimination is shown in Figure 10.1a. Obtain IR and NMR spectra of your starting material and product. Compare your spectra with those of authentic samples of 2-bromo-2-methylbutane (Figs. 10.2 and 10.3) and of the two possible alkenes (Figs. 10.4–10.7). The ^1H NMR spectrum of the product mixture obtained from a representative experiment is given in Figure 10.8.

B ▪ *Elimination with Potassium* Tert-*Butoxide*

MINISCALE PROCEDURE

Preparation Sign in at **www.cengage.com/login** to answer Pre-Lab Exercises, access videos, and read the MSDSs for the chemicals used or produced in this procedure. Review Sections 2.2, 2.4, 2.9, 2.10, 2.11, 2.14, 2.22, 2.27, and 6.4.

Apparatus A *dry* 50-mL round-bottom flask, drying tube, apparatus for fractional distillation, magnetic stirring, and *flameless* heating.

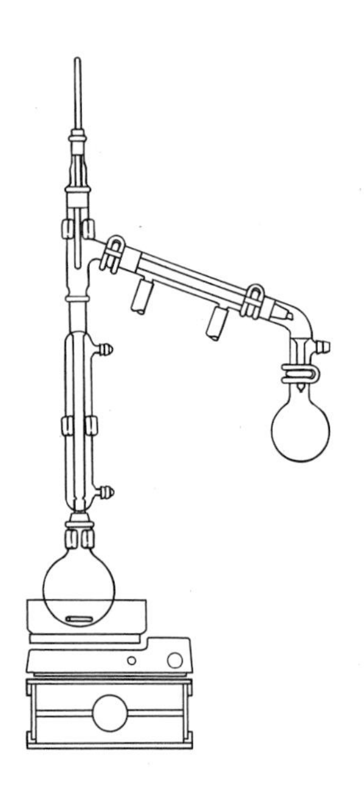

Setting Up Equip the flask with a stirbar. Working quickly, add 25 mL of a 1 *M* solution of potassium *tert*-butoxide in *tert*-butyl alcohol. This reagent is *very* moisture-sensitive, so exposure to the atmosphere *must* be minimized. Then add 2.5 mL of 2-bromo-2-methylbutane to the solution and equip the flask for fractional distillation. Complete set-up according to the directions in the Miniscale Procedure of Part **A.**

Elimination and Isolation Follow the procedure given in the Miniscale Procedure of Part **A** for completing the reaction.

Analysis Weigh the distillate and calculate the yield of products. Then put the container in an ice-water bath until all tests on the distillate have been completed. Test the distillate for unsaturation using the bromine and Baeyer tests (Secs. 25.8A and B, respectively). Analyze your distillate by GLC or submit a sample of it for analysis. After obtaining the results, calculate the relative percentages of the two isomeric alkenes formed; assume that the response factors are identical for the two alkenes. A typical GLC trace of the products from this elimination is shown in Figure 10.1b. Obtain IR and NMR spectra of your starting material and product. Compare your spectra with those of authentic samples of 2-bromo-2-methylbutane (Figs. 10.2 and 10.3) and of the two possible alkenes (Figs. 10.4–10.7). The ^1H NMR spectrum of the product mixture obtained from a representative experiment is given in Figure 10.9.

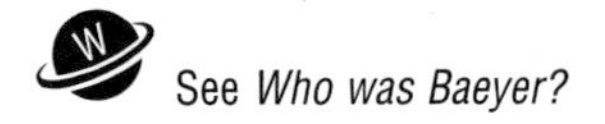

See *Who was Baeyer?*

Discovery Experiment *Elimination of Alternate Non-Terminal Alkyl Halides*

Obtain a sample of another non-terminal bromoalkane such as 2-bromoheptane from your instructor. Team up with another student, with one of you performing a base-promoted dehydrobromination using alcoholic potassium hydroxide according to the procedure in Part **A** and the other using potassium *tert*-butoxide according to the procedure in Part **B.** Calculate your yield and analyze the products by GLC to identify and to determine the ratio of the three possible alkenes. Compare the ratio of products obtained using the two different bases and explain any differences. Be sure to consult with your instructor before undertaking any experimental procedures.

WRAPPING IT UP

Neutralize the *residue remaining in the stillpot* with 10% aqueous hydrochloric acid; then flush it down the drain. Place the *dichloromethane solution* from the bromine test for unsaturation in a container for halogenated organic liquids; put the *manganese dioxide* from the Baeyer test for unsaturation in a container for heavy metals.

EXERCISES

1. Create a reaction profile (potential energy diagram) analogous to that in Figure 13.1, making it consistent with the information provided regarding the relative amounts of 7 and 8 (Eq. 10.5) and the relative energies of 9 and 10.
2. Demonstrate that the amount of 4 *M* potassium hydroxide solution used for the elimination provides an excess of base relative to the alkyl bromide.
3. How might the ratio of alkenes obtained in the *tert*-butoxide-promoted elimination change if the *tert*-butyl alcohol contains water?
4. Why does the excess of base used in these eliminations favor the E2 over the E1 mechanism for elimination?
5. What is the solid material that precipitates as the eliminations proceed?
6. If all the elimination reactions in the experimental section had proceeded by the E1 mechanism, would the results have been different from those actually obtained? Why?
7. From the results of the experiments that were performed and/or from the GLC data in Figure 10.1, what conclusions can be drawn concerning the effect of relative base size upon product distribution?
8. In the miniscale procedures, what purpose is served by packing the Hempel column with Raschig rings or other packing materials?
9. Draw the Newman projections for the three staggered conformations about the C2–C3 bond of 2-bromoheptane and indicate which of these possesses an anti-periplanar relationship between the β-hydrogen atom and the bromine atom.
10. If the steric demands of the leaving group in the 2-methyl-2-butyl system were greater than those of a methyl group, why would 2-methyl-1-butene be expected to be formed in greater amounts than if the leaving group were smaller

than methyl, regardless of which base is used? Use Newman projections to support your explanation.

11. It has been shown that there is an important stereoelectronic requirement for an anti-periplanar relationship between the leaving group and the β-hydrogen during an E2 elimination. Predict the structure of the main product of the dehydrobromination of *trans*-2-bromomethylcyclohexane using alcoholic potassium hydroxide. (*Hint:* The elimination may not occur via the lowest energy chair conformer.)

12. What is the main side reaction that competes with elimination when a primary alkyl halide is treated with alcoholic potassium hydroxide, and why does this reaction compete with elimination of a primary alkyl halide but not a tertiary alkyl halide?

Sign in at **www.cengage.com/login** and use the spectra viewer and Tables 8.1–8.8 as needed to answer the blue-numbered questions on spectroscopy.

13. Consider the NMR data for 2-bromo-2-methylbutane (Fig. 10.3).

a. In the ^1H NMR spectrum, assign the various resonances to the hydrogen nuclei responsible for them.

b. For the ^{13}C NMR data, assign the various resonances to the carbon nuclei responsible for them.

14. Consider the spectral data for 2-methyl-1-butene (Figs. 10.4 and 10.5).

a. In the functional group region of the IR spectrum, specify the absorptions associated with the carbon-carbon double bond and the hydrogens attached to the C1 carbon atom.

b. In the ^1H NMR spectrum, assign the various resonances to the hydrogen nuclei responsible for them.

c. For the ^{13}C NMR data, assign the various resonances to the carbon nuclei responsible for them.

15. Consider the spectral data for 2-methyl-2-butene (Figs. 10.6 and 10.7).

a. In the functional group region of the IR spectrum, specify the absorptions associated with the carbon-carbon double bond and the hydrogen attached to the C3 carbon atom.

b. In the ^1H NMR spectrum, assign the various resonances to the hydrogen nuclei responsible for them.

c. For the ^{13}C NMR data, assign the various resonances to the carbon nuclei responsible for them.

16. Discuss the differences observed in the IR and NMR spectra of 2-bromo-2-methylbutane and of 2-methyl-1-butene and 2-methyl-2-butene that are consistent with dehydrohalogenation occurring in this experiment.

17. Referring to the spectra on the website for Figures 10.5 and 10.7–10.9, calculate the percentage compositions of the mixtures of isomeric methylbutenes obtained from the reaction of 2-bromo-2-methylbutane with potassium hydroxide and potassium *tert*-butoxide. In the spectra of the mixtures, the integration of the resonances in the region of δ 5.0 ppm has been electronically amplified in the upper stepped line so that the relative areas of the two multiplets in that region can be more accurately measured.

SPECTRA

Starting Materials and Products

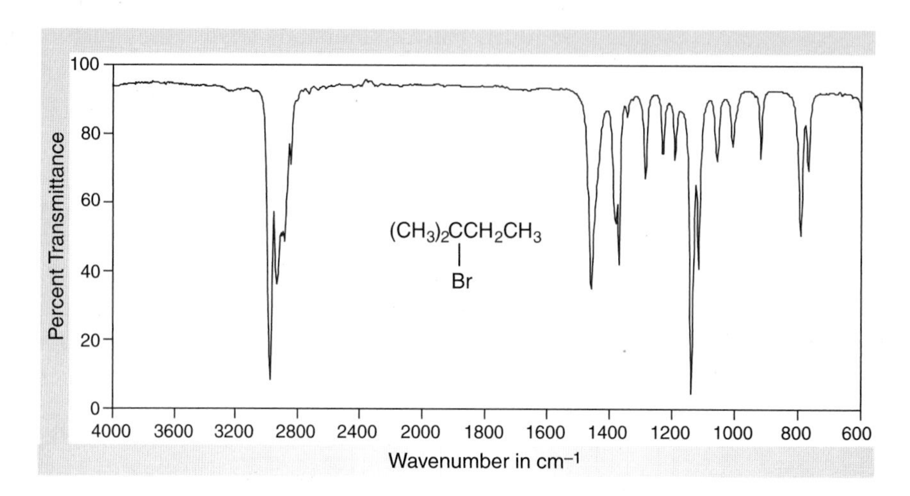

Figure 10.2
IR spectrum of 2-bromo-2-methylbutane (neat).

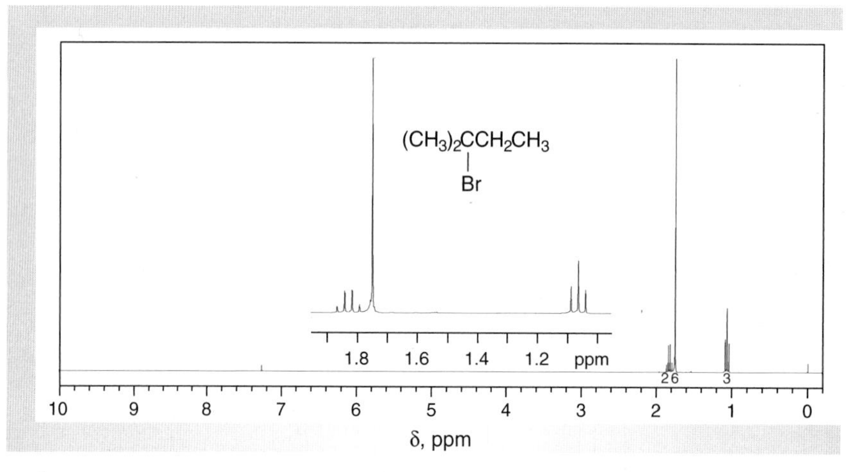

Figure 10.3
NMR data for 2-bromo-2-methylbutane ($CDCl_3$).

(a) ^1H NMR spectrum (300 MHz).
(b) ^{13}C data: δ 10.7, 33.8, 40.3, 69.2.

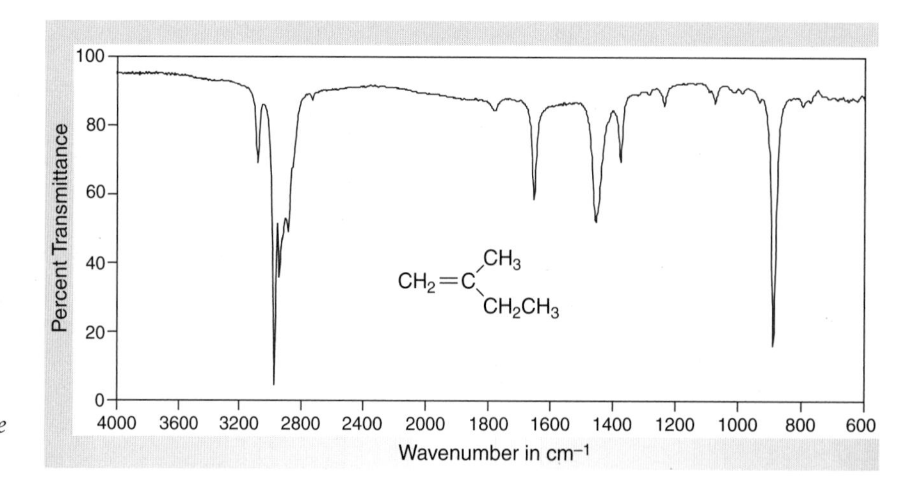

Figure 10.4
IR spectrum of 2-methyl-1-butene (neat).

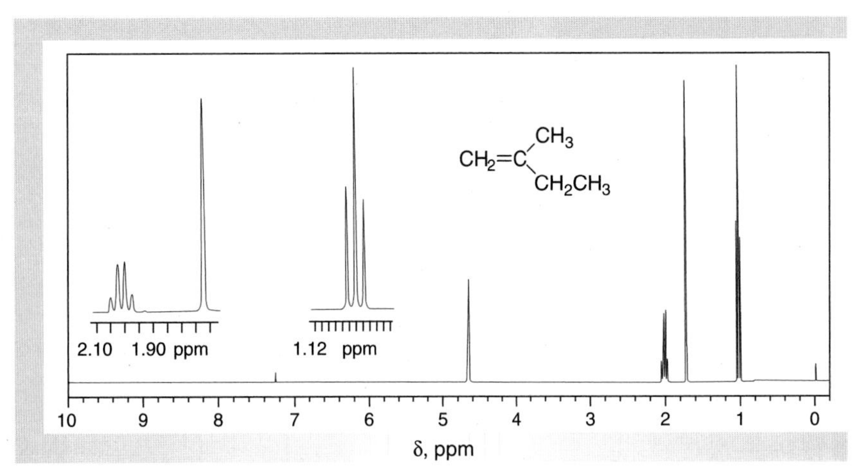

Figure 10.5
NMR data for 2-methyl-1-butene ($CDCl_3$).

(a) ^{1}H NMR spectrum (300 MHz).
(b) ^{13}C data: δ 12.5, 22.3, 31.0, 108.8, 147.5.

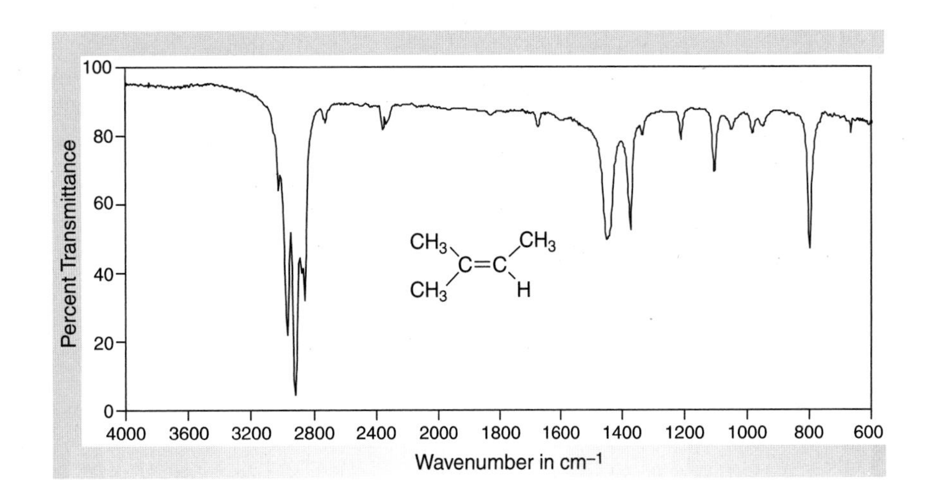

Figure 10.6
IR spectrum of 2-methyl-2-butene (neat).

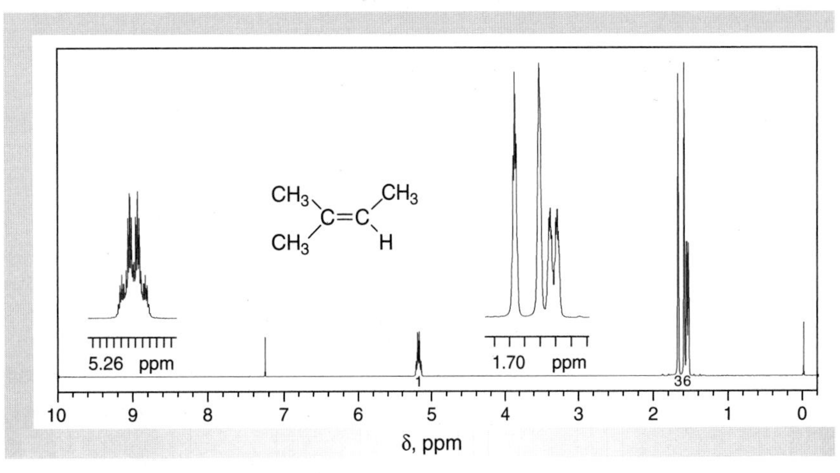

Figure 10.7
NMR data for 2-methyl-2-butene ($CDCl_3$).

(a) ^{1}H NMR spectrum (300 MHz).
(b) ^{13}C data: δ 13.4, 17.3, 25.6, 118.8, 132.0.

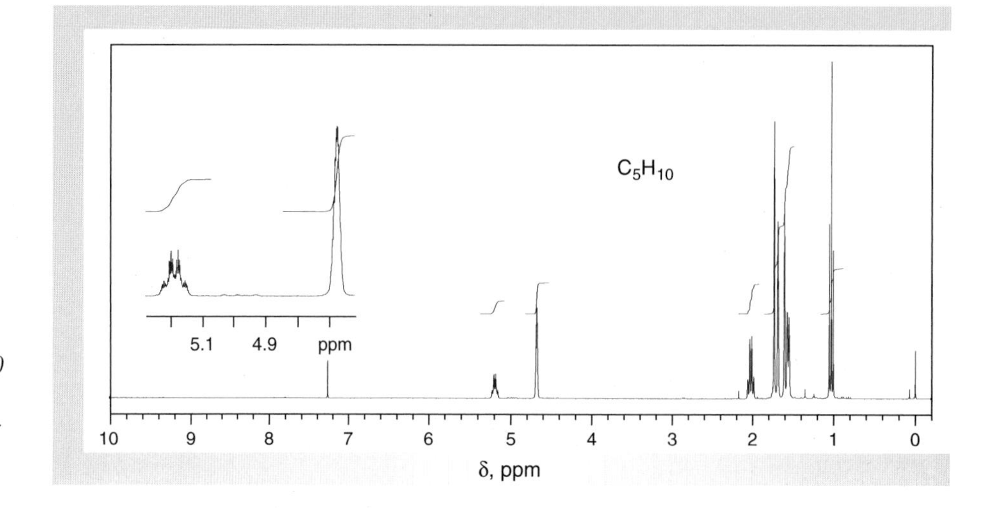

Figure 10.8
^{1}H NMR spectrum ($CDCl_3$, 300 MHz) of the product mixture from the elimination of 2-bromo-2-methylbutane with potassium hydroxide.

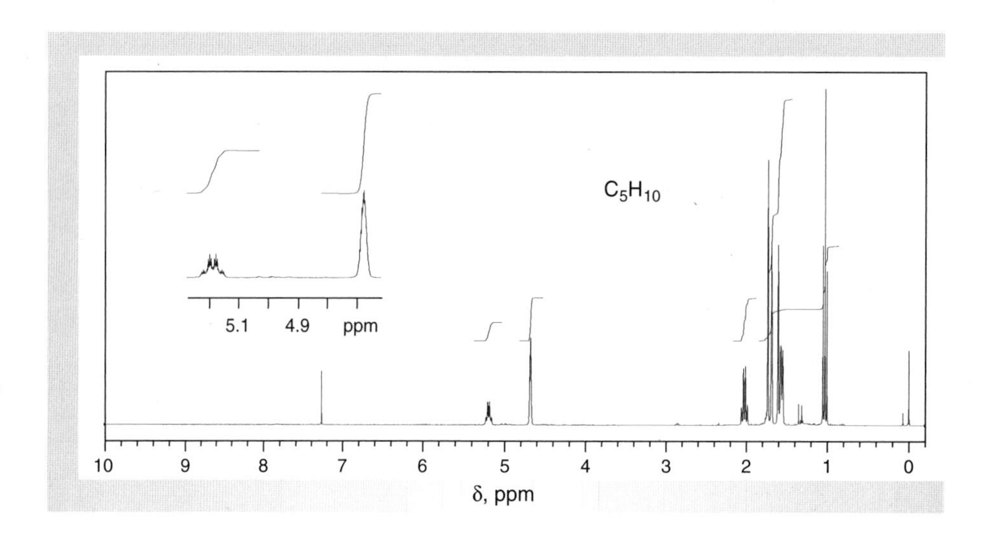

Figure 10.9
^{1}H NMR spectrum ($CDCl_3$, 300 MHz) of the product mixture from the elimination of 2-bromo-2-methylbutane with potassium tert-*butoxide.*

10.3 DEHYDRATION OF ALCOHOLS

See *Who was Brønsted?*

See *Who was Lewis?*

See *Who was Lowry?*

The nonbonding lone pairs of electrons on the oxygen atom of the hydroxyl group, which is the functional group of alcohols, serve as **Lewis bases** that may complex with **Brønsted-Lowry** and **Lewis acids,** which are defined as proton-donating and electron pair-accepting acids, respectively. For example, the protonation of an alcohol, **11,** is rapid and reversible and produces an **oxonium ion, 12,** in which the positively charged oxygen atom further polarizes the carbon-oxygen single bond. In contrast to the hydroxide ion, a strong base and a poor leaving group, water is a weak base and an excellent leaving group. Hence, **12** may undergo facile elimination and substitution reactions upon reaction with a base or **nucleophile** $Nu:^-$ (Eq. 10.6).

$$\mathbf{11} \text{ (An alcohol)} + H\text{—}\overset{+}{O}H_2 \rightleftharpoons \mathbf{12} \text{ (An oxonium ion)} \xrightarrow[\text{(rds)}]{Nu:^-} \mathbf{13} + H_2O \text{ (Elimination)}; \quad \mathbf{12} \xrightarrow[\text{(rds)}]{Nu:^-} \mathbf{14} + H_2O \text{ (Substitution)} \qquad (10.6)$$

The loss of water that effects **dehydration** of the oxonium ion **12** to produce an alkene, **13,** may occur by mechanistic pathways like those shown in Equations 10.7 and 10.8. With *secondary* and *tertiary* alcohols, **12** undergoes **endothermic** heterolysis via the transition state **12**‡ to produce the **carbocation 15** and water. Rapid transfer of a proton from an adjacent carbon atom to a weak base such as water then gives **13** (Eq. 10.7). The fragmentation of **12** to give **15** is kinetically a **first-order** process and is the **rate-determining step** (rds) of the overall reaction. Consequently, the reaction is classified as **E1,** which stands for *elimination unimolecular*. Because the formation of the carbocation **15** is an endothermic process, the transition state **12**‡ is productlike and is characterized by extensive heterolysis of the carbon-oxygen bond with simultaneous development of partial positive charge on the carbon atom. Thus, the ease of dehydration of alcohols by the E1 mechanism is in the order $3° > 2° \gg 1°$, paralleling the relative stabilities of carbocations. In the case of *primary* alcohols, ionization of the oxonium ion **12** would produce a highly unstable 1° carbocation (**12,** $R^3 = R^4 = H$), so the alternate E2 (*elimination bimolecular*) mechanism of Equation 10.8 applies.

See more on *Kinetics*

$$\mathbf{12} \xrightleftharpoons{\text{rds}} [\mathbf{12}^‡]^‡ \text{ (Transition state)} \rightleftharpoons \mathbf{15} + H_2O \rightleftharpoons \mathbf{13} + R\text{—}\overset{+}{O}H_2 \qquad (10.7)$$

$$\mathbf{12} \xrightleftharpoons{\text{rds}} R^1R^2C{=}CH_2 + H_2O + R\text{—}\overset{+}{O}H_2 \qquad (10.8)$$

As illustrated in Equations 10.6–10.8, each of the steps along the reaction pathway is reversible, so an alkene may undergo acid-catalyzed hydration to form an alcohol. In practice, reversal of the dehydration may be avoided by removing the alkene, whose boiling point is always *lower* than the parent alcohol, from the reaction

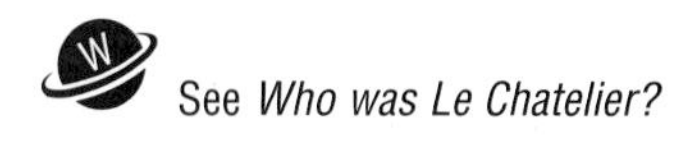

See *Who was Le Chatelier?*

mixture by distillation. This technique shifts the equilibrium to the right and maximizes the yield of alkene. You may recall that shifting an equilibrium of a reversible reaction to the right by removing one of the products as it is formed follows from the **Le Chatelier principle.**

According to Equation 10.6, elimination and substitution reactions can be competing pathways (see also Sec. 14.3), but the relative importance of each may often be controlled through proper choice of reaction conditions and reagents. In the case of the acid-catalyzed dehydration of secondary and tertiary alcohols, these competing reactions involve combining the intermediate carbocation with the **conjugate base,** A^-, of the acid catalyst A–H or with some other nucleophile such as solvent. For example, if hydrochloric acid were chosen as the catalyst for dehydration, the chloride ion could trap the carbocation to give an alkyl chloride (Eq. 10.9). This side reaction thus not only produces an undesired by-product but also consumes the acid catalyst. Consequently, the desired dehydration is terminated because formation of the alkyl chloride is irreversible under the conditions normally used for dehydration.

$$R_3C{-}OH \xrightarrow[-H_2O]{HCl} R_3C^+ \;(+\,Cl{:}^-) \longrightarrow R_3C{-}Cl \qquad (10.9)$$

The side reaction depicted in Equation 10.9 is avoided when the formation of the substitution product is reversible. For this reason, sulfuric acid is often used to effect the dehydration of alcohols in the undergraduate laboratory because the intermediate alkyl bisulfate ($Nu{:}^- = HSO_3O^-$) substitution product readily reionizes to the intermediate carbocation under the reaction conditions (Eq. 10.10). Subsequent loss of a proton then gives the desired alkene.

$$R^1{-}C(R^2)(H){-}CH(R^3){-}OH \underset{-H_2O}{\overset{H_3O^+}{\rightleftharpoons}} R^1{-}C(R^2)(H){-}\overset{+}{C}H(R^3) \overset{-H^+}{\rightleftharpoons} (R^2)(R^1)C{=}CHR^3 \qquad (10.10)$$

Elimination product

$$R^1{-}C(R^2)(H){-}\overset{+}{C}H(R^3) \overset{Nu{:}^-}{\rightleftharpoons} R^1{-}C(R^2)(H){-}CH(R^3)(Nu)$$

Substitution product

Two or more isomeric alkenes may be formed in E1 dehydration reactions. For example, consider the unsymmetrically substituted carbocation **17,** formed from alcohol **16** by protonation and subsequent loss of water (Eq. 10.11). Loss of H_A produces **18,** whereas loss of H_B yields **19.** The relative amounts of the two alkenes produced depends upon the relative free energies of the transition states for the loss of the H_A and H_B. In these **transition states,** the C–H bond is partially broken, and the C–C double bond is partially formed, as depicted in **20** and **21.** Since the major difference between these two transition states is the substitution on the incipient carbon-carbon double bond, their relative energies parallel the relative stabilities of the product alkenes. Consequently the lower-energy transition state is the one leading to the more highly substituted and more stable alkene, which is **19** in this

See more on *Transition State*

See more on *Steric Hindrance*

See *Who was Zaitsev?*

generalized case. This is the mechanistic basis for Zaitsev's rule. Furthermore, as a general rule, the (*E*)-isomer of a pair of diastereomers (geometric isomers) predominates over the (*Z*)-isomer for steric reasons.

$$R^1\text{—}\underset{H_A}{CH}\text{—}\overset{OH}{CH}\text{—}\overset{R^2}{\underset{H_B}{C}}\text{—}R^3 \underset{}{\overset{-H_2O}{\rightleftharpoons}} R^1\text{—}\underset{H_A}{CH}\text{—}\overset{+}{CH}\text{—}\overset{R^2}{\underset{H_B}{C}}\text{—}R^3 \quad \begin{cases} \xrightarrow{-H_A^+} R^1\text{—}CH{=}CH\text{—}\overset{R^2}{\underset{H_B}{C}}\text{—}R^3 \;\; \mathbf{18} \\ \xrightarrow{-H_B^+} R^1\text{—}\underset{H_A}{CH}\text{—}CH{=}\overset{R^2}{C}\text{—}R^3 \;\; \mathbf{19} \end{cases} \tag{10.11}$$

16 **17**

$$\left[R^1\text{—}\underset{\delta^+ H_A}{CH}{=\!=}\overset{\delta^+}{CH}\text{—}\overset{R^2}{\underset{H_B}{C}}\text{—}R^3\right]^{\ddagger} \quad \left[R^1\text{—}\underset{H_A}{CH}\text{—}\overset{\delta^+}{CH}{=\!=}\overset{R^2}{\underset{\delta^+ H_B}{C}}\text{—}R^3\right]^{\ddagger}$$

20 **21**

The distribution of alkenes from the acid-catalyzed dehydration of alcohols is not always predictable from the orientational factors just described. Rather, the intermediate carbocation may rearrange through the migration of either a **hydride,** H:$^-$, or an **alkyl group,** R:$^-$, from a carbon atom adjacent to the cationic center. The rearrangement is more likely if the new carbocation is more stable than the original ion. The subsequent loss of a proton from the rearranged carbocation according to the orientation effects already discussed then leads to the additional alkenes.

$$(CH_3)_2CHCH_2\overset{OH}{CH}\text{—}CH_3 \xrightarrow[\Delta]{H_3O^+} (CH_3)_2CHCH_2CH{=}CH_2 + \underset{H}{\overset{(CH_3)_2CH}{}}C{=}C\underset{CH_3}{\overset{H}{}}$$

22 4-Methyl-2-pentanol; **23** 4-Methyl-1-pentene; **24** *trans*-4-Methyl-2-pentene

$$\underset{H}{\overset{(CH_3)_2CH}{}}C{=}C\underset{H}{\overset{CH_3}{}} + \underset{H_3C}{\overset{H_3C}{}}C{=}C\underset{H}{\overset{CH_2CH_3}{}} + \underset{H}{\overset{H}{}}C{=}C\underset{CH_2CH_2CH_3}{\overset{CH_3}{}} \tag{10.12}$$

25 *cis*-4-Methyl-2-pentene; **26** 2-Methyl-2-pentene; **27** 2-Methyl-1-pentene

The acid-catalyzed dehydration of 4-methyl-2-pentanol (**22**), which is one of the experiments in this section, is shown in Equation 10.12 and nicely illustrates the preceding principles. Dehydration of **22** produces a complex mixture of isomeric alkenes, including 4-methyl-1-pentene (**23**), *trans*-4-methyl-2-pentene (**24**), *cis*-4-methyl-2-pentene (**25**), 2-methyl-2-pentene (**26**), and 2-methyl-1-pentene (**27**).

A pathway to **23–25** involves deprotonating the intermediate carbocation **28.** However, forming **26** and **27** requires intervention of carbocations **29** and **30** (see Exercise 13).

$$(CH_3)_2CH{-}CH_2{-}\overset{+}{C}HCH_3 \quad \textbf{28} \qquad (CH_3)_2CH{-}\overset{+}{C}H{-}CH_2CH_3 \quad \textbf{29} \qquad (CH_3)_2\overset{+}{C}{-}CH_2{-}CH_2CH_3 \quad \textbf{30}$$

In contrast to **22,** cyclohexanol (**31**) undergoes acid-catalyzed dehydration *without* rearrangement to yield a single product (Eq. 10.13). This reaction may be accompanied by acid-catalyzed polymerization of the desired cyclohexene, however.

$$\text{Cyclohexanol } (\textbf{31}) \xrightarrow[\Delta]{H^+} \text{Cyclohexene} + H_2O \qquad (10.13)$$

EXPERIMENTAL PROCEDURES

Dehydration of Alcohols

Purpose To determine the product distribution in the acid-catalyzed dehydration of alcohols to alkenes.

SAFETY ALERT

1. **The majority of materials, particularly the product alkenes, that will be handled during this experiment are highly flammable, so use *flameless* heating and be certain that there are no open flames in your vicinity.**
2. **Several experimental operations require pouring, transferring, and weighing chemicals and reagents that cause burns on contact with your skin. Wear latex gloves when handling the strongly acidic catalysts and during the work-up and washing steps in which a separatory funnel is used. Should acidic solutions accidentally come in contact with your skin, immediately flood the affected area with water, then wash it with 5% sodium bicarbonate solution.**

A ■ *Dehydration of 4-Methyl-2-Pentanol*

MINISCALE PROCEDURE

Preparation Sign in at **www.cengage.com/login** to answer Pre-Lab Exercises, access videos, and read the MSDSs for the chemicals used or produced in this procedure. Review Sections 2.4, 2.9, 2.10, 2.11, 2.13, 2.14, 2.2 2, 2.25, 2.27, and 6.4.

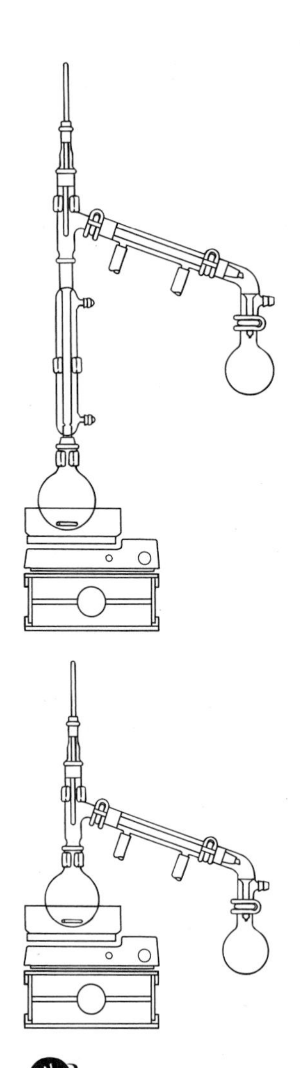

Apparatus A 25-mL and two 10-mL round-bottom flasks, drying tube, ice-water bath, apparatus for simple and fractional distillation, magnetic stirring, and *flameless* heating.

Setting Up Place a stirbar and 4.0 mL of 4-methyl-2-pentanol in the 25-mL round-bottom flask. Add 2.5 mL of 9 *M* sulfuric acid and thoroughly mix the liquids by gently swirling the flask. Equip the flask for fractional distillation and use an ice-water bath to cool the receiving flask.

Elimination Heat the reaction mixture and collect all distillates while maintaining the head temperature below 90 °C. If the reaction mixture is not heated too strongly, the head temperature will stay below 80 °C for most of the reaction. When about 2.5 mL of liquid remains in the reaction flask, discontinue heating. Transfer the distillate to an Erlenmeyer flask and add several spatula-tips full of anhydrous potassium carbonate to neutralize any acid and to dry the distillate.★ Occasionally swirl the mixture during a period of 10–15 min to hasten drying; add additional portions of anhydrous potassium carbonate if the liquid remains cloudy.

Isolation Transfer all of the dried organic mixture into a dry 10-mL round-bottom flask by decantation or by using a Pasteur pipet. Add a stirbar and equip the flask for simple distillation (Fig. 2.37a). Use a tared 10-mL flask as the receiver. Protect the receiver from atmospheric moisture by equipping the vacuum adapter with a drying tube and cool the receiver in an ice-water bath.

Isolate the alkenes by simple distillation. The expected products of the reaction are 4-methyl-1-pentene, *cis*-4-methyl-2-pentene, *trans*-4-methyl-2-pentene, 2-methyl-1-pentene, and 2-methyl-2-pentene. Collect the fraction boiling between 53 and 69 °C (760 torr) in a pre-weighed, dry 10-mL receiving flask.

See *Who was Baeyer?*

Analysis Weigh the distillate and determine the yield of products. Test the distillate for unsaturation using the bromine and Baeyer tests (Secs. 25.8A and B, respectively). Analyze your distillate by GLC or submit a sample of it for analysis. After obtaining the results, calculate the relative percentages of the isomeric alkenes formed; assume that the response factors are identical for the isomers. A typical GLC trace of the alkenes from this elimination is shown in Figure 10.10. Obtain IR and NMR spectra of your starting material and compare your spectra with those of an authentic sample (Figs. 10.11 and 10.12).

B ■ *Dehydration of Cyclohexanol*

MINISCALE PROCEDURE

Preparation Sign in at **www.cengage.com/login** to answer Pre-Lab Exercises, access videos, and read the MSDSs for the chemicals used or produced in this procedure. Review Sections 2.9, 2.10, 2.11, 2.13, 2.14, 2.22, 2.27, and 6.4.

Apparatus A 25-mL and two 10-mL round-bottom flasks, drying tube, ice-water bath, apparatus for simple and fractional distillation, magnetic stirring, and *flameless* heating.

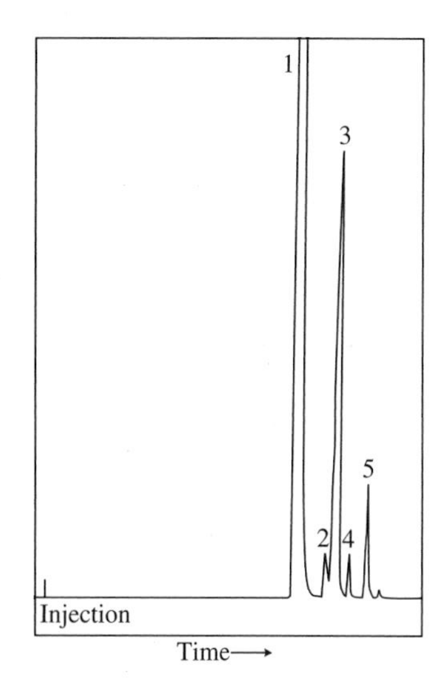

Figure 10.10
GLC trace of the product mixture from the dehydration of 4-methyl-2-pentanol. Assignments and peak areas (in parentheses): peak 1: diethyl ether (solvent for analysis); peak 2: 4-methyl-1-pentene (3030); peak 3: cis- *and* trans-*4-methyl-2-pentene (51693); peak 4: 2-methyl-1-pentene (2282); peak 5: 2-methyl-2-pentene (9733). Column and conditions: 0.25-mm × 30-m APT-Hep-Tex; 37 °C, 35 mL/min.*

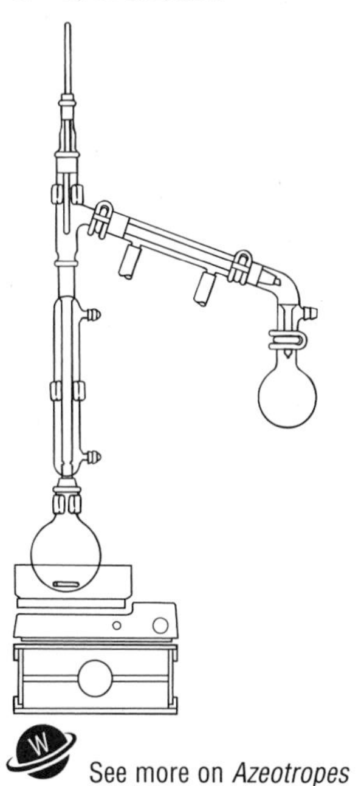

Setting Up Place a stirbar and 5.0 mL of cyclohexanol in the 25-mL round-bottom flask. Add 2.5 mL of 9 *M* sulfuric acid and thoroughly mix the liquids by gently swirling the flask. Equip the flask for fractional distillation and use an ice-water bath to cool the receiving flask.

Elimination and Isolation Follow the same protocol as provided in Part **A**, with the *single exception* that the product should be collected over the range of 80–85 °C (760 torr). Note that water and cyclohexene form a **minimum-boiling azeotrope,** which is a mixture of the two compounds that distills *below* the boiling point of either pure compound (see Figure 4.4). This makes it vital that the crude product be dried *thoroughly* prior to the simple distillation.

See more on *Azeotropes*

Analysis Weigh the distillate and, assuming that it is pure cyclohexene, determine the yield of product. Test the distillate for unsaturation using the bromine and Baeyer tests (Secs. 25.8A and B, respectively). Analyze your distillate by GLC or submit a sample of it for analysis. Obtain IR and ^{1}H NMR spectra of your starting material and product and compare them with those of authentic samples (Figs. 10.23–10.26).

MICROSCALE PROCEDURE

Preparation Sign in at **www.cengage.com/login** to answer Pre-Lab Exercises, access videos, and read the MSDSs for the chemicals used or produced in this procedure. Review Sections 2.4, 2.9, 2.13, 2.22, 2.25, and 6.4.

Apparatus A 3-mL conical vial, apparatus for magnetic stirring, simple distillation, and *flameless* heating.

Setting Up Place a spinvane in the conical vial. Using a calibrated Pasteur pipet, add 1 mL of cyclohexanol and 0.5 mL of 9 *M* sulfuric acid to the vial. Thoroughly mix the contents by briefly stirring or swirling the liquid for a few seconds. Equip the vial for microscale distillation and circulate cold water through the condenser.

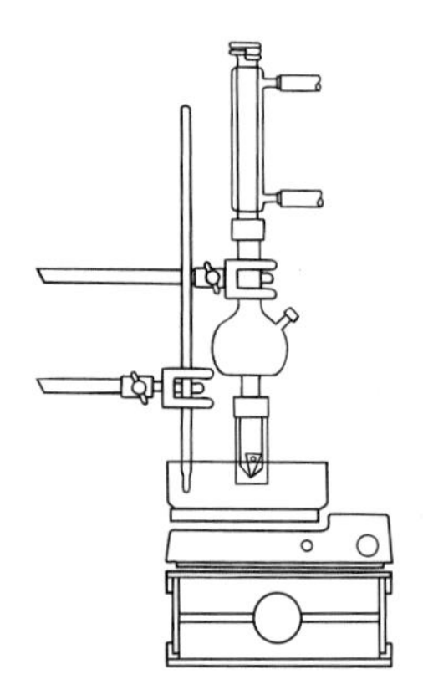
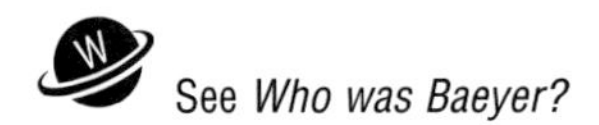

Elimination and Isolation Heat the reaction vial in a bath that has been preheated to about 120 °C (bath temperature). The vial should be positioned so that the levels of the top of the bath and of the liquid in the reaction vial are equal. Slowly raise the temperature of the bath as necessary to maintain distillation of the product and continue distilling until about 0.5–0.7 mL of liquid remains in the vial. It should not be necessary to heat the bath higher than about 130–135 °C. Dismantle the apparatus, and remove the distillate from the Hickman stillhead using a Pasteur or filter-tip pipet. Transfer the distillate to a dry sample vial or test tube, and add several microspatula-tips full of anhydrous potassium carbonate to neutralize any acid and to dry the distillate. Occasionally swirl the mixture for 5–10 min to hasten drying and add additional portions of anhydrous potassium carbonate if the liquid remains cloudy.

See *Who was Baeyer?*

Analysis Test the distillate for unsaturation using the bromine and Baeyer tests (Secs. 25.8A and B, respectively). Analyze your distillate by GLC or submit a sample of it for analysis. Obtain IR and ^{1}H NMR spectra of your starting material and product and compare them with those of authentic samples (Figs. 10.23–10.26).

Discovery Experiment

Elimination of Stereoisomeric Alcohols

Obtain a sample of another alcohol such as 2-methylcyclohexanol, which is typically available as a mixture of *cis*- and *trans*-isomers, from your instructor. Perform the acid-catalyzed dehydration of 2-methylcyclohexanol according to the procedure in Part **B.** Calculate your yield and analyze the products by capillary GLC to identify and to determine the ratio of the three possible alkenes. Consult with your instructor before undertaking any experiments.

WRAPPING IT UP

Dilute the *residue remaining in the stillpot* with water, carefully neutralize it with sodium carbonate, and flush it down the drain with large quantities of water. Dry the *potassium carbonate* on a tray in the hood and flush it down the drain or place it in a container for nonhazardous solids. Pour the *dichloromethane solution* from the bromine test for unsaturation in a container for halogenated organic liquids, and put the *manganese dioxide* from the Baeyer test for unsaturation in a container for heavy metals.

EXERCISES

General Questions

1. Create a reaction profile (potential energy diagram) analogous to that in Figure 13.1, making it consistent with the information provided regarding the relative amounts of **18** and **19** (Eq. 10.11) and the relative energies of **20** and **21**.
2. Define the Le Chatelier principle.
3. Why is the boiling point of the parent alcohol higher than that of the product alkene?
4. Why are the distillates obtained in the initial step of the dehydration reaction dried over anhydrous potassium carbonate?
5. Why is the head temperature kept below 90 °C in the initial step of the dehydration reaction?
6. Why is it necessary to separate a dried organic solution from the drying agent prior to distilling the solution?

7. In principle, the equilibrium in the dehydration of an alcohol could be shifted to the right by removal of water. Why is this tactic not a good option for the dehydration of 4-methyl-2-pentanol and cyclohexanol?
8. The loss of a proton from suitably substituted carbocations can provide both the *trans*- and *cis*-isomers of the resulting alkene, but the *trans*-isomer normally predominates. For example, deprotonation at C3 of the 2-pentyl carbocation produces mainly *trans*-pentene. By analyzing the relative energies of the conformational isomers of the carbocation that lead to the two isomeric 2-pentenes, explain why the *trans*-isomer is formed preferentially. This analysis is aided by the use of Newman projections based on the partial structure shown.

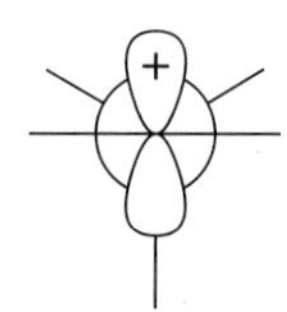

9. Why is it proposed that the alcohol functional group is protonated by acid before dehydration can occur via either an E1 or an E2 mechanism?
10. List two of the major differences in the dehydration of an alcohol by an E1 and E2 reaction mechanism.
11. What is the driving force for the rearrangement of intermediate carbocations during the dehydration of alcohols via the E1 mechanism?
12. Give a detailed mechanism for the dehydration reaction shown. Use curved arrows to symbolize the flow of electrons.

Me, OH, Me $\xrightarrow[-H_2O]{H_2SO_4\text{(cat.)}}$ Me, Me

Questions for Dehydration of 4-Methyl-2-pentanol

13. Give a detailed mechanism explaining the formation of **23–25** from the dehydration of 4-methyl-2-pentanol (**22**). Use curved arrows to symbolize the flow of electrons.
14. A mechanism for formation of **26** and **27** can be developed from rearrangement of the carbocation **28** by way of hydride shifts.
    - **a.** Provide a mechanism by which **26** may be formed from **28** through a hydride shift followed by deprotonation. Use curved arrows to symbolize the flow of electrons.
    - **b.** Show how your mechanism in Part **a** can be elaborated to provide for formation of **27.**
    - **c.** An alternative pathway for formation of the alkene **26** involves addition of a proton to **24** or **25** and deprotonation of the resulting carbocation. Similarly, **27** could be produced by the addition of a proton to **26** followed by deprotonation of the intermediate carbocation. Write out mechanisms for these transformations using curved arrows to symbolize the flow of electrons.
    - **d.** How might you experimentally distinguish the two types of mechanisms for the formation of **26** and **27**?

15. The dehydration of 4-methyl-2-pentanol (**22**) gives a mixture of the five isomeric alkenes **23–27**.

   a. Which of the alkenes **23–27** would you expect to be the major product? Explain the basis for your answer.

   b. Provide a rationale for why only small amounts of 4-methyl-1-pentene (**23**) and 2-methyl-1-pentene (**27**) are produced in this reaction.

16. Near the end of the dehydration of 4-methyl-2-pentanol, a white solid may precipitate from the reaction mixture. What is the solid likely to be?

17. Give the structure, including stereochemistry, of the product of adding bromine (refer to Sec. 10.6) to *trans*-4-methyl-2-pentene (**24**) and to *cis*-4-methyl-2-pentene (**25**). Relative to one another, are these products identical, enantiomeric, or diastereomeric (Sec. 7.1)? Give your reasoning.

Sign in at **www.cengage.com/login** and use the spectra viewer and Tables 8.1–8.8 as needed to answer the blue-numbered questions on spectroscopy.

18. Consider the spectral data for 4-methyl-2-pentanol (Figs. 10.11 and 10.12).

   a. In the functional group region of the IR spectrum, specify any absorptions associated with the alcohol function.

   b. In the ^{1}H NMR spectrum, assign the various resonances to the hydrogen nuclei responsible for them.

   c. For the ^{13}C NMR data, assign the various resonances to the carbon nuclei responsible for them.

19. Consider the spectral data for 4-methyl-1-pentene (Figs. 8.25, 10.13, and 10.14).

   a. In the IR spectrum, specify the absorptions associated with the carbon-carbon double bond and the vinylic hydrogen atoms.

   b. In the ^{1}H NMR spectrum, assign the various resonances to the hydrogen nuclei responsible for them.

   c. For the ^{13}C NMR data, assign the various resonances to the carbon nuclei responsible for them.

20. Consider the spectral data for *trans*-4-methyl-2-pentene (Figs. 10.15 and 10.16).

   a. In the IR spectrum, specify the absorptions, if any, associated with the carbon-carbon double bond and with the *trans* relationship of the vinylic hydrogen atoms.

   b. In the ^{1}H NMR spectrum, assign the various resonances to the hydrogen nuclei responsible for them.

   c. For the ^{13}C NMR data, assign the various resonances to the carbon nuclei responsible for them.

21. Consider the spectral data for *cis*-4-methyl-2-pentene (Figs. 10.17 and 10.18).

   a. In the IR spectrum, specify the absorptions, if any, associated with the carbon-carbon double bond and with the *cis* relationship of the vinylic hydrogen atoms.

   b. In the ^{1}H NMR spectrum, assign the various resonances to the hydrogen nuclei responsible for them.

   c. For the ^{13}C NMR data, assign the various resonances to the carbon nuclei responsible for them.

22. Consider the spectral data for 2-methyl-1-pentene (Figs. 10.19 and 10.20).

- **a.** In the IR spectrum, specify the absorptions associated with the carbon-carbon double bond and with the hydrogens attached to the C1 carbon atom.
- **b.** In the ^{1}H NMR spectrum, assign the various resonances to the hydrogen nuclei responsible for them.
- **c.** For the ^{13}C NMR data, assign the various resonances to the carbon nuclei responsible for them.

23. Consider the spectral data for 2-methyl-2-pentene (Figs. 10.21 and 10.22).

- **a.** In the IR spectrum, specify the absorptions associated with the carbon-carbon double bond and with the vinylic hydrogen atom.
- **b.** In the ^{1}H NMR spectrum, assign the various resonances to the hydrogen nuclei responsible for them.
- **c.** For the ^{13}C NMR data, assign the various resonances to the carbon nuclei responsible for them.

24. Discuss the differences observed in the IR and NMR spectra of 4-methyl-2-pentanol and the various methylpentenes that are consistent with dehydration occurring in this experiment.

Questions for Dehydration of Cyclohexanol

25. Define the term *minimum-boiling azeotrope* and explain why it is impossible to separate an azeotropic mixture completely by distillation.

26. Why is it particularly important that the crude cyclohexene be dry prior to its distillation?

27. Give the structure, including stereochemistry, of the product of addition of bromine (refer to Sec. 10.6) to cyclohexene. Should it be possible, at least in principle, to resolve this dibromide into separate enantiomers? Explain your answer.

28. Provide a detailed mechanism for the acid-catalyzed dehydration of cyclohexanol. Use curved arrows to symbolize the flow of electrons.

29. Cyclohexyl carbocation undergoes attack by nucleophiles. With this in mind, propose a mechanism whereby cyclohexene might form the dimer **31** in the presence of H_3O^+. Note that this type of reaction leads to polymerization of cyclohexene.

31

Sign in at **www.cengage.com/login** and use the spectra viewer and Tables 8.1–8.8 as needed to answer the blue-numbered questions on spectroscopy.

30. Consider the spectral data for cyclohexanol (Figs. 10.23 and 10.24).

- **a.** In the functional group region of the IR spectrum, specify any absorptions associated with the alcohol function.
- **b.** In the ^{1}H NMR spectrum, assign the various resonances to the hydrogen nuclei responsible for them.
- **c.** For the ^{13}C NMR data, assign the various resonances to the carbon nuclei responsible for them.

31. Consider the spectral data for cyclohexene (Figs. 10.25 and 10.26).

a. In the IR spectrum, specify the absorptions associated with the carbon-carbon double bond and with the vinylic hydrogen atoms.

b. In the ^{1}H NMR spectrum, assign the various resonances to the hydrogen nuclei responsible for them.

c. For the ^{13}C NMR data, assign the various resonances to the carbon nuclei responsible for them.

32. Discuss the differences observed in the IR and NMR spectra of cyclohexanol and cyclohexene that are consistent with dehydration occurring in this experiment.

SPECTRA

Starting Materials and Products

The ^{1}H NMR spectrum of 4-methyl-1-pentene is provided in Figure 8.25.

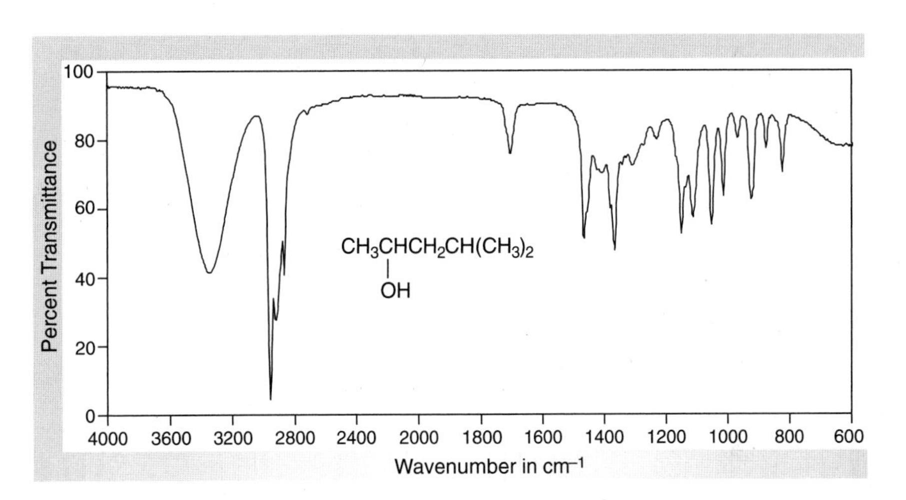

Figure 10.11
IR spectrum of 4-methyl-2-pentanol (neat).

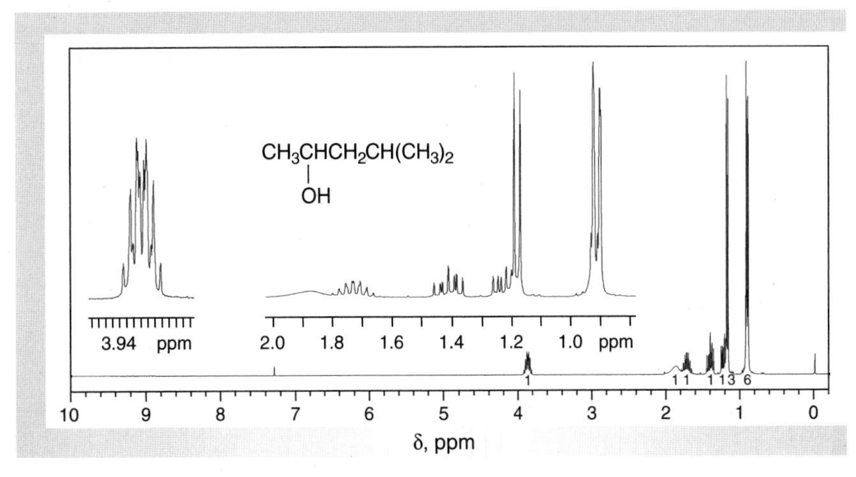

Figure 10.12
NMR data for 4-methyl-2-pentanol ($CDCl_3$).

(a) ^{1}H NMR spectrum (300 MHz).
(b) ^{13}C data: δ 22.5, 23.2, 23.9, 24.9, 48.8, 65.8.

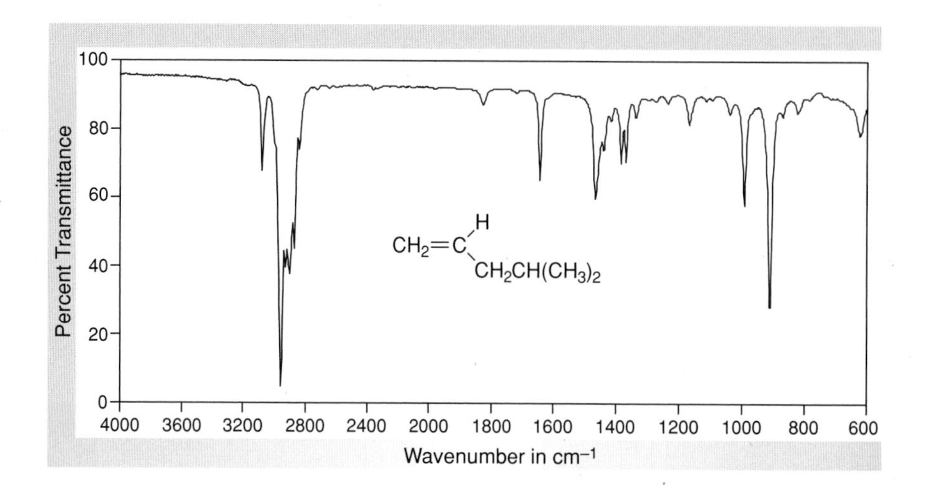

Figure 10.13
IR spectrum of 4-methyl-1-pentene (neat).

Figure 10.14
^{13}C NMR data for 4-methyl-1-pentene ($CDCl_3$).
Chemical shifts: δ 22.3, 28.1, 43.7, 115.5, 137.8.

$(CH_3)_2CHCH_2CH{=}CH_2$

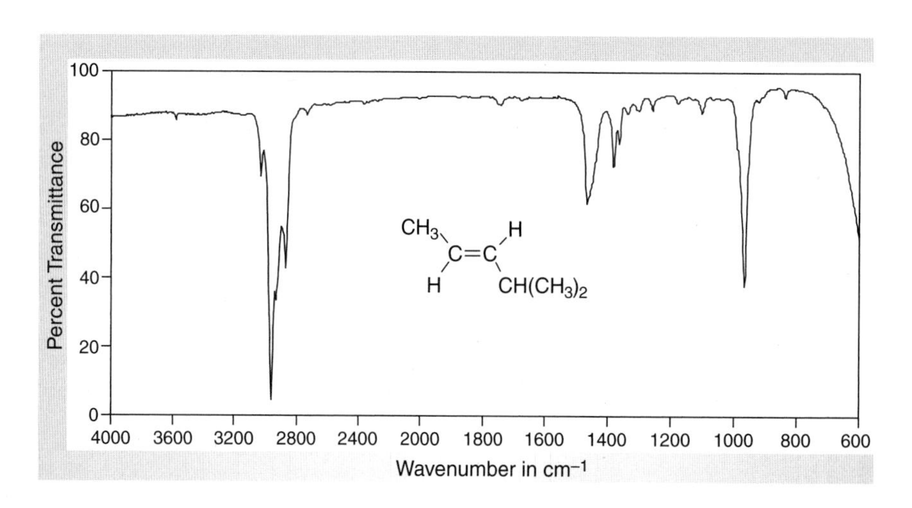

Figure 10.15
IR spectrum of trans-*4-methyl-2-pentene (neat).*

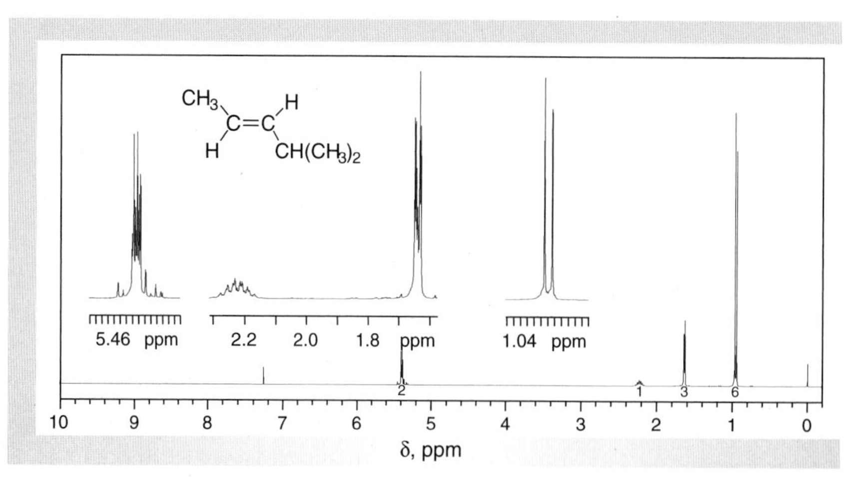

Figure 10.16
NMR data for trans-*4-methyl-2-pentene ($CDCl_3$).*

(a) ^1H NMR spectrum (300 MHz).
(b) ^{13}C data: δ 17.6, 22.7, 31.5, 121.6, 139.4.

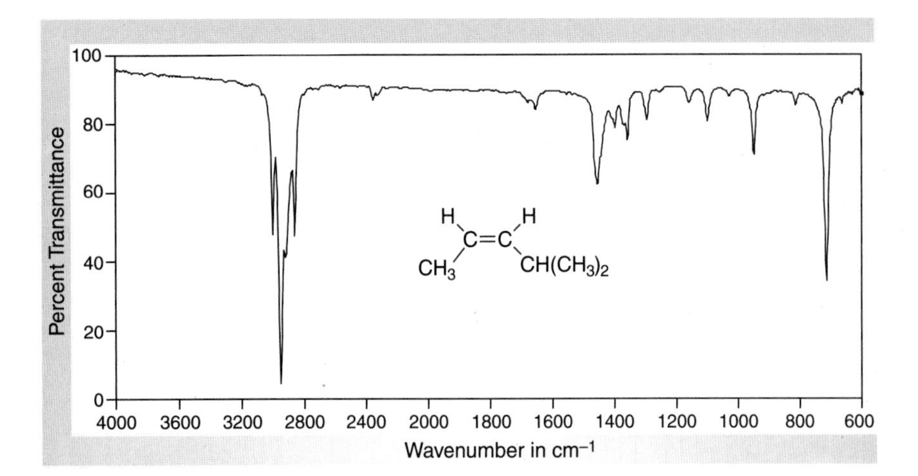

Figure 10.17
IR spectrum of cis-*4-methyl-2-pentene (neat).*

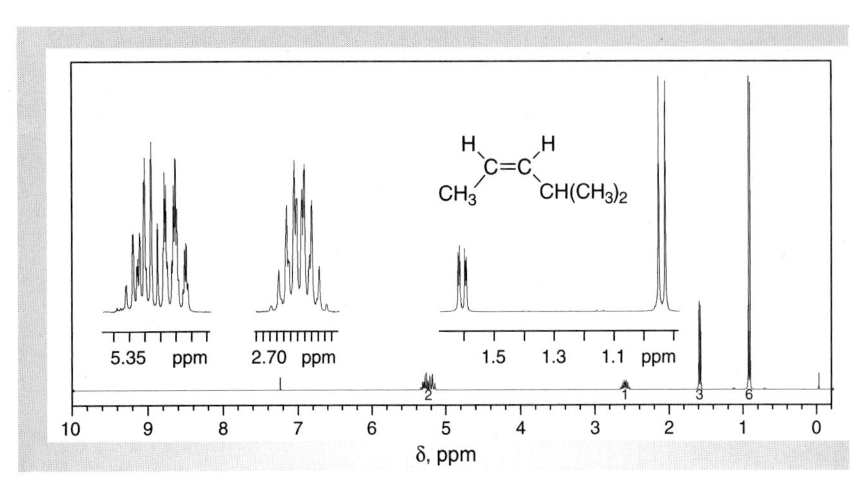

Figure 10.18
NMR data for cis-*4-methyl-2-pentene ($CDCl_3$).*

(a) ^{1}H NMR spectrum (300 MHz).
(b) ^{13}C data: δ 12.7, 23.1, 26.4, 121.4, 138.6.

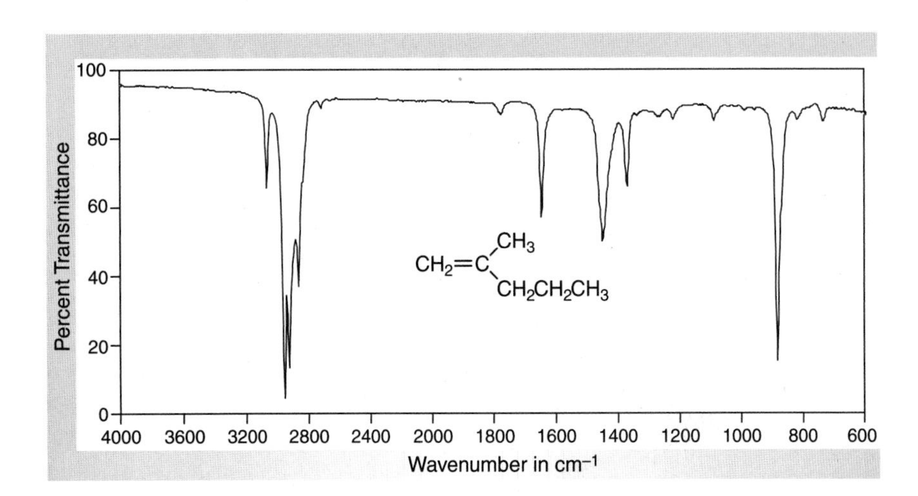

Figure 10.19
IR spectrum of 2-methyl-1-pentene (neat).

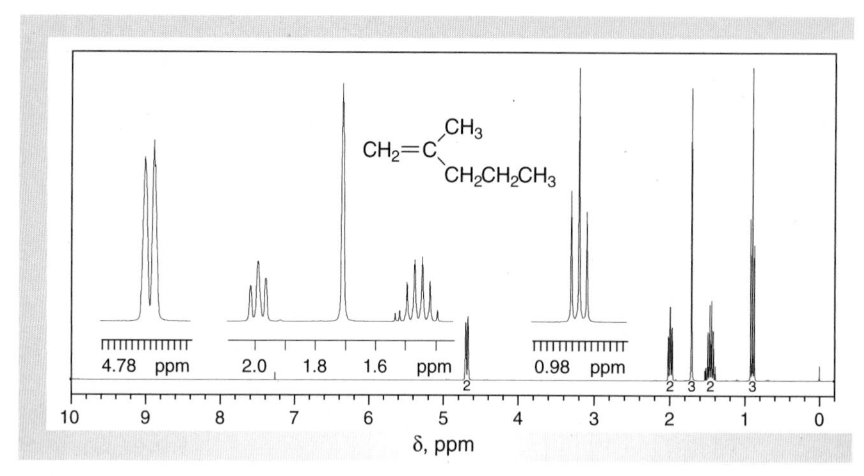

Figure 10.20
NMR data for 2-methyl-1-pentene ($CDCl_3$).

(a) ^1H NMR spectrum (300 MHz).
(b) ^{13}C data: δ 13.8, 20.8, 22.3, 40.0, 109.7, 146.0.

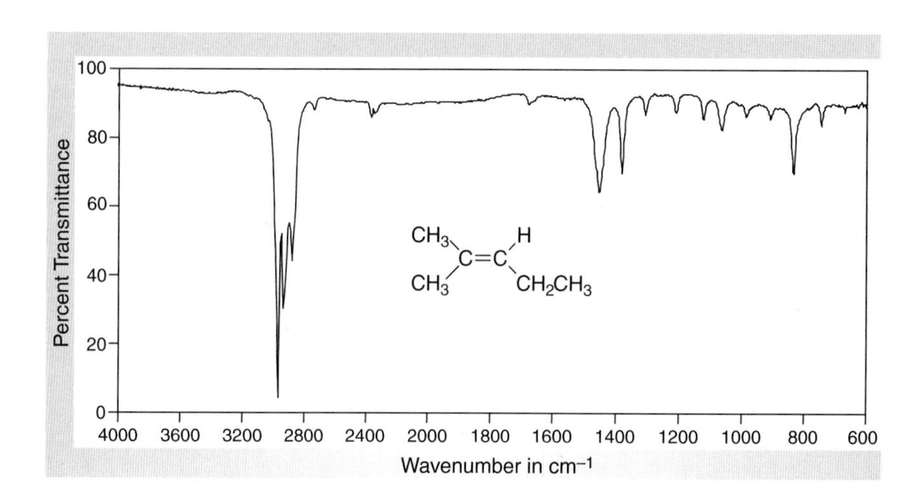

Figure 10.21
IR spectrum of 2-methyl-2-pentene (neat).

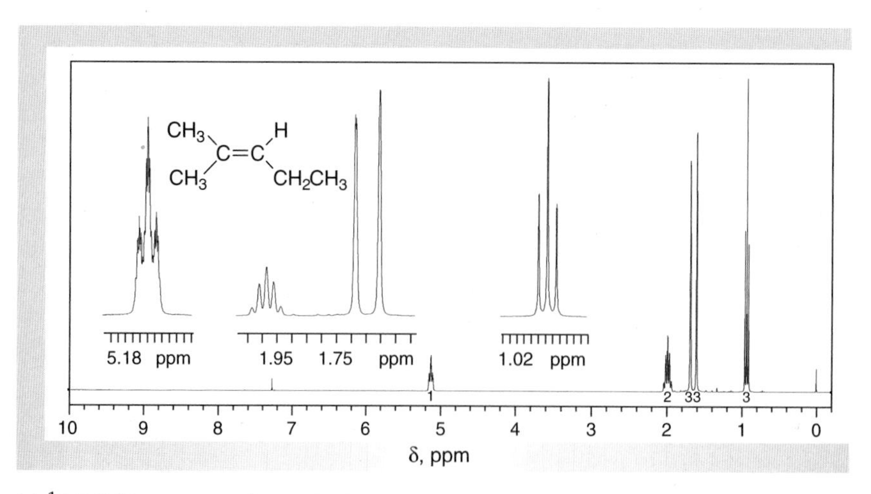

Figure 10.22
NMR data for 2-methyl-2-pentene ($CDCl_3$).

(a) ^1H NMR spectrum (300 MHz).
(b) ^{13}C data: δ 14.5, 17.5, 21.6, 25.7, 126.9, 130.6.

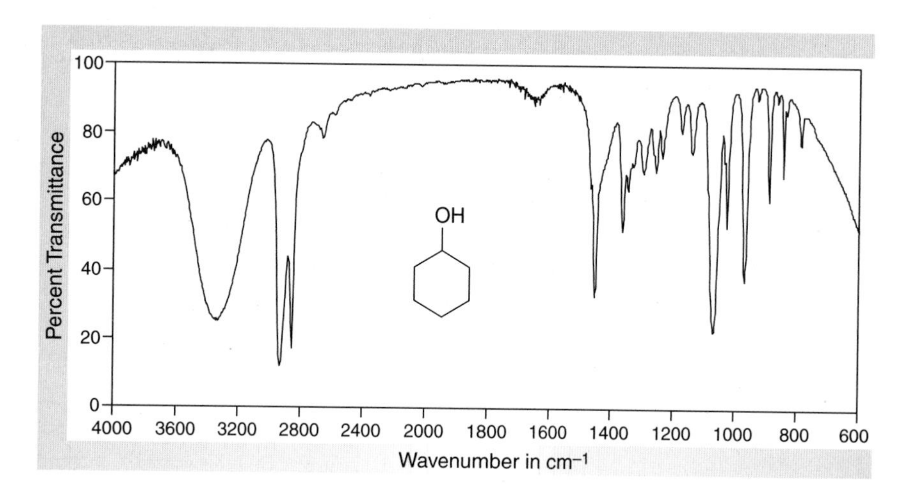

Figure 10.23
IR spectrum of cyclohexanol (neat).

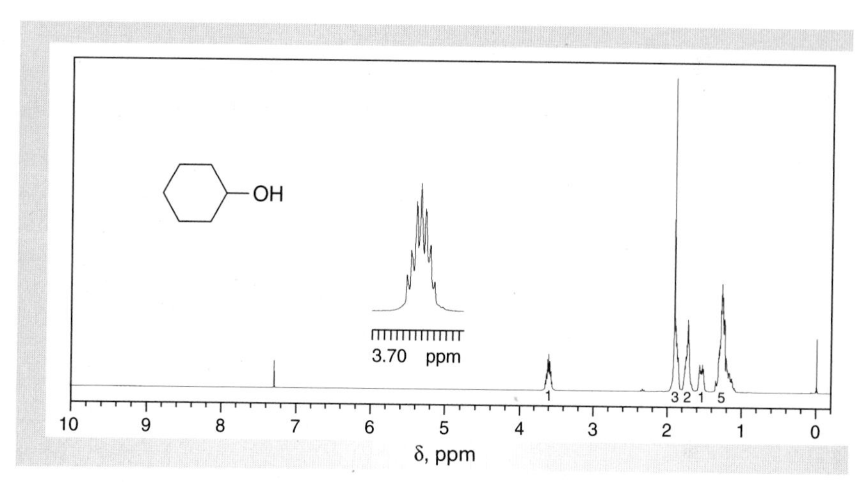

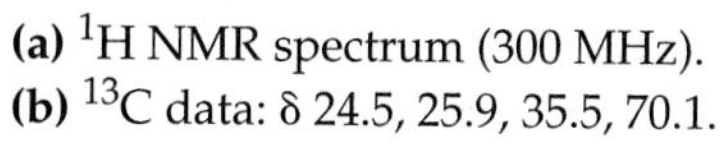

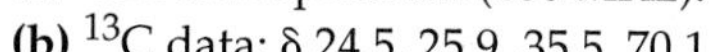

Figure 10.24
NMR data for cyclohexanol ($CDCl_3$).

(a) ^1H NMR spectrum (300 MHz).
(b) ^{13}C data: δ 24.5, 25.9, 35.5, 70.1.

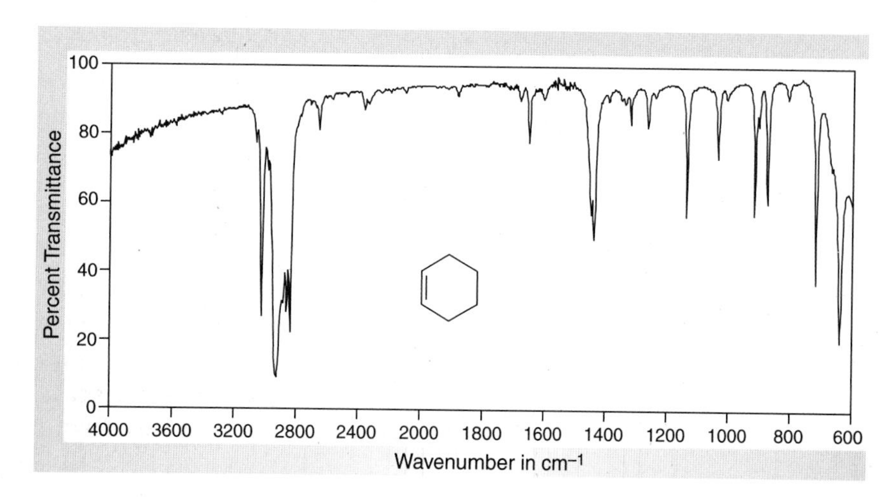

Figure 10.25
IR spectrum of cyclohexene (neat).

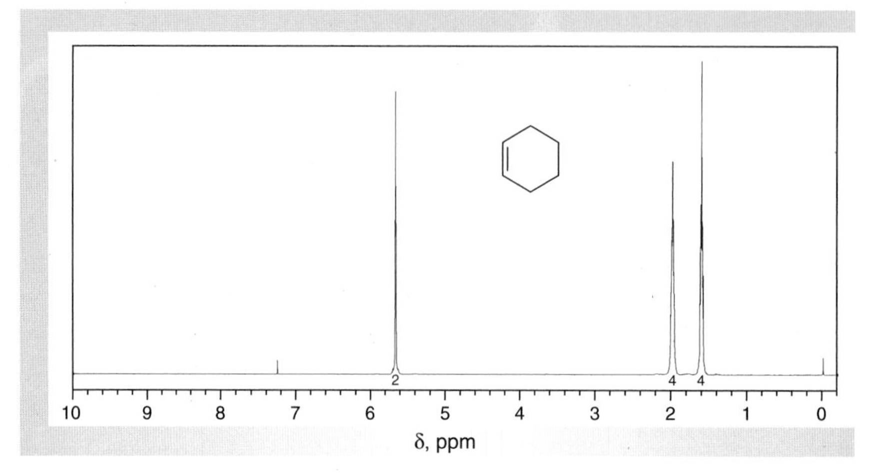

Figure 10.26
NMR data for cyclohexene ($CDCl_3$).

(a) ^1H NMR spectrum (300 MHz).
(b) ^{13}C data: δ 22.9, 25.2, 127.3.

10.4 ADDITION REACTIONS OF ALKENES: OVERVIEW

Alkenes are useful starting materials for organic syntheses because they undergo a large variety of reactions involving their functional group, the carbon-carbon double or π-bond. One of the typical reactions of alkenes is the **addition** of a reagent X–Y across the π-bond (Eq. 10.14). Such additions occur because of two principal factors. First, the strength of the carbon-carbon π-bond is 60–65 kcal/mol, a value substantially lower than the 80–100 kcal/mol that is typical for the σ-bond strengths of other atoms bound to carbon. Adding a reagent across the double bond is therefore usually *exothermic,* because one π-bond is being replaced by two σ-bonds. Second, the π-electrons of the double bond are more loosely held than σ-electrons, so they are more polarizable. Consequently, the π-electron "cloud" is readily distorted through electrostatic interaction with an electron-deficient reagent, and this polarization enhances the reactivity of the alkene toward attack.

$$R^1R^2C{=}CR^3R^4 + X{-}Y \longrightarrow R^2{-}C(R^1)(X){-}C(Y)(R^4){-}R^3 \qquad (10.14)$$

The most common mode of addition of X–Y across the π-bond of alkenes involves an *ionic* **stepwise mechanism** in which an **electrophile,** E^+, first attacks the π-bond, functioning as a nucleophile, to produce a carbocation. In such cases, X–Y may be rewritten as E–Nu, wherein E is the *less* electronegative of the two atoms. The **nucleophile,** $Nu{:}^-$, then attacks the carbocation to produce the final product. The overall process is termed **electrophilic addition.**

In order to understand the mechanism of electrophilic addition, it is helpful to consider this type of addition in terms of sequential reactions of Lewis acids and Lewis bases. In the first step, the electrons of the polarizable π-bond, a Lewis base, attack the electrophile E^+, an electron-deficient species and a Lewis acid, to give either an acyclic planar carbocation **32** or a cyclic cation **33** (Eq. 10.15). If carbocation **32** is formed as an intermediate in the electrophilic addition to an *unsymmetrically*

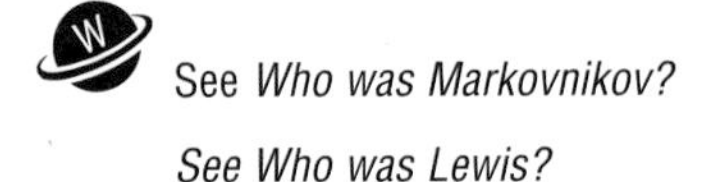

See *Who was Markovnikov?*

See *Who was Lewis?*

substituted alkene, the electrophile E^+ generally adds to the terminus of the double bond that produces the *more stable carbocation* **32** according to **Markovnikov's rule** (see the Historical Highlight at the end of this chapter). Subsequent reaction of this carbocation, a Lewis acid, with the nucleophile $Nu:^-$, which is electron-rich and a Lewis base, then gives the addition product. When the positively charged carbon of **32** is substituted with three different groups, the nucleophile may attack from *both* faces of the planar ion by paths **a** and **b** to give a mixture of **34** and **35,** which are stereoisomers (Sec. 7.1). Acyclic carbocations like **32** may also rearrange via 1,2-hydride or 1,2-alkyl migration to form a more stable carbocation (Sec. 10.3); subsequent reaction of this isomeric carbocation with the nucleophilic partner produces adducts isomeric with **34** and **35.** Some electrophilic reagents add to a carbon-carbon double bond to form a cyclic cation **33;** such cations tend *not* to rearrange. The nucleophile $Nu:^-$ may then attack **33** by paths **c** and **d** to give mixtures of isomers **36** and **37.**

$$R^1R^2C{=}CR^3R^4 + E^+ + Nu:^- \rightleftharpoons \left[R^2{-}C(R^1)(E){-}C^+R^3R^4 \quad Nu:^-\right]_{\mathbf{32}} \text{ or } \left[R^1R^2C{-}CR^3R^4 \text{ bridged by } E^+, \; Nu:^-\right]_{\mathbf{33}} \qquad (10.15)$$

$$R^1R^2C{=}CR^3R^4 + E{-}Nu \text{ (equilibrium with } E^+ + Nu:^-)$$

$$\mathbf{32} \xrightarrow{\mathbf{a}} R^2{-}C(R^1)(E){-}C(Nu)R^3R^4 \;(\mathbf{34}) \qquad \mathbf{32} \xrightarrow{\mathbf{b}} R^2{-}C(R^1)(E){-}C(R^3)(R^4)Nu \;(\mathbf{35})$$

$$\mathbf{33} \xrightarrow{\mathbf{c}} R^2{-}C(R^1)(Nu){-}C(E)(R^4){-}R^3 \;(\mathbf{36}) \qquad \mathbf{33} \xrightarrow{\mathbf{d}} R^2{-}C(R^1)(E){-}C(Nu)(R^4){-}R^3 \;(\mathbf{37})$$

The general principles set forth in Equation 10.15 may be illustrated by considering several examples. The addition of the unsymmetrical reagent hydrogen bromide, H–Br, to an alkene is shown in Equation 10.16, and the acid-catalyzed addition of water, or **hydration,** of an alkene is depicted in Equation 10.17. In both reactions, the π-bond is protonated to form the more stable intermediate carbocation according to Markovnikov's rule. This carbocation may then rearrange to a more stable carbocation (Sec. 10.3), or, as shown in the present case, may undergo direct nucleophilic attack from *both* faces of the planar carbocation to give the observed alkyl bromide or alcohol. These two examples of electrophilic additions to alkenes are **regioselective** processes because the two unsymmetrical reactants combine *predominantly* in one orientational sense to give one **regioisomer** preferentially.

$$R^1{-}C(R^2){=}CH(R^3) + H{-}Br \longrightarrow \left[R^1{-}\overset{+}{C}(R^2){-}CH(R^3)(H)\right] \xrightarrow{Br:^-} R^1{-}C(R^2)(Br){-}CH(R^3)(H) \qquad (10.16)$$

As mixture of stereoisomers

(10.17)

As mixture of stereoisomers

Since the ionic addition of hydrogen bromide and the acid-catalyzed addition of water to an alkene proceed via planar carbocations, the formation of products derived from rearranged carbocations is relatively common. In contrast, the bromination of alkenes, which is illustrated by the addition of bromine to cyclopentene (Eq. 10.18), occurs via a **cyclic bromonium ion 38** related to **33**, so skeletal rearrangements of such cations are *not* generally observed. Moreover, the intervention of this cyclic ion dictates that the stereochemistry of the addition is completely in the *anti* sense, as shown. The reddish color of bromine is discharged upon addition to an alkene, making this reaction a useful qualitative test for unsaturation (Sec. 25.8A).

(10.18)

Cyclopentene (*colorless*) + Bromine (*red-brown*) $\xrightarrow{CH_2Cl_2}$ **38** A cyclic bromonium ion ⟶ *trans*-1,2-Dibromocyclopentene (*colorless*)

See *Who was Baeyer?*

The addition of reagents X–Y to carbon-carbon π-bonds may also proceed via a **concerted mechanism** in which each new σ-bond is formed simultaneously on the *same* face of the π-bond. The stereochemistry of such reactions is necessarily *syn*. For example, the reaction of potassium permanganate, which is purple, with an alkene such as cyclohexene proceeds via *syn*-addition of permanganate ion across the π-bond to give **39**, which is colorless. Subsequent decomposition of **39** gives a *cis*-1,2-diol and manganese dioxide, the brown precipitate that is observed as the other product of the reaction (Eq. 10.19). This decoloration of potassium permanganate by alkenes forms the basis of the Baeyer qualitative test for the presence of carbon-carbon π-bonds (Sec. 25.8B).

(10.19)

Cyclohexene (*colorless*) + $KMnO_4$ Potassium permanganate (*purple*) $\xrightarrow{H_2O}$ **39** ⟶ *cis*-Cyclohexane-1,2-diol (*colorless*) + MnO_2 Manganese dioxide (*brown*)

10.5 ADDITION OF HYDROBROMIC ACID TO ALKENES

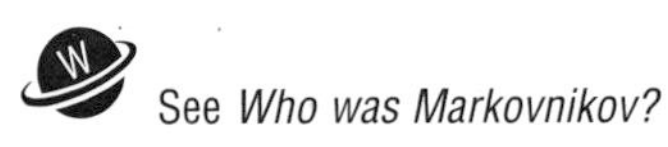

See *Who was Markovnikov?*

In this experiment, you will study the **ionic addition** of hydrobromic acid, H–Br, to 1-hexene to ascertain whether 2-bromohexane (**41**) is produced preferentially according to Markovnikov's rule (Eq. 10.20). This type of reaction is normally rather difficult to perform in the undergraduate laboratory for several reasons. First, common alkenes are immiscible with concentrated aqueous hydrobromic acid, and the reaction is sluggish if the layers are not mixed efficiently. Second, H–Br is a *strong* acid that protonates water extensively to give the hydronium ion, H_3O^+. Because a hydronium ion is a *weaker* acid than undissociated hydrobromic acid, it is unable to protonate the alkene rapidly under mild reaction conditions. Moreover, the presence of water in the reaction mixture introduces the possibility of competing acid-catalyzed addition of water to the alkene (Eq. 10.17, Sec. 10.4). These problems are reduced by using anhydrous hydrogen bromide, but the highly corrosive nature of this gas makes it difficult to handle and use.

$$\underset{\text{1-Hexene}}{\underset{\mathbf{40}}{CH_3(CH_2)_3CH{=}CH_2}} + H{-}Br \longrightarrow \underset{\text{2-Bromohexane}}{\underset{\mathbf{41}}{CH_3(CH_2)_3\underset{Br}{\underset{|}{C}}H{-}\underset{H}{\underset{|}{C}}H_2}} \qquad (10.20)$$

A convenient solution to these experimental difficulties entails the addition of a catalytic amount of a quaternary ammonium salt such as methyltrioctylammonium chloride, $CH_3(n\text{-}C_8H_{17})_3N^+Cl^-$, also known as Aliquat 336®, to the heterogeneous mixture of the aqueous acid and the alkene. The tetraalkyl ammonium salt partitions between the aqueous and organic phases because of the amphoteric nature of the catalyst: It is **lipophilic** or **nonpolar-loving** due to the alkyl groups and **hydrophilic** or **polar-loving** because of the ionic ammonium function, respectively. By forming a complex such as **42** with the H–Br, the quaternary ammonium salt extracts H–Br from the aqueous phase, transporting it into the organic phase and the presence of the alkene (Scheme 10.1). The quaternary ammonium salt repartitions into the aqueous phase to complete the catalytic cycle. The transfer of the H–Br into the organic phase essentially dehydrates the acid, making it more reactive toward the alkene so that the addition becomes possible.

Scheme 10.1

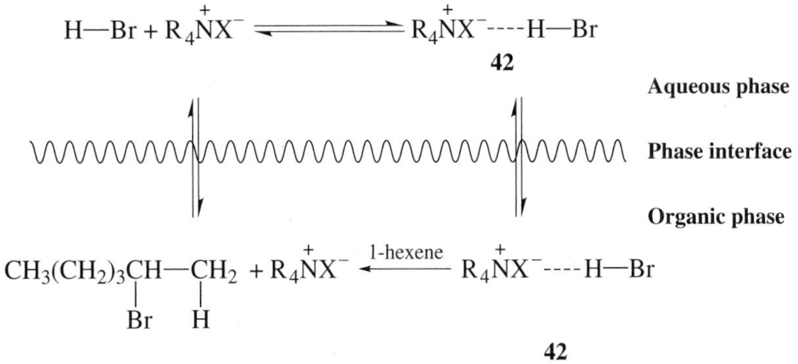

Aqueous phase:

$$H{-}Br + R_4\overset{+}{N}X^- \rightleftharpoons \underset{\mathbf{42}}{R_4\overset{+}{N}X^-{\text{----}}H{-}Br}$$

Phase interface (both species transfer reversibly across)

Organic phase:

$$CH_3(CH_2)_3\underset{Br}{\underset{|}{C}}H{-}\underset{H}{\underset{|}{C}}H_2 + R_4\overset{+}{N}X^- \xleftarrow{\text{1-hexene}} \underset{\mathbf{42}}{R_4\overset{+}{N}X^-{\text{----}}H{-}Br}$$

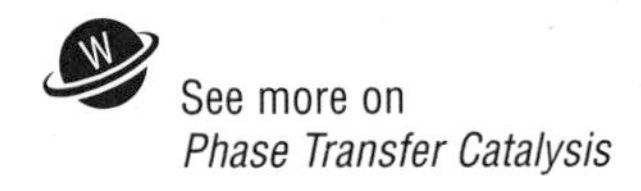

See more on *Phase Transfer Catalysis*

Compounds that promote the transport of reagents between immiscible layers by means of **ion pairs** like **42** are called **phase-transfer catalysts.** Their presence can have dramatic effects on the rates of bimolecular reactions between reagents contained in immiscible phases: Rate accelerations of 10^4 to 10^9 are common. One factor that determines the overall rate of a reaction involving phase-transfer catalysis is the efficiency of the partitioning of the reagents and reactants between the two phases. This is a function of the total surface area at the interface of the phases. To increase this area, the reaction mixture must be *vigorously* agitated by stirring or shaking to promote emulsification and the formation of tiny droplets of the immiscible layers.

As depicted in Scheme 10.2, the mechanism of the reaction presumably involves protonation of 1-hexene to afford the secondary carbocation **43.** Attack of bromide ion on this ion then leads to 2-bromohexane (**41**). The carbocation may also rearrange by way of a hydride shift (Sec. 10.3) to provide a different secondary carbocation, **44,** which would provide 3-bromohexane (**45**) upon reaction with bromide ion. Alternatively, it may deprotonate to form 2-hexene (**46**), addition of H–Br to which could afford both **41** and **45.** In this experiment, you will determine the regiochemistry of the addition of H–Br to the unsymmetrical alkene 1-hexene (**40**) and thereby assess whether this ionic reaction proceeds according to Markovnikov's rule.

Scheme 10.2

$CH_3(CH_2)_3CH{=}CH_2 + H{-}Br{---}X^-\ {}^+NR_4 \longrightarrow CH_3(CH_2)_2CH(H){-}\overset{+}{C}HCH_2{-}H \xrightarrow{Br^-}$ **41**

43; $\sim H^-$ (to **44**); $-H^+$ (to **46**)

$CH_3(CH_2)_2CH(Br){-}CH_2CH_3 \xleftarrow{Br^-} CH_3CH_2\overset{+}{C}H{-}CH(H){-}CH_3 \underset{}{\overset{Br{-}H}{\rightleftharpoons}} CH_3(CH_2)_2CH{=}CHCH_3$

45 3-Bromohexane; **44**; **46** 2-Hexene

Addition of Hydrogen Bromide to 1-Hexene

Purpose To determine the regiochemistry of the polar addition of H–Br to an alkene.

SAFETY ALERT

1. **Concentrated hydrobromic acid (47–49%) is a *highly corrosive* and *toxic* material. Working at a hood, measure out the amount of acid required and avoid inhaling its vapors. Use latex gloves when transferring the acid between containers. If the acid comes in contact with your skin, flood the affected area immediately and thoroughly with water and rinse it with 5% sodium bicarbonate solution.**

2. **Quaternary ammonium salts are *toxic* substances and can be absorbed through the skin. Should they accidentally come in contact with your skin, wash the affected area immediately with large amounts of water.**
3. ***If* a flame must be used in the distillation step, be certain that all joints in the apparatus are well lubricated and tightly mated because 1-hexene and the solvents used, particularly petroleum ether, are highly flammable. *Flameless* heating is preferred.**

MINISCALE PROCEDURE

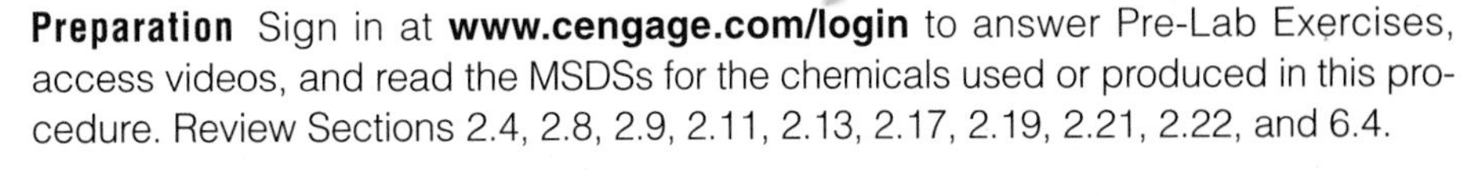

Preparation Sign in at **www.cengage.com/login** to answer Pre-Lab Exercises, access videos, and read the MSDSs for the chemicals used or produced in this procedure. Review Sections 2.4, 2.8, 2.9, 2.11, 2.13, 2.17, 2.19, 2.21, 2.22, and 6.4.

Apparatus A 25- and a 50-mL round-bottom flask, separatory funnel, apparatus for heating under reflux, simple distillation, magnetic stirring, and *flameless* heating.

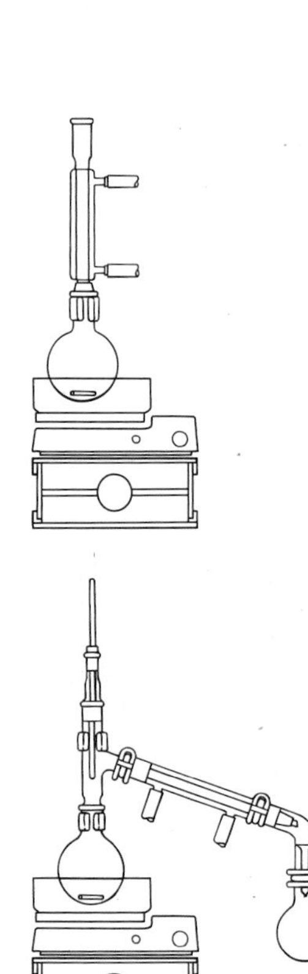

Setting Up Combine 3.0 mL of 1-hexene, 14 mL of 48% aqueous hydrobromic acid, and 1.0 g of methyltrioctylammonium chloride in the 50-mL round-bottom flask. Equip the flask with a water-cooled condenser and set up the apparatus heating under reflux.

Addition and Work-Up With *rapid* stirring, heat the heterogeneous reaction mixture under gentle reflux for 2 h and then allow the mixture to cool to room temperature.★

Carefully transfer the two-phase mixture to a separatory funnel, rinse the reaction flask with 15 mL of petroleum ether (bp 60–80 °C, 760 torr), and add the rinse to the separatory funnel. Shake the funnel thoroughly and allow the layers to separate. Verify that the lower one is the aqueous phase and remove it. Sequentially wash the organic phase with two 15-mL portions of 10% sodium bicarbonate solution. Vent the funnel *frequently* because gas is evolved and excessive pressure must *not* develop in the funnel. Transfer the organic layer to an Erlenmeyer flask and dry it with swirling over several spatula-tips full of anhydrous sodium sulfate for at least 0.5 h.★ Add additional portions of anhydrous sodium sulfate if the liquid remains cloudy.

Isolation Decant or gravity-filter the dried solution into a 25-mL round-bottom flask, equip the flask for simple distillation, and distill the product. Carefully control the rate of heating throughout the course of this distillation because severe foaming can occur. *Do not attempt to remove the solvent too rapidly,* as this may result in excessive loss of product due to foaming.

After the first fraction containing solvent and unreacted 1-hexene is removed over a range of room temperature to 80 °C (760 torr),★ change the receiving flask and collect the bromohexane as a single fraction in a tared container. Because the volume of product is likely to be only 1–2 mL, an accurate boiling point may be difficult to obtain, and all material that distills above 110–115 °C (760 torr) should be collected in order to obtain a reasonable yield.

Analysis Weigh the distillate and determine the yield of crude product. Determine the structure of the principal product of the reaction using IR and/or NMR spectroscopic methods or by GLC, assuming authentic samples of the isomers are available.

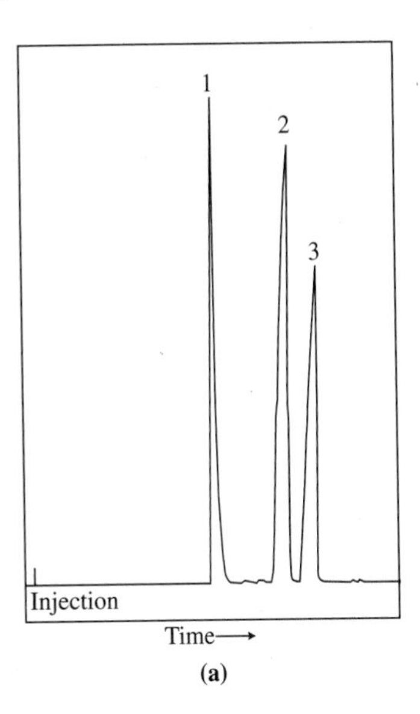

(a)

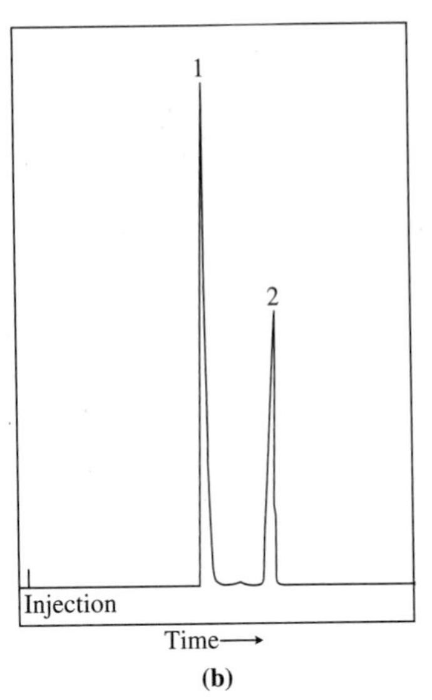

(b)

Figure 10.27
GLC traces of the bromohexanes. (a) Mixture of petroleum ether (peak 1), 2-bromohexane (peak 2), and 1-bromohexane (peak 3). (b) Distillate from addition of HBr to 1-hexene containing petroleum ether (peak 1). Column and conditions: 0.25-mm × 25-m AT-1 Helifex 100% dimethylsiloxane; 110 °C, 40 mL/min.

An important difference in the ^1H NMR spectra for 2-bromohexane and 3-bromohexane is that the multiplets for the hydrogen atom on the same carbon atom as the bromine are centered at δ 4.13 and δ 4.00, respectively. A GLC trace of a mixture of 1- and 2-bromohexanes is given in Figure 10.27a, along with a chromatogram of the product of the reaction (Fig. 10.27b); note that 2- and 3-bromohexane are *not* separated under the GLC conditions used. Obtain IR and ^1H NMR spectra of your starting material and product and compare them with those of authentic samples (Figs. 10.28–10.33). Alternatively, determine whether or not Markovnikov addition has occurred by subjecting the product to the silver nitrate or sodium iodide/acetone tests for classifying alkyl halides (Secs. 25.9A and 25.9B, respectively).

Discovery Experiment

Analysis of Bromohexanes

Using authentic samples of 2- and 3-bromohexane and after consulting with your instructor, attempt to find GLC conditions that allow separation of these isomers. This may require the use of other types of capillary columns than that employed to obtain the traces of Figure 10.27. If your investigation is successful, analyze your distillate for the presence of 3-bromohexane.

MICROSCALE PROCEDURE

Preparation Sign in at **www.cengage.com/login** to answer Pre-Lab Exercises, access videos, and read the MSDSs for the chemicals used or produced in this procedure. Review Sections 2.4, 2.8, 2.9, 2.11, 2.13, 2.17, 2.19, 2.21, 2.22, and 6.4.

Apparatus A 5-mL conical vial, screw-cap centrifuge tube, apparatus for heating under reflux, simple distillation, magnetic stirring, and *flameless* heating.

Setting Up and Addition Equip the conical vial with a spinvane and add to it 500 μL (0.5 mL) of 1-hexene, 2 mL of 48% aqueous hydrobromic acid, and 150 mg of methyltrioctylammonium chloride. Equip the vial with a water-cooled condenser and assemble the apparatus for heating under reflux. With *rapid* stirring, heat the

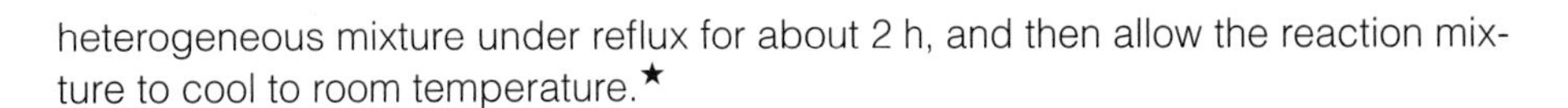

heterogeneous mixture under reflux for about 2 h, and then allow the reaction mixture to cool to room temperature.★

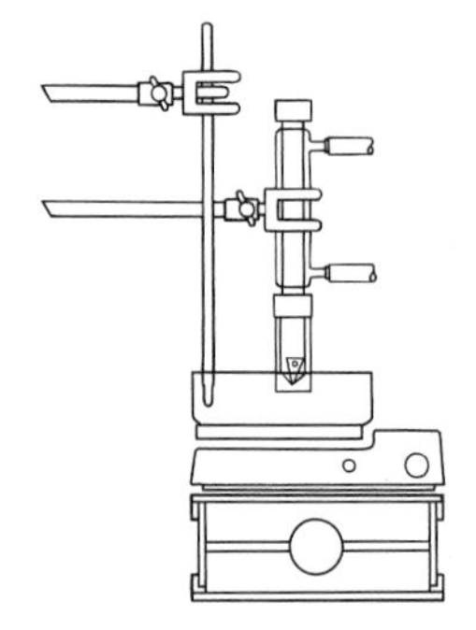

Work-Up Add 1 mL of petroleum ether (bp 60–80 °C, 760 torr) to the reaction mixture and shake the vial, venting as necessary to relieve pressure. Allow the layers to separate. Remove the aqueous layer with a Pasteur pipet. Do *not* discard this liquid until you are certain that it is indeed the aqueous layer!

Add 0.5 mL of 10% sodium bicarbonate solution, stir the mixture (*Caution:* gas evolution!), and remove the aqueous layer after allowing separation. Using pHydrion paper, check whether this layer is basic. If it is not, repeat washing the organic layer with aqueous sodium carbonate until the aqueous washes are basic. Finally, dry the reaction mixture over several microspatula-tips full of anhydrous sodium sulfate for about 15 min; swirl the vial occasionally to hasten drying. Add additional portions of anhydrous sodium sulfate if the liquid remains cloudy.

Distillation Using a filter-tip pipet, transfer the dried organic layer into a clean 5-mL conical vial equipped with a stirbar. Attach the vial to a Hickman stillhead equipped with a water-cooled condenser (Fig. 2.38) and perform a microscale distillation. The first fraction should contain solvent and unchanged 1-hexene and should distill at a bath temperature below 110 °C. Remove this fraction from the still, and collect a second fraction, using a bath temperature of 165 °C. This fraction should mainly consist of the desired product. Transfer this distillate to a tared screw-cap vial.

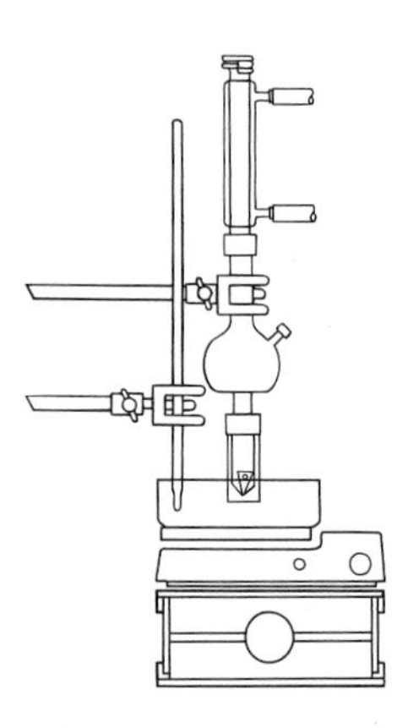

Analysis Weigh the distillate and determine the yield of crude product. Determine the structure of the principal product of the reaction using IR and/or NMR spectroscopic methods or by GLC, assuming authentic samples of the isomers are available. An important difference in the ^{1}H NMR spectra for 2-bromohexane and 3-bromohexane is that the multiplets for the hydrogen atom on the same carbon atom as the bromine are centered at δ 4.13 and δ 4.00, respectively. A GLC trace of a mixture of 1- and 2-bromohexanes is given in Figure 10.27a, along with a chromatogram of the product of the reaction (Fig. 10.27b); note that 2- and 3-bromohexane are *not* separated under the GLC conditions used. Obtain IR and ^{1}H NMR spectra of your starting material and product and compare them with those of authentic samples (Figs. 10.28–10.33). Alternatively, determine whether or not Markovnikov addition has occurred by subjecting the product to the silver nitrate or sodium iodide/acetone tests for classifying alkyl halides (Secs. 25.9A and 25.9B, respectively).

Discovery Experiment

Analysis of Bromohexanes

Using authentic samples of 2- and 3-bromohexane and after consulting with your instructor, attempt to find GLC conditions that allow separation of these isomers. This may require the use of other types of capillary columns than that employed to obtain the traces of Figure 10.27. If your investigation is successful, analyze your distillate for the presence of 3-bromohexane.

WRAPPING IT UP

Carefully (foaming may occur!) combine the *aqueous layers and washes,* neutralize if necessary with sodium carbonate, and flush them down the drain with excess water. Since the *first distillation fraction* may contain some halogenated product, pour it in a container for halogenated organic solvents. Evaporate the petroleum ether from the *sodium sulfate* used as drying agent in the hood, and place the solid in a container for nonhazardous solids. Dilute the *silver nitrate test solution* with water and flush it down the drain with excess water unless instructed to do otherwise.

EXERCISES

1. Define the following terms:
   a. lipophilic
   b. hydrophobic
   c. hydrophilic
2. What structural feature of an alkene makes it nucleophilic?
3. It was stated that aqueous H–Br exists predominantly as H_3O^+ Br^-. Confirm the validity of this statement by calculating the equilibrium constant for the process shown below. The values in parentheses are the pK_as of the indicated acids. (*Hint:* See Section 5.3.)

$$\underset{(-8)}{\text{H—Br}} + H_2O \rightleftharpoons \underset{(-1.7)}{H_3O^+} + Br^-$$

4. The yield of 2-bromohexane is much higher when the reaction mixture is stirred than when it is not. Why is this?
5. Account mechanistically for the fact that polar addition of H–Br to 1-hexene preferentially forms 2- rather than 1-bromohexane.
6. Provide an explanation for the order of carbocation stability: 3°>2°>1°.
7. Specify those structural characteristics of a phase-transfer catalyst that allow it to be soluble in both organic and aqueous solvents.
8. Show how a phase transfer catalyst can interact with H–Br via electrostatic, hydrogen bonding, dipole-dipole, and van der Waals interactions.
9. When methyltrioctylammonium chloride is used as the phase-transfer catalyst, formation of 2-*chloro*hexane as a by-product is possible. Why would contamination of the desired 2-bromohexane by the chloro compound be a minor concern in this experimental procedure?
10. Why is methyltrioctylammonium chloride only partially soluble in 1-hexene? In aqueous hydrobromic acid?
11. Outline a procedure that would allow monitoring of the course of this addition reaction as a function of time. In other words, how might the reaction mixture be analyzed periodically so that you could determine when the reaction is complete?
12. In the work-up of this procedure, why is it necessary to wash the organic phase with a solution of sodium bicarbonate? What gas is evolved?
13. Devise an experiment that would demonstrate that a phase-transfer catalyst accelerates the rate of reaction between 1-hexene and aqueous hydrobromic acid. Why would it be difficult to perform the ionic addition of hydrogen bromide to 1-pentene using the procedure outlined in this experiment?
14. Refer to Figure 10.27a and determine whether the stationary phase used for the GLC analysis shown separates the bromohexanes on the basis of their relative boiling points or of some other property. Explain your answer.

15. What is the mechanistic basis (a) of the silver nitrate and (b) of the sodium iodide in acetone tests for differentiating 1° and 2° alkyl bromides from each other?

16. Write the structure of the major product expected from the ionic addition of H–Br to each of the alkenes shown.

a. Me, Me   b. Me, Me   c. Me, Me, Me

17. Using curved arrows to symbolize the flow of electrons, provide a mechanism for the reaction of hydrogen bromide with each of the alkenes in Exercise 16. Provide a rationale for any reactions expected to proceed with skeletal rearrangement.

Sign in at **www.cengage.com/login** and use the spectra viewer and Tables 8.1–8.8 as needed to answer the blue-numbered questions on spectroscopy.

18. Consider the spectral data for 1-hexene (Figs. 10.28 and 10.29).

    a. In the IR spectrum, specify the absorptions associated with the carbon-carbon double bond and the vinylic hydrogen atoms.

    b. In the ^1H NMR spectrum, assign the various resonances to the hydrogen nuclei responsible for them.

    c. For the ^{13}C NMR data, assign the various resonances to the carbon nuclei responsible for them.

19. Consider the NMR data for 1-bromohexane (Fig. 10.31).

    a. In the ^1H NMR spectrum, assign the various resonances to the hydrogen nuclei responsible for them.

    b. For the ^{13}C NMR data, assign the various resonances to the carbon nuclei responsible for them.

20. Consider the NMR data for 2-bromohexane (Fig. 10.33).

    a. In the ^1H NMR spectrum, assign the various resonances to the hydrogen nuclei responsible for them.

    b. For the ^{13}C NMR data, assign the various resonances to the carbon nuclei responsible for them.

21. Discuss the differences observed in the IR and NMR spectra of 1-hexene and 2-bromohexane that are consistent with addition of HBr occurring in this experiment.

22. Assume you obtained the ^1H NMR spectrum of a mixture of 1- and 2-bromohexane and found that the integrations for the multiplets centered at δ 3.4 and 4.1 were 20 units and 40 units, respectively. What is the relative ratio of these two compounds in the sample?

23. The ^1H NMR spectrum of the crude bromohexane isolated by distillation in one experiment reveals a multiplet centered at δ 4.00, which is slightly upfield of the multiplet centered at δ 4.13 for 2-bromohexane.

    a. Determine whether this multiplet may signal the presence of 3-bromohexane by looking up the ^1H NMR spectrum of this isomer in either the Aldrich or Sadtler catalogs of reference spectra.

    b. The integrated area of the multiplet centered at δ 4.00 is 45 units, whereas that of the multiplet centered at δ 4.13 is 200 units. Assuming that the upfield multiplet represents 3-bromohexane, what is its percentage in the mixture of bromohexanes?

SPECTRA

Starting Materials and Products

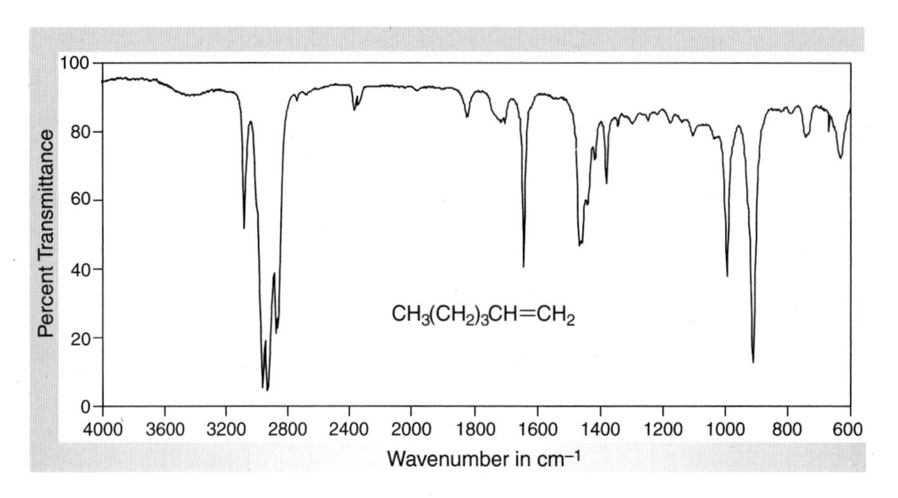

Figure 10.28
IR spectrum of 1-hexene (neat).

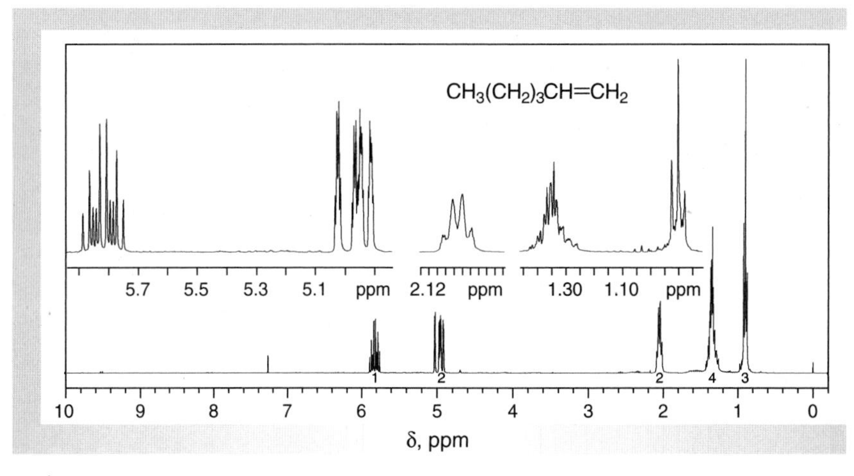

Figure 10.29
NMR data for 1-hexene ($CDCl_3$).

(a) ^{1}H NMR spectrum (300 MHz).
(b) ^{13}C data: δ 14.0, 22.5, 31.6, 33.8, 114.3, 139.2.

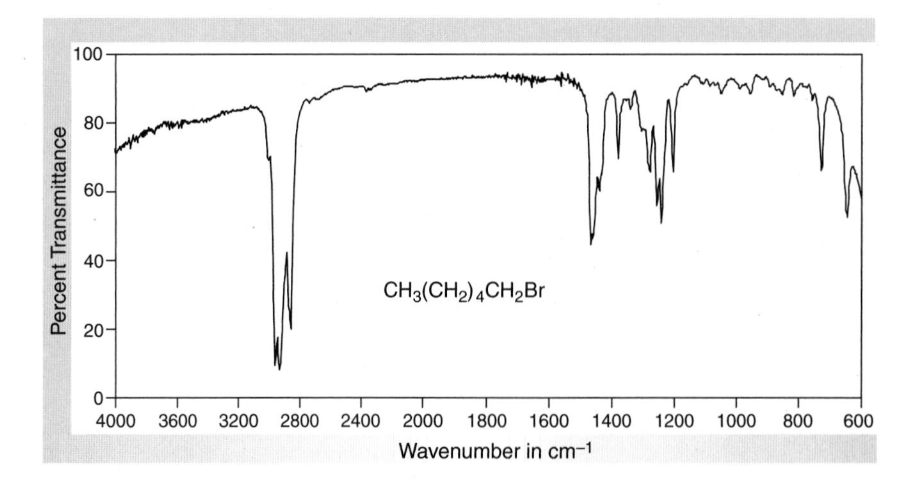

Figure 10.30
IR spectrum of 1-bromohexane (neat).

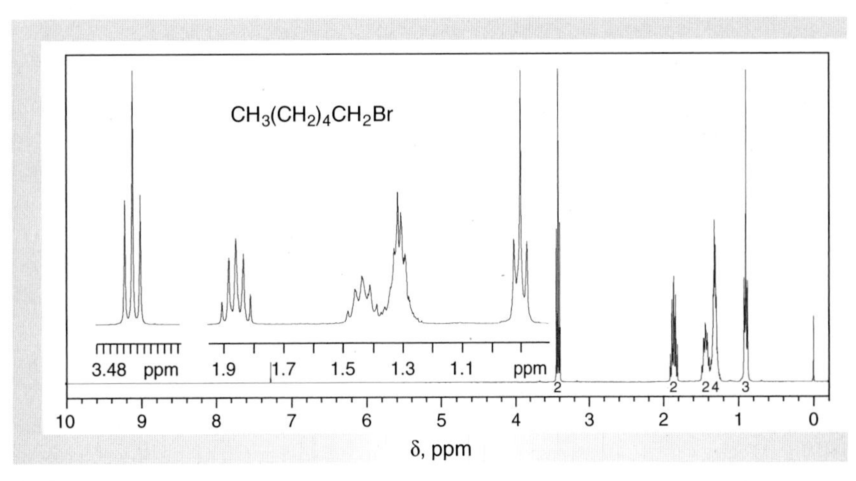

Figure 10.31
NMR data for 1-bromohexane ($CDCl_3$).

(a) ^{1}H NMR spectrum (300 MHz).
(b) ^{13}C data: δ 14.0, 22.7, 28.1, 31.2, 33.1, 33.4.

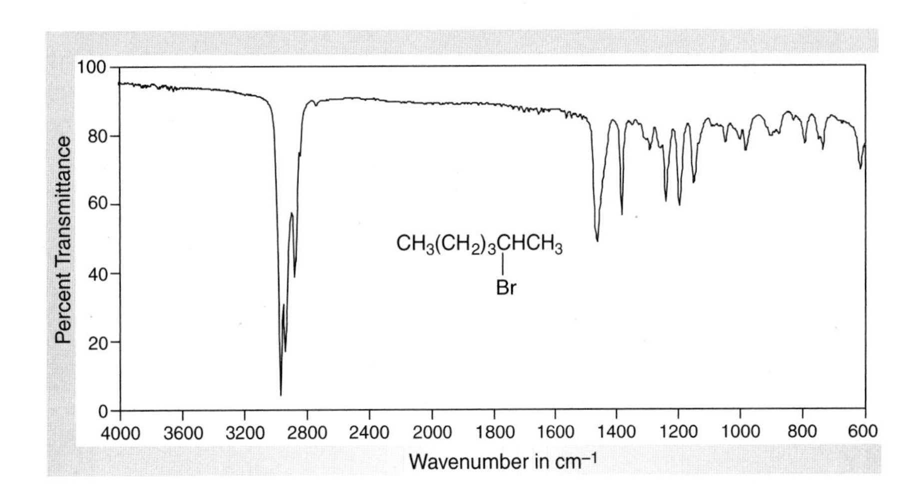

Figure 10.32
IR spectrum of 2-bromohexane (neat).

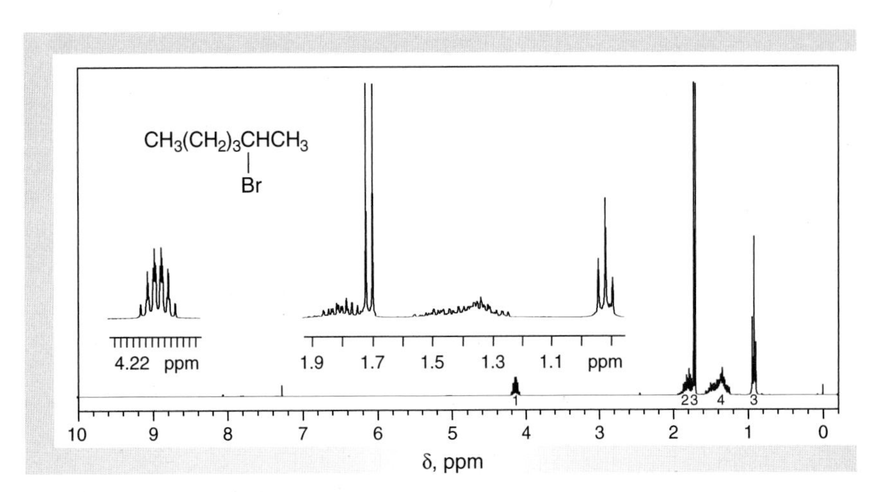

Figure 10.33
NMR data for 2-bromohexane ($CDCl_3$).

(a) ^{1}H NMR spectrum (300 MHz).
(b) ^{13}C data: δ 13.9, 22.1, 26.5, 29.9, 41.0, 51.0.

10.6 BROMINATION OF ALKENES

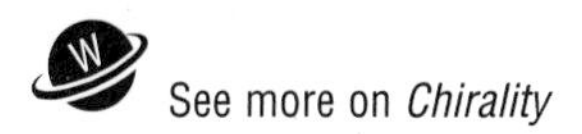

The reaction of bromine with (*E*)-stilbene (**47**) to give *meso*-stilbene dibromide (**48**) as the major product (Eq. 10.21) is another example of an electrophilic addition reaction of alkenes. The addition of bromine to many alkenes is a **stereospecific** reaction that proceeds by *anti* addition to the double bond. However, the addition of bromine to **47** is *not* stereospecific because small amounts of *dl*-stilbene dibromide (**49**) are also formed in this reaction. The formation of *meso*-stilbene dibromide presumably occurs via the nucleophilic attack of bromide on the intermediate cyclic bromonium ion, **50.** The possible interconversion of **50** and the acyclic carbocation **51** (Eq. 10.22) is one possible way to account for the presence of *dl*-stilbene dibromide in the product.

The bromination of (*E*)-stilbene using a solution of bromine in dichloromethane involves a chlorinated solvent. This protocol violates a principle of "green chemistry" (see Historical Highlight in Chapter 16) because halogenated compounds are environmental hazards as a general rule. Thus, alternative methods have been developed for producing bromine in nonhalogenated solvents. One of these involves the *in situ* formation of bromine by oxidation of hydrobromic acid with hydrogen peroxide (Eq. 10.23). Because stilbene is basically insoluble in water, a product of the reaction and the solvent for the hydrogen peroxide, an organic solvent that is environmentally more benign than a halogenated liquid and dissolves both the alkene and water is needed. Ethanol or other alcohols serve this purpose in the green experiment described in this section.

The dibromides **48** and **49** are *diastereomers* and thus have different physical and chemical properties (Sec. 7.1). Indeed, their melting points differ by 100 °C. This makes it very easy to identify the major product of the bromination.

47 (*E*)-Stilbene (*colorless*) $\xrightarrow[(red)]{Br_2/CH_2Cl_2}$ **48** *meso*-Stilbene dibromide (*colorless*) + **49** *dl*-Stilbene dibromide (*colorless*) (+ enantiomer) (10.21)

50 ⇌ **51** (10.22)

$$HBr + H_2O_2 \longrightarrow Br_2 + H_2O \qquad (10.23)$$

EXPERIMENTAL PROCEDURES

Bromination of (E)-*Stilbene*

Purpose To determine the stereochemistry of the electrophilic addition of bromine to an alkene.

SAFETY ALERT

1. ***Bromine is a hazardous chemical that may cause serious chemical burns.* Do not breathe its vapors or allow it to come into contact with the skin. Perform all operations involving the transfer of the pure liquid or its solutions at a hood and wear latex gloves. If you get bromine on your skin, wash the area immediately with soap and warm water and soak the affected area in 0.6 *M* sodium thiosulfate solution, for up to 3 h if the burn is particularly serious.**
2. **Dispense the 1 *M* bromine in dichloromethane solution from burets or similar devices fitted with Teflon stopcocks and located in hoods.**
3. **Bromine reacts with acetone to produce the powerful lachrymator α-bromoacetone, $BrCH_2COCH_3$. Do *not* rinse glassware containing residual bromine with acetone!**
4. **All parts of this experiment should be conducted in a hood if possible.**

See more on *Lachrymators*

MINISCALE PROCEDURE

Preparation Sign in at **www.cengage.com/login** to answer Pre-Lab Exercises, access videos, and read the MSDSs for the chemicals used or produced in this procedure. Review Sections 2.7, 2.11, and 2.17.

Apparatus A 25-mL round-bottom flask, apparatus for magnetic stirring, and vacuum filtration.

Setting Up Equip the flask with a stirbar and add to it 0.9 g of (*E*)-stilbene and 10 mL of dichloromethane. Stir or swirl the mixture to effect dissolution.

Bromination and Isolation Measure 5 mL of freshly prepared 1 *M* bromine in dichloromethane directly from the dispenser into the round-bottom flask. Swirl the flask gently during the addition to mix the contents. After the addition of the bromine solution is complete, stopper the reaction flask loosely and stir the mixture for 15 min. Isolate the product by vacuum filtration. Wash the product with one or two 1-mL portions of *cold* dichloromethane until it is white. Transfer the product to a watchglass or a piece of filter or weighing paper and allow it to air-dry. Recrystallize the product from xylenes.

Analysis Weigh the product and determine the yield. Measure the melting point to determine the stereochemistry of the bromination. *Caution:* Do *not* use mineral oil as the heating fluid for this determination! Obtain IR and ^1H NMR spectra of your starting material and product and compare them with those of authentic samples (Figs. 10.34–10.37).

MICROSCALE PROCEDURE

Preparation Sign in at **www.cengage.com/login** to answer Pre-Lab Exercises, access videos, and read the MSDSs for the chemicals used or produced in this procedure. Review Sections 2.7, 2.11, and 2.17.

Apparatus A 5-mL conical vial, apparatus for magnetic stirring, vacuum filtration, and Craig tube filtration.

Setting Up Equip the vial with a spinvane and add to it 180 mg of (*E*)-stilbene and 2 mL of dichloromethane. Stir or swirl the mixture to effect dissolution.

See *Who was Hirsch?*

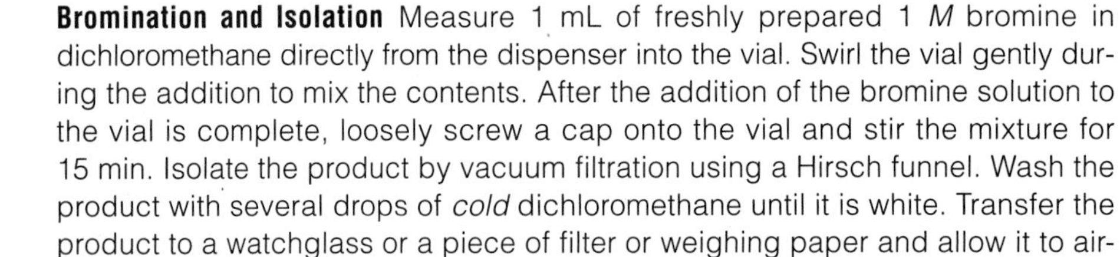

Bromination and Isolation Measure 1 mL of freshly prepared 1 *M* bromine in dichloromethane directly from the dispenser into the vial. Swirl the vial gently during the addition to mix the contents. After the addition of the bromine solution to the vial is complete, loosely screw a cap onto the vial and stir the mixture for 15 min. Isolate the product by vacuum filtration using a Hirsch funnel. Wash the product with several drops of *cold* dichloromethane until it is white. Transfer the product to a watchglass or a piece of filter or weighing paper and allow it to air-dry. Using a Craig tube, recrystallize the product from xylenes.

Analysis Weigh the product and determine the yield. Measure the melting point to determine the stereochemistry of the bromination. *Caution:* Do *not* use mineral oil as the heating fluid for this determination! Obtain IR and ^1H NMR spectra of your starting material and product and compare them with those of authentic samples (Figs. 10.34–10.37).

Discovery Experiment

Green Experiment

Bromination of (E)*-Stilbene: The Green Approach*

Purpose To demonstrate the stereochemistry of the electrophilic addition of bromine, generated *in situ*, to an alkene using an environmentally more benign process.

SAFETY ALERT

1. **Concentrated hydrobromic acid (47–49%) is a *highly corrosive* and *toxic* material. Working at a hood, measure out the amount of acid required and avoid inhaling its vapors. Use latex gloves when transferring the acid between containers. If the acid comes in contact with your skin, flood the affected area immediately and thoroughly with water and rinse it with 5% sodium bicarbonate solution.**
2. **The 30% hydrogen peroxide used is a *strong oxidant* and may blister the skin on contact. If you accidentally spill some on your skin, wash the affected area with copious amounts of water.**

MINISCALE PROCEDURE

Preparation Sign in at **www.cengage.com/login** to answer Pre-Lab Exercises, access videos, and read the MSDSs for the chemicals used or produced in this procedure. Review Sections 2.5, 2.9, 2.11, 2.17, and 2.21.

Apparatus A 50-mL round-bottom flask, 1- and 2-mL graduated pipets, apparatus for heating under reflux, magnetic stirring, vacuum filtration, and *flameless* heating.

Setting Up Place 0.5 g of (*E*)-stilbene and 12 mL of 95% ethanol in the round-bottom flask containing a stirbar and set up the apparatus for heating under reflux. Heat the mixture under reflux until most of the solid has dissolved.

Bromination and Isolation Using a graduated pipet, add 1.2 mL of *concentrated* hydrobromic acid dropwise to the boiling mixture; this may cause some precipitation. Continue heating and add 0.8 mL of 30% hydrogen peroxide dropwise, using a graduated pipet, and note the change in color of the mixture. Continue stirring and heating under reflux until the color is discharged. Allow the reaction mixture to cool to room temperature and adjust its pH to 5–7 by adding saturated aqueous sodium bicarbonate. Cool the resulting mixture in an ice-water bath to complete precipitation of the product, and collect it by vacuum filtration. Wash the filter cake with water, transfer the solid to a watch glass, a piece of filter paper, or weighing paper, and allow it to air-dry. Recrystallize the crude product from xylenes.

Analysis Weigh the product and determine the yield. Measure the melting point to determine the stereochemistry of the bromination. *Caution:* Do *not* use mineral oil as the heating fluid for this determination! Obtain IR and ^1H NMR spectra of your starting material and product and compare them with those of authentic samples (Figs. 10.34–10.37).

MICROSCALE PROCEDURE

Preparation Sign in at **www.cengage.com/login** to answer Pre-Lab Exercises, access videos, and read the MSDSs for the chemicals used or produced in this procedure. Review Sections 2.5, 2.9, 2.11, 2.17, and 2.21.

Apparatus A 5-mL conical vial, 1-mL graduated pipet, apparatus for heating under reflux, magnetic stirring, vacuum filtration, Craig tube filtration, and *flameless* heating.

Setting Up Place 0.13 g of (*E*)-stilbene and 3 mL of 95% ethanol in the conical vial containing a stirbar and set up the apparatus for heating under reflux. Heat the mixture under reflux until most of the solid has dissolved.

Bromination and Isolation Using the pipet, add 0.3 mL of *concentrated* hydrobromic acid dropwise to the refluxing mixture; this may cause some precipitation to occur. Continue heating and add 0.2 mL of 30% hydrogen peroxide dropwise, noting the change in color of the mixture. Continue stirring and heating under reflux until the color is discharged. Allow the reaction mixture to cool to room temperature and adjust its pH to 5–7 by addition of saturated aqueous sodium bicarbonate. Cool the resulting mixture in an ice-water bath to complete precipitation of product, and collect it by vacuum filtration using a Hirsch funnel. Wash the filter cake with drops of water, transfer it to a watch glass or a piece of filter or weighing paper, and allow it to air-dry. Using a Craig tube, recrystallize the crude product from xylenes.

Analysis Weigh the product and determine the yield. Measure the melting point to determine the stereochemistry of the bromination. *Caution:* Do *not* use mineral oil as the heating fluid for this determination! Obtain IR and ^{1}H NMR spectra of your starting material and product and compare them with those of authentic samples (Figs. 10.34–10.37).

Discovery Experiment

Bromination of (Z)-Stilbene

Apply the procedure given for bromination of (*E*)-stilbene to study this reaction with (*Z*)-stilbene. Develop a protocol whereby you could demonstrate the stereochemical outcome of the dibromide(s) produced. Consult with your instructor before undertaking any experimental procedures.

Discovery Experiment

Solvent Effects on the Stereochemistry of Bromination

According to the literature (Buckles, R. E.; Bader, J. M.; Thurmaier, R. J. *J. Org. Chem.* **1962**, *27,* 4523–4527), the stereochemical outcome of the bromination of (*Z*)-stilbene is dependent upon the polarity of the solvent used in the reaction. Develop a protocol for confirming this report, using NMR spectroscopy (see Hartshorn, M. P.; Opie, M. C. A.; Vaughan, J. *Aust. J. Chem.* **1973**, *26,* 917–920) as the analytical tool for analyzing the stereochemistry of the reaction as a function of solvents of differing polarities. Consult with your instructor before undertaking any experimental procedures for evaluating the reported solvent effect.

Discovery Experiment

Substituent Effects on the Stereochemistry of Bromination

Apply the microscale procedure given for bromination of (*E*)-stilbene to study this reaction with one or more substituted (*E*)-stilbenes bearing electron-withdrawing or electron-donating substituents. Develop a protocol whereby you could demonstrate the stereochemical outcome of the dibromide(s) produced. Guidance for determining the stereochemical result is available in the following references: Buckles, R. E.; Bader, J. M.; Thurmaier, R. J. *J. Org. Chem.* **1962**, *27,* 4523–4527; Hartshorn, M. P.; Opie, M. C. A.; Vaughan, J. *Aust. J. Chem.* **1973**, *26,* 917–920. Compare your result(s) with those of others who have explored the reaction with substituted stilbenes different from yours. Consult with your instructor before undertaking any experimental procedures.

WRAPPING IT UP

Decolorize any *solutions* in which the color of bromine is visible by the dropwise addition of cyclohexene; then discard the resulting solutions together with *all other non-aqueous solutions* in a container for halogenated organic liquids. Pour all aqueous solutions down the drain.

EXERCISES

1. What does the "*E*" in (*E*)-stilbene stand for and what does it mean?
2. Is *meso*-stilbene dibromide optically active? What symmetry element is present in this molecule that precludes the existence of an enantiomer?
3. Explain why, in contrast to *meso*-stilbene dibromide, the dibromide **49** exists in two enantiomeric forms.
4. For the procedure in which a solution of bromine is used, why does solid not separate immediately after you begin adding the solution to that containing (*E*)-stilbene?
5. In the procedure involving *in situ* generation of bromine, what is the solid that appears when the *concentrated* hydrogen bromide is added?
6. What is the source of color that is produced upon addition of aqueous hydrogen peroxide to the reaction mixture?
7. Why should mineral oil *not* be used as the heating fluid for determining the melting point of *meso*-stilbene dibromide?
8. Write the structure of the product obtained when cyclohexene is used to decolorize bromine-containing solutions.
9. Using suitable stereochemical structures, write the mechanism for the addition of bromine to (*E*)-stilbene to give *meso*-stilbene dibromide via the intermediate cyclic bromonium ion **50.** Use curved arrows to symbolize the flow of electrons.
10. To what general class of mechanisms does the attack of bromide ion, Br^-, on the cyclic bromonium ion **50** belong? Does this ring-opening step follow first- or second-order kinetics?
11. How many centers of chirality are present in the bromonium ion **50** and in the isomeric carbocation **51**?
12. Draw suitable three-dimensional structures of the two products obtained by attack of bromide ion at both of the carbon atoms in the cyclic bromonium ion **50** and show that they are identical.
13. Based upon the mechanism you provided in Exercise 9, predict the major product of the addition of bromine to (*Z*)-stilbene.
14. Demonstrate that attack of bromide ion, Br^-, on the carbocation **51** can provide *both dl*- and *meso*-stilbene dibromide.
15. The reaction of bromine with cyclopentene according to Equation 10.18 is stereospecific and proceeds by *anti* addition. On the other hand, the addition of bromine to (*E*)-stilbene gives both *meso*-stilbene dibromide (**48**) and *dl*-stilbene dibromide (**49**). Rationalize this difference by comparing the relative stabilities of the carbocations resulting from ring-opening of the cyclic bromonium ions **38** and **50.**
16. How do the results of this experiment support the hypothesis that the addition of bromine to (*E*)-stilbene proceeds primarily through **50** rather than **51?**

17. Write the expected products for the addition of bromine to the following alkenes.

a. $CH_3CH_2CH{=}CH_2$

b. (CH_3 and H on one alkene carbon; H and CH_3 on the other, trans)

c. (cyclohexene with D on each alkene carbon)

Sign in at **www.cengage.com/login** and use the spectra viewer and Tables 8.1–8.8 as needed to answer the blue-numbered questions on spectroscopy.

18. Consider the spectral data for (*E*)-stilbene (Figs. 10.34 and 10.35).
    a. In the IR spectrum, identify the absorption(s) consistent with the *trans* relationship of the vinylic hydrogen atoms and explain the absence of a significant absorption in the range of 1600–1700 cm^{-1} normally characteristic of alkenes.
    b. In the ^1H NMR spectrum, assign the various resonances to the hydrogen nuclei responsible for them.
    c. For the ^{13}C NMR data, assign the various resonances to the carbon nuclei responsible for them.

19. Consider the ^1H NMR spectrum of *meso*-stilbene dibromide (Fig. 10.37a).
    a. In the ^1H NMR spectrum, assign the various resonances to the hydrogen nuclei responsible for them.
    b. For the ^{13}C NMR data, assign the various resonances to the carbon nuclei responsible for them.

20. Discuss the differences observed in the IR (Fig. 10.36) and NMR spectra of (*E*)-stilbene and *meso*-stilbene dibromide that are consistent with addition of Br_2 occurring in this experiment.

SPECTRA

Starting Materials and Products

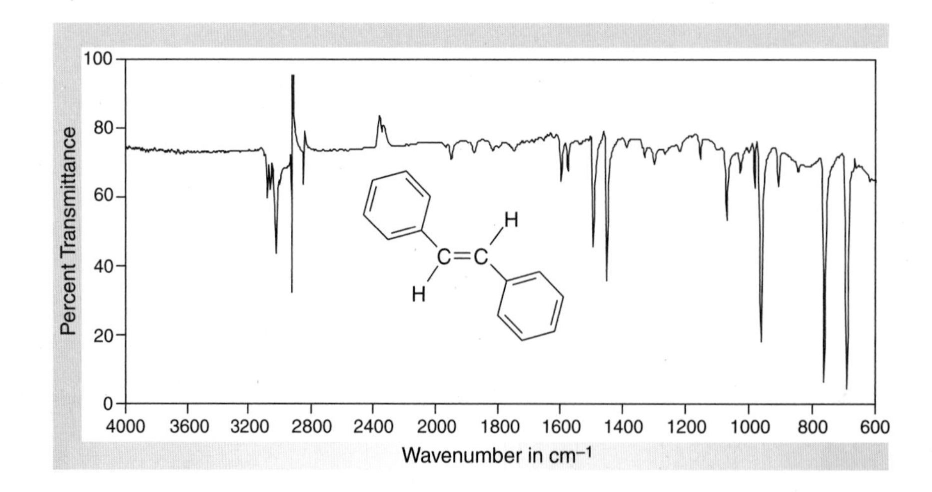

Figure 10.34
IR spectrum of (E)-*stilbene (IR card).*

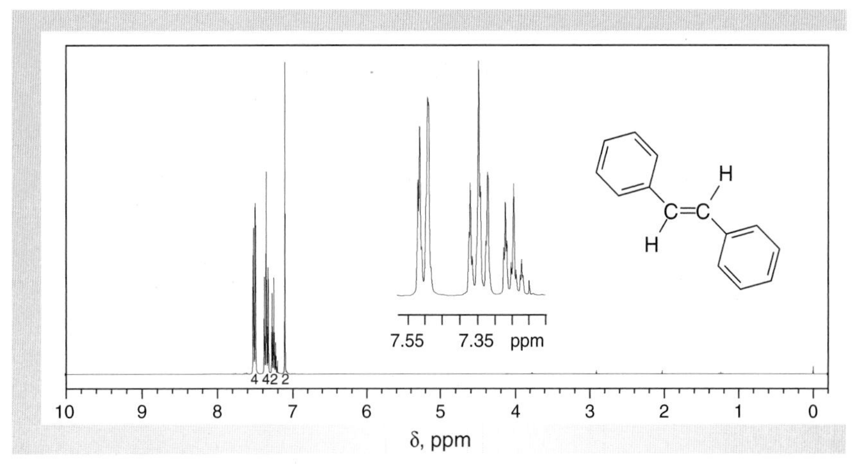

Figure 10.35
NMR data for (E)-*stilbene* ($CDCl_3$).

(a) ^{1}H NMR spectrum (300 MHz).
(b) ^{13}C data: δ 126.3, 127.8, 128.9, 129.0, 137.6.

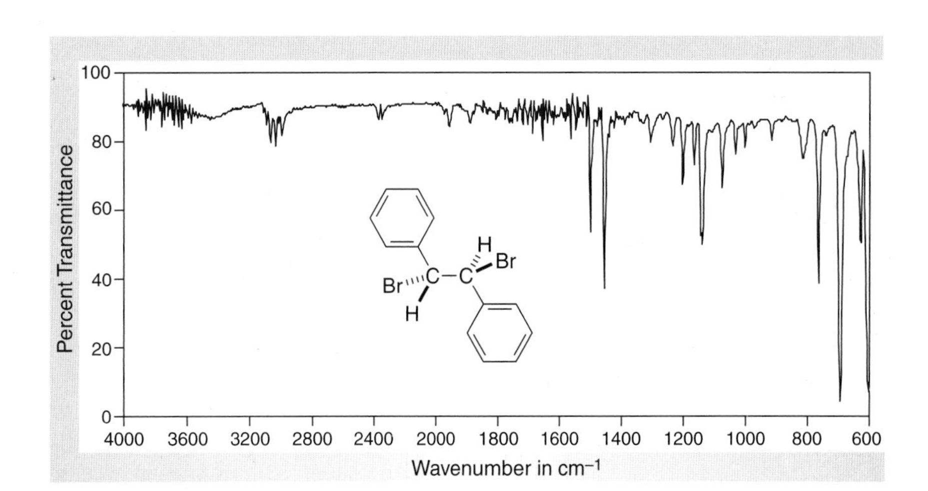

Figure 10.36
IR spectrum of meso-*stilbene dibromide (IR card).*

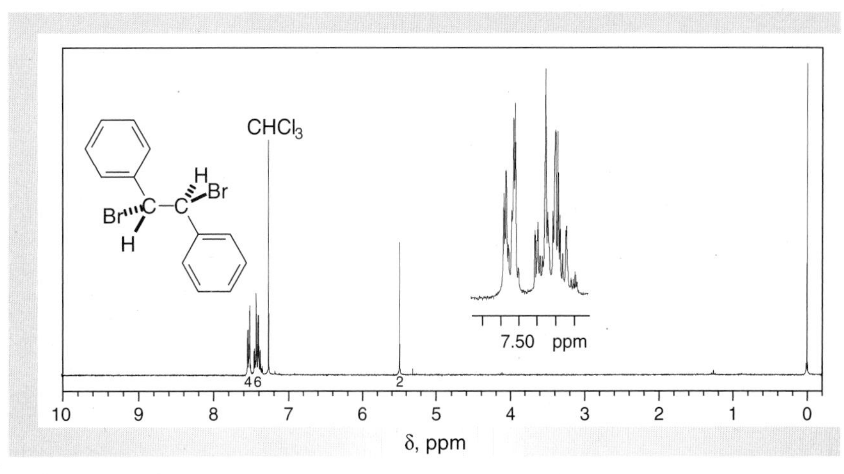

Figure 10.37
NMR data for meso-*stilbene dibromide* ($CDCl_3$).

(a) ^{1}H NMR spectrum (300 MHz).
(b) ^{13}C data: δ 57.0, 129.1, 129.6, 129.7, 141.7.

10.7 ACID-CATALYZED HYDRATION OF ALKENES

The acid-catalyzed addition of the elements of water across a carbon-carbon π-bond to give an alcohol is referred to as **hydration** of an alkene (Eq. 10.17). Mechanistically, this process is simply the reverse of the acid-catalyzed **dehydration** of alcohols (Sec. 10.3). The position of the equilibrium for these two competing processes depends upon the reaction conditions. Hydration of a double bond requires the presence of excess water to drive the reaction to completion, whereas the dehydration of an alcohol requires removing water to complete the reaction. In the experiment that follows, you will examine the acid-catalyzed hydration of norbornene (**52**) to give *exo*-norborneol (**53**) (Eq. 10.24) as the exclusive product.

H_3O^+ OH   (10.24)

52 Norbornene   **53** *exo*-Norborneol

The mechanism of this reaction and the structure of the intermediate carbocation were once the subject of intense controversy. When the double bond of norbornene is protonated, a carbocation is produced (Eq. 10.25). The question of whether this cation is **54a** and is in rapid equilibrium with the isomeric cation **54b,** or whether **54a** and **54b** are simply two contributing resonance structures to the resonance hybrid **55,** has been the focus of the debate. Note that the contributing structures **54a** and **54b** *differ in the location of an σ-bond,* not in the location of π-bonds as is more commonly encountered in **resonance structures.** The delocalized structure **55** is thus referred to as a **nonclassical carbocation,** and most chemists now accept this formulation as the more likely representation of this intermediate. It is evident from empirical results that the sterically more accessible side of the carbocation is the one away from or *exo* to the bridging carbon atom as shown. Nucleophilic attack of water solely from this side leads to *exo*-norborneol (**53**) rather than *endo*-norborneol (**56**).

See more on *Nonclassical Ions*

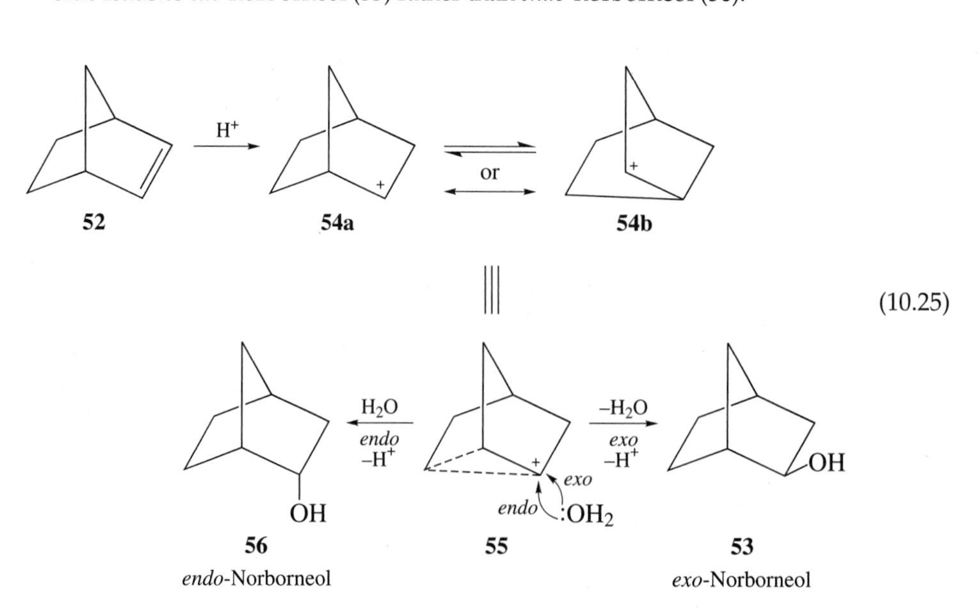

EXPERIMENTAL PROCEDURE

Hydration of Norbornene

Purpose To study the acid-catalyzed addition of water to an alkene.

SAFETY ALERT

1. **Norbornene is a volatile solid. Perform all weighings in a hood if possible.**
2. **If the acidic solution used in the first part of the experiment comes in contact with your skin, immediately flood the affected area with water and thoroughly rinse it with 5% sodium bicarbonate solution.**
3. **If the basic solution used in this experiment comes in contact with your skin, immediately flood the affected area with water and thoroughly rinse the area with a dilute solution (1%) of acetic acid.**

MINISCALE PROCEDURE

Preparation Sign in at **www.cengage.com/login** to answer Pre-Lab Exercises, access videos, and read the MSDSs for the chemicals used or produced in this procedure. Review Sections 2.7, 2.20, 2.21, and 2.29.

Apparatus A 25-mL Erlenmeyer flask, separatory funnel, ice-water bath, apparatus for simple distillation, magnetic stirring, sublimation, and *flameless* heating.

Setting Up Slowly add 2 mL of concentrated sulfuric acid to 1 mL of water in a 25-mL Erlenmeyer flask containing a stirbar and cool the solution to room temperature.

Hydration Working at the hood, add 1.0 g of norbornene in small pieces to the stirred solution of aqueous sulfuric acid. Continue stirring the mixture until all of the norbornene dissolves. Cool the flask *briefly* in an ice-water bath if the mixture becomes noticeably warm, but *do not cool the contents of the flask below room temperature.* Stir this solution for 15–30 min. Prepare a solution of 1.5 g of potassium hydroxide in 7.5 mL of water.

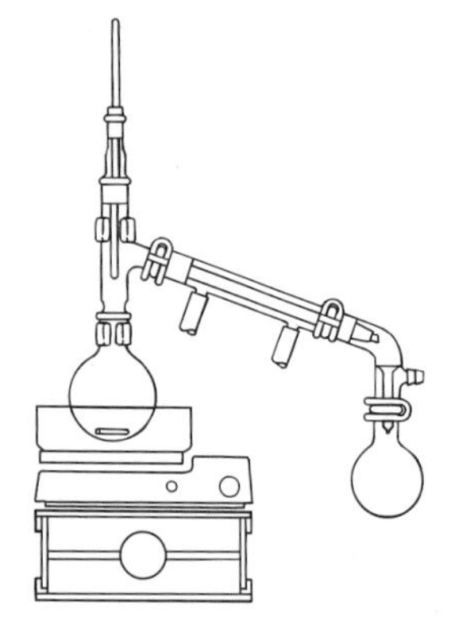

Work-Up and Isolation Cool both solutions in an ice-water bath and slowly add the base to the acidic reaction mixture to partially neutralize the acid. Be sure that the resulting mixture is at room temperature or below, then transfer it to a small separatory funnel, and add 15 mL of diethyl ether. If any solid is present in the bottom of the separatory funnel, add 1–2 mL of water to dissolve it. After shaking the funnel, *periodically venting it to release pressure,* separate the aqueous layer (*Save!*) and pour the ethereal solution into an Erlenmeyer flask. Return the aqueous layer to the separatory funnel, and extract again with a fresh 10-mL portion of diethyl ether. Separate the aqueous layer, and combine the two ethereal extracts in the separatory funnel.

Wash the combined extracts with 5 mL of water. Separate the aqueous layer, and wash the ethereal solution sequentially with 5-mL portions of saturated sodium bicarbonate solution and saturated sodium chloride solution; *periodically vent the separatory funnel* to release pressure that may develop. Separate the layers, place the ethereal solution in a dry 50-mL Erlenmeyer flask, and dry it over several spatula-tips full of anhydrous sodium sulfate with occasional swirling for 10–15 min.★ Add additional portions of anhydrous sodium sulfate if the liquid remains cloudy.

Transfer the dried ethereal solution to a 50-mL round-bottom flask equipped for simple distillation and concentrate the solution by distillation. Alternatively, rotary evaporation or other techniques may be used to concentrate the solution. Transfer the residue to a 25- or 50-mL filter flask.

Sublimation Equip the filter flask with a filter adapter or rubber stopper containing a centrifuge tube or a 13-mm × 120-mm test tube that extends to within about 2 cm of the bottom of the flask (Fig. 2.58b). The adapter or stopper should be lightly greased to prevent water from entering the sublimation chamber. Evacuate the flask using a water aspirator equipped with a trap and *then* half-fill the centrifuge tube or test tube with tightly packed chipped ice. *Gently* heat the flask using a steam bath or the sweeping motion of a heat gun or small flame. Add ice as necessary to the cold-finger on which the sublimed alcohol collects. When the alcohol has completely sublimed, cool the flask, break the vacuum to the aspirator, and disconnect the flask from the vacuum hose. Carefully decant the water out of the cold finger. Carefully remove the cold finger from the flask, while maintaining the cold finger in a horizontal or slightly inverted position. Scrape the soft crystalline product from the test tube.

Analysis Weigh the purified solid and determine the yield. Measure the melting point by placing the sample in a capillary tube that has been sealed about 2 cm from the open end. Obtain IR and ^{1}H NMR spectra of your starting material and product and compare them with those of authentic samples (Figs. 10.38–10.41).

WRAPPING IT UP

Slowly combine the *aqueous layers and washes*. If necessary, neutralize them with sodium carbonate, and flush them down the drain with excess water. Pour any *diethyl ether* obtained during the concentration step into a container for flammable organic solvents. After the diethyl ether has evaporated from the *sodium sulfate* in the hood, place the solid in a container for nonhazardous solids.

EXERCISES

1. What purpose is served by washing the ethereal solution of product with bicarbonate solution? Saturated brine?
2. How would the yield of *exo*-norborneol be affected if this alcohol were left under vacuum for an excessive period of time to remove all traces of solvent?
3. Why is it necessary to seal the capillary tube before taking the melting point of *exo*-norborneol?
4. Using curved arrows to symbolize the flow of electrons, show the mechanism by which the classical cations **54a** and **54b** can be interconverted.
5. What is the significance of the dashed bonds in the structure of the nonclassical cation **55**? What is "nonclassical" about this cation (i.e., how is **55** different from a classical carbocation)?

6. *Exo*-Norborneol (**53**) contains three centers of chirality, but the product obtained from the hydration of norbornene (**52**) in this experiment is optically inactive.
   a. Is norbornene chiral? Explain your answer by showing why it is or is not.
   b. Identify the three chiral centers in **53** with asterisks.
   c. Write the chemical structure for the stereoisomer of **53** that is produced by the hydration of **52** and show how it is formed from intermediate **55.** Are **53** and this stereoisomer enantiomers or diastereomers?
7. Why is sulfuric acid a superior acid to HCl for effecting the acid-catalyzed hydration of alkenes?
8. *Exo*-norborneol undergoes reaction with concentrated HBr to form a single diastereomer. Write the structure of this bromide and propose a stepwise mechanism for its formation, using curved arrows to symbolize the flow of electrons. Why is only one diastereomer produced?
9. For each of the alkenes **a–e,** write the structure of the alcohol that should be the major product of acid-catalyzed hydration.

a. Me, Me, Me **b.** Me **c.** Me

d. Me **e.** NO_2, MeO

Sign in at **www.cengage.com/login** and use the spectra viewer and Tables 8.1–8.8 as needed to answer the blue-numbered questions on spectroscopy.

10. Consider the spectral data for norbornene (Figs. 10.38 and 10.39).
    a. In the functional group region of the IR spectrum, identify any absorptions associated with the carbon-carbon double bond.
    b. In the ^1H NMR spectrum, assign the various resonances to the hydrogen nuclei responsible for them.
    c. For the ^{13}C NMR data, assign the various resonances to the carbon nuclei responsible for them. Explain why there are only four resonances in the spectrum.
11. Consider the spectral data for *exo*-norborneol (Figs. 10.40 and 10.41).
    a. In the functional group region of the IR spectrum, specify any absorptions associated with the alcohol function.
    b. In the ^1H NMR spectrum, assign the various resonances to the hydrogen nuclei responsible for them.
    c. For the ^{13}C NMR data, assign the various resonances to the carbon atom responsible for them.
12. Discuss the differences observed in the IR and NMR spectra of norbornene and *exo*-norborneol that are consistent with the addition of water occurring in this experiment.

SPECTRA

Starting Materials and Products

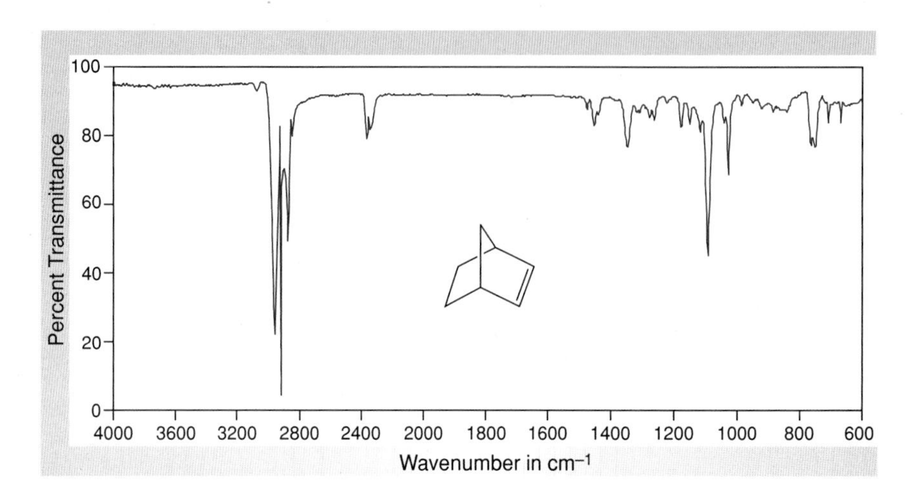

Figure 10.38
IR spectrum of norbornene (IR card).

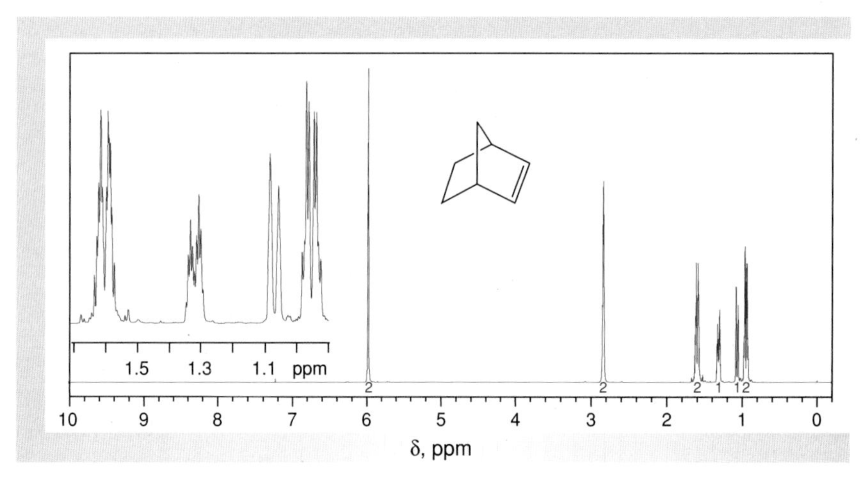

Figure 10.39
NMR data for norbornene ($CDCl_3$).

(a) ^{1}H NMR spectrum (300 MHz).
(b) ^{13}C data: δ 24.8, 42.0, 48.8, 135.4.

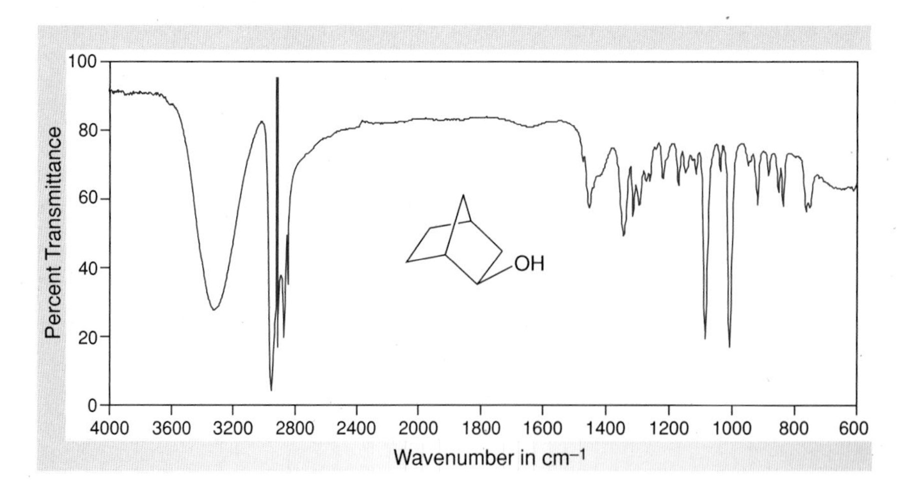

Figure 10.40
IR spectrum of exo-norborneol (IR card).

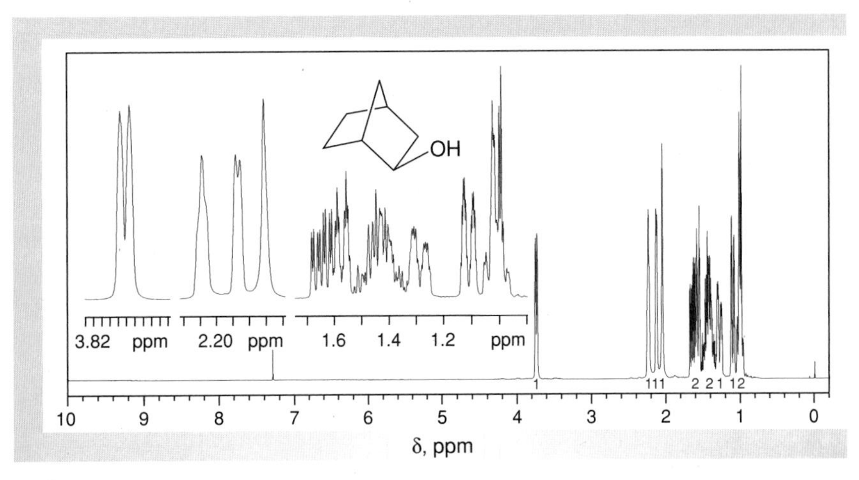

Figure 10.41
NMR data for exo-norborneol ($CDCl_3$).

(a) ^{1}H NMR spectrum (300 MHz).
(b) ^{13}C data: δ 24.7, 28.4, 34.5, 35.5, 42.2, 44.2, 74.6.

10.8 HYDROBORATION-OXIDATION OF ALKENES

See *Who was Markovnikov?*

See more on *Hydroboration-Oxidation*

The **regiochemistry** of acid-catalyzed hydration of alkenes to give alcohols is that predicted by Markovnikov's rule (Sec. 10.4) because the more stable intermediate carbocation is preferentially formed. Sometimes it is desirable to add the elements of water across a carbon-carbon π-bond in the opposite regiochemical sense to provide the **anti-Markovnikov** product (see the Historical Highlight at the end of this chapter). In order to accomplish this goal, a process termed **hydroboration-oxidation** was developed that involves the reaction of an alkene sequentially with diborane, B_2H_6, and basic hydrogen peroxide (Eq. 10.26).

$$R^1R^2C{=}CHR^3 \xrightarrow[\text{2. } H_2O_2/HO^-]{\text{1. } B_2H_6} R^1{-}C(R^2)(H){-}CH(R^3)(OH) \qquad (10.26)$$

Diborane is a dimer of borane, BH_3. The bonding in diborane is unusual because the hydrogen atoms bridge the two boron atoms with the two monomeric BH_3 subunits being bound by *two-electron, three-center bonds.* Because the boron atom in borane possesses an empty *p*-orbital, borane is a Lewis acid, and it forms stable complexes upon reaction with tetrahydrofuran (THF) and other ethers, which function as Lewis bases, as illustrated by the formation of a borane-THF complex (Eq. 10.27).

$$\underset{\text{Diborane}}{B_2H_6} + 2\ \underset{\text{Tetrahydrofuran}}{C_4H_8O} \rightleftharpoons \underset{\text{A borane-THF complex}}{2\ H_3\overset{-}{B}{-}\overset{+}{O}C_4H_8} \qquad (10.27)$$

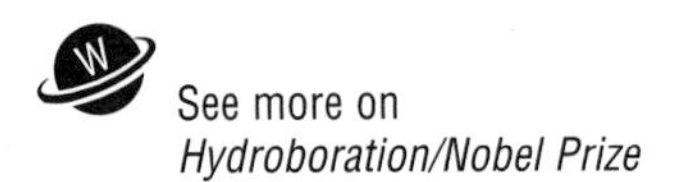

See more on *Hydroboration/Nobel Prize*

As you learned from reading Section 10.4, alkenes may serve as Lewis bases. Hence, alkenes can react with borane by the net addition of the boron-hydrogen bond across the carbon-carbon π-bond to give a monoalkylborane as the initial product. This process, known as **hydroboration,** is illustrated for 1-butene in Scheme 10.3. This intermediate monoalkylborane, **57,** then adds successively to additional molecules of 1-butene to form a dialkylborane **58** or a trialkylborane **59.**

The extent to which dialkyl- and trialkylboranes will be formed is dependent upon steric factors and the degree of substitution on the double bond.

Scheme 10.3

$$\underset{\text{1-Butene}}{CH_3CH_2CH{=}CH_2} \;(H{-}BH_2) \longrightarrow \underset{\text{A monoalkylborane}}{\underset{\mathbf{57}}{CH_3CH_2\overset{H}{C}H{-}\overset{BH_2}{C}H_2}} \xrightarrow{\text{1-butene}} \underset{\text{A dialkylborane}}{\underset{\mathbf{58}}{[CH_3(CH_2)_2CH_2]_2BH}} \xrightarrow{\text{1-butene}} \underset{\text{A trialkylborane}}{\underset{\mathbf{59}}{[CH_3(CH_2)_2CH_2]_3B}}$$

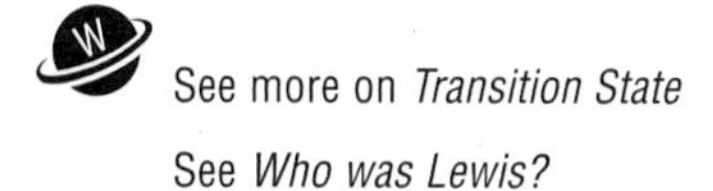

See more on *Transition State*

See *Who was Lewis?*

As observed in the electrophilic additions to alkenes depicted in Equations 10.16 and 10.17, the addition of borane to alkenes is *regioselective* (Sec. 10.4) and occurs so the *electrophilic boron atom* preferentially attaches to the *less substituted carbon atom* of the carbon-carbon π-bond. This selectivity may be understood by considering the mechanism of hydroboration. As borane approaches the double bond, a Lewis acid–Lewis base interaction occurs in which the π-electrons of the double bond begin to interact with the empty *p*-orbital on the Lewis acidic boron atom, resulting in the development of a partial positive charge on the more highly substituted carbon atom. The boron atom then assumes a partial negative charge that facilitates the transfer of the hydride ion, H:$^-$, from the boron to the carbon atom bearing the partial positive charge. This process occurs through the four-centered **transition state 60,** in which all four atoms simultaneously undergo changes in bonding.

$$\left[\begin{array}{c} \delta^- \\ H{-}{-}BH_2 \\ \delta^+ \\ C_2H_5 \end{array}\right]^{\ddagger}$$

60

Thus, the regiochemistry of hydroboration is predicted by the same general rule that applies to all electrophilic additions to alkenes: *The reaction of an electrophile with a carbon-carbon double bond occurs preferentially via the transition state in which a partial positive charge develops on that carbon atom better able to accommodate it.* Geometric constraints inherent in the cyclic transition state **60** require that the addition of borane to the alkene proceed so that both the boron and the hydrogen add from the same face of the double bond, a process called ***syn*-addition.**

Treating a trialkylborane with alkaline hydrogen peroxide promotes oxidative cleavage to form three moles of an alcohol and boric acid, as exemplified in the conversion of **59** into 1-butanol (**64**), shown in Scheme 10.4. The first step of the reaction involves nucleophilic attack of the hydroperoxide anion on the Lewis acidic boron atom of **59** to give **61.** In the next step, migration of the alkyl group from boron to oxygen proceeds with displacement of hydroxide ion and cleavage of the weak oxygen-oxygen bond to give the monoalkoxyborane **62.** An important consequence of the *intramolecularity* of this rearrangement is that it proceeds with *retention* of configuration at the carbon atom. The sequence is repeated until all of the boron-carbon bonds have been converted into oxygen-carbon bonds, whereupon the intermediate trialkoxyborane **63** undergoes alkaline hydrolysis to give 1-butanol and boric acid.

Scheme 10.4

$$[CH_3(CH_2)_2CH_2]_3B + {}^-\ddot{O}\text{—}OH \longrightarrow [CH_3(CH_2)_2CH_2]_2\overset{-}{B}\begin{matrix}CH_2(CH_2)_2CH_3\\ O\text{—}OH\end{matrix} \xrightarrow{-HO^-}$$

59 **61**

$$[CH_3(CH_2)_2CH_2]_2B\text{—}O\text{—}CH_2(CH_2)_2CH_3 \xrightarrow{{}^-\ddot{O}\text{—}OH} CH_3(CH_2)_2CH_2B\begin{matrix}O(CH_2)_3CH_3\\ O(CH_2)_3CH_3\end{matrix}$$

62
A monoalkoxyborane

A dialkoxyborane

$$\xrightarrow{{}^-\ddot{O}\text{—}OH} [CH_3(CH_2)_3O]_3B \xrightarrow[H_2O]{HO^-} 3\ CH_3(CH_2)_2CH_2\text{—}OH + B(OH)_3$$

63 A trialkoxyborane; **64** 1-Butanol; Boric acid

By considering the reactions depicted in Schemes 10.3 and 10.4, you can see how the hydroboration-oxidation of an alkene gives an alcohol that is the product of overall anti-Markovnikov addition of the elements of H–OH to the carbon-carbon double bond. Remember, however, that the key step determining the regiochemistry of the reaction is the Markovnikov addition of the hydrogen-boron bond across the π-bond.

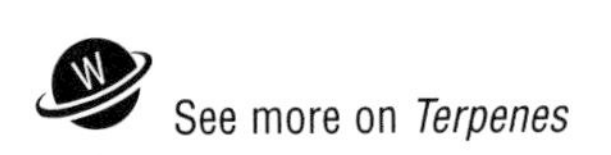

See more on *Terpenes*

In this experiment, you will examine the regiochemistry and stereochemistry of the sequence of hydroboration-oxidation using (+)-α-pinene (**65**), a terpene, as the alkene and a solution of borane-tetrahydrofuran complex as the hydroborating reagent (Eq. 10.28). Using this complex avoids the need to handle diborane, B_2H_6, a potentially hazardous reagent. Since the double bond in **65** is trisubstituted and sterically hindered, only *two* molecules of this alkene react with each molecule of borane, thus forming the intermediate dialkylborane **66.** Subsequent oxidation of **66** with alkaline hydrogen peroxide then gives (−)-isopinocampheol (**67**).

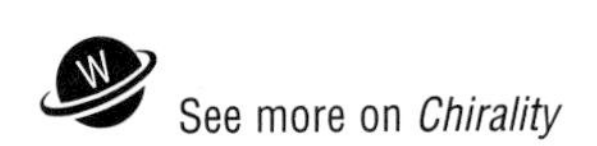

See more on *Chirality*

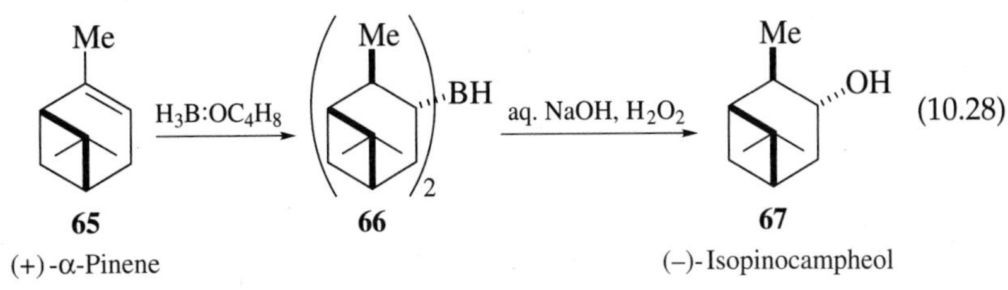

You will be able to verify that (−)-isopinocampheol (**67**) is the sole product of the hydroboration-oxidation of (+)-α-pinene (**65**) by determining the melting points of the product of your reaction and that of a suitable derivative, such as an α-naphthylurethane (Sec. 24.11D). Comparing these data with those in Table 10.1 for the alternative products **68–70,** which could also arise from adding borane in a Markovnikov sense, and the possible products **71** and **72,** which would be obtained from an anti-Markovnikov hydroboration, provides compelling evidence that the structure of the product is indeed **67.** Hence, by performing the hydroboration-oxidation of (+)-α-pinene (**65**), you will find that the addition of borane to alkenes occurs in a highly regioselective fashion with exclusive *syn*-addition.

Table 10.1 *Melting Points of Isomeric Campheols and Pinanols*

Structure	°C	°C, *of α-Naphthylurethane Derivative*
67	51–53 (55–56)	87.5–88
68	45–47 (48)	88
69	27	148
70	67	91
71	78–79	Not formed
72	58–59	Not formed

Moreover, you will discover that the hydroboration of (+)-α-pinene is an example of a **stereoselective** reaction because only one of several possible stereoisomeric products is formed by addition of the borane reagent to only one face of the π-bond of **65.**

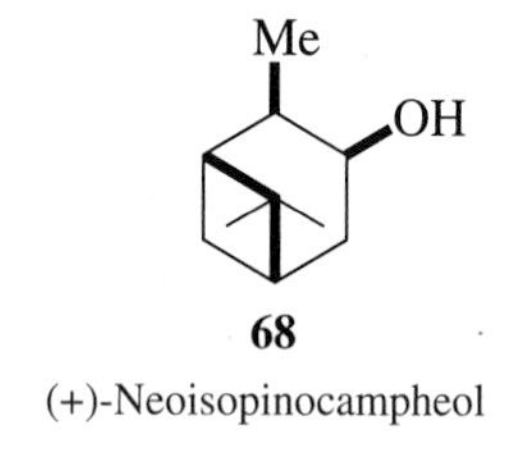

68
(+)-Neoisopinocampheol

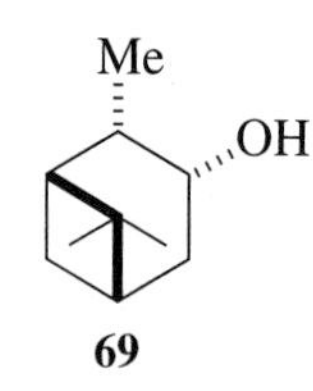

69
(–)-Neopinocampheol

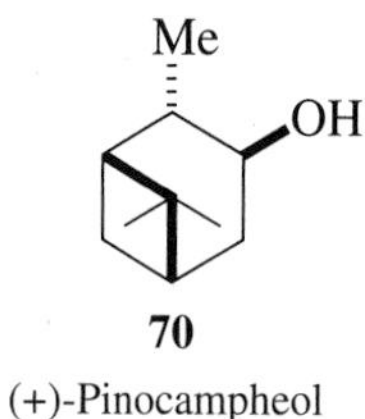

70
(+)-Pinocampheol

71
(+)-*cis*-2-Pinanol

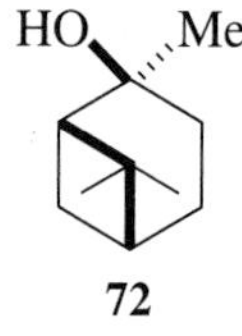

72
(+)-*trans*-2-Pinanol

EXPERIMENTAL PROCEDURES

Hydroboration-Oxidation of (+)-α-Pinene

Purpose To study the hydration of an alkene by means of hydroboration-oxidation and to discover the regiochemistry and stereochemistry of the process.

SAFETY ALERT

1. **Tetrahydrofuran is a flammable and potentially toxic solvent. Do not use this solvent near open flames.**
2. **The 30% hydrogen peroxide used is a *strong oxidant* and may blister the skin on contact. If you accidentally spill some on your skin, wash the affected area with copious amounts of water.**
3. **The complex of borane and tetrahydrofuran reacts *vigorously* with water. Use care when disposing of any excess reagent and when quenching the reaction.**

MINISCALE PROCEDURE

Preparation Sign in at **www.cengage.com/login** to answer Pre-Lab Exercises, access videos, and read the MSDSs for the chemicals used or produced in this procedure. Review Sections 2.5, 2.10, 2.11, 2.13, 2.21, 2.27, 2.28, and 2.29.

Apparatus A 50-mL round-bottom flask, Claisen adapter, separatory funnel, two 5-mL glass syringes with needles, drying tube, water bath, apparatus for simple distillation, magnetic stirring, and *flameless* heating.

Setting Up Equip the round-bottom flask with the Claisen adapter, and prepare a drying tube filled with calcium chloride for attachment to the *sidearm* of the adapter. Fit a rubber septum onto the joint of the adapter that is directly in line with the lower male joint (Fig. 10.42). Dry the glass parts of the syringes in an oven at 110 °C for at least 15 min. Using a small flame or a heat gun, gently heat the flask to remove traces of water from the apparatus; do *not* overheat the joint with the septum. After the flask has been heated for a few minutes, discontinue the heating, add a stirbar, attach the drying tube, and allow the flask to *cool to room temperature.* Remove the glass syringes from the oven and allow them to *cool to room temperature,* preferably in a desiccator. Prepare a solution of 1.6 mL of (+)-α-pinene in 2 mL of dry tetrahydrofuran.

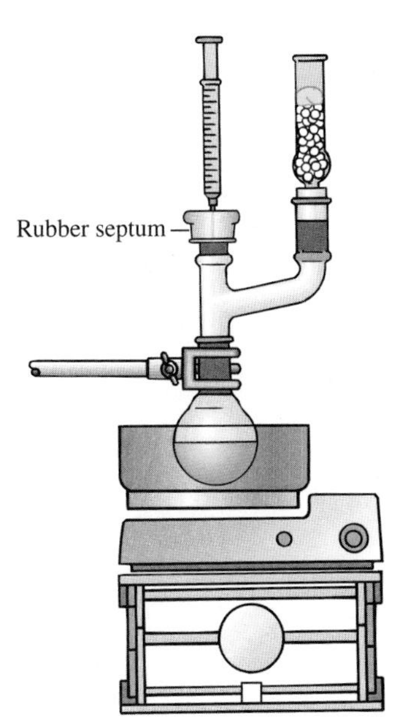

Figure 10.42
Miniscale hydroboration apparatus.

Hydroboration Draw 5 mL of a 1.0 *M* solution of borane-tetrahydrofuran complex in tetrahydrofuran into the *dry* 5-mL glass syringe. Pierce the rubber septum with the needle of the syringe, being careful not to press down on the plunger, and add the solution directly to the round-bottom flask. Place the apparatus in a water bath made from cold tap water and located on a magnetic stirrer. Use another dry 5-mL glass syringe to transfer the solution of (+)-α-pinene slowly and dropwise to the stirred solution of borane-tetrahydrofuran complex; this addition should require about 10–15 min. Remove the water bath and stir the reaction mixture at room temperature for about 1.5 h. Remove the septum from the reaction flask, and place the flask in an ice-water bath on a magnetic stirrer. Slowly add 0.5 mL of water with stirring to the reaction mixture to decompose any unreacted borane-tetrahydrofuran complex.

Oxidation Prepare 3 mL of an alkaline solution of hydrogen peroxide by mixing equal parts of 30 weight % aqueous hydrogen peroxide and 3 *M* aqueous NaOH. Using a Pasteur pipet, *slowly* add this solution to the stirred reaction mixture. Perform the addition *cautiously,* as vigorous gas evolution may occur. Once the addition is complete, stir the resulting mixture for 5–10 min.

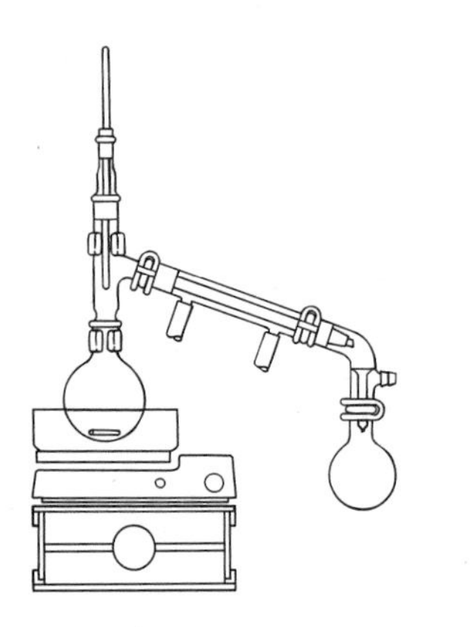

Work-Up and Isolation Transfer the reaction mixture to a small separatory funnel. Add 5 mL of water and separate the layers. Extract the aqueous layer with two separate 10-mL portions of diethyl ether. Wash the combined organic layers sequentially with two 10-mL portions of water and two 10-mL portions of saturated brine, and dry them over several spatula-tips full of anhydrous sodium sulfate.★ Add additional portions of anhydrous sodium sulfate if the liquid remains cloudy. Filter or decant the dried solution into a dry 50-mL round-bottom flask and remove the volatile solvents by simple distillation. Alternatively, rotary evaporation or other techniques may be used to remove the solvents. The final traces of solvents may be removed by attaching the flask to a vacuum source and gently swirling the contents as the vacuum is applied.★ Transfer the solid residue to a tared container.

Analysis Weigh your product and determine the yield. Obtain IR and NMR spectra of your starting material and product and compare them with those of authentic samples (Figs. 10.44–10.47). You may also verify the identity of the product by comparing its properties and the melting point of its α-naphthylurethane derivative, prepared according to the procedure in Section 25.11D, with the data summarized in Table 10.1.

Discovery Experiment

Regiochemistry of Hydroboration/Oxidation of an Acyclic Alkene

Devise an experimental procedure for hydroboration/oxidation of an alkene, the results of which will test the net "anti-Markovnikov" orientation of the overall reaction. One possible substrate would be 1-hexene. Be certain the spectral data are available in the literature for the possible isomers that could be formed, so that you can analyze the product(s) you obtain. Check with your instructor before undertaking the procedure.

Discovery Experiment

Regio- and Stereochemistry of Hydroboration/ Oxidation of a Cyclopentene

Devise an experimental procedure for hydroboration/oxidation of 1-methylcyclopentene, the results of which will test the net "anti-Markovnikov" orientation of the overall reaction and its stereochemistry. Be certain the spectral data are available in the literature for the possible isomers that could be formed, so that you can analyze the product(s) you obtain. Check with your instructor before undertaking the procedure.

MICROSCALE PROCEDURE

Preparation Sign in at **www.cengage.com/login** to answer Pre-Lab Exercises, access videos, and read the MSDSs for the chemicals used or produced in this procedure. Review Sections 2.5, 2.10, 2.11, 2.13, 2.27, 2.28, and 2.29.

Apparatus A 3- and 5-mL conical vial, two 1-mL glass syringes with needles, screw-cap centrifuge tube, Claisen adapter, drying tube, water bath, apparatus for magnetic stirring, simple distillation, and *flameless* heating.

Setting Up Dry the 5-mL conical vial, the Claisen adapter, and the glass parts of the two syringes in an oven at 110 °C for at least 15 min. Remove the vial and adapter from the oven, and after they have cooled to room temperature, preferably in a desiccator, place a spinvane in the vial, attach the adapter, and fit its sidearm with a drying tube. Seal the other opening of the adapter with a rubber septum (Fig. 10.43). Place the apparatus in a water bath made from cold tap water and located on a magnetic stirrer. Remove the glass syringes from the oven and allow them to *cool to room temperature,* preferably in a desiccator. Prepare a solution of 0.3 mL of (+)-α-pinene in 0.5 mL of dry tetrahydrofuran.

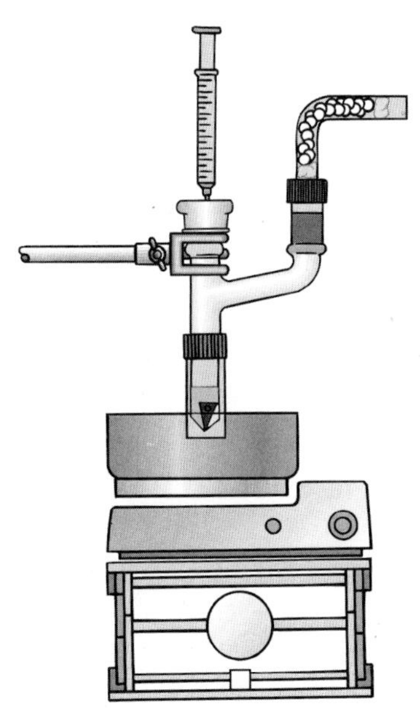

Figure 10.43
Microscale hydroboration apparatus.

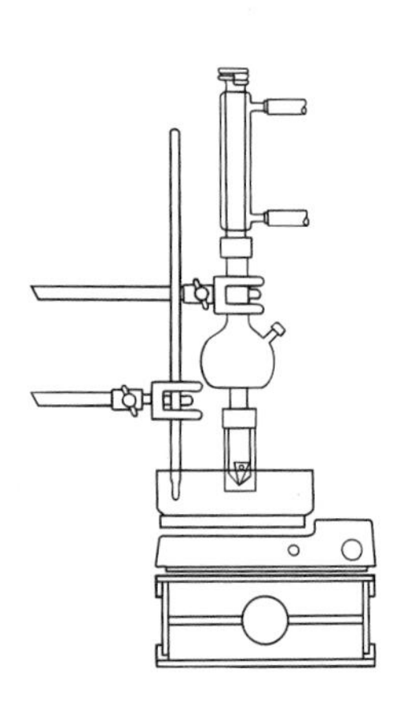

Hydroboration Draw 1 mL of a 1.0 *M* solution of borane-tetrahydrofuran complex in tetrahydrofuran into a *dry* 1-mL glass syringe. Pierce the rubber septum with the needle of the syringe, being careful not to press down on the plunger, and add the solution directly to the conical vial. Use the other dry 1-mL glass syringe to transfer the solution of (+)-α-pinene slowly and *dropwise* to the stirred solution of borane-tetrahydrofuran complex; this addition should require about 5–10 min. The rate of addition may be controlled by slowly twisting the plunger down the body of the syringe. Remove the water bath and stir the reaction mixture at room temperature for about 1.5 h. Remove the septum from the reaction vial and place the vial in an ice-water bath on a magnetic stirrer. Slowly add 2–3 drops of water with stirring to the reaction mixture to decompose any unreacted borane-tetrahydrofuran complex.

Oxidation Prepare 0.6 mL of an alkaline solution of hydrogen peroxide by mixing equal parts of 30 weight % aqueous hydrogen peroxide and 3 *M* aqueous NaOH. Using a Pasteur pipet, *slowly* add this solution to the stirred reaction mixture. Perform this addition *cautiously,* as vigorous gas evolution may occur. Once the addition is complete, stir the resulting mixture for about 5 min and then add 1 mL of water followed by 1 mL of diethyl ether.

Work-Up and Isolation Stir the mixture in the vial for about 5 min in order to mix the liquid and solid materials. Remove the lower aqueous layer using a Pasteur pipet and transfer it to a 3-mL conical vial; transfer the organic layer to a screw-cap centrifuge tube. Extract the aqueous layer three times with separate 1-mL-portions of diethyl ether, each time transferring the organic layer to the centrifuge tube. Vent the tube as necessary to relieve pressure. Wash the combined organic layers sequentially with two 2-mL portions each of water and saturated brine. Transfer the organic layer to a dry test tube using a Pasteur pipet, and add several microspatula-tips full of anhydrous sodium sulfate. Allow the solution to stand for 10–15 min, with occasional swirling to hasten drying.★ Add additional portions of anhydrous sodium sulfate if the liquid remains cloudy.

Using a Pasteur pipet, transfer the organic solution to a clean 5-mL conical vial and add a stirbar to the vial. Remove the volatile solvents by simple distillation. Alternatively, rotary evaporation or other techniques may be used to remove the solvent. Following removal of solvents, transfer the solid residue to a tared container.

Analysis Weigh your product and determine the yield. Obtain IR and NMR spectra of your starting material and product and compare them with those of authentic samples (Figs. 10.44–10.47). You may also verify the identity of the product by comparing its properties and the melting point of its α-naphthylurethane derivative, prepared according to the procedure in Section 25.11, with the data summarized in Table 10.1.

Discovery Experiment

Regiochemistry of Hydroboration/Oxidation of an Acyclic Alkene

Devise an experimental procedure for hydroboration/oxidation of an alkene, the results of which will test the net "anti-Markovnikov" orientation of the overall reaction. One possible substrate would be 1-hexene. Be certain the spectral data are available in the literature for the possible isomers that could be formed, so that you can analyze the product(s) you obtain. Check with your instructor before undertaking the procedure.

Discovery Experiment

Regio- and Stereochemistry of Hydroboration/Oxidation of a Cyclopentene

Devise an experimental procedure for hydroboration/oxidation of 1-methylcyclopentene, the results of which will test the net "anti-Markovnikov" orientation of the overall reaction and its stereochemistry. Be certain the spectral data are available in the literature for the possible isomers that could be formed, so that you can analyze the product(s) you obtain. Check with your instructor before undertaking the procedure.

WRAPPING IT UP

Dilute the *aqueous layer* with water, neutralize it with acetic acid, and then flush it down the drain with excess water. Pour the *distillate* containing THF and ether in the container for nonhalogenated organic solvents.

EXERCISES

1. Why is it necessary to maintain the pH of the reaction mixture above 7 during the oxidation of an alkyl borane with hydrogen peroxide?
2. Why is it necessary to use dry apparatus and solvents when using a solution of borane-THF complex?
3. The hydroborating agent used in this procedure is a complex of borane and tetrahydrofuran.
   a. Why do you think this reagent was used rather than diborane, B_2H_6, itself?
   b. Provide a Lewis dot structure for BH_3•THF showing the direction of each bond dipole with the symbol $\mapsto$.
   c. Why is uncomplexed BH_3 an unstable compound?
4. (+)-α-Pinene is the limiting reagent in the hydroboration reaction in this procedure. How many molar equivalents of BH_3 and of hydride ion, H^-, are used?
5. Label each center of chirality in (+)-α-pinene with an asterisk. Draw the chemical structure for (−)-α-pinene.
6. Why is the hydroboration-oxidation of an alkene considered an *addition* reaction? What small molecule has been "added" to the alkene in the overall reaction?
7. Provide explanations based on the mechanism of the hydroboration-oxidation process that would account for why compounds **68–72** are not formed from (+)-α-pinene.

8. Predict the structure of the major product that would be formed upon hydroboration-oxidation of each of the alkenes **a–e.** Include the correct stereochemistry wherever appropriate.

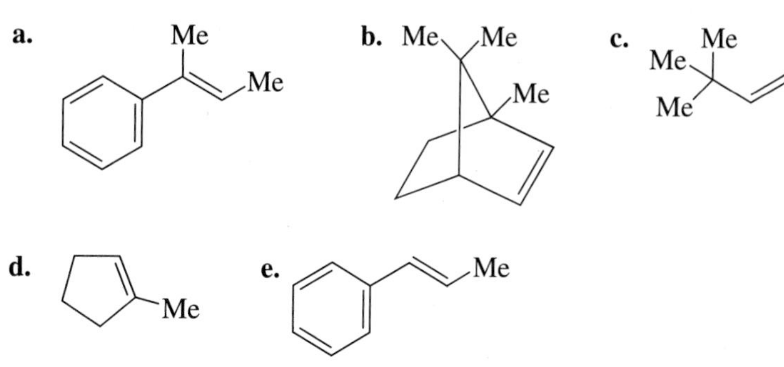

9. For each of the alkenes in Exercise 8, write the structure of the alcohol that should be the major product upon *acid-catalyzed* hydration.

10. An important advantage of hydroboration-oxidation for hydration of alkenes is that rearrangements of the carbon skeleton do not occur, whereas they do in acid-catalyzed hydration. For example, acid-catalyzed hydration of (+)-α-pinene (**65**) proceeds with skeletal rearrangements to produce a number of isomeric alcohols.
    - **a.** Draw the structure(s) of the product(s) you would expect to obtain upon acid-catalyzed hydration of (+)-α-pinene if no skeletal rearrangements occurred.
    - **b.** Draw the structures of the isomeric alcohols having the formula $C_{10}H_{18}O$ that you might expect to obtain upon treatment of (+)-α-pinene with aqueous acid.

11. The hydroboration of **65** is a stereoselective reaction in which borane approaches from only one face of the carbon-carbon double bond to give the dialkylborane **66.** Provide a rationale for this observed stereoselectivity.

12. Write the structure of the α-naphthylurethane derivative of **67.**

13. Neither of the alcohols **71** or **72** forms an α-naphthylurethane derivative, whereas the isomers **67–70** each do. Explain this difference in reactivities.

Sign in at **www.cengage.com/login** and use the spectra viewer and Tables 8.1–8.8 as needed to answer the blue-numbered questions on spectroscopy.

14. Consider the spectral data for (+)-α-pinene (Figs. 10.44 and 10.45).
    - **a.** In the functional group region of the IR spectrum, identify any absorptions associated with the carbon-carbon double bond.
    - **b.** In the ^1H NMR spectrum, assign the various resonances to the hydrogen nuclei responsible for them.
    - **c.** For the ^{13}C NMR data, assign the resonances at δ 116.2 and 144.1 to the carbon nuclei responsible for them.

15. Consider the spectral data for (−)-isopinocampheol (Figs. 10.46 and 10.47).
    - **a.** In the functional group region of the IR spectrum, specify any absorptions associated with the alcohol function.
    - **b.** In the ^1H NMR spectrum, assign the various resonances to the hydrogen nuclei responsible for them.
    - **c.** For the ^{13}C NMR data, assign the resonance at δ 71.7 to the carbon atom responsible for it.

16. Discuss the differences observed in the IR and NMR spectra of (+)-α-pinene and (−)-isopinocampheol that are consistent with overall addition of water to the double bond occurring in this experiment.

SPECTRA

Starting Materials and Products

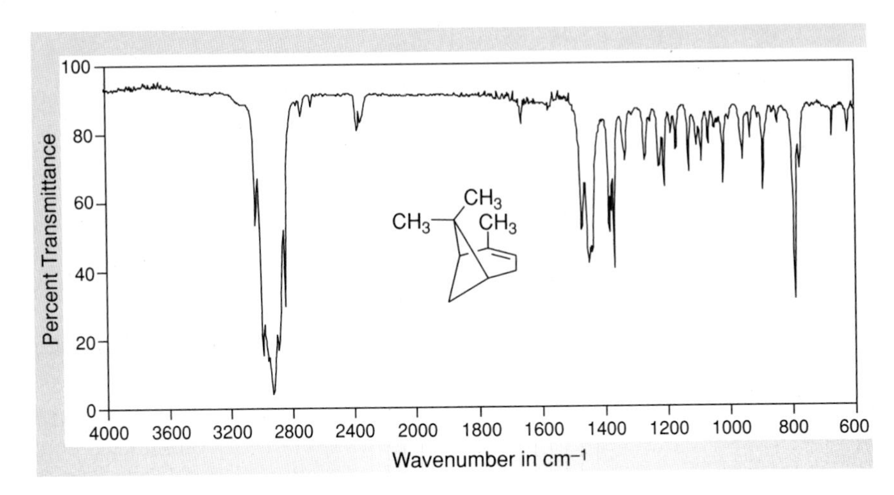

Figure 10.44
IR spectrum of (+)-α-pinene (neat).

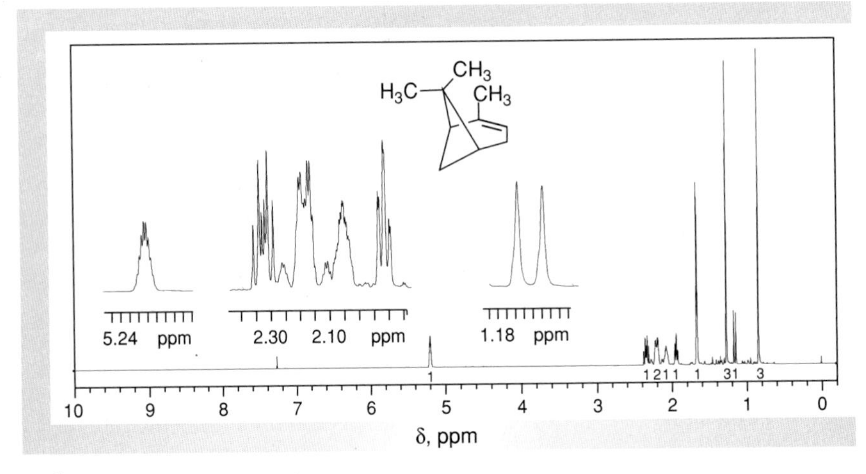

Figure 10.45
NMR data for (+)-α-pinene ($CDCl_3$).

(a) ^{1}H NMR spectrum (300 MHz).
(b) ^{13}C data: δ 20.8, 22.8, 26.8, 31.4, 31.5, 38.0, 41.5, 42.2, 116.2, 144.1.

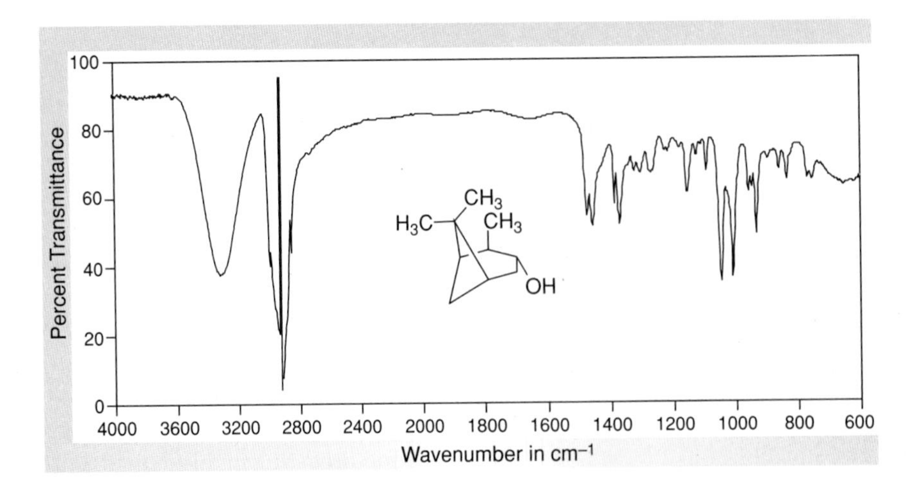

Figure 10.46
IR spectrum of (−)-isopinocampheol (IR card).

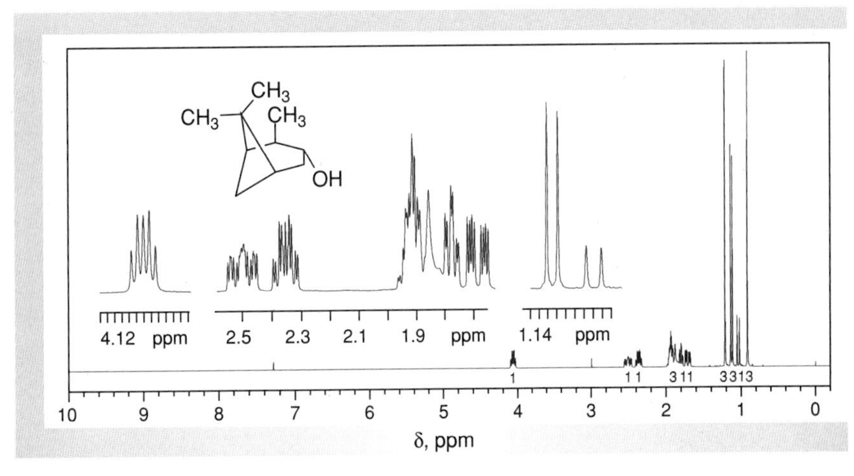

Figure 10.47
NMR data for (−)-isopinocampheol ($CDCl_3$).

(a) ^{1}H NMR spectrum (300 MHz).
(b) ^{13}C data: δ 20.7, 23.7, 27.7, 34.4, 38.1, 39.1, 41.8, 47.8, 47.9, 71.7.

HISTORICAL HIGHLIGHT

Additions Across Carbon-Carbon π-Bonds

A characteristic reaction of molecules containing carbon-carbon multiple bonds is the addition of a generalized reagent X–Y across the π-bond. If both this reagent and the π-bond to which it is adding are unsymmetrical, two orientations of addition are possible, as shown below for the possible modes of addition of H–Br to 1-methylcyclohexene. The mode of addition providing the 1-bromo-1-methyl isomer is known as "Markovnikov addition" in recognition of the contributions of V. Markovnikov, who was the first chemist to recognize that when a reagent H–Y adds to an unsymmetrically substituted alkene, the proton becomes attached to the carbon atom already bearing the larger number of hydrogen atoms.

1-Methylcyclohexene + HBr ⟶ 1-Bromo-1-methylcyclohexane

1-Methylcyclohexene + HBr ⟶ 1-Bromo-2-methylcyclohexane

Markovnikov, a Russian chemist, was on the faculty of Kazan State University in the second half of the 19th century, and it was during this period that he performed the research that led to his "rule." Although not a well-known academic institution in the West, this university sired a long line of chemists whose names are familiar to us, among which are Zaitsev, Arbuzov, and Reformatsky, all of whom have a rule or reaction bearing their name. The university also has the distinction of having expelled Lenin in 1887 for taking part in an antigovernment students' rally!

The theoretical basis for Markovnikov's rule was unknown at the time of its formulation, and indeed there were important apparent exceptions to it. Thus in 1929, Frank Mayo, a graduate student of M. S. Kharasch at the University of Chicago, found that adding H–Br to allyl bromide provided *either* 1,2-dibromopropane, the "Markovnikov adduct," *or* 1,3-dibromopropane, the "anti-Markovnikov adduct," as the principal product, depending on the particular reaction conditions used. Thus, bubbling nitrogen through a commercial sample of allyl bromide, followed by the addition of anhydrous H–Br, afforded 1,2-dibromopropane almost exclusively, whereas substituting oxygen for nitrogen provided mainly 1,3-dibromopropane. Indeed, treating various samples of allyl

(Continued)

HISTORICAL HIGHLIGHT *Additions Across Carbon-Carbon π-Bonds (Continued)*

bromide under seemingly identical reaction conditions provided widely differing ratios of the two adducts: sealing mixtures of allyl bromide and anhydrous H–Br under vacuum provided ratios of 1,2-dibromopropane to 1,3-dibromopropane ranging from 4:1 to 1:5. Such variations were confusing, to say the least, and Kharasch was led on one occasion to comment, regarding the variation in percentages, "You know, Mayo, maybe it's the phase of the moon!"

$$CH_2{=}CHCH_2Br + H{-}Br \rightarrow CH_2(H){-}CH(Br)CH_2Br$$

Allyl bromide → 1,2-Dibromopropane

$$CH_2{=}CHCH_2Br + H{-}Br \rightarrow CH_2(Br){-}CH(H)CH_2Br$$

1,3-Dibromopropane

Not believing the phase of the moon was relevant, Mayo continued his efforts to define the variables dictating the orientation of addition of H–Br to allyl bromide. He performed nearly 500 different trials, some of which lasted for over a year. He systematically varied a number of parameters, such as temperature and the presence of the water, air, hydrogen, and light, to determine their effects on the ratio of the two isomers. His persistence ultimately paid off, as he was able to show that the anti-Markovnikov mode of addition that provided 1,3-dibromopropane was due to contamination of the allyl bromide with small amounts of peroxides, although a mechanistic understanding of the "peroxide effect" was lacking at the time he and Kharasch published their results (Kharasch, M. S.; Mayo, F. R. *J. Am. Chem. Soc.* 1933, *55*, 2468). It is not surprising that they failed to propose free radicals as being responsible for the peroxide effect: it was only in 1929 that a report from another research group showed that radicals played a role in organic reactions, so free-radical chemistry was in its infancy at the time of Mayo's experiments. Indeed, in the opinion of many chemists of that era, a unified basis for the whole of organic chemistry would be founded on "quasi-ionic" or "heterolytic" reaction mechanisms.

Based on modern concepts of reaction mechanisms, the peroxide effect arises from decomposition of a peroxide to peroxy radicals, which then react with hydrogen bromide to produce bromine atoms, as shown below. This initiates free-radical chain addition rather than polar addition of H–Br to the π-bond. Consequently, the product-determining step in free-radical chain addition of H–Br to an alkene is attack of Br• on the π-bond, whereas that for polar addition of this H–Br is electrophilic attack of a proton. This results in opposite regiochemistries of the overall addition, since in both cases the more stable tertiary species is produced.

$$RO{-}OR \longrightarrow 2\ RO^{\bullet}$$

A peroxide → Peroxy radicals

$$RO^{\bullet} + H{-}Br \longrightarrow RO{-}H + Br^{\bullet}$$

Radical Addition of HBr to an Alkene

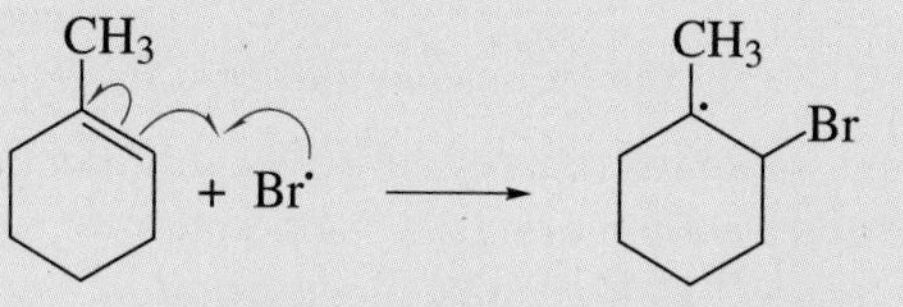

1-methylcyclohexene (CH3) + Br• → A tertiary radical (CH3, Br) → 1-Bromo-2-methylcyclohexane

Polar Addition of HBr to an Alkene

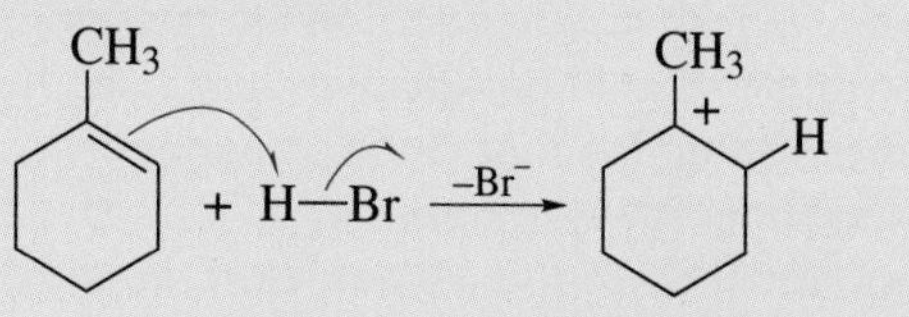

1-methylcyclohexene (CH3) + H—Br —(−Br⁻)→ A tertiary carbocation (CH3, +, H) → 1-Bromo-1-methylcyclohexane

Although the experimental findings of Mayo revealed the "peroxide effect" on the regiochemistry of addition across carbon-carbon double bonds, the insights and guidance of his mentor were critical to the success of the project. For example, Prof. Kharasch had considerable experience with chemicals used by the rubber industry, which had been undergoing rapid development as the automobile became increasingly popular and affordable. Among these chemicals were

(Continued)

HISTORICAL HIGHLIGHT *Additions Across Carbon-Carbon π-Bonds (Continued)*

antioxidants, a class of molecules that increased the durability of tires. Kharasch had Mayo study the effect of several of these substances on the addition of H–Br to allyl bromide, and Mayo found that they suppressed the formation of 1,3-dibromopropane, the anti-Markovnikov product. We now know that this is because antioxidants function by trapping free radicals and thereby stopping free-radical chain reactions. By continuing their investigations of the role of peroxides and antioxidants on additions to alkenes, Prof. Kharasch and his students made many contributions to the understanding of free-radical chain reactions. Their accomplishments in this area of research are considered by many scientists to have had the most influence of any group on the development of the organic chemistry of radicals.

Interestingly, Kharasch, whose research ultimately led to the understanding of anti-Markovnikov addition to a carbon-carbon double bond, was born in the Ukraine and thus shares a Russian heritage with Markovnikov himself. These two eminent chemists laid the foundation for the understanding of addition reactions to carbon-carbon π-bonds.

Relationship of Historical Highlight to Experiments

Several of the procedures in this chapter involve addition reactions characteristic of alkenes. Two of them, the addition of hydrogen bromide to 1-hexene (Sec. 10.5) and of borane to α-pinene (Sec. 10.8), represent examples in which both the reagent and the alkene to which it is adding are unsymmetrical. Consequently, identifying the products from these reactions provides a means of testing whether a Markovnikov or an anti-Markovnikov mode of addition has occurred.

See *Who was Markovnikov?*
See *Who was Kharasch?*
See *Who was Zaitsev?*
See *Who was Arbuzov?*
See *Who was Reformatsky?*

Alkynes

When you see this icon, sign in at this book's premium website at **www.cengage.com/login** to access videos, Pre-Lab Exercises, and other online resources.

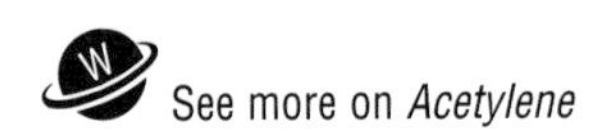

You may have been introduced to the simplest alkyne, acetylene, because it is frequently used as a fuel in welding. Indeed, when acetylene is burned with oxygen, the flame reaches temperatures of about 3300 °C (6000 °F), hotter than all but two or three other mixtures of combustible gases. You already know that flames and organic labs can be a dangerous combination, so we will explore tamer properties of alkynes, avoiding studies of their combustibility. Like the alkenes discussed in Chapter 10, alkynes are unsaturated hydrocarbons, and their chemistry is also dominated by the presence of carbon-carbon multiple or π-bonds. Whereas alkenes have a double bond, alkynes are characterized by a triple bond, which is composed of two orthogonal carbon-carbon π-bonds. Hence, if you understand the reactions that lead to the formation of alkenes, you will be able to apply this knowledge to preparing alkynes. Similarly, you will be able to extend your knowledge of the reactions of alkenes to predicting products of reactions of alkynes. Because alkynes have two double bonds, however, they basically just do everything twice. Well, it's *almost* that simple.

11.1 INTRODUCTION

Unsaturated organic compounds that contain a carbon-carbon triple bond as the **functional group** are called **alkynes.** Acetylene (ethyne), H–C≡C–H, is the simplest alkyne and is widely used in industry as a fuel and as a chemical feedstock for the preparation of other organic compounds such as acetic acid (CH_3CO_2H), vinyl chloride (CH_2=CHCl), a monomer used in the manufacture of polyvinyl chloride, and chloroprene (CH_2=CCl–CH=CH_2), which polymerizes to give neoprene.

The carbon-carbon triple bond in an alkyne is formed by the overlap of two *orthogonal* pairs of *p*-orbitals on adjacent *sp*-hybridized carbon atoms. Because the functional group in alkynes is related to the carbon-carbon double bond in alkenes (Chap. 10), the chemistry of these two classes of compounds is similar. For example, in much the same manner as one **elimination reaction** may be used to form the double bond in alkenes, *two* sequential elimination reactions yield the triple bond in alkynes (Eq. 11.1).

$$\underset{\text{A 1,2-dihaloalkane}}{\text{R—CHX—CHX—R}} \xrightarrow[\Delta]{\text{strong base}} \underset{\text{An alkyne}}{\text{R—C}\equiv\text{C—R}} \qquad (11.1)$$

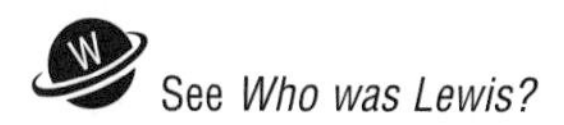
See *Who was Lewis?*

The π-electrons in both alkenes and alkynes provide a Lewis-base site for interaction with electrophilic reagents, which are Lewis acids, so one of the typical reactions of alkenes *and* alkynes is **electrophilic addition** as illustrated in Equations 11.2 and 11.3. Such reactions of alkynes can be stopped after the addition of one equivalent of a reagent, but the use of excess reagent leads to the formation of saturated products in which a second equivalent has been added. Some aspects of the chemistry of alkynes will be explored in the experiments in this chapter.

$$\underset{\text{An alkyne}}{R{-}C{\equiv}C{-}R} + \underset{\text{A hydrogen halide}}{2\,H{-}X} \longrightarrow \underset{\text{A 1,1-dihaloalkane}}{R{-}CX_2{-}CH_2{-}R} \quad (11.2)$$

$$\underset{\text{An alkyne}}{R{-}C{\equiv}C{-}R} + \underset{\text{A halogen}}{2\,X_2} \longrightarrow \underset{\text{A 1,1,2,2-tetrahaloalkane}}{R{-}CX_2{-}CX_2{-}R} \quad (11.3)$$

11.2 DEHYDROHALOGENATION OF 1,2-DIHALOALKANES

The dehydrohalogenation of alkyl halides (Sec. 10.2) is a general method to produce a double bond (Eq. 11.4). By analogy, the sequential elimination of two molecules of a hydrogen halide will lead to the formation of a triple bond as shown by the general reaction in Equations 11.5 and 11.6. The first stage of the reaction involves the base-induced elimination of a hydrogen halide to give a haloalkene, also referred to as a vinyl halide. This step proceeds by a **concerted E2 mechanism,** with the base abstracting the proton on the carbon atom β to the leaving halide ion. This reaction involves a **transition state** in which the hydrogen atom and the departing halogen atom are at a dihedral angle of 180° to one another, the **anti-periplanar** orientation, as shown in Equation 11.5. The second step of the sequence (Eq. 11.6) requires a strong base and more forcing conditions such as heating. Although the relationship between the hydrogen and the leaving halide ion in the reaction shown in Equation 11.6 is anti-periplanar, elimination may also occur when the orientation of these two groups on the vinyl halide is **syn-periplanar,** wherein the dihedral angle between the groups is 180°. However, under these circumstances, even higher temperatures are required to effect dehydrohalogenation.

See more on *Transition State*

$$\underset{\text{A haloalkane}}{RCH(H){-}CHR(X)} \xrightarrow{\text{base}} \underset{\text{An alkene}}{RCH{=}CHR} \quad (11.4)$$

$$\underset{\text{A 1,2-dihaloalkane}}{R(X)(H)C{-}C(X)(R)(H)} \xrightarrow{B:^-} \underset{\text{A haloalkene}}{(X)(R)C{=}C(R)(H)} + B{-}H + X:^- \quad (11.5)$$

$$\underset{\text{A haloalkene}}{X(R)C{=}C(R)H} + B:^- \xrightarrow{\Delta} \underset{\text{An alkyne}}{R{-}C{\equiv}C{-}R} + B{-}H + X:^- \qquad (11.6)$$

In the experiment that follows, you will perform the dehydrobromination of *meso*-stilbene dibromide (**1**), which can be prepared by the addition of bromine to (*E*)-stilbene (Sec. 10.6), to give diphenylacetylene (**2**) according to Equation 11.7. Thus, the combination of the bromination of an alkene followed by the double dehydrobromination of the intermediate *vicinal* dibromide represents a useful means of converting an alkene into an alkyne. Because the temperature required to effect elimination of hydrogen bromide from the intermediate vinyl bromide is nearly 200 °C, it is necessary to use a high-boiling solvent such as triethylene glycol so the reaction may be carried out without special apparatus. However, if a microwave apparatus is available for heating (Sec. 2.9), the reaction can be effected in a pressure-rated tube at 150 °C using methanol as solvent. It is instructive to compare the reaction times required and the results obtained using conventional versus microwave heating for promoting the reaction.

$$\underset{\substack{\mathbf{1}\\ meso\text{-Stilbene dibromide}}}{C_6H_5CHBr{-}CHBrC_6H_5} + 2\,KOH \xrightarrow[190\ ^\circ C]{HO(CH_2CH_2O)_2CH_2CH_2OH} \underset{\substack{\mathbf{2}\\ \text{Diphenylacetylene}}}{C_6H_5{-}C{\equiv}C{-}C_6H_5} \qquad (11.7)$$

EXPERIMENTAL PROCEDURES

Dehydrobromination of Meso-*Stilbene Dibromide*

Purpose To demonstrate the preparation of an alkyne by a double dehydrohalogenation.

SAFETY ALERT

The potassium hydroxide solution used in this experiment is *highly* caustic. *Do not allow it to come in contact with your skin.* If this should happen, flood the affected area with water and then thoroughly rinse the area with a solution of dilute acetic acid. Wear latex gloves when preparing and transferring all solutions.

MINISCALE PROCEDURE

Preparation Sign in at **www.cengage.com/login** to answer Pre-Lab Exercises, access videos, and read the MSDSs for the chemicals used or produced in this procedure. Review Sections 2.9, 2.10, and 2.17.

Apparatus A sand bath, hot plate, 25-mL Erlenmeyer flask, apparatus for vacuum filtration.

Setting Up Preheat the sand bath to about 190–200 °C. Place 800 mg of *meso*-stilbene dibromide and 5 pellets (about 400 mg) of commercial potassium hydroxide in the Erlenmeyer flask. Add 4 mL of triethylene glycol and a *carborundum* boiling stone to the flask.

Dehydrobromination and Isolation Place the flask in the sand bath and heat the mixture. After potassium bromide begins to separate from solution, heat the mixture for an additional 5 min. Remove the flask from the sand bath and allow it to cool to room temperature. Add 10 mL of water and place the flask in an ice-water bath for 5 min. Collect the diphenylacetylene that precipitates by vacuum filtration. Wash the solid with about 2–3 mL of cold water. Recrystallize the product from a small quantity of 95% ethanol or an ethanol-water mixture. If the solution is allowed to cool slowly undisturbed, you should obtain large, sparlike, colorless crystals.

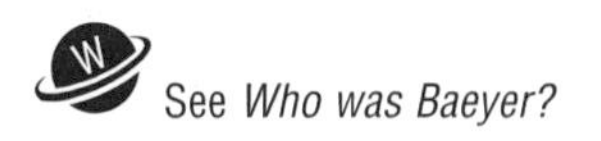

Analysis Determine the melting point, weight, and percent yield of the recrystallized product. Test the product for unsaturation using the bromine and Baeyer tests (Secs. 25.8A and B, respectively). Obtain IR and ^{1}H NMR spectra of your starting material and product and compare them with those of authentic samples (Figs. 10.36, 10.37, 11.1, and 11.2).

MICROSCALE PROCEDURE

Preparation Sign in at **www.cengage.com/login** to answer Pre-Lab Exercises, access videos, and read the MSDSs for the chemicals used or produced in this procedure. Review Sections 2.9, 2.10, and 2.17.

Apparatus A sand bath, hot plate, 13-mm × 120-mm Pyrex test tube, apparatus for vacuum filtration.

Setting Up Preheat the sand bath to about 190–200 °C. Place 150 mg of *meso*-stilbene dibromide and 1 pellet (about 80 mg) of commercial potassium hydroxide in the Pyrex test tube. Add 1 mL of triethylene glycol and a carborundum boiling stone to the test tube.

Dehydrobromination and Isolation Place the test tube in the sand bath and heat the mixture. After potassium bromide begins to separate from solution, heat the mixture for an additional 5 min. Remove the test tube from the sand bath and allow it to cool to room temperature. Add 2 mL of water and place the test tube in an ice-water bath for 5 min. Collect the diphenylacetylene that precipitates by vacuum filtration. Wash the solid with about 1 mL of cold water. Recrystallize the product from a small quantity of 95% ethanol or an ethanol-water mixture.

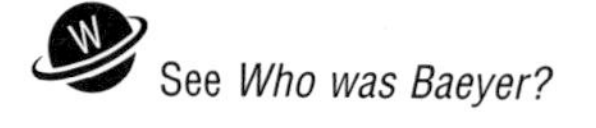
See *Who was Baeyer?*

Analysis Determine the melting point, weight, and percent yield of the recrystallized product. Test the product for unsaturation using the bromine and Baeyer tests (Secs. 25.8A and B, respectively). Obtain IR and ^{1}H NMR spectra of your starting material and product and compare them with those of authentic samples (Figs. 10.36, 10.37, 11.1, and 11.2).

MINISCALE PROCEDURE FOR MICROWAVE OPTION

Preparation Sign in at **www.cengage.com/login** to answer Pre-Lab Exercises, access videos, and read the MSDSs for the chemicals used or produced in this procedure. Review Sections 2.9, 2.10, and 2.17.

Apparatus A 10-mL pressure-rated tube with cap, stirbar, apparatus for microwave heating with magnetic stirring and vacuum filtration, ice-water bath.

Setting Up Equip the 10-mL pressure-rated tube with the stirbar and add 0.40 g of *meso*-stilbene dibromide, about 0.2 g (two pellets) of solid potassium hydroxide, and 2 mL of methanol. Cap the pressure-rated tube and gently shake it or place it on a magnetic stirrer to facilitate initial mixing of its contents. Place the tube in the cavity of the microwave apparatus.

Dehydrobromination and Isolation Program the unit to heat the reaction mixture with stirring according to the directions provided by your instructor. Generally, the reaction temperature should be set at 150 °C and the power set at a maximum of 25 W with a 1-min ramp time and a 5-min hold time; the pressure limit should be set at 275 psi. Allow the mixture to cool to room temperature and remove the tube from the microwave apparatus. Add 5 mL of water and place the tube in an ice-water bath for 5 min. Collect the diphenylacetylene that precipitates by vacuum filtration. Wash the solid with about 2 mL of cold water. Recrystallize the product from a small quantity of 95% ethanol or an ethanol-water mixture. If the solution is allowed to cool slowly undisturbed, you should obtain large, sparlike, colorless crystals.

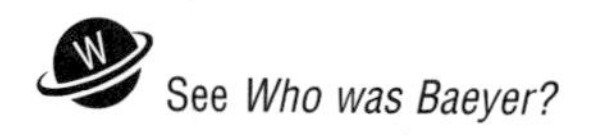
See *Who was Baeyer?*

Analysis Determine the melting point, weight, and percent yield of the recrystallized product. Test the product for unsaturation using the bromine and Baeyer tests (Secs. 25.8A and B, respectively). Obtain IR and ^{1}H NMR spectra of your starting material and product and compare them with those of authentic samples (Figs. 10.36, 10.37, 11.1, and 11.2).

WRAPPING IT UP

Combine the *filtrate* from the reaction mixture with the *mother liquor* from the recrystallization, dilute with water, and neutralize with 10% aqueous hydrochloric acid; then flush the solution down the drain. Place the *dichloromethane solution* from the bromine test for unsaturation in a container for halogenated organic liquids; put the *manganese dioxide* from the Baeyer test for unsaturation in a container for heavy metals.

EXERCISES

1. Write three-dimensional structures of *meso*-stilbene dibromide and one enantiomer of *dl*-stilbene dibromide.
2. The solvent used in the preparation of diphenylacetylene (**2**) is triethylene glycol.
   a. Write the structure of triethylene glycol; circle and label each of the functional groups in this molecule.
   b. What structural features account for the high boiling point of this solvent?
3. The functional group in alkynes is the carbon-carbon triple bond.
   a. Using a suitable drawing, show how the atomic *p*-orbitals in acetylene overlap to form π-molecular orbitals.
   b. What is the angle between these two π-orbitals?
   c. What is the angle between each of the π-orbitals and the carbon-carbon σ-bond?
4. Give a stepwise mechanism showing the base-induced formation of diphenylacetylene from *meso*-stilbene dibromide. Use curved arrows to symbolize the flow of electrons.
5. The E2 reaction of most compounds is known to proceed preferentially by removal of a proton anti-periplanar to the leaving group. Based upon this generalization, predict the geometry of the 1-bromo-1,2-diphenylethylene that is produced as the intermediate in the double dehydrobromination of *meso*-stilbene dibromide.
6. Why do you think the enthalpy of activation, $\Delta H^{\ddagger}$, for a syn-periplanar elimination is higher than that for an anti-periplanar elimination?
7. If *meso*-stilbene dibromide is treated with KOH in ethanol, it is possible to isolate the 1-bromo-1,2-diphenylethylene that is formed from the first dehydrobromination. The E2 elimination of the second molecule of hydrogen bromide from this intermediate alkene to give diphenylacetylene has a higher activation enthalpy than the first elimination and thus requires a higher reaction temperature. Explain.
8. Provide an explanation for why ethanol or ethylene glycol are not suitable solvents for the second dehydrobromination.

Sign in at **www.cengage.com/login** and use the spectra viewer and Tables 8.1–8.8 as needed to answer the blue-numbered questions on spectroscopy.

9. Consider the NMR spectral data for *meso*-stilbene dibromide (Fig. 10.37).
   a. In the ^{1}H NMR spectrum, assign the various resonances to the hydrogen nuclei responsible for them.
   b. For the ^{13}C NMR data, assign the various resonances to the carbon nuclei responsible for them.
10. Consider the spectral data for diphenylacetylene (Figs. 11.1 and 11.2).
   a. In the functional group region of the IR spectrum, identify the absorptions associated with the aromatic rings. Why is there no absorption for the carbon-carbon triple bond?
   b. In the ^{1}H NMR spectrum, assign the various resonances to the hydrogen nuclei responsible for them.
   c. For the ^{13}C NMR data, assign the various resonances to the carbon nuclei responsible for them.

11. Discuss the differences observed in the NMR spectra of *meso*-stilbene dibromide and diphenylacetylene that are consistent with the double dehydrobromination in this experiment.

SPECTRA

Starting Material and Product

The IR and NMR spectra of meso-stilbene dibromide are provided in Figures 10.36 and 10.37, respectively.

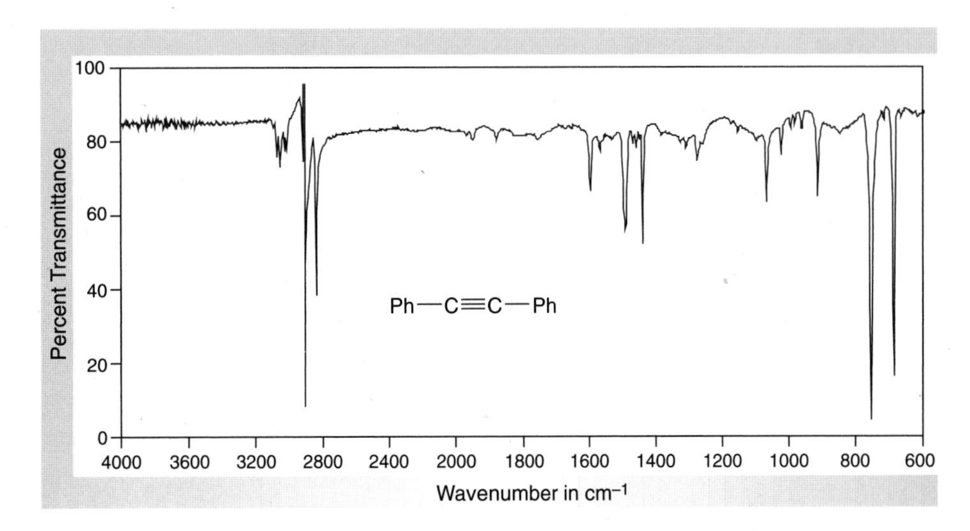

Figure 11.1
IR spectrum of diphenylacetylene (IR card).

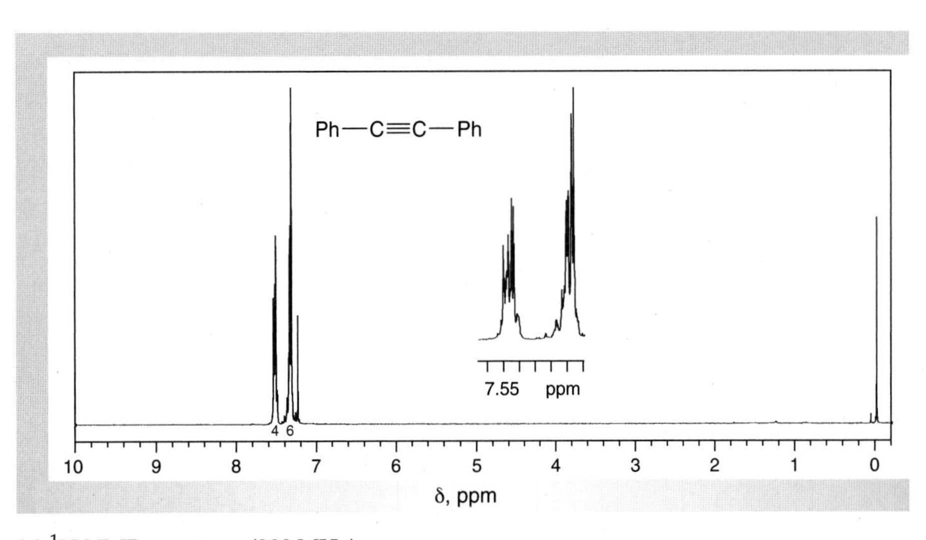

Figure 11.2
NMR data for diphenylacetylene ($CDCl_3$).

(a) ^{1}H NMR spectrum (300 MHz).
(b) ^{13}C NMR data: δ 89.6, 123.3, 128.1, 128.2, 131.6.

11.3 ADDITION REACTIONS OF ALKYNES

One of the general classes of reactions that characterizes the chemistry of alkynes as well as alkenes (Sec. 10.4) is the **electrophilic addition** of reagents generalized as **E–Nu** (Eq. 11.8) to the carbon-carbon multiple bond. This reaction typically proceeds in two stages. Under carefully controlled reaction conditions in which only one equivalent of electrophilic reagent **E–Nu** is used, it is sometimes possible to stop the addition reaction at the alkene stage. However, if more than one equivalent of the reagent is present, further addition occurs, and substituted alkanes are formed (Sec. 11.1). Typical reagents that add to alkynes include the halogens chlorine and bromine and acids such as hydrogen chloride and hydrogen bromide.

$$R^1{-}C{\equiv}C{-}R^2 \xrightarrow[\text{1 mole}]{\text{E–Nu}} (R^1)(E)C{=}C(Nu)(R^2) \xrightarrow[\text{1 mole}]{\text{E–Nu}} R^1{-}C(E)(E){-}C(Nu)(Nu){-}R^2 \quad (11.8)$$

An alkyne — A substituted alkene — A substituted alkane

See *Who was Markovnikov?*

The *regiochemistry* of the electrophilic addition of unsymmetrical reagents to terminal alkynes follows *Markovnikov's rule* in the same fashion as observed for alkenes (Sec. 10.4). Thus, the Lewis acid component, E^+, of the reagent E–Nu adds to the electron-rich triple bond to form the more stable, or more substituted, of the two possible intermediate vinyl carbocations. This carbocation then undergoes nucleophilic attack by the Lewis base component, Nu^-, of the reagent (Eq. 11.9). The addition of the second mole of E–Nu to the resulting alkene occurs with the same regiochemistry as the first (Eqs. 11.8 and 11.9).

$$R{-}C{\equiv}C{-}H \xrightarrow[\text{1 mole}]{\text{E–Nu}} (R)(Nu)C{=}C(E)(H) \xrightarrow[\text{1 mole}]{\text{E–Nu}} R{-}C(Nu)(Nu){-}C(E)(E){-}H \quad (11.9)$$

A terminal alkyne — A substituted alkene — A substituted alkane

Alkynes undergo electrophilic addition reactions with hydrogen halides and bromine, but these reactions have limited synthetic utility. However, one reaction of alkynes that is commonly used in organic chemistry is **hydration** of the carbon-carbon triple bond to give a **ketone,** a transformation that is catalyzed by mercuric ion in the presence of sulfuric acid (Eq. 11.10).

$$R{-}C({=}O){-}CH_3 \quad (11.10)$$

The role that the mercuric ion, a polarizable Lewis acid, plays in this reaction is not completely understood. However, a reasonable hypothesis is that it first coordinates with the triple bond to generate a vinyl cation that may be stabilized by conversion to a cyclic mercurinium ion (Eq. 11.11). Attack by water on the cyclic ion then occurs at the more substituted carbon. Following several proton transfers, mercuric ion is lost to give a substituted vinyl alcohol, which is unstable and tautomerizes to give the **carbonyl group,** C=O, of the product ketone (Eq. 11.12).

$$\underset{\text{An alkyne}}{R{-}C{\equiv}C{-}H} \xrightarrow{Hg^{2+}} \underset{\text{A vinyl cation}}{R{-}\overset{+}{C}{=}C(H)\ddot{H}g^{+}} \rightleftharpoons \underset{\text{A cyclic mercurinium ion}}{R(C{=}C)H\ \text{bridged by}\ Hg^{2+}} \quad (11.11)$$

$$R(C{=}C)H\ \text{with}\ Hg^{2+} + H_2O{:} \longrightarrow R(HO^{+}H)C{=}C(H)Hg^{+} \xrightarrow{-H^{+}}$$

$$R(HO)C{=}C(H)Hg^{+} \xrightarrow[-Hg^{2+}]{+H^{+}} \underset{\substack{\text{A vinyl alcohol (enol)}\\ \textit{(unstable; not isolated)}}}{R(HO)C{=}CH_2} \rightleftharpoons \underset{\text{A ketone}}{R{-}C({=}O){-}CH_3} \quad (11.12)$$

In the experiment that follows, the hydration of a terminal alkyne is illustrated by the conversion of 2-methyl-3-butyn-2-ol (**3**) to 3-hydroxy-3-methyl-2-butanone (**4**), as shown in Equation 11.13. The presence of a hydroxyl group in **3** has little effect on the chemical properties of the carbon-carbon triple bond. Rather, the main effect of the polar hydroxyl group is on the physical properties of the molecule, with the boiling point of **3** being considerably higher than those of other acetylenic hydrocarbons having the same molecular weight.

$$\underset{\substack{\mathbf{3}\\ \text{2-Methyl-3-butyn-2-ol}}}{CH_3{-}C(OH)(CH_3){-}C{\equiv}C{-}H} + H_2O \xrightarrow[Hg^{2+}]{H_2SO_4} \underset{\substack{\mathbf{4}\\ \text{3-Hydroxy-3-methyl-2-butanone}}}{CH_3{-}C(OH)(CH_3){-}C({=}O){-}CH_3} \quad (11.13)$$

$$\underset{\substack{\mathbf{5}\\ \text{(not formed)}}}{CH_3{-}C(OH)(CH_3){-}CH_2{-}C({=}O){-}H}$$

This experiment provides an opportunity to verify experimentally that the hydration of a terminal alkyne occurs in accordance with Markovnikov's rule (see Historical Highlight in Chap. 10). For example, that the structure of the product of the hydration of **3** is **4** rather than **5** may be confirmed by comparing the IR, ^{1}H, and ^{13}C NMR spectra of the product of the reaction with known standard samples of **4** and **5**, the spectra of which are clearly distinct. When *all* of the spectral properties of a compound are superimposable with those of a standard sample, the two are identical.

Another common method of assigning a structure to the product of a reaction involves preparing a solid derivative and comparing its melting point with the melting point of the same derivative of a series of known compounds. If the melting points of the derivatives of the unknown and a known sample are identical, then a mixed melting point (Sec. 3.3) is taken to confirm the identity of the two samples. In the present case, a solid derivative of the product of the reaction can be made for comparison with the same derivatives of authentic samples of **4** and **5.** For example, forming the semicarbazone **6** from **4** proceeds smoothly, but converting the aldehyde **5** to its semicarbazone occurs with concomitant dehydration to give **7,** which is the semicarbazone of the unsaturated aldehyde $(CH_3)_2C{=}CHCHO$.

$$CH_3-C(OH)(CH_3)-C(CH_3){=}NNHCONH_2 \quad \textbf{6}$$

$$(CH_3)_2C{=}CH-C(H){=}NNHCONH_2 \quad \textbf{7}$$

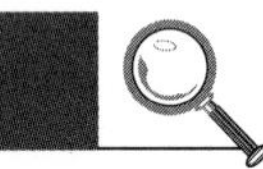

EXPERIMENTAL PROCEDURE

Preparation of 3-Hydroxy-3-methyl-2-Butanone

Purpose To demonstrate the hydration of a terminal alkyne and to determine the regiochemistry of the addition.

SAFETY ALERT

1. **Concentrated sulfuric acid may cause severe chemical burns, so do not allow it to come into contact with your skin. Wear latex gloves when transferring this reagent. Wash any affected area immediately with water and then with 5% sodium bicarbonate solution.**
2. **Be sure that you add the concentrated acid slowly *to water* and *not* the reverse.**

MINISCALE PROCEDURE

Preparation Sign in at **www.cengage.com/login** to answer Pre-Lab Exercises, access videos, and read the MSDSs for the chemicals used or produced in this procedure. Review Sections 2.9, 2.10, 2.11, 2.13, 2.16, 2.21, 2.22, and 2.29.

Apparatus A 10-mL, a 50-mL, and a 100-mL round-bottom flask, separatory funnel, ice-water bath, 13-mm × 120-mm test tube, apparatus for heating under reflux, simple distillation, magnetic stirring, steam distillation with internal steam generation, vacuum filtration, and *flameless* heating.

Setting Up *Carefully* add 3 mL of *concentrated* sulfuric acid *to* 20 mL of water contained in the 100-mL round-bottom flask equipped with a stirbar. Dissolve 0.2 g of reagent-grade mercuric oxide in the resulting warm solution and cool the flask to about 50 °C. Attach a reflux condenser to the flask, and add 3.6 mL of 2-methyl-3-butyn-2-ol in one portion through the top of the condenser. The cloudy white precipitate that forms is presumably the mercury complex of the alkyne.

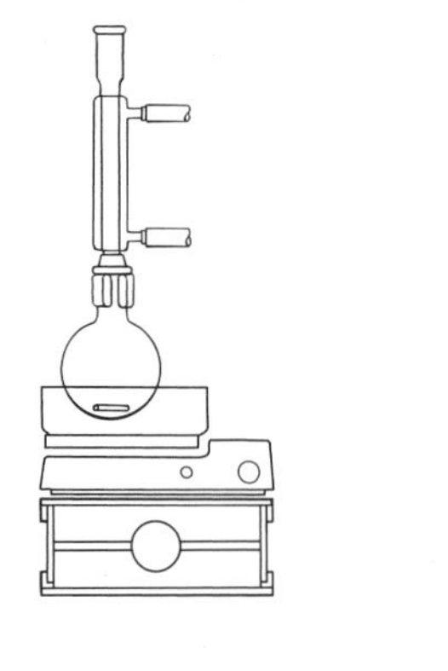

Reaction Stir the reaction flask to mix the contents thoroughly, whereupon a mildly exothermic reaction may ensue. Allow the reaction to proceed by itself for about 2 min. Heat the mixture under reflux for 30 min, and then allow it to cool to room temperature.★

Isolation While the reaction mixture is cooling, prepare an apparatus for steam distillation *with internal steam generation.* Add about 35–40 mL of water to the reaction flask. Heat the flask and steam-distill the product until about 30–35 mL of distillate has been collected.★

Transfer the distillate to a small separatory funnel, add 2–3 g of potassium carbonate, and then carefully saturate the solution with sodium chloride; avoid adding more sodium chloride than will dissolve readily. If a second layer forms at this point, do not separate it, but continue with the extraction. Extract the mixture sequentially with two 10-mL portions of dichloromethane and combine the extracts. Dry the organic extracts over several spatula-tips full of anhydrous sodium sulfate. Swirl the flask occasionally for a period of 10–15 min to facilitate drying; add further portions of anhydrous sodium sulfate if the solution does not become clear.

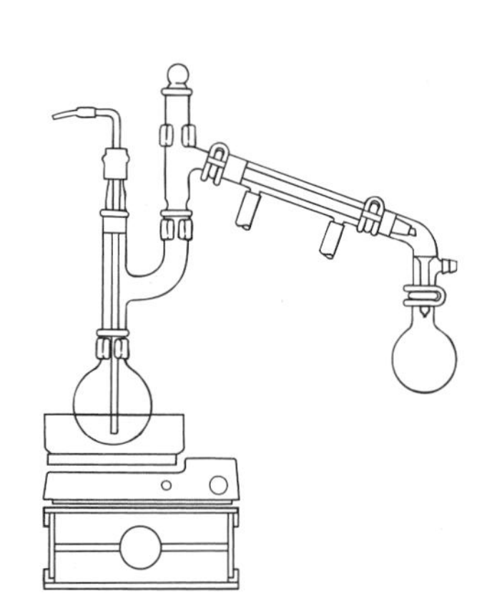

Decant the solution into a 50-mL round-bottom flask and remove most of the dichloromethane by simple shortpath distillation; cool the receiving flask in an ice-water bath. Alternatively, use rotary evaporation or other techniques to remove most of the dichloromethane. Using a Pasteur pipet to minimize material loss, transfer the crude product into a 10-mL round-bottom flask. Remove the last traces of dichloromethane from the crude product by connecting the flask to a vacuum source. Then continue the distillation using the shortpath apparatus fitted with a clean, tared receiving flask. Since the volume of product is likely to be only 1–2 mL, an accurate boiling point may be difficult to obtain, and all material that distills above 110–115 °C should be collected in order to obtain a reasonable yield.

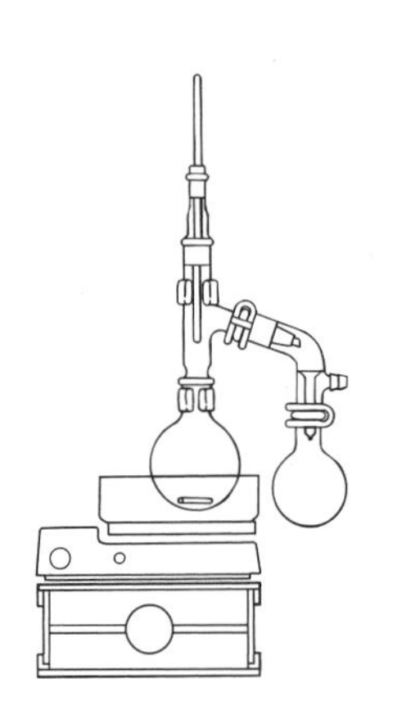

Derivatization Add 0.5 mL of **4** to a solution prepared from 0.5 g of semicarbazide hydrochloride and 0.8 g of sodium acetate dissolved in 2.5 mL of water contained in a test tube. Shake the mixture vigorously. Collect the crude solid semicarbazone that forms by vacuum filtration; use a glass rod to scratch at the air-liquid interface if necessary to initiate crystallization. Dissolve the solid product in a minimum volume of hot 2-propanol, cool the mixture to room temperature and then in an ice-water bath, and collect the crystals by vacuum filtration. Air-dry the crystals.

Analysis Determine the weight and percent yield of the distilled product. Determine the melting point of the semicarbazone of your product. Use this information to determine the identity of the product of hydration of 2-methyl-3-butyn-2-ol. The melting point of **6** is 162–163 °C, whereas **7** melts at 222–223 °C. Obtain IR and ^{1}H NMR spectra of your starting material and product and compare them with those of authentic samples (Figs. 11.3–11.6).

WRAPPING IT UP

Since the *residue in the stillpot* from the steam distillation contains soluble mercury salts, place it in a special container for solutions of mercuric salts; do *not* pour it down the drain. If directed to do so by your instructor, you may convert the *soluble mercuric salts* into insoluble mercuric sulfide by reaction with sodium sulfide or hydrosulfide; collect the *mercuric sulfide* by vacuum filtration for recycling. Flush the *other aqueous solutions* down the drain with water. Pour the *dichloromethane* recovered as the forerun in the distillation into a container for halogenated organic solvents. Dilute the *filtrate* from the preparation of the semicarbazone derivative and the *mother liquor* from the recrystallization with water, and flush the solutions down the drain.

EXERCISES

1. 3-Hydroxy-3-methyl-2-butanone (**4**) was produced by the hydration of 2-methyl-3-butyn-2-ol (**3**) in the experiment you performed.
   a. Does the formation of **4** from **3** follow Markovnikov's rule?
   b. Draw the structure of **4** and a tautomer of **4.**
   c. Draw a resonance structure of **4.**
   d. Using your answers to Parts **b** and **c,** describe how two tautomeric structures and two resonance structures differ.
2. Using curved arrows to symbolize the flow of electrons, give a stepwise mechanism for the hydration of 2-methyl-3-butyn-2-ol with sulfuric acid in the presence of mercuric oxide.
3. How is the ketone **4** isolated from the reaction mixture, and what advantages does this technique offer over direct extraction?
4. Give the structures for the products that you would expect to obtain on hydration of 1-octyne and of 2-octyne.
5. Consider the compounds hexane, 1-hexene, and 1-hexyne (see Chapters 9 and 10 for a review of reactions of alkanes and alkenes). What similarities and what differences would you expect upon treatment of these compounds with
   a. bromine in dichloromethane at room temperature in the dark?
   b. an aqueous solution of potassium permanganate?
   c. an aqueous solution of sulfuric acid?

   Give the structures of the products, if any, that would be obtained from each of these reactions.
6. The reaction of 1-butyne with hydrogen bromide gives 2,2-dibromobutane, as shown in Equation 11.14. Write a stepwise mechanism for this process, using curved arrows to symbolize flow of electrons. Also provide a rationale for the observed regioselectivity for addition of *each* mole of hydrogen bromide.

$$CH_3CH_2C{\equiv}C{-}H + 2\,H{-}Br \longrightarrow CH_3CH_2CBr_2CH_3 \qquad (11.15)$$

Sign in at **www.cengage.com/login** and use the spectra viewer and Tables 8.1–8.8 as needed to answer the blue-numbered questions on spectroscopy.

7. Consider the spectral data for 2-methyl-3-butyn-2-ol (Figs. 11.3 and 11.4).
   a. In the functional group region of the IR spectrum, identify the absorptions associated with the carbon-carbon triple bond and the hydroxyl group.
   b. In the ^1H NMR spectrum, assign the various resonances to the hydrogen nuclei responsible for them.
   c. For the ^{13}C NMR data, assign the various resonances to the carbon nuclei responsible for them.
8. Consider the spectral data for 3-hydroxy-3-methyl-2-butanone (Figs. 11.5 and 11.6).
   a. In the functional group region of the IR spectrum, identify the absorptions associated with the carbonyl and hydroxyl groups.

b. In the ^{1}H NMR spectrum, assign the various resonances to the hydrogen nuclei responsible for them.

c. For the ^{13}C NMR data, assign the various resonances to the carbon nuclei responsible for them.

9. ^{1}H NMR spectroscopy could be used to determine whether **4** or **5** is produced by the hydration of **3.** Sketch the ^{1}H NMR spectrum you would expect from **5,** and discuss the differences that you would expect in the ^{1}H NMR spectra of **4** and **5.**

10. IR spectroscopy may also be used to determine whether **4** or **5** is produced by the hydration of **3.** Describe the important difference(s) you would expect in the IR spectra of the two compounds.

SPECTRA

Starting Material and Product

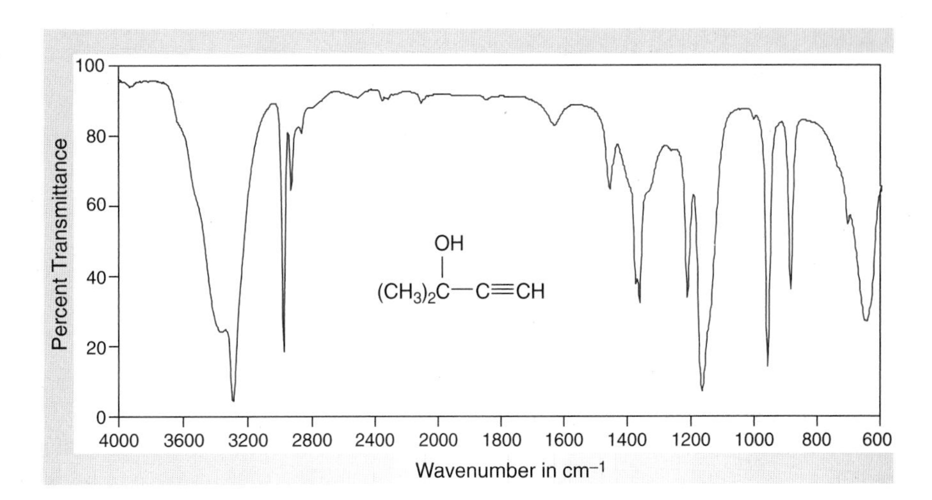

Figure 11.3
IR spectrum of 2-methyl-3-butyn-2-ol (neat).

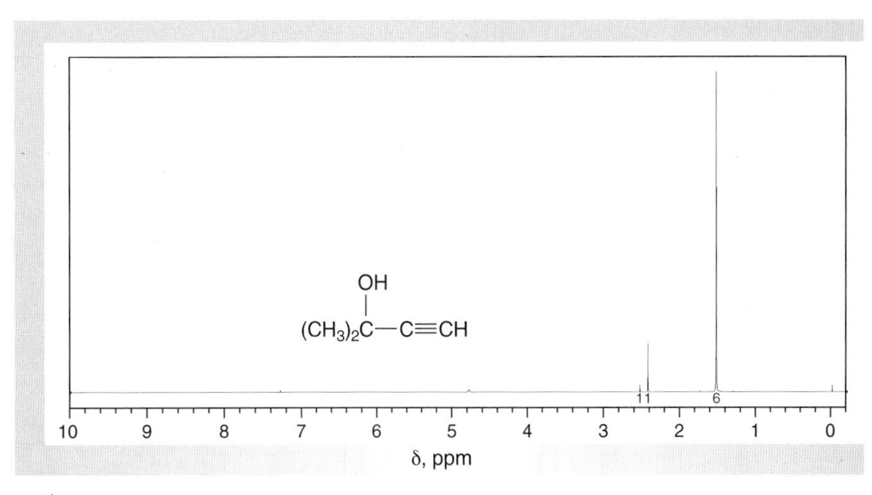

Figure 11.4
NMR data for 2-methyl-3-butyn-2-ol ($CDCl_3$).

(a) ^{1}H NMR spectrum (300 MHz).
(b) ^{13}C NMR data: δ 31.1, 64.8, 70.4, 89.1.

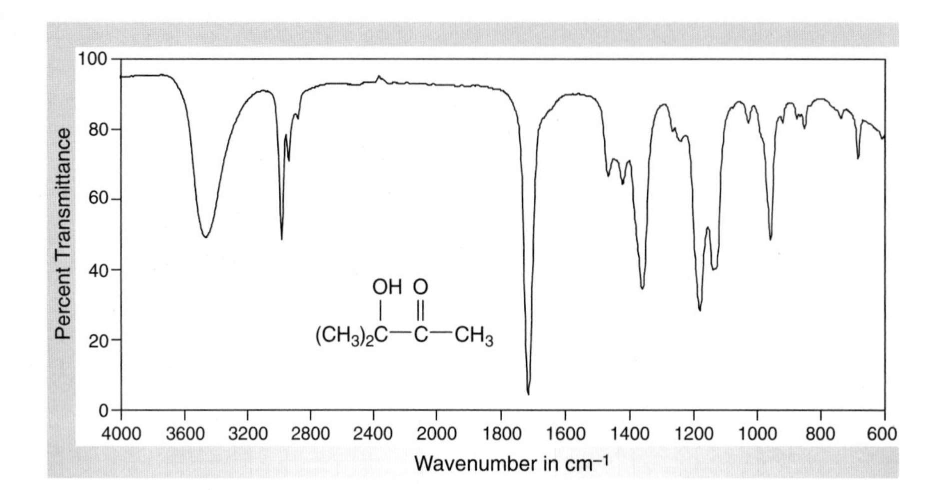

Figure 11.5
IR spectrum of 3-hydroxy-3-methyl-2-butanone (neat).

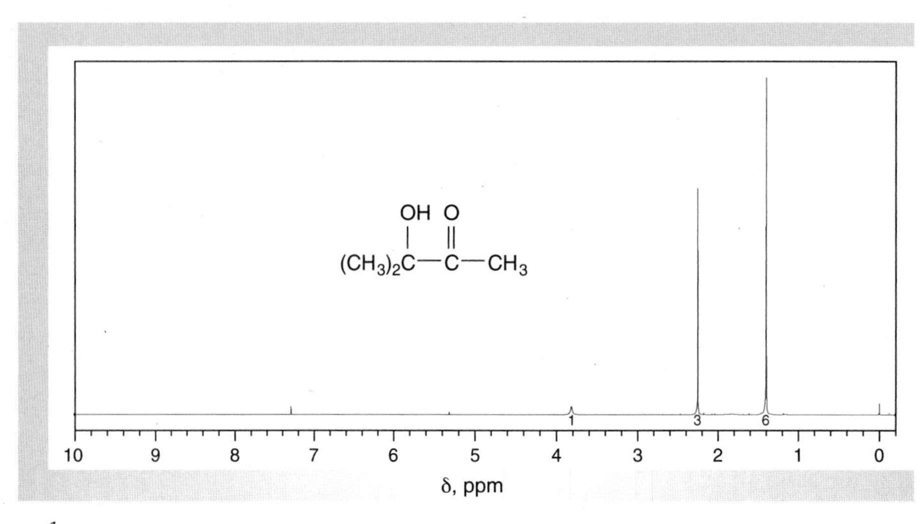

Figure 11.6
NMR data for 3-hydroxy-3-methyl-2-butanone ($CDCl_3$).

(a) ^{1}H NMR spectrum (300 MHz).
(b) ^{13}C NMR data: δ 23.8, 26.4, 76.7, 213.2.

11.4 TERMINAL ALKYNES AS ACIDS

The acidity of the acetylenic hydrogen of terminal alkynes (R–C≡C–H), p$K_a \approx 25$, provides the basis of a simple and specific qualitative test for such compounds. Treating a terminal alkyne with a solution containing silver ammonia complex, which is prepared by dissolving silver nitrate in ammonia, provides a solid silver salt, as illustrated by the general reaction shown in Equation 11.15. This test differentiates terminal alkynes from alkenes, internal alkynes, dienes, and allenes.

$$\underset{\text{A terminal alkyne}}{R{-}C{\equiv}C{-}H} + [Ag(NH_3)_2]^{+} \longrightarrow \underset{\text{A silver acetylide}}{\underline{R{-}C{\equiv}C{-}Ag}} + NH_3 + \overset{+}{N}H_4 \qquad (11.16)$$

This unique reaction of terminal alkynes may also be used to separate them from nonterminal alkynes, which do *not* form insoluble silver salts and remain in solution. The silver acetylide salts can be reconverted to the terminal alkynes on treatment with hydrochloric acid, as shown in Equation 11.16. *Dry silver salts of this type are quite sensitive to shock and tend to decompose explosively.* Therefore, they should *never* be allowed to dry out before being decomposed by hydrochloric acid.

$$\underline{R{-}C{\equiv}C{-}Ag} + HCl \longrightarrow R{-}C{\equiv}C{-}H + \underline{AgCl} \qquad (11.16)$$

EXPERIMENTAL PROCEDURE

Formation of a Silver Acetylide and Its Decomposition

Purpose To demonstrate the formation and decomposition of the silver acetylide of a terminal alkyne.

SAFETY ALERT

1. **Avoid spilling solutions of silver salts on your skin. If you do come into contact with the solutions, wash the affected area immediately with copious amounts of water.**
2. **Take special care to destroy *all* of the silver salt acetylide with hydrochloric acid *before* discarding it.**

Procedure

Preparation Sign in at **www.cengage.com/login** to answer Pre-Lab Exercises, access videos, and read the MSDSs for the chemicals used or produced in this procedure. Review Section 2.17.

Apparatus Two 13-mm × 120-mm test tubes, apparatus for vacuum filtration.

Setting Up In a 10-mL Erlenmeyer flask, prepare a solution of silver ammonia complex from 2.5 mL of 0.1 *M* silver nitrate solution by adding ammonium hydroxide solution *dropwise*. Brown silver oxide forms first; add *just enough* ammonium hydroxide to dissolve the silver oxide. Dilute the solution by adding 1.5 mL of water.

Formation and Decomposition Add 3 mL of the diluted silver ammonia complex solution to about 0.1 mL of 2-methyl-3-butyn-2-ol in a test tube, and note the formation of the silver acetylide salt. Collect the silver salt by vacuum filtration of the aqueous solution; *be careful not to let the silver salt dry, because the dry salt is explosive.* Transfer the *wet* silver salt to a clean test tube, and add a small amount of dilute hydrochloric acid. Observe what changes occur, especially in the color and form of the precipitate.

WRAPPING IT UP

Dilute the remaining *silver ammonia complex test solution* with water and flush it down the drain with water. Destroy all *solid silver acetylide salt* by treatment with hydrochloric acid. Collect the *silver chloride* and put it in a container for heavy metals, and then flush the *filtrate* down the drain.

EXERCISES

1. The pK_a of a terminal alkyne is about 25 and the pK_a of ammonium ion, $\overset{+}{N}H_4$, is about 9.

   a. Use these values to predict the K_{eq} for Equation 11.17 to prove that the acid-base equilibrium lies to the left as shown.

$$R{-}C{\equiv}C{-}H + NH_3 \rightleftharpoons R{-}C{\equiv}\bar{C} + \overset{+}{N}H_4 \qquad (11.18)$$

   b. Provide an explanation for the observation that the equilibrium for the reaction of a terminal alkyne with silver ammonia complex (Eq. 11.15) favors the formation of silver acetylide.

2. How might you separate, and obtain in pure form, 1-octyne and 2-octyne from a mixture containing both?

3. Explain the dramatic decrease in acidity of the boldfaced "terminal" hydrogen atom in the following series of hydrocarbons. (*Hint:* Consider the difference in the nature of the respective carbon-hydrogen σ-bonds and the anticipated relative stabilities of the anions resulting from deprotonation.)

$$H_3C{-}C{\equiv}C{-}\mathbf{H} > (H)(H_3C)C{=}C(\mathbf{H})(H) > H{-}C(H)(H_3C){-}C(\mathbf{H})(H){-}H$$

4. Consider the acetylide anion $CH_3C{\equiv}C^-$, which can be prepared by deprotonation of propyne.

   a. Explain why this anion is a good nucleophile.

   b. Write the structure of the product of the reaction of the anion $CH_3C{\equiv}C^-$ with ethyl bromide and indicate to what mechanistic class this reaction belongs.

   c. The reaction of the anion $CH_3C{\equiv}C^-$ with a tertiary halide does *not* yield a substitution product. Write the structure of the organic product(s) formed from the reaction of $CH_3C{\equiv}C^-$ and a tertiary alkyl halide of your choosing and explain why substitution does not occur.

5. Carbon-carbon bond-forming reactions are very important in synthetic organic chemistry. Explain how you could prepare heptane from 1-butyne using any other organic and inorganic reagents you desire.

HISTORICAL HIGHLIGHT

Acetylene: A Valuable Small Molecule

A budding organic chemist might wonder why a simple two-carbon molecule such as acetylene could be of so much chemical interest and value, beyond its use as a combustible fuel for devices such as cutting torches and carbide lamps. The answer is that acetylene originally served as a source of a variety of commercially important basic organic chemicals. Among them are acetaldehyde, acetic acid, acetone, vinyl

(Continued)

HISTORICAL HIGHLIGHT *Acetylene: A Valuable Small Molecule (Continued)*

chloride, 1,3-butadiene, and 1,4-butanediol. You can see that some of these substances are the result of carbon-carbon bond-forming reactions, whereby the two-carbon atoms of acetylene become part of more complex molecules. Other precursors derived from coal tar, crude oil, or natural gas have now replaced it for producing some of these compounds, but uses of acetylene remain of historical interest. A particularly interesting example is associated with World War I and II.

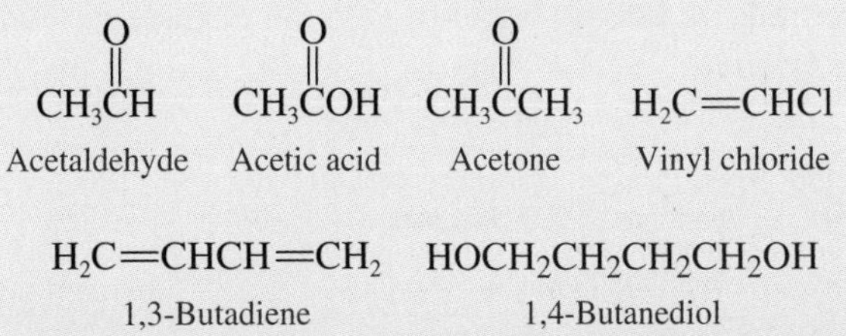

$CH_3CH{=}O$ (Acetaldehyde) $CH_3C({=}O)OH$ (Acetic acid) $CH_3C({=}O)CH_3$ (Acetone) $H_2C{=}CHCl$ (Vinyl chloride)

$H_2C{=}CHCH{=}CH_2$ (1,3-Butadiene) $HOCH_2CH_2CH_2CH_2OH$ (1,4-Butanediol)

Though they may seem mundane, tires are critical to armed forces because of the inevitable dependence on trucks, vehicles, and the like for transporting personnel and material. Rubber, then, is a particularly critical commodity during wartime. Latex is the source of natural rubber, a polymer of isoprene (2-methyl-1,3-butadiene), and it is contained in the sap produced by rubber trees, which are grown on plantations in East and Southeast Asia. During both of the World Wars, naval blockades cut Germany off from materials from this part of the world, so the country was forced to depend on synthetic rubber. The original such material was "methyl rubber," a polymer of 2,3-dimethyl-1,3-butadiene, that was produced from acetic acid, which could be derived from acetylene according to the following sequence:

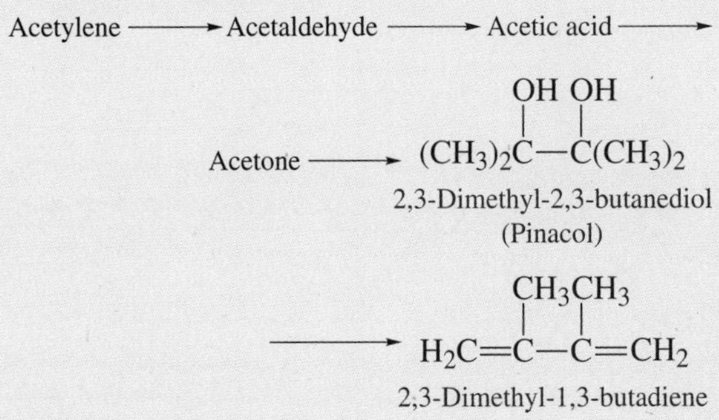

Acetylene ⟶ Acetaldehyde ⟶ Acetic acid ⟶

Acetone ⟶ $(CH_3)_2C(OH){-}C(OH)(CH_3)_2$
2,3-Dimethyl-2,3-butanediol (Pinacol)

⟶ $H_2C{=}C(CH_3){-}C(CH_3){=}CH_2$
2,3-Dimethyl-1,3-butadiene

Acetylene was the key building block from which methyl rubber was built. This synthetic rubber proved to be an unsatisfactory substitute for the natural material, however, because it was sticky and was readily degraded by oxygen. Nonetheless, it was the best that the Germans could get during World War I.

How did the Germans obtain the acetylene that is the starting material for this diene? This was easy, as it turns out, because Germany has vast quantities of coal (C) and limestone ($CaCO_3$) that can be used to produce acetylene according to the following reaction scheme:

$$\underset{\text{Calcium carbonate}}{CaCO_3} \xrightarrow{-CO_2} \underset{\text{Calcium oxide}}{CaO} \xrightarrow[-CO]{3\,C}$$

$$\underset{\text{Calcium carbide}}{CaC_2} \xrightarrow[-Ca(OH)_2]{2\,H_2O} \underset{\text{Acetylene}}{HC{\equiv}CH}$$

It was one thing to be able to make acetylene, but it was quite another to handle it safely. Because acetylene is a highly explosive substance, its production in large quantities is fraught with danger. Indeed, up until the 1930s it was only produced in small quantities, which, when combined with the risks in handling it, greatly limited its industrial utility. Acetylene is so dangerous that many countries actually prohibited working with it under pressure, a desirable and necessary option for many of its reactions. Germany, for example, legally forbade subjecting it to pressures of more than 1.5 bar (750 torr, less than one atmosphere). It was only in the late 1930s when J. Walter Reppe, a German chemist working at BASF, invented the so-called "Reppe glasses"—sealable stainless steel spheres or cylinders that allow high-pressure reactions to be performed with acetylene—that its use as a chemical building block became more widespread. A broad set of commercially important reactions, dubbed "Reppe Chemistry," ultimately resulted.

Reppe's timing was fortuitous with respect to the war effort of Nazi Germany—the "Third Reich"—in the 1930s. As the country launched World War II, it again became dependent on synthetic rubber, but

(Continued)

HISTORICAL HIGHLIGHT *Acetylene: A Valuable Small Molecule (Continued)*

by this time a better non-natural rubber had been invented. It was a copolymer formed by the sodium-promoted (**Na**) polymerization of 1,3-**Bu**tadiene and **S**tyrene, and hence the name, **Buna-S.** As mentioned earlier, the Germans were able to make 1,3-butadiene from acetylene, and they could produce styrene by dehydrogenation of ethylbenzene. Ethylbenzene was available from a Friedel-Crafts reaction (see Chapter 15.2) between benzene and ethylene. Benzene can be prepared from coal tar, which is derived from coal, and ethylene results from partial reduction of acetylene, as summarized below. In short, the role played by acetylene was critical to Germany's war effort. Although we naturally decry the very existence of warfare, the German chemists must be recognized for their ingenuity and creativity in making acetylene, a very hazardous substance, a valuable chemical feedstock during and after the worldwide conflict.

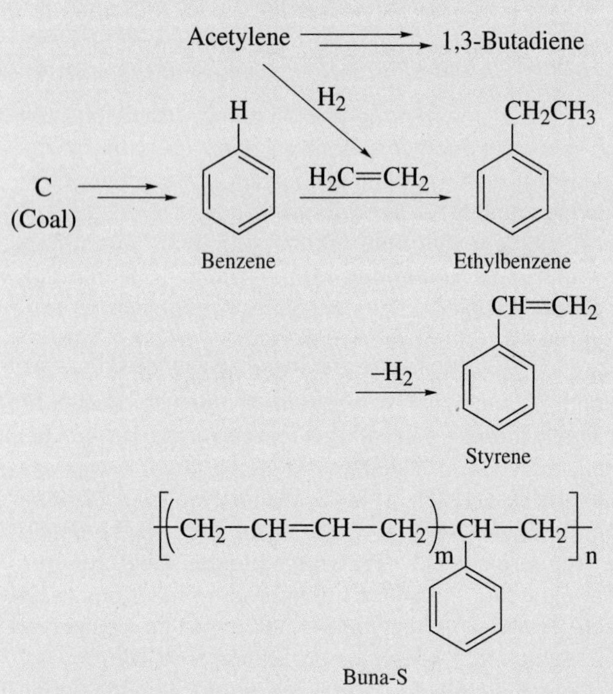

Acetylene $\longrightarrow$ 1,3-Butadiene

Acetylene $\xrightarrow{H_2}$ $H_2C{=}CH_2$

C (Coal) $\longrightarrow$ Benzene $\xrightarrow{H_2C{=}CH_2}$ Ethylbenzene (CH_2CH_3) $\xrightarrow{-H_2}$ Styrene ($CH{=}CH_2$)

$$\left[\left(CH_2{-}CH{=}CH{-}CH_2\right)_m CH(C_6H_5){-}CH_2\right]_n$$

Buna-S

Relationship of Historical Highlight to Experiments

The procedures presented in this chapter represent basic reactions of alkynes. That involving hydration of 2-methyl-3-butyn-2-ol through electrophilic addition of water to the π-system is a reaction analogous to the conversion of acetylene to acetaldehyde, a precursor to acetic acid and acetone. In addition, the formation of an alkyne via an elimination reaction illustrates an alternate approach to forming a carbon-carbon triple bond, although one that is not nearly so easy experimentally as adding water to calcium carbide to make acetylene!

See more on *1,3-Butadiene*
See more on *Acetic Acid*
See more on *Vinyl Chloride*
See more on *1,4-Butanediol*
See more on *Acetaldehyde*
See more on *Carbide Lamps*
See more on *Acetylene*
See more on *Latex*
See more on *Rubber*
See more on *Benzene*
See more on *Ethylene*
See *Who was Reppe?*
See *Who was Friedel?*
See *Who was Crafts?*

CHAPTER 12

Dienes: The Diels-Alder Reaction

When you see this icon, sign in at this book's premium website at **www.cengage.com/login** to access videos, Pre-Lab Exercises, and other online resources.

See more on *Diels-Alder Reaction*

In Chapters 10 and 11, you can discover that both alkenes and alkynes undergo a number of different electrophile addition reactions. These transformations typically proceed by stepwise mechanisms, involving reactions of a π-bond first with an electrophile to give an intermediate cation that then reacts with a nucleophile. In this chapter, you will discover that alkenes and alkynes also undergo a unique cycloaddition reaction with conjugated 1,3-dienes in an interesting process known as the Diels-Alder reaction. Because two new carbon-carbon bonds *and* a new six-membered ring are formed in a single step, the Diels-Alder reaction is one of the most powerful reactions in organic chemistry, and it has been widely used as a key step in the preparation of complex molecules of biological interest. Indeed, there is now good evidence that the Diels-Alder reaction is even used to prepare compounds that occur in living organisms.

12.1 INTRODUCTION

See more on *Diels-Alder/Nobel Prize*

In synthetic organic chemistry, reactions that produce carbocyclic rings from acyclic precursors are of great importance because they lead to the formation of complex molecular structures. Rings may be produced from the cyclization of an acyclic starting material by forming a *single* carbon-carbon bond. However, one of the most useful methods for constructing six-membered rings from acyclic starting materials involves the reaction of a **1,3-diene** with an alkene to give a derivative of cyclohexene by forming *two* new carbon-carbon bonds (Eq. 12.1). In this reaction, the alkene partner is generally referred to as a **dienophile.** This **cycloaddition** reaction is called the **Diels-Alder reaction** in honor of Otto Diels and Kurt Alder, the two German chemists who recognized its importance and shared the 1950 Nobel Prize in Chemistry for their extensive development of this reaction. A more detailed account of their elegant work is included in the Historical Highlight in this chapter. The Diels-Alder reaction also belongs to a class of reactions termed **1,4-additions,** because the two new carbon-carbon σ-bonds are formed between the 1- and 4-carbon atoms of the diene and the two π-bonded carbon atoms of the dienophile.

1 2 3 4 — 1 2 3 4 (12.1)

A 1,3-diene A dienophile A cycloadduct

The scope of the Diels-Alder reaction is broad, and many combinations of dienes and dienophiles are known to furnish cycloadducts in good yields. The presence of electron-withdrawing substituents such as cyano, C≡N, and carbonyl, C=O, groups on the dienophile increases the rate and yield of the reaction. The Diels-Alder reaction is remarkably free of complicating side reactions, and yields of the desired product are often high. Probably the single most important side reaction that may be encountered is dimerization or polymerization of the diene and/or dienophile.

12.2 MECHANISTIC AND STEREOCHEMICAL ASPECTS

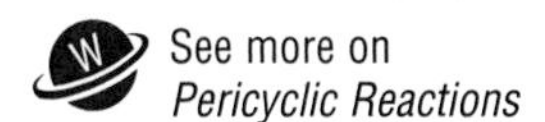
See more on *Pericyclic Reactions*

The mechanism of the Diels-Alder reaction is now fairly well understood. The accumulated evidence favors a picture in which reorganization of the π-electrons in the transition state for the reaction occurs so the two new σ-bonds are formed in a simultaneous, **concerted** fashion; little charge or free-radical character is developed at any of the terminal carbon atoms (Eq. 12.2). The reaction is one of a number of **pericyclic reactions** that are controlled by **orbital symmetry.** Detailed discussions of orbital symmetry and its ramifications may be found in all modern organic chemistry textbooks.

‡ (12.2)

Reactants Six-membered transition state Product

A consequence of the concerted nature of the Diels-Alder reaction is that the diene must be able to adopt a *planar conformation* in which the dihedral angle between the two double bonds is 0°. Such a conformation is designated as *s-cis,* meaning **cisoid** about the single bond. Dienes that have difficulty achieving this conformation are expected to undergo Diels-Alder reactions slowly or not at all. A concerted reaction of a dienophile with a diene in its more stable *s-trans* conformation would lead to a six-membered ring containing a highly strained *trans* double bond, an improbable structure.

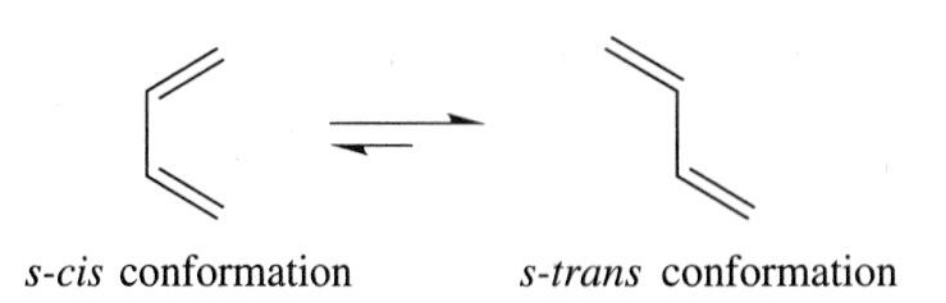

s-cis conformation *s-trans* conformation

A consequence of the highly ordered relationship between the diene and dienophile in the transition state depicted in Equation 12.2 is that Diels-Alder reactions are highly **stereoselective,** with the formation of the new carbon-carbon σ-bonds occurring on the *same* face of *both* the diene and the dienophile. Such addition reactions are termed ***syn*-additions.** Thus, the cycloadduct resulting from reaction of *trans, trans*-2,4-hexadiene and a dienophile is *exclusively cis*-3,6-dimethylcyclohexene (Eq. 12.3), whereas the product obtained from *cis, trans*-2,4-hexadiene and a dienophile is *trans*-3,6-dimethylcyclohexene. Both of these reactions proceed by *syn*-1,4-addition to the diene, with the stereochemical relationship between the two methyl groups on the diene being maintained. Similarly, the reaction between 1,3-butadiene and dimethyl maleate gives solely the *cis*-diester **1** (Eq. 12.4). This example illustrates that the Diels-Alder reaction proceeds by exclusive *syn*-1,2-addition to the double bond of the dienophile; no isomerization to the more stable *trans*-diester is observed. The results of the reactions depicted in Equations 12.3 and 12.4 are in accord with the hypothesis that the Diels-Alder reaction is concerted, with both new carbon-carbon bonds forming simultaneously, although not necessarily at the same rate.

CH_3 ... CH_3 + (alkene) ⟶ H, Me ... H, Me (12.3)

trans, trans-2,4-Hexadiene An alkene A *cis*-3,6-dimethylcyclohexene

(diene) + H, CO_2Me, H, CO_2Me ⟶ H, CO_2Me, CO_2Me, H **1** (12.4)

s-cis-1,3-Butadiene Dimethyl maleate Dimethyl 4-cyclohexene-*cis*-1,2-dicarboxylate

There is a second type of stereoselectivity that is characteristic of the Diels-Alder reaction. The addition of a dienophile such as maleic anhydride to a *cyclic* diene like 1,3-cyclopentadiene could provide two products, the *endo*-adduct **2** and the *exo*-adduct **3** (Eq. 12.5). However, only the *endo*-cycloadduct **2,** in which the two bold-faced hydrogens are *syn* to the one-carbon bridge, is observed experimentally, and its preferential formation follows what is now commonly termed the **Alder rule.** The basis for this result is believed to be stabilization of the **transition state 4** by **secondary orbital interactions** that occur through space between the *p*-orbitals on the internal carbons of the diene and the carbonyl carbon atoms of the dienophile, as shown by the dashed lines in **4.** Analogous stabilization is not possible in transition state **5.** Structure **4** is thus characterized as the one being stabilized by *maximum orbital overlap.* It should be noted that not all Diels-Alder reactions are as stereoselective as the one between 1,3-cyclopentadiene and maleic anhydride; mixtures of *endo*- and *exo*-products are sometimes obtained.

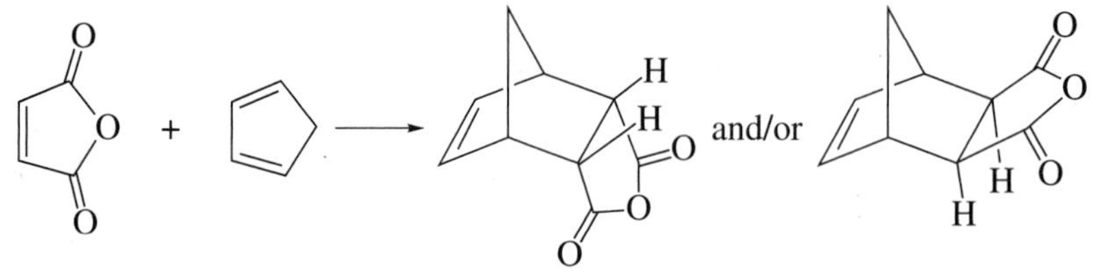

(12.5)

2 | 3

Maleic anhydride | 1,3-Cyclopentadiene | Bicyclo[2.2.1]hept-5-ene-*endo*-2,3-dicarboxylic anhydride | Bicyclo[2.2.1]hept-5-ene-*exo*-2,3-dicarboxylic anhydride

4 | **5**

12.3 APPLICATIONS OF DIELS-ALDER REACTIONS

Two experimental procedures using maleic anhydride, a highly reactive dienophile, nicely illustrate various features of the Diels-Alder reaction. In Part **A**, the Diels-Alder reaction of 1,3-butadiene with maleic anhydride to give 4-cyclohexene-*cis*-1,2-dicarboxylic anhydride (**6**) is performed (Eq. 12.6). Since 1,3-butadiene is a gas at room temperature (bp –4.5 °C, 760 torr), Diels-Alder reactions involving this diene are normally performed in sealed steel pressure vessels into which the diene is introduced under pressure. However, 1,3-butadiene can be conveniently generated *in situ* by the thermal decomposition of 3-sulfolene that occurs with the extrusion of sulfur dioxide (Eq. 12.7), and this strategy is used in this experiment. The decomposition can be effected by heating a solution of 3-sulfolene under reflux in the presence of the diene. This process can also be promoted through microwave heating (Sec. 2.9) if the appropriate apparatus is available. It is instructive to compare the reaction times required and the results obtained using conventional versus microwave heating for promoting the reaction.

(12.6)

6

1,3-Butadiene | Maleic anhydride | 4-Cyclohexene-*cis*-1,2-dicarboxylic anhydride

(12.7)

3-Sulfolene | 1,3-Butadiene

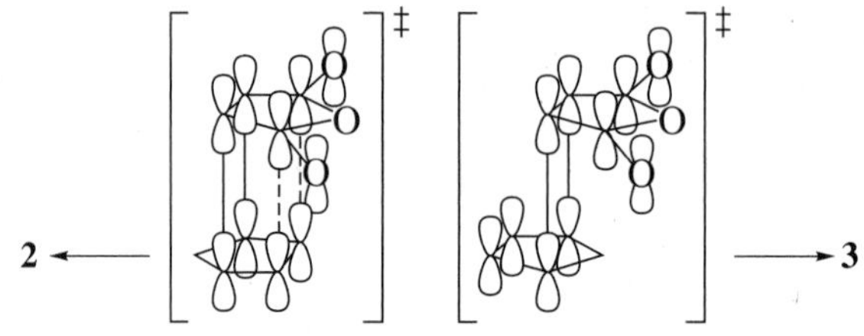

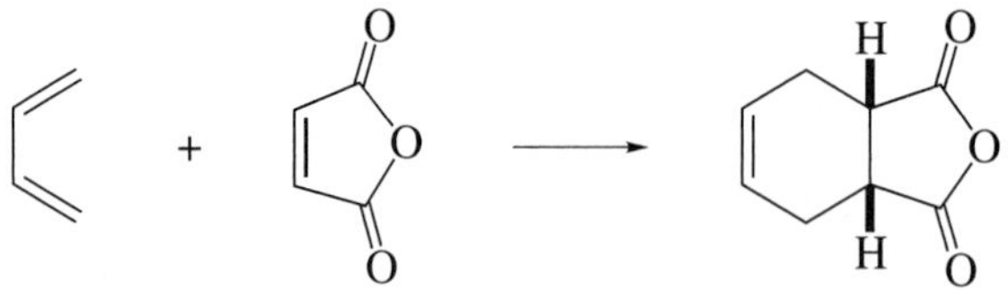

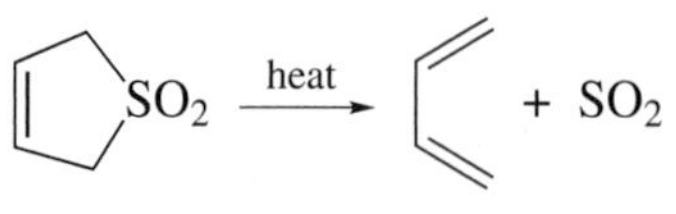

In Part **B**, the cycloaddition of 1,3-cyclopentadiene with maleic anhydride is examined (Eq. 12.5). This experiment provides an opportunity to investigate the stereoselectivity of the Diels-Alder reaction. Monomeric 1,3-cyclopentadiene cannot be purchased because it readily dimerizes at room temperature by a Diels-Alder reaction to give dicyclopentadiene (Eq. 12.8), which is commercially available. Fortunately, the equilibrium between the monomer and the dimer can be established at the boiling point of the dimer (170 °C, 760 torr) by a process commonly called **cracking,** and the lower-boiling 1,3-cyclopentadiene may then be isolated by fractional distillation. The diene *must* be kept cold in order to prevent its redimerization via a Diels-Alder reaction prior to use in the desired Diels-Alder reaction.

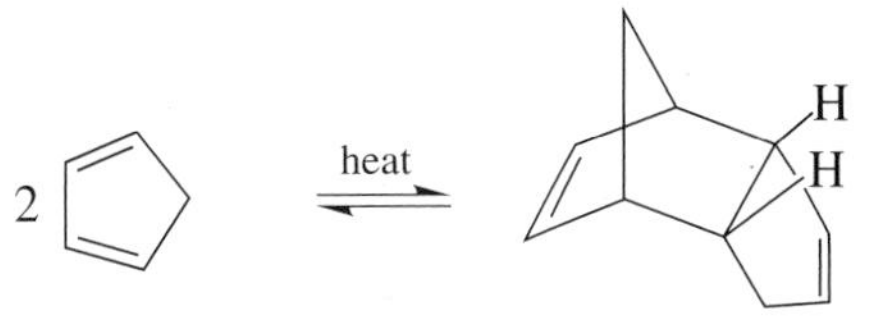

2 1,3-Cyclopentadiene ⇌ (heat) Dicyclopentadiene (12.8)

The Diels-Alder adducts **2** and **6** retain the anhydride function that was originally present in the dienophile, maleic anhydride. As a rule, this functionality is unstable in the presence of water even at pH 7 and hydrolyzes to a dicarboxylic acid according to the general sequence outlined in Scheme 12.1. This process is initiated by nucleophilic attack of water on the carbonyl function, followed by cleavage of a C–O bond of the anhydride. Because anhydrides are reactive toward water, it is important that the apparatus and reagents you use are dry to maximize the yield of cycloadduct. You may be instructed to prepare the dicarboxylic acids **7** and **8** by intentional hydrolysis of the corresponding anhydrides. Protocols for doing so are included in Part **C** of the experimental procedures.

Scheme 12.1

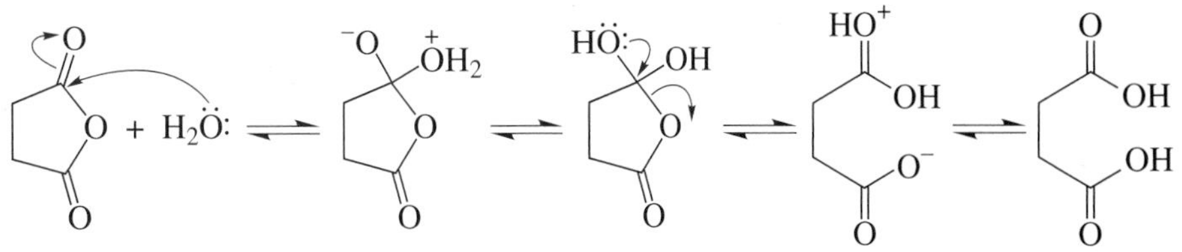

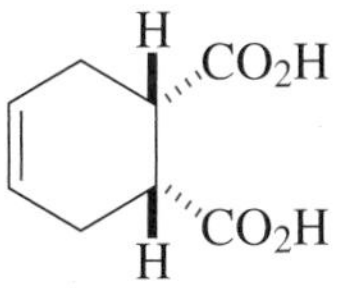

4-Cyclohexene-*cis*-1,2-dicarboxylic acid
7

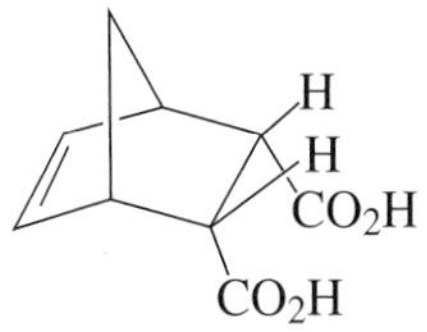

Bicyclo[2.2.1]hept-5-ene-*endo*-2,3-dicarboxylic acid
8

EXPERIMENTAL PROCEDURES

Diels-Alder Reaction

Purpose To demonstrate the formation of six-membered rings by a cycloaddition reaction.

SAFETY ALERT

1. **Be certain that all joints in the apparatus are tight and well lubricated before heating the reaction mixture so that sulfur dioxide, a toxic and foul-smelling gas, does not escape into the laboratory.**
2. **The organic solvents are highly flammable, so use *flameless* heating.**

A ■ *Reaction of 1,3-Butadiene and Maleic Anhydride*

MINISCALE PROCEDURE

Preparation Sign in at **www.cengage.com/login** to answer Pre-Lab Exercises, access videos, and read the MSDSs for the chemicals used or produced in this procedure. Review Sections 2.9, 2.11, 2.17, 2.22, and 2.23.

Apparatus A 25-mL round-bottom flask, gas trap, apparatus for heating under reflux, magnetic stirring, vacuum filtration, and *flameless* heating.

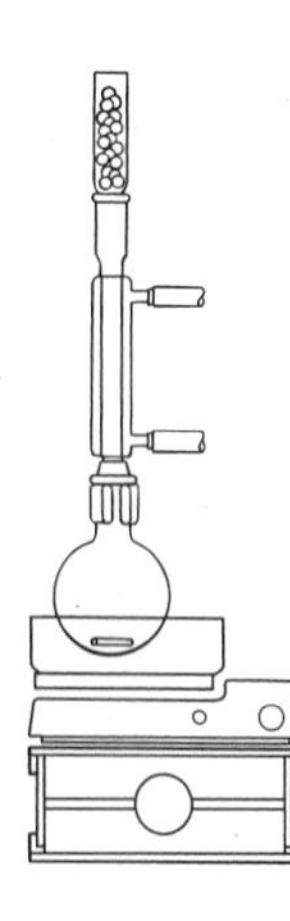

Setting Up Place 2.5 g of 3-sulfolene, 1.5 g of finely pulverized maleic anhydride, and 1 mL of dry xylenes in the flask and equip it with a stirbar. Assemble the apparatus for heating under reflux. Attach the condenser to the gas trap specified by your instructor.

Cycloaddition and Isolation Gently warm the flask with stirring to effect *complete* dissolution of all solids; failure to do so will lower the yield of the reaction *significantly.* Then heat the mixture under gentle reflux for about 0.5 h.★ Allow the reaction mixture to cool below the boiling point and add 10 mL of xylenes. Reheat the stirred mixture to dissolve all of the solids, and then pour the hot solution into an Erlenmeyer flask. If any product crystallizes during this operation, heat the mixture gently until all of the solid redissolves.★ Carefully add petroleum ether (bp 60–80 °C, 760 torr) to the hot solution until cloudiness develops, and then set the solution aside to cool to room temperature.★ Collect the crystals by vacuum filtration, transfer them to a piece of filter or weighing paper, and air-dry them thoroughly.

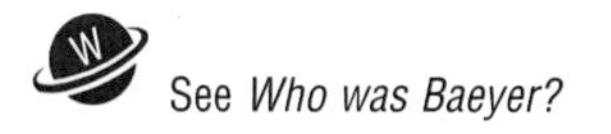

Analysis Weigh your product to determine the yield and measure its melting point. Test the product for unsaturation using the bromine in dichloromethane (Sec. 25.8A) and the Baeyer tests (Sec. 25.8B). Obtain IR and ^{1}H NMR spectra of your starting materials and product and compare them with those of authentic samples (Figs. 12.1–12.6).

MICROSCALE PROCEDURE

Preparation Sign in at **www.cengage.com/login** to answer Pre-Lab Exercises, access videos, and read the MSDSs for the chemicals used or produced in this procedure. Review Sections 2.9, 2.11, 2.17, 2.22, and 2.23.

Apparatus A 3-mL conical vial, gas trap, apparatus for heating under reflux, magnetic stirring, Craig tube filtration, and *flameless* heating.

Setting Up Equip the conical vial with a spinvane and add 250 mg of 3-sulfolene, 150 mg of finely pulverized maleic anhydride, and 0.5 mL of dry xylenes. Assemble the apparatus for heating under reflux using the gas trap specified by your instructor.

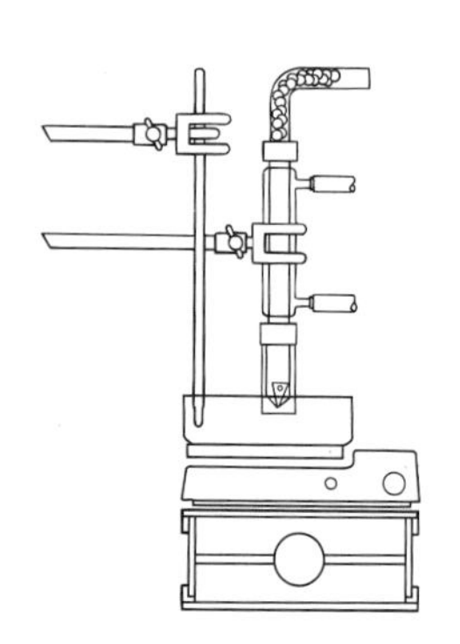

Cycloaddition and Isolation Gently warm the vial with stirring to effect *complete* dissolution of all solids; failure to do so will lower the yield of the reaction *significantly.* Then heat the mixture under gentle reflux for about 0.5 h. Allow the reaction mixture to cool below the boiling point and add 1 mL of xylenes.★ Reheat the mixture to effect dissolution of all solids and then rapidly transfer the contents of the vial to a Craig tube. If any product crystallizes during this operation, heat the mixture gently until all of the solid redissolves. Carefully add petroleum ether (bp 60–80 °C, 760 torr) to the hot solution until cloudiness develops, and then set the solution aside to cool to room temperature.★ Complete the crystallization by cooling the mixture in an ice-water bath, equip the Craig tube for filtration, and isolate the product by centrifugation.

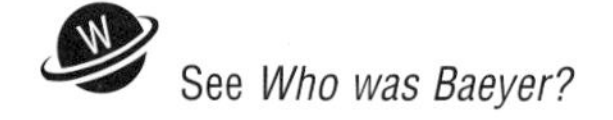
See *Who was Baeyer?*

Analysis Weigh your product to determine the yield and measure its melting point. Test the product for unsaturation using the bromine in dichloromethane (Sec. 25.8A) and the Baeyer tests (Sec. 25.8B). Obtain IR and ^{1}H NMR spectra of your starting materials and product and compare them with those of authentic samples (Figs. 12.1–12.6).

MINISCALE PROCEDURE FOR MICROWAVE OPTION

Preparation Sign in at **www.cengage.com/login** to answer Pre-Lab Exercises, access videos, and read the MSDSs for the chemicals used or produced in this procedure. Review Sections 2.9, 2.11, 2.17, 2.22, and 2.23.

Apparatus A 25-mL round-bottom flask, stirbar, condenser, gas trap, and apparatus for microwave heating with magnetic stirring and vacuum filtration.

See more on *Microwave Heating*

Setting Up Place 2.5 g of 3-sulfolene, 1.5 g of pulverized maleic anhydride, and 1 mL of dry xylenes in the flask and equip it with a stirbar. Place the flask in the cavity of the microwave apparatus on top of the vessel stand, secure the microwave door, and attach the condenser and gas trap to the flask.

Cycloaddition and Isolation Program the microwave apparatus for heating open reaction vessels with stirring according to the directions provided by your instructor. Generally, the reaction temperature should be set at 150 °C and the power set at 300 W with a 1-min ramp time and a 2-min hold time; the pressure limit should be set at 250 psi. Allow the mixture to cool to room temperature, remove the flask from the microwave apparatus, and add 10 mL of xylenes. Return the apparatus to the microwave and, using the same directions as before, heat the solution to dissolve all solids. Transfer the hot solution to an Erlenmeyer flask. If any product crystallizes during this operation, heat the mixture gently until all of the solid redissolves.★ Carefully add petroleum ether (bp 60–80 °C, 760 torr) to the hot solution until cloudiness develops, and then set the solution aside to cool to room temperature. Collect the crystals by vacuum filtration, transfer them to a piece of filter or weighing paper, and air-dry them thoroughly.

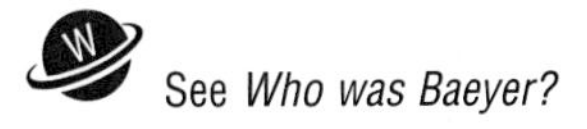
See *Who was Baeyer?*

Analysis Weigh your product to determine the yield and measure its melting point. Test the product for unsaturation using the bromine in dichloromethane (Sec. 25.8A) and the Baeyer tests (Sec. 25.8B). Obtain IR and ^{1}H NMR spectra of your starting materials and product and compare them with those of authentic samples (Figs. 12.1–12.6).

WRAPPING IT UP

Pour the *mixture of xylenes and petroleum ether* recovered from the reaction into a container for nonhalogenated organic solvents. Pour the *dichloromethane solution* from the bromine test for unsaturation in a container for halogenated organic liquids, and put the *manganese dioxide* from the Baeyer test for unsaturation in a container for heavy metals.

B ■ *Reaction of 1,3-Cyclopentadiene and Maleic Anhydride*

SAFETY ALERT

1. **1,3-Cyclopentadiene is a mildly toxic, foul-smelling, and volatile substance. Prepare and use it in a hood if possible; avoid inhaling its vapors. Keep the diene cold at all times to minimize vaporization and dimerization.**
2. **Be certain that all joints in the apparatus are tight before heating dicyclopentadiene to produce the monomer.**
3. **The 1,3-cyclopentadiene and organic solvents used are highly flammable, so use *flameless* heating.**

MINISCALE PROCEDURE

Preparation Sign in at **www.cengage.com/login** to answer Pre-Lab Exercises, access videos, and read the MSDSs for the chemicals used or produced in this procedure. Review Sections 2.9, 2.10, 2.11, 2.14, and 2.17.

Apparatus A 10- and a 25-mL round-bottom flask, 25-mL Erlenmeyer flask, drying tube, ice-water bath, apparatus for fractional distillation, magnetic stirring, vacuum filtration, and *flameless* heating.

Setting Up Place a stirbar and 7 mL of dicyclopentadiene in a 25-mL round-bottom flask, and attach the flask to a fractional distillation apparatus that is equipped with an *unpacked* fractionating column and the 10-mL round-bottom flask as the receiver. Cool the receiver in an ice-water bath and fit the vacuum adapter with a drying tube.

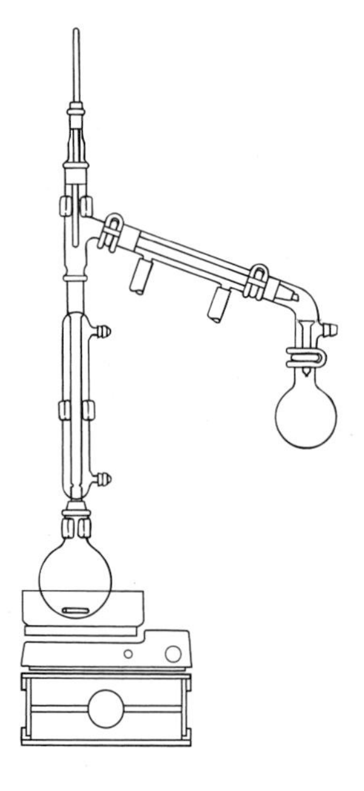

Preparing Reactants Heat the dimer until the solution is briskly refluxing (*Caution*: occasional foaming) and the monomer begins to distill in the range 40–42 °C. Distill the monomer as *rapidly* as possible, but do *not* permit the head temperature to exceed 43–45 °C. About 2.5 mL of 1,3-cyclopentadiene should be obtained from the distillation, which will require about 0.5 h. If the distillate is cloudy because of condensation of moisture in the cold receiver, add about 0.5 g of *anhydrous* calcium chloride.

While the distillation is in progress, dissolve 1.5 g of maleic anhydride in 5 mL of ethyl acetate contained in a 25-mL Erlenmeyer flask. Gently heat the mixture to hasten dissolution and then cool the solution thoroughly in an ice-water bath.

Cycloaddition and Isolation Add 1.5 mL of dry 1,3-cyclopentadiene in one portion to the cooled solution of maleic anhydride, and swirl the resulting solution gently until the exothermic reaction subsides. Add 5 mL of petroleum ether (bp 60–80 °C, 760 torr) to cause precipitation of the cycloadduct as a white solid.★ Heat the mixture until the solid has redissolved.

To prevent the desired product from oiling out of solution rather than crystallizing, the reaction mixture *must* cool slowly. To accomplish this, place the Erlenmeyer flask in a water bath warmed to about 60 °C, and let the bath and flask cool to room temperature.★ Seeding the solution to effect crystallization may be required and can be performed according to the procedure described in Section 3.2. Once the solution has cooled to room temperature and no more crystallization is apparent, cool the flask in an ice-water bath for a few minutes to complete the process. Then isolate the product by vacuum filtration, and air-dry it.

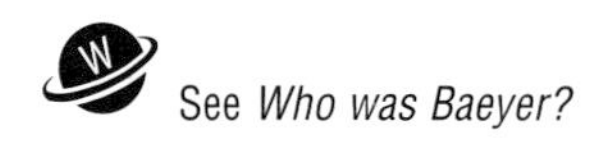
See *Who was Baeyer?*

Analysis Weigh your product to determine the yield and measure its melting point. Test the product for unsaturation using the bromine in dichloromethane (Sec. 25.8A) and the Baeyer tests (Sec. 25.8B). Obtain IR and ^{1}H NMR spectra of your starting materials and product and compare them with those of authentic samples (Figs. 12.1, 12.2, and 12.7–12.10).

WRAPPING IT UP

Pour the *pot residue of dicyclopentadiene*, any unused *1,3-cyclopentadiene*, and the *mother liquors* from the filtration into a container for recovered dicyclopentadiene or a container for nonhalogenated organic solvents. Allow any volatiles on the *calcium chloride* that was used to dry the 1,3-cyclopentadiene to evaporate in the hood, and then place the *residual solid* in a container for nonhazardous solids. Pour the *dichloromethane solution* from the bromine test for unsaturation in a container for halogenated organic liquids, and put the *manganese dioxide* from the Baeyer test for unsaturation in a container for heavy metals.

MICROSCALE PROCEDURE

Preparation Sign in at **www.cengage.com/login** to answer Pre-Lab Exercises, access videos, and read the MSDSs for the chemicals used or produced in this procedure. Review Sections 2.9, 2.10, and 2.17.

Apparatus Water bath and apparatus for Craig tube filtration.

Setting Up Place 100 mg of maleic anhydride and 0.4 mL of ethyl acetate in the Craig tube and mix the contents of the tube carefully to effect dissolution; warming may be helpful. Then add 0.4 mL of petroleum ether (bp 60–90 °C, 760 torr), and mix the contents of the tube by gentle swirling.

Cycloaddition and Isolation In one portion, add 0.1 mL of dry 1,3-cyclopentadiene, which may be obtained from your instructor, to the Craig tube and mix the contents by swirling the tube until a homogeneous solution results. The exothermicity of the reaction may keep the desired product in solution, but if not, heat the tube gently to dissolve any solids that form.

To prevent the desired product from oiling out of solution rather than crystallizing, the reaction mixture *must* cool slowly. To accomplish this, place the Craig tube in a water bath warmed to about 60 °C, and let the bath and tube cool to room temperature.★ Seeding the solution to effect crystallization may be required and can be performed according to the procedure described in Section 3.2. Once the solution has cooled to room temperature and no more crystallization is apparent, cool the Craig tube in an ice-water bath for a few minutes to complete the process. Then equip the Craig tube for filtration and isolate the product by centrifugation.

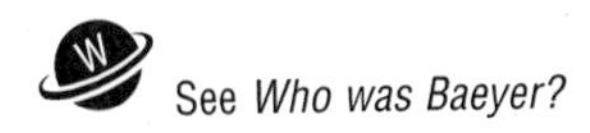

Analysis Weigh your product to determine the yield and measure its melting point. The reported melting point is 164–165 °C. Test the product for unsaturation using the bromine in dichloromethane (Sec. 25.8A) and the Baeyer tests (Sec. 25.8B). Obtain IR and ^{1}H NMR spectra of your starting materials and product, and compare them with those of authentic samples (Figs. 12.1, 12.2, and 12.7–12.10).

WRAPPING IT UP

Pour the *mother liquors* from the filtration into a container for nonhalogenated organic solvents. Pour the *dichloromethane solution* from the bromine test for unsaturation in a container for halogenated organic liquids, and put the *manganese dioxide* from the Baeyer test for unsaturation in a container for heavy metals.

C ■ Hydrolysis of Anhydrides

1. 4-Cyclohexene-*cis*-1,2-dicarboxylic Acid

MINISCALE PROCEDURE

Preparation Sign in at **www.cengage.com/login** to answer Pre-Lab Exercises, access videos, and read the MSDSs for the chemicals used or produced in this procedure. Review Sections 2.9, 2.10, and 2.17.

Apparatus A 25-mL Erlenmeyer flask, ice-water bath, apparatus for vacuum filtration, and *flameless* heating.

Setting Up Place 1.0 g of anhydride **6** and 5 mL of distilled water in the Erlenmeyer flask and add a stirbar.

Hydrolysis and Isolation Heat the mixture to boiling and continue heating until all the oil that forms initially dissolves. Allow the solution to cool to room temperature and then induce crystallization by scratching the flask at the air-liquid interface.★ After crystallization begins, cool the flask in an ice-water bath to complete the process, collect the solid by vacuum filtration, and air-dry it. Recrystallize the diacid from water.

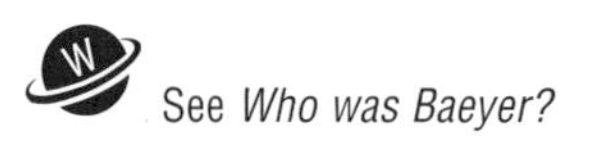

Analysis Weigh the product to determine the yield, and measure its melting point. Test the product for unsaturation using the bromine in dichloromethane (Sec. 25.8A) and the Baeyer tests (Sec. 25.8B) to determine whether hydrolysis has affected the carbon-carbon π-bond. Also test a saturated aqueous solution of the diacid with pHydrion paper and record the result. Obtain IR and ^{1}H NMR spectra of your product and compare them with those of an authentic sample (Figs. 12.11 and 12.12).

Optional Procedure Hydrolysis of the anhydride **6** to the diacid **8** without isolating **6** is possible as follows. Modify the procedure in Part **A** such that after the initial 0.5 h period of reflux the reaction mixture is allowed to cool below the boiling point. Then add about 5 mL of water to the flask through the top of the condenser and heat the resulting mixture under gentle reflux with stirring for at least 0.5 h. Allow the mixture to cool to room temperature.★ If no crystals form, acidify the solution with about 0.3 mL (6 drops) of concentrated sulfuric acid, and cool the resulting mixture in an ice-water bath. Isolate and analyze the diacid according to the directions given above.

MICROSCALE PROCEDURE

Preparation Sign in at **www.cengage.com/login** to answer Pre-Lab Exercises, access videos, and read the MSDSs for the chemicals used or produced in this procedure. Review Sections 2.9, 2.10, and 2.17.

Apparatus A 10-mm × 75-mm test tube, ice-water bath, apparatus for Craig tube filtration, and *flameless* heating.

Setting Up, Hydrolysis, and Isolation Place 100 mg of the anhydride **6** in the test tube and add 0.5 mL of water and a boiling stone. Heat the mixture to boiling and continue heating until all the oil that forms initially dissolves. Using a preheated Pasteur pipet, transfer the hot solution to a Craig tube. Allow the solution to cool to room temperature. If necessary, induce crystallization by gently scratching the tube at the air-liquid interface.★ Complete crystallization by cooling the mixture in an ice-water bath.

Equip the Craig tube for filtration, isolate the crystalline product by centrifugation, and transfer it to a piece of filter or weighing paper for air-drying. Recrystallize the diacid from water.

Analysis Weigh the product to determine the yield and measure its melting point. Test the product for unsaturation using the bromine in dichloromethane (Sec. 25.8A) and the Baeyer tests (Sec. 25.8B) to determine whether hydrolysis has affected the carbon-carbon π-bond. Also test a saturated aqueous solution of the diacid with pHydrion paper and record the result. Obtain IR and ^{1}H NMR spectra of your product and compare them with those of an authentic sample (Figs. 12.11 and 12.12).

2. Bicyclo[2.2.1]hept-5-ene-*endo*-2,3-dicarboxylic Acid

MINISCALE PROCEDURE

Preparation Sign in at **www.cengage.com/login** to answer Pre-Lab Exercises, access videos, and read the MSDSs for the chemicals used or produced in this procedure. Review Sections 2.9, 2.10, and 2.17.

Apparatus A 25-mL Erlenmeyer flask, ice-water bath, apparatus for vacuum filtration, and *flameless* heating.

Setting Up Place 1.0 g of anhydride **2** and 5 mL of distilled water in the Erlenmeyer flask and add a stirbar.

Hydrolysis and Isolation Follow the Miniscale Procedure of Part **1** to effect hydrolysis of **2** and isolation of the diacid **8**.

Analysis Perform the same analysis as that described in the Miniscale Procedure of Part **1**. The reported melting point of bicyclo[2.2.1]hept-5-en-*endo*-2,3-dicarboxylic acid is 180–182 °C. Obtain IR and ^{1}H NMR spectra of your product, and compare them with those of an authentic sample (Figs. 12.13 and 12.14).

MICROSCALE PROCEDURE

Preparation Sign in at **www.cengage.com/login** to answer Pre-Lab Exercises, access videos, and read the MSDSs for the chemicals used or produced in this procedure. Review Sections 2.9, 2.10, and 2.17.

Apparatus A 10-mm × 75-mm test tube, ice-water bath, apparatus for Craig tube filtration, and *flameless* heating.

Setting Up, Hydrolysis, and Isolation Follow the Microscale Procedure of Part **1** to effect hydrolysis of **2** and isolation of the diacid **8**.

Analysis Perform the same analysis as that described in the Microscale Procedure of Part **1**. The reported melting point of bicyclo[2.2.1]hept-2-ene-*endo*-5,6-dicarboxylic acid is 180–182 °C. Obtain IR and ^{1}H NMR spectra of your product, and compare them with those of an authentic sample (Figs. 12.13 and 12.14).

Discovery Experiment

Hydrolysis of Anhydrides

Investigate the possible catalysis of the hydrolysis of anhydrides by designing a protocol for determining whether the hydrolysis of an anhydride may be accelerated by performing the reaction at pH > 7 and pH < 7 as well as at pH 7. The results test whether the hydrolysis is subject to base and acid catalysis. Consult with your instructor before undertaking any experimental procedures.

WRAPPING IT UP

Using sodium carbonate, neutralize the *aqueous mother liquors* from hydrolysis of the anhydride, dilute the solution, and flush it down the drain. Pour the *dichloromethane solution* from the bromine test for unsaturation in a container for halogenated organic liquids, and put the *manganese dioxide* from the Baeyer test for unsaturation in a container for heavy metals.

EXERCISES

General Questions

1. Explain why a *cyclic* 1,3-diene like 1,3-cyclopentadiene dimerizes much more readily than does an *acyclic* one like *trans*-1,3-pentadiene.
2. Write structures for the products expected in the following possible Diels-Alder reactions. If no reaction is anticipated, write "N.R."

a. 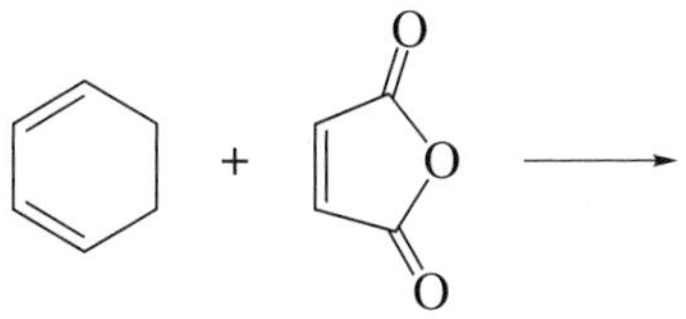 (1,3-cyclohexadiene) + (maleic anhydride) ⟶

b. 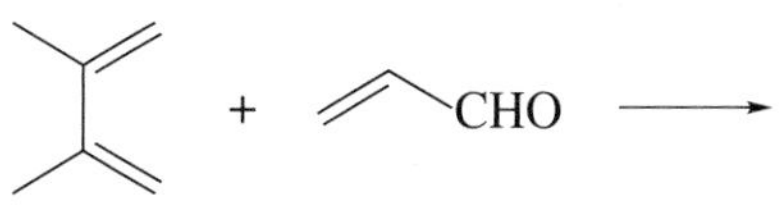 (2,3-dimethyl-1,3-butadiene) + $CH_2{=}CH{-}CHO$ ⟶

c. 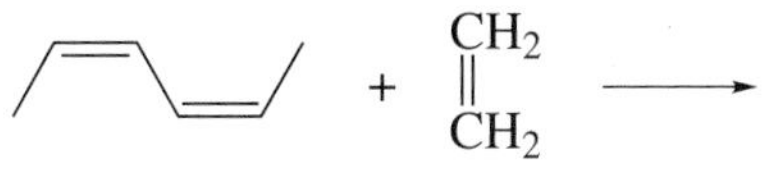 (2,4-hexadiene) + $CH_2{=}CH_2$ ⟶

d. 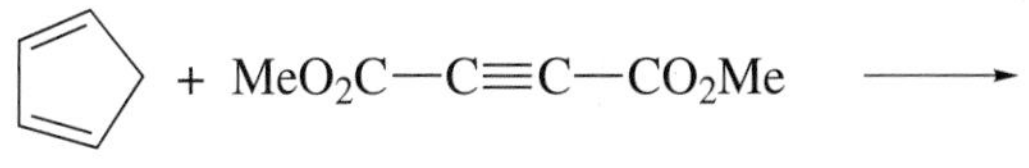 (1,3-cyclopentadiene) + $MeO_2C{-}C{\equiv}C{-}CO_2Me$ ⟶

3. When producing Diels-Alder adducts using maleic anhydride as the dienophile, why is it important that the reagents and apparatus be dry?
4. Write the structure of the product from the Diels-Alder reaction between dimethyl maleate (Sec. 7.3) and 1,3-cyclopentadiene.
5. Provide an explanation for why the stereoisomers shown below would not be produced in the Diels-Alder reaction of Exercise 4.

a. 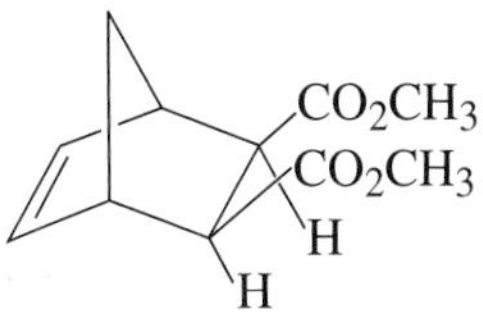 (bicyclic structure with CO_2CH_3, CO_2CH_3, H, H)

b. 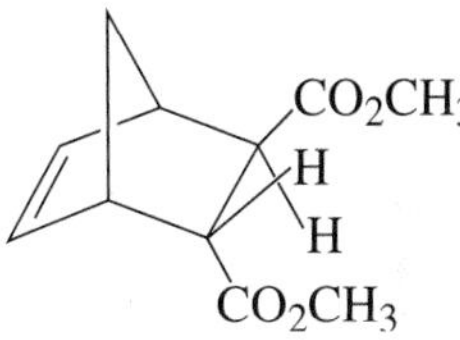 (bicyclic structure with CO_2CH_3, H, H, CO_2CH_3)

6. Consider the structure of bicyclo[2.2.1]hept-4-ene-*endo*-2,3-dicarboxylic anhydride (**2**), which is produced by the Diels-Alder reaction of maleic anhydride and 1,3-cyclopentadiene.
   a. Draw the structure of **2** and label the bridgehead carbons with an asterisk.
   b. Highlight the two new carbon-carbon bonds that were formed in the Diels-Alder reaction.
   c. What structural features of **2** lead to its classification as a bridged bicyclic compound?
   d. The nomenclature terms *endo* and *exo* are used to indicate the relative stereochemistry of a substituent on a bridged bicyclic structure. Circle the larger of the two unsubstituted carbon bridges in **2** and define the term *endo* as it relates to the orientation of substituents relative to the larger carbon bridge of **2.**
   e. Circle the carbon atoms in **2** that are chiral and determine whether **2** is optically active.
7. The primary orbital interactions in the transition state of normal Diels-Alder reactions occur between the highest occupied molecular orbital (the HOMO)

of the diene and the lowest unoccupied molecular orbital (the LUMO) of the dienophile; however, in so-called inverse demand Diels-Alder reactions, the primary orbital interactions occur between the LUMO of the diene and the HOMO of the dienophile.

a. The four π-molecular orbitals for 1,3-cyclopentadiene are shown below. Show how each of the orbitals is occupied with electrons and label the HOMO and the LUMO.

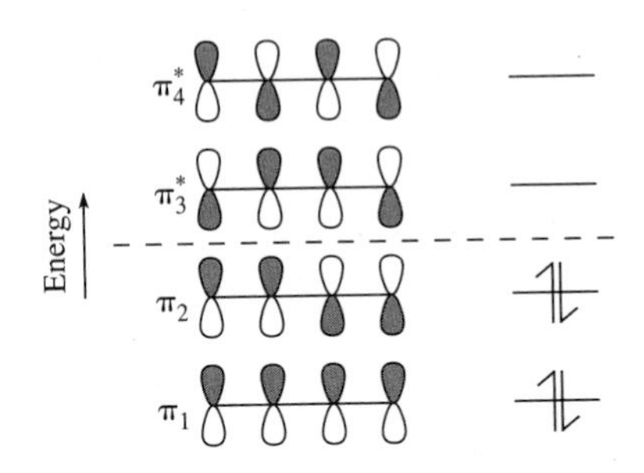

b. Write the two π-molecular orbitals for ethylene, $CH_2{=}CH_2$, and label the HOMO and the LUMO.

c. Is the HOMO or the LUMO an antibonding orbital?

d. Show the two possible HOMO and LUMO primary orbital interactions between ethylene and 1,3-butadiene.

e. Which of the HOMO-LUMO interactions in Part **d** are the orbital interactions between the carbon atoms in the bonds that are being formed in-phase (i.e., constructive orbital overlap)?

8. Write the structures of the diene and the dienophile that would form each of the products shown via a Diels-Alder reaction.

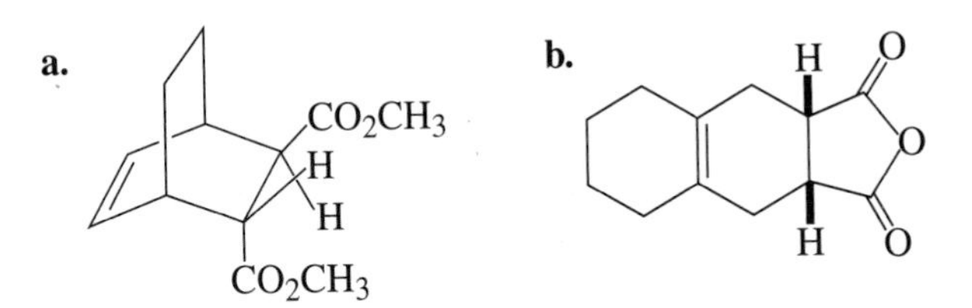

Sign in at **www.cengage.com/login** and use the spectra viewer and Tables 8.1–8.8 as needed to answer the blue-numbered questions on spectroscopy.

9. Consider the spectral data for maleic anhydride (Figs. 12.1 and 12.2).

a. In the functional group region of the IR spectrum, specify the absorptions associated with the carbon-carbon and carbon-oxygen double bonds.

b. The ^1H NMR spectrum shows the chemical shift of the vinylic protons to be δ 7.10. Explain why this resonance is downfield of the normal range observed for vinyl hydrogen atoms.

c. For the ^{13}C NMR data, assign the various resonances to the carbon nuclei responsible for them.

Questions for Part A

10. In this experiment, heating 3-sulfolene produces 1,3-butadiene and a gas.

a. What gas is evolved and is it toxic?

b. How did you avoid releasing this gas into the laboratory?

c. Using curved arrows to symbolize the flow of electrons, give a mechanism for this reaction and indicate whether it is a concerted reaction or not.

d. Is this reaction reversible under the conditions you used? If not, can you describe a set of conditions under which the reaction might be reversible?

11. Which of the two reactants in this procedure was the dienophile and which was the diene?

12. Draw the structures of the three isomeric xylenes and give two reasons why a mixture of these is a good solvent for this reaction.

13. Why should 3-sulfolene and maleic anhydride be *completely* dissolved in xylenes before heating the mixture to effect reaction?

14. Write the structure, including stereochemistry, of the expected addition product of bromine to the Diels-Alder adduct obtained by this procedure.

Sign in at **www.cengage.com/login** and use the spectra viewer and Tables 8.1–8.8 as needed to answer the blue-numbered questions on spectroscopy.

15. Consider the spectral data for 3-sulfolene (Figs. 12.3 and 12.4).

a. In the IR spectrum, specify the absorption associated with the carbon-carbon double bond.

b. In the ^{1}H NMR spectrum, assign the various resonances to the hydrogen nuclei responsible for them.

c. For the ^{13}C NMR data, assign the various resonances to the carbon nuclei responsible for them.

16. Consider the spectral data for 4-cyclohexene-*cis*-1,2-dicarboxylic anhydride (Figs. 12.5 and 12.6).

a. In the functional group region of the IR spectrum, specify the absorptions associated with the carbon-carbon double bond and the carbonyl groups of the anhydride function.

b. In the ^{1}H NMR spectrum, assign the various resonances to the hydrogen nuclei responsible for them.

c. For the ^{13}C NMR data, assign the various resonances to the carbon nuclei responsible for them.

17. a. Given that sulfones typically show two strong absorptions near 1300 and 1150 cm^{-1}, discuss the differences observed in the IR and NMR spectra of 3-sulfolene and of 4-cyclohexene-*cis*-1,2-dicarboxylic anhydride that are consistent with the loss of sulfur dioxide in this experiment.

b. Discuss the differences observed in the IR and NMR spectra of maleic anhydride and of 4-cyclohexene-*cis*-1,2-dicarboxylic anhydride that are consistent with the involvement of the carbon-carbon π-bond of maleic anhydride in this experiment.

Questions for Part B

18. Why should the head temperature be maintained below about 45 °C when dicyclopentadiene is being cracked?

19. Why is it good technique to induce crystallization of the diacid derived from the Diels-Alder adduct *before* cooling the solution to 0 °C?

20. Why does 1,3-cyclopentadiene react more rapidly with maleic anhydride than with another molecule of itself?

21. The "*cracking*" of dicyclopentadiene to two moles of 1,3-cyclopentadiene (Eq. 12.8) is an example of a retro-Diels-Alder reaction. Predict the products to be anticipated from an analogous reaction with the compounds **a–d.**

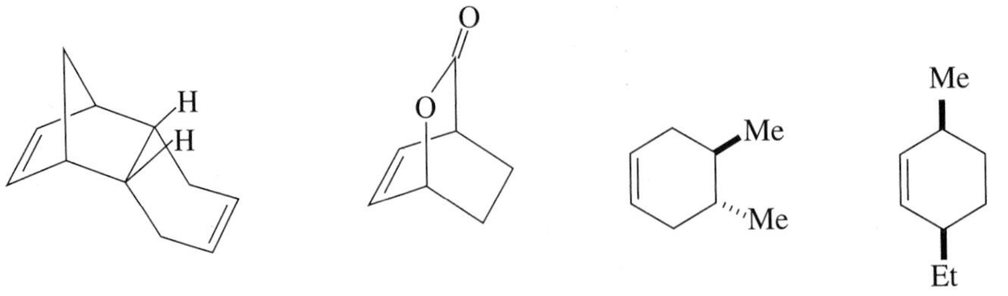

H H O O Me Me Me Et

a. **b.** **c.** **d.**

22. Write an equation that shows the equilibrium process involving dimerization of 1,3-cyclopentadiene to give dicyclopentadiene and cracking of dicyclopentadiene to give 1,3-cyclopentadiene. Using curved arrows to symbolize the flow of electrons, give a mechanism for each reaction.

23. Describe the process of "seeding" a solution and explain how it can facilitate crystallization of a solid.

24. What experimental evidence do you have that the Diels-Alder reaction of 1,3-cyclopentadiene with maleic anhydride is a stereoselective reaction?

25. Explain why maleic anhydride is a much better dienophile than ethylene, $CH_2{=}CH_2$.

26. Consider the spectral data for 1,3-cyclopentadiene (Figs. 12.7 and 12.8).
    - **a.** In the functional group region of the IR spectrum, specify the absorption associated with the carbon-carbon π-bond.
    - **b.** In the ^{1}H NMR spectrum, assign the various resonances to the hydrogen nuclei responsible for them. Explain why the resonance associated with the two C5 hydrogen atoms is not a triplet.
    - **c.** For the ^{13}C NMR data, assign the various resonances to the carbon nuclei responsible for them.

Sign in at **www.cengage.com/login** and use the spectra viewer and Tables 8.1–8.8 as needed to answer the blue-numbered questions on spectroscopy.

27. Consider the spectral data for bicyclo[2.2.1]hept-5-ene-*endo*-2,3-dicarboxylic anhydride (Figs. 12.9 and 12.10).
    - **a.** In the functional group region of the IR spectrum, specify the absorption associated with the carbon-carbon and carbon-oxygen double bonds.
    - **b.** In the ^{1}H NMR spectrum, assign the various resonances to the hydrogen nuclei responsible for them.
    - **c.** For the ^{13}C NMR data, assign the various resonances to the carbon nuclei responsible for them.

28. **a.** Discuss the differences observed in the IR and NMR spectra of 1,3-cyclopentadiene and of bicyclo[2.2.1]hept-5-en-*endo*-2,3-dicarboxylic anhydride that are consistent with the loss of the 1,3-diene function in this experiment.
    - **b.** Discuss the differences observed in the IR and NMR spectra of maleic anhydride and of bicyclo[2.2.1]hept-2-ene-*endo*-5,6-dicarboxylic anhydride that are consistent with the involvement of the carbon-carbon π-bond of maleic anhydride in this experiment.

Questions for Part C

29. In this experiment, you prepared a 1,4-dicarboxylic acid by hydrolyzing an anhydride with water. However, it is known that this hydrolysis is facilitated by the addition of a catalytic amount of acid. Using curved arrows to symbolize the flow of electrons, give a mechanism for the acid-catalyzed hydrolysis of 4-cyclohexene-*cis*-1,2-dicarboxylic anhydride (**6**) and explain why acid increases the rate of the reaction.

30. Suggest a way to prepare 4-cyclohexene-*cis*-1,2-dicarboxylic anhydride (**6**) from the corresponding dicarboxylic acid **7.**

31. Amides and esters, which are also derivatives of carboxylic acids, are stable in water under neutral conditions, whereas anhydrides react readily with water by nucleophilic acyl substitution. Explain the basis for the difference in reactivity of anhydrides, amides, and esters toward water at neutral pH.

Sign in at **www.cengage.com/login** and use the spectra viewer and Tables 8.1–8.8 as needed to answer the blue-numbered questions on spectroscopy.

32. Consider the spectral data for 4-cyclohexene-*cis*-1,2-dicarboxylic acid (Figs. 12.11 and 12.12).

a. In the functional group region of the IR spectrum, specify the absorptions associated with the carbon-carbon double bond and the carboxylic acid functions.

b. In the ^{1}H NMR spectrum, assign the various resonances to the hydrogen nuclei responsible for them.

c. For the ^{13}C NMR data, assign the various resonances to the carbon nuclei responsible for them. Does the number of resonances observed in the ^{13}C NMR spectrum allow you to prove that the *cis*- rather than the *trans*-dicarboxylic acid was produced by the hydrolysis of anhydride **6**? Explain.

33. Discuss the differences observed in the IR and NMR spectra of 4-cyclohexene-*cis*-1,2-dicarboxylic anhydride (Figs. 12.5 and 12.6) and of 4-cyclohexene-*cis*-1,2-dicarboxylic acid that are consistent with the conversion of an anhydride to a diacid function in this experiment.

34. Consider the spectral data for bicyclo[2.2.1]hept-5-ene-*endo*-2,3-dicarboxylic acid (Figs. 12.13 and 12.14).

a. In the functional group region of the IR spectrum, specify the absorptions associated with the carbon-carbon double bond and the carboxylic acid functions.

b. In the ^{1}H NMR spectrum, assign the various resonances to the hydrogen nuclei responsible for them.

c. For the ^{13}C NMR data, assign the various resonances to the carbon nuclei responsible for them. Does the number of resonances observed in the ^{13}C NMR spectrum allow you to prove that the *cis*- rather than the *trans*-dicarboxylic acid was produced by the hydrolysis of anhydride **2**? Explain.

35. Discuss the differences observed in the IR and NMR spectra of bicyclo[2.2.1]hept-5-ene-*endo*-2,3-dicarboxylic anhydride and of bicyclo[2.2.1]hept-2-ene-*endo*-5,6-dicarboxylic acid that are consistent with the conversion of an anhydride to a diacid function in this experiment.

SPECTRA

Starting Materials and Products

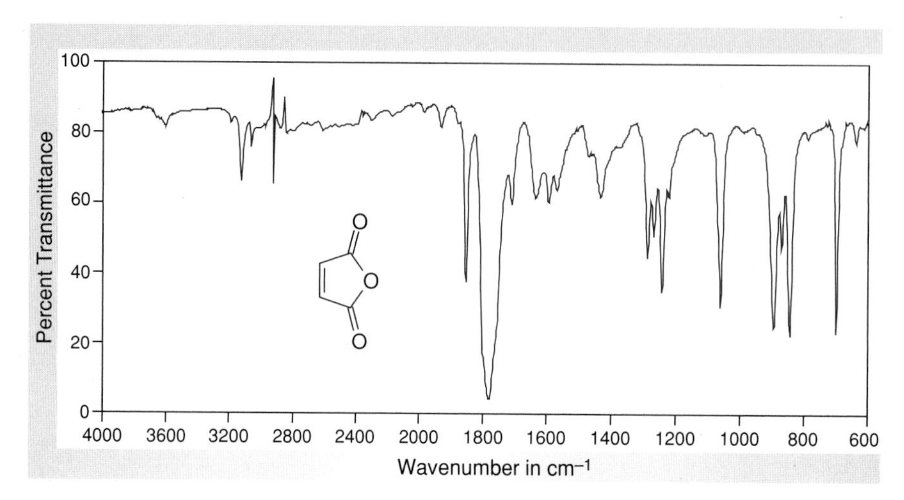

Figure 12.1
IR spectrum of maleic anhydride (IR card).

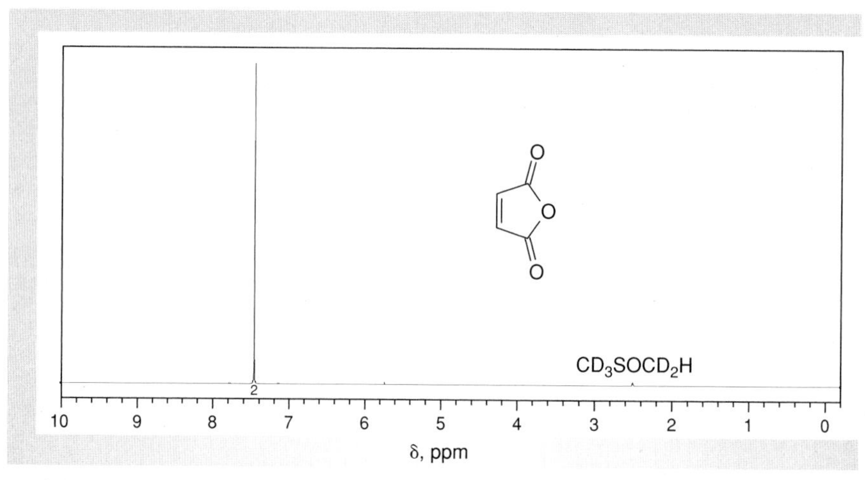

Figure 12.2
NMR data for maleic anhydride (DMSO-d_6).

(a) ^{1}H NMR spectrum (300 MHz).
(b) ^{13}C NMR data: δ 137.0, 165.0.

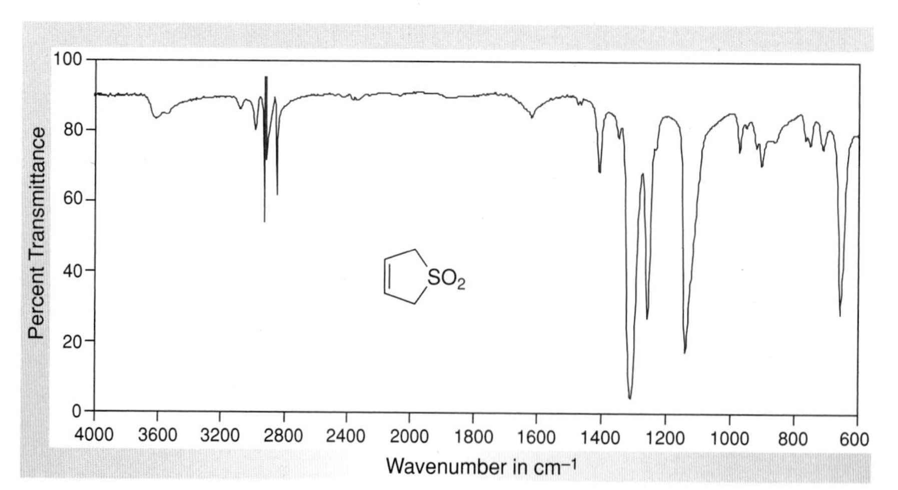

Figure 12.3
IR spectrum of 3-sulfolene (IR card).

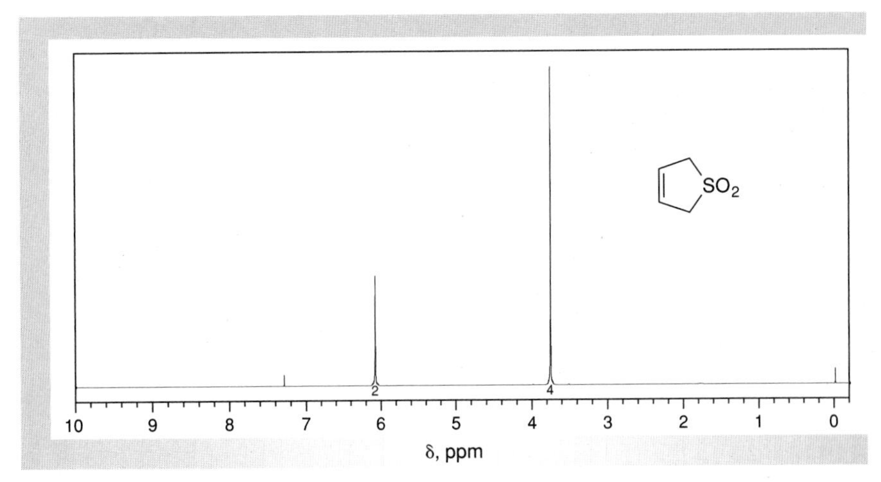

Figure 12.4
NMR data for 3-sulfolene ($CDCl_3$).

(a) ^1H NMR spectrum (300 MHz).
(b) ^{13}C NMR data: δ 55.7, 124.7.

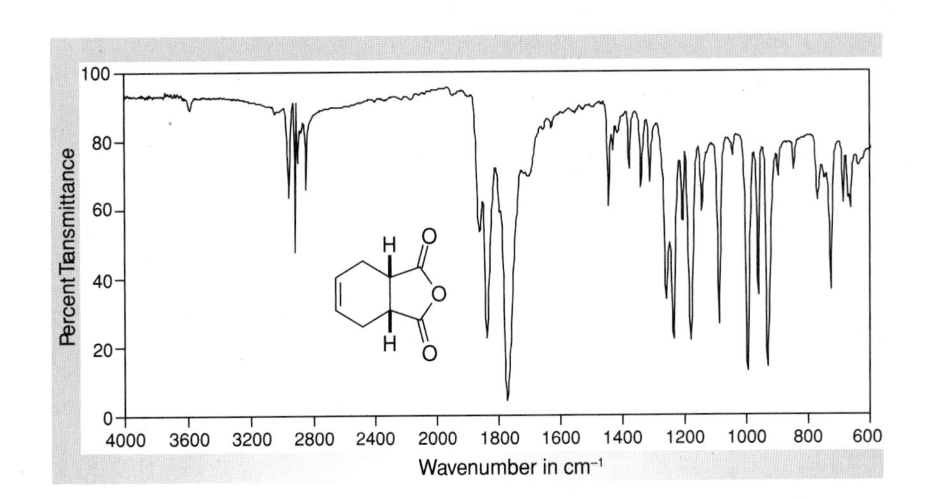

Figure 12.5
IR spectrum of 4-cyclohexene-cis-dicarboxylic anhydride (IR card).

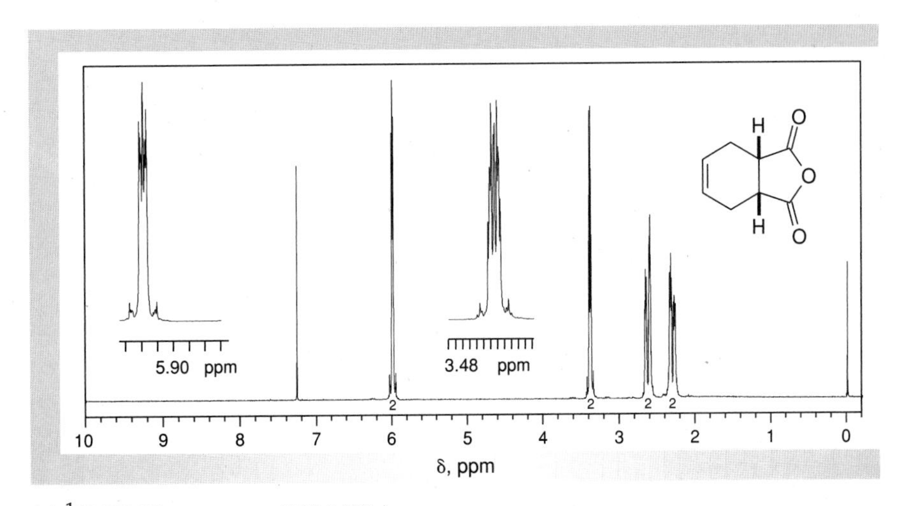

Figure 12.6
NMR data for 4-cyclohexene-cis-dicarboxylic anhydride ($CDCl_3$).

(a) ^1H NMR spectrum (300 MHz).
(b) ^{13}C NMR data. Chemical shifts: δ 26.1, 39.3, 125.5, 174.7.

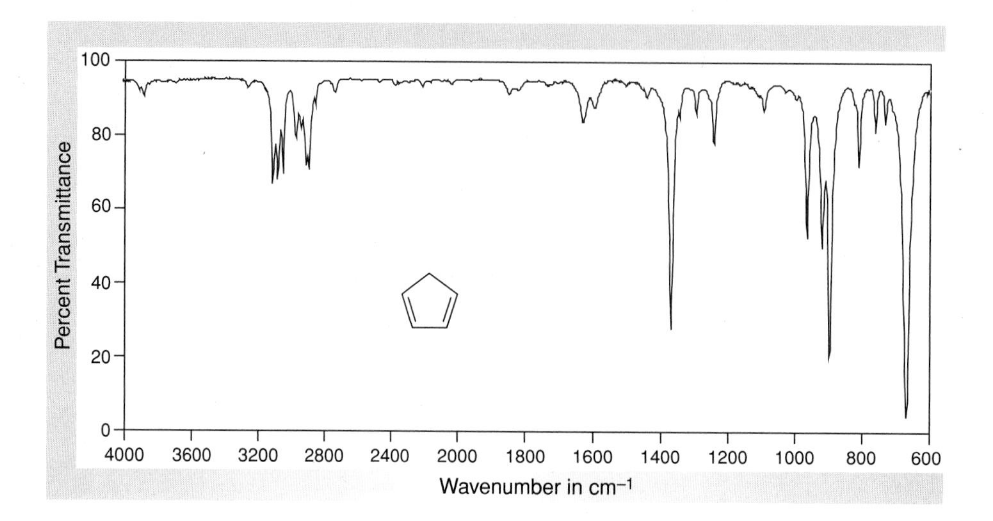

Figure 12.7
IR spectrum of 1,3-cyclopentadiene (neat).

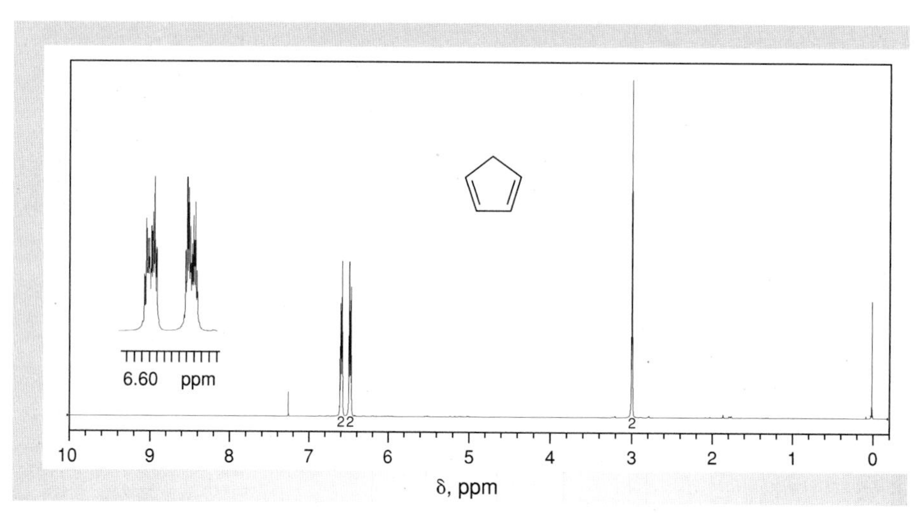

Figure 12.8
NMR data for 1,3-cyclopentadiene ($CDCl_3$).

(a) ^{1}H NMR spectrum (300 MHz).
(b) ^{13}C NMR data. Chemical shifts: δ 42.2, 133.0, 134.4.

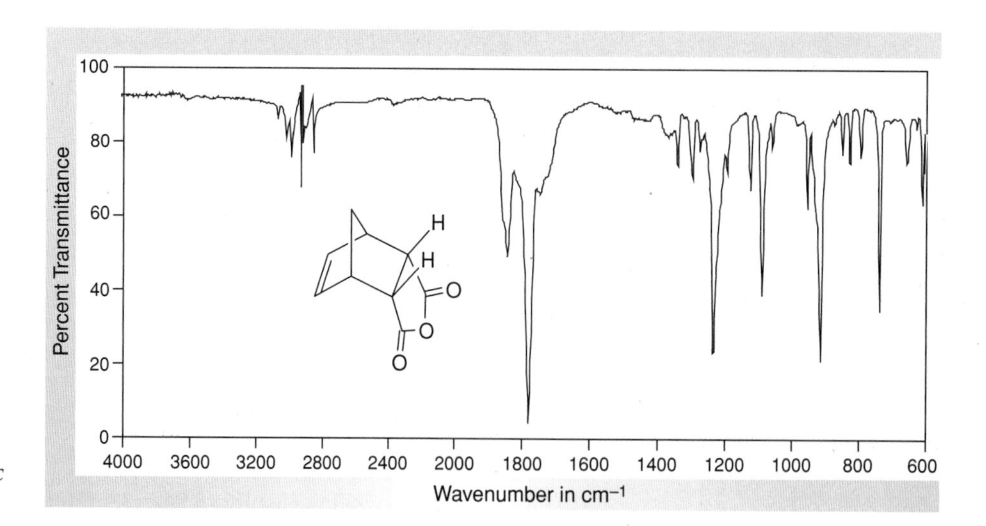

Figure 12.9
*IR spectrum of bicyclo[2.2.1]hept-5-ene-*endo*-2,3-dicarboxylic anhydride (IR card).*

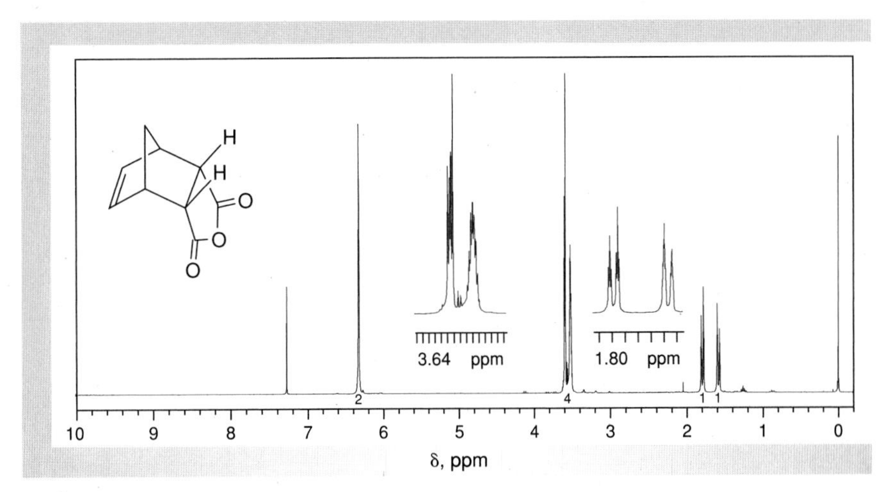

Figure 12.10
NMR data for bicyclo[2.2.1]hept-5-ene-endo-2,3-dicarboxylic anhydride ($CDCl_3$).

(a) ^{1}H NMR spectrum (300 MHz).
(b) ^{13}C NMR data: δ 47.6, 53.5, 136.1, 171.9.

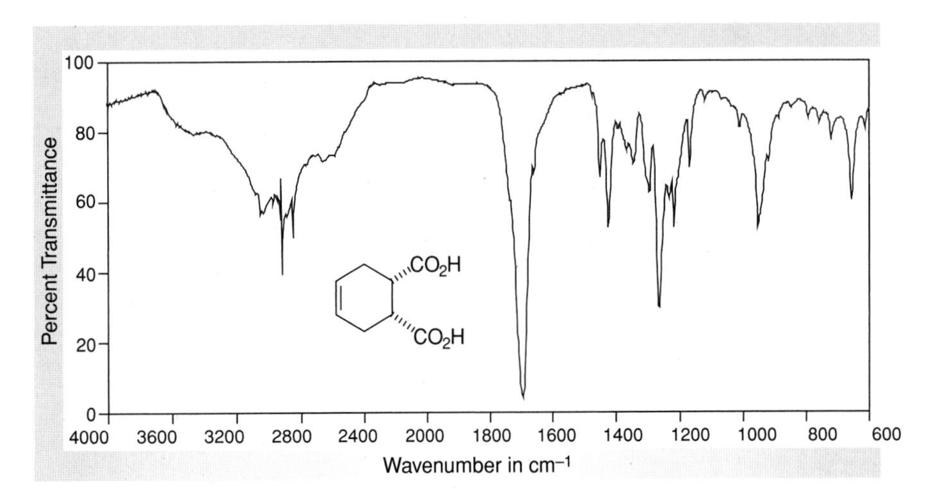

Figure 12.11
IR spectrum of 4-cyclohexene-cis-1,2-dicarboxylic acid (IR card).

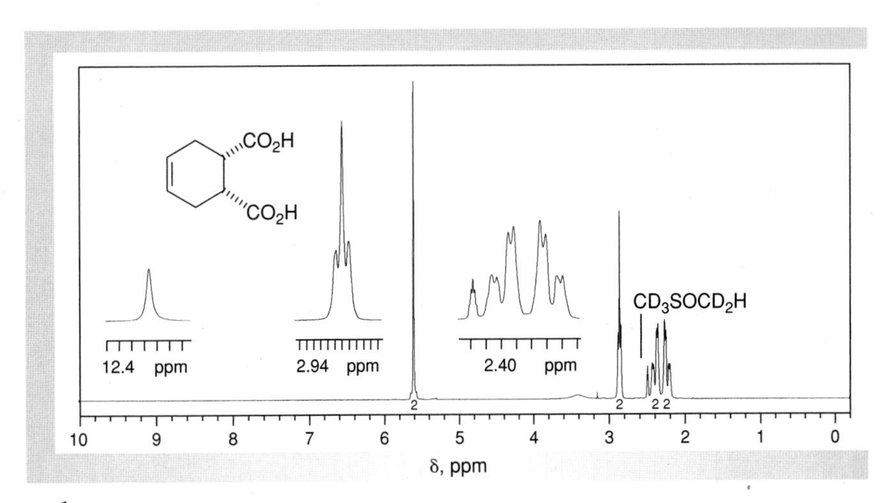

Figure 12.12
NMR data for 4-cyclohexene-cis-1,2-dicarboxylic acid (DMSO-d_6).

(a) ^{1}H NMR spectrum (300 MHz).
(b) ^{13}C NMR data: δ 25.9, 39.2, 125.2, 174.9.

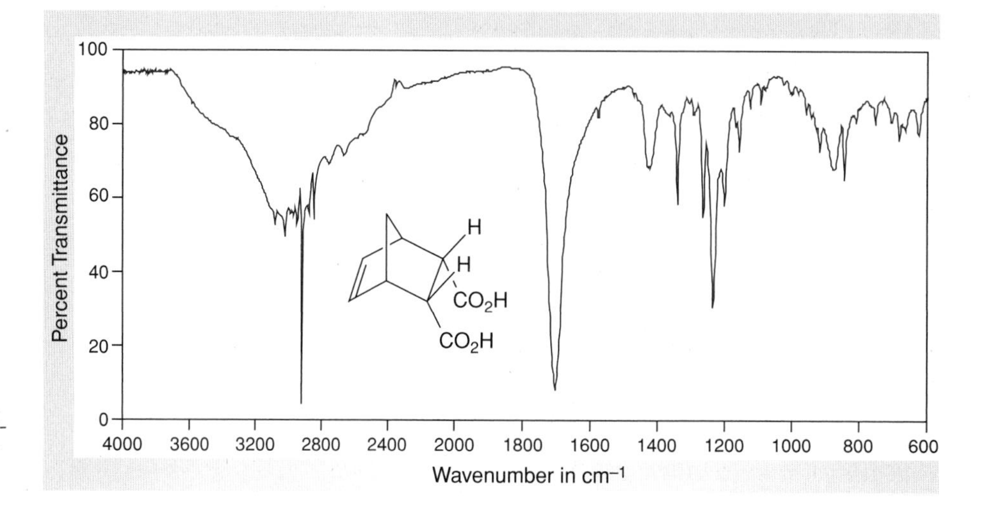

Figure 12.13
*IR spectrum of bicyclo[2.2.1]hept-5-ene-*endo*-2,3-dicarboxylic acid (IR card).*

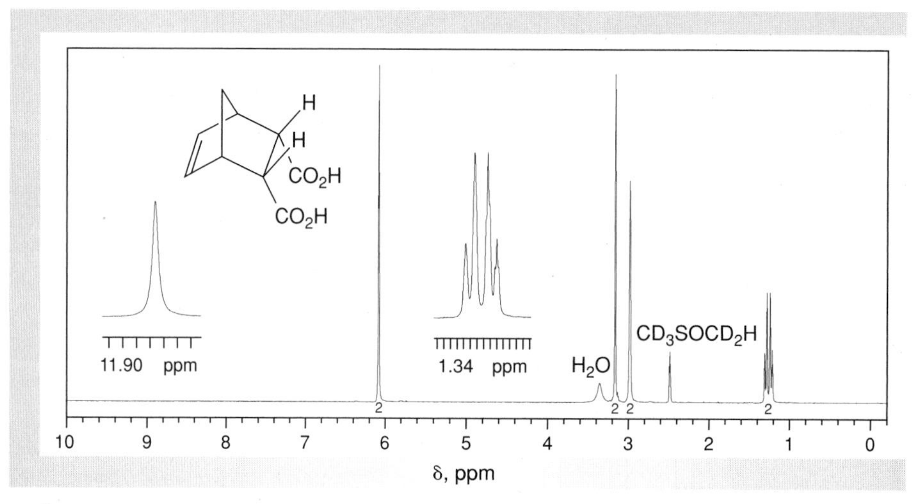

Figure 12.14
*NMR data for bicyclo[2.2.1]hept-5-ene-*endo*-2,3-dicarboxylic acid (DMSO-d_6).*

(a) ^{1}H NMR spectrum (300 MHz).
(b) ^{13}C NMR data: δ 46.8, 48.7, 49.1, 135.6, 174.4.

HISTORICAL HIGHLIGHT

Discovery of the Diels-Alder Reaction

The Diels-Alder reaction was discovered by the two German chemists whose names it bears. Otto Diels was a professor at the University of Kiel, and Kurt Alder was a postdoctoral student working with him. Diels, who was born in Hamburg in 1866, studied at the University of Berlin, where he received a Ph.D.; he later worked as an assistant to the world-famous chemist Emil Fischer (see Historical Highlight in Chap. 23). Alder, born in Upper Silesia in 1902, studied chemistry at the University of Berlin and then at the University of Kiel, where he became associated with Diels and received his Ph.D. in 1926.

One of the subjects that intrigued Diels and Alder was a novel addition of one or two molecules of 1,3-cyclopentadiene to one molecule of *p*-benzoquinone, a reaction that had been investigated in

(Continued)

HISTORICAL HIGHLIGHT *Discovery of the Diels-Alder Reaction (Continued)*

other laboratories for over twenty years. For example, H. Albrecht, another German chemist, was apparently the first to observe this reaction, and in 1906 he proposed that the structures of the products were **1** and **2.** Between 1906 and 1928, this reaction was studied in several laboratories by a number of renowned chemists, and several other structures were proposed for the products. Diels and Alder conducted extensive investigations of this and similar reactions, and in a classic paper published in 1928, they questioned the assignment of the structures **1** and **2.** In a series of reports in succeeding years, they showed conclusively that the structures proposed by Albrecht and others were incorrect and that the products from the reaction of 1,3-cyclopentadiene with *p*-benzoquinone were in fact **3** and **4.**

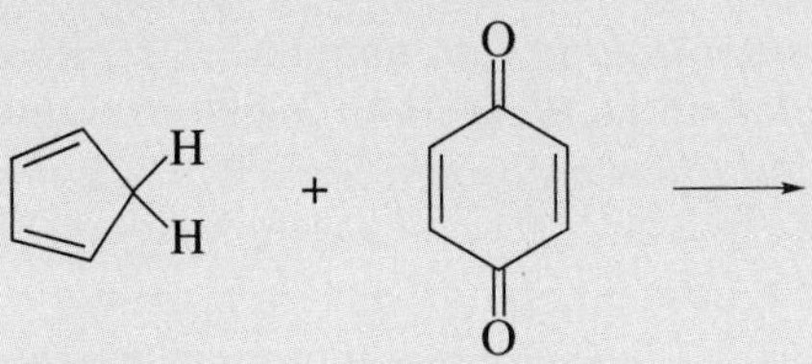

1,3-Cyclopentadiene *p*-Benzoquinone

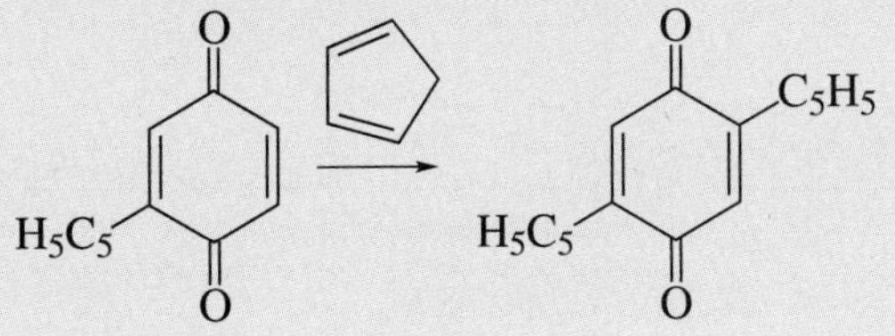

H_5C_5 **1** C_5H_5 H_5C_5 **2**

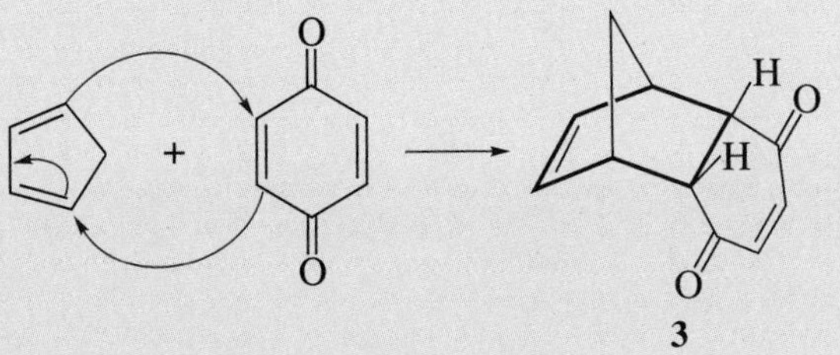

3

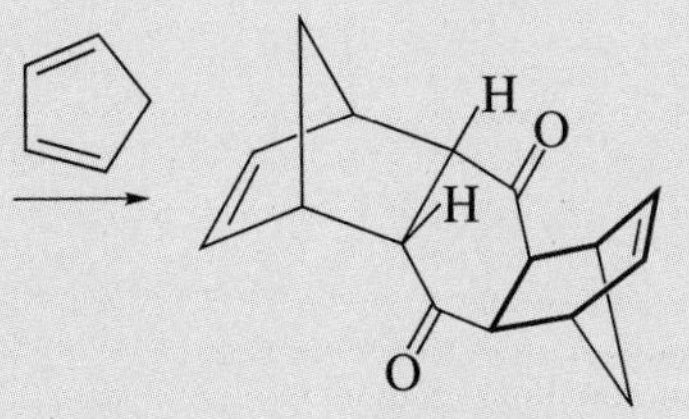

4

As a result of their experiments, Diels and Alder suggested that this reaction and numerous other related processes belonged to a common reaction type in which a 1,3-diene undergoes cycloaddition to a π-bond of a dienophile to form a product containing a new six-membered ring, which is highlighted by the bonds in the structures of **3** and **4.** At the time of its discovery, this was an entirely new and important type of cycloaddition reaction for the construction of six-membered rings.

The Diels-Alder reaction is considered to be one of the few truly new organic reactions of the 20th century, and it ranks in synthetic importance with the Wittig and Grignard reactions (see Historical Highlights for Chaps. 18 and 19, respectively). In recognition of their research and development of this process, Diels and Alder were awarded the Nobel Prize in 1950. In his Nobel Prize lecture, Diels commented that he thought it plausible that nature might use this type of reaction to fabricate some of the complicated polycyclic compounds that are so important to living organisms. The Diels-Alder reaction often does not require temperatures that lie outside of the range at which life exists, and it frequently proceeds in almost quantitative yield without using any special catalysts, reagents, or solvents. Despite Diels's suggestion that the Diels-Alder reaction would be a common transformation in nature, the biosyntheses of only a few natural products appear to involve such a process. On the other hand, this reaction has been extensively used as a key step in the laboratory syntheses of a number of important naturally occurring substances, including alkaloids, terpenes, dyes, vitamins, and hormones.

The Diels-Alder reaction has also been a key reaction in industrial-scale syntheses of substances that are *not* naturally occurring. Prominent among these are certain bicyclic polychlorinated compounds, such as aldrin, dieldrin, and chlordane, that are potent insecticides. The synthesis of each of these materials relies on perchloro-1,3-cyclopentadiene as the diene for the cycloaddition. Unfortunately, the beneficial uses of members of this class of compounds are often outweighed by their negative environmental impact. Like DDT, whose use is now banned in the United States although not throughout the world, they tend to persist for months, if not years, after being applied. The resulting long-term, low-level exposure of many generations

(Continued)

HISTORICAL HIGHLIGHT *Discovery of the Diels-Alder Reaction (Continued)*

of the target pest to the insecticide may allow the species to develop biological resistance to it. Moreover, organisms that feed on the insects may accumulate the compound in their own tissues and suffer a variety of harmful effects as a result. Consequently, alternative chemical insect controls are constantly being sought, an important criterion for such alternatives being that they are readily biodegradable.

Chlordane

Aldrin

Dieldrin

DDT

A sense of the growth in importance of the Diels-Alder reaction is gained by looking at the Cumulative Indexes of *Chemical Abstracts.* Only two entries were listed under the name "Diels-Alder reaction" in the ten-year period from 1927 to 1936, whereas such entries had increased to 180 by the decade spanning 1947 to 1956, during which time Diels and Alder shared the Nobel Prize in Chemistry. In the five-year period from 1987 to 1991, a total of about 2400 entries appeared under this heading, and the number of citations presently exceeds 800 per year.

An informative and more extensive description of the history of the Diels-Alder reaction may be found in Berson, J. *Tetrahedron* 1992, *48,* 3–17.

Relationship of Historical Highlight to Experiments

The procedures in this chapter illustrate the basic principles of the Diels-Alder reaction, many of which were first unveiled by the extensive research efforts of Diels and Alder. In all cases, the reactions form products containing new, unsaturated six-membered rings, which is the fundamental structural outcome of the Diels-Alder process.

See more on *Diels-Alder/Nobel Prize*

See more on *Diels-Alder Reaction*

See more on *Chlordane*

See more on *Aldrin*

See more on *Dieldrin*

See more on *DDT*

CHAPTER 13

Kinetic and Thermodynamic Control of a Reaction

When you see this icon, sign in at this book's premium website at **www.cengage.com/login** to access videos, Pre-Lab Exercises, and other online resources.

Only one product is formed in many of the reactions you will study in organic chemistry. However, sometimes two (or more) different products may be formed from competing reactions. If two products are formed, the inquiring scientist wants to know which is formed faster and which is more stable. While the product that forms faster is frequently also the more stable, this is not always the case. Sometimes it is possible to favor the formation of one product or another by modifying the conditions under which the reaction is conducted. The two experimental parameters that are most commonly varied to favor formation of one product over another are reaction time and the temperature at which the reaction is conducted. By performing the experimental procedures in this chapter, you will have an unusual opportunity to discover different means of controlling the course of chemical reactions using some of the same techniques that are practiced in industry for the production of compounds of practical interest.

13.1 INTRODUCTION

See more on *Free Energy*

See more on *Enthalpy*

See more on *Entropy*

Organic chemists frequently need to predict which of several possible competing reactions will predominate when a specified set of reactants is used. The problem is illustrated by considering the general situation where reagents **W** and **X** can combine to produce either **Y** (Eq. 13.1) or **Z** (Eq. 13.2). Predicting which product will be formed preferentially under a particular set of conditions requires developing an **energy diagram** for the two reactions. Strictly speaking, such a diagram should be based on the **free energy changes,** ΔG, that occur during the conversion of reactants to products. To generate the needed diagram would require knowledge of both the **enthalpic,** ΔH, and **entropic,** ΔS, changes involved. If the reactions being considered are similar, as we assume is the case for the processes of Equations 13.1 and 13.2, entropic changes are not too important, and using values for ΔH gives a close approximation to the changes occurring in ΔG. This is the basis on which our potential energy diagram will be constructed.

$$W + X \rightleftharpoons Y \qquad (13.1)$$

$$W + X \rightleftharpoons Z \qquad (13.2)$$

We will start our analysis by assuming that both of the reactions are *exothermic and* that **Y** is *thermodynamically less stable* than **Z.** These assumptions mean that (1) the energy levels of both products are below those of the starting materials and (2) the heat of reaction for formation of **Z** is greater than that for **Y,** as symbolized by ΔH_Z and ΔH_Y in Figure 13.1. The activation barriers for converting the reactants to the two transition states **Y**‡ and **Z**‡ for forming **Y** and **Z,** respectively, then correspond to $\Delta H^\ddagger_Y$ and $\Delta H^\ddagger_Z$. The latter barrier has been set *lower* on the diagram, so $\Delta H^\ddagger_Z < \Delta H^\ddagger_Y$. This ordering of the relative energies of the two transition states is arbitrary but is consistent with the *empirical* observation that when comparing two *similar* organic reactions, the *more exothermic process often has the lower enthalpy of activation,* $\Delta H^\ddagger$. In other words, the relative energies of the transition states reflect the relative energies of the products that are produced from those transition states. This generalization is useful because it allows us to complete the potential energy diagram and to predict that because less energy is required for reactants **W** and **X** to reach the transition state leading to **Z,** this product will be formed faster than **Y.**

For all practical purposes, most organic reactions are either irreversible or are performed under conditions such that equilibrium between products and starting materials is *not* attained. The ratio of products obtained from reactions run under these circumstances is said to be subject to **kinetic control.** This means that the relative amounts of the various products are determined by the relative rates of their formation rather than by their relative thermodynamic stabilities. The major product of such a reaction is often called the **kinetically controlled product** even though other compounds may be formed. In the example of Figure 13.1, **Z** would represent the kinetically controlled product; note that $\Delta H^\ddagger_Z < \Delta H^\ddagger_Y$.

Organic reactions that are performed under conditions in which *equilibrium* is established give ratios of products, and perhaps reactants too, that are subject to **thermodynamic control.** In such cases, the predominant product is called the **thermodynamically controlled product** and would be the more stable **Z** under the circumstances illustrated in Figure 13.1. Note that the ratio **Y:Z** would probably *not* be the same under both sets of reaction conditions because it is unlikely that the

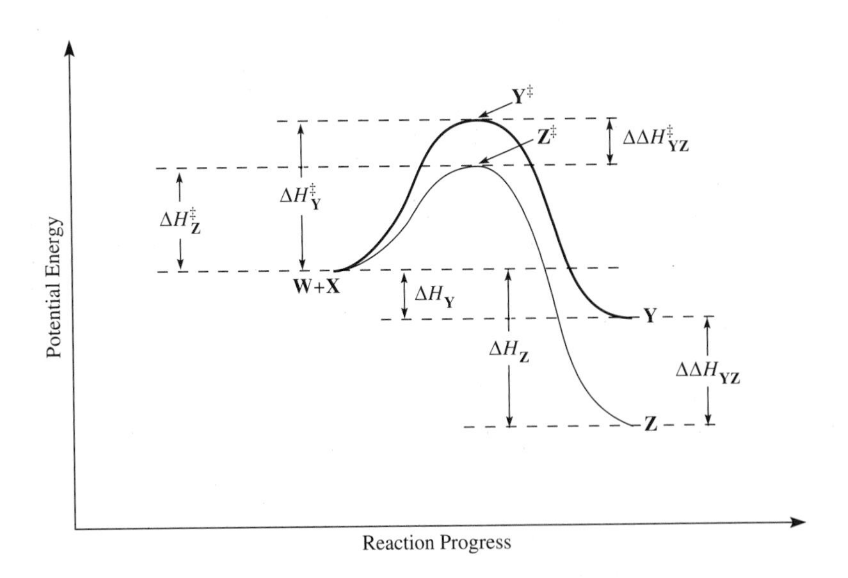

Figure 13.1
Typical reaction profile for competing reactions.

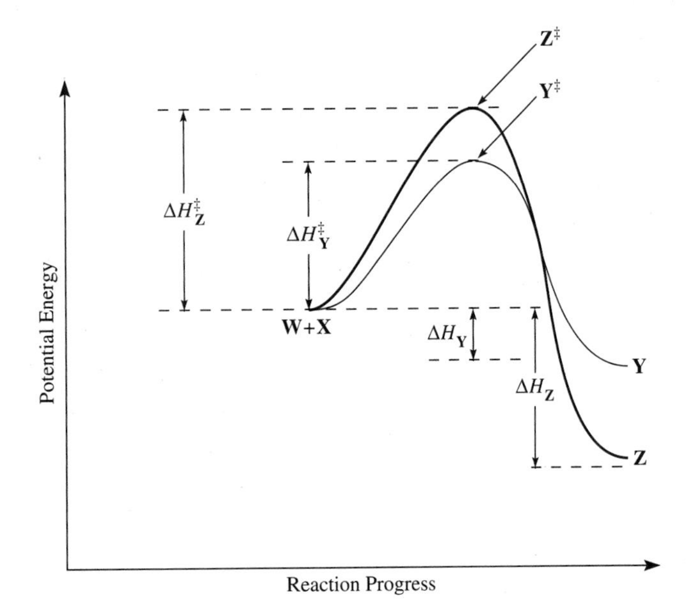

Figure 13.2
Reaction profile that predicts different products from kinetic and thermodynamic control of competing reactions.

difference in enthalpies of activation, $\Delta\Delta H^{\ddagger}_{YZ}$, between the two transition states would be equal to the enthalpy difference, $\Delta\Delta H_{YZ}$, between the two ground states of the products.

The situation shown in Figure 13.1 is the one most frequently encountered in organic chemistry; namely, the major product of kinetic control and thermodynamic control is the same. There are exceptions to this, however, and the reaction profiles of Figure 13.2 depict just such a case. To generate these profiles, it is again assumed that the reactions of Equations 13.1 and 13.2 are exothermic and that **Z** is more stable than **Y,** but the activation barrier, $\Delta H^{\ddagger}_{Y}$, for forming **Y** is now *less* than that for forming **Z,** $\Delta H^{\ddagger}_{Z}$. This lower barrier means that **Y** would now be the product of kinetic control, although **Z** would remain the product of thermodynamic control.

Note that in developing Figures 13.1 and 13.2, conclusions about the preferred products of kinetic control and thermodynamic control do not depend on the exothermicity or endothermicity of the overall reactions. *All that matters are the relative energies of the transition states leading to the products, if the reaction is under kinetic control, or the relative energies of the products themselves, if the conditions are those of thermodynamic control.* Whether different major products are produced under reversible or irreversible reaction conditions is determined by measuring the ratios of products formed under the two different circumstances. If the products of a reaction can be separated, there is another way to ascertain whether the reaction was performed under conditions of kinetic or thermodynamic control. Each product is resubjected separately to the original reaction conditions. If each is recovered unchanged, the original reaction was conducted under conditions of kinetic control, but if each product gives a mixture of products identical to that of the original reaction, then the reaction was subject to thermodynamic control. In the procedures that follow, you will have the opportunity to determine whether reactions are being conducted under thermodynamic or kinetic control using several experimental techniques.

13.2 FORMATION OF SEMICARBAZONES UNDER KINETIC AND THERMODYNAMIC CONTROL

The principle of kinetic and thermodynamic control in organic chemistry may be illustrated by studying the competing reactions of semicarbazide (**1**) with two carbonyl compounds, cyclohexanone (**2**) and 2-furaldehyde (**4**), as shown in Equations 13.3 and 13.4, respectively. In this case, one compound, semicarbazide, reacts with two different compounds, cyclohexanone and 2-furaldehyde, to give two different **semicarbazones, 3** and **5**, respectively. Both products are crystalline solids that have distinctive melting points by which they may be easily identified. Thus, it is easy to determine experimentally which compound is the product of kinetic control and which is the product of thermodynamic control.

$$H_2NC(=O)NHNH_2 + O{=}C_6H_{10} \rightleftharpoons H_2NC(=O)NHN{=}C_6H_{10} + H_2O \quad (13.3)$$

1 Semicarbazide; **2** Cyclohexanone; **3** Cyclohexanone semicarbazone

$$H_2NC(=O)NHNH_2 + O{=}C(H){-}C_4H_3O \rightleftharpoons H_2NC(=O)NHN{=}C(H){-}C_4H_3O + H_2O \quad (13.4)$$

1; **4** 2-Furaldehyde; **5** 2-Furaldehyde semicarbazone

The rates and equilibrium constants of reactions of carbonyl compounds with nucleophiles such as semicarbazide are affected significantly by variations in the pH of the reaction medium. This may be understood by examining the results of protonating the reactants.

First, consider the effect that protonation of the carbonyl function has on the rate of reaction of a carbonyl compound with nucleophiles. The addition of a proton to the oxygen atom of a carbonyl group produces **6** (Eq. 13.5). The carbonyl carbon atom in **6** is much more *electrophilic* than in the unprotonated compound because the partial positive charge on the carbon atom is increased, as shown by consideration of the two resonance structures for **6.** Since the rate-determining step for the formation of semicarbazones is the nucleophilic attack of semicarbazide (**1**) on the carbonyl carbon atom of the substrate, the enhanced electrophilicity of **6** *increases* the rate of formation of the semicarbazone.

$$R_2C{=}\ddot{O} + H^+ \rightleftharpoons \left[R_2C{=}\overset{+}{\ddot{O}}H \longleftrightarrow R_2\overset{+}{C}{-}\ddot{O}H\right] \quad (13.5)$$

Carbonyl compound; **6**

Now assess what happens if the nucleophile is protonated. Adding a proton to **1** produces its conjugate acid **7** (Eq. 13.6), which is *not* nucleophilic and does not add to a carbonyl group. Hence, if the medium in which the reaction is performed is sufficiently acidic, essentially all of the **1** will be converted to **7,** thereby lowering the concentration of the nucleophile **1.** The formation of the semicarbazone would then become extremely slow or might not occur at all. Thus both the rate and equilibrium constant for the reaction of Equation 13.7 are affected significantly by the pH at which the process is conducted.

$$H_2NC(=O)NH\ddot{N}H_2 + H^+ \rightleftharpoons H_2NC(=O)NH\overset{+}{N}H_3 \quad (13.6)$$

1 → **7**

$$H_2NC(=O)NH\ddot{N}H_2 \;\; R_2C{=}\overset{+}{\ddot{O}}H \rightleftharpoons \left[H_2NC(=O)NH\overset{+}{N}H{-}CR_2(OH)\right] \rightleftharpoons \left[H_2NC(=O)NH\ddot{N}{-}CR_2(\overset{+}{O}H_2)\right] \quad (13.7)$$

1 **6**

$$H_3O^+ + H_2NC(=O)NH\ddot{N}{=}CR_2 \underset{}{\overset{H_2\ddot{O}:}{\rightleftharpoons}} H_2NC(=O)NH\overset{+}{N}(H){=}CR_2 + H_2O$$

The preceding discussion makes clear that with respect to Equation 13.7, the concentration of the nucleophile **1** is reduced at low pH whereas that of the activated electrophile **6** is reduced at high pH. It follows that there is an optimum pH at which the mathematical product of the concentrations of the nucleophilic reagent and conjugate acid of the carbonyl compound, [**1**] × [**6**], is maximized.

C_6H_5–NH$\ddot{N}H_2$ (**8**) H$\ddot{O}$NH$_2$ (**9**)

For reactions of aldehydes and ketones with reagents such as semicarbazide (**1**), phenylhydrazine (**8**), and hydroxylamine (**9**), the desired pH is produced and maintained by **buffer solutions.** A buffer solution resists changes in its pH. In general, a buffer is comprised of a pair of substances in the solution, one that is a weak acid and neutralizes hydroxide ions and another that is a weak base and neutralizes protons. For example, the $H_2PO_4^-/HPO_4^{2-}$ **buffer system** is produced by the addition of dibasic potassium phosphate, K_2HPO_4, a weak base, to semicarbazide hydrochloride. The weakly acidic $H_2PO_4^-$ component of the buffer system is produced as shown in Equation 13.8. The function of $H_2PO_4^-$ and HPO_4^{2-} in neutralizing hydroxide and hydrogen ions is illustrated in Equations 13.9 and 13.10, respectively.

Each different buffer system functions to maintain the pH within a rather narrow range characteristic of the weak acid and weak base of that particular system. In the case of the phosphate system, the range is pH 6.1–6.2. By comparison, a carbonate/bicarbonate buffer system (CO_3^{2-}/HCO_3^-) maintains a pH of 7.1–7.2.

$$H_2NC(=O)NH\overset{+}{N}H_3\,Cl^- + HPO_4^{2-} \rightleftharpoons H_2NC(=O)NH\ddot{N}H_2 + H_2PO_4^- + Cl^- \quad (13.8)$$

1

$$H_2PO_4^- + HO^- \rightleftharpoons HPO_4^{2-} + H_2O \quad (13.9)$$

$$HPO_4^{2-} + H_3O^+ \rightleftharpoons H_2PO_4^- + H_2O \quad (13.10)$$

The maximum rates of the reactions of most aldehydes and ketones with semicarbazide occur in the pH range of 4.5–5.0. For the purpose of making derivatives of carbonyl compounds (Sec. 25.7), semicarbazide is best used in an acetate buffer ($CH_3CO_2H/CH_3CO_2^-$) solution, which maintains a pH in the maximum rate range of 4.5–5.0. However, to demonstrate the principle of kinetic and thermodynamic control of reactions, buffers that maintain higher pHs, and thus produce lower rates, are more desirable. Parts A–C of the experimental procedure involve a phosphate buffer system, whereas the bicarbonate system is used in Part D. It is then possible to compare how the difference in rates in the two buffer systems affects the product ratio. Analysis of the products from the various parts of these experiments provides strong clues as to which of the semicarbazones is the product of kinetic control and which is the product of thermodynamic control.

In Section 13.1, it was noted that implicit in the theory of kinetic and thermodynamic control is the fact that the *kinetic product* (**Z** in Fig. 13.1 and **Y** in Fig. 13.2), which is produced more rapidly, is also reconverted to starting material more rapidly. The *thermodynamic product*, on the other hand, being more stable, is not so readily reconverted to starting material. Experimental Procedure E provides tests of the relative stabilities of the two semicarbazone products toward the reverse reaction. The results of these experiments should provide additional evidence as to which of the semicarbazones is the product of kinetic control and which is the product of thermodynamic control.

Kinetic and Thermodynamic Control of a Reaction

Purpose To study the phenomenon of kinetic and thermodynamic control of competing reactions by determining which product is formed faster and which is more stable.

SAFETY ALERT

Because the melting point of one of the compounds is near 200 °C, do *not* use a liquid-filled apparatus such as a Thiele tube *unless* the heating medium is silicone oil. A metal block melting-point apparatus is satisfactory for determining the melting points of the semicarbazones in this experiment.

Preparation Sign in at **www.cengage.com/login** to answer Pre-Lab Exercises, access videos, and read the MSDSs for the chemicals used or produced in this procedure. Review Sections 2.9, 2.10, and 2.17.

You may be assigned only some parts of these procedures. If so, plan to share results with other students who perform the other parts. This will enable you to draw your own conclusions and report them accordingly.

Apparatus Two 13-mm × 100-mm test tubes, two 25-mL, seven 5-mL, and two 125-mL Erlenmeyer flasks, 1-mL graduated pipet, ice-water bath, apparatus for vacuum filtration and *flameless* heating.

A ■ *Preparation of Cyclohexanone Semicarbazone*

Dissolve 0.5 g of semicarbazide hydrochloride and 1.0 g of dibasic potassium phosphate (K_2HPO_4) in 6 mL of water contained in a 25-mL Erlenmeyer flask. Using a 1-mL graduated pipet, deliver 0.5 mL of cyclohexanone into a test tube containing 2.5 mL of 95% ethanol. Pour the ethanolic solution into the aqueous semicarbazide solution, and swirl or stir the mixture immediately. Allow 5–10 min for crystallization of the semicarbazone to reach completion. Collect the crystals by vacuum filtration and wash them with a little cold water. *Thoroughly* air-dry the crystals★ and determine their weight and melting point. Obtain IR and NMR spectra of your starting material and product. Compare your spectra with those of authentic samples (Figs. 13.3–13.6).

B ■ *Preparation of 2-Furaldehyde Semicarbazone*

For best results, the 2-furaldehyde should be freshly distilled immediately before use. Prepare the semicarbazone of 2-furaldehyde by following the procedure of Part A exactly, except use 0.4 mL of 2-furaldehyde instead of 0.5 mL of cyclohexanone.★ Recrystallization of the product is unnecessary. Obtain IR and NMR spectra of your starting material and product. Compare your spectra with those of authentic samples (Figs. 13.7–13.10).

C ■ *Reactions of Semicarbazide with Cyclohexanone and 2-Furaldehyde in Phosphate Buffer Solution*

Dissolve 3.0 g of semicarbazide hydrochloride and 6.0 g of dibasic potassium phosphate in 75 mL of water. Because this is an aqueous solution, it is referred to as *solution W.* Prepare a solution of 3.0 mL of cyclohexanone and 2.5 mL of 2-furaldehyde in 15 mL of 95% ethanol. Because this is an ethanolic solution, it is referred to as *solution E.*★

1. Cool a 25-mL portion of solution W and a 5-mL portion of solution E separately in an ice-water bath to 0–2 °C. Add solution E to solution W and swirl the mixture; crystals should form almost immediately. Place the mixture in an ice-water bath for 3–5 min; then collect the crystals by vacuum filtration and wash them on the filter with 2–3 mL of cold water. *Thoroughly* air-dry the crystals★ and determine their weight and melting point.
2. Add a 5-mL portion of solution E to a 25-mL portion of solution W at room temperature; crystals should be observed in 1–2 min. Allow the mixture to stand at room temperature for 5 min, cool it in an ice-water bath for about 5 min, and then collect the crystals by vacuum filtration and wash them on the filter with 2–3 mL of cold water. *Thoroughly* air-dry the crystals★ and determine their weight and melting point.
3. Warm a 25-mL portion of solution W and a 5-mL portion of solution E *separately* to 80–85 °C; add solution E to solution W and swirl the mixture. Continue to heat the solution for 10–15 min, cool it to room temperature, and then

place it in an ice-water bath for about 5–10 min. Collect the crystals by vacuum filtration, and wash them on the filter with 2–3 mL of cold water. *Thoroughly* air-dry the crystals★ and determine their weight and melting point.

D ■ Reactions of Semicarbazide with Cyclohexanone and 2-Furaldehyde in Bicarbonate Buffer Solution

Dissolve 2.0 g of semicarbazide hydrochloride and 4.0 g of sodium bicarbonate in 50 mL of water. Prepare a solution of 2.0 mL of cyclohexanone and 1.6 mL of 2-furaldehyde in 10 mL of 95% ethanol. Divide each of these solutions into two equal portions.

1. Mix half of the aqueous solution and half of the ethanolic solution and allow the mixture to stand at room temperature for 5 min. Collect the crystals by vacuum filtration and wash them on the filter with 2–3 mL of cold water. *Thoroughly* air-dry the crystals★ and determine their weight and melting point.
2. Warm the other portions of the aqueous and ethanolic solutions *separately* to 80–85 °C. Combine them and continue heating the mixture for 10–15 min. Cool the solution to room temperature, and place it in an ice-water bath for 5–10 min. Collect the crystals by vacuum filtration and wash them on the filter with 2–3 mL of cold water. *Thoroughly* air-dry the crystals★ and determine their weight and melting point.

E ■ Tests of Reversibility of Semicarbazone Formation

1. Place 0.3 g of cyclohexanone semicarbazone, prepared in Part A, 0.3 mL of 2-furaldehyde, 2 mL of 95% ethanol, and 10 mL of water in a 25-mL Erlenmeyer flask. Warm the mixture until a homogeneous solution is obtained (about 1 or 2 min should suffice), and continue warming for an additional 3 min. Cool the mixture to room temperature and then place it in an ice-water bath. Collect the crystals on a filter, and wash them with 2–3 mL of cold water. *Thoroughly* air-dry the crystals★ and determine their weight and melting point.
2. Repeat the preceding experiment, but use 0.3 g of 2-furaldehyde semicarbazone (prepared in Part B) and 0.3 mL of cyclohexanone in place of the cyclohexanone semicarbazone and 2-furaldehyde.

On the basis of your results from experiments C, D, and E, deduce which semicarbazone is the product of kinetic control and which is the product of thermodynamic control. To do this, first use the observed melting points of the crystals produced in parts C1, C2, and C3 to deduce whether the product in each part is the semicarbazone of cyclohexanone, of 2-furaldehyde, or of a mixture of the two. Note that in C1 the crystals of product separate almost immediately, in C2 after 1 or 2 min, and in C3 only after 10–15 min at a higher temperature; thus the reaction time is shortest in C1, intermediate in C2, and longest in C3. Then compare the product of D1 with that of C2 and the product of D2 with that of C3.

When analyzing the results of the experiments in Part E, remember that the thermodynamic product, being the more stable, is not easily transformed into the less stable kinetic product. In contrast, the kinetic product is readily converted into the more stable thermodynamic product. These circumstances arise because the reverse reaction from the less stable kinetic product has a lower activation enthalpy than the reverse reaction of the thermodynamic product.

Your completed laboratory report should include the diagram specified in Exercise 2 and answers to Exercises 3 and 4 as well, unless you were instructed to omit some parts of the experiment.

Discovery Experiment

Effect of pH on Kinetic vs. Thermodynamic Control

Repeat experiments C1–3 using a buffer that will provide a pH different from that obtained using dibasic potassium phosphate and sodium bicarbonate. You may be instructed to prepare a buffer having a specific pH, or you may be instructed to use sodium citrate, potassium hydrogen phosphate, sodium carbonate, potassium phosphate, or a salt of another acid to serve as the buffer for the reaction. Use the same quantities of semicarbazide hydrochloride, 2-furaldehyde, cyclohexanone, water, and 95% ethanol as were used in experiments C1–3; the molar ratio of the salt to semicarbazide hydrochloride should be about 2:1. Depending upon the buffer you use, crystals may form faster or slower than under the conditions of experiments C1–3; record the length of time required for crystals to form with the different buffers under the different conditions. On the basis of your results from these experiments, deduce which semicarbazone is the product of kinetic control and which is the product of thermodynamic control at different pHs. Do this by using the observed melting points of the crystals produced in Parts C1, C2, and C3 to deduce the nature of your product(s). If possible, compare the rates of forming the kinetic or thermodynamic product at different pHs. Your instructor may have you work in groups with several students using different buffers, so you may compare your results with theirs. As always, consult with your instructor before undertaking any experimental work.

WRAPPING IT UP

Neutralize the various *filtrates* and then flush them down the drain.

EXERCISES

1. Why is it particularly important in this experiment that the semicarbazones be dry before their melting points are determined?
2. On the basis of the results from experiments **C, D,** and **E,** draw a diagram similar to Figure 13.2 and clearly label the products corresponding to those obtained in experiments **A** and **B.**
3. On the basis of the results from the experiments of Part **D,** explain the effect of the higher pH on the reactions between semicarbazide and the two carbonyl compounds.
4. What results would be expected if sodium acetate buffer, which provides a pH of approximately 5, were used in experiments analogous to those of Part **C?** Explain.
5. What different result, if any, is expected if the heating period at 80–85 °C is extended to one hour in experiment D2?
6. There are two NH_2 groups in semicarbazide (**1**), yet only one of them reacts with the carbonyl group to form a product. Explain.

7. Based upon your observations, describe the impact that temperature has upon whether a reaction is subject to kinetic or thermodynamic control.
8. The change in the free energy, ΔG, for a chemical reaction is mathematically related to the change in enthalpy, ΔH, and the change in entropy, ΔS. What is this mathematical relationship?
9. For a chemical reaction at equilibrium, the difference in the free energy, ΔG_{eq}, between the reactants and the products is mathematically related to the equilibrium constant, K_{eq}, for the reaction. What is this mathematical relationship?
10. In order to draw the energy diagrams in Figures 13.1 and 13.2 and in Exercise 2, we assumed that entropic changes were not important.
    a. Define entropy.
    b. Explain why it was appropriate to ignore entropy changes in comparing the reactions of cyclohexanone and 2-furaldehyde with semicarbazide.
11. Using the energy diagram in Figure 13.1 to illustrate your answer, describe how the terms below are related.
    a. ΔH and K_{eq}
    b. $\Delta H^{\ddagger}$ and k
12. Formation of a semicarbazone is the basis for a qualitative test to determine whether an aldehyde or a ketone functional group is present in a molecule (Sec. 25.7). For the compounds listed below, indicate whether they will yield a positive or negative semicarbazone test. For those compounds giving a positive test result, write the structure of the semicarbazone produced.

a. 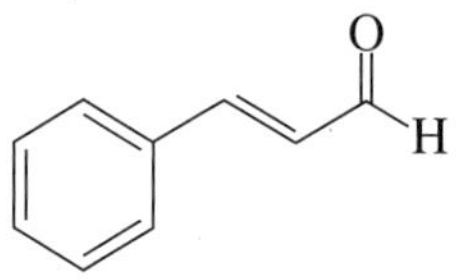 H O b. 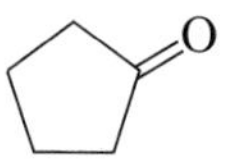 O c. 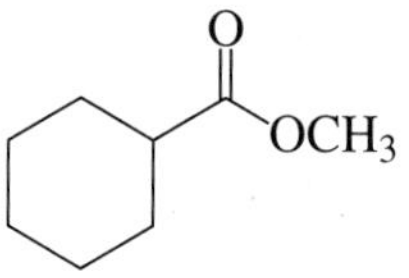 O OCH_3

13. Using curved arrows to symbolize the flow of electrons, provide a stepwise mechanism showing all proton transfer steps for the reaction between cyclohexanone and semicarbazide at pH < 7.
14. Which step of the mechanism for semicarbazone formation occurs at a reduced rate when the solution pH is
    a. very high?
    b. very low?
15. Explain how a strong acid and a weak acid differ in terms of their respective K_as and pK_as.
16. Which of the following pairs contain the prerequisites for a good buffer? (More than one answer may be correct.)
    a. strong acid/conjugate base
    b. weak acid/conjugate base
    c. weak base/conjugate acid
    d. strong base/conjugate acid
17. The carbonate/bicarbonate buffer system is an important biological buffer that is critical in maintaining blood pH.

a. Write the acid-base reaction that shows how this buffer prevents *acidosis*, or a decrease in blood pH.

b. Write the acid-base reaction that shows how this buffer system prevents an increase in blood pH.

18. The Henderson-Hasselbach equation, shown below, is used by biochemists and chemists to calculate the pH of a buffer solution.

$$pH = pK_a + \log\left(\frac{[A^-]_{initial}}{[HA]_{initial}}\right)$$

a. Calculate the pH of a buffer solution that is 0.02 *M* CH_3CO_2Na and 0.04 *M* CH_3CO_2H. The pK_a of acetic acid is 4.75.

b. Using the Henderson-Hasselbach equation, show how $pH = pK_a$ when the concentration of the acid and the conjugate base are the same.

19. Protonation of an enolate ion can occur via two competing exothermic reactions, as shown. Protonation on the oxygen atom of the enolate ion to form the enol is faster than protonation on the carbon atom to form the ketone; however, the ketone is thermodynamically more stable than the enol.

Assuming that the ΔSs for the two reactions are similar, draw an energy diagram similar to that shown in Figure 13.1 or 13.2 for the protonation of an enolate ion to give an enol and a ketone. Clearly label the curve with the structures of the reactants and the products. Your curve should show the relative enthalpies of the reactants and products as well as the relative enthalpies of activation, $\Delta H^{\ddagger}$, and the relative enthalpies, ΔH, for the two reactions.

Sign in at **www.cengage.com/login** and use the spectra viewer and Tables 8.1–8.8 as needed to answer the blue-numbered questions on spectroscopy.

20. Consider the spectral data for cyclohexanone (Figs. 13.3 and 13.4).

a. In the functional group region of the IR spectrum, specify the absorption associated with the carbon-oxygen double bond.

b. In the ^1H NMR spectrum, assign the various resonances to the hydrogen nuclei responsible for them and explain why the multiplet integrating for four protons is downfield of that integrating for six.

c. For the ^{13}C NMR data, assign the various resonances to the carbon nuclei responsible for them.

21. Consider the NMR data for cyclohexanone semicarbazone (Fig. 13.6).

a. In the ^1H NMR spectrum, assign the various resonances to the hydrogen nuclei responsible for them.

b. For the ^{13}C data, assign the various resonances to the carbon nuclei responsible for them.

22. Consider the spectral data for 2-furaldehyde (Figs. 13.7 and 13.8).
    a. In the functional group region of the IR spectrum, specify the absorptions associated with the carbon-oxygen and carbon-carbon double bonds and with the aldehydic C–H bond.
    b. In the ^{1}H NMR spectrum, assign the various resonances to the hydrogen nuclei responsible for them.
    c. For the ^{13}C NMR data, assign the various resonances to the carbon nuclei responsible for them.
23. a. The IR spectra portrayed in Figures 13.5 and 13.9 contain a sharp absorption at approximately 3500 cm^{-1}. Specify the functional group responsible.
    b. Both of these spectra have complex absorptions in the range of 3100–3400 cm^{-1}. What functional group is responsible?
    c. In Figure 13.9, what functionality is responsible for the absorption near 1700 cm^{-1}?
24. Consider the NMR data for 2-furaldehyde semicarbazone in Figure 13.10.
    a. In the ^{1}H NMR spectrum, assign the various resonances to the hydrogen nuclei responsible for them.
    b. For the ^{13}C data, assign the various resonances to the carbon nuclei responsible for them.
25. Discuss the differences observed in the IR and NMR spectra of cyclohexanone and of its semicarbazone that are consistent with the replacement of the carbonyl function by a carbon-nitrogen double bond in this experiment.
26. Discuss the differences observed in the IR and NMR spectra of 2-furaldehyde and of its semicarbazone that are consistent with the replacement of the carbonyl function by a carbon-nitrogen double bond in this experiment.

SPECTRA

Starting Materials and Products

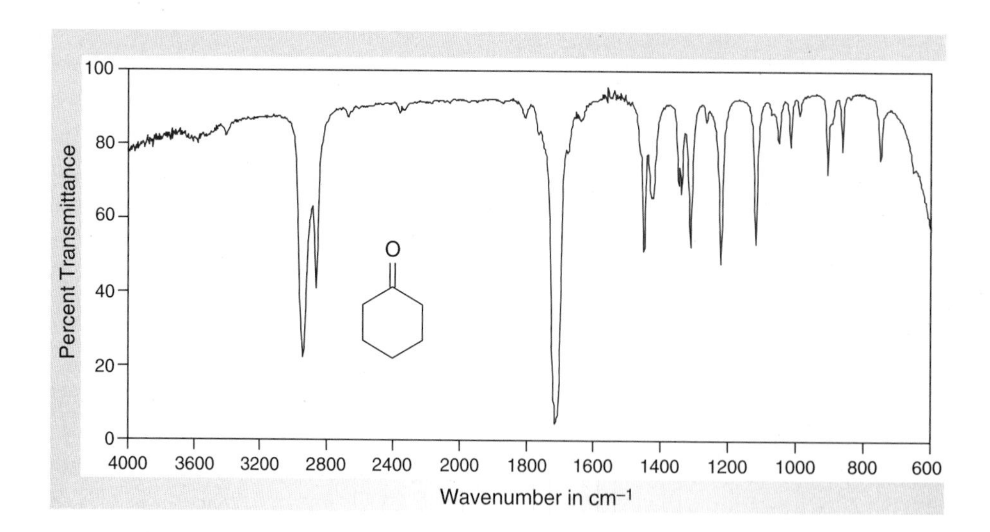

Figure 13.3
IR spectrum of cyclohexanone (neat).

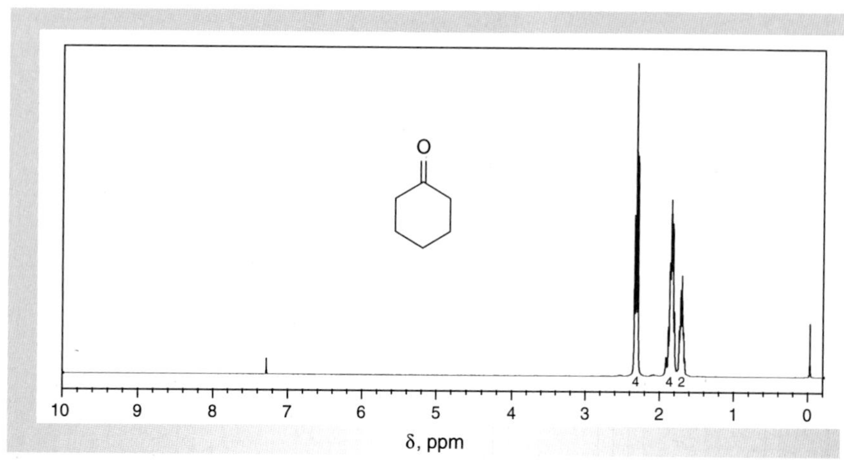

Figure 13.4
NMR data for cyclohexanone ($CDCl_3$).

(a) ^{1}H NMR spectrum (300 MHz).
(b) ^{13}C NMR data: δ 25.1, 27.2, 41.9, 211.2.

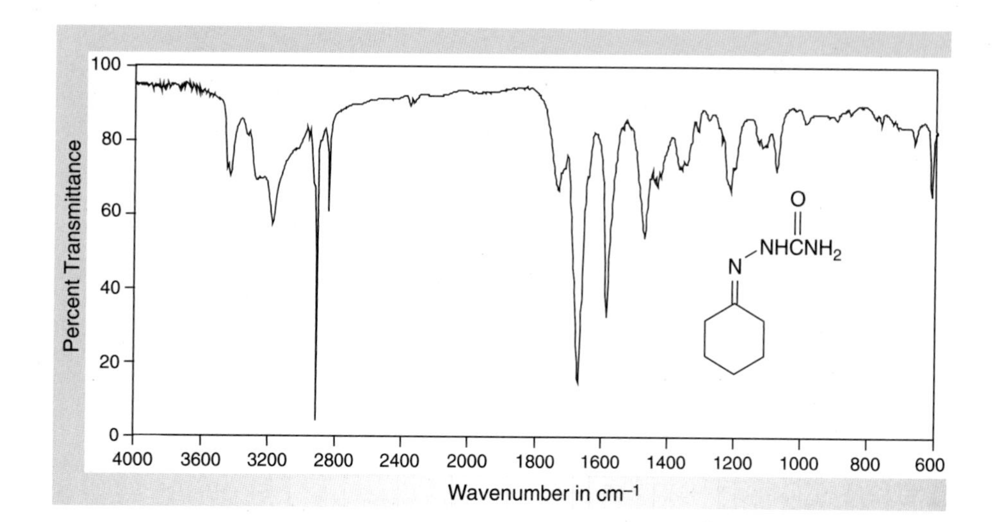

Figure 13.5
IR spectrum of cyclohexanone semicarbazone (IR card).

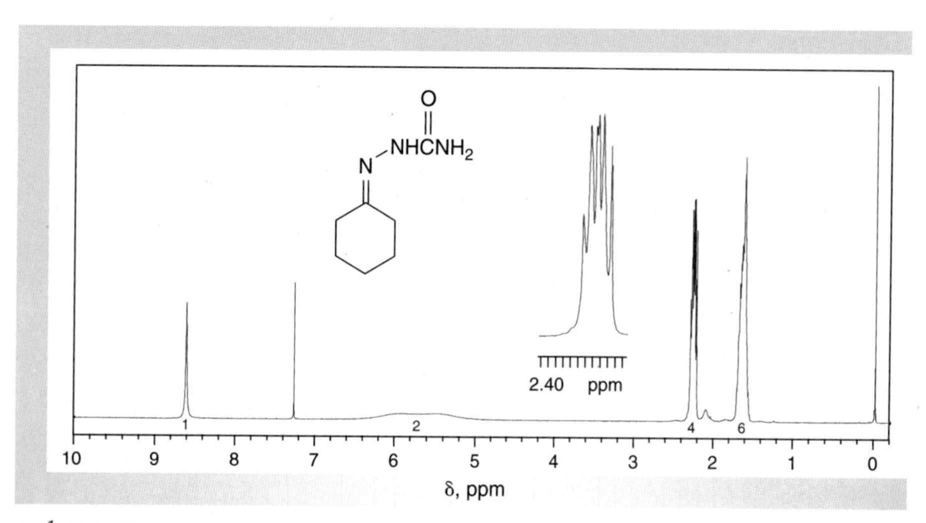

Figure 13.6
NMR data for cyclohexanone semicarbazone ($CDCl_3$).

(a) ^{1}H NMR spectrum (300 MHz).
(b) ^{13}C NMR data: δ 25.6, 25.8, 26.5, 27.0, 35.4, 153.4, 158.8.

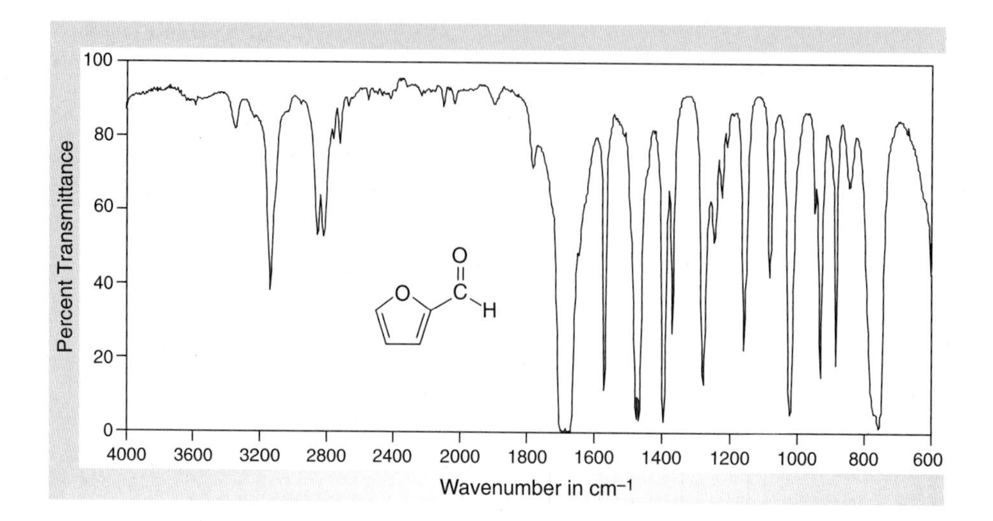

Figure 13.7
IR spectrum of 2-furaldehyde (neat).

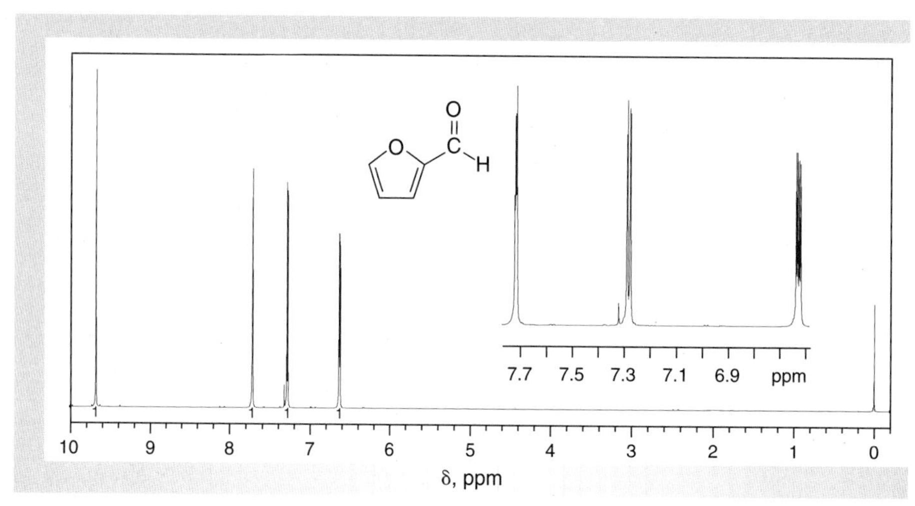

Figure 13.8
NMR data for 2-furaldehyde ($CDCl_3$).

(a) ^{1}H NMR spectrum (300 MHz).
(b) ^{13}C NMR data: δ 112.9, 121.6, 148.6, 153.3, 178.1.

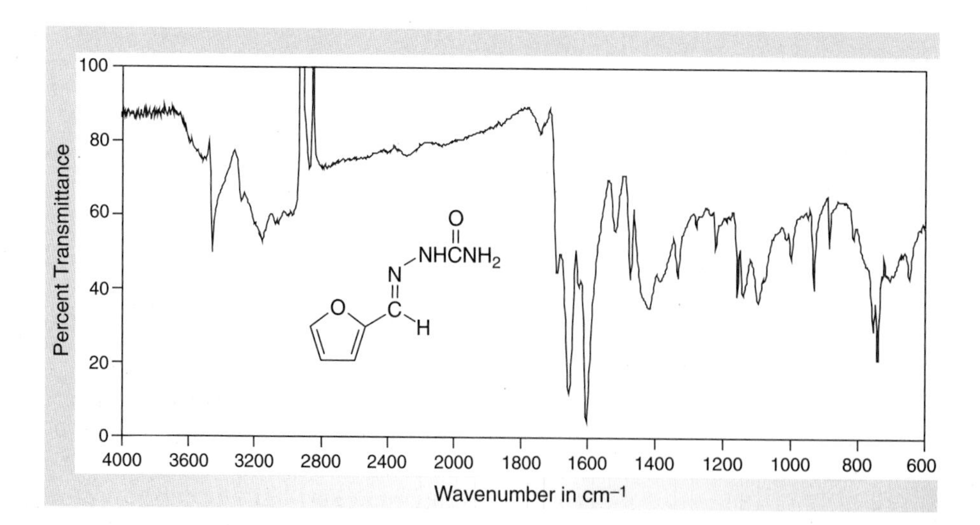

Figure 13.9
IR spectrum of 2-furaldehyde semicarbazone (IR card).

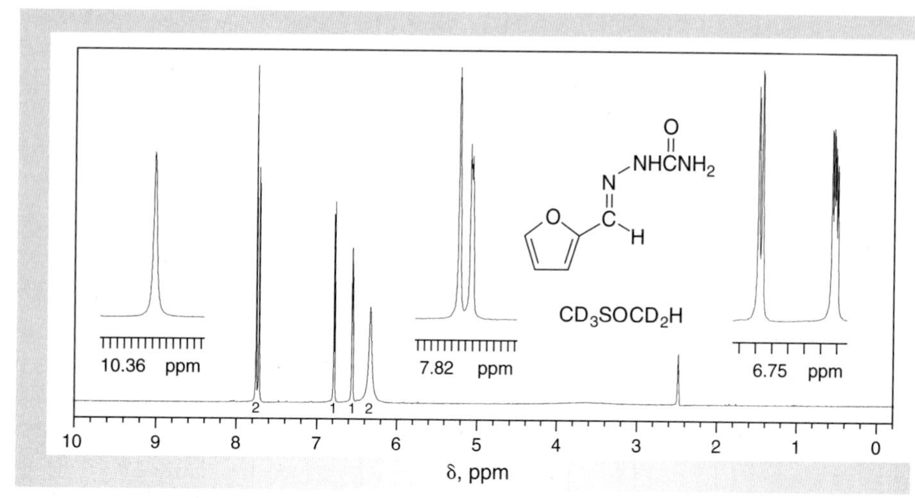

Figure 13.10
NMR data for 2-furaldehyde semicarbazone (DMSO-d_6).

(a) ^{1}H NMR spectrum (300 MHz).
(b) ^{13}C NMR data: δ 110.6, 111.8, 129.8, 143.8, 149.9, 156.4.

Nucleophilic Aliphatic Substitution

Preparation of Alkyl Halides

When you see this icon, sign in at this book's premium website at **www.cengage.com/login** to access videos, Pre-Lab Exercises, and other online resources.

Nucleophiles, Nu:, may effect substitutions at aliphatic carbon atoms by a process that reveals many interesting principles of organic chemistry, including the roles steric and solvent effects play in substitutions. For the synthetic chemist, these types of reactions are very important because they frequently allow a wide variety of functionalities to be prepared from a single starting material simply by varying the nature of the nucleophilic partner. You might say that "variety is the spice of life," and this certainly applies to the range of transformations available through substitution reactions. In this chapter we'll review the general features of these types of reactions, and you will conduct experiments to study the conversion of alcohols into alkyl halides, a representative substitution reaction.

14.1 GENERAL CONCEPTS

The substitution of one group for another at a saturated, sp^3-hybridized carbon atom is a reaction commonly used to interconvert different functional groups, and the process is called a **nucleophilic aliphatic substitution.** This conversion is exemplified in Equation 14.1, where **Nu:** is a symbol for a generalized **nu**cleophile and represents a neutral molecule or an anion that has Lewis basic character, and **L:** represents a **l**eaving group. The reaction may be considered to be a Lewis acid-base reaction because the carbon atom bearing the electronegative leaving group has Lewis acidic character. This results from the polarization of the C–L bond, which produces a partial positive charge on the carbon atom.

$$\text{Nu:}^- + \text{R}-\overset{|}{\underset{|}{\text{C}}}{}^{\delta+}-\text{L}^{\delta-} \longrightarrow \text{Nu}-\overset{|}{\underset{|}{\text{C}}}-\text{R} + \text{L:}^- \qquad (14.1)$$

Nucleophiles have the property in common of bearing *at least one pair of nonbonding electrons* and being either *neutral or negatively charged.* During a substitution reaction, the nonbonding pair of electrons of the nucleophile is donated to an electrophilic, Lewis acidic carbon atom with concomitant formation of a new covalent bond. Examples of some typical nucleophiles include Cl^-, Br^-, I^-, HO^-, $N{\equiv}C^-$, $R{-}C{\equiv}C^-$, H_2O, N_3^-, R_3N, and RS^-.

The leaving group, L:, which may also be neutral or negatively charged, must accept the pair of bonding electrons from the carbon atom as the C–L bond breaks. The ease with which various groups leave in nucleophilic substitution reactions has been determined experimentally by studying the rates of reaction of a specific nucleophile with substrates having different leaving groups. These investigations show that the leaving ability of a particular group L: correlates with the strength of its conjugate acid: the better leaving groups are those that are **conjugate bases** of the stronger acids. For example, a leaving group such as Cl^- is the conjugate base of hydrochloric acid, a strong acid, so Cl^- is a good leaving group. On the other hand, HO^- is a poor leaving group because it is the conjugate base of the weak acid water; however, H_2O itself is a good leaving group because it is the conjugate base of hydronium, H_3O^+, another strong acid.

14.2 CLASSIFICATION OF NUCLEOPHILIC SUBSTITUTION REACTIONS

Nucleophilic substitution is a general reaction for aliphatic compounds in which the leaving group is attached to an sp^3-hybridized carbon (Eq. 14.1). However, the mechanism for a given transformation depends upon the structure of the alkyl group bearing the leaving group. The two different mechanistic pathways that apply to such substitutions are designated by the symbols $\mathbf{S_N1}$ (**S** for substitution, **N** for nucleophilic, and **1** for unimolecular) and $\mathbf{S_N2}$ (**2** for bimolecular). These two mechanisms are depicted in general form in Equations 14.2 and 14.3.

S_N1 Mechanism

$$R_3C{-}L \underset{\text{slow}}{\overset{\text{rds}}{\rightleftharpoons}} L:^- + R_3C^+ \xrightarrow[\text{fast}]{Nu:^-} R_3C{-}Nu \quad (14.2)$$

S_N2 Mechanism

$$Nu:^- + R^1(H)(R^2)C{-}L \xrightarrow{\text{rds}} \left[\overset{\delta^-}{Nu}{:}{\cdots}C(R^1)(H)(R^2){\cdots}\overset{\delta^-}{L} \right]^{\ddagger} \longrightarrow Nu{-}C(R^1)(H)(R^2) + L:^- \quad (14.3)$$

Transition state

When nucleophilic substitution occurs by an S_N1 mechanism, the reaction proceeds in two successive steps, as illustrated in Equation 14.2. The first one involves the **heterolytic cleavage,** or **ionization,** of the bond between the carbon atom and the leaving group. This step is assisted by polar interactions between solvent molecules and the incipient cationic and anionic centers. Because the leaving group acquires the pair of bonding electrons, the organic fragment becomes a **carbocation,** with the carbon atom formerly bound to L now bearing a positive charge. The intermediate carbocation may then undergo the usual reactions of carbocations: (1) rearrangement to a more stable carbocation (Sec. 10.3); (2) loss of a proton to give an alkene by net **elimination** of the elements of **H–L** (Sec. 10.3); or (3) combination with a nucleophile to form the substitution product (Eq. 14.2). Normally, the concentration of the nucleophile, Nu:$^-$, is high compared to that of the L:$^-$ that has been produced, so the reaction of the carbocation with L:$^-$ to give the starting material is relatively unimportant. If the nucleophile is the solvent, the reaction is known as a **solvolysis.**

The first step of an S_N1 reaction is much slower than the second because it involves breaking the C–L bond to generate an unstable carbocation, an endothermic process. The second step is a fast, exothermic process involving bond formation. Thus, the first step of an S_N1 reaction is the **rate-determining step (rds)** of the reaction, and the rate of the reaction depends *only* upon the concentration of the substrate, R–L. Such a reaction is termed **unimolecular.** This is expressed mathematically in Equation 14.4, where k_1 is the **first-order rate constant.**

$$\text{Rate} = k_1[\text{R}-\text{L}] \tag{14.4}$$

When substitution occurs by an S_N2 mechanism, the nucleophile directly attacks the substrate, with the angle of approach being 180° to the C–L bond. This is called "backside attack," and the reaction proceeds with inversion of stereochemistry, the so-called "Walden inversion." The C–L bond is being broken concurrently with the formation of the C–Nu bond, so both the substrate, R–L, *and* the nucleophile are involved in the transition state of the rate-determining step. Reactions in which two reactants are involved in the transition state of the rate-determining step are termed **bimolecular,** and the rate of such processes depends on the concentration of the substrate and the nucleophile, as shown in Equation 14.5, where k_2 is the **second-order rate constant.**

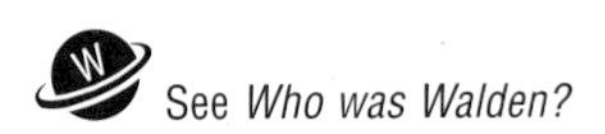

See *Who was Walden?*

$$\text{Rate} = k_2[\text{R}-\text{L}][\text{Nu:}] \tag{14.5}$$

We can summarize some of the important factors that dictate whether a particular substrate undergoes substitution preferentially by an S_N1 or S_N2 mechanism as follows.

1. As more alkyl groups are attached to the carbon atom **C–L** undergoing substitution, it becomes sterically more difficult for the nucleophile to attack from the backside because of the bulk of these groups, thereby decreasing the ease with which the S_N2 process can occur.
2. With increasing substitution of alkyl groups on the carbon atom **C–L,** the incipient carbocation in the S_N1 reaction becomes more stable, thereby increasing its ease of formation along the S_N1 pathway.

These two effects reinforce one another and yield the following general trends:

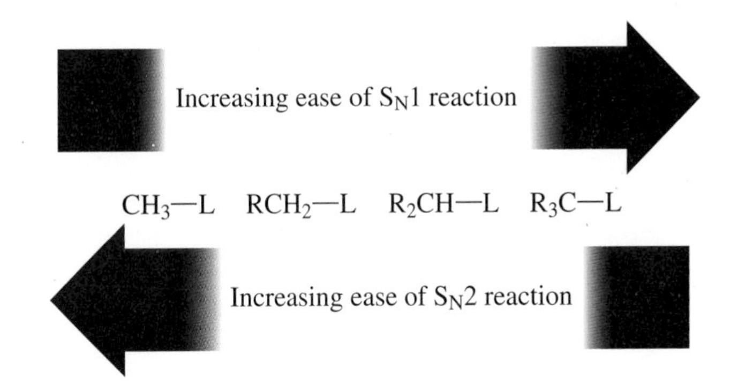

Primary substrates (1°, RCH_2–L) undergo nucleophilic substitution predominantly by an S_N2 mechanism, whereas tertiary substrates (3°, R_3C–L) react via an S_N1 mechanism. Secondary substrates (2°, R_2CH–L) may react by both mechanisms, and the specific pathway is dictated by various factors such as solvent, reaction conditions, and nature of the nucleophile. Such generalizations must always be applied cautiously, and the mechanism by which a particular reaction occurs *must* be confirmed *experimentally*. You may be able to deduce the mechanism by which a particular substitution occurred if you know the structures of the starting substrate and the product(s). For example, since carbocations are involved in the S_N1 mechanism, the formation of a product arising from a rearranged carbocation points to this mechanism; the lack of rearrangement does *not* necessarily exclude an S_N1 process, however. Determining whether the role of the substitution reaction follows a first- or a second-order rate law (Eqs. 14.4 and 14.5, respectively) also allows you to characterize the process as S_N1 or S_N2.

14.3 COMPETITION BETWEEN SUBSTITUTION AND ELIMINATION

Elimination reactions to produce alkenes may compete in reactions in which nucleophilic aliphatic substitution is the desired process. Unimolecular elimination reactions, **E1,** compete with S_N1 substitutions, and bimolecular elimination processes, **E2** (E stands for elimination and 2 for bimolecular), compete with S_N2 transformations. These competitions are shown in Equations 14.6 and 14.7. The nature of E1 reactions is discussed in detail in Section 10.3 and that of E2 processes in Section 10.2.

S_N2 Versus E2

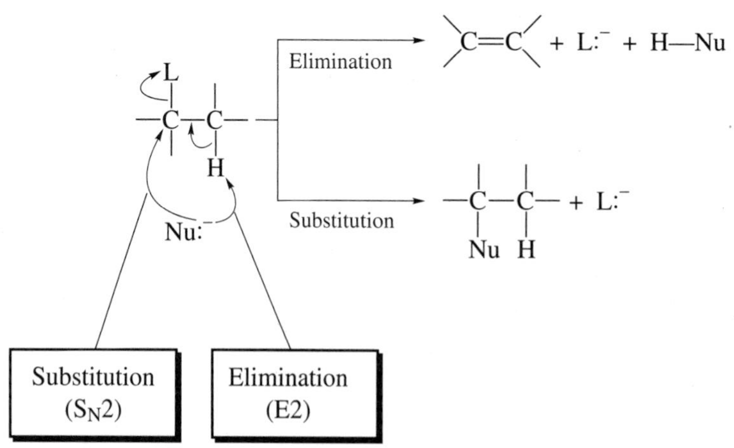

Elimination → C=C + L:⁻ + H—Nu

Substitution → —C(Nu)—C(H)— + L:⁻ (14.6)

Nu:⁻

Substitution (S_N2)

Elimination (E2)

S_N1 *Versus* *E1*

slow, $-L:^-$; Elimination → $C{=}C$ + H—Nu ; Substitution → —C—C— (Nu, H) ; $Nu:^-$

Substitution (S_N1) | Elimination (E1) (14.7)

The course of the reaction may be influenced by the nature of the nucleophile that is present. *Substitution is favored* with weakly basic and highly polarizable nucleophiles such as I^-, Br^-, Cl^-, H_2O, and $CH_3CO_2^-$, whereas *elimination is favored* when strongly basic and only slightly polarizable nucleophiles such as RO^-, H_2N^-, H^-, and HO^- are used. **Polarizability** is a measure of the ease with which the electron cloud of the Lewis base is distorted by a nearby center that bears a partial or full positive charge. Furthermore, bulky nucleophiles tend to favor elimination because the hydrogen atom is more sterically accessible than is the carbon atom bearing the leaving group.

14.4 PREPARATION OF 1-BROMOBUTANE: AN S_N2 REACTION

A common technique for converting a primary alcohol to an alkyl halide involves treating the alcohol with a hydrogen halide H–X (X = Cl, Br, or I) as shown in Equation 14.8. This reaction is *reversible,* and displacing the equilibrium to the right normally involves using a large excess of the acid, a strategy in accord with the **Le Chatelier principle.**

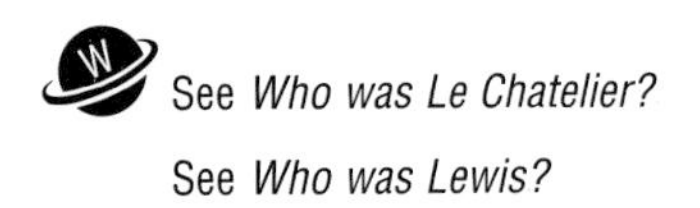

See *Who was Le Chatelier?*

See *Who was Lewis?*

$$\text{H—X} + \text{R—OH} \underset{}{\overset{\Delta}{\rightleftharpoons}} \text{R—X} + \text{H}_2\text{O} \quad (14.8)$$

where X = Cl, Br, I

For example, 1-bromobutane may be prepared by heating 1-butanol with hydrobromic acid, H–Br, in the presence of sulfuric acid (Eq. 14.9). The mechanism for this reaction is known to occur in two steps, the first being protonation of the alcohol to give the oxonium ion **1** via a Lewis acid-base reaction (Eq. 14.10). This oxonium ion then undergoes displacement by the bromide ion to form 1-bromobutane and water. This process is an S_N2 reaction in which *water is the leaving group and bromide ion is the nucleophile.* The sulfuric acid serves two important purposes: (1) It is a *dehydrating* agent that reduces the activity of water and shifts the position of equilibrium to the right, and (2) it provides an added source of hydrogen ions to increase the concentration of oxonium ion **1**. The use of *concentrated* hydrobromic acid also helps to establish a favorable equilibrium.

$$CH_3CH_2CH_2CH_2{-}OH + HBr \xrightleftharpoons[\Delta]{H_2SO_4} CH_3CH_2CH_2CH_2{-}Br + H_2O \quad (14.9)$$

1-Butanol (*n*-butyl alcohol) → 1-Bromobutane (*n*-butyl bromide)

$$n\text{-}C_3H_7CH_2{-}\ddot{O}H + H^+ \rightleftharpoons n\text{-}C_3H_7CH_2{-}\overset{+}{O}H_2 \;(\mathbf{1}) \xrightleftharpoons{Br:^-} n\text{-}C_3H_7CH_2{-}Br + H_2O \quad (14.10)$$

No reaction occurs between 1-butanol and NaBr in the absence of strong acid because leaving groups in nucleophilic substitution reactions must be *weakly* basic (Sec. 14.1), as is water in Equation 14.10. If the reaction of 1-butanol and NaBr were to occur (Eq. 14.11), the leaving group would necessarily be the *strongly* basic hydroxide ion, a poor leaving group. Thus the forward reaction depicted in Equation 14.11 does *not* occur. On the other hand, the reverse reaction between 1-bromobutane and hydroxide proceeds readily because the leaving group in this reaction is the weakly basic bromide ion.

$$CH_3CH_2CH_2CH_2{-}OH + NaBr \not\rightleftharpoons CH_3CH_2CH_2CH_2{-}Br + NaOH \quad (14.11)$$

The mixture of hydrobromic acid and sulfuric acid may be prepared by either adding concentrated sulfuric acid to concentrated hydrobromic acid or by generating the hydrobromic acid *in situ* by adding concentrated sulfuric acid to aqueous sodium bromide (Eq. 14.12). Both of these methods work well and give good yields of the alkyl bromide from low-molar-mass alcohols. The method of generating H–Br *in situ* is not effective with higher-molar-mass alcohols because of their low solubility in concentrated salt solutions, so concentrated (48%) hydrobromic acid is used instead.

$$NaBr + H_2SO_4 \rightleftharpoons HBr + NaHSO_4 \quad (14.12)$$

Although the presence of concentrated sulfuric acid promotes the formation of the alkyl bromide, several side reactions involving sulfuric acid and the alcohol can occur. One of these is esterification of the alcohol by sulfuric acid to form an alkyl hydrogen sulfate **2** (Eq. 14.13).

$$RO{-}H + H_2SO_4 \rightleftharpoons RO{-}SO_3H + H_2O \quad (14.13)$$

2
An alkyl hydrogen sulfate

This reaction is reversible, and the position of the equilibrium is shifted to the left, regenerating the alcohol from **2** as the alkyl bromide is produced. The formation of **2** itself does not directly decrease the yield of alkyl bromide; rather, **2** undergoes other reactions to give undesired by-products. For example, it suffers elimination on heating to give a mixture of alkenes (Eq. 14.14). It may also react with another molecule of alcohol to give a dialkyl ether by an S_N2 reaction in which the nucleophile is ROH (Eq. 14.15). Both of these side reactions consume alcohol, so the yield of alkyl bromide is decreased. Fortunately, these side reactions may be minimized for *primary* alcohols by controlling the temperature of the reaction and the concentration of sulfuric acid.

$$RO{-}SO_3H \xrightarrow{\text{heat}} \text{alkenes} + H_2SO_4 \tag{14.14}$$

$$RO{-}SO_3H + ROH \xrightarrow{\text{heat}} R{-}O{-}R + H_2SO_4 \tag{14.15}$$

It is necessary to use different procedures to prepare secondary alkyl bromides from *secondary* alcohols because such alcohols are easily dehydrated by concentrated sulfuric acid to give alkenes by way of Equations 14.13 and 14.14. In fact, the acid-catalyzed dehydration of secondary and tertiary alcohols is a common method for synthesizing alkenes (Sec. 10.3). This problem may be circumvented by using concentrated hydrobromic acid; however, it is better to prepare secondary alkyl bromides by the reaction of secondary alcohols with phosphorus tribromide, PBr_3 (Eq. 14.16).

$$3\,R{-}OH + \underset{\text{Phosphorus tribromide}}{PBr_3} \longrightarrow 3\,R{-}Br + H_3PO_3 \tag{14.16}$$

EXPERIMENTAL PROCEDURES

Preparation of 1-Bromobutane

Purpose To demonstrate the conversion of a primary alcohol to a primary bromoalkane using hydrobromic acid.

SAFETY ALERT

1. **Examine your glassware for cracks and chips. This procedure involves heating *concentrated* acids, and defective glassware could break under these conditions, spilling hot corrosive chemicals on you and those working around you.**
2. **Wear latex gloves while performing this experiment.**
3. **Be very careful when handling concentrated sulfuric acid. If any concentrated sulfuric acid comes in contact with your skin, immediately wash it off with copious amounts of cold water and then with dilute sodium bicarbonate solution.**
4. ***Concentrated* sulfuric acid and water mix with the evolution of lots of heat. *Always add the acid to the water,* a technique that disperses the heat through warming of the water. Add the acid slowly and with swirling to ensure continuous and thorough mixing.**

MINISCALE PROCEDURE

Preparation Sign in at **www.cengage.com/login** to answer Pre-Lab Exercises, access videos, and read the MSDSs for the chemicals used or produced in this procedure. Review Sections 2.9, 2.10, 2.11, 2.13, 2.21, and 2.22.

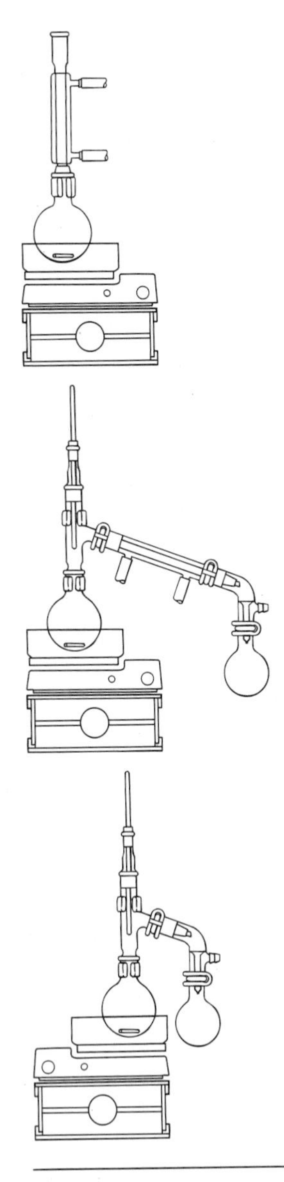

Apparatus A 25-mL and a 100-mL round-bottom flask, separatory funnel, ice-water bath, apparatus for heating under reflux, simple distillation, magnetic stirring, and *flameless* heating.

Setting Up Place 11.1 g of sodium bromide in the 100-mL round-bottom flask, and add about 10 mL of water and 10 mL of 1-butanol. Mix the contents of the flask thoroughly by swirling; then cool the flask in an ice-water bath. *Slowly* add 10 mL of *concentrated* sulfuric acid to the cold mixture with swirling and continuous cooling. Remove the flask from the ice-water bath, add a stirbar, and set up the apparatus for heating under reflux.

Reaction, Work-Up, and Isolation Warm the flask gently until most of the salts have dissolved and then heat the mixture under gentle reflux for about 45 min.★ Equip the flask for simple distillation. Distill the mixture rapidly and collect the distillate of water and 1-bromobutane in an ice-cooled receiver. Continue distilling until the distillate is clear; the head temperature should be around 115 °C at this point because sulfuric acid and hydrobromic acid are now co-distilling with water.★

Transfer the distillate to a separatory funnel and add about 10 mL of water. Shake the funnel gently with venting. Separate the layers and determine which of these is the organic layer. Wash the organic layer sequentially with two 5-mL portions of cold 2 *M* aqueous sodium hydroxide solution, 10 mL of water, and 10 mL of saturated aqueous sodium chloride. Transfer the cloudy 1-bromobutane layer to an Erlenmeyer flask, and dry it over several spatula-tips full of anhydrous sodium sulfate.★ Swirl the flask occasionally for a period of 10–15 min until the crude 1-bromobutane is clear; add further small portions of anhydrous sodium sulfate if the solution appears cloudy.

Using a Pasteur pipet, carefully transfer the crude 1-bromobutane to a clean, dry 25-mL round-bottom flask. Add a stirbar, equip the flask for shortpath distillation, and carefully distill the product into a tared receiver. Because of the relatively small quantity of product expected, it may be difficult to obtain an accurate boiling point, so you should collect the fraction having a boiling point greater than about 90 °C (760 torr).

Analysis Weigh the product and compute the yield. Perform the alcoholic silver nitrate and sodium iodide/acetone classification tests on your product (Sec. 25.9). Obtain IR and ^{1}H NMR spectra of your starting material and product and compare them with those of authentic samples (Figs. 8.28 and 14.1–14.4).

MICROSCALE PROCEDURE

Preparation Sign in at **www.cengage.com/login** to answer Pre-Lab Exercises, access videos, and read the MSDSs for the chemicals used or produced in this procedure. Review Sections 2.9, 2.10, 2.11, 2.13, 2.21, and 2.22.

Apparatus A 3-mL and a 5-mL conical vial, ice-water bath, apparatus for heating under reflux, simple distillation, magnetic stirring, and *flameless* heating.

Setting Up Place 1.1 g of sodium bromide in the 5-mL conical vial and add 1.0 mL of water, 1.0 mL of 1-butanol, and a spinvane. Mix the contents of the vial thoroughly by stirring and then cool the vial in an ice-water bath. *Slowly* add 1.0 mL of *concentrated* sulfuric acid to the cold mixture with stirring and continuous cooling. Remove the vial from the bath and set up the apparatus for heating under reflux.

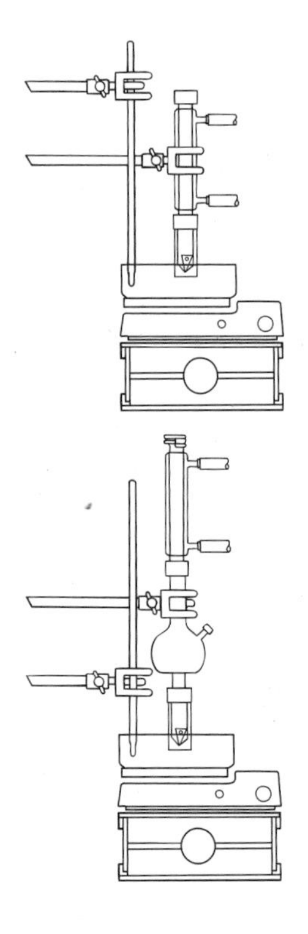

Reaction, Work-Up, and Isolation Warm the vial gently until most of the salts have dissolved and then heat the mixture under gentle reflux for 45 min.★ Equip the vial for simple distillation. Distill the mixture rapidly and collect the distillate water and 1-bromobutane in the Hickman stillhead. Continue the distillation, increasing the temperature of the heating source as necessary until there is about 1.5–2.0 mL remaining in the vial and the distillate is clear.★

Using a Pasteur pipet, transfer the distillate to the 3-mL conical vial and add 1.0 mL of water. Cap the vial and shake it gently with venting to mix the contents. Separate the layers and determine which of these is the organic layer. Wash the organic layer sequentially with two 0.5-mL portions of cold 2 *M* aqueous sodium hydroxide solution, 1.0 mL of water, and 1.0 mL of saturated aqueous sodium chloride. Transfer the cloudy 1-bromobutane layer to a screw-cap centrifuge tube, and dry it over several microspatula-tips full of anhydrous sodium sulfate.★ Swirl the contents in the centrifuge tube occasionally for a period of 10–15 min until the crude 1-bromobutane is clear; add further small portions of anhydrous sodium sulfate if the solution appears cloudy. Using a Pasteur pipet, carefully transfer the crude 1-bromobutane to a tared sample vial.

Analysis Weigh the product and compute the yield. Perform the alcoholic silver nitrate and sodium iodide/acetone classification tests on your product (Sec. 25.9). Obtain IR and ^{1}H NMR spectra of your starting material and product and compare them with those of authentic samples (Figs. 8.28 and 14.1–14.4).

Discovery Experiment

Analysis of S_N Reactions as a Function of Substrate

In Section 14.2, we state that the relative rate of an S_N reaction for a substrate R–L is a function of the degree of the carbon atom bearing the leaving group. This proposition is subject to experimental verification using simple qualitative tests described in Chapter 25. By referring to the procedures described under "Alkyl Halides" in Section 25.6, propose and execute experiments that explore the relative rates of S_N1 and S_N2 reactions for substrates in which the leaving group is on carbon atoms of various degrees. You may consider using both alkyl and aryl halides in your study, but be certain to consult with your instructor before undertaking any experiments.

WRAPPING IT UP

Carefully dilute the *stillpot residue from the reaction* with water and then slowly combine this with the *water and the sodium hydroxide washes.* Neutralize the *combined aqueous mixture* with sodium carbonate and flush the solution down the drain with water. The *sodium sulfate* used as the drying agent is contaminated with product, so place it in the container for halogenated solids. Place the *residue in the stillpot from the distillation in the Miniscale Procedure* into the container for halogenated liquids. Dilute the *silver nitrate test solution* with water and flush it down the drain unless instructed otherwise. Place the *sodium iodide/acetone test solution* in the container for halogenated liquids.

EXERCISES

1. Some water was added to the initial reaction mixture in the procedure you performed.
   a. How might the yield of 1-bromobutane be affected if the water were not added, and what product(s) would be favored?
   b. How might the yield of product be affected by adding twice as much water as is specified, while keeping the quantities of the other reagents the same?
2. In the purification process, the organic layer is washed sequentially with 2 *M* NaOH, water, and saturated aqueous sodium chloride. What is the purpose of each of these washes and why is *cold* 2 *M* NaOH recommended?
3. After the washes described in Exercise 2, the 1-bromobutane is treated with anhydrous sodium sulfate.
   a. Why is this done?
   b. Could solid sodium hydroxide or potassium hydroxide be used for this purpose? Explain.
4. The final step of the purification process in the Miniscale Procedure involves a simple distillation. What impurities are removed by this distillation?
5. Using curved arrows to symbolize the flow of electrons, propose a mechanism for formation of the by-product(s) from elimination that could be produced when 1-butanol is treated with HBr. Specify whether your mechanism is E1, E2, or both.
6. List two factors, one in the substrate R–L and the other in the nucleophile/base Nu: being used, that should increase the yield of E2 relative to S_N2 products in the reaction of R–L with Nu:.
7. Treating 1-butanol with sulfuric acid establishes an equilibrium with 1-butyl bisulfate and water as shown.

$$\underset{\text{1-Butanol}}{CH_3(CH_2)_2CH_2OH} + H_2SO_4 \rightleftharpoons \underset{\text{1-Butyl bisulfate}}{CH_3(CH_2)_2CH_2OSO_3H} + H_2O$$

   a. Using curved arrows to symbolize the flow of electrons, propose a mechanism for the conversion of 1-butanol to 1-butyl bisulfate.
   b. 1-Butyl bisulfate may serve as the precursor to an ether and an alkene, as well as the desired 1-bromobutane.
      i. Provide the structures expected for the ether and the alkene that might be produced from 1-butyl bisulfate.
      ii. Using curved arrows to symbolize the flow of electrons, provide reaction mechanisms for the formation of each of these by-products and specify whether the mechanisms you write are S_N1, S_N2, E1, E2, or none of these.
      iii. Explain why you would expect bisulfate to be a good leaving group for substitution or elimination reactions.
8. Consider the mechanistic step in the conversion of 1-butanol to 1-bromobutane in which bromide ion displaces water from the oxonium ion **1** (Eq. 14.10).
   a. What evidence, if any, is there that this step involves backside attack by the nucleophile?

b. Propose an experiment that would allow you to prove that backside attack of the nucleophile on the substrate was indeed the mechanism for the reaction.

9. How would doubling the concentration of the nucleophile affect the rate of an S_N2 reaction?

10. 1-Butanol does not undergo S_N2 reactions in the absence of acids. If the alcohol were converted to the corresponding *p*-toluenesulfonate ester shown below, would you expect this ester to undergo an S_N2 reaction with NaBr in the absence of acid? Give your reasoning.

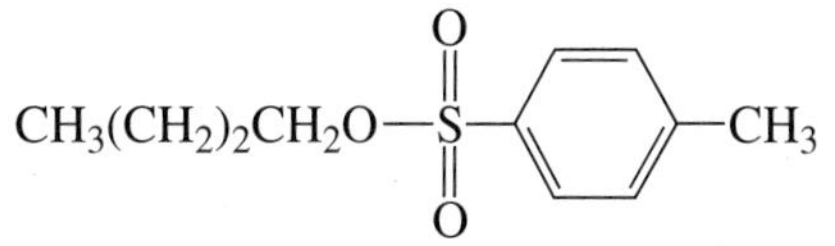

$$CH_3(CH_2)_2CH_2O{-}\overset{O}{\underset{O}{\overset{\|}{\underset{\|}{S}}}}{-}C_6H_4{-}CH_3$$

1-Butyl *p*-toluenesulfonate

11. Consider the structure of 1-butanol and explain why this alcohol does not undergo S_N1 reactions.

12. Specify the most electrophilic carbon atom in 1-butanol and rationalize your choice.

13. Neopentyl chloride, $CH_3C(CH_3)_2CH_2Cl$, is a primary alkyl halide that undergoes S_N2 reactions at extremely slow rates. Offer an explanation for this fact that is consistent with the mechanism for this reaction. Draw a suitable illustration of the transition state for the rate-determining step as part of your explanation.

14. A student attempted to prepare *tert*-butyl ethyl ether by a substitution reaction between 2-chloro-2-methylpropane (*tert*-butyl chloride) and sodium ethoxide. None of the desired ether was obtained. What product was formed and why did this route to the ether fail? Suggest a different substitution reaction that should yield the desired product.

$$(CH_3)_3C{-}Cl + NaOCH_2CH_3 \not\longrightarrow (CH_3)_3C{-}O{-}CH_2CH_3$$

2-Chloro-2-methylpropane; Sodium ethoxide; *tert*-Butyl ethyl ether

15. The reaction of 3-methyl-2-butanol with concentrated hydrobromic acid gives the two isomeric bromoalkanes shown.

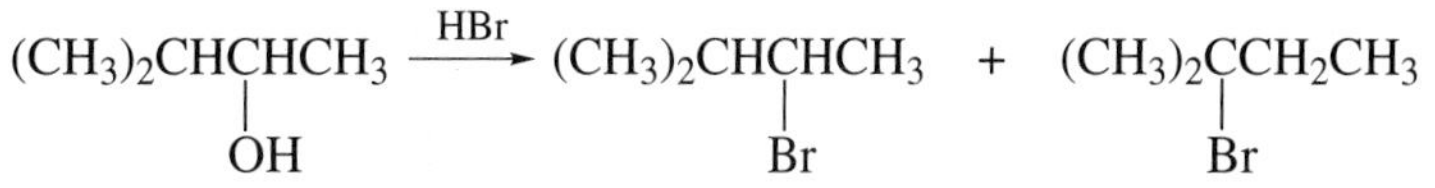

$$(CH_3)_2CHCH(OH)CH_3 \xrightarrow{HBr} (CH_3)_2CHCH(Br)CH_3 + (CH_3)_2C(Br)CH_2CH_3$$

3-Methyl-2-butanol; 2-Bromo-3-methylbutane 10%; 2-Bromo-2-methylbutane 90%

Using curved arrows to symbolize the flow of electrons, suggest two reasonable mechanisms for this reaction that account for the formation of each of the products. Indicate whether each is an S_N1 or an S_N2 reaction.

Sign in at **www.cengage.com/login** and use the spectra viewer and Tables 8.1–8.8 as needed to answer the blue-numbered questions on spectroscopy.

16. Consider the spectral data for 1-butanol (Figs. 8.28, 14.1, and 14.2).

a. In the functional group region of the IR spectrum, specify the absorption associated with the hydroxyl group, and explain why this peak is broad.

b. In the ^1H NMR spectrum, assign the various resonances to the hydrogen nuclei responsible for them.

c. If the NMR sample of 1-butanol in $CDCl_3$ were shaken with D_2O *prior* to taking the ^1H NMR spectrum, what differences might you expect to observe?

d. For the ^{13}C NMR data, assign the various resonances to the carbon nuclei responsible for them.

17. Consider the NMR spectral data for 1-bromobutane (Fig. 14.4).

a. In the ^1H NMR spectrum, assign the various resonances to the hydrogen nuclei responsible for them.

b. For the ^{13}C NMR data, assign the various resonances to the carbon nuclei responsible for them.

18. Discuss the differences in the IR and NMR spectra of 1-butanol and 1-bromobutane that are consistent with the conversion of an alcohol into a bromoalkane in this procedure.

19. Discuss the differences observed in the ^1H and ^{13}C NMR spectra of 1-bromobutane, which appear in Figure 14.4a and b, with those of 1-chlorobutane, which appear in Figure 9.3a and b. To what factor do you attribute these differences?

SPECTRA

Starting Material and Product

The ^1H NMR spectrum of 1-butanol is provided in Figure 8.28.

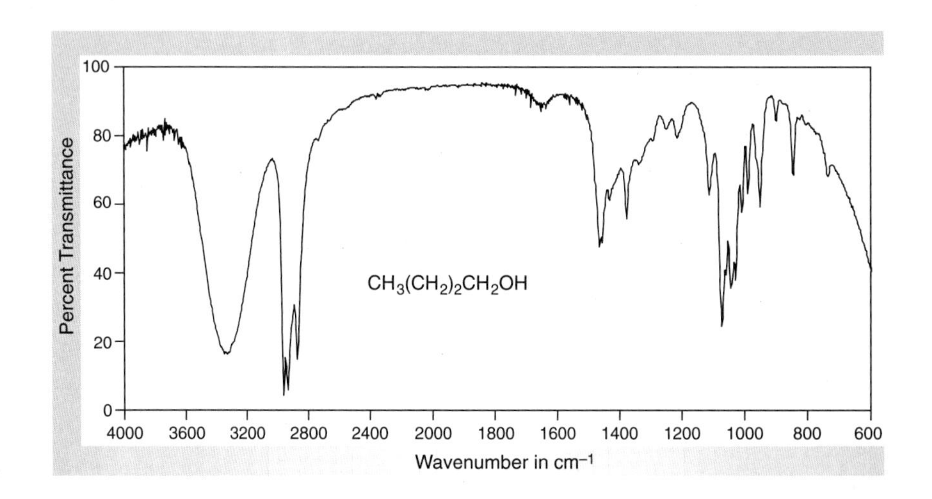

Figure 14.1
IR spectrum of 1-butanol (neat).

Figure 14.2
^{13}C NMR data for 1-butanol ($CDCl_3$).
Chemical shifts: Δ 13.9, 19.2, 35.0, 62.2.

$CH_3(CH_2)_2CH_2OH$

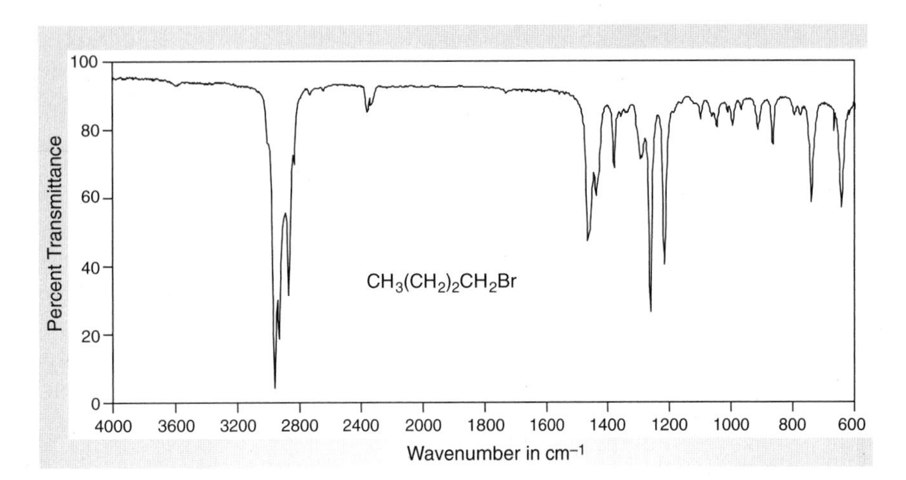

Figure 14.3
IR spectrum of 1-bromobutane (neat).

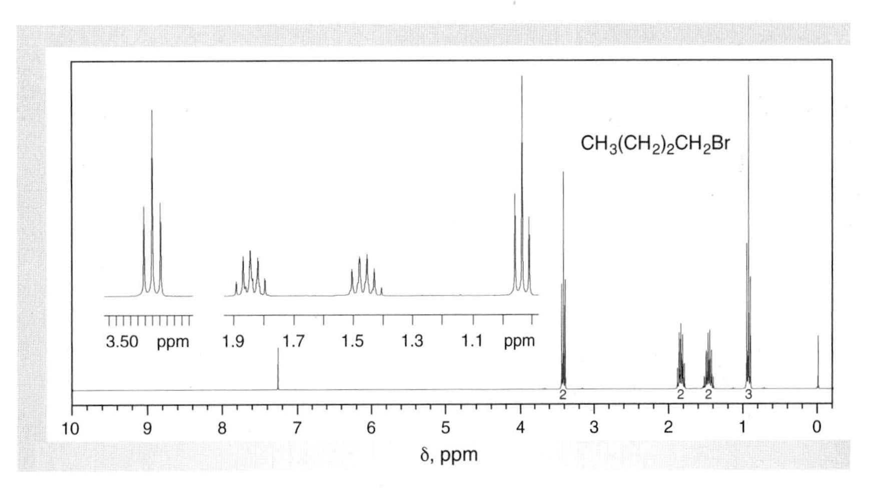

Figure 14.4
NMR data for 1-bromobutane ($CDCl_3$).

(a) ^{1}H NMR spectrum (300 MHz).
(b) ^{13}C NMR data: δ 13.2, 21.5, 33.1, 35.0.

14.5 PREPARATION OF 2-CHLORO-2-METHYLBUTANE: AN S_N1 REACTION

Different reagents such as HX and PX_3 may be used to prepare alkyl halides from primary and secondary alcohols. However, because elimination reactions predominate when tertiary alcohols are treated with phosphorous trihalides, preparing tertiary alkyl halides from tertiary alcohols proceeds with good yields only if *concentrated* hydrogen halides, HX, are used. The reaction of 2-methyl-2-butanol with hydrochloric acid to produce 2-chloro-2-methylbutane (Eq. 14.17) illustrates this transformation.

$$CH_3CH_2-\underset{CH_3}{\overset{CH_3}{\underset{|}{\overset{|}{C}}}}-OH + HCl \rightleftharpoons CH_3CH_2-\underset{CH_3}{\overset{CH_3}{\underset{|}{\overset{|}{C}}}}-Cl + H_2O \quad (14.17)$$

2-Methyl-2-butanol → 2-Chloro-2-methylbutane

The conversion of tertiary alcohols to the corresponding tertiary alkyl halide using concentrated hydrogen halides proceeds via an S_N1 mechanism (Eqs. 14.18–14.20). Mechanistically, the first step of the process involves protonation of the hydroxyl group of the alcohol, a Lewis acid-base reaction (Eq. 14.18). Ionization then occurs in the second step, and a molecule of water is lost (Eq. 14.19). Owing to steric hindrance at the tertiary carbon atom, attack of chloride on **3** in an S_N2 process (Sec. 14.4) does not occur. This difference in reactivity between the oxonium ions derived from tertiary alcohols and those of secondary or primary alcohols reflects the relative stabilities of the three types of carbocations (3° > 2° > 1°) that would be formed upon the loss of a molecule of water. The ionization is the slow, rate-determining (rds) step in the sequence. In the final step of the reaction, chloride ion attacks the intermediate carbocation to give 2-chloro-2-methylbutane (Eq. 14.20).

$$CH_3CH_2{-}C(CH_3)_2{-}\ddot{O}{-}H + H^+ \overset{\text{fast}}{\rightleftharpoons} CH_3CH_2{-}C(CH_3)_2{-}O^+H_2 \quad (14.18)$$

3

$$CH_3CH_2{-}C(CH_3)_2{-}O^+H_2 \overset{\text{rds}}{\rightleftharpoons} CH_3CH_2{-}C^+(CH_3)_2 + H_2O \quad (14.19)$$

Relatively stable tertiary carbocation

$$CH_3CH_2{-}C^+(CH_3)_2 + :\ddot{Cl}:^- \overset{\text{fast}}{\rightleftharpoons} CH_3CH_2{-}C(CH_3)_2{-}Cl \quad (14.20)$$

The principal side reaction in S_N1 reactions of this type is E1 elimination, which results from the loss of a proton from the tertiary carbocation to give 2-methyl-1-butene together with 2-methyl-2-butene (Eqs. 14.21 and 14.22), as discussed in more detail in Section 10.3. Under the reaction conditions that you will use, however, elimination is reversible through Markovnikov addition of HCl to the 2-methyl-1-butene or the 2-methyl-2-butene that is produced, and this addition gives the desired 2-chloro-2-methylbutane (Eq. 14.23). A more extensive discussion of the ionic addition of hydrogen halides to alkenes is presented in Sections 10.4 and 10.5.

$$CH_3CH_2{-}C^+(CH_3){-}CH_2{-}H + :\ddot{Cl}:^- \longrightarrow CH_3CH_2{-}C(=CH_2){-}CH_3 + H{-}Cl \quad (14.21)$$

2-Methyl-1-butene

$$CH_3CH(H){-}C^+(CH_3)_2 + :\ddot{Cl}:^- \longrightarrow CH_3CH{=}C(CH_3)_2 + H{-}Cl \quad (14.22)$$

2-Methyl-2-butene

$$CH_3CH_2{-}C({=}CH_2)CH_3 + H{-}Cl \longrightarrow CH_3CH_2{-}\overset{CH_3}{\underset{CH_3}{C^+}} + :\ddot{Cl}:^- \longrightarrow CH_3CH_2{-}\overset{CH_3}{\underset{CH_3}{C}}{-}Cl \qquad (14.23)$$

or

$$CH_3CH{=}C(CH_3)CH_3 + H{-}Cl \longrightarrow CH_3CH_2{-}\overset{CH_3}{\underset{CH_3}{C^+}}$$

EXPERIMENTAL PROCEDURES

Preparation of 2-Chloro-2-methylbutane

Purpose To demonstrate the conversion of a tertiary alcohol into a tertiary alkyl chloride using hydrochloric acid.

SAFETY ALERT

Wear latex gloves throughout the experiment, because *concentrated* hydrochloric acid is being used. If any acid spills on your skin, wash it off with large volumes of water and then with dilute sodium bicarbonate solution.

MINISCALE PROCEDURE

Preparation Sign in at **www.cengage.com/login** to answer Pre-Lab Exercises, access videos, and read the MSDSs for the chemicals used or produced in this procedure. Review Sections 2.10, 2.13, and 2.21.

Apparatus A 25-mL round-bottom flask, separatory funnel, ice-water bath, apparatus for simple distillation, magnetic stirring, and *flameless* heating.

Setting Up Place 10 mL of 2-methyl-2-butanol and 25 mL of *concentrated* (12 *M*) hydrochloric acid in the separatory funnel.

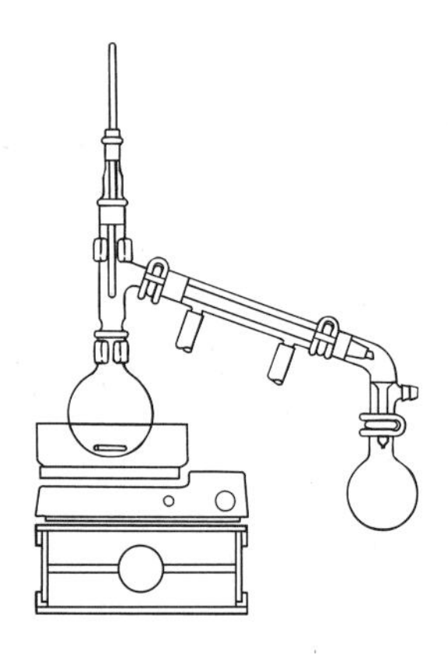

Reaction, Work-Up, and Isolation Swirl the contents of the separatory funnel gently *without* the stopper on the funnel to mix the reactants. After swirling the funnel for about 1 min, stopper and then carefully invert it. Release the excess pressure by opening the stopcock with the funnel inverted. *Do not shake the funnel until the pressure has been equalized.* Now close the stopcock and shake the funnel for several minutes, with intermittent venting. Allow the mixture to separate into two distinct layers.

Separate the layers and determine which is the organic layer. Wash the organic layer sequentially with 10-mL portions of saturated aqueous sodium chloride and *cold* saturated aqueous sodium bicarbonate. On initial addition of the bicarbonate solution, vigorous gas evolution will normally occur; gently swirl the *unstoppered* separatory funnel until this stops. Stopper the funnel and *carefully* invert it; vent the funnel immediately to release gas pressure. Shake the funnel gently with *frequent* venting and then shake it vigorously with frequent venting. Separate the organic layer, and again wash it sequentially with 10-mL portions of water and saturated aqueous sodium chloride. Carefully remove the aqueous layer, transfer the 2-chloro-2-methylbutane to an Erlenmeyer flask, and dry the product over several spatula-tips full of anhydrous sodium sulfate.★ Swirl the flask occasionally for 10–15 min until the product is dry; add further small portions of anhydrous sodium sulfate if the liquid appears cloudy.

Using a Pasteur or a filter-tip pipet, carefully transfer the crude product to the round-bottom flask, add a stirbar, and equip the flask for shortpath distillation. Distill the product, collecting it in a tared receiver cooled in an ice-water bath. Because of the relatively small quantity of product, it may be difficult to obtain an accurate boiling point, so you should collect the fraction having a boiling point greater than about 75 °C (760 torr).

Analysis Determine the mass of 2-chloro-2-methylbutane isolated and calculate the percent yield. Perform the alcoholic silver nitrate and sodium iodide/acetone classification tests on your product (Sec. 25.9). Obtain IR and ^{1}H NMR spectra of your starting material and product and compare them with those of authentic samples (Figs. 14.5–14.8).

MICROSCALE PROCEDURE

Preparation Sign in at **www.cengage.com/login** to answer Pre-Lab Exercises, access videos, and read the MSDSs for the chemicals used or produced in this procedure. Review Sections 2.10, 2.13, and 2.21.

Apparatus A 5-mL conical vial.

Setting Up Place 1.0 mL of 2-methyl-2-butanol and 2.5 mL of *concentrated* (12 *M*) hydrochloric acid in the conical vial.

Reaction, Work-Up, and Isolation Using a small glass rod, stir the contents of the conical vial gently to mix the reactants. After about 1 min of stirring, cap and invert the vial, and then return it to its original upright position. Release the excess pressure in the vial by *carefully* venting. *Do not shake the vial until the pressure has been equalized.* Replace the screw-cap and shake the vial for several minutes with intermittent venting. Allow the mixture to separate into two distinct layers.

Separate the layers and determine which is the organic layer. Wash the organic layer sequentially with 1-mL portions of saturated aqueous sodium chloride and *cold* saturated aqueous sodium bicarbonate. On initial addition of the bicarbonate solution, vigorous gas evolution will normally occur. Using a glass rod, gently stir the contents of the vial until this stops. Cap and invert the vial and return it to its original upright position; release the excess pressure by venting *carefully.* Shake the vial gently with *frequent* venting and then shake it vigorously with frequent venting. Separate the organic layer, and again wash it sequentially with 1-mL portions of water and saturated sodium chloride solution. Carefully remove the aqueous layer.

Using a filter-tip pipet, transfer the 2-chloro-2-methylbutane to a centrifuge tube, and dry the liquid over several microspatula-tips full of anhydrous sodium sulfate.★ Swirl the tube occasionally for 10–15 min until the product is dry. Add several small portions of anhydrous sodium sulfate if the liquid is cloudy. Using a Pasteur or a filter-tip pipet, transfer the crude product to a tared sample vial.

Analysis Determine the mass of the 2-chloro-2-methylbutane isolated and calculate the percent yield. Perform the alcoholic silver nitrate and sodium iodide/acetone classification tests on your product (Sec. 25.9). Obtain IR and ^1H NMR spectra of your starting material and product and compare them with those of authentic samples (Figs. 14.5–14.8).

Discovery Experiment

Analysis of S_N Reactions as a Function of Substrate

In Section 14.2, we stated that the relative rate of an S_N reaction for a substance R–L is a function of the degree of the carbon atom bearing the leaving group. This proposition is subject to experimental verification using simple qualitative tests described in Chapter 25. By referring to the procedures described under "Alkyl Halides" in Section 25.6, propose and execute experiments that explore the relative rates of S_N1 and S_N2 reactions for substrates in which the leaving group is on carbon atoms of various degrees. You may consider using both alkyl and aryl halides in your study, but be certain to consult with your instructor before undertaking any experiments.

WRAPPING IT UP

Carefully dilute the *aqueous layer* from the first separation with water and then neutralize it with sodium carbonate. Combine this solution with the other *aqueous washes (water, saturated sodium bicarbonate, and saturated sodium chloride) and layers*, and flush them down the drain with water. The *sodium sulfate* used as the drying agent is contaminated with product, so place it in the container for halogenated solids. Pour the *residue in the stillpot from the Miniscale Procedure* into the container for halogenated liquids. Dilute the *silver nitrate test solution* with water and flush it down the drain unless instructed otherwise. Place the *sodium iodide/acetone test solution* in the container for halogenated liquids.

EXERCISES

1. The work-up procedure calls for washing the crude 2-chloro-2-methylbutane with *cold* sodium bicarbonate solution.
   a. What purpose does this wash serve?
   b. This washing procedure is accompanied by vigorous gas evolution, which increases the difficulty of handling and requires considerable caution. Alternatively, one might consider using a dilute solution of sodium hydroxide instead of the sodium bicarbonate. Discuss the relative advantages and disadvantages of using these two basic solutions in the work-up.
   c. On the basis of your answer to **b**, why were you instructed to use sodium bicarbonate, even though it is more difficult to handle?
2. What is the purpose of the final wash of the organic layer with saturated sodium chloride solution in the purification process?
3. The 2-chloro-2-methylbutane is dried with anhydrous sodium sulfate in this procedure. Could solid sodium hydroxide or potassium hydroxide be used for this purpose? Explain your answer.
4. Draw a reaction profile to illustrate the conversion of 2-methyl-2-butanol and HCl to 2-chloro-2-methylbutane and water (Eqs. 14.18–14.20). Label enthalpy of activation, $\Delta H^{\ddagger}$, for the rate-determining step and the enthalpy, ΔH_{rxn}, for the overall reaction.

5. Explain why polar solvents would be expected to increase the rate of the reaction in Exercise 4.

6. Consider the by-products that might be formed by E1 and/or E2 processes in the reaction of 2-methyl-2-butanol with HCl.

$$CH_3C(CH_3)(OH)CH_2CH_3 + HCl_{aq} \longrightarrow ?$$

2-Methyl-2-butanol

   a. Provide structures for these products.

   b. Which alkene would you expect to be favored and why?

7. Draw the structures of all the alcohols that are isomeric with 2-methyl-2-butanol.

   a. Arrange these alcohols in order of increasing reactivity toward concentrated hydrochloric acid.

   b. Which, if any, of these other alcohols would you expect to give a reasonable yield of the corresponding alkyl chloride under such reaction conditions?

8. Consider the reaction of (*S*)-3-methyl-3-hexanol with *concentrated* HCl.

$$CH_3CH_3C(CH_3)(OH)CH_2CH_2CH_3 + HCl_{conc} \longrightarrow CH_3CH_3C(CH_3)(Cl)CH_2CH_2CH_3$$

3-Methyl-3-hexanol → 3-Chloro-3-methylhexane

   a. Draw a three-dimensional representation for this enantiomer.

   b. Predict whether the 3-chloro-3-methylhexane formed in the reaction would be *R*-, *S*-, or racemic, and explain your prediction.

9. Why does ionization of a tertiary substrate $R_3C{-}L$ proceed at a rate faster than that of a secondary substrate $R_2CH{-}L$?

10. How would doubling the concentration of the nucleophile affect the rate of an S_N1 reaction?

11. Benzyl chloride, $C_6H_5CH_2Cl$, readily undergoes an S_N1 reaction, yet it is a primary substrate. Explain this seemingly anomalous result.

12. For each of the following pairs, predict which one will undergo an SN_1 reaction faster. Explain your predictions.

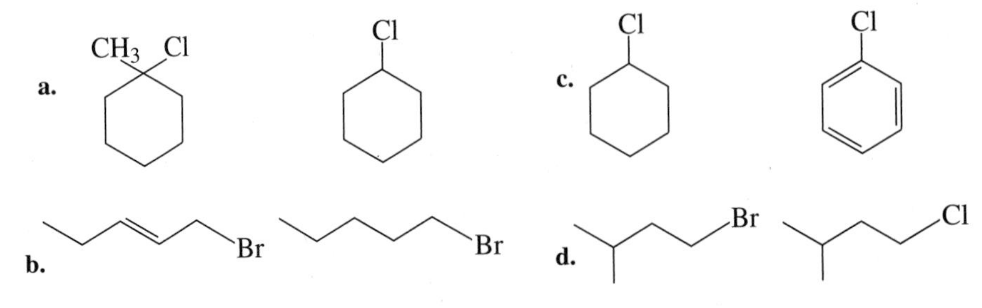

13. On prolonged heating with concentrated hydrochloric acid, 2,2-dimethyl-1-propanol, a primary alcohol, reacts to give 2-chloro-2-methylbutane. Provide a stepwise mechanism for this reaction, using curved arrows to symbolize the

flow of electrons. Also account for the fact that the observed product is a tertiary alkyl chloride even though the starting alcohol is primary.

$$CH_3-C(CH_3)_2-CH_2-OH \xrightarrow{HCl} CH_3-C(CH_3)(Cl)-CH_2CH_3 + H_2O$$

2,2-Dimethyl-1-propanol → 2-Chloro-2-methylbutane

Sign in at **www.cengage.com/login** and use the spectra viewer and Tables 8.1–8.8 as needed to answer the blue-numbered questions on spectroscopy.

14. Consider the spectral data for 2-methyl-2-butanol (Figs. 14.5 and 14.6).

a. In the functional group region of the IR spectrum, identify the absorption associated with the hydroxyl group, and explain why this peak is broad.

b. In the ^1H NMR spectrum, assign the various resonances to the hydrogen nuclei responsible for them.

c. If the NMR sample of 2-methyl-2-butanol in $CDCl_3$ were shaken with D_2O *prior* to taking the ^1H NMR spectrum, what differences might you expect to observe?

d. For the ^{13}C NMR data, assign the various resonances to the carbon nuclei responsible for them.

15. Consider the NMR spectral data for 2-chloro-2-methylbutane (Fig. 14.8).

a. In the ^1H NMR spectrum, assign the various resonances to the hydrogen nuclei responsible for them.

b. For the ^{13}C NMR data, assign the various resonances to the carbon nuclei responsible for them.

16. Discuss the differences in the **IR** and **NMR** spectra of 2-methyl-2-butanol and 2-chloro-2-methylbutane that are consistent with the conversion of an alcohol into an alkyl chloride in this procedure.

SPECTRA

Starting Material and Product

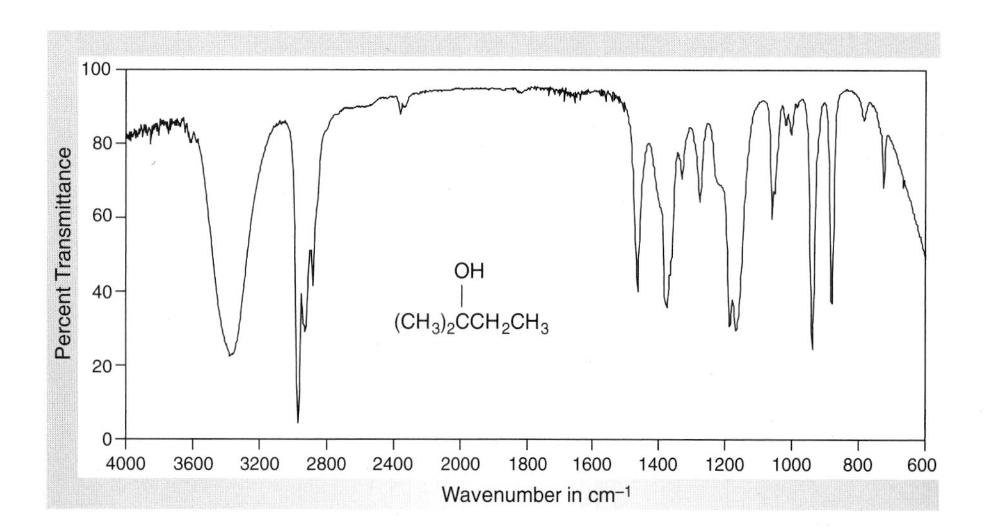

Figure 14.5
IR spectrum of 2-methyl-2-butanol (neat).

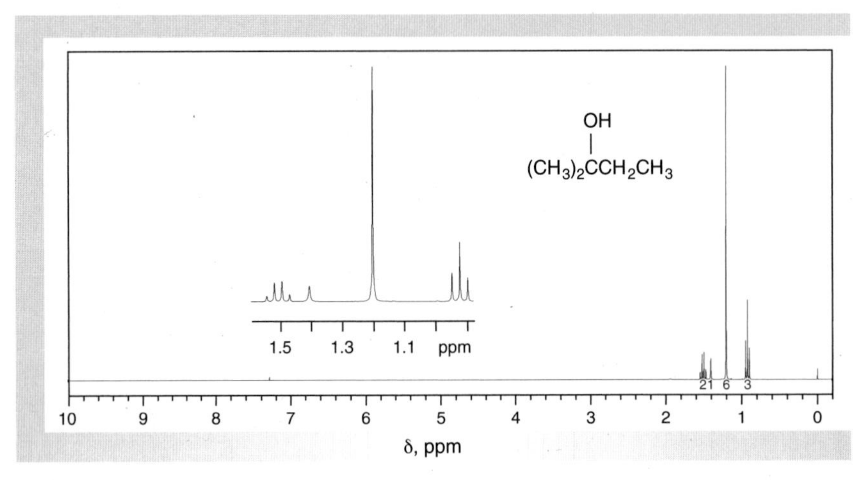

Figure 14.6
NMR data for 2-methyl-2-butanol ($CDCl_3$).

(a) ^1H NMR spectrum (300 MHz).
(b) ^{13}C NMR data: δ 8.7, 28.6, 36.5, 71.1.

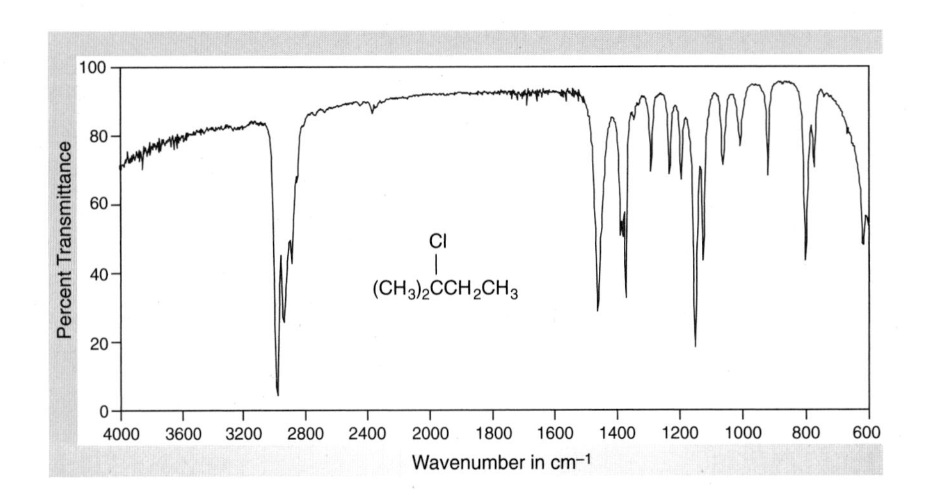

Figure 14.7
IR spectrum of 2-chloro-2-methylbutane (neat).

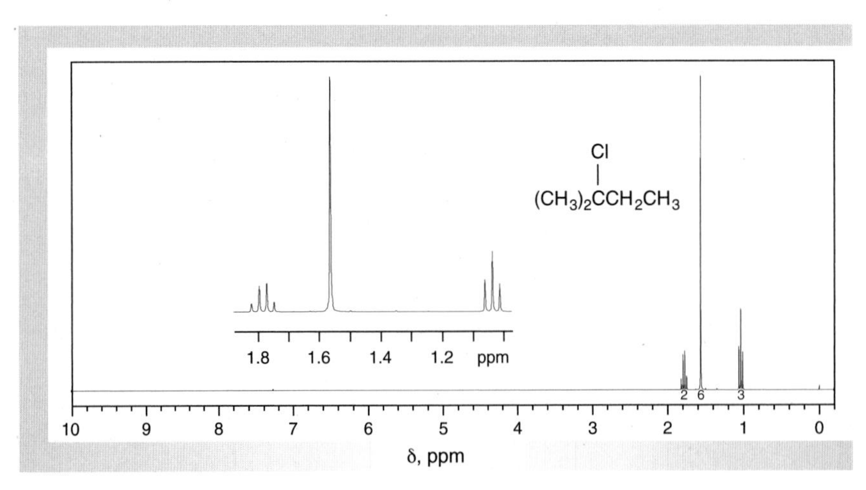

Figure 14.8
NMR data for 2-chloro-2-methylbutane ($CDCl_3$).

(a) ^1H NMR spectrum (300 MHz).
(b) ^1C NMR data: δ 9.5, 32.0, 39.0, 70.9.

14.6 CHEMICAL KINETICS: EVIDENCE FOR NUCLEOPHILIC SUBSTITUTION MECHANISMS

Introduction

Continuous changes in the concentrations of reactants and products are observed during the course of chemical reactions. Thus, as an *irreversible* reaction proceeds, we find that the concentration of each reactant decreases until that of the limiting reagent becomes zero, at which point the reaction stops. Simultaneously, the concentration of the product increases from zero to its maximum value when the reaction is complete. How fast these concentrations change *as a function of time* is determined by the **reaction rate** for a chemical transformation.

The field of **chemical kinetics** involves investigation of the interplay of factors and variables influencing the rates of reactions. Because chemical kinetics provide great insight into the nature and details of reaction mechanisms, it is hard to overstate the importance of this area of research. Indeed, kinetic studies have provided some of the most important evidence in support of the S_N1 and S_N2 mechanisms (Sec. 14.2).

To simplify our discussion of kinetics, let's first assume that the reactions are *irreversible* and then consider the mathematical expressions or **rate laws** governing these classes of reaction. To do so, we need only examine the *rate-determining step* for the particular type of substitution. In the case of an S_N1 reaction, this step is the formation of a carbocation from the precursor R–L (Eq. 14.2). The rate of the overall reaction is then proportional only to the concentration of substrate, as expressed in Equation 14.24. We see that the rate is **first order** in the concentration of R–L, expressed as $[\text{R–L}]^1$, and **zeroeth order** in that of Nu:, that is, $[\text{Nu:}]^0$, which means that the rate is *independent* of its concentration. Adding the two exponents for the concentration gives the overall order of the S_N1 reaction, which is seen to be **first order.** A simplified version of the rate law is seen in Equation 14.25, and this is the form in which it is normally written.

$$\text{Rate of reaction} = k_1[\text{R–L}]^1[\text{Nu:}]^0 \quad (14.24)$$

$$\text{Rate of reaction} = k_1[\text{R–L}] \quad (14.25)$$

For S_N2 reactions, the rate-determining step involves attack by the nucleophile, Nu:, on the substrate R–L (Eq. 14.3). The corresponding rate law is shown in Equation 14.26, wherein the rate is seen to be proportional to the concentrations of R–L *and* Nu:. Both concentrations are taken to the first power, so the rate of the reaction is said to be **second order** overall.

$$\text{Rate of reaction} = k_1[\text{R–L}][\text{Nu:}] \quad (14.26)$$

Both k_1 and k_2 (Eqs. 14.24–14.26) are **rate constants** or proportionality factors that relate the rate of reaction and the concentrations of reactants; the subscripts "1" and "2" indicate that they are for unimolecular and bimolecular processes, respectively. Rate constants may have different units: that for k_1 is $(\text{time})^{-1}$, whereas that for k_2 is $(\text{concentration})^{-1}$ $(\text{time})^{-1}$.

As compared to their S_N2 relatives, S_N1 reactions are somewhat easier to study experimentally, and calculations of their rate constants are easier. Consequently, you will be performing quantitative measurements of the kinetics of an S_N1 rather than an S_N2 reaction in the procedure of this section.

Kinetic Study of an S_N1 Reaction

Because an S_N1 reaction is a first-order reaction, the rate of the reaction is linearly dependent on the concentration of the reactant R–L (Eq. 14.24). For example, doubling the concentration doubles the rate. A graph of *rate* versus *concentration* thus yields a straight line whose slope is k_1.

When the R–L is being consumed in a first-order reaction, its concentration decreases exponentially with time. If C_0 is the *initial* concentration of the substrate at time $t = 0$ (t_0) and C is its concentration at any elapsed time t, where t is measured in a unit of time, typically seconds or minutes, their relationship is given by Equation 14.27. This equation may be rewritten as shown in Equations 14.28 and 14.29.

$$C_t = C_0 e^{-k/t} \tag{14.27}$$

$$k_1 t = \ln (C_0/C_t) \tag{14.28}$$

$$k_1 t = 2.303 \log (C_0/C_t) \tag{14.29}$$

If the initial concentration, C_0, of R–L is known and if its concentration, C_t, is measured at various time intervals, t, while the reaction is proceeding, the rate constant may be determined in the following ways.

1. The values of C_0, C_t, and t measured at each point during the reaction are substituted into Equation 14.28 or 14.29, which is then solved for k_1. This produces several values of k_1, which may then be averaged. The *correct* rate constant is not easily obtained by this method because the average value will be determined without bias, that is, without compensating for any measurements that may be incorrect due to experimental error.
2. A better method for calculating k_1 from experimental data is to plot either (C_0/C_t) or $\log (C_0/C_t)$ versus t. A straight line is then drawn so that it lies closest to the largest number of points on the graph and is called the "best fit." This line is drawn *with bias* in that it purposely gives more "weight" to the majority of points lying close to the line; those lying farther from the line are taken to be less reliable and thus are given less "weight." The slope of the line is the rate constant k_1 if natural logarithms are used (Eq. 14.28) or $k_1/2.303$ if $\log_{10}$ are used (Eq. 14.29). Alternatively, a least-squares analysis may be performed on the data using commercially available software such as *Cricket Graph*™ to obtain the slope of the line; this eliminates the human element in defining where the line should be drawn. You should consult with your instructor if this technique is to be used.

The discussion and experiments presented below illustrate methods of studying chemical kinetics and determining the effects of structure on reactivity, as exemplified by the **solvolysis** of tertiary alkyl halides. The term "solvolysis" describes a substitution reaction in which the solvent, HOS, functions as the nucleophile (Eq. 14.30). In principle, solvolyses may be performed in any nucleophilic solvent such as water (hydrolysis), alcohols (alcoholysis), and carboxylic acids (for example, acetolysis with acetic acid). However, a practical limitation in choosing a solvent is the solubility of the substrate in the solvent because the reaction mixture must be homogeneous; if it is not, surface effects at the interface of the phases will make the kinetic results difficult to interpret and probably nonreproducible as well. In the experiment described here, you will explore solvolyses in mixtures of 2-propanol and water.

$$R\text{–}X + H\text{–}OS \rightarrow R\text{–}OS + H\text{–}X \tag{14.30}$$

You will need to measure the quantities C_0, C_i, and t in order to determine k_1. The procedure we'll use to measure C_t depends on the fact that a molecule of acid, H–X, is produced for each molecule of alkyl halide that reacts (Eq. 14.30). Thus, you may monitor the progress of the reaction by determining the concentration of hydrogen ion, $[H^+]$, produced as a function of time: at any time t, $[H^+]_t = C_0 - C_I$, so that C_t is defined by Equation 14.31.

$$C_t = C_0 - [H^+]_t \tag{14.31}$$

The value of $[H^+]_t$ is determined experimentally by withdrawing an accurately measured sample, referred to as an **aliquot,** from the reaction mixture and "quenching" it in a quantity of 98% 2-propanol sufficient to prevent further solvolysis; the solvolysis does not proceed at a measurable rate when so little water is present. The elapsed time t is simply the time that the aliquot was removed minus the time t_0 that the reaction was started. The aliquot is titrated with base to yield $[H^+]$. If C_0 is known, C_t may then be calculated from Equation 14.31.

The value of C_0 can be determined in two different ways. (1) A solution of defined molarity of the alkyl halide in the solvent mixture to be used may be prepared by accurately measuring the mass of alkyl halide and the volume of solvent. (2) A more reliable and easier procedure is to allow the solvolysis to go to completion, at which time all the alkyl halide will have reacted. Accurate titration of an aliquot of this mixture is the "infinity point," the point at which $[H^+]_\infty = C_0$. Equation 14.31 may then be transformed to Equation 14.32.

$$C_t = [H^+]_\infty - [H^+] \tag{14.32}$$

It is *not* necessary to use a standardized solution of base for the titrations so long as the *same* basic solution is used for *all* the titrations you perform and the volumes of the aliquots withdrawn are *identical.* Furthermore, it is unnecessary to calculate $[H^+]$ for each aliquot, because the value is directly proportional to the *volume* of basic solution required to titrate the acid that has formed according to Equation 14.30. Using this fact, and Equation 14.32, allows Equation 14.29 to be rewritten as Equation 14.33.

$$2.303 \log \frac{(\text{mL of NaOH})_\infty}{(\text{mL of NaOH})_\infty - (\text{mL of NaOH})_t} = k_1 t \tag{14.33}$$

A variety of factors influence the rate of S_N1 reactions, and the experimental procedures you perform will allow you to investigate some of them. These include the influence of solvent and the structure of the alkyl halide undergoing solvolysis, as discussed below. Additional factors affecting reaction rates include temperature and the nature of L:, the leaving group; exploration of these is possible in the **Optional Discovery Experiments** provided.

1. **Solvent Composition.** The nature of the solvent is expected to affect the rate of an S_N1 reaction because formation of ions occurs in the rate-determining step (Eq. 14.2). Based on your understanding of the principles of solvation, you would predict that more polar solvents would accelerate the rate of the reaction. You will test this hypothesis with experiments that use various mixtures of 2-propanol and water for the solvolysis.

2. **Alkyl Group.** Tertiary alkyl halides may be used in solvolyses with confidence that the reactions are proceeding by the S_N1 mechanism. Two compounds that allow you to examine the possible effects of the structure of the alkyl group on reaction rates are 2-chloro-2-methylpropane (*tert*-butyl chloride) and 2-chloro-2-methylbutane (*tert*-pentyl chloride).

The composition of the reaction mixture formed in an S_N1 reaction affects neither the rate of the reaction nor the value of the rate constant because the products all arise from a common intermediate, namely a carbocation, that is formed in the rate-determining step (Eq. 14.2). Thus, the four products that are likely to be formed in the solvolysis of 2-chloro-2-methylpropane in aqueous 2-propanol may individually be formed at different rates because of the differing values of the rate constants, $k_a - k_d$, and of the concentrations of the reagents that react with the carbocation. Nevertheless, the overall rate of the reaction is defined by the ionization that produces the carbocation and chloride ion and has the rate constant k_1 (Scheme 14.1).

Scheme 14.1

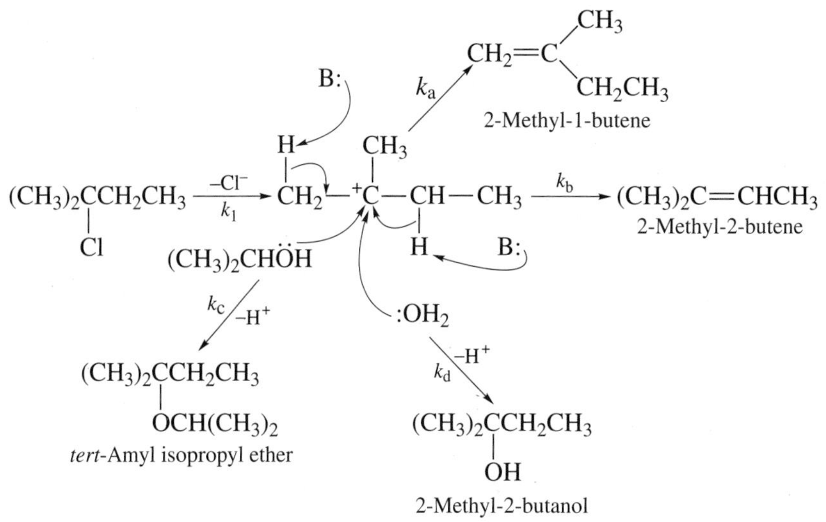

$(CH_3)_2C(Cl)CH_2CH_3 \xrightarrow[k_1]{-Cl^-} CH_2(H)\text{—}{}^+C(CH_3)\text{—}CH(H)\text{—}CH_3$

k_a (B:): $CH_2{=}C(CH_3)CH_2CH_3$ 2-Methyl-1-butene

k_b (B:): $(CH_3)_2C{=}CHCH_3$ 2-Methyl-2-butene

k_c, $(CH_3)_2CH\ddot{O}H$, $-H^+$: $(CH_3)_2C(OCH(CH_3)_2)CH_2CH_3$ *tert*-Amyl isopropyl ether

k_d, $:OH_2$, $-H^+$: $(CH_3)_2C(OH)CH_2CH_3$ 2-Methyl-2-butanol

EXPERIMENTAL PROCEDURES

Kinetics of Solvolysis of 2-Chloro-2-methylbutane

Purpose To investigate the effect of various factors on the rates of S_N1 reactions.

SAFETY ALERT

Avoid contact of the aqueous solutions with your skin, as they may dry it out and/or cause chemical burns. Wash any affected areas with copious amounts of water.

MINISCALE PROCEDURE

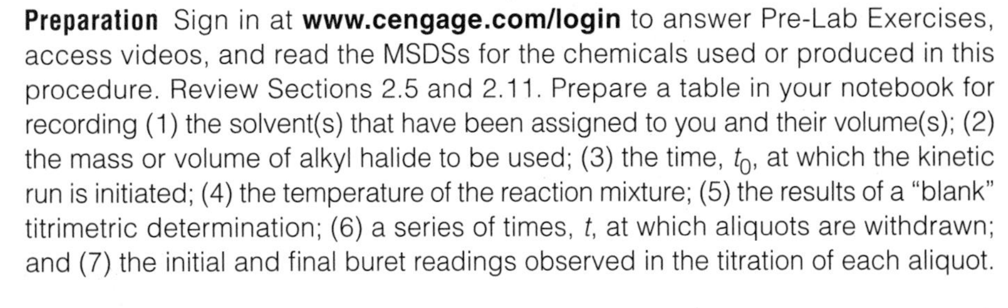

Preparation Sign in at **www.cengage.com/login** to answer Pre-Lab Exercises, access videos, and read the MSDSs for the chemicals used or produced in this procedure. Review Sections 2.5 and 2.11. Prepare a table in your notebook for recording (1) the solvent(s) that have been assigned to you and their volume(s); (2) the mass or volume of alkyl halide to be used; (3) the time, t_0, at which the kinetic run is initiated; (4) the temperature of the reaction mixture; (5) the results of a "blank" titrimetric determination; (6) a series of times, *t*, at which aliquots are withdrawn; and (7) the initial and final buret readings observed in the titration of each aliquot.

Apparatus A 10-mL and a 100-mL graduated cylinder, 50-mL buret, a 25-mL beaker, 10-mL volumetric pipet, thermometer, one 125-mL and three 250-mL Erlenmeyer flasks, a 10-mm × 75-mm test tube, disposable pipet, timer, apparatus for magnetic stirring.

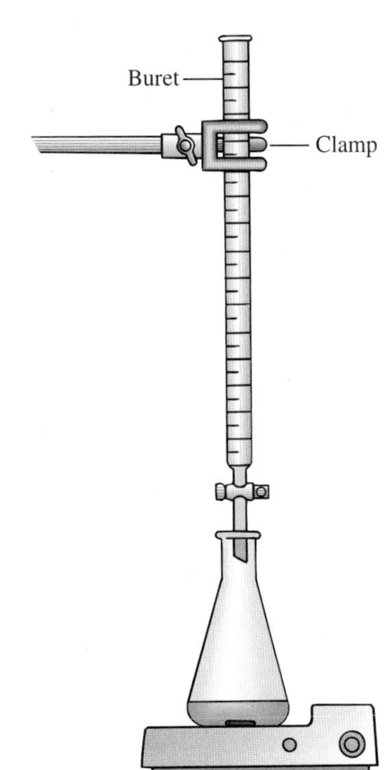

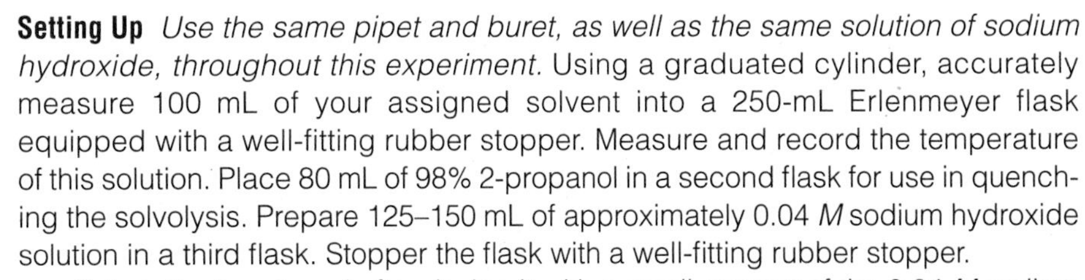

Setting Up *Use the same pipet and buret, as well as the same solution of sodium hydroxide, throughout this experiment.* Using a graduated cylinder, accurately measure 100 mL of your assigned solvent into a 250-mL Erlenmeyer flask equipped with a well-fitting rubber stopper. Measure and record the temperature of this solution. Place 80 mL of 98% 2-propanol in a second flask for use in quenching the solvolysis. Prepare 125–150 mL of approximately 0.04 *M* sodium hydroxide solution in a third flask. Stopper the flask with a well-fitting rubber stopper.

Set up the buret, and after rinsing it with a small amount of the 0.04 *M* sodium hydroxide solution, fill it with this solution, ensuring that all air bubbles are out of the tip, and cover the top with a test tube or beaker to minimize absorption of carbon dioxide from the air.

Put about 2 mL of phenolphthalein indicator solution in the test tube and have it and a dropper available for use in each titration. Connect a short length of rubber tubing to the nearest aspirator or vacuum line. This will be used to draw air through the volumetric pipet for a minute or two after each sampling in order to dry it before taking the next sample. Have a timer available that may be read to at least the nearest minute.

This experiment requires that a series of quantitative measurements be performed in a relatively short time. You should be prepared to work *rapidly* and *carefully* in order to maintain a high standard of accuracy. Buret readings should be made to the nearest 0.02 mL if possible, although precision within 0.05 mL will normally be satisfactory. Time measurements should be made *at least* to the nearest minute.

Reaction Initiate the kinetic run by adding about 1 g of the alkyl halide to the solvent mixture. The alkyl halide may either be weighed out or be measured with a 1-mL pipet. Swirl the mixture gently to obtain homogeneity. Record the time of addition as t_0. Keep the flask tightly stoppered to avoid evaporation that would change the concentration of the mixture.

While waiting to make the first measurement, determine a "blank" correction for the solvent as follows. Using a graduated cylinder, measure a 10-mL portion of your assigned solvent into the 125-mL Erlenmeyer flask. Next add 10 mL of 98% 2-propanol and 4–5 drops of phenolphthalein to the flask and titrate the stirred solution with aqueous base until a faint pink color persists for 30 sec. In this and all other titrations, use a white background below the titration flask to assist you in detecting color. You may also wish to accentuate the lower edge of the meniscus in the buret by holding dark paper or some other dark object just below it to make the graduations on the buret easier to read. The blank correction will probably be no more than 0.05–0.15 mL. Dispose of the titrated solution as specified in "Wrapping It Up." Rinse the flask with water and then with a small amount of 98% 2-propanol so it is ready for further use.

At regular intervals, use the volumetric pipet to remove a 10-mL aliquot from the reaction mixture, and quench the reaction by adding the aliquot to 10 mL of 98% 2-propanol contained in the 125-mL Erlenmeyer flask. Be sure to note the time of addition of the aliquot, probably best taken as the time at which one-half of it has been added. Titrate the solution with base to the phenolphthalein endpoint just as you did in the blank determination.

The suggested *approximate* times for taking aliquots under various conditions are as follows:

1. 50% 2-propanol/water and 2-chloro-2-methylpropane: 10, 20, 35, 50, 75, and 100 min.
2. 55% 2-propanol/water and 2-chloro-2-methylpropane: 15, 30, 50, 75, 100, and 135 min.
3. 60% 2-propanol/water and 2-chloro-2-methylpropane: 20, 40, 70, 100, 130, and 170 min.
4. 50% 2-propanol/water and 2-chloro-2-methylbutane: 10, 20, 30, 40, 50, and 60 min.
5. 55% 2-propanol/water and 2-chloro-2-methylbutane: 15, 30, 45, 60, 80, 110, and 140 min.
6. 60% 2-propanol/water and 2-chloro-2-methylbutane: 20, 40, 60, 80, 100, and 120 min.

At room temperature, the *fastest* of these solvolyses requires about 4 h to reach 99.5% completion, and the slowest requires over 12 h. Therefore it is easiest to wait until the next laboratory period to perform the infinity titration that is necessary to obtain C_0. Stopper the reaction flask *tightly* to avoid evaporation and store it in your desk. *Be sure to save at least 30 mL of the sodium hydroxide solution in a tightly stoppered flask so that it will be available for the infinity titration.*

Analysis Obtain IR and NMR spectra of the starting materials and compare them with those of authentic samples (Figs. 14.7–14.10). Compute the desired rate constants, k_1, by the following sequence of steps.

1. Using the buret readings, determine by difference the number of milliliters of sodium hydroxide solution used in each titration. Apply the blank correction to all values by subtracting it from each volume and use these corrected values in your calculations. Determine the elapsed time at which each aliquot was withdrawn from the reaction mixture. Apply Equation 14.33 by calculating the log term for each kinetic point, multiplying that value by 2.303, and *plotting* the resulting number (ordinate) vs. elapsed time t (abscissa) in hours. Draw the best straight line through the points. Determine the slope of the line; this slope is the rate constant k_1. Alternatively, analyze your experimental data using least-squares techniques to obtain the rate constant.
2. Using the same data and Equation 14.31, calculate the value of k_1 separately for each kinetic point. Compare the *average* of these values with the rate constant obtained graphically. Also compare this average with each of the values that were averaged. In your laboratory report for this experiment, specify

which procedure, graphical or averaging, allows you most easily to identify a point that is likely in error.

3. Equation 14.34 provides the half-life, $t_{1/2}$, the time necessary for one-half of the original alkyl halide to react. Calculate the half-life of your reaction using the value of k_1 obtained from the graph. Then reexamine your experimental data to determine whether about one-half of the total volume of NaOH used in the infinity titration had been consumed by the end of the first half-life of the reaction. If not, an error has been made in the calculations, and they should be rechecked.

$$t_{1/2} = 0.693/k_1 \quad (14.34)$$

Discovery Experiment

Effect of Temperature on Rates of Solvolysis

Temperature changes affect the rates of reactions, the *rough* rule of thumb being that the rate doubles for each rise of 10 °C. You may explore this effect by comparing solvolysis rates at room temperature and at 0 °C. Based on the experimental procedure of this section, design and execute a protocol for measuring the rate of solvolysis of either 2-chloro-2-methylpropane or 2-chloro-2-methylbutane at 0 °C. You should consider the effect of solvent polarity on the rate of the reaction so that the conditions you propose to use will allow measurable amounts of solvolysis to occur during the time period you have available for removing aliquots for titration. Consult with your instructor before performing your proposed procedure.

Discovery Experiment

Effect of Leaving Group on Rates of Solvolysis

Because the bond between the carbon atom and the leaving group, L, in R–L is broken in the rate-determining step of an S_N reaction, you might expect that the rate of the reaction would be dependent on the nature of L. You may investigate this possibility by comparing the rates of solvolysis of alkyl bromides with those of alkyl chlorides. Based on the experimental procedure of this section, design and execute a protocol for measuring the rate of solvolysis of 2-bromo-2-methylpropane at room temperature. Consult with your instructor before performing your proposed procedure.

WRAPPING IT UP

Although you normally should not dispose of solutions containing organic halides by pouring them down the drain, it is acceptable to do so in this case because of the small quantities of the alkyl halides being used and their facile solvolysis. Make the solvolysis solution strongly basic and allow it to remain in your desk until the next laboratory period. Neutralize the basic solution and flush it down the drain using copious amounts of water.

EXERCISES

1. Why does the titration endpoint color fade after 30–60 sec?
2. You covered the top of the buret with a test tube or beaker to protect its contents from air. Why was a rubber or cork stopper not used instead?

3. In the kinetics experiment you performed, individual aliquots were quenched by adding them to 98% 2-propanol. Why does this stop the solvolysis?

4. Suppose the reaction flask was left unstoppered until the following laboratory period when the infinity titration was performed. Would the calculated rate constant have been larger or smaller than the "correct" value if some evaporation of solvent had occurred during this time?

5. Derive Equation 14.33 from Equation 14.29. (*Hint:* Set up the concentrations C_0 and C_t in terms of moles/liter for the volume of NaOH used in each titration, and recognize the fact that $[OH^-] = [H^+]$. Substitute these values into Equation 14.29, and cancel out constants that appear in the new equation.)

6. Give the possible advantages for using an infinity titration and Equation 14.33 for analyzing your data instead of calculating concentrations of the alkyl halide and using Equation 14.29 for the analysis. Be sure to consider the alternatives for obtaining C_0.

7. Equation 14.34 gives the *half-life*, $t_{1/2}$, of a reaction, which is defined as the time required for a reaction to reach 50% completion. This equation applies to any first-order reaction, including radioactive disintegration. Show how Equation 14.34 may be derived from Equation 14.29. (*Hint:* When the reaction is 50% complete, $C_t = 1/2\ C_0$.)

8. List the possible errors involved in the determination of rate constants by the procedure you used and state the relative importance of each.

9. Refer to Scheme 14.1 and predict the ratio of 2-methyl-2-butanol:*tert*-amyl isopropyl ether expected if the solvent for the reaction were equimolar in water and 2-propanol and $k_c = 5k_d$.

Sign in at **www.cengage.com/login** and use the spectra viewer and Tables 8.1–8.8 as needed to answer the blue-numbered questions on spectroscopy.

10. Consider the NMR spectral data for 2-chloro-2-methylbutane (Fig. 14.8).
    a. In the ^1H NMR spectrum, assign the various resonances to the hydrogen nuclei responsible for them.
    b. For the ^{13}C NMR data, assign the various resonances to the carbon nuclei responsible for them.

11. Consider the NMR spectral data for 2-chloro-2-methylpropane (Fig. 14.10).
    a. In the ^1H NMR spectrum, assign the resonance to the hydrogen nuclei responsible for it.
    b. For the ^{13}C NMR data, assign the resonance to the carbon nuclei responsible for it.

12. Consider the spectral data for 2-methyl-2-butanol (Figs. 14.5 and 14.6).
    a. In the functional group region of the IR spectrum, identify the absorption associated with the hydroxyl group, and explain why this peak is broad.
    b. In the ^1H NMR spectrum, assign the various resonances to the hydrogen nuclei responsible for them.
    c. If the NMR sample of 2-methyl-2-butanol in $CDCl_3$ were shaken with D_2O *prior* to taking the ^1H NMR spectrum, what differences might you expect to observe?
    d. For the ^{13}C NMR data, assign the various resonances to the carbon nuclei responsible for them.

13. Consider the spectral data for 2-methyl-1-butene (Figs. 10.4 and 10.5).

  a. In the functional group region of the IR spectrum, specify the absorptions associated with the carbon-carbon double bond and the hydrogens attached to the C1 carbon atom.

  b. In the ^1H NMR spectrum, assign the various resonances to the hydrogen nuclei responsible for them.

  c. For the ^{13}C NMR data, assign the various resonances to the carbon nuclei responsible for them.

14. Consider the spectral data for 2-methyl-2-methylbutene (Figs. 10.6 and 10.7).

  a. In the functional group region of the IR spectrum, specify the absorptions associated with the carbon-carbon double bond and the hydrogens attached to the C3 carbon atom.

  b. In the ^1H NMR spectrum, assign the various resonances to the hydrogen nuclei responsible for them.

  c. For the ^{13}C NMR data, assign the various resonances to the carbon nuclei responsible for them.

SPECTRA

Starting Materials and Products

The IR spectrum and NMR spectral data for 2-chloro-2-methylbutane, 2-methyl-2-butanol, 2-methyl-1-butene, and 2-methyl-2-methylbutene are provided in Figures 14.7, 14.8, 14.5, 14.6, and 10.4–10.7, respectively.

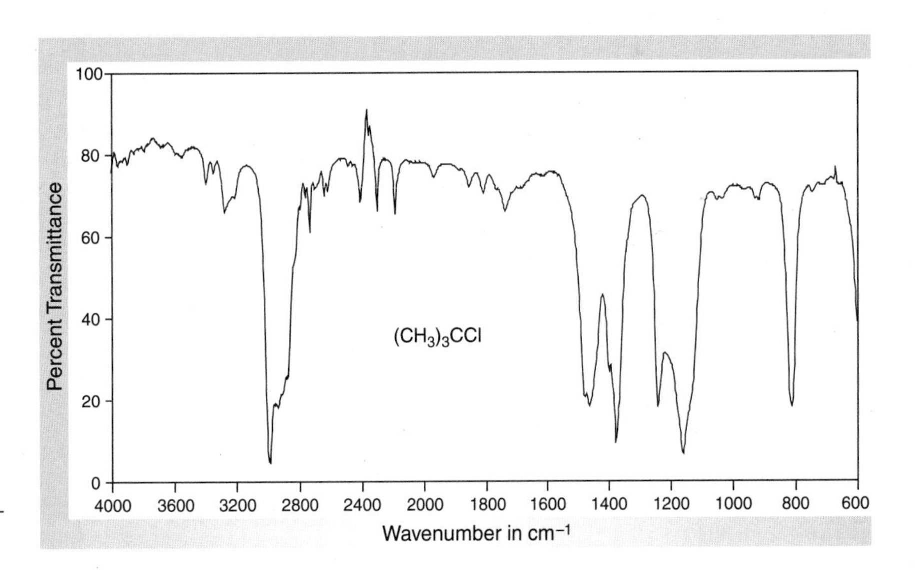

Figure 14.9
IR spectrum of 2-chloro-2-methylpropane (neat).

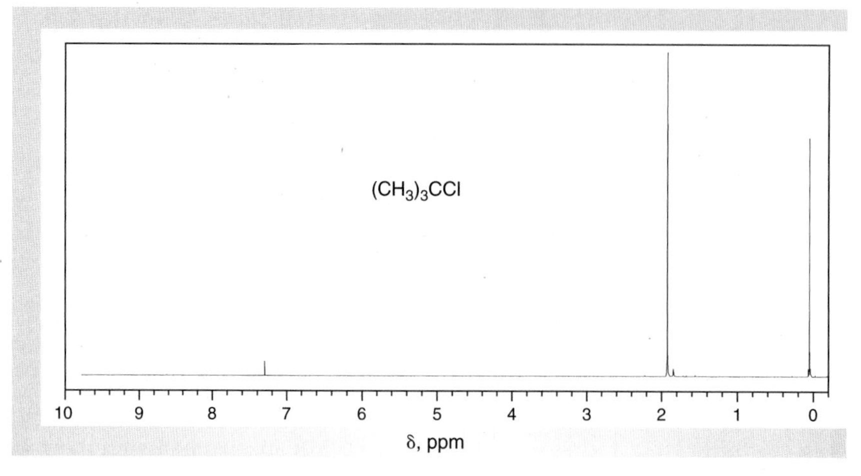

Figure 14.10
NMR data for 2-chloro-2-methylpropane ($CDCl_3$).

(a) ^1H NMR spectrum (300 MHz).
(b) ^{13}C NMR data: δ 34.5, 66.5.

CHAPTER 15

Electrophilic Aromatic Substitution

When you see this icon, sign in at this book's premium website at **www.cengage.com/login** to access videos, Pre-Lab Exercises, and other online resources.

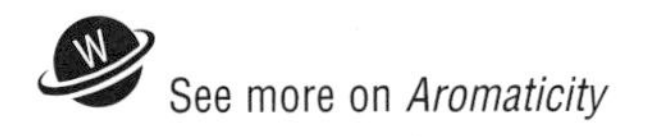

See more on *Aromaticity*

Something chemically remarkable happens when three conjugated double bonds are incorporated into a six-membered ring to produce a so-called **aromatic compound.** In sharp contrast to the addition reactions that typify alkenes (Secs. 10.4 and 17.2 and Chap. 12) and alkynes (Sec. 11.3), the chemistry of aromatic molecules is characterized by substitutions. Namely, aromatic compounds undergo chemical transformations in which a proton on the ring is replaced by some other atom or functional group. We'll explore these fascinating and useful reactions in this chapter. You will have the opportunity not only to execute reactions in which functional groups are introduced onto the aromatic ring but also to study both qualitatively and quantitatively the effects that substituents on the ring have on the rates at which these substitutions occur. These experiments will help you understand the extraordinary chemistry of aromatic compounds.

15.1 INTRODUCTION

Electrophilic aromatic substitution is an important reaction that allows the introduction of many different functional groups onto an aromatic ring. A general form of the reaction is given by Equation 15.1, where Ar–H is an aromatic compound, an **arene,** and E^+ represents an **electrophile** that replaces an H on the ring. This equation is oversimplified because the electrophile is usually generated during the reaction, and a Lewis base assists in the removal of H^+.

See *Who was Lewis?*

$$\underset{\text{An arene}}{\text{Ar—H}} + \underset{\text{An electrophile}}{E^+} \longrightarrow \underset{\text{A substituted arene}}{\text{Ar—E}} + H^+ \quad (15.1)$$

The rates of electrophilic aromatic substitutions are generally *second-order* overall, being *first-order* in both the aromatic component *and* the electrophile (Eq. 15.2). A general mechanism that is consistent with this kinetic expression is depicted in Equations 15.3–15.5.

$$\text{Rate} = k_2\,[\text{Ar—H}][E^+] \quad (15.2)$$

Step 1: Formation of electrophile:

$$\text{E—Nu} \underset{}{\overset{\text{catalyst}}{\rightleftharpoons}} \text{E}^+ + \text{Nu}^- \quad (15.3)$$

Step 2: Reaction with electrophile:

$$\text{Ar—H} + \text{E}^+ \overset{\text{rds}}{\rightleftharpoons} \left[\overset{+}{\text{Ar}} \begin{matrix} \text{H} \\ \text{E} \end{matrix} \right] \quad (15.4)$$

Step 3: Loss of proton to give product:

$$\left[\overset{+}{\text{Ar}} \begin{matrix} \text{H} \\ \text{E} \end{matrix} \right] \overset{\text{fast}}{\rightleftharpoons} \text{Ar—E} + \text{H}^+ \quad (15.5)$$

The electrophile is usually produced by the reaction between a catalyst and a compound containing a potential electrophile (Eq. 15.3). The second-order nature of the reaction arises from the step shown in Equation 15.4 in which one molecule each of arene and electrophile react to give a cationic intermediate. The formation of this cation is the **rate-determining step (rds)** in the overall reaction; the subsequent deprotonation of the cation (Eq. 15.5) is fast. The bimolecular nature of the transition state for the rate-limiting step and the fact that an electrophile is involved in attacking the aromatic substrate classifies the reaction as **S_E2** (**S**ubstitution **E**lectrophilic **Bi**molecular). Experiments involving four different such reactions are given in this chapter: Friedel-Crafts alkylation and acylation, nitration, and bromination.

15.2 FRIEDEL-CRAFTS ALKYLATION OF *p*-XYLENE WITH 1-BROMOPROPANE

See *Who were Friedel and Crafts?*

The **Friedel-Crafts alkylation reaction** using alkyl halides (Eq. 15.6) is a classic example of electrophilic aromatic substitution (see the Historical Highlight at the end of this chapter). As a versatile method for directly attaching alkyl groups to aromatic rings, it is a process of great industrial importance. The three main limitations to this reaction as a synthetic tool are: (1) the substitution fails if the aromatic ring carries strongly **ring-deactivating groups,** such as NO_2, R_3N^+, C(O)R, and CN; (2) more than one alkyl group may be introduced onto the aromatic ring, a process termed polyalkylation; and (3) mixtures of isomeric products may be formed if the alkyl group rearranges. The first restriction is associated with the fact that aromatic rings bearing electron-withdrawing substituents are not sufficiently nucleophilic, or Lewis basic, to react with the electrophiles, or Lewis acids, generated under the reaction conditions. The second problem arises because addition of the first alkyl group activates the ring toward further substitution. This side reaction may be minimized by using a large excess of the arene, as is done in our experiment. The final limitation results from structural rearrangements within the alkyl group, R.

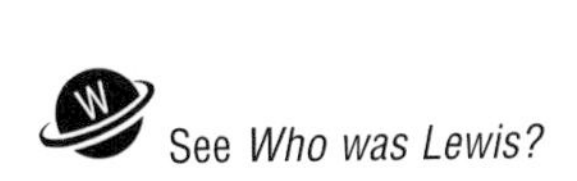

See *Who was Lewis?*

$$\underset{\text{An arene}}{\text{Ar—H}} + \underset{\substack{\text{An alkyl halide} \\ (\text{X = halogen})}}{\text{R—X}} \overset{\text{catalyst}}{\rightleftharpoons} \underset{\text{An alkylarene}}{\text{Ar—R}} + \text{HX} \quad (15.6)$$

A generally accepted mechanism for Friedel-Crafts alkylation using alkyl halides is illustrated in Equations 15.7–15.9. In the first step, the active electrophile, written as R^+ for the present purpose, is produced from an alkyl halide by reaction with a Lewis acid such as $AlCl_3$ (Eq. 15.7). The rate-determining step involves attack of the arene, which functions as a Lewis base, on the electrophile to yield a positively charged, resonance-stabilized **σ-complex** (Eq. 15.8). Finally, a base such as a halide ion from $XAlCl_3^-$ deprotonates this complex to give HX, re-forming the aromatic system and regenerating the catalyst (Eq. 15.9).

$$R{-}X + AlCl_3 \rightleftharpoons R^+ + X\bar{A}lCl_3 \quad (15.7)$$

$$R^+ + C_6H_6 \overset{\text{rds}}{\rightleftharpoons} [\sigma\text{-complex resonance structures}] \quad (15.8)$$

σ-Complex

$$Cl_3\bar{A}l{-}X + \sigma\text{-complex} \xrightarrow{\text{fast}} C_6H_5R + HX + AlCl_3 \quad (15.9)$$

Representing the electrophile as a carbocation, as in Equation 15.7, is reasonable when *secondary* and *tertiary* alkyl halides are used. However, this is probably *not* the case with the electrophile derived from a *primary* alkyl halide, where the reactive species is better written as a *polarized complex* such as $R{-}X^+{-}\bar{A}lCl_3$.

Molecular rearrangements involving hydride, H^-, and alkyl, R^-, shifts become possible when a carbocation or a polarized cation-like complex is an intermediate in a reaction. Thus, the aluminum chloride-promoted reaction of 1-bromopropane with benzene gives *n*-propylbenzene (**1**) and isopropylbenzene (**2**) in a ratio of about 1:2 (Eq. 15.10). As shown in Scheme 15.1, the formation of **2** is explained by a 1,2-hydride rearrangement of the *primary carbocation-like* complex **3** to produce the *secondary carbocation* **4.** The formation of comparable amounts of **1** and **2** in the reaction means that the rate of addition of **3** to benzene followed by deprotonation to give **1** must be competitive with that for isomerization of **3** to the secondary ion **4,** which is the precursor of **2.** In mathematical terms, k_2[**3**][C_6H_6] must be similar to k_1[**3**], since these expressions measure the rates of formation of **1** and **2,** respectively.

$$C_6H_6 + CH_3CH_2CH_2Br \xrightarrow{AlCl_3} C_6H_5CH_2CH_2CH_3 + C_6H_5CH(CH_3)_2 \quad (15.10)$$

Benzene; 1-Bromopropane; **1** *n*-Propylbenzene; **2** Isopropylbenzene

Scheme 15.1

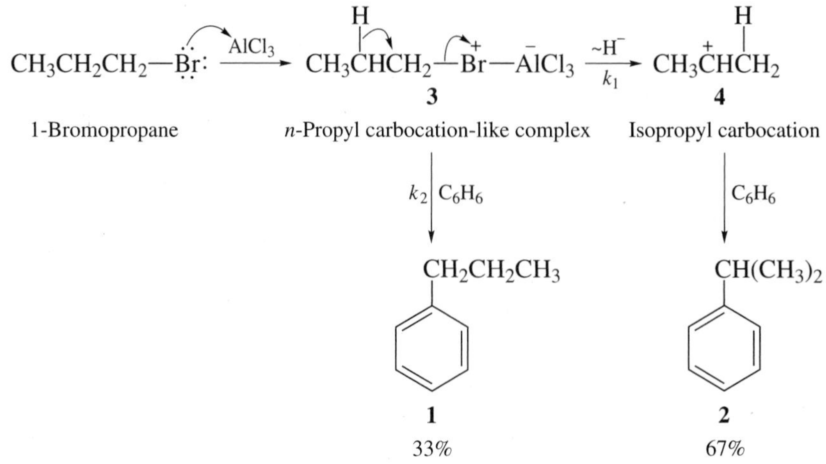

$$CH_3CH_2CH_2{-}\ddot{B}r: \xrightarrow{AlCl_3} CH_3\overset{H}{C}HCH_2{-}\overset{+}{B}r{-}\overset{-}{A}lCl_3 \ (\mathbf{3}) \xrightarrow[k_1]{\sim H^-} CH_3\overset{+}{C}H\overset{H}{C}H_2 \ (\mathbf{4})$$

1-Bromopropane; *n*-Propyl carbocation-like complex (3); Isopropyl carbocation (4)

3 → (k_2, C_6H_6) → $C_6H_5CH_2CH_2CH_3$ (1), 33%

4 → (C_6H_6) → $C_6H_5CH(CH_3)_2$ (2), 67%

The present experiment demonstrates Friedel-Crafts alkylation and allows you to examine the following hypothesis: The amount of alkyl rearrangement that occurs during a Friedel-Crafts alkylation of a series of arenes depends on the *nucleophilicity* of the arene, Ar–R. This hypothesis is derived as follows. The rate constant, k_1, is for a *unimolecular* process of **3** that does *not* involve the arene (Scheme 15.1), so its value is *independent* of the nucleophilicity of the arene. In contrast, k_2 is the rate constant for a *bimolecular* reaction between **3** *and* the arene, so its value is *dependent* on the nucleophilicity of the arene. Substituents, R, that make the arene more nucleophilic yield higher values of the bimolecular rate constant k_2, thereby increasing the overall rate of the electrophilic substitution. Consequently, formation of the product derived from rearrangement becomes a relatively less important reaction pathway.

The importance of the nucleophilicity of the arene in determining the ratio of unrearranged products to rearranged products may be studied by comparing the Friedel-Crafts reactions of benzene and *p*-xylene with 1-bromo-propane (Eq. 15.11). Because methyl groups activate the ring toward the S_E2 reaction and make k_2 (*p*-xylene) > k_2 (benzene), we expect from our hypothesis that the ratio of unrearranged product **5** to its rearranged isomer **6** is *greater* than that of 1:2. The experiment you will perform allows this prediction to be tested.

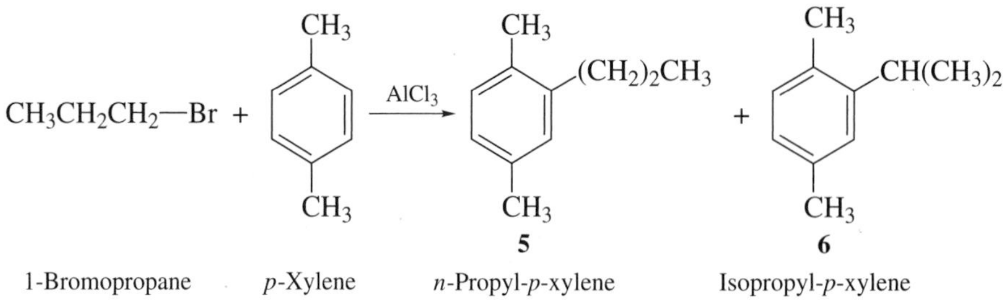

$$CH_3CH_2CH_2{-}Br + p\text{-}(CH_3)_2C_6H_4 \xrightarrow{AlCl_3} \underset{\mathbf{5}}{2,5\text{-}(CH_3)_2C_6H_3(CH_2)_2CH_3} + \underset{\mathbf{6}}{2,5\text{-}(CH_3)_2C_6H_3CH(CH_3)_2} \quad (15.11)$$

1-Bromopropane; *p*-Xylene; *n*-Propyl-*p*-xylene (5); Isopropyl-*p*-xylene (6)

Friedel-Crafts Alkylation of p*-Xylene*

Purpose To demonstrate alkylation by electrophilic aromatic substitution and to assess the relationship between the extent of carbocation rearrangement and arene nucleophilicity.

SAFETY ALERT

1. **Anhydrous aluminum chloride is *extremely* hygroscopic and reacts rapidly with water, even the moisture on your hands, producing highly corrosive fumes of hydrogen chloride. *Do not allow aluminum chloride to come in contact with your skin.* If it does, flush the affected area with copious amounts of water. *Minimize exposure of this chemical to the atmosphere!***
2. **You may be provided with a weighed amount of aluminum chloride. If not, *quickly* weigh it into a dry vial in the hood and then transfer it to the reaction flask. The success of this experiment is highly dependent on the quality of the aluminum chloride that is used, so obtain it from a *freshly* opened bottle. In handling this chemical, be aware that it is a powdery solid that easily becomes airborne.**
3. ***p*-Xylene is flammable. Assemble the apparatus carefully and be sure that all joints are tightly mated. Have your instructor inspect your set-up before you begin the distillation and use *flameless* heating.**

MINISCALE PROCEDURE

Preparation Sign in at **www.cengage.com/login** to answer Pre-Lab Exercises, access videos, and read the MSDSs for the chemicals used or produced in this procedure. Review Sections 2.9, 2.10, 2.11, 2.13, 2.14, 2.23, 2.27, 4.4, and 6.4.

Apparatus A 25- and a 50-mL round-bottom flask, drying tube, Claisen adapter, separatory funnel, gas trap, apparatus for simple distillation, fractional distillation, magnetic stirring, and *flameless* heating.

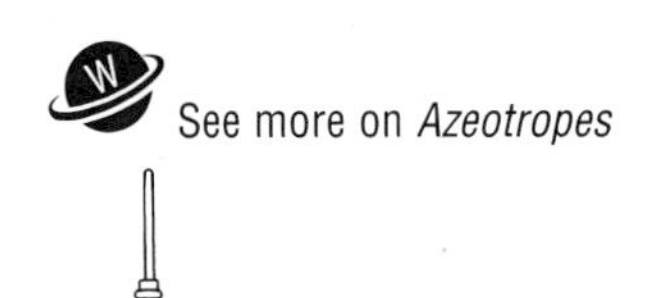

Setting Up Equip the 50-mL round-bottom flask for simple distillation and add a stirbar and 15 mL of *p*-xylene. Dry the *p*-xylene through azeotropic distillation by distilling it into a receiver until the distillate is not cloudy; this should occur after no more than 3 mL of distillate has been obtained.

Allow the residual *p*-xylene in the stillpot to cool to room temperature, and equip the flask with a Claisen adapter, a water-cooled condenser, the gas trap specified by your instructor for removing fumes of hydrogen chloride, and a separatory funnel (Fig. 2.66b); fit the funnel with a drying tube rather than a stopper. *Briefly* remove the condenser from the reaction flask, *quickly* pour 0.7 g of *anhydrous* powdered aluminum chloride into the flask, and *immediately* reassemble the apparatus.

Using a graduated cylinder, measure 8.5 mL of 1-bromopropane and pour it into the funnel (stopcock *closed!*). Prepare an ice-water bath in case it is needed to cool the reaction mixture.

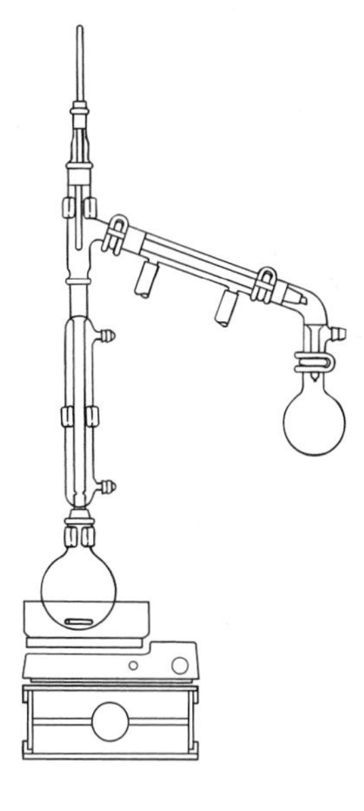

Addition and Work-Up Add the 1-bromopropane dropwise to the stirred mixture of *p*-xylene and aluminum chloride. If the evolution of gas becomes too vigorous, reduce the rate of addition and raise the cooling bath to lower the temperature of the reaction mixture. The addition should take about 10 min. After completing the addition, continue stirring the reaction mixture for about 0.5 h. *Working at the hood* because of the hydrogen chloride that will be evolved, pour the mixture into a beaker containing about 10 g of crushed ice. Stir the mixture until all the ice has melted, pour it into a separatory funnel, and remove the aqueous layer. Pour the organic layer into an Erlenmeyer flask containing anhydrous sodium sulfate and swirl it for about 1 min; if the solution appears cloudy, add additional sodium sulfate with swirling.★

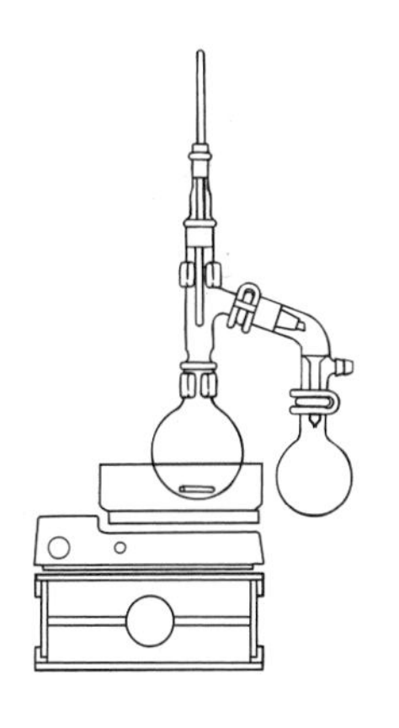

Isolation Decant the dried solution into the 25-mL round-bottom flask containing a stirbar and equip the flask for fractional distillation using an *air-cooled* condenser. Make sure that all connections are tight and then distill. Collect the excess *p*-xylene and forerun in a tared, labeled receiver until the head temperature reaches 180 °C (760 torr). Discontinue heating and allow any liquid in the column to drain into the stillpot.★ Refit the stillpot for shortpath distillation and resume the distillation, collecting any additional distillate boiling below 180 °C (760 torr) in the original receiver. Use a second tared, labeled receiver to collect the fraction boiling between 180 and 207 °C (760 torr). Isomers **5** and **6** have bp 204 °C and 196 °C (760 torr), respectively.

Analysis Weigh the two fractions and analyze each by GLC; a typical trace of the second fraction is provided in Figure 15.1. Use the GLC analysis to calculate the yield of propylxylenes obtained and their proportion to each other; assume that *p*-xylene and the propylxylenes have the same density and GLC response factor. Obtain IR and ^{1}H NMR spectra of your starting materials and products and compare them with those of authentic samples (Figs. 15.4–15.11).

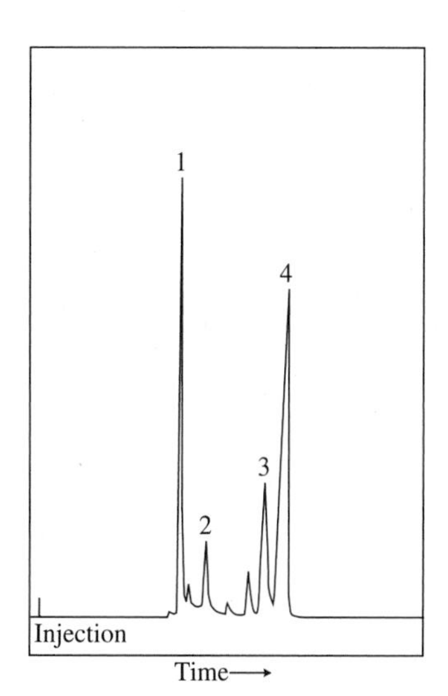

Figure 15.1
GLC analysis of reaction mixture from miniscale alkylation of p-xylene *with 1-bromopropane. Numbers in parentheses are retention times in minutes and peak areas, respectively. Peak 1: solvent; peak 2:* p-xylene *(1.7, 14417); peak 3: isopropyl-p-xylene (1.8, 71721); peak 4:* n-propyl-p-xylene *(2.0, 229275). Column and conditions: 0.25-mm × 25-m AT-1 with 100% dimethylsiloxane as stationary phase; flowrate 1.2 mL/min; temperature 185 °C.*

MICROSCALE PROCEDURE

Preparation Sign in at **www.cengage.com/login** to answer Pre-Lab Exercises, access videos, and read the MSDSs for the chemicals used or produced in this procedure. Review Sections 2.9, 2.11, 2.13, 2.17, 2.23, 4.4, and 6.4.

See more on *Azeotropes*

Apparatus A 3- and a 5-mL conical vial, 1-mL syringe, Claisen adapter, gas trap, apparatus for distillation, magnetic stirring, and *flameless* heating.

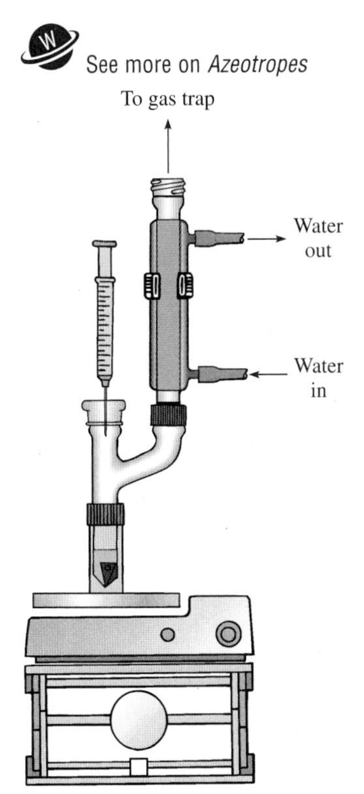

Figure 15.2
Microscale apparatus for alkylation of p-*xylene.*

Setting Up Equip the 5-mL vial for simple distillation and add a spinvane and 2 mL of *p*-xylene. Dry the *p*-xylene through azeotropic distillation by distilling it until the distillate is not cloudy; this should occur after no more than 0.5 mL of distillate has been obtained.

Allow the residual *p*-xylene in the vial to cool to room temperature. Replace the condenser on the vial with a Claisen adapter fitted with a rubber septum on the opening directly above the vial; on the other arm of the adapter, put a water-cooled condenser equipped with the gas trap specified by your instructor for removing fumes of hydrogen chloride (Fig. 15.2). *Briefly* remove the vial from the adapter, *quickly* pour 0.1 g of *anhydrous* powdered aluminum chloride into it, and *immediately* reassemble the apparatus. Draw 0.8 mL of 1-bromopropane into the syringe and pierce the rubber septum with its needle.

Addition and Work-Up Add the 1-bromopropane dropwise to the stirred mixture of *p*-xylene and aluminum chloride over a period of 5 to 10 min. If the evolution of gas becomes too vigorous, reduce the rate of addition. After completing the addition, continue stirring the reaction mixture for about 20 min. *Working at the hood* because of the hydrogen chloride that will be evolved, pour the mixture into a beaker containing about 1 g of crushed ice. Stir the mixture until all the ice has melted, pour it into a centrifuge tube, and remove the aqueous layer. Add a microspatula-tip full of anhydrous sodium sulfate to the organic layer and swirl it for about 1 min; add additional sodium sulfate with swirling if the solution appears cloudy.★

Isolation Using a filter-tip pipet, transfer the dried solution into the 3-mL conical vial containing a spinvane and equip the vial for distillation, with internal monitoring of the temperature. Make sure that all connections are tight and then distill. Collect the excess *p*-xylene and forerun until the bath temperature reaches 180 °C (760 torr). Discontinue heating and allow the apparatus to cool for a few minutes.★ Transfer the distillate to a tared, labeled screw-cap vial and resume the distillation, collecting as one fraction all volatiles that distill with a bath temperature of up to 240 °C. Isomers **5** and **6** have bp 204 °C and 196 °C (760 torr), respectively.

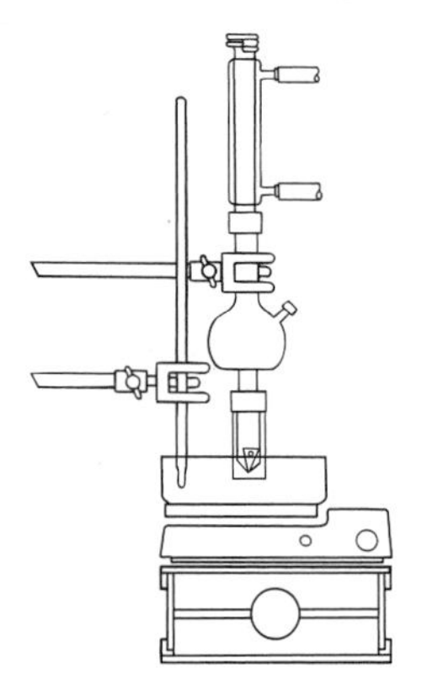

Analysis Weigh the two fractions and analyze each by GLC; a typical trace of the second fraction is provided in Figure 15.3. Obtain IR and ^{1}H NMR spectra of your starting materials and products and compare them with those of authentic samples (Figs. 15.4–15.11).

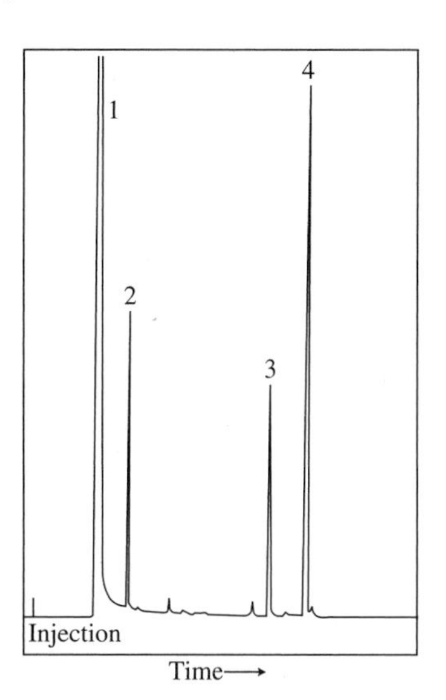

Figure 15.3
GLC trace of reaction mixture from microscale alkylation of p-xylene *with 1-bromopropane (numbers in parentheses are retention times in minutes and peak areas, respectively). Peak 1: solvent; peak 2:* p-xylene *(1.8, 74565); peak 3: isopropyl-*p-xylene *(4.3, 92699); peak 4:* n-propyl-p-xylene *(5.0, 236258). Column and conditions: 0.25-mm × 25-m AT-1 column with 100% dimethylpolysiloxane as stationary phase; flowrate 1.2 mL/min; temperature 110 °C.*

WRAPPING IT UP

Neutralize the *aqueous layer* from hydrolysis of the reaction mixture with sodium carbonate and flush it down the drain. Spread the *sodium sulfate* used for drying on a tray in the hood to allow evaporation of the volatiles; then put spent *drying agent* in the container for nontoxic solids. Place the *liquid collected* as the *low-boiling fraction* and the *pot residue* from the distillation in the container for nonhalogenated organic liquids.

EXERCISES

1. What purpose is served by distilling a small quantity of *p*-xylene prior to combining the reagents in this experiment?
2. What compound(s) might be present in the higher-boiling residue remaining in the stillpot after you collect the isomeric propylxylenes?
3. Calculate the molar ratio of 1-bromopropane to *p*-xylene used in the experiment. What would be the effect on the experimental results of using a ratio twice that actually used? Consider this in connection with your answer to Exercise 2.
4. Use either your own GLC analysis or that given in Figure 15.1 or 15.3 to calculate the yield of propylxylenes obtained and their proportion to each other; assume that *p*-xylene and the propylxylenes have the same densities and GLC response factor. Discuss whether these calculations support the hypothesis regarding the relative amounts of unrearranged product versus rearranged product to be expected from *p*-xylene as compared to benzene.
5. Account for the higher boiling point of **5** as compared to that of **6**. Are the boiling points consistent with the relative retention times of the two isomers as seen in the GLC trace of Figure 15.1 or 15.3? Explain.
6. Why do you think it is recommended that the distillation fraction containing the mixture of **5** and **6** be taken over such a wide temperature range?
7. Consider aluminum chloride, $AlCl_3$, the catalyst used in this procedure.
   a. Provide a Lewis dot structure for $AlCl_3$.
   b. Define the term *Lewis acid*.

c. What aspect of the Lewis structure is consistent with this compound being a Lewis acid?

d. How does addition of $AlCl_3$ promote formation of carbocations from alkyl halides?

e. Provide an example of a Lewis acid that does not contain aluminum.

8. Using curved arrows to symbolize the flow of electrons, write the stepwise mechanism for the reaction between *p*-xylene and *n*-propyl cation. Show all resonance structures where applicable.

9. On the assumption that relative cost is not an overriding consideration, give a reason why *p*-xylene was used in preference to *o*-xylene as the arene in this experiment.

10. Outline a synthesis of *p*-xylene starting from benzene. Assume that the isomeric xylenes that might be formed could be separated from one another through a purification procedure.

11. Nitrobenzene is sometimes used as the solvent for Friedel-Crafts alkylation of other arenes. Why is this possible?

12. Alkylation of toluene, $C_6H_5CH_3$, with 1-bromopropane gives a mixture of four isomeric propyltoluenes. Write formulas for these compounds, and predict the relative amounts in which they would be formed, based on the results of the alkylation of *p*-xylene.

13. Consider the Friedel-Crafts alkylation of toluene and anisole, $C_6H_5OCH_3$, with 1-chloropropane/$AlCl_3$. Given your understanding of substituent effects on S_E2 reactions, how would you expect the ratio of *n*-propyl alkylation to isopropyl alkylation to differ for these two arenes? Explain your answer.

Sign in at **www.cengage.com/login** and use the spectra viewer and Tables 8.1–8.8 as needed to answer the blue-numbered questions on spectroscopy.

14. Consider the spectral data for *p*-xylene (Figs. 15.4 and 15.5).

a. In the functional group region of the IR spectrum, specify the absorption associated with the π-bonds of the aromatic ring. Indicate with what structural feature the strong band at about 800 cm^{-1} is associated.

b. In the ^1H NMR spectrum, assign the various resonances to the hydrogen nuclei responsible for them.

c. For the ^{13}C NMR data, assign the various resonances to the carbon nuclei responsible for them.

15. Consider the NMR spectral data for 1-bromopropane (Fig. 15.7).

a. In the ^1H NMR spectrum, assign the various resonances to the hydrogen nuclei responsible for them.

b. For the ^{13}C NMR data, assign the various resonances to the carbon nuclei responsible for them.

16. Consider the spectral data for *n*-propyl-*p*-xylene (Figs. 15.8 and 15.9).

a. In the functional group region of the IR spectrum, specify the absorption associated with the π-bonds of the aromatic ring. Indicate with what structural feature the strong band at about 810 cm^{-1} is associated.

b. In the ^1H NMR spectrum, assign the various resonances to the hydrogen nuclei responsible for them.

c. For the ^{13}C NMR data, assign the various resonances to the carbon nuclei responsible for them.

17. Consider the spectral data for isopropyl-*p*-xylene (Figs. 15.10 and 15.11).

a. In the functional group region of the IR spectrum, specify the absorption associated with the π-bonds of the aromatic ring. Indicate with what structural feature the strong band at about 800 cm^{-1} is associated.

b. In the ^1H NMR spectrum, assign the various resonances to the hydrogen nuclei responsible for them.

c. For the ^{13}C NMR data, assign the various resonances to the carbon nuclei responsible for them.

18. What differences in the IR and NMR spectra of *p*-xylene and of **5** and **6** are consistent with the incorporation of a C_3H_7 group on the ring in this experiment?

19. What differences in the IR and NMR spectra of *p*-xylene, and of **5** and **6**, are consistent with *mono*substitution rather than *di*substitution of *p*-xylene having occurred in the Friedel-Crafts alkylation you performed?

20. *Ipso* substitution, as given in the equation below, is sometimes observed in electrophilic aromatic substitution reactions. Explain how the data provided in Figures 15.8–15.11 are inconsistent with the structures of the *ipso* products shown.

CH_3 / CH_3 (*p*-Xylene) + $CH_3CH_2CH_2Br$ $\xrightarrow{AlCl_3}$ $CH_2CH_2CH_3$ / CH_3 (1-Methyl-4-*n*-propylbenzene) + $CH(CH_3)_2$ / CH_3 (1-Methyl-4-isopropylbenzene)

p-Xylene

1-Methyl-4-*n*-propylbenzene 1-Methyl-4-isopropylbenzene

Ipso substitution products

SPECTRA

Starting Materials and Products

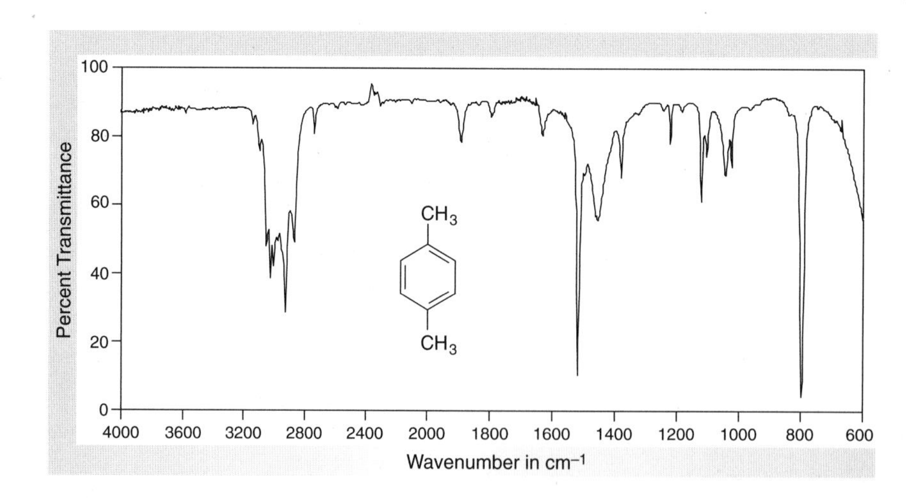

Figure 15.4
IR spectrum of p*-xylene (neat).*

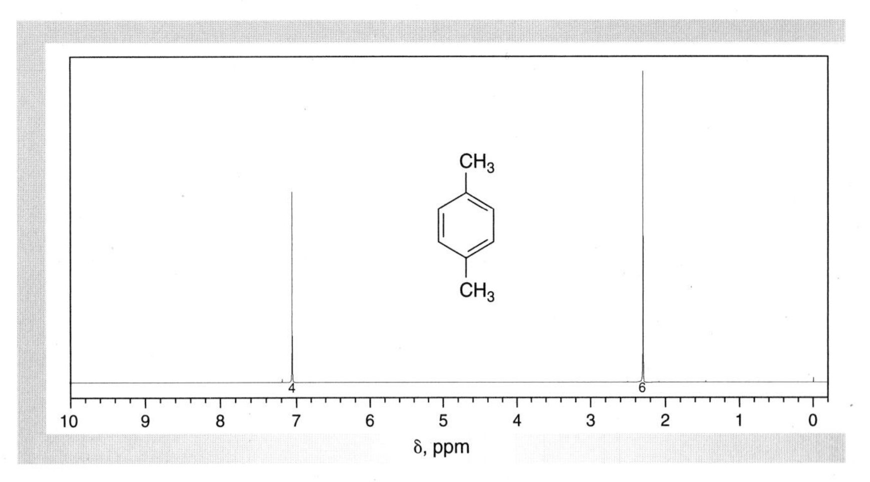

Figure 15.5
NMR data for p-xylene ($CDCl_3$).

(a) ^{1}H NMR spectrum (300 MHz).
(b) ^{13}C NMR data: δ 20.9, 129.0, 134.6.

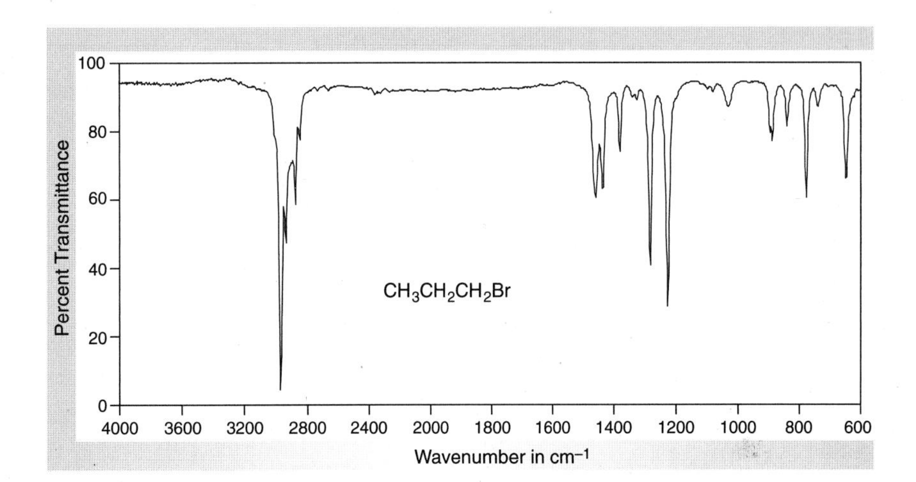

Figure 15.6
IR spectrum of 1-bromopropane (neat).

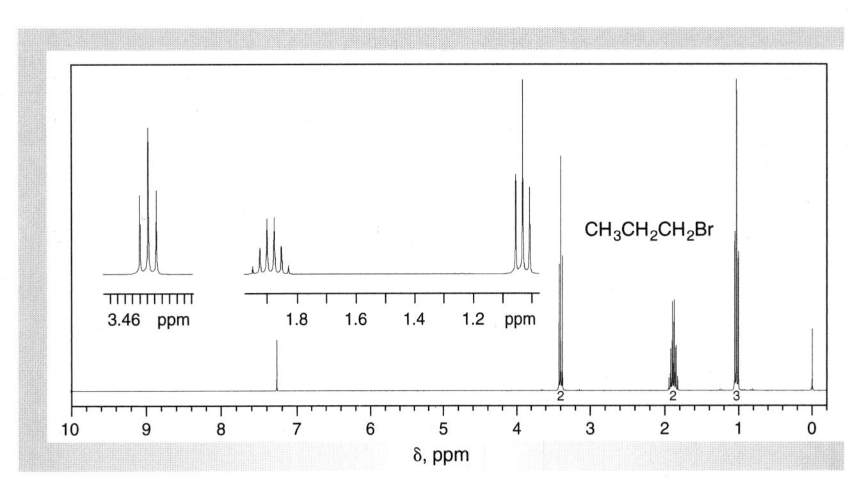

Figure 15.7
NMR data for 1-bromopropane ($CDCl_3$).

(a) ^{1}H NMR spectrum (300 MHz).
(b) ^{13}C NMR data: δ 13.0, 26.2, 35.9.

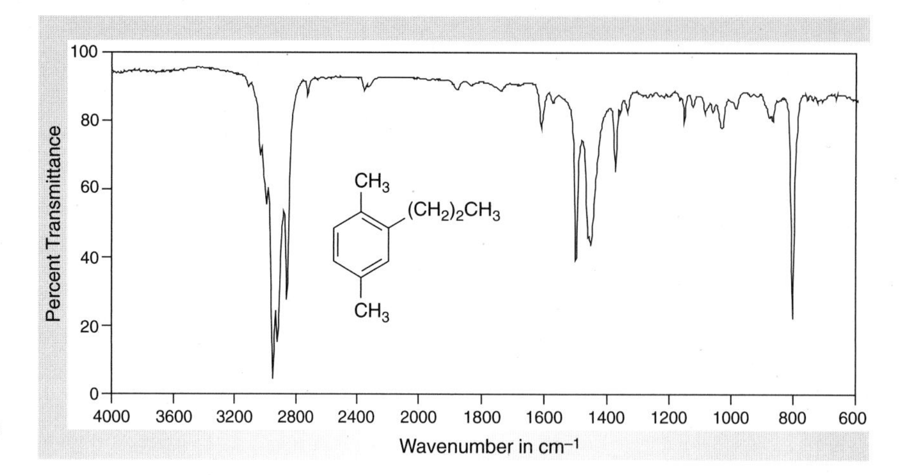

Figure 15.8
IR spectrum of n-*propyl*-p-*xylene (neat).*

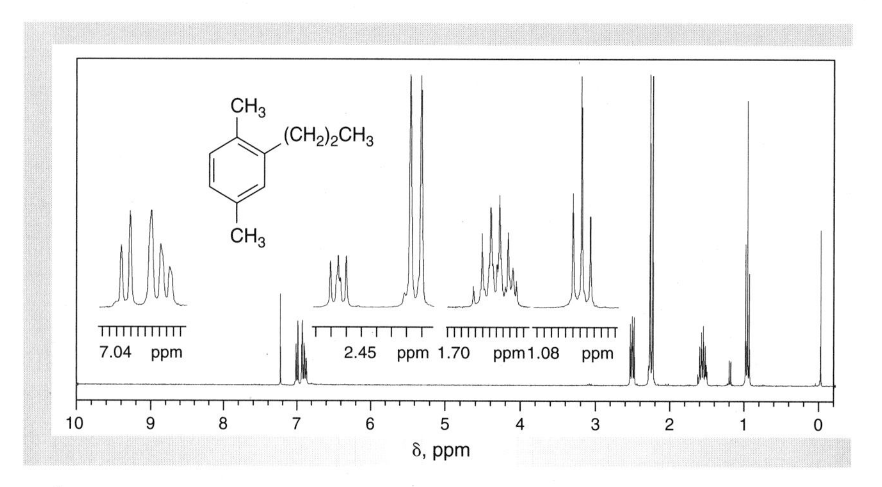

Figure 15.9
NMR data for n-*propyl*-p-*xylene (contains 4% of isopropyl isomer) (CDCl*$_3$*).*

(a) ^{1}H NMR spectrum (300 MHz).
(b) ^{13}C NMR data: δ 14.2, 18.8, 21.0, 23.5, 35.4, 126.4, 129.7, 129.9, 132.7, 135.1, 140.7.

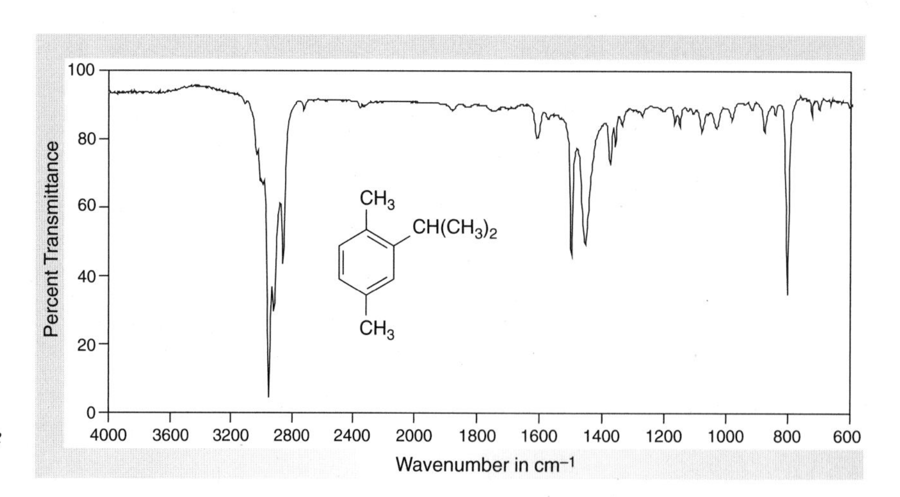

Figure 15.10
IR spectrum of isopropyl-p-*xylene (neat).*

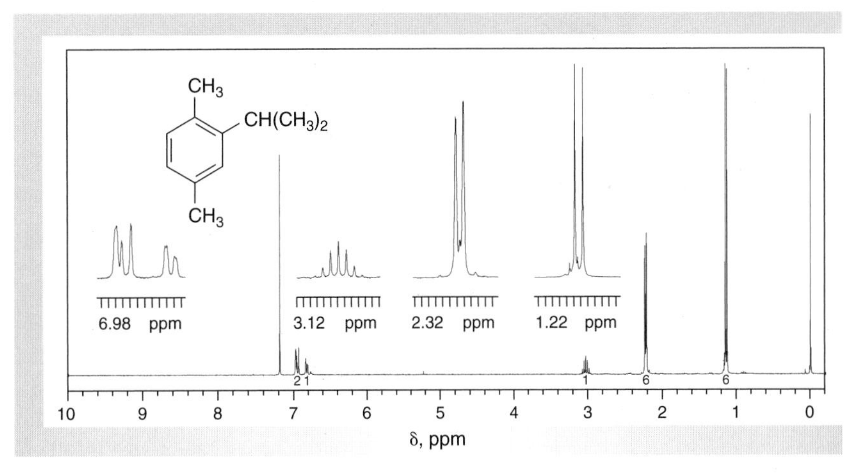

Figure 15.11
NMR data for isopropyl-p-xylene ($CDCl_3$).

(a) ^{1}H NMR spectrum (300 MHz).
(b) ^{13}C NMR data: δ 18.8, 21.2, 23.2, 29.1, 125.4, 126.2, 130.1, 131.8, 135.4, 146.6.

15.3 FRIEDEL-CRAFTS ACYLATION OF *m*-XYLENE

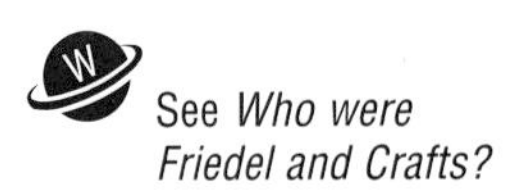

See *Who were Friedel and Crafts?*

As described in Section 15.2 and in the Historical Highlight at the end of this chapter, Friedel and Crafts discovered that alkyl groups could be substituted onto the aromatic ring by reaction of arenes with alkyl halides in the presence of aluminum chloride, $AlCl_3$ (Eq. 15.7). The role of aluminum chloride, a strong Lewis acid, is to convert the alkyl halide into a reactive electrophilic intermediate, in the form of a carbocation or a highly polarized carbon-halogen bond. This electrophile then undergoes attack by an arene (Sec. 15.1), which functions as a Lewis base, resulting in aromatic substitution. They also explored the reaction of aluminum chloride with acid chlorides, **7**, and anhydrides, **8**, to produce **acylonium ions, 9,** which function as electrophiles, just as alkyl carbocations do. Upon reaction of **9** with an arene, an acyl group is introduced onto the ring to provide an aryl ketone, **10** (Eq. 15.12). The overall reaction is named the **Friedel-Crafts acylation.**

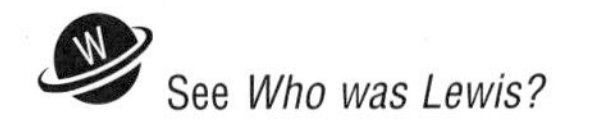

See *Who was Lewis?*

The resonance structures in Equation 15.12 symbolize distribution of the positive charge in acylonium ions. This charge delocalization stabilizes the cation so that acylonium ions such as **9** do *not* undergo structural rearrangements as do alkyl carbocations (Sec. 15.2). The increased stability of acylonium ions also makes them less electrophilic and thus less reactive than alkyl carbocations. Nevertheless, cation **9** is sufficiently reactive to undergo the S_E2 reaction (Sec. 15.1) with arenes, *provided* the aromatic ring does not bear strongly deactivating groups such as NO_2, R_3N^+, C(O)R, and CN.

An important difference between the Friedel-Crafts alkylation and acylation reactions is that the latter process requires use of a *stoichiometric* rather than a catalytic amount of aluminum chloride. This is because the product aryl ketones **10** undergo a Lewis acid-base reaction with aluminum chloride to form a strong 1:1 complex, **11;** the aluminum chloride involved in the complex no longer fosters formation of acylonium ions and in essence is "consumed" in the reaction. Thus, although generation of acylonium ions needed for the Friedel-Crafts acylation is catalytic in aluminum chloride, complexation with the product makes aluminum chloride a stoichiometric reagent.

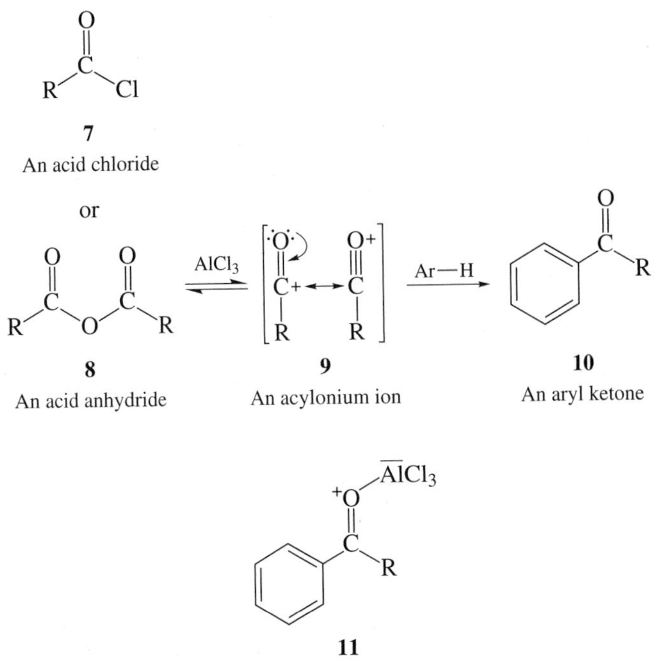

7
An acid chloride

or

8
An acid anhydride

$\xrightleftharpoons{AlCl_3}$

9
An acylonium ion

$\xrightarrow{Ar-H}$

10
An aryl ketone

(15.12)

11

The acylation performed in this section involves reaction of phthalic anhydride (**12**), a cyclic anhydride, with *m*-xylene (**13**) in the presence of aluminum chloride to afford 2-(2′,4′-dimethylbenzoyl)benzoic acid (**15**) (Scheme 15.2), a ketoacid. The initial product of the acylation is **14**, in which one mole of aluminum chloride has reacted with the acid function to form the salt RCO_2AlCl_2, and a second mole of aluminum chloride is complexed to the carbonyl group. Adding ice and hydrochloric acid decomposes the complex **14** to produce **15** and water-soluble aluminum salts.

Scheme 15.2

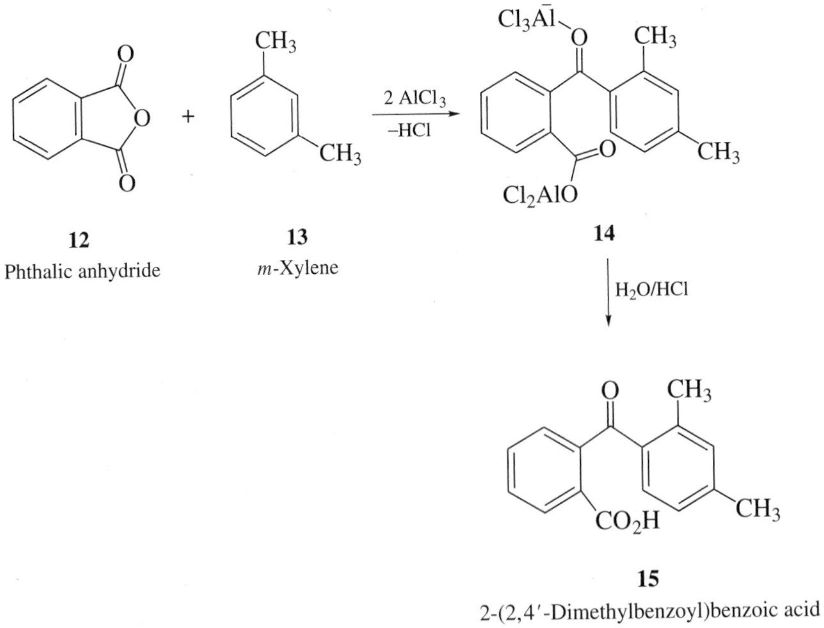

12
Phthalic anhydride

13
m-Xylene

$\xrightarrow[-HCl]{2\ AlCl_3}$

14

H_2O/HCl

15
2-(2,4′-Dimethylbenzoyl)benzoic acid

Formation of **14** involves reaction of aluminum chloride with anhydride **12** to give the electrophile **16** (Eq. 15.13), which then reacts with *m*-xylene (**13**). An issue in this step of the sequence is which of the three different positions of **13** undergoes attack by **16**. Because methyl groups are *o,p*-directors, reaction should occur preferentially at C2 and C4 of the ring. Steric hindrance impedes attack at C2, so *m*-xylene reacts selectively at C4 to give **14**, which provides the ketoacid **15** upon hydrolysis, as shown in Scheme 15.2.

12 + $AlCl_3$ ⇌ [intermediate: $\ddot{O}^{+}$–$\bar{A}lCl_3$] ⇌ **16** [$C{\equiv}O^{+}$; O–$\bar{A}lCl_3$] (15.13)

EXPERIMENTAL PROCEDURES

Friedel-Crafts Acylation of m-*Xylene with Phthalic Anhydride*

Purpose To demonstrate acylation by electrophilic aromatic substitution.

SAFETY ALERT

1. **Anhydrous aluminum chloride is *extremely* hygroscopic and reacts rapidly with water, even the moisture on your hands, producing fumes of hydrogen chloride, which are highly corrosive. *Do not allow aluminum chloride to come in contact with your skin.* If it does, flush the affected area with copious amounts of water. *Minimize exposure of this chemical to the atmosphere!***
2. **You may be provided with a weighed amount of aluminum chloride. If not, *quickly* weigh it into a dry vial or test tube *in the hood* and then transfer it to the reaction flask. The success of this experiment is highly dependent on the quality of the aluminum chloride that is used, so obtain it from a *freshly* opened bottle. In handling this chemical, be aware that it is a powdery solid that easily becomes airborne.**
3. ***m*-Xylene is flammable. Assemble the apparatus carefully and be sure that all joints are tightly mated. Use *flameless* heating.**

MINISCALE PROCEDURES

Preparation Sign in at **www.cengage.com/login** to answer Pre-Lab Exercises, access videos, and read the MSDSs for the chemicals used or produced in this procedure. Review Sections 2.9, 2.10, 2.11, 2.13, 2.23, 2.25, and 2.29.

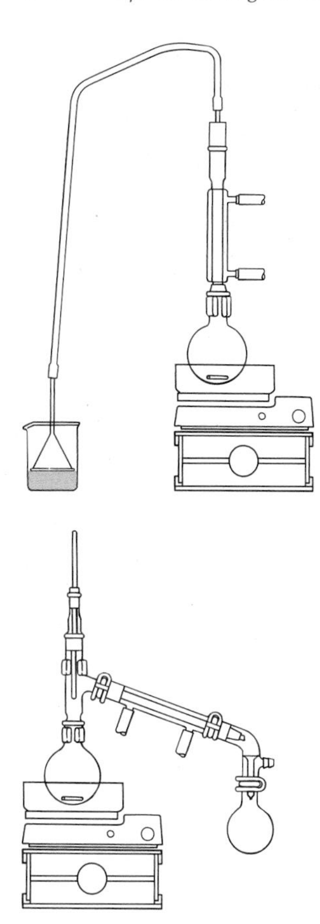

Apparatus A dry 25-mL round-bottom flask, gas trap, ice-water bath, separatory funnel, apparatus for magnetic stirring, heating under reflux, vacuum filtration, and *flameless* heating.

Setting Up *Working at the hood,* weigh 2 g of *anhydrous* aluminum chloride into a dry test tube or screw-cap vial and *immediately* close the vessel. Equip the round-bottom flask with a stirbar and a dry reflux condenser fitted with the gas trap specified by your instructor for removing fumes of hydrogen chloride. Place 1 g of phthalic anhydride and 6 mL of *m*-xylene in the flask, and cool the mixture to 0 °C in an ice-water bath.

Acylation Remove the condenser from the flask, and, in one portion, *quickly* add the anhydrous aluminum chloride. *Immediately* replace the condenser and stir the reaction mixture for a minute or two while maintaining its temperature at 0 °C and then allow it to warm to room temperature. Initiation of the reaction will be evidenced by evolution of bubbles. If this does not occur, gently warm the flask until gas evolution begins. Be prepared to immerse the flask into an ice-water bath if the reaction becomes too vigorous. Continue gently heating the mixture with stirring until the reaction is proceeding smoothly, and then heat it under reflux for about 0.5 h.

Work-Up, Isolation, and Purification Remove the flask from the apparatus and stopper it. *Working at the hood,* pour the cold reaction mixture onto 8–10 g of ice contained in a beaker while stirring the mixture thoroughly. With continued vigorous stirring, slowly add 2 mL of *concentrated* hydrochloric acid and then 8–10 mL of ice-cold water to the beaker; if necessary, cool the mixture in an ice-water bath to keep it near room temperature during the additions. Add an additional 4 mL of ice-cold water with stirring and cool the resulting mixture to room temperature or below.★

Add 5 mL of diethyl ether to the beaker. With the aid of a stirring rod, scrape any solid from the neck and walls of the beaker and carefully break up any lumps at the bottom. Stir the mixture vigorously for several minutes to complete decomposition of the aluminum salt of the product, extraction of the organic product into diethyl ether, and dissolution of the inorganic aluminum salts.★

Transfer the two-phase mixture into a separatory funnel, rinse the beaker with a few milliliters of diethyl ether, and add the rinse to the funnel. Shake the mixture, frequently venting the funnel, and allow the layers to separate. A grayish, fluffy precipitate may appear below the organic layer. If so, dissolve the precipitate before separating the layers by adding 5 mL of 6 *M* HCl and shaking the mixture for a few minutes, occasionally venting the funnel. Should any precipitate or grayish emulsion remain with the ethereal solution, remove it by vacuum filtration through a pad of a filter-aid. After separating the layers, extract the aqueous layer sequentially with two additional 10-mL portions of diethyl ether, venting the funnel occasionally during the extraction. Combine all of the ethereal solutions in an Erlenmeyer flask and add several spatula-tips full of anhydrous sodium sulfate, swirling the mixture occasionally over a period of about 15 min; if the solution appears cloudy, add additional sodium sulfate with swirling.★

Decant the ethereal solution into a 50-mL round-bottom flask and equip the flask for simple distillation. Concentrate the solution to about half its original volume. Alternatively, use rotary evaporation or other techniques to concentrate the solution. Transfer the concentrated solution to an Erlenmeyer flask and cool it to room temperature. Once crystallization has begun, further cool the solution in an ice-water bath and then collect **15** by vacuum filtration. Recrystallize it from 50% aqueous ethanol and air-dry the purified product.

Analysis Weigh your product and calculate the yield. Measure the melting point of **15,** which is reported to be 142.5–142.8 °C. Obtain IR and NMR spectra of your starting materials and product and compare them with those of authentic samples (Figs. 15.12–15.17).

MICROSCALE PROCEDURE

Preparation Sign in at **www.cengage.com/login** to answer Pre-Lab Exercises, access videos, and read the MSDSs for the chemicals used or produced in this procedure. Review Sections 2.9, 2.10, 2.11, 2.13, 2.23, 2.25, and 2.29.

Apparatus A dry 5-mL conical vial, gas trap, ice-water bath, 15-mL screw-cap centrifuge tube, apparatus for magnetic stirring, heating under reflux, vacuum and Craig-tube filtration, and *flameless* heating.

Setting Up *Working at the hood,* weigh 0.7 g of *anhydrous* aluminum chloride into a dry test tube or vial and *immediately* stopper the vessel. Equip the conical vial with a spinvane and a dry reflux condenser fitted with the gas trap specified by your instructor for removing fumes of hydrogen chloride. Add 0.3 g of phthalic anhydride and 1.8 mL of *m*-xylene to the vial and cool the mixture to 0 °C in an ice-water bath.

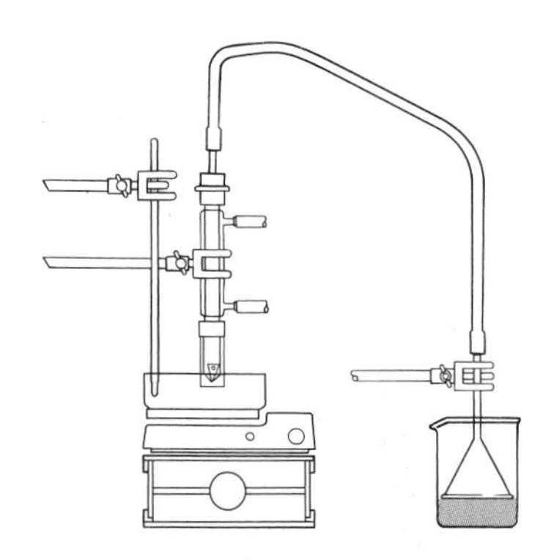

Acylation Remove the condenser from the conical vial, and, in one portion, quickly add the *anhydrous* aluminum chloride. *Immediately* replace the condenser, stir the reaction mixture for a minute or two, and then allow it to warm to room temperature. The start of the reaction will be evidenced by evolution of bubbles. If this does not occur, gently warm the flask in a sand or oil bath until gas evolution begins. Be prepared to immerse the vial into an ice-cold water bath if the reaction becomes too vigorous. Continue heating the mixture gently with stirring until the reaction appears to be proceeding smoothly and then heat it under reflux for about 20 min. At the end of the reflux period, cool the reaction mixture to 0 °C with stirring.

Work-Up, Isolation, and Purification Remove the conical vial from the apparatus and stopper it. *Working at the hood,* pour the cold reaction mixture into a 15-mL screw-cap centrifuge tube or screw-cap vial containing 2 mL of ice-water and stir the mixture thoroughly. With continued vigorous stirring, slowly add 0.5 mL of *concentrated* hydrochloric acid and then 2 mL of ice-cold water. Cool the resulting mixture to room temperature or below.★

Add 1 mL of diethyl ether and break up any lumps of solid with the aid of a stirring rod. Cap the centrifuge tube and shake it vigorously for several minutes to complete decomposition of aluminum salts, extraction of the product into the organic phase, and dissolution of inorganic aluminum salts.★ Vent the system occasionally to relieve any pressure that may develop.

Using a Pasteur or filter-tip pipet, transfer the organic phase to a test tube or vial with a 15-mL capacity. Extract the aqueous layer sequentially with three 2-mL portions of diethyl ether, combining the organic layers with the original ethereal extract. Combine all of the ethereal solutions in an Erlenmeyer flask, and add several microspatula-tips full of anhydrous sodium sulfate, swirling the mixture occasionally over a period of about 15 min; if the solution appears cloudy, add additional sodium sulfate with swirling.

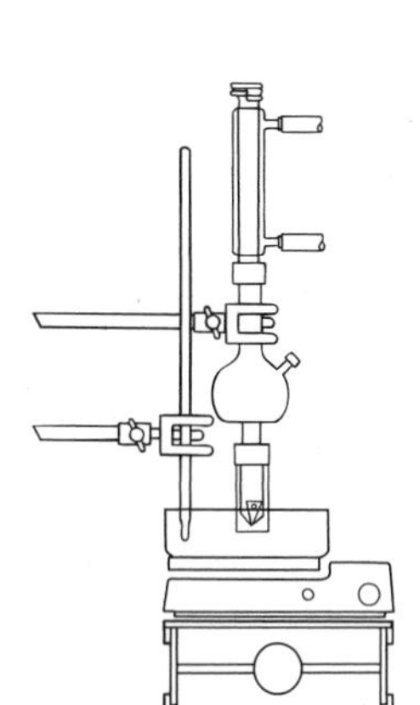

Decant about one-half of the ethereal solution into a 5-mL conical vial equipped for simple distillation and concentrate the solution to about half its original volume, removing the distillate from the Hickman stillhead as necessary. Allow the solution to cool below the boiling point, add the remainder of the dried ethereal extracts, and again concentrate the solution to about one-half its original volume. Alternatively, use rotary evaporation or other techniques to concentrate the solution. Cool the concentrated solution to room temperature. Once crystallization has begun, further cool the solution in an ice-water bath, and then collect the product **15** by vacuum filtration. Recrystallize it from 50% aqueous ethanol and air-dry the purified product.

Analysis Weigh the product and calculate the yield. Measure the melting point of **15,** which is reported to be 142.5–142.8 °C. Obtain IR and NMR spectra of your starting materials and product and compare them with those of authentic samples (Figs. 15.12–15.17).

WRAPPING IT UP

Discard the *grayish solid* obtained by gravity filtration in the container for nonhazardous solids. Pour the *diethyl ether* obtained by distillation in a container for nonhalogenated organic liquids. Flush the *filtrate from the recrystallization* and the *aqueous layer containing aluminum salts* down the drain.

EXERCISES

1. Explain why a catalytic amount of $AlCl_3$ is insufficient to promote the Friedel-Crafts reaction of **12** and **13.**
2. Why is loss of carbon monoxide (decarbonylation) from the acylonium ion **16** an unlikely reaction?
3. Why is it important that the apparatus used for the Friedel-Crafts acylation be dry?

4. Write the products of the reaction of $AlCl_3$ and excess H_2O.

5. Consider aluminum chloride, $AlCl_3$, the catalyst used in this procedure.
   a. Provide a Lewis dot structure for $AlCl_3$.
   b. Define the term *Lewis acid*.
   c. What aspect of the Lewis structure is consistent with this compound being a Lewis acid?
   d. How does addition of $AlCl_3$ promote formation of carbocations from alkyl halides?
   e. Provide an example of a Lewis acid that does not contain aluminum.

6. Write the structure of the product from reaction between phthalic anhydride and water.

7. What gas is responsible for the bubbles observed in the early stages of the reaction between *m*-xylene, phthalic anhydride, and aluminum chloride?

8. In which of the two solvents, water or ethanol, should the product first be dissolved for the mixed solvent recrystallization? Explain.

9. Why does the acylonium ion **16** react more rapidly with **13** than with **12?**

10. Using curved arrows to symbolize the flow of electrons, write the stepwise mechanism for the reaction between *m*-xylene and the acylonium ion **16** to give the complex **14**. Write the contributing resonance structures for the s-complex formed by attack of **16** on *m*-xylene and provide a three-dimensional representation of the complex.

11. In the Friedel-Crafts acylation of *m*-xylene with phthalic anhydride, explain why each of the isomers shown below is *not* produced.

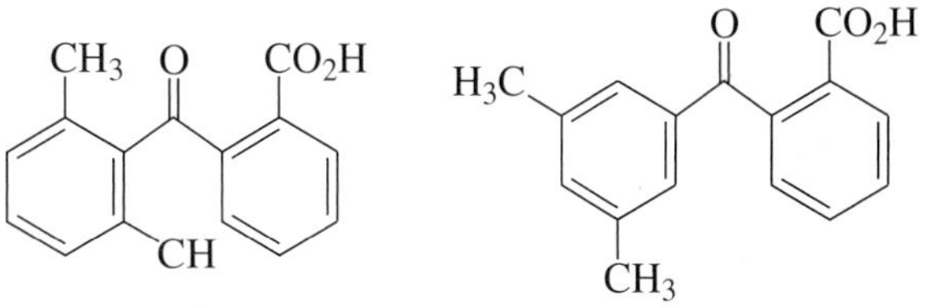

12. Suggest a sequence of reactions for synthesizing *m*-xylene from any substituted aromatic precursor of your choice. (*Hint:* More than one step will be necessary.)

13. Provide two reasons why intramolecular acylation involving the carboxy carbonyl group of **14** does not occur to produce the compound shown below.

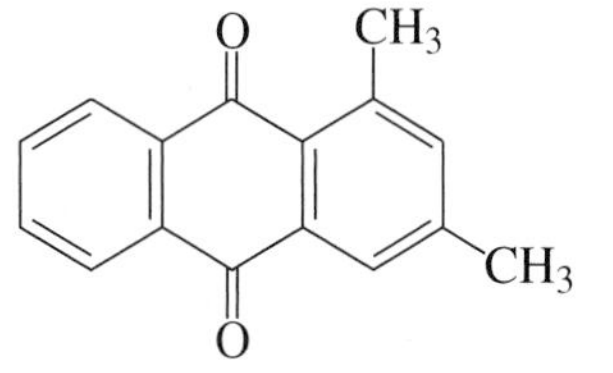

14. What evidence do you have from this experiment that substitution of phthalic acid for phthalic anhydride would not provide acylonium ion **16** required for the acylation reaction? (*Hint:* Write the product of reaction of aluminum chloride with one of the carboxylic acid groups of phthalic acid and compare it to the carboxylate **14.**)

Sign in at **www.cengage.com/login** and use the spectra viewer and Tables 8.1–8.8 as needed to answer the blue-numbered questions on spectroscopy.

15. Consider the spectral data for *m*-xylene (Figs. 15.12 and 15.13).

a. In the functional group region of the IR spectrum, specify the absorption associated with the π-bonds of the aromatic ring. Indicate the bands in the fingerprint region that characterize the *meta* orientation of the methyl groups.

b. In the ^{1}H NMR spectrum, assign the various resonances to the hydrogen nuclei responsible for them.

c. For the ^{13}C NMR data, assign the various resonances to the carbon nuclei responsible for them.

d. Explain how ^{13}C spectroscopy allows differentiation of the *meta* isomer from the *ortho* and *para* isomers.

16. Consider the spectral data for phthalic anhydride (Figs. 15.14 and 15.15).

a. In the functional group region of the IR spectrum, specify the absorptions associated with the anhydride function and the aromatic ring.

b. In the ^{1}H NMR spectrum, assign the various resonances to the hydrogen nuclei responsible for them.

c. For the ^{13}C NMR data, assign the various resonances to the carbon nuclei responsible for them.

17. Consider the spectral data for 2-(2′,4′-dimethylbenzoyl)benzoic acid (Figs. 15.16 and 15.17).

a. In the functional group region of the IR spectrum, specify the absorptions associated with the ketone carbonyl group, the aromatic rings, and the carboxylic acid function. Explain why the latter absorption is broad.

b. In the ^{1}H NMR spectrum, assign the various resonances to the hydrogen nuclei responsible for them.

c. For the ^{13}C NMR data, assign the various resonances to the carbon nuclei responsible for them.

18. What differences in the IR and NMR spectra of phthalic anhydride and 2-(2′,4′-dimethylbenzoyl)benzoic acid are consistent with the conversion of the anhydride group to ketone and carboxylic acid groups in this experiment?

19. Explain how ^{1}H NMR and ^{13}C NMR spectra could be used to differentiate between **15** and the isomer of it shown below.

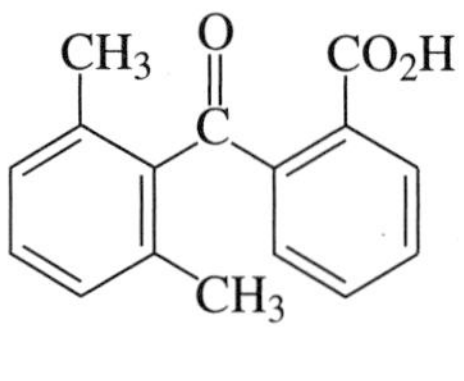

CH_3 O CO_2H
C
CH_3

SPECTRA

Starting Materials and Products

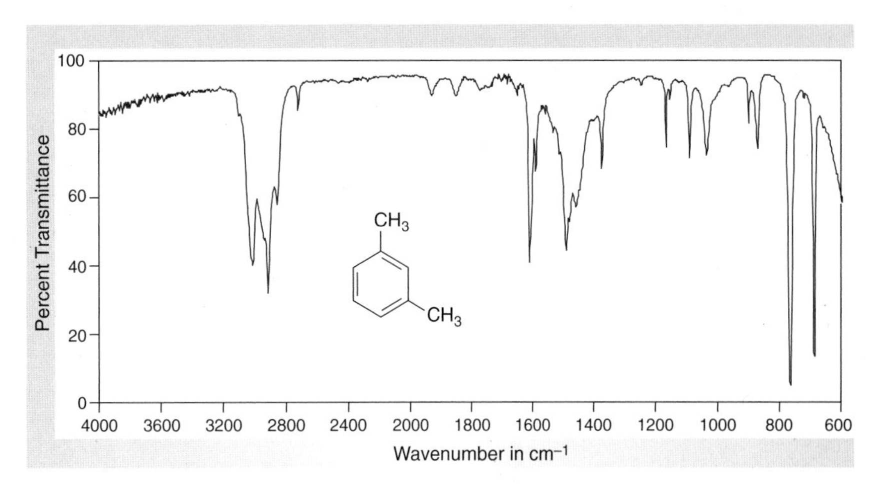

Figure 15.12
IR spectrum of m*-xylene (neat).*

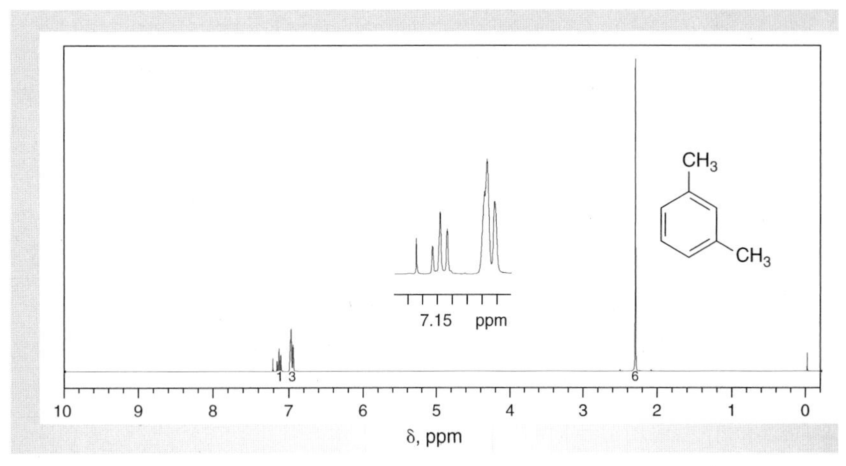

Figure 15.13
NMR data for m*-xylene ($CDCl_3$).*

(a) ^{1}H NMR spectrum (300 MHz).
(b) ^{1}C NMR data: δ 21.3, 126.2, 128.2, 130.0, 137.6.

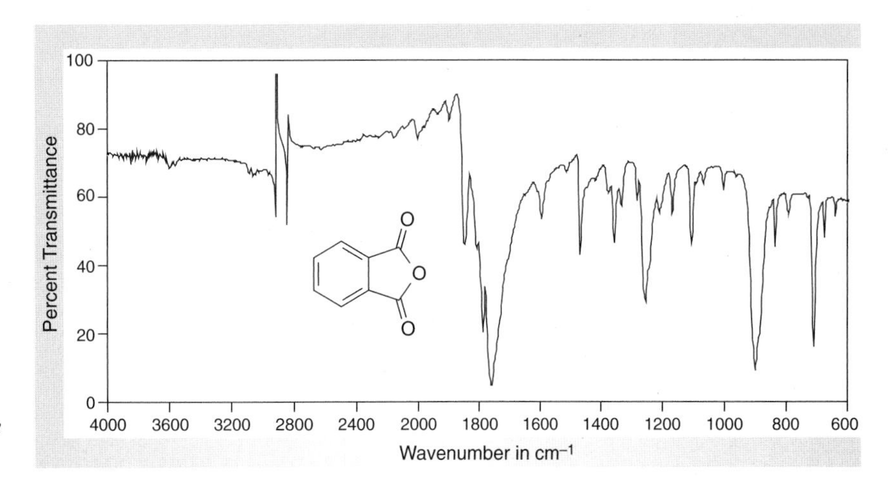

Figure 15.14
IR spectrum of phthalic anhydride (IR card).

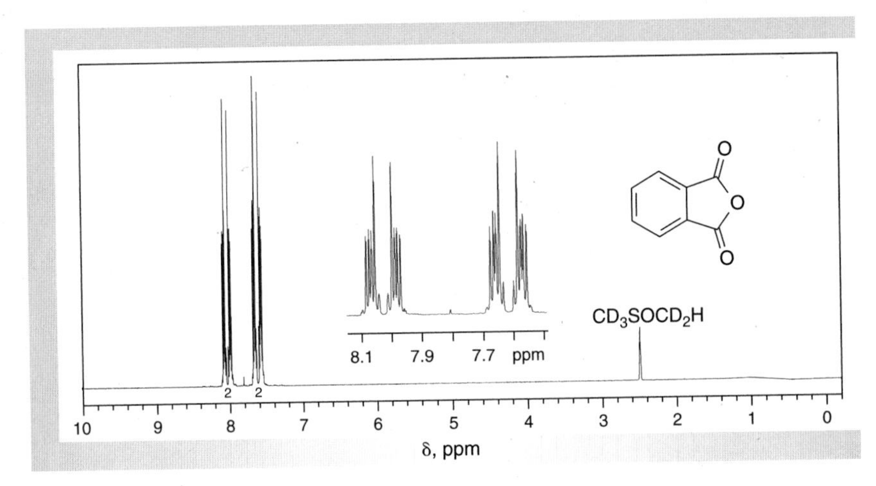

Figure 15.15
NMR data for phthalic anhydride (DMSO-d_6).

(a) ^{1}H NMR spectrum (300 MHz).
(b) ^{13}C NMR data: δ 125.3, 131.1, 136.1, 163.1.

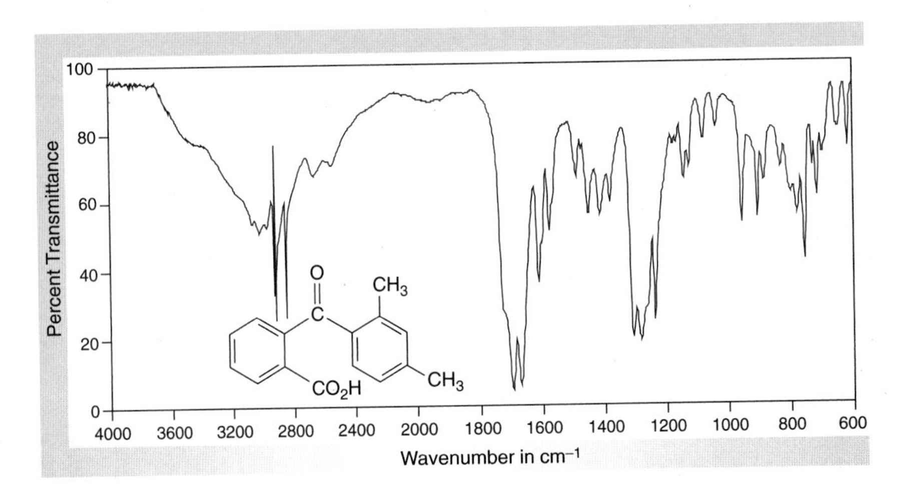

Figure 15.16
IR spectrum of 2-(2′,4′-dimethylbenzoyl)benzoic acid (IR card).

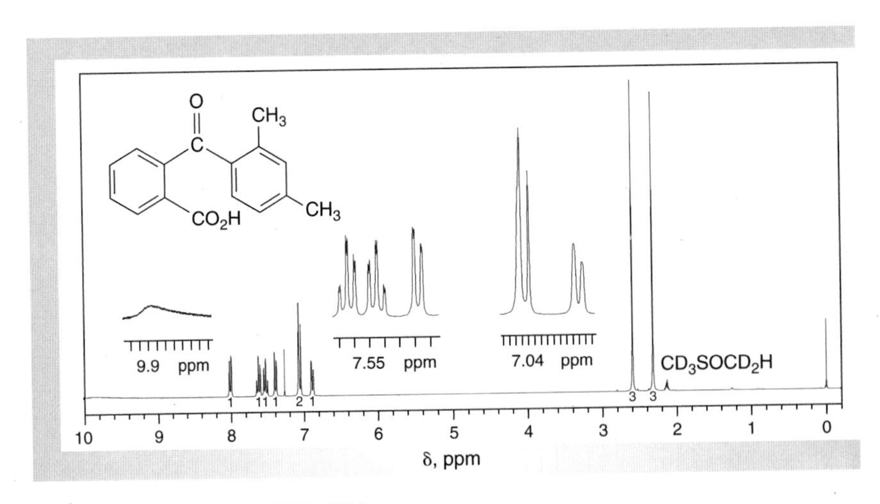

Figure 15.17
NMR data for 2-(2′,4′-dimethylbenzoyl)benzoic acid ($CDCl_3$).

(a) ^{1}H NMR spectrum (300 MHz).
(b) ^{13}C NMR data: δ 21.3, 21.4, 125.7, 128.0, 128.1, 129.4, 130.6, 131.8, 132.7, 132.8, 133.8, 140.3, 142.4, 143.9, 171.2, 198.4.

15.4 NITRATION OF BROMOBENZENE

Reaction of an aromatic compound such as benzene with a mixture of concentrated sulfuric and nitric acids introduces a nitro group on the ring by way of electrophilic aromatic substitution (Sec. 15.1) as depicted in Equation 15.14. The electrophilic species is the **nitronium ion,** NO_2^+, which is produced by reaction of sulfuric acid with nitric acid (Eq. 15.15). It is interesting to note that nitric acid, the *weaker* of these two strong mineral acids, is serving as a *base* rather than as an acid in this equilibrium.

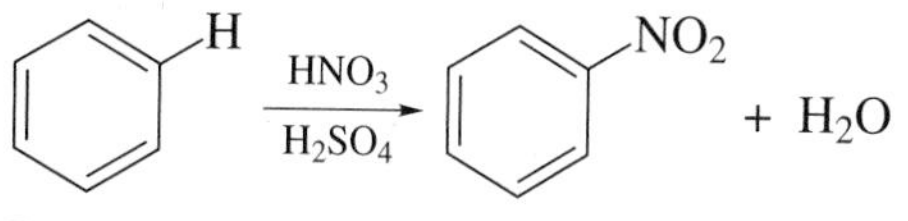

$$\text{Benzene} \xrightarrow[H_2SO_4]{HNO_3} \text{Nitrobenzene} + H_2O \qquad (15.14)$$

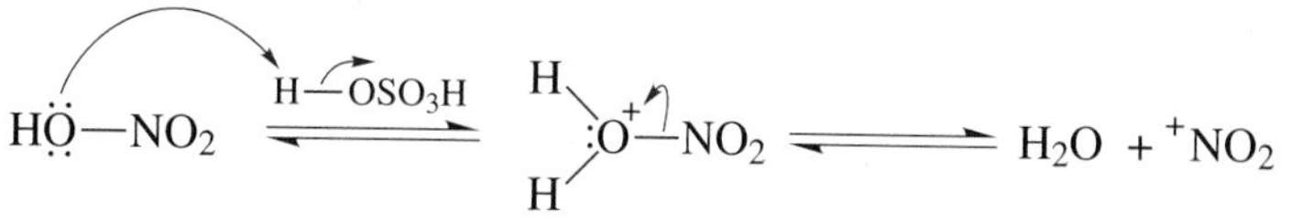

$$\underset{\text{Nitric acid}}{H\ddot{O}{-}NO_2} \underset{}{\overset{H{-}OSO_3H}{\rightleftharpoons}} H_2\overset{+}{O}{-}NO_2 \rightleftharpoons H_2O + \underset{\text{Nitronium ion}}{^{+}NO_2} \qquad (15.15)$$

The rate-determining step (rds) in the nitration reaction involves nucleophilic attack of the aromatic ring, a Lewis base, on the nitronium ion, a Lewis acid, to form the delocalized σ-complex **17** (Eq. 15.16). In the final step of the reaction, a base such as water or bisulfate deprotonates this complex to regenerate the aromatic ring (Eq. 15.17).

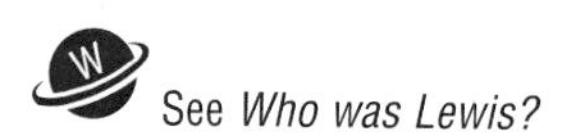

See *Who was Lewis?*

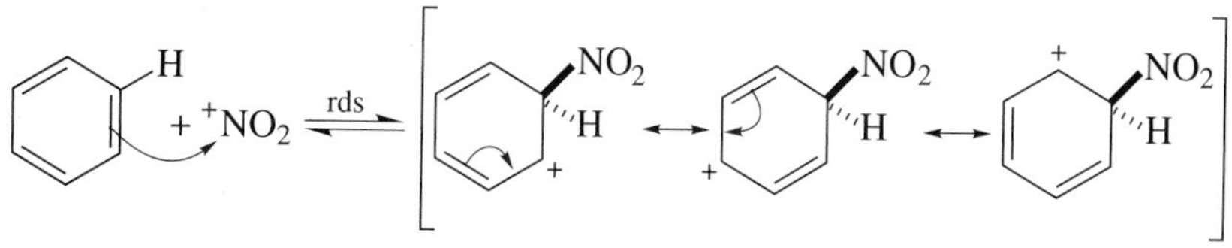

$$C_6H_6 + {}^{+}NO_2 \overset{\text{rds}}{\rightleftharpoons} \left[\text{resonance structures}\right] \qquad (15.16)$$

17
σ-Complex

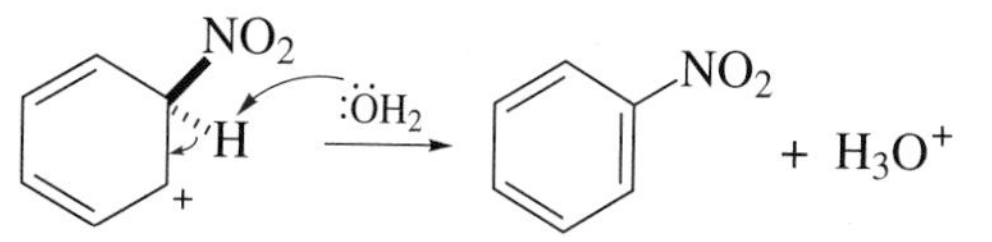

$$\mathbf{17} \xrightarrow{:\ddot{O}H_2} C_6H_5NO_2 + H_3O^+ \qquad (15.17)$$

The mechanism of the reaction of a nitronium ion with a *substituted* benzene such as bromobenzene (**18**), the arene used in this experiment, is similar to that with benzene. Now, however, *three* different products can be formed because of the presence of the bromine atom on the ring (Eq. 15.18). The activation enthalpies, $\Delta H^{\ddagger}$, for attack at the positions *ortho* and *para* to the substituent are lower than that for attack

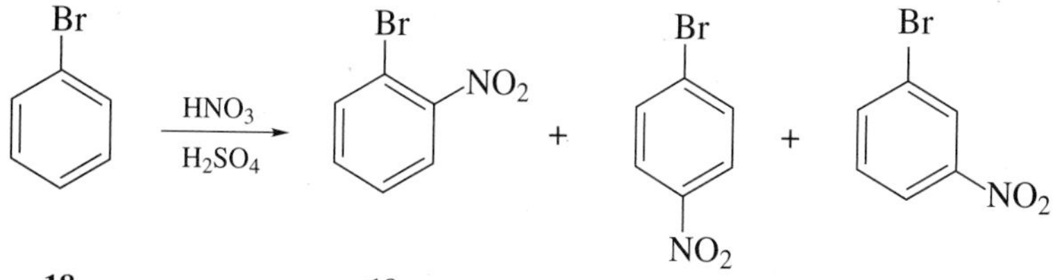

$$\underset{\text{Bromobenzene}}{\mathbf{18}} \xrightarrow[H_2SO_4]{HNO_3} \underset{\text{2-Bromo-nitrobenzene}}{\mathbf{19}} + \underset{\text{4-Bromo-nitrobenzene}}{\mathbf{20}} + \underset{\text{3-Bromo-nitrobenzene}}{\mathbf{21}} \qquad (15.18)$$

at the *meta* position. This is because the σ-complexes **22** and **23**, which are the precursors to the *ortho* and *para* products **19** and **20**, respectively, are more stable than complex **24**, which gives the *meta* compound **21**. The increased stability of **22** and **23** relative to **24** is associated with the participation of nonbonding electrons on the bromine atom in delocalizing the positive charge in the complexes **22** and **23**. Thus, like an alkyl group, a bromine atom is an ***ortho-para director***, because it favors formation of these isomers over the *meta* isomer. For this nitration reaction the preference is found to be about 100:1.

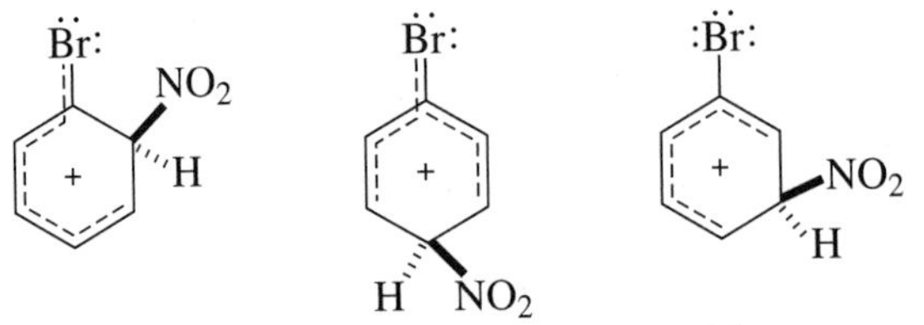

22 **23** **24**

The *o:p* ratio itself is of interest because isolating pure disubstituted products requires separation of these isomers. Assuming that attack of the electrophile at the two positions occurs on a purely statistical basis, this ratio is predicted to be 2:1, because there are two *ortho* carbon atoms and only one *para* carbon atom in bromobenzene. This prediction neglects the possibility that the steric bulk and the inductive electron withdrawing effect of the bromine atom might inhibit approach of the electrophile to the *ortho* relative to the *para* position. By determining the relative amounts of **19** and **20** produced, you will be able to evaluate this hypothesis.

Another product may result if further nitration of **19** or **20** occurs. Consideration of the directing effects of the bromo and nitro substituents leads to the prediction that the major product from both isomers would be 1-bromo-2,4-dinitrobenzene (**25**), as shown in Equation 15.19. The possible formation of 1-bromo-2,6-dinitrobenzene (**26**) from **19** is less likely because of the steric and inductive effects of the bromo substituent. Fortunately, it is relatively easy to minimize dinitration by performing the reaction at temperatures below 60 °C. Controlling the temperature suppresses dinitration because the strongly deactivating nitro substituent in **19** and **20** raises the activation enthalpy for their nitration above that of **18**. Another factor decreasing the dinitration of **20** is its precipitation during the course of the reaction, thereby removing it from the nitrating medium.

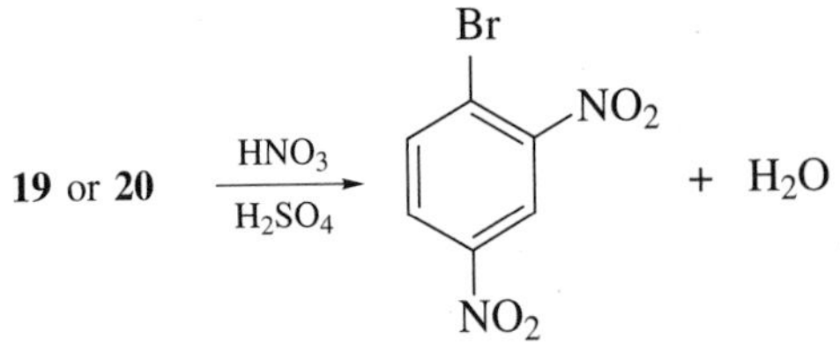

19 or **20** $\xrightarrow[H_2SO_4]{HNO_3}$ **25** + H_2O (15.19)

25
1-Bromo-2,4-dinitrobenzene

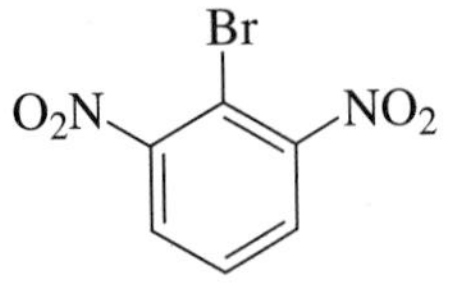

26
1-Bromo-2,6-dinitrobenzene

Based on the preceding discussion, the major obstacle to isolating pure 4-bromonitrobenzene (**20**) is separating it from 2-bromonitrobenzene (**19**). The strategy for doing this relies on the greater polarity of **19,** which makes this isomer more soluble in polar solvents than **20.** This difference is dramatically demonstrated by their relative solubilities in ethanol. At room temperature, the *ortho* isomer **19** is very soluble, whereas the *para* isomer is only slightly soluble, dissolving to the extent of only 1.2 g/100 mL. This large difference in solubilities allows separation of the isomers by the technique of **fractional crystallization.** The mixture of products obtained from nitration of bromobenzene is dissolved in hot 95% ethanol, and the solution is allowed to cool. The less soluble *para* isomer selectively crystallizes from solution and is isolated by filtration. Concentration of the filtrate allows isolation of a second crop of **20.**

In a fractional crystallization, it is commonly possible to induce crystallization of the more soluble component, in this case the *ortho* isomer **19,** once the less soluble material has been mostly removed. However, the low melting point of **19** means that this isomer is difficult to crystallize in the presence of impurities. Consequently, isolating this isomer involves column chromatography (Sec. 6.3).

EXPERIMENTAL PROCEDURES

Nitration of Bromobenzene

Purpose To demonstrate nitration by electrophilic aromatic substitution and to test the directing effects of a bromo substituent.

SAFETY ALERT

Because *concentrated* sulfuric and nitric acids may cause severe chemical burns, *do not allow them to contact your skin.* Wear latex gloves when handling these reagents. Wipe off any drips and runs on the outside surface of reagent bottles and graduated cylinders *before* picking them up. Wash any affected area immediately and thoroughly with cold water, and apply 5% sodium bicarbonate solution.

A ▪ *Nitration*

MINISCALE PROCEDURE

Preparation Sign in at **www.cengage.com/login** to answer Pre-Lab Exercises, access videos, and read the MSDSs for the chemicals used or produced in this procedure. Review Sections 2.9, 2.10, 2.11, and 2.29.

Apparatus A 25-mL round-bottom flask, Claisen adapter, thermometer, water-cooled condenser, ice-water bath, apparatus for magnetic stirring and *flameless* heating.

Setting Up Prepare a solution of 4.0 mL of concentrated nitric acid and 4.0 mL of concentrated sulfuric acid in the round-bottom flask and cool it to room temperature

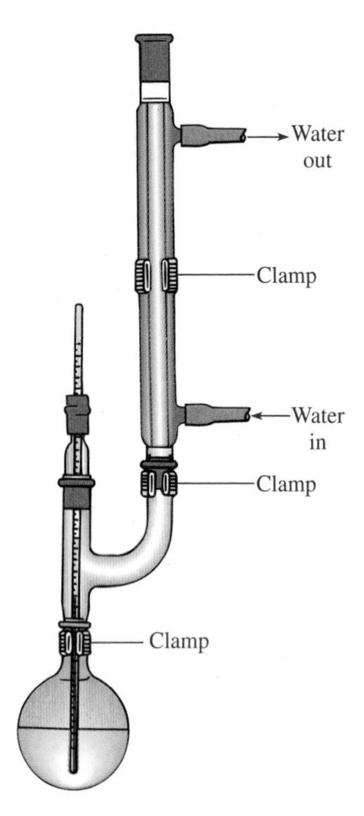

Figure 15.18
Apparatus for brominating nitrobenzene.

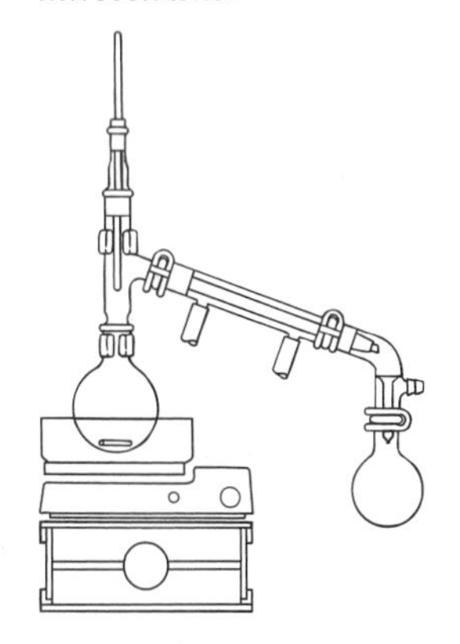

with a water bath. Equip the flask with a stirbar and a Claisen adapter fitted with the condenser and a thermometer that extends into the flask (Fig. 15.18).

Reaction and Work-Up In portions of approximately 0.5 mL, add 4.5 mL of bromobenzene to the stirred mixture through the top of the condenser over a period of about 10 min. Do *not* allow the temperature of the reaction mixture to exceed 50–55 °C during the addition. Control the temperature by allowing more time between the addition of successive portions of bromobenzene and by cooling the reaction flask with an ice-water bath.

After the addition is complete and the exothermic reaction has subsided, heat the stirred mixture for 15 min, keeping its temperature below 60 °C.★ Cool the reaction mixture to room temperature and then pour it carefully and with stirring into 40 mL of cold water contained in a beaker.

Isolation and Purification Isolate the mixture of crude bromonitrobenzenes by vacuum filtration. Wash the filter cake thoroughly with cold water until the washes are neutral to pHydrion paper; allow the solid to drain under vacuum until nearly dry.★

Transfer the filter cake to an Erlenmeyer flask and recrystallize the crude product from 95% ethanol. Allow the residual solution to cool slowly to room temperature; then cool it to 0 °C in an ice-water bath. Isolate the crystalline product by vacuum filtration. Wash the product with a little *ice-cold* 95% ethanol, allowing the washes to drain into the filter flask with the mother liquors. Transfer the product to a watchglass or a piece of filter paper for air-drying.

Concentrate the mother liquors to a volume of about 10 mL by simple distillation. Perform this operation in a hood to prevent release of vapors into the laboratory. Alternatively, use rotary evaporation or other techniques to concentrate the solution. Allow the residual solution to cool to room temperature to produce a second crop of 4-bromonitrobenzene. Isolate it by vacuum filtration and, after air-drying, put it in a separate vial from the first crop.

Further concentrate the mother liquors from the second crop to a volume of 3–4 mL. The resulting oil contains crude 2-bromonitrobenzene. Separate the oil from the two-phase mixture by means of a Pasteur pipet and weigh and reserve it for chromatographic analysis (Parts B and C).

Analysis Weigh both crops of product and calculate the yield. Measure the melting points of both crops. Obtain IR and NMR spectra of your starting material and product and compare them with those of authentic samples (Figs. 8.31, 15.19–15.22).

MICROSCALE PROCEDURES

Preparation Sign in at **www.cengage.com/login** to answer Pre-Lab Exercises, access videos, and read the MSDSs for the chemicals used or produced in this procedure. Review Sections 2.9, 2.10, 2.11, and 2.29.

Apparatus A 3-mL conical vial, water-cooled condenser, ice-water bath, apparatus for magnetic stirring, vacuum and Craig tube filtration, and *flameless* heating.

Setting Up Prepare a solution of 0.5 mL of concentrated nitric acid and 0.5 mL of concentrated sulfuric acid in the vial, and cool it to room temperature with a water bath. Equip the vial with a spinvane and the condenser.

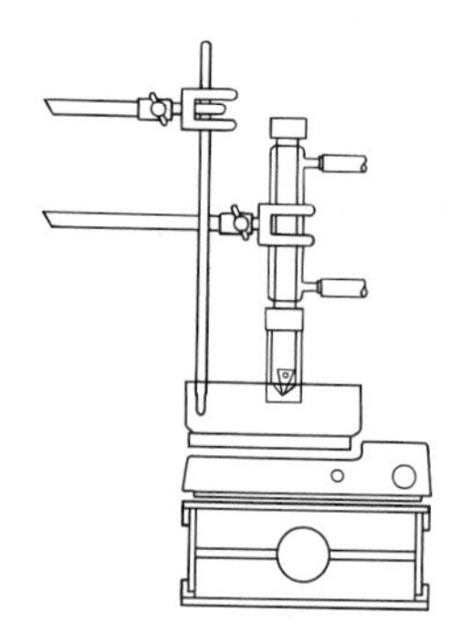

Reaction and Work-Up In portions of approximately 0.1 mL, add 0.5 mL of bromobenzene to the stirred mixture through the top of the condenser over a period of about 5 min. Although the temperature of the reaction mixture may become *warm* to the touch, do not allow the reaction mixture to become *hot* to the touch during the addition. Control the temperature by allowing more time between the addition of successive portions of bromobenzene and by cooling the vial with an ice-water bath if necessary.

After the addition is complete and the exothermic reaction has subsided, heat the stirred reaction mixture for 15 min, keeping the bath temperature below 70 °C.★ Cool the vial to room temperature, and then pour the reaction mixture carefully and with stirring into 5 mL of cold water contained in a beaker.

Isolation and Purification Isolate the mixture of crude bromonitrobenzenes by vacuum filtration. Wash the filter cake with small portions of cold water until the washes are neutral, and allow the solid to drain under vacuum until nearly dry.★

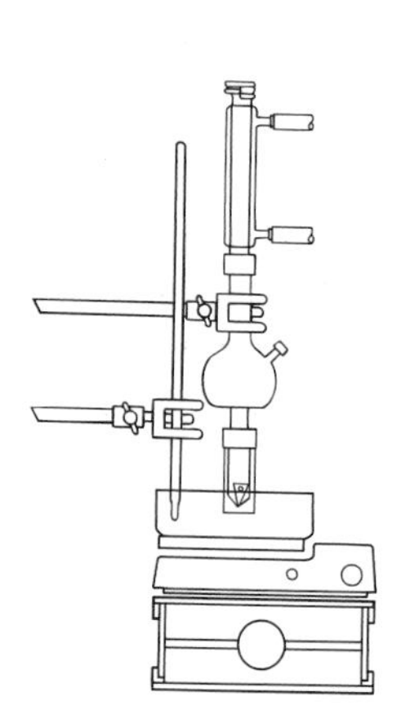

Transfer the filter cake to a Craig tube and recrystallize the crude product from 95% ethanol. Allow the solution to cool slowly to room temperature; then cool it to 0 °C in an ice-water bath. Isolate the crystalline product by vacuum or Craig tube filtration.

Concentrate the mother liquors to a volume of about 1 mL by simple distillation. Perform this operation in a hood to prevent release of vapors into the laboratory. Alternatively, use rotary evaporation or other techniques to concentrate the solution. Allow the residual solution to cool to room temperature to produce a second crop of 4-bromonitrobenzene. Isolate it by Craig tube filtration and, after air-drying, put it in a separate vial from the first crop.

Further concentrate the mother liquors from the second crop to a volume of about 0.5 mL. The resulting oil contains crude 2-bromonitrobenzene. Separate the oil from the two-phase mixture by means of a Pasteur pipet and weigh and reserve it for TLC analysis (Part B).

Analysis Weigh both crops of product and calculate the yield. Measure the melting points of both crops. Obtain IR and NMR spectra of your starting material and product and compare them with those of authentic samples (Figs. 8.31, 15.19–15.22).

WRAPPING IT UP

Neutralize the *aqueous filtrates* with sodium carbonate and flush them down the drain. Pour any *oil* obtained by concentration of the *ethanolic filtrate* into the container for halogenated liquids, *unless* the oil is to be subjected to chromatography.

B ■ *Thin-Layer Chromatography*

Preparation Sign in at **www.cengage.com/login** to answer Pre-Lab Exercises, access videos, and read the MSDSs for the chemicals used or produced in this procedure. Review Section 6.3.

Apparatus Capillary pipets, developing chamber.

Setting Up In two small vials, separately prepare solutions of 4-bromonitrobenzene and of the oil containing 2-bromonitrobenzene in about 0.5 mL of dichloromethane. Apply spots of each solution to the TLC plate and allow the spots to dry.★

Chromatography and Analysis Develop the chromatogram in a TLC chamber with 9:1 (*v:v*) hexane:ethyl acetate as the eluting solvent. When the solvent is within about 0.5 cm of the top of the plate, remove the developed chromatogram from the chamber, *quickly* mark the solvent front with a pencil, and allow the plate to dry.

Make the spots visible by placing the dry plate under an ultraviolet (UV) lamp or in a chamber whose atmosphere is saturated with iodine vapor. Calculate the R_f-values of the spots observed, and identify them as either 2- or 4-bromonitrobenzene. A small orange spot may be observed very near the origin of the chromatogram; this spot is 1-bromo-2,4-dinitrobenzene.

WRAPPING IT UP

Put the unused *eluant* in the container for nonhalogenated organic liquids. Discard the dry *TLC plates* in the container for halogenated organic compounds.

C ■ *Column Chromatography*

MINISCALE PROCEDURE

Preparation Sign in at **www.cengage.com/login** to answer Pre-Lab Exercises, access videos, and read the MSDSs for the chemicals used or produced in this procedure. Review Sections 6.2 and 6.3.

Apparatus Silica gel chromatography column, eight 13-mm × 100-mm test tubes, capillary pipets, developing chamber, *flameless* heating.

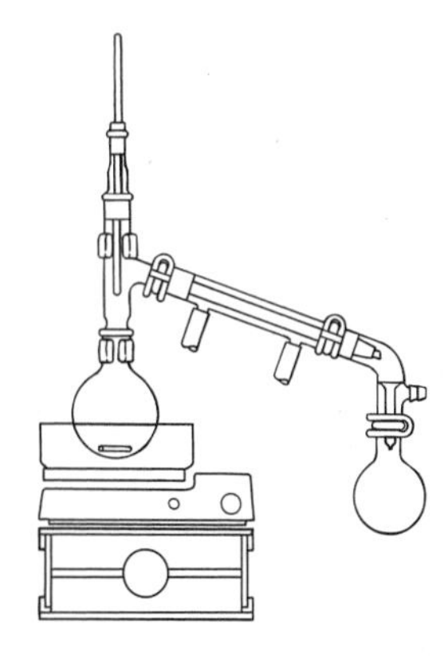

Setting Up Obtain a 25- or 50-mL buret and 5 g of silica gel. Pack the column according to the procedure described in the experimental procedure of Section 6.2 with the exception that a 9:1 (*v:v*) solution of hexane and ethyl acetate rather than petroleum ether is to be used for packing the column. With a Pasteur pipet, apply a 0.2-g sample of the oil containing the 2-bromonitrobenzene to the head of the column, and rinse the inside of the buret with 1 mL of ethyl acetate. Open the stopcock and allow the liquid to drain just to the top of the sand. Close the stopcock and fill the buret with 9:1 (*v:v*) hexane:ethyl acetate.

Chromatography and Analysis Elute the column until a total of 40 mL of the solvent has passed through the column. Do not allow the level of liquid to drain below the sand at the top of the column. Collect the eluate in 5- to 8-mL fractions in separate, labeled test tubes.

Analyze each fraction by TLC using 9:1 (*v:v*) hexane:ethyl acetate as the eluting solvent. With careful spotting, it should be possible to analyze at least three different fractions on each 3-cm × 8-cm strip of a silica gel chromatogram. In order to introduce sufficient sample onto the strip, re-spot two or three times.

Combine the fractions containing only 2-bromonitrobenzene and then remove the solvent by simple distillation. Alternatively, use rotary evaporation or other techniques to concentrate the solution. Characterize the residue obtained by either melting point determination, TLC, or spectroscopic analysis. 2-Bromonitrobenzene has a low reported melting point and thus may be isolated as an oil. Obtain IR and NMR spectra of your product and compare them with those of an authentic sample (Figs. 15.23 and 15.24).

WRAPPING IT UP

After first allowing any residual solvent to evaporate from the silica gel in the hood, put the *silica gel* in the container for nontoxic solids. Pour all *eluates containing bromonitrobenzenes* into the container for halogenated organic compounds; put *all other eluates* and any unused eluant in the container for nonhalogenated solvents.

EXERCISES

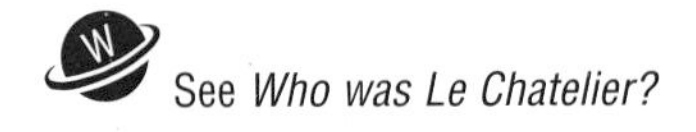
See *Who was Le Chatelier?*

1. Precipitation of the mononitration product of bromobenzene prevents dinitration from occurring. Explain how this experimental result is an application of the Le Chatelier principle.
2. How does maintaining the reaction temperature below 60 °C help suppress formation of dinitration by-products?
3. The pK_as of sulfuric and nitric acids are −3 (estimated value) and −1.3, respectively.
   a. Which of the two acids is stronger?
   b. Given your prediction in Part **a,** write a chemical equation for the equilibrium involving reaction of sulfuric and nitric acids.
   c. What is the value of K_{eq} for the acid-base reaction in Part **b?** Assume that the pK_a of the conjugate acid of nitric acid is −13.
4. Explain why 4-bromonitrobenzene (**20**) is *less* polar than 2-bromonitrobenzene (**19**).
5. Explain why the melting point of 4-bromonitrobenzene (**20**) is considerably higher than that of the 2-isomer **19.**
6. a. Provide the resonance structures that contribute to the σ-complexes **23** and **24.**
   b. Use these resonance structures to explain why the formation of 3-bromonitrobenzene (**21**) is disfavored relative to 4-bromonitrobenzene.
7. Why does 4-bromonitrobenzene (**20**) have a larger R_f-value in the TLC analysis than does the 2-isomer **19?**
8. Which isomer, 4-bromonitrobenzene or 2-bromonitrobenzene, will elute from a silica gel chromatography column first, and what physical property accounts for this order?
9. Why should a chromatography column never be allowed to go dry?
10. It is hard to measure the yield of 2-bromonitrobenzene accurately in the reaction you performed because this isomer is not purified in the procedure. However, the yield may be crudely approximated by assuming that the oil obtained from concentrating the mother liquors from recrystallization is comprised entirely of this isomer.
    a. Using the assumption that the oil is the 2-bromonitrobenzene, determine the *o*:*p* ratio of isomers formed in the mononitration of bromobenzene.
    b. Explain whether the result of your determination in Part **a** supports the hypothesis that the steric effect of the bromo substituent suppresses or augments the formation of the *o*-isomer relative to that expected statistically.
    c. A better way to assess the *o*:*p* ratio of the two bromonitro isomers would be to analyze the crude reaction mixture prior to any purification steps.

Propose a way by which you might perform such an analysis. Be specific in your answer.

d. The experimentally observed *o:p* ratio is reported to be 38:62.

i. What does this ratio indicate with regard to the steric effect of the bromo substituent?

ii. Use this ratio and the amount of 4-bromonitrobenzene actually isolated to estimate the experimental yield of mononitration in the reaction.

iii. What errors might attend using this method to calculate the extent of mononitration?

11. The *o:p* ratio in the mononitration of bromobenzene has been reported to be 38:62. Use this ratio and the amount of 4-bromonitrobenzene actually isolated to estimate the experimental yield of mononitration in the reaction. What errors are there in using this method to calculate the extent of mononitration?

12. Explain why 4-bromonitrobenzene cannot be prepared efficiently by the bromination of nitrobenzene.

13. Would nitration of bromobenzene or nitration of 4-bromonitrobenzene be expected to have the higher enthalpy of activation? Explain your answer.

Sign in at **www.cengage.com/login** and use the spectra viewer and Tables 8.1–8.8 as needed to answer the blue-numbered questions on spectroscopy.

14. Consider the spectral data for bromobenzene (Figs. 15.19 and 15.20).

a. In the IR spectrum, specify the absorption associated with the π-bonds of the aromatic ring. Indicate with what structural feature the strong absorptions at about 740 and 690 cm^{-1} are associated.

b. In the ^1H NMR spectrum, assign the various resonances to the hydrogen nuclei responsible for them.

c. For the ^{13}C NMR data, assign the various resonances to the carbon nuclei responsible for them.

15. Consider the spectral data for 4-bromonitrobenzene (Figs. 8.31, 15.21, and 15.22).

a. In the functional group region of the IR spectrum, specify the absorption associated with the π-bonds of the aromatic ring and indicate with what structural feature the strong absorption at 825 cm^{-1} is associated.

b. In the ^1H NMR spectrum, assign the various resonances to the hydrogen nuclei responsible for them.

c. For the ^{13}C NMR data, assign the various resonances to the carbon nuclei responsible for them.

16. Consider the spectral data for 2-bromonitrobenzene (Figs. 15.23 and 15.24).

a. In the functional group region of the IR spectrum, specify the absorption associated with the π-bonds of the aromatic ring and indicate with what structural feature the strong absorption at about 750 cm^{-1} is associated.

b. In the ^1H NMR spectrum, assign the various resonances to the hydrogen nuclei responsible for them.

c. For the ^{13}C NMR data, assign the various resonances to the carbon nuclei responsible for them.

17. What differences in the IR and NMR spectra of bromobenzene and 4-bromonitrobenzene are consistent with the introduction of a nitro group onto the ring in this experiment?

SPECTRA

Starting Materials and Products

The ^1H NMR spectrum of 4-bromonitrobenzene is provided in Figure 8.30.

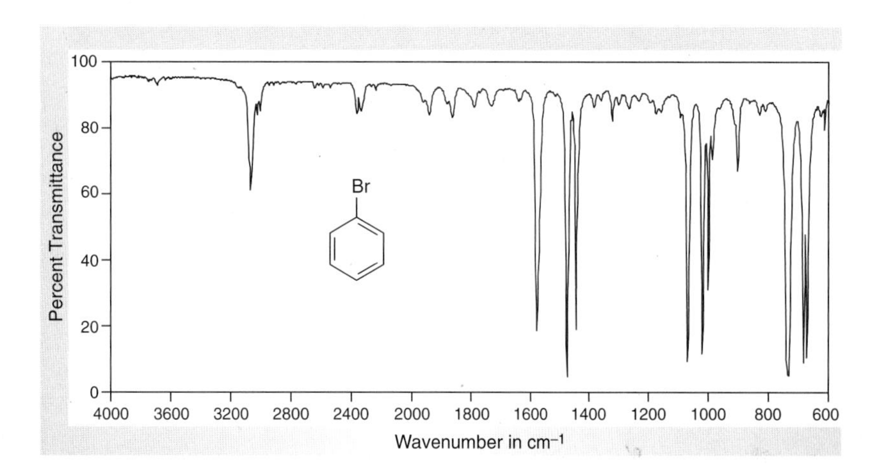

Figure 15.19
IR spectrum of bromobenzene (neat).

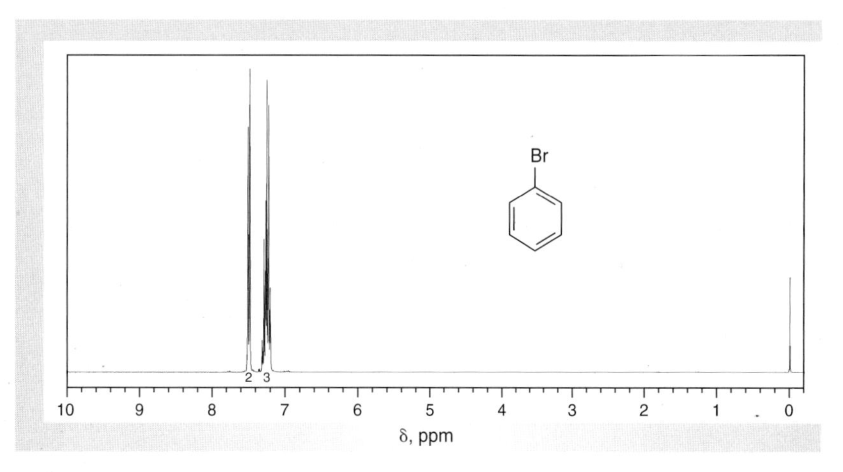

Figure 15.20
NMR data for bromobenzene ($CDCl_3$).

(a) ^1H NMR spectrum (300 MHz).
(b) ^{13}C NMR data: δ 122.5, 126.7, 129.8, 131.4.

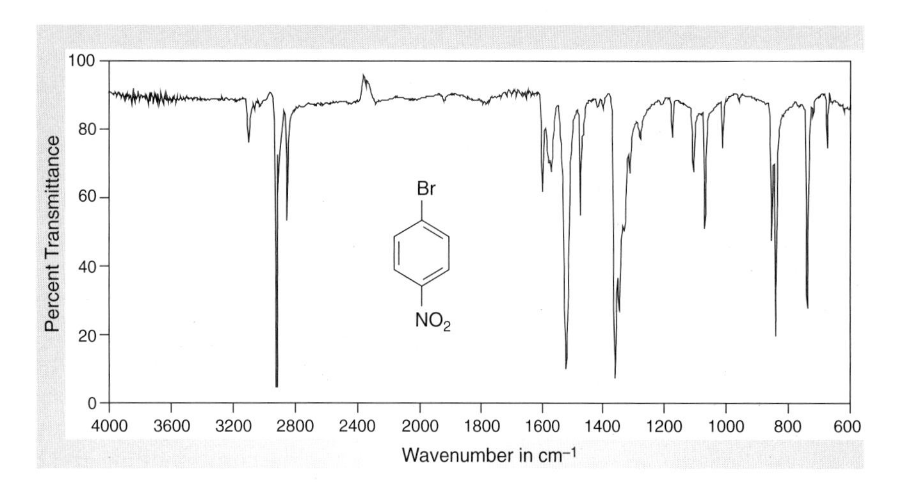

Figure 15.21
IR spectrum of 4-bromonitrobenzene (IR card).

Figure 15.22
NMR data for 4-bromonitrobenzene.

^{13}C NMR data: δ 124.9, 129.8, 132.5, 147.0.

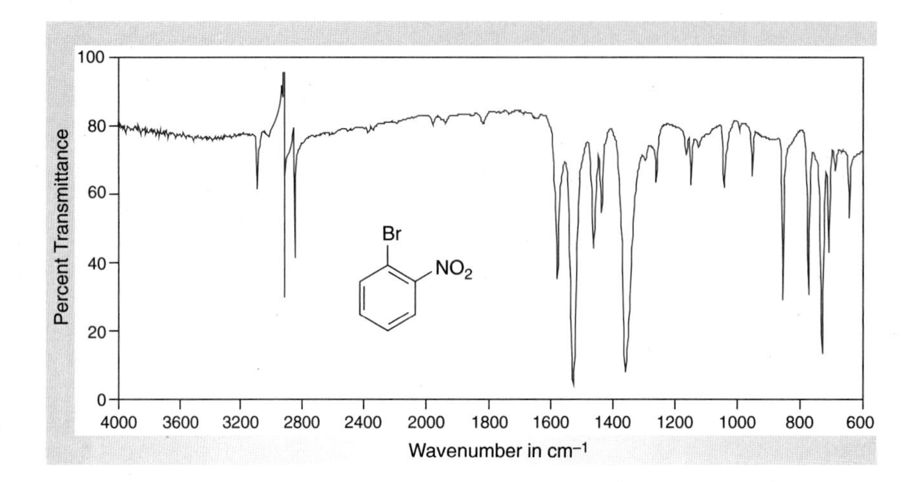

Figure 15.23
IR spectrum of 2-bromonitrobenzene (neat).

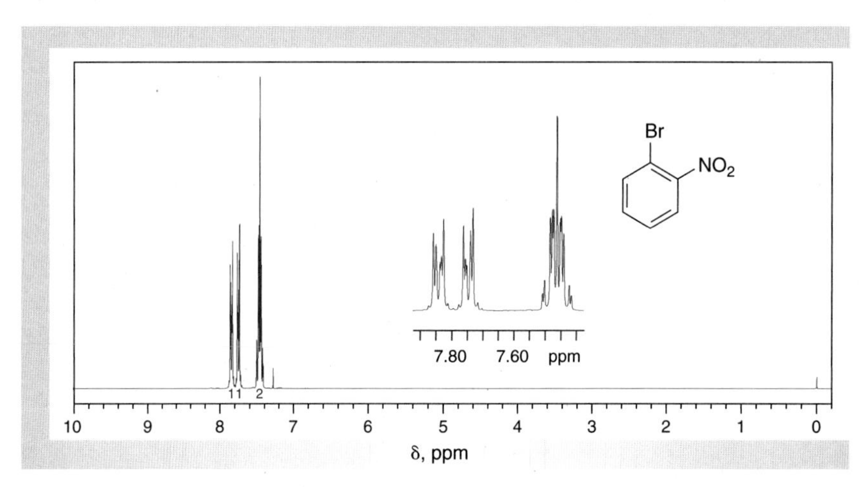

Figure 15.24
NMR data for 2-bromonitrobenzene ($CDCl_3$).

(a) ^{1}H NMR spectrum (300 MHz).
(b) ^{13}C NMR data: δ 114.1, 125.4, 128.2, 133.2, 134.8, 136.4.

15.5 RELATIVE RATES OF ELECTROPHILIC AROMATIC SUBSTITUTION

Electrophilic aromatic substitution reactions (Sec. 15.1) are among the best understood of all organic reactions. The *qualitative* aspects of the reactions that are discussed in textbooks include the effect substituents have on the reactivity of arenes toward electrophiles and the orientation, *ortho, meta,* or *para,* of their attack on the ring. However, relatively little information is given in textbooks about the *quantitative* differences in rates and reactivities of substituted aromatic compounds. The experimental procedures of this section provide both *semiquantitative* and *quantitative* measures of the differences in reactivity of a series of arenes toward the bromonium ion, Br^+, to produce the corresponding aryl bromides (Eq. 15.20).

$$\text{Ar—H} + \text{Br}_2 \xrightarrow{\text{HOAc}} \text{Ar—Br} + \text{HBr} \qquad (15.20)$$

An arene — Bromine (*red-brown*) — An aryl bromide (*colorless*)

Studies of electrophilic substitutions on arenes are reported in which the experimental conditions allow a direct comparison of the relative reaction rates. For example, the relative reactivities of benzene and toluene toward halogenation, acetylation, sulfonation, nitration, and methylation have been determined. In all cases, electrophilic aromatic substitution was more rapid with toluene. For example, bromination of toluene is some 600 times faster than that of benzene. Such studies have led to the classification of substituents as **ring activators** or **deactivators,** depending on whether the substituted arene reacts *faster* or *slower* than benzene itself. Thus, the methyl group of toluene is a ring activator.

The effect substituents have on the relative rate of a specific type of S_E2 reaction may exceed several orders of magnitude, and the range of reactivity when comparing substituents may be enormous. For example, phenol reacts 10^{12} times faster and nitrobenzene reacts 10^5 times *slower* than benzene in electrophilic bromination. The hydroxy group is thus a powerful *ring activator,* and the nitro group is a potent *ring deactivator.* The range of difference in reactivity due to these two substituents is a remarkable 10^{17}!

You will be measuring the relative rates of bromination of six monosubstituted arenes **27–32.** This particular electrophilic aromatic substitution is selected because the relative rates may be measured both qualitatively and quantitatively. The results should allow you to determine the order of ring activation associated with the various substituents that you will investigate.

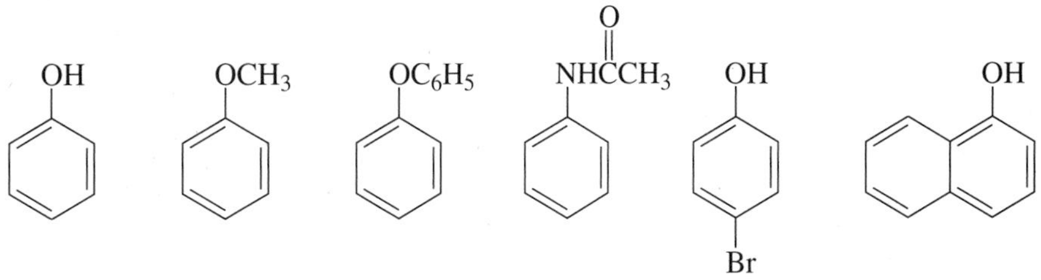

OH — **27** Phenol; OCH_3 — **28** Anisole; OC_6H_5 — **29** Diphenyl ether; $NHCCH_3$ (C=O) — **30** Acetanilide; OH, Br — **31** 4-Bromophenol; OH — **32** 1-Naphthol

The qualitative measurements are easier to perform because they simply rely on your observing the disappearance of the characteristic reddish color of molecular bromine as a function of time. You should be able to rank-order the substituent effects, but the results will be subject to the vagaries of human judgment with respect to knowing when all of the color has been discharged from the reaction mixture.

Quantitative measurements also depend on determining the rate of disappearance of the color of bromine, but a spectrophotometer is used rather than the human eye to measure changes in the visible portion of the electromagnetic spectrum (Sec. 8.1). None of the substrates **28–30** that we are studying absorb at this wavelength; only molecular bromine does, so a decrease in light absorption reflects a decrease in the concentration of molecular bromine. It is not possible to extend quantitative analyses to **27, 31,** and **32** because they react so fast that obtaining accurate measurements of absorbance versus time requires use of more expensive spectrophotometers than are normally available in undergraduate laboratories.

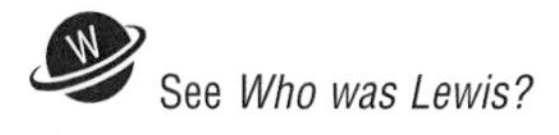
See *Who was Lewis?*

In order to analyze the data from the quantitative measurements, we start by considering the rate expression for an S_E2 reaction that is given in Equation 15.21; this equation is identical to Equation 15.2 (Sec. 15.1) except that $[Br_2]$, the precursor to Br^+, has replaced $[E^+]$. This expression could be more complex if a Lewis acid catalyst were required to promote formation of the electrophile, which is the case when the arene is less reactive toward electrophilic aromatic substitution than the ones we've selected. The rate is seen to be first-order in both reagents and therefore is second-order overall (Sec. 14.6).

$$\text{Rate} = k_2[\text{ArH}][\text{Br}_2] \tag{15.21}$$

Calculations for second-order reactions are more complex than those for first-order processes, so we've designed our experimental protocol so that the kinetics will be pseudo-first-order by using a sufficiently large excess of the aromatic substrate so that its concentration does not change appreciably during the reaction. Equation 15.21 then may be simplified to Equation 15.22, in which the concentration of the arene has been incorporated into the rate constant, k_1; k_1 is then called a *pseudo-first-order rate constant.*

$$\text{Rate} = k_1[\text{Br}_2] \tag{15.22}$$

The *integrated* form of Equation 15.22 is given in Equation 15.23, which can be rearranged to provide Equations 15.24 and 15.25, where $[Br_2]_0$ and $[Br_2]_t$ are the initial concentration of bromine and the concentration of bromine at any time t, respectively. By plotting the natural log, ln, or $\log_{10}$ of the ratio of concentrations versus time, a straight line should result if the reaction is indeed pseudo-first-order; the slope of the line is k_1 if the natural log is used, but is $k_1/2.303$ if the common log is used.

$$[\text{Br}_2]_t = [\text{Br}_2]_0 e^{-k/t} \tag{15.23}$$

$$k_1 t = \ln \frac{[\text{Br}_2]_0}{[\text{Br}_2]_t} \tag{15.24}$$

$$k_1 t = 2.303 \log \frac{[\text{Br}_2]_0}{[\text{Br}_2]_t} \tag{15.25}$$

You will be using an ultraviolet-visible (UV-Vis) spectrophotometer to measure the concentration of molecular bromine as a function of time. Fortunately, a relatively inexpensive nonrecording instrument is sufficient for our purposes because all measurements will be made at a single wavelength of 400 nm. You personally will be serving as the recorder, noting the absorbance, A (Eq. 15.26), of Br_2 at this wavelength as a function of time. Equation 15.26 was originally discussed in Section 8.4 (Eq. 8.14), and you should review that discussion. Because the same cell will be used for all measurements, l is constant. The molar extinction coefficient, ε, is also constant because only molecular bromine is absorbing at the selected wavelength. Consequently, Equations 15.25 and 15.26 may be combined to give Equation 15.27, since $A_o = \varepsilon[Br_2]_0 l$ and $A_t = \varepsilon[Br_2]_t l$. This is a useful form of Equation 15.24 because it relates absorbance to the rate constant k_1.

$$A = \log \frac{I_o}{I} = \varepsilon c l \tag{15.26}$$

$$k_1 t = 2.303 \log \frac{A_o}{A_t} \tag{15.27}$$

EXPERIMENTAL PROCEDURES

Relative Rates of Electrophilic Aromatic Bromination

Purpose To evaluate the relative reactivities of different arenes in electrophilic aromatic bromination.

SAFETY ALERT

1. ***Bromine is a hazardous chemical that may cause serious chemical burns.*** **Do not breathe its vapors or allow it to come into contact with your skin. Perform all operations involving the transfer of the pure liquid or its solutions in a hood and wear latex gloves for these manipulations. Even though the solutions used in the experiments are dilute, take proper precautions. If you get bromine on your skin, wash the area *immediately* with soap and warm water and soak any affected area in 0.6 *M* sodium thiosulfate solution, for up to 3 h if the burn is particularly serious.**
2. **Do not use your mouth to fill pipets when transferring any of the solutions in this experiment; this applies to both the bromine solutions and the acetic acid solutions of the substrates.**

A ■ *Qualitative Measurements*

Note It is convenient to work in pairs for this experiment.

MINISCALE PROCEDURES

Preparation Sign in at **www.cengage.com/login** to answer Pre-Lab Exercises, access videos, and read the MSDSs for the chemicals used or produced in this procedure. Review Section 2.5.

Apparatus A 1-L beaker, three calibrated Pasteur pipets, six 13-mm × 100-mm test tubes, and *flameless* heating.

Setting Up As accurately as possible, calibrate several Pasteur pipets to contain 1.5 mL. In separate labeled test tubes, place 1.5 mL of a 0.2 *M* stock solution of each of the following substrates in 15 *M* (90%) acetic acid: (a) phenol, (b) anisole, (c) diphenyl ether, (d) acetanilide, (e) 4-bromophenol, and (f) 1-naphthol. Use special care to ensure that one stock solution is not contaminated by another during any transfer operations you perform.

Use the 1-L beaker to make a water bath and establish a bath temperature of 35 ± 2 °C. Suspend the test tubes partially in the water bath by looping a piece of copper wire around the neck of the test tube and over the rim of the beaker. Transfer about 25 mL of a solution containing 0.05 *M* bromine in 15 *M* acetic acid to an Erlenmeyer flask, and equilibrate this solution in the water bath for a few minutes.

Bromination Add 1.5 mL of the bromine solution to a test tube containing one of the substrates. Make the addition *rapidly,* mix the solution *quickly,* and *record the* *exact time* of addition. Monitor the reaction mixture and note the elapsed time for the bromine color to become faint yellow or to disappear. If decoloration has not occurred within 5 min, which can happen because some substrates react slowly, proceed to a second substrate while waiting for the endpoint of the first to be reached. Record the reaction times. Repeat this procedure with each of the arenes, using the same endpoint color for each one.

To confirm your conclusions regarding the relative reactivities of the series of arenes, repeat the experiment at 0 °C in an ice-water bath, using only those substrates that were not clearly differentiated at 35 °C.

Analysis On the basis of your observations, list the arenes in order of *decreasing* reactivity toward bromine. Obtain IR and NMR spectra of the starting materials and compare them with those of authentic samples (Figs. 8.7 and 15.25–15.35).

B ■ *Quantitative Measurements*

Note It is convenient to work in pairs for this experiment. In preparing for and executing the kinetic runs, you must be well organized so that you can work quickly. You may need to perform several kinetic runs before you are able to work rapidly enough to get acceptable results.

MINISCALE PROCEDURES

Preparation Sign in at **www.cengage.com/login** to answer Pre-Lab Exercises, access videos, and read the MSDSs for the chemicals used or produced in this procedure. Review Section 2.5.

Apparatus UV-Vis spectrophotometer or colorimeter, three capped cuvettes, three calibrated Pasteur pipets, glass stirring rod, timer, and thermometer.

Setting Up Have at your disposal 0.5 *M* stock solutions of anisole, diphenyl ether, and acetanilide in 15 *M* (90%) aqueous acetic acid and a 0.02 *M* solution of bromine in 15 *M* acetic acid. Calibrate the spectrophotometer to zero absorbance at 400 nm using a solution of the arene substrate in acetic acid. Prepare this solution by using a 2-mL pipet to combine 2 mL of 15 *M* acetic acid and 2 mL of the stock solution of the substrate arene in a clean and dry cuvette. Stir the solution to ensure homogeneity and place the cuvette in the spectrophotometer. Adjust the absorbance reading of the spectrometer to zero. Record the laboratory temperature.

Reaction Once you have zeroed the spectrophotometer for a particular substrate, clean and dry the cuvette used for the calibration, and continue with the following operations. (1) Add 2 mL of 0.02 *M* bromine solution to the cuvette; (2) add 2 mL of the 0.5 *M* solution of substrate; (3) quickly stir the solution one time only, and have your partner record the exact time of mixing; (4) place the cuvette in the spectrophotometer; and (5) record the absorbance of the solution at 400 nm as a function of time. Obtain as many readings as possible before the absorbance drops below about 0.5, although readings may be taken below 0.5. Record the time at which the needle on the meter crosses a line on the absorbance scale rather than attempting interpolation between those lines.

Repeat the above procedure for each of the other substrates. For those runs that are particularly fast, it is advisable to perform duplicate runs.

Analysis Obtain IR and NMR spectra of the starting materials and compare them with those of authentic samples (Figs. 8.7, 15.27–15.31). Compute the desired rate constants, k_l, according to the following procedure.

1. The data recorded above constitute a series of values A_t, and it is necessary to extrapolate these data to the initial time, t_o, of mixing to obtain A_o. Plot log A_t as the ordinate versus time in seconds as the abscissa and extrapolate the straight line obtained to zero time to obtain log A_o. Determine A_o for each substrate.
2. Using the value of A_o and the recorded values of A_t, calculate a series of values of 2.303 log A_o/A_t for each substrate. Plot these values as the ordinate versus time in seconds as the abscissa. Draw the best straight line possible through these points. Calculate the slope of that line to obtain k_1 (sec^{-1}) for that substrate. You may also wish to determine the slope of the line using a least-squares analysis.
3. Divide all values of k_1 by the *smallest* value of this rate constant to obtain the *relative* reactivities of the substrate toward electrophilic aromatic bromination.

WRAPPING IT UP

Neutralize the combined *acetic acid solutions* by adding saturated aqueous sodium bicarbonate and filter the resulting mixture under vacuum. Flush the *filtrate* down the drain; put the *filter cake,* which contains aryl bromides, in the container for halogen-containing solid waste.

EXERCISES

General Questions

1. Write the structures for the major monobromination products that would be formed from each of the substrates used. Explain your predictions.
2. Using curved arrows to symbolize the flow of electrons, write the stepwise mechanism for the bromination of anisole to provide the major product you proposed in Exercise 1. Write the resonance structures for any charged, delocalized intermediates that are involved.
3. Explain how a nonpolar molecule like bromine can act as an electrophile.
4. Why does performing some of the S_E2 reactions at 0 °C allow you to identify the order of reactivity more definitively?
5. What problem might there be if you attempted to perform the kinetic experiment at −20 °C?
6. Would you expect the relative rate of S_E2 reactions to be faster or slower if molecular chlorine, Cl_2, were used instead of molecular bromine, Br_2? Explain your reasoning.
7. Why is the first step in an electrophilic aromatic bromination reaction so much more endothermic than the first step in an electrophilic addition to an alkene?
8. Do any of the substrates in this experiment contain substituents that deactivate the aromatic ring toward electrophilic aromatic substitution?
9. Which compound would you expect to undergo bromination at a faster rate, aniline, $C_6H_5NH_2$, or acetanilide? Anilinium ion, $C_6H_5NH_3^+$, or aniline? Explain your reasoning.
10. Explain why a methyl group is a mildly activating group, whereas a methoxy group is strongly activating.

11. Using inductive and/or resonance effects, explain the reactivity order you observed.

12. a. Determine the *initial* concentrations of bromine and of substrate arene in each reaction mixture.

    b. Which reagent, bromine or arene, is present in excess in each case?

    c. Explain why the experimental procedure calls for an excess of the specified reactant in **b** over the other.

    d. Discuss whether you would be able to determine the rates of reaction if the concentrations of arene and of bromine had been reversed over what is specified in the experimental procedure.

Questions for Part A

13. Provide a mechanistically based explanation for the relative order of reactivity that you obtained.

14. What might be a major source of error in determining relative rates of reaction by the procedure that you used?

Questions for Part B

15. Derive Equation 15.27 from Equations 15.25 and 15.26. On the basis of this derivation, explain why it is unnecessary to use either the extinction coefficient, ε, or the cell length, l, for analyzing the data.

16. Specify sources of error in this procedure for determining the relative rates of electrophilic aromatic substitution.

Sign in at **www.cengage.com/login** and use the spectra viewer and Tables 8.1–8.8 as needed to answer the blue-numbered questions on spectroscopy.

17. Consider the spectral data for phenol (Figs. 15.25 and 15.26).

    a. In the functional group region of the IR spectrum, specify with what structural feature the broad band centered at about 3280 cm^{-1} is associated, and explain why the band is broad.

    b. In the ^1H NMR spectrum, assign the various resonances to the hydrogen nuclei responsible for them.

    c. For the ^{13}C NMR data, assign the various resonances to the carbon nuclei responsible for them.

18. Consider the spectral data for anisole (Figs. 15.27 and 15.28).

    a. In the functional group region of the IR spectrum, specify the absorption associated with the π-bonds of the aromatic ring.

    b. In the ^1H NMR spectrum, assign the various resonances to the hydrogen nuclei responsible for them.

    c. For the ^{13}C NMR data, assign the various resonances to the carbon nuclei responsible for them.

19. Consider the spectral data for diphenyl ether (Figs. 15.29 and 15.30).

    a. In the functional group region of the IR spectrum, specify the absorption associated with the π-bonds and with the C–H bonds of the aromatic rings.

    b. In the ^1H NMR spectrum, assign the various resonances to the hydrogen nuclei responsible for them.

    c. For the ^{13}C NMR data, assign the various resonances to the carbon nuclei responsible for them.

20. Consider the spectral data for acetanilide (Figs. 8.7 and 15.31).

    a. In the functional group region of the IR spectrum, specify the absorptions associated with the N–H bond, with the carbon-oxygen double bond, and with the aromatic ring.

b. In the ^1H NMR spectrum, assign as many as possible of the various resonances to the hydrogen nuclei responsible for them.

c. For the ^{13}C NMR data, assign the various resonances to the carbon nuclei responsible for them.

21. Consider the spectral data for 4-bromophenol (Figs. 15.32 and 15.33).

a. In the functional group region of the IR spectrum, specify the absorption associated with the π-bonds of the aromatic ring; indicate with what structural feature the broad band centered at about 3400 cm^{-1} is associated, and explain why the band is broad.

b. In the ^1H NMR spectrum, assign the various resonances to the hydrogen nuclei responsible for them.

c. For the ^{13}C NMR data, assign the various resonances to the carbon nuclei responsible for them.

22. Consider the spectral data for 1-naphthol (Figs. 15.34 and 15.35).

a. In the functional group region of the IR spectrum, specify the absorption associated with the π-bonds of the aromatic ring. Indicate with what structural feature the broad band centered at about 3300 cm^{-1} is associated, and explain why the band is broad.

b. In the ^1H NMR spectrum, assign the various resonances to the hydrogen nuclei responsible for them.

c. For the ^{13}C NMR data, assign the various resonances to the carbon nuclei responsible for them.

23. When the NMR solutions of phenol, 4-bromophenol, and 1-naphthol were each shaken with a drop of D_2O, the resonances assigned to the hydrogen of the hydroxyl group disappeared. Explain this observation.

SPECTRA

Starting Materials and Products

The IR spectrum of acetanilide is provided in Figure 8.7.

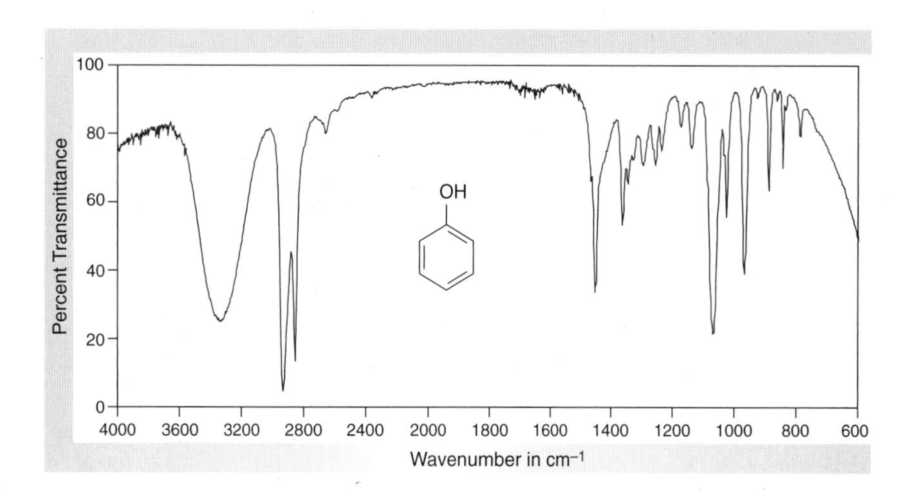

Figure 15.25
IR spectrum of phenol (neat).

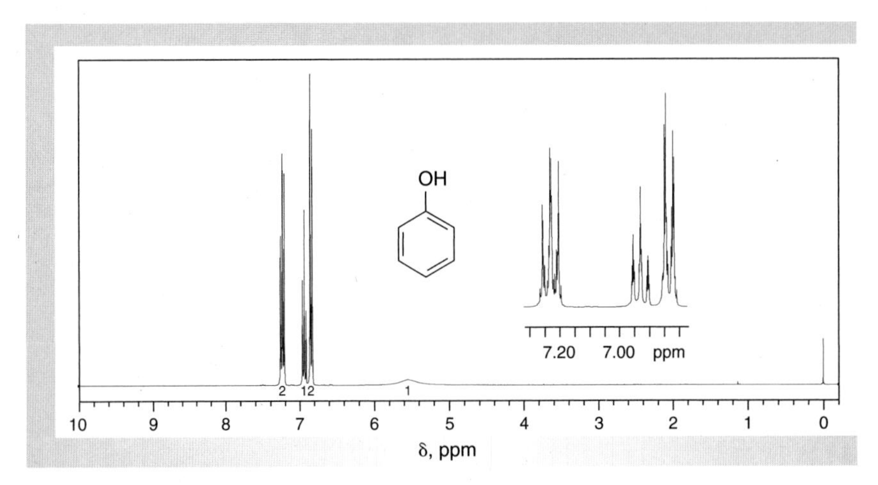

Figure 15.26
NMR data for phenol ($CDCl_3$).

(a) ^1H NMR spectrum (300 MHz).
(b) ^{13}C NMR data: δ 115.4, 121.0, 129.7, 154.9.

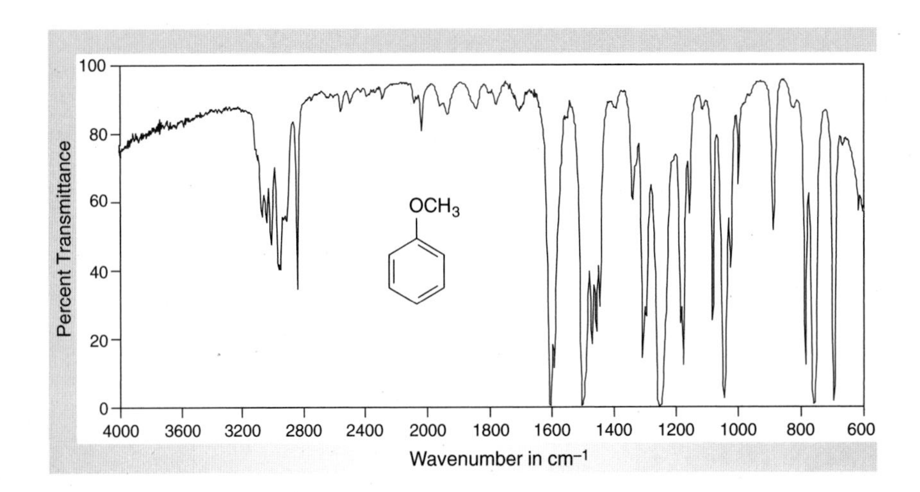

Figure 15.27
IR spectrum of anisole (neat).

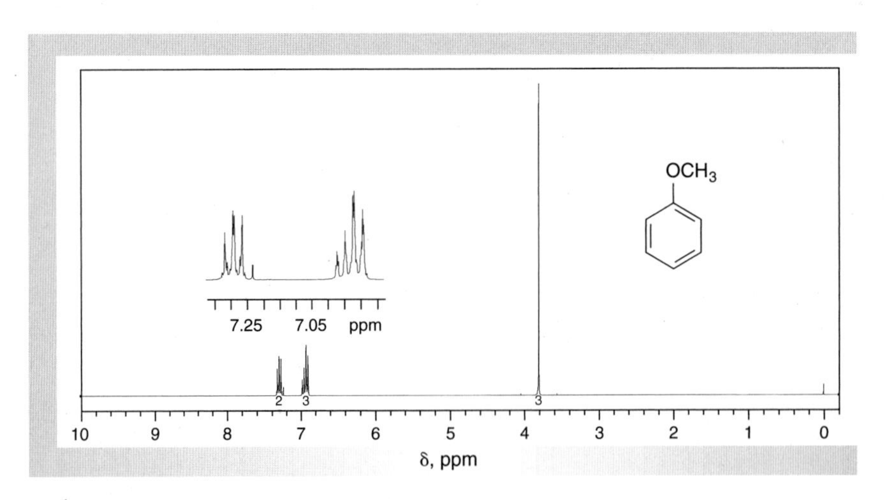

Figure 15.28
NMR data for anisole ($CDCl_3$).

(a) ^1H NMR spectrum (300 MHz).
(b) ^{13}C NMR data: δ 54.8, 114.1, 120.7, 129.5, 159.9.

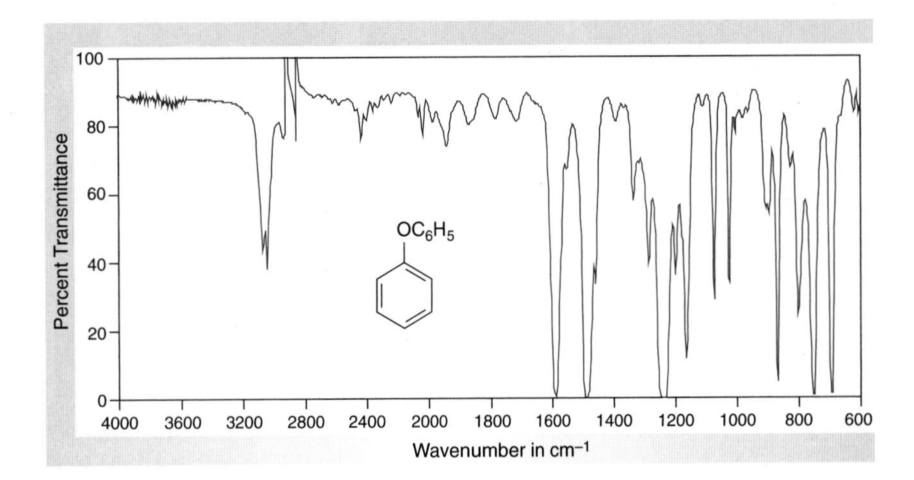

Figure 15.29
IR spectrum of diphenyl ether (IR card).

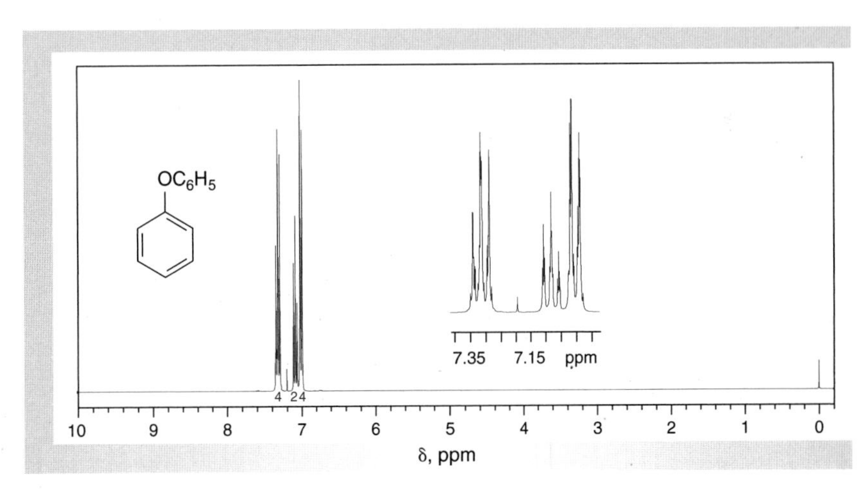

Figure 15.30
NMR data for diphenyl ether ($CDCl_3$).

(a) ^1H NMR spectrum (300 MHz).
(b) ^{13}C NMR data: δ 119.0, 123.2, 129.8, 157.6.

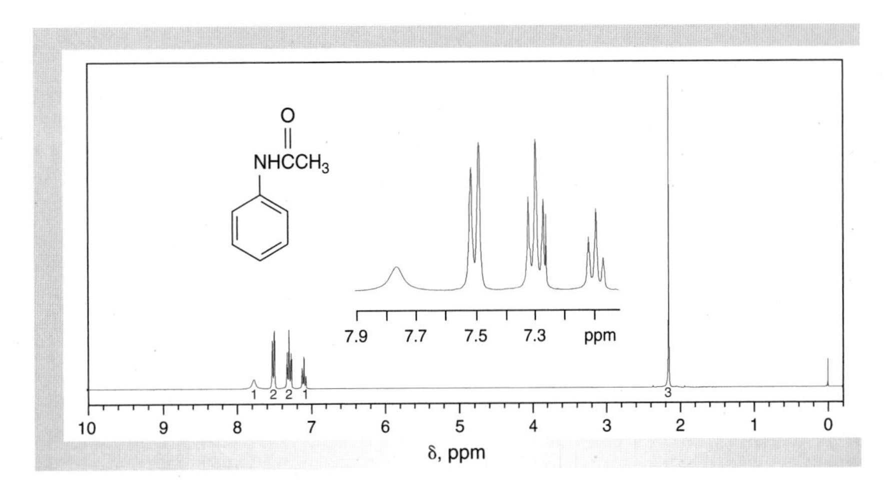

Figure 15.31
NMR data for acetanilide ($CDCl_3$).

(a) ^1H NMR spectrum (300 MHz).
(b) ^{13}C NMR data: δ 24.0, 119.7, 123.3, 128.5, 139.2, 168.7.

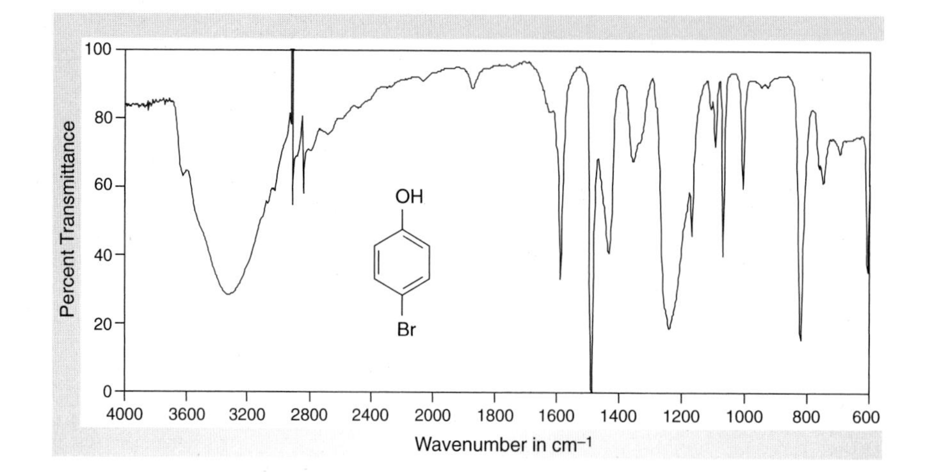

Figure 15.32
IR spectrum of 4-bromophenol (IR card).

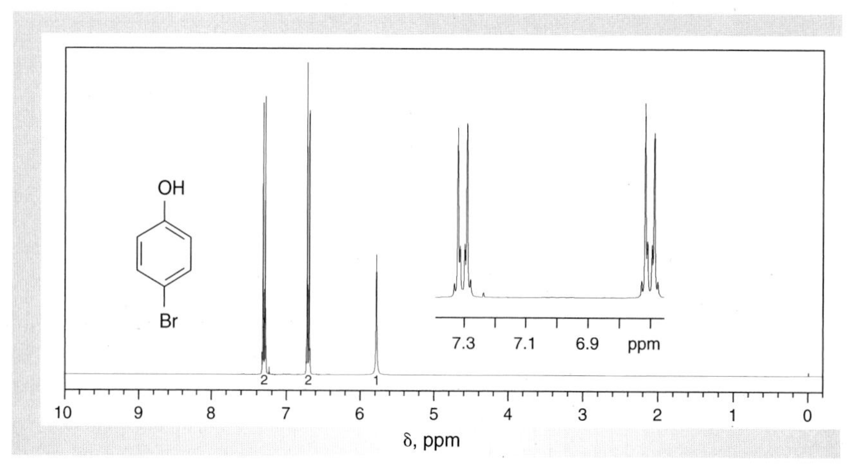

Figure 15.33
NMR data for 4-bromophenol ($CDCl_3$).

(a) ^{1}H NMR spectrum (300 MHz).
(b) ^{13}C NMR data: δ 113.2, 117.2, 132.5, 153.9.

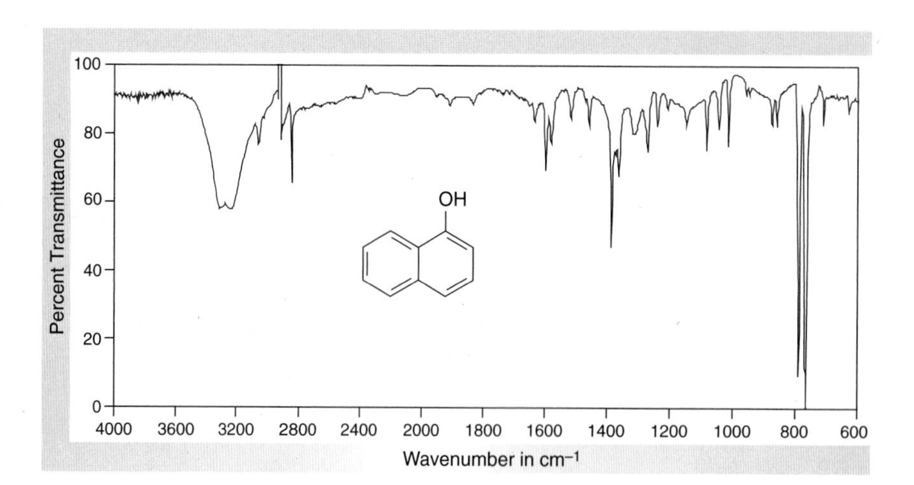

Figure 15.34
IR spectrum of 1-naphthol (IR card).

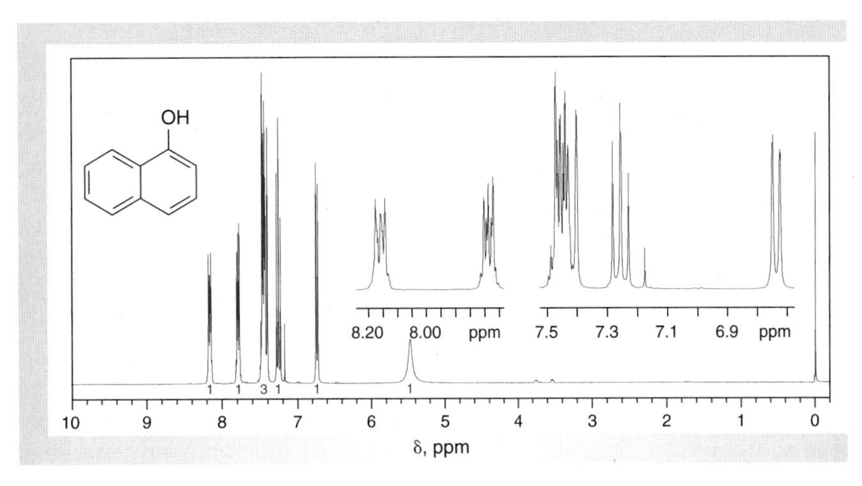

Figure 15.35
NMR data for 1-naphthol ($CDCl_3$).

(a) ^{1}H NMR spectrum (300 MHz).
(b) ^{13}C NMR data: δ 108.8, 119.8, 122.1, 124.9, 126.2, 127.6, 134.9, 152.4.

HISTORICAL HIGHLIGHT

Discovery of the Friedel-Crafts Reaction

The Friedel-Crafts reaction bears the names of two chemists, Charles Friedel, a Frenchman, and James Crafts, an American, who accidentally discovered this reaction in Friedel's laboratory in Paris in 1877. Friedel and Crafts quickly recognized the potential significance of their discovery, and they immediately secured patents in France and England on procedures for preparing hydrocarbons and ketones using the reaction. Their judgment was accurate. Probably no other organic reaction has been of more practical value. Major industrial processes for producing high-octane gasoline, synthetic rubber, plastics, and synthetic detergents are representative applications of Friedel-Crafts chemistry. All of this came from an unexpected laboratory result and the brilliance of the two scientists who observed, interpreted, and extended it.

Crafts was born in 1839 in Boston. After graduating from Harvard University at the age of 19, he spent a year studying mining engineering and then decided to go to Europe for further education. Crafts became fascinated with chemistry while studying metallurgy in Freiburg, Germany, and he subsequently secured positions in the laboratories of Bunsen in Heidelberg and Wurtz in Paris. It was in Wurtz's laboratory that he met Friedel, and, owing to their similar scientific interests, the two began a collaboration in 1861.

Crafts returned to the United States in 1865 and, after a brief tenure as a mining inspector in Mexico and California, accepted a position as Professor of Chemistry at Cornell University shortly after this institution was founded. Three years later, he moved to M.I.T. Because of poor health, he decided to return to Europe in 1874 and resume his collaboration with Friedel in Wurtz's laboratory in Paris.

When Crafts left M.I.T., he had expected to return in a short time, presumably after recovering his health. Owing to the change of climate or, perhaps, the excitement of the discovery he shared with Friedel in 1877, his health improved dramatically, yet he remained in Paris for another 17 years. Friedel and Crafts conducted an energetic research collaboration and by 1888 had produced over 50 publications, including patents, related to the reactions of aluminum chloride with organic compounds.

What was the accidental research discovery that engraved the names of Friedel and Crafts in scientific and industrial annals? They were attempting to prepare 1-iodopentane by treating 1-chloropentane with aluminum and iodine. They observed that the reaction took an entirely different course from that expected, producing large amounts of hydrogen chloride and, unexpectedly, hydrocarbons. Upon further investigation, they

(Continued)

HISTORICAL HIGHLIGHT *Discovery of the Friedel-Crafts Reaction (Continued)*

found that using aluminum chloride in place of aluminum gave the same results. Other researchers had earlier reported somewhat similar results from reactions of organic chlorides with certain metals but had not explained them or implicated the metal chloride as a reactant or catalyst. It was the work of Friedel and Crafts that first showed that the presence of the metal chloride was essential.

They subsequently performed a reaction of 1-chloropentane with aluminum chloride using benzene as the solvent. Once again, they observed evolution of hydrogen chloride, but this time they found that 1-pentylbenzene ("amylbenzene") was a major product! In describing their discovery to the Chemical Society of France in 1877, they reported, "With a mixture of [alkyl] chloride and hydrocarbon [an arene], the formation is established, in good yield, of hydrocarbons from the residues of the hydrocarbon less H and from the chloride less Cl. It is thus that ethylbenzene, amylbenzene, benzophenone, etc., are obtained." A general representation of the transformation is shown in the following equation.

$$C_6H_5\text{—}H + R\text{—}Cl \xrightarrow{AlCl_3} C_6H_5\text{—}R + HCl$$

Friedel and Crafts recognized that their unexpected result promised the possibility of synthesizing a wide variety of hydrocarbons and ketones, and they immediately proved this by experiment. In the ensuing years, the voluminous research papers and patents that came from Friedel and Crafts established a whole new area of research and practice in organic chemistry and laid the foundation for some important modern industrial chemical processes.

Friedel-Crafts chemistry has touched our lives in many important but perhaps unrecognized ways. For example, Winston Churchill, referring to winning the air war over Britain and the role of fighter pilots, once said, "Never in the field of human conflict was so much owed by so many to so few." What was not recognized by many at the time, perhaps even Churchill, was that the victory in the air war was due not only to the skill and daring of the British pilots but to the superiority of their aviation gasoline over that of the Germans. It has been generally acknowledged that the German fighter planes were mechanically superior to the British planes, but their fuel was not. The gasoline used in the British planes was of higher quality because it contained toluene and other alkylated aromatic hydrocarbons, which provided higher octane ratings. Production of these types of hydrocarbons was a direct outgrowth of Friedel-Crafts chemistry.

Similarly, "synthetic rubber" was vital to the ground-war effort in World War II, after the Japanese cut off the Allies from sources of natural rubber in Southeast Asia (see Historical Highlight in Chapter 11). A synthetic rubber was developed in an amazingly short time in a remarkable display of cooperation between government officials and industrial and academic scientists. This rubber was a copolymer (Sec. 22.1) that contained styrene as a key component. Styrene is made from ethylbenzene, which is prepared industrially by the Friedel-Crafts reaction of benzene and ethylene, as shown below.

Styrene can also be homopolymerized to produce polystyrene (Sec. 22.2). This material is one of the most versatile plastics ever invented. For example, it can be molded into rigid cases for radios and batteries or formulated in a more flexible form useful in the manufacture of toys and all kinds of containers. Polystyrene can also be produced in the form of a lightweight foam known as Styrofoam®. This foam is used for insulation in buildings and is molded into ice chests and disposable cups for hot and cold drinks.

The Friedel-Crafts reaction is a key part of the manufacture of many synthetic detergents, cleaning agents

$$\underset{\text{Benzene}}{C_6H_5\text{—}H} + \underset{\text{Ethylene}}{CH_2{=}CH_2} \xrightarrow[\text{catalyst}]{\text{Acid}} \underset{\text{Ethylbenzene}}{C_6H_5\text{—}CH_2CH_3} \xrightarrow{\text{Dehydrogenation}} \underset{\text{Styrene}}{C_6H_5\text{—}CH{=}CH_2}$$

(Continued)

HISTORICAL HIGHLIGHT *Discovery of the Friedel-Crafts Reaction (Continued)*

that have revolutionized the way we live today. We wash dishes and clothes with them, and, in contrast to soap, they work as well in hard water as in soft. A typical example of a biodegradable synthetic detergent is sodium dodecylbenzenesulfonate, in which the 12-carbon side chain is attached to the benzene ring by a Friedel-Crafts alkylation reaction, as illustrated here.

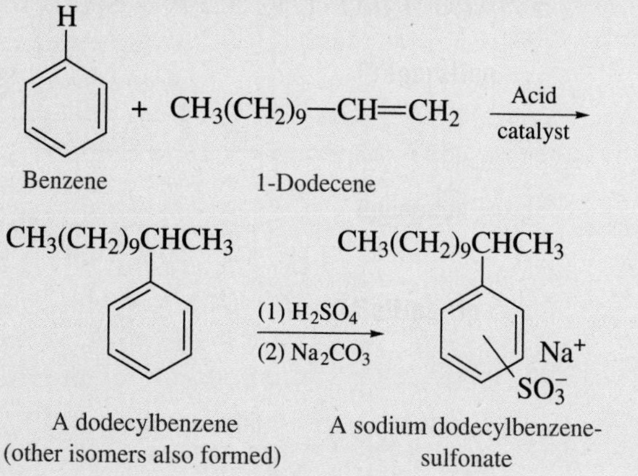

Benzene + $CH_3(CH_2)_9{-}CH{=}CH_2$ (1-Dodecene) $\xrightarrow{\text{Acid catalyst}}$ $CH_3(CH_2)_9CHCH_3$ (attached to benzene ring)

A dodecylbenzene (other isomers also formed) $\xrightarrow{(1)\ H_2SO_4;\ (2)\ Na_2CO_3}$ $CH_3(CH_2)_9CHCH_3$ (ring bearing SO_3^- Na^+)

A sodium dodecylbenzene-sulfonate

These are just a few of the practical applications that have been developed in the years since Friedel and Crafts made their initial observation of an unexpected experimental result. By proper interpretation of the result, they have earned credit for a major discovery.

Relationship of Historical Highlight to Experiments

The experiments in Sections 15.2 and 15.3 illustrate the Friedel-Crafts alkylation and acylation of aromatic hydrocarbons, respectively. A complication of Friedel-Crafts reactions is apparent in the alkylation experiment, wherein rearrangements of the carbocations generated from the alkyl halide provide mixtures of substitution products. The acylation reaction of Section 15.3 provides an example of how a combination of electronic and steric effects can affect the orientation of electrophilic attack on an aromatic ring.

See *Who were Friedel and Crafts?*

See *Who was Wurtz?*

See *Who was Bunsen?*

See *Who was Churchill?*

See more on *Detergents*

Oxidation of Alcohols and Carbonyl Compounds

When you see this icon, sign in at this book's premium website at **www.cengage.com/login** to access videos, Pre-Lab Exercises, and other online resources.

In this and the following chapter, we'll explore oxidation and reduction reactions in organic chemistry. Such reactions are extremely important as they are commonly used to convert one functional group into another during the course of preparing more complex materials from simpler ones. In the simplest sense, oxidation in organic chemistry involves increasing the number of carbon-oxygen bonds. For example, the complete combustion of hydrocarbons or other organic compounds to produce carbon dioxide, water, and heat is an oxidation, but because such oxidations destroy the organic molecule, they are not useful in synthesis. Hence, in the experiments that follow, we'll explore the controlled oxidations of alcohols and aldehydes to give carbonyl compounds and carboxylic acids, respectively, as these transformations are widely used in contemporary synthetic organic chemistry.

16.1 INTRODUCTION

Oxidation, often represented by the symbol [O], is a fundamental type of reaction in chemistry and is the opposite of **reduction.** The general concept of oxidation is typically introduced in beginning chemistry courses as an electron transfer process involving a *loss* of electrons from an ion or a neutral atom. Such definitions are difficult to apply to reactions in organic chemistry, however, as carbon forms covalent bonds and does not normally lose electrons. Nevertheless, oxidations of organic compounds usually do involve a loss of electron density at carbon as a consequence of forming a new bond between a carbon atom and a more electronegative atom such as nitrogen, oxygen, or a halogen. Reactions that result in breaking carbon-hydrogen bonds are also oxidations.

Oxidations are frequently used in organic chemistry to effect functional group transformations. For example, oxidations involving the conversions of an alkene into a 1,2-diol (Eq. 16.1 and Sec. 10.4), of a primary or secondary alcohol into an aldehyde or ketone, respectively (Eq. 16.2), and of an aldehyde into a carboxylic acid (Eq. 16.3) are common laboratory reactions. In processes involving the degradation of organic molecules, oxidations may be used to cleave carbon-carbon double bonds, as illustrated by the ozonolysis of an alkene to give the corresponding carbonyl compounds (Eq. 16.4).

$$\underset{\text{An alkene}}{R^1\text{-}CH{=}CH_2} \xrightarrow{KMnO_4} \underset{\text{A 1,2-diol}}{R^1\text{-}CH(OH)\text{-}CH_2OH} \tag{16.1}$$

$$\underset{\text{A primary alcohol}}{R^1\text{—}CH_2\text{—}OH} \text{ or } \underset{\text{A secondary alcohol}}{R^1\text{—}CH(R^2)\text{—}OH} \xrightarrow{[O]} \underset{\text{An aldehyde}}{R^1\text{-}C(=O)\text{-}H} \text{ or } \underset{\text{A ketone}}{R^1\text{-}C(=O)\text{-}R^2} \tag{16.2}$$

$$\underset{\text{An aldehyde}}{R^1\text{-}C(=O)\text{-}H} \xrightarrow{[O]} \underset{\text{A carboxylic acid}}{R^1\text{-}C(=O)\text{-}OH} \tag{16.3}$$

$$\underset{\text{An alkene}}{R^1\text{-}CH{=}C(R^2)R^3} \xrightarrow[\text{(2) Zn, HOAc}]{\text{(1) }O_3} \underset{\text{An aldehyde}}{R^1\text{—}CHO} + \underset{\text{A ketone}}{R^2\text{-}C(=O)\text{-}R^3} \tag{16.4}$$

See more on *Metabolism*

The processes of life also depend on oxidation of organic substrates, whereby metabolic energy is derived from the overall oxidation of carbohydrates, fats, and proteins to carbon dioxide and water, among other products, as illustrated in Equation 16.5. Potentially poisonous substances are commonly detoxified by biological oxidation to more benign substances. Thus, nicotine, which is toxic to humans if present in sufficiently high concentration, is oxidized in the liver to cotinine (Eq. 16.6), a substance of low toxicity.

$$\underset{\text{A carbohydrate}}{C_6H_{12}O_6} \xrightarrow{[O]} 6\,CO_2 + 6\,H_2O + \text{energy} \tag{16.5}$$

$$\text{Nicotine} \xrightarrow[\text{Liver}]{[O]} \text{Cotinine} \tag{16.6}$$

As noted previously, oxidation does *not* necessarily result in introducing oxygen into an organic molecule. In a general sense, and in analogy to concepts of oxidation as applied to inorganic compounds, oxidation of an organic substance involves an increase in the **oxidation number,** or **oxidation state,** of carbon. There are several ways of defining the oxidation numbers of carbon atoms, but the method that follows is a useful one.

1. *Select* the carbon atom whose oxidation number is to be determined.
2. *Assign* oxidation numbers to the atoms attached to this carbon atom using the following values:
   a. +1 for hydrogen
   b. –1 for halogen, nitrogen, oxygen, and sulfur
   c. 0 for carbon

   Simply put, with respect to the atoms bound to the carbon atom, those that are *more* electronegative are assigned a value of –1, whereas those that are *less* electronegative are given a value of +1.
3. *Sum* the oxidation numbers of these atoms. If a heteroatom (see 2b) is multiply bound to a carbon atom, multiply its oxidation number by the number of bonds linking it to the carbon atom involved.
4. *Determine* that the sum of the number from step 3 and the oxidation number of the carbon atom under consideration equals the charge on the carbon atom, which is zero unless it bears a positive or negative charge.

Application of this method shows that the conversion of an alkene to the corresponding dibromide (Eq. 16.7) and that of an alkane to an alkyl halide (Eq. 16.8), reactions discussed in Section 10.6 and Chapter 9, respectively, are oxidations even though oxygen is not incorporated in these reactions. Determining the changes in the oxidation number of the various carbon atoms during the conversion of acetic acid to carbon dioxide and water (Eq. 16.9) illustrates how the method is used when heteroatoms are multiply bound to a carbon atom.

$$\text{Cyclopentene (C-1, C-2: } -1, -1) \xrightarrow[CCl_4]{Br_2} \textit{trans}\text{-1,2-Dibromocyclopentane (C-1, C-2: } 0, 0) \quad (16.7)$$

$$\underset{\text{Ethane}}{CH_3\overset{-3}{C}H_2{-}H} \xrightarrow{Cl_2} \underset{\text{1-Chloroethane}}{CH_3\overset{-1}{C}H_2{-}Cl} \quad (16.8)$$

Net increase in oxidation number of 2 (oxidation at carbon atom)

$$\underset{\text{An alkyne}}{R{-}C{\equiv}C{-}R(H)} \xrightarrow[H_2SO_4,\ H_2O]{HgSO_4} \underset{\text{A ketone}}{R{-}\overset{O}{\overset{\|}{C}}{-}CH_2{-}R\ (H)} \quad (16.9)$$

16.2 PREPARATION OF ALDEHYDES AND KETONES BY OXIDATION OF ALCOHOLS

Aldehydes and ketones have a central role in organic synthesis, and efficient procedures for their preparation are of great importance. Such compounds are synthesized in a number of ways, including hydration or hydroboration-oxidation of alkynes (Eqs. 16.10 and 16.11, respectively, and Chap. 11) and reaction of carboxylic acids or their derivatives with organometallic reagents or reducing agents

(Eqs. 16.12 and 16.13). The latter two reactions involve net **reduction** of the carbonyl carbon atom. However, the transformations shown in Equations 16.10 and 16.11 are neither *net* oxidation nor reduction of the alkyne. You should determine the net change in oxidation state at the carbon atoms involved in these transformations to convince yourself that this is true.

$$\underset{\text{An alkyne}}{R{-}C{\equiv}C{-}R(H)} \xrightarrow[\substack{H_2SO_4 \\ H_2O}]{HgSO_4} \underset{\text{A ketone}}{R{-}\overset{\overset{\large O}{\|}}{C}{-}CH_2{-}R\ (H)} \tag{16.10}$$

$$R{-}C{\equiv}C{-}R(H) \xrightarrow[\substack{(2)\ H_2O_2/ \\ HO^-}]{(1)\ B_2H_6} R{-}CH_2{-}\overset{\overset{\large O}{\|}}{C}{-}R\ (H) \tag{16.11}$$

$$\underset{\text{A carboxylic acid}}{R^1{-}\overset{\overset{\large O}{\|}}{C}{-}OH} \xrightarrow[(2)\ H_3O^+]{(1)\ 2\ R^2Li} R^1{-}\overset{\overset{\large O}{\|}}{C}{-}R^2 \tag{16.12}$$

$$\underset{\text{An ester}}{R^1{-}\overset{\overset{\large O}{\|}}{C}{-}OR} \xrightarrow[(2)\ H_3O^+]{(1)\ (i\text{-}Bu)_2AlH} R^1{-}\overset{\overset{\large O}{\|}}{C}{-}H \tag{16.13}$$

One of the most common synthetic methods for preparing aldehydes and ketones is the oxidation of primary and secondary alcohols, a reaction involving a two-electron change in the oxidation number of the functionalized carbon atom (Eq. 16.14). Chromic ion, Cr^{6+}, frequently in the form of chromic acid, H_2CrO_4, is a common oxidant, although other oxidants, such as halonium ion, X^+, as found in hypohalous acids, HOX, and permanganate ion, Mn^{7+}, as in potassium permanganate, $KMnO_4$, can be used.

$$R{-}\overset{\overset{\large OH}{|}}{CH}{-}R(H) \xrightarrow[\substack{\text{or HOX} \\ \text{or } KMnO_4}]{H_2CrO_4} R{-}\overset{\overset{\large O}{\|}}{C}{-}R(H) \tag{16.14}$$

Oxidation number 0 (2 alcohol)   +2 (ketone)
−1 (1 alcohol)   +1 (aldehyde)

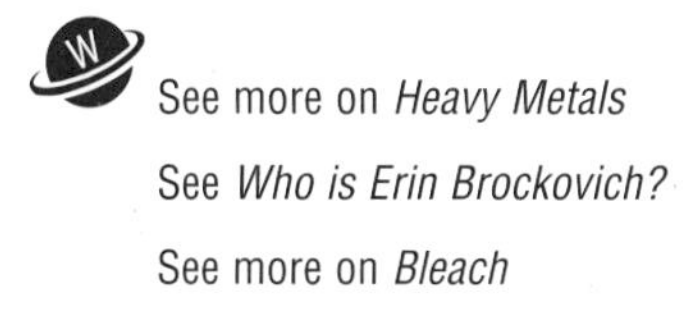

See more on *Heavy Metals*
See *Who is Erin Brockovich?*
See more on *Bleach*

Chromic acid and potassium permanganate are excellent oxidizing agents, but these oxidants are derived from **heavy metals**, a class of elements that are commonly toxic and therefore environmentally hazardous, as exemplified in the film "Erin Brockovich." The safe disposal of these metals and their derivatives is of considerable concern, and avoiding their use is desirable from an environmental perspective. Fortunately, an alternative to chromate and permanganate ions for oxidizing alcohols are the hypohalites, XO^- (X = halogen), which are made by reaction of the corresponding halogen with aqueous base (Eq. 16.15). The most familiar sources of this type of oxidizing agent are household bleach, which is an aqueous solution of sodium hypochlorite, NaOCl, and granular $Ca(OCl)_2$, which is a chlorinating agent for swimming pools.

$$\text{X—X} + \text{HO}^- \rightleftharpoons \text{X}^- + \text{XO—H} \overset{\text{HO}^-}{\rightleftharpoons} \text{XO}^- + \text{H}_2\text{O} \qquad (16.15)$$

(X = halogen) Hypohalous acid; Hypohalite ion

The mechanism by which hypohalites oxidize alcohols probably involves initial formation of an alkyl hypohalite (Eq. 16.16). This product arises from reaction of the alcohol with the hypohalous acid that is in equilibrium with hypohalite ion in aqueous medium (Eq. 16.15). Base-promoted E2 elimination of the elements of H–X from the alkyl hypochlorite leads directly to either an aldehyde or ketone. The advantage of using hypohalite as an oxidant is immediately obvious upon examining Equation 16.16. The inorganic by-product derived from the oxidant is a halide salt that can be safely flushed down the drain. Reactions that do not produce toxic by-products are environmentally friendly and are now commonly referred to as "Green Chemistry." (See the Historical Highlight at the end of this chapter.)

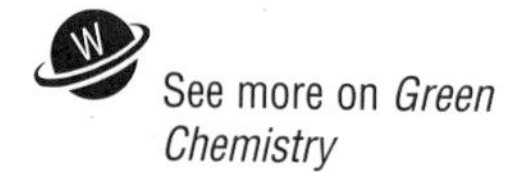
See more on *Green Chemistry*

$$\text{R}_2\overset{0}{\text{C}}\text{H—}\ddot{\text{O}}\text{H} + \overset{+1}{\text{X}}\text{—OH} \overset{-\text{H}_2\text{O}}{\rightleftharpoons} \text{R}_2\text{CH—O—X} \xrightarrow{^-\text{OAc}} \text{R}_2\overset{+2}{\text{C}}\text{=O} + \overset{-1}{\text{X}^-} \qquad (16.16)$$

Alkyl hypohalite; Halide ion

The stoichiometry of the reaction may be determined from changes in oxidation numbers in the reactants and products. During the oxidation, the carbon atom of the alcohol undergoes a two-electron change in oxidation state, from 0 to +2 in the case of a secondary alcohol, and the halogen also undergoes a two-electron change in oxidation number from +1 to –1 (Eq. 16.16). Consequently, a one-to-one ratio of the two reactants is required stoichiometrically.

Some side reactions may complicate the oxidation of a primary alcohol to an aldehyde using hypohalous acid. For example, an aldehyde may undergo reaction with an additional equivalent of hypohalite to form a carboxylic acid (Eq. 16.17), a process that may be initiated by acid-catalyzed formation of the hydrate of the intermediate aldehyde as shown in Equation 16.18. Subsequent steps in the oxidation are then analogous to those for converting an alcohol to an aldehyde or ketone (Eq. 16.16). It may be difficult to suppress this further oxidation, so unless carboxylic acids are the desired products, the use of hypohalite as an oxidant is limited to the conversion of *secondary* alcohols to ketones.

$$\text{R}^1\text{—C(=O)—H} + \text{X—OH} + \text{H}_2\text{O} \longrightarrow \text{R}^1\text{—C(=O)—OH} \qquad (16.17)$$

An aldehyde; A carboxylic acid

$$\text{R—C(=O)—H} \overset{\text{H}^+}{\rightleftharpoons} \text{R—C(=O}^+\text{H)—H} \overset{:\ddot{\text{O}}\text{H}_2}{\rightleftharpoons} \text{R—CH(OH)—}^+\text{OH}_2 \overset{-\text{H}^+}{\rightleftharpoons} \text{R—CH(OH)—OH} \qquad (16.18)$$

An aldehyde; A hydrate

Another side reaction that may occur when hypohalite is used as an oxidizing agent is the α-halogenation of the desired carbonyl compound (Eq. 16.19). This process involves the rapid, *base-catalyzed* formation of an enolate ion, followed by its reaction with hypohalous acid or some other source of halonium ion, such as X_2 (X = halogen). The replacement of the α-hydrogens of a ketone or aldehyde is the basis for the **haloform test**, which is described in Section 25.7E. When oxidations of alcohols to give carbonyl compounds are conducted in acidic media, α-halogenation of the product is not a significant problem for two reasons: the rate of enolization of a ketone is slower in acidic than in basic media, and the *enols* that are produced in acid are less reactive nucleophiles than the enolates that are formed in base.

$$\underset{\text{A carbonyl compound}}{R{-}C(=O){-}CH(H)R} \underset{}{\overset{HO^-}{\rightleftharpoons}} \underset{\text{An enolate}}{R{-}C(O^-){=}CHR \longleftrightarrow R{-}C(=O){-}\bar{C}HR} \xrightarrow{X{-}OH} \underset{\text{An α-halocarbonyl compound}}{R{-}C(=O){-}CH(X)R} \quad (16.19)$$

In the experiments that follow, several representative oxidations of alcohols using hypochlorous acid will be performed. In the first experiment, cyclododecanol (**1**), a secondary alcohol, is oxidized to cyclododecanone (**2**) using sodium hypochlorite or commercial household bleach (Eq. 16.20). In the second experiment, 4-chlorobenzyl alcohol (**3**) is oxidized directly to 4-chlorobenzoic acid (**4**) using calcium hypochlorite (Eq. 16.21).

Because the concentration of sodium hypochlorite in bleach depends on the age of the solution, it is important to monitor the oxidation of **1** with starch/iodide test paper to ensure that excess oxidant is present. The test paper should turn blue-black owing to the formation of the complex of starch and iodine that is produced upon oxidation of iodide ion by the hypochlorite ion. On the other hand, commercially available calcium hypochlorite is quite stable and may be stored at room temperature without decomposition. Therefore, the amount of oxidant used is easy to determine, but it is still good practice to ensure that an excess of oxidant is present by applying the starch/iodide test, especially if the reaction mixture is heated. Excess oxidant may be reduced at the end of the experiment using sodium bisulfite, $NaHSO_3$.

$$\underset{\textbf{1}\ \text{Cyclododecanol}}{C_{12}H_{23}OH} \xrightarrow[CH_3CO_2H]{NaOCl,\ H_2O} \underset{\textbf{2}\ \text{Cyclododecanone}}{C_{12}H_{22}O} \quad (16.20)$$

$$\underset{\textbf{3}\ \text{4-Chlorobenzyl alcohol}}{Cl{-}C_6H_4{-}CH_2OH} \xrightarrow[CH_3CO_2H]{Ca(OCl)_2,\ H_2O} \underset{\textbf{4}\ \text{4-Chlorobenzoic acid}}{Cl{-}C_6H_4{-}CO_2H} \quad (16.21)$$

EXPERIMENTAL PROCEDURES

Oxidation of Alcohols

A ■ *Oxidation of Cyclododecanol to Cyclododecanone*

Green Experiment

Purpose To demonstrate the oxidation of a secondary alcohol to the corresponding ketone using hypochlorous acid.

SAFETY ALERT

Do not allow the solution of sodium hypochlorite to come in contact with your skin or eyes. If it does, flush the affected area immediately with copious amounts of water. Sodium hypochlorite will also bleach clothing.

MINISCALE PROCEDURE

Preparation Sign in at **www.cengage.com/login** to answer Pre-Lab Exercises, access videos, and read the MSDSs for the chemicals used or produced in this procedure. Review Sections 2.9, 2.11, 2.13, 2.21, 2.22, and 2.29.

Apparatus A 25-mL round-bottom flask, separatory funnel, apparatus for heating under reflux, magnetic stirring, simple distillation, and *flameless* heating.

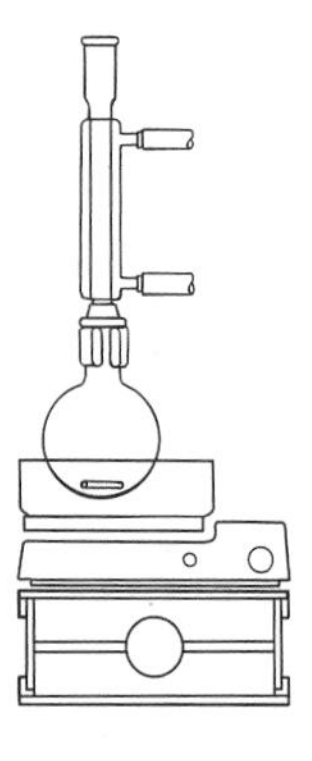

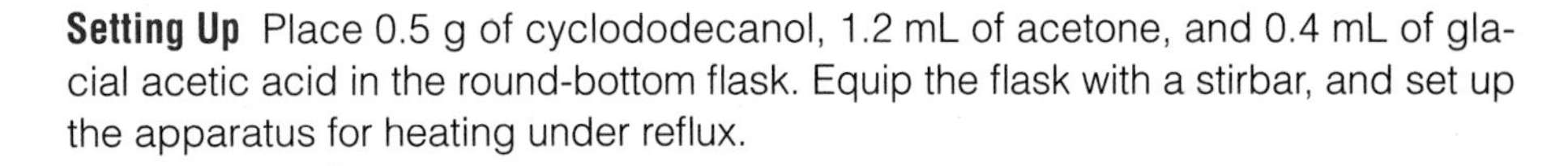

Setting Up Place 0.5 g of cyclododecanol, 1.2 mL of acetone, and 0.4 mL of glacial acetic acid in the round-bottom flask. Equip the flask with a stirbar, and set up the apparatus for heating under reflux.

Oxidation Stir the mixture and warm it to approximately 45 °C; maintain this temperature within ±5 °C throughout the course of the reaction. You may wish to monitor the temperature by suspending a thermometer through the top of the condenser using a copper wire to hold the thermometer in place. Using a Pasteur pipet, add 4.5 mL of commercial bleach (ca. 5.3% sodium hypochlorite) dropwise to the stirred mixture through the top of the condenser over a period of about 0.5 h. Upon completing the addition, stop stirring and heating the mixture so the layers may separate. Using a Pasteur pipet, remove a small portion of the aqueous layer, and place a drop or two of this solution on a dampened piece of starch/iodide test paper to determine whether sufficient hypochlorite has been added. The indicator paper immediately turns blue-black in color if sufficient bleach has been added. If this color does not develop, add an additional 0.4 mL of bleach to the reaction mixture. Stir the resulting mixture with heating for 2–3 min and repeat the test for excess hypochlorite. Add additional 0.4-mL portions of bleach until a positive test for oxidant is observed. Then stir the reaction mixture with heating for an additional 10 min and retest for hypochlorite. If the test is negative, add a final 0.4 mL of bleach. Whether this last test is positive or negative, stir the mixture with heating for 10 min more to complete the reaction.

Work-Up Allow the reaction mixture to cool to room temperature, and transfer it to a separatory funnel using a Pasteur pipet. Rinse the round-bottom flask with 5 mL of diethyl ether, and use a filter-tip pipet to transfer this wash to the separatory funnel. Shake the two-phase mixture, and separate the layers. Extract the aqueous layer with an additional 5-mL portion of diethyl ether, and add this extract to the original one. Wash the combined organic extracts with 5 mL of saturated sodium bicarbonate. Before shaking this mixture, swirl the *unstoppered* funnel until the evolution of carbon dioxide ceases. Shake the mixture, venting the funnel frequently to relieve any pressure that might develop. Wash the organic solution sequentially with 5-mL portions of saturated aqueous sodium bisulfite and saturated aqueous sodium chloride. Transfer the organic solution to an Erlenmeyer flask, and dry it over several spatula-tips full of anhydrous sodium sulfate.★ Swirl the flask occasionally for a period of 10–15 min to facilitate drying; add further small portions of anhydrous sodium sulfate if the solution does not become clear.

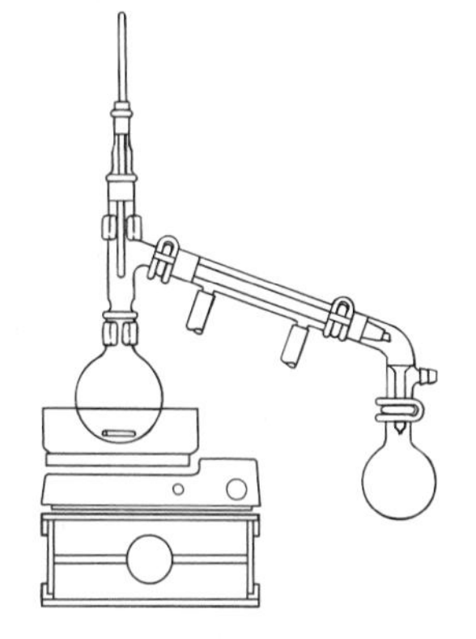

Isolation and Purification Using a filter-tip pipet, transfer the dried ethereal solution to a *tared* 25-mL round-bottom flask, and equip it for simple distillation. Remove the diethyl ether by simple distillation. Alternatively, use rotary evaporation or other techniques to concentrate the solution. The final traces of solvent may be removed by attaching the flask to a vacuum source and gently swirling the contents as the vacuum is applied.★ The oil that is initially formed after removal of the solvents should solidify. Recrystallize the cyclododecanone from aqueous methanol.

Analysis Weigh the flask and calculate the yield of solid cyclododecanone. Determine the melting point of the product. Prepare the semicarbazone (mp 218–219 °C) or oxime (mp 131–132 °C) according to the procedures given in Sections 25.7G and 25.7H. If necessary, recrystallize the derivatives from methanol. Obtain IR and ^{1}H NMR spectra of your starting material and product, and compare them with those of authentic samples (Figs. 16.1–16.4).

WRAPPING IT UP

Flush all *aqueous solutions* down the drain. Spread the *sodium sulfate* on a tray in the hood to evaporate residual solvent from it, and then put the used drying agent in the nonhazardous solid waste container. Place the *diethyl ether* recovered by distillation in the container for nonhalogenated organic solvents.

EXERCISES

1. Prove that no net oxidation occurs at the carbon atoms in the transformations of Equations 16.10 and 16.11 by comparing the oxidation numbers for carbon in the reactants and products.
2. Consider the oxidation of cyclododecanol (**1**) to cyclododecanone (**2**) using hypochlorous acid to answer the following questions.
   a. What is the oxidation number of the carbon atom bearing the alcohol functional group in **1**?
   b. What is the oxidation number of this carbon atom in **2**?

c. What is reduced in this reaction, and to what is it reduced?

d. Suggest a reagent other than bleach that would oxidize **1** into **2.**

e. Write the chemical structure of a side-product that might be formed by the α-halogenation of **2.**

3. Would it be possible to convert cyclododecanone (**2**) to cyclododecanol (**1**)? If so, what type of reaction would this be and what reagent(s) might you use?
4. Why is glacial acetic acid used in this reaction?
5. What is the function of sodium bisulfite in the procedure for isolating cyclododecanone?
6. Would cyclododecanone be expected to give a positive iodoform test (Sec. 25.7E)? Explain.
7. Determine whether the conversion of iodide ion to iodine is an oxidation or a reduction.
8. What was the purpose of testing the reaction mixture with starch/iodide paper? Illustrate your answer by showing the chemical reaction that transforms iodide ion into iodine in this test.
9. For each of the reactions shown below, indicate whether it is a net reduction, an oxidation, or neither and calculate the change in oxidation number for any carbon being reduced or oxidized.

a. (cyclohexanecarboxylic acid, C(=O)OH) → (cyclohexanecarboxamide, C(=O)NH_2)

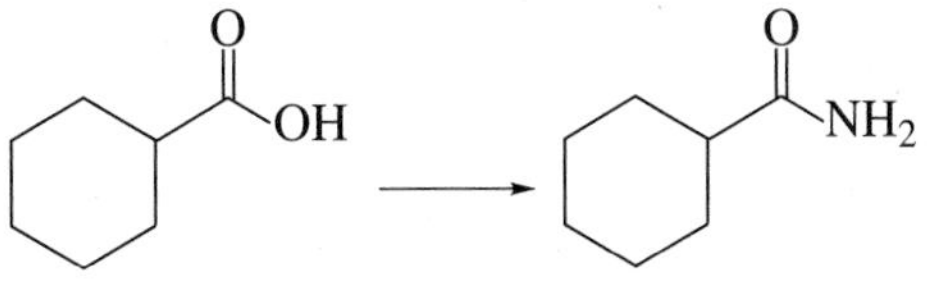

b. (cyclopentanol, –OH) → (cyclopentanone, =O)

c. (cyclohexene) → (cyclohexane)

d. (cyclopentane) → (chlorocyclopentane, –Cl)

e. (1-pentene) → (1,2-dibromopentane, Br, Br)

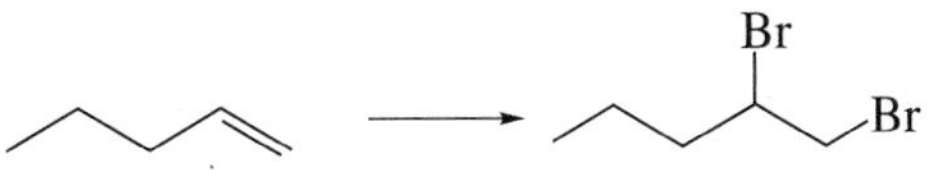

Sign in at **www.cengage.com/login** and use the spectra viewer and Tables 8.1–8.8 as needed to answer the blue-numbered questions on spectroscopy.

10. Consider the spectral data for cyclododecanol (Figs. 16.1 and 16.2).

a. In the IR spectrum, there is a broad absorption at about 3250 cm^{-1}. What functional group is responsible for this absorption, and why is the absorption broad?

b. In the ^1H NMR spectrum, assign the various resonances to the hydrogen nuclei responsible for them.

c. For the ^{13}C NMR data, assign the various resonances to the carbon nuclei responsible for them.

11. Consider the spectral data for cyclododecanone (Figs. 16.3 and 16.4).
    a. In the functional group region of the IR spectrum, specify the absorption associated with the carbon-oxygen double bond.
    b. In the ^{1}H NMR spectrum, assign the various resonances to the hydrogen nuclei responsible for them.
    c. For the ^{13}C NMR data, assign the various resonances to the carbon nuclei responsible for them.
12. Discuss the differences observed in the IR and NMR spectra of cyclododecanol and cyclododecanone that are consistent with the conversion of an alcohol to a ketone in this experiment.

SPECTRA

Starting Material and Product

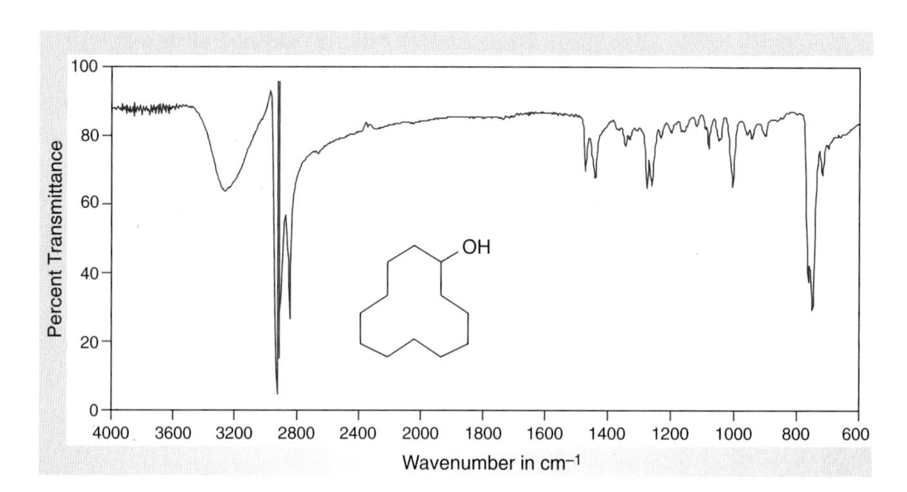

Figure 16.1
IR spectrum of cyclododecanol (IR card).

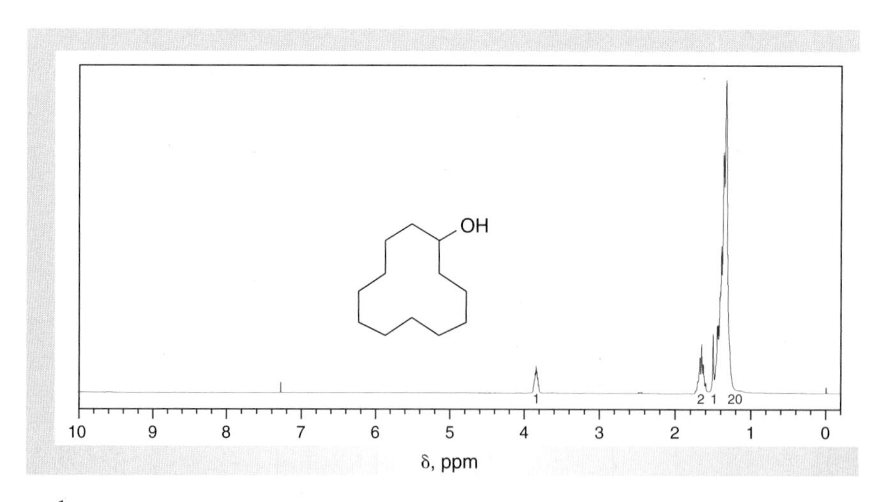

Figure 16.2
NMR data for cyclododecanol ($CDCl_3$).

(a) ^{1}H NMR spectrum (300 MHz).
(b) ^{13}C NMR data: δ 21.0, 23.5, 24.0, 24.4, 32.5, 69.1.

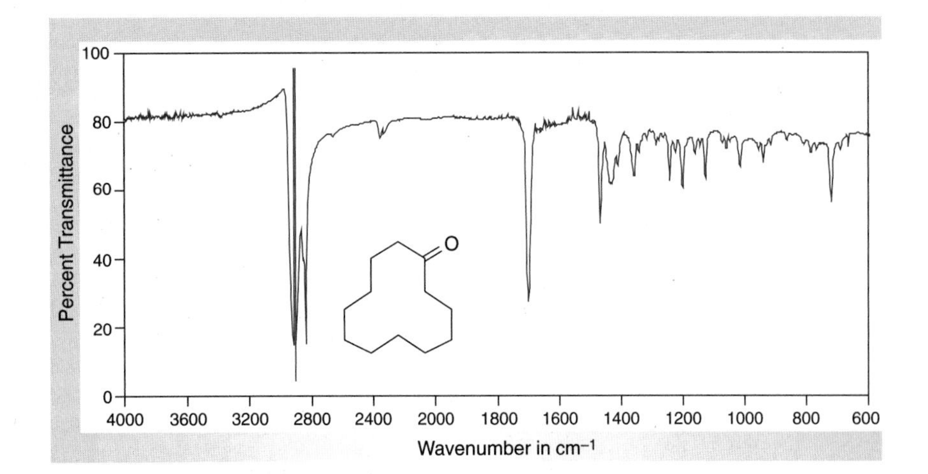

Figure 16.3
IR spectrum of cyclododecanone (IR card).

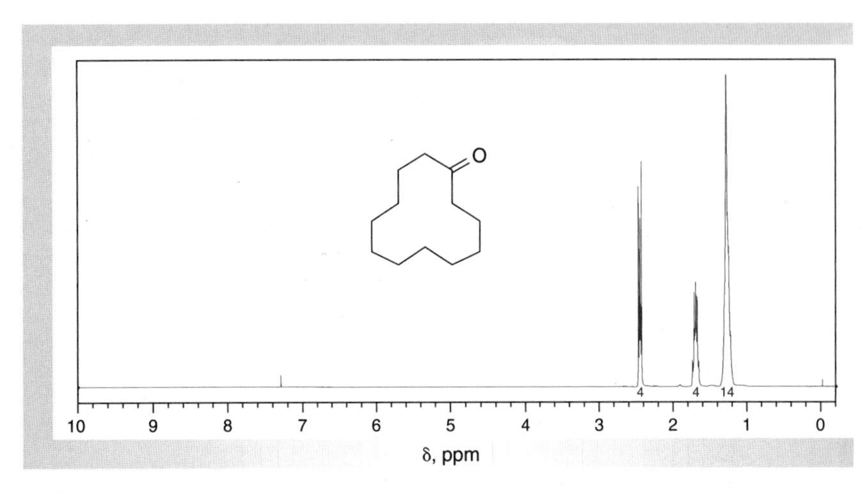

Figure 16.4
NMR data for cyclododecanone ($CDCl_3$).

(a) ^{1}H NMR spectrum (300 MHz).
(b) ^{13}C NMR data: δ 22.7, 24.4, 24.9, 40.2, 211.4.

B ■ *Oxidation of 4-Chlorobenzyl Alcohol to 4-Chlorobenzoic Acid*

Green Experiment **Purpose** To demonstrate the oxidation of an alcohol to the corresponding carboxylic acid using hypochlorous acid.

SAFETY ALERT

Do not allow the solution of calcium hypochlorite to come in contact with your skin or eyes. If it does, flush the affected area immediately with copious amounts of water. Calcium hypochlorite will also bleach clothing.

MINISCALE PROCEDURE

Preparation Sign in at **www.cengage.com/login** to answer Pre-Lab Exercises, access videos, and read the MSDSs for the chemicals used or produced in this procedure. Review Sections 2.9, 2.11, 2.17, and 2.21.

Apparatus A 50-mL round-bottom flask, separatory funnel, apparatus for heating under reflux, magnetic stirring, vacuum filtration, and *flameless* heating.

Setting Up Place 2.4 g of commercial calcium hypochlorite (65%) in the round-bottom flask containing a stirbar. Add 20 mL of water to the flask and stir the mixture while adding 2 mL of glacial acetic acid dropwise. Weigh 0.5 g of 4-chlorobenzyl alcohol, and add it to a small Erlenmeyer flask containing 5 mL of acetonitrile; swirl the contents to dissolve the alcohol. Set up the apparatus for heating under reflux.

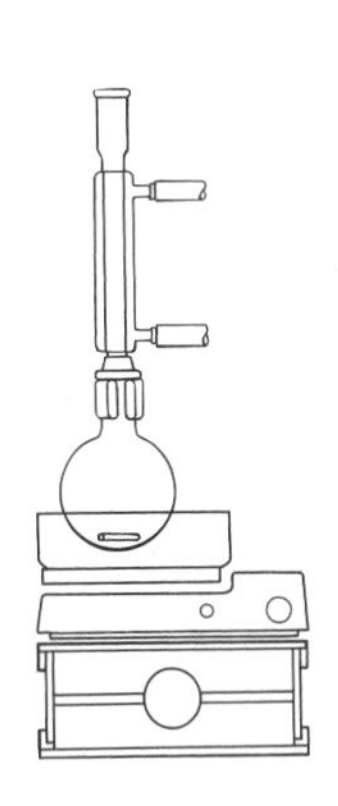

Oxidation Using a Pasteur pipet, transfer the solution of 4-chlorobenzyl alcohol in acetonitrile dropwise to the stirred solution of calcium hypochlorite through the top of the condenser. Stir the mixture vigorously for 1.5–2 h at 45–50 °C; *do not heat above 50 °C.* You may wish to monitor the temperature by suspending a thermometer through the top of the condenser using a copper wire to hold the thermometer in place. Remove a small aliquot of the reaction mixture using a Pasteur pipet, and place a drop or two of this solution on a dampened piece of starch/iodide paper to determine whether sufficient hypochlorite has been added. The indicator paper immediately turns blue-black in color if excess oxidizing agent is present. If this color does not develop, add an additional 150 mg of commercial calcium hypochlorite to the reaction mixture. Whether or not additional calcium hypochlorite is required, continue stirring the mixture for an additional 1 h at 45–50 °C, and then cool it to room temperature.★

Work-Up, Isolation, and Purification Transfer the cooled reaction mixture to the separatory funnel and add 10 mL of water. Rinse the reaction flask with 10 mL of diethyl ether, and transfer this rinse to the separatory funnel. Shake the funnel and separate the layers. Extract the aqueous layer sequentially with two additional 10-mL portions of diethyl ether. Extract the *combined* diethyl ether layers with two separate 10-mL portions of saturated aqueous sodium bicarbonate. Slowly add *concentrated* hydrochloric acid dropwise to the *combined* bicarbonate washings until the solution is slightly acidic (pH = 3). Collect the precipitate by vacuum filtration, wash the filter cake with 5–10 mL of water, and air-dry the crude product. 4-Chlorobenzoic acid may be recrystallized from methanol.

Analysis Weigh the product and calculate the percent yield. Determine the melting point of the recrystallized 4-chlorobenzoic acid. Obtain IR and ^1H NMR spectra of your starting material and product, and compare them with those of authentic samples (Figs. 16.5–16.8).

MICROSCALE PROCEDURE

Preparation Sign in at **www.cengage.com/login** to answer Pre-Lab Exercises, access videos, and read the MSDSs for the chemicals used or produced in this procedure. Review Sections 2.7, 2.9, 2.11, 2.17, and 2.21.

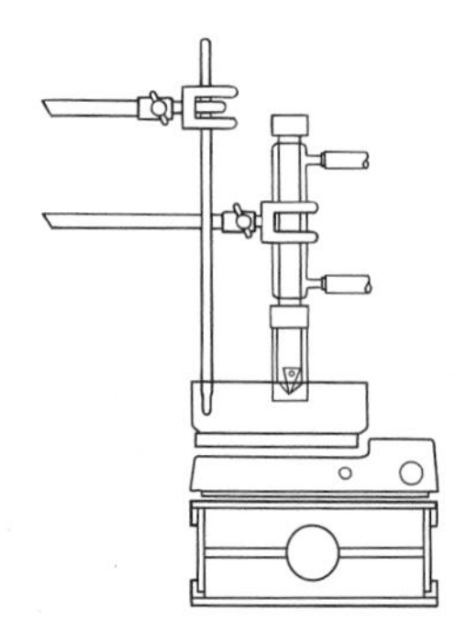

Apparatus A 5-mL conical vial, apparatus for heating under reflux, Pasteur pipet, magnetic stirring, vacuum filtration, Craig tube filtration, and *flameless* heating.

Setting Up Weigh 0.5 g of commercial calcium hypochlorite (65%), and transfer it to the conical vial containing a spinvane. Add 3.0 mL of water to the vial, and stir the mixture while adding 0.3 mL of glacial acetic acid dropwise. Weigh 0.1 g of 4-chlorobenzyl alcohol, and add it to a small test tube containing 1 mL of acetonitrile; swirl the contents to dissolve the alcohol. Set up the apparatus for heating under reflux.

Oxidation Using a Pasteur pipet, transfer the solution of 4-chlorobenzyl alcohol in acetonitrile dropwise to the stirred solution of calcium hypochlorite through the top of the condenser. Stir the mixture vigorously for 1.5–2 h at 45–50 °C; *do not heat above 50 °C*. You may wish to monitor the temperature by suspending a thermometer through the top of the condenser using a copper wire to hold the thermometer in place. Remove a small aliquot of the reaction mixture using a Pasteur pipet, and place a drop or two of this solution on a dampened piece of starch/iodide paper to determine whether sufficient hypochlorite has been added. The indicator paper immediately turns blue-black in color if there is excess oxidizing agent present. If this color does not develop, add an additional 50 mg of commercial calcium hypochlorite to the reaction mixture. Whether or not additional calcium hypochlorite is required, continue stirring the mixture for an additional 1 h and then cool it to room temperature.★

Work-Up, Isolation, and Purification Transfer the cooled reaction mixture to a screw-cap centrifuge tube and add 1.5 mL of water. Rinse the reaction vial with 2 mL of diethyl ether, and add this rinse to the centrifuge tube. Cap the centrifuge tube, and shake the contents to mix the layers. Separate the layers and extract the aqueous layer sequentially with two additional 2-mL portions of diethyl ether. Extract the *combined* diethyl ether layers with two separate 2-mL portions of saturated aqueous sodium bicarbonate. Slowly add *concentrated* hydrochloric acid dropwise to the *combined* bicarbonate washings until the solution is slightly acidic (pH = 3). Collect the precipitate by vacuum filtration, wash the filter cake with 1–2 mL of water, and air-dry the crude product. 4-Chlorobenzoic acid may be recrystallized from methanol in a Craig tube.

Analysis Weigh the product and calculate the percent yield. Determine the melting point of the recrystallized 4-chlorobenzoic acid. Obtain IR and ^{1}H NMR spectra of your starting material and product, and compare them with those of authentic samples (Figs. 16.5–16.8).

WRAPPING IT UP

Neutralize the *combined aqueous solutions* with solid sodium carbonate and flush them down the drain. Because there may be some residual starting alcohol or intermediate aldehyde in the *diethyl ether extract,* transfer it to the container for halogenated organic liquids.

EXERCISES

1. Prove that no net oxidation occurs at the carbon atoms in the transformations of Equations 16.10 and 16.11 by comparing the oxidation numbers for carbon in the reactants and products.
2. Consider the oxidation of 4-chlorobenzyl alcohol (**3**) to 4-chlorobenzoic acid (**4**) using hypochlorous acid to answer the following questions.
   a. What is the oxidation number of the carbon atom bearing the alcohol functional group in **3**?
   b. What is the oxidation number of this carbon atom in **4**?
   c. What reactant is reduced in this reaction, and what is the reduced product?
3. Why is glacial acetic acid used in this reaction?
4. Write a balanced equation for the oxidation of **3** into **4** to determine how many equivalents of hypochlorous acid must be used.
5. Why is 4-chlorobenzaldehyde not produced in this reaction?
6. There are reagents that may be used to convert 4-chlorobenzoic acid (**4**) into 4-chlorobenzyl alcohol (**3**). Suggest one such reagent, and indicate whether this reaction is an oxidation, a reduction, or neither.
7. What was the purpose of testing the reaction mixture with starch/iodide paper? Illustrate your answer by showing the chemical reaction that transforms iodide ion into iodine in this test.
8. Determine whether the conversion of iodide ion to molecular iodine is an oxidation or a reduction.
9. What is the reduced form of $Ca(OCl)_2$ that remains at the end of the experiment?
10. For each of the reactions shown below, indicate whether it is a reduction, an oxidation, or neither.

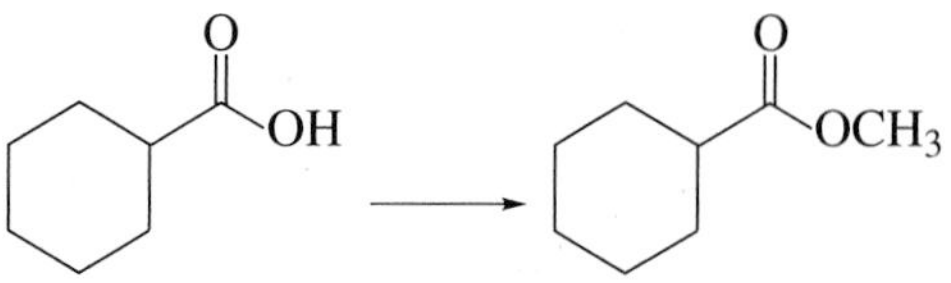

a. [cyclohexyl–C(=O)OH] ⟶ [cyclohexyl–CH₂OH]

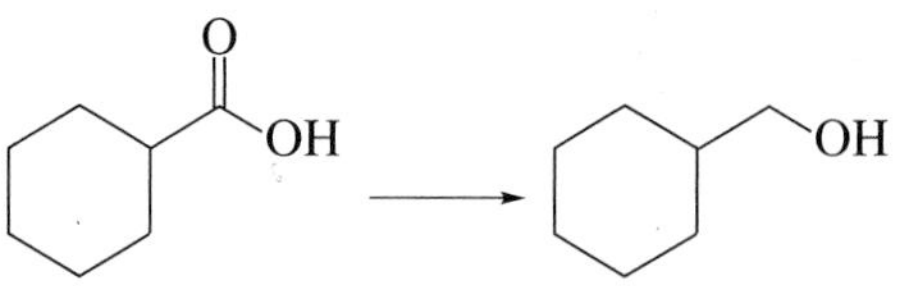

b. [cyclopentyl–OH] ⟶ [cyclopentanone, =O]

c. [cyclohexyl–C(=O)OH] ⟶ [cyclohexyl–C(=O)OCH₃]

d.

e.

Br

11. The equilibrium between an aldehyde and its hydrate greatly favors the aldehyde, as shown.

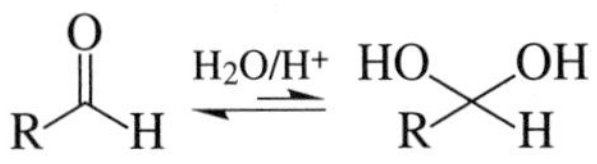

$$\mathrm{RCHO} \underset{}{\overset{H_2O/H^+}{\rightleftharpoons}} \mathrm{RCH(OH)_2}$$

a. Using curved arrows to symbolize the flow of electrons, show the mechanism for the acid-catalyzed conversion of the hydrate into the aldehyde.

b. If only the hydrate is capable of undergoing oxidation, explain why the aldehyde is oxidized efficiently to the carboxylic acid, even though only small amounts of the hydrate are present at any given time.

c. Based upon the principle that an alcohol such as methanol, CH_3OH, is a derivative of water, what compound might be formed upon mixing an aldehyde, RCHO, with methanol?

d. What product would be formed if the intermediate in Part **c** were oxidized?

12. One of the side reactions that may occur when hypohalite is used as an oxidizing agent for benzylic alcohols is halogenation of the aromatic ring. Although this side reaction is a problem when benzyl alcohol is oxidized with calcium hypochlorite, it is not observed in the oxidation of 4-chlorobenzyl alcohol. Explain this observation.

Sign in at **www.cengage.com/login** and use the spectra viewer and Tables 8.1–8.8 as needed to answer the blue-numbered questions on spectroscopy.

13. Consider the spectral data for 4-chlorobenzyl alcohol (Figs. 16.5 and 16.6).

a. In the IR spectrum, there is a broad absorption at about 3400 cm^{-1}. What functional group is responsible for this absorption, and why is the absorption broad?

b. In the ^1H NMR spectrum, assign the various resonances to the hydrogen nuclei responsible for them.

c. For the ^{13}C NMR data, assign the various resonances to the carbon nuclei responsible for them.

14. Consider the spectral data for 4-chlorobenzoic acid (Figs. 16.7 and 16.8).

a. In the functional group region of the IR spectrum, specify the absorptions associated with the carbonyl component of the carboxyl group. What functional group is responsible for the broad absorption in the region of about 2800 cm^{-1}, and why is it broad?

b. In the ^1H NMR spectrum, assign the various resonances to the hydrogen nuclei responsible for them.

c. For the ^{13}C NMR data, assign the various resonances to the carbon nuclei responsible for them.

15. Discuss the differences observed in the IR and NMR spectra of 4-chlorobenzyl alcohol and 4-chlorobenzoic acid that are consistent with the oxidation of an alcohol to a carboxylic acid rather than to an aldehyde in this experiment.

SPECTRA

Starting Material and Product

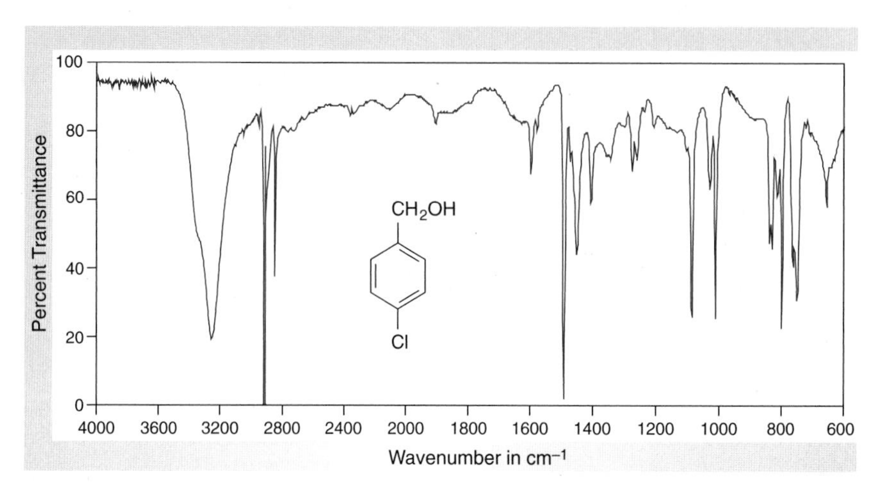

Figure 16.5
IR spectrum of 4-chlorobenzyl alcohol (IR card).

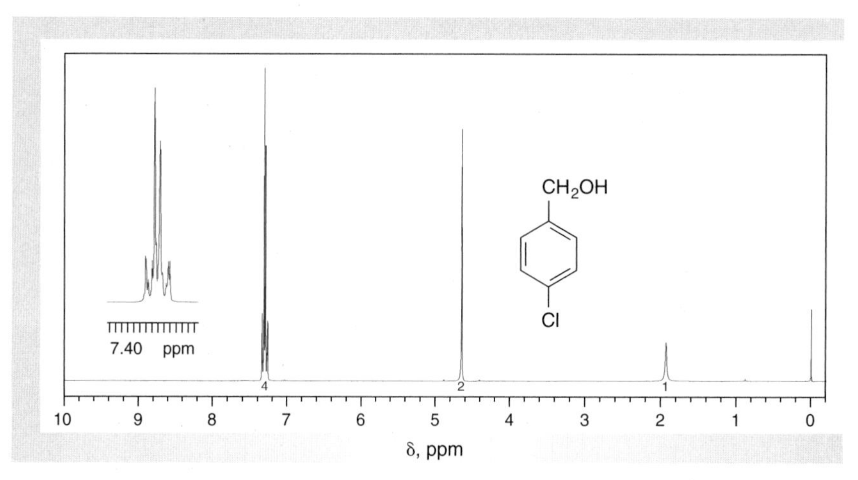

Figure 16.6
NMR data for 4-chlorobenzyl alcohol ($CDCl_3$).

(a) ^{1}H NMR spectrum (300 MHz).
(b) ^{13}C NMR data: δ 62.5, 128.0, 131.6, 141.4.

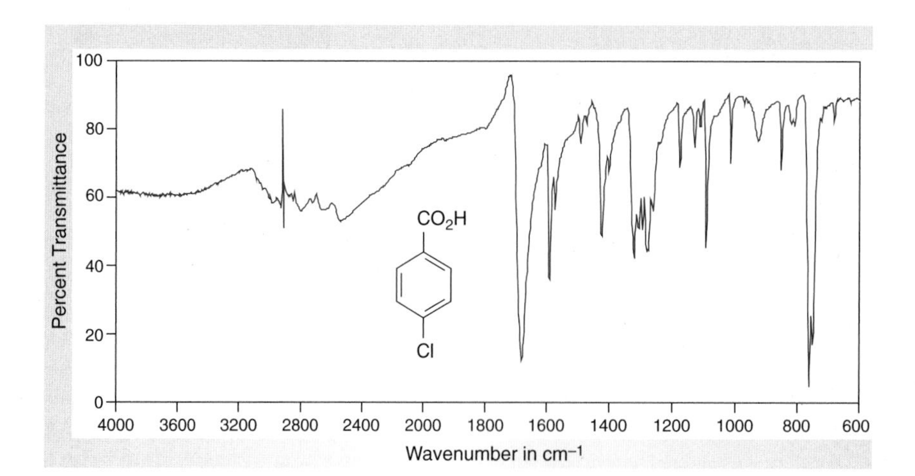

Figure 16.7
IR spectrum of 4-chlorobenzoic acid (IR card).

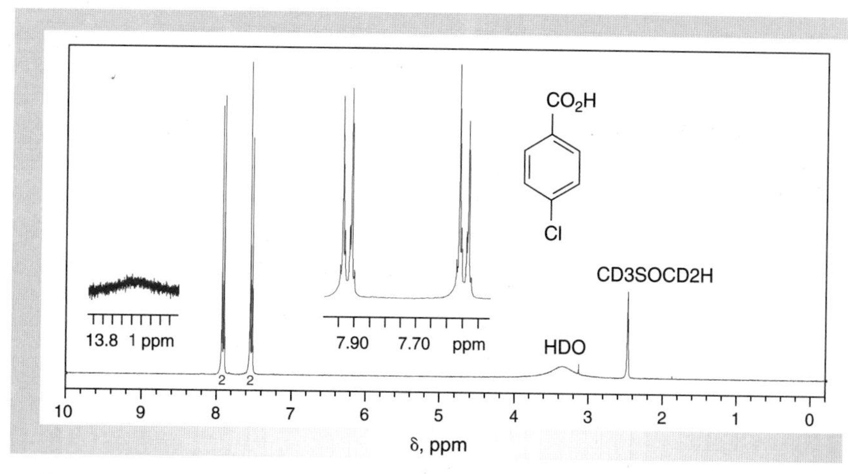

Figure 16.8
NMR data for 4-chlorobenzoic acid ($CDCl_3$).

(a) ^1H NMR spectrum (300 MHz).
(b) ^{13}C NMR data: δ 128.5, 129.8, 131.1, 138.3, 166.8.

16.3 BASE-CATALYZED OXIDATION-REDUCTION OF ALDEHYDES: THE CANNIZZARO REACTION

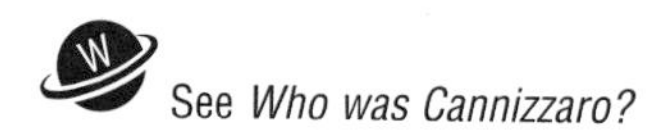

Aldehydes having no hydrogens on the **α-carbon atom,** the carbon atom adjacent to the carbonyl group, undergo mutual oxidation and reduction in the presence of strong alkali (Eq. 16.22). In contrast, aldehydes with hydrogen atoms on the **α-carbon atom** preferentially undergo other types of base-promoted reactions such as the **aldol condensation** (Eq. 16.23), which is described in Section 18.3. The mutual oxidation-reduction, called the **Cannizzaro reaction,** is a consequence of the fact that an aldehyde is intermediate in oxidation state between an alcohol and a carboxylic acid. It may be converted into either one by a decrease or gain of 2, respectively, in the oxidation number (Sec. 16.1) of its carbonyl carbon atom, as given in Equation 16.22.

$$2\,R_3C-\overset{+1}{C}(=O)-H \xrightarrow[(2)\ H_3O^+]{(1)\ HO^-} R_3C-\overset{-1}{C}H_2-OH + R_3C-\overset{+3}{C}(=O)-OH \qquad (16.22)$$

An aldehyde → An alcohol + A carboxylic acid

$$2\,R-CH_2-C(=O)-H \xrightarrow[(2)\ H_3O^+]{(1)\ HO^-} R-CH_2-CH(OH)-CH(R)-C(=O)-H \qquad (16.23)$$

An aldehyde → An aldol (β-hydroxycarbonyl compound)

The accepted mechanism of the reaction follows logically from the ease with which nucleophiles add to the carbonyl group, particularly that of an aldehyde. The first step is attack of hydroxide ion on the carbonyl group to give a tetrahedral intermediate (Eq. 16.24), and this is followed by transfer of a hydride ion to the carbonyl group of another aldehyde function (Eq. 16.25). The step depicted in Equation 16.25 is the one in which both oxidation and reduction occur. You should verify this fact by determining the oxidation numbers of the carbon atoms involved. The remaining steps (Eqs. 16.26 and 16.27) simply illustrate the types of acid-base chemistry expected to occur in the strongly basic medium in which the reaction is performed. Summing of Equations 16.24–16.27 gives Equation 16.22.

$$RCHO + {}^{-}OH \longrightarrow R{-}C(O^{-})(OH){-}H \tag{16.24}$$

$$R{-}C(O^{-})(OH){-}H + RCHO \longrightarrow RCOOH + R{-}C(O^{-})(H){-}H \tag{16.25}$$

$$RCO{-}O{-}H + {}^{-}OH \longrightarrow RCO_2^{-} + H{-}OH \tag{16.26}$$

$$R{-}CH_2{-}O^{-} + H{-}OH \longrightarrow R{-}CH_2{-}O{-}H + {}^{-}OH \tag{16.27}$$

Aromatic aldehydes, formaldehyde, and α-trisubstituted acetaldehydes all undergo the Cannizzaro reaction. In the experiment of this section, 4-chlorobenzaldehyde, an aromatic aldehyde, is converted to 4-chlorobenzyl alcohol and potassium 4-chlorobenzoate, which gives 4-chlorobenzoic acid upon acidification of the reaction mixture (Eq. 16.28).

$$4\text{-}ClC_6H_4CHO + KOH \longrightarrow 4\text{-}ClC_6H_4CH_2OH + 4\text{-}ClC_6H_4CO_2^{-}K^{+} \xrightarrow{H_3O^{+}} 4\text{-}ClC_6H_4CO_2H \tag{16.28}$$

4-Chlorobenzaldehyde; 4-Chlorobenzyl alcohol; Potassium 4-chlorobenzoate; 4-Chlorobenzoic acid

EXPERIMENTAL PROCEDURES

Base-Catalyzed Oxidation-Reduction of Aldehydes by the Cannizzaro Reaction

Purpose To demonstrate the oxidation-reduction of an aromatic aldehyde to give the corresponding alcohol and carboxylic acid salt by the Cannizzaro reaction.

SAFETY ALERT

The concentrated solution of potassium hydroxide is *highly* corrosive and caustic. *Do not allow it to come into contact with skin.* Should it do so, flood the affected area immediately with water and then thoroughly rinse it with 1% acetic acid. Wear latex gloves when preparing and transferring solutions of potassium hydroxide.

MINISCALE PROCEDURE

Preparation Sign in at **www.cengage.com/login** to answer Pre-Lab Exercises, access videos, and read the MSDSs for the chemicals used or produced in this procedure. Review Sections 2.9, 2.11, 2.13, 2.17, 2.21, 2.22, and 2.29.

Apparatus A 25-mL round-bottom flask, thermometer, separatory funnel, apparatus for heating under reflux, magnetic stirring, simple distillation, vacuum filtration, and *flameless* heating.

Setting Up Dissolve 5.0 g of potassium hydroxide in 5 mL of distilled water contained in an Erlenmeyer flask. Weigh 1.0 g of 4-chlorobenzaldehyde and transfer it to the round-bottom flask containing a stirbar. Add 2.5 mL of methanol and stir to dissolve the 4-chlorobenzaldehyde. Using a Pasteur pipet, transfer 1.5 mL of the 50% aqueous potassium hydroxide solution to the round-bottom flask; do not get any of this solution on the ground-glass joint. Assemble the apparatus for heating under reflux; *be sure that the joint of the flask is well greased so that it does not freeze.*

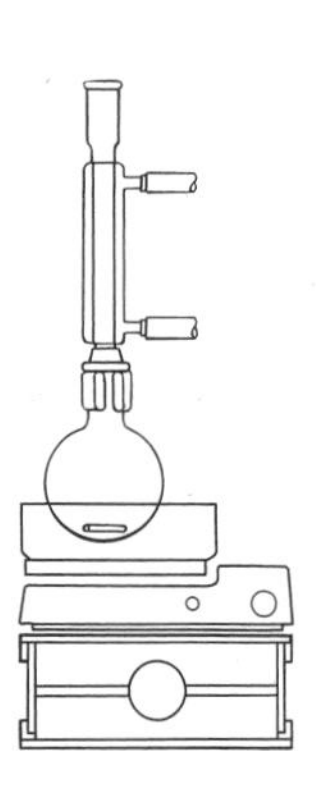

Reaction and Work-Up Place the round-bottom flask in a heating bath that has been preheated to about 75 °C, and heat the reaction mixture with stirring at this temperature for 1.5 h. Remove the flask from the bath, cool the mixture to room temperature, and add 15 mL of water. Transfer the mixture to a separatory funnel. Extract the aqueous mixture with three separate 5-mL portions of dichloromethane; shake the funnel *gently* to avoid forming an emulsion. Test the lower layer to verify that it is the organic layer, and combine all of the organic extracts. *Save both the organic and the aqueous layers, as each must be worked up separately to isolate the two products.*

Isolation and Purification Wash the combined *dichloromethane extracts* with two separate 5-mL portions of saturated aqueous sodium chloride. Dry the organic layer over several spatula-tips full of anhydrous sodium sulfate.★ Swirl the flask occasionally for a period of 10–15 min to facilitate drying; add further small portions of anhydrous sodium sulfate if the solution does not become clear. Filter or

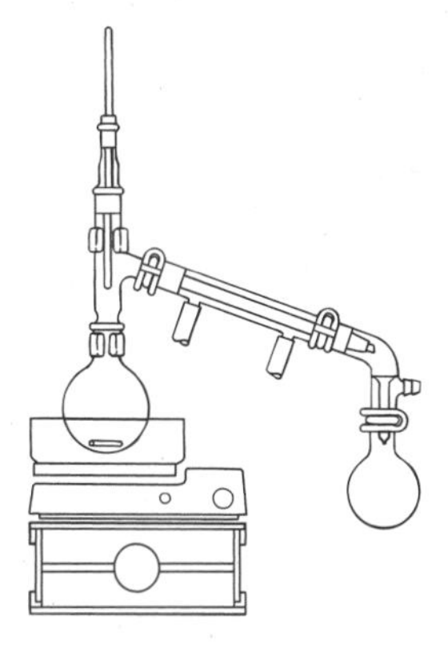

decant the dried solution into a clean, tared round-bottom flask, rinse the sodium sulfate with about 1 mL of fresh dichloromethane, and transfer the rinse to the round-bottom flask. Remove the dichloromethane by simple distillation. Alternatively, use rotary evaporation or other techniques to concentrate the solution. The final traces of solvent may be removed by attaching the flask to a vacuum source and gently swirling the contents as the vacuum is applied.★ The crude 4-chlorobenzyl alcohol may be recrystallized by mixed solvent techniques from acetone and hexane (Sec. 3.2). Air-dry the purified product.

Cool the *aqueous phase* saved from the initial dichloromethane extraction in an ice-water bath. Acidify the solution by slowly adding 2.5 mL of *concentrated* hydrochloric acid. If the solution is not acidic (pH ~3), continue adding concentrated hydrochloric acid *dropwise* until it is. Collect the precipitate by vacuum filtration, wash the filter cake with 5–10 mL of water, and air-dry the product. The crude 4-chlorobenzoic acid may be recrystallized from methanol.

Analysis Weigh the products and calculate percent yields. Determine the melting points of the recrystallized products. Obtain IR and ^{1}H NMR spectra of your starting material and product and compare them with those of authentic samples (Figs. 16.5–16.10).

MICROSCALE PROCEDURE

Preparation Sign in at **www.cengage.com/login** to answer Pre-Lab Exercises, access videos, and read the MSDSs for the chemicals used or produced in this procedure. Review Sections 2.9, 2.11, 2.13, 2.17, 2.21, 2.22, and 2.29.

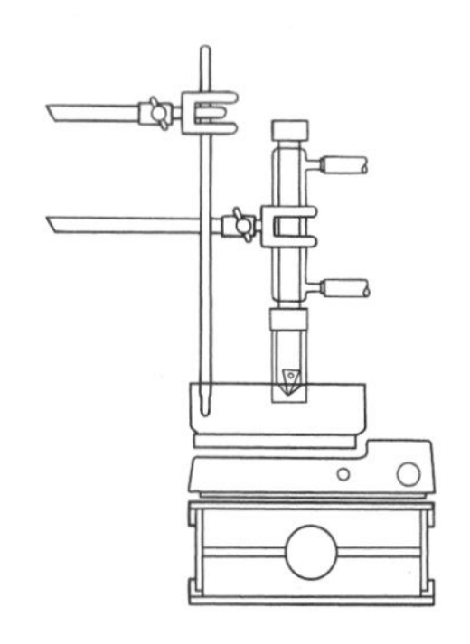

Apparatus A 5-mL conical vial, thermometer, screw-cap centrifuge tubes, apparatus for heating under reflux, magnetic stirring, simple distillation, vacuum filtration, Craig tube filtration, and *flameless* heating.

Setting Up Dissolve 1.0 g of potassium hydroxide in 1 mL of distilled water contained in a test tube. Weigh 200 mg of 4-chlorobenzaldehyde and transfer it to the 5-mL conical vial containing a spinvane. Add 0.5 mL of methanol and stir to dissolve the 4-chlorobenzaldehyde. Using a Pasteur pipet, transfer 0.3 mL of the 50% aqueous potassium hydroxide solution to the vial; do not get any of this solution on the ground-glass joint. Assemble the apparatus for heating under reflux; *be sure that the joint of the vial is well greased so it does not freeze.*

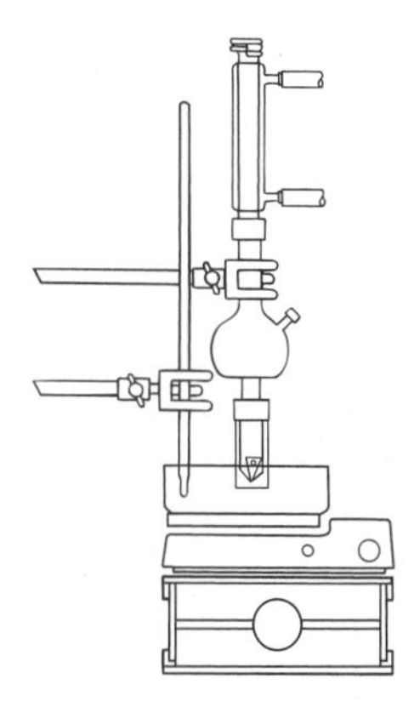

Reaction and Work-Up Place the base of the 5-mL conical vial in a heating bath that has been preheated to about 75 °C, and heat the reaction mixture with stirring at this temperature for 1.5 h. Remove the vial from the bath, cool the mixture to room temperature, and add 2.5 mL of water. Extract the aqueous mixture with three separate 0.7-mL portions of dichloromethane. For each extraction, you should cap the vial and shake it gently to avoid forming an emulsion; carefully vent the vial by slowly loosening the cap. Remove the lower layer using a Pasteur pipet and verify that it is the organic layer; combine all of the organic extracts in a centrifuge tube. *Save both the organic and the aqueous layers, as each must be worked up separately to isolate the two products.*

Isolation and Purification Wash the combined *dichloromethane extracts* with two separate 0.5-mL portions of saturated aqueous sodium chloride. Transfer the dichloromethane layer to a dry centrifuge tube, and add several microspatula-tips full of anhydrous sodium sulfate.★ Swirl the tube occasionally for a period of 10–15 min to facilitate drying; add further small portions of anhydrous sodium sulfate if the solution does not become clear. Using a filter-tip pipet, transfer the dried solution into a clean 5-mL conical vial, rinse the sodium sulfate with about 0.2 mL of dichloromethane, and transfer the rinse to the conical vial. Remove the dichloromethane by simple distillation. Alternatively, use rotary evaporation or other techniques to concentrate the solution. The final traces of solvent may be removed by attaching the vial to a vacuum source and gently swirling the contents as the vacuum is applied.★ The crude 4-chlorobenzyl alcohol may be recrystallized from 4% acetone in hexane in a Craig tube. Air-dry the purified product.

Cool the *aqueous phase* saved from the initial dichloromethane extraction in an ice-water bath. Acidify the solution by slowly adding 0.5 mL of *concentrated* hydrochloric acid. If the solution is not acidic (pH ~3), continue adding concentrated hydrochloric acid *dropwise* until it is. Collect the precipitate by vacuum filtration, wash the filter cake with 1–2 mL of water, and air-dry the product. The crude 4-chlorobenzoic acid may be recrystallized from methanol in a Craig tube.

Analysis Weigh the products and calculate the percent yields. Determine the melting points of the recrystallized products. Obtain IR and ^1H NMR spectra of your starting material and product, and compare them with those of authentic samples (Figs. 16.5–16.10).

WRAPPING IT UP

Flush the *saturated sodium chloride solution* used during the extraction process to wash the dichloromethane down the drain. Neutralize the *aqueous filtrate* from isolation of 4-chlorobenzoic acid with solid sodium carbonate, and flush it down the drain. Place the *dichloromethane* obtained by distillation in the container for halogenated organic liquids. Dispose of the *sodium sulfate* in the container for nontoxic solids after allowing residual solvent to evaporate from it in the hood.

EXERCISES

1. Determine the oxidation number for the carbonyl carbon in 4-chlorobenzaldehyde. Then confirm that oxidation and reduction of the aldehyde group occurs in the Cannizzaro reaction by calculating the oxidation numbers for the corresponding carbon atoms in 4-chlorobenzyl alcohol and 4-chlorobenzoic acid.
2. After completion of the Cannizzaro reaction of 4-chlorobenzaldehyde, the mixture was transferred into a separatory funnel and extracted with CH_2Cl_2. Write the structures of the products that were dissolved in the organic and aqueous layers, respectively.
3. After separating the organic and aqueous layers, the dichloromethane layer is first washed with two portions of saturated aqueous sodium chloride prior to drying with anhydrous sodium sulfate. What is the purpose of these washes?
4. What is the maximum yield of 4-chlorobenzoic acid that can be expected in this experiment?

5. When the Cannizzaro reaction is performed on 4-chlorobenzaldehyde in D_2O solution, no deuterium is found on the benzylic carbon atom in the 4-chlorobenzyl alcohol formed. How does this support the mechanism given in Equations 16.24–16.27?
6. What is the solid that is formed after 4-chlorobenzaldehyde has been allowed to react with aqueous potassium hydroxide?
7. The Cannizzaro reaction occurs much more slowly in dilute than in concentrated potassium hydroxide solution. Why is this?
8. How would 2,2-dimethylpropanal react with potassium hydroxide solution under the conditions of this experiment?
9. A *crossed* Cannizzaro reaction involving two different aldehydes may be performed if neither of the aldehydes contains α-hydrogens. Write an equation for the crossed Cannizzaro reaction of a mixture of 4-chlorobenzaldehyde and formaldehyde with concentrated potassium hydroxide solution, followed by acidification of the reaction mixture. Show all organic products.
10. Heterolytic cleavage of a C–H bond is typically a very endothermic step. Explain why hydride transfer can occur in the Cannizzaro reaction.
11. The Cannizzaro reaction is an example of a general class of reactions known as ***disproportionation*** reactions. Explain how this term applies to the Cannizzaro reaction.
12. A carbon-based hydride reducing reagent in biological systems is nicotinamide adenine dinucleotide, NADH, which reduces carbonyl compounds by a mechanism related to the Cannizzaro reaction. In these reactions, NADH transfers a hydride to a carbonyl compound to yield an alcohol and NAD^+, as shown.

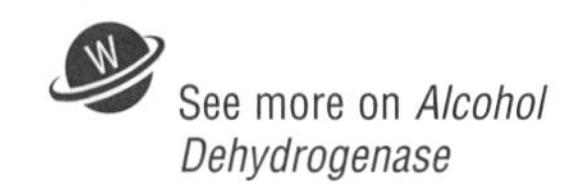

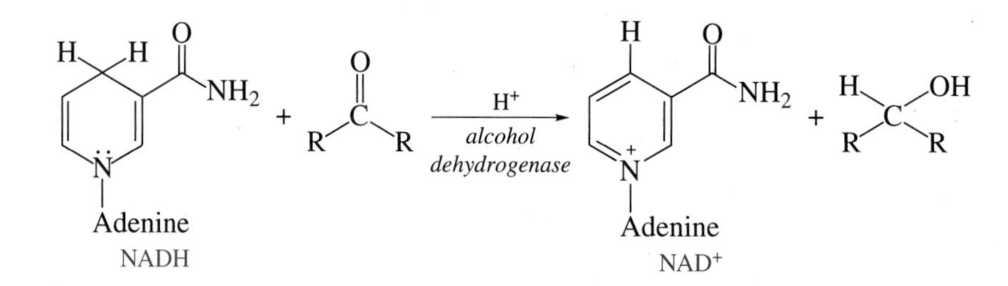

   a. Using curved arrows to symbolize the flow of electrons, write the mechanism for the hydride transfer from NADH to a carbonyl compound. Your mechanism should show how NADH becomes NAD^+.
   b. What is the driving force for this reaction? (*Hint:* Consider why NAD^+ is more stable than NADH.)
   c. Does the conversion of NADH to NAD^+ represent an oxidation or a reduction of NADH?

Sign in at **www.cengage.com/login** and use the spectra viewer and Tables 8.1–8.8 as needed to answer the blue-numbered questions on spectroscopy.

13. Consider the spectral data for 4-chlorobenzaldehyde (Figs. 16.9 and 16.10).
   a. In the functional group region of the IR spectrum, specify the absorptions associated with the carbonyl group and with the aromatic ring.
   b. In the ^1H NMR spectrum, assign the various resonances to the hydrogen nuclei responsible for them.
   c. For the ^{13}C NMR data, assign the various resonances to the carbon nuclei responsible for them.

14. Consider the spectral data for 4-chlorobenzyl alcohol (Figs. 16.5 and 16.6).

a. In the functional group region of the IR spectrum, there is a broad absorption at about 3400 cm^{-1}. What functional group is responsible for this absorption, and why is it broad?

b. In the ^{1}H NMR spectrum, assign the various resonances to the hydrogen nuclei responsible for them.

c. For the ^{13}C NMR data, assign the various resonances to the carbon nuclei responsible for them.

15. Consider the spectral data for 4-chlorobenzoic acid (Figs. 16.7 and 16.8).

a. In the functional group region of the IR spectrum, specify the absorptions associated with the carbonyl component of the carboxyl group. What functional group is responsible for the broad absorption in the region of about 2800 cm^{-1}, and why is the absorption broad?

b. In the ^{1}H NMR spectrum, assign the various resonances to the hydrogen nuclei responsible for them.

c. For the ^{13}C NMR data, assign the various resonances to the carbon nuclei responsible for them.

16. Discuss the differences observed in the IR and NMR spectra of 4-chlorobenzaldehyde, 4-chlorobenzyl alcohol, and 4-chlorobenzoic acid that are consistent with the formation of the two products from 4-chlorobenzaldehyde by the Cannizzaro reaction.

SPECTRA

Starting Material and Products

The IR and NMR spectra of 4-chlorobenzyl alcohol are given in Figures 16.5 and 16.6, respectively; those of 4-chlorobenzoic acid are presented in Figures 16.7 and 16.8, respectively.

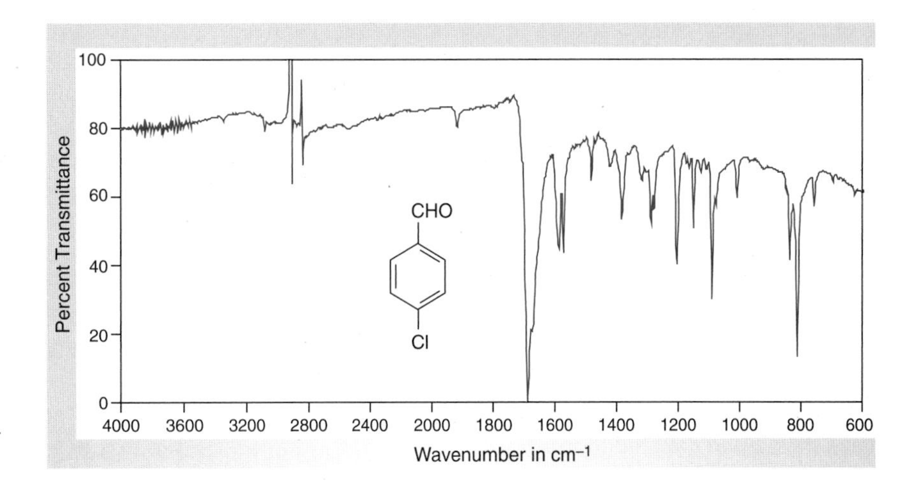

Figure 16.9
IR spectrum of 4-chlorobenzaldehyde (IR card).

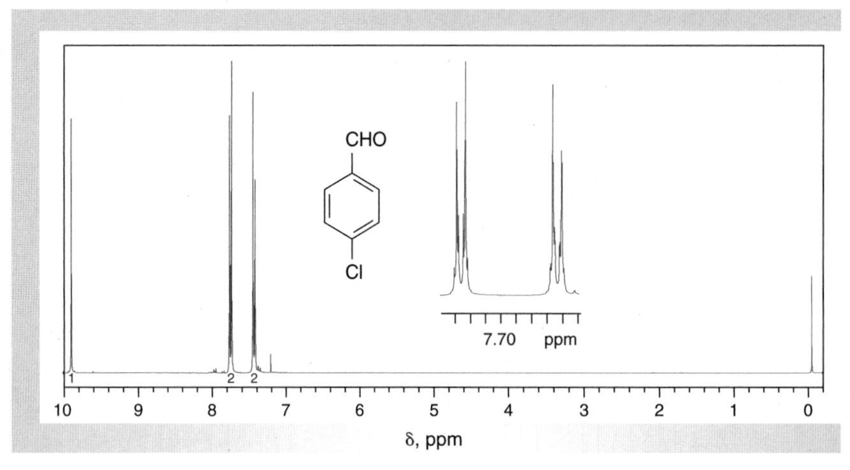

Figure 16.10
NMR data for 4-chlorobenzaldehyde ($CDCl_3$).

(a) ^1H NMR spectrum (300 MHz).
(b) ^{13}C NMR data: δ 129.4, 130.8, 134.9, 140.7, 190.5.

HISTORICAL HIGHLIGHT

Green Chemistry

What is "green chemistry"? An oxymoron? Absolutely not! Ever since the fateful lessons of Love Canal, chemists, environmentalists, and politicians have sought environmentally friendly, or "green," ways to manufacture the chemicals, pharmaceuticals, and agrochemicals that have become so much an integral part of our society. In the twenty-first century, the chemical industry will certainly become "greener" as we seek to protect the world in which we live. The ultimate goals of green chemistry are to use renewable raw materials, to eliminate waste, and to avoid use and generation of toxic and/or hazardous materials at all stages of chemical manufacture and application. Hence, true green chemistry is directed toward prevention of primary pollution rather than clean-up and remediation of waste.

Billions of kilograms of various compounds are produced annually in the chemical industry, and reducing the impact of such large-scale production upon the environment is a major concern. The importance of making chemistry green is reflected in the fact that an increasing number of international conferences have been devoted to this topic, and there is now even a journal, *Green Chemistry*, where recent developments in this field are published. To encourage research focusing on environmentally friendly technologies in this country, the Presidential Green Chemistry Challenge Awards were established in 1995.

There are a number of ways in which chemistry can be made greener. Recycling certainly ranks as one of the important strategies, but this has its limit because not all of the unwanted chemicals associated with a given process can be recycled cost-effectively. Indeed, cost will always be a major factor in determining whether a green process will be commercialized. Another way to make chemical processes more environmentally friendly is to minimize the use of potentially hazardous organic solvents for reactions. This may be accomplished by employing alternative solvents such as water, which can be purified and returned to lakes and streams, or supercritical carbon dioxide, which simply vaporizes and is returned to the atmosphere. It is also important to avoid those reactions and processes in which toxic reagents are used or toxic by-products are generated.

The tools of biology are playing an increasingly important role in green chemistry. For example, one of the objectives of developing genetically engineered

(Continued)

HISTORICAL HIGHLIGHT *Green Chemistry (Continued)*

or modified plants was to reduce the need for the numerous pesticides and herbicides used in food production. However, the application of this technology is currently the subject of considerable debate, and whether genetically modified foods will eventually be accepted for human consumption remains an open question. Yet another biological strategy is to use biocatalytic processes that exploit enzymes or genetically modified microbes to develop alternative synthetic routes to chemicals.

This latter approach to making industrial processes greener is nicely exemplified by the synthesis of adipic acid from D-glucose. Nearly 2 billion kg of adipic acid are produced annually for the manufacture of the polymer Nylon-6,6 (see the Historical Highlight at the end of Chapter 22). The principal starting material in the commercial synthesis of adipic acid is benzene, a known carcinogen that is derived from nonrenewable fossil fuels. The final step in the synthesis is the oxidation of a mixture of cyclohexanol and cyclohexanone using nitric acid, but a by-product of this reaction is nitrous oxide, which is involved in the greenhouse effect and ozone depletion. Hence, the present commercial synthesis of adipic acid can hardly be construed as being green.

An environmentally benign route to adipic acid was invented by K. Draths and J. Frost (*J. Am. Chem. Soc.* 1994, *116*, 399–400). In this procedure, a genetically engineered microbe is used to convert D-glucose, which may be obtained as plant starch and cellulose from abundant and renewable sources, into *cis,cis*-muconic acid via a biosynthetic pathway that is not known to occur in nature. The *cis,cis*-muconic acid thus produced is then reduced by catalytic hydrogenation to give adipic acid. Although this process has not yet been commercialized, it is promising and is representative of the advances that chemists are making to improve the quality of the environment.

Relationship of Historical Highlight to Experiments

Some chemical processes require the use of hazardous materials or heavy metals. Oxidations of alcohols into carbonyl compounds or carboxylic acids were frequently performed using either chromic acid or potassium permanganate. However, these two oxidants are derived from heavy metals, which are toxic and hence environmentally hazardous. Safe disposal of salts of these metals thus presents a considerable environmental concern. Fortunately, hypohalites may also be used to oxidize alcohols into carbonyl compounds and carboxylic acids, as exemplified in the procedures in Section 16.2. This oxidation procedure exemplifies green chemistry because the hypohalites are reduced to water and halide ion, which has little environmental impact.

HO, HO, HO, OH, OH (D-Glucose) → HO_2C … CO_2H (*cis, cis*-Muconic acid) → HO_2C … CO_2H (Adipic acid)

D-Glucose — *cis, cis*-Muconic acid — Adipic acid

See more on *Green Chemistry*

See more on *Love Canal*

See more on *Supercritical Carbon Dioxide*

See more on *Genetic Engineering*

See more on *Heavy Metals*

See more on *Ozone Hole*

See more on *Greenhouse Effect*

See more on *Genetically Modified Foods*

CHAPTER 17

Reduction Reactions of Double Bonds

Alkenes, Carbonyl Compounds, and Imines

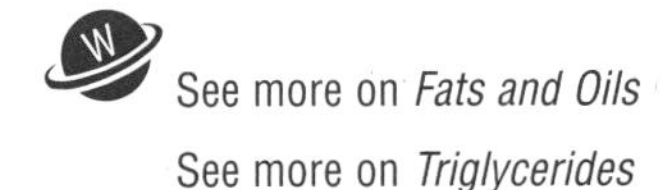

When you see this icon, sign in at this book's premium website at **www.cengage.com/login** to access videos, Pre-Lab Exercises, and other online resources.

See more on *Fats and Oils*

See more on *Triglycerides*

In this chapter, we'll examine the reductions of a number of different functional groups. A reduction in organic chemistry frequently involves increasing the number of carbon-hydrogen bonds by simply adding a molecule of hydrogen across a double or triple bond. For example, a reduction that is commonly performed in the food industry is the catalytic hydrogenation of the polyunsaturated triglycerides that are found in peanut, soybean, cottonseed, and other oils. In this process, which is commonly known as hardening, oils are converted into fats like shortening, margarine, and butter substitutes. You will have an opportunity to conduct a catalytic hydrogenation similar to the reaction used to make fats from oils in one of the experiments in this chapter. In other experiments, you will be able to use reductions to convert one functional group selectively into another, a common transformation in modern research laboratories.

17.1 INTRODUCTION

Reduction, often represented by the symbol [H], is a basic type of chemical reaction and is the opposite of **oxidation** (Chap. 16). You were probably introduced to reductions as electron-transfer processes in which an ion or a neutral atom *gained* electrons. In the reactions that are typically encountered in organic chemistry, however, carbon usually forms covalent bonds and hence does not really *gain* electrons. Nevertheless, reductions of organic compounds often do involve an increase in the electron density at a carbon atom as a consequence of replacing a bond between the carbon atom and a more electronegative atom, for example, nitrogen, oxygen, or halogen, with a carbon-hydrogen bond. Reactions in which new carbon-hydrogen bonds are formed by adding one or more hydrogen atoms to a functional group are also reductions.

The term **oxidation number** was defined in Section 16.1, and an organic reaction was considered an oxidation whenever the oxidation number on a carbon atom *increased.* This same concept may be applied to ascertain if an organic reaction is a reduction, because a reduction results in a *decrease* in the oxidation number of a carbon atom. For example, from the discussion in Section 16.1, we know that the oxidation number of each carbon atom in ethylene, $CH_2{=}CH_2$, is –2; however, in ethane, $CH_3{-}CH_3$, the corresponding number is –3. Thus, conversion of ethylene to ethane involves a *decrease* in the oxidation number of each carbon atom by 1 and a net decrease in the oxidation numbers of both carbon atoms of 2.

Organic functional groups containing double and triple bonds undergo reduction by the *net* addition of the elements of one or two molecules of hydrogen across the π-bond(s). The starting compounds in these processes are termed **unsaturated,** whereas the final products are **saturated** with respect to hydrogen because they contain only single bonds and can absorb no more hydrogen. Equations 17.1–17.4 illustrate some types of reductions that are commonly performed by organic chemists. In addition to the functional groups shown in these equations, many other functional groups containing π-bonds can be reduced to form products containing only single σ-bonds. These include carboxylic acids and their derivatives (esters, amides, anhydrides, and nitriles), nitro groups, and aromatic rings.

$$\underset{\text{An alkene}}{>C{=}C<} \xrightarrow{[H]} \underset{\text{An alkane}}{-\overset{H}{\underset{|}{\overset{|}{C}}}-\overset{H}{\underset{|}{\overset{|}{C}}}-} \qquad (17.1)$$

$$\underset{\text{An alkyne}}{-C{\equiv}C-} \xrightarrow[\text{1 mol}]{[H]} \underset{\text{An alkene}}{\overset{\diagdown\quad\diagup}{\underset{H\quad\;\; H}{C{=}C}}} \xrightarrow[\text{1 mol}]{[H]} \underset{\text{An alkane}}{-\overset{H}{\underset{H}{\overset{|}{\underset{|}{C}}}}-\overset{H}{\underset{H}{\overset{|}{\underset{|}{C}}}}-} \qquad (17.2)$$

$$\underset{\text{A carbonyl compound}}{>C{=}O} \xrightarrow{[H]} \underset{\text{An alkohol}}{-\underset{H}{\underset{|}{\overset{|}{C}}}-\underset{H}{\underset{|}{O}}} \qquad (17.3)$$

$$\underset{\text{An imine}}{>C{=}N\diagdown} \xrightarrow{[H]} \underset{\text{An amine}}{-\underset{H}{\underset{|}{\overset{|}{C}}}-\underset{H}{\underset{|}{N}}\diagup} \qquad (17.4)$$

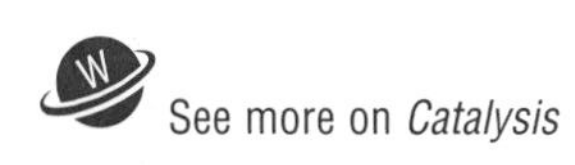
See more on *Catalysis*

Reductions like those shown in Equations 17.1–17.4 are commonly effected by either **catalytic** or **chemical** methods. Sometimes both techniques can be used, but one of them will usually be preferred for a given substrate, depending upon what other functional groups might be present in the molecule. A brief discussion of these methods will help enable you to understand the basic differences.

See more on *Catalytic Hydrogenation/Nobel Prize*

The technique of **catalytic hydrogenation,** which is discussed in more detail in Section 17.2, is widely used in organic chemistry to reduce the π-bonds in a variety of functional groups. This reaction involves the addition of a molecule of hydrogen across a carbon-carbon or carbon-heteroatom double or triple bond in the presence of a metal **catalyst** such as platinum, palladium, nickel, or rhodium that is often adsorbed on an inert solid support like carbon. These catalysts have different properties and selectivities toward reducible groups, so it is sometimes necessary to try several in order to optimize a particular catalytic hydrogenation. Preparative catalytic hydrogenation is frequently used in research and industrial laboratories to synthesize compounds of interest, and analytical methods have been developed in which the amount of hydrogen consumed and the heat, ΔH_{hydrog}, evolved may be accurately measured. You will have an opportunity to perform the catalytic hydrogenation of a compound containing a carbon-carbon double bond in the experimental procedure in Section 17.2.

The **chemical reduction** of a number of organic functional groups is commonly effected using **metal hydride** reducing agents that have different reactivities toward specific functional groups. This property is of much practical importance because one of the challenges in contemporary synthetic organic chemistry is developing new reducing reagents that react *selectively* with only one type of functional group in the presence of other reducible groups. In this context, it is significant that hydride reagents may be used to reduce a carbon-oxygen or a carbon-nitrogen double bond in the presence of a carbon-carbon double bond, a reaction that is not easily accomplished by catalytic hydrogenation. One such selective reduction is performed in the experimental procedure in Section 17.3.

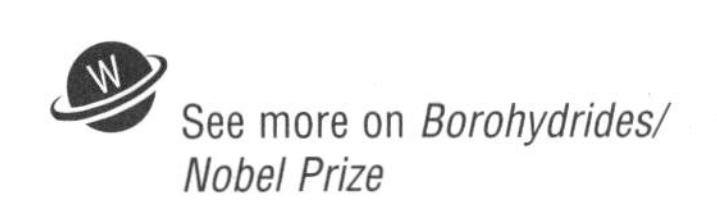

There are a large number of hydride-containing reducing agents that exhibit a broad range of selectivities toward various unsaturated functional groups. For example, lithium aluminum hydride, $LiAlH_4$, is a very reactive, and hence unselective, hydride donor that rapidly reduces aldehydes, ketones, and esters with ease. Sodium borohydride, $NaBH_4$, on the other hand, is less reactive and thus more selective. It reduces aldehydes and ketones but *not* esters, for example. Not only is chemical selectivity a concern in choosing the appropriate reagent, but experimental factors such as ease of handling and safety often play important roles. Thus, if lithium aluminum hydride and sodium borohydride could both be used for a specific reduction, the latter would be preferred because it is safer and easier to handle. Sections 17.3 and 17.4 include experiments for several chemical reductions that utilize sodium borohydride.

The nitro group, $-NO_2$, is reduced to the amino group, $-NH_2$, by catalytic hydrogenation or by using lithium aluminum hydride. However, this conversion may also be accomplished by an alternative method of chemical reduction known as **dissolving metal reduction**. These reductions are typically performed either using finely divided iron or tin with hydrochloric acid or an alkaline earth metal such as lithium, sodium, and potassium in liquid ammonia. The reduction of nitrobenzene, $C_6H_5NO_2$, to aniline, $C_6H_5NH_2$, using tin and hydrochloric acid (Sec. 20.2) is an example of such a reduction.

17.2 CATALYTIC HYDROGENATION OF THE CARBON-CARBON DOUBLE BOND

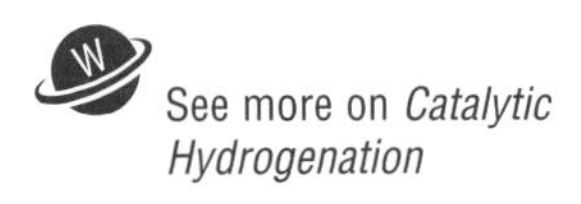

Catalytic hydrogenation of alkenes belongs to the general class of reactions known as **addition reactions**, which were introduced in Section 10.4. In these reactions, one mole of hydrogen is *stereospecifically* added to the *same* side of the carbon-carbon π-bond, as exemplified by the hydrogenation of 1,2-dimethylcyclopentene (**1**) to give *cis*-1,2-dimethylcyclopentane (**2**), as shown in Equation 17.5. Such processes are termed ***syn*-additions**. Although the detailed mechanism by which this reaction occurs is not fully understood, it is generally believed that hydrogen gas is adsorbed on the surface of the finely divided metal catalyst to produce hydrogen atoms. The π-electrons of the double bond also complex with the surface of the catalyst, and the hydrogen atoms are transferred from the surface to the carbon atoms to form the saturated product. Catalytic hydrogenation is the method of choice for reducing alkenes to alkanes and alkynes to *cis*-alkenes.

CH_3 CH_3 **1** 1,2-Dimethylcyclopentene $\xrightarrow{H_2/Pt}$ CH_3 H H CH_3 **2** *cis*-1,2-Dimethylcyclopentane (17.5)

Catalytic hydrogenation is commonly performed by first suspending the appropriate metallic catalyst in a solution containing the reactant in a suitable solvent. The resulting mixture is then stirred or shaken under an atmosphere of hydrogen gas at pressures typically ranging from 15 to 60 psi. Hydrogen, a highly flammable gas, is normally purchased in pressurized gas cylinders, and special apparatus and precautions must be used to deliver the hydrogen from the cylinder to the apparatus. However, it is more convenient in simple experiments to generate hydrogen gas *in situ*. Combining sodium borohydride and concentrated hydrochloric acid provides a ready source of hydrogen via the reaction of H^+ from HCl with the hydride ions, $H:^-$, contained in $NaBH_4$ (Eq. 17.6). The catalyst, platinum on carbon, may be prepared in the same flask by reducing chloroplatinic acid, H_2PtCl_6, with hydrogen in the presence of decolorizing carbon, which serves as a solid support for the finely divided metallic platinum produced (Eq. 17.7).

$$NaBH_4 + HCl + 3\,H_2O \rightarrow 3\,H_2 + B(OH)_3 + NaCl \quad (17.6)$$

$$H_2PtCl_6 + 2\,H_2 \rightarrow Pt^0 + 6\,HCl \quad (17.7)$$

The experiment that follows involves the conversion of 4-cyclohexene-*cis*-1,2-dicarboxylic acid (**3**) to cyclohexane-*cis*-1,2-dicarboxylic acid (**4**), as shown in Equation 17.8. The catalytic hydrogenation of **3** is considered to be selective because the carboxylic acid groups, $-CO_2H$, are *not* reduced under the reaction conditions. The starting material for this reduction may be readily prepared by the hydrolysis of 4-cyclohexene-*cis*-1,2-dicarboxylic acid anhydride, which is the product of the Diels-Alder reaction between butadiene and maleic anhydride, according to the procedure outlined in Section 12.3, Part C1.

H H CO_2H CO_2H **3** 4-Cyclohexene-*cis*-1,2-dicarboxylic acid $\xrightarrow{H_2,\ Pt(C)}$ H H H H CO_2H CO_2H **4** Cyclohexane-*cis*-1,2-dicarboxylic acid (17.8)

EXPERIMENTAL PROCEDURE

Hydrogenation of 4-Cyclohexene-cis-1,2-dicarboxylic Acid

Purpose To demonstrate the selective reduction of an alkene by catalytic hydrogenation.

SAFETY ALERT

1. ***Hydrogen gas is extremely flammable.* Use no flames in this experiment, and make certain that there are no open flames nearby.**
2. **The platinum catalyst prepared in this experiment may be *pyrophoric* and combust spontaneously in air. Do *not* allow it to become dry!**
3. **Use *flameless* heating when removing the diethyl ether.**
4. **Wear latex gloves throughout the experiment to avoid skin contamination by the solutions you are using. If you do get these solutions on your hands, wash them thoroughly with water. In the case of contact with the sodium borohydride solution, rinse the affected areas with 1% acetic acid solution. In the case of acid burns, apply a paste of sodium bicarbonate to the area for a few minutes and then rinse with large amounts of water.**

MINISCALE PROCEDURE

Preparation Sign in at **www.cengage.com/login** to answer Pre-Lab Exercises, access videos, and read the MSDSs for the chemicals used or produced in this procedure. Review Sections 2.9, 2.11, 2.13, 2.17, 2.21, and 2.29.

Apparatus A 50-mL filter flask, a 2-mL plastic syringe, separatory funnel, a 50-mL round-bottom flask, apparatus for magnetic stirring, simple distillation, vacuum filtration, and *flameless* heating.

Setting Up Prepare a reaction vessel for hydrogenation by using rubber bands or wire to attach a balloon to the sidearm of the filter flask. Place a stirbar in the flask. Prepare a 1 *M* aqueous solution of sodium borohydride by dissolving 0.2 g of sodium borohydride in 5 mL of 1% aqueous sodium hydroxide. Place 5 mL of water, 0.5 mL of a 5% solution of chloroplatinic acid ($H_2PtCl_6{\cdot}6\ H_2O$), and 0.2 g of decolorizing carbon in the reaction flask, and slowly add 1.5 mL of the 1 *M* sodium borohydride solution with stirring. Stir the resulting slurry for 5 min to allow the active catalyst to form. During this time, dissolve 0.5 g of 4-cyclohexene-*cis*-1,2-dicarboxylic acid (Sec. 12.3, Part C1) in 5 mL of water by heating. Pour 2 mL of *concentrated* hydrochloric acid into the reaction flask containing the catalyst, and add the hot aqueous solution of the diacid to the flask. Seal the flask with a rubber septum that is securely wired in place, and stir the aqueous suspension.

Reaction Draw 0.8 mL of the 1 *M* sodium borohydride solution into the syringe, carefully push the needle of the syringe through the rubber septum, and inject the solution dropwise into the stirred solution of reactants in the filter flask. If the balloon on the sidearm of the flask begins to inflate *rapidly*, add the solution of sodium borohydride more slowly. When the syringe is empty, remove it from the flask, refill it with an additional 0.8 mL of 1 *M* sodium borohydride solution, and add this second portion of sodium borohydride to the stirred reaction mixture. The balloon should inflate somewhat, indicating a positive pressure of hydrogen in the system. Remove the syringe from the rubber septum, and continue stirring for 5 min. Then heat the flask at 70–80 °C for 1.5–2 h with continued stirring to complete the reaction.

Work-Up and Isolation Release the pressure from the reaction flask by pushing a syringe needle through the rubber septum. *Do not allow the reaction mixture to cool.* Remove the catalyst by vacuum filtration of the *hot* reaction mixture using a Hirsch funnel, and *immediately* treat the catalyst according to the protocol given in the Wrapping It Up section.★ Cool the filtrate and use pHydrion paper to determine its pH. If the filtrate is not acidic, add *concentrated* hydrochloric acid dropwise until the pH is below 5. Transfer the filtrate to the separatory funnel. Rinse the filter flask with 15 mL of diethyl ether to dissolve the solid product that may have separated upon cooling, and add this rinse to the separatory funnel. Based upon the amount of water used in the experiment, calculate the weight of sodium chloride that would be needed to saturate the filtrate, and add this amount of solid sodium chloride to the separatory funnel. Shake the separatory funnel, venting as necessary to relieve pressure. Allow the layers to separate; any insoluble solid should be sodium chloride. Separate the layers, and extract the aqueous layer sequentially with two 10-mL portions of diethyl ether. Dry the *combined* ethereal extracts by swirling them over several spatula-tips full of anhydrous sodium sulfate.★ Swirl the flask occasionally for a period of 10–15 min to facilitate drying; add further small portions of anhydrous sodium sulfate if the solution remains cloudy.

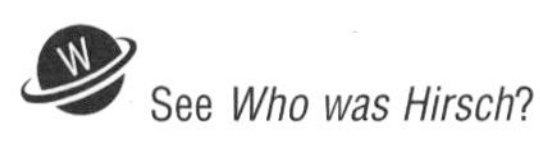

See *Who was Hirsch?*

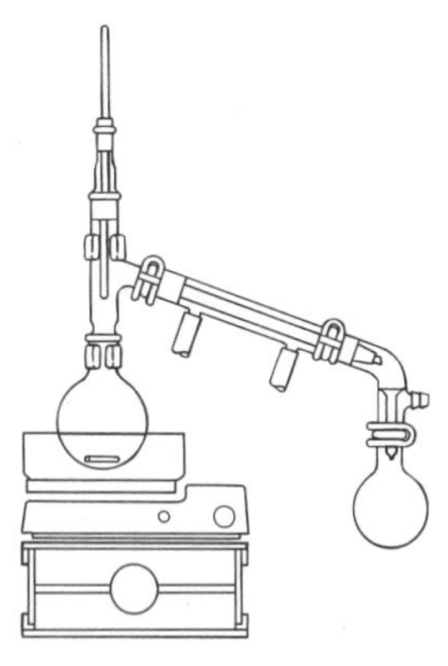

Carefully decant the dried solution into a tared round-bottom flask containing a stirbar. Equip the flask for simple distillation and remove the diethyl ether by simple distillation. Alternatively, use rotary evaporation or other techniques to concentrate the solution. The final traces of solvent may be removed by attaching the flask to a vacuum source and gently swirling the contents as the vacuum is applied. The product remains as a solid residue in the flask.★ Air-dry the crude product and determine its weight and melting point.

Purification The crude product may be recrystallized from a *minimum* amount of water as follows. Carefully scrape the bulk of the crude diacid from the round-bottom flask into a *small* Erlenmeyer flask. Add *no more* than 1 mL of water to the round-bottom flask, heat the contents to dissolve any residual diacid, and pour the hot solution into the Erlenmeyer flask. Heat the aqueous mixture to boiling. If necessary, *slowly* add water *dropwise* to the boiling mixture to dissolve all of the diacid. Do *not* add more than an additional 1 mL of water, because the diacid is rather soluble in water. Determine the pH of the aqueous solution to ascertain whether hydrogenation has affected the acidic nature of the molecule. When all of the diacid has dissolved, add 2 drops of *concentrated* hydrochloric acid to decrease its solubility. Allow the mixture to cool to room temperature, and then place the flask in an ice-water bath to complete crystallization of the product. Collect the product by vacuum filtration and air-dry it completely.★ The reported melting point of cyclohexane-*cis*-1,2-dicarboxylic acid is 192–193 °C.

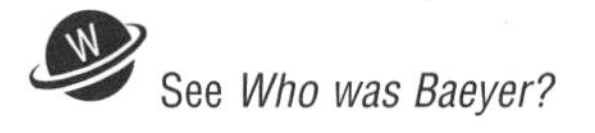
See *Who was Baeyer?*

Analysis Weigh the product and calculate the percent yield; determine its melting point. Perform the qualitative tests for unsaturation using the bromine and Baeyer tests (Secs. 25.8A and B, respectively) on your product, and record your observations. Obtain IR and ^{1}H NMR spectra of your starting material and product, and compare them with those of authentic samples (Figs. 12.11, 12.12, 17.1, and 17.2).

WRAPPING IT UP

Wet the *catalyst* and the *filter paper* immediately with water, and remove them from the Hirsch funnel; place them in the container reserved for recovered catalyst, which should be kept wet with water at all times. Neutralize the *combined aqueous filtrates and washes* with sodium carbonate, and then flush them down the drain with excess water. Place the *recovered ethereal distillate* in the container for nonhalogenated organic solvents. After allowing the diethyl ether to evaporate from the *sodium sulfate* on a tray in the hood, place the used drying agent in the container for nonhazardous solids. Place the *dichloromethane solution* from the bromine test for unsaturation in a container for halogenated organic liquids; put the *manganese dioxide* from the Baeyer test for unsaturation in a container for heavy metals.

EXERCISES

1. Why is the sodium borohydride solution prepared in basic rather than acidic aqueous media?
2. Why is decolorizing carbon used in the procedure for the catalytic hydrogenation of 4-cyclohexene-*cis*-1,2-dicarboxylic acid (**3**)?
3. Why does adding hydrochloric acid to an aqueous solution of a carboxylic acid decrease its solubility in water?
4. What is the purpose of swirling the ethereal extracts over sodium sulfate?
5. How did you determine that the two carboxylic acid groups were not reduced by catalytic hydrogenation of **3**?
6. Determine the oxidation numbers (Sec. 16.1) of the carbon atoms undergoing change in the following reactions, and indicate the net increase or decrease of oxidation number in each.
   a. $CH_3Cl \rightarrow CH_4$
   b. $CH_3CHO \rightarrow CH_3CH_2OH$
   c. acetylene → ethylene
   d. $CH_3CO_2H \rightarrow CH_3CHO$
   e. $CH_3CH = NH \rightarrow CH_3CH_2NH_2$
   f. $H_2C = CH_2 \rightarrow BrCH_2CH_2Br$
7. Draw the structure of the product of catalytic hydrogenation of:
   a. 1,2-dimethylcyclohexene
   b. *cis*-2,3-dideutero-2-butene
   c. *trans*-2,3-dideutero-2-butene
   d. *cis*-3,4-dimethyl-3-hexene
8. Consider the conversion of nitrobenzene to aniline as shown.

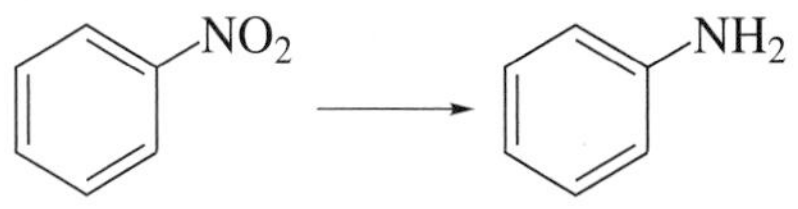

NO_2 → NH_2

Nitrobenzene   Aniline

a. Determine the oxidation number of the nitrogen atom in the starting material and the product. What is the net change in the oxidation number of this atom?

b. Is this reaction an oxidation or a reduction?

c. Suggest a method, including necessary reagents, that may be used to accomplish this transformation.

9. The heats of hydrogenation, ΔH_{hydrog}, for three isomeric butenes have been determined quantitatively by catalytic hydrogenation.

1-butene	$\Delta H_{hydrog} = -30.3$ kcal/mol
cis-2-butene	$\Delta H_{hydrog} = -28.6$ kcal/mol
trans-2-butene	$\Delta H_{hydrog} = -27.6$ kcal/mol

a. Give the reagent(s) that would effect these catalytic hydrogenations.

b. Write the structure(s) of the products(s) formed from each of these reductions.

c. Rank these butenes in order of *decreasing* stability.

d. Draw a reaction profile for each of the three hydrogenations on the same diagram. For each reaction, label the starting alkene, the product, and ΔH_{hydrog}, and show the relative stabilities of the starting material and the product.

Sign in at **www.cengage.com/login** and use the spectra viewer and Tables 8.1–8.8 as needed to answer the blue-numbered questions on spectroscopy.

10. Consider the spectral data for 4-cyclohexene-*cis*-1,2-dicarboxylic acid (Figs. 12.11 and 12.12).

a. In the functional group region of the IR spectrum, specify the absorptions associated with the hydroxyl and the carbonyl components of the carboxyl group.

b. In the ^{1}H NMR spectrum, assign the various resonances to the hydrogen nuclei responsible for them.

c. For the ^{13}C NMR data, assign the various resonances to the carbon nuclei responsible for them.

11. Consider the spectral data for cyclohexane-*cis*-1,2-dicarboxylic acid (Figs. 17.1 and 17.2).

a. In the functional group region of the IR spectrum, specify the absorptions associated with the hydroxyl and the carbonyl components of the carboxyl group.

b. In the ^{1}H NMR spectrum, assign the various resonances to the hydrogen nuclei responsible for them.

c. For the ^{13}C NMR data, assign the various resonances to the carbon nuclei responsible for them.

12. Discuss the differences observed in the IR and NMR spectra of 4-cyclohexene-*cis*-1,2-dicarboxylic acid and cyclohexane-*cis*-1,2-dicarboxylic acid that are consistent with the selective reduction of the carbon-carbon double bond in this experiment.

SPECTRA

Starting Material and Product

*The IR and NMR spectra of 4-cyclohexene-*cis*-1,2-dicarboxylic acid are given in Figures 12.11 and 12.12, respectively.*

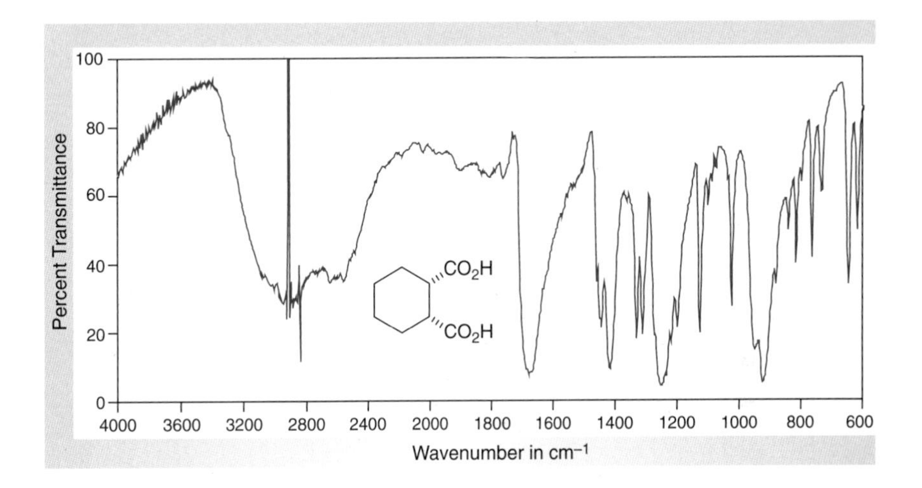

Figure 17.1
*IR spectrum of cyclohexane-*cis*-1,2-dicarboxylic acid (IR card).*

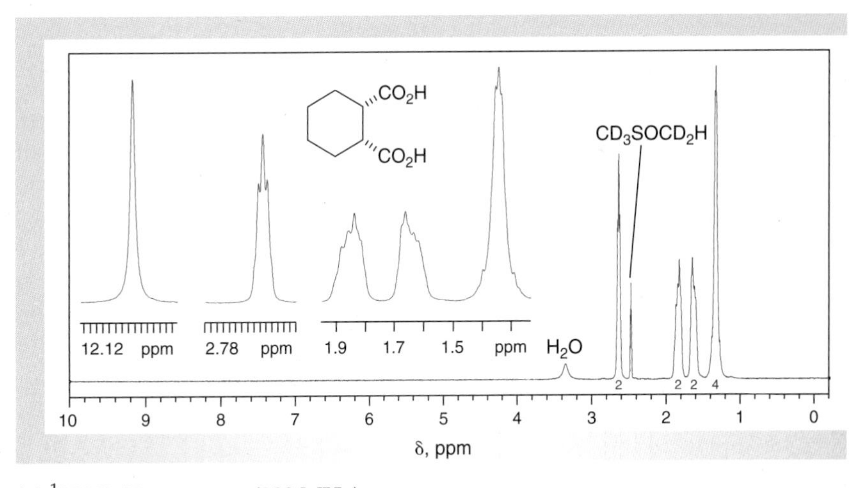

Figure 17.2
*NMR data for cyclohexane-*cis*-1,2-dicarboxylic acid (DMSO-d_6).*

(a) ^{1}H NMR spectrum (300 MHz).
(b) ^{13}C NMR data: δ 23.8, 26.3, 42.3, 175.6.

17.3 REDUCTION OF IMINES; PREPARATION OF AMINES

Imines are readily formed by the condensation of primary amines with carbonyl compounds such as aldehydes and ketones according to Equation 17.9. The stability of an imine depends upon the nature of the substituents, R^1–R^3, on the carbon and nitrogen atoms of the imine functional group. For example, imines that bear only alkyl groups are less stable toward hydrolysis and polymerization than those having one or more aromatic substituents on the carbon-nitrogen double bond; the most stable imines are those having only aryl substituents.

$$R^1R^2C{=}O + R^3{-}NH_2 \longrightarrow R^1R^2C{=}N{\sim}R^3 + H_2O \qquad (17.9)$$

An aldehyde or a ketone — A primary amine — An imine

Imines undergo addition of one mole of hydrogen in the presence of a catalyst such as palladium, platinum, or nickel to produce secondary amines (Eq. 17.10). This process is analogous to the hydrogenation of alkenes to alkanes and of carbonyl compounds to alcohols.

$$R^1R^2C{=}N{\sim}R^3 + H_2 \xrightarrow{\text{Catalyst}} R^1R^2C(H){-}N(H)R^3 \qquad (17.10)$$

An imine — A secondary amine

Other functional groups, such as C=C, C≡C, N=O (as in NO_2), and C–X, undergo catalytic hydrogenation under the same conditions required to reduce imines. Hence, it may not be possible to reduce an imine selectively by catalytic hydrogenation in the presence of one or more of these functional groups. This problem may be readily overcome, however, by using a metal hydride, such as sodium borohydride, $NaBH_4$, to reduce the imine, as depicted in Equation 17.11. Sodium borohydride also reacts with methanol, which is frequently used as the solvent, but it reacts much faster with the imine. Thus, when performing small-scale reactions, it is generally more convenient to use an excess of $NaBH_4$ to allow for its reaction with the solvent than to use a less-reactive solvent in which this reagent is less soluble.

$$R^1R^2C{=}N{\sim}R^3 \xrightarrow[(2)\ H_2O]{(1)\ NaBH_4,\ CH_3OH} R^1R^2C(H){-}N(H){-}R^3 \qquad (17.11)$$

An imine — A secondary amine

In this section, you will prepare *N*-cinnamyl-*m*-nitroaniline (**9**) by a sequence beginning with the condensation of cinnamaldehyde (**5**) with *m*-nitroaniline (**6**), followed by reduction of the intermediate imine **7** with sodium borohydride, as shown in Equations 17.12–17.14. The formation of the imine is reversible, but the reaction is driven to completion by **azeotropic distillation**. Because cyclohexane and water form a **minimum-boiling azeotrope** (Sec. 4.4), the water generated by the condensation of **5** and **6** is continuously removed by distilling the cyclohexane-water azeotrope throughout the course of the reaction.

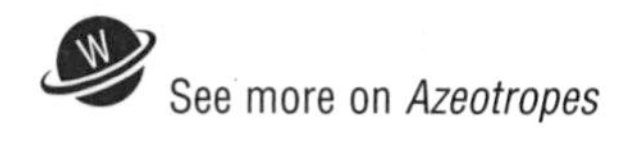

See more on *Azeotropes*

$$\underset{\substack{\mathbf{5}\\ \text{Cinnamaldehyde}}}{C_6H_5CH{=}CH{-}\underset{\substack{|\\H}}{C}{=}O} + \underset{\substack{\mathbf{6}\\ m\text{-Nitroaniline}}}{H_2N{-}C_6H_4{-}NO_2} \underset{}{\overset{\text{Cyclohexane}}{\rightleftharpoons}} \underset{\substack{\mathbf{7}\\ N\text{-Cinnamylidene-}m\text{-nitroaniline}}}{C_6H_5CH{=}CH{-}\underset{\substack{|\\H}}{C}{=}N{-}C_6H_4{-}NO_2} + H_2O \quad (17.12)$$

$$4\ \underset{\mathbf{7}}{C_6H_5CH{=}CH{-}\underset{\substack{|\\H}}{C}{=}N{-}C_6H_4{-}NO_2} + NaBH_4 \longrightarrow \underset{\mathbf{8}}{\left(C_6H_5CH{=}CH{-}\overset{\substack{H\\|}}{C}H{-}\underset{\substack{|\\C_6H_4NO_2}}{N}\right)_4}B^-\,Na^+ \quad (17.13)$$

$$\underset{\mathbf{8}}{\left(C_6H_5CH{=}CH{-}CH_2{-}\underset{\substack{|\\C_6H_4NO_2}}{N}\right)_4}B^-\,Na^+ + 3\ H_2O \longrightarrow 4\ \underset{\substack{\mathbf{9}\\ N\text{-Cinnamyl-}m\text{-nitroaniline}}}{C_6H_5CH{=}CH{-}CH_2{-}\underset{\substack{|\\C_6H_4NO_2}}{NH}} + NaH_2BO_3 \quad (17.14)$$

The reaction of sodium borohydride with the imine **7** (Eq. 17.13) is analogous to the addition of sodium borohydride (Sec. 17.4) or a Grignard reagent (Secs. 19.1 and 19.3) to a carbonyl compound. A nucleophilic hydride ion (H:) is transferred from the borohydride anion (BH_4^-) to the electrophilic carbon atom of the carbon-nitrogen double bond, and the electron-deficient boron atom becomes attached to nitrogen. All four hydrides of the borohydride anion may be transferred to the imine carbon to produce the organoborate anion **8**, which is subsequently decomposed with water to yield the secondary amine **9** as shown in Equation 17.14. In this experiment, you will have an opportunity to determine whether sodium borohydride selectively reduces the imine group in **7** but not the carbon-carbon double bond, the nitro group, or the benzene rings that are also present.

Formation and Reduction of N-*Cinnamylidene*-m-*nitroaniline*

Purpose To demonstrate the reductive alkylation of an amine by the sequence of forming and reducing an imine and to evaluate the selectivity of a hydride reducing agent.

SAFETY ALERT

1. **Avoid contact of methanolic solutions of sodium borohydride with your skin; they are *highly caustic.* If possible, wear latex gloves when handling these solutions. If these solutions accidentally get on your skin, wash the affected area first with a 1% acetic acid solution and then with copious quantities of water.**
2. **Do *not* stopper flasks containing methanolic sodium borohydride; the solution slowly evolves hydrogen gas, and a dangerous buildup of pressure could occur in a stoppered flask.**
3. **Cyclohexane is flammable; use *flameless* heating in this experiment.**

MINISCALE PROCEDURE

Preparation Sign in at **www.cengage.com/login** to answer Pre-Lab Exercises, access videos, and read the MSDSs for the chemicals used or produced in this procedure. Review Sections 2.9, 2.11, 2.13, 2.17, and 2.22.

Apparatus A 25-mL round-bottom flask, apparatus for heating under reflux, simple distillation, magnetic stirring, vacuum filtration, and *flameless* heating.

Setting Up Place 0.6 g of *freshly distilled* cinnamaldehyde, 0.6 g of *m*-nitroaniline, 5 mL of cyclohexane, and a stirbar in the round-bottom flask. Assemble the apparatus for simple distillation.

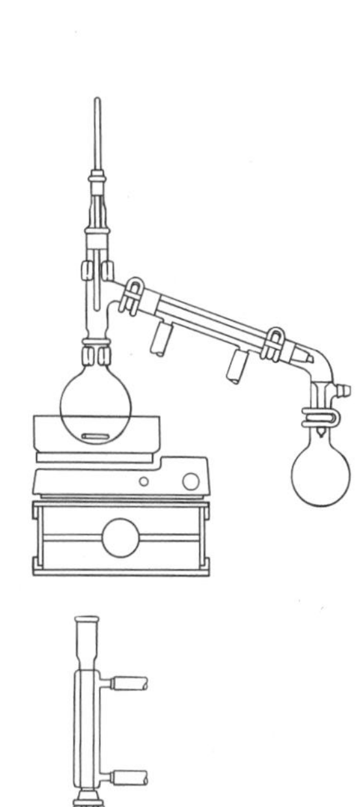

Imine Formation Heat the reaction flask with stirring, and distill until most of the cyclohexane is removed; about 4 mL of distillate should be obtained in approximately 5–10 min. *Discontinue* heating when the distillation rate decreases. Add another 5 mL of cyclohexane, and resume the distillation. Stop heating when the rate of distillation again decreases, and remove the heat source.

Isolation and Characterization of the Imine (Optional) Take about 0.2 mL of the residual liquid from the stillpot with a Pasteur pipet. Transfer this aliquot to a small test tube, and add 1 mL of methanol. Swirl the tube to effect solution and place the test tube in an ice-water bath. Collect any crystals that separate, air-dry them, and determine their melting point. The crude *N*-cinnamylidene-*m*-nitroaniline may be purified by recrystallization from a small volume of methanol. Pure *N*-cinnamylidene-*m*-nitroaniline has a melting point of 92–93 °C.

Reduction Add 4 mL of methanol to the stillpot, stir to dissolve the crude imine, and set up the apparatus for heating under reflux. Prepare a solution of 0.15 g of sodium borohydride in 3 mL of methanol. Because sodium borohydride reacts slowly with methanol to evolve hydrogen gas, this solution should be prepared in an *unstoppered* vessel *immediately* before use. Using a Pasteur pipet or a syringe, transfer the methanolic solution of sodium borohydride dropwise through the top of the reflux condenser to the stirred solution of the imine; the addition should be completed within 1 min. Try to add the solution *directly* into the flask without touching the walls of the condenser. After all of the borohydride solution is added, heat the reaction mixture under reflux for 15 min.

Work-Up, Isolation, and Purification Cool the reaction mixture to room temperature and add 10 mL of water.★ Stir the mixture for 10–15 min. Collect the crystals that separate, and wash them with water.★ Recrystallize the crude *N*-cinnamyl-*m*-nitroaniline from 95% ethanol. Air-dry the product.

Analysis Weigh the product and calculate the percent yield; determine its melting point. The melting point of pure *N*-cinnamyl-*m*-nitroaniline is 106–107 °C. Perform the qualitative test for unsaturation using the bromine test (Sec. 25.8A) on your product, and record your observations. Obtain IR and ^1H NMR spectra of your starting materials, the intermediate imine if isolated, and the product, and compare them with those of authentic samples (Figs. 17.3–17.10).

MICROSCALE PROCEDURE

Preparation Sign in at **www.cengage.com/login** to answer Pre-Lab Exercises, access videos, and read the MSDSs for the chemicals used or produced in this procedure. Review Sections 2.9, 2.11, 2.13, 2.17, and 2.22.

Apparatus A 5-mL conical vial, 1-mL syringe, apparatus for heating under reflux, simple distillation, vacuum filtration, magnetic stirring, Craig tube filtration, and *flameless* heating.

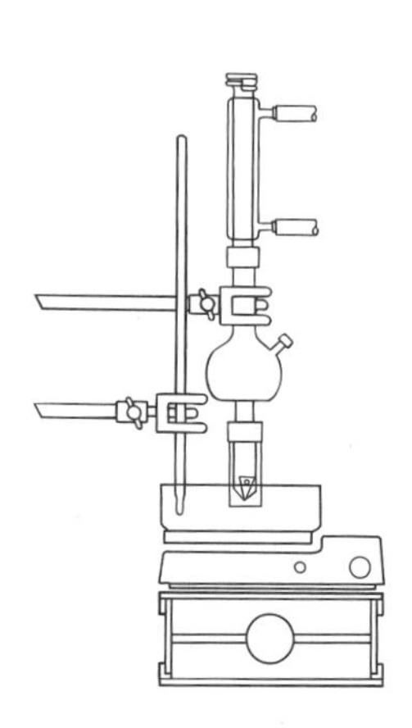

Setting Up Place 120 mg of *freshly distilled* cinnamaldehyde, 120 mg of *m*-nitroaniline, 1 mL of cyclohexane, and a spinvane in the conical vial. Assemble the apparatus for simple distillation.

Imine Formation Heat the reaction vial with stirring and distill until most of the cyclohexane is removed; this should require about 3–5 min. *Discontinue* heating when the distillation rate decreases. Add another 1 mL of cyclohexane, and resume the distillation; stop heating when the rate of distillation again decreases, and remove the heat source.

Reduction Add 0.8 mL of methanol to the conical vial, stir to dissolve the crude imine, and set up the apparatus for heating under reflux. Prepare a solution of 30 mg of sodium borohydride in 0.6 mL of methanol. Because sodium borohydride slowly reacts with methanol to evolve hydrogen gas, this solution should be prepared in an *unstoppered* vessel *immediately* before use. Using a Pasteur pipet or a syringe, transfer the methanolic solution of sodium borohydride *dropwise* through the top of the reflux condenser to the stirred solution of the imine; the addition should be completed within 1 min. Try to add the solution *directly* into the vial without touching the walls of the condenser. After all of the borohydride solution is added, heat the reaction mixture under reflux for 15 min.

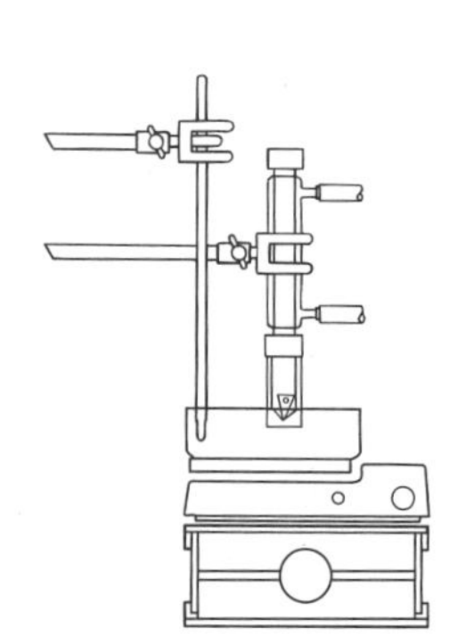

Work-Up, Isolation, and Purification Cool the reaction mixture to room temperature and add 2 mL of water.★ Stir the mixture for 10–15 min. Collect the crystals that separate, and wash them with water.★ Recrystallize the crude *N*-cinnamyl-*m*-nitroaniline from 95% ethanol in a Craig tube. Air-dry the product.

Analysis Weigh the product and calculate the percent yield; determine its melting point. The melting point of pure *N*-cinnamyl-*m*-nitroaniline is 106–107 °C. Perform the qualitative test for unsaturation using the bromine test (Sec. 25.8A) on your product, and record your observations. Obtain IR and ^{1}H NMR spectra of your starting materials and product, and compare them with those of authentic samples (Figs. 17.3–17.6, 17.9, and 17.10).

WRAPPING IT UP

Pour the *cyclohexane distillates* in the container for nonhalogenated organic solvents. Dilute the *aqueous methanolic filtrate* with water, neutralize it with acetic acid to destroy any excess sodium borohydride, and flush the mixture down the drain with excess water. Place the *dichloromethane solution* from the bromine test for unsaturation in a container for halogenated organic liquids.

EXERCISES

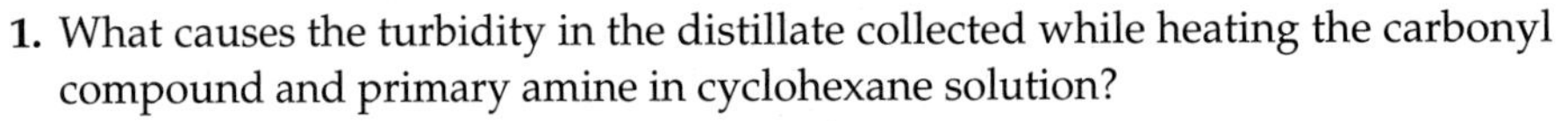

1. What causes the turbidity in the distillate collected while heating the carbonyl compound and primary amine in cyclohexane solution?
2. Why was it not necessary to use *dry* cyclohexane?
3. How does the fact that water forms a minimum-boiling azeotrope with cyclohexane allow water to be removed during the course of the reaction to produce **7**?

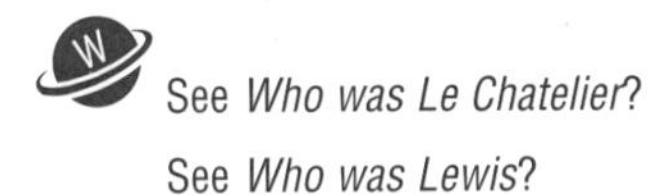

See *Who was Le Chatelier?*

See *Who was Lewis?*

4. Explain how Le Chatelier's principle is being applied in this experiment to give **7**.
5. Draw a Lewis dot structure for $NaBH_4$, and indicate the direction of the B–H bond dipoles.
6. Determine the molar ratio of $NaBH_4$ to imine that you actually used in the experiment. Why is it necessary to use a greater molar ratio than theoretical?
7. After the reaction between sodium borohydride and the imine is complete, the reaction mixture is treated with water to produce the desired secondary amine. Explain this reaction by indicating the source of the hydrogen that ends up on the nitrogen atom.
8. Although sodium borohydride is fairly unreactive toward methanol, adding a mineral acid to this solution results in the rapid destruction of the $NaBH_4$. Explain this observation.
9. Suppose that each of the following pairs of compounds were subjected to the reactions provided in the experimental procedures. Draw the structures of the intermediate imines that would be formed, as well as those of the final amine product.
   - **a.** cinnamaldehyde and aniline
   - **b.** benzaldehyde and *m*-nitroaniline
   - **c.** benzaldehyde and aniline
   - **d.** cyclohexanone and aniline
10. Propose a mechanism to show how imine **7** can be hydrolyzed by water at pH 6 to produce **5** and **6**. Symbolize the flow of electrons with curved arrows.
11. Explain why aliphatic imines are much less stable than those having at least one aromatic group attached to the carbon-nitrogen double bond.
12. Write the structure(s) of the product(s) that you would expect from reducing *N*-cinnamylidene-*m*-nitroaniline by *catalytic* hydrogenation.

13. A nitro group, a carbon-carbon double bond, a carbon-nitrogen double bond, and an aromatic ring may each be reduced under certain conditions. Rank these functional groups in the order of *increasing* reactivity toward hydride reducing agents such as sodium borohydride.

14. Why do you think an imine bearing an aryl group on the nitrogen atom is more stable than one bearing an alkyl group?

15. The optimal pH for condensing an amine with a carbonyl group to give an imine according to Equation 17.9 is about 5–6.

- **a.** Using curved arrows to symbolize the flow of electrons, provide a stepwise mechanism showing all proton transfers for this reaction.
- **b.** In the mechanism you proposed in Part **a**, what step(s) would be slower at *higher* pH? Explain your reasoning.
- **c.** In the mechanism you proposed in Part **a**, what steps(s) would be slower at *lower* pH? Explain your reasoning.

16. What fundamental characteristic of alkenes makes them unreactive toward hydride reagents?

Sign in at **www.cengage.com/login** and use the spectra viewer and Tables 8.1–8.8 as needed to answer the blue-numbered questions on spectroscopy.

17. Consider the spectral data for cinnamaldehyde (Figs. 17.3 and 17.4).

- **a.** In the functional group region of the IR spectrum, specify the absorptions associated with the α, β-unsaturated aldehyde and with the aromatic ring.
- **b.** In the ^{1}H NMR spectrum, assign the various resonances to the hydrogen nuclei responsible for them.
- **c.** For the ^{13}C NMR data, assign the various resonances to the carbon nuclei responsible for them.

18. Consider the spectral data for *m*-nitroaniline (Figs. 17.5 and 17.6).

- **a.** In the functional group region of the IR spectrum, specify the absorptions associated with the nitro group and the aromatic ring. There is a sharp absorption at about 3400 cm^{-1}. What functional group is responsible for this absorption?
- **b.** In the ^{1}H NMR spectrum, assign the various resonances to the hydrogen nuclei responsible for them.
- **c.** For the ^{13}C NMR data, assign the various resonances to the carbon nuclei responsible for them.

19. Consider the spectral data for *N*-cinnamylidene-*m*-nitroaniline (Figs. 17.7 and 17.8).

- **a.** In the functional group region of the IR spectrum, specify the absorptions associated with the nitro group, the carbon-nitrogen double bond, and the carbon-carbon π-bonds of the aromatic ring and alkene.
- **b.** In the ^{1}H NMR spectrum, assign the various resonances to the hydrogen nuclei responsible for them.
- **c.** For the ^{13}C NMR data, assign the various resonances to the carbon nuclei responsible for them.

20. Consider the spectral data for *N*-cinnamyl-*m*-nitroaniline (Figs. 17.9 and 17.10).

- **a.** In the functional group region of the IR spectrum, specify the absorptions associated with the amino group, the nitro group, and the aromatic ring.

b. In the ^{1}H NMR spectrum, assign the various resonances to the hydrogen nuclei responsible for them.

c. For the ^{13}C NMR data, assign the various resonances to the carbon nuclei responsible for them.

21. Discuss the differences observed in the IR and NMR spectra of cinnamaldehyde and *N*-cinnamylidene-*m*-nitroaniline that are consistent with the formation of the carbon-nitrogen double bond of the imine in this experiment.

22. Discuss the differences observed in the IR and NMR spectra of *N*-cinnamylidene-*m*-nitroaniline and *N*-cinnamyl-*m*-nitroaniline that are consistent with the *selective* reduction of the carbon-nitrogen double bond in this experiment.

SPECTRA

Starting Materials and Products

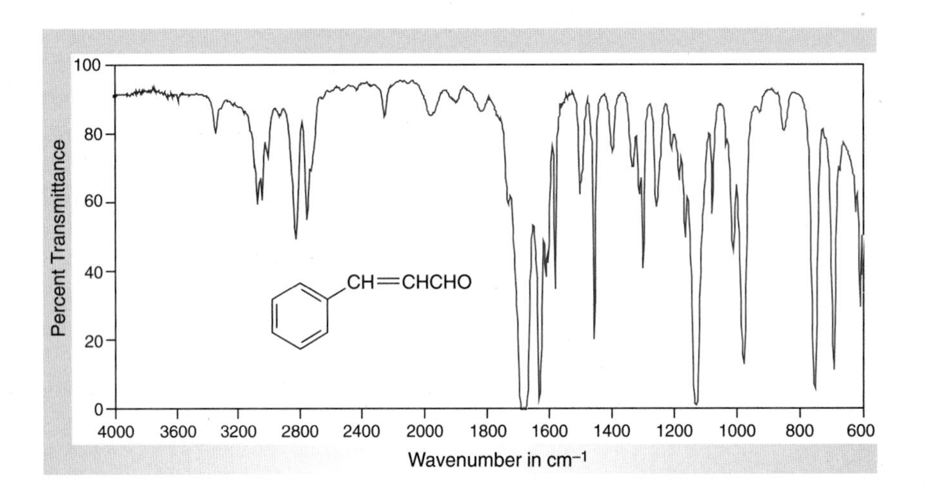

Figure 17.3
IR spectrum of cinnamaldehyde (neat).

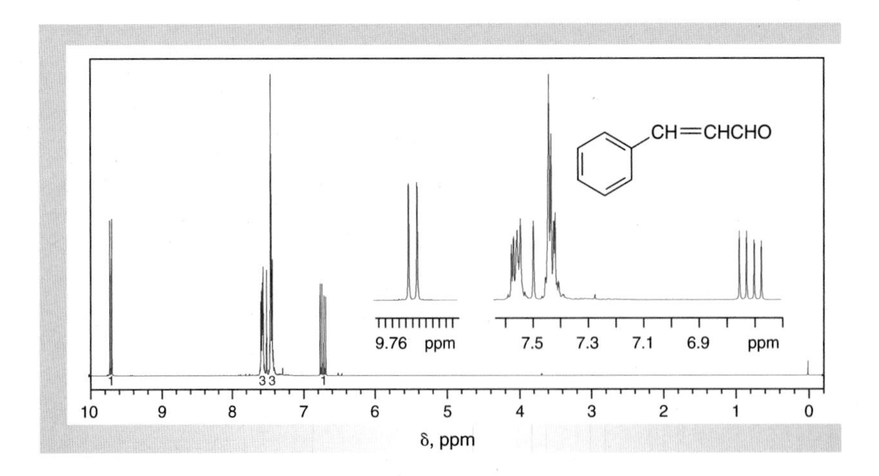

Figure 17.4
NMR data for cinnamaldehyde ($CDCl_3$).

(a) ^{1}H NMR spectrum (300 MHz).
(b) ^{13}C NMR data: δ 128.5, 129.0, 131.1, 134.1, 152.3, 193.2.

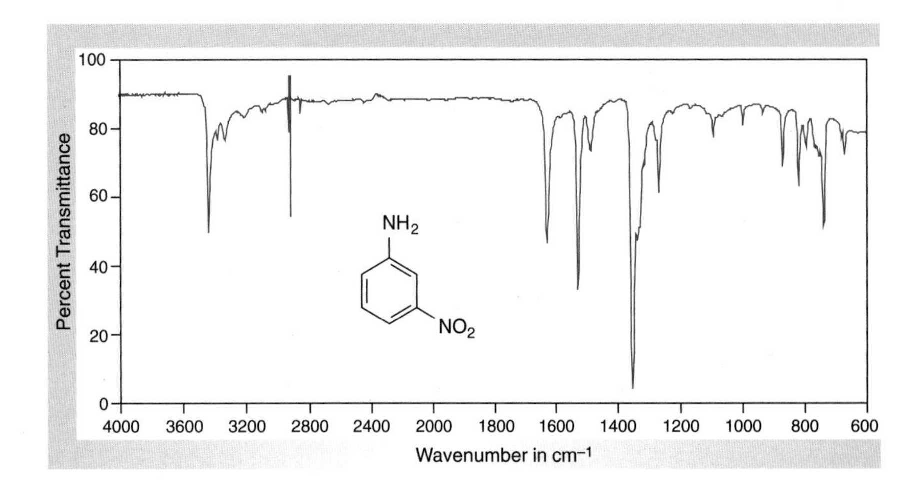

Figure 17.5
IR spectrum of m-*nitroaniline (IR card).*

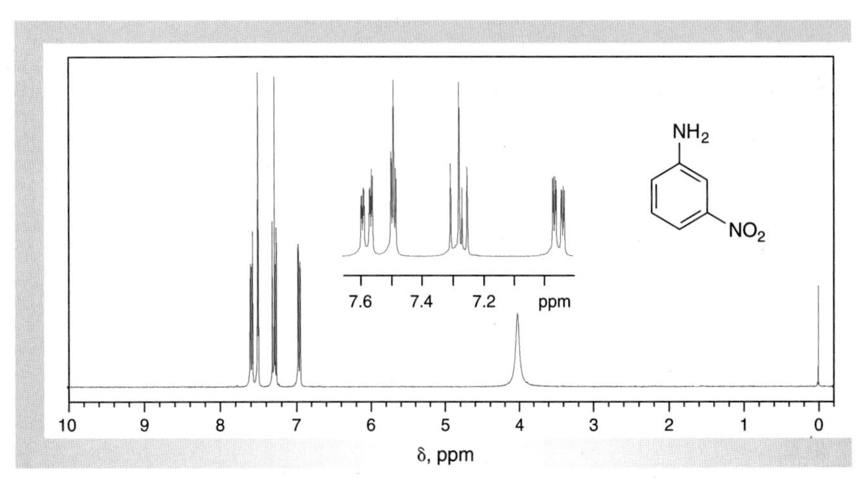

Figure 17.6
NMR data for m-*nitroaniline ($CDCl_3$).*

(a) ^1H NMR spectrum (300 MHz).
(b) ^{13}C NMR data: δ 107.8, 110.5, 120.2, 129.6, 149.1, 149.8.

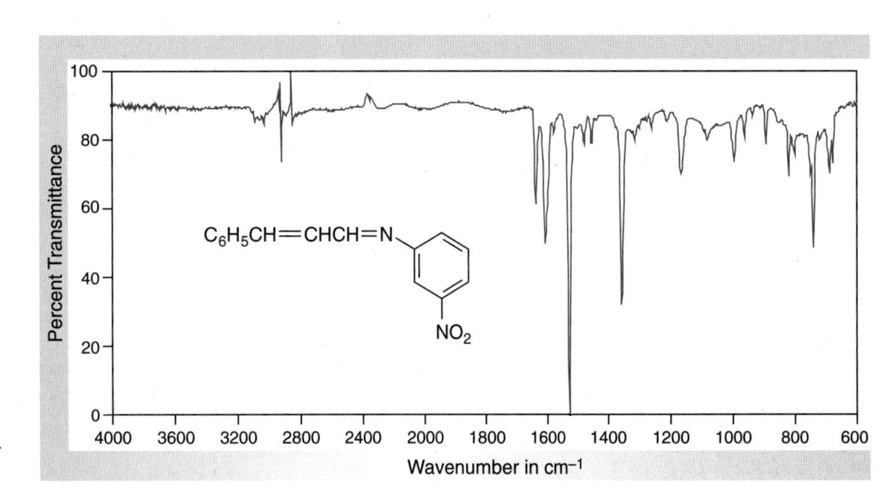

Figure 17.7
IR spectrum of N-*cinnamylidene-* m-*nitroaniline (IR card).*

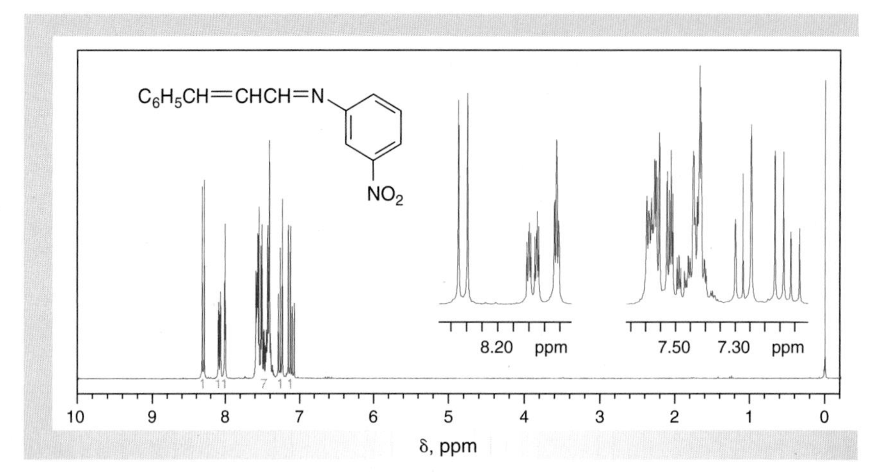

Figure 17.8
NMR data for N-*cinnamylidene*-m-*nitroaniline* ($CDCl_3$).

(a) ^{1}H NMR spectrum (300 MHz).
(b) ^{13}C NMR data: δ 114.8, 120.1, 127.4, 128.6, 129.5, 129.8, 129.9, 130.2, 134.8, 145.7, 148.6, 152.6, 163.4.

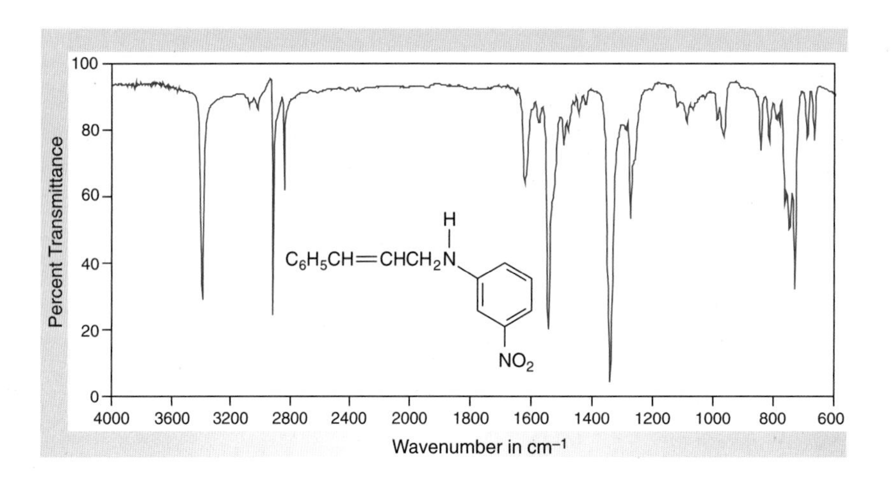

Figure 17.9
IR spectrum of N-*cinnamyl*-m-*nitroaniline (IR card).*

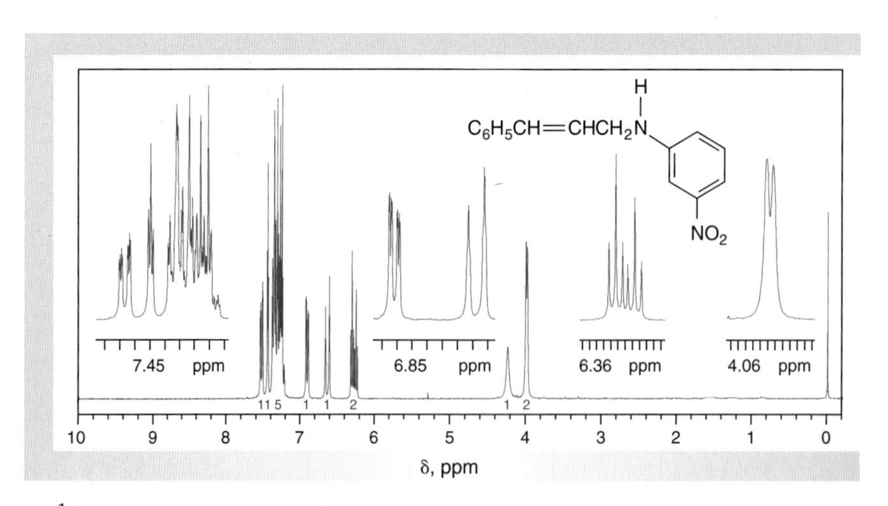

Figure 17.10
NMR data for N-*cinnamyl*-m-*nitroaniline* ($CDCl_3$).

(a) ^{1}H NMR spectrum (300 MHz).
(b) ^{13}C NMR data: δ 45.7, 106.5, 111.9, 118.8, 125.4, 126.3, 127.7, 128.6, 129.6, 132.1, 136.4, 148.7, 149.3.

17.4 REDUCTION OF CARBONYL COMPOUNDS; PREPARATION OF ALCOHOLS

Carbonyl compounds are commonly reduced to alcohols by catalytic hydrogenation or with metal hydrides. When applied to aldehydes, reduction, represented by the symbol [H], provides a convenient route to primary alcohols (Eq. 17.15), whereas the reduction of ketones gives secondary alcohols (Eq. 17.16). Although catalytic hydrogenation of carbonyl groups is frequently the method of choice in industrial processes, lithium aluminum hydride, sodium borohydride, and their derivatives are generally used in the research laboratory. Sodium borohydride may be used in alcoholic and even aqueous solutions, because it reacts much more rapidly with the carbonyl group than with the solvent. On the other hand, lithium aluminum hydride reacts rapidly with protic solvents, so it *must* be used in *anhydrous* ethereal solvents such as diethyl ether or tetrahydrofuran.

$$\underset{\text{An aldehyde}}{RC(=O)H} \xrightarrow{[H]} \underset{\text{A primary alcohol}}{R{-}CH_2OH} \qquad (17.15)$$

R = alkyl or aryl

$$\underset{\text{A ketone}}{R^1C(=O)R^2} \xrightarrow{[H]} \underset{\text{A secondary alcohol}}{R^1{-}CH(OH){-}R^2} \qquad (17.16)$$

R^1 and R^2 = alkyl or aryl

In this experiment, you will examine the reduction of 9-fluorenone (**10**) using sodium borohydride to give 9-fluorenol (**12**), as shown in Equation 17.17. This reaction is mechanistically analogous to the reduction of imines with sodium borohydride (Sec. 17.3) and involves the transfer of hydride ion ($H:^-$) from borohydride ion, BH_4^-, to the electrophilic carbonyl carbon with concomitant transfer of the electron-deficient boron atom to the carbonyl oxygen. Theoretically, all four of the hydrogen atoms attached to boron may be transferred in this way to produce the intermediate borate salt **11**, which is decomposed upon addition of water and acid to yield 9-fluorenol (**12**).

$$4\ \underset{\substack{\textbf{10}\\ \text{9-Fluorenone}\\ \textit{(yellow)}}}{\text{C}_{12}\text{H}_8\text{C=O}} + NaBH_4 \xrightarrow{MeOH} \underset{\textbf{11}}{\left(\text{C}_{12}\text{H}_8\text{CH–O}\right)_4 B^- Na^+} \xrightarrow{H_3O^+/\Delta} 4\ \underset{\substack{\textbf{12}\\ \text{9-Fluorenol}\\ \textit{(colorless)}}}{\text{C}_{12}\text{H}_8\text{CH(OH)}} \qquad (17.17)$$

EXPERIMENTAL PROCEDURES

Reduction of 9-Fluorenone

Purpose To demonstrate the reduction of a ketone to an alcohol using sodium borohydride.

SAFETY ALERT

1. **Avoid contact of methanolic solutions of sodium borohydride with your skin; they are *highly caustic.* If possible, wear latex gloves when handling these solutions. If these solutions accidentally get on your skin, wash the affected area first with a 1% acetic acid solution and then with copious quantities of water.**
2. **Do *not* stopper flasks containing methanolic sodium borohydride; the solution slowly evolves hydrogen gas, and a dangerous buildup of pressure could occur in a stoppered flask.**

MINISCALE PROCEDURE

Preparation Sign in at **www.cengage.com/login** to answer Pre-Lab Exercises, access videos, and read the MSDSs for the chemicals used or produced in this procedure. Review Sections 2.9 and 2.17.

Apparatus A 25-mL Erlenmeyer flask, apparatus for vacuum filtration, and *flameless* heating.

Setting Up Add 0.6 g of 9-fluorenone to the Erlenmeyer flask containing 6 mL of methanol, and swirl the flask with slight warming to dissolve the ketone. Allow the solution to cool to room temperature. *Quickly* weigh 0.05 g of sodium borohydride into a dry test tube, and stopper the test tube *immediately* to avoid undue exposure of the hygroscopic reagent to atmospheric moisture.

Reduction Add the sodium borohydride in *one* portion to the solution of 9-fluorenone in methanol, and swirl the mixture vigorously to dissolve the reagent. After all of the sodium borohydride dissolves, allow the solution to stand at room temperature for 20 min; swirl the solution occasionally. If the solution does not become colorless during this time, add an additional small portion of sodium borohydride with swirling to complete the reaction.★

Work-Up Add 2 mL of 3 *M* sulfuric acid to the reaction mixture. Heat the contents in the flask gently and intermittently for 5–10 min. Stir the mixture occasionally with a glass rod to help dissolve the solid; maintain the internal temperature just below the reflux point to *minimize the loss of solvent.* If all of the solids do not dissolve, *gradually* add methanol in about 0.5- to 1-mL portions with continued heating until a solution is obtained. When all of the precipitated solids redissolve, allow the solution to cool to room temperature, and then cool in an ice-water bath for 10–15 min.

Isolation and Purification Collect the solid product by vacuum filtration, *wash it thoroughly* with water until the filtrate is *neutral,* and air-dry the product.★ Recrystallize the 9-fluorenol by a mixed solvent recrystallization (Sec. 3.2) using methanol and water. After cooling the solution first to room temperature and then in an ice-water bath for 10–15 min, collect and air-dry the crystals.

Analysis Weigh the product and calculate the percent yield; determine its melting point. Obtain IR and ^{1}H NMR spectra of your starting material and product, and compare them with those of authentic samples (Figs. 17.11–17.14).

MICROSCALE PROCEDURE

Preparation Sign in at **www.cengage.com/login** to answer Pre-Lab Exercises, access videos, and read the MSDSs for the chemicals used or produced in this procedure. Review Sections 2.9 and 2.17.

Apparatus A 5-mL conical vial, apparatus for magnetic stirring, heating under reflux, vacuum filtration, Craig tube filtration, and *flameless* heating.

Setting Up Add 100 mg of 9-fluorenone to the conical vial containing a spinvane and 1 mL of methanol, and stir the contents of the vial with slight warming to dissolve the ketone. Allow the solution to cool to room temperature. *Quickly* weigh 10 mg of sodium borohydride into a dry test tube, and stopper the test tube *immediately* to avoid undue exposure of the hygroscopic reagent to atmospheric moisture.

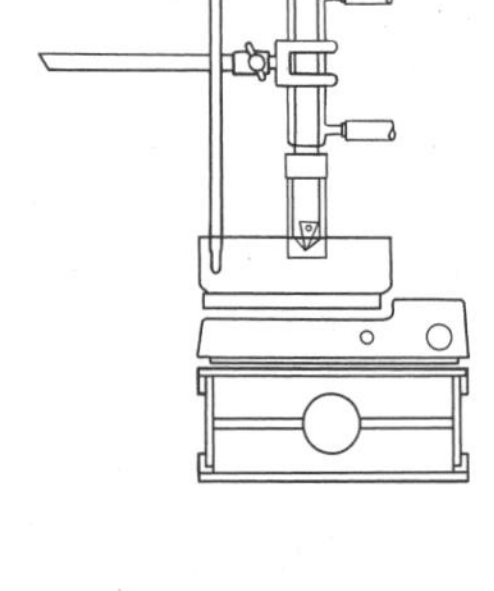

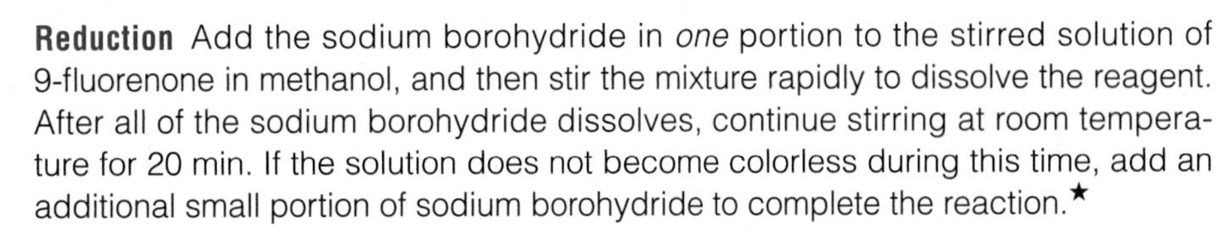

Reduction Add the sodium borohydride in *one* portion to the stirred solution of 9-fluorenone in methanol, and then stir the mixture rapidly to dissolve the reagent. After all of the sodium borohydride dissolves, continue stirring at room temperature for 20 min. If the solution does not become colorless during this time, add an additional small portion of sodium borohydride to complete the reaction.★

Work-Up Add 0.35 mL of 3 *M* sulfuric acid to the reaction mixture. Set up the apparatus for heating under reflux, and heat the vial gently with stirring for 5–10 min to dissolve the solid; adjust the internal temperature to maintain a *gentle* reflux. If all of the solids do not dissolve, *gradually* add methanol *dropwise* to the vial with continued heating and stirring until a solution is obtained. When all of the precipitated solids redissolve, allow the solution to cool to room temperature, and then cool it in an ice-water bath for 10–15 min.

Isolation and Purification Collect the solid product by vacuum filtration, *wash it thoroughly* with water until the filtrate is *neutral,* and air-dry the product.★ Recrystallize the 9-fluorenol by a mixed solvent recrystallization (Sec. 3.2) using methanol and water in a Craig tube. After cooling the solution first to room temperature and then in an ice-water bath for 10–15 min, collect and air-dry the crystals.

Analysis Weigh the product and calculate the percent yield; determine its melting point. Obtain IR and ^{1}H NMR spectra of your starting material and product, and compare them with those of authentic samples (Figs. 17.11–17.14).

Discovery Experiment

Reduction of 4-tert-Butylcyclohexanone

Consult with your instructor before performing this experiment, in which you will determine the stereoselectivity of the reduction of 4-*tert*-butylcyclohexanone with sodium borohydride. You might also be asked to use other hydride reducing agents and compare their stereoselectivities with that of sodium borohydride. Follow either the Miniscale or the Microscale Procedure for the reduction of 9-fluorenone to reduce 4-*tert*-butylcyclohexanone using sodium borohydride. Modify the work-up by using diethyl ether to extract the aqueous mixture obtained after acidification. Dry the extracts over anhydrous sodium sulfate, and remove the solvent. Obtain the ^{1}H NMR spectrum of your product and compare it with the ^{1}H NMR spectra of authentic samples of *cis*- and *trans*-4-*tert*-butylcyclohexanol. Determine the ratio of *cis*- and *trans*-4-*tert*-butylcyclohexanols in your product, using the integrations of the multiplets for the methine protons at 4.0 and 3.5 ppm, respectively, on the carbon atoms bearing the hydroxyl groups. Perform a GLC analysis of your reaction mixture to determine the ratio of *cis*- and *trans*-4-*tert*-butylcyclohexanols in your product, and compare the results of these two determinations.

Discovery Experiment

Reduction of Benzoin

Consult with your instructor before performing this experiment, in which you will determine the stereoselectivity of the reduction of benzoin with sodium borohydride. Follow either the Miniscale or Microscale Procedure described for reducing 9-fluorenone to reduce benzoin with sodium borohydride, but use ethanol rather than methanol as the solvent. After slowly adding 3 *M* HCl to decompose the excess borohydride, add enough water to adjust the solvent composition to 50% (*v:v*) ethanol and water. You may recrystallize the crude product from 50% (*v:v*) ethanol and water. Obtain the melting point and the IR, ^{1}H, and ^{13}C NMR spectra of the purified product for characterization. Compare these data with those for racemic and *meso*-hydrobenzoin to determine the identity of the product and the stereochemistry of the reduction. If authentic samples of racemic and *meso*-hydrobenzoin are available, determine mixed melting points to support your assignment.

WRAPPING IT UP

Dilute the combined *aqueous methanol filtrates* with water, neutralize the resulting solution with sodium carbonate, and flush the mixture down the drain with excess water.

EXERCISES

1. Determine the molar ratio of sodium borohydride to 9-fluorenone that you used in the experiment. Why is it necessary to use a greater molar ratio than theoretical?
2. After the reaction between sodium borohydride and the ketone is complete, the reaction mixture is treated with water and acid to produce the desired secondary alcohol. Explain this reaction by indicating the source of the hydrogen atom that ends up on the oxygen atom.
3. Sodium borohydride is fairly unreactive toward methanol, but adding a mineral acid to this solution results in the rapid destruction of the sodium borohydride. Explain.
4. How many molar equivalents of hydride does sodium borohydride contain?
5. Using curved arrows to symbolize the flow of electrons, write the mechanism for the steps involved in the conversion of 9-fluorenone to 9-fluorenol with sodium borohydride followed by aqueous acid according to Equation 17.17.
6. Suggest a structure for the white precipitate formed in the reaction of 9-fluorenone with sodium borohydride.

7. What gas is evolved when sulfuric acid is added to the reaction mixture?
8. 9-Fluorenone is colored, but 9-fluorenol is not. What accounts for this difference?
9. Draw the structure of the product that results from complete reduction of the following compounds by sodium borohydride.
   a. cyclohexanone
   b. 3-cyclohexen-1-one
   c. 1,4-butanediol
   d. 4-oxohexanal
   e. acetophenone
10. Draw the structure of the product that would be formed from allowing each of the compounds in Exercise 9 to react with excess hydrogen gas in the presence of a nickel catalyst.
11. A nitro group, a carbon-carbon double bond, a carbon-nitrogen double bond, and an aromatic ring may each be reduced under certain conditions. Rank these functional groups in order of *increasing* reactivity toward hydride reducing agents such as sodium borohydride.
12. Esters are normally unreactive toward sodium borohydride, but they react readily with lithium aluminum hydride to produce alcohols.
   a. Propose a rationale for the lack of reactivity of esters toward sodium borohydride.
   b. Propose a rationale for the greater reactivity of lithium aluminum hydride over that of sodium borohydride.
   c. Write the structures of the organic products that will be formed when methyl benzoate is treated with lithium aluminum hydride and the reaction mixture is worked up with aqueous acid.

CO_2CH_3

Methyl benzoate

Sign in at **www.cengage.com/login** and use the spectra viewer and Tables 8.1–8.8 as needed to answer the blue-numbered questions on spectroscopy.

13. Consider the spectral data for 9-fluorenone (Figs. 17.11 and 17.12).
   a. In the functional group region of the IR spectrum, specify the absorptions associated with the carbonyl group and with the aromatic rings.
   b. In the ^1H NMR spectrum, assign the various resonances to the hydrogen nuclei responsible for them.
   c. For the ^{13}C NMR data, assign the various resonances to the carbon nuclei responsible for them.
14. Consider the spectral data for 9-fluorenol (Figs. 17.13 and 17.14).
   a. In the functional group region of the IR spectrum, specify the functional group that is associated with the broad absorption at about 3100–3250 cm^{-1}; why is this absorption broad?
   b. In the ^1H NMR spectrum, assign the various resonances to the hydrogen nuclei responsible for them.
   c. For the ^{13}C NMR data, assign the various resonances to the carbon nuclei responsible for them.
15. Discuss the differences observed in the IR and NMR spectra of 9-fluorenone and 9-fluorenol that are consistent with the reduction of the ketone functional group in this experiment.

SPECTRA

Starting Materials and Products

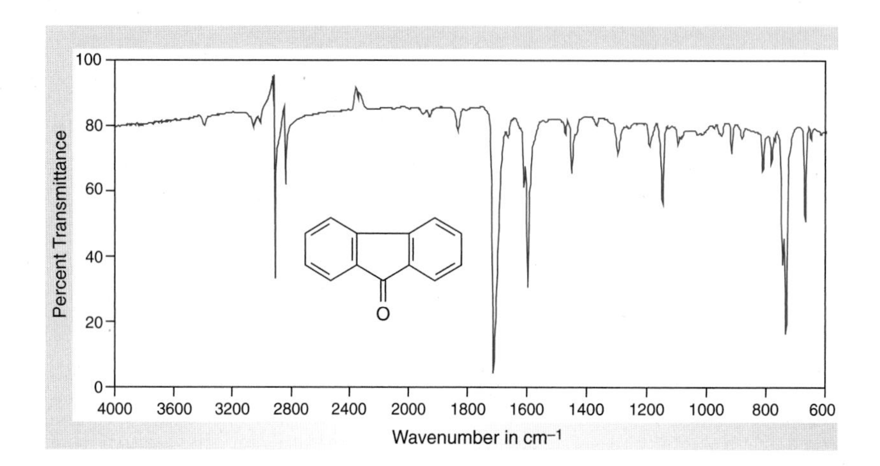

Figure 17.11
IR spectrum of 9-fluorenone (IR card).

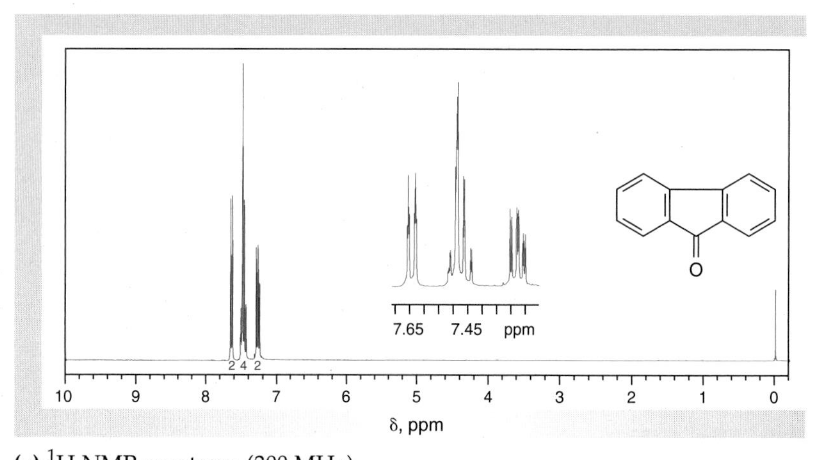

(a) ^{1}H NMR spectrum (300 MHz).
(b) ^{13}C NMR data: δ 120.1, 123.8, 128.8, 133.9, 134.4, 144.1, 193.1.

Figure 17.12
NMR data for 9-fluorenone ($CDCl_3$).

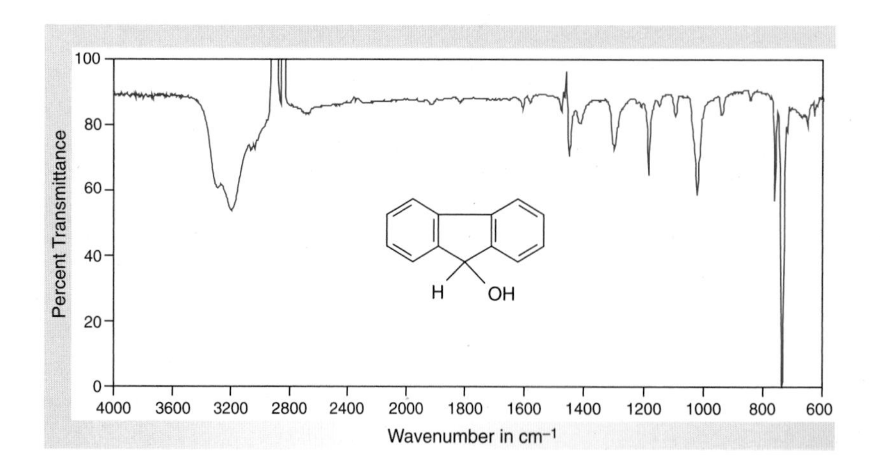

Figure 17.13
IR spectrum of 9-fluorenol (IR card).

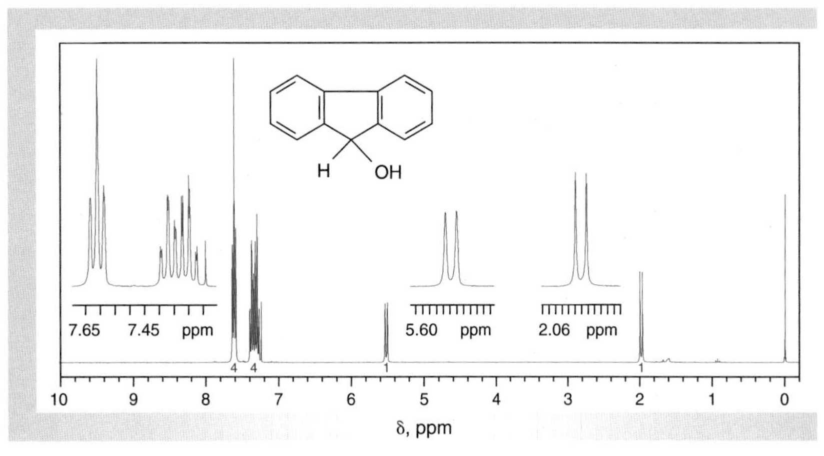

Figure 17.14
NMR data for 9-fluorenol ($CDCl_3$).

(a) ^{1}H NMR spectrum (300 MHz).
(b) ^{13}C NMR data: δ 73.8, 119.6, 125.0, 127.2, 128.2, 139.5, 146.8.

17.5 ENZYMATIC REDUCTION: A CHIRAL ALCOHOL FROM A KETONE

In modern synthetic organic chemistry, it is often necessary to use optically pure starting materials to prepare a target molecule, such as a drug, in enantiomerically pure form. However, reduction by catalytic hydrogenation or with hydride reducing agents such as sodium borohydride or lithium aluminum hydride of a **prochiral** ketone **13** that bears two *different* alkyl or aryl residues on the carbonyl carbon atom gives equal amounts of the chiral alcohol **14** and its enantiomer ***ent*-14** (Eq. 17.18). This **racemate** is produced because these *achiral* reducing reagents attack *both* faces *a* and *b* of the planar prochiral carbonyl function with *equal* probability.

R^1, R^2 C=O (**13**) —[H], face *a*→ R^1, R^2, H, OH C (**14**); —[H], face *b*→ R^1, R^2, OH, H C (***ent*-14**) (17.18)

However, if the reducing agent is chiral, it is possible to reduce a prochiral ketone to give a chiral alcohol in enantiomerically pure or enriched form. For example, the reduction of such ketones by catalytic hydrogenation in the presence of chiral catalysts or by the use of chiral hydride reducing agents may produce secondary alcohols with high levels of **enantiomeric purity**. There are also certain enzymes, which are polypeptides composed of L-amino acids and thus have chiral active sites, that can perform such **enantioselective** reductions. Reactions resulting in preferential formation of one enantiomer are examples of **asymmetric synthesis**.

See more on *Enantioselective Hydrogenation/Nobel Prizes*

See more on *Enzymes as Catalysts in Organic Synthesis*

See more on *Asymmetric Synthesis*

When a chiral product is formed from an achiral starting material, the **optical yield** or **optical purity**, not just the chemical yield, of the product is important.

Optical purity is commonly evaluated by calculating the **enantiomeric excess (ee)**. For example, reduction of a prochiral ketone **13** with sodium borohydride produces 50% (*R*)- and 50% (*S*)-alcohol **14** and ***ent*-14**; the ee for this process is 0%. If 90% of the (*S*)- and 10% of the (*R*)-isomer were produced by the reaction, the ee would be 80%.

In the experiment that follows, methyl acetoacetate (**15**) is reduced to methyl (*S*)-(+)-3-hydroxybutanoate (**16**) using one of the reducing enzymes found in baker's yeast (Eq. 17.19). The enzyme is one of many that are involved in the metabolism of D-glucose to ethanol. Enantiomeric excesses ranging from 70% to 97% have been reported for this reaction, with higher optical purities being reported when oxygen is excluded during the fermentation. Such fermentations are termed **anaerobic**. The hydroxy ester **16** is a building block that has been used as a starting material in a number of syntheses of optically pure natural products having useful biological activities.

$$\underset{\substack{\textbf{15}\\ \text{Methyl acetoacetate}}}{CH_3COCH_2CO_2CH_3} \xrightarrow[\text{sucrose}]{\text{baker's yeast}} \underset{\substack{\textbf{16}\\ \text{Methyl (}S\text{)-(+)-3-hydroxybutanoate}}}{CH_3CH(OH)CH_2CO_2CH_3} \quad (17.19)$$

EXPERIMENTAL PROCEDURE

Enzymatic Reduction of Methyl Acetoacetate

Purpose To demonstrate the enantioselective reduction of a ketone with an enzyme.

MINISCALE PROCEDURE

Preparation Sign in at **www.cengage.com/login** to answer Pre-Lab Exercises, access videos, and read the MSDSs for the chemicals used or produced in this procedure. Review Sections 2.11, 2.17, 2.21, and 2.29.

Apparatus A 250-mL Erlenmeyer flask, separatory funnel, apparatus for magnetic stirring, anaerobic fermentation, vacuum filtration, simple distillation, and *flameless* heating.

Setting Up Dissolve 40 g of sucrose and 0.25 g of disodium hydrogen phosphate, which buffers the mixture to maintain an optimal pH, in 75 mL of warm (40 °C) tap water contained in an Erlenmeyer flask. Add a stirbar and one packet (8 g) of dry, *active* baker's yeast, and assemble the apparatus for anaerobic fermentation shown in Figure 17.15. The 3% barium hydroxide solution should be protected from atmospheric carbon dioxide by a layer of mineral oil. A precipitate of barium carbonate will form as carbon dioxide is produced during the course of the fermentation. Stir the mixture *vigorously* for about 1 h in a warm location to suspend all of the yeast and to initiate the fermentation.

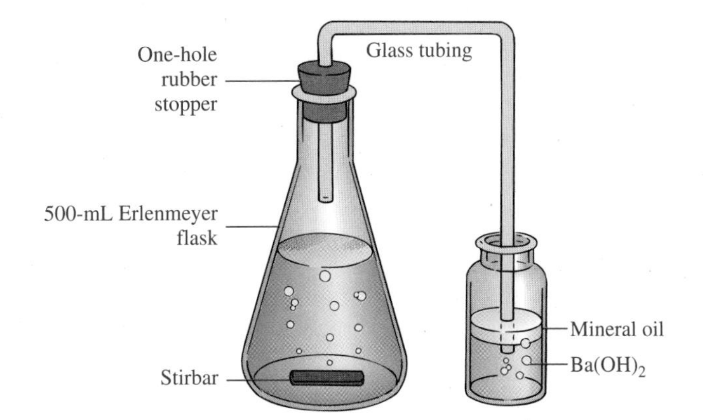

Figure 17.15
Apparatus for the anaerobic fermentation of methyl acetoacetate.

Reduction Remove the stopper in the Erlenmeyer flask, and add 2.5 mL of methyl acetoacetate. Replace the stopper and continue to stir the contents of the flask *vigorously*, under anaerobic conditions, and in a warm place, ideally at 30–35 °C, for at least 48 h, although one week is better.★

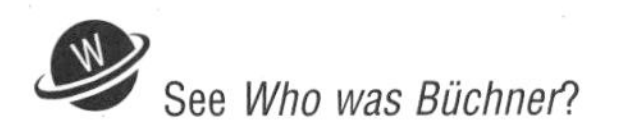
See *Who was Büchner?*

Work-Up and Isolation Prepare a uniform bed of a filter-aid by filtering a slurry made from 10 g of filter-aid in 50 mL of water through a 10-cm Büchner funnel containing a piece of filter paper; do not allow the bed to dry completely. Discard the filtrate according to the procedure described in Wrapping It Up. Add another 10 g of filter-aid to the reaction mixture containing the yeast cells, and swirl the mixture thoroughly to mix the filter-aid with the cells. Allow the solids to settle for a few minutes. Then *carefully* decant the clear supernatant fluid through the bed of filter-aid while applying *gentle* suction to remove the suspended yeast cells. After filtering the supernatant liquid, pour the residual solids and liquid through the bed of filter-aid. Because the tiny yeast particles tend to clog the filter-bed, it will be necessary to increase the vacuum during the filtration.

Wash the filter-aid with 25 mL of water and saturate the combined filtrates with solid sodium chloride. Extract the resulting solution three times with 25-mL portions of dichloromethane by *gently* shaking the separatory funnel to mix the layers. Do *not* shake the funnel too vigorously, because a bad emulsion will form. If this occurs, pass the contents of the separatory funnel through a *fresh* bed of filter-aid by vacuum filtration; this filtration will remove most of the fine particles that cause the emulsion. Wash the bed with 15 mL of dichloromethane. Transfer the filtrate and washing to the separatory funnel, and continue the extractions; any fine emulsion that remains may be included with the aqueous phase. Dry the combined dichloromethane layers with several spatula-tips full of anhydrous sodium sulfate.★ Swirl the flask occasionally for a period of 10–15 min to facilitate drying; add further small portions of anhydrous sodium sulfate if the solution remains cloudy.

Decant the dry dichloromethane solution into a tared distilling flask, and remove most of the dichloromethane by simple distillation. Alternatively, use rotary evaporation or other techniques to concentrate the solution.★ Cool the stillpot to room temperature and remove the last traces of solvent by attaching the flask to a vacuum source and gently swirling the contents as the vacuum is applied.★

Analysis Weigh the product and determine the percent yield of crude methyl (*S*)-(+)-3-hydroxybutanoate. Use the ferric ion test (Sec. 25.12C) to ascertain whether starting ketoester remains; record the result. Also analyze the product by

TLC using dichloromethane as the eluting solvent. If starting material is present, the optical yield of methyl (*S*)-(+)-3-hydroxybutanoate as measured by specific rotation will be lower. Obtain IR and ^{1}H NMR spectra of your starting material and product, and compare them with those of authentic samples (Figs. 17.16–17.19).

WRAPPING IT UP

Flush the *filtrate* from preparing the filter-aid bed down the drain. Dilute the *aqueous layer* with water, and flush it down the drain. Dry the *filter-aid and yeast cells*, and place them in the container for nonhazardous solids. After the dichloromethane evaporates from the *sodium sulfate* in the hood, place the solid in the container for nonhazardous solids. In the container for halogenated organic solvents, place the *recovered dichloromethane* that was used as the extraction solvent and as the eluting solvent for the TLC.

EXERCISES

1. Define the term *enantioselective.*
2. For applying the rules for naming centers of chirality as *S* or *R*, which substituent in **16** has the highest priority? Which has the second highest?
3. The keto group of methyl acetoacetate may also be reduced selectively with sodium borohydride. Describe how the product of this reaction would differ from the product of the enzymatic reduction of methyl acetoacetate with baker's yeast.
4. What nucleophile is supplied by the yeast?
5. What is the purpose of using sucrose in the baker's yeast reduction of methyl acetoacetate?
6. What gas is evolved during the reduction of methyl acetoacetate with baker's yeast?
7. What is an emulsion, and what substances present in the reaction mixture could be responsible for the formation of an emulsion during the work-up?
8. *Alcohol dehydrogenase* is one of the active enzymes in yeast. The active site in *alcohol dehydrogenase* contains a zinc ion, Zn^{2+}, that is coordinated to the sulfur atoms of two cysteine residues of the enzyme. The hydride reducing reagent in alcohol dehydrogenase is nicotinamide adenine dinucleotide, NADH, which transfers a hydride ion to a carbonyl compound to yield an alcohol and NAD^+, in a mechanism that is related to the Cannizzaro reaction (Sec. 16.3).

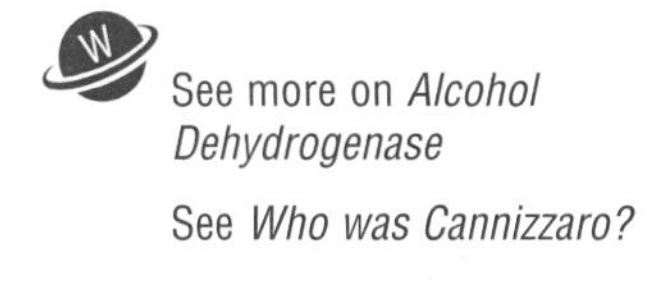

See more on *Alcohol Dehydrogenase*

See *Who was Cannizzaro?*

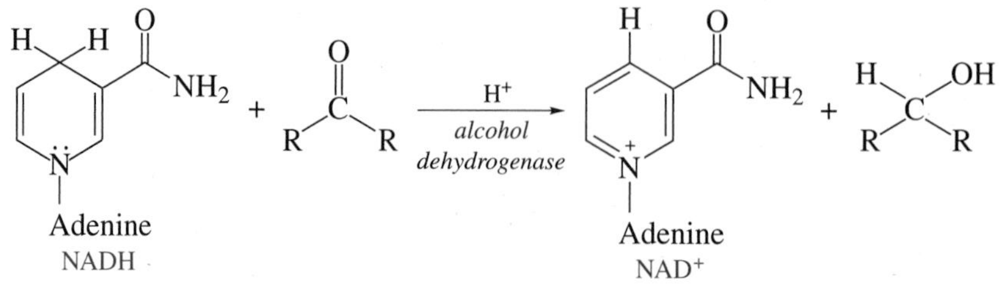

H H O NH$_2$ + O C R R $\xrightarrow[\text{alcohol dehydrogenase}]{H^+}$ H O NH$_2$ + H OH C R R

Adenine NADH

Adenine NAD^+

a. Speculate how the zinc ion might assist the addition of hydride ion to a ketone such as methyl acetoacetate.

b. Using curved arrows to symbolize the flow of electrons, write the mechanism for the hydride transfer from NADH to a carbonyl compound. Your mechanism should show how NADH becomes NAD^+.

c. Describe what general structural features are characteristic of enzymes that enable them to catalyze enantioselective reactions.

9. Why is it necessary to saturate the aqueous mixture with sodium chloride before extracting the product with dichloromethane?

10. Based upon the stereochemistry of the reduction of **15**, predict the structure of the alcohols obtained by reducing the ketones **a–c** with baker's yeast.

a. O, $CO_2C_2H_5$ **b.** O **c.** O, O, O—$P(OCH_3)_2$

Sign in at **www.cengage.com/login** and use the spectra viewer and Tables 8.1–8.8 as needed to answer the blue-numbered questions on spectroscopy.

11. Consider the spectral data for methyl acetoacetate (Figs. 17.16 and 17.17).

a. In the functional group region of the IR spectrum, specify the absorptions associated with the two carbonyl groups.

b. In the ^1H NMR spectrum, assign the various resonances to the hydrogen nuclei responsible for them.

c. For the ^{13}C NMR data, assign the various resonances to the carbon nuclei responsible for them.

12. Consider the spectral data for methyl (*S*)-(+)-3-hydroxybutanoate (Figs. 17.18 and 17.19).

a. In the functional group region of the IR spectrum, specify the absorptions associated with the hydroxyl and ester groups. Why is the absorption at about 3200–3600 cm^{-1} broad?

b. In the ^1H NMR spectrum, assign the various resonances to the hydrogen nuclei responsible for them.

c. For the ^{13}C NMR data, assign the various resonances to the carbon nuclei responsible for them.

13. Discuss the differences observed in the IR and NMR spectra of methyl acetoacetate and methyl (*S*)-(+)-3-hydroxybutanoate that are consistent with the reduction of the carbonyl group in this experiment.

SPECTRA

Starting Materials and Products

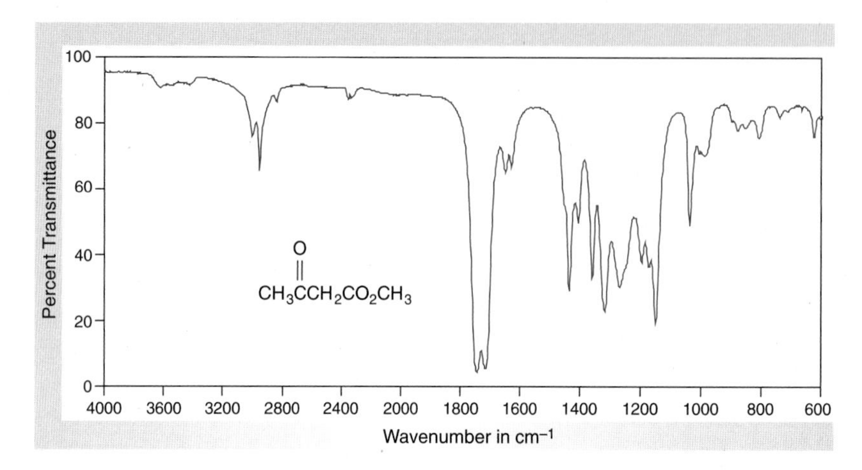

Figure 17.16
IR spectrum of methyl acetoacetate (neat).

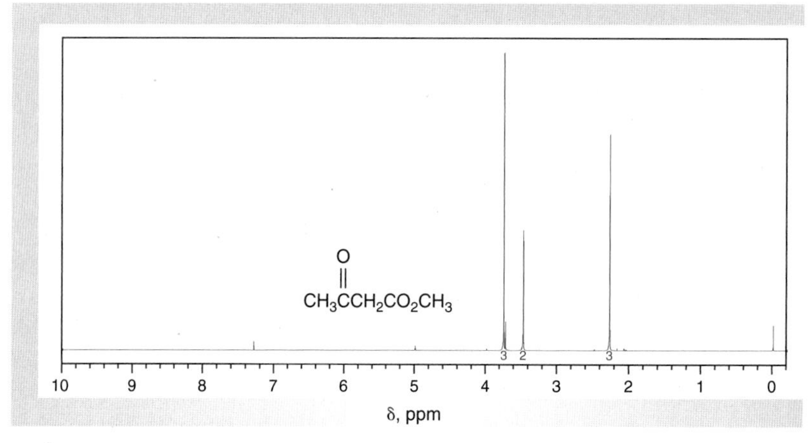

Figure 17.17
NMR data for methyl acetoacetate ($CDCl_3$).

(a) ^{1}H NMR spectrum (300 MHz).
(b) ^{13}C NMR data: δ 29.9, 49.8, 52.1, 168.0, 200.8.

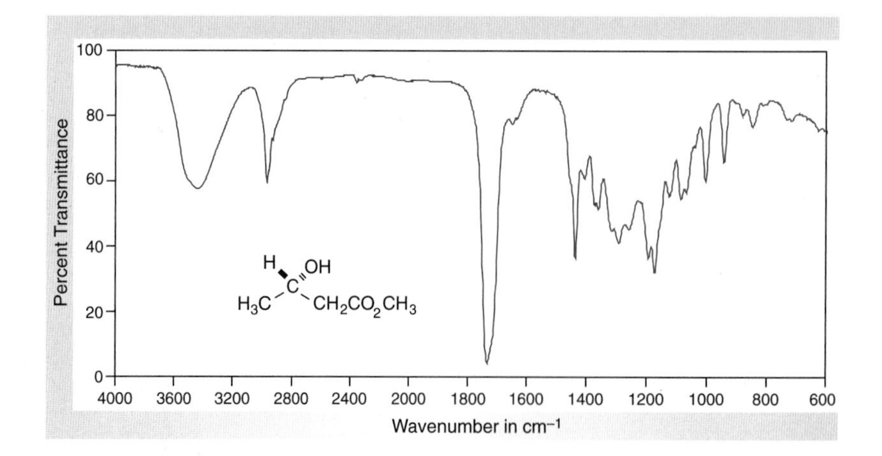

Figure 17.18
IR spectrum of methyl (S)-(+)-3-hydroxybutanoate (neat).

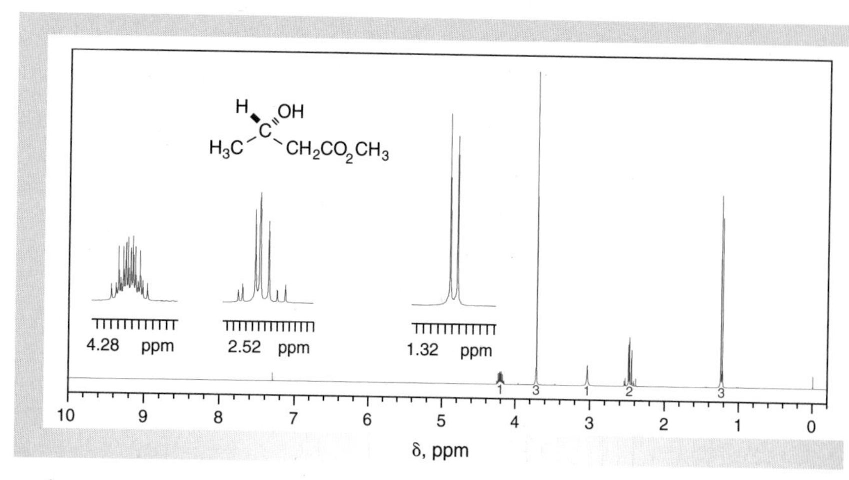

Figure 17.19
NMR data for methyl (S)-(+)-3-hydroxybutanoate ($CDCl_3$).

(a) ^{1}H NMR spectrum (300 MHz).
(b) ^{13}C NMR data: δ 22.6, 42.8, 51.7, 64.3, 173.2.

17.6 DETERMINING OPTICAL PURITY

There are several ways to determine the optical purity of a compound obtained as the product of an enantioselective synthesis. The simplest technique involves measuring the optical rotation using a polarimeter (Sec. 7.5). The reported specific rotation, $[\alpha]^{25}{}_{D}$, of pure methyl (*S*)-(−)-3-hydroxybutanoate is +38.5° (*c* = 1.80, chloroform), but the specific rotation of the product you obtain might range from +27° to +36°. For example, a measured specific rotation of 32.7° (*c* = 1.3, chloroform) for the product of this enzymatic reduction would correspond to an ee of 85%.

When only small quantities of a material are available, the use of ^{1}H NMR and a chiral shift reagent is a more accurate method for determining optical purity. In an *achiral* environment, the ^{1}H NMR spectra of methyl (*S*)-(+)-3-hydroxybutanoate and methyl (*R*)-(−)-3-hydroxybutanoate are *identical*, and each corresponds to that shown in Figure 17.19a. However, a Lewis acid having chiral ligands can form *diastereomeric complexes* with the two enantiomers, and these complexes often give different ^{1}H NMR spectra. In the present case, the Lewis acid is the chiral shift reagent *tris*[3-(heptafluoropropylhydroxymethylene)-(+)-camphorato]europium (III) or Eu(hfc)$_3$ (**17**); the europium atom coordinates with the hydroxyl group in methyl 3-hydroxybutanoate and its camphor-derived ligand to form diastereomeric complexes, each of which gives a different ^{1}H NMR spectrum.

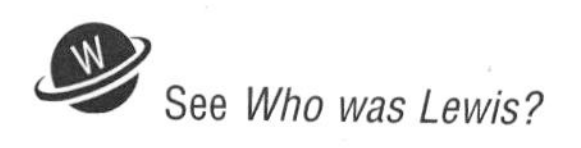

See *Who was Lewis?*

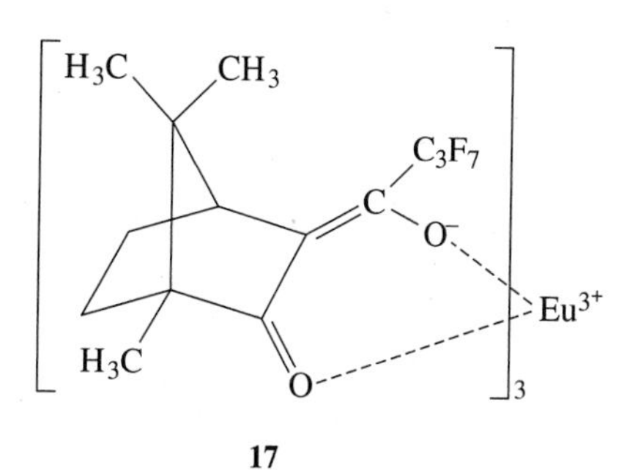

17

An important property of europium shift reagents is that they spread the peaks in the ^1H NMR spectrum over a wider range, generally moving all of the peaks *downfield*. The magnitude of the shift of each peak depends on the concentration of the shift reagent and the proximity of the europium ion to the protons corresponding to that peak. The complexes that are formed between **17** and methyl (*S*)-(+)-3-hydroxybutanoate and methyl (*R*)-(+)-3-hydroxybutanoate are *diastereomeric*, so the peaks for the protons on one enantiomer move downfield to different extents than the corresponding peaks for the other. Thus, the two diastereomeric complexes may be easily distinguished in the ^1H NMR spectrum, as shown in Figure 17.20. It is then possible to integrate the peak areas corresponding to each of the enantiomers and determine the enantiomeric excess. Determining the ee using this NMR method gives results that are accurate to within approximately 2–3%.

In this experiment, you will use the chiral europium shift reagent **17** to determine the enantiomeric excess of the product that you obtained upon enantioselective reduction of methyl acetoacetate with baker's yeast.

EXPERIMENTAL PROCEDURE

Determining Optical Purity of Methyl (S)-(+)-3-Hydroxybutanoate

Purpose To determine the optical purity of methyl (*S*)-(+)-3-hydroxybutanoate by ^1H NMR spectroscopy.

Procedure

Preparation Sign in at **www.cengage.com/login** to answer Pre-Lab Exercises, access videos, and read the MSDSs for the chemicals used or produced in this procedure.

Apparatus Two NMR tubes.

Preparing the Samples Dissolve 25–30 mg of racemic methyl 3-hydroxybutanoate in 0.75 mL of deuterochloroform and transfer the solution to an NMR tube using a Pasteur or a filter-tip pipet. Dissolve 25–30 mg of enantiomerically enriched methyl (*S*)-(+)-3-hydroxybutanoate, prepared by the enzymatic reduction in the first part of this experiment, in 0.75 mL of deuterochloroform and transfer the solution to another NMR tube, using a Pasteur or a filter-tip pipet.

Measurements Measure the ^1H NMR spectrum of the racemic methyl 3-hydroxybutanoate. Add 20 mg of *tris*[3-(heptafluoropropylhydroxymethylene)-(+)-camphorato]europium shift reagent to this solution, shake the mixture to effect dissolution, and allow the solution to stand for about 20 min. Again measure the ^1H NMR spectrum. The peaks of primary interest for determining the optical purity are the methoxy hydrogens of the ester. Continue the procedure of adding shift reagent and measuring the spectrum until these peaks are separated and accurate integrals may be obtained as shown in Figure 17.20; no more than about 80 mg of shift reagent should be required.

Now perform the same series of measurements on the sample of enantiomerically enriched methyl (*S*)-(+)-3-hydroxybutanoate. A typical spectrum is shown in Figure 17.21. Determine the ratio of the two enantiomers in both of the samples by comparing the heights of the singlets that correspond to the methoxy protons of the ester, and calculate the percentages of each enantiomer in the two samples based upon this ratio. Calculate the enantiomeric excess (ee) of methyl (*S*)-(+)-3-hydroxybutanoate.

WRAPPING IT UP

Place the *solution of deuterochloroform and shift reagent* in the container for halogenated organic liquids.

EXERCISES

1. What are the advantages of using a chiral shift reagent instead of polarimetry to determine the optical purity of a sample?
2. Which of the following terms may be applied to the nature of the complex formed between Eu^{3+} and the alcohol functional group in methyl 3-hydroxybutanoate: Covalent? Ionic? Lewis acid/Lewis base? Van der Waals?
3. What is the meaning of the word "shift" in the term "chiral shift reagent"?
4. What properties of a chiral shift reagent allow you to determine the ratio of enantiomeric alcohols formed by an enantioselective reduction?
5. The ^1H NMR spectrum in Figure 17.21 is of a mixture of the chiral shift reagent **17** and a sample of methyl (*S*)-(+)-3-hydroxybutanoate that was produced by the enzymatic reduction of methyl acetoacetate using baker's yeast. Determine the enantiomeric excess of the methyl (*S*)-(+)-3-hydroxybutanoate contained in this sample.
6. Why is it necessary to obtain a ^1H NMR spectrum of *racemic* methyl 3-hydroxybutanoate in the presence of **17**?
7. Which of the following compounds will form diastereomeric complexes with the chiral europium shift reagent **17**?

a. 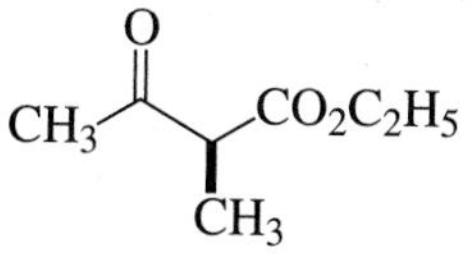 CH_3 O $CO_2C_2H_5$ CH_3 **b.** 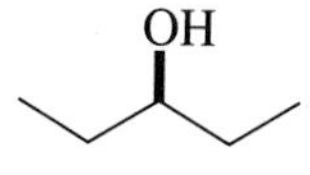 OH **c.** 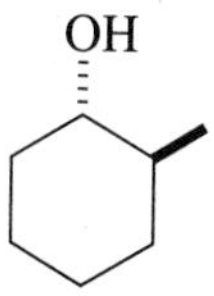 OH

d. 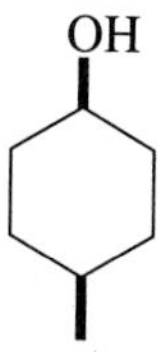 OH **e.** 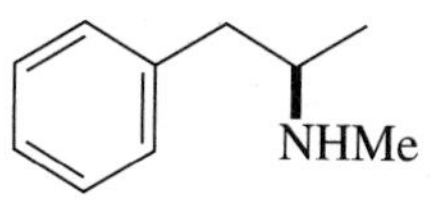 NHMe **f.**

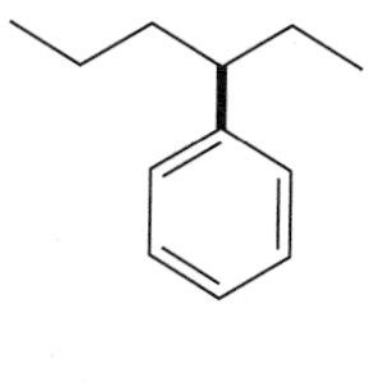

SPECTRA

Chiral Complexes of Products

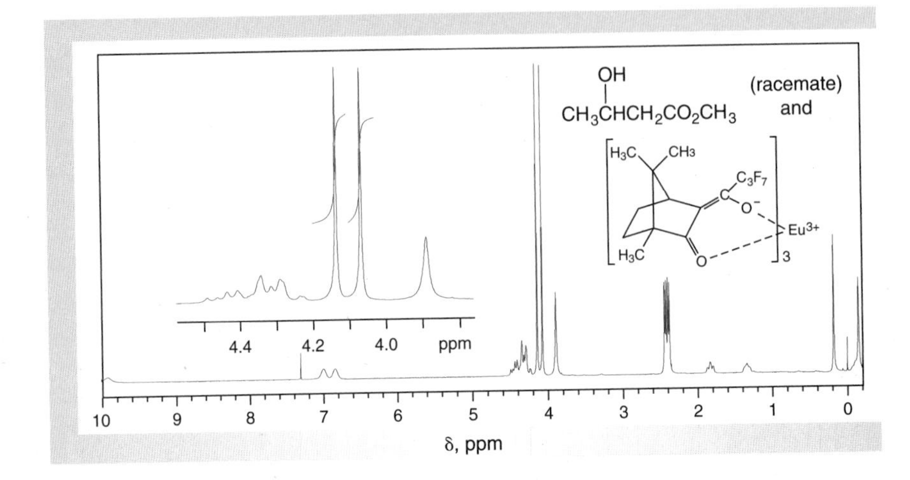

Figure 17.20
*^1H NMR spectrum (300 MHz, $CDCl_3$) of racemic methyl 3-hydroxybutanoate in the presence of chiral shift reagent **17**.*

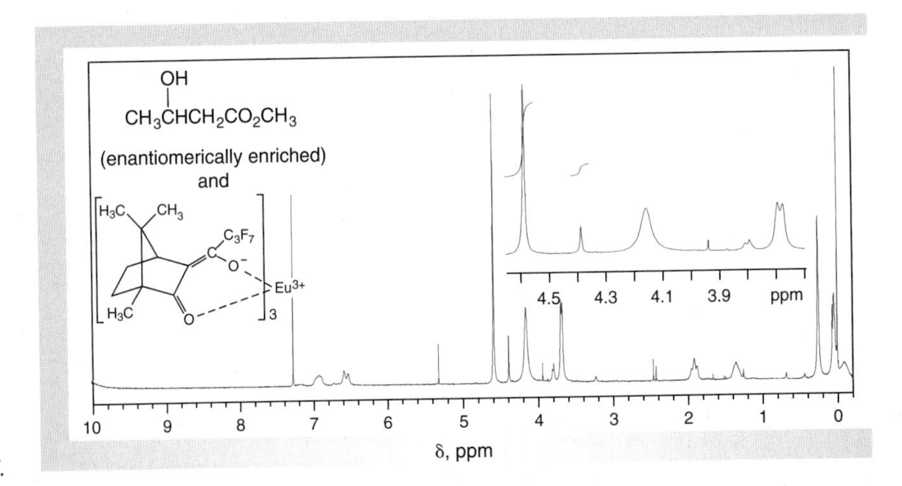

Figure 17.21
*^1H NMR spectrum (300 MHz, $CDCl_3$) of enriched methyl (S)-(+)-3-hydroxybutanoate in the presence of chiral shift reagent **17**.*

HISTORICAL HIGHLIGHT

Chiral Drugs

An ever-increasing number of the drugs discovered and developed by medicinal chemists in the pharmaceutical industry are chiral (see the Historical Highlight at the end of Chapter 7). This isn't surprising, given that the enzymes, receptors, and nucleic acids that are typically the targets of these drugs are chiral, as well. One enantiomer of a drug will interact differently with a particular biological molecule than the other enantiomer because the two interactions are diastereoisomeric in nature. Many chiral drugs were once sold as racemates (Sec. 7.1), but it's becoming more and more important to be able to prepare these compounds as single enantiomers by enantioselective synthesis. The motivation for preparing a single enantiomer of a drug is that one enantiomer is frequently more potent and/or more biologically available than its mirror-image isomer; sometimes one enantiomer may have side effects, even devastating ones, as is the case for thalidomide (see the Historical Highlight at the end of Chapter 7). Because the

(Continued)

HISTORICAL HIGHLIGHT *Chiral Drugs (Continued)*

enzymes responsible for clearing drugs from the body by oxidation or other chemical reactions are chiral, it is not unexpected that the enantiomers of a drug may be metabolized at different rates. Moreover, the metabolic products of one enantiomer may be different from the other, and this can be of significance because a metabolite of a drug has biological activity that may or may not be beneficial.

Owing to the dissimilar biological activities of the two enantiomers of a potential drug, the Food and Drug Administration (FDA) now requires that the biological profiles and activities of each enantiomer as well as the racemate be studied in detail. Hence, there is an incentive for pharmaceutical companies to develop single enantiomers owing to the reduced costs for testing and clinical trials of the drug. Companies sometimes exploit the chirality of some drugs to extend their patent life, and hence their life cycle as a marketable product. For example, a company might have obtained its initial patents and begun marketing a chiral drug as its racemate. If one enantiomer is subsequently found to be more efficacious or less prone to inducing side effects, it becomes possible for the same or another company to file a new patent on the single enantiomer. The firm filing this patent would then have exclusive rights to market the enantiomerically pure drug, even as generic forms of the racemic drug are marketed.

Let's consider a few examples of chiral drugs to help you understand the potential importance of using single enantiomers. Do you know anyone who suffers from depression? If so, one therapy associated with treating this condition may include taking drugs that target serotonin pathways. Serotonin is a neurotransmitter that is involved in modulating anger, mood, aggression, sleep, and a number of other important activities in the brain. It has been found that compounds that selectively inhibit the reuptake of serotonin into nerve presynapses may be useful therapeutic agents for treating a variety of psychiatric disorders. Indeed, two well-known drugs for treating depression are Prozac™ (fluoxetine) and Zoloft™ (sertraline), both of which are inhibitors of serotonin reuptake. The racemic form of fluoxetine was marketed first, but it was later discovered that the antidepressant activity actually resided in the (*S*)-enantiomer. It was thus possible to obtain an independent patent on the single enantiomeric form, although most of the sales of Prozac™ today are still of the racemic form. Interestingly, sertraline has always been sold as a single enantiomer.

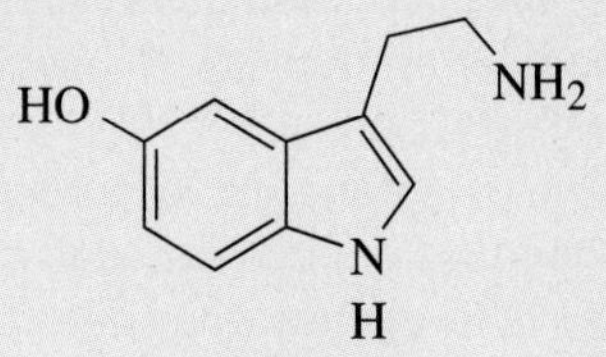

HO NH$_2$ N H

Serotonin

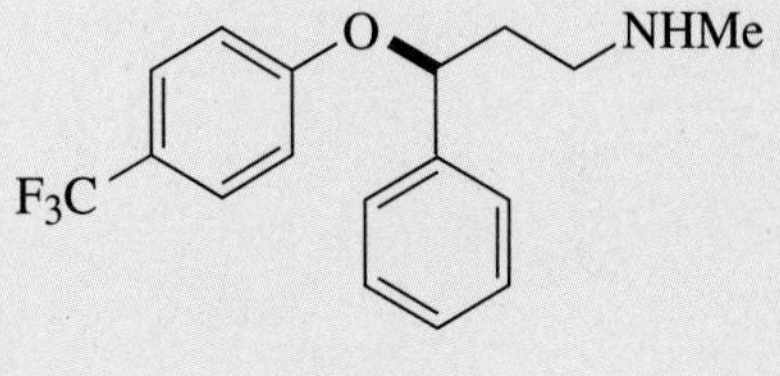

O NHMe F_3C

(*S*)-Fluoxetine

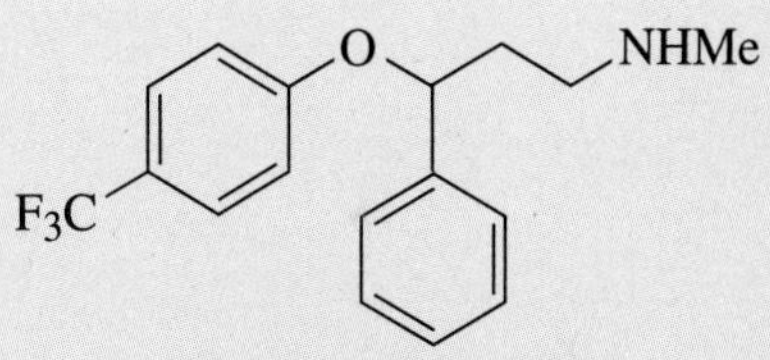

O NHMe F_3C

Fluoxetine (Prozac™)

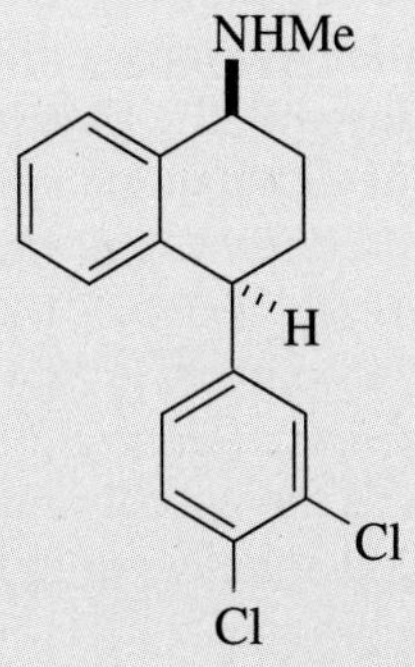

NHMe H Cl Cl

Sertraline (Zoloft™)

Have you ever had an asthma attack? Perhaps you used an inhalant to relieve the symptoms and to make it easier to breathe. One important drug for asthma

(Continued)

HISTORICAL HIGHLIGHT *Chiral Drugs (Continued)*

sufferers is racemic albuterol (Ventolin™), also known as salbutamol. Albuterol is a short-acting β_2-adrenergic agonist that exhibits activity as a bronchodilator and is thus used as an aerosol inhalant to treat asthmatic bronchospasms. As was the case with Prozac, Ventolin™ was first marketed in its racemic form, but it has been suggested that the *R*-enantiomer, levalbuterol (Xopenex™), produces fewer side effects, especially on the β_1-adrenergic receptors in the heart. This has not been well documented in clinical trials, however, so use of levalbuterol has been limited, owing perhaps in part to its higher cost.

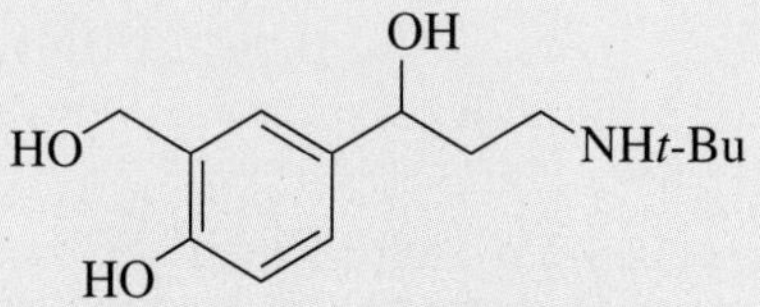

Albuterol (Ventolin™)

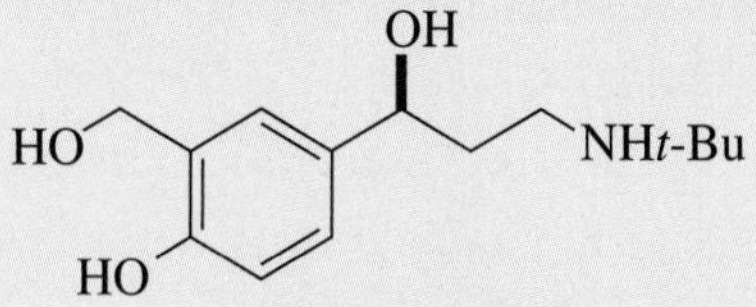

Levalbuterol (Xopenex™)

You probably know someone who suffers from attention-deficit hyperactivity disorder, ADHD. Students who have this condition may have difficulty doing well in school. Perhaps surprisingly, this condition may be treated using psychostimulants that are inhibitors of catecholamine amine uptake. Catecholamines, which derive their name from the fact that they contain a catechol or a 1,2-dihydroxyphenyl group, are sympathomimetic "fight-or-flight" hormones released by the adrenal glands and include compounds such as the neurotransmitters dopamine and adrenaline. One effective drug for treating ADHD is racemic methylphenidate (Ritalin™), which is a psychostimulant that increases *trans*-synaptic concentrations of dopamine and adrenaline. Because methylphenidate has two stereocenters, there are four possible stereoisomers, but only the racemic *threo* isomer has been used clinically. The D-*threo* isomer of methylphenidate, which is marketed as dexmethylphenidate (Focalin®), is 13 times more active than the racemic form of the drug, so lower doses are possible with formulations containing only this enantiomer.

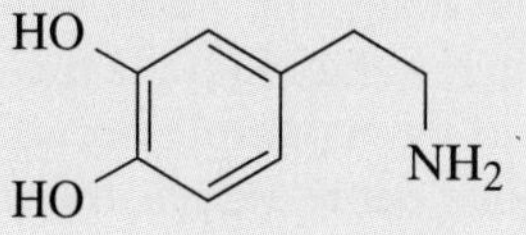

Dopamine

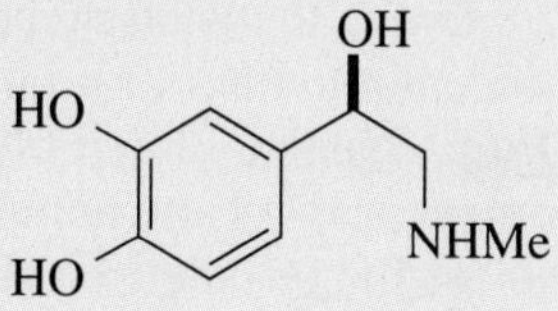

Adrenaline (Epinephrine)

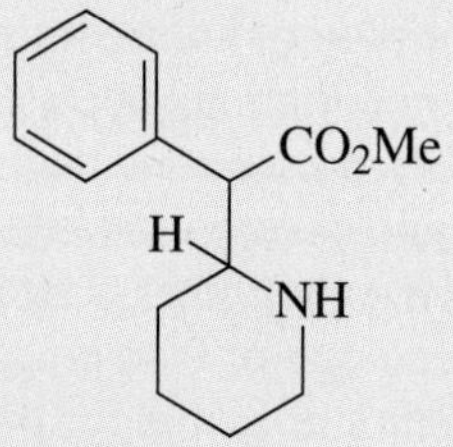

Methylphenidate (Ritalin™)

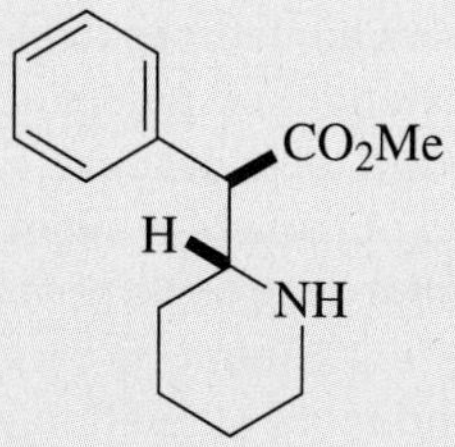

Dexmethylphenidate (Focalin®)

From these few examples, you can readily understand the value of using a single enantiomer of drugs containing one or even more stereocenters. Undesirable pharmacological activities in the form of side effects and harmful metabolites can be minimized, and a lesser amount of the drug may be required to achieve the desired medicinal effect. The challenge for the synthetic chemist, then, is to develop cost-effective ways to prepare these compounds in enantiomerically pure form. One possible method to solve this problem is to resolve the enantiomers by preparing diastereoisomeric salts or other derivatives that may be separated by crystallization or chromatography (Sec. 7.6). Enantiomers may also be separated by column chromatography using stationary phases that are themselves chiral. However, perhaps the best way to prepare enantiomerically pure compounds is by asymmetric synthesis using chiral catalysts or chiral auxiliaries.

(Continued)

HISTORICAL HIGHLIGHT *Chiral Drugs (Continued)*

Relationship of Historical Highlight to Experiments

The challenge of preparing chiral compounds in enantiomerically pure form is routinely faced by chemists in the pharmaceutical industry as they work toward the development of improved therapeutic agents. Indeed, an article describing recent approaches to performing such syntheses on a large scale is provided in *Chemical and Engineering News*, August 6, 2007, pages 11–19. One of the steps commonly encountered during the course of synthesizing potential drugs is the reduction of double bonds. The use of enzymes as catalysts to effect such reactions in an enantioselective manner, as demonstrated by the reduction of methyl acetoacetate using baker's yeast (Sec.17.5), is an important technique that may be employed on an industrial scale.

See more on *Serotonin*
See more on *Prozac™*
See more on *Zoloft™*
See more on *Adrenergic Receptors*
See more on *Albuterol*
See more on *Levalbuterol*
See more on *ADHD*
See more on *Catecholamines*
See more on *Ritalin*
See more on *Dexmethylphenidate*
See more on *Chiral Column Chromatography*
See more on *Asymmetric Synthesis*
See more on *Chiral Catalysts/Nobel Prize*
See more on *Chiral Auxiliaries*

CHAPTER 18

Reactions of Carbonyl Compounds

When you see this icon, sign in at this book's premium website at **www.cengage.com/login** to access videos, Pre-Lab Exercises, and other online resources.

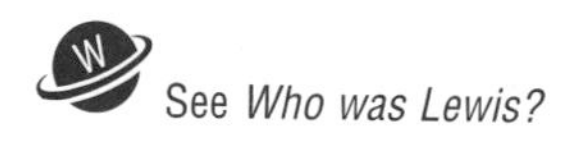

See *Who was Lewis?*

The chemistry of carbon-carbon multiple bonds is illustrated by the experiments with alkenes (Chaps. 10 and 12, and Sec. 17.2), alkynes (Sec. 11.3), and arenes (Chaps. 15 and 21). The reactions of these classes of organic compounds typically involve electrophilic processes, with the transformations being initiated by addition of an electrophile or a Lewis acid such as a proton, a bromonium ion, and so forth, to the π-system. What you will *not* find in your study of alkenes and alkynes are reactions in which a nucleophile or a Lewis base attacks the carbon-carbon π-bond, although examples of such processes may be found when a carbon atom involved in the π-bond is appropriately substituted. The picture changes dramatically when you examine the chemistry of carbonyl compounds, which are characterized by the presence of a carbon-oxygen double bond. Now you will find that the carbonyl carbon atom is highly susceptible to attack by nucleophiles and that this leads to reactions having broad applications, including the formation of new carbon-carbon bonds. In this chapter, we can only give you a small sample of the diversity of reactions available to aldehydes and ketones, a class of compounds in which the carbonyl carbon atom, in addition to being doubly bound to an oxygen atom, is singly bound to other carbon or hydrogen atoms. Nonetheless, we think that this "taste" of carbonyl chemistry will suffice to whet your appetite for more.

18.1 INTRODUCTION

The **carbonyl group, 1,** is a rich source of many important reactions in organic chemistry, and two fundamental properties of this functionality are primarily responsible for its diverse chemistry. The first is the polarization of the carbon-oxygen π-bond, owing to the relatively high electronegativity of the oxygen atom. In terms of resonance theory, this polarization arises from the contribution of the dipolar resonance structure **1b** to the resonance hybrid for this functional group.

1a $\longleftrightarrow$ **1b**

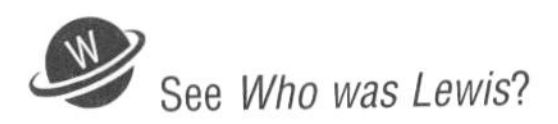
See *Who was Lewis?*

Consequently, the carbonyl group undergoes a variety of reactions in which the electrophilic carbonyl carbon atom is attacked by **nucleophiles,** which are Lewis bases; the oxygen atom, in turn, reacts with **electrophiles** or **Lewis acids.** The net effect is *addition* of a reagent, symbolized as Nu–E, across the π-bond of the carbonyl function. The sequence by which the elements of Nu–E add to the carbonyl group varies according to the particular reagents and reaction conditions used but, in a general sense, follows the pathway of Equation 18.1 in basic media and that of Equation 18.2 in acidic media.

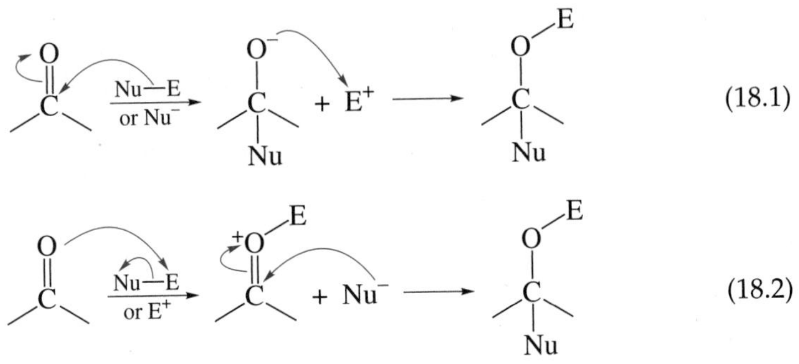

(18.1)

(18.2)

The second property of a carbonyl moiety is to increase the acidity of the **α-hydrogen atoms,** which are those on the carbon atoms directly attached to the carbonyl carbon atom. A result of the enhanced acidity of these α-hydrogens is that the **α-carbon atoms** can become **nucleophilic** either through deprotonation to form an **enolate ion** (Eq. 18.3), or by a keto–enol equilibration, called **tautomerization,** to give an **enol** (Eq. 18.4). As shown in Equation 18.3, an enolate, which is the resonance hybrid of the two contributing resonance structures **2a** and **2b,** can react with electrophiles, E^+, at an α-carbon atom to give net substitution of the electrophile for an α-hydrogen atom. A similar result attends the reaction of an enol **3** with an electrophile, E^+ (Eq. 18.4).

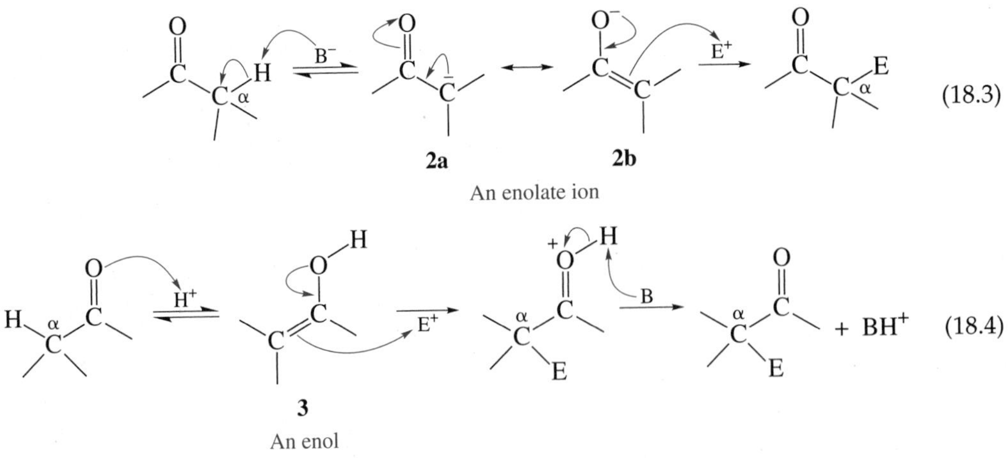

(18.3)

2a **2b**
An enolate ion

(18.4)

3
An enol

The experiments described in this chapter illustrate some of the representative reactions of carbonyl compounds according to the principles outlined in Equations 18.1–18.4. For example, you may study reactions in which the electrophilic carbonyl carbon atom of an aldehyde undergoes attack by a carbon nucleophile that is stabilized by an electron-withdrawing group such as a phosphorus-containing substituent, as found in the **Wittig reaction** and the **Horner-Wadsworth-Emmons modification** thereof (Sec. 18.2), or a carbonyl function, as found in the **aldol condensation** (Sec. 18.3). The **1,4-** or **conjugate addition** of the enol form of an aldehyde to an α,β-unsaturated ketone to give a dicarbonyl compound that cyclizes by an *intramolecular* aldol reaction is explored in Section 18.4.

See *Who was Wittig?*

See more on *Horner-Wadsworth-Emmons Reaction*

18.2 THE WITTIG AND RELATED REACTIONS

Introduction

As noted in the preceding section, the ability of the carbonyl group to increase the acidity of the α-hydrogens makes it possible to generate enolate ions, carbon nucleophiles that are stabilized by resonance (Eq. 18.3). Certain phosphorus-containing groups are also known to increase the acidity of the hydrogen atoms bound to the same carbon atom, and deprotonation gives rise to another type of carbon nucleophile. For example, the bold-faced hydrogen atom of the phosphonium salt **4** exhibits enhanced acidity, and treatment of **4** with a strong base produces the nucleophilic species **5** (Eq.18.5), a **phosphorane.**

$$(C_6H_5)_3\overset{+}{P}-CH(\mathbf{H})C_6H_5\ Cl^- \xrightarrow{^-B} \left[(C_6H_5)_3\overset{+}{P}-\overset{-}{C}HC_6H_5 \longleftrightarrow (C_6H_5)_3P{=}CHC_6H_5\right] \quad (18.5)$$

4 Benzyltriphenylphosphonium chloride; **5a**, **5b** Benzylidenetriphenylphosphorane

See more on *Wittig Reagent/Nobel Prize*

Phosphoranes like **5** are frequently referred to as **Wittig reagents** in recognition of their inventor, Georg Wittig, who showed that they react with aldehydes and ketones to produce alkenes. Wittig won the Nobel Prize in Chemistry in 1979 for this discovery. (See the Historical Highlight at the end of this chapter.) An example of a Wittig reaction is illustrated in Scheme 18.1 for the synthesis of (*Z*)-stilbene (**9**) and (*E*)-stilbene (**10**) by the reaction of the ylide **5** with benzaldehyde (**6**). The reaction commences with the addition of the nucleophilic carbon atom of **5** to the electrophilic carbonyl carbon atom of **6** to give the betaine **7** as a mixture of two diastereomers. The betaine **7** collapses to the **oxaphosphetane 8,** which then undergoes fragmentation to give a mixture of the isomeric alkenes **9** and **10,** together with triphenylphosphine oxide (**11**). Although phosphorane **5** may be written as the neutral species **5b,** the contribution of resonance structure **5a** imparts dipolar character to it, so it is commonly referred to as a phosphorus **ylide.**

Scheme 18.1

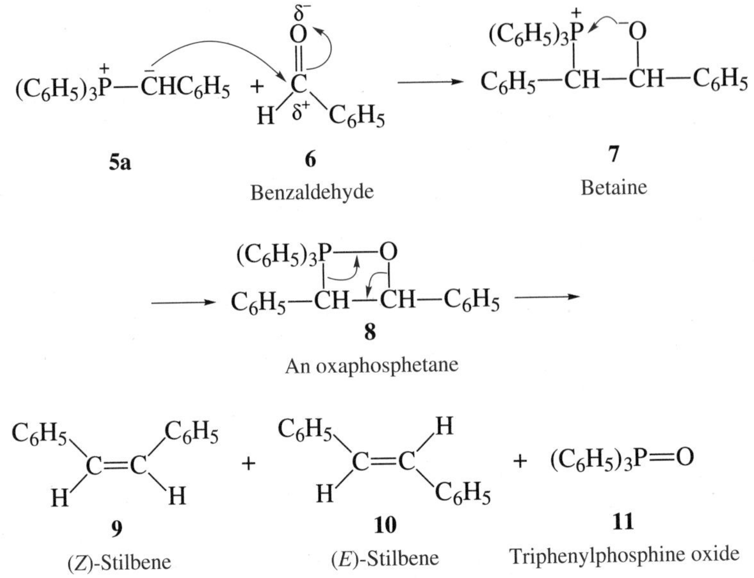

$$(C_6H_5)_3\overset{+}{P}-\overset{-}{C}HC_6H_5 + \overset{\delta^-}{O}=\overset{\delta^+}{C}(H)C_6H_5 \longrightarrow (C_6H_5)_3\overset{+}{P}-\overset{-}{O} \; / \; C_6H_5-CH-CH-C_6H_5$$

5a — **6** Benzaldehyde — **7** Betaine

$$\longrightarrow (C_6H_5)_3P-O \; / \; C_6H_5-CH-CH-C_6H_5 \longrightarrow$$

8 An oxaphosphetane

$$(Z)\text{-}C_6H_5CH=CHC_6H_5 + (E)\text{-}C_6H_5CH=CHC_6H_5 + (C_6H_5)_3P=O$$

9 (*Z*)-Stilbene — **10** (*E*)-Stilbene — **11** Triphenylphosphine oxide

The net effect of the Wittig reaction is the conversion of a *carbon-oxygen* double bond to a *carbon-carbon* double bond. This transformation represents a general method for preparing alkenes that has two important advantages over other methods for preparing this functional group: (1) the carbonyl group is replaced *specifically* by a carbon-carbon double bond without forming isomeric alkenes having the π-bond at other positions, and (2) the reactions are carried out under mild conditions.

See more on *Phase-Transfer Catalysis*

The Wittig reaction typically involves highly reactive phosphoranes and is therefore not normally suited for the introductory laboratory. For example, the very strong bases such as *n*-butyllithium that are normally used to produce the intermediate ylides require special handling techniques, and it is generally necessary to perform all experimental operations under a dry, inert atmosphere. However, it has recently been shown that the Wittig reaction of **5** with aromatic aldehydes such as benzaldehyde can be performed under **phase-transfer** conditions in which the phosphonium salt serves as the phase-transfer catalyst (Sec. 10.5). This simple expedient is used to prepare a mixture of (Z)-stilbene (**9**) and (*E*)-stilbene (**10**) in one of the experimental procedures in this section. You will be able to determine the ratio of **9** and **10** by examining an ^{1}H NMR spectrum of the mixture. It is also possible to isomerize (Z)-stilbene to (*E*)-stilbene by irradiation of a solution of (Z)-stilbene containing a small amount of molecular iodine (Eq. 18.6).

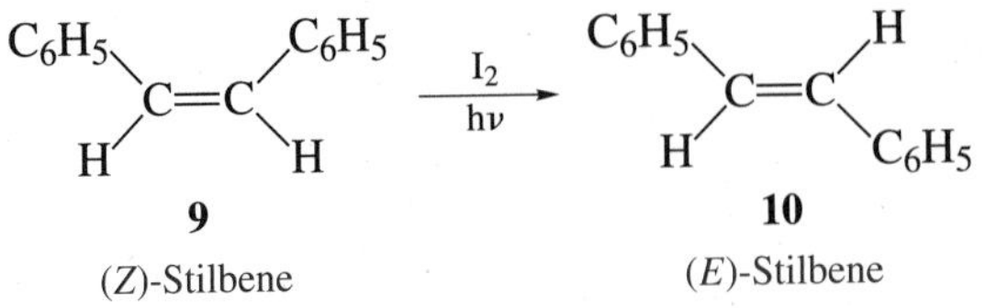

$$(Z)\text{-}C_6H_5CH=CHC_6H_5 \xrightarrow[h\nu]{I_2} (E)\text{-}C_6H_5CH=CHC_6H_5 \quad (18.6)$$

9 (*Z*)-Stilbene — **10** (*E*)-Stilbene

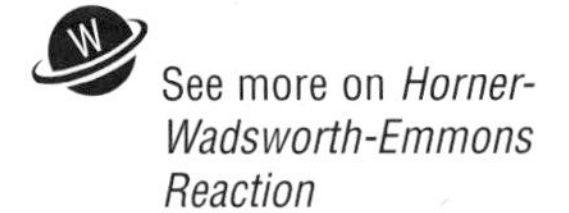

See more on *Horner-Wadsworth-Emmons Reaction*

The **Horner-Wadsworth-Emmons reaction** is an important variant of the Wittig reaction and involves using a **phosphonate ester** in place of a phosphonium salt. Like the phase-transfer Wittig reaction just discussed, these reactions may be easily performed in the undergraduate laboratory. In one of the procedures that follows, the phosphonate ester **12** is deprotonated with potassium *tert*-butoxide in the polar, *aprotic* solvent *N,N*-dimethylformamide, $(CH_3)_2NCHO$ (DMF), to provide the resonance-stabilized, nucleophilic phosphonate anion **13** (Eq. 18.7).

$$(C_2H_5O)_2P(=O)-CHC_6H_5(H) + {}^-OC(CH_3)_3 \longrightarrow \left[(C_2H_5O)_2P(=O)-\bar{C}HC_6H_5 \longleftrightarrow (C_2H_5O)_2P(-O^-)=CHC_6H_5\right] \quad (18.7)$$

12 Diethyl benzylphosphonate; *tert*-Butoxide ion; **13a**, **13b** Potassium diethyl benzylphosphate

As shown in Scheme 18.2, the anion **13** reacts with benzaldehyde to give one of the stilbenes **9** or **10** with high selectivity by a sequence of reactions similar to that observed for the reaction of the ylide **5** with **6** (Scheme 18.1). The intermediate adduct **14** undergoes cyclization to form **15,** which then fragments to give the stilbene and potassium diethyl phosphate (**16**).

By obtaining spectral data for your product, you will be able to determine which of the stilbenes is produced. An explanation of why the Wadsworth-Emmons reaction mainly yields a single diastereomer of stilbene is beyond the scope of this discussion.

Scheme 18.2

$$(C_2H_5O)_2P(=O)-\bar{C}HC_6H_5\ K^+ + O^{\delta-}=C^{\delta+}(H)C_6H_5 \longrightarrow (C_2H_5O)_2P(=O)(-O^-)\cdots C_6H_5-CH-CH-C_6H_5$$

13a **6** **14**

$$\longrightarrow \underset{\mathbf{15}}{(C_2H_5O)_2P(=O)-O \text{ ring: } C_6H_5-CH-CH-C_6H_5} \longrightarrow$$

$$C_6H_5CH=CHC_6H_5 + (C_2H_5O)_2P(=O)-O^-\ K^+$$

9 and/or **10** Stilbene; **16** Potassium diethyl phosphate

Wittig and Horner-Wadsworth-Emmons Reactions

A ■ *Preparation of (Z)- and (E)-Stilbenes by a Wittig Reaction*

Purpose To synthesize a mixture of stilbenes by a phase-transfer Wittig reaction, to determine the major product, and to perform their geometric isomerization.

SAFETY ALERT

The 50% (by mass) aqueous sodium hydroxide solution is *highly* caustic. *Do not allow it to come in contact with your skin.* If this should happen, wash the affected area with dilute acetic acid and then copious amounts of water. Wear latex gloves when transferring the solution.

MINISCALE PROCEDURE

Preparation Sign in at **www.cengage.com/login** to answer Pre-Lab Exercises, access videos, and read the MSDSs for the chemicals used or produced in this procedure. Review Sections 2.9, 2.11, 2.13, 2.17, 2.21, 2.22, and 2.29.

Apparatus A 25- and a 50-mL round-bottom flask, separatory funnel, apparatus for heating under reflux, magnetic stirring, simple distillation, vacuum filtration, and *flameless* heating.

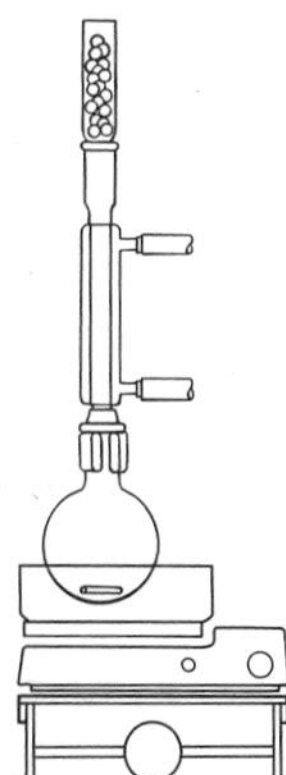

Setting Up Prepare a 50% (by mass) sodium hydroxide solution by dissolving 5.0 g of sodium hydroxide in 5 mL of water. The mixture may be heated gently to facilitate dissolution, but be sure to cool the solution to room temperature before proceeding. Add 3.8 g of benzyltriphenylphosphonium chloride and 1 mL of benzaldehyde to the 50-mL round-bottom flask containing 10 mL of dichloromethane and a stirbar. Set up the apparatus for heating under reflux. Heat the mixture with stirring until it refluxes *gently.*

Reaction Using a Pasteur pipet, add 5 mL of the 50% aqueous sodium hydroxide solution dropwise through the top of the reflux condenser to the solution, which should be stirred as *vigorously* as possible to ensure complete mixing of the phases. Try to add the solution *directly* into the flask without allowing it to touch the walls of the condenser. After the addition is complete, continue stirring under gentle reflux for 30 min.

Work-Up and Isolation of Isomeric Stilbenes Allow the reaction mixture to cool to room temperature and transfer it to a separatory funnel. Rinse the round-bottom flask with a 5-mL portion of dichloromethane and transfer the wash to the separatory funnel. Separate the layers and wash the organic layer sequentially with 10 mL of water and 15 mL of saturated aqueous sodium bisulfite. Finally, wash the organic layer with 10-mL portions of water until the pH of the wash is neutral. Transfer the organic solution to a dry Erlenmeyer flask and add several spatula-tips full of anhydrous sodium sulfate. Occasionally swirl the mixture over a period of 10–15 min to facilitate drying; add further small portions of anhydrous sodium sulfate if the solution remains cloudy.★

To determine the ratio of (*Z*)- and (*E*)-stilbene in the *crude* reaction mixture by ^{1}H NMR spectroscopy, use a filter-tip pipet to transfer a 0.5-mL portion of the dry organic solution to a small round-bottom flask and evaporate the dichloromethane by one of the techniques described in Section 2.29. The final traces of solvent may be removed by attaching the flask to a vacuum source and gently swirling the contents as vacuum is applied.★

Isomerization Filter or decant the remainder of the dry dichloromethane solution into the 25-mL round-bottom flask containing a stirbar, rinse the sodium sulfate with about 1 mL of fresh dichloromethane, and transfer this wash to the round-bottom flask. Add about 75 mg of iodine to the dichloromethane solution, fit the flask with a reflux condenser, and irradiate the solution with stirring for 1 h with a 150-W lightbulb.★

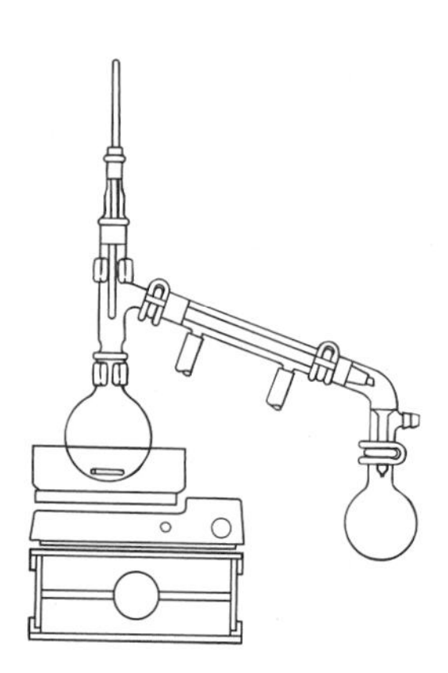

Work-Up and Isolation of Stilbene Decant the solution into a separatory funnel, and wash the dichloromethane solution with a 5-mL portion of saturated aqueous sodium bisulfite; shake the mixture vigorously to decolorize the dichloromethane layer. Wash the organic solution with a 5-mL portion of saturated aqueous sodium chloride, transfer it to a dry Erlenmeyer flask, and add several spatula-tips full of anhydrous sodium sulfate. Occasionally swirl the mixture over a period of 10–15 min to facilitate drying; add further small portions of anhydrous sodium sulfate if the solution remains cloudy.★

Filter or decant the dried solution into a tared round-bottom flask, equip the flask for simple distillation, and distil the dichloromethane. Alternatively, use rotary evaporation or other techniques to concentrate the solution. The final traces of solvent may be removed by attaching the flask to a vacuum source and gently swirling the contents as vacuum is applied.★

Remove about 50 mg of the residue for analysis by ^{1}H NMR spectroscopy. Dissolve the remaining residue in 10–12 mL of hot 95% ethanol. Cool the resulting solution first to room temperature and then in an ice-water bath for 15–20 min to complete crystallization. Collect the precipitate and air-dry it. Recrystallize the crude stilbene from 95% ethanol.

See *Who was Baeyer?*

Analysis Weigh the product and calculate the percent yield; determine its melting point. Perform the tests for unsaturation on the product using the bromine and Baeyer tests (Secs. 25.8A and B, respectively). Obtain IR and ^{1}H NMR spectra of your starting materials, the final product, and the crude product mixtures before and after irradiation with iodine. Compare these spectra with those of authentic samples (Figs. 10.34, 10.35, 18.1–18.6) to determine the ratio of the isomeric stilbenes **9** and **10** in the crude mixture and which one is the product after irradiation.

MICROSCALE PROCEDURE

Preparation Sign in at **www.cengage.com/login** to answer Pre-Lab Exercises, access videos, and read the MSDSs for the chemicals used or produced in this procedure. Review Sections 2.9, 2.11, 2.13, 2.17, 2.21, 2.22, and 2.29.

Apparatus A 5-mL conical vial, screw-cap centrifuge tube, apparatus for heating under reflux, magnetic stirring, simple distillation, vacuum filtration, Craig tube filtration, and *flameless* heating.

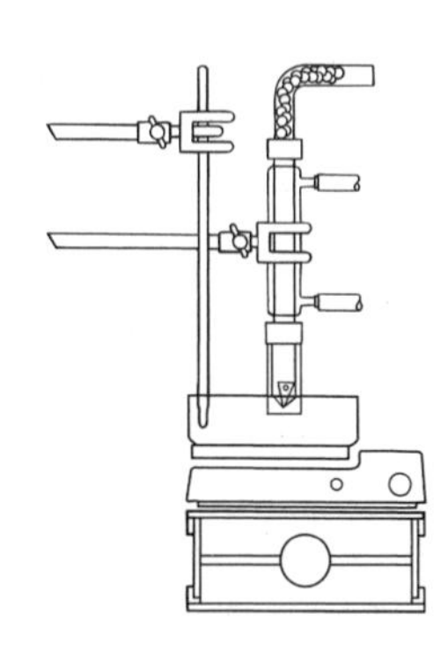

Setting Up Prepare a 50% (by mass) sodium hydroxide solution by dissolving 1.0 g of sodium hydroxide in 1 mL of water. The mixture may be heated gently to facilitate dissolution, but be sure to cool the solution to room temperature before proceeding. Add 0.76 g of benzyltriphenylphosphonium chloride and 0.2 mL of benzaldehyde to the conical vial containing 2 mL of dichloromethane and a spinvane. Set up the apparatus for heating under reflux. Heat the solution with stirring until it refluxes *gently*.

Wittig Reaction Using a Pasteur pipet, add 1 mL of the 50% aqueous sodium hydroxide solution dropwise through the top of the reflux condenser to the solution, which should be stirred as *vigorously* as possible to ensure adequate mixing of the phases. Try to add the solution *directly* into the flask without allowing it to touch the walls of the condenser. After the addition is complete, continue stirring under *gentle* reflux for 30 min.

Work-Up and Isolation of Isomeric Stilbenes Allow the reaction mixture to cool to room temperature, and transfer it to a screw-cap centrifuge tube. Rinse the conical vial with 0.5 mL of dichloromethane, and transfer the rinse to the centrifuge tube. Separate the layers and wash the organic layer sequentially with 1.5 mL of water and then with 2.5 mL of saturated aqueous sodium bisulfite. Finally, wash the organic layer with 1.5-mL portions of water until the pH of the wash is neutral. Transfer the organic solution to a dry centrifuge tube and add several microspatula-tips full of anhydrous sodium sulfate. Occasionally shake the mixture over a period of 10–15 min to facilitate drying; add further small portions of anhydrous sodium sulfate if the solution remains cloudy.★ Consult with your instructor to determine whether you should obtain a ^{1}H NMR spectrum of the crude product mixture as in the Miniscale Procedure.

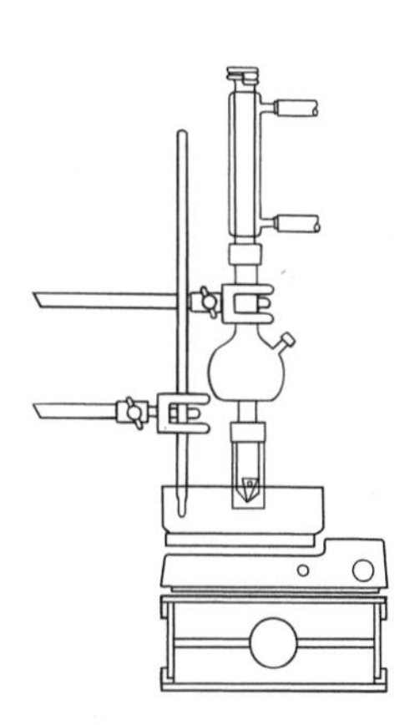

Isomerization Using a filter-tip pipet, transfer the dried dichloromethane solution into the 5-mL conical vial equipped with a spinvane. Rinse the sodium sulfate with 0.5 mL of dichloromethane and transfer the rinse to the conical vial. Add a *very small* crystal of iodine to the dichloromethane solution and irradiate the solution with stirring for 1 h with a 150-W lightbulb.★

Work-Up and Isolation of Stilbene Remove the spinvane, add 1 mL of saturated aqueous sodium bisulfite, and shake the mixture vigorously to decolorize the dichloromethane layer. Separate the layers and wash the organic layer with 1 mL of saturated aqueous sodium chloride. Transfer the organic solution to a dry centrifuge tube and add several microspatula-tips full of anhydrous sodium sulfate. Occasionally swirl the mixture over a period of 10–15 min to facilitate drying; add further small portions of anhydrous sodium sulfate if the solution remains cloudy.★ Using a filter-tip pipet, transfer the dried dichloromethane solution into a clean, dry 5-mL conical vial equipped with a spinvane. Remove the dichloromethane by simple distillation; withdraw solvent from the Hickman stillhead as needed. Alternatively, use rotary evaporation or other techniques to concentrate the solution. The final traces of solvents may be removed by attaching the vial to a vacuum source.★

Remove about 25 mg of the residue for analysis by ^{1}H NMR. Dissolve the remaining thick, cloudy residue in 2 mL of hot 95% ethanol. Cool the resulting solution first to room temperature and then in an ice-water bath for 15–20 min to complete crystallization. Collect the precipitate and air-dry it. Recrystallize the crude stilbene from 95% ethanol.

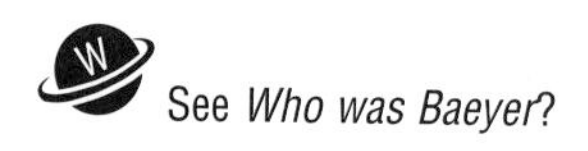

See *Who was Baeyer?*

Analysis Weigh the product and calculate the percent yield; determine its melting point. Perform the tests for unsaturation on the product using the bromine and Baeyer tests (Secs. 25.8A and B, respectively). Obtain IR and ^{1}H NMR spectra of your starting materials and the final product, and compare them with those of authentic samples (Figs. 10.34, 10.35, 18.1–18.6) to determine which one of the isomeric stilbenes **9** and **10** has been formed.

Discovery Experiment

Wittig Reaction of 9-Anthraldehyde

The Wittig reaction of benzyltriphenylphosphorane (**5**) with benzaldehyde (**6**) produces a mixture of (*E*)- and (*Z*)-stilbene (Scheme 18.1). These isomers are expected to have differing thermodynamic stabilities owing to steric interactions that should make the *Z*-isomer **9** less stable than the *E*-isomer **10.** You might then predict that increasing the steric interactions would foster a larger preference for forming the thermodynamically favored diastereomer in a Wittig reaction. This hypothesis is testable by executing a base-promoted Wittig reaction of benzyltriphenylphosphonium chloride with 9-anthraldehyde rather than benzaldehyde (Eq. 18.8). Even if you don't perform the experiment, you may explore the proposal that the difference in thermodynamic stabilities is less for **9** and **10** than it is for the corresponding isomers derived from 9-anthraldehyde if you have access to a molecular mechanics program for calculating energies. Consult with your instructor to ascertain whether this is possible.

$$\text{9-Anthraldehyde (CHO)} \xrightarrow[\text{Base}]{(C_6H_5)_3\overset{+}{P}CH_2C_6H_5\ Cl^-} \text{9-Anthryl–CH=CH–}C_6H_5\ (Z)\ \text{and/or}\ \text{9-Anthryl–CH=CH–}C_6H_5\ (E) \quad (18.8)$$

Design and implement an experimental protocol for performing the reaction of Equation 18.8. Because 9-anthraldehyde is relatively expensive, you should plan to carry out the reaction at the microscale level unless instructed to do otherwise. You should be sure that your proposal includes a means to analyze for the possible presence of both isomeric products in the reaction mixture. Appropriate quantities of reagents are 0.97 g of benzyltriphenylphosphonium chloride (**4**), 0.57 g of 9-anthraldehyde, and 1.5 mL of 50% (by mass) of aqueous sodium hydroxide solution. Use 2-propanol as the solvent for recrystallizing the crude product.

WRAPPING IT UP

Neutralize the combined *aqueous layers and washes* with dilute hydrochloric acid, and flush them down the drain. Place the *recovered dichloromethane*, the *dichloromethane solution* from the bromine test for unsaturation, and any NMR solutions in a container for halogenated organic liquids; put the *manganese dioxide* from the Baeyer test for unsaturation in a container for heavy metals. Flush the *ethanolic filtrate* down the drain.

B ■ *Preparation of a Stilbene by the Horner-Wadsworth-Emmons Reaction*

Purpose To demonstrate the synthesis of a stilbene by the Horner-Wadsworth-Emmons modification of the Wittig reaction.

SAFETY ALERT

Solutions of potassium *tert*-butoxide are *highly* caustic. *Do not allow them to come in contact with your skin.* If this should happen, wash the affected area with dilute acetic acid and then copious amounts of water. Wear latex gloves when transferring the solution.

MINISCALE PROCEDURE

Preparation Sign in at **www.cengage.com/login** to answer Pre-Lab Exercises, access videos, and read the MSDSs for the chemicals used or produced in this procedure. Review Sections 2.10, 2.11, 2.17, 2.27, and 2.28.

Apparatus A 25-mL round-bottom flask, Claisen adapter, rubber septum, drying tube, two 1-mL syringes and a 5-mL syringe with needles, cold-water bath, apparatus for magnetic stirring, vacuum filtration, and *flameless* heating.

Setting Up Dry the glass parts of the syringes in an oven at 110 °C for 15 min. Equip the round-bottom flask with a stirbar and the Claisen adapter. Fit the rubber septum onto the joint of the Claisen adapter that is directly *in line* with the lower male joint (see Fig. 10.42). Using a small flame or a heat gun, gently heat the flask and Claisen adapter to remove traces of water from the apparatus; do *not* overheat the joint with the septum. After the flask has been heated for a few minutes, discontinue the heating, attach the drying tube to the *sidearm* of the adapter, and allow the apparatus to cool to room temperature. Remove the glass syringes from the oven and allow them to cool to room temperature, preferably in a desiccator.

Draw 5 mL of a 1 *M* solution of potassium *tert*-butoxide in dry *N,N*-dimethylformamide into the 5-mL syringe. Pierce the rubber septum with the needle of the syringe, being careful not to press down on the plunger, and add the solution directly to the round-bottom flask. Cool the flask below 20 °C in a cold-water bath. Using a 1-mL syringe, transfer 1 mL of diethyl benzylphosphonate through the rubber septum into the stirred solution of potassium *tert*-butoxide and continue stirring the solution for 5 min.

Reaction Using a 1-mL syringe, slowly add 0.5 mL of benzaldehyde through the rubber septum while continuing to stir the cooled solution. Once the addition is complete, remove the apparatus from the cooling bath and stir the mixture at room temperature for 30 min.★

Work-Up and Isolation Remove the Claisen adapter from the flask and add 5 mL of water with stirring. Collect the precipitated product by vacuum filtration, wash it with cold 1:1 methanol-water, and air-dry it. Reserve about 50 mg of the crude stilbene for analysis by ^{1}H NMR spectroscopy, and recrystallize the remainder from 95% ethanol.

See *Who was Baeyer?*

Analysis Weigh the product and calculate the percent yield; determine its melting point. Perform the tests for unsaturation on the product using bromine in dichloromethane (Sec. 25.8A) and the Baeyer test (Sec. 25.8B). Obtain IR and ^{1}H NMR spectra of your starting materials and product, and compare them with those of authentic samples (Figs. 10.34, 10.35, 18.1–18.4, 18.7, and 18.8). In the ^{1}H NMR spectrum of the crude product, determine whether there is any of the other diastereomeric stilbene present; if so, calculate the isomeric ratio.

MICROSCALE PROCEDURE

Preparation Sign in at **www.cengage.com/login** to answer Pre-Lab Exercises, access videos, and read the MSDSs for the chemicals used or produced in this procedure. Review Sections 2.10, 2.11, 2.17, 2.27, and 2.28.

Apparatus A 5-mL conical vial, Claisen adapter, rubber septum, drying tube, three 1-mL syringes with needles, cold-water bath, apparatus for magnetic stirring, vacuum filtration, Craig tube filtration, and *flameless* heating.

Setting Up Dry the conical vial, the Claisen adapter, and the glass parts of the syringes in an oven at 110 °C for 15 min. Remove the vial and adapter from the oven, and after they have cooled, preferably in a desiccator, place a spinvane in the vial and attach the Claisen adapter. Fit the rubber septum onto the joint of the Claisen adapter that is directly *in line* with the lower male joint and attach the drying tube to the *sidearm* of the adapter (see Fig. 10.43). Remove the glass syringes from the oven and allow them to cool to room temperature, preferably in a desiccator.

Draw 1 mL of a 1 *M* solution of potassium *tert*-butoxide in dry *N,N*-dimethylformamide into one of the 1-mL syringes. Pierce the rubber septum with the needle of the syringe, being careful not to press down on the plunger, and add the solution directly to the vial. Cool the vial below 20 °C in a cold-water bath. Using another 1-mL syringe, transfer 0.2 mL of diethyl benzylphosphonate through the rubber septum into the stirred solution of potassium *tert*-butoxide, and continue stirring the solution for 5 min.

Reaction Using a 1-mL syringe, slowly add 0.1 mL of benzaldehyde through the rubber septum while continuing to stir the cooled solution. Once the addition is complete, remove the apparatus from the cooling bath and stir the mixture at room temperature for 30 min.★

Work-Up and Isolation Remove the Claisen adapter from the vial, and add 1 mL of water with stirring. Collect the precipitated product by vacuum filtration, wash it with cold 1:1 methanol-water, and air-dry it. Recrystallize the crude stilbene from 95% ethanol.

See *Who was Baeyer?*

Analysis Weigh the product and calculate the percent yield; determine its melting point. Perform the tests for unsaturation on the product using the bromine and Baeyer tests (Secs. 25.8A and B, respectively). Obtain IR and ^{1}H NMR spectra of your starting materials and product, and compare them with those of authentic samples (Figs. 10.34, 10.35, 18.1–18.4, 18.7, and 18.8). In the ^{1}H NMR spectrum of the product, determine whether there is any of the other diastereomeric stilbene present; if so, calculate the isomeric ratio and identify the major isomer.

WRAPPING IT UP

Place the *recovered dichloromethane* and the *dichloromethane solution* from the bromine test for unsaturation in a container for halogenated organic liquids; put the *manganese dioxide* from the Baeyer test for unsaturation in a container for heavy metals. Flush all *aqueous filtrates* down the drain.

EXERCISES

General Questions

1. Compare the mechanism of aldol addition (Sec. 18.3) to that of the Wittig synthesis, pointing out similarities and differences.
2. Why should the aldehydes used as starting materials in Wittig syntheses be free of contamination by carboxylic acids?
3. Considering the mechanism of the Wittig reaction, speculate what the driving force for the decomposition of the oxaphosphetane intermediate might be.
4. Ylides react readily with aldehydes and ketones but slowly or not at all with esters. Explain this difference in reactivity.
5. Explain why you would expect an anion of type **5** to be more stable if one of the R groups is cyano, C≡N, rather than alkyl.
6. An ylide like **5** is a stabilized carbanion. In what way(s) does the phosphorus atom provide stability to the carbanion?
7. Write equations for the preparation of the following alkenes by the Wittig reaction or the Horner-Wadsworth-Emmons modification of the Wittig reaction. Start with any carbonyl compound and Wittig or Horner-Wadsworth-Emmons reagent together with any other organic or inorganic reagents that you require.
   a. $C_6H_5CH{=}C(CH_3)C_6H_5$
   b. $CH_2{=}CH{-}CH{=}CH{-}C_6H_5$
   c. $(CH_3)_2C{=}CH{-}CO_2C_2H_5$
8. Suggest *two* different routes for preparing $C_6H_5CH{=}CHCH_3$ via a Wittig reaction, using a phosphorane as the ylide precursor in both cases.

Sign in at **www.cengage.com/login** and use the spectra viewer and Tables 8.1–8.8 as needed to answer the blue-numbered questions on spectroscopy.

9. Consider the spectral data for benzaldehyde (Figs. 18.1 and 18.2).
   a. In the functional group region of the IR spectrum, specify the absorptions associated with the carbonyl group and the aromatic ring.
   b. In the ^{1}H NMR spectrum, assign the various resonances to the hydrogen nuclei responsible for them.
   c. For the ^{13}C NMR data, assign the various resonances to the carbon nuclei responsible for them.
10. Consider the spectral data for (*E*)-stilbene (Figs. 10.34 and 10.35).
    a. In the functional group region of the IR spectrum, specify the absorptions associated with the aromatic rings.

b. In the ^{1}H NMR spectrum, assign the various resonances to the hydrogen nuclei responsible for them.

c. For the ^{13}C NMR data, assign the various resonances to the carbon nuclei responsible for them.

11. Consider the spectral data for (*Z*)-stilbene (Figs. 18.3 and 18.4).

a. In the functional group region of the IR spectrum, specify the absorptions associated with the aromatic rings.

b. In the ^{1}H NMR spectrum, assign the various resonances to the hydrogen nuclei responsible for them.

c. For the ^{13}C NMR data, assign the various resonances to the carbon nuclei responsible for them.

12. Discuss the differences observed in the IR and NMR spectra of (*Z*)-stilbene and (*E*)-stilbene that enable you to distinguish between the two isomers. What differences in the IR and NMR spectra of benzaldehyde, (*Z*)-stilbene, and (*E*)-stilbene are consistent with the conversion of benzaldehyde to a mixture of the isomeric stilbenes in this experiment?

Questions for Part A

13. Suggest a method for preparing benzyltriphenylphosphonium chloride (**4**) via an S_N2 reaction.

14. Define the term *phase-transfer catalyst.*

15. Consider the step in which you washed the crude reaction mixture obtained *immediately* following the Wittig reaction with saturated aqueous sodium bisulfite.

a. What potential contaminant of the final product is removed in this step?

b. Write the chemical equation by which the contaminant is removed and specify whether it is being oxidized, reduced, or neither in the process.

16. Propose a mechanism for the isomerization of (*Z*)-stilbene into (*E*)-stilbene using a trace of iodine and light (Eq. 18.6). (*Hint:* The iodine-iodine bond undergoes homolysis upon irradiation with light; see also Section 7.3.)

17. Based on your experimental results, what are you able to conclude about the iodine/light-promoted isomerization of (*E*)-stilbene? Which of the stilbene isomers appears to be the more stable, and how do you rationalize this difference in stabilities?

18. By analysis of the ^{1}H NMR spectra of the crude product before *and* after irradiation in the presence of iodine, determine the approximate ratios of (*Z*)-stilbene and (*E*)-stilbene.

Sign in at **www.cengage.com/login** and use the spectra viewer and Tables 8.1–8.8 as needed to answer the blue-numbered questions on spectroscopy.

19. Consider the NMR spectral data for benzyltriphenylphosphonium chloride (Fig. 18.6). Note that ^{31}P has $I_z = 1/2$ and therefore couples with neighboring hydrogen and carbon and carbon nuclei having $I_z = 1/2$.

a. In the ^{1}H NMR spectrum, assign the various resonances to the hydrogen nuclei responsible for them.

b. For the ^{13}C NMR data, assign the various resonances to the carbon nuclei responsible for them.

Questions for Part B

20. Account for the fact that diethyl benzylphosphonate (**12**) is a stronger acid than benzyltriphenylphosphonium chloride (**4**).

21. Which phosphorus compound, **5** or **13,** is more nucleophilic? Explain your answer.

22. Why should the exposure of the solution of potassium *tert*-butoxide to the atmosphere be minimized?
23. *N,N*-Dimethylformamide (DMF) is very water-soluble. Why?
24. Consider the NMR spectral data for diethyl benzylphosphonate (Fig. 18.8). Note that ^{31}P has $I_z = 1/2$ and therefore couples with neighboring hydrogen and carbon and carbon nuclei having $I_z = 1/2$.
    a. In the ^{1}H NMR spectrum, assign the various resonances to the hydrogen nuclei responsible for them.
    b. For the ^{13}C NMR data, assign the various resonances to the carbon nuclei responsible for them.

Sign in at **www.cengage.com/login** and use the spectra viewer and Tables 8.1–8.8 as needed to answer the blue-numbered questions on spectroscopy.

SPECTRA

Starting Materials and Products

The IR and NMR spectra of (E)-stilbene are provided in Figures 10.34 and 10.35, respectively.

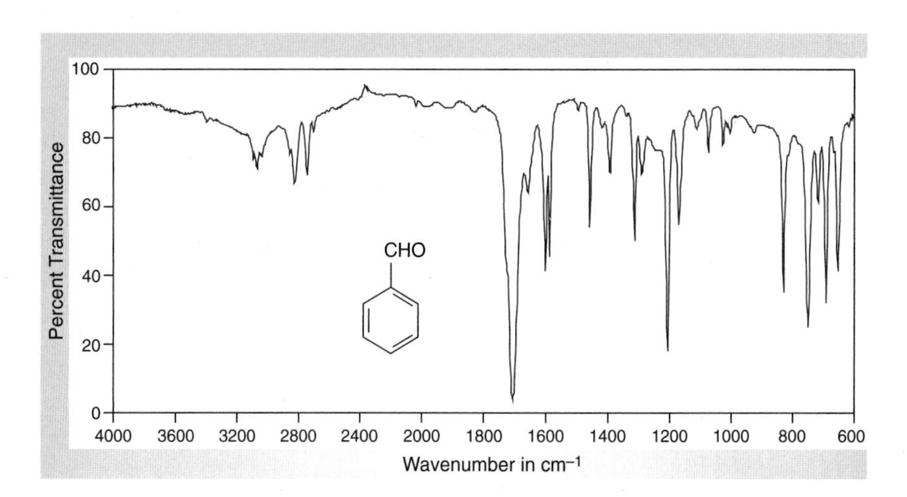

Figure 18.1
IR spectrum of benzaldehyde (neat).

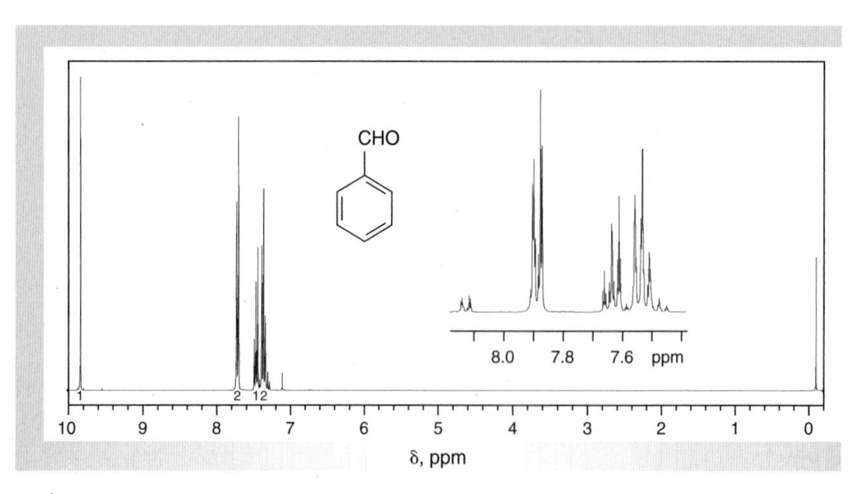

Figure 18.2
NMR data for benzaldeyde ($CDCl_3$).

(a) ^{1}H NMR spectrum (300 MHz).
(b) ^{13}C NMR data: δ 129.0, 129.7, 134.4, 136.6, 192.0.

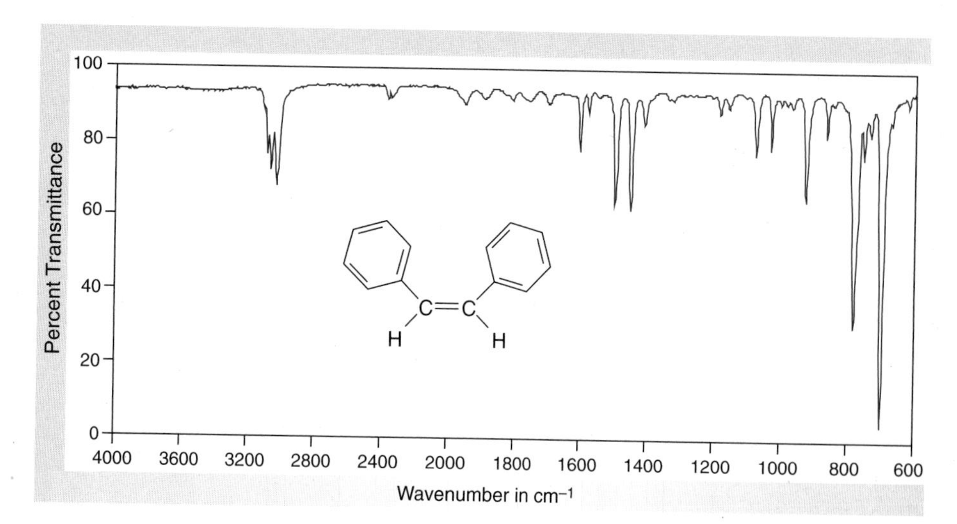

Figure 18.3
IR spectrum of (Z)-stilbene (neat).

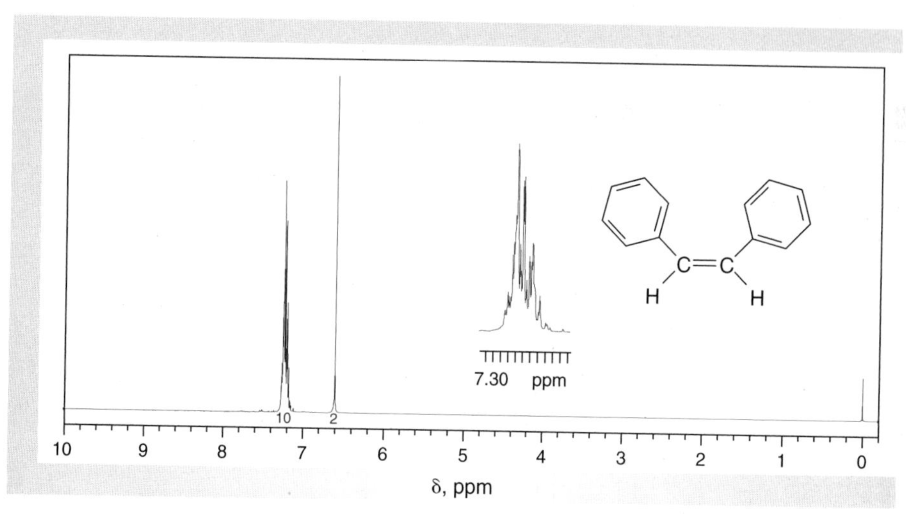

Figure 18.4
NMR data for (Z)-stilbene ($CDCl_3$).

(a) ^1H NMR spectrum (300 MHz).
(b) ^{13}C NMR data: δ 127.3, 128.4, 129.1, 130.5, 137.5.

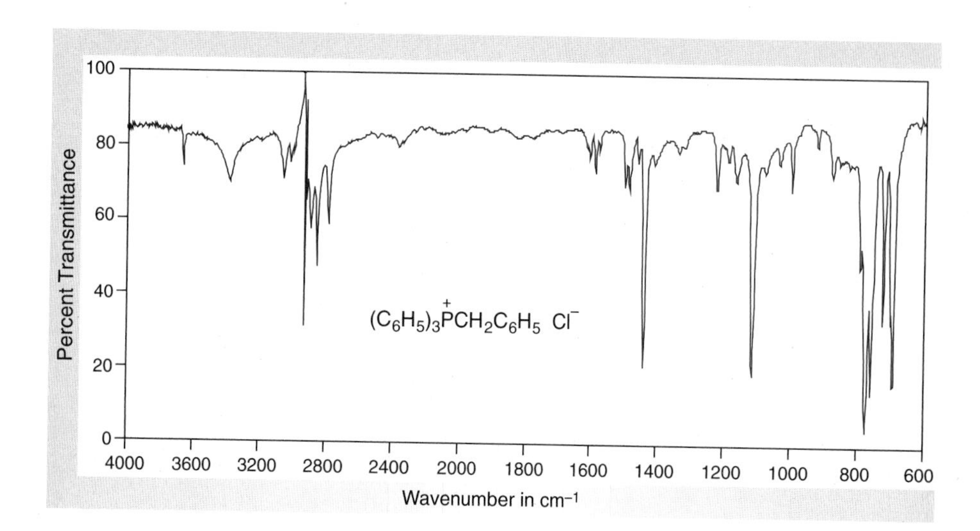

Figure 18.5
IR spectrum of benzyltri-phenylphosphonium chloride (IR card).

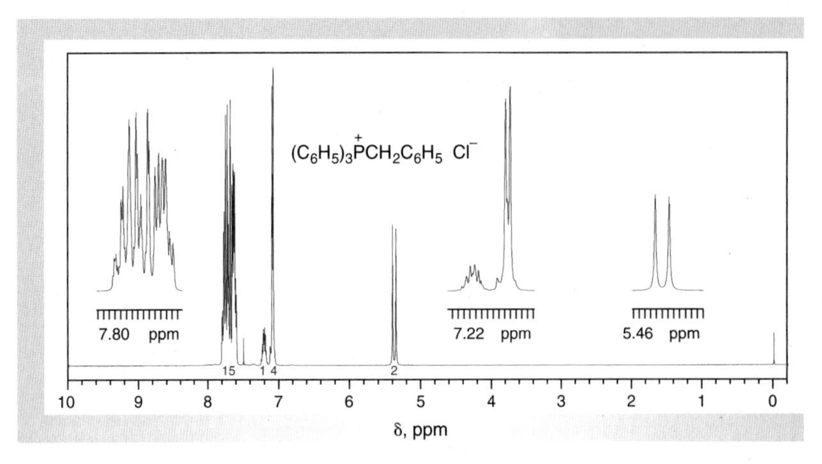

Figure 18.6
NMR data for benzyltriphenylphosphonium chloride ($CDCl_3$).

(a) ^{1}H NMR spectrum (300 MHz).
(b) ^{13}C NMR data (^{31}P-coupled): δ 29.8, 30.6, 116.6, 118.0, 126.8, 127.0, 127.9, 128.3, 128.4, 129.6, 129.8, 13.10, 131.1, 133.8, 133.9, 134.6.

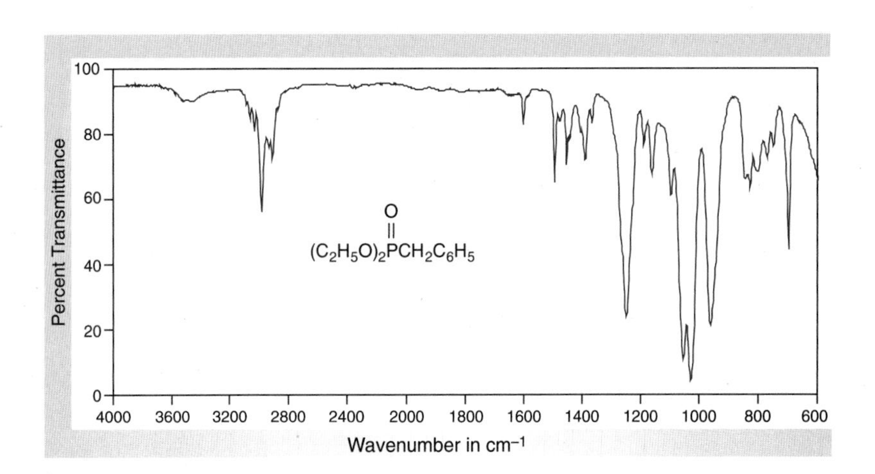

Figure 18.7
IR spectrum of diethyl benzylphosphonate (neat).

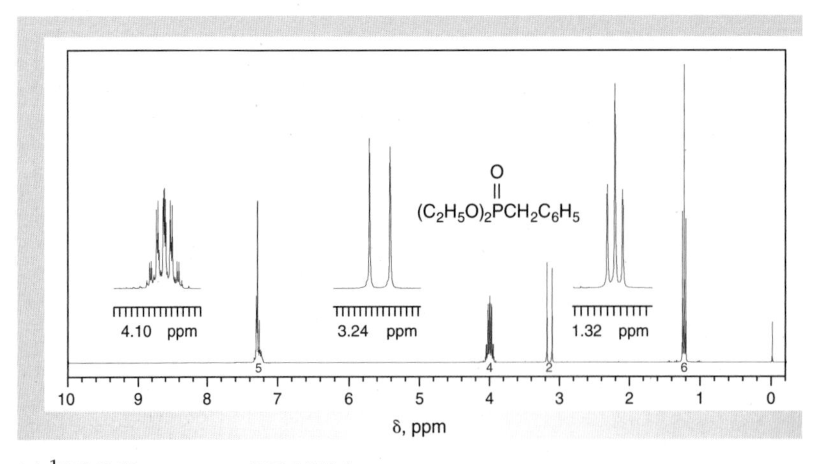

Figure 18.8
NMR data for diethyl benzylphosphonate ($CDCl_3$).

(a) ^{1}H NMR spectrum (300 MHz).
(b) ^{13}C NMR data (^{31}P-coupled): δ 15.9, 16.0, 32.2, 34.4, 61.5, 61.6, 126.3, 126.4, 128.0, 128.1, 129.3, 129.4, 131.1, 131.3.

18.3 REACTIONS OF STABILIZED CARBANIONS FROM CARBONYL COMPOUNDS

The presence of the carbonyl function in a molecule allows the α-carbon atom to become nucleophilic through formation of an **enolate ion** (Eq. 18.3) or an **enol** (Eq. 18.4). The nucleophilic property of an enol is the basis for the experiment in Section 18.4. The corresponding property of enolate ions, which are considerably more reactive nucleophiles than enols, is the subject of this section.

Formation of enolate ions by deprotonation (Eq. 18.3) involves the use of a base, and the strength of the base required will depend on the pK_a of the hydrogen atom attached to the α-carbon atom. The pK_a is defined as the negative log of K_a, the equilibrium constant for the ionization shown in Equation 18.9. Thus, $pK_a = -\log K_a$. Water is the usual reference solvent for reporting values of pK_a, which are known to be solvent-dependent. Experimentally, the pK_a values of aldehydes and ketones are in the range of 18–20. This is a remarkably low value when compared to those for saturated hydrocarbons, which are in excess of 50!

(18.9)

17

Enolate ion

The dramatic acidifying effect of the carbonyl group is due to its inductive effect and its ability to delocalize and thereby stabilize the negative charge in the enolate ion, as reflected in **17** (Eq. 18.9), which is the **resonance hybrid** of the **resonance structures 3a** and **3b** (Sec. 18.1). Because the pK_a values of aldehydes and ketones fall in the range of water (pK_a 15.7) and alcohols (pK_a 15.5–18), it is possible to generate enolate ions using anions such as hydroxide or alkoxide (Eq. 18.3, $B^- = HO^-$ and RO^-, respectively); these are the bases utilized in the experimental procedures of this section.

Aldol Condensations

An important general reaction of enolate ions involves **nucleophilic addition** to the electrophilic carbonyl carbon atom of the aldehyde or ketone from which the enolate is derived. A dimeric anion **18** results, which may then be neutralized by abstraction of a proton to produce a β-hydroxycarbonyl compound, **19** (Eq. 18.10). If the reaction is performed in hydroxylic solvents such as water or an alcohol, the source of the proton may be the solvent, whose deprotonation will regenerate the base required for forming the enolate ion. Thus, the overall process is *catalytic* in the base that is used.

(18.10)

3b 3a

Enolate ion

18

19

An aldol

(β-hydroxycarbonyl compound)

The β-hydroxycarbonyl compound **19** is commonly called an **aldol,** and the reaction leading to its formation is referred to as an **aldol addition.** The term "aldol" reflects the fact that use of an aldehyde in the reaction provides a product containing an ***ald*ehyde function** and an **alcoh*ol* group.** The ability to isolate such addition products depends on the reaction conditions used, because dehydration to an α,β-unsaturated carbonyl compound **20** may occur *if* **19** still has a hydrogen atom on the α-carbon atom that served as the nucleophile in the addition reaction (Eq. 18.11). The overall reaction described by Equations 18.10 and 18.11 is called an **aldol condensation.**

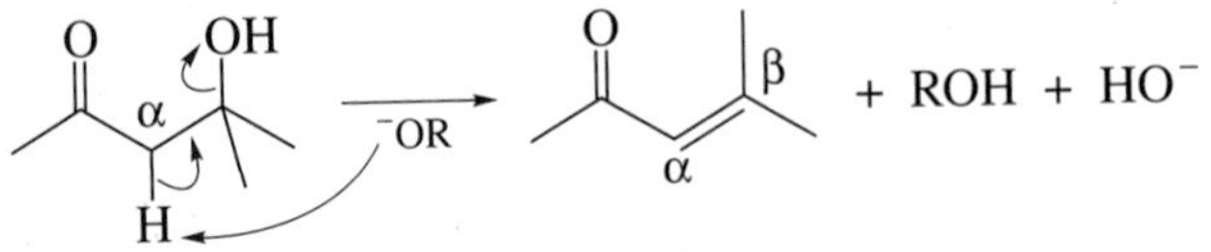

19 → ($^-$OR) **20** α,β-Unsaturated compound + ROH + HO$^-$ (18.11)

The experiment presented in this section is an example of a **mixed-**or **crossed-aldol condensation.** This term describes cases in which two *different* carbonyl compounds are the reactants. Such reactions are synthetically practical under certain circumstances, selectively producing a single major condensation product. For example, a ketone may preferentially condense with an aldehyde rather than undergoing self-addition with another molecule of itself (Scheme 18.3). This is because the carbonyl carbon atom of ketones is sterically and electronically not as susceptible to nucleophilic attack as is that of aldehydes. The aldehydic partner in such a reaction generally has no α-hydrogen atoms, so that it is unable to undergo an aldol reaction.

Scheme 18.3

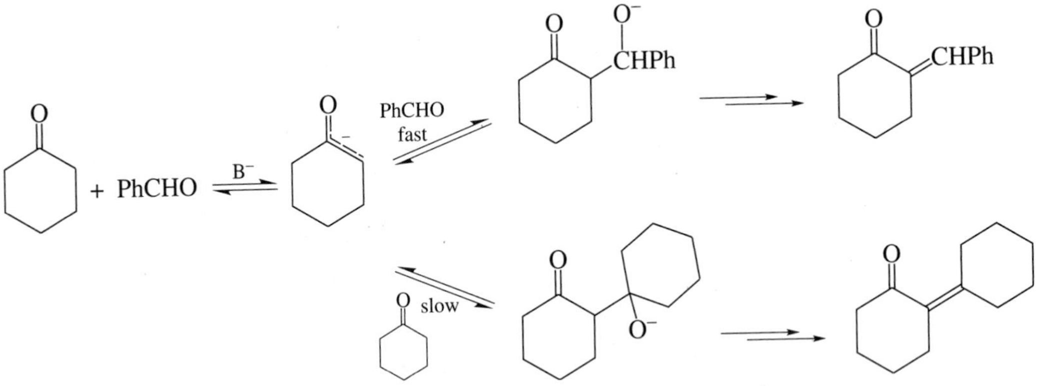

+ PhCHO (B$^-$) PhCHO fast CHPh CHPh slow

You will perform the crossed-aldol condensation of acetophenone (**21**) with *p*-anisaldehyde (4-methoxybenzaldehyde, **23**) to give *trans-p*-anisalacetophenone (**25**) according to the sequence outlined in Scheme 18.4. For steric and electronic reasons, the enolate ion **22** reacts preferentially with **23** rather than **21,** leading to the aldol **24.** This product then dehydrates in the presence of base, in the manner outlined in Equation 18.11, to yield the condensation product **25.** Part of the thermodynamic driving force for this dehydration is associated with forming a new carbon-carbon π-bond that is **conjugated** with the aromatic ring as well as with the carbonyl function. As a general rule, formation of an extended conjugated system, as occurs here, increases the stability of a molecule.

Scheme 18.4

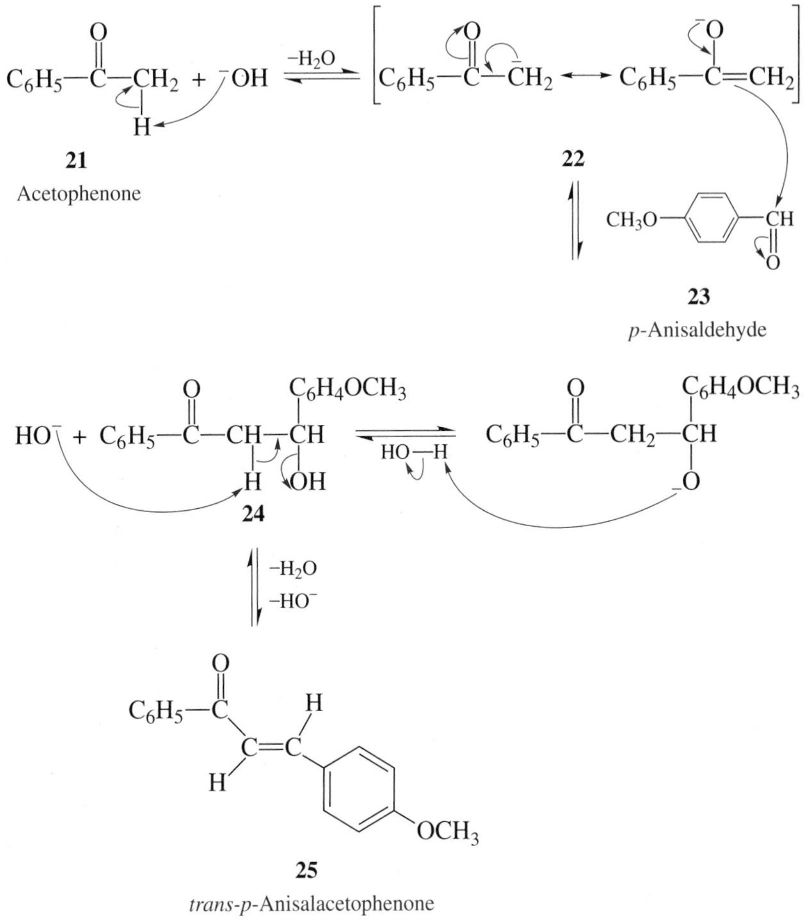

$C_6H_5-C(=O)-CH_2-H$ + ^-OH $\xrightarrow{-H_2O}$ [$C_6H_5-C(=O)-\bar{C}H_2 \longleftrightarrow C_6H_5-C(-O^-)=CH_2$]

21
Acetophenone

22

$CH_3O-C_6H_4-CH=O$

23
p-Anisaldehyde

HO^- + $C_6H_5-C(=O)-CH(H)-CH(OH)(C_6H_4OCH_3)$ $\rightleftharpoons$ (HO—H) $C_6H_5-C(=O)-CH_2-CH(O^-)(C_6H_4OCH_3)$

24

$-H_2O$
$-HO^-$

$C_6H_5-C(=O)-CH=CH-C_6H_4-OCH_3$

25
trans-p-Anisalacetophenone

EXPERIMENTAL PROCEDURES

Preparation of trans-p-*Anisalacetophenone*

Purpose To demonstrate the synthesis of an α,β-unsaturated carbonyl compound by a crossed-aldol condensation.

SAFETY ALERT

The 50% (by mass) aqueous sodium hydroxide solution is *highly* caustic. *Do not allow it to come in contact with your skin.* If this should happen, wash the affected area with dilute acetic acid and then copious amounts of water. Wear latex gloves when transferring the solution.

MINISCALE PROCEDURE

Preparation Sign in at **www.cengage.com/login** to answer Pre-Lab Exercises, access videos, and read the MSDSs for the chemicals used or produced in this procedure. Review Sections 2.9, 2.10, 2.11, and 2.17.

Apparatus A 13-mm × 100-mm test tube, two 1-mL syringes, apparatus for vacuum filtration and *flameless* heating.

Setting Up Prepare a 50% (by mass) sodium hydroxide solution by dissolving 1.0 g sodium hydroxide in 1 mL of water. You may heat the mixture gently to hasten dissolution, but be sure to cool the solution to room temperature before proceeding. Place 1.0 mL of *p*-anisaldehyde and 1.0 mL of acetophenone in the test tube and add 3.0 mL of 95% ethanol. Shake the test tube gently to mix and dissolve the reactants.

Aldol Reaction, Work-Up, and Isolation Using a Pasteur pipet, transfer 5 drops of the 50% sodium hydroxide solution into the ethanolic solution of the carbonyl compounds, shake the mixture for a minute or two until a homogeneous solution results, and allow it to stand with occasional shaking at room temperature for 15 min.★ Cool the reaction mixture in an ice-water bath. If crystals do not form, scratch at the liquid-air interface to induce crystallization. Collect the product by vacuum filtration, wash the product with 1–2 mL of *cold* 95% ethanol, and air-dry the crystals.★ Recrystallize the crude *trans-p*-anisalacetophenone from methanol.

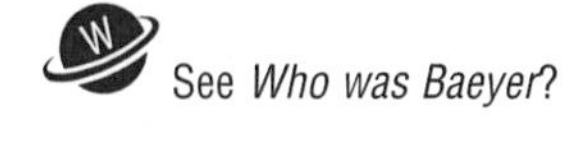
See *Who was Baeyer?*

Analysis Weigh the product and calculate the percent yield; determine its melting point. Prepare the 2,4-dinitrophenylhydrazones of the product and of the two starting materials according to the procedure in Section 25.7F; compare the melting points of these derivatives. Perform the tests for unsaturation on the product using the bromine and Baeyer tests (Secs. 25.8A and B, respectively). Obtain IR and ^{1}H NMR spectra of your starting materials and product and compare them with those of authentic samples (Figs. 18.9–18.14).

MICROSCALE PROCEDURE

Preparation Sign in at **www.cengage.com/login** to answer Pre-Lab Exercises, access videos, and read the MSDSs for the chemicals used or produced in this procedure. Review Sections 2.9, 2.10, 2.11, and 2.17.

Apparatus A 3-mL conical vial, two 1-mL syringes, apparatus for magnetic stirring, vacuum filtration, Craig tube filtration, and *flameless* heating.

Setting Up Prepare a 50% (by mass) sodium hydroxide solution by dissolving 0.5 g of sodium hydroxide in 0.5 mL of water. You may heat the mixture gently to hasten dissolution, but be sure to cool the solution to room temperature before proceeding. Place 0.2 mL of *p*-anisaldehyde and 0.2 mL of acetophenone in the conical vial containing a spinvane. Add 0.7 mL of 95% ethanol and stir the contents of the vial to mix and dissolve the reactants.

Aldol Reaction, Work-Up, and Isolation Using a Pasteur pipet, transfer 2 drops of the 50% NaOH solution into the ethanolic solution of the carbonyl compounds and stir the reaction mixture at room temperature for 15 min.★ Cool the reaction mixture in an ice-water bath. If crystals do not form, scratch at the liquid-air interface to induce crystallization.★ Collect the product by vacuum filtration, wash the product with about 5–10 drops of *cold* 95% ethanol, and air-dry the crystals. Recrystallize the crude *trans-p*-anisalacetophenone from methanol.

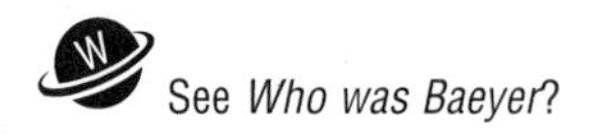

See *Who was Baeyer?*

Analysis Weigh the product and calculate the percent yield; determine its melting point. Prepare the 2,4-dinitrophenylhydrazones of the product and the two starting materials according to the procedure in Section 25.7F; compare the melting points of these derivatives. Perform the tests for unsaturation on the product using the bromine and Baeyer tests (Secs. 25.8A and B, respectively). Obtain IR and ^{1}H NMR spectra of your starting materials and product, and compare them with those of authentic samples (Figs. 18.9–18.14).

Discovery Experiment

Synthesis of trans,trans-*Dibenzylideneacetone*

Consult with your instructor before performing this experiment, in which you will prepare *trans,trans*-dibenzylideneacetone by the crossed-aldol condensation of benzaldehyde with acetone. Design an experiment using either a miniscale or microscale procedure to produce the product. Determine the relative amounts of benzaldehyde and acetone that should be used by considering the stoichiometry of the reaction and the potential problems that might be associated with side reactions and the use of large excesses of reactants. Freshly distilled benzaldehyde should be used. The product may be recrystallized from an 85% (*v:v*) ethanol-water mixture. Perform the tests for unsaturation on the product using the bromine and Baeyer tests (Secs. 25.8A and B, respectively). Obtain the melting point and the IR, ^{1}H, and ^{13}C NMR spectra of the purified product for purposes of characterization, and compare these with the literature data. Determine the λ_{max} and ε_{max} in the UV spectrum for the product, and compare to the literature values.

Discovery Experiment

Solvent-Free Aldol Condensation

Some but not all organic reactions proceed well in the absence of solvent(s). Develop a solvent-free protocol for the aldol reaction using solid sodium or potassium hydroxide as the base. The reaction can be successfully performed on a 5–10 mmol scale, using an equivalent of base. Bear in mind that the base should be finely divided, so it should first be crushed with a mortar and pestle. In addition, the condensation partners should be mixed thoroughly before the base is added. The reaction should take no more than 20–25 minutes, and the mixture should be acidified to pH ~5 with 10% HCl prior to isolation of the product. Characterize the product by its melting point and using spectroscopic techniques.

Possible combinations that you might use include acetophenone with either 4-methylbenzaldehyde or 4-chlorobenzaldehyde, 1-indanone with either 4-phenylcyclohexanone or 3,4-dimethoxybenzaldehyde, 4-phenylcyclohexanone with 3,4-dimethoxybenzaldehyde, and 3,4-dimethoxybenzaldehyde with 3,4-dimethoxyacetophenone. Self-condensation of 4-methylbenzaldehyde is also a possibility. Consult with your instructor before undertaking any experiments.

WRAPPING IT UP

Neutralize the *ethanolic filtrates* with dilute hydrochloric acid, and flush them down the drain. Place the *dichloromethane solution* from the bromine test for unsaturation in a container for halogenated organic liquids.

EXERCISES

1. Compute the equilibrium constant, K_{eq}, for the reaction of equimolar amounts of acetophenone, $C_6H_5COCH_3$, and hydroxide ion to generate the enolate ion. The pK_a values of the ketone and of water are 19.0 and 15.7, respectively.

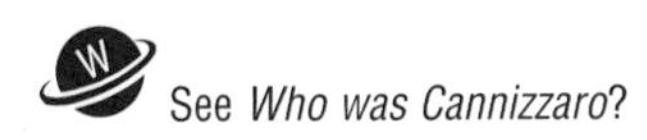
See *Who was Cannizzaro?*

2. Explain why the main reaction between acetophenone and *p*-anisaldehyde is the mixed-aldol reaction rather than (a) self-condensation of acetophenone or (b) the Cannizzaro reaction (Sec. 16.3) of *p*-anisaldehyde.
3. Identify the nucleophile and electrophile involved in the rate-determining step of the synthesis of *trans-p*-anisalacetophenone.
4. Explain why *trans-p*-anisalacetophenone would be expected to be more stable than the corresponding *cis* isomer.
5. Propose a synthesis of cinnamaldehyde using a mixed-aldol condensation reaction.

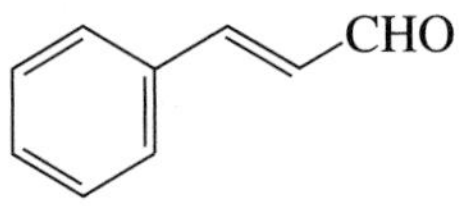

CHO

Cinnamaldehyde

6. Write structures for the various aldol condensation products you expect from the aldol self-condensation of 2-butanone, $CH_3C(=O)CH_2CH_3$.
7. Predict the product of the *intramolecular* aldol reaction of the diketone shown below.

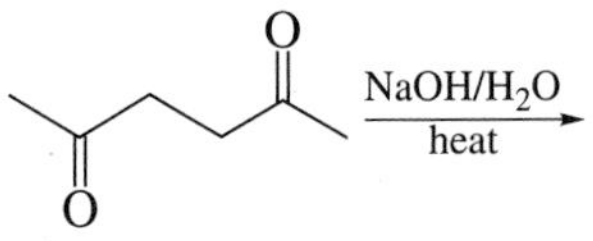

$NaOH/H_2O$, heat

2,4-Pentanedione

8. α,β-Unsaturated carbonyl compounds such as *trans-p*-anisalacetophenone (**25**) undergo reactions that are typical of both carbonyl compounds and alkenes. Illustrate this by showing the reaction of **25** with bromine (Sec. 10.6) and 2,4-dinitrophenylhydrazine (Sec. 25.7F) to give the corresponding products.
9. The addition of bromine to *trans-p*-anisalacetophenone (**25**) gives a single diastereomer.
   a. Using curved arrows to symbolize the flow of electrons and suitable stereochemical drawings, show the mechanism of this reaction (*Hint:* See Sec. 10.6).
   b. How does the mechanism you proposed in Part **a** support the formation of the racemic form of a single diastereomer?

Sign in at **www.cengage.com/login** and use the spectra viewer and Tables 8.1–8.8 as needed to answer the blue-numbered questions on spectroscopy.

10. Consider the spectral data for acetophenone (Figs. 18.9 and 18.10).
    a. In the functional group region of the IR spectrum, specify the absorptions associated with the carbonyl group and the aromatic ring.
    b. In the ^{1}H NMR spectrum, assign the various resonances to the hydrogen nuclei responsible for them.
    c. For the ^{13}C NMR data, assign the various resonances to the carbon nuclei responsible for them.

11. Consider the spectral data for *p*-anisaldehyde (Figs. 18.11 and 18.12).

a. In the functional group region of the IR spectrum, specify the absorptions associated with the carbonyl group and the aromatic ring. Also specify the absorption in the fingerprint region that is associated with the *para* substitution on the aromatic ring.

b. In the ^{1}H NMR spectrum, assign the various resonances to the hydrogen nuclei responsible for them.

c. For the ^{13}C NMR data, assign the various resonances to the carbon nuclei responsible for them.

12. Consider the spectral data for *trans*-*p*-anisalacetophenone (Figs. 18.13 and 18.14).

a. In the functional group region of the IR spectrum, specify the absorptions associated with the unsaturated ketone group and the aromatic rings. Also specify the absorptions in the fingerprint region that are associated with the *para* substitution on the aromatic ring and the *trans* carbon-carbon double bond.

b. In the ^{1}H NMR spectrum, assign the various resonances to the hydrogen nuclei responsible for them.

c. For the ^{13}C NMR data, assign the various resonances to the carbon nuclei responsible for them.

13. Discuss the differences observed in the IR and NMR spectra of acetophenone, *p*-anisaldehyde, and *trans*-*p*-anisalacetophenone that are consistent with the crossed-aldol condensation occurring in this experiment.

SPECTRA

Starting Materials and Products

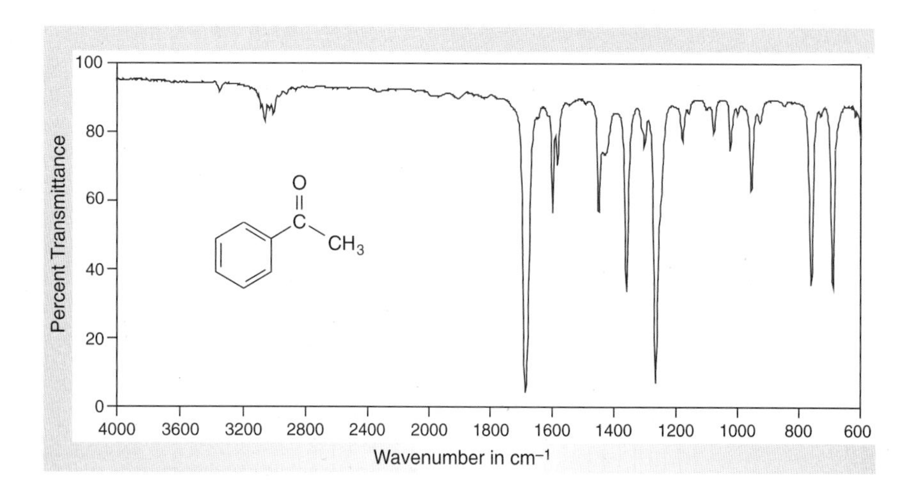

Figure 18.9
IR spectrum of acetophenone (neat).

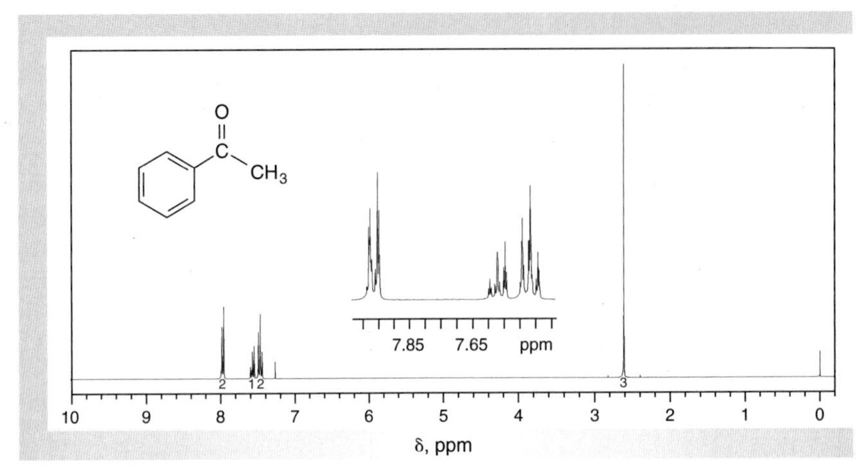

Figure 18.10
NMR data for acetophenone ($CDCl_3$).

(a) ^1H NMR spectrum (300 MHz).
(b) ^{13}C NMR data: δ 26.3, 128.3, 128.8, 133.0, 137.3, 197.4.

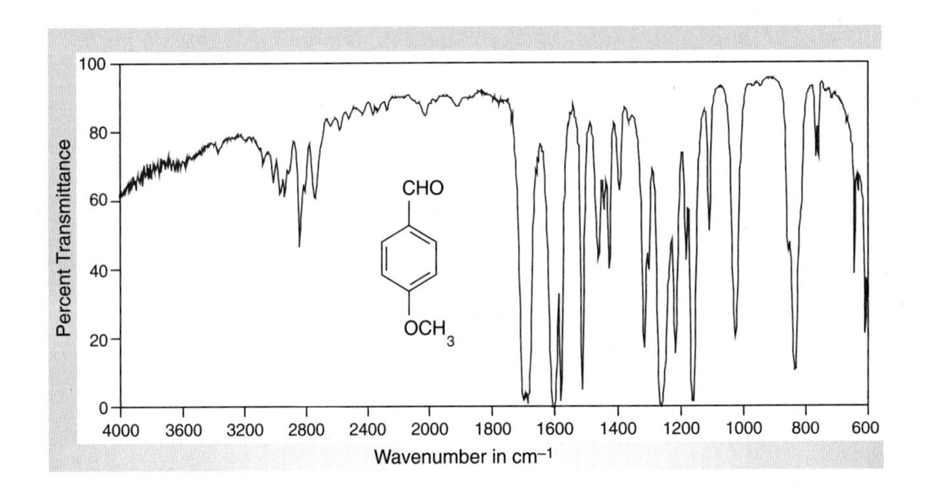

Figure 18.11
IR spectrum of p-anisaldehyde (neat).

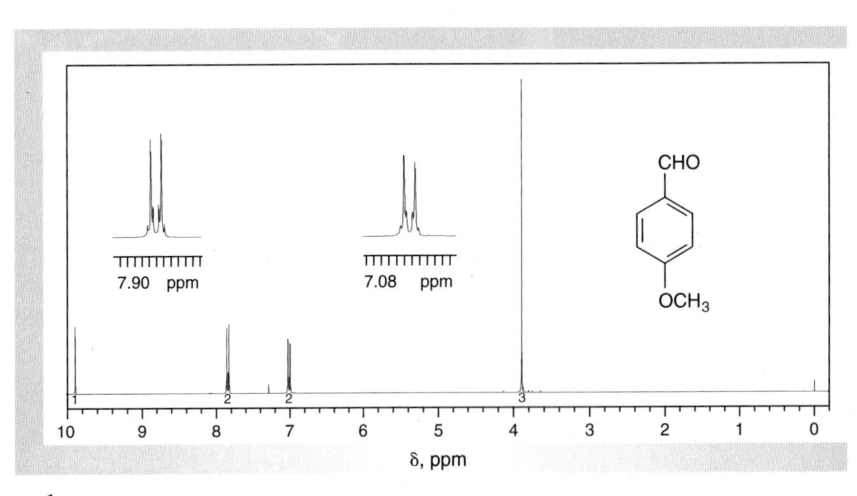

Figure 18.12
NMR data for p-anisaldehyde ($CDCl_3$).

(a) ^1H NMR spectrum (300 MHz).
(b) ^{13}C NMR data: δ 55.5, 114.5, 130.2, 131.9, 164.6, 190.5.

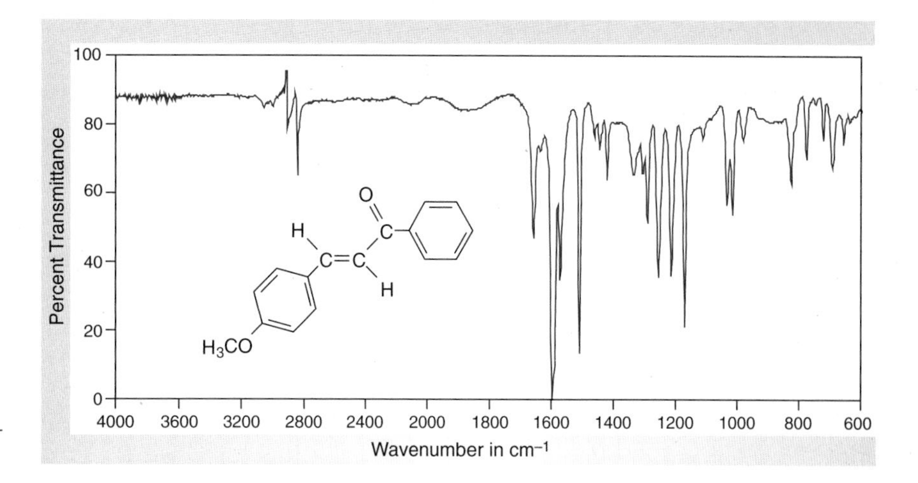

Figure 18.13
IR spectrum of trans-p-*anisalacetophenone (IR card).*

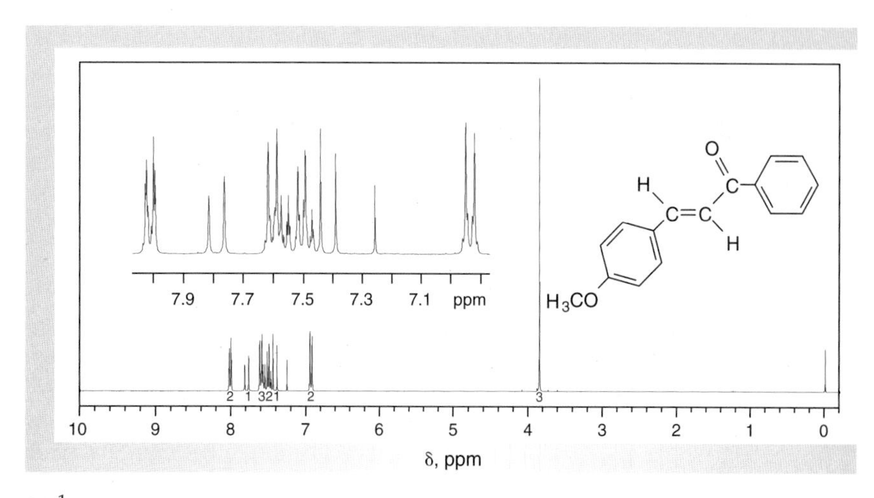

Figure 18.14
NMR data for trans-p-*anisalacetophenone* ($CDCl_3$).

(a) ^{1}H NMR spectrum (300 MHz).
(b) ^{13}C NMR data: δ 55.3, 114.4, 119.7, 127.6, 128.4, 128.5, 130.2, 132.5, 138.5, 144.6, 161.7, 190.4.

18.4 CONJUGATE ADDITION TO AN α, β-UNSATURATED KETONE

A carbon atom *alpha* to a carbonyl function may become nucleophilic either by forming an *enolate* (Eq. 18.3) or an *enol* (Eq. 18.4). In the crossed-aldol reaction described in the previous section (Scheme 18.4), the nucleophilic character of the carbon atom *alpha* to a ketone function was provided by enolate formation. However, in the experiment that follows, nucleophilicity at the α-carbon atom is promoted by enol formation. The overall reaction is the *acid-catalyzed* conversion of 2-methylpropanal (**26**) and 3-buten-2-one (**27**) into 4,4-dimethyl-2-cyclohexen-1-one (**29**), as shown in Equation 18.12. For the first stage of the reaction, the enol form of **26** serves as the nucleophile in the **1,4-** or **conjugate addition** to the α,β-unsaturated ketone **27** to give **28.** In the second stage, an enol is the nucleophile in the cyclization of **28** by an **intramolecular aldol condensation** that is acid-catalyzed. The conjugate addition to give **28** is an example of a

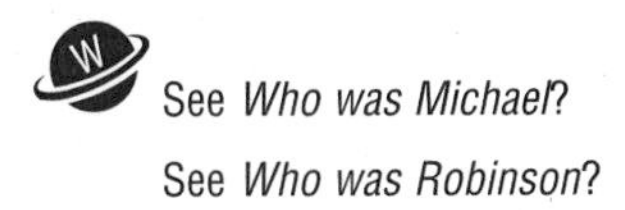
See *Who was Michael?*
See *Who was Robinson?*

Michael-type addition, whereas the combination of this addition followed by an aldol reaction to form the six-membered ring of **29** is called the **Robinson annulation.** The names honor the chemists who discovered the two types of reactions.

$$\underset{\textbf{26}}{H{-}C({=}O){-}CH(CH_3)_2} + \underset{\textbf{27}}{CH_3{-}C({=}O){-}CH{=}CH_2} \xrightarrow[\Delta]{H_3O^+} \underset{\textbf{28}}{OHC{-}C(CH_3)_2{-}CH_2CH_2{-}C({=}O){-}CH_3} \xrightarrow[\Delta\ (-H_2O)]{H_3O^+} \underset{\textbf{29}}{\text{4,4-dimethylcyclohex-2-en-1-one}} \qquad (18.12)$$

26 2-Methylpropanal (isobutyraldehyde); **27** 3-Buten-2-one (methyl vinyl ketone); **28** 2,2-Dimethylhexanal-5-one; **29** 4,4-Dimethyl-2-cyclohexen-1-one

A plausible overall mechanism for the reactions leading to **29** is shown in Scheme 18.5. 2-Methylpropanal (**26**) first undergoes acid-catalyzed tautomerization to its enol form **30,** in which the α-carbon atom is now nucleophilic. Protonation of the 3-buten-2-one (**27**) converts it into the delocalized cation **31,** which is more electrophilic than the unprotonated form. Because **30** is a weak nucleophile relative to an enolate ion, the formation of **31** facilitates the next stage of the reaction, which results in a new carbon-carbon bond between **26** and **27** to give the enol **32.** The boldfaced atoms in structure **32** show that the α-C–H bond of **26** has added in a conjugate-, or 1,4-, manner to **27**. Acid-catalyzed tautomerization of **32** leads to the thermodynamically more stable keto form **28.**

Although **28** is potentially isolable, it is not observed under our reaction conditions but rather undergoes an acid-catalyzed *intramolecular aldol addition.* This process is initiated by the acid-catalyzed tautomerization of **28** to the isomeric enol form **33.** Activating the remaining carbonyl function by protonation sets the stage for cyclization to give the β-hydroxycarbonyl compound **34** after deprotonation. The fact that **28** undergoes efficient cyclization rather than intermolecular condensation with the enolic form of 2-methylpropanal or other potential nucleophiles is mainly attributable to *entropic* considerations that favor *intra*molecular processes over *inter*molecular ones. In the final stage of the reaction, acid-catalyzed dehydration of **34** to give the desired **29** occurs; the mechanism of this reaction is left as an exercise at the end of this section.

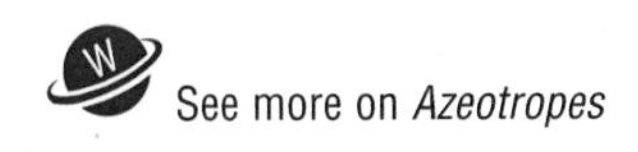
See more on *Azeotropes*

The dehydration of **34** to **29** is a reversible process that is driven to completion by removing the water from the reaction mixture. This is conveniently done by using toluene as the solvent for the reaction. Water and toluene form an **azeotrope** (Sec. 4.4), so azeotropic distillation allows the continuous separation of water as dehydration occurs. To minimize the amount of solvent that is required for distillations of this type, a Dean-Stark trap (Fig. 18.15) is commonly used. Because such traps are often not available in the undergraduate laboratory, an operational equivalent may be devised by assembling the apparatus in a way such that water, but not toluene, can be prevented from returning to the reaction flask (Fig. 18.16).

The choice of an aldehyde such as 2-methylpropanal (**26**) having only *one* α-hydrogen atom is important to the success of this crossed-aldol condensation because its dimerization is of no consequence. For example, aldehyde **26** can readily react with another molecule of itself to give **35** (Eq. 18.13), but this process is *reversible* under the reaction conditions. Moreover, unlike the aldol product **34, 35** is incapable of dehydrating to give an α,β-unsaturated product and water, which would drive the equilibrium leading to the dimerization product. It is left to you to consider the possibility of self-condensation of 3-buten-2-one (**27**) and of the reaction of enol **30** with the carbon atom of the protonated carbonyl function in **31** (see Exercises 12 and 13 at the end of this section).

Scheme 18.5

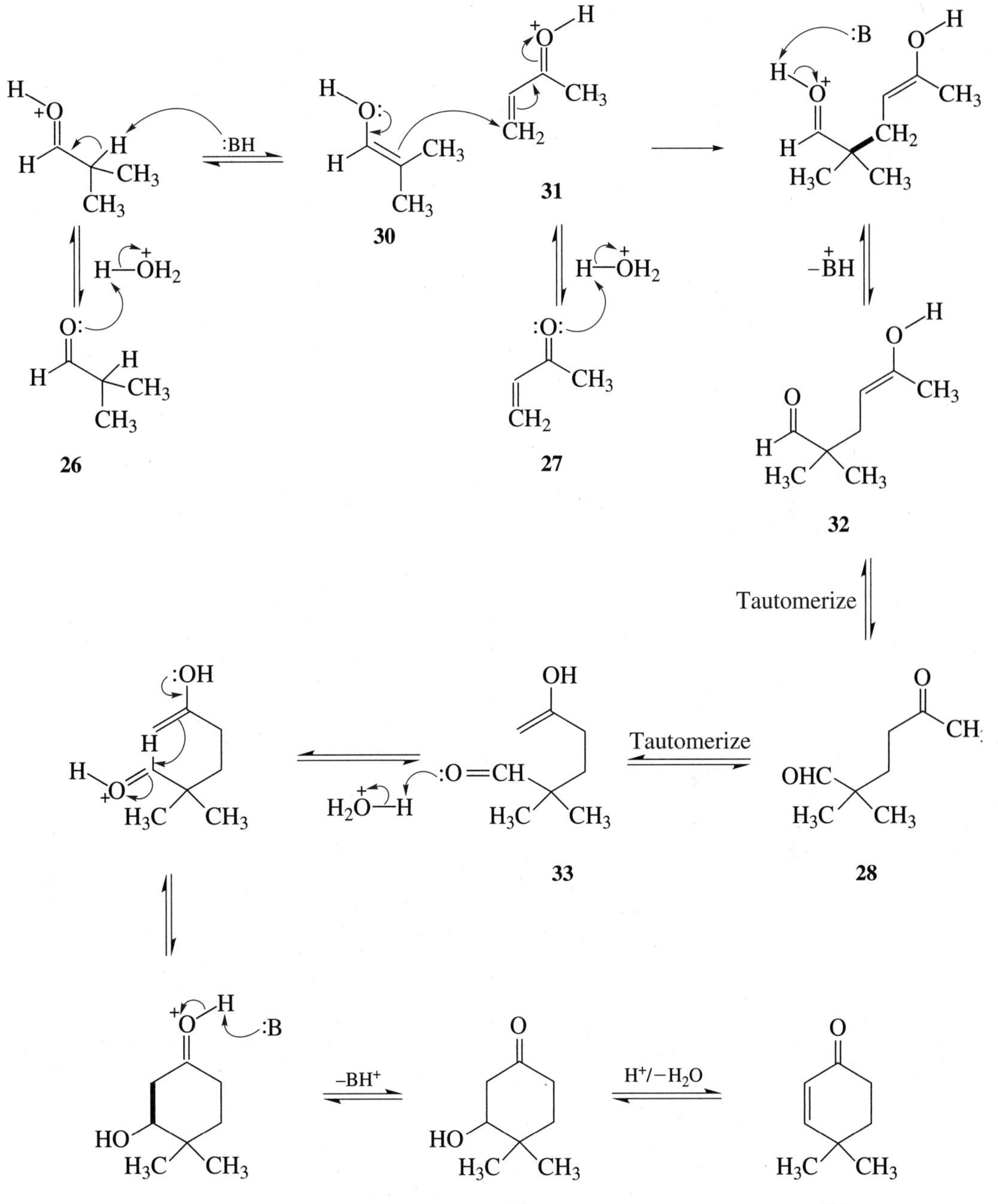

H
+O
H
H
CH3
CH3
:BH
H
O:
H
CH3
CH3
+O
H
CH3
CH2
30
31
:B
H
O
H
H
+O
H
CH2
H3C
CH3
CH3
H—OH2
H—OH2
−BH
O:
H
H
CH3
CH3
:O:
CH3
CH2
O
H
O
H
H3C
CH3
CH3
26
27
32
Tautomerize
:OH
H
H
O
H3C
CH3
H2O—H
:O=CH
OH
H3C
CH3
Tautomerize
O
CH
OHC
H3C
CH3
33
28
H
O
:B
HO
H3C
CH3
−BH+
O
HO
H3C
CH3
H+/−H2O
O
H3C
CH3
34
29

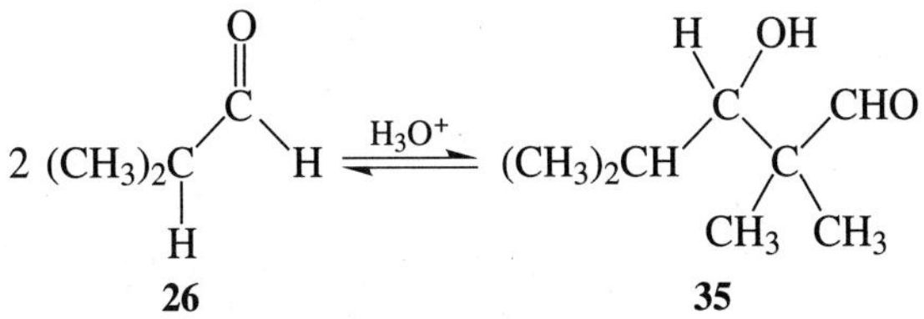

O
C
2 (CH3)2C
H
H
H3O+
H OH
C
CHO
(CH3)2CH
C
CH3 CH3

(18.13)

26 **35**

2,2,4-Trimethyl-3-hydroxypentanal

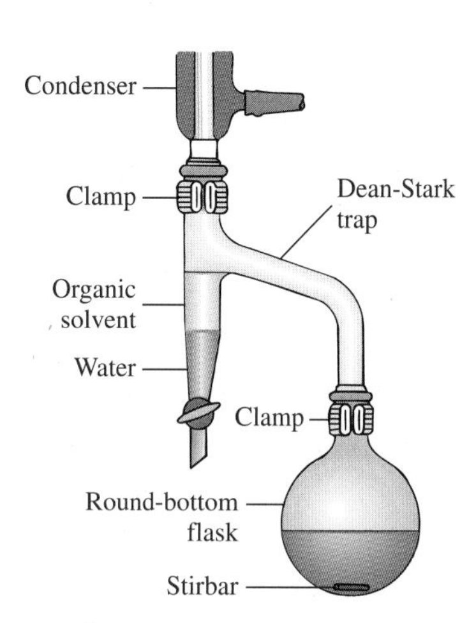

Figure 18.15
Dean-Stark trap for removal of water by azeotropic distillation.

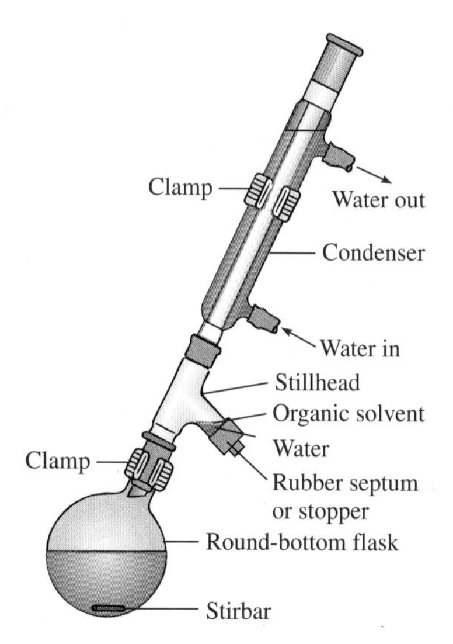

Figure 18.16
Apparatus for reaction of 2-methylpropanal with 3-buten-2-one.

EXPERIMENTAL PROCEDURES

Preparation of 4,4-Dimethyl-2-cyclohexen-1-one

Purpose To demonstrate the preparation of a cyclohexenone via sequential conjugate addition and an intramolecular aldol condensation.

SAFETY ALERT

1. **Because volatile and flammable solvents are used in this experiment, *flameless* heating should be used for heating the reaction under reflux and for distilling the product. However, if it is necessary to use a small microburner to distill the product, be sure that no one is using flammable solvents in the laboratory.**
2. **Be certain that all joints in the apparatus are properly lubricated and firmly mated in order to keep vapors from escaping.**
3. **2-Methylpropanal and 3-buten-2-one are highly volatile irritants with penetrating odors. Do *not* breathe their vapors. If possible, these chemicals should be handled in a hood and measured using syringes to minimize their release into the laboratory.**

MINISCALE PROCEDURE

Preparation Sign in at **www.cengage.com/login** to answer Pre-Lab Exercises, access videos, and read the MSDSs for the chemicals used or produced in this procedure. Review Sections 2.9, 2.11, 2.13, 2.21, 2.22, 2.29, and 6.4.

Apparatus A 125-mL and a 25-mL round-bottom flask, two 5-mL syringes, stillhead, reflux condenser, separatory funnel, microburner, apparatus for magnetic stirring, simple distillation, and *flameless* heating.

Setting Up Equip the 125-mL round-bottom flask with a stirbar and add 0.1 g of 2-naphthalenesulfonic acid and 25 mL of toluene. Fit the stillhead with a rubber septum or stopper firmly seated in the male joint that is normally used for connecting to a condenser. Secure the rubber septum or stopper to the joint with wire. Working at the hood, use two different syringes to measure 5.0 mL of 2-methylpropanal and 3.5 mL of 3-buten-2-one into an Erlenmeyer flask containing 10 mL of toluene. Transfer this solution quickly, but carefully to avoid spillage, to the round-bottom flask. Attach the stillhead to the flask and fit the stillhead with a reflux condenser. Finish assembling the apparatus shown in Figure 18.16 for heating under reflux with azeotropic removal of water. Tilt the flask at an angle of about 30° from the vertical so that the condensate will drip into the sidearm of the distillation head.

Robinson Annulation Bring the solution to a brisk reflux, and continue heating under reflux for 2–2.5 h. The reflux ring must rise above the sidearm of the stillhead so that water collects in the plugged sidearm as the reaction proceeds. Following the period of reflux, allow the reaction mixture to cool to room temperature. Carefully disconnect the apparatus, *making certain that the collected water is not spilled back into the reaction flask.* Pour the water and solvent contained in the sidearm into a small graduated cylinder to determine what percentage of the theoretical amount of water has been collected.★

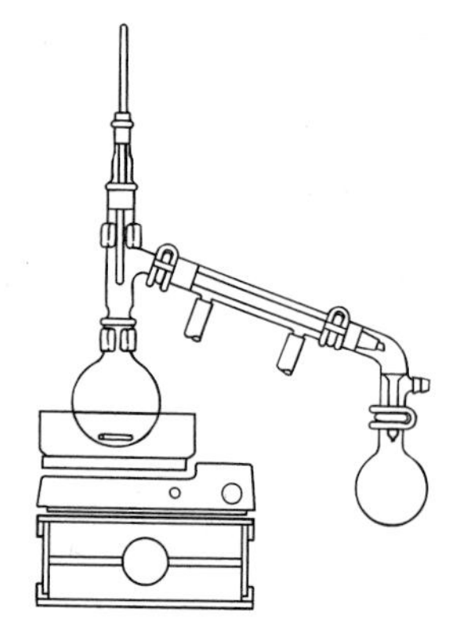

Work-Up and Isolation Transfer the reaction mixture to a separatory funnel, and wash it with a 20-mL portion of saturated aqueous sodium bicarbonate. Transfer the organic solution to a dry Erlenmeyer flask, and add several spatula-tips full of anhydrous sodium sulfate. Occasionally shake the mixture over a period of 10–15 min to facilitate drying; add further small portions of anhydrous sodium sulfate if the solution remains cloudy.★

Filter or decant the dry organic solution into a dry 125-mL round-bottom flask and equip the flask for simple distillation. Carefully remove the solvent and any unreacted starting materials by simple distillation; (*Caution: foaming*) collect a single fraction boiling up to 110–115 °C (760 torr). Alternatively, use rotary evaporation or other techniques to concentrate the solution. The final traces of solvent may be removed by attaching the flask to a vacuum source and gently swirling the contents as the vacuum is applied.★

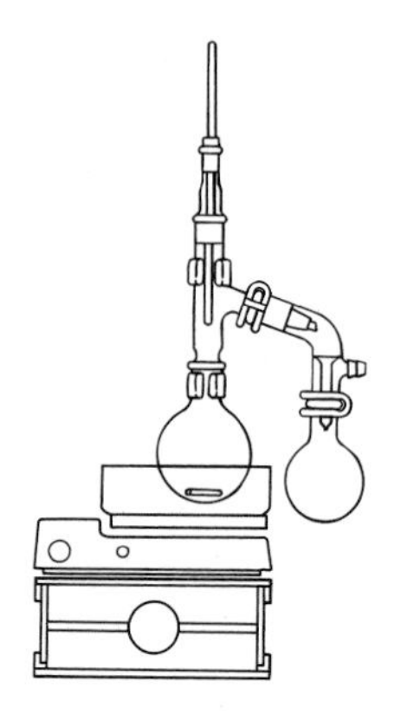

Purification Using a Pasteur pipet, transfer the *cooled* pot residue to a 25-mL round-bottom flask containing a stirbar. Rinse the original stillpot with 1–2 mL of diethyl ether and add the rinse to the 25-mL round-bottom flask. Remove the diethyl ether by attaching the flask to a vacuum source and gently swirling the contents as the vacuum is applied. Fit this flask for a shortpath distillation, insulating the top of the distillation flask and stillhead with cotton or glasswool wrapped with aluminum foil. Using a microburner, distill the 4,4-dimethyl-2-cyclohexen-1-one into a *tared* receiver. Collect it as a single fraction boiling above 130 °C (760 torr), and record the temperature range. Note that the head temperature may never reach the reported boiling point, owing to the high boiling temperature and the small amount of sample being distilled. Terminate heating when the dark pot residue becomes viscous and starts to evolve fumes.

Analysis Weigh the product and calculate the percent yield. The product may be contaminated with residual toluene, as illustrated in Figures 18.17 and 18.18, the amount of which will be dependent upon the efficiency with which toluene was removed in the first stage of the distillation. Determine the purity of your product by GLC or ^1H NMR spectroscopic analysis. Prepare a 2,4-dinitrophenylhydrazone derivative from a portion of your product according to the procedure in Section 25.7F; this derivative may be recrystallized from 95% ethanol. Obtain IR and ^1H NMR spectra of your starting materials and product, and compare them with those of authentic samples (Figs. 18.19–18.24).

MICROSCALE PROCEDURE

Preparation Sign in at **www.cengage.com/login** to answer Pre-Lab Exercises, access videos, and read the MSDSs for the chemicals used or produced in this procedure. Review Sections 2.9, 2.11, 2.13, 2.21, 2.22, 2.25, 2.29, and 6.4.

Apparatus A 5-mL conical vial, two 1-mL syringes, microburner, apparatus for magnetic stirring, simple distillation, and *flameless* heating.

Setting Up Equip the conical vial with a spinvane and add 10 mg of 2-naphthalenesulfonic acid and 2.5 mL of toluene. Working at the hood, use two different syringes to measure first 0.35 mL of 3-buten-2-one and then 0.50 mL of 2-methylpropanal directly into the conical vial. Set up the apparatus for simple distillation. Using a Pasteur pipet, fill the well of the Hickman stillhead with toluene, being careful not to allow any toluene to flow back into the conical vial.

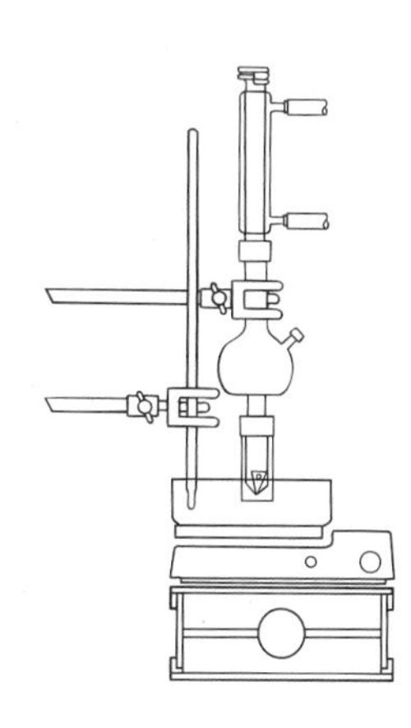

Robinson Annulation Bring the solution to a brisk reflux with stirring and continue heating under reflux for 2–2.5 h. To remove water from the reaction mixture, it is important that the reflux ring rise *into the bulb* of the Hickman stillhead. Allow the reaction mixture to cool to room temperature.★

Work-Up and Isolation Remove the spinvane and add 0.5 mL of saturated aqueous sodium bicarbonate to the conical vial. Cap the vial, invert it, and return it to its original upright position; release the pressure by *carefully* venting the screw-cap. Now replace the cap and shake the vial with intermittent venting to relieve pressure. Separate the layers and transfer the organic layer to a dry, screw-cap centrifuge tube. Add several microspatula-tips full of anhydrous sodium sulfate. Shake the centrifuge tube occasionally for a period of 10–15 min to facilitate drying; add further small portions of anhydrous sodium sulfate if the solution remains cloudy.★

Using a Pasteur pipet, transfer the organic solution to a dry 5-mL conical vial and equip the vial for simple distillation. Carefully remove the solvent by simple distillation (*Caution:* foaming); withdraw solvent from the Hickman stillhead as needed. Alternatively, use rotary evaporation or other techniques to concentrate the solution. The final traces of solvents may be removed by attaching the vial to a vacuum source and gently swirling the contents as the vacuum is applied.★

Purification (Optional) Remove the source of heat and dissemble the apparatus. Clean and dry the Hickman stillhead. Reassemble the apparatus for simple distillation, insulating the upper portion of the conical vial and the lower portion

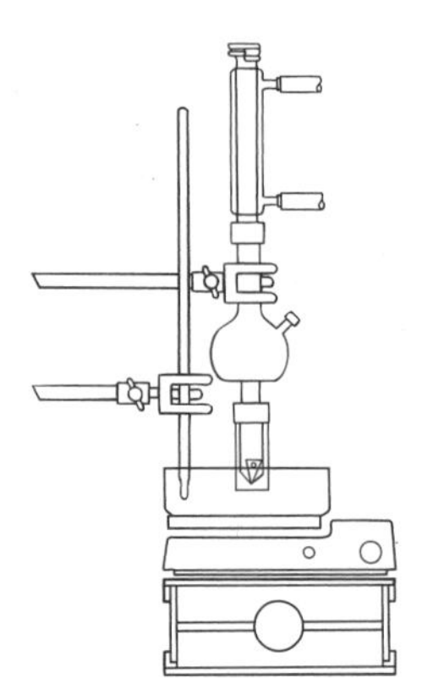

of the stillhead with cotton or glasswool wrapped with aluminum foil. Using a microburner, distil the 4,4-dimethyl-2-cyclohexen-1-one. Collect it as a single fraction. Terminate heating when the dark pot residue becomes viscous and starts to evolve fumes. Transfer the product to a tared sample vial.

Analysis Weigh the product and calculate the percent yield. The product may be contaminated with residual toluene, as illustrated in Figures 18.17 and 18.18, the amount of which will be dependent upon the efficiency with which toluene was removed. Determine the purity of your product by GLC or ^{1}H NMR spectroscopic analysis. Prepare a 2,4-dinitrophenylhydrazone derivative of a portion of your product according to the procedure in Section 25.7F; this derivative may be recrystallized from 95% ethanol. Obtain IR and ^{1}H NMR spectra of your starting materials and product, and compare them with those of authentic samples (Figs. 18.19–18.24).

WRAPPING IT UP

Place the *recovered toluene* that was obtained from the azeotropic distillation and from removal from the reaction mixture into the container for nonhalogenated organic liquids. Flush the *sodium bicarbonate wash* down the drain. Use a small amount of *recovered toluene* to rinse the *pot residue* from the simple distillation of the product into the container for nonhalogenated organic liquids. After residual solvent has evaporated from the *sodium sulfate* on a tray in the hood, place the drying agent in a container for nonhazardous waste.

EXERCISES

1. Although benzene could be used as the solvent for this experiment, toluene is the solvent of choice because it is not carcinogenic. Furthermore, the reaction appears to go faster in toluene than in benzene, as judged by the rate at which water is formed. How might you account for this rate acceleration?
2. What is the structure of 2-naphthalenesulfonic acid, and what function does it serve in the preparation of 4,4-dimethyl-2-cyclohexen-1-one?
3. Why is it important that water *not* return to the reaction vessel during the reflux period?
4. Why is the final product of this reaction washed with aqueous sodium bicarbonate?
5. **a.** Figure 18.23 is the GLC trace of a sample of product obtained in this experiment. Calculate the ratio of toluene to **29** that is present, assuming that the response factor of the detector (Sec. 6.4) is the same for the two compounds. Note that the GLC column used in this analysis separates substances on the basis of their relative boiling points.
   b. Figure 18.24 is the ^{1}H NMR spectrum of this same sample. Calculate the ratio of toluene to **29** that is present according to this method of analysis.
   c. Account for the discrepancy, if any, between the two ratios that you obtain.
   d. The distillate from which the samples for these analyses were drawn weighed a total of 2.8 g. Based on the ^{1}H NMR analysis of the mixture,

calculate how many grams of the desired **29** were present in the distillate. Calculate the yield of the reaction, assuming that it was run on the scale described in the Miniscale Procedure of this section.

6. Why is an enolate more nucleophilic than an enol?
7. Write two resonance structures for the α,β-unsaturated ketone **27**, and discuss how these structures show that both the carbonyl carbon atom and the β-carbon atom have electrophilic character.
8. Refer to Equation 18.12 and write the stepwise mechanism for the self-condensation of **26** to form **35**, using curved arrows to symbolize the flow of electrons.
9. Why does compound **35** *not* undergo dehydration?
10. Why is compound **31** less likely to form an enol than compound **26**?
11. Product **36** could potentially be produced by aldol condensation of **28** but is not. Provide a stepwise reaction mechanism for the possible formation of **36** in the presence of an acid catalyst. Explain why this alternative product is not observed.

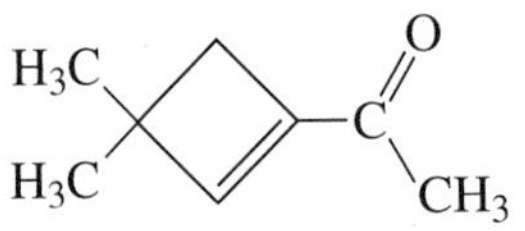

H_3C, H_3C, O, C, CH_3

36

12. Write a product of self-condensation of 3-buten-2-one (**27**), and explain why this process might be less favorable than the observed reaction between **26** and **27**.
13. Write the product expected to result from attack of enol **30** on the carbonyl carbon atom of **31**. Explain why formation of this product should be thermodynamically *less* favorable than formation of **28**.
14. Why is protonation of the carbonyl group in **33** likely to be involved in the conversion of **33** to **34**?
15. Using curved arrows to symbolize the flow of electrons, write a stepwise reaction mechanism for the acid-catalyzed dehydration of **34** to **29**.
16. Attempted acid-catalyzed reaction of propanal, CH_3CH_2CHO, with 3-buten-2-one (**27**) fails to give a good yield of the desired product, **37**. Explain why this might be.

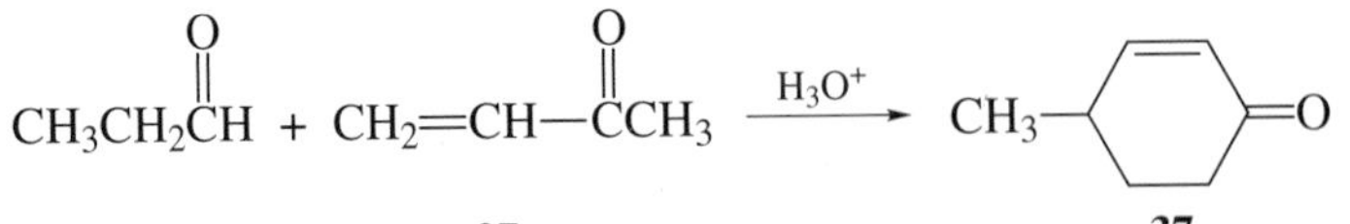

$$CH_3CH_2\overset{O}{\overset{\|}{C}}H + CH_2{=}CH{-}\overset{O}{\overset{\|}{C}}CH_3 \xrightarrow{H_3O^+} CH_3{-}\text{(cyclohexenone)}{=}O$$

Propanal — **27** — **37** 4-Methyl-2-cyclohexen-1-one

17. In the Robinson ring annulation reaction, the formation of five- and six-membered rings is favored over formation of rings having other sizes or bicyclic rings. Given this information, predict the product of the following Robinson ring annulation.

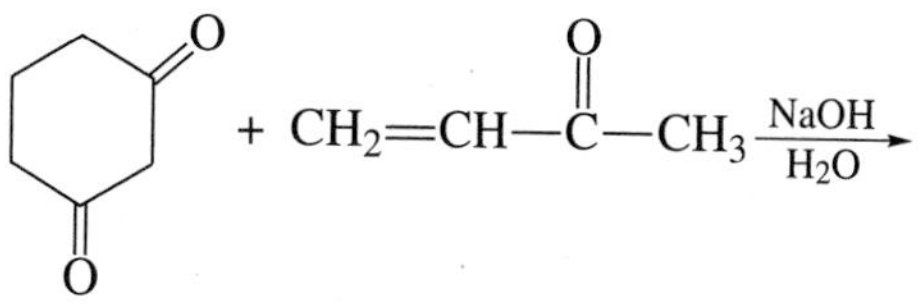

O=C1CCCC(=O)C1 + $CH_2{=}CH{-}\overset{O}{\overset{\|}{C}}{-}CH_3 \xrightarrow[H_2O]{NaOH}$

1,3-Cyclohexanedione 3-Buten-2-one

18. Outline a synthesis of the compound below via a Michael reaction using cyclopentanone as one of your starting materials.

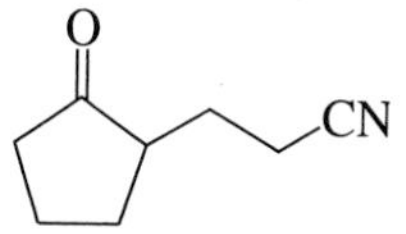

(2-oxocyclopentyl)–CH_2CH_2–CN

Sign in at **www.cengage.com/login** and use the spectra viewer and Tables 8.1–8.8 as needed to answer the blue-numbered questions on spectroscopy.

19. Consider the spectral data for 2-methylpropanal (Figs. 18.17 and 18.18).

a. In the functional group region of the IR spectrum, specify the absorptions associated with the carbonyl component and the carbon-hydrogen bond of the aldehyde group.

b. In the ^1H NMR spectrum, assign the various resonances to the hydrogen nuclei responsible for them.

c. For the ^{13}C NMR data, assign the various resonances to the carbon nuclei responsible for them.

20. Consider the spectral data for 3-buten-2-one (Figs. 18.19 and 18.20).

a. In the functional group region of the IR spectrum, specify the absorptions associated with the carbonyl group and the carbon-carbon double bond. Also specify the absorption in the fingerprint region that is characteristic for the terminal vinyl group.

b. In the ^1H NMR spectrum, assign the various resonances to the hydrogen nuclei responsible for them.

c. For the ^{13}C NMR data, assign the various resonances to the carbon nuclei responsible for them.

21. Consider the spectral data for 4,4-dimethyl-2-cyclohexen-1-one (Figs. 18.21 and 18.22).

a. In the functional group region of the IR spectrum, specify the absorptions associated with the carbonyl group and the carbon-carbon double bond.

b. In the ^1H NMR spectrum, assign the various resonances to the hydrogen nuclei responsible for them.

c. For the ^{13}C NMR data, assign the various resonances to the carbon nuclei responsible for them.

22. Discuss the differences observed in the IR and NMR spectra of 2-methylpropanal, 3-buten-2-one, and 4,4-dimethyl-2-cyclohexen-1-one that are consistent with the formation of the latter by the Robinson annulation in this experiment.

SPECTRA

Starting Materials and Products

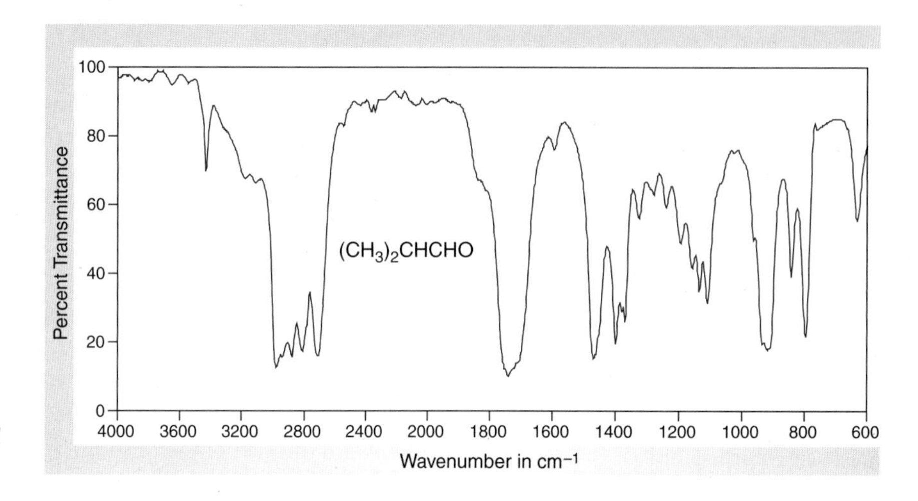

Figure 18.17
IR spectrum of 2-methylpropanal (neat).

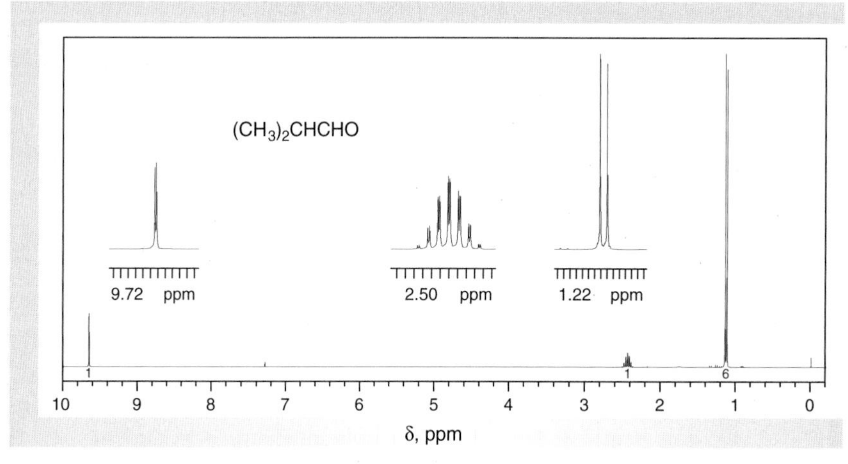

Figure 18.18
NMR data for 2-methylpropanal ($CDCl_3$).

(a) ^{1}H NMR spectrum (300 MHz).
(b) ^{1}C NMR data: δ 15.5, 41.2, 204.7.

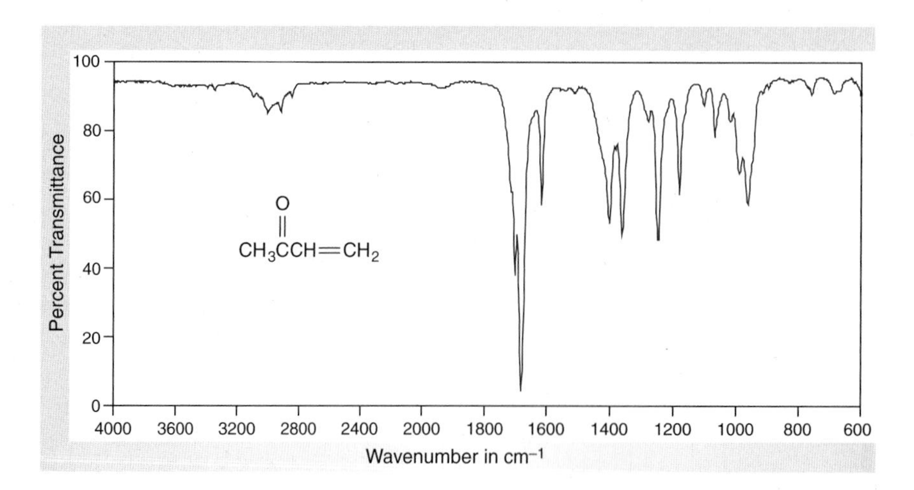

Figure 18.19
IR spectrum of 3-buten-2-one (neat).

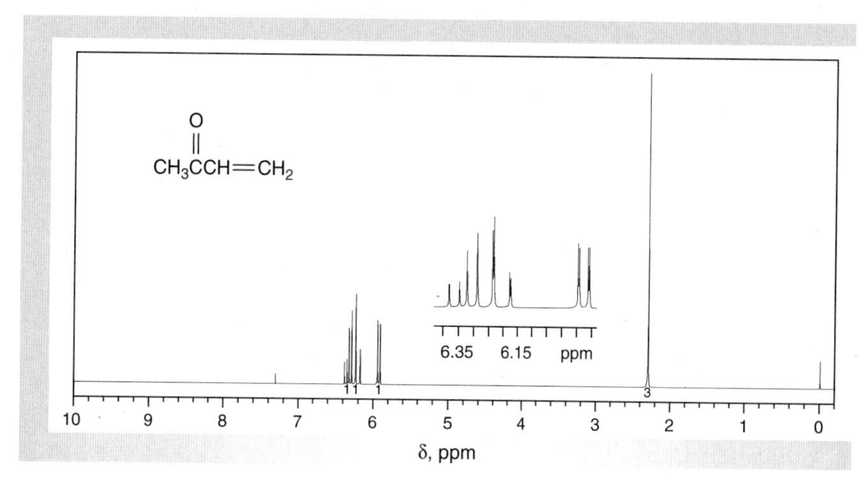

Figure 18.20
NMR data for 3-buten-2-one ($CDCl_3$).

(a) ^1H NMR spectrum (300 MHz).
(b) ^{13}C NMR data: δ 26.3, 129.0, 137.5, 198.8.

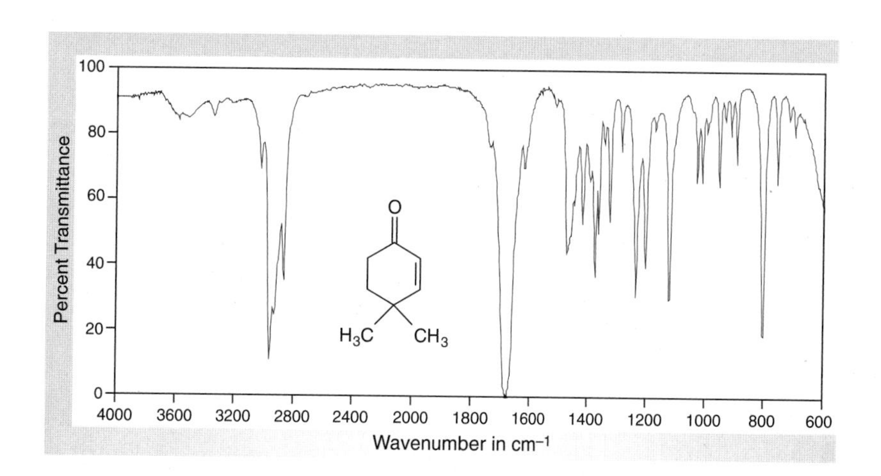

Figure 18.21
IR spectrum of 4,4-dimethyl-2-cyclohexen-1-one (neat).

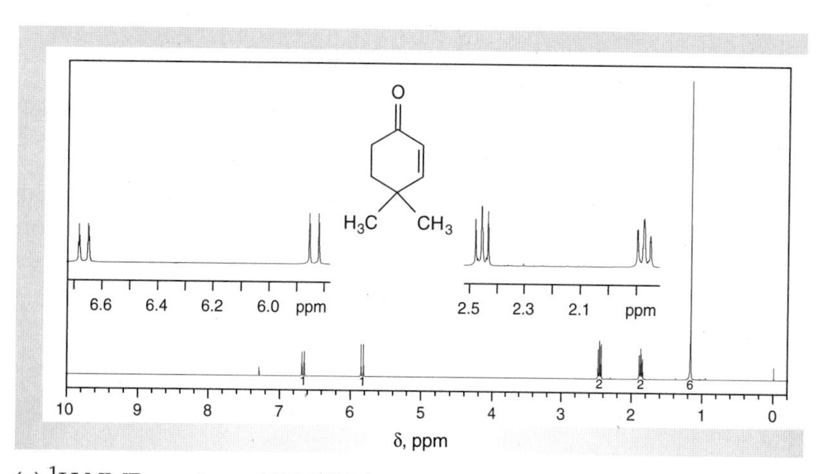

Figure 18.22
NMR data for 4,4-dimethyl-2-cyclohexen-1-one ($CDCl_3$).

(a) ^1H NMR spectrum (300 MHz).
(b) ^{13}C NMR data: δ 27.7, 32.8, 34.4, 36.1, 126.8, 159.6, 199.0.

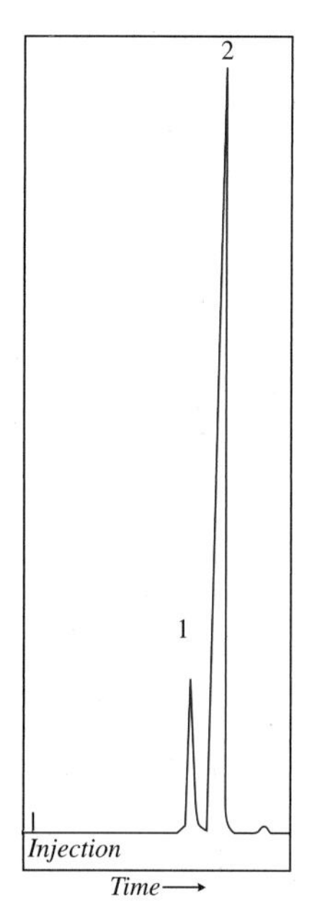

Figure 18.23
GLC analysis of reaction mixture for Exercise 5a. The retention times in minutes and the peak areas, respectively, are: Peak 1: 1.7, 10250; peak 2: 1.8, 37513.

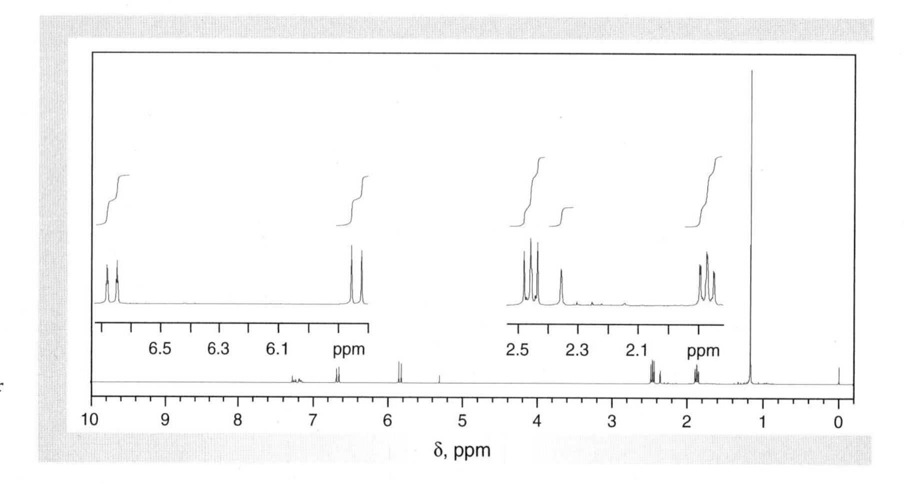

Figure 18.24
^1H NMR spectrum (300 MHz) of 4,4-dimethyl-2-cyclohexen-1-one and toluene for Exercise 5b.

HISTORICAL HIGHLIGHT

The Wittig Reaction

In the early 1950s, the synthesis of alkenes with complete control of the position of the carbon-carbon double bond was not possible. Mixtures of alkenes, sometimes having rearranged carbon frameworks, were often formed upon base-induced dehydrohalogenation of alkenes and acid-catalyzed dehydration of alcohols (Secs. 10.2 and 10.3). This situation was soon to change, however, because in 1953 Georg Wittig observed the unexpected formation of 1,1-diphenylethylene from the reaction of methylene-triphenylphosphorane with

(Continued)

HISTORICAL HIGHLIGHT *The Wittig Reaction (Continued)*

benzophenone. At the time, Wittig and Georg Geissler, one of his students, were studying the chemistry of pentavalent organophosphorus compounds, and they published this unexpected result as a minor part of a paper summarizing other aspects of the chemistry of these organophosphorus compounds. Because Wittig and Geissler were unaware of a related reaction that had been reported 30 years earlier by Staudinger, they were astonished by their result. Nevertheless, Wittig recognized the importance of the discovery. He and another student, Ullrich Schöllkopf, quickly published a second paper on the reactions of methylenetri phenylphosphorane with a variety of aldehydes and ketones to give alkenes under very mild conditions and with no isomerization. For the first time, it was possible to prepare alkenes with predictable structure, because the carbon-carbon double bond that was produced by the reaction *always* replaced the carbon-oxygen bond of the carbonyl-containing reagent.

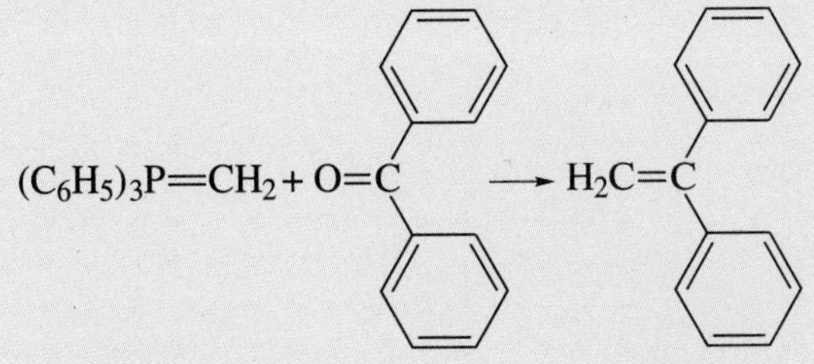

$(C_6H_5)_3P{=}CH_2 + O{=}C(C_6H_5)_2 \longrightarrow H_2C{=}C(C_6H_5)_2$

Methylenetriphenylphosphorane *An ylide* | Benzophenone | 1,1-Diphenylethylene

Soon the reaction was being widely used in research laboratories throughout the world, and today the Wittig reaction and variants thereof stand as some of the most important reactions in organic chemistry for the efficient synthesis of a diverse array of substituted alkenes. In research laboratories, the Wittig synthesis of alkenes is commonly applied to the preparation of small quantities of alkenes, but it may also be carried out on a large scale in industry, where it has been used to prepare tonnage quantities of vitamin A!

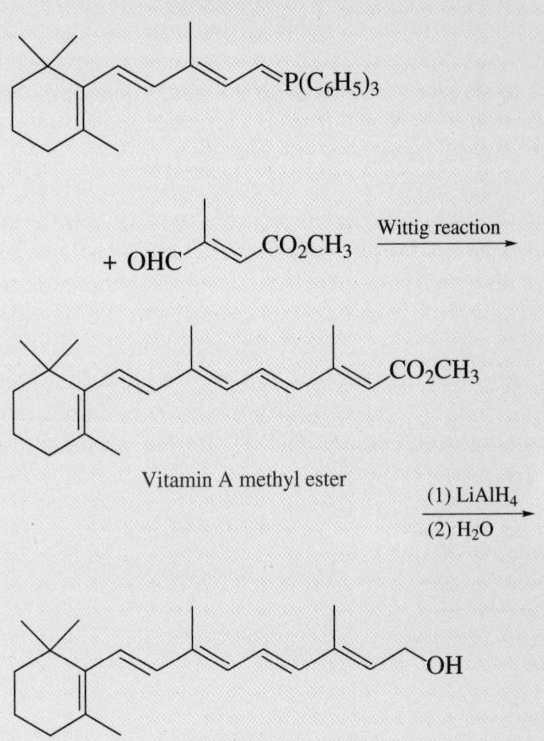

$=P(C_6H_5)_3$ + OHC ... CO_2CH_3 $\xrightarrow{\text{Wittig reaction}}$

... CO_2CH_3

Vitamin A methyl ester

$\xrightarrow[(2)\ H_2O]{(1)\ LiAlH_4}$

... OH

Vitamin A

Georg Friedrich Wittig was born in Berlin in 1897. Early in his life, he became keenly interested in science, music, and mountain climbing, and he pursued these passions for his entire life. He began his studies in chemistry at the University of Tübingen, but his education was interrupted by his military service during World War I. After the war, he finished his studies at the University of Marburg, where he also obtained his Ph.D. He remained at Marburg until 1932, when he went to the University of Braunschweig and then to the University of Freiburg. In 1944, he moved to Tübingen as professor and head of the Institute of Chemistry. Finally, in 1956, he moved to the University of Heidelberg as head of the department, where he became Professor Emeritus in 1967. He died in 1987 at the age of 90.

Wittig was involved in many areas of research, including the concept of ring strain, valence tautomerism, diradicals, the formation and reactions of

(Continued)

HISTORICAL HIGHLIGHT *The Wittig Reaction (Continued)*

dehydrobenzenes, carbanions, and organophosphorus chemistry. His interest in the latter two subjects gave rise to his work with the **ylides** that were formed upon deprotonation of alkyltriphenylphosphonium salts. In due course, this research resulted in his discovery of the Wittig synthesis of alkenes, work for which he shared the Nobel Prize in Chemistry in 1979 with Herbert Brown of Purdue University. His studies of carbanions also led to important discoveries with metalated derivatives of Schiff bases or imines, which are termed metalloenamines; these reactive intermediates have proven to be very useful in directed aldol reactions (Sec. 18.3) and for effecting the monoalkylations of aldehydes and ketones. But it is for his serendipitous discovery of the reaction that bears his name that Wittig is best remembered.

Relationship of Historical Highlight to Experiments

The Wittig synthesis of alkenes and variations of it, such as the Horner-Wadsworth-Emmons reaction, are now regarded as among the classic reactions of organic chemistry. Such reactions are of enormous importance because they produce alkenes of predictable structure under mild conditions. In performing some of the experiments in this chapter, you may be executing a functional group transformation that earned its discoverer a Nobel Prize in Chemistry (Sec. 18.2).

See *Who was Wittig?*

CHAPTER 19

Organometallic Chemistry

When you see this icon, sign in at this book's premium website at **www.cengage.com/login** to access videos, Pre-Lab Exercises, and other online resources.

It should by now be no surprise to you that carbon atoms are not the least bit "happy" about bearing a charge. This reluctance to form ions is seen in the high energies of carbocations, for example, which are not formed at all at primary carbon atoms and are possible for tertiary carbon atoms only if a highly polar solvent is present to assist in stabilizing the ion (Chap. 14). The same aversion to accommodate a charge attends the formation of unstabilized carbanions, that is, anions that are not delocalized through electronegative substitutents like carbonyl groups (Chap. 18). Nonetheless, it's possible to generate chemical species that behave as though they are unstabilized carbanions. Their creation and reactions are the subject of this chapter.

19.1 INTRODUCTION

See *Who was Grignard?*

See more on *Organometallic Compounds*

Organometallic compounds are substances that contain carbon-metal bonds, and they may be generally represented by the structure **1,** in which the metal, **M,** may be Li, Na, Mg, Cu, Hg, Pd, or other transition elements. Organomagnesium compounds **1,** M = MgBr, which are commonly called **Grignard reagents,** were the first organometallic substances to be extensively studied, and they are still among the most important. (See the Historical Highlight at the end of this chapter.) The polarization of the bond between the carbon atom and the electropositive metal in these reagents renders the carbon atom electron-rich, or Lewis basic, and the carbon atom bears a partial negative charge, δ^-, as shown. With this polarization in mind, you should not consider the metallated carbon atom as a true anion, because it retains some covalent bonding to its metallic partner, although the extent of this bonding remains a point of discussion among chemists.

One of the characteristic properties of organometallic reagents **1** is that the carbon atom serves as a **nucleophile** in chemical reactions. In contrast, when a carbon atom is bonded to more electronegative elements such as the halogen atom in the alkyl halide **2** (X = Cl, Br, or I) or oxygen atom in the carbonyl compound **3,** it is electron-deficient or Lewis acidic and possesses a partial positive charge. Such carbon atoms then serve as **electrophiles** in chemical reactions.

$R^1R^2R^3C^{\delta-}\text{—}M^{\delta+}$ | $R^1R^2R^3C^{\delta-}\text{—}X^{\delta+}$ | $R^1R^2C^{\delta+}\text{=}O^{\delta-}$

1
An organometallic reagent

2
An alkyl halide

3
A carbonyl compound

Because of their nucleophilic character, organometallic compounds are widely utilized as reagents in reactions that produce new carbon-carbon bonds. For example, two typical reactions of organometallic reagents **1** with carbon electrophiles such as alkyl halides **2** and carbonyl compounds **3** are illustrated by the general transformations shown in Equations 19.1 and 19.2. In each of these reactions, the nucleophilic carbon atom of one reactant becomes attached to the electrophilic carbon atom of the other reactant with the resulting formation of a new carbon-carbon bond. Thus, like many bond-forming processes, these reactions may be viewed in the simple context of *combinations of Lewis bases with Lewis acids.*

See *Who was Lewis?*

$$\underset{\text{An organometallic reagent}}{R^1\overset{\delta^-}{C}H_2{-}\overset{\delta^+}{M}} + \underset{\text{An alkyl halide}}{\overset{\delta^-}{X}{-}\overset{\delta^+}{C}H_2R^2} \longrightarrow \underset{\text{An alkane}}{R^1CH_2{-}CH_2R^2} + MX \qquad (19.1)$$

$$\underset{\text{An organometallic reagent}}{R^1\overset{\delta^-}{C}H_2{-}\overset{\delta^+}{M}} + \underset{\text{A carbonyl compound}}{R^2R^3\overset{\delta^+}{C}{=}\overset{\delta^-}{O}} \longrightarrow \underset{\text{A metal alkoxide}}{R^1CH_2{-}C(R^2)(R^3){-}O^-M^+} \xrightarrow{H_3O^+} \underset{\text{An alcohol}}{R^1CH_2{-}C(R^2)(R^3){-}OH} \qquad (19.2)$$

The following discussions focus on the preparation and reactions of one important class of organometallic compounds, the Grignard reagents **1** (M = Mg). However, many of the principles that are presented may be applied to the chemistry of other organometallic reagents. The experiments that you'll be performing involve the initial preparation of Grignard reagents from aryl and alkyl bromides, followed by their typical reactions with (1) an ester to produce a tertiary alcohol, (2) carbon dioxide to produce a carboxylic acid, and (3) an aldehyde to produce a secondary alcohol.

19.2 GRIGNARD REAGENTS: PREPARATION

Reactions of Organic Halides with Magnesium Metal

Grignard reagents, R–MgX or Ar–MgX, are typically prepared by the reaction of an alkyl halide, R–X, or an aryl halide, Ar–X, with magnesium metal in an *anhydrous* ethereal solvent (Eq. 19.3); the organometallic reagent dissolves as it is formed. You may note that carbon is transformed from an *electrophilic center* in the starting material R–X or Ar–X into a *nucleophilic center* in the product R–MgX or Ar–MgX in this process.

$$\underset{\text{An alkyl halide}}{\overset{\delta^+}{R}{-}\overset{\delta^-}{X}} \text{ or } \underset{\text{An aryl halide}}{\overset{\delta^+}{Ar}{-}\overset{\delta^-}{X}} + Mg^\circ \xrightarrow[\text{or THF (solvent)}]{\text{dry diethyl ether}} \underset{\text{An alkyl magnesium halide}}{\overset{\delta^-}{R}{-}\overset{\delta^+}{Mg}X} \text{ or } \underset{\text{An aryl magnesium halide}}{\overset{\delta^-}{Ar}{-}\overset{\delta^+}{Mg}X} \qquad (19.3)$$

Although it is customary to represent the Grignard reagent by the formula R–MgX or Ar–MgX, the structure of the organometallic species in solution is rather more complex. For example, with alkyl magnesium halides, there is an equilibrium between RMgX, R_2Mg, and MgX_2 that depends on the solvent, the halide ion, and the nature of the alkyl group. Moreover, the various organometallic species form aggregates in solution.

Grignard reagents are readily prepared from alkyl and aryl chlorides, bromides, and iodides but are rarely synthesized from organofluorides. The ease of formation of Grignard reagents from alkyl halides follows the order R–I > R–Br > R–Cl. Aryl halides are less reactive than their alkyl counterparts and aryl bromides and chlorides are comparable in reactivity.

In the experiments in this chapter, you will prepare one or both of the Grignard reagents derived from bromobenzene (**4**) and 1-bromobutane (**6**) in an ethereal solvent, according to Equations 19.4 and 19.5. The preparation of these reagents theoretically requires equivalent amounts of the organic halide and magnesium, but a slight excess of magnesium is normally used.

$$C_6H_5\text{–Br} + Mg \xrightarrow{\text{diethyl ether}} C_6H_5\text{–MgBr} \quad (19.4)$$

4 Bromobenzene → **5** Phenylmagnesium bromide

$$CH_3CH_2CH_2CH_2\text{—Br} + Mg \xrightarrow{\text{diethyl ether}} CH_3CH_2CH_2CH_2\text{—MgBr} \quad (19.5)$$

6 1-Bromobutane → **7** 1-Butylmagnesium bromide

Using an ethereal solvent is critical for the efficient preparation of the Grignard reagent because the basic oxygen atom of the ether complexes with the electropositive magnesium atom to help stabilize the organometallic species. The ethereal solvents most commonly used in this reaction are diethyl ether, $(C_2H_5)_2O$, and tetrahydrofuran, THF, $(CH_2)_4O$. Diethyl ether is often the solvent of choice because it is less expensive, may be purchased in anhydrous form, and is easily removed from the reaction mixture owing to its low boiling point (bp 36 °C, 760 torr). Tetrahydrofuran is a stronger Lewis base than diethyl ether and it also has better solvating ability; it may be used when the Grignard reagent does not form readily in diethyl ether.

You should be careful to use freshly opened containers of anhydrous diethyl ether or THF, because opening the container exposes its contents to atmospheric oxygen, which promotes the formation of **hydroperoxides.** For example, diethyl ether can be converted to the hydroperoxide, $CH_3CH(OOH)OCH_2CH_3$. Such peroxides are *explosive,* and large volumes of anhydrous ether or THF suspected to contain peroxides should *not* be evaporated to dryness. Moreover, emptied containers should be *thoroughly* rinsed with water to remove any peroxides before being discarded. Peroxides in anhydrous diethyl ether or THF may be conveniently destroyed by distillation from alkali metals such as sodium or potassium metal or from lithium aluminum hydride, $LiAlH_4$. The presence of peroxides in diethyl ether and THF may be detected by placing a drop of the solvent on a piece of *moistened* starch/iodide test paper. If the paper turns dark violet, owing to formation of a starch-iodine complex, then the ether contains peroxides.

The magnesium metal used for the preparation of the Grignard reagent is normally in the form of turnings—thin shavings that have a high surface area relative to chunks of the metal. This type of magnesium is generally suitable for the preparation of most Grignard reagents. However, if the turnings have been repeatedly exposed to atmospheric oxygen, their surface may be covered by a coating of magnesium oxide, which decreases their reactivity. Magnesium ribbon rather than turnings may also be used *after* the oxide coating on it has been scraped off with the edge of a spatula.

Initiating the reaction between the organic halide and magnesium may be difficult, especially with unreactive halides. In these cases, the reaction may often be initiated by adding a small crystal of iodine, I_2, to the reaction mixture. The iodine facilitates the reaction either by activating the metal through removal of some of its oxide coating or by converting a small amount of the R–X halide to the corresponding iodide, which is more reactive toward magnesium. Alternatively, the mixture can be placed in the bath of an ultrasonicator to initiate the reaction.

The formation of a Grignard reagent is an **exothermic** process. Since side reactions may occur if the reaction is allowed to proceed uncontrolled, it is important to regulate the rate at which the reaction proceeds. Grignard reactions often require an **induction period,** so it is *important* to be sure that the reaction of the alkyl or aryl halide with the magnesium has initiated *before* adding large quantities of the halide. Once the reaction is in progress, the halide should be added *dropwise and slowly* to the stirred suspension of the magnesium metal in the ethereal solvent; the halide may be added either neat or as a solution in the ethereal solvent. Adding the halide slowly keeps its concentration low, thus enabling better control of the rate of the reaction and the evolution of heat. The heat that is generated usually brings the solvent to its reflux temperature, so that the heat of the reaction is ultimately transferred to the cooling water in the condenser. Should it appear that the reaction is getting out of control, as evidenced by vapors escaping from the top of the condenser, the reaction mixture must be immediately cooled with an ice-water bath. Indeed, as a general rule, it is prudent always to have such a cooling bath prepared when performing an exothermic reaction.

Side Reactions

Side reactions may be encountered when forming Grignard reagents. For example, as the **conjugate bases** of the exceedingly weak organic acids R–H (R = aryl and alkyl), Grignard reagents are *very* strong bases that react rapidly with water according to Equation 19.6. This reaction results in the destruction of the Grignard reagent and the formation of the corresponding hydrocarbon, RH, and a basic magnesium salt. Grignard reagents also cannot be prepared when carboxyl (CO_2H), hydroxyl (OH), or amino (NH_2) groups are present in the alkyl or aryl halide; the acidic hydrogens of these functional groups will simply protonate the highly basic carbon atom of the Grignard reagent, thus destroying it as it forms.

$$R{-}MgX + H_2O \longrightarrow R{-}H + HO{-}Mg{-}X \qquad (19.6)$$

The reaction of Grignard reagents with water (Eq. 19.6) dictates that all reagents, solvents, and apparatus used for their preparation must be *thoroughly* dry. Consequently, *anhydrous* ethereal solvents that typically contain less than 0.01% water must be used. Anhydrous diethyl ether rapidly absorbs atmospheric moisture, so opening a container of it a number of times over a period of several days renders the diethyl ether unsuitable for use in preparing a Grignard reagent. Only freshly opened cans should be used, and the cans should always be tightly sealed *immediately* after the necessary volume of ether has been removed. These operations will also minimize the formation of dangerous hydroperoxides.

Other side reactions that may occur during formation of a Grignard reagent are shown in Equations 19.7–19.9, but these can be minimized by taking certain precautions. For example, reaction of the Grignard reagent with oxygen and carbon dioxide (Eqs. 19.7 and 19.8) is typically avoided in the research laboratory by performing the reaction under an inert atmosphere such as nitrogen (N_2) or argon. However, this precaution is not essential in the undergraduate laboratory, because when diethyl ether is used as the solvent, its very high vapor pressure effectively excludes most of the air from the reaction vessel.

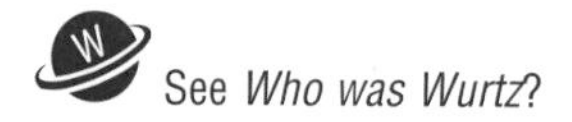
See *Who was Wurtz?*

The coupling reaction (Eq. 19.9) is an example of a Wurtz-type reaction. Although this can be a useful process for preparing symmetrical hydrocarbons, R–R, it is normally desirable to minimize this side reaction by using dilute solutions, thereby avoiding high localized concentrations of the halide. This is accomplished by efficient stirring and by slowly adding the halide to the suspension of magnesium in the ethereal solvent.

$$RMgX + O_2 \longrightarrow R{-}O{-}O^{-}\overset{+}{M}gX \tag{19.7}$$

$$RMgX + CO_2 \longrightarrow RCO_2^{-}\overset{+}{M}gX \tag{19.8}$$

$$RMgX + RX \longrightarrow R{-}R + MgX_2 \tag{19.9}$$

In the experiments that follow, phenylmagnesium bromide (**5**) and 1-butylmagnesium bromide (**7**) are prepared according to Equations 19.4 and 19.5, respectively. The most important side reaction in these experiments involves the Wurtz-type coupling of the Grignard reagent with the organic halide. For example, during the preparation of phenylmagnesium bromide, small quantities of biphenyl (**8**) are formed according to Equation 19.10; however, the presence of **8** does not interfere with the subsequent reactions of phenylmagnesium bromide. Although coupling also occurs during the preparation of 1-butylmagnesium bromide, the *n*-octane that is produced is volatile and easily removed.

$$C_6H_5MgBr\ (\mathbf{5}) + C_6H_5Br\ (\mathbf{4}) \longrightarrow C_6H_5{-}C_6H_5\ (\mathbf{8}) \tag{19.10}$$

5 Phenylmagnesium bromide; **4** Bromobenzene; **8** Biphenyl

EXPERIMENTAL PROCEDURES

Preparation of Grignard Reagents

Purpose To execute techniques required to prepare Grignard reagents from aryl and alkyl halides.

SAFETY ALERT

1. **Diethyl ether is *extremely* flammable and volatile, and its vapors can easily travel several feet along the bench top or the floor and then be ignited. Consequently, *be certain there are no open flames anywhere in the laboratory***

whenever you are working with ether. Use a *flameless* heating source whenever heating is required.

2. **The *anhydrous* diethyl ether used in this experiment is contained in metal cans, and the screw or plastic cap should *always* be in place when the can is not in use to prevent evaporation, absorption of atmospheric moisture and oxygen, and accidental fires.**
3. ***Open containers of diethyl ether must not be kept at your laboratory bench or stored in your laboratory drawer.* Estimate the total volume of ether you will need and measure it in the hood into a container that is *loosely* stoppered.**
4. **You should use ovens to dry your glassware if possible. However, if drying ovens are not available and it is necessary to dry the glass apparatus with a flame or a heat gun, be *certain* that no one in the laboratory is working with diethyl ether. *Consult with your instructor before using any open flame.* Avoid excessive heating in the vicinity of the ring seals in the condenser and near the stopcock in the addition funnel, particularly if the stopcock is made of plastic or Teflon.**
5. **Lubricate all ground-glass joints in the apparatus carefully and mate them tightly to prevent the escape of diethyl ether during the reaction.**
6. **On the small scale of these experiments, the exothermic formation and reaction of Grignard reagents rarely causes a problem. Nevertheless, it is still good laboratory practice for you to have an ice-water bath ready if the reaction proceeds too rapidly, as evidenced by an *excessively* rapid rate of reflux and the emission of vapors from the top of the condenser.**

MINISCALE PROCEDURE

Preparation Sign in at **www.cengage.com/login** to answer Pre-Lab Exercises, access videos, and read the MSDSs for the chemicals used or produced in this procedure. Review Sections 2.9, 2.10, 2.11, 2.22, 2.27, and 2.28.

Apparatus A 50-mL round-bottom flask, 5-mL syringe, separatory funnel with a ground-glass joint, condenser, Claisen adapter, drying tube, ice-water bath, and apparatus for magnetic stirring and *flameless* heating.

Setting Up Weigh 0.5 g of magnesium turnings that have been freshly crushed with a spatula into the round-bottom flask, and add a stirbar. Place this flask and its contents, the separatory funnel, condenser, Claisen adapter, and drying tube in an oven at 110 °C for at least 30 min. If the separatory funnel has a plastic or Teflon stopcock and stopper, do *not* put the stopcock, its plastic retaining nut, and the stopper in the oven, as they may melt or soften. Using gloves or tongs, remove the glassware from the oven and let it cool, preferably in a desiccator. After the glassware is cool enough to handle, lubricate all the joints and quickly assemble the apparatus shown in Figure 2.66b. Attach the drying tube to the top of the condenser and place the stopper and stopcock in the separatory funnel. Allow the apparatus to cool to room temperature.

Optional Measures If an oven is not available, it will be necessary to dry the apparatus with a microburner or a heat gun. Assemble the apparatus as described above. *Be sure that no one in the laboratory is working with diethyl ether,* and then dry the assembled apparatus. Do not overheat any plastic parts of the apparatus. Allow the apparatus to cool to room temperature.

Verify that there are no *flames* in the laboratory before continuing. Prepare a solution of *either* 2.4 mL of bromobenzene or 2.5 mL of 1-bromobutane in 5 mL of anhydrous diethyl ether in a small, *dry* Erlenmeyer flask. Swirl the solution to achieve homogeneity. Add 5 mL of *anhydrous* diethyl ether to the round-bottom flask *through* the separatory funnel; *close the stopcock*. Be sure that water is running through the condenser. Transfer the ethereal solution of halide to the separatory funnel.

Reaction Add a 0.5-mL portion of the ethereal solution from the separatory funnel onto the magnesium turnings and stir the resulting mixture. If small bubbles form at the surface of the magnesium turnings or if the mixture becomes slightly cloudy or chalky, the reaction has started. The flask should become *slightly* warm. If the reaction has started, *disregard* the optional instructions in the next paragraph.

Optional Measures If the reaction does not start spontaneously, warm the mixture gently for several minutes and observe whether the mixture becomes slightly cloudy or chalky. If it does not, then obtain one or two additional magnesium turnings and crush them *thoroughly* with a heavy spatula or the end of a clamp. Remove the separatory funnel just long enough to add these broken pieces of magnesium to the flask and quickly replace the funnel. The clean, unoxidized surfaces of magnesium that are exposed should aid in initiating the reaction. If the reaction still has not started after an additional 3–5 min of warming, consult your instructor. The best remedy at this point is to add a small crystal of iodine to the mixture. Alternatively, a small amount of the preformed Grignard reagent may be added if it is available.

Once the reaction has started, *gently* heat the reaction mixture so the solvent refluxes smoothly. Add another 5-mL portion of anhydrous diethyl ether to the reaction mixture through the top of the *condenser* and continue heating and stirring until the solvent is again refluxing. Add the remainder of the ethereal solution of the halide dropwise to the stirred reaction mixture at a rate that is just fast enough to maintain a gentle reflux. If the reaction becomes too vigorous, reduce the rate of addition and discontinue heating the flask if necessary. If the spontaneous boiling of the mixture slows, increase the rate of addition slightly. If the rate of reflux still does not increase, heat the mixture as necessary to maintain gentle reflux during the remainder of addition. *It is important that reflux be maintained throughout the addition of the ethereal solution.* The addition should take about 5–10 min. Upon completing the addition, continue heating the mixture under gentle reflux for 15 min. If necessary, add anhydrous diethyl ether to the reaction flask so that there is no less than about 15 mL of solution. At the end of the reaction, the solution normally has a tan to brown, chalky appearance, and most of the magnesium will have disappeared, although residual bits of metal usually remain. Discontinue heating and allow the mixture to cool to room temperature.

Use the Grignard reagent as soon as possible after preparing it. Phenylmagnesium bromide is used in Parts A and B of Section 19.4, and 1-butylmagnesium bromide is used in Part C of that section.

MICROSCALE PROCEDURE

Preparation Sign in at **www.cengage.com/login** to answer Pre-Lab Exercises, access videos, and read the MSDSs for the chemicals used or produced in this procedure. Review Sections 2.9, 2.10, 2.11, 2.22, 2.27, and 2.28.

Apparatus A 3-mL and 5-mL conical vial, screw-cap centrifuge tube, two 1-mL plastic (or glass) syringes, Pasteur pipet with 0.5- and 1.0-mL calibration marks, Claisen adapter, drying tube, ice-water bath, and apparatus for magnetic stirring and *flameless* heating.

Setting Up Dry the conical vials, Claisen adapter, centrifuge tube, drying tube, and calibrated Pasteur pipet in an oven at 110 °C for at least 30 min. Do *not* put any plastic connectors or rubber O-rings in the oven, as they may melt or soften. Using gloves or tongs, remove the glassware from the oven and let it cool, preferably in a desiccator. Lubricate the joints and assemble the apparatus shown in Figure 19.1 by adding a spinvane to the 5-mL conical vial and then fitting the Claisen adapter to the vial. Finally, place a rubber septum and the drying tube on the Claisen adapter.

Optional Measures If an oven is not available, it will be necessary to dry the apparatus with a microburner or a heat gun. Assemble the apparatus as described above. *Be sure that no one in the laboratory is working with diethyl ether,* and then dry the assembled apparatus. Be careful heating the conical vial, as this heavy-walled vessel can easily crack from thermal shock. Also, do not overheat any plastic parts of the apparatus. Allow the apparatus to cool to room temperature.

Verify that there are no *flames* in the laboratory before continuing. Transfer about 3–4 mL of anhydrous diethyl ether to a screw-cap centrifuge tube and cap the tube. Use this during the remainder of the experiment whenever anhydrous diethyl ether is required. Weigh 0.05 g of magnesium turnings that have been freshly crushed with a heavy spatula or the end of a clamp. Remove the rubber septum and transfer the turnings to the reaction vial. Add one *small* crystal of iodine to the vial and replace the rubber septum. Add 0.2 mL of anhydrous diethyl ether to the vial using a dry 1-mL syringe inserted through the rubber septum. Prepare a solution of 0.24 mL of bromobenzene in 0.5 mL of anhydrous diethyl ether in a dry 3-mL conical vial. Swirl the solution to achieve homogeneity.

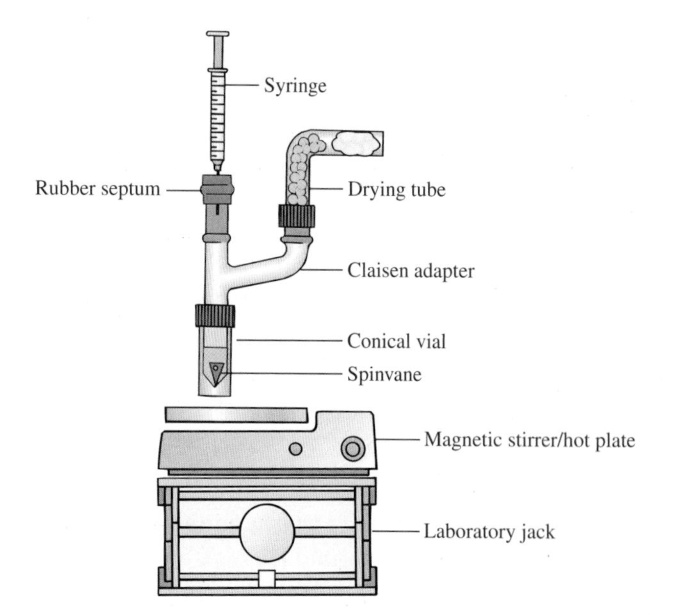

Figure 19.1
Microscale apparatus for preparing and reaction of Grignard reagents.

Reaction Stir the contents of the conical vial while warming them gently. Draw the ethereal solution of bromobenzene into the plastic syringe and cap the conical vial containing the bromobenzene solution. Insert the syringe needle through the rubber septum, and add a 0.1-mL portion of this solution onto the magnesium turnings; stir the mixture. *Hold* the plunger of the syringe to control the amount of solution added. If small bubbles form at the surface of the magnesium turnings or if the mixture becomes slightly cloudy or chalky, the reaction has started.

Optional Measures If the reaction does not start spontaneously, consult your instructor. The best remedy at this point is to continue warming the vial and add an additional crystal of iodine to the mixture. You may also add a small sample of phenylmagnesium bromide from a stock solution if it is available.

Once the reaction has started, continue heating the reaction mixture gently so the solvent refluxes slightly. Using a second plastic syringe inserted through the rubber septum, add another 0.5-mL portion of anhydrous diethyl ether to the reaction mixture and continue heating and stirring until the solvent is again refluxing. Add the remainder of the bromobenzene-ether solution *dropwise* to the stirred reaction mixture at a rate that is just fast enough to maintain a gentle reflux. If the reaction becomes too vigorous, reduce the rate of adding the ethereal solution of aryl halide and, if necessary, discontinue heating the vial. If the spontaneous boiling of the mixture slows, increase the rate of addition slightly. If the rate of reflux still does not increase, increase heating and maintain gentle reflux during the remainder of addition. *It is important that reflux be maintained throughout the addition of the bromobenzene-ether solution.* The addition should take about 3–5 min. Upon completion of the addition, place about 0.5 mL of anhydrous diethyl ether in the vial that contained the bromobenzene solution. Draw this solution into a syringe and add it in one portion to the reaction mixture. Continue heating the mixture with stirring under gentle reflux for 15 min. If necessary, add anhydrous diethyl ether to maintain a level *no lower* than the 2-mL mark on the conical vial. At the end of the reaction, the solution normally has a tan to brown, chalky appearance, and most of the magnesium will have disappeared, although residual bits of metal usually remain. Discontinue heating and allow the mixture to cool to room temperature.

Use the Grignard reagent as soon as possible after preparing it, following one of the procedures given in Section 19.4.

EXERCISES

1. Answer the following questions for a Grignard reagent, R_3C–MgX.
   a. Indicate the direction of the dipole in the C–Mg bond.
   b. Is the carbon atom associated with magnesium electrophilic or nucleophilic?
   c. Write the conjugate acid of a Grignard reagent, and explain why Grignard reagents are strong bases.
2. In the reaction of bromobenzene with magnesium to form phenylmagnesium bromide, C_6H_5MgBr, which reagent undergoes oxidation and which undergoes reduction? Write the half-reactions for the two processes.
3. Why should the glassware not be flame-dried in the vicinity of diethyl ether, the solvent used for preparing the Grignard reagent?

4. Why were you cautioned not to heat excessively in the vicinity of plastic parts, rubber O-rings, and the ring seals of the condenser when *flame*-drying the apparatus for this experiment?
5. Ethanol is often present in solvent-grade diethyl ether. If this grade rather than anhydrous were used, what effect would the ethanol have on the formation of the Grignard reagent?
6. Why is it necessary to lubricate the joints of the apparatus used to prepare a Grignard reagent?
7. What is the purpose of adding an additional portion of diethyl ether after the Grignard reaction is initiated and before the addition of halide is continued?
8. Why is it important to add the ethereal solution of the halide *slowly* to the magnesium turnings?
9. Why is it possible to use solvent-grade diethyl ether in the work-up procedure, whereas the more expensive *anhydrous* form must be used to prepare the Grignard reagent?
10. Explain how the undesired coupling product **8** is formed and how its formation can be minimized.
11. Why is it unwise to allow the solution of the Grignard reagent to remain exposed to air, even if it is protected from moisture by drying tubes?
12. Give a three-dimensional structure for the complex $RMgBr \cdot 2(C_2H_5)_2O$. What is the nature of the bonding between the diethyl ether molecules and the Grignard reagent?
13. The intermediates in the mechanism for the conversion of an aryl halide into a Grignard reagent are shown below. Use curved arrows to symbolize the flow of electrons in the three steps given.

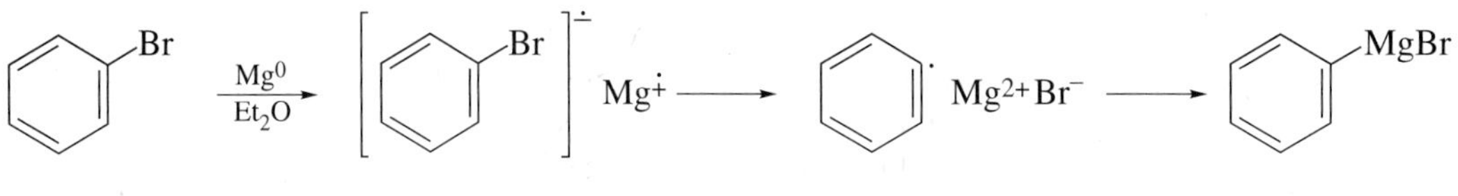

$$C_6H_5Br \xrightarrow[Et_2O]{Mg^0} [C_6H_5Br]^{\dot{-}}\ Mg^{\dot{+}} \longrightarrow C_6H_5^{\cdot}\ Mg^{2+}Br^- \longrightarrow C_6H_5MgBr$$

14. What problems might be encountered if bromocyclohexane were used as an additive to help initiate the formation of phenylmagnesium bromide and 1-butylmagnesium bromide? How does the use of 1,2-dibromoethane avoid such experimental difficulties?
15. The reaction of 5-bromo-1-pentanol with magnesium in anhydrous ether did not give the expected Grignard reagent, 5-hydroxy-1-pentylmagnesium bromide. Provide an explanation for the failure of this reaction. What organic product is formed instead?

$$H{-}OCH_2(CH_2)_3CH_2{-}Br + Mg^0 \xrightarrow[\text{dry } Et_2O]{} H{-}OCH_2(CH_2)_3CH_2{-}MgBr$$

5-Bromo-1-pentanol → 5-Hydroxy-1-pentylmagnesium bromide

Sign in at **www.cengage.com/login** and use the spectra viewer and Tables 8.1–8.8 as needed to answer the blue-numbered questions on spectroscopy.

16. Consider the spectral data for bromobenzene (Figs. 15.19 and 15.20).
    a. In the functional group region of the IR spectrum, specify the absorptions associated with the aromatic ring.

b. In the ^{1}H NMR spectrum, assign the various resonances to the hydrogen nuclei responsible for them.

c. For the ^{13}C NMR data, assign the various resonances to the carbon nuclei responsible for them.

17. Consider the NMR spectral data for 1-bromobutane (Fig. 14.4).

a. In the ^{1}H NMR spectrum, assign the various resonances to the hydrogen nuclei responsible for them.

b. For the ^{13}C NMR data, assign the various resonances to the carbon nuclei responsible for them.

SPECTRA

Starting Materials and Products

The IR and NMR spectra of 1-bromobutane and bromobenzene are given in Figures 14.3, 14.4, 15.19, and 15.20, respectively.

19.3 GRIGNARD REAGENTS: REACTIONS

Reaction of Phenylmagnesium Bromide with Methyl Benzoate

The aryl carbon atom bearing the magnesium in phenylmagnesium bromide is nucleophilic and reacts readily with electrophiles (Sec. 19.1). In Part A of the experimental procedures that follow, two equivalents of phenylmagnesium bromide (**5**, Eq. 19.4) are allowed to react with methyl benzoate (**9**), an ester, in an exothermic reaction to give triphenylmethanol (**11**) via the sequence of reactions depicted in Scheme 19.1. In the first step of the reaction, nucleophilic attack of phenylmagnesium bromide on the electrophilic carbon atom of the ester group of **9** leads to benzophenone (**10**) as an intermediate. Subsequent reaction of **10** with an additional mole of **5** then produces an alkoxide salt, which is converted to **11** upon protonation with acid. The reaction is worked up with acid rather than water alone to avoid precipitation of the basic magnesium salt, HOMgX.

Scheme 19.1

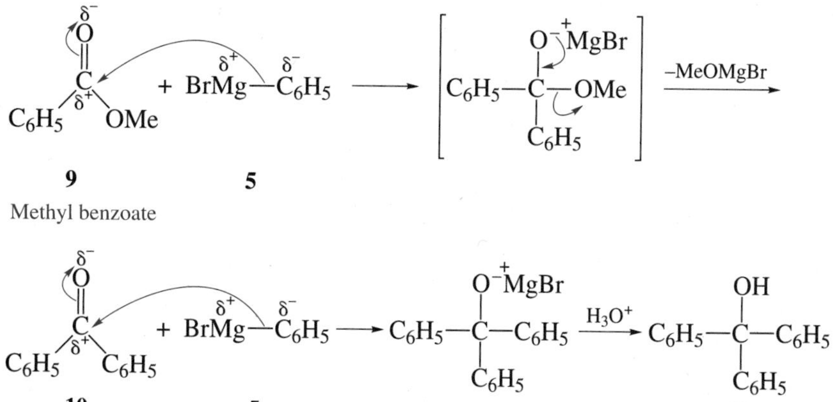

$$C_6H_5C(=O^{\delta-})^{\delta+}OMe + {}^{\delta+}BrMg{-}C_6H_5^{\delta-} \longrightarrow \left[C_6H_5{-}C(O^-\,{}^+MgBr)(C_6H_5){-}OMe \right] \xrightarrow{-MeOMgBr}$$

9 Methyl benzoate; **5**

$$C_6H_5C(=O^{\delta-})^{\delta+}C_6H_5 + {}^{\delta+}BrMg{-}C_6H_5^{\delta-} \longrightarrow C_6H_5{-}C(O^-\,{}^+MgBr)(C_6H_5){-}C_6H_5 \xrightarrow{H_3O^+} C_6H_5{-}C(OH)(C_6H_5){-}C_6H_5$$

10 Benzophenone; **5**; **11** Triphenylmethanol

The principal organic products present after the aqueous work-up are the desired triphenylmethanol (**11**), benzene from any unreacted **5**, and biphenyl (**8**). Fortunately, it is possible to separate **11** and **8** easily owing to their relative solubilities in nonpolar hydrocarbon solvents. Biphenyl is considerably more soluble in cyclohexane than is triphenylmethanol, so recrystallization of the crude product mixture gives pure triphenylmethanol.

Reaction of Phenylmagnesium Bromide with Carbon Dioxide

Nucleophilic addition of phenylmagnesium bromide (**5**, Eq. 19.4) to carbon dioxide produces an intermediate carboxylate salt that may be converted into benzoic acid (**12**) by acidification of the reaction mixture (Eq. 19.11). This transformation, which is the subject of Part B of the following experimental procedures, may be accompanied by side reactions that lead to the formation of benzophenone (**10**) and/or triphenylmethanol (**11**), as shown in Equation 19.12. However, formation of these by-products can be minimized by controlling the reaction conditions. Thus, the bromomagnesium salt of benzoic acid is only slightly soluble in diethyl ether, so the salt precipitates from solution and cannot undergo further reaction when the reaction is performed in this solvent. A large excess of carbon dioxide is also used to increase the likelihood that **5** will react with carbon dioxide rather than with the magnesium salt of benzoic acid. If the phenylmagnesium bromide is added slowly *to* the dry ice, carbon dioxide is always present in excess, thus favoring the desired reaction. Finally, when dry ice is used as the source of carbon dioxide, the temperature of the reaction is maintained at −78 °C until all of the Grignard reagent has been consumed. At this low temperature, the reaction of phenylmagnesium bromide with the bromomagnesium salt of benzoic acid is slow.

$$\overset{\delta^-}{O}{=}\overset{\delta^+}{C}{=}O + \overset{\delta^+}{BrMg}{-}\overset{\delta^-}{C_6H_5} \longrightarrow C_6H_5C(=O)O^-\,\overset{+}{M}gBr \xrightarrow{H_3O^+} C_6H_5C(=O)OH \tag{19.11}$$

5 — Bromomagnesium benzoate — **12** Benzoic acid

$$C_6H_5C(=O)O^-\,\overset{+}{M}gBr \xrightarrow{C_6H_5-MgBr} C_6H_5C(=O)C_6H_5 \xrightarrow{C_6H_5-MgBr} (C_6H_5)_3C{-}O^-\,\overset{+}{M}gBr \xrightarrow{H_3O^+} (C_6H_5)_3C{-}OH \tag{19.12}$$

10 — **11**

After the addition of the Grignard reagent to carbon dioxide is complete, the excess dry ice is allowed to evaporate, and the mixture is acidified. Extraction of the aqueous mixture with diethyl ether then gives a solution containing benzoic acid, benzophenone, triphenylmethanol, benzene, and biphenyl. The benzoic acid is readily separated from the neutral side products in the crude mixture by extracting the solution with dilute aqueous sodium hydroxide, whereby benzoic acid is converted to its water-soluble sodium salt; the neutral by-products remain in the organic layer. Acidification of the aqueous layer regenerates the benzoic acid, which can be purified by recrystallization.

Reaction of 1-Butylmagnesium Bromide with 2-Methylpropanal

The reaction of 1-butylmagnesium bromide (**7**, Eq. 19.5) with 2-methylpropanal (**13**) provides an excellent example of the Grignard synthesis of secondary alcohols (Eq. 19.13). In this reaction, the Grignard reagent **7** adds to 2-methylpropanal to give the magnesium salt of 2-methyl-3-heptanol, and the alcohol **14** may be isolated after acidification of the mixture. This transformation is found in Part C of the following experimental procedures.

$$\underset{\substack{\textbf{7}\\ \text{1-Butylmagnesium bromide}}}{CH_3CH_2CH_2\overset{\delta^-}{C}H_2-\overset{\delta^+}{Mg}Br} + \underset{\substack{\textbf{13}\\ \text{2-Methylpropanal}}}{H-\overset{\delta^+}{\underset{}{C}}(=\overset{\delta^-}{O})-CH(Me)Me} \longrightarrow \tag{19.13}$$

$$CH_3(CH_2)_3-CH(O^-\overset{+}{Mg}Br)-CH(CH_3)CH_3 \xrightarrow{H_3O^+} \underset{\substack{\textbf{14}\\ \text{2-Methyl-3-heptanol}}}{CH_3(CH_2)_3-CH(OH)-CH(CH_3)CH_3}$$

19.4 SPECIAL EXPERIMENTAL TECHNIQUES

Some of the side reactions that are encountered during the preparation of Grignard reagents and during their reaction with electrophiles can be controlled by prudent choice of solvent and reaction temperature. Another experimental technique that can be used to minimize undesired side reactions involves the **order of addition** of the Grignard reagent and the electrophilic reactant. For example, the electrophiles methyl benzoate and 2-methylpropanal are added *to* a solution of the appropriate Grignard reagent, which is a highly reactive nucleophile, to prepare triphenylmethanol (**11**) and 2-methyl-3-heptanol (**14**), respectively. This mode of addition is referred to as **normal addition.** On the other hand, in the preparation of benzoic acid, the nucleophilic Grignard reagent is added *to* the carbon dioxide, the electrophile. This procedure is called **inverse addition.**

Whether an addition is performed according to the *normal* or *inverse* mode is dictated by the specific nature of the reactants and the potential side reactions that may occur. For example, if carbon dioxide gas were bubbled into a solution of phenylmagnesium bromide (**5**) to prepare benzoic acid (normal addition), the by-products **10** and/or **11** (Eq. 19.12) would become significant, perhaps predominant. There would always be an excess of the Grignard reagent **5** present relative to carbon dioxide, so further reaction between **5** and the bromomagnesium salt of benzoic acid would be favored. In the preparation of triphenylmethanol (**11**), the combination of the Grignard reagent with methyl benzoate could be done equally well in either the normal or inverse fashion because two equivalents of **5** are necessary to complete the reaction and the magnesium alkoxide salt of **11** is unreactive toward **5.** Inverse addition is less convenient in this case, however, because it requires transferring the Grignard reagent **5** from one flask to a dropping funnel and then adding it to the reaction flask containing the methyl benzoate. Similarly, the reaction of 1-butylmagnesium bromide with 2-methylpropanal (Eq. 19.13) may be performed using either normal or inverse addition, but normal addition is again more convenient.

EXPERIMENTAL PROCEDURES

Reactions of Grignard Reagents

A ■ *Preparation of Triphenylmethanol*

Purpose To demonstrate the preparation of a tertiary alcohol by the reaction of a Grignard reagent with an ester.

SAFETY ALERT

Review the Safety Alert for Preparation of Grignard Reagents (Sec. 19.2).

MINISCALE PROCEDURE

Preparation Sign in at **www.cengage.com/login** to answer Pre-Lab Exercises, access videos, and read the MSDSs for the chemicals used or produced in this procedure. Review Sections 2.10, 2.11, 2.13, 2.17, 2.21, 2.22, and 2.29.

Apparatus Glass apparatus from the miniscale experimental procedure of Section 19.2, separatory funnel, ice-water bath, and apparatus for magnetic stirring, simple distillation, vacuum filtration, and *flameless* heating.

Setting Up While the reaction mixture for the preparation of phenylmagnesium bromide (Sec. 19.2) is cooling to room temperature, dissolve 1.2 mL of methyl benzoate in about 5 mL of *anhydrous* diethyl ether, and place this solution in the separatory funnel with the *stopcock closed.* Cool the reaction flask containing the phenylmagnesium bromide in the ice-water bath.

Reaction Begin the *slow, dropwise* addition of the solution of methyl benzoate to the *stirred* solution of phenylmagnesium bromide. This reaction is *exothermic*, so you should control the rate of reaction by adjusting the rate of addition *and* by occasionally cooling the reaction flask as needed with the ice-water bath. The ring of condensate should be allowed to rise no more than one-third of the way up the reflux condenser. A white solid may form during the reaction, but this is normal. After the addition is complete and the exothermic reaction subsides, you may complete the reaction in one of two ways. Consult with your instructor to determine whether you should (1) heat the reaction mixture at reflux for 30 min or (2) stopper the flask after cooling the contents to room temperature and place it in the *hood* until the next laboratory period (no reflux required).★

Work-Up, Isolation, and Purification Place about 10 mL of cold 6 *M* sulfuric acid and about 5–10 g of crushed ice in a beaker. If the reaction mixture solidified upon cooling, add a small quantity of solvent-grade diethyl ether to the reaction flask. Pour the reaction mixture gradually with stirring into the ice-acid mixture. Rinse the round-bottom flask with 2–3 mL of solvent-grade diethyl ether and add this wash to the beaker. Continue stirring until the heterogeneous mixture is completely free of undissolved solids. It may be necessary to add a small portion of solvent-grade diethyl ether to dissolve all the organic material; the total volume of ether should be about 15–20 mL. Verify that the aqueous layer is acidic; if it is not, add cold 6 *M* sulfuric acid dropwise until the layer is acidic. If necessary, sequentially add 2- to 3-mL portions of solvent-grade diethyl ether and then water to dissolve all of the solids.

Transfer the entire mixture to a separatory funnel. Shake the funnel vigorously with venting to relieve pressure; separate the aqueous layer.★ Wash the organic layer sequentially with about 5 mL of 3 *M* sulfuric acid, two 5-mL portions of saturated aqueous sodium bicarbonate (*vent!*), and finally with one 5-mL portion of saturated sodium chloride solution. Dry the organic layer using several spatula-tips full of anhydrous sodium sulfate. Swirl the flask occasionally for a period of 10–15 min to facilitate drying; add further small portions of anhydrous sodium sulfate if the solution remains cloudy.★

Filter or decant the solution into a 50-mL round-bottom flask and equip the flask for simple distillation. Remove the diethyl ether by simple distillation. Alternatively, use rotary evaporation or other techniques to concentrate the solution. The final traces of solvent may be removed by attaching the flask to a vacuum source and gently swirling the contents as the vacuum is applied. After the crude solid residue has dried, determine its melting range, which may be wide.★

Purify the triphenylmethanol by dissolving it in a *minimum* amount of boiling cyclohexane (ca. 10 mL/g product). Perform this operation at the hood or use a funnel that is attached to a vacuum source and inverted over the flask (Fig. 2.71b). Once all the material is in solution, evaporate the solvent *slowly* until small crystals of triphenylmethanol start to form. Allow the crystallization to continue at room temperature and then in an ice-water bath until no more crystals form. Isolate the product by vacuum filtration and air-dry it.

Analysis Weigh the triphenylmethanol and calculate the percent yield; determine its melting point. Obtain IR and ^{1}H NMR spectra of your starting materials and product, and compare them with those of authentic samples (Figs. 15.19, 15.20, and 19.2–19.5). If possible, analyze your product by GC-MS to determine if it is contaminated with benzophenone (**10**).

MICROSCALE PROCEDURE

Preparation Sign in at **www.cengage.com/login** to answer Pre-Lab Exercises, access videos, and read the MSDSs for the chemicals used or produced in this procedure. Review Sections 2.10, 2.11, 2.13, 2.17, 2.21, 2.22, and 2.29.

Apparatus Glass apparatus from the microscale experimental procedure of Section 19.2, 3-mL conical vial, two screw-cap centrifuge tubes, 1-mL plastic syringe, Pasteur pipet with 0.5- and 1.0-mL calibration marks, ice-water bath, and apparatus for magnetic stirring, simple distillation, vacuum filtration, Craig tube filtration, and *flameless* heating.

Setting Up While the reaction mixture for the preparation of phenylmagnesium bromide (Sec. 19.2) is cooling to room temperature, dissolve 120 μL of methyl benzoate in 0.5 mL of *anhydrous* diethyl ether contained in the 3-mL conical vial. Cool the conical vial containing the phenylmagnesium bromide in the ice-water bath.

Reaction Draw the solution of methyl benzoate into the syringe and recap the vial. Insert the syringe needle through the rubber septum and begin the *slow, dropwise* addition of the solution of methyl benzoate to the *stirred* solution of phenylmagnesium bromide. This reaction is *exothermic,* so you should control the rate of reaction by adjusting the rate of addition *and* by occasionally cooling the reaction vial with the ice-water bath if necessary. The reflux ring should not be allowed to rise into the Claisen adapter. A white solid may form during the reaction, but this is normal. Upon completion of the addition, place about 0.2–0.3 mL of anhydrous diethyl ether in the vial that contained the methyl benzoate solution. Draw this solution into a syringe and add it in one portion to the reaction mixture. After the exothermic reaction subsides, you may complete the reaction in one of two ways. Consult with your instructor to determine whether you should (1) heat the reaction mixture at *gentle* reflux for 30 min or (2) cap the vial after cooling the contents to room temperature and place it in the *hood* until the next laboratory period (no reflux required). The mixture may solidify on cooling.★

Work-Up, Isolation, and Purification Remove the Claisen adapter and drying tube, and place the conical reaction vial in an ice-water bath. Using a Pasteur pipet, slowly add 1.5 mL of cold 3 *M* sulfuric acid dropwise to neutralize the reaction mixture. *Be careful*, as the addition of acid may be accompanied by frothing. Use a small glass stirring rod to break up the solid during this addition. You may need to cap the vial tightly and shake it with venting to facilitate dissolution of the solids. Continue shaking the vial until the heterogeneous mixture is free of undissolved solids. You should obtain two distinct layers as the solid gradually dissolves. It may be necessary to add solvent-grade diethyl ether to dissolve all the organic material; the total volume of ether should be about 2.0–2.5 mL. Verify that the aqueous layer is acidic; if it is not, add cold 3 *M* sulfuric acid dropwise until the layer is acidic.

Optional Measures If some solid material remains after the addition of aqueous acid is complete, the entire mixture may be transferred to a screw-cap centrifuge tube. Sequentially add solvent-grade diethyl ether and water in 0.5-mL portions to the centrifuge tube. After each addition, cap the tube and shake it to dissolve the solids. Continue the extraction as described, using appropriately sized conical vials or screw-cap centrifuge tubes.

Remove the spinvane with forceps and, holding it above the vial, rinse it with several drops of solvent-grade ether. Cap the vial tightly, invert it gently several times with venting to mix the layers, and allow the layers to separate. Using a Pasteur pipet, transfer the lower aqueous layer to a conical vial. Wash this layer with one 0.5-mL portion of solvent-grade ether and combine this ethereal layer with the first.★ Wash the combined organic layers sequentially with a 0.5-mL portion of 3 *M* sulfuric acid, two 0.5-mL portions of saturated aqueous sodium bicarbonate (*vent!*), and finally with one 0.5-mL portion of saturated aqueous sodium chloride. Using a clean filter-tip pipet, transfer the ether layer to a screw-cap centrifuge tube containing several microspatula-tips full of anhydrous sodium sulfate. Swirl the tube occasionally for a period of 10–15 min to facilitate drying; add further small portions of anhydrous sodium sulfate if the solution remains cloudy.★

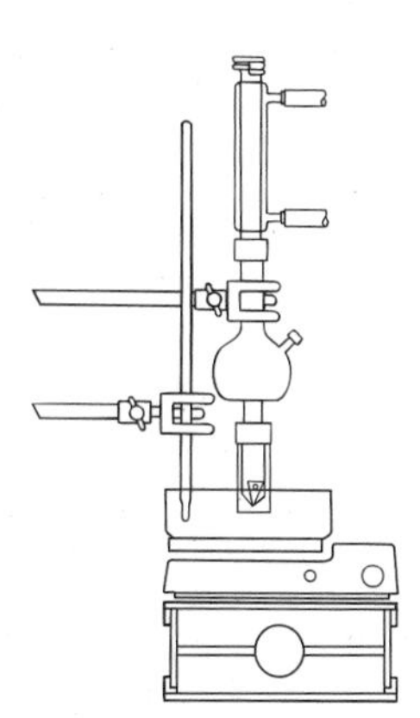

With a filter-tip pipet, transfer the dried organic solution into a clean 5.0-mL conical vial, rinse the sodium sulfate with about 0.2 mL of diethyl ether, and add this rinse to the vial. Equip the vial for simple distillation. Remove the solvent by simple distillation; withdraw the distillate from the Hickman stillhead as needed. Alternatively, use rotary evaporation or other techniques to concentrate the solution. The final traces of solvents may be removed by attaching the vial to a vacuum source. After the crude solid residue has dried, determine its melting range, which may be wide.★

Purify the triphenylmethanol by dissolving it in a *minimum* amount of boiling cyclohexane (ca. 1 mL/0.1 g product) in a Craig tube. Perform this operation at the hood. Once all the material is in solution, evaporate the solvent *slowly* until small crystals of triphenylmethanol start to form. Allow the crystallization to continue at room temperature and then in an ice-water bath until no more crystals form. Isolate and air-dry the product.

Analysis Weigh the triphenylmethanol and calculate the percent yield; determine its melting point. Obtain IR and ^{1}H NMR spectra of your starting materials and product and compare them with those of authentic samples (Figs. 15.19, 15.20, and 19.2–19.5). If possible, analyze your product by GC-MS to determine if it is contaminated with benzophenone (**10**).

WRAPPING IT UP

Dilute the *combined aqueous layers and washes* with water, neutralize the solution if necessary, and flush it down the drain with excess water. Place the *ether distillate* and the *cyclohexane mother liquor* in the container for nonhalogenated organic solvents. Spread the *calcium chloride* from the drying tube and the *sodium sulfate* on a tray in the hood and, after the ether has evaporated, place them and the *filter paper* in the container for nonhazardous solids.

B ■ *Preparation of Benzoic Acid*

Purpose To demonstrate the preparation of a carboxylic acid by the reaction of a Grignard reagent with carbon dioxide.

SAFETY ALERT

1. **Review the Safety Alert for Preparation of Grignard Reagents (Sec. 19.2).**
2. **Use tongs or gloves to handle the solid carbon dioxide (dry ice) because contact with skin can cause severe frostbite. Crush the dry ice by wrapping larger pieces in a clean, dry towel and striking them with a mallet or a wooden block. Use the dry ice *immediately* after crushing, to minimize contact with atmospheric moisture.**

MINISCALE PROCEDURE

Preparation Sign in at **www.cengage.com/login** to answer Pre-Lab Exercises, access videos, and read the MSDSs for the chemicals used or produced in this procedure. Review Sections 2.17 and 2.21.

Apparatus Glass apparatus from the miniscale experimental procedure of Section 19.2, separatory funnel, apparatus for vacuum filtration and *flameless* heating.

Setting Up After the reaction mixture for the preparation of phenylmagnesium bromide (Sec.19.2) has cooled to room temperature, place about 10 g of coarsely crushed dry ice in a 125-mL Erlenmeyer flask. The dry ice should be crushed *immediately* before use and protected as much as possible from atmospheric moisture.

Reaction Slowly pour the phenylmagnesium bromide solution onto the dry ice with vigorous swirling; the mixture normally becomes rather viscous. Rinse the original round-bottom reaction flask with 2–3 mL of anhydrous diethyl ether and add this wash to the Erlenmeyer flask. Continue swirling the contents of the flask for a few minutes. After the addition is complete, cover the flask with a small watchglass or piece of filter paper and allow the excess carbon dioxide to sublime by letting the flask stand in the hood, properly labeled with your name, until the next laboratory period.★

You may expedite the removal of the excess carbon dioxide by shaking or swirling the flask while warming it *very slightly* in a warm-water bath or by adding small amounts of warm water to the reaction mixture and then shaking or swirling the flask. Be careful, because both of these methods may cause a *sudden* evolution of carbon dioxide gas and splash the contents of the flask onto the floor or bench top. The flask should *never* be stoppered.

Work-Up, Isolation, and Purification After the excess dry ice has completely sublimed, most of the ether will have also evaporated, so add 20 mL of solvent-grade diethyl ether to the reaction flask. Combine about 5 g of ice and 10 mL of cold 3 *M* sulfuric acid in an Erlenmeyer flask, and add this cold acid to the reaction mixture *slowly* to avoid excessive foaming. If the ether evaporates appreciably during this operation, more solvent-grade diethyl ether should be added so that the total volume of ether is about 20–25 mL. Verify that the aqueous layer is acidic; if it is not, add cold 3 *M* sulfuric acid dropwise until the layer is acidic.

Swirl the mixture and transfer it to a separatory funnel. Rinse the flask with a small portion of solvent-grade diethyl ether and add the rinse to the separatory funnel. Shake the funnel cautiously with venting and separate the layers. Extract the aqueous layer with a 10-mL portion of solvent-grade diethyl ether and add this to the original ether layer. Extract the *combined* ethereal fractions sequentially with two 10-mL portions of a 1 *M* solution of sodium hydroxide; vent the funnel frequently during the extractions. Transfer the two alkaline extracts to an Erlenmeyer flask and slowly add 6 *M* hydrochloric acid until precipitation of the benzoic acid is complete and the aqueous mixture is *acidic*. Cool the solution in an ice-water bath, isolate the solid benzoic acid by vacuum filtration, wash it with water, and air-dry it.★ Recrystallize the benzoic acid from water.

Analysis Weigh the benzoic acid and calculate the percent yield; determine its melting point. Obtain IR and ^{1}H NMR spectra of your starting material and product and compare them with those of authentic samples (Figs. 15.19, 15.20, 19.6, and 19.7).

Discovery Experiment

Exploring the Influence of Mode of Addition

Develop a protocol for investigating how using the "normal" mode addition of phenylmagnesium bromide and carbon dioxide affects the outcome of the reaction. You will need access to gaseous carbon dioxide to perform the reaction. Consult with your instructor before undertaking any experiments.

MICROSCALE PROCEDURE

Preparation Sign in at **www.cengage.com/login** to answer Pre-Lab Exercises, access videos, and read the MSDSs for the chemicals used or produced in this procedure. Review Sections 2.17 and 2.21.

Apparatus Glass apparatus from the microscale experimental procedure of Section 19.2, a 3-mL and a 5-mL conical vial, screw-cap centrifuge tube, Pasteur pipet with 0.5- and 1.0-mL calibration marks, apparatus for vacuum filtration, and Craig tube filtration.

Setting Up After the reaction mixture for the preparation of phenylmagnesium bromide (Sec.19.2) has cooled to room temperature, place about 1 g of coarsely crushed dry ice in a small beaker. The dry ice should be crushed immediately before use and protected as much as possible from atmospheric moisture.

Reaction Using a filter-tip pipet, slowly transfer the phenylmagnesium bromide solution onto the dry ice. Stir the mixture with a glass rod during the addition, as the mixture normally becomes rather viscous. Rinse the conical vial with 0.5 mL of anhydrous diethyl ether and add this wash to the beaker. After the addition is complete, cover the beaker with a small watchglass or piece of filter paper, and allow the excess carbon dioxide to sublime by letting the beaker stand in the hood, properly labeled with your name, until the next laboratory period.★

You may expedite the removal of the excess carbon dioxide by stirring the contents of the beaker while warming it *very slightly* in a warm-water bath or by adding small amounts of warm water to the reaction mixture with stirring. Because both of these methods may cause a *sudden* evolution of carbon dioxide gas and splash the contents of the beaker onto the floor or bench top, be careful using them.

Work-Up, Isolation, and Purification After the excess dry ice has completely sublimed, most of the ether will have also evaporated, so add 2.5 mL of solvent-grade diethyl ether. Then *slowly* add 1.5 mL of cold 3 *M* sulfuric acid to the reaction mixture to avoid excessive foaming. If the ether evaporates appreciably during this operation, more solvent-grade diethyl ether should be added so that the total volume of ether is about 2.0–2.5 mL. Verify that the aqueous layer is acidic; if it is not, add cold 3 *M* sulfuric acid dropwise until the layer is acidic.

Optional Measures If some solid material remains after the addition of aqueous acid is complete, the entire mixture may be transferred to a screw-cap centrifuge tube. Sequentially add solvent-grade diethyl ether and water in 0.5-mL portions to the centrifuge tube. After each addition, cap the tube and shake it to dissolve the solids. Continue the extraction as described using appropriately sized conical vials or screw-cap centrifuge tubes.

Using a Pasteur pipet, transfer the entire mixture to a 5-mL conical vial.★ Rinse the beaker with a small portion of solvent-grade diethyl ether and add the rinse to the conical vial. Cap the vial tightly, invert it gently several times with venting to mix the layers thoroughly, and allow the layers to separate. Transfer the aqueous layer to a 3-mL conical vial, extract the aqueous layer with a 0.5-mL portion of solvent-grade diethyl ether, and add this extract to the original ether layer. Extract the *combined* ethereal fractions sequentially with two 1-mL portions of a 1 *M* solution of sodium hydroxide, venting the vial frequently during the extractions. Transfer the two alkaline extracts to a small Erlenmeyer flask and warm the flask for about 5 min to remove the ether that is dissolved in the aqueous phase; a saturated solution of diethyl ether in water contains 7% diethyl ether. Cool the alkaline solution and slowly add 6 *M* hydrochloric acid until precipitation of the benzoic acid is complete and the aqueous mixture is acidic. Cool the solution in an ice-water bath, isolate the solid benzoic acid by vacuum filtration, and air-dry the sample.★ Recrystallize the benzoic acid from water in a Craig tube.

Analysis Weigh the benzoic acid and calculate the percent yield; determine its melting point. Obtain IR and ^{1}H NMR spectra of your starting material and product and compare them with those of authentic samples (Figs. 15.19, 15.20, 19.6, and 19.7).

Discovery Experiment ***Exploring the Influence of Mode of Addition***

Develop a protocol for investigating how using the "normal" mode addition of phenylmagnesium bromide and carbon dioxide affects the outcome of the reaction. You will need access to gaseous carbon dioxide to perform the reaction. Consult with your instructor before undertaking any experiments.

WRAPPING IT UP

Dilute the *combined aqueous layers and washes* with water, neutralize the solution if necessary, and flush it down the drain with excess water. Place the *combined ethereal layers* in the container for nonhalogenated organic solvents. Spread the *calcium chloride* from the drying tube on a tray in the hood to allow the ether to evaporate; then place it and the *filter paper* in the container for nonhazardous solids.

C ■ *Preparation of 2-Methyl-3-Heptanol*

Purpose To demonstrate the preparation of a secondary alcohol by the reaction of a Grignard reagent with an aldehyde.

SAFETY ALERT

Review the Safety Alert for Preparation of Grignard Reagents (Sec. 19.2).

MINISCALE PROCEDURE

Preparation Sign in at **www.cengage.com/login** to answer Pre-Lab Exercises, access videos, and read the MSDSs for the chemicals used or produced in this procedure. Review Sections 2.10, 2.11, 2.13, 2.17, 2.21, and 2.29.

Apparatus Glass apparatus from the miniscale experimental procedure of Section 19.2, a 25-mL round-bottom flask, separatory funnel, ice-water bath, and apparatus for magnetic stirring, simple distillation, shortpath distillation, and *flameless* heating.

Setting Up While the reaction mixture for the preparation of 1-butylmagnesium bromide (Sec.19.2) is cooling to room temperature, dissolve 1.8 mL of *freshly distilled* 2-methylpropanal in 5 mL of *anhydrous* diethyl ether, and place this solution in the separatory funnel with the *stopcock closed.* Cool the reaction flask containing the 1-butylmagnesium bromide in an ice-water bath and begin stirring.

Reaction Over a period of about 5 min, add the solution of 2-methylpropanal *dropwise* to the stirred solution of 1-butylmagnesium bromide. Control the resulting exothermic reaction by adjusting the rate of addition so that the ring of condensate rises no more than one-third of the way up the reflux condenser. After completing the addition, allow the reaction mixture to stand for about 15 min. The reaction mixture may be stored, loosely stoppered, *in the hood* until the next laboratory period.★

Work-Up, Isolation, and Purification Place about 10 mL of cold 6 *M* sulfuric acid and 5–10 g of crushed ice in a beaker. Pour the reaction mixture *slowly* with stirring into the ice-acid mixture. After the addition is complete, transfer the cold mixture, which may contain some precipitate, to a separatory funnel and shake it gently. The precipitate should dissolve. Separate the layers. Extract the aqueous layer sequentially with two 5-mL portions of solvent-grade diethyl ether, and add these extracts to the main ethereal layer. Transfer the combined ethereal extracts into the separatory funnel. Venting the funnel frequently to relieve pressure, wash the solution sequentially with 5 mL of saturated aqueous sodium bisulfite, two 5-mL portions of saturated aqueous sodium bicarbonate, and finally with 5 mL of saturated aqueous sodium chloride. Dry the ethereal solution over several spatula-tips full of anhydrous sodium sulfate.★ Swirl the flask occasionally for a period of 10–15 min to facilitate drying; add further small portions of anhydrous sodium sulfate if the solution remains cloudy.

Filter or carefully decant the dried organic solution into a round-bottom flask equipped for simple distillation, and remove the diethyl ether by simple distillation. Alternatively, use rotary evaporation or other techniques to concentrate the solution. The final traces of solvent may be removed by attaching the flask to a vacuum source and gently swirling the contents as the vacuum is applied.★

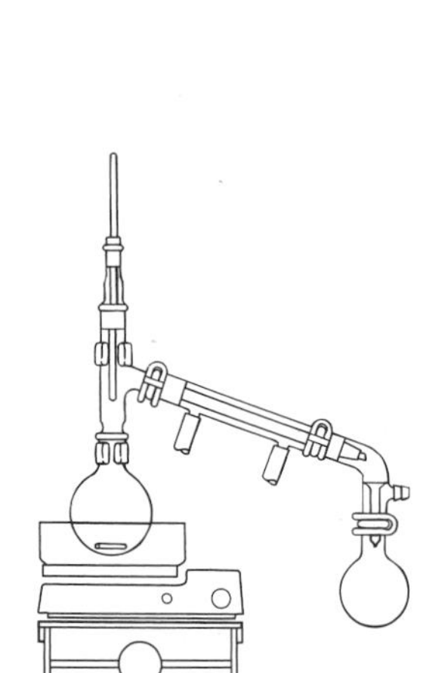

Using a Pasteur pipet, transfer the cooled pot residue to a 25-mL round-bottom flask containing a stirbar. Rinse the original stillpot with 1–2 mL of diethyl ether, and add the rinse to the 25-mL round-bottom flask. Remove the remaining diethyl ether by attaching the flask to a vacuum source and gently swirling the contents as the vacuum is applied. Fit this flask for a shortpath distillation, insulating the top of the distilling flask and the stillhead with cotton or glasswool wrapped with aluminum foil. Distill the 2-methyl-3-heptanol into a *tared* receiver, collecting all distillate that boils above 130 °C (760 torr). Note that the head temperature may never reach the reported boiling point, owing to the high boiling temperature and the small amount of sample being distilled. Terminate heating when the dark pot residue becomes viscous and starts to evolve fumes.

Analysis Weigh the product and calculate the yield. Obtain IR and ^{1}H NMR spectra of your starting materials and product and compare them with those of authentic samples (Figs. 14.3, 14.4, 18.17, 18.18, 19.8, and 19.9).

Discovery Experiment

Preparation and Characterization of a 3° Alcohol

Following the general procedures outlined in Sections 19.2 and 19.4C for the preparation and reaction of 1-butylmagnesium bromide with 2-methylpropanal, perform a Grignard synthesis with 30 mmol of a 1° alkyl bromide and 28 mmol of a ketone, which are unknowns that you obtain from your instructor. Use 33 mmol of magnesium turnings. The possible alkyl halide and ketone unknowns are listed below in alphabetical order, as are the 3° alcohols that are products of the various combinations of reagents. Characterize your product by its boiling point, IR spectrum, and, if possible,

1H NMR spectrum and use the data to determine which alcohol you have produced. Be certain to consult with your instructor before undertaking any procedures.

Alkyl Halide	Ketone	Alcohol
1-Bromobutane	Acetone	3-Ethyl-3-heptanol
Bromoethane	Butanone	4-Ethyl-4-heptanol
1-Bromopentane	2-Pentanone	2-Methyl-2-heptanol
1-Bromopropane	3-Pentanone	3-Methyl-3-heptanol
		4-Methyl-4-heptanol
		3-Ethyl-3-hexanol
		2-Methyl-2-hexanol
		3-Methyl-3-hexanol
		2-Methyl-2-octanol
		3-Methyl-3-octanol
		4-Methyl-4-octanol
		3-Ethyl-3-pentanol
		2-Methyl-2-pentanol
		3-Methyl-3-pentanol

WRAPPING IT UP

Dilute the *combined aqueous layers and washes* with water, neutralize the solution if necessary, and flush it down the drain with excess water. Place the *ether distillate* in the container for nonhalogenated organic solvents. Spread the *calcium chloride* from the drying tube and the *sodium sulfate* on a tray in the hood, and then place them in the container for nonhazardous solids after the volatiles have evaporated.

EXERCISES

General Questions

1. Arrange the following compounds in order of increasing reactivity toward attack of a Grignard reagent at the carbonyl carbon atom: methyl benzoate, benzoic acid, benzaldehyde, and benzophenone. Explain the basis for your decision, making use of mechanisms where needed.
2. What is (are) the product(s) of reaction of each of the carbonyl-containing compounds in Exercise 1 with *excess* Grignard reagent, RMgBr?
3. How might primary, secondary, and tertiary alcohols be prepared from a Grignard reagent and a suitable carbonyl-containing compound? Write chemical reactions for these preparations using any starting materials you wish; indicate stoichiometry where important.

Questions for Part A

4. Why is it unwise to begin addition of the solution of methyl benzoate to the Grignard reagent before the latter has cooled to room temperature and then been placed in an ice-water bath?
5. Why should anhydrous rather than solvent-grade diethyl ether be used to prepare the solution of methyl benzoate that is added to the Grignard reagent?
6. What is the solid that forms during the addition of the ester to the Grignard reagent?
7. Why is it necessary to acidify the mixture obtained after the reaction of methyl benzoate with phenylmagnesium bromide?

8. Cyclohexane is used as the recrystallization solvent to purify the triphenylmethanol by removing the biphenyl impurity. Why is this a better choice of solvent than a solvent such as isopropyl alcohol?

9. Comment on the use of steam distillation (Sec. 4.4) as a possible alternative procedure for purifying crude triphenylmethanol. Consider what possible starting materials, products, and by-products might be present, and indicate which of these should steam-distill and which should not. Would this method of purification yield pure triphenylmethanol? Give your reasoning.

Sign in at **www.cengage.com/login** and use the spectra viewer and Tables 8.1–8.8 as needed to answer the blue-numbered questions on spectroscopy.

10. Consider the spectral data for methyl benzoate (Figs. 19.2 and 19.3).
    a. In the functional group region of the IR spectrum, identify the absorptions associated with the ester functional group and the aromatic ring.
    b. In the ^1H NMR spectrum, assign the various resonances to the hydrogen nuclei responsible for them.
    c. For the ^{13}C NMR data, assign the various resonances to the carbon nuclei responsible for them.

11. Consider the spectral data for triphenylmethanol (Figs. 19.4 and 19.5).
    a. In the functional group region of the IR spectrum, specify the absorptions due to the aromatic ring. There is a broad absorption at about 3450 cm^{-1}. What functional group is responsible for this absorption and why is the absorption broad?
    b. In the ^1H NMR spectrum, assign the various resonances to the hydrogen nuclei responsible for them.
    c. For the ^{13}C NMR data, assign the various resonances to the carbon nuclei responsible for them.

12. Discuss the differences observed in the IR and NMR spectra of methyl benzoate and triphenylmethanol that are consistent with the conversion of an ester into a tertiary alcohol in this experiment.

Questions for Part B

13. What molecular feature or features allow carbon dioxide to function as an electrophile?

14. Define the term *inverse addition* as it applies to a Grignard reaction and explain why this is the preferred mode of addition for the reaction between C_6H_5MgBr and CO_2.

15. Why is it necessary to acidify the mixture obtained after the reaction of phenylmagnesium bromide with carbon dioxide?

16. Consider the acid-base reaction that occurs when benzoic acid is extracted with aqueous sodium hydroxide.
    a. Write the equation for this reaction, labeling the acid, base, conjugate acid, and conjugate base.
    b. What is the anticipated relative solubility of benzoic acid and its conjugate base in diethyl ether?
    c. The pK_a of benzoic acid is 4.2, whereas that of carbonic acid, H_2CO_3, is 6.4. On the basis of this information, would aqueous sodium bicarbonate be sufficiently basic to deprotonate benzoic acid? Explain your reasoning.
    d. The pK_a of hydrochloric acid is about −6. On the basis of this information, would aqueous sodium chloride be sufficiently basic to deprotonate benzoic acid? Explain your reasoning.

17. What is the purpose of using diethyl ether to extract the aqueous mixture obtained after acidification of the reaction? What products are in the aqueous and organic layers?

18. What function does extracting the ethereal solution of organic products with aqueous base have in the purification of benzoic acid? What remains in the ether layer?

19. The yield of benzoic acid obtained when only enough acid is added to the aqueous solution of sodium benzoate to bring the pH to 7 is lower than that obtained if the pH is brought below 5. Explain.

20. Explain why neither triphenylmethanol, $(C_6H_5)_3COH$, nor benzophenone, $C_6H_5COC_6H_5$, was produced in the reaction between C_6H_5MgBr and CO_2.

Sign in at **www.cengage.com/login** and use the spectra viewer and Tables 8.1–8.8 as needed to answer the blue-numbered questions on spectroscopy.

21. Consider the spectral data for benzoic acid (Figs. 19.6 and 19.7).
    a. In the functional group region of the IR spectrum, specify the absorptions associated with the carbonyl component of the carboxyl group and the aromatic ring. There is a broad absorption in the region of about 2800–3050 cm^{-1}. What functional group is responsible for this absorption, and why is the absorption broad?
    b. In the ^1H NMR spectrum, assign the various resonances to the hydrogen nuclei responsible for them.
    c. For the ^{13}C NMR data, assign the various resonances to the carbon nuclei responsible for them.

22. Discuss the differences observed in the IR and NMR spectra of bromobenzene and benzoic acid that are consistent with the conversion of bromobenzene into benzoic acid in this experiment.

Questions for Part C

23. Which reagent functions as the electrophile in the reaction of 1-butylmagnesium bromide and 2-methylpropanal, and why should it be freshly distilled before use?

24. Why is it unwise to begin addition of the solution of 2-methylpropanal to the Grignard reagent before the latter has cooled to room temperature and then been placed in an ice-water bath?

25. What is the solid that forms upon reaction of 1-butylmagnesium bromide with 2-methylpropanal?

26. How may the exothermicity associated with the addition of the Grignard reagent to a solution of 2-methypropanal be controlled?

27. Why is it necessary to acidify the mixture obtained after the reaction of 1-butylmagnesium bromide with 2-methylpropanal?

28. Consumption of unreacted magnesium metal occurs when the reaction mixture is quenched with sulfuric acid. Write the balanced equation for the reaction between Mg^0 and sulfuric acid.

29. The work-up in this reaction calls for successive washes of an ethereal solution of the product with aqueous sodium bisulfite, sodium bicarbonate, and sodium chloride. What is the purpose of each of these steps?

30. Another organometallic reagent that reacts in a manner similar to the Grignard reagent is an alkyllithium, R–Li.
    a. Predict the addition product formed after acidification of the mixture resulting from reaction of $CH_2{=}CH{-}CH2{-}Li$ with 2-methylpropanal.
    b. Suppose $CH_3CHCH{=}CH{-}CH_2{-}Li$ had been used for the reaction with 2-methylpropanal. Write the addition product(s) expected after acidification of the reaction mixture.

Sign in at **www.cengage.com/login** and use the spectra viewer and Tables 8.1–8.8 as needed to answer the blue-numbered questions on spectroscopy.

31. Consider the spectral data for 2-methylpropanal (Figs. 18.17 and 18.18).
    a. In the functional group region of the IR spectrum, specify the absorptions associated with the carbonyl component and the carbon-hydrogen bond of the aldehyde group.
    b. In the ^{1}H NMR spectrum, assign the various resonances to the hydrogen nuclei responsible for them.
    c. For the ^{13}C NMR data, assign the various resonances to the carbon nuclei responsible for them.
32. Consider the spectral data for 2-methyl-3-heptanol (Figs. 19.8 and 19.9).
    a. In the IR spectrum, specify the absorptions associated with the hydroxyl group.
    b. In the ^{1}H NMR spectrum, assign the various resonances to the hydrogen nuclei responsible for them.
    c. For the ^{13}C NMR data, assign the various resonances to the carbon nuclei responsible for them. Explain why there are eight peaks in the ^{13}C NMR spectrum.
33. Discuss the differences observed in the IR and NMR spectra of 2-methylpropanal and 2-methyl-3-heptanol that are consistent with the conversion of an aldehyde into a secondary alcohol in this experiment.
34. The IR spectrum of 2-methyl-3-heptanol prepared by the procedure of Part **C** sometimes shows a band at 1720 cm^{-1}. Give a possible source for this absorption.

SPECTRA

Starting Materials and Products

The ^{13}C NMR spectrum of methyl benzoate is shown in Figure 8.47. The IR and NMR spectra for 1-bromobutane, bromobenzene, and 2-methylpropanal are presented in Figures 14.2, 14.4, 15.19, 15.20, 18.17, and 18.18, respectively.

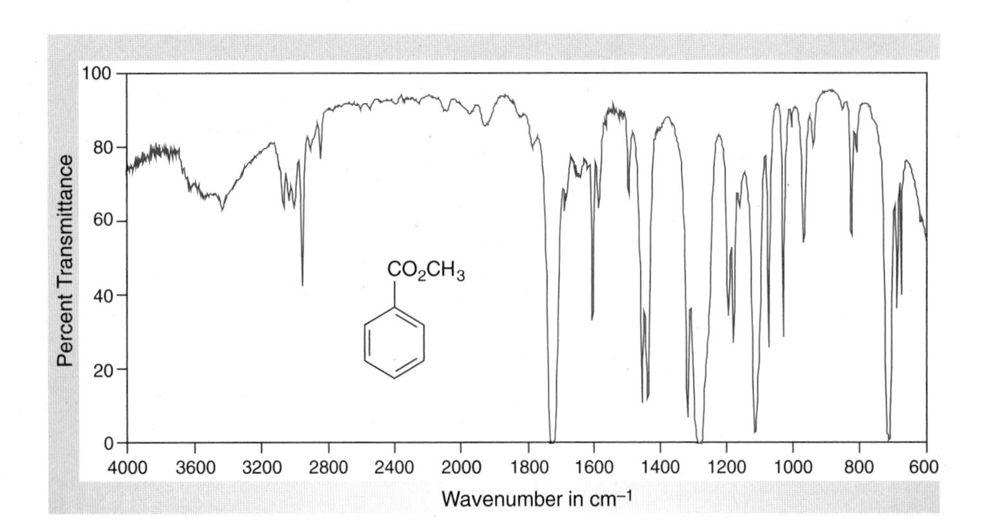

Figure 19.2
IR spectrum of methyl benzoate (neat).

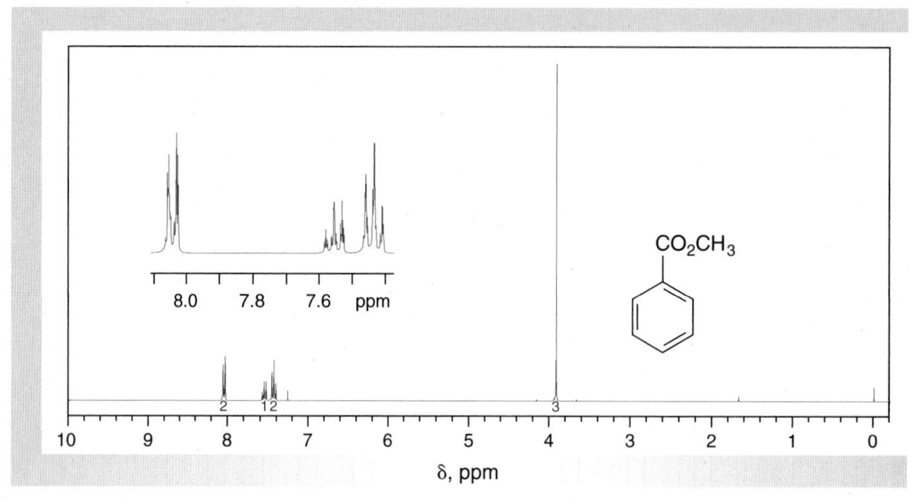

Figure 19.3
NMR data for methyl benzoate ($CDCl_3$).

(a) ^{1}H NMR spectrum (300 MHz).
(b) ^{13}C NMR data: δ 51.8, 128.5, 129.7, 130.5, 132.9, 166.8 (see Figure 8.47 for spectrum).

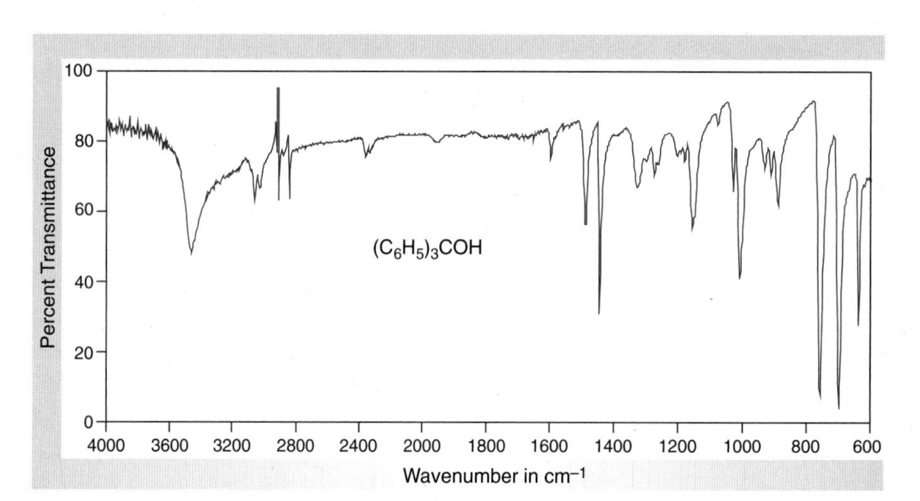

Figure 19.4
IR spectrum of triphenylmethanol (IR card).

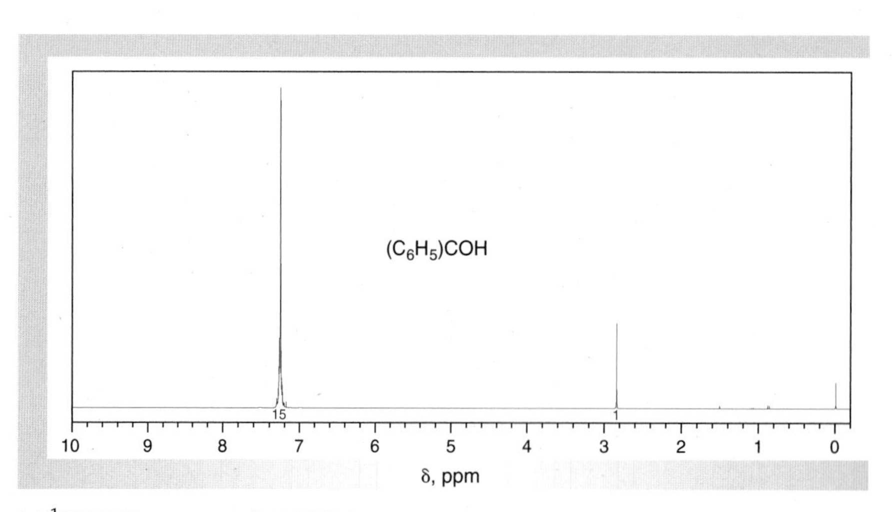

Figure 19.5
NMR data for triphenylmethanol ($CDCl_3$).

(a) ^{1}H NMR spectrum (300 MHz).
(b) ^{13}C NMR data: δ 82.0, 127.2, 127.9, 146.9.

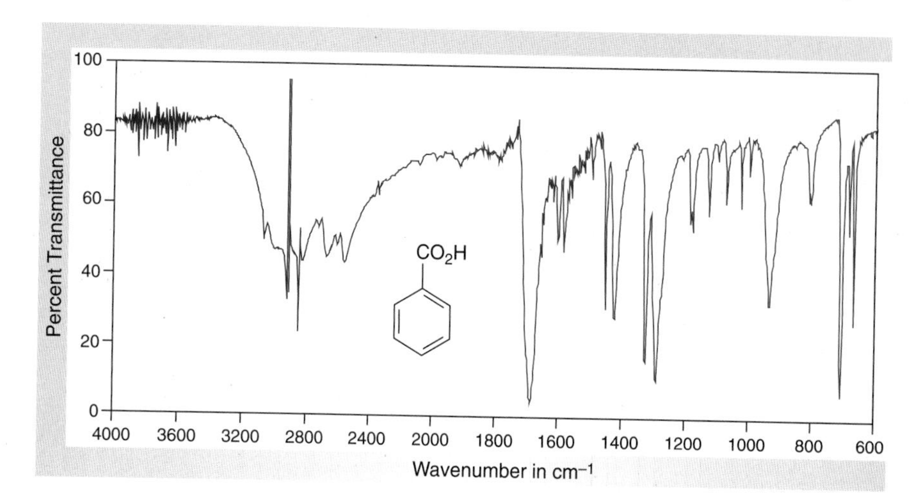

Figure 19.6
IR spectrum of benzoic acid (IR card).

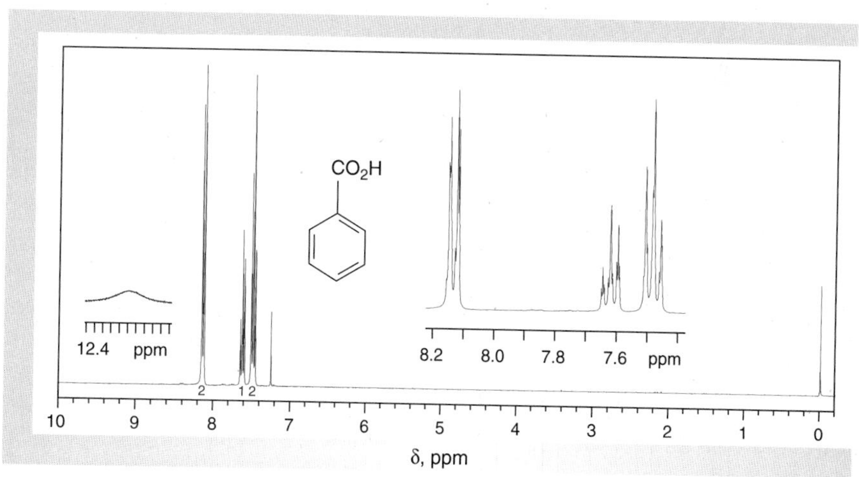

Figure 19.7
NMR data for benzoic acid ($CDCl_3$).

(a) ^1H NMR spectrum (300 MHz).
(b) ^{13}C NMR data: δ 128.5, 129.5, 130.3, 133.8, 172.7.

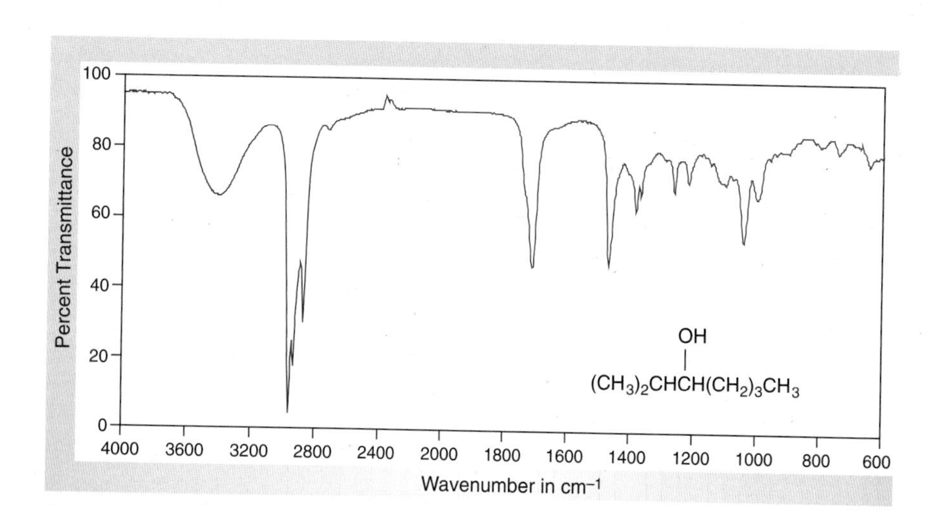

Figure 19.8
IR spectrum of 2-methyl-3-heptanol (neat).

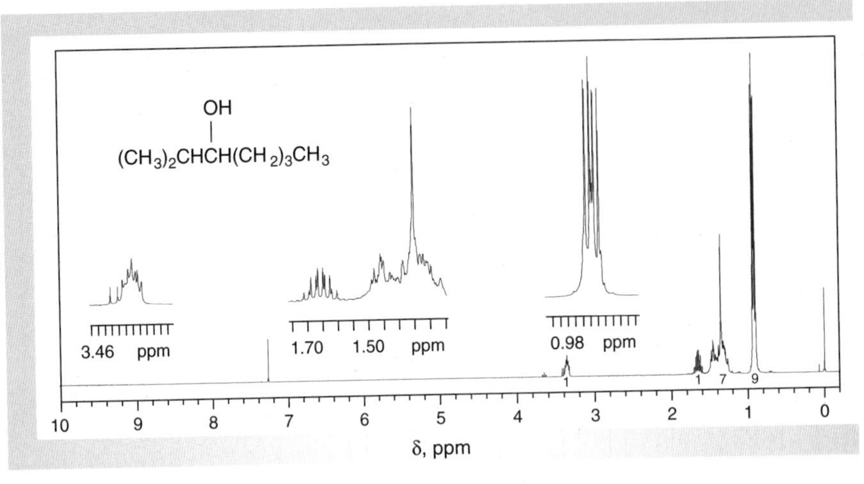

Figure 19.9
NMR data for 2-methyl-3-heptanol ($CDCl_3$).

(a) ^1H NMR spectrum (300 MHz).
(b) ^{13}C NMR data: δ 14.2, 17.5, 19.3, 23.1, 28.6, 33.8, 34.1, 76.6.

HISTORICAL HIGHLIGHT

Grignard and the Beginnings of Modern Organometallic Chemistry

François Auguste Victor Grignard was born in Cherbourg, France, in 1871. He began his graduate studies at the University of Lyons under Barbier, one of the pioneers in developing the chemistry of compounds that contain a carbon-metal bond. In 1899 Barbier had discovered that magnesium could be used in place of zinc to promote the reaction of methyl iodide with a ketone to give a tertiary alcohol. This represented a useful advance in organometallic chemistry because the zinc reagents that were being used at the time were pyrophoric and difficult to handle. Given this exciting finding, Barbier suggested to the young Grignard that he conduct further exploratory studies of the reactions of magnesium with alkyl halides for his doctoral research project. Every young chemist should be given such a golden opportunity, and Grignard made the most of his!

Grignard quickly discovered that a wide variety of alkyl and aryl halides reacted readily with magnesium in anhydrous ether to give the corresponding alkyl- and arylmagnesium halides that are now known as Grignard reagents. His initial results were reported in 1900, and he published seven additional papers within the next year on the preparation of organomagnesium compounds and their reactions with carbonyl compounds to give alcohols. After receiving his Ph.D. in 1901, Grignard continued his research at the University of Lyons and then the University of Nancy, where he explored the scope and limitations of the reactions of Grignard reagents. During this time, he discovered that the reactions that now bear his name could be used to prepare primary, secondary, and tertiary alcohols, ketones, esters, and carboxylic acids.

The discovery and development of the chemistry of Grignard reactions were especially important for applications to the synthesis of complex molecules because a new carbon-carbon bond is formed by combining two simpler building blocks, namely an alkyl or aryl halide and a carbonyl compound. The products of Grignard reactions also contain functionality that may then be used in subsequent transformations to fashion even larger molecules.

Grignard was awarded the Nobel Prize in Chemistry in 1912 for his extensive work with organomagnesium compounds. He returned to the University of Lyons in 1919, where he succeeded Barbier as head of the department and continued his research. Grignard also conducted an extensive research program in the fields of terpenoid natural products, ozonolysis, aldol reactions, and catalytic hydrogenation and dehydrogenation. However, it is for his

(Continued)

HISTORICAL HIGHLIGHT *Grignard and the Beginnings . . . (Continued)*

development of the chemistry of Grignard reagents that he is best remembered. Indeed, at the time of his death in 1935, more than 6000 papers had appeared in which applications of Grignard reactions were described. Tens of thousands more have been published since then.

Relationship of Historical Highlight to Experiments

Grignard reagents belong to the larger class of substances known as organometallic compounds. The reactivity of these compounds is determined by the polarized nature of the carbon-metal bond in which carbon bears a partial negative charge. In the experiments in this chapter, you will conduct some of the same reactions that Grignard performed while a graduate student. For example, you will find that the reaction of an alkyl or aryl halide with magnesium in anhydrous ether occurs rapidly. You will also discover that these reagents are highly reactive toward electrophiles and that the carbon atom of the reagent serves as a nucleophile in its chemical reactions. Although you will not win a Nobel Prize for your observations in these experiments, there are other reactions that remain to be discovered. Perhaps your training in the organic laboratory will enable you to find one of these.

See *Who was Grignard?*
See *Who was Barbier?*
See more on *Pyrophoricity*

Carboxylic Acids and Their Derivatives

When you see this icon, sign in at this book's premium website at **www.cengage.com/login** to access videos, Pre-Lab Exercises, and other online resources.

The chemistry of organic compounds containing the carbon-oxygen double bond is arguably richer than that of any other class of organic molecules. In the experiments in Chapter 18, there is an opportunity to learn that the chemistry of aldehydes and ketones is determined by the presence of the carbonyl functional group. In this chapter you will be able to study the chemistry associated with carboxylic acids and some of their derivatives. These compounds contain a carbon-oxygen double bond, in which the carbonyl carbon atom bears an additional heteroatom such as oxygen, nitrogen, or halogen. The presence of this heteroatom dramatically changes the chemical properties of the carbonyl group, but in ways that you will learn are predictable. Derivatives of carboxylic acids are commonplace in materials we encounter daily and include fabrics, antibiotics, and insecticides, to name a few. In the experiments in this chapter, you will be introduced to the varied chemistry and the importance of carboxylic acid derivatives by preparing compounds having use as an anesthetic and as an insect repellent. You will also have the opportunity to study a fascinating chemical reaction that produces light, resulting from decomposition of a compound you have synthesized.

20.1 INTRODUCTION

Carboxylic acids are characterized by the presence of the **carboxyl functional group,** $-CO_2H$, which is composed of a **carb**onyl group and a hydr**oxyl** group, and have the general structural formula **1** in which R is a hydrogen or an alkyl or an aryl substituent. Their derivatives arise by replacing the hydroxyl group in **1** with other heteroatom substituents such as Cl in an **acid chloride,** $OCOR^1$ in an **acid anhydride,** OR^2 in an **ester,** and NR^3R^4 in an **amide.** These derivatives may be summarized by the general formula **2** in which the **acyl group,** R–C(=O)–, is bonded to an electronegative substituent Z, which may represent a single atom, as in acid chlorides, or a group of atoms, as in anhydrides, esters, and amides. Although they will not be included in the discussion that follows, **nitriles,** R–CN, are generally classified as derivatives of carboxylic acids because they may be hydrolyzed to produce these acids.

$$\underset{\mathbf{1}:\ Z = OH \quad \mathbf{2}:\ Z = Cl,\ OCOR^1,\ OR^2,\ NR^3R^4}{R{-}\overset{\delta^+}{C}(=O^{\delta^-}){-}Z^{\delta^-}}$$

The acyl group defines the chemistry of carboxylic acids and their derivatives. Like the carbonyl group, C=O, in aldehydes and ketones (Chap. 18), the carbon-oxygen π-bond in carboxylic acids and their derivatives is polarized because of the electronegativity of the oxygen atom. The *electrophilicity* of the **acyl carbon atom** induced by this polarization is further increased by the presence of the other electronegative substituent, Z. This combination of an electronegative substituent attached to the polarized carbon-oxygen double bond has important consequences on the reactivity of carboxylic acid derivatives toward nucleophiles. Indeed, the general type of reaction that is most characteristic of these compounds is **nucleophilic acyl substitution reaction** (Eq. 20.1).

$$R{-}C(=O){-}Z + Nu^- \longrightarrow R{-}C(=O){-}Nu + Z^- \tag{20.1}$$

A number of mechanistic pathways exist for nucleophilic acyl substitution. In the simplest of these, a negatively charged nucleophile, Nu^-, attacks the electrophilic acyl carbon atom of **3** to give the tetrahedral intermediate **4**. This then collapses to regenerate the carbon-oxygen double bond with loss of the **leaving group,** Z^-, to provide a substitution product **5,** which is also a carboxylic acid derivative (Eq. 20.2). The first step in this reaction may be considered to be a Lewis acid-Lewis base reaction in which the acyl carbon atom is the Lewis acid and the nucleophile is the Lewis base.

See *Who was Lewis?*

$$\underset{\substack{\mathbf{3}\\ \text{A carboxylic}\\ \text{acid derivative}}}{R{-}\overset{\delta^+}{C}(=O^{\delta^-}){-}Z^{\delta^-}} + Nu^- \longrightarrow \underset{\mathbf{4}}{R{-}C(O^-)(Nu){-}Z} \longrightarrow \underset{\substack{\mathbf{5}\\ \text{A nucleophilic}\\ \text{substitution product}}}{R{-}C(=O){-}Nu} + Z^- \tag{20.2}$$

When weaker, typically neutral, nucleophiles or less reactive carboxylic acid derivatives are involved, the reaction may be catalyzed by acid as outlined in Scheme 20.1. The oxygen atom of the acyl group in **3** is first protonated to give the cation **6** in which the carbon-oxygen bond is even more highly polarized and the acyl carbon atom more electrophilic than in **3.** Nucleophilic attack by a neutral nucleophile, H–Nu, then gives the tetrahedral intermediate **7,** which will transfer a proton to a base, B^-, in the reaction mixture to give **8.** The details of how the tetrahedral intermediate **8** is converted to the substitution product **5** will depend upon the nature of the leaving group, Z^-, the nucleophile, Nu^-, and the reaction conditions. For example, if Z^- is not a good leaving group relative to Nu^-, it will likely be protonated to give **9** prior to breaking of the C–Z bond; the cation **9** will then collapse to give **5.**

Scheme 20.1

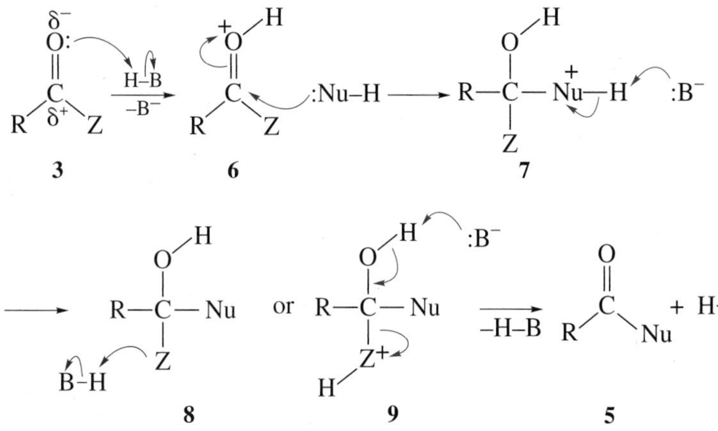

$$\underset{\mathbf{3}}{R\overset{\delta^+}{C}(=\overset{\delta^-}{O}{:})Z} \xrightarrow[-B^-]{H\text{–}B} \underset{\mathbf{6}}{R\text{–}C(=\overset{+}{O}H)\text{–}Z} \xrightarrow{:Nu\text{–}H} \underset{\mathbf{7}}{R\text{–}C(OH)(Z)\text{–}\overset{+}{Nu}\text{–}H}\;\; :B^-$$

$$\longrightarrow \underset{\mathbf{8}}{R\text{–}C(OH)(Z)\text{–}Nu}\;\; B\text{–}H \;\; \text{or} \;\; \underset{\mathbf{9}}{R\text{–}C(OH)(\overset{+}{Z}H)\text{–}Nu}\;\; :B^- \xrightarrow{-H\text{–}B} \underset{\mathbf{5}}{R\text{–}C(=O)\text{–}Nu} + H\text{–}Z$$

Comparison of the reactions of carboxylic acid derivatives with those of carbonyl compounds (Chap. 18) reveals some similarities as well as some important differences. The reactions of carbonyl compounds, whether they are acid-catalyzed or not, lead to the formation of *nucleophilic addition products* according to Equation 20.3, whereas carboxylic acid derivatives react with nucleophiles to give **substitution products** (Eq. 20.1). In both cases the first step is nucleophilic attack on the electrophilic carbon atom of the carbon-oxygen double bond. However, aldehydes and ketones do not have an electronegative group, Z, that can form a stable anion and thus serve as a good leaving group. A good leaving group must be a relatively weak base such as Cl^-, RO^-, H_2O, and ROH, and not a very strong base like H^- or R^- (Sec. 14.1).

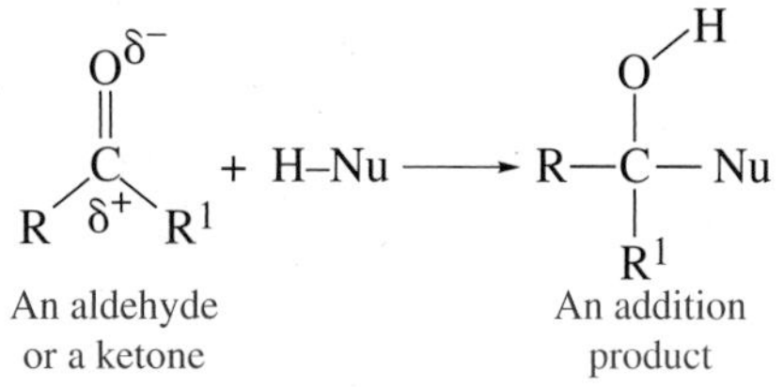

$$\underset{\text{An aldehyde or a ketone}}{R\overset{\delta^+}{C}(=O^{\delta^-})R^1} + H\text{–}Nu \longrightarrow \underset{\text{An addition product}}{R\text{–}C(OH)(R^1)\text{–}Nu} \qquad (20.3)$$

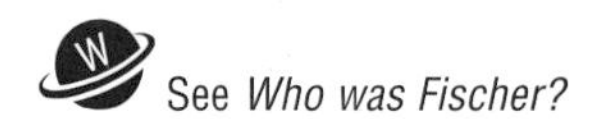

The experiments described in this chapter illustrate some of the representative reactions of carboxylic acids and their derivatives. For example, you will perform a **Fischer esterification,** one of the classic reactions in organic chemistry that was discovered by the Nobel laureate Emil Fischer (see the Historical Highlight at the end of Chapter 23 for an account of the life of this famous chemist), to prepare the anesthetic agent benzocaine from *p*-aminobenzoic acid (Sec. 20.2). In another experiment, you will synthesize the mosquito repellent *N,N*-diethyl-*m*-toluamide (DEET) by a two-step process that involves the conversion of a carboxylic acid into an acid chloride and subsequent reaction with an amine to produce the desired amide (Sec. 20.3).

Finally, you will prepare a hydrazide, a close relative of an amide, and study its chemical decomposition in a process that produces visible light.

20.2 ESTERS AND THE FISCHER ESTERIFICATION

The direct conversion of a carboxylic acid and an alcohol into an ester is usually catalyzed by a mineral acid such as concentrated sulfuric acid, as outlined in Scheme 20.2. The catalyst promotes the addition of the weakly nucleophilic alcohol to the carbon-oxygen double bond by protonating the carbonyl oxygen of the acid **10** to give **11.** Consequently, **11** is activated toward nucleophilic attack and hence reacts rapidly with an alcohol to give the tetrahedral intermediate **12.** Two proton transfer steps follow to give **13,** which collapses by loss of water, a good leaving group, to provide first **14** and then the ester **15.**

Scheme 20.2

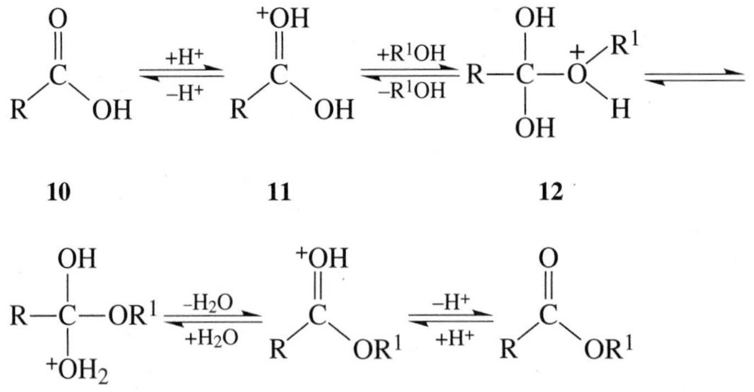

$$\underset{\mathbf{10}}{\mathrm{R{-}C({=}O){-}OH}} \underset{-\mathrm{H^+}}{\overset{+\mathrm{H^+}}{\rightleftharpoons}} \underset{\mathbf{11}}{\mathrm{R{-}C({=}\overset{+}{O}H){-}OH}} \underset{-\mathrm{R^1OH}}{\overset{+\mathrm{R^1OH}}{\rightleftharpoons}} \underset{\mathbf{12}}{\mathrm{R{-}C(OH)_2{-}\overset{+}{O}(H)R^1}} \rightleftharpoons$$

$$\underset{\mathbf{13}}{\mathrm{R{-}C(OH)(\overset{+}{O}H_2){-}OR^1}} \underset{+\mathrm{H_2O}}{\overset{-\mathrm{H_2O}}{\rightleftharpoons}} \underset{\mathbf{14}}{\mathrm{R{-}C({=}\overset{+}{O}H){-}OR^1}} \underset{+\mathrm{H^+}}{\overset{-\mathrm{H^+}}{\rightleftharpoons}} \underset{\mathbf{15}}{\mathrm{R{-}C({=}O){-}OR^1}}$$

As you can see from the multiple equilibria of Scheme 20.2, the acid-catalyzed hydrolysis of an ester **15** to the corresponding acid **10** is simply the reverse of the esterification. Special strategies are therefore required to effect esterification of a carboxylic acid in good yield. These may involve using a large excess of one of the reactants or removing one of the products as it is formed. Both of these experimental tactics make use of the Le Chatelier principle (Sec. 10.3). When the alcohol is inexpensive and volatile, it is frequently used as the solvent, thus forcing the equilibrium to the right and increasing the yield of the ester. In other cases, it may be possible to remove the product water from the reaction using a drying agent, like activated molecular sieves (Sec. 2.24), or by azeotropic distillation (Sec. 4.4). Often, ester formation by the Fischer esterification is inefficient because of the existence of the equilibration, and esters are more generally prepared by the reaction of alcohols with acid chlorides or acid anhydrides.

See *Who was Le Chatelier?*
See more on *Molecular Sieves*
See more on *Azeotropes*
See more on *Benzocaine*

In the procedure described in this section, you will perform the Fischer esterification of *p*-aminobenzoic acid (**16**) with ethanol to give ethyl *p*-aminobenzoate (**17**), more commonly known as benzocaine, according to Equation 20.4. Benzocaine is a useful topical anesthetic, and the discovery of its anesthetic properties represents an interesting example of "rational drug design" as described in the Historical Highlight at the end of this chapter.

$$p\text{-}H_2NC_6H_4CO_2H \;(\mathbf{16}) + CH_3CH_2OH \underset{\text{(catalyst)}}{\overset{\text{conc. } H_2SO_4}{\rightleftharpoons}} p\text{-}H_2NC_6H_4CO_2CH_2CH_3 \;(\mathbf{17}) + H_2O \quad (20.4)$$

16 *p*-Aminobenzoic acid

17 Ethyl *p*-aminobenzoate (benzocaine)

The equlibrium between the starting materials and products is readily established by heating a mixture of **16** and ethanol under reflux in the presence of a catalytic amount of sulfuric acid to give **17** and water (Eq. 20.4). However, if a microwave apparatus is available for heating (Sec. 2.9), the reaction can be effected in a pressurized vial. It is instructive to compare the reaction times required and the results obtained using conventional versus microwave heating for promoting the reaction.

EXPERIMENTAL PROCEDURES

Preparation of Benzocaine

Purpose To demonstrate the acid-catalyzed esterification of a carboxylic acid with an alcohol.

SAFETY ALERT

***Concentrated* sulfuric acid is *very corrosive* and may cause serious chemical burns if allowed to come into contact with your skin. Wear latex gloves when handling this chemical. When it is poured from the reagent bottle, some may run down the outside of the bottle. If any concentrated sulfuric acid comes into contact with your skin, immediately flood the affected area with cold water and then with 5% sodium bicarbonate solution.**

MINISCALE PROCEDURE

Preparation Sign in at **www.cengage.com/login** to answer Pre-Lab Exercises, access videos, and read the MSDSs for the chemicals used or produced in this procedure. Review Sections 2.9, 2.10, 2.11, 2.17, and 2.22.

Apparatus A 25-mL round-bottom flask, apparatus for heating under reflux, magnetic stirring, vacuum filtration, and *flameless* heating.

Setting Up Add 1.0 g of *p*-aminobenzoic acid and 10 mL of *absolute* ethanol to the round-bottom flask containing a spinvane. Stir the mixture until the solid is *completely* dissolved. Add 1 mL of *concentrated* sulfuric acid *dropwise* to the ethanolic

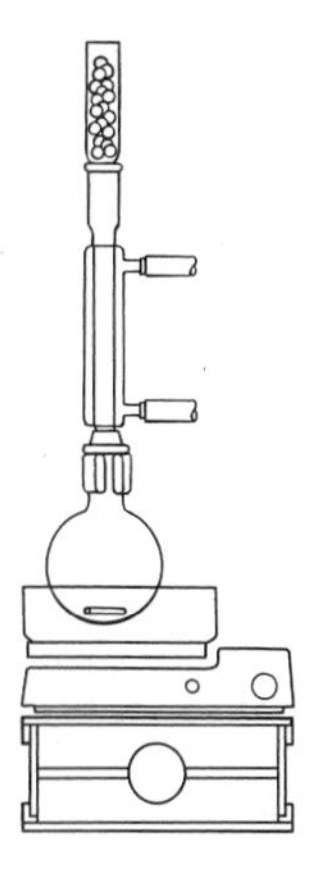

solution of *p*-aminobenzoic acid, equip the flask with a condenser, and set up the apparatus for heating under reflux.

Reaction Heat the mixture under gentle reflux for 30 min. If any solid remains in the flask at this time, remove the heat source and allow the mixture to cool for 2–3 min. Add 3 mL of ethanol and 0.5 mL of concentrated H_2SO_4 to the reaction flask and resume heating under reflux. After the reaction mixture becomes homogeneous, continue heating it under gentle reflux for another 30 min. *For the experiment to succeed, it is important that all of the solids dissolve during the period of reflux.*★

Work-Up and Isolation Allow the reaction mixture to cool to room temperature, and then pour it into a beaker containing 30 mL of water. Bring the mixture to a pH of about 8 by *slowly* adding 10% aqueous sodium carbonate with stirring. Be careful in this step, as frothing occurs during the neutralization. Beforehand, you should calculate the approximate volume of 10% aqueous sodium carbonate that will be required to neutralize the *total* amount of sulfuric acid you used. By vacuum filtration, collect the crude benzocaine that precipitates. Use three 10-mL portions of cold water to rinse the solid from the beaker and wash the filter cake, and then air-dry the product.★

Purification Weigh the crude product and transfer it to a 50-mL Erlenmeyer flask containing a stirbar and 20 mL of water. Heat the mixture with stirring to about 60 °C and then add just enough methanol to dissolve the solid (5–10 mL); do *not* add more methanol than necessary. When the solid has dissolved, allow the solution to cool to room temperature and then cool it in an ice-water bath for 10–15 min to complete crystallization. Isolate the crystals by vacuum filtration, wash them with 5–10 mL of cold water, and allow them to air-dry.

Analysis Weigh the recrystallized product and calculate the percent yield; determine its melting point. Obtain IR and ^1H NMR spectra of your starting materials and product, and compare them with those of authentic samples (Figs. 20.1–20.6).

MINISCALE PROCEDURE FOR MICROWAVE OPTION

Preparation Sign in at **www.cengage.com/login** to answer Pre-Lab Exercises, access videos, and read the MSDSs for the chemicals used or produced in this procedure. Review Sections 2.9, 2.10, 2.11, 2.17, and 2.22.

Apparatus A 10-mL pressure-rated tube with cap, stirbar, apparatus for microwave heating with magnetic stirring and vacuum filtration, ice-water bath.

Setting Up Place 0.50 g of *p*-aminobenzoic acid and a stirbar in the 10-mL tube. Add 5 mL of *absolute* ethanol. Add about 0.3 mL (6 drops) of *concentrated* sulfuric acid, seal the vial with the cap, and place it in the cavity of the microwave apparatus.

Reaction Program the unit to heat the reaction mixture with stirring according to the directions provided by your instructor. Generally, the reaction temperature should be set at 140 °C and the power set at a maximum of 300 W with a 1-min ramp time and a 5-min hold time; the pressure limit should be set at 250 psi.

Work-Up and Isolation Allow the reaction mixture to cool to room temperature and then pour it into a beaker containing 15 mL of water. Bring the mixture to a pH of about 8 by *slowly* adding 10% aqueous sodium carbonate with stirring. Be careful in this step, as frothing occurs during the neutralization. Beforehand, you should calculate the approximate volume of 10% aqueous sodium carbonate that will be required to neutralize the amount of sulfuric acid that you used. By vacuum filtration, collect the crude benzocaine that precipitates. Use three 5-mL portions of cold water to rinse the solid from the beaker and wash the filter cake, and then air-dry product.★

Purification Weigh the crude product and transfer it to a 25-mL Erlenmeyer flask containing a stirbar and 20 mL of water. Heat the mixture with stirring to about 60 °C and then add just enough methanol to dissolve the solid (3–5 mL); do *not* add more methanol than necessary. When the solid has dissolved, allow the solution to cool to room temperature and then cool it in an ice-water bath for 10–15 min to complete crystallization. Isolate the crystals by vacuum filtration, wash them with 3–5 mL of cold water, and allow them to air-dry.

Analysis Weigh the recrystallized product and calculate the percentage yield; determine the melting point. Obtain IR and ^{1}H NMR spectra of your starting materials and product, and compare them with those of authentic samples (Figs. 20.1–20.6).

MICROSCALE PROCEDURE

Preparation Sign in at **www.cengage.com/login** to answer Pre-Lab Exercises, access videos, and read the MSDSs for the chemicals used or produced in this procedure. Review Sections 2.9, 2.10, 2.11, 2.17, and 2.22.

Apparatus A 5-mL conical vial, apparatus for heating under reflux, magnetic stirring, vacuum filtration, and *flameless* heating.

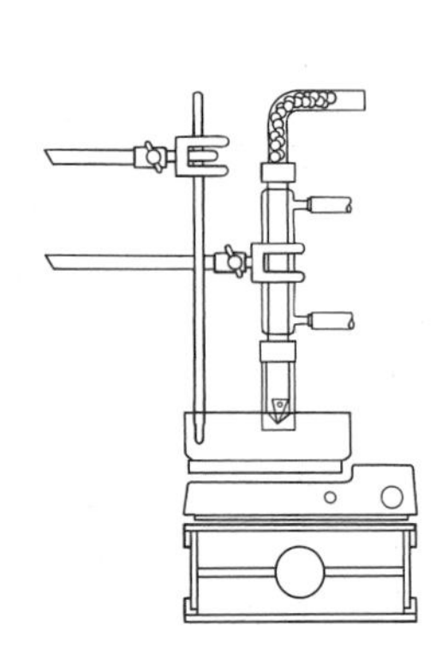

Setting Up Add 0.2 g of *p*-aminobenzoic acid and 2.0 mL of *absolute* ethanol to the conical vial containing a spinvane. Stir the mixture until the solid is *completely* dissolved. Add 0.2 mL of concentrated sulfuric acid *dropwise* to the ethanolic solution of *p*-aminobenzoic acid, equip the vial with a condenser, and set up the apparatus for heating under reflux.

Reaction Heat the mixture under gentle reflux for 20 min. If any solid remains in the vial at this time, remove the heat source and allow the mixture to cool for 2–3 min. Then add 0.5 mL of *absolute* ethanol and 0.1 mL of concentrated H_2SO_4 to the reaction vial and resume heating under gentle reflux. After the reaction mixture becomes homogeneous, continue heating it under gentle reflux for another 30 min. *For the experiment to succeed, it is important that all of the solids dissolve during the period of reflux.*★

Work-Up and Isolation Allow the reaction mixture to cool to room temperature and then pour it into a beaker containing 5 mL of water. Bring the mixture to a pH of about 8 by *slowly* adding 10% aqueous sodium carbonate with stirring. Be careful in this step, as frothing occurs during the neutralization. Beforehand, you should calculate the approximate volume of 10% aqueous sodium carbonate that will be required to neutralize the *total* amount of sulfuric acid you used. By vacuum filtration, collect the crude benzocaine that precipitates. Use three 2-mL portions of cold water to rinse the solid from the beaker and wash the filter cake, and then air-dry the product.★

Purification Weigh the crude product and transfer it to a 10-mL Erlenmeyer flask containing a magnetic stirbar and 4 mL of water. Heat the mixture with stirring to about

60 °C and then add just enough methanol to dissolve the solid (1–2 mL); do *not* add more methanol than necessary. When the solid has dissolved, allow the solution to cool to room temperature and then cool it in an ice-water bath for 10–15 min to complete crystallization. Isolate the crystals by vacuum filtration, wash them with 1–2 mL of cold water, and allow them to air-dry.

Analysis Weigh the recrystallized product and calculate the percent yield; determine its melting point. Obtain IR and ^1H NMR spectra of your starting materials and product, and compare them with those of authentic samples (Figs. 20.1–20.6).

WRAPPING IT UP

Flush all *filtrates* down the drain.

EXERCISES

Sign in at **www.cengage.com/login** and use the spectra viewer and Tables 8.1–8.8 as needed to answer the blue-numbered questions on spectroscopy.

1. Why is absolute ethanol rather than 95% ethanol used in this experiment?
2. Propose a structure for the solid that separates when concentrated sulfuric acid is added to the solution of *p*-aminobenzoic acid in ethanol.
3. Why is it important that the solid referred to in Exercise 2 dissolve during the reaction in order to obtain a good yield of ethyl *p*-aminobenzoate?
4. Sodium carbonate is used to adjust the pH to 8 during the work-up of the reaction.
   a. Why is it necessary to adjust to this pH prior to extracting the aqueous mixture with diethyl ether?
   b. What undesired reaction might occur if you made the solution strongly basic with aqueous sodium hydroxide?
   c. Using curved arrows to symbolize the flow of electrons, write the stepwise mechanism for the reaction of Part **b.**
5. Consider the equilibrium for the esterification of *p*-aminobenzoic acid (**16**) to give ethyl *p*-aminobenzoate (**17**) shown in Equation 20.4.
   a. Write the mathematical expression for the equilibrium constant, K_{eq}, as a function of the concentrations of the products and reactants for this reaction.
   b. Provide two specific ways whereby the Le Chatelier principle may be applied to this equilibrium to drive the reaction completely to the ester **17.**
   c. Which one of the techniques given in Part **b** is used in this experiment?
6. Using curved arrows to symbolize the flow of electrons, write the stepwise mechanism for the acid-catalyzed esterification of **16** to give **17.**
7. A strong acid is generally used to catalyze the Fischer esterification of carboxylic acids. What *two* steps in the reaction are accelerated by the presence of strong acid, and what function does the acid play in each of these steps?
8. Consider the spectral data for *p*-aminobenzoic acid (Figs. 20.1 and 20.2).
   a. In the functional group region of the IR spectrum, specify the absorption associated with the carbonyl component of the carboxyl group. What

functional group is responsible for the broad absorption in the region of about 2800 cm^{-1}, and why is the absorption broad?

b. In the ^1H NMR spectrum, assign the various resonances to the hydrogen nuclei responsible for them.

c. For the ^{13}C NMR data, assign the various resonances to the carbon nuclei responsible for them.

9. Consider the spectral data for ethanol (Figs. 20.3 and 20.4).

   a. What functional group accounts for the broad absorption centered at about 3350 cm^{-1} in the IR spectrum, and why it is broad?

   b. In the ^1H NMR spectrum, assign the various resonances to the hydrogen nuclei responsible for them.

   c. In the ^{13}C NMR spectrum, assign the various resonances to the carbon nuclei responsible for them.

10. Consider the spectral data for ethyl *p*-aminobenzoate (Figs. 20.5 and 20.6).

    a. In the functional group region of the IR spectrum, specify the absorption associated with the carbonyl component of the ester group. What functional group is responsible for the broad absorption in the region of about 3200 cm^{-1}, and why is the absorption broad?

    b. In the ^1H NMR spectrum, assign the various resonances to the hydrogen nuclei responsible for them.

    c. For the ^{13}C NMR data, assign the various resonances to the carbon nuclei responsible for them.

11. Discuss the differences observed in the IR and NMR spectra of *p*-aminobenzoic acid and ethyl *p*-aminobenzoate that are consistent with the formation of the latter in this procedure.

SPECTRA

Starting Materials and Products

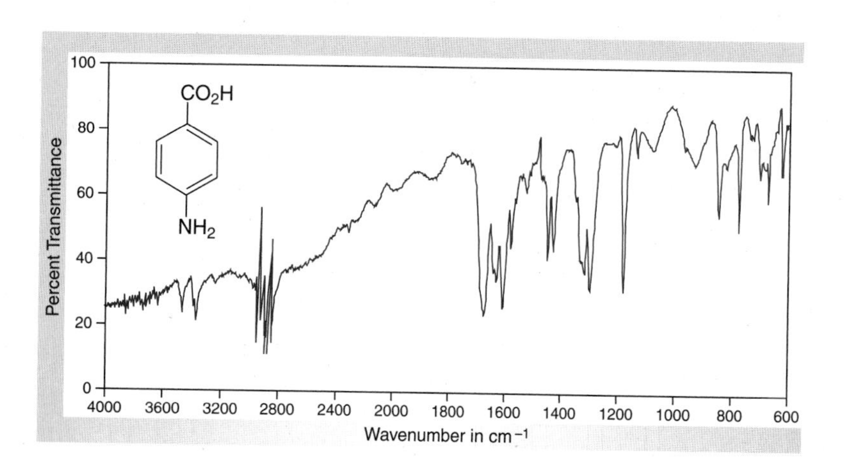

Figure 20.1
IR spectrum of p-*aminobenzoic acid (IR card).*

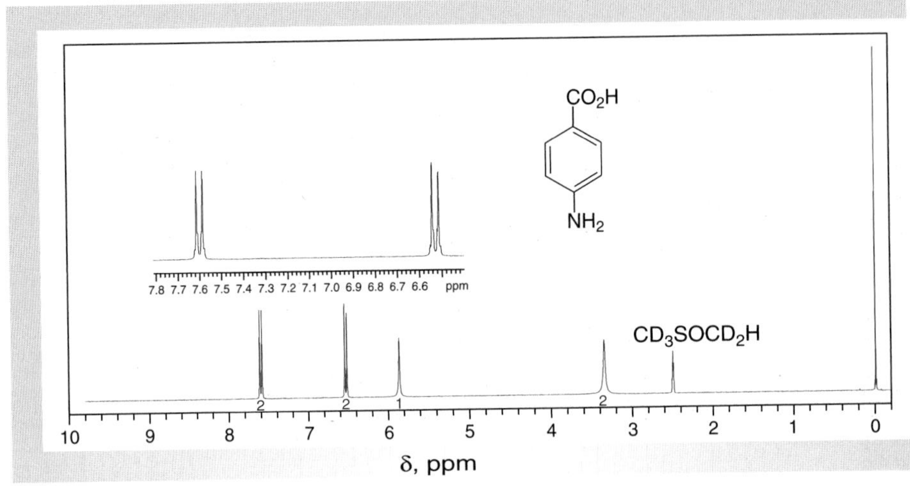

Figure 20.2
NMR data for p-aminobenzoic acid (DMSO-d_6).

(a) ^1H NMR spectrum (300 MHz).
(b) ^{13}C NMR data: δ 112.6, 116.9, 131.2, 153.1, 167.5.

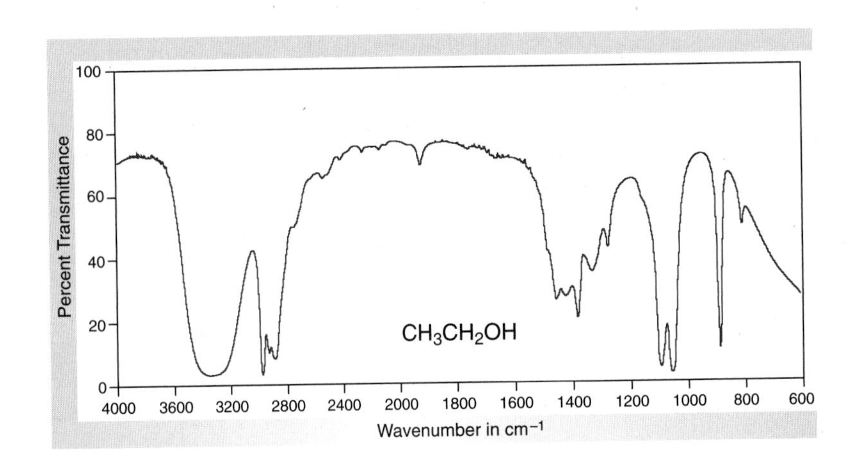

Figure 20.3
IR spectrum of ethanol (neat).

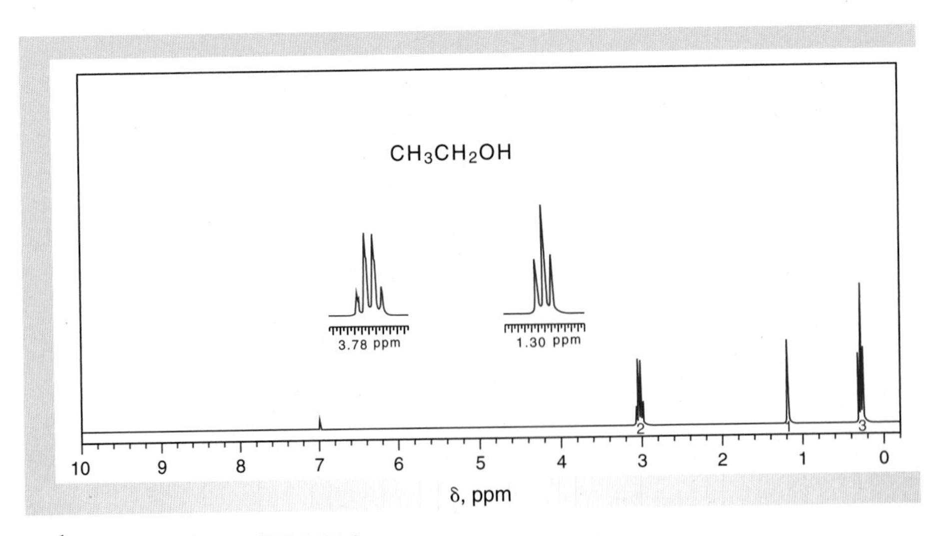

Figure 20.4
NMR data for ethanol ($CDCl_3$).

(a) ^1H NMR spectrum (300 MHz).
(b) ^{13}C NMR data: 18.1, 57.8.

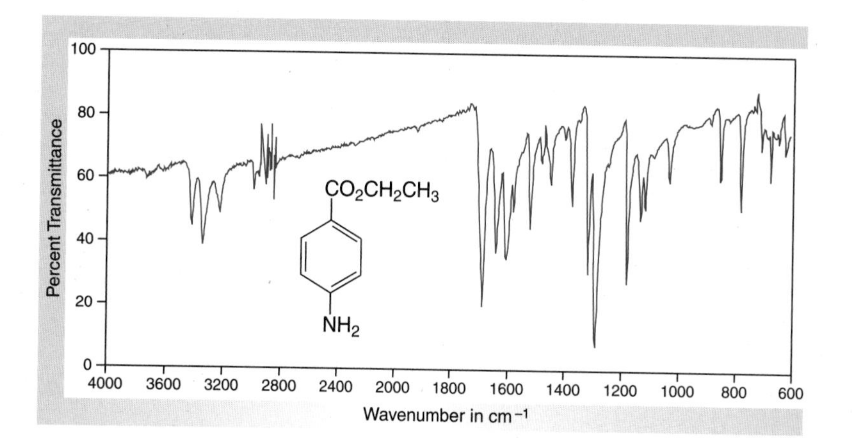

Figure 20.5
IR spectrum of ethyl p-*amino-benzoate (IR card).*

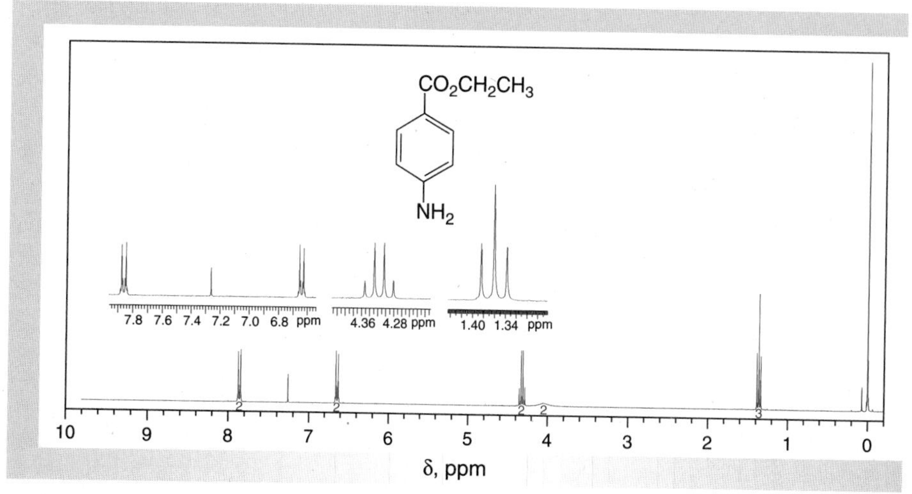

Figure 20.6
NMR data for ethyl p-*aminobenzoate ($CDCl_3$).*

(a) ^{1}H NMR spectrum (300 MHz).
(b) ^{13}C NMR data: δ 14.4, 60.9, 113.8, 120.2, 131.5, 150.7, 166.7.

20.3 AMIDES AND INSECT REPELLENTS

Amides cannot be easily prepared by the direct reaction of ammonia or amines with carboxylic acids because amines function as bases to convert carboxylic acids into their unreactive ammonium carboxylate salts (Eq. 20.5). Although these salts may be converted into amides by heating to high temperatures (>200 °C), the forcing nature of these conditions severely limits the applicability of this method (see Sec. 20.4).

$$\underset{\text{A carboxylic acid}}{\mathrm{R{-}C({=}O)OH}} + \underset{\text{An amine}}{\mathrm{R^1NH_2}} \longrightarrow \underset{\text{An ammonium carboxylate}}{\mathrm{R{-}C({=}O)O^-\;\; R^1\overset{+}{N}H_3}} \xrightarrow{>200°\,C} \underset{\text{An amide}}{\mathrm{R{-}C({=}O)NHR^1}} + H_2O \qquad (20.5)$$

A more general method for preparing amides involves a two-step procedure in which the carboxylic acid is first converted to an acid chloride that is then treated with an amine to give the desired amide according to Equation 20.6. Acid chlorides are among the most reactive derivatives of carboxylic acids and thus are often used to prepare other organic compounds having an acyl group, including esters and aryl alkyl ketones (Sec. 15.3).

$$\text{R–C(=O)OH} \xrightarrow{SOCl_2} \text{R–C(=O)Cl} \xrightarrow{2\,R^1NH_2} \text{R–C(=O)NHR}^2 + R^1\overset{+}{N}H_3\,Cl^- \quad (20.6)$$

A carboxylic acid — An acid chloride — An amide — An ammonium salt

The reagent commonly used to prepare an acid chloride from a carboxylic acid is thionyl chloride, $SOCl_2$. The reaction probably proceeds via the intermediate *mixed anhydride* **18** (Eq. 20.7), which is very reactive. Consequently, it undergoes rapid attack by chloride ion via a nucleophilic acyl substitution in which sulfur dioxide and chloride ion are lost and the acid chloride is produced.

$$\text{R–C(=O)–OH} + \text{Cl–S(=O)–Cl} \xrightarrow{-HCl} \text{R–C(=O)–O–S(=O)–Cl} \;(\mathbf{18}) \xrightarrow[-Cl^-]{-SO_2} \text{R–C(=O)Cl} \quad (20.7)$$

A carboxylic acid — Thionyl chloride — **18** — An acid chloride

The acid chlorides prepared in this fashion are often not isolated and purified, but rather are used directly in a subsequent reaction. This is possible because both of the side products, HCl and SO_2, are gases and readily lost or removed from the mixture. For example, if an amide is the desired product, then the crude acid chloride is simply allowed to react with an excess of an amine. An excess of the amine is used to neutralize the HCl that is generated during the preparation of the acid chloride (Eq. 20.7) *and* by the reaction of the acid chloride with the amine (Eq. 20.6). Like **18,** acid chlorides are highly reactive, and they tend to undergo nucleophilic acyl substitution according to the general mechanism shown in Equation 20.2.

The experimental procedure for the synthesis of *N,N*-diethyl-*m*-toluamide (**21**) nicely illustrates a common method for preparing amides. In the first step, *m*-toluic acid (**19**) reacts with thionyl chloride to give the acid chloride **20** (Eq. 20.8). The acid chloride is treated directly with an excess of diethylamine to give *N,N*-diethyl-*m*-toluamide and diethylammonium hydrochloride (Eqs. 20.9, 20.10).

$$m\text{-}CH_3C_6H_4COOH \;(\mathbf{19}) + SOCl_2 \xrightarrow[-SO_2]{-HCl} m\text{-}CH_3C_6H_4COCl \;(\mathbf{20}) \quad (20.8)$$

19 *m*-Toluic acid — **20** *m*-Toluoyl chloride

$$m\text{-}CH_3C_6H_4COCl + (CH_3CH_2)_2NH \longrightarrow m\text{-}CH_3C_6H_4CON(CH_2CH_3)_2 + HCl \qquad (20.9)$$

20 *m*-Toluoyl chloride; **21** *N,N*-Diethyl-*m*-toluamide

$$HCl + (CH_3CH_2)_2NH \longrightarrow (CH_3CH_2)_2\overset{+}{N}H_2\ Cl^- \qquad (20.10)$$

Diethylammonium chloride

See more on *DEET*

N,N-Diethyl-*m*-toluamide (**21**), commonly known as DEET, is widely used as an insect repellent and is the active ingredient in the commercial repellent OFF™, for example. Whereas other repellents are useful toward only one or two types of insects, DEET is unusual in that it is effective against a broad spectrum of insects, including mosquitoes, fleas, chiggers, ticks, deerflies, and gnats. Although the exact mode of action of *N,N*-diethyl-*m*-toluamide as a repellent is not known, at least for mosquitoes, it appears to interfere with the receptors or sensors that enable the insect to detect moisture gradients. Hence, the female mosquito is unable to find a suitable host or target, like you! So this is an experiment you can do without fear of being bit by a stray mosquito that might wander into the lab, no matter where you live.

EXPERIMENTAL PROCEDURES

Preparation of N,N*-Diethyl-*m*-toluamide*

Purpose To demonstrate the synthesis of an amide from a carboxylic acid via the corresponding acid chloride.

Note: The entire procedure up to the first stopping point should be completed in one laboratory period in order to obtain a good yield of product.

SAFETY ALERT

1. **Thionyl chloride is a volatile and *corrosive* chemical that undergoes rapid reaction with water and atmospheric moisture to produce sulfur dioxide and hydrogen chloride, both of which are also corrosive. Wear latex gloves when handling this chemical and do not breathe it or allow it to come in contact with your skin. If contact occurs, immediately flood the affected area with cold water and then rinse with 5% sodium bicarbonate solution. Dispense thionyl chloride only at a hood and minimize its exposure to atmospheric moisture by keeping the bottle tightly closed when not in use.**

2. **Diethyl ether is *extremely* flammable and volatile, and its vapors can easily travel several feet along the bench top or the floor and then be ignited. Consequently, *be certain there are no open flames anywhere in the laboratory* whenever you are working with it. Use a *flameless* heating source *whenever* heating is required.**
3. **Diethylamine is a noxious-smelling and corrosive chemical. Wear latex gloves when handling this chemical and dispense it only at a hood.**
4. **Make sure your glassware is dry, lubricate the ground-glass joints of the apparatus carefully, and be certain that the joints are intimately mated. Otherwise, noxious gases will escape.**
5. **If possible, this experiment should be performed in a hood.**
6. **Heptane is extremely flammable; use *flameless* heating to concentrate solutions of heptane.**

MINISCALE PROCEDURE

Preparation Sign in at **www.cengage.com/login** to answer Pre-Lab Exercises, access videos, and read the MSDSs for the chemicals used or produced in this procedure. Review Sections 2.9, 2.10, 2.11, 2.13, 2.21, 2.22, 2.23, 2.25, 2.29, and 6.3.

Apparatus A 100-mL round-bottom flask, a 50-mL Erlenmeyer flask, Claisen adapter, gas trap, separatory funnel, separatory funnel with standard-taper glass joint, ice-water bath, reflux condenser, apparatus for magnetic stirring, column chromatography, simple distillation, and *flameless* heating.

Setting Up Oven-dry the apparatus prior to assembling it. Because an *airtight* seal is required for all connections, *carefully grease all joints* when assembling the apparatus. Place 2.0 g of *dry m*-toluic acid in the round-bottom flask and equip the flask with a stirbar and the Claisen adapter. Fit the *sidearm* of the adapter with the reflux condenser and fit the top of the condenser with the gas trap specified by your instructor; one example is illustrated in Figure 20.7. *Working at the hood* and using a graduated pipet, measure 2.2 mL of thionyl chloride into the separatory funnel; be certain that the stopcock of the funnel is firmly seated and closed prior to this transfer. Stopper the funnel and place it on the straight arm of the Claisen adapter above the reaction flask (Fig. 20.7). Prepare a solution of 5.0 mL of diethylamine in 10 mL of *anhydrous* diethyl ether in a small Erlenmeyer flask, stopper the flask, and place it in an ice-water bath. *Be sure the flask is secured so it does not tip over while you are making the acid chloride.*

Reaction Open the stopcock of the funnel fully so that the thionyl chloride is added as rapidly as possible to the flask containing the *m*-toluic acid; it may be necessary to lift the stopper on the funnel *briefly* to equalize the pressure in the system if the flow of thionyl chloride becomes erratic or slow. When the addition has been completed, *close the stopcock* and heat the mixture under gentle reflux for 15 min. The bubbling that is initially observed should slow during the reflux period and the reaction mixture should become clear.

Cool the mixture to room temperature with continued stirring and add 30 mL of *anhydrous* ether to the separatory funnel. Open the stopcock of the funnel fully

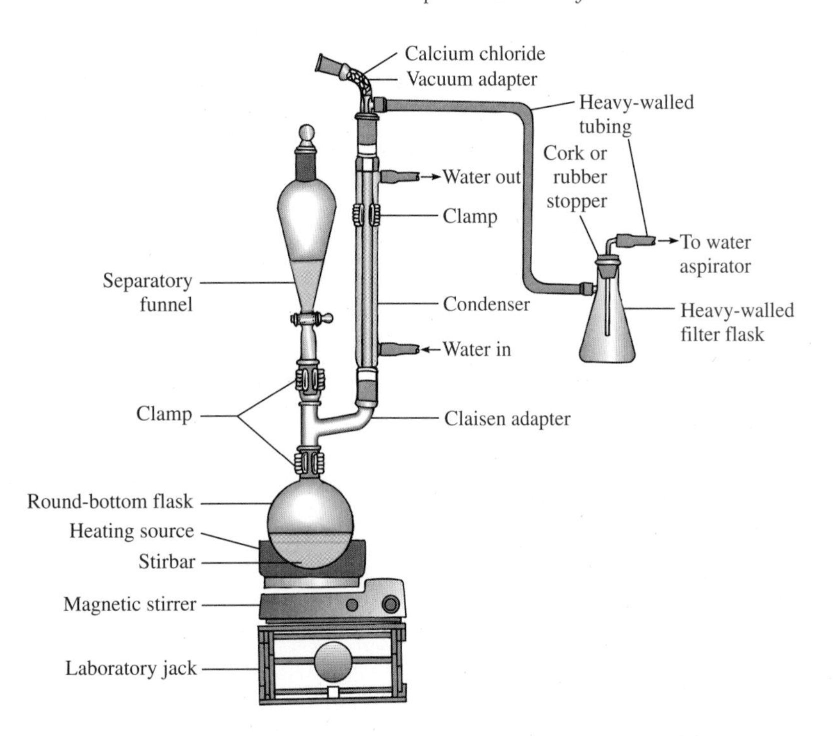

Figure 20.7
Miniscale apparatus for heating reaction under reflux, equipped with a gas trap and a separatory funnel for adding a solution.

so that the ether drains as rapidly as possible into the flask. *Close the stopcock*, transfer the *cold* ethereal solution of diethylamine to the funnel, and stopper the funnel. Some white smoke may form in the funnel during this transfer. Add the ethereal solution of diethylamine dropwise to the stirred solution of acid chloride over a period of about 10–15 min. While this solution is being added to the flask, quantities of white smoke will be produced. After the addition is complete, continue stirring for 15 min to ensure completion of the reaction.

Work-Up and Isolation Remove the gas trap, separatory funnel, condenser, and Claisen adapter from the reaction flask. Add 15 mL of 2.5 *M* aqueous sodium hydroxide to the stirred solution and continue stirring for 15 min; swirl the flask if necessary to dissolve all of the white solid. Transfer the reaction mixture to a separatory funnel, separate the layers, and remove the lower aqueous layer. Wash the organic layer with 15 mL of *cold* 3 *M* hydrochloric acid, *venting as necessary*, and then with 15 mL of cold water. Dry the organic layer over several spatula-tips full of anhydrous sodium sulfate.★ Filter or carefully decant the solution into a *tared* round-bottom flask and remove most of the ether by simple distillation. Alternatively, use rotary evaporation or other techniques to concentrate the solution. The final traces of solvent may be removed by connecting the cool stillpot to a vacuum source. Determine the yield of crude product.★

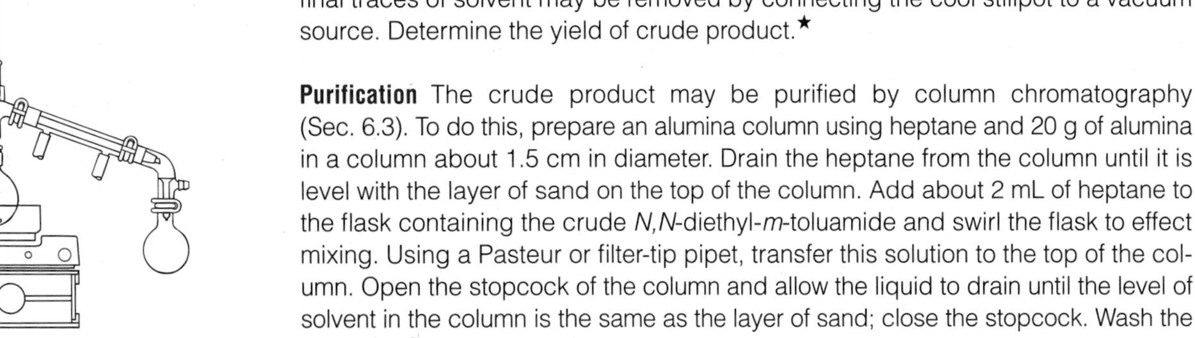

Purification The crude product may be purified by column chromatography (Sec. 6.3). To do this, prepare an alumina column using heptane and 20 g of alumina in a column about 1.5 cm in diameter. Drain the heptane from the column until it is level with the layer of sand on the top of the column. Add about 2 mL of heptane to the flask containing the crude *N,N*-diethyl-*m*-toluamide and swirl the flask to effect mixing. Using a Pasteur or filter-tip pipet, transfer this solution to the top of the column. Open the stopcock of the column and allow the liquid to drain until the level of solvent in the column is the same as the layer of sand; close the stopcock. Wash the flask with 2–3 mL of heptane, transfer this rinse to the top of the column, and drain the column as before. Elute the column with about 50 mL of heptane, collecting the eluant as a single fraction. In order to determine whether all of the product has eluted, collect several drops on a watchglass and evaporate the liquid. If there is no residue, the amide is no longer eluting.★ Transfer the eluant into a *tared* round-bottom flask, and remove most of the heptane by simple distillation. Alternatively, use rotary evaporation or other techniques to concentrate the solution. The final traces of heptane may be removed by connecting the cool stillpot to a vacuum source.

Analysis Weigh the product and calculate the percent yield. Obtain IR and ^{1}H NMR spectra of your starting material and product, and compare them with those of authentic samples (Figs. 20.9–20.12).

MICROSCALE PROCEDURE

Preparation Sign in at **www.cengage.com/login** to answer Pre-Lab Exercises, access videos, and read the MSDSs for the chemicals used or produced in this procedure. Review Sections 2.5, 2.9, 2.10, 2.11, 2.13, 2.21, 2.22, 2.23, 2.25, 2.29, and 6.3.

Apparatus A 3- and 5-mL conical vial, 2-mL glass syringe, Claisen adapter, gas trap, ice-water bath, reflux condenser, a screw-cap centrifuge tube, apparatus for magnetic stirring, column chromatography, simple distillation, and *flameless* heating.

Setting Up This reaction should be set up at a hood. Oven-dry the apparatus prior to assembling it. Because an *airtight* seal is required for all connections, *carefully grease all joints* when assembling the apparatus. Place 0.2 g of *dry m*-toluic acid in the 5-mL conical vial containing a spinvane. Equip the vial with a Claisen adapter, and fit the *sidearm* of the adapter with the reflux condenser. Fit the top of the condenser with the gas trap specified by your instructor; one example is illustrated in Figure 20.8. If you use the gas trap shown, be sure that no water from the moistened cotton runs into the condenser. Prepare a solution of 0.5 mL of diethylamine in 2.0 mL of *anhydrous* diethyl ether in a screw-cap centrifuge tube, screw on the cap, and place the tube in an ice-water bath. *Be sure the tube is secured so it does not tip over while you are making the acid chloride.*

Reaction *Working at the hood*, transfer 0.2 mL of thionyl chloride into the vial using a graduated pipet. Seal the joint of the *straight portion* of the adapter with a rubber septum. The apparatus may now be moved to your desk if desired. Heat the mixture under gentle reflux for 15 min. The bubbling that is initially observed should slow during the reflux period and the reaction mixture should become clear.

Cool the mixture to room temperature with continued stirring. Using the syringe, add 1.0 mL of *anhydrous* diethyl ether through the rubber septum to the vial. Add the ethereal solution of diethylamine dropwise to the stirred solution of acid chloride over a period of about 2–3 min. While this solution is being added to the flask, quantities of white smoke will be produced. After the addition is complete, continue stirring for 15 min to ensure completion of the reaction.

Work-Up and Isolation Remove the rubber septum from the Claisen adapter, add 1.5 mL of 2.5 *M* aqueous sodium hydroxide to the stirred solution, and continue stirring for 15 min. If necessary, screw a cap onto the vial and shake it briefly to dissolve all of the white solid. Remove the lower aqueous layer with a Pasteur pipet and

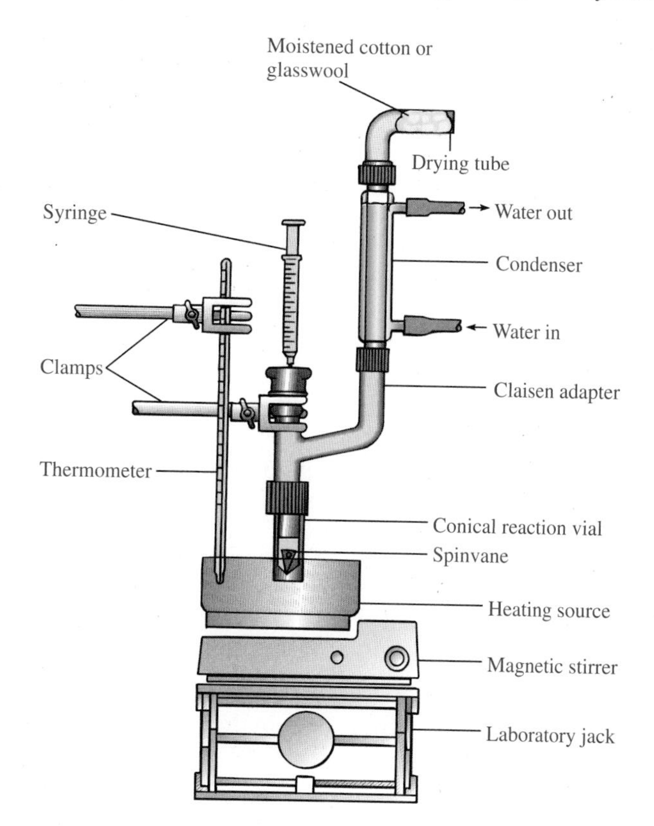

Figure 20.8
Microscale apparatus for heating under reflux, equipped with a gas trap and a syringe for adding a solution.

wash the organic layer with 1.5 mL of *cold* 3 *M* hydrochloric acid, *venting as necessary*, and then with 1.5 mL of cold water. Transfer the organic layer to a dry test tube using a Pasteur or filter-tip pipet and add several microspatula-tips full of anhydrous sodium sulfate. Allow the solution to stand for 10–15 min, with occasional swirling to hasten drying.★ Using a Pasteur or filter-tip pipet, transfer the organic solution to a *tared* dry 5-mL conical vial, and add a spinvane to the vial. Remove most of the ether by simple distillation. Alternatively, use rotary evaporation or other techniques to concentrate the solution. The final traces of solvent may be removed by connecting the cool stillpot to a vacuum source. Determine the yield of crude product.★

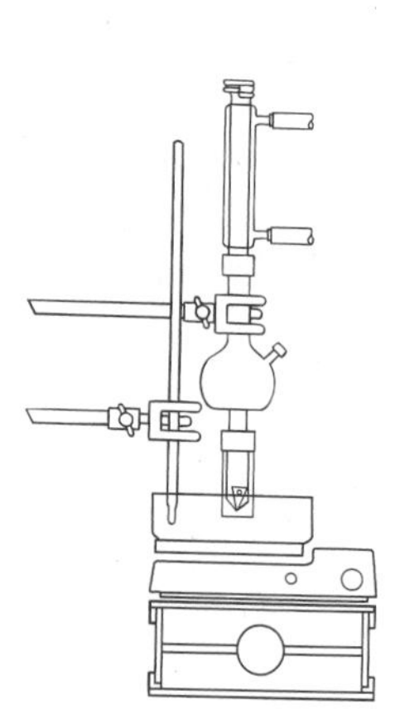

Purification The crude product may be purified by column chromatography (Sec. 6.3). To do this, prepare an alumina column using heptane and 2 g of alumina in a Pasteur filtering pipet (see Fig. 2.9b). Dissolve your crude product in about 0.2 mL of heptane and transfer the solution to the top of the alumina column. Wash the column with about 5 mL of heptane, collecting the eluant as a single fraction. In order to determine whether all of the product has eluted, collect several drops on a watchglass and evaporate the liquid. If there is no residue, the amide is no longer eluting.★ Transfer the eluant into a *tared* round-bottom flask and remove most of the heptane by simple distillation. Alternatively, use rotary evaporation or other techniques to concentrate the solution. The final traces of heptane may be removed by connecting the cool stillpot to a vacuum source. The pure product should be a colorless liquid.

Analysis Weigh the product and calculate the percent yield. Obtain IR and ^{1}H NMR spectra of your starting material and product and compare them with those of authentic samples (Figs. 20.9–20.12).

WRAPPING IT UP

Combine all of the *aqueous layers* and determine the pH with pHydrion paper. Adjust the pH to approximately 7 with either solid sodium carbonate or 3 *M* HCl. Flush all *filtrates* down the drain. Pour the recovered *diethyl ether* and the *heptane* in a container for nonhalogenated organic liquids. Spread the *sodium sulfate* and the *alumina* on a tray in the hood to evaporate residual solvent, and then put them in the container for nonhazardous solids.

EXERCISES

1. Write the balanced equation for the reaction of thionyl chloride with water.
2. Write the balanced equation for the reaction of *m*-toluoyl chloride with water.
3. What is the white solid that is formed when diethylamine is added to the reaction mixture containing *m*-toluoyl chloride?
4. What is the white smoke that is formed when diethylamine is added to the reaction mixture containing *m*-toluoyl chloride?
5. The thionyl chloride remaining after the formation of the acid chloride also reacts with diethylamine. Write a balanced equation for this reaction.
6. In the work-up of the reaction, the organic layer is extracted with 2.5 *M* sodium hydroxide solution. What is removed from the organic layer in this extraction?
7. What is the purpose of the next step of the work-up in which the organic layer is washed with 3 *M* hydrochloric acid?
8. Using curved arrows to symbolize the flow of electrons, write a stepwise mechanism for the formation of *m*-toluoyl chloride from *m*-toluic acid using thionyl chloride according to Equation 20.8.
9. Using curved arrows to symbolize the flow of electrons, write a stepwise mechanism for the formation of *N,N*-diethyl-*m*-toluamide from *m*-toluoyl chloride with diethylamine according to Equation 20.9.

Sign in at **www.cengage.com/login** and use the spectra viewer and Tables 8.1–8.8 as needed to answer the blue-numbered questions on spectroscopy.

10. Consider the spectral data for *m*-toluic acid (Figs. 20.9 and 20.10).
    - **a.** In the functional group region of the IR spectrum, specify the absorption associated with the carbonyl component of the carboxyl group and with the π-bonds of the aromatic ring.
    - **b.** In the ^{1}H NMR spectrum, assign the various resonances to the hydrogen nuclei responsible for them.
    - **c.** For the ^{13}C NMR data, assign the various resonances to the carbon nuclei responsible for them.
11. Consider the spectral data for *N,N*-diethyl-*m*-toluamide (Figs. 20.11 and 20.12).
    - **a.** In the functional group region of the IR spectrum, specify the absorption associated with the carbonyl component of the amide group and with the π-bonds of the aromatic ring.

b. In the ^{1}H NMR spectrum, assign the various resonances to the hydrogen nuclei responsible for them.

c. For the ^{13}C NMR data, assign the various resonances to the carbon nuclei responsible for them.

12. Discuss the differences observed in the IR and NMR spectra of *m*-toluic acid and *N,N*-diethyl-*m*-toluamide that are consistent with the formation of the latter in this procedure.

SPECTRA

Starting Materials and Products

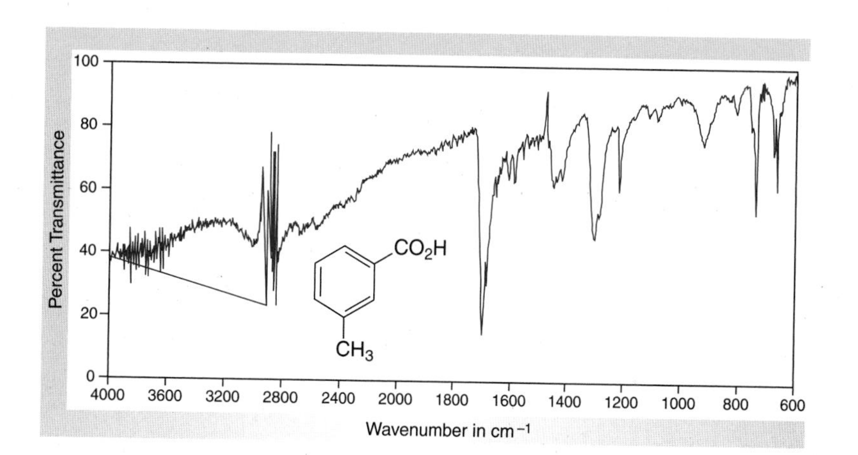

Figure 20.9
IR spectrum of m-toluic acid *(IR card).*

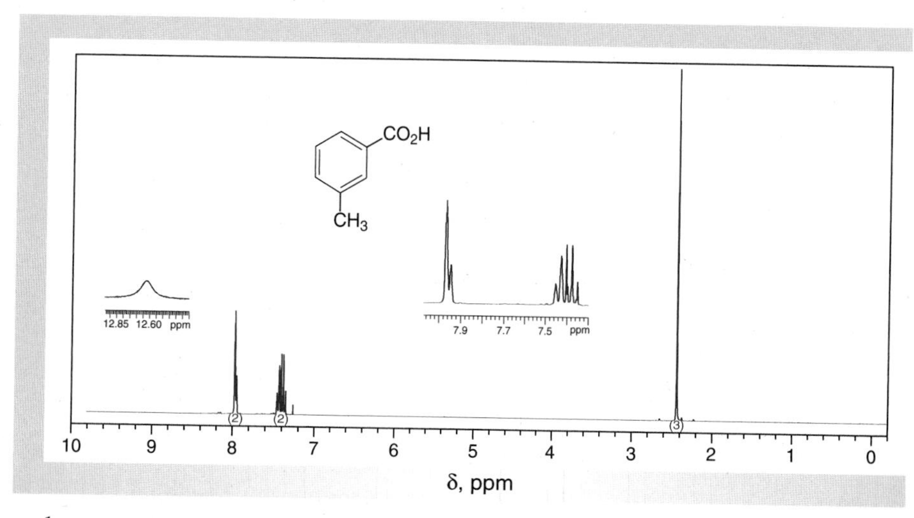

Figure 20.10
NMR data for m-toluic acid *($CDCl_3$).*

(a) ^{1}H NMR spectrum (300 MHz).
(b) ^{13}C NMR data: δ 21.2, 127.4, 128.4, 129.2, 130.7, 134.6, 138.3, 172.8.

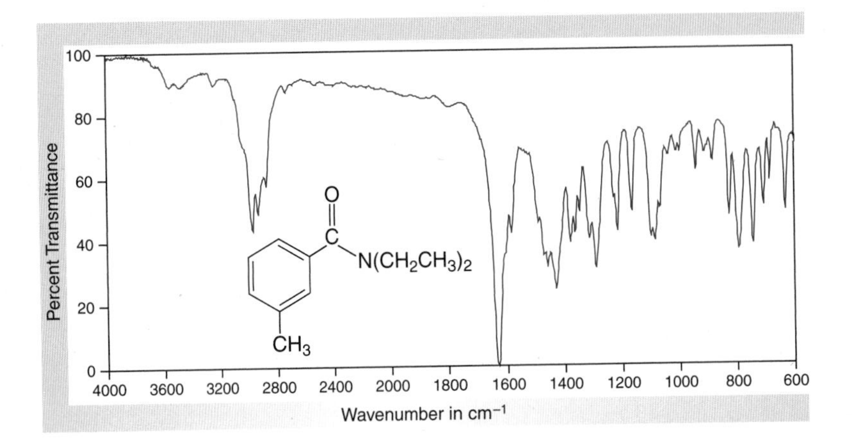

Figure 20.11
IR spectrum of N,N-*diethyl*-m-*toluamide (neat).*

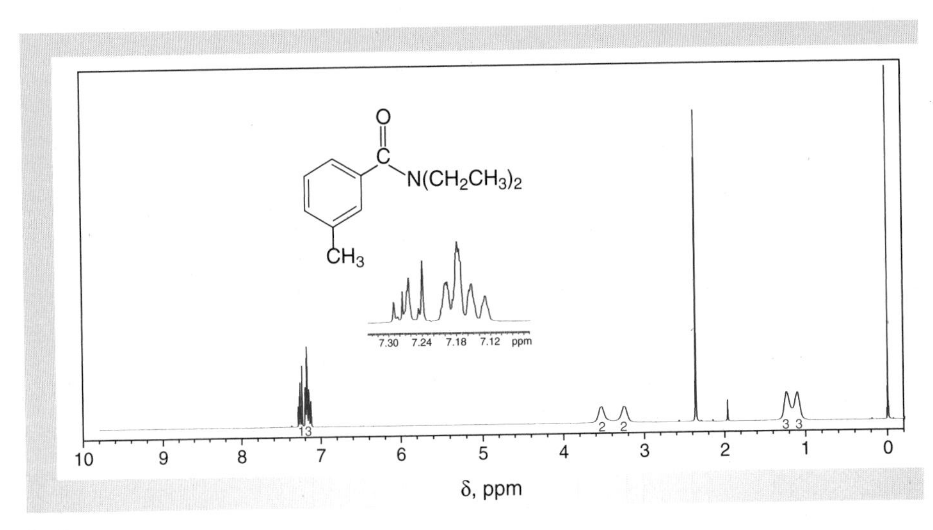

Figure 20.12
NMR data for N,N-*diethyl*-m-*toluamide ($CDCl_3$).*

(a) ^{1}H NMR spectrum (300 MHz).
(b) ^{13}C NMR data: δ 14.2, 21.4, 39.1, 123.1, 126.9, 128.2, 129.8, 137.2, 138.2, 171.5.

20.4 AMIDES AND CHEMILUMINESCENCE

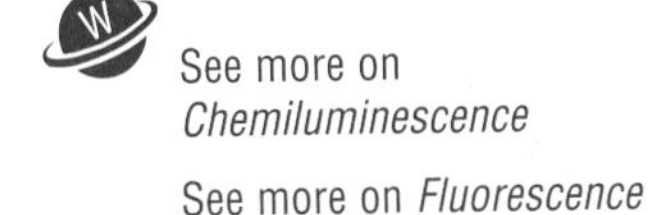

See more on *Fluorescence*

See more on *Phosphorescence*

A phenomenon related to the fundamental principles associated with UV-Vis spectroscopy (Sec. 8.4) is that of **chemiluminescence,** which is a process whereby light is generated by a chemical reaction. This occurs when the reaction produces a product in an *excited* electronic state that ultimately relaxes to the *ground* electronic state, with concomitant emission of energy in the form of a photon (light). In UV-Vis spectroscopy (Sec. 8.4), of course, exposing a sample to light of the proper wavelength fosters just the opposite behavior, namely, transforming a ground electronic state to an excited electronic state by promotion of an electron from one orbital to another (Fig. 8.49). In any case, the excited state may relax to the ground state through **radiationless** loss of energy or through emission of visible light, as seen when **fluorescence** and **phosphorescence** occur (Fig. 20.13). These last two modes of relaxation involve the **singlet** and **triplet** electronic states, respectively (see below).

Each electronic state comprises a number of vibrational energy states, labeled as $\nu_0 - \nu_5$ (Fig. 20.13). For simplicity's sake, we've shown the relaxation phenomena as originating solely from the lowest vibrational energy state, ν_0, of the electronically

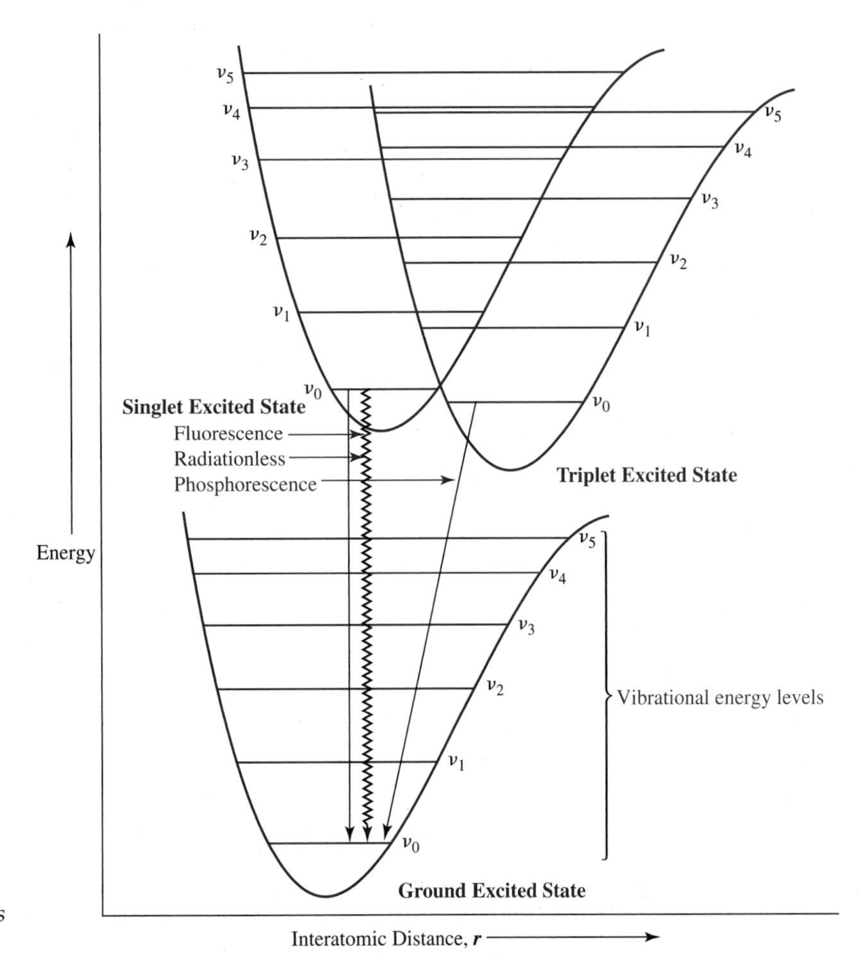

Figure 20.13
Potential energy diagram for singlet and triplet electronic states of a molecule.

excited states. In this simplified diagram, you can see that the energy released upon conversion of a triplet excited state to the ground electronic state is less than that for corresponding transformation of a singlet excited state. A consequence of this difference is that light of longer wavelength, and thus lower in energy (Eq. 8.1), results from phosphorescence as compared to fluorescence.

A number of living organisms emit light through chemiluminescent reactions involving naturally occurring compounds. If you have ever watched fireflies on a summer evening, you have witnessed one of the more familiar examples of **bioluminescence**, an example of chemiluminescence in which a biochemical process results in the emission of light. When in search of a mate, the male firefly emits flashes of visible light that result from the reaction of *luciferin* with molecular oxygen in a process catalyzed by the enzyme luciferase (Eq. 20.11).

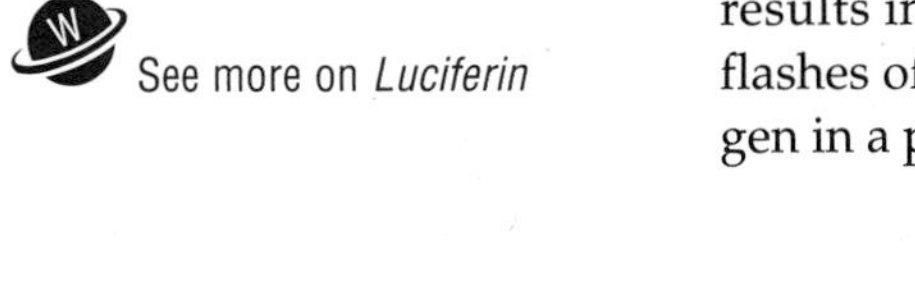
See more on *Luciferin*

$$\text{Luciferin} \xrightarrow[-CO_2]{\text{Luciferase}/O_2} \text{Decarboxyketoluciferin} + \text{Light} \qquad (20.11)$$

See more on *Luminol*

Unlike luciferin, 3-aminophthalhydrazide (**15**), commonly called luminol (Scheme 20.3) is not a natural product, but it also undergoes chemiluminescence, although the mechanism by which this occurs is not fully understood. However, it is thought to involve initial deprotonation of **15** to form a dianion that leads to the formation of a peroxide in the presence of an iron catalyst. The peroxide decomposes to form the dianion of 3-diaminophthalic acid in an electronically excited *triplet state* **16T**, a species having two unpaired electrons with the same spin quantum number, i.e., $+1/2$ or $-1/2$. As seen in Figure 20.14, **16T** undergoes intersystem crossing (ISC) to an excited *singlet state* **16S***, which has two unpaired electrons but with *opposite* spin quantum numbers. Relaxation (decay) to the *singlet* ground electronic state **16S** of the dianion then results in fluorescence in the form of bluish-green light.

The chemiluminescent properties of luminol have a number of practical applications. A forensic assay that is used to identify the presence of blood at crime scenes is one of these. Thus, when an alkaline solution of luminol and hydrogen peroxide is exposed to blood, the characteristic bluish-green light associated with the chemiluminescence of luminol is observed; the iron catalyst required for the reaction is provided by the hemoglobin present in blood.

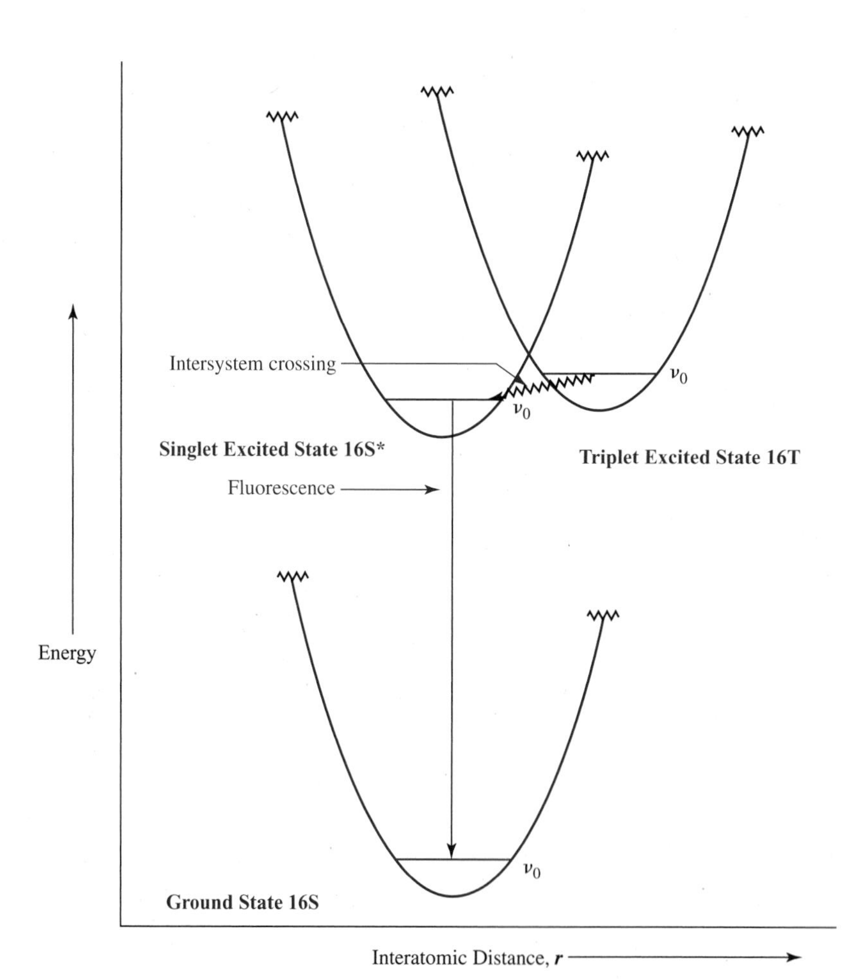

Figure 20.14
Potential energy diagram illustrating intersystem crossing and fluorescence.

Scheme 20.3

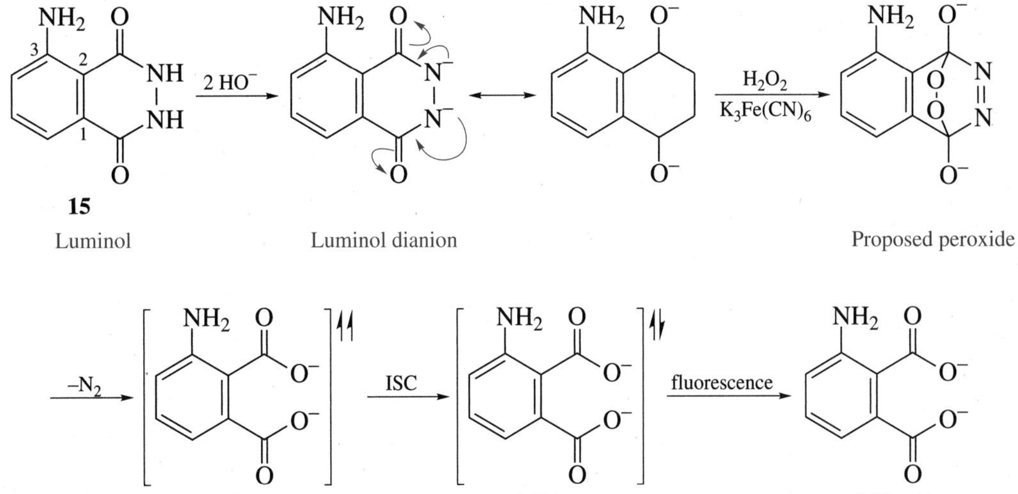

15 Luminol → ($2\ HO^-$) Luminol dianion ↔ → (H_2O_2, $K_3Fe(CN)_6$) Proposed peroxide

→ ($-N_2$) **16T** Triplet dianion → (ISC) **16S*** Excited-state singlet dianion → (fluorescence) **16S** Ground-state singlet dianion

Note that the half-headed arrows (⇂) represent the orientation of the spin of an electron.

In the procedure described below, luminol is synthesized by a sequence of reactions that involves the condensation of 3-nitrophthalic acid (**16**) with hydrazine (**18**) to produce 3-nitrophthalhydrazide (**20**), which is then reduced by sodium dithionite to form luminol (Eq. 20.12). The chemiluminescent reaction will then be demonstrated by mixing luminol with hydrogen peroxide and potassium ferricyanide, $K_2Fe(CN)_6$ (Scheme 20.3).

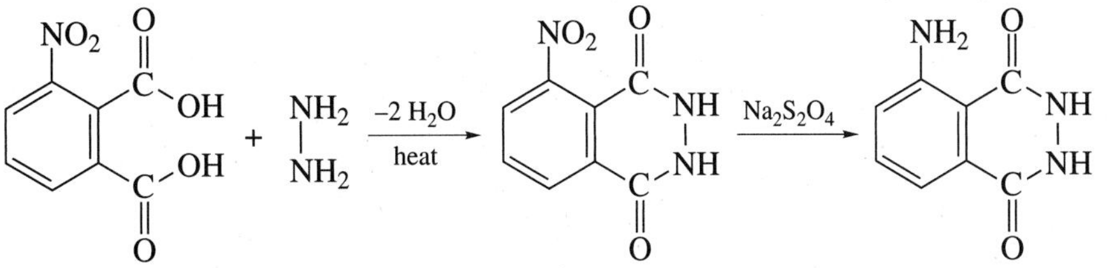

17 3-Nitrophthalic acid + **19** Hydrazine → ($-2\ H_2O$, heat) **20** 3-Nitrophthalhydrazide → ($Na_2S_2O_4$) **15** Luminol (20.12)

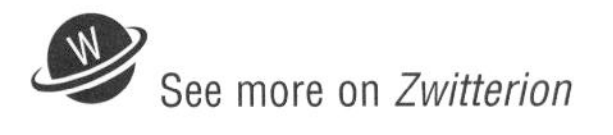

See more on *Zwitterion*

The formation of **20** is basically a reaction in which a 1,2-dicarboxylic acid and a 1,2-diamine react to form a cyclic diamide or hydrazide. The mechanistic steps by which an acid and amine are converted to an amide are portrayed in Scheme 20.4. The acid and amine are in rapid equilibrium with the corresponding ammonium salt and the carboxylate ion, the latter of which is not subject to nucleophilic attack at the carbonyl carbon atom. The acid itself may undergo nucleophilic attack, however, so the small amount of it that exists at equilibrium (see Exercise 4) reacts with free amine to produce the tetrahedral zwitterionic intermediate **21.** Subsequent proton transfers and loss of water afford the amide **22.** Repetition of this series of steps generates **20.**

Scheme 20.4

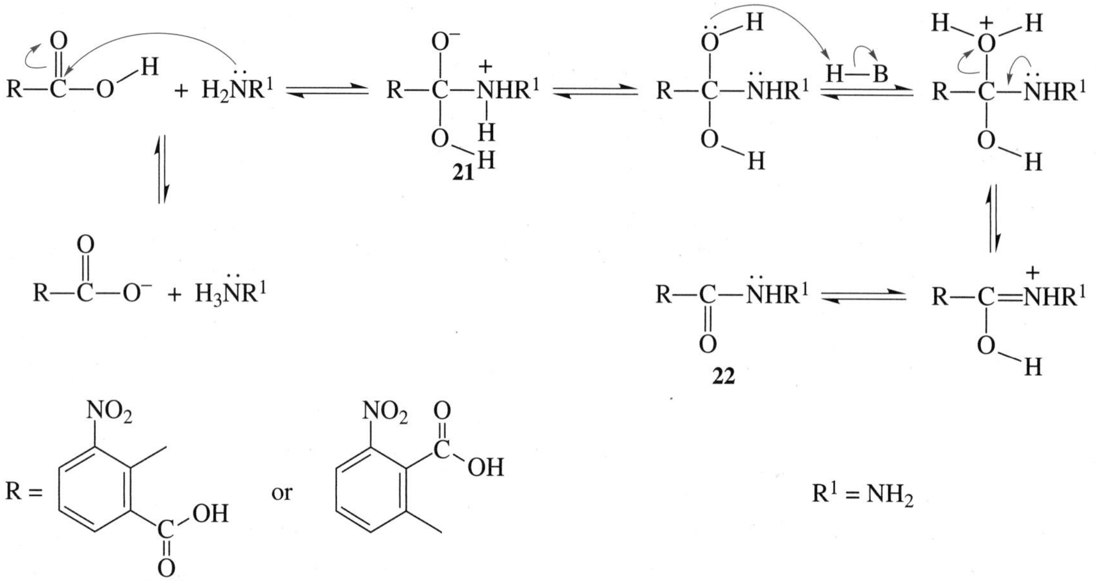

R—C(=O)—O—H + $H_2\ddot{N}R^1$ ⇌ R—C(O^-)($\overset{+}{N}H_2R^1$)—OH (**21**) ⇌ R—C(OH)(OH)—$\ddot{N}HR^1$ ⇌(H—B) R—C($\overset{+}{O}H_2$)(OH)—$\ddot{N}HR^1$

R—C(=O)—O^- + $H_3\ddot{N}R^1$

R—C(=O)—$\ddot{N}HR^1$ (**22**) ⇌ R—C(OH)=$\overset{+}{N}HR^1$

R = 2-nitro-6-methylbenzoic acid residue (NO$_2$, C(=O)OH) or (NO$_2$, C(=O)OH)

$R^1 = NH_2$

The reduction of the nitro group of 3-nitrophthalhydrazide (**20**) with dithionate is mechanistically complex, but a possible sequence of events is illustrated in Scheme 20.5, with $ArNO_2$ representing **20.** The reaction presumably is initiated by transfer of an electron from dithionate to produce the anion radical **23.** Subsequent steps involving protonation and addition of a second electron afford the *N,N*-dihydroxy intermediate **24,** which can dehydrate to produce the nitroso compound **25.** Further addition of electrons from dithionate, protonation, and loss of water from the hydroxylamine **26** leads to the reduced product **15.**

Scheme 20.5

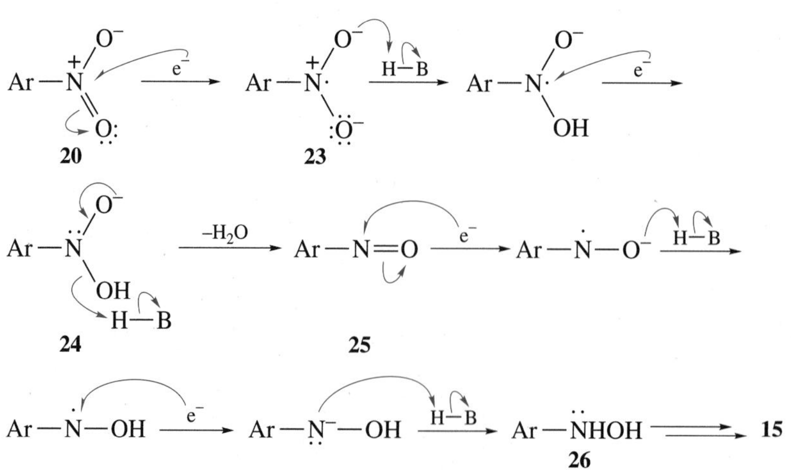

Ar—$\overset{+}{N}$(O^-)=O (**20**) —e^-→ Ar—$\overset{+}{N}$·(O^-)—O^- (**23**) —H—B→ Ar—N·(O^-)—OH —e^-→

Ar—$\ddot{N}$(O^-)—OH (**24**) (H—B) —$-H_2O$→ Ar—N=O (**25**) —e^-→ Ar—$\dot{N}$—O^- —H—B→

Ar—$\dot{N}$—OH —e^-→ Ar—$\ddot{N}^-$—OH —H—B→ Ar—$\ddot{N}$HOH (**26**) ⟶ **15**

EXPERIMENTAL PROCEDURES

Preparation and Chemiluminescence of Luminol

Purpose To demonstrate the preparation of luminol and its chemiluminescence.

SAFETY ALERT

Wear latex gloves when handling solutions of hydrazine, as it is readily absorbed through the skin. If hydrazine comes into contact with your skin, immediately wash the area with soap and water and flush it with copious amounts of water.

A ■ *Preparation of Luminol*

MINISCALE PROCEDURE

Preparation Sign in at **www.cengage.com/login** to answer Pre-Lab Exercises, access videos, and read the MSDSs for the chemicals used or produced in this procedure. Review Sections 2.9, 2.10, and 2.17.

Apparatus A 20-mm × 150-mm test tube, 25-mL Erlenmeyer flask, thermometer, and apparatus for vacuum filtration and *flameless* heating.

Setting Up Combine 1 g of 3-nitrophthalic acid and 2 mL of an 8% aqueous solution of hydrazine in the test tube and carefully heat the mixture until the solid dissolves. Add 3 mL of triethylene glycol and a boiling stone and clamp the vial in a vertical position. Insert a thermometer into the solution, securing the thermometer with a clamp.

Reaction Bring the solution to a vigorous boil to remove excess water. During this time, the temperature should be around 110–120 °C. After the water has evaporated, the temperature should rise to 215 °C in a 3- to 4-min period. Maintain the temperature at 215–220 °C for 2 min and then remove the test tube from the heat source and allow the solution to cool to about 100 °C. While the test tube and its contents are cooling, bring about 15 mL of water to boiling in the Erlenmeyer flask. Slowly add the hot water to the test tube, stir the contents with a glass stirring rod, cool the mixture to room temperature, and collect the solid nitrohydrazide by vacuum filtration.

Return the damp solid to the uncleaned test tube, add 5 mL of 3 *M* aqueous NaOH, and mix the contents until the solid is dissolved. Add 3 g of fresh sodium hydrosulfite dihydrate to the solution, rinsing any solid adhering to the walls of the test tube into the solution with a few drops of water, and heat the resulting mixture to just below boiling for 5 min, taking care to avoid bumping.

Work-Up and Isolation Add 2 mL of *glacial* acetic acid to the reaction mixture, cool the test tube in a beaker of cold water, and collect the crude luminol by vacuum filtration. A second crop of product may separate from the filtrate upon standing, but do not combine it with the first crop, which is to be used in the chemiluminescence experiment.

Analysis If instructed to do so, recrystallize the 3-nitrophthalhydrazide and luminol, and obtain their IR and ^{1}H NMR spectra and those of 3-nitrophthalic acid; compare these spectra with those of authentic samples (Figs. 20.15–20.20).

MICROSCALE PROCEDURE

Preparation Sign in at **www.cengage.com/login** to answer Pre-Lab Exercises, access videos, and read the MSDSs for the chemicals used or produced in this procedure. Review Sections 2.9, 2.10, and 2.17.

Apparatus A 5-mL conical vial, 25-mL Erlenmeyer flask, thermometer, and apparatus for vacuum filtration and *flameless* heating.

Setting Up Combine 200 mg of 3-nitrophthalic acid and 0.4 mL of an 8% aqueous solution of hydrazine in the conical vial and heat the mixture until the solid dissolves. Add 0.6 mL of triethylene glycol and a boiling stone, and clamp the vial in a vertical position. Insert a thermometer into the solution, securing the thermometer with a clamp.

Reaction Bring the solution to a vigorous boil to remove excess water. During this time, the temperature should be around 110–120 °C. After the water has evaporated, the temperature should rise to 215 °C in a 3- to 4-min period. Maintain the temperature at 215–220 °C for about 2 min, and then remove the vial from the heat source and allow the solution to cool to about 100 °C. While the tube and its contents are cooling, bring about 10 mL of water to boiling in the Erlenmeyer flask. Slowly add 3 mL of boiling water to the reaction mixture, stir the contents of the conical vial with a glass stirring rod, cool the resulting mixture to room temperature, and collect the solid nitrohydrazide by vacuum filtration.

Return the damp solid to the conical vial, add 1 mL of 3 *M* aqueous NaOH, and mix the contents until the solid is dissolved. Add 0.6 g of fresh sodium hydrosulfite dihydrate to the solution, and heat the resulting mixture to just below boiling for 5 min, taking care not to cause bumping.

Work-Up and Isolation Add 0.4 mL of *glacial* acetic acid to the reaction mixture, cool the vial in a beaker of cold water, and collect the luminol by vacuum filtration.

Analysis If instructed to do so, recrystallize the 3-nitrophthalhydrazide and luminol, and obtain their IR and ^{1}H NMR spectra and those of 3-nitrophthalic acid; compare these spectra with those of authentic samples (Figs. 20.15–20.20).

WRAPPING IT UP

Flush the *aqueous filtrates* down the drain with excess water. Put the *filter papers* in the container for nontoxic waste.

B ▪ *Chemiluminescence*

MINISCALE PROCEDURE

Preparation Sign in at **www.cengage.com/login** to answer Pre-Lab Exercises, access videos, and read the MSDSs for the chemicals used or produced in this procedure. Review Section 2.11.

Apparatus Two 50-mL, one 125-mL, and one 250-mL Erlenmeyer flask.

Setting Up Prepare stock solution A by placing the crude luminol (40–60 mg dry weight, but which need not be dry) in a 50-mL Erlenmeyer flask and dissolving it in 2 mL of 3 *M* aqueous NaOH and 18 mL of water. Prepare solution B in the second 50-mL Erlenmeyer flask by combining 4 mL of 3% aqueous potassium ferricyanide, 4 mL of 3% aqueous hydrogen peroxide, and 32 mL of water. Combine 5 mL of solution A with 35 mL of water in the 125-mL Erlenmeyer flask.

Reaction Working in a low-light environment, simultaneously pour the diluted portion of solution A and all of solution B into the 250-mL Erlenmeyer flask. Swirl the flask and record what you observe. Test what happens if you add additional small portions of the aqueous NaOH solution and crystals of potassium ferricyanide to the reaction mixture.

MICROSCALE PROCEDURE

Preparation Sign in at **www.cengage.com/login** to answer Pre-Lab Exercises, access videos, and read the MSDSs for the chemicals used or produced in this procedure. Review Section 2.11.

Apparatus Two 50-mL and one 250-mL Erlenmeyer flask.

Setting Up Working with a partner, prepare solution A by combining two samples of crude luminol, which need not be dry, in a 50-mL Erlenmeyer flask and dissolving the solids in 2 mL of 3 *M* aqueous NaOH and 18 mL of water. Prepare solution B in the second 50-mL Erlenmeyer flask by combining 4 mL of 3% aqueous potassium ferricyanide, 4 mL of 3% aqueous hydrogen peroxide, and 32 mL of water in the other small Erlenmeyer flask.

Reaction Working in a low-light environment, simultaneously pour solution A and solution B into the 250-mL Erlenmeyer flask. Swirl the flask and record what you observe.

WRAPPING IT UP

Neutralize the *luminol solution* and flush it down the drain with excess water.

EXERCISES

1. Define the following terms.
   a. fluorescence
   b. phosphorescence
   c. chemiluminescence
   d. intersystem crossing
2. What is the difference in electronic configuration of a singlet and a triplet state?

3. There are two carboxylic acid functions in 3-nitrophthalic acid (**17**). The pK_a of the acid function at the 2-position is approximately 2, whereas that at the 1-position is about 3. Which of the two is the more acidic and why?
4. Given the approximate pK_as provided, compute K_{eq} for the reaction below.

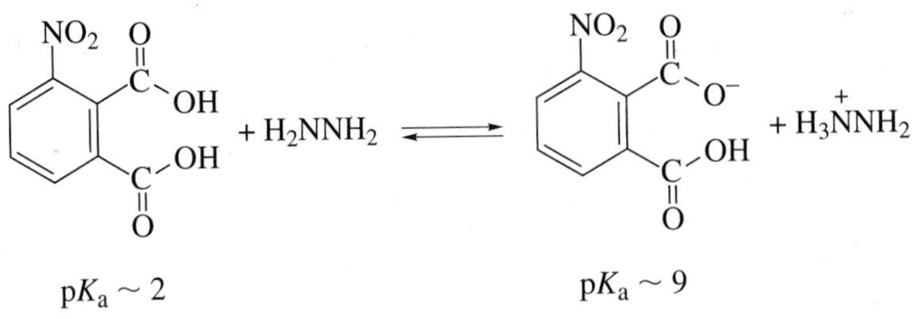

$pK_a \sim 2$ $\qquad$ $+ H_2NNH_2 \rightleftharpoons$ $\qquad$ $pK_a \sim 9$ $\qquad$ $+ H_3\overset{+}{N}NH_2$

5. Using curved arrows to symbolize the flow of electrons, propose a mechanism for the transformation shown below.

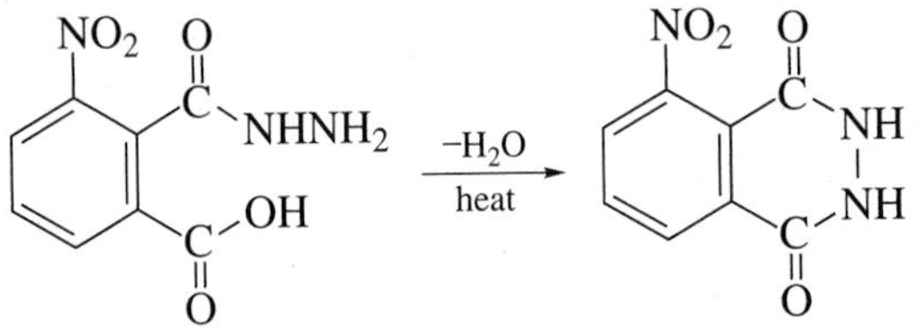

$\xrightarrow[\text{heat}]{-H_2O}$

20

6. What factor(s) make the hydrogen atoms of the hydrazide moiety sufficiently acidic so that hydroxide can convert luminol to the dianion portrayed in Scheme 20.3?

Sign in at **www.cengage.com/login** and use the spectra viewer and Tables 8.1–8.8 as needed to answer the blue-numbered questions on spectroscopy.

7. Consider the spectral data for 3-nitrophthalic acid (Figs. 20.15 and 20.16).
   a. In the functional group region of the IR spectrum, specify the absorptions associated with the O–H bonds and the carbonyl and nitro functions of the molecule.
   b. In the ^{1}H NMR spectrum, assign the various resonances to the hydrogen nuclei responsible for them.
   c. In the ^{13}C NMR spectrum, assign the various resonances to the carbon nuclei responsible for them.
8. Consider the spectral data for 3-nitrophthalhydrazide (Figs. 20.17 and 20.18).
   a. In the functional group region of the IR spectrum, specify the absorptions associated with the N–H bonds and carbonyl and nitro functions of the molecule.
   b. In the ^{1}H NMR spectrum, assign the various resonances to the hydrogen nuclei responsible for them.
   c. In the ^{13}C NMR spectrum, assign the various resonances to the carbon nuclei responsible for them.
9. Consider the spectral data for luminol (3-aminophthalhydrazide) (Figs. 20.19 and 20.20).
   a. In the functional group region of the IR spectrum, specify the absorptions associated with the N–H bonds of the hydrazide and of the amino group and that of the carbonyl functions of the molecule.

b. In the ^{1}H NMR spectrum, assign the various resonances to the hydrogen nuclei responsible for them.

c. In the ^{13}C NMR spectrum, assign the various resonances to the carbon nuclei responsible for them.

10. Discuss the differences observed in the IR and NMR spectra of 3-nitrophthalic acid and 3-nitrophthalhydrazide that are consistent with replacement of the carboxylic acid functions with the hydrazide moiety in this experiment.

11. Discuss the differences observed in the IR and NMR spectra of 3-nitrophthalhydrazide and luminol that are consistent with reduction of the nitro function to an amino group in this experiment.

SPECTRA

Starting Material and Products

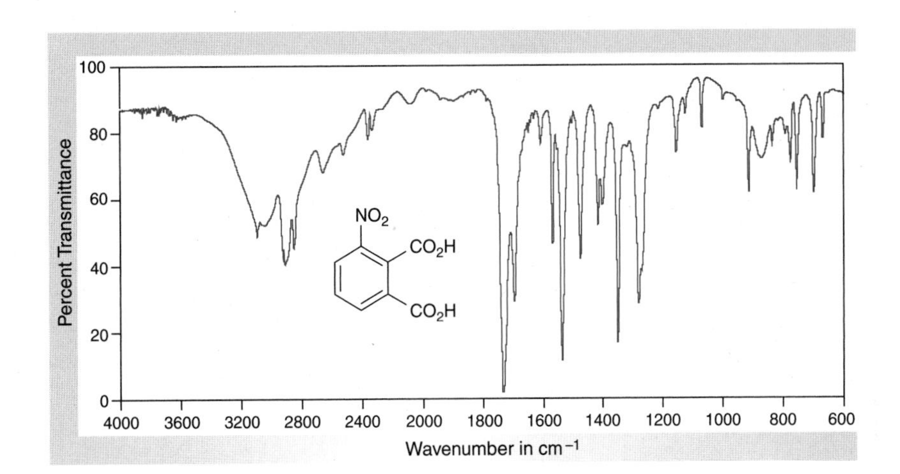

Figure 20.15
IR spectrum of 3-nitrophthalic acid (IR card).

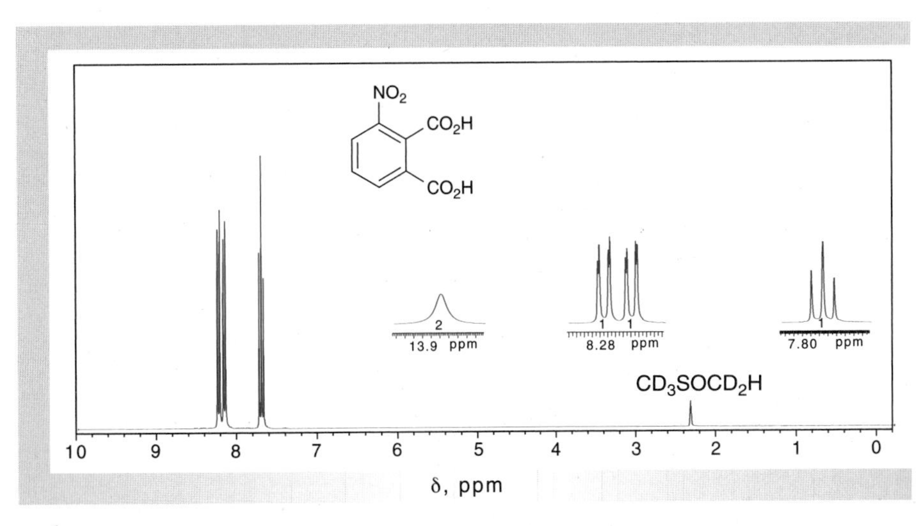

Figure 20.16
NMR data for 3-nitrophthalic acid (DMSO-d_6).

(a) ^{1}H NMR spectrum (300 MHz).
(b) ^{13}C NMR data: δ 127.4, 130.4, 130.7, 131.3, 134.9, 146.5, 165.7, 165.8.

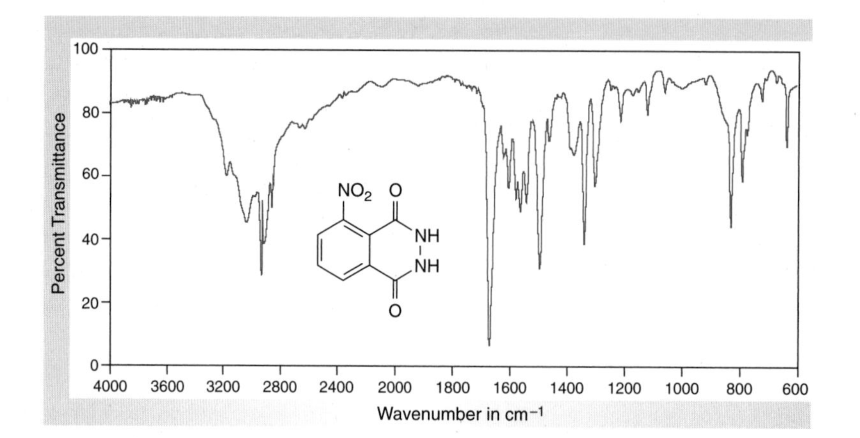

Figure 20.17
IR spectrum of 3-nitrophthalhydrazide (IR card).

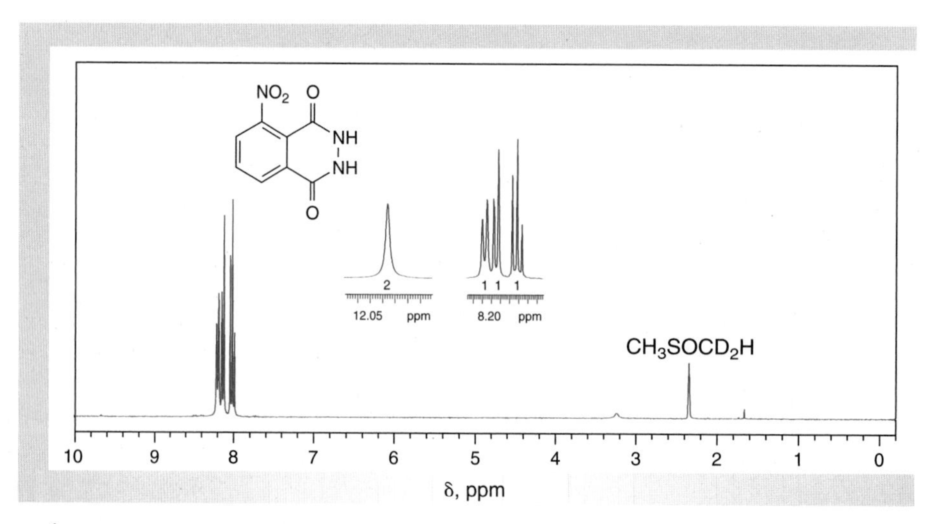

Figure 20.18
NMR data for 3-nitrophthalhydrazide (DMSO-d_6).

(a) ^{1}H NMR spectrum (300 MHz).
(b) ^{13}C NMR data: δ 118.4, 126.1, 127.6, 127.8, 133.9, 147.7, 151.9, 152.7.

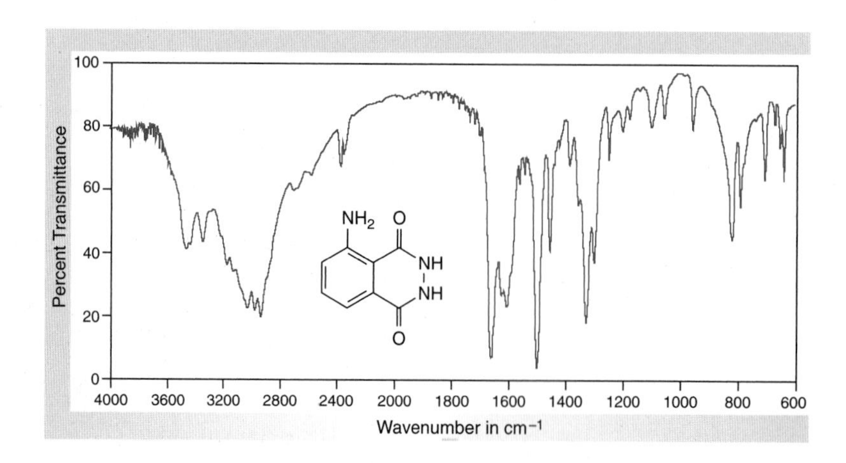

Figure 20.19
IR spectrum of luminol (3-aminophthalhydrazide) (KBr pellet).

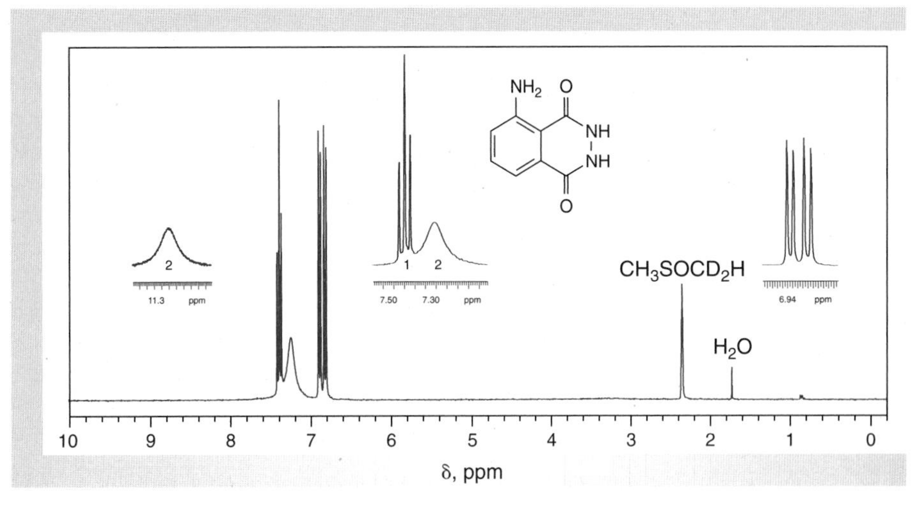

Figure 20.20
NMR data for luminol (3-aminophthalhydrazide) (DMSO-d_6).

(a) ^{1}H NMR spectrum (300 MHz).
(b) ^{13}C NMR data: δ 109.8, 110.6, 116.7, 126.5, 133.9, 150.2, 151.5, 161.2.

HISTORICAL HIGHLIGHT

Evolution of Synthetic Analgesics

Medicinal chemists are constantly searching for new organic compounds that have useful biological activity. Such substances are sometimes identified by broad-screen testing of large numbers of synthetic compounds. However, biologically active compounds are frequently discovered in the folk medicines used by the indigenous people of an area. These natural compounds often serve as an inspiration to imaginative organic chemists, who examine the structure of the new compound and develop a hypothesis regarding what structural features might be responsible for the biological activity. An analog of the natural product that incorporates these features, the so-called **pharmacophore,** is then prepared and evaluated for biological activity. Often, simpler molecules having the biological activity of the original compound are found, and further structural modifications are then made to optimize the desired pharmacological activity and eliminate any undesired side effects. Indeed, the separation of the beneficial from the undesired properties of a drug remains one of the biggest challenges that medicinal chemists face today.

The development of benzocaine as a topical anesthetic, or painkiller, follows this pattern of "rational drug design," and the story of its discovery is a fascinating one that begins with the coca shrub found in the Andes Mountains of Peru. The native Andean Indians had long known about the stimulant effects of the leaves of the coca shrub. Chewing the leaves led to a sense of mental and physical well-being that perhaps enabled them to endure the harsh conditions encountered in their daily lives. We now know that the organic compound responsible for the biological properties of the coca leaves is the **alkaloid** cocaine. Alkaloids are natural products that contain a basic nitrogen atom and are derived biosynthetically from amino acids. Perhaps the Andean Indians were fortunate that only small amounts of cocaine were consumed by chewing the leaves, as prolonged consumption and/or overindulgence causes mental and physical debilitation and may even cause death. The structural similarities of benzocaine and cocaine are highlighted by the boldfaced lines in their structures.

(Continued)

HISTORICAL HIGHLIGHT *Evolution of Synthetic Analgesics (Continued)*

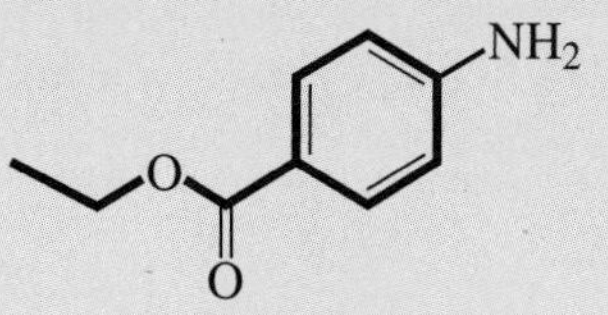

Benzocaine

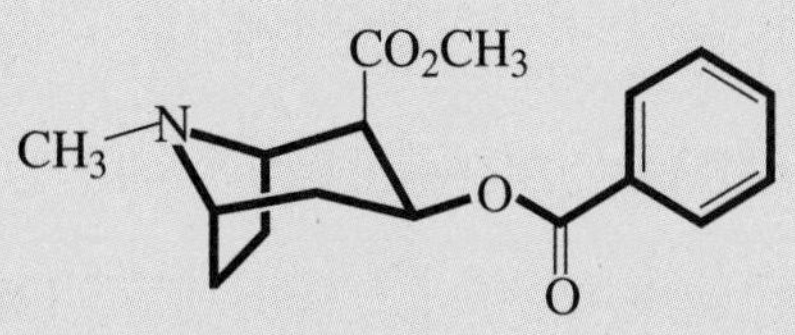

Cocaine

Pure cocaine was first isolated by Niemann in 1862. A common practice of chemists at that time, but one that is certainly no longer recommended, was to taste the compounds they made or isolated. Niemann found that cocaine, like many alkaloids, had a bitter taste, but more interestingly, he discovered that cocaine numbed the tongue to the point that it was almost devoid of sensation. Nearly twenty years later, Von Anrep discovered that subcutaneous injection of cocaine made the surrounding skin numb and insensitive to the prick of a pin. Sigmund Freud, better known for his contributions to psychoanalysis, and his assistant Karl Koller also played a role in the development of cocaine as a local anesthetic. They showed that putting a few drops of a solution of cocaine onto the eye numbed the eye muscles, causing cessation of the involuntary reflex movements that had made eye surgery difficult. Freud was also interested in the effect that cocaine had on the central nervous system and discovered its addictive properties in the mid-1880s.

Because of its ability to block pain, cocaine became widely used as a local anesthetic to deaden nerves during various medical procedures, but it was not an ideal anesthetic owing to its undesired side effects. For example, in eye surgery it produces **mydriasis**, or dilation of the pupil; it also has powerful addictive properties and exhibits dangerous effects on the central nervous system. Medicinal chemists thus began a search for substances related to cocaine that would retain its valuable anesthetic properties but would not produce its side effects.

Eucaine, the first unnatural analog of cocaine to be synthesized, was reported in 1918. As shown by the bold lines in its structure, eucaine is very similar to cocaine. Eucaine was found to function as a local anesthetic, and it did not produce mydriasis, nor was it addictive; however, it was not the "Holy Grail" of local anesthetics because it was highly toxic. Another analog of cocaine is piperocaine, which, as shown by the boldfaced lines, retains some of the basic elements of cocaine but is structurally more different from cocaine than is eucaine. Piperocaine has the same beneficial characteristics as eucaine but is only one-third as toxic as cocaine. A further iteration of the structure is seen in the form of procaine, which is commonly known as Novocain, perhaps the most successful anesthetic agent derived from cocaine. Novocain is about one-fourth as toxic as cocaine and is not habit-forming.

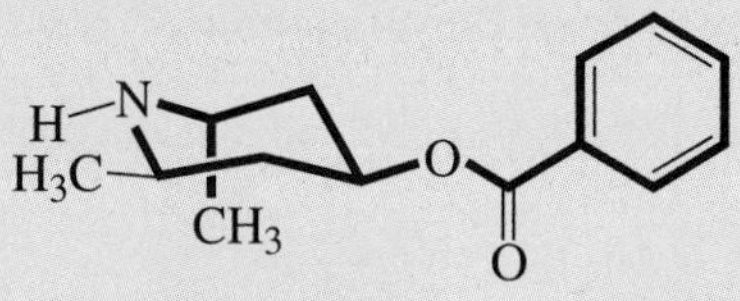

Eucaine

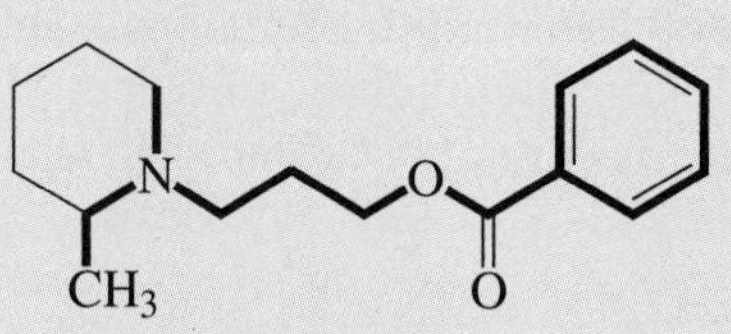

Piperocaine

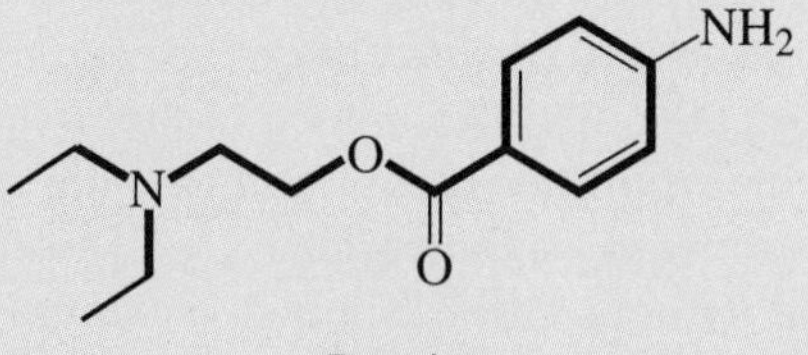

Procaine

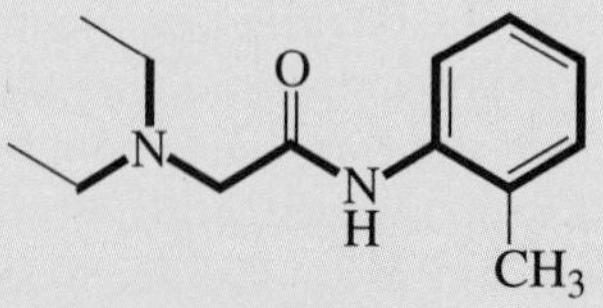

Lidocaine

(Continued)

HISTORICAL HIGHLIGHT *Evolution of Synthetic Analgesics (Continued)*

Hundreds of other new local anesthetics have been synthesized and tested over the years, one of which is lidocaine (see Sec. 21.4); most of these are not generally used, and the search for the perfect local anesthetic agent continues. All of these compounds have a pharmacophore containing an aromatic ring, and most have a tertiary or secondary amine function linked to this aromatic ring by an ester group and a chain of up to three other atoms. The enzymatic hydrolysis of the ester moiety in the bloodstream is a key step in the detoxification of these compounds, and compounds lacking this functionality are typically longer-lasting and more toxic; lidocaine is an exception to this generalization. The amino group is important for solubility, and most of these compounds are used as their hydrochloride salts, which are water-soluble and thus injectable. Benzocaine, which lacks this amino group, is not used for injection and is not water-soluble. Rather, it is usually incorporated into an ointment or salve for direct application and is a common ingredient of many sunburn preparations. Interestingly, it is made by esterification of *p*-aminobenzoic acid (PABA), which frequently is also a component of such preparations.

Relationship of Historical Highlight to Experiments

The preparation of derivatives of carboxylic acids, especially esters and amides, is a key step in the syntheses of many biologically active compounds, as these functional groups are common to many drugs and other substances used in over-the-counter preparations used in health care. The preparation of benzocaine in Section 20.2 using a Fischer esterification illustrates one important route to simple esters, a functional group found in the pharmacophore of many anesthetic agents. A general method for producing amides from carboxylic acids via an intermediate acid chloride is found in the synthesis of the insect repellent *N,N*-diethyl-*m*-toluamide. A related procedure is used in the synthesis of lidocaine (Sec. 21.4).

See more on *Pharmacophore*
See more on *Cocaine*
See more on *Eucaine*
See more on *Procaine*
See more on *Lidocaine*
See more on *PABA*
See *Who was Fischer?*

Multistep Organic Synthesis

When you see this icon, sign in at this book's premium website at **www.cengage.com/login** to access videos, Pre-Lab Exercises, and other online resources.

An everyday application of organic chemistry is the synthesis of compounds that enable us to lead better and more enjoyable lives. Such compounds may be drugs, pesticides, insecticides, or polymers, or they may be used as additives in fragrances, flavors, foods, cosmetics, adhesives, or detergents. The syntheses of these substances from simple starting materials typically involve numerous chemical steps and hence require careful planning and extensive experimentation in order to optimize the overall processes. It's usually necessary to marshal the combined skills of expert synthetic and mechanistic organic chemists to solve the problems that arise. In the experiments in this chapter, you will become better acquainted with the practical aspects of organic chemistry by preparing a useful compound by a multistep sequence. In so doing, perhaps you will further understand the importance of studying organic chemistry.

21.1 INTRODUCTION

Many organic compounds that we use daily are prepared in large quantities, especially by companies in the pharmaceutical and agricultural industries. Hence, solving the problems associated with the efficient and cost-effective syntheses of complex, multifunctional molecules from simple starting materials is a challenge that must be routinely addressed by organic chemists. In these synthetic processes, it is necessary to perform each transformation selectively and to avoid all side reactions that may lead to reduced yields and impure products. It is perhaps equally important to minimize the number of individual steps that are required because of the "arithmetic demon" that haunts the overall yield in sequential reactions that are not quantitative. For example, even a five-step synthesis in which each step occurs in a respectable 80% yield gives an overall yield of only about 33% [(0.8)5 × 100%]. In view of the manifold problems, it is noteworthy that there are presently economical routes to biologically important molecules that require as many as 20 steps.

The *optimal route* to a particular compound *is not always the shortest*, as is illustrated by the simple problem of preparing 2-chloropropane on an industrial scale, shown in Scheme 21.1. Although 2-chloropropane is produced from propane in one step by direct chlorination, it must be separated from the 1-chloropropane formed concurrently; this is *not* an easy separation. However, if the mixture of 1- and 2-chloropropane is treated with a base, both isomers undergo dehydrochlorination to give propane, and subsequent addition of hydrogen chloride produces 2-chloropropane as the *sole* product. In this case, a three-step process is the preferred way to obtain pure 2-chloropropane. As this simple example demonstrates, an important consideration in designing a synthesis is the ease of separating the desired product from unwanted by-products and purifying it. Other considerations include the availability and cost of starting materials, simplicity of equipment and instrumentation, energy costs, activity of catalysts, selectivity of reactions in polyfunctional molecules, and stereochemical control.

Scheme 21.1

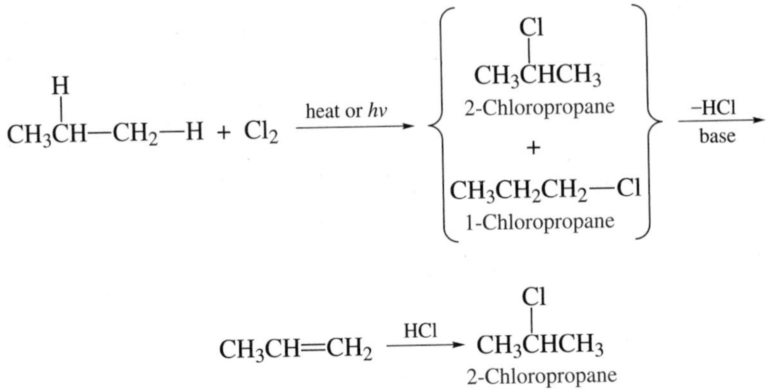

$$CH_3CH(H){-}CH_2{-}H + Cl_2 \xrightarrow{\text{heat or } h\nu} \left\{ \begin{array}{c} CH_3CH(Cl)CH_3 \\ \text{2-Chloropropane} \\ + \\ CH_3CH_2CH_2{-}Cl \\ \text{1-Chloropropane} \end{array} \right\} \xrightarrow[\text{base}]{-HCl}$$

$$CH_3CH{=}CH_2 \xrightarrow{HCl} \underset{\text{2-Chloropropane}}{CH_3CH(Cl)CH_3}$$

Because so many variables must be considered in planning the synthesis of complex molecules, it is not surprising that organic chemists sometimes use computers to help design and analyze multistep syntheses. Computers can handle an enormous amount of information, and it may eventually be possible to design optimal sequences of reaction steps using sophisticated software. In considering this fascinating prospect, however, it must be remembered that the information is provided to the computer by humans, specifically chemists, who will be vital to the success of such a project. Chemists still have at least one significant competitive advantage over computers—they are more creative!

Appreciation of and insight into some of the problems that may be encountered in planning a multistep synthesis may be gained from preparing a compound that requires a fairly small number of steps. In this chapter, the syntheses of sulfanilamide (Sec. 21.2), 1-bromo-3-chloro-5-iodobenzene (Sec. 21.3), and lidocaine (Sec. 21.4) are used to illustrate the fundamental principles. In these experiments, the product of one reaction is used in a subsequent step, and if you are to be successful, you *must* use good experimental technique. You may either isolate and purify the intermediates by distillation or recrystallization, or you may use the crude material directly in the next step. This is a choice the practicing synthetic organic chemist constantly faces. In general, extensive purification of each intermediate in a sequence is avoided *if* the impurities can eventually be removed *and* their presence does not interfere with the course of the desired reactions. However, even if the entire quantity of an intermediate is not purified, it is good scientific practice to purify a small sample of it for *complete* characterization by spectroscopic (Chap. 8) and physical (bp, mp, etc.) methods.

21.2 SULFANILAMIDE: DISCOVERY AND SYNTHESIS OF THE FIRST ANTIBIOTIC

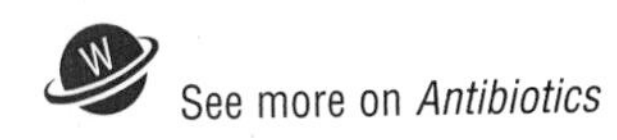

An important area of research in the pharmaceutical industry is the discovery and development of new orally active **antibiotics** to treat bacterial infections. Commonly used antibiotics fall into several important classes, including **β-lactams,** the

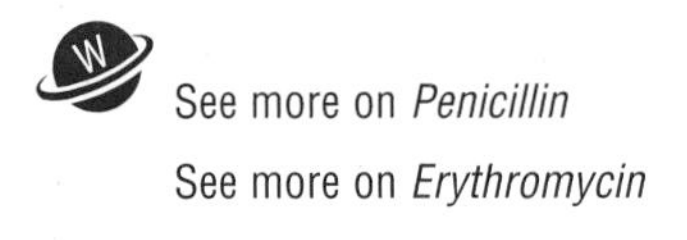
See more on *Penicillin*
See more on *Erythromycin*

Sulfamethoxazole
A sulfa drug

Terramycin
A tetracycline antibiotic

Ampicillin
A β-lactam antibiotic

Erythromycin A
A macrolide antibiotic

Spectinomycin
An aminocyclitol antibiotic

Figure 21.1
Classes of antibiotics.

macrolides, the **tetracyclines,** the **aminocyclitols,** and the **sulfa drugs** (Fig. 21.1). Each antibiotic has a different profile of biological activity, and all of them are used in modern medicine to treat various infections.

Background

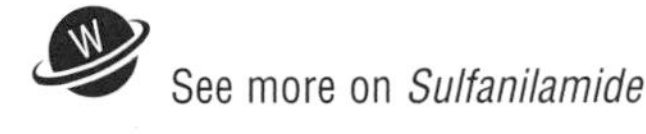
See more on *Sulfanilamide*

See more on *Sunscreens*

The broad-spectrum antibacterial activity of sulfanilamide (**1**) was first revealed in the mid-1930s by serendipity, as are many important discoveries in science, and an interesting account of this discovery can be found in the Historical Highlight at the end of this chapter. Over one thousand derivatives of sulfanilamide have been synthesized and tested as potential antibiotics, and some of these are still used today. The mode of action of the sulfa drugs is another interesting story because it provides some insights into strategies that might be generally exploited to design biologically active compounds.

Early in the development of sulfa drugs as antibiotics, it was found that *p*-aminobenzoic acid (PABA, **2**), which is now used in sunscreens and sunblocks, inhibits the antibacterial action of sulfanilamide. Since *p*-aminobenzoic acid and sulfanilamide are structurally similar, this discovery led to the speculation that the two compounds competed with each other in some biological process that was essential for bacterial growth. This speculation was eventually supported by experimentation. *p*-Aminobenzoic acid is used by bacteria in the synthesis of the essential enzyme cofactor folic acid (**3**). When sulfanilamide is present, it successfully competes with *p*-aminobenzoic acid for the active site in the enzyme that

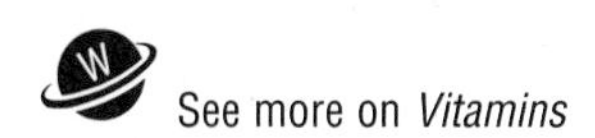
See more on *Vitamins*

incorporates *p*-aminobenzoic acid into folic acid. By functioning as a competitive inhibitor of this enzyme, sulfanilamide blocks the biosynthesis of folic acid, and without folic acid, the bacteria cannot grow. On the other hand, animal cells cannot synthesize folic acid, which is an **essential vitamin,** and it therefore must be part of the diet. Because only bacteria rely on the biosynthesis of folic acid from *p*-aminobenzoic acid, the sulfa compounds are "ideal" drugs, as they kill only the bacteria and not the animal host. Of course, they are not truly ideal, since some people have allergic reactions to sulfa drugs and because bacteria develop resistance to them over time.

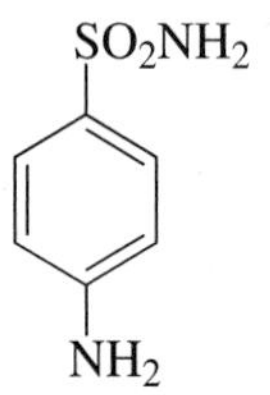

1 Sulfanilamide (SO_2NH_2, NH_2)

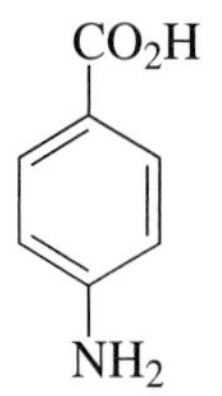

2 *p*-Aminobenzoic acid (PABA) (CO_2H, NH_2)

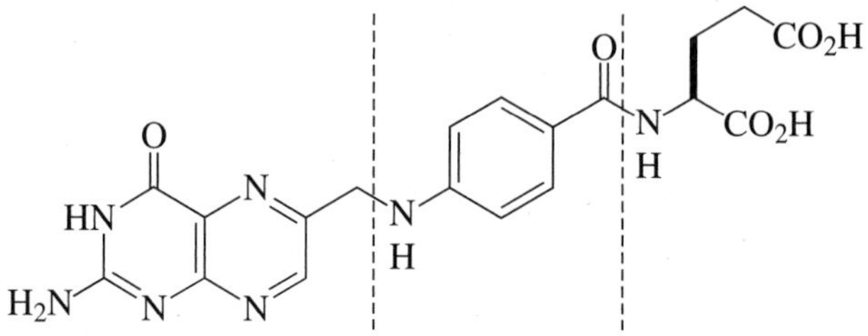

3 Folic acid

The preparation of sulfanilamide (**1**) from benzene (**4**) (Scheme 21.2) serves as an excellent example of a multistep synthesis that produces a biologically active, nonnatural molecule. Owing to the toxicity of benzene, the experimental sequence that follows will commence with the second step, the reduction of nitrobenzene (**5**), which is considerably less toxic. Each of the intermediate compounds in the synthesis may either be isolated or used directly in the subsequent step without further purification. A discussion of the reactions involved in each step follows.

Scheme 21.2

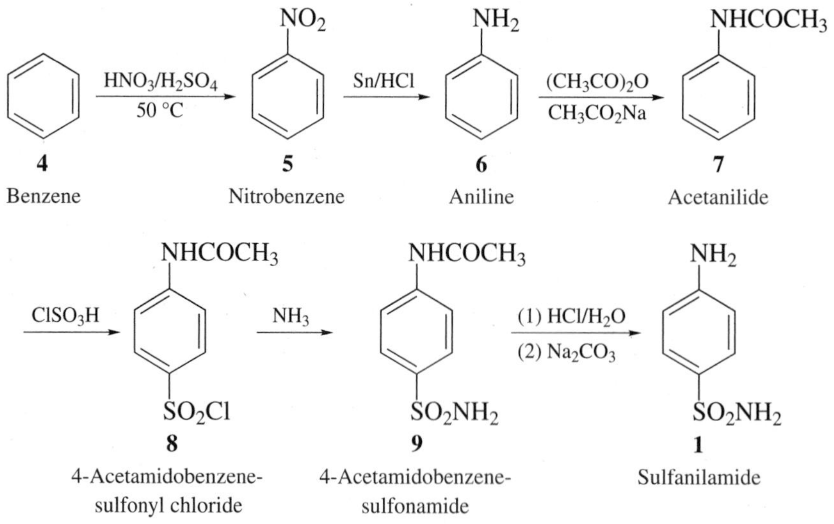

4 Benzene → (HNO_3/H_2SO_4, 50 °C) → **5** Nitrobenzene (NO_2) → (Sn/HCl) → **6** Aniline (NH_2) → ($(CH_3CO)_2O$, CH_3CO_2Na) → **7** Acetanilide ($NHCOCH_3$)

→ ($ClSO_3H$) → **8** 4-Acetamidobenzenesulfonyl chloride ($NHCOCH_3$, SO_2Cl) → (NH_3) → **9** 4-Acetamidobenzenesulfonamide ($NHCOCH_3$, SO_2NH_2) → ((1) HCl/H_2O, (2) Na_2CO_3) → **1** Sulfanilamide (NH_2, SO_2NH_2)

Reduction of Aromatic Nitro Compounds: Preparation of Aniline (6)

Aromatic nitro compounds, which are readily prepared by electrophilic aromatic nitration (Sec. 15.1), are reduced to the corresponding aromatic amines by a variety of methods. Chemical reduction is commonly used in the research laboratory, but the most important commercial method is catalytic hydrogenation (Chap. 17). Although the precise mechanism of the reduction of nitro compounds is not fully understood, various intermediates in this stepwise process have been isolated, as shown in Equation 21.1. The symbol **[H]** over the arrows represents reduction without specifying the reagent. The product that is ultimately isolated from the reduction of an aromatic nitro compound can be controlled to a significant degree by the nature of the reducing agent. For example, the reduction of nitrobenzene (**5**) with zinc metal and ammonium chloride gives only *N*-phenylhydroxylamine (**11**), whereas the use of tin metal and hydrochloric acid gives aniline (**6**).

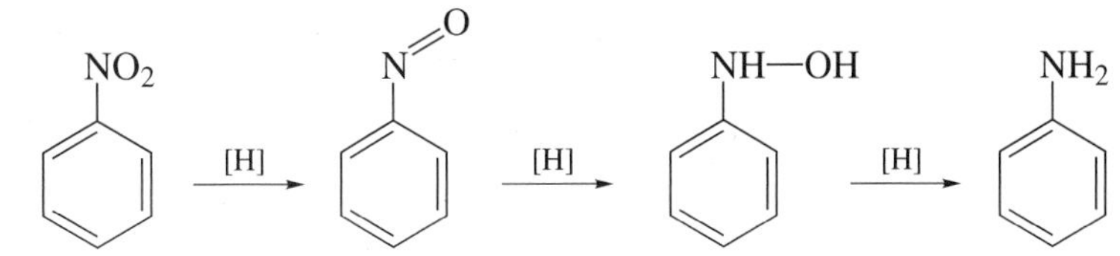

$$C_6H_5NO_2 \xrightarrow{[H]} C_6H_5N{=}O \xrightarrow{[H]} C_6H_5NH{-}OH \xrightarrow{[H]} C_6H_5NH_2 \quad (21.1)$$

5 Nitrobenzene; **10** Nitrosobenzene; **11** *N*-Phenylhydroxylamine; **6** Aniline

The reduction of nitrobenzene to aniline is a typical oxidation-reduction reaction in which tin metal, Sn°, is oxidized to stannic ion, Sn^{4+}, in the form of stannic chloride, $SnCl_4$; hydrochloric acid serves as the source of protons. A plausible mechanism of this reaction is outlined in Scheme 21.3. Generally, the reduction occurs by a sequence of steps in which an electron is first transferred from a tin atom to the organic substrate to give an intermediate radical ion that is then protonated. The oxygen atoms on the nitro group are eventually removed as water molecules. It is left as an exercise to write a *balanced* equation for the overall reaction and to provide a mechanism for the reduction of *N*-phenylhydroxylamine (**11**) into aniline (see Exercises 13 and 14 at the end of this section).

Scheme 21.3

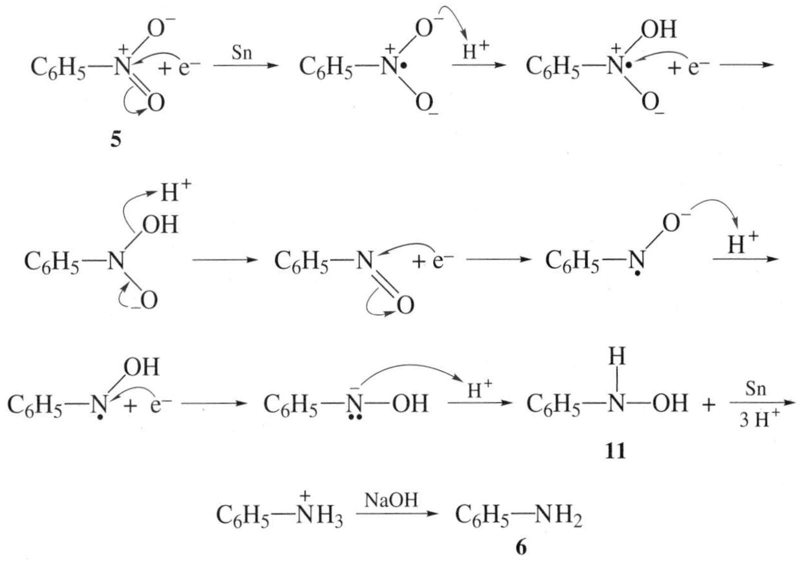

$$C_6H_5{-}\overset{+}{N}(O^-){=}O\ (\mathbf{5}) + e^- \xrightarrow{Sn} C_6H_5{-}\overset{+}{N}^{\bullet}(O^-)(O^-) \xrightarrow{H^+} C_6H_5{-}\overset{+}{N}^{\bullet}(OH)(O^-) + e^- \longrightarrow$$

$$C_6H_5{-}N(OH)(O^-) + H^+ \longrightarrow C_6H_5{-}N{=}O + e^- \longrightarrow C_6H_5{-}\dot{N}{-}O^- \xrightarrow{H^+}$$

$$C_6H_5{-}\dot{N}{-}OH + e^- \longrightarrow C_6H_5{-}\ddot{N}^-{-}OH \xrightarrow{H^+} C_6H_5{-}NH{-}OH\ (\mathbf{11}) + \xrightarrow[3\,H^+]{Sn}$$

$$C_6H_5{-}\overset{+}{N}H_3 \xrightarrow{NaOH} C_6H_5{-}NH_2\ (\mathbf{6})$$

The procedure for isolating and purifying the aniline formed in this experiment represents an excellent example of how the physical and chemical properties of a component in a mixture of organic substances can be exploited to isolate it *without* using chromatographic techniques. For example, after completing the procedure for the chemical reduction of nitrobenzene, it is necessary to remove aniline from its principal impurities, which are unchanged nitrobenzene and two by-products, benzidine (**12**), and 4-aminophenol (**13**).

Steam distillation (Secs. 2.16 and 4.5) is a good technique for separating volatile organic compounds from nonvolatile organic and inorganic substances. However, the reaction mixture is *acidic,* and it must be made *basic* prior to performing the steam distillation so aniline is present as the free base, and the 4-aminophenol is converted to its *water-soluble* sodium phenoxide salt. The aniline and nitrobenzene are then removed from the reaction mixture by steam distillation. The nonvolatile salt of 4-aminophenol and the nonvolatile benzidine remain in the aqueous phase.

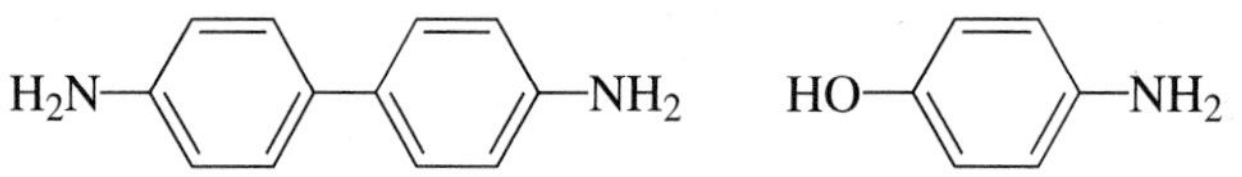

H_2N–C_6H_4–C_6H_4–NH_2 $\qquad$ HO–C_6H_4–NH_2

12
Benzidine

13
4-Aminophenol

Aniline may be separated from nitrobenzene by first acidifying the steam distillate with hydrochloric acid to convert aniline to its water-soluble hydrochloride salt **14.** The acidic solution is then extracted with diethyl ether to remove nitrobenzene. The aqueous layer, which contains **14,** is made basic, and the aniline that is formed is isolated from the mixture by extraction with diethyl ether.

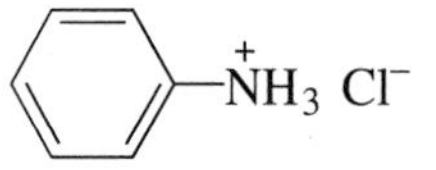

C_6H_5–$\overset{+}{N}H_3$ Cl^-

14
Anilinium hydrochloride

Protecting Groups: Preparation of Acetanilide (7)

The amino function on aniline is an electron-donating group that *activates* the aromatic ring toward electrophilic aromatic substitution reactions (Chap. 15). However, if an aryl amino group is protonated or complexed with a Lewis acid, the resulting ammonium group *deactivates* the ring toward electrophilic aromatic substitution. Electrophilic reagents or other functional groups that may be present on the ring can also react directly with the amino group. Thus, because of its basic and nucleophilic properties, it is necessary to protect the amino group on aniline to ensure that the desired reaction will take place. In general, placing a **protecting group** on a reactive functional group alters the chemical properties of that group, thereby rendering it unreactive toward certain reagents. The protecting group must be carefully selected, however, because it must be stable to the reaction conditions employed in the various transformations, *and* it must be removable under conditions that do not adversely affect other functional groups in the molecule.

In our synthesis of sulfanilamide (**1**), adding a sulfonamido, $-SO_2NH_2$, group to the aromatic ring requires the use of a protecting group for the amino function. This is necessary because introducing a sulfonamido function onto an arene, Ar–H, commonly involves sequential chlorosulfonation, an electrophilic aromatic substitution reaction, followed by treatment of the intermediate aryl sulfonyl chloride, $Ar–SO_2Cl$, with ammonia to give the sulfonamide, $Ar–SO_2NH_2$ (Eq. 21.2). If aniline itself were treated with chlorosulfonic acid, the amino group could react with either hydrogen chloride or SO_3, which are generated *in situ*, to provide ammonium salts such as **14** or **15;** these groups would then deactivate the ring toward electrophilic aromatic substitution. Moreover, a chlorosulfonyl group cannot be generated in the presence of an amino group since the chlorosulfonyl group of one molecule would react with the amino group of another, thereby forming a *polymeric* material containing sulfonamide linkages.

$$\underset{\text{An arene}}{Ar—H} \xrightarrow[(2)\ NH_3]{(1)\ ClSO_3H} \underset{\text{An aryl sulfonamide}}{Ar—SO_2NH_2} \qquad (21.2)$$

$$C_6H_5—\overset{+}{N}H_2\ SO_3^-$$

15

In the experiment you will perform, the amino group is protected as an amide by acetylation. The free amine is later regenerated by removing the acetyl group via controlled hydrolysis *after* the chlorosulfonyl moiety has been introduced onto the aromatic ring and converted into the sulfonamido group. The selective hydrolysis of the amido group of **9** in the presence of the sulfonamido group is possible because the sulfur atom of the sulfonamido group is *tetrahedral* and hence more sterically hindered toward nucleophilic attack by water than the *trigonal* carbonyl carbon atom (Scheme 21.2).

Aniline (**6**) is converted to acetanilide (**7**) by acetylation according to the sequence shown in Equation 21.3. In this reaction, aniline is first converted to its hydrochloride salt **14.** Acetic anhydride and sodium acetate are then added to give a mixture in which sodium acetate and **14** are in equilibrium with acetic acid and aniline. As free aniline is produced by this acid-base reaction, it rapidly undergoes acetylation by acetic anhydride to give acetanilide. This particular method for acetylating amines is general, and the yields are usually high.

$$\underset{\mathbf{6}}{C_6H_5NH_2} \xrightarrow{HCl} \underset{\mathbf{14}}{C_6H_5\overset{+}{N}H_3\ Cl^-} \xrightarrow[CH_3CO_2Na]{(CH_3CO)_2O} \underset{\mathbf{7}}{C_6H_5NHCOCH_3} + CH_3CO_2H + NaCl \qquad (21.3)$$

It is always important to consider possible side reactions that may occur. One example in this procedure is diacetylation (Eq. 21.4). Although the acetylation of aniline in pure acetic anhydride does give substantial quantities of the diacetyl compound **16,** this side reaction is minimized in aqueous solution.

$$\text{7 } (C_6H_5NHCOCH_3) \xrightarrow{(CH_3CO)_2O} \text{16 } (C_6H_5N(COCH_3)_2) + CH_3CO_2H \quad (21.4)$$

Chlorosulfonation: Preparation of 4-Acetamidobenzenesulfonyl Chloride (8)

The sulfonyl group, $-SO_2Cl$, can be introduced *para* to the *N*-acetyl group of acetanilide by a one-step process called chlorosulfonation (Eq. 21.5). Although an electron-donating group can direct an incoming electrophile to either the *ortho* or *para* position, the acetamido group orients the incoming group predominantly *para,* presumably as a consequence of the steric bulk of the acetamido group; virtually none of the *ortho* isomer is observed. The reaction is known to proceed through the intermediate sulfonic acid **17,** which is converted to the sulfonyl chloride **8** by reaction with chlorosulfonic acid. At least two equivalents of chlorosulfonic acid are thus required per equivalent of acetanilide (**7**). The attacking electrophile is probably SO_3, which is generated *in situ* from chlorosulfonic acid according to Equation 21.6.

$$\text{7 } (C_6H_5NHCOCH_3) \xrightarrow{ClSO_3H\ (1\ mol)} \text{17 } (4\text{-}CH_3CONHC_6H_4SO_3H) + HCl \xrightarrow{ClSO_3H\ (1\ mol)} \text{8 } (4\text{-}CH_3CONHC_6H_4SO_2Cl) + H_2SO_4 \quad (21.5)$$

17: 4-Acetamidobenzene-sulfonic acid

$$ClSO_3H \rightleftharpoons SO_3 + HCl \quad (21.6)$$

The reaction is worked up by pouring it into ice-water, whereupon the 4-acetamidobenzenesulfonyl chloride (**8**) separates as a white solid. The excess chlorosulfonic acid is hydrolyzed by water (Eq. 21.7). In general, sulfonyl chlorides are much less reactive toward water than are carboxylic acid chlorides, but you should not expose them to water for extended periods of time because they hydrolyze slowly to give the corresponding sulfonic acids according to Equation 21.8. It is unnecessary to dry or purify the 4-acetamidobenzenesulfonyl chloride (**8**) in this sequence; rather, **8** is simply treated directly with aqueous ammonia.

$$ClSO_3H + H_2O \longrightarrow HCl + H_2SO_4 \quad (21.7)$$

$$\text{8 } (4\text{-}CH_3CONHC_6H_4SO_2Cl) + H_2O \longrightarrow \text{17 } (4\text{-}CH_3CONHC_6H_4SO_3H) + HCl \quad (21.8)$$

Ammonolysis and Hydrolysis: Preparation of Sulfanilamide (1)

The final steps in the preparation of sulfanilamide (**1**) involve treating 4-acetamidobenzenesulfonyl chloride (**8**) with an excess of aqueous ammonia to give 4-acetamidobenzenesulfonamide (**9**), followed by selective hydrolytic removal of the *N*-acetyl protecting group (Eq. 21.9). To avoid hydrolysis of the sulfonyl chloride group (Eq. 21.8) before its reaction with ammonia, it is *imperative* to treat **8** with ammonia *immediately* after its isolation in the previous experiment. If the conversion of acetanilide (**7**) to **9** is not completed within a *single* laboratory period, the overall yield for the sequence may be significantly reduced.

Hydrolysis of the acetamido moiety of **9** may be effected using either aqueous acid or aqueous base without affecting the sulfonamido group, which hydrolyzes slowly; an acid-catalyzed hydrolysis is performed in this experiment. Because the amine group will form a hydrochloride salt under the conditions of the hydrolysis, it is necessary to neutralize the solution with a base, such as sodium carbonate, in order to isolate sulfanilamide, which may then be purified by recrystallization from water.

$$\underset{\mathbf{8}}{4\text{-}CH_3CONH\text{-}C_6H_4\text{-}SO_2Cl} \xrightarrow{NH_4OH} \underset{\mathbf{9}}{4\text{-}CH_3CONH\text{-}C_6H_4\text{-}SO_2NH_2} \xrightarrow[\Delta]{HCl/H_2O} 4\text{-}\overset{+}{N}H_3\,Cl^-\text{-}C_6H_4\text{-}SO_2NH_2 \xrightarrow[H_2O]{Na_2CO_3} \underset{\mathbf{1}}{4\text{-}H_2N\text{-}C_6H_4\text{-}SO_2NH_2} \qquad (21.9)$$

EXPERIMENTAL PROCEDURES

Synthesis of Sulfanilamide

Purpose To demonstrate principles of multistep synthesis, protecting groups, and electrophilic aromatic substitution.

Note Throughout the sequence of reactions, quantities of reagents must be adjusted according to the amount of starting material that is available from the previous reaction, but you should *not* run the reactions on a larger scale than is indicated. If the amount of your starting material varies from that indicated, do *not* change the reaction times unless directed to do so by your instructor. You should be able to obtain sufficient quantities of product from each reaction to complete the entire sequence starting with the amount of acetanilide provided in the first step.

A ▪ *Preparation of Aniline*

SAFETY ALERT

1. **Nitrobenzene is toxic; avoid inhaling its vapors or allowing it to come into contact with your skin. Wear latex gloves when transferring this chemical.**

2. ***Concentrated* hydrochloric acid can cause burns if it comes in contact with the skin. Should contact occur, flood the affected area with water and rinse it thoroughly with dilute aqueous sodium bicarbonate solution. Wear latex gloves when transferring this acid.**
3. **The solution of 12 *M* sodium hydroxide is highly caustic and can cause burns and loss of hair; avoid contact of this solution with your skin. Should contact occur, flood the affected area with water and rinse it thoroughly with dilute aqueous acetic acid solution. Wear latex gloves when transferring this solution.**
4. **Use *flameless* heating when concentrating ethereal solutions.**

MINISCALE PROCEDURE

Preparation Sign in at **www.cengage.com/login** to answer Pre-Lab Exercises, access videos, and read the MSDSs for the chemicals used or produced in this procedure. Review Sections 2.9, 2.10, 2.11, 2.13, 2.16, 2.21, 2.22, 2.25, and 2.29.

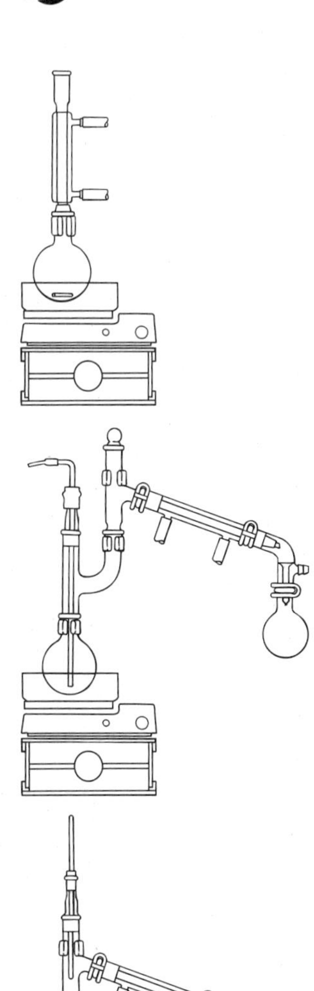

Apparatus A 250-mL round-bottom flask, thermometer, ice-water bath, separatory funnel, apparatus for heating under reflux, magnetic stirring, steam distillation, simple distillation, and *flameless* heating.

Setting Up Add 13.1 g of finely divided tin powder and 5.2 mL of nitrobenzene to the round-bottom flask equipped with a stirbar.

Reduction Add 28 mL of *concentrated* hydrochloric acid to the flask, insert a thermometer, and stir the contents of the flask to mix the three-phase system. Monitor the temperature and cool the flask as necessary in an ice-water bath to maintain the temperature below 60 °C. The initial exothermic portion of the reaction should be complete after about 15 min. Attach a reflux condenser and heat the contents of the flask under reflux for about 20 min with continued stirring. During this time, the color, which is due to the intermediate reduction product, and any droplets of nitrobenzene in the condenser should disappear.

Work-Up When the reflux period is complete, cool the acidic solution in an ice-water bath and slowly add 50 mL of 12 *M* sodium hydroxide solution directly to the reaction mixture. Check the pH of the mixture to ensure that it is basic; if it is not, add additional base. Equip the flask for steam distillation and steam-distill the mixture until the condensate no longer appears oily, although it may still be slightly cloudy; about 35–40 mL of distillate should be collected.★

Add 4.5 mL of *concentrated* hydrochloric acid to the distillate and transfer the mixture to a separatory funnel. Check the pH of the mixture to be certain it is acidic, and then extract it sequentially with two 15-mL portions of diethyl ether to remove any unreacted nitrobenzene. Transfer the *aqueous* layer to an Erlenmeyer flask. Cool the contents of the flask in an ice-water bath and make the aqueous solution *basic* by slowly adding a *minimum* volume of 12 *M* sodium hydroxide solution. Saturate the aqueous solution with solid sodium chloride, cool the mixture to room temperature, and transfer it to a *clean* separatory funnel. Extract the aqueous mixture sequentially with two 15-mL portions of diethyl ether, using the first portion to rinse the flask in which the neutralization was done. Separate the aqueous layer from the organic layer as thoroughly as possible each time. Dry the *combined* organic extracts over several spatula-tips full anhydrous sodium sulfate.★ Decant the dried solution into a round-bottom flask, and remove most of the diethyl ether by simple distillation. Alternatively,

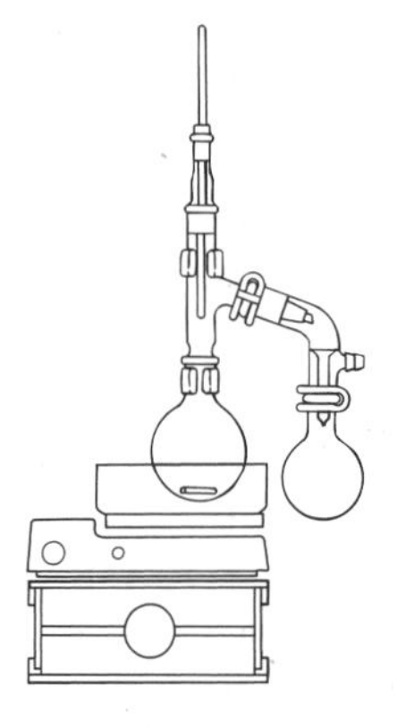

use rotary evaporation or other techniques to concentrate the solution. Remove the final traces of solvents by connecting the cool stillpot to a vacuum source.

The aniline remaining in the stillpot may be used directly in the next step of the sequence. Alternatively, you may purify the aniline by shortpath distillation. Wrap the stillhead with a layer of glasswool and then aluminum foil to minimize heat losses, thereby facilitating the distillation. Collect three fractions having approximate boiling ranges (760 torr) of 35–90 °C, 90–180 °C, and 180–185 °C, respectively.

Analysis Weigh the product and calculate its yield. Obtain IR and ^{1}H NMR spectra of your starting material and product, and compare them with those of authentic samples (Figs. 21.2–21.5). Pure aniline is colorless but may darken soon following distillation, owing to air-oxidation.

WRAPPING IT UP

Vacuum-filter the *pot residue from the steam distillation* through a bed of filter-aid, and place the *tin salts* together with the *filter-aid* and *filter paper* in the container for nonhazardous solids. Neutralize the *filtrate* and the other *aqueous layers* with dilute hydrochloric acid and flush them down the drain with excess water. Pour the combined *ethereal extracts* and *distillate* into a container for nonhalogenated organic liquids. Put the *sodium sulfate* on a tray in the hood and, after the ether evaporates, pour it into the container for nonhazardous solids.

B ■ *Preparation of Acetanilide*

SAFETY ALERT

Acetic anhydride is a lachrymator. Transfer this material in a hood and do not inhale its vapors.

MINISCALE PROCEDURE

Preparation Sign in at **www.cengage.com/login** to answer Pre-Lab Exercises, access videos, and read the MSDSs for the chemicals used or produced in this procedure. Review Sections 2.9, 2.10, 2.11, 2.17, and 2.26.

Apparatus A 250-mL Erlenmeyer flask, thermometer, ice-water bath, apparatus for magnetic stirring, vacuum filtration, and *flameless* heating.

Setting Up Add 3.6 mL of aniline to the Erlenmeyer flask containing 100 mL of 0.4 *M* hydrochloric acid and equipped with a stirbar. Stir the mixture to effect solution and warm it to about 50 °C. Prepare a solution containing 6.0 g of sodium acetate trihydrate dissolved in 20 mL of water, and, in a *separate* container, measure out 4.4 mL of acetic anhydride.

Reaction and Isolation Add the acetic anhydride in *one* portion to the warm solution of anilinium hydrochloride. Stir the solution *vigorously* and then add the solution of sodium acetate *immediately and in one portion.* Cool the reaction mixture to 5 °C in an ice-water bath, and continue to stir the mixture until the crystalline product completely precipitates. Collect the acetanilide by vacuum filtration, wash it with a small portion of ice-cold water, and air-dry it.★ If impure or slightly colored acetanilide is obtained, recrystallize it from a minimum volume of hot water, using decolorizing carbon to give a colorless product.

Analysis Weigh the product and calculate its yield. Determine its melting point. Obtain IR and ^{1}H NMR spectra of your starting material and product and compare them with those of an authentic sample (Figs. 8.7 and 15.31).

WRAPPING IT UP

After neutralizing the *aqueous filtrate* with sodium carbonate, flush it and the *mother liquor* from crystallization down the drain with a large excess of water.

C ■ *Preparation of 4-Acetamidobenzenesulfonyl Chloride*

Note This entire procedure as well as the next one, up to the first stopping point, must be completed in a *single* laboratory period.

SAFETY ALERT

1. **This experiment should be performed in a hood, if possible. Otherwise, use the gas trap prescribed by your instructor to prevent escape of hydrogen chloride, SO_2, and SO_3 into the laboratory.**
2. **Chlorosulfonic acid is *highly corrosive* and may cause serious burns if it is allowed to come in contact with your skin. Wear latex gloves when handling or transferring this reagent, and use *extreme* care when working with it. Should any chlorosulfonic acid come in contact with your skin, *immediately* flood the affected area with cold water and then rinse it with 5% sodium bicarbonate solution.**
3. **Chlorosulfonic acid reacts *violently* with water. Open containers of chlorosulfonic acid will fume upon exposure to atmospheric moisture owing to the reaction of chlorosulfonic acid with water to give HCl and SO_3, both of which are noxious and corrosive gases. Measure and transfer this acid only *in the hood.* Several graduated cylinders should be kept in the hood for all students to use for measuring the volume of chlorosulfonic acid needed.**
4. **To destroy residual chlorosulfonic acid in graduated cylinders and other glassware that has contained it, add cracked ice to the glassware *in the hood* and let the glassware remain there until the ice has melted. Then rinse the apparatus with large amounts of cold water.**
5. **Make sure your glassware is dry, lubricate the ground-glass joints of the apparatus carefully, and be certain that the joints are intimately mated. Otherwise, noxious gases will escape.**

MINISCALE PROCEDURE

Preparation Sign in at **www.cengage.com/login** to answer Pre-Lab Exercises, access videos, and read the MSDSs for the chemicals used or produced in this procedure. Review Sections 2.9, 2.10, 2.11, 2.17, 2.23, and 2.28.

Apparatus A 50-mL round-bottom flask, Claisen adapter, gas trap, separatory funnel with ground-glass joint, thermometer, ice-water bath, apparatus for magnetic stirring, vacuum filtration, and *flameless* heating.

Setting Up Oven-dry the glass apparatus prior to assembling it. Because an *airtight* seal is required for all connections, *carefully grease all joints* when assembling the apparatus. Place 2.7 g of *dry* acetanilide in the round-bottom flask and equip the flask with a stirbar and the Claisen adapter. Fit the *sidearm* of the adapter with the gas trap specified by your instructor; some examples are illustrated in Figure 2.68. *Working at the hood,* measure 8.0 mL of chlorosulfonic acid into a small separatory funnel having a standard-taper ground-glass joint; be certain that the stopcock of the funnel is firmly seated and closed *prior* to this transfer. Stopper the funnel and place it on the straight arm of the Claisen adapter above the reaction flask so that the chlorosulfonic acid will drip *directly* onto the acetanilide. Cool the flask in a water bath that is maintained at 10–15 °C; do *not* allow the bath temperature to drop below 10 °C.

Reaction Open the stopcock of the funnel fully so that the chlorosulfonic acid is added as rapidly as possible to the flask containing the acetanilide. It may be necessary to *briefly* lift the stopper on the funnel to equalize the pressure in the system if the flow of chlorosulfonic acid becomes erratic or slow. When the addition has been completed, stir the mixture to hasten dissolution of the acetanilide; maintain the temperature of the water bath below 20 °C. After most of the solid has dissolved and the initial reaction has subsided, remove the cooling bath, and allow the reaction mixture to warm to room temperature with stirring. To complete the reaction, warm the mixture at about 70–80 °C with continued stirring until there is no longer any increase in the rate of gas evolution; about 10–20 min of heating will be required.

Work-Up and Isolation Cool the mixture to room temperature or slightly below using an ice-water bath. *Working at the hood,* place about 150 g of crushed ice in a beaker and pour the reaction mixture slowly onto the ice with stirring; stir the mixture thoroughly with a glass rod during the addition. *Be careful not to add the mixture too quickly to avoid splattering the chlorosulfonic acid.* Rinse the reaction flask with a little ice-water and transfer this to the beaker. The remainder of the work-up procedure may be performed at the bench. The precipitate that forms is crude 4-acetamidobenzenesulfonyl chloride and may be white to pink in color. If it is allowed to stand without stirring, a hard mass may form. Any lumps that form should be thoroughly broken up with a stirring rod. Collect the crude material by vacuum filtration. Wash the solid thoroughly with 15-mL portions of ice-water *until the filtrate is neutral;* press the filter cake as dry as possible with a clean cork. The crude sulfonyl chloride should be used *immediately* in the following procedure.

Purification Purify about 50 mg of the product by dissolving it in a minimum amount of boiling dichloromethane in a test tube. Using a Pasteur pipet, remove the upper, aqueous layer as quickly as possible. Be careful when separating the layers. Cool the organic layer, collect the purified product by vacuum filtration, and air-dry it.

Analysis Determine the melting point of the recrystallized solid. Obtain its IR and ^{1}H NMR spectra and compare them with those of an authentic sample (Figs. 21.6 and 21.7).

WRAPPING IT UP

Neutralize the combined *aqueous solutions* with sodium carbonate and flush them down the drain with a large excess of water. Cover any spilled droplets of *chlorosulfonic acid* with sodium carbonate and flush the resulting powder down the drain. If a gas trap that contains a drying agent has been used, put the *drying agent* in the container for nonhazardous solids.

D ▪ *Preparation of 4-Acetamidobenzenesulfonamide*

SAFETY ALERT

1. **Ammonia is evolved in this procedure. If possible, perform this reaction in a hood. Otherwise, use an inverted funnel attached to a vacuum source (Fig. 2.71b) to minimize escape of vapors of ammonia into the laboratory.**
2. **Concentrated ammonium hydroxide is caustic and may cause burns if it is allowed to come in contact with your skin. Wear latex gloves when handling or transferring this reagent, and use care when working with it. Should any of the liquid come in contact with your skin, *immediately* flood the affected area with cold water.**

MINISCALE PROCEDURE

Preparation Sign in at **www.cengage.com/login** to answer Pre-Lab Exercises, access videos, and read the MSDSs for the chemicals used or produced in this procedure. Review Sections 2.9, 2.10, 2.11, and 2.17.

Apparatus A 125-mL Erlenmeyer flask, ice-water bath, apparatus for vacuum filtration, and *flameless* heating.

Reaction, Work-Up, and Isolation Transfer the crude 4-acetamidobenzenesulfonyl chloride obtained in the previous experiment to the Erlenmeyer flask, and add 15 mL of concentrated (28%) ammonium hydroxide. A rapid exothermic reaction may occur if the crude 4-acetamidobenzenesulfonyl chloride contains acidic contaminants that were not removed by the aqueous washings in the previous step. Use a stirring rod to break up any lumps of solid that may remain; the reaction mixture should be thick but homogeneous. Heat the mixture at 70–80 °C for about 0.5 h. Cool the reaction mixture in an ice-water bath and collect the product by vacuum filtration. Wash the crystals with cold water and air-dry them.★

Analysis Weigh the product and calculate its yield. Determine its melting point. Obtain IR and ^1H NMR spectra of your product and compare them with those of an authentic sample (Figs. 21.8 and 21.9).

WRAPPING IT UP

Neutralize the combined *aqueous solutions* and *filtrates* with 10% hydrochloric acid and flush them down the drain with a large excess of water.

E ■ *Preparation of Sulfanilamide*

SAFETY ALERT

Concentrated hydrochloric acid is corrosive and may cause serious burns if it is allowed to come in contact with your skin. Wear latex gloves when handling or transferring this reagent and use care when working with it. Should any acid come in contact with your skin, *immediately* flood the affected area with cold water and then rinse it with 5% sodium bicarbonate solution.

MINISCALE PROCEDURE

Preparation Sign in at **www.cengage.com/login** to answer Pre-Lab Exercises, access videos, and read the MSDSs for the chemicals used or produced in this procedure. Review Sections 2.9, 2.10, 2.11, 2.17, and 2.22.

Apparatus A 50-mL round-bottom flask, ice-water bath, apparatus for heating under reflux, magnetic stirring, vacuum filtration, and *flameless* heating.

Setting Up Prepare 20 mL of a solution of *dilute* hydrochloric acid by adding 10 mL of *concentrated* hydrochloric acid to 10 mL of water; swirl to mix. Equip the round-bottom flask with a stirbar and add the 4-acetamidobenzenesulfonamide and an amount of *dilute* hydrochloric acid that is about *twice* the weight of the crude 4-acetamidobenzenesulfonamide. Assemble the apparatus for heating under reflux.

Reaction, Work-Up, and Isolation Heat the reaction mixture with stirring under gentle reflux for 45 min and then allow the mixture to cool to room temperature. If any solid, which is unreacted starting material, appears on cooling, reheat the mixture under reflux for an additional 15 min. Add an equal volume of water to the cooled solution and transfer the resulting mixture to a beaker. Neutralize the excess hydrochloric acid by adding *small* quantities of *solid* sodium carbonate until the solution is slightly alkaline.

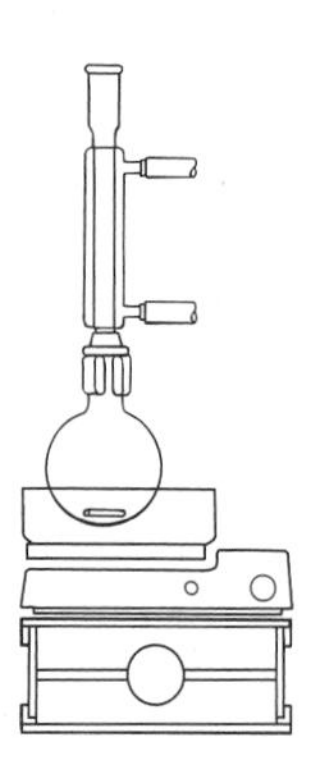

A precipitate should form during neutralization; scratching at the air-liquid interface may be necessary to initiate crystallization. Cool the mixture in an ice-water bath to complete the precipitation of product.★ Collect the crystals by vacuum filtration, wash them with a small amount of cold water, and air-dry them.

Purification Purify the crude product by recrystallization from the minimum volume of hot water. If necessary, decolorize the hot solution and use a preheated funnel for the hot-filtration step to prevent premature crystallization of the product in the funnel. Cool the aqueous solution of sulfanilamide in an ice-water bath, isolate the crystalline product by vacuum filtration, and air-dry it.★

Analysis Weigh the product and calculate its yield. Determine its melting point. Test the solubility of sulfanilamide in 1.5 *M* hydrochloric acid solution and in 1.5 *M* sodium hydroxide solution. Obtain IR and ^{1}H NMR spectra of your product and compare them with those of an authentic sample (Figs. 21.10 and 21.11).

WRAPPING IT UP

Neutralize the combined *aqueous solutions and filtrates* with sodium carbonate and flush them down the drain with a large excess of water.

EXERCISES

General Questions

1. What biological properties of sulfanilamide enable it to interfere with bacterial but not human cell growth?
2. If each step in the synthesis of sulfanilamide from nitrobenzene proceeded in 90% yield, how much nitrobenzene would be needed to obtain 1.0 kg of sulfanilamide according to the sequence of reactions in Scheme 21.2?
3. Using benzene as the starting material together with any organic or inorganic reagents you desire, outline a synthesis of *p*-aminobenzoic acid (PABA) (**2**).
4. Outline a possible synthesis of the sulfanilamide derivative, **18**, using benzene as the only source of an aromatic ring. Use any needed aliphatic or inorganic reagents.

H_2N—C_6H_4—SO_2NH—C_6H_5

18

5. Two hypothetical sequences have been developed for converting an arbitrary compound A into E in the same overall yield.

$$A \xrightarrow{25\%} B \xrightarrow{49\%} C \xrightarrow{60\%} D \xrightarrow{57\%} E$$

$$A \xrightarrow{58\%} B \xrightarrow{57\%} C \xrightarrow{51\%} D \xrightarrow{30\%} E$$

   a. What is the overall yield for each sequence?
   b. As the production manager responsible for selecting the more economical of the two routes, what factor(s) would you consider in reaching a decision? Assume that no new capital investment would be required for either sequence.

6. A student proposed the following alternative sequence of reactions for preparing **1** from sulfanilic acid. Discuss this modified approach, commenting on any possible side reactions that might be encountered in either of the two steps. Propose a modification of this sequence that could be used to prepare sulfanilamide.

NH_2 / SO_3H —PCl_5→ NH_2 / SO_2Cl —NH_3→ NH_2 / SO_2NH_2 **1**

Sulfanilic acid   *p*-Aminosulfonyl chloride

Questions for Part A

7. Why must an organic compound be *immiscible* with water in order to be purified by steam distillation?
8. Use Dalton's law and Raoult's law to explain why the steam distillation of aniline occurs at a lower temperature than the boiling point of aniline (Sec. 4.5).
9. Outline in a flow diagram the procedure for purifying aniline. Indicate the importance of each step in the procedure, and give reasons for doing the steam distillation with a *basic* solution. What is the purpose of performing the subsequent extractions from aqueous acid and then base? Write the equation(s) for the reactions that occur when base and acid are added.
10. What gas is evolved upon adding concentrated hydrochloric acid to tin powder in this reduction?
11. Suppose a student decided that a suitable stopping point in the procedure might be immediately after the initially exothermic reaction between tin and hydrochloric acid. What is the potential problem with storing the reaction mixture at this stage?
12. Why is sodium chloride added to the steam distillate *after* it has been made basic but *before* extraction with diethyl ether?
13. Write the balanced half-reactions for the reduction of nitrobenzene (**5**) into aniline (**6**) and for the oxidation of tin metal into stannic ion, Sn^{4+}. Sum these equations and write the balanced equation for the overall process.
14. Write a mechanism for the reduction of *N*-phenylhydroxylamine (**11**) into aniline (**6**). Show all electron and proton transfer steps, using curved arrows to symbolize the flow of electrons.
15. Protection and deprotection of the aryl amino group might be avoided if it were possible to reduce the nitro group as the last step in the synthesis. Explain why this is not an available option.

Sign in at **www.cengage.com/login** and use the spectra viewer and Tables 8.1–8.8 as needed to answer the blue-numbered questions on spectroscopy.

16. Consider the spectral data for nitrobenzene (Figs. 21.2 and 21.3).
    a. In the functional group region of the IR spectrum, specify the absorptions associated with the nitro group and with the aromatic ring.
    b. In the ^1H NMR spectrum, assign the various resonances to the hydrogen nuclei responsible for them.
    c. For the ^{13}C NMR data, assign the various resonances to the carbon nuclei responsible for them.
17. Consider the spectral data for aniline (Figs. 21.4 and 21.5).
    a. In the functional group region of the IR spectrum, specify the absorptions associated with the amino group and with the aromatic ring.

b. In the ^1H NMR spectrum, assign the various resonances to the hydrogen nuclei responsible for them.

c. For the ^{13}C NMR data, assign the various resonances to the carbon nuclei responsible for them.

18. Discuss the differences observed in the IR and NMR spectra of nitrobenzene and aniline that are consistent with the formation of the latter in this procedure.

Questions for Part B

19. Why is it necessary to *N*-acetylate aniline to give acetanilide (Part B) when the *N*-acetyl group (CH_3CO–) will be removed at the end of the synthesis (Part E)?
20. Why should the acetic anhydride not be allowed to stay in contact with the aqueous solution of anilinium hydrochloride (**14**) for an extended period of time before the solution of sodium acetate is added?
21. Why is aqueous sodium acetate preferred to aqueous sodium hydroxide for the conversion of anilinium hydrochloride (**14**) to aniline?
22. Why is aniline (**6**) soluble in aqueous hydrochloric acid, whereas acetanilide (**7**) is not?
23. Using curved arrows to symbolize the flow of electrons, give a stepwise reaction mechanism for the reaction of aniline (**6**) with acetic anhydride.

Sign in at **www.cengage.com/login** and use the spectra viewer and Tables 8.1–8.8 as needed to answer the blue-numbered questions on spectroscopy.

24. Consider the spectral data for acetanilide (Figs. 8.7 and 15.31).

a. In the functional group region of the IR spectrum, specify the absorptions associated with the nitrogen-hydrogen bonds, with the carbon-oxygen double bond, and with the aromatic ring.

b. In the ^1H NMR spectrum, assign the various resonances to the hydrogen nuclei responsible for them.

c. For the ^{13}C NMR data, assign the various resonances to the carbon nuclei responsible for them.

25. Discuss the differences observed in the IR and NMR spectra of aniline and acetanilide that are consistent with the formation of the latter in this procedure.

Questions for Part C

26. Why does the sulfonation of acetanilide (**7**) only occur in the *para* position?
27. Explain why 4-acetamidobenzenesulfonyl chloride (**8**) is much *less* susceptible to hydrolysis of the acid chloride function than is 4-acetamidobenzoyl chloride.

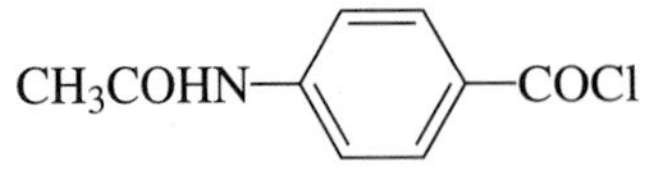

CH_3COHN–C_6H_4–COCl

4-Acetamidobenzoyl chloride

28. What materials, organic or inorganic, may contaminate the crude sulfonyl chloride **8** prepared in this reaction? Which of them are likely to react with the ammonia used in the next reaction step of the sequence?
29. Using curved arrows to symbolize the flow of electrons, write a stepwise mechanism for the conversion of acetanilide (**7**) into 4-acetamidobenzenesulfonic acid (**17**).

30. Write the structure of the material containing sulfonamide linkages that might be obtained upon polymerization of 4-aminobenzenesulfonyl chloride.

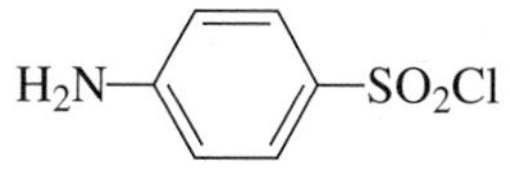

H_2N–⟨benzene⟩–SO_2Cl

4-Aminobenzenesulfonyl chloride

31. Predict the product of monochlorosulfonation of the following compounds.

a. H_3CO–⟨benzene⟩–NO_2 **b.** ⟨benzene⟩–C(=O)–O–⟨benzene⟩

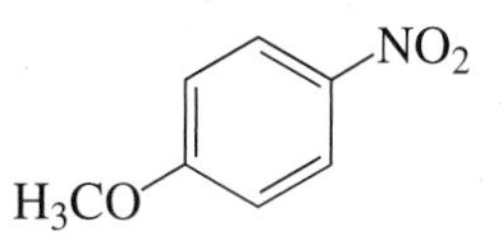

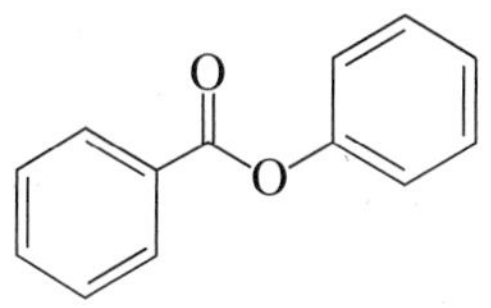

Sign in at **www.cengage.com/login** and use the spectra viewer and Tables 8.1–8.8 as needed to answer the blue-numbered questions on spectroscopy.

32. Consider the spectral data for 4-acetamidobenzenesulfonyl chloride (Figs. 21.6 and 21.7).

a. In the functional group region of the IR spectrum, specify the absorptions associated with the nitrogen-hydrogen bond, with the carbon-oxygen double bond, and with the aromatic ring.

b. In the ^{1}H NMR spectrum, assign the various resonances to the hydrogen nuclei responsible for them.

c. For the ^{13}C NMR data, assign the various resonances to the carbon nuclei responsible for them.

33. Discuss the differences observed in the IR and NMR spectra of acetanilide and 4-acetamidobenzenesulfonyl chloride that are consistent with the formation of the latter in this procedure.

Questions for Part D

34. What acids might be present in the crude sulfonyl chloride that would cause an exothermic reaction with ammonia?

35. Why would you expect the yield of 4-acetamidobenzenesulfonamide (**9**) to be lowered if the crude sulfonyl chloride **8** were not combined with ammonia until the laboratory period following its preparation?

36. Concentrated ammonia contains 72% water, yet its reaction with 4-acetamidobenzenesulfonyl chloride (**8**) produces 4-acetamidobenzenesulfonamide (**9**) rather than 4-acetamidobenzenesulfonic acid (**17**). Why is ammonia more reactive than water toward **9**?

37. Using curved arrows to symbolize the flow of electrons, write a stepwise mechanism for the aminolysis of 4-acetamidobenzenesulfonyl chloride (**8**) to give 4-acetamidobenzenesulfonamide (**9**).

Sign in at **www.cengage.com/login** and use the spectra viewer and Tables 8.1–8.8 as needed to answer the blue-numbered questions on spectroscopy.

38. Consider the spectral data for 4-acetamidobenzenesulfonamide (Figs. 21.8 and 21.9).

a. In the functional group region of the IR spectrum, specify the absorptions associated with the nitrogen-hydrogen bonds, with the carbon-oxygen double bond, and with the aromatic ring.

b. In the ^{1}H NMR spectrum, assign the various resonances to the hydrogen nuclei responsible for them.

c. For the ^{13}C NMR data, assign the various resonances to the carbon nuclei responsible for them.

39. Discuss the differences observed in the IR and NMR spectra of 4-acetamidobenzenesulfonyl chloride and 4-acetamidobenzenesulfonamide that are consistent with the formation of the latter in this procedure.

Questions for Part E

40. In the preparation of sulfanilamide (**1**) from 4-acetamidobenzenesulfonamide (**9**), only the acetamido group is hydrolyzed. Explain this difference in reactivity of the acetamido and sulfonamido groups toward aqueous acid.

41. Following hydrolysis of 4-acetamidobenzenesulfonamide (**9**) with aqueous acid, the reaction mixture is homogeneous, whereas 4-acetamidobenzenesulfonamide is insoluble in aqueous acid. Explain the change in solubility that occurs as a result of the hydrolysis.

42. Explain the results obtained when the solubility of sulfanilamide was determined in 1.5 *M* hydrochloric acid and in 1.5 *M* sodium hydroxide. Write equations for any reaction(s) that occurred.

43. Why might there be a preference for using solid sodium carbonate instead of solid sodium hydroxide for basifying the acidic hydrolysis solution obtained in this experiment?

44. What would be observed if 4-acetamidobenzenesulfonamide were subjected to vigorous hydrolysis conditions, such as concentrated hydrochloric acid and heat, for a long period of time? Write an equation for the reaction that might occur.

45. Calculate the overall yield of sulfanilamide obtained in the sequence of reactions that you performed.

46. An amide derivative of a carboxylic acid is hydrolyzed in aqueous acid or base more slowly than an ester, anhydride, or acid chloride of the same acid.

a. Provide a rationale for this observation, using derivatives of acetic acid to illustrate your answer.

b. Would you expect the amide function in 4-acetamidobenzenesulfonamide to be more or less readily hydrolyzed than the amide group of an aliphatic amide? Explain your answer.

47. Using curved arrows to symbolize the flow of electrons, write a stepwise mechanism for the hydrolysis of 4-acetamidobenzenesulfonamide (**9**) in aqueous acid to give sulfanilamide (**1**).

Sign in at **www.cengage.com/login** and use the spectra viewer and Tables 8.1–8.8 as needed to answer the blue-numbered questions on spectroscopy.

48. Consider the spectral data for sulfanilamide (Figs. 21.10 and 21.11).

a. In the functional group region of the IR spectrum, specify the absorptions associated with the nitrogen-hydrogen bonds and with the aromatic ring.

b. In the ^{1}H NMR spectrum, assign the various resonances to the hydrogen nuclei responsible for them.

c. For the ^{13}C NMR data, assign the various resonances to the carbon nuclei responsible for them.

49. Discuss the differences observed in the IR and NMR spectra of 4-acetamidobenzenesulfonyl chloride and 4-acetamidobenzenesulfonamide that are consistent with the formation of the latter in this procedure.

SPECTRA

Starting Materials and Products

The IR and NMR spectra of acetanilide are provided in Figures 8.7 and 15.31, respectively.

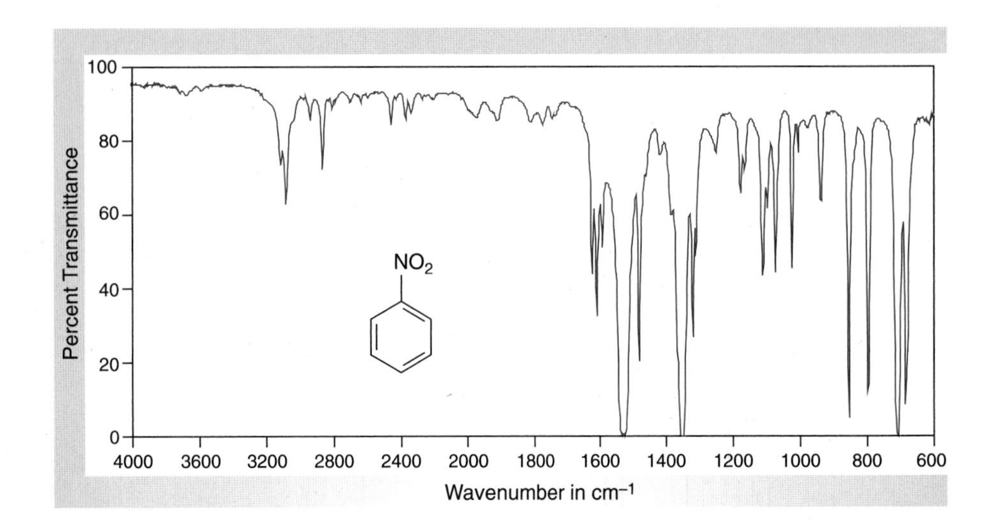

Figure 21.2
IR spectrum of nitrobenzene (neat).

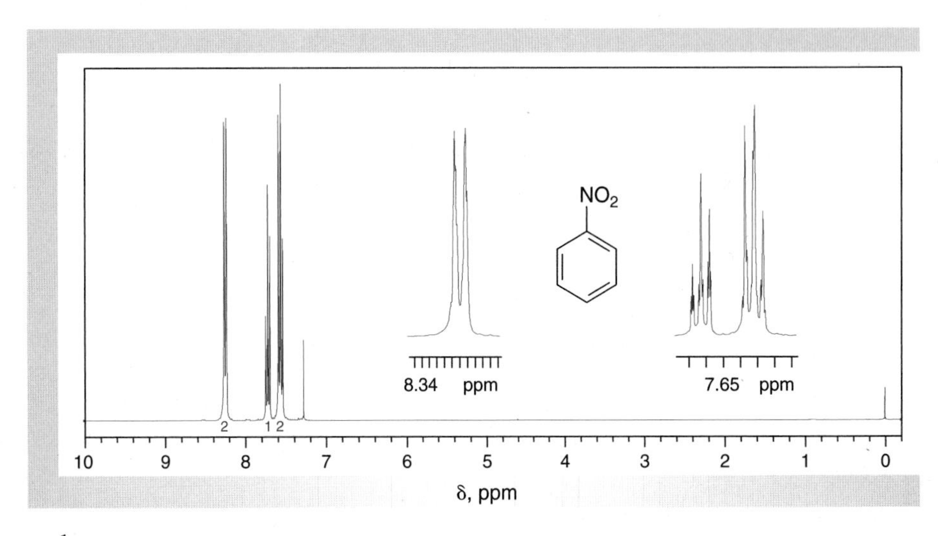

Figure 21.3
NMR data for nitrobenzene ($CDCl_3$).

(a) ^{1}H NMR spectrum (300 MHz).
(b) ^{13}C NMR data: δ 123.5, 129.5, 134.8, 148.3.

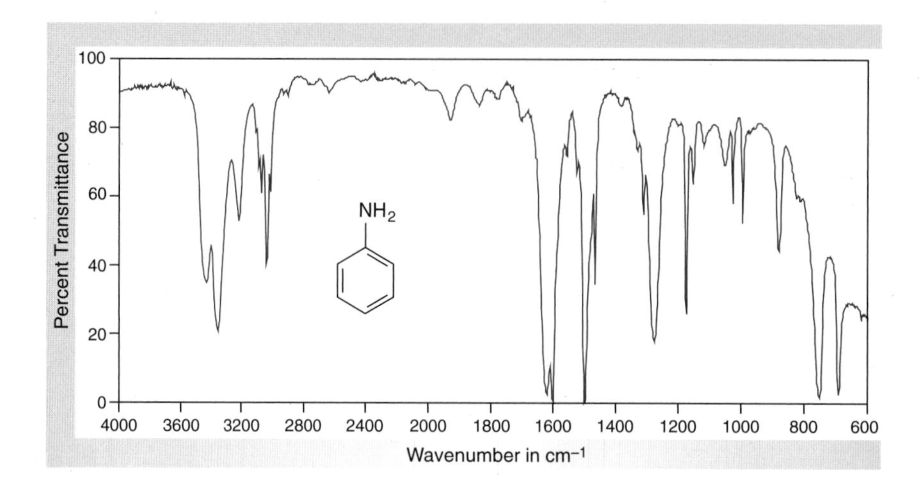

Figure 21.4
IR spectrum of aniline (neat).

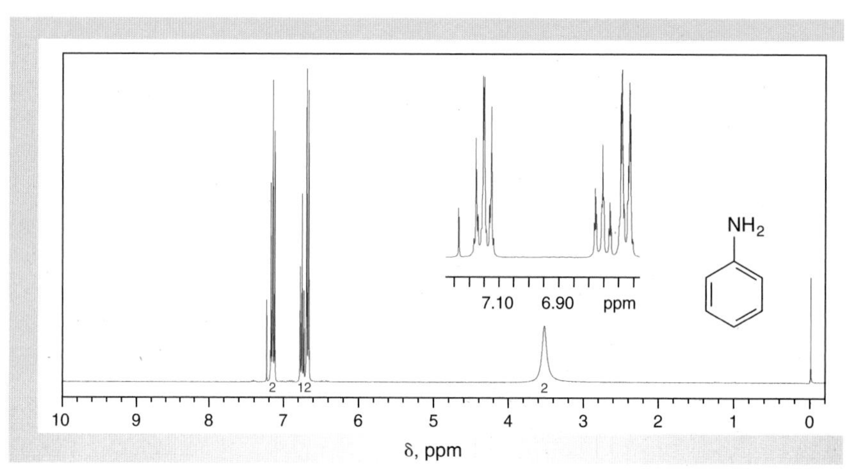

Figure 21.5
NMR data for aniline ($CDCl_3$).

(a) ^{1}H NMR spectrum (300 MHz).
(b) ^{13}C NMR data: δ 115.1, 118.2, 129.2, 146.7.

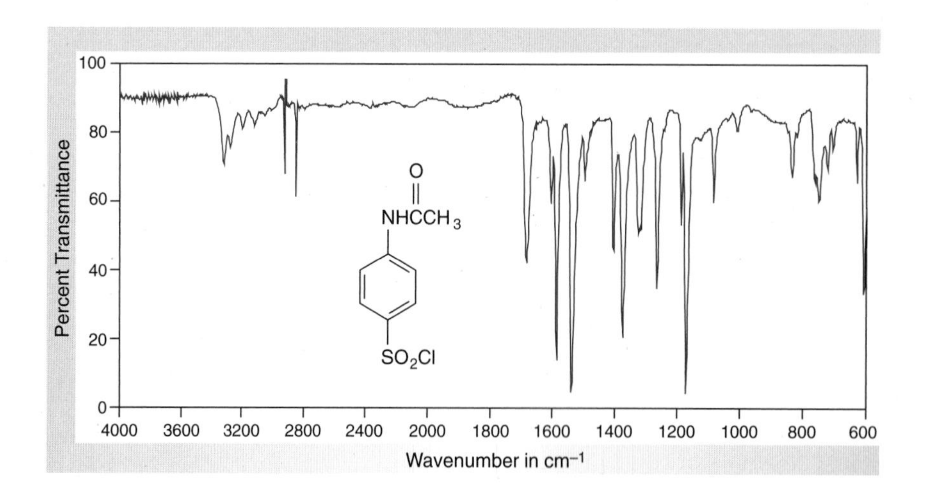

Figure 21.6
IR spectrum of 4-acetamidobenzenesulfonyl chloride (IR card).

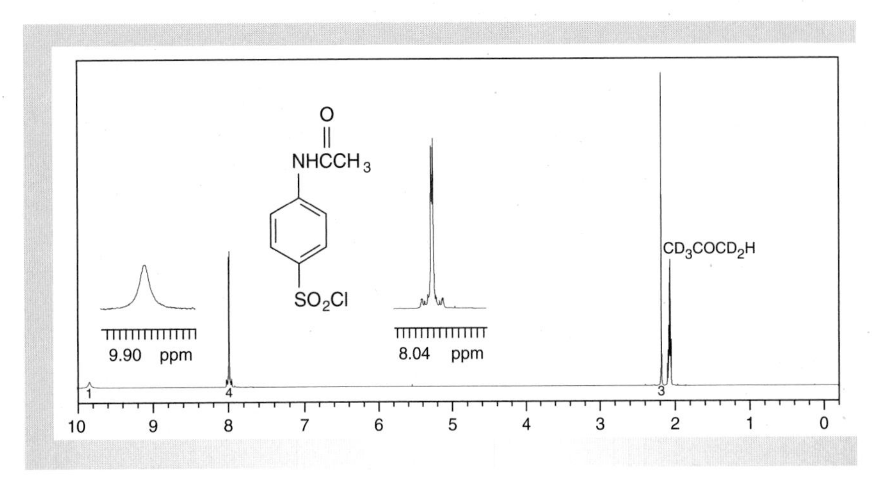

Figure 21.7
NMR data for 4-acetamidobenzenesulfonyl chloride (DMSO-d_6).

(a) ^{1}H NMR spectrum (300 MHz).
(b) ^{13}C NMR data: δ 25.3, 119.7, 127.5, 141.6, 141.7, 170.3.

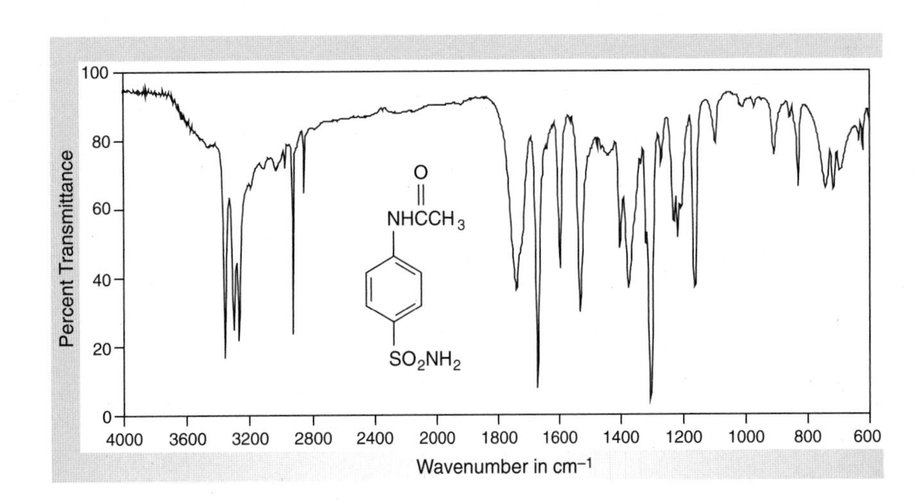

Figure 21.8
IR spectrum of 4-acetamidobenzenesulfonamide (IR card).

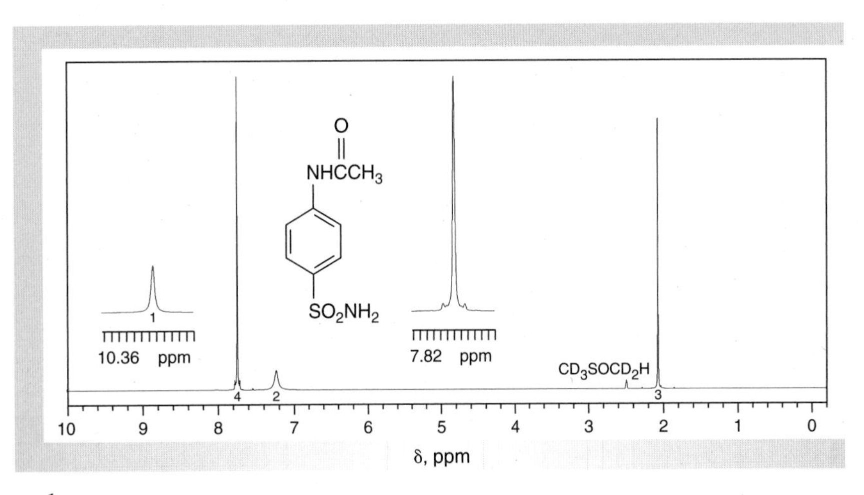

Figure 21.9
NMR data for 4-acetamidobenzenesulfonamide (DMSO-d_6).

(a) ^{1}H NMR spectrum (300 MHz).
(b) ^{1}C NMR data: δ 24.2, 118.5, 126.7, 138.1, 142.3, 169.0.

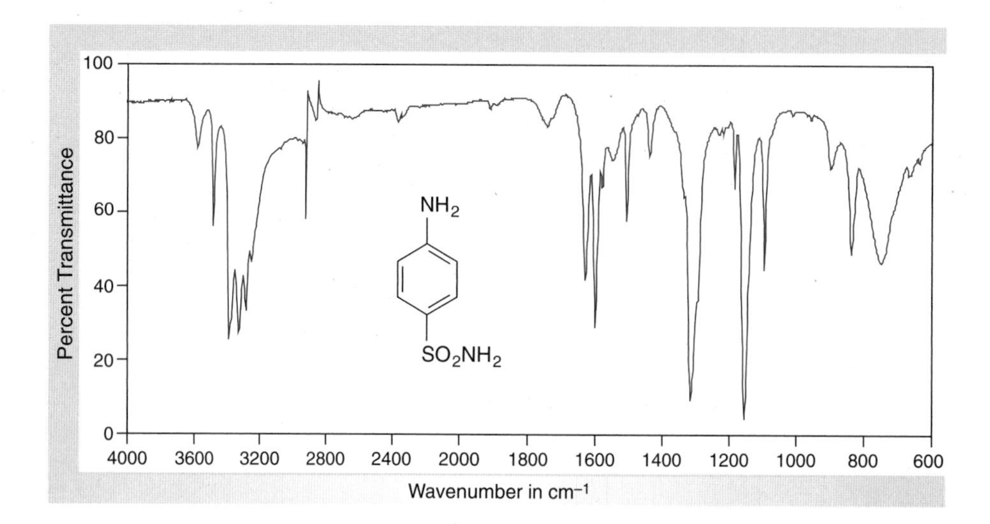

Figure 21.10
IR spectrum of sulfanilamide (IR card).

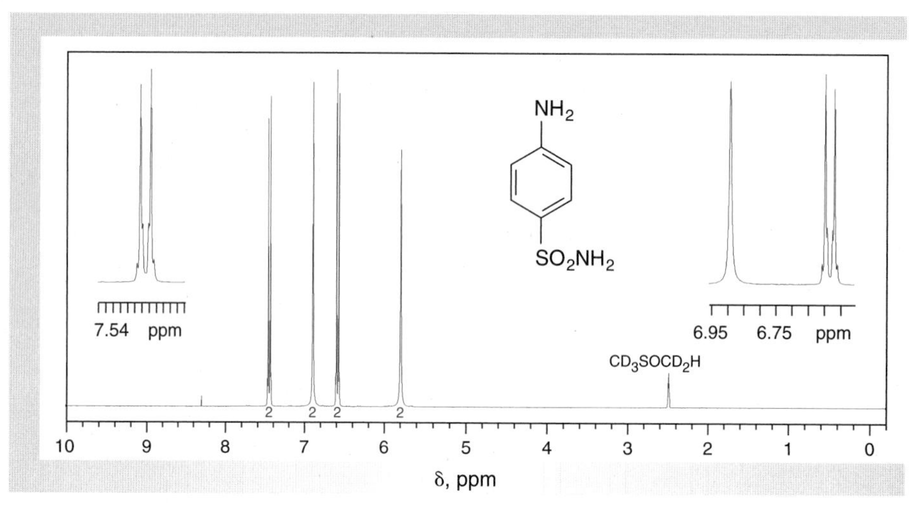

Figure 21.11
NMR data for sulfanilamide (DMSO-d_6).

(a) ^{1}H NMR spectrum (300 MHz).
(b) ^{13}C NMR data: δ 113.1, 127.6, 130.2, 151.6.

21.3 SYNTHESIS OF 1-BROMO-3-CHLORO-5-IODOBENZENE

The preparation of 1-bromo-3-chloro-5-iodobenzene (**24**), as illustrated in Scheme 21.4, is a fascinating excursion into the role of **protecting groups** and **substituent effects** as important factors in multistep synthesis. The approach used here involves a sequence of **electrophilic aromatic substitutions** and illustrates two important general strategies in synthesis: (1) protecting a reactive functionality to prevent undesired side reactions and (2) using substituents to control the regiochemical course of a reaction. In the present example, the first principle is represented by the conversion of the *amino* group of **6** into the *amido* group of **7**, and the second is illustrated by the directing and ring-activating properties of the amino and amido functions. The yields of the individual steps are good to high, ranging from 60% to 96%, an important consideration in designing multistep syntheses, as was noted in Section 21.1.

Although benzene (**4**) is a potential starting point of our synthesis (Scheme 21.4), it is an alleged carcinogen, so the sequence of reactions begins with nitrobenzene (**5**). Discussion of and experimental procedures for the reduction of nitrobenzene and the conversion of aniline (**6**) to acetanilide (**7**) are found in Section 21.2, so the present description starts with the bromination of acetanilide.

Scheme 21.4

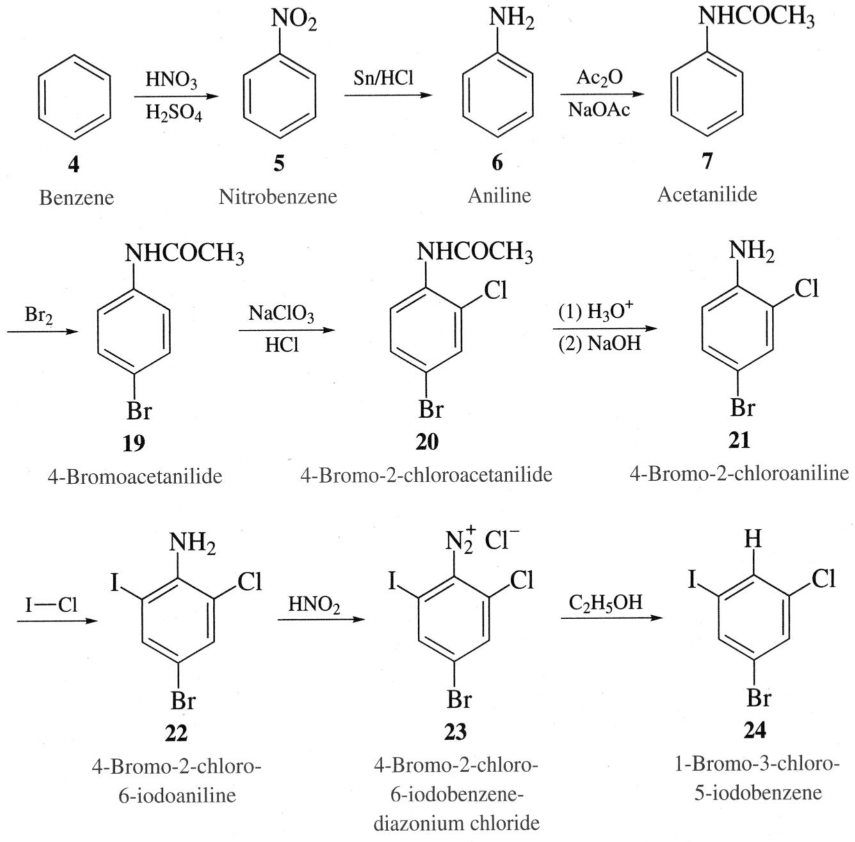

4 Benzene $\xrightarrow[H_2SO_4]{HNO_3}$ **5** Nitrobenzene (NO_2) $\xrightarrow{Sn/HCl}$ **6** Aniline (NH_2) $\xrightarrow[NaOAc]{Ac_2O}$ **7** Acetanilide ($NHCOCH_3$)

$\xrightarrow{Br_2}$ **19** 4-Bromoacetanilide $\xrightarrow[HCl]{NaClO_3}$ **20** 4-Bromo-2-chloroacetanilide $\xrightarrow[(2)\ NaOH]{(1)\ H_3O^+}$ **21** 4-Bromo-2-chloroaniline

$\xrightarrow{I-Cl}$ **22** 4-Bromo-2-chloro-6-iodoaniline $\xrightarrow{HNO_2}$ **23** 4-Bromo-2-chloro-6-iodobenzene-diazonium chloride ($N_2^+\ Cl^-$) $\xrightarrow{C_2H_5OH}$ **24** 1-Bromo-3-chloro-5-iodobenzene

Acetylation and Bromination: Preparation of 4-Bromoacetanilide (19)

The introduction of bromine by electrophilic aromatic substitution is not performed directly on aniline (**6**) because its amino group so greatly activates the ring toward electrophiles that even mild brominating agents such as an aqueous solution of bromine lead to trisubstitution (Eq. 21.10). Acetylating the amino group of aniline attentuates the activating effect of the nonbonded pair of electrons on the nitrogen atom. Like the amino group, the acetamido group, $CH_3CONH–$, of acetanilide (**7**) is an activating group, so selective monobromination of **7** to give **19** occurs under mild conditions, such as bromine in acetic acid *without* a Lewis acid catalyst. The acetamido group is *ortho-* and *para-directing,* so both 2-bromo- and 4-bromoacetanilide are produced. However, the steric bulk of the acetamido substituent hinders attack at the 2-position, thereby strongly favoring formation of the 4-bromo isomer **19** (Eq. 21.11).

NH_2-benzene (6) $\xrightarrow[H_2O]{3\ Br_2}$ 2,4,6-Tribromoaniline + 3 HBr (21.10)

$NHCOCH_3$-benzene (7) $\xrightarrow[CH_3CO_2H]{Br_2}$ 4-bromoacetanilide (19, major) + 2-Bromoacetanilide (minor) (21.11)

As a general rule, separating *ortho*- and *para*-isomers from a mixture containing both is fairly simple. Because *para*-isomers are more symmetrical than *ortho*-isomers, they are *less* soluble in a given solvent, as noted in Section 15.4. 4-Bromoacetanilide (**19**), for example, is separable from the 2-bromo isomer by a single recrystallization of the crude mixture. A second consequence of the greater molecular symmetry of *para*-isomers is that they typically have higher melting points than their *ortho* relatives. In the case of the *ortho*- and *para*-isomers of bromonitrobenzene, this difference is approximately 70 °C.

Chlorination: Preparation of 4-Bromo-2-Chloroacetanilide (20)

4-Bromoacetanilide is slightly less reactive toward further electrophilic substitution than acetanilide, but it is still sufficiently reactive to be chlorinated under mild conditions *without* a strong Lewis acid catalyst. Thus, molecular chlorine dissolved in acetic acid chlorinates 4-bromoacetanilide to produce 4-bromo-2-chloroacetanilide (**20**) as shown in Equation 21.12. The formation of the 2- rather than the 3-chloro isomer demonstrates the more powerful directing effect of the acetamido group as compared to the bromo substituent.

4-bromoacetanilide (19) $\xrightarrow[CH_3CO_2H]{Cl_2}$ 4-bromo-2-chloroacetanilide (20) (21.12)

To avoid the hazards associated with using chlorine gas, chlorine may be easily generated *in situ* by reaction of concentrated hydrochloric acid and sodium chlorate ($NaClO_3$). In this reaction, an oxidation-reduction process occurs by which chloride ion is oxidized to chlorine (Eq. 21.13) and chlorate ion is reduced to chlorine (Eq. 21.14); balancing of these equations is left to the student. Not only is this method for generating chlorine easy and safe, but it also allows for the preparation of precise amounts of molecular chlorine.

$$Cl^- \rightarrow 1/2\ Cl_2 + e^- \qquad (21.13)$$

$$6H^+ + ClO_3^- + 5e^- \rightarrow 1/2\ Cl_2 + 3H_2O \qquad (21.14)$$

Hydrolysis: Preparation of 4-Bromo-2-Chloro-6-Iodoaniline (22)

The acetamido group of 4-bromo-2-chloroacetanilide (**20**) is the most powerful directing group of the three substituents on the aromatic ring, so further electrophilic substitution will preferentially occur at the 6-position of the ring. However, attempted iodination of **20** fails because of the deactivating effects of the halogen substituents and because the iodonium ion, I^+, is a weak and unreactive electrophile compared to bromonium, Br^+, or chloronium, Cl^+, ions. The solution to this problem is to regenerate the amino group, a more powerful ring activator, by hydrolyzing the amide (Eq. 21.15). This reaction is effected using a solution of concentrated hydrochloric acid in ethanol.

NHCOCH$_3$ / Cl / Br **20** —(conc. HCl / CH_3CH_2OH)→ NH$_2$ / Cl / Br **21** (21.15)

The iodinating agent in our procedure is iodine monochloride, ICl. Because iodine is the less electronegative of the two halogens, this reagent is polarized to make the iodine atom electrophilic, and it may dissociate to give Cl^- and I^+. The electrophile for the reaction is thus either polarized molecular I–Cl or iodonium ion (Eq. 21.16).

$\overset{\delta^-}{Cl}-\overset{\delta^+}{I} \rightleftharpoons Cl^- + I^+$ —(**21** / CH_3CO_2H)→ **22** (I, NH$_2$, Cl, Br) (21.16)

Diazotization and Protonolysis: Preparation of 1-Bromo-3-Chloro-5-Iodobenzene (24)

The final step in the synthesis is the removal of the amino group from 2-chloro-4-bromo-6-iodoaniline (**22**). After the amino group has been diazotized with nitrous acid at 0 °C to produce **23,** the diazo group may be replaced by a hydrogen atom upon treatment with ethanol or hypophosphorous acid, H_3PO_2, as depicted in Equation 21.17. Although the latter is the more general reagent for replacing a diazonium group by a hydrogen atom, ethanol works well with certain types of aromatic compounds, in particular those containing halogen, and it will be used in our procedure. The transformation presumably involves an oxidation-reduction reaction in which ethanol is oxidized to acetaldehyde or acetic acid. The process is mechanistically ill-defined, but donation of a hydride ion to reduce the diazo compound is a likely possibility (Eq. 21.18).

$$\textbf{22} \xrightarrow[0\,^\circ C]{NaNO_2/conc.H_2SO_4} \textbf{23}\ (N_2^+\,HSO_4^-) \xrightarrow[warm]{CH_3CH_2OH} \textbf{24} \quad (21.17)$$

$$CH_3CH_2OH + \textbf{23} \xrightarrow{-N_2} \textbf{24} + CH_3CHO + H_2SO_4 \quad (21.18)$$

A potentially important side reaction that might occur is dissociation of the diazonium salt **23** to a carbocation and molecular nitrogen. The carbocation could then react with ethanol to give an aryl ethyl ether as a by-product (Eq. 21.19). Fortunately, ether formation is a minor process in the present reaction, and any ether that is produced is removed by recrystallization of the product.

$$ArN_2^+\,HSO_4^- \xrightarrow{-N_2} Ar^+ \xrightarrow[-H^+]{H\ddot{O}CH_2CH_3} ArOCH_2CH_3 \quad (21.19)$$

Overview

The logic underlying the specific sequence of steps for making **24** is summarized as follows. Introducing the amino group as a director on the benzene ring provides a way to control the locations at which the various halogen substituents are introduced. When the directive effects of the amino group are no longer needed, this functionality is replaced with a hydrogen atom, the final step of the present synthesis. Before introducing a halogen atom onto the ring, the ring-activating power of the amino group is lessened by converting it to an acetamido group, which makes monohalogenation possible. The reason the bromo substituent is added first is that the acetamido group preferentially directs an electrophile to the *para* position. The chloro substituent was introduced next because chlorination of 4-bromoacetanilide is possible, whereas iodination is not. If the amino function had been regenerated *prior* to chlorinating and iodinating the ring, undesired polyhalogenation would have occurred. The iodo group is added last because the aromatic ring must be highly activated for this reaction to succeed, and there is only one activated position remaining where substitution can occur, making polyiodination improbable.

EXPERIMENTAL PROCEDURES

Synthesis of 1-Bromo-3-Chloro-5-Iodobenzene

Purpose To demonstrate principles of multistep synthesis, protecting groups, and electrophilic aromatic substitution.

Note Throughout the sequence of reactions, quantities of reagents must be adjusted according to the amount of starting material that is available from the previous reaction, but you should *not* run the reactions on a larger scale than is indicated. If the amount of your starting material varies from that indicated, do *not* change the reaction *times* unless directed to do so by your instructor. You should be able to obtain sufficient quantities of product from each reaction to complete the entire sequence starting with the amount of acetanilide used in Step B.

A ■ *Preparation of Aniline and Acetanilide*

See Section 21.2 for Pre-Lab Exercises and Experimental Procedures for the preparation of aniline and acetanilide and Figures 21.4, 21.5, 8.7, and 15.31, respectively, for their spectra.

B ■ *Preparation of 4-Bromoacetanilide*

SAFETY ALERT

1. **Bromine is a hazardous chemical that may cause serious chemical burns. Do not breathe its vapors or allow it to come into contact with the skin. Perform all operations involving the transfer of solutions of bromine at a hood, and wear latex gloves when handling bromine or solutions of it. If you get bromine on your skin, immediately wash the affected area with soap and warm water and soak the skin in 0.6 *M* sodium thiosulfate solution for up to 3 h if the burn is particularly serious.**
2. **Do *not* use acetone to rinse glassware containing residual bromine! This prevents formation of α-bromoacetone, a severe *lachrymator.* First rinse the glassware with aqueous sodium bisulfite.**
3. **Glacial acetic acid is a corrosive liquid. Wear latex gloves when handling this chemical and thoroughly wash with large amounts of water any areas of your skin that may come in contact with it.**

MINISCALE PROCEDURE

Preparation Sign in at **www.cengage.com/login** to answer Pre-Lab Exercises, access videos, and read the MSDSs for the chemicals used or produced in this procedure. Review Sections 2.9, 2.11, and 2.17.

Apparatus A 250-mL round-bottom flask, apparatus for magnetic stirring, vacuum filtration, and *flameless* heating.

Setting Up Working at the hood, prepare a solution of 3.2 mL of bromine in 6 mL of glacial acetic acid in an Erlenmeyer flask. Dissolve 8.1 g of acetanilide in 30 mL of glacial acetic acid in the round-bottom flask containing a stirbar.

Bromination, Work-Up, and Isolation Add the bromine solution to the rapidly stirred solution of acetanilide over a period of 1–2 min; continue stirring the reaction mixture for about 10 min after completing the addition. Slowly add 100 mL of ice-cold water with stirring and then add just enough ice-cold saturated aqueous sodium bisulfite to discharge the color of the mixture that results. Cool the mixture in an ice-water bath and collect the product by vacuum filtration. If the crude product appears yellow, wash it with aqueous sodium bisulfite. In any case, wash it well with cold water and press it as dry as possible on the filter. The product may be recrystallized from methanol.★

Analysis Weigh the product and calculate its yield. Determine its melting point. Obtain IR and ^1H NMR spectra of your starting material and product and compare them with those of authentic samples (Figs. 8.7, 15.31, 21.12, and 21.13).

MICROSCALE PROCEDURE

Preparation Sign in at **www.cengage.com/login** to answer Pre-Lab Exercises, access videos, and read the MSDSs for the chemicals used or produced in this procedure. Review Sections 2.9, 2.11, and 2.17.

Apparatus A 5-mL conical vial, apparatus for magnetic stirring, vacuum filtration, and *flameless* heating.

Setting Up Working at the hood, prepare a solution of 0.15 mL of bromine and 0.3 mL of glacial acetic acid in a test tube. Prepare a solution of 375 mg of acetanilide and 3 mL of glacial acetic acid in the conical vial equipped with a spinvane.

Bromination, Work-Up, and Isolation In one portion, add the bromine solution to the rapidly stirred solution of acetanilide and continue stirring for about 10 min at room temperature. Pour the reaction mixture into 25 mL of ice-cold water contained in a beaker that is equipped with a stirbar. Add just enough ice-cold saturated aqueous sodium bisulfite to discharge the color of the mixture. Stir the resulting mixture in an ice-water bath for 15 min and isolate the precipitated product by vacuum filtration.★ If the crude product appears yellow, wash it with aqueous sodium bisulfite. In any case, wash it well with cold water and press it as dry as possible on the filter. The product may be recrystallized from methanol.★

Analysis Weigh the product and calculate its yield. Determine its melting point. Obtain IR and ^1H NMR spectra of your starting material and product and compare them with those of authentic samples (Figs. 8.7, 15.31, 21.12, and 21.13).

WRAPPING IT UP

If the *filtrate* and *washings* contain residual bromine, as evidenced by a yellow color, add saturated aqueous sodium bisulfite until the solution is colorless. Flush the *decolorized solution* down the drain.

C ■ *Preparation of 4-Bromo-2-Chloroacetanilide*

SAFETY ALERT

1. **Chlorine gas is evolved in this procedure. To minimize exposure to it, perform this reaction in the hood, if possible. Alternatively, invert a funnel over the reaction flask and attach the funnel to the water aspirator to remove the gaseous chlorine (Fig. 2.71b).**
2. ***Concentrated* hydrochloric acid and glacial acetic acid are corrosive liquids. Wear latex gloves when handling these chemicals and thoroughly wash any areas of your skin that may come in contact with them.**

MINISCALE PROCEDURE

Preparation Sign in at **www.cengage.com/login** to answer Pre-Lab Exercises, access videos, and read the MSDSs for the chemicals used or produced in this procedure. Review Sections 2.9, 2.10, 2.11, and 2.17.

Apparatus A 250-mL round-bottom flask, ice-water bath, apparatus for magnetic stirring, vacuum filtration, and *flameless* heating.

Setting Up Suspend 10.7 g of 4-bromoacetanilide in a stirred solution of 23 mL of *concentrated* hydrochloric acid and 28 mL of glacial acetic acid contained in the round-bottom flask. Heat the mixture gently with stirring until it becomes homogeneous and then cool the solution to 0 °C. Prepare a solution of 2.8 g of sodium chlorate in about 7 mL of water.

Chlorination, Work-Up, and Isolation Add the aqueous sodium chlorate solution dropwise to the stirred cold mixture over a period of about 5 min; some chlorine gas is evolved. After completing the addition, stir the reaction mixture at room temperature for 1 h and then collect the precipitate by vacuum filtration.★ Wash the crude product *thoroughly* with ice-cold water until the washings are neutral or nearly so. The product may be recrystallized from methanol.

Analysis Weigh the product and calculate its yield. Determine its melting point. Obtain its IR and ^{1}H NMR spectra and compare them with those of an authentic sample (Figs. 21.14 and 21.15).

MICROSCALE PROCEDURE

Preparation Sign in at **www.cengage.com/login** to answer Pre-Lab Exercises, access videos, and read the MSDSs for the chemicals used or produced in this procedure. Review Sections 2.9, 2.10, 2.11, and 2.17.

Apparatus A 5-mL conical vial, ice-water bath, apparatus for magnetic stirring, vacuum filtration, and *flameless* heating.

Setting Up Suspend 500 mg of 4-bromoacetanilide in a stirred solution of 1 mL of *concentrated* hydrochloric acid and 1.3 mL of glacial acetic acid contained in the conical vial. If necessary, heat the mixture gently with stirring until it becomes homogeneous and then cool the solution to 0 °C. Prepare a solution of 150 mg of sodium chlorate in about 0.5 mL of water.

Chlorination, Work-Up, and Isolation Add the aqueous sodium chlorate solution dropwise to the stirred cold mixture over a period of about 2 min; some chlorine gas is evolved. After completing the addition, stir the reaction mixture at room temperature for 1 h and then collect the precipitate by vacuum filtration.★ Wash the crude product *thoroughly* with ice-cold water until the washings are neutral or nearly so. The product may be recrystallized from methanol.

Analysis Weigh the product and calculate its yield. Determine its melting point. Obtain its IR and ^1H NMR spectra and compare them with those of an authentic sample (Figs. 21.14 and 21.15).

WRAPPING IT UP

If the *filtrate* and *washings* contain residual chlorine, as evidenced by formation of a blue-black color when a drop of the liquid is placed on starch/iodide paper, add saturated aqueous sodium bisulfite until the solution gives a negative test to starch/iodide paper. Flush the *decolorized solution* down the drain.

D ■ *Preparation of 4-Bromo-2-Chloroaniline*

SAFETY ALERT

***Concentrated* hydrochloric acid is a corrosive liquid, and 14 *N* aqueous sodium hydroxide is highly caustic. Wear latex gloves when handling these chemicals and thoroughly wash any areas of your skin that may come in contact with them.**

MINISCALE PROCEDURE

Preparation Sign in at **www.cengage.com/login** to answer Pre-Lab Exercises, access videos, and read the MSDSs for the chemicals used or produced in this procedure. Review Sections 2.9, 2.11, 2.17, and 2.22.

Apparatus A 250-mL round-bottom flask, apparatus for heating under reflux, magnetic stirring, vacuum filtration, and *flameless* heating.

Setting Up Mix 11.2 g of crude 4-bromo-2-chloroacetanilide with 20 mL of 95% ethanol and 13 mL of *concentrated* hydrochloric acid in the flask. Add a stirbar and assemble the apparatus for heating under reflux.

Hydrolysis and Isolation Heat the mixture under reflux for *ca.* 0.5 h. At the end of this period, add 90 mL of *hot* water, swirl the mixture to dissolve any solids completely,

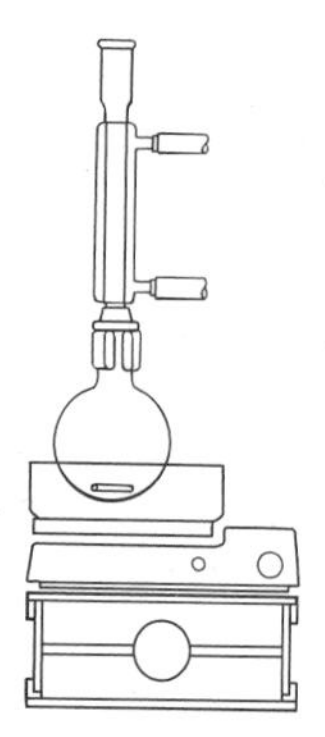

and pour the hot solution onto about 150 g of ice contained in a beaker. Slowly add 12 mL of 14 *M* sodium hydroxide solution to the resulting mixture, stirring the mixture well during the addition. Add more ice if it has all melted before completing addition of the aqueous base. Check the pH of the reaction mixture to ensure that it is basic; if not, add more aqueous base. Isolate the solid by vacuum filtration, wash it thoroughly with ice-cold water, and press it as dry as possible on the filter.★

Purification The crude product may be recrystallized by first dissolving it in boiling methanol, allowing the solution to cool to below 40 °C, and adding cold water dropwise until crystallization begins. Should the product begin to oil out rather than crystallize, add a few drops of methanol to redissolve the oil. Cool the mixture thoroughly in an ice-water bath and collect the product by vacuum filtration, washing it with a *small* volume of ice-cold 1:1 methanol-water.

Analysis Weigh the product and calculate its yield. Determine its melting point. Obtain its IR and ^1H NMR spectra and compare them with those of an authentic sample (Figs. 21.16 and 21.17).

MICROSCALE PROCEDURE

Preparation Sign in at **www.cengage.com/login** to answer Pre-Lab Exercises, access videos, and read the MSDSs for the chemicals used or produced in this procedure. Review Sections 2.9, 2.11, 2.17, and 2.22.

Apparatus A 5-mL conical vial, apparatus for heating under reflux, magnetic stirring, vacuum filtration, Craig tube filtration, and *flameless* heating.

Setting Up Mix 450 mg of crude 4-bromo-2-chloroacetanilide with 1.5–2 mL of 95% ethanol and 0.5 mL of *concentrated* hydrochloric acid in the conical vial. Add a spinvane and assemble the apparatus for heating under reflux.

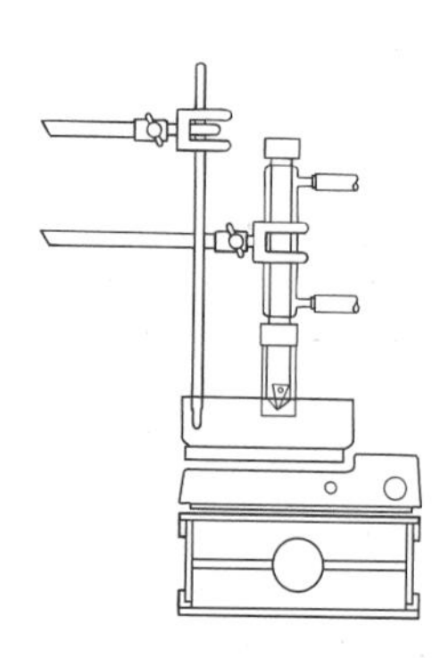

Hydrolysis and Isolation Heat the mixture under reflux for *ca.* 0.5 h. At the end of this period, add 2 mL of *hot* water, stir the mixture to dissolve any solids completely, and pour the hot solution onto 3–4 g of ice contained in a beaker. Slowly add 0.5 mL of 14 *M* sodium hydroxide solution to the resulting mixture, stirring the mixture well during the addition. Add more ice if it has all melted before completing addition of the aqueous base. Check the pH of the reaction mixture to ensure that it is basic; if not, add more aqueous base. Isolate the solid by vacuum filtration, and thoroughly dry it.★

Purification The crude product may be recrystallized by first dissolving it in boiling methanol, allowing the solution to cool to below 40 °C, and adding cold water dropwise until crystallization begins. Should the product begin to oil out rather than crystallize, add a few drops of methanol to redissolve the oil. Cool the mixture thoroughly in an ice-water bath and collect the product by Craig tube or vacuum filtration, washing it with a *few drops* of ice-cold 1:1 methanol-water.

Analysis Weigh the product and calculate its yield. Determine its melting point. Obtain its IR and ^1H NMR spectra and compare them with those of an authentic sample (Figs. 21.16 and 21.17).

WRAPPING IT UP

Neutralize the *filtrate* and flush it down the drain.

E ■ *Preparation of 4-Bromo-2-Chloro-6-Iodoaniline*

SAFETY ALERT

Glacial acetic acid is a corrosive liquid, can cause burns, and is harmful if inhaled. Wear latex gloves when handling this chemical and thoroughly wash with water any areas of your skin that may come in contact with it.

MINISCALE PROCEDURE

Preparation Sign in at **www.cengage.com/login** to answer Pre-Lab Exercises, access videos, and read the MSDSs for the chemicals used or produced in this procedure. Review Sections 2.9, 2.10, 2.11, and 2.17.

Apparatus A 250-mL round-bottom flask, thermometer, ice-water bath, apparatus for magnetic stirring, vacuum filtration, and *flameless* heating.

Setting Up With magnetic stirring, dissolve 2.5 g of recrystallized 4-bromo-2-chloroaniline in 40 mL of glacial acetic acid in the round-bottom flask and add 10 mL of water to the mixture. In an Erlenmeyer flask, prepare a solution of 2.5 g of technical iodine monochloride in 10 mL of glacial acetic acid.

Iodination, Work-Up, and Isolation Over a period of 8–10 min, add the iodine monochloride solution to the stirred solution of the aniline. Heat the resulting mixture to 90 °C with continued stirring and then allow the solution to cool to about 50 °C. Add just enough saturated aqueous sodium bisulfite to turn the color of the mixture bright yellow; record the volume of sodium bisulfite solution required. Dilute the reaction mixture with enough extra water such that the *combined* volume of the aqueous sodium bisulfite *plus* the volume of added water is about 15 mL. Cool the reaction mixture in an ice-water bath and isolate the crude product that results by vacuum filtration. Wash the crystals with a small amount of ice-cold 33% aqueous acetic acid and then at least 50 mL of ice-cold water. Press the filter cake as dry as possible.★

Purification Recrystallize the crude product from acetic acid-water as follows. Mix the product with glacial acetic acid in the ratio of about 20 mL of glacial acetic acid per gram of product. Heat the mixture and slowly add 5 mL of water per gram of product to the mixture as it is heating. Terminate heating once dissolution has occurred and allow the solution to cool slowly to room temperature. Complete the crystallization by cooling the mixture in an ice-water bath and isolate the product by vacuum filtration. Wash the product with cold water and air-dry it.

Analysis Weigh the product and calculate its yield. Determine its melting point. Obtain its IR and ^{1}H NMR spectra and compare them with those of an authentic sample (Figs. 21.18 and 21.19).

MICROSCALE PROCEDURE

Preparation Sign in at **www.cengage.com/login** to answer Pre-Lab Exercises, access videos, and read the MSDSs for the chemicals used or produced in this procedure. Review Sections 2.9, 2.10, 2.11, and 2.17.

Apparatus A 25-mL Erlenmeyer flask, thermometer, ice-water bath, apparatus for magnetic stirring, vacuum filtration, and *flameless* heating.

Setting Up With magnetic stirring, dissolve 250 mg of recrystallized 4-bromo-2-chloroaniline in 4 mL of glacial acetic acid in the Erlenmeyer flask, and add 1 mL of water to the mixture. In a test tube, prepare a solution of 250 mg of technical iodine monochloride in 1 mL of glacial acetic acid.

Iodination, Work-Up, and Isolation Over a period of 3–5 min, add the iodine monochloride solution to the stirred solution of the aniline. Heat the resulting mixture to 90 °C with continued stirring and then allow the solution to cool to about 50 °C. Add just enough saturated aqueous sodium bisulfite to turn the color of the mixture bright yellow; record the volume of sodium bisulfite solution required. Dilute the reaction mixture with enough extra water such that the *combined* volume of the aqueous sodium bisulfite *plus* the volume of added water is about 3 mL. Cool the reaction mixture in an ice-water bath and isolate the crude product that results by vacuum filtration. Wash the crystals with a small amount of ice-cold 33% aqueous acetic acid and then with at least 5 mL of ice-cold water. Press the filter cake as dry as possible.★

Purification Using a test tube, recrystallize the crude product from acetic acid-water as follows. Mix the product with glacial acetic acid in the ratio of about 2 mL of glacial acetic acid per 100 mg of product. Heat the mixture and slowly add 0.5 mL of water per 100 mg of product to the mixture as it is heating. Terminate heating once dissolution has occurred and allow the solution to cool slowly to room temperature. Complete the crystallization by cooling the mixture in an ice-water bath and isolate the product by vacuum filtration. Wash the product with cold water and air-dry it.

Analysis Weigh the product and calculate its yield. Determine its melting point. Obtain its IR and ^{1}H NMR spectra and compare them with those of an authentic sample (Figs. 21.18 and 21.19).

WRAPPING IT UP

Flush the *aqueous filtrates* down the drain. Use solid sodium carbonate to neutralize the *filtrate* from the recrystallization before flushing it down the drain.

F ■ *Preparation of 1-Bromo-3-Chloro-5-Iodobenzene*

SAFETY ALERT

***Concentrated* sulfuric acid is a corrosive liquid. Wear latex gloves when handling this chemical and thoroughly wash any areas of your skin that may come in contact with it.**

MINISCALE PROCEDURE

Preparation Sign in at **www.cengage.com/login** to answer Pre-Lab Exercises, access videos, and read the MSDSs for the chemicals used or produced in this procedure. Review Sections 2.9, 2.10, 2.11, 2.13, 2.16, 2.17, 2.22, and 2.29.

Apparatus A 250-mL and a 50-mL round-bottom flask, thermometer, ice-water bath, apparatus for heating under reflux, magnetic stirring, steam distillation, vacuum filtration, simple distillation, and *flameless* heating.

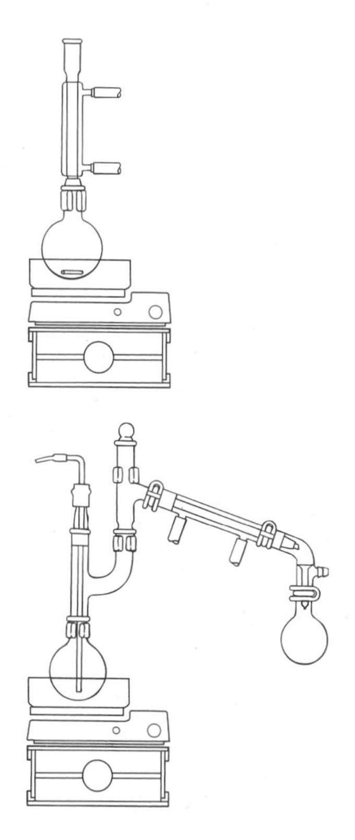

Setting Up Suspend 2.0 g of 4-bromo-2-chloro-6-iodoaniline in about 10 mL of *absolute* ethanol in the 250-mL round-bottom flask containing a stirbar. While stirring the mixture, add 4.0 mL of *concentrated* sulfuric acid dropwise. Equip the flask for heating under reflux.

Diazotization, Isolation, and Purification Add 0.7 g of powdered sodium nitrite in small portions through the top of the condenser. When the addition is complete, use a few drops of *absolute* ethanol to rinse any solid that may stick to the walls of the condenser into the flask and then heat the mixture under gentle reflux for about 10 min. Add 125 mL of hot water to the flask and equip it for steam distillation. Steam-distill the mixture, collecting about 50 mL of distillate, which should be clear and not contain any organic product. The desired product should form as a solid in the condenser, so monitor the condenser closely to ensure that it does not become plugged during the distillation. Should the condenser become clogged with solid, open it by running hot water or a slow stream of steam through the condenser jacket.

Upon completing the distillation, extract the distillate with two 15-mL portions of dichloromethane and put the extract in the 50-mL round-bottom flask. Use a few milliliters of dichloromethane to rinse any solid in the condenser into the flask containing the dichloromethane extract.★ Remove the dichloromethane by simple distillation. Alternatively, use rotary evaporation or other techniques to remove the solvent. Recrystallize the residue from methanol, and isolate the product by vacuum filtration.

Analysis Weigh the product and calculate its yield. Determine its melting point. Obtain its IR and ^{1}H NMR spectra, and compare them with those of an authentic sample (Figs. 21.20 and 21.21).

MICROSCALE PROCEDURE

Preparation Sign in at **www.cengage.com/login** to answer Pre-Lab Exercises, access videos, and read the MSDSs for the chemicals used or produced in this procedure. Review Sections 2.9, 2.10, 2.11, 2.13, 2.16, 2.17, 2.22, and 2.29.

Apparatus A 5-mL conical vial, ice-water bath, apparatus for heating under reflux, magnetic stirring, simple distillation, Craig tube filtration, and *flameless* heating.

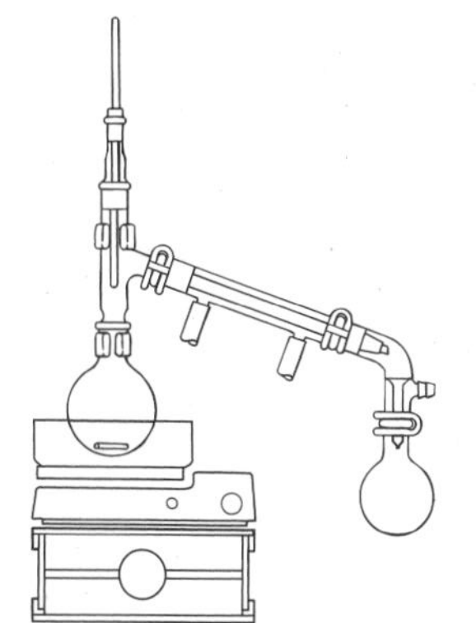

Setting Up Suspend 200 mg of 4-bromo-2-chloro-6-iodoaniline in about 1.5 mL of absolute ethanol in the conical vial containing a spinvane. While stirring the mixture, add 0.4 mL of *concentrated* sulfuric acid dropwise.

Diazotization, Isolation, and Purification Add 0.07 g of powdered sodium nitrite in small portions to the vial. When the addition is complete, use a drop or two of *absolute* ethanol to rinse any solid that may stick to the sides of the vial into the solution. Equip the vial with a condenser and heat the mixture under gentle reflux for about 10 min. Add 2 mL of hot water to the vial and equip it for simple distillation. Steam-distill the mixture until about 1 mL of liquid remains in the vial.

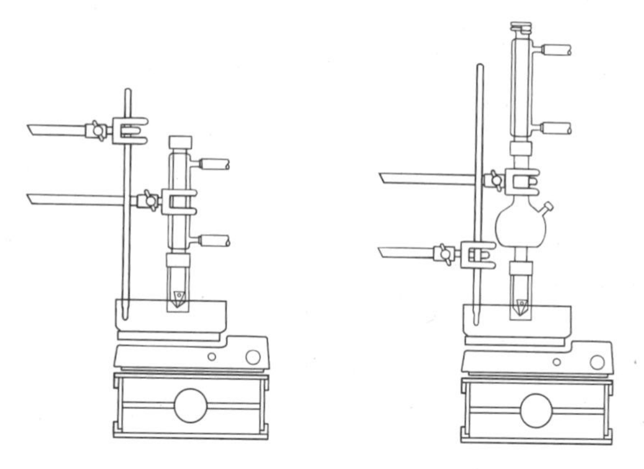

Remove distillate from the Hickman stillhead as necessary and place it in a conical test tube or vial.

Upon completing the distillation, combine all distillates in the conical vessel. Rinse any solid contained in the Hickman stillhead into the conical vessel with 1–2 mL of dichloromethane and separate the layers. Extract the aqueous layer once with 2 mL of dichloromethane and combine the dichloromethane solutions.★ Remove the dichloromethane by simple distillation. Alternatively, use rotary evaporation or other techniques to remove the solvent. Recrystallize the residue from methanol in a Craig tube and isolate the product by Craig tube filtration.

Analysis Weigh the product and calculate its yield. Determine its melting point. Obtain its IR and ^{1}H NMR spectra and compare them with those of authentic samples (Figs. 21.20 and 21.21).

WRAPPING IT UP

Neutralize the *pot residue* from the steam distillation of the *reaction mixture* and flush the resulting *aqueous layer* and that obtained from the *steam distillate* down the drain. Pour the *organic layer* from the *pot residue*, the recovered *dichloromethane*, and the *filtrate* from recrystallization into the container for halogenated compounds.

EXERCISES

General Questions

1. Polyhalogenation is a potential problem at several stages of the synthesis of 1-bromo-3-chloro-5-iodobenzene. Describe what tactics are used to avoid polyhalogenation at each of these stages.
2. Which step(s) in the multistep synthesis of 1-bromo-3-chloro-5-iodobenzene from benzene is/are:
   - **a.** a protection?
   - **b.** a deprotection?
   - **c.** a reduction?
   - **d.** an electrophilic aromatic substitution reaction?

3. Using the concepts developed in the synthesis of 1-bromo-3-chloro-5-iodobenzene, outline synthetic procedures, starting from acetanilide, for preparing each of the following compounds:
   a. 1,3,5-tribromobenzene
   b. 2-bromo-4-chloro-6-iodophenol
   c. 2-bromo-4,6-dichloroaniline
   d. 3-bromotoluene
4. Why is the amino group a more powerful ring activator than the amido group?
5. In the sequence used, which steps demonstrate that the amido group is a more powerful *ortho-*, *para-*director than a bromo or chloro substituent?
6. Hydroxylic solvents, such as water or low-molar-mass alcohols, are often used to purify amides such as the acetanilides produced in this experiment. Briefly explain why these are used in preference to hydrocarbon solvents like petroleum ether.

Questions for Part A

See *Questions 9–25 of Section 21.2.*

Questions for Part B

7. Sodium bisulfite is used to discharge the yellow color produced in the electrophilic aromatic bromination of acetanilide.
   a. What is the source of the yellow color?
   b. Write the chemical equation for the reaction in which this yellow color is discharged.
8. Explain why electrophilic bromination of acetanilide produces 4- rather than 2-bromoacetanilide as the major product.
9. Compared to $Br_2/HOAc$, $Br_2/FeBr_3$ is a more powerful combination of reagents for effecting electrophilic aromatic bromination. What difficulty might arise if this combination were used for brominating acetanilide?
10. Account for the color change you observed upon reaction of bromine with acetanilide.
11. Using curved arrows to symbolize the flow of electrons, provide a stepwise mechanism for the bromination of acetanilide with $Br_2/HOAc$.
12. Consider the spectral data for 4-bromoacetanilide (Figs. 21.12 and 21.13).
   a. In the functional group region of the IR spectrum, specify the absorptions associated with the nitrogen-hydrogen bond, carbon-oxygen double bond, and the aromatic ring.
   b. In the ^1H NMR spectrum, assign the various resonances to the hydrogen nuclei responsible for them.
   c. For the ^{13}C NMR data, assign the various resonances to the carbon nuclei responsible for them.
13. How many resonances would you expect for the ^1H-decoupled ^{13}C spectrum of 2-bromoacetanilide?
14. Discuss the differences observed in the IR and NMR spectra of acetanilide and 4-bromoacetanilide that are consistent with the formation of the latter in this procedure.

Sign in at **www.cengage.com/login** and use the spectra viewer and Tables 8.1–8.8 as needed to answer the blue-numbered questions on spectroscopy.

Questions for Part C

15. What is the oxidation number of the chlorine atom in the chlorate ion, ClO_3, and in molecular chlorine, Cl_2, in Equation 21.14?

16. Why does an amide group activate an aromatic ring toward electrophilic aromatic substitution more than a halogen atom?

17. How might acetic acid interact with molecular chlorine to make it more electrophilic in character?

18. Would 4-bromoacetanilide or 4-bromo-2-chloroacetanilide be more reactive toward electrophilic substitution? Explain.

19. 4-Bromo-3-chloroacetanilide is a possible product from chlorination of 4-bromoacetanilide. Why is this isomer *not* produced to any significant extent?

Sign in at **www.cengage.com/login** and use the spectra viewer and Tables 8.1–8.8 as needed to answer the blue-numbered questions on spectroscopy.

20. Consider the spectral data for 4-bromo-2-chloroacetanilide (Figs. 21.14 and 21.15).

a. In the functional group region of the IR spectrum, specify the absorptions associated with the nitrogen-hydrogen bond, the carbon-oxygen double bond, and the aromatic ring.

b. In the ^{1}H NMR spectrum, assign the various resonances to the hydrogen nuclei responsible for them.

c. For the ^{13}C NMR data, assign the various resonances to the carbon nuclei responsible for them.

21. Discuss the differences observed in the IR and NMR spectra of 4-bromoacetanilide and 4-bromo-2-chloroacetanilide that are consistent with the formation of the latter in this procedure.

Questions for Part D

22. What is the white precipitate that forms during the hydrolysis of 4-bromo-2-chloroacetanilide?

23. Why must aqueous base be added in order to isolate 4-bromo-2-chloroaniline from the aqueous acidic mixture?

24. Using curved arrows to symbolize the flow of electrons, provide a stepwise mechanism for the acid-catalyzed hydrolysis of 4-bromo-2-chloroacetanilide.

25. Why is it necessary to remove the protecting group on nitrogen at this stage of the synthesis rather than as the last step?

Sign in at **www.cengage.com/login** and use the spectra viewer and Tables 8.1–8.8 as needed to answer the blue-numbered questions on spectroscopy.

26. Consider the spectral data for 4-bromo-2-chloroaniline (Figs. 21.16 and 21.17).

a. In the functional group region of the IR spectrum, specify the absorptions associated with the amino group and the aromatic ring.

b. In the ^{1}H NMR spectrum, assign the various resonances to the hydrogen nuclei responsible for them.

c. For the ^{13}C NMR data, assign the various resonances to the carbon nuclei responsible for them.

27. Discuss the differences observed in the IR and NMR spectra of 4-bromo-2-chloroacetanilide and 4-bromo-2-chloroaniline that are consistent with the formation of the latter in this procedure.

Questions for Part E

28. The combination of reagents used in this experiment to iodinate 4-bromo-2-chloroaniline (**21**) is I–Cl/CH_3CO_2H because I_2/CH_3CO_2H is not as effective. Provide a mechanistic rationale to explain this fact.

29. a. Using curved arrows to symbolize the flow of electrons, provide a stepwise mechanism for the reaction that occurs between iodine monochloride and 4-bromo-2-chloroaniline.

b. Suppose that this same reaction were carried out using bromine monochloride, BrCl. What electrophilic substitution reaction might occur were this reagent, rather than iodine monochloride, allowed to react with 4-bromo-2-chloroaniline? Explain.

30. Iodonium ion, I^+, is a less reactive electrophile than bromonium ion, Br^+. Explain why.

31. Halogen atoms deactivate the aromatic ring toward electrophilic substitution.

a. Provide a mechanistic rationale to explain this observation.

b. Which halogen is the most deactivating? Explain your answer.

32. Explain why halogen atoms direct electrophilic aromatic substitution to the *ortho*- and *para*-positions rather than the *meta*-position.

33. Why is iodine the last halogen substituent introduced onto the aromatic ring?

34. The pK_a of acetic acid is about 4.8, whereas that of 4-bromo-2-chloro-6-iodoanilinium ion is estimated to be about 3.

a. Using these values, calculate K_{eq} for the acid-base reaction given below.

b. Given that the anilinium ion should be soluble in polar media such as the combination of HOAc and H_2O used in recrystallizing 4-bromo-2-chloro-6-iodoaniline (**22**), explain how your calculated K_{eq} is consistent with the fact that **22** precipitates from aqueous acetic acid.

$$CH_3CO_2H + \text{Br–C}_6\text{H}_2\text{(I)(Cl)–NH}_2 \rightleftharpoons CH_3CO_2^- + \text{Br–C}_6\text{H}_2\text{(I)(Cl)–}\overset{+}{N}H_3$$

22

4-Bromo-2-chloro-6-iodoanilinium ion

Sign in at **www.cengage.com/login** and use the spectra viewer and Tables 8.1–8.8 as needed to answer the blue-numbered questions on spectroscopy.

35. Consider the spectral data for 4-bromo-2-chloro-6-iodoaniline (Figs. 21.18 and 21.19).

a. In the functional group region of the IR spectrum, specify the absorptions associated with the amino group and the aromatic ring.

b. In the ^{1}H NMR spectrum, assign the various resonances to the hydrogen nuclei responsible for them.

c. For the ^{13}C NMR data, assign the various resonances to the carbon nuclei responsible for them.

36. Discuss the differences observed in the IR and NMR spectra of 4-bromo-2-chloroaniline and 4-bromo-2-chloro-6-iodoaniline that are consistent with the formation of the latter in this procedure.

Questions for Part F

37. Using curved arrows to symbolize the flow of electrons, provide a stepwise mechanism for the conversion of 4-bromo-2-chloro-6-iodoaniline (**22**) to the diazonium salt **23.**

38. The reduction of the diazonium salt **23** may involve transfer of hydrogen from ethanol, which is oxidized to acetaldehyde. Propose an experiment that would allow you to determine whether the hydrogen comes from the hydroxyl group or from Cl of ethanol.

39. Calculate the overall yield of 1-bromo-3-chloro-5-iodobenzene obtained in the sequence of reactions that you performed.

40. Arene diazonium salts are relatively stable in aqueous solutions below about 10 °C. At higher temperatures, or if allowed to dry, they may decompose explosively.

a. Write the products of this decomposition reaction, using **23** as an example.

b. What is the driving force for decomposition?

41. In addition to reduction, diazonium salts can be converted directly to various other useful aromatic functional groups. Predict the product formed when the diazonium salt **23** is treated with the following reagents:

a. H_2O

b. CuCl

c. CuCN

d. HBF_4

Sign in at **www.cengage.com/login** and use the spectra viewer and Tables 8.1–8.8 as needed to answer the blue-numbered questions on spectroscopy.

42. Consider the spectral data for 1-bromo-3-chloro-5-iodobenzene (Figs. 21.20 and 21.21).

a. In the functional group region of the IR spectrum, specify the absorptions associated with the aromatic ring.

b. In the ^1H NMR spectrum, assign as many as possible of the various resonances to the hydrogen nuclei responsible for them.

c. For the ^{13}C NMR data, assign the various resonances to the carbon nuclei responsible for them.

43. Discuss the differences observed in the IR and NMR spectra of 1-bromo-3-chloro-5-iodobenzene and 4-bromo-2-chloro-6-iodobenzene that are consistent with the formation of the latter in this procedure.

SPECTRA

Starting Materials and Products

The IR and NMR spectra of acetanilide are provided in Figures 8.7 and 15.31, respectively.

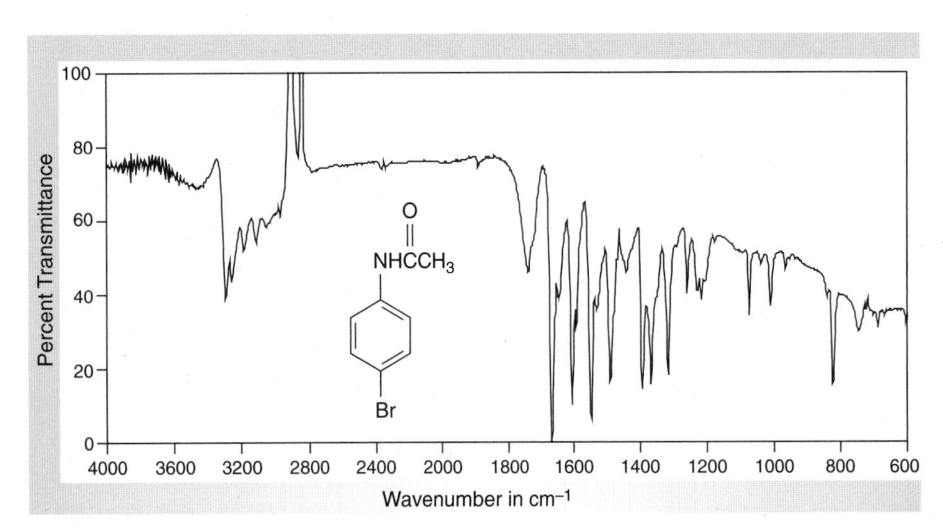

Figure 21.12
IR spectrum of 4-bromoacetanilide (IR card).

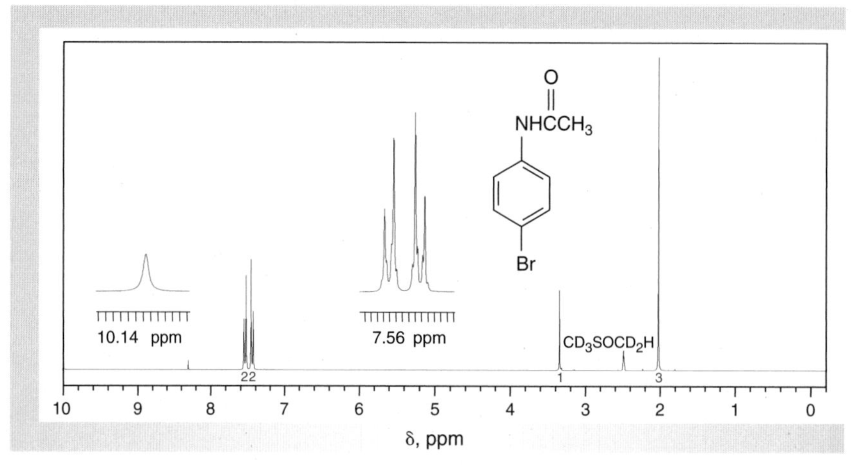

Figure 21.13
NMR data for 4-bromoacetanilide (DMSO-d_6).

(a) ^{1}H NMR spectrum (300 MHz).
(b) ^{13}C NMR data: δ 25.1, 115.5, 121.9, 132.5, 139.7, 169.5.

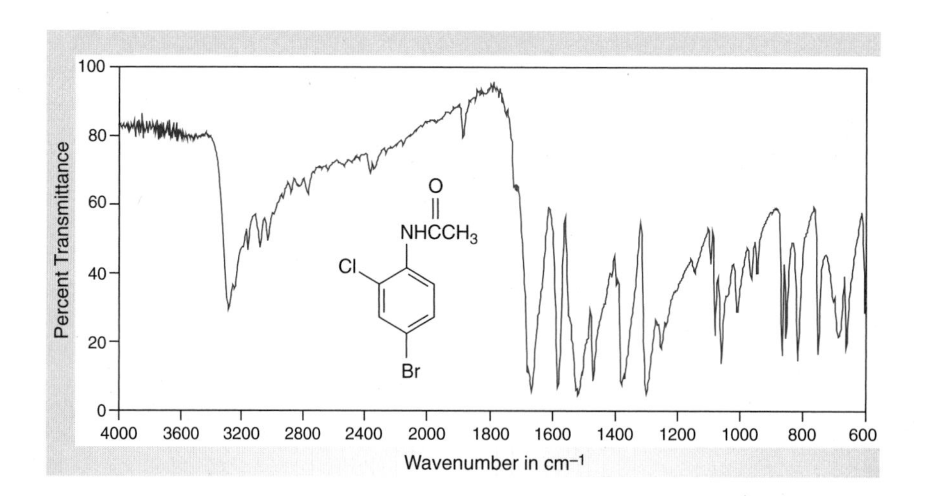

Figure 21.14
IR spectrum of 4-bromo-2-chloroacetanilide (IR card).

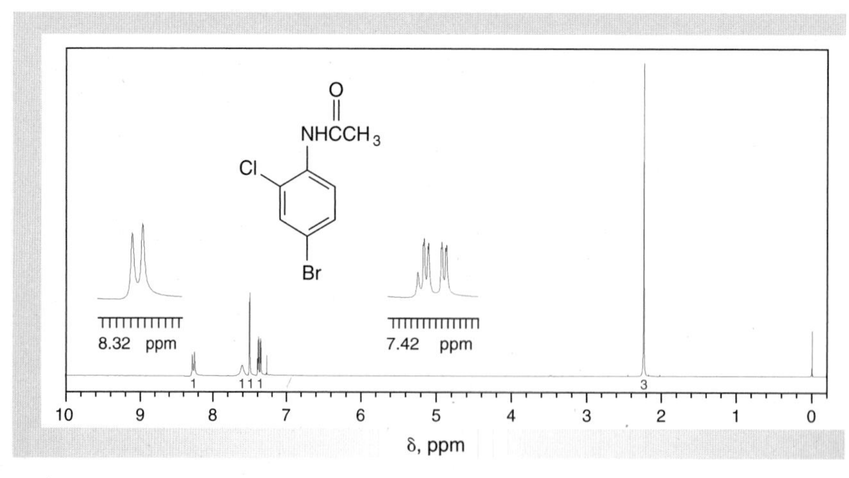

Figure 21.15
NMR data for 4-bromo-2-chloroacetanilide ($CDCl_3$).

(a) ^{1}H NMR spectrum (300 MHz).
(b) ^{13}C NMR data: δ 24.8, 116.2, 122.6, 123.2, 130.7, 131.3, 133.7, 168.2.

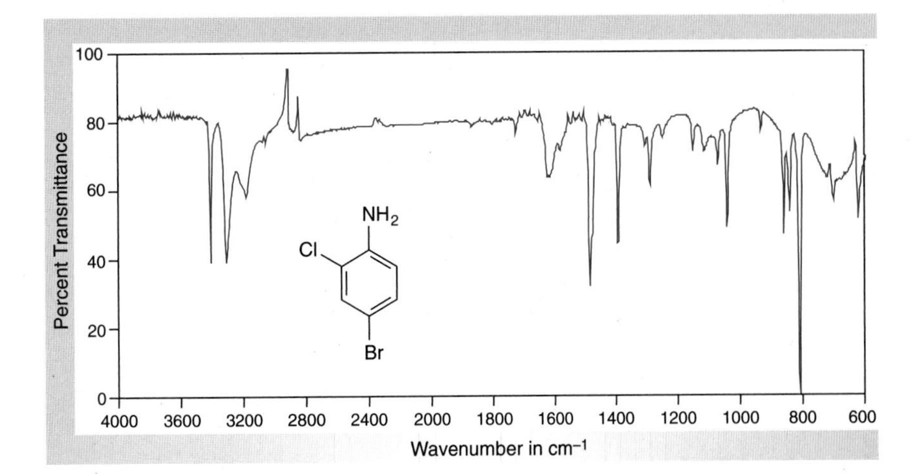

Figure 21.16
IR spectrum of 4-bromo-2-chloroaniline (IR card).

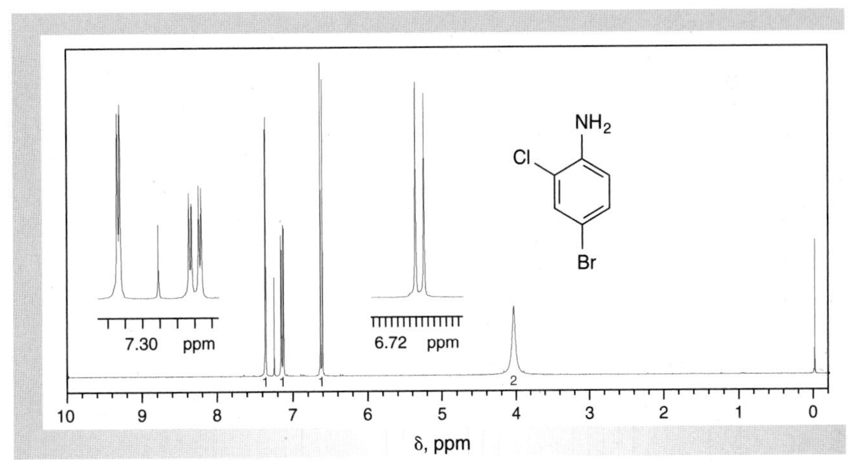

Figure 21.17
NMR data for 4-bromo-2-chloroaniline ($CDCl_3$).

(a) ^{1}H NMR spectrum (300 MHz).
(b) ^{13}C NMR data: δ 109.2, 116.8, 119.8, 130.4, 131.5, 142.1.

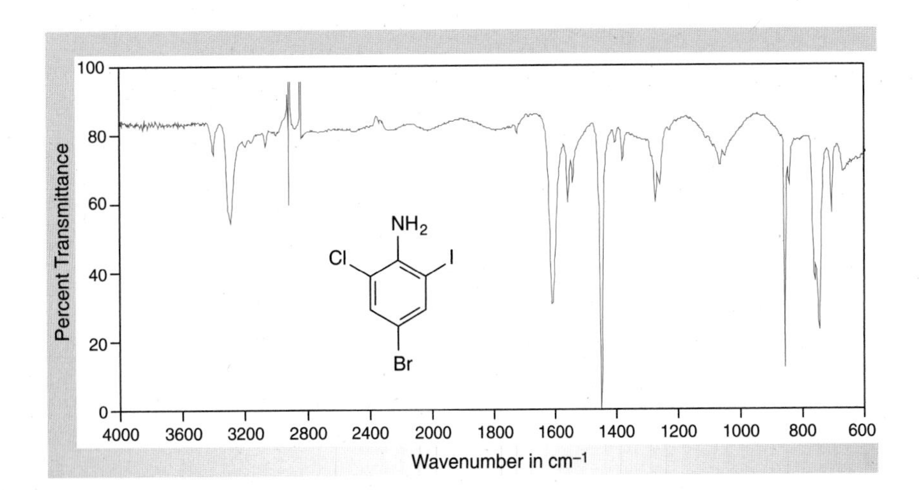

Figure 21.18
IR spectrum of 4-bromo-2-chloro-6-iodoaniline (IR card).

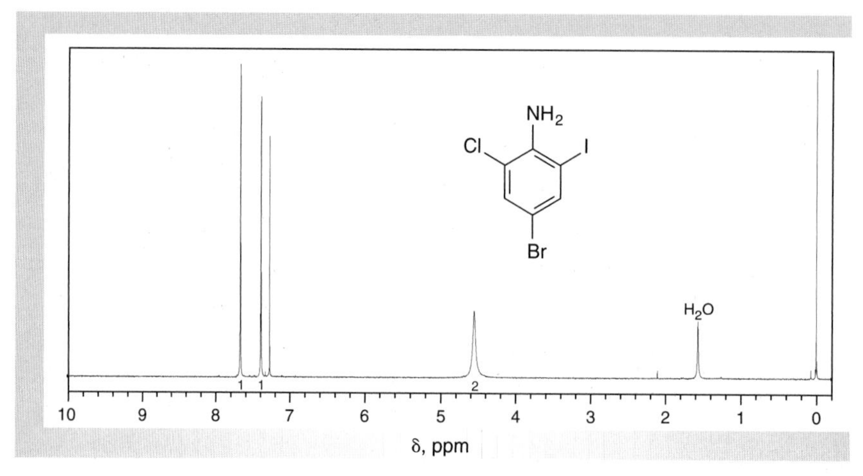

Figure 21.19
NMR spectrum of 4-bromo-2-chloro-6-iodoaniline ($CDCl_3$).

(a) ^1H NMR spectrum (300 MHz).
(b) ^{13}C NMR data: δ 84.0, 109.7, 118.6, 132.4, 139.7, 143.2.

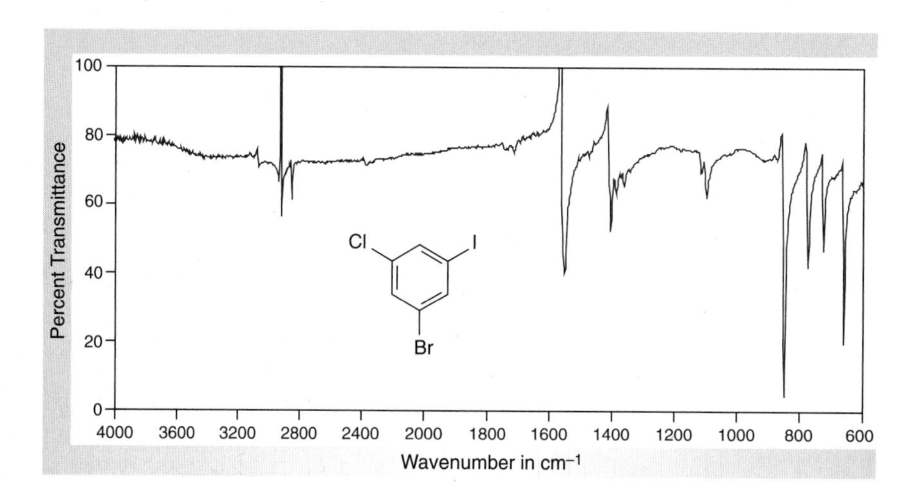

Figure 21.20
IR data for 1-bromo-3-chloro-5-iodobenzene (IR card).

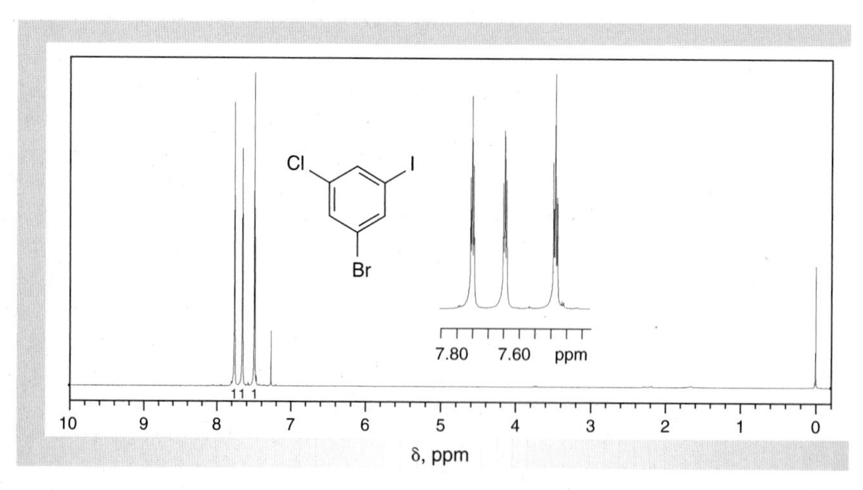

Figure 21.21
NMR data for 1-bromo-3-chloro-5-iodobenzene ($CDCl_3$).

(a) ^1H NMR spectrum (300 MHz).
(b) ^{13}C NMR data: δ 94.7, 123.7, 131.6, 136.3, 136.5, 138.7.

21.4 LIDOCAINE: SYNTHESIS OF AN ANESTHETIC AGENT

See more on *Lidocaine*
See more on *Anesthetics*
See more on *Procaine*
See more on *Benzocaine*
See more on *Procainamide*

Lidocaine (**25**) is the *generic* name of an important member of a category of useful drugs that exhibit local anesthetic activity. One of the common trade names for this compound is Xylocaine, but its chemical name is 2-diethylamino-*N*-2′,6′-dimethylphenylacetamide. Two other prominent members of this family of anesthetic agents are procaine (**26**), known more commonly by the trade name Novocain, and benzocaine (**27**) (Sec. 20.2). In addition to having applications as topical anesthetics, lidocaine and procaine are effective in treating arrhythmia, a condition involving erratic beating of the heart. Procaine has a very short half-life in the body because the ester group is readily hydrolyzed by enzymes called lipases. However, the amide derivative procainamide (**28**) is more stable *in vivo* and serves as a useful cardiac depressant and antiarrhythmic agent. Interestingly, the antiarrhythmic properities of these compounds were discovered accidentally by cardiologists who used these drugs as anesthetics during surgical procedures—a nice example of serendipitous discovery in science.

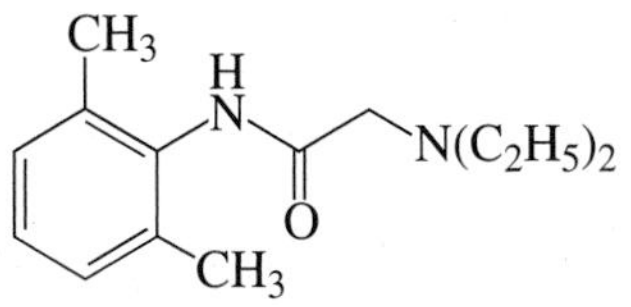

25
Lidocaine

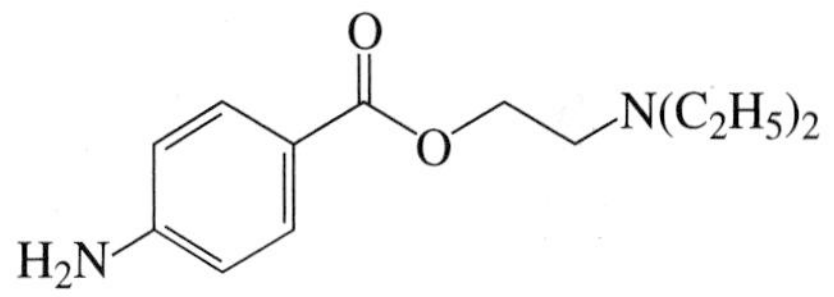

26
Procaine

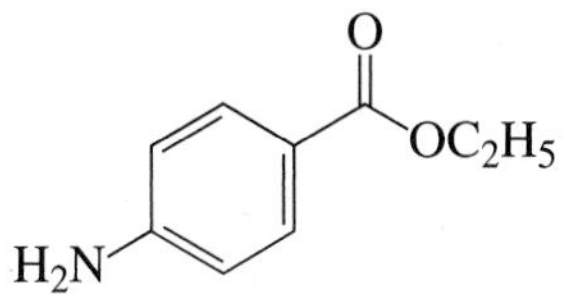

27
Benzocaine

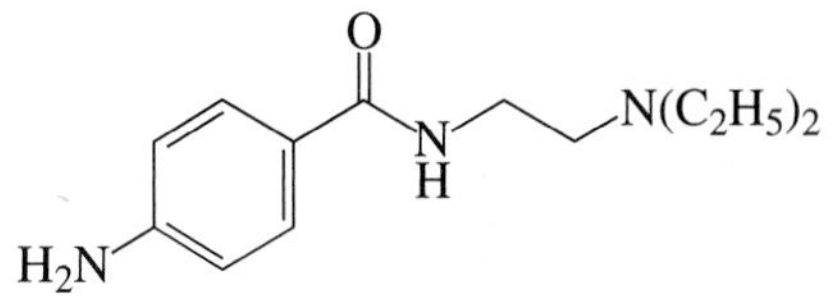

28
Procainamide

Examination of biologically active compounds often reveals certain structural or functional subunits that are common to other naturally and nonnaturally occurring substances having a broad spectrum of important medicinal applications. For example, compounds **25, 26,** and **28** each possess a dialkylamino moiety, a group that is also found in the *tranquilizer* perazine (**29**), the *antidepressant* imipramine (**30**), the *antihistamine* chlorpheniramine (**31**), the *psychotomimetic* psilocybin (**32**), and the *antimalarial* chloroquine (**33**).

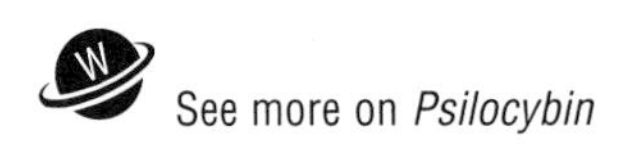

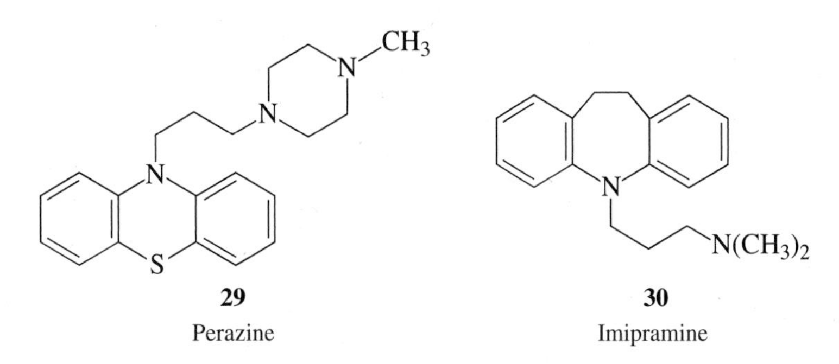

29
Perazine

30
Imipramine

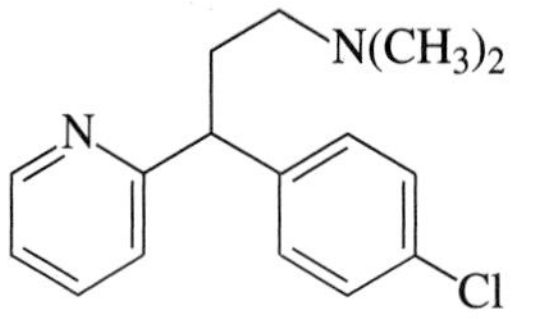

$N(CH_3)_2$ N Cl

31
Chlorpheniramine

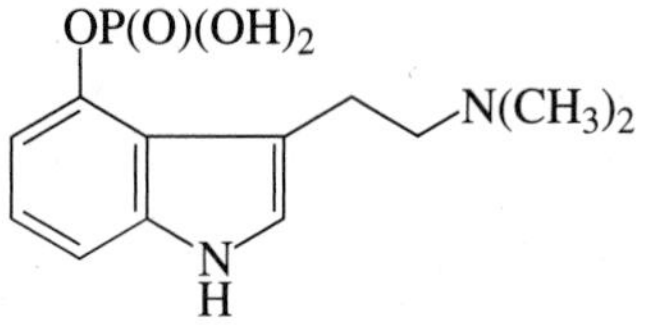

$OP(O)(OH)_2$ $N(CH_3)_2$ N H

32
Psilocybin

See more on *Malaria*

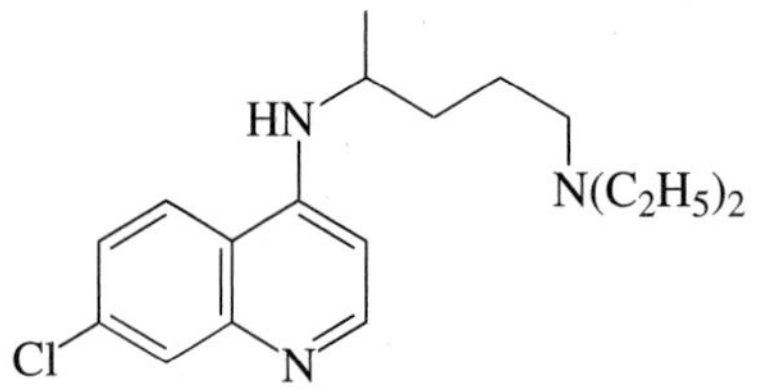

HN $N(C_2H_5)_2$ Cl N

33
Chloroquine

H_3C N CO_2CH_3 O O

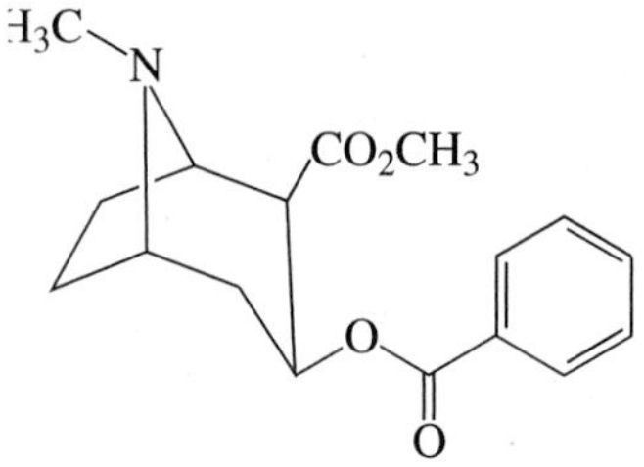

34
Cocaine

With the sole exception of psilocybin, compounds **25–33** are all synthetic, or man-made. Indeed, many of the drugs commonly used today to treat diseases are the result of the ingenuity and synthetic skills of organic chemists who work in the pharmaceutical industry. Sometimes the inspiration for a new drug is based upon analogy with other known drugs, but new drugs are frequently developed from lead compounds found in nature. For example, compounds with beneficial pharmacological and medicinal properties are often found in folk medicines that are used by native populations in various parts of the world. Indeed, the anesthetic agents **25–27** were discovered during the course of developing structural mimics of cocaine (**34**), an alkaloid found in the coca plant indigenous to South America. Cocaine has addictive as well as anesthetic properties, so there were compelling reasons to develop a synthetic analog lacking addicting potential. An interesting account of the evolution of **25–27** from cocaine can be found in the Historical Highlight at the end of Chapter 20.

See more on *Cocaine*

The present synthesis of lidocaine (**25**) from 2,6-dimethylnitrobenzene (**35**) requires three steps, as shown in Scheme 21.5. Some of the details of each step in the sequence are discussed in the following paragraphs.

Scheme 21.5

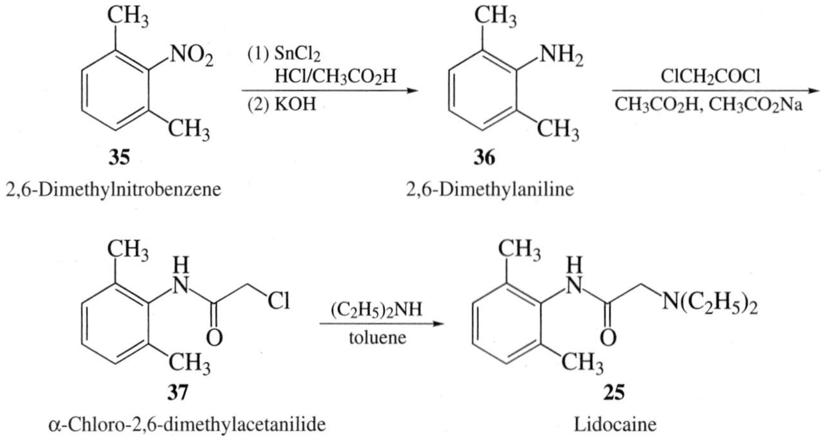

CH_3 NO_2 CH_3 —(1) $SnCl_2$, HCl/CH_3CO_2H; (2) KOH→ CH_3 NH_2 CH_3 —$ClCH_2COCl$; CH_3CO_2H, CH_3CO_2Na→

35
2,6-Dimethylnitrobenzene

36
2,6-Dimethylaniline

CH_3 H N Cl O CH_3 —$(C_2H_5)_2NH$; toluene→ CH_3 H N $N(C_2H_5)_2$ O CH_3

37
α-Chloro-2,6-dimethylacetanilide

25
Lidocaine

Reduction: Preparation of 2,6-Dimethylaniline (36)

The preparation of lidocaine starts with the chemical reduction of 2,6-dimethylnitrobenzene (**35**) to give 2,6-dimethylaniline (**36**) using stannous chloride as a reducing agent (Eq. 21.20). In this reaction, stannous chloride, $SnCl_2$, is oxidized to stannic chloride, $SnCl_4$. This reduction is similar to the reduction of nitrobenzene to aniline, which was discussed in detail in Section 21.2 (Scheme 21.3), and the side reactions associated with this process are discussed in that section.

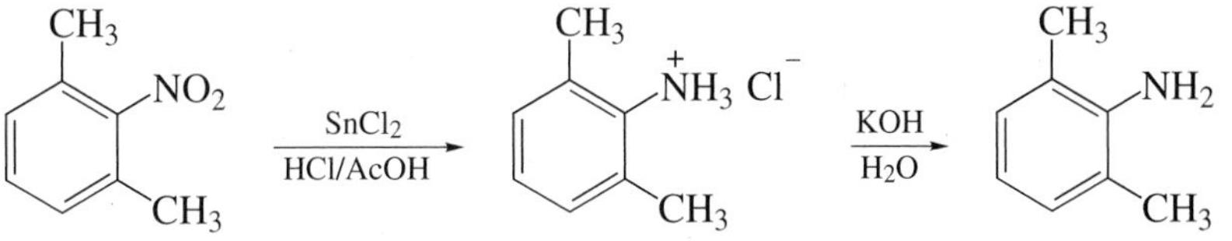

$$\textbf{35} \xrightarrow[HCl/AcOH]{SnCl_2} \textbf{38} \xrightarrow[H_2O]{KOH} \textbf{36} \quad (21.20)$$

35 2,6-Dimethyl-nitrobenzene; **38** 2,6-Dimethylanilinium hydrochloride; **36** 2,6-Dimethylaniline

The hydrochloride salt **38** that is formed upon reduction of the nitro group precipitates from the reaction mixture. Isolating this salt by filtration makes purification easy, as this operation frees **38** from contaminants such as unchanged **35** and all by-products that are soluble in the reaction medium. The **36** that is obtained after treating **38** with aqueous potassium hydroxide is sufficiently pure for use in the next step of the sequence. On the other hand, purification of aniline (**6**) from the chemical reduction of nitrobenzene is more tedious (Sec. 21.2, Part A).

Acylation: Preparation of α-Chloro-2,6-Dimethylacetanilide (37)

The substituted aniline **36** is next converted into the amide **37** by reaction with the difunctional reagent α-chloroacetyl chloride, $ClCH_2COCl$ (Eq. 21.21). Selective substitution at the acyl carbon atom in this step reflects the substantially greater reactivity of nucleophiles with acid chlorides as compared to alkyl chlorides. This results from the difference in electrophilicities and steric environments of the two possible sites for nucleophilic attack. Therefore, reaction at the carbon atom alpha to the carbonyl carbon atom to give **39** or the disubstitution product **40** is a minor competing reaction at best.

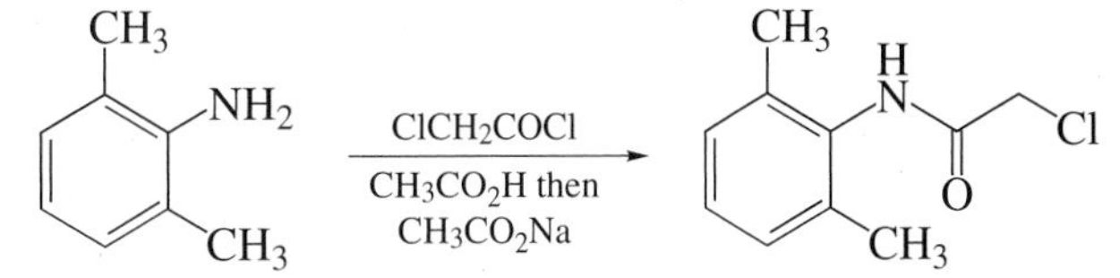

$$\textbf{36} \xrightarrow[CH_3CO_2H \text{ then } CH_3CO_2Na]{ClCH_2COCl} \textbf{37} \quad (21.21)$$

36 2,6-Dimethylaniline; **37** α-Chloro-2,6-dimethylacetanilide

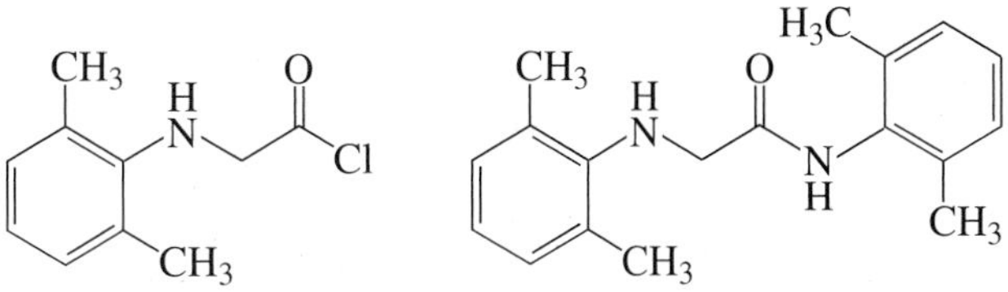

39 **40**

This reaction is performed in glacial acetic acid, so **36** will be in equilibrium with the salt **41** (Eq. 21.22). As the reaction proceeds, hydrochloric acid is liberated, so **36** is also partially converted to its hydrochloride salt **38.** Any **38** remaining at the end of the reaction period would contaminate the precipitated **37** because **38** is also insoluble in cold acetic acid. Hence, to avoid coprecipitation of the desired **37** and the hydrochloride acid salt **38,** aqueous sodium acetate is added to the warm reaction mixture to reestablish the process shown in Equation 21.22 as the only significant acid-base equilibrium involving **36.** The acetate salt **41** *is* soluble in cold aqueous acetic acid, so that filtration allows isolation of crystalline **37.** Any unreacted **36,** or its acetate salt **41,** will remain in solution and appear in the filtrate.

CH_3, NH_2, CH_3 — **36** (2,6-Dimethylaniline) $\xrightleftharpoons{CH_3CO_2H}$ CH_3, $\overset{+}{N}H_3\ \overset{-}{O_2}CCH_3$, CH_3 — **41** (21.22)

2,6-Dimethylaniline

Nucleophilic Substitution: Preparation of Lidocaine (25)

The reaction of the alkyl chloride **37** with diethylamine, which is an S_N2 process (Sec. 14.2), completes the synthetic sequence and is another example of a selective reaction. In this case, nucleophilic attack at the carbonyl function of the amido group is disfavored relative to reaction at the α-carbon atom because of the disruption of amide resonance that would accompany attack at the carbonyl group. The diethylamine serves the dual roles of acting as a nucleophile and as a base in this step. Namely, it displaces a chloride ion from **37,** and it reacts with the hydrogen chloride formed in the reaction (Eq. 21.23). Because of this latter reaction, it is necessary to use an excess of diethylamine in order to obtain high yields of lidocaine (**25**).

37 (α-Chloro-2,6-dimethylacetanilide) $\xrightarrow{2\ (C_2H_5)_2NH}$ **25** (Lidocaine) + $(C_2H_5)_2\overset{+}{N}H_2\ \overset{-}{Cl}\downarrow$ (Diethylammonium hydrochloride) (21.23)

α-Chloro-2,6-dimethylacetanilide

Lidocaine

Diethylammonium hydrochloride

Isolation of **25** involves filtering the reaction mixture to remove the diethylammonium hydrochloride, followed by extraction of the basic **25** into aqueous hydrochloride. All nonbasic contaminants, such as unchanged **37,** remain in the toluene solution. Lidocaine is liberated from its hydrochloride acid salt with aqueous base and then extracted into diethyl ether. After removal of the diethyl ether, the lidocaine is isolated as a low-melting solid or oil, depending upon its purity. In order to facilitate purification of lidocaine in this procedure, it is converted into its solid bisulfate salt **42** by reaction with sulfuric acid (Eq. 21.24). Many drugs that contain a basic site, like lidocaine, are sold in the form of their hydrochloride or sulfate salts because these salts are typically more stable and more soluble in water than the free base.

25 (Lidocaine) + H_2SO_4 $\longrightarrow$ **42** (Lidocaine bisulfate; $\overset{+}{N}H(C_2H_5)_2\ HSO_4^-$) (21.24)

Lidocaine

Lidocaine bisulfate

EXPERIMENTAL PROCEDURES

Synthesis of Lidocaine

Purpose To demonstrate the principles of multistep synthesis and substitution reactions.

Note Quantities of reagents must be adjusted according to the amount of starting material that is available from the previous reaction, but you should *not* run the reactions on a larger scale than is indicated. If the amount of your starting material varies from that indicated, do *not* change the reaction *times* unless directed to do so by your instructor. You should be able to obtain sufficient quantities of product from each reaction to complete the entire sequence starting with the amount of 2,6-dimethylnitrobenzene used in the first step.

A ■ *Preparation of 2,6-Dimethylaniline*

SAFETY ALERT

1. ***Concentrated* hydrochloric acid and glacial acetic acid are corrosive liquids that may cause burns if allowed to come in contact with your skin. Wear latex gloves when handling these chemicals. Should contact occur, flood the affected area with water and then rinse it thoroughly with dilute aqueous sodium bicarbonate solution.**
2. **The solution of 8 *M* sodium hydroxide is highly caustic and can cause burns and loss of hair; avoid contact of this solution with your skin. Wear latex gloves when transferring this solution. Should contact occur, flood the affected area with water and then rinse it thoroughly with dilute aqueous acetic acid solution.**
3. **Be certain there are no open flames in the vicinity when you are working with diethyl ether. Use *flameless* heating when concentrating ethereal solutions.**

MINISCALE PROCEDURE

Preparation Sign in at **www.cengage.com/login** to answer Pre-Lab Exercises, access videos, and read the MSDSs for the chemicals used or produced in this procedure. Review Sections 2.9, 2.11, 2.13, 2.17, 2.21, 2.25, and 2.29.

Apparatus A 125- and 250-mL Erlenmeyer flask, 50-mL round-bottom flask, water bath, separatory funnel, apparatus for vacuum filtration, simple distillation, and *flameless* heating.

Setting Up Dissolve 33.9 g of $SnCl_2 \cdot 2\ H_2O$ in 40 mL of *concentrated* hydrochloric acid contained in the 125-mL Erlenmeyer flask. Heating may be required for complete dissolution and the solution may appear milky. Dissolve 5.0 g of 2,6-dimethylnitrobenzene in 50 mL of glacial acetic acid contained in the 250-mL Erlenmeyer flask.

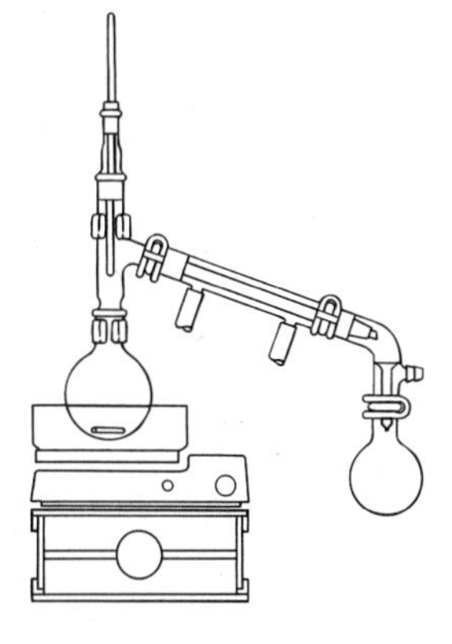

Reduction Add the solution of $SnCl_2 \cdot 2\ H_2O$ in concentrated hydrochloric acid *in one portion* to the solution of 2,6-dimethylnitrobenzene in glacial acetic acid. Swirl the resulting mixture briefly and let the resulting warm solution stand for 15 min.

Work-Up and Isolation Cool the reaction mixture to room temperature using a water bath and collect the precipitate that has formed by vacuum filtration.★ Transfer the damp solid to an Erlenmeyer flask containing 25 mL of water. Add 40–50 mL of 8 *M* aqueous KOH solution to the flask and check the pH of the mixture to ensure that it is strongly basic (pH >10). Cool the warm mixture to room temperature using a water bath and transfer it to a separatory funnel. Extract the aqueous mixture sequentially with two 15-mL portions of diethyl ether, using the first portion to rinse the Erlenmeyer flask in which the aqueous mixture was made basic. Wash the *combined* organic extracts sequentially with two 10-mL portions of water and then dry the organic layer over several spatula-tips full of anhydrous sodium sulfate.★ Filter or carefully decant the solution into the *tared* round-bottom flask and remove most of the diethyl ether by simple distillation. Alternatively, use rotary evaporation or other techniques to concentrate the solution. Remove the final traces of diethyl ether by connecting the cool stillpot to a vacuum source.

Analysis Weigh the product and calculate its yield. Obtain IR and ^{1}H NMR spectra of your starting material and product, and compare them with those of authentic samples (Figs. 21.22–21.25).

MICROSCALE PROCEDURE

Preparation Sign in at **www.cengage.com/login** to answer Pre-Lab Exercises, access videos, and read the MSDSs for the chemicals used or produced in this procedure. Review Sections 2.9, 2.11, 2.13, 2.17, 2.21, 2.25, and 2.29.

Apparatus A 25-mL Erlenmeyer flask, a 5-mL conical vial, a 13-mm × 100-mm test tube, water bath, separatory funnel, apparatus for vacuum filtration, simple distillation, and *flameless* heating.

Setting Up Dissolve 3.4 g of $SnCl_2 \cdot 2\ H_2O$ in 4 mL of *concentrated* hydrochloric acid contained in the test tube. Heating may be required for complete dissolution and the solution may appear milky. Dissolve 0.5 g of 2,6-dimethylnitrobenzene in 5 mL of glacial acetic acid contained in the Erlenmeyer flask.

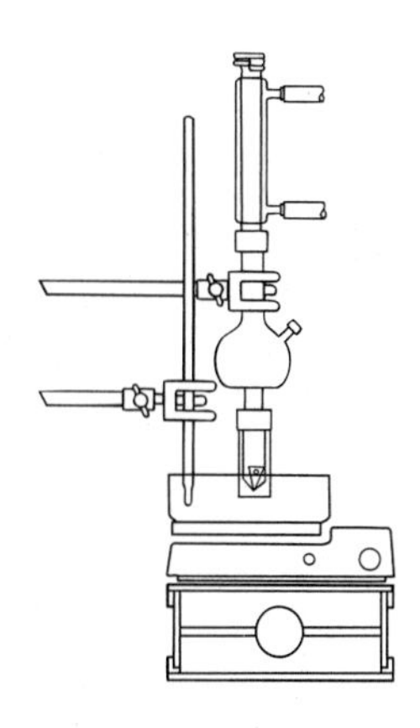

Reduction Add the solution of $SnCl_2 \cdot 2\ H_2O$ in concentrated hydrochloric acid *in one portion* to the solution of 2,6-dimethylnitrobenzene in acetic acid. Swirl the resulting mixture briefly and let the resulting warm solution stand for 15 min.

Work-Up and Isolation Cool the reaction mixture to room temperature using a water bath and collect the precipitate that has formed by vacuum filtration.★ Transfer the damp solid to an Erlenmeyer flask containing 2.5 mL of water. Add 4–5 mL of 8 *M* aqueous KOH solution to the flask and check the pH of the mixture to ensure that it is strongly basic (pH >10). Cool the warm mixture to room temperature using a water bath and transfer the mixture to a separatory funnel. Extract the aqueous mixture sequentially with two 1.5-mL portions of diethyl ether, using the first portion to rinse the Erlenmeyer flask in which the aqueous mixture was made basic. Wash the *combined* organic extracts sequentially with two 1-mL portions of water and then dry the organic layer over several microspatula-tips full of anhydrous sodium sulfate.★ Filter or carefully decant the solution into the *tared* conical vial, and remove most of the diethyl ether by simple distillation. Alternatively, use rotary evaporation or other techniques to concentrate the solution. Remove the final traces of diethyl ether by connecting the cool stillpot to a vacuum source.

Analysis Weigh the product and calculate its yield. Obtain IR and ^1H NMR spectra of your starting material and product and compare them with those of authentic samples (Figs. 21.22–21.25).

WRAPPING IT UP

Combine all of the *aqueous solutions* and neutralize them with either dilute sodium hydroxide or dilute hydrochloric acid. Flush the combined solutions down the drain with excess water. Place the *ethereal distillate* in a container for nonhalogenated organic solvents. Put the *sodium sulfate* on a tray in the hood, and after the diethyl ether evaporates, pour it into the container for nonhazardous solids.

B ■ *Preparation of α-Chloro-2,6-Dimethylacetanilide*

SAFETY ALERT

1. **Glacial acetic acid is a corrosive liquid. Wear latex gloves when handling this chemical. Should contact occur, flood the affected area with water and rinse it thoroughly with dilute aqueous sodium bicarbonate solution.**
2. **α-Chloroacetyl chloride is irritating to the mucous membranes and to the skin. Wear latex gloves when handling and transferring containers of this material, and work at the ventilation hood. Should contact occur, flood the affected area with water and rinse it thoroughly with dilute aqueous sodium bicarbonate solution.**

MINISCALE PROCEDURE

Preparation Sign in at **www.cengage.com/login** to answer Pre-Lab Exercises, access videos, and read the MSDSs for the chemicals used or produced in this procedure. Review Sections 2.9, 2.10, 2.11, 2.17, and 2.26.

Apparatus Two 250-mL Erlenmeyer flasks, ice-water bath, apparatus for vacuum filtration, and *flameless* heating.

Setting Up Dissolve 3.0 g of 2,6-dimethylaniline in 20 mL of glacial acetic acid contained in the Erlenmeyer flask. Dissolve 4.3 g of sodium acetate trihydrate in 80 mL of water in the other Erlenmeyer flask.

Reaction Add 2.8 g of α-chloroacetyl chloride to the solution of 2,6-dimethylaniline in acetic acid and swirl the solution to mix the reactants. Allow the reaction to proceed for 15 min with occasional swirling. Add the aqueous solution of sodium acetate in one portion with swirling.

Work-Up and Isolation Cool the resulting mixture in an ice-water bath and collect the product by vacuum filtration. Rinse the filter cake with water until the odor of acetic acid can no longer be detected. Press the solid with a clean spatula while the vacuum source is still attached to dry it as completely as possible, and then air-dry it for at least 24 h.

Analysis Weigh the product and calculate its yield. Determine its melting point (reported mp, 145–146 °C). The product may be recrystallized from a small volume of toluene. Obtain IR and ^{1}H NMR spectra of your product and compare them with those of an authentic sample (Figs. 21.26 and 21.27).

MICROSCALE PROCEDURE

Preparation Sign in at **www.cengage.com/login** to answer Pre-Lab Exercises, access videos, and read the MSDSs for the chemicals used or produced in this procedure. Review Sections 2.9, 2.10, 2.11, 2.17, and 2.26.

Apparatus Two 25-mL Erlenmeyer flasks, ice-water bath, apparatus for vacuum filtration, and *flameless* heating.

Setting Up Dissolve 0.3 g of 2,6-dimethylaniline in 2 mL of glacial acetic acid contained in an Erlenmeyer flask. Dissolve 0.45 g of sodium acetate trihydrate in 8 mL of water in the other Erlenmeyer flask.

Reaction Add 0.28 g of α-chloroacetyl chloride to the solution of 2,6-dimethylaniline in acetic acid and swirl the solution to mix the reactants. Allow the reaction to proceed for 15 min with occasional swirling. Add the aqueous solution of sodium acetate in one portion with swirling.

Work-Up and Isolation Cool the resulting mixture in an ice-water bath and collect the product by vacuum filtration. Rinse the filter cake with water until the odor of acetic acid can no longer be detected. Press the solid with a clean spatula while the vacuum source is still attached to dry it as completely as possible, and then air-dry it for at least 24 h.

Analysis Weigh the product and calculate its yield. Determine its melting point (reported mp, 145–146 °C). Obtain IR and ^{1}H NMR spectra of your product and compare them with those of an authentic sample (Figs. 21.26–21.27).

WRAPPING IT UP

Neutralize the *filtrate* with dilute sodium hydroxide and flush it down the drain with excess water.

C ■ *Preparation of Lidocaine*

SAFETY ALERT

1. **Diethylamine is an unpleasant-smelling and corrosive liquid. Measure it out in a hood. Should any of this chemical come in contact with the skin, flood the affected area with cold water.**
2. **The solution of 8 *M* potassium hydroxide is highly caustic and can cause burns and loss of hair; avoid contact of this solution with your skin. Wear latex gloves when transferring this solution. Should contact occur, flood the affected area with water and rinse it thoroughly with dilute aqueous acetic acid solution.**
3. **Be certain there are no open flames in the vicinity when you are working with diethyl ether. Use *flameless* heating when concentrating diethyl ether solutions.**

MINISCALE PROCEDURE

Preparation Sign in at **www.cengage.com/login** to answer Pre-Lab Exercises, access videos, and read the MSDSs for the chemicals used or produced in this procedure. Review Sections 2.7, 2.9, 2.10, 2.11, 2.17, 2.21, 2.22, 2.28, and 2.29.

Apparatus A 125-mL round-bottom flask, 250-mL Erlenmeyer flask, an ice-water bath, separatory funnel, apparatus for heating under reflux, magnetic stirring, vacuum filtration, simple distillation, and *flameless* heating.

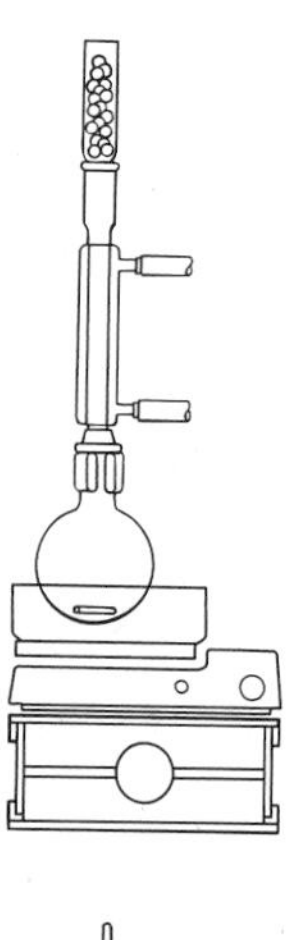

Setting Up Oven-dry the glass apparatus for heating under reflux. Dissolve 2.2 g of dry α-chloro-2,6-dimethylacetanilide in 30 mL of toluene in the dry round-bottom flask containing a stirbar. Assemble the apparatus for heating under reflux.

Reaction Add 2.4 g of diethylamine to the solution of α-chloro-2,6-dimethylacetanilide with stirring, and then heat the reaction mixture under reflux for 90 min. Allow the mixture to cool to room temperature, and then cool it briefly in an ice-water bath.

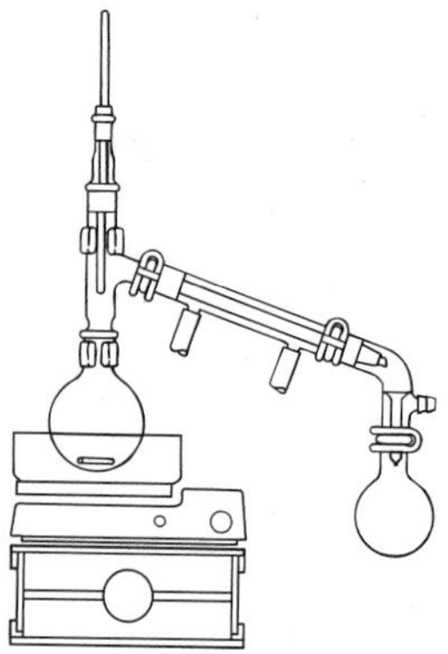

Work-Up and Isolation Filter the reaction mixture by vacuum filtration and rinse the filter cake with about 5 mL of toluene. Transfer the combined filtrate and washing to a separatory funnel and extract it with two 20-mL portions of 3 *M* HCl. Shake the funnel vigorously with venting. Combine the two acidic aqueous extracts in an Erlenmeyer flask and add about 25 mL of 8 *M* KOH solution to make the mixture strongly basic to pHydrion paper. Cool the alkaline mixture thoroughly in an ice-water bath. Transfer the chilled alkaline solution to a separatory funnel. Extract the aqueous mixture sequentially with two 15-mL portions of diethyl ether, using the first portion to rinse the Erlenmeyer flask in which the aqueous mixture was basified. The extractions should be carried out with vigorous shaking and frequent venting. Wash the *combined* organic extracts with 25 mL of water and dry the organic layer over several spatula-tips full of anhydrous sodium sulfate.★ Filter or carefully decant the solution into a dry, *tared* round-bottom flask, and remove most of the diethyl ether by simple distillation. Alternatively, use rotary evaporation or other techniques to concentrate the solution. Remove the final traces of solvent by connecting the cool stillpot to a vacuum source.

Salt Formation Dissolve the crude lidocaine in diethyl ether (10 mL of solvent per g of solute; *no flames!*) and then add a solution of 2 mL of 2.2 *M* sulfuric acid in ethanol per g of solute. Mix the solutions thoroughly and scratch at the air-liquid interface with a glass rod to induce crystallization. Dilute the mixture with an equal volume of reagent-grade acetone and isolate the precipitated salt by vacuum filtration. Rinse the filter cake with a few milliliters of cold reagent-grade acetone and then air-dry and weigh the product. The lidocaine bisulfate may be recrystallized by dissolving it in an equal weight of hot water and then adding *in one portion* a volume of reagent-grade acetone equal to 20 times the volume of this aqueous solution. Swirl the mixture briefly and then allow the solution to stand until crystallization is complete; you may need to scratch at the air-liquid interface with a glass rod to induce crystallization. Cool in an ice-bath for 15 min to complete crystallization.

Analysis Weigh the lidocaine bisulfate and calculate its yield. Determine its melting point (reported mp, 210–212 °C). (*Caution:* Do *not* use a liquid-bath melting-point apparatus to determine the melting point; use a metal block apparatus, as illustrated in Figure 2.19, Section 2.7.) Obtain IR and ^{1}H NMR spectra of your product and compare them with those of an authentic sample (Figs. 21.28–21.29).

MICROSCALE PROCEDURE

Preparation Sign in at **www.cengage.com/login** to answer Pre-Lab Exercises, access videos, and read the MSDSs for the chemicals used or produced in this procedure. Review Sections 2.7, 2.9, 2.10, 2.11, 2.17, 2.21, 2.22, 2.28, and 2.29.

Apparatus A 5-mL conical vial, 25-mL Erlenmeyer flask, two screw-cap centrifuge tubes, an ice-water bath, apparatus for heating under reflux, magnetic stirring, vacuum filtration, simple distillation, and *flameless* heating.

Setting Up Oven-dry the glass apparatus for heating under reflux. Dissolve 0.2 g of dry α-chloro-2,6-dimethylacetanilide in 2.5 mL of toluene in the dry conical vial containing a spinvane. Assemble the apparatus for heating under reflux.

Reaction Add 0.22 g of diethylamine to the solution of α-chloro-2,6-dimethylacetanilide with stirring, and then heat the reaction mixture under reflux for 90 min. Allow the mixture to cool to room temperature and then cool it briefly in an ice-water bath.

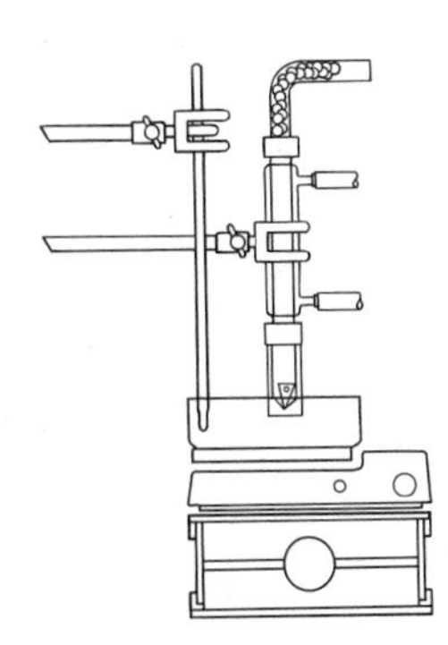

Work-Up and Isolation Filter the reaction mixture by vacuum filtration and rinse the filter cake with about 0.5 mL of toluene. Transfer the combined filtrate and washing to a screw-cap centrifuge tube and extract it with two 1.5-mL portions of 3 *M* HCl. Shake the tube vigorously with venting. Combine the two acidic aqueous extracts in the Erlenmeyer flask and add about 2.0 mL of 8 *M* KOH solution to make the mixture strongly basic to pHydrion paper. Cool the alkaline mixture thoroughly in an ice-water bath. Transfer the chilled alkaline solution to a screw-cap centrifuge tube. Extract the aqueous mixture sequentially with two 1.5-mL portions of diethyl ether, using the first portion to rinse the Erlenmeyer flask in which the aqueous mixture was basified. The extractions should be carried out with vigorous shaking and frequent venting. Wash the *combined* organic extracts with 2 mL of water and dry the organic layer over several microspatula-tips full of anhydrous sodium sulfate.★ Filter or carefully decant the solution into a dry, *tared* conical vial, and

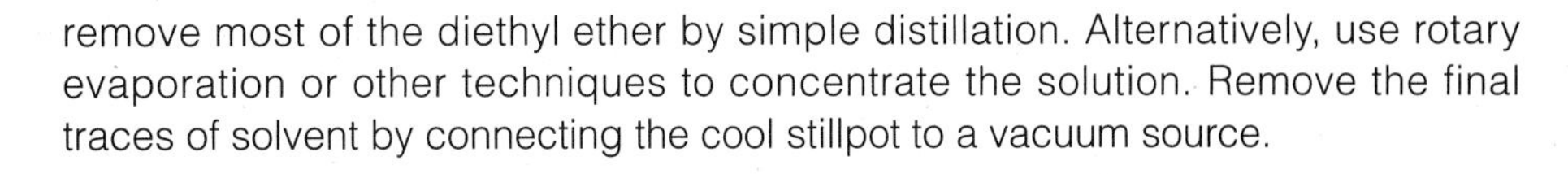

remove most of the diethyl ether by simple distillation. Alternatively, use rotary evaporation or other techniques to concentrate the solution. Remove the final traces of solvent by connecting the cool stillpot to a vacuum source.

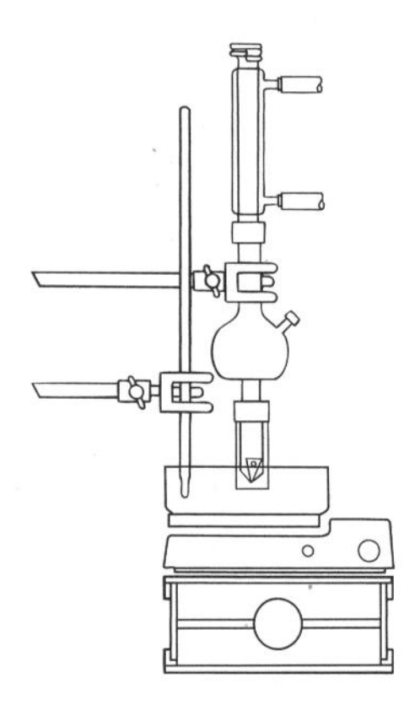

Salt Formation Dissolve the crude lidocaine in diethyl ether (1 mL of solvent per 0.1 g of solute; *no flames!*) and then add a solution of 0.2 mL of 2.2 *M* sulfuric acid in ethanol per 0.1 g of solute. Mix the solutions thoroughly and scratch at the air-liquid interface with a glass rod to induce crystallization. Dilute the mixture with an equal volume of reagent-grade acetone and isolate the precipitated salt by vacuum filtration. Rinse the filter cake with a few milliliters of cold reagent-grade acetone and then air-dry and weigh the product. The lidocaine bisulfate may be recrystallized by dissolving it in an equal weight of hot water and then adding *in one portion* a volume of reagent-grade acetone equal to 20 times the volume of this aqueous solution. Swirl the mixture briefly and then allow the solution to stand until crystallization is complete; you may need to scratch at the air-liquid interface with a glass rod to induce crystallization. Cool in an ice-bath for 15 min to complete crystallization.

Analysis Weigh the lidocaine bisulfate and calculate its yield. Determine its melting point (reported mp, 210–212 °C). (*Caution:* Do *not* use a liquid-bath melting-point apparatus to determine the melting point; use a metal block apparatus, as illustrated in Figure 2.19, Section 2.7.) Obtain IR and ^{1}H NMR spectra of your product and compare them with those of an authentic sample (Figs. 21.28–21.29).

WRAPPING IT UP

Neutralize the *aqueous solution* with dilute hydrochloric acid and flush it down the drain with excess water. Place the *combined diethyl ether distillate and filtrates* in a container for nonhalogenated organic solvents. Put the *sodium sulfate* on a tray in the hood, and after the diethyl ether evaporates, pour it and the *diethylammonium hydrochloride* into the container for nonhazardous solids.

EXERCISES

General Questions

1. 2,6-Dimethylnitrobenzene (**35**) may be prepared in one step from 1,3-dimethylbenzene. What reagents would be necessary to effect this transformation, and would this reaction give **35** as the exclusive product? Explain your answer.
2. Given the yields that you obtained in the laboratory, calculate the number of moles of 2,6-dimethylnitrobenzene (**35**) that would be needed to prepare one mole of lidocaine bisulfate (**42**).

Questions for Part A

3. Write the balanced equation for the reduction of 2,6-dimethylnitrobenzene (**35**) to give **36** using stannous chloride, $SnCl_2$, in concentrated hydrochloric acid.
4. What is the precipitate that is originally collected in the reduction of 2,6-dimethylnitrobenzene (**35**) by stannous chloride?
5. What would be the consequence of failing to make the aqueous solution basic prior to extracting the 2,6-dimethylaniline (**36**) into ether?

6. Write structures for at least two organic by-products expected to be produced by the reduction of 2,6-dimethylnitrobenzene (**35**) with stannous chloride.
7. Write a mechanism for the reduction of **35** into **36.** Show all electron and proton transfer steps using curved arrows to symbolize the flow of electrons.

Sign in at **www.cengage.com/login** and use the spectra viewer and Tables 8.1–8.8 as needed to answer the blue-numbered questions on spectroscopy.

8. Consider the spectral data for 2,6-dimethylnitrobenzene (Figs. 21.22 and 21.23).
   a. In the IR spectrum, specify the absorptions in the functional group region associated with the nitro group and with the aromatic ring.
   b. In the ^1H NMR spectrum, assign the various resonances to the hydrogen nuclei responsible for them.
   c. For the ^{13}C NMR data, assign the various resonances to the carbon nuclei responsible for them.
9. Consider the spectral data for 2,6-dimethylaniline (Figs. 21.24 and 21.25).
   a. In the IR spectrum, specify the absorptions in the functional group region associated with the nitrogen-hydrogen bonds and with the aromatic ring.
   b. In the ^1H NMR spectrum, assign the various resonances to the hydrogen nuclei responsible for them.
   c. For the ^{13}C NMR data, assign the various resonances to the carbon nuclei responsible for them.
10. Discuss the differences observed in the IR and NMR spectra of 2,6-dimethylnitrobenzene and 2,6-dimethylaniline that are consistent with the formation of the latter in this procedure.

Questions for Part B

11. Why would ethanol be a poor choice of a solvent for the reaction between 2,6-dimethylaniline (**36**) and α-chloroacetyl chloride?
12. Why is the anilide **37** much less basic than 2,6-dimethylaniline (**36**)?
13. What organic by-product(s) might be formed if **36** were allowed to react with two moles of α-chloroacetyl chloride?
14. What factor(s) favor attack of 2,6-dimethylaniline (**36**) at the carbonyl carbon atom rather than the α-carbon atom of α-chloroacetyl chloride?
15. Using curved arrows to symbolize the flow of electrons, write a stepwise mechanism for the reaction of **36** with α-chloroacetyl chloride.

Sign in at **www.cengage.com/login** and use the spectra viewer and Tables 8.1–8.8 as needed to answer the blue-numbered questions on spectroscopy.

16. Consider the spectral data for α-chloro-2,6-dimethylacetanilide (Figs. 21.26 and 21.27).
    a. In the IR spectrum, specify the absorptions in the functional group region associated with the nitrogen-hydrogen bond, with the carbon-oxygen double bond, and with the aromatic ring.
    b. In the ^1H NMR spectrum, assign the various resonances to the hydrogen nuclei responsible for them.
    c. For the ^{13}C NMR data, assign the various resonances to the carbon nuclei responsible for them.
17. Discuss the differences observed in the IR and NMR spectra of 2,6-dimethylaniline and α-chloro-2,6-dimethylacetanilide that are consistent with the formation of the latter in this procedure.

Questions for Part C

18. What would be the expected effect on yield if only one mole of diethylamine per mole of **37** were used in the final step of the preparation of lidocaine? Explain.

19. What side reaction(s) would be expected if **37** were wet when the reaction with diethylamine was performed?

20. What is the solid that is isolated initially in the reaction between **37** and diethylamine?

21. Why is **37** less susceptible to nucleophilic attack at the carbonyl carbon atom than is α-chloroacetyl chloride?

22. Why does sulfuric acid protonate the nitrogen atom of the diethylamine group of lidocaine preferentially to that of the amido group, as shown by formation of **42**?

Sign in at **www.cengage.com/login** and use the spectra viewer and Tables 8.1–8.8 as needed to answer the blue-numbered questions on spectroscopy.

23. Consider the spectral data for lidocaine bisulfate (Figs. 21.28 and 21.29).

    a. In the IR spectrum, specify the absorptions in the functional group region associated with the nitrogen-hydrogen bond, with the carbon-oxygen double bond, and with the aromatic ring.

    b. In the ^{1}H NMR spectrum, assign the various resonances to the hydrogen nuclei responsible for them.

    c. For the ^{13}C NMR data, assign the various resonances to the carbon nuclei responsible for them.

24. Discuss the differences observed in the IR and NMR spectra of α-chloro-2, 6-dimethylacetanilide and lidocaine bisulfate that are consistent with the formation of the latter in this procedure.

SPECTRA

Starting Materials and Products

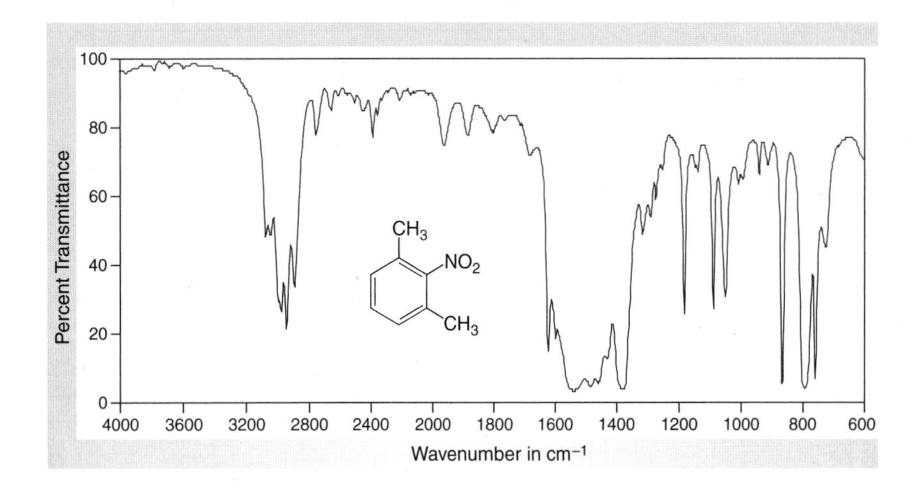

Figure 21.22
IR spectrum of 2,6-dimethylnitrobenzene (neat).

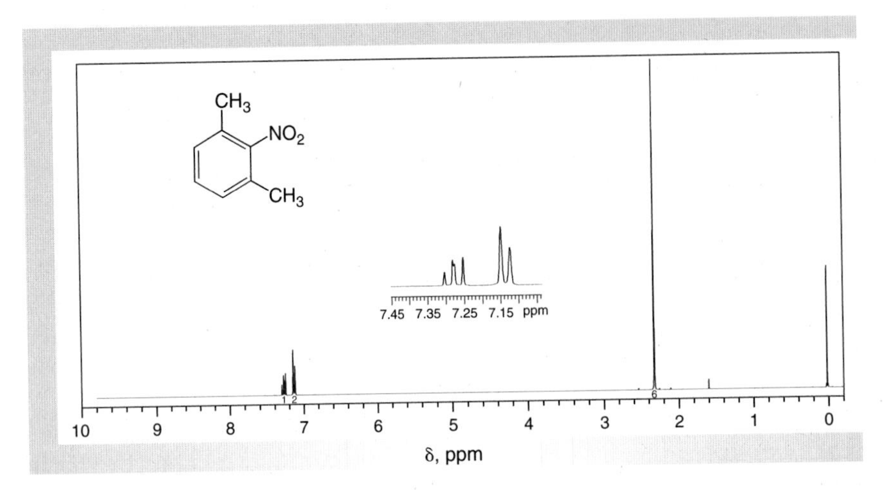

Figure 21.23
NMR data for 2,6-dimethylnitrobenzene ($CDCl_3$).

(a) ^{1}H NMR spectrum (300 MHz).
(b) ^{13}C NMR data: δ 17.3, 128.8, 129.4, 129.9, 153.4.

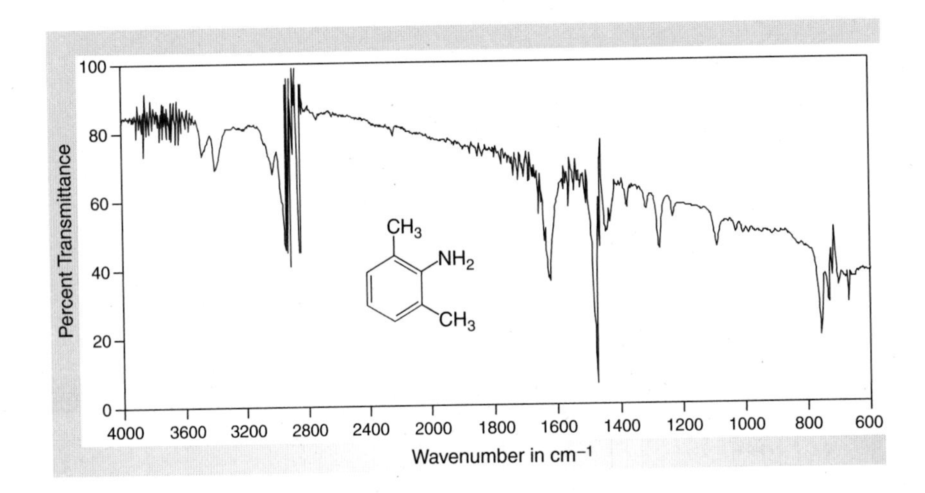

Figure 21.24
IR spectrum of 2,6-dimethylaniline (IR card).

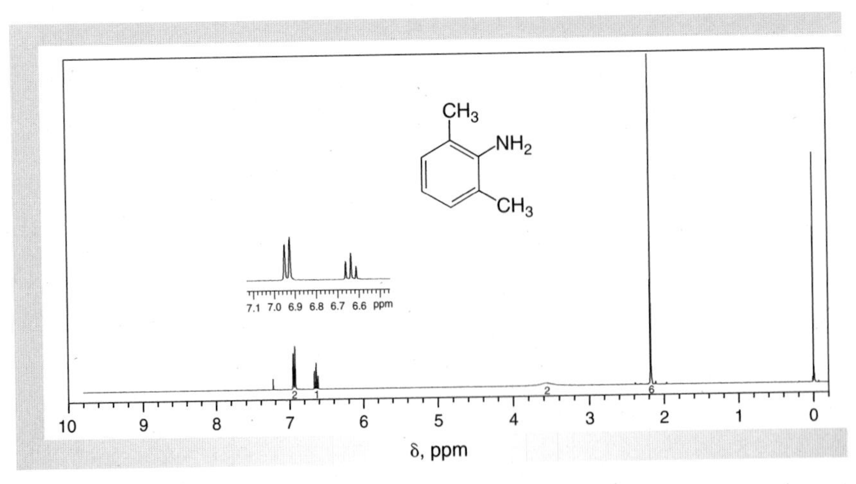

Figure 21.25
NMR data for 2,6-dimethylaniline ($CDCl_3$).

(a) ^{1}H NMR spectrum (300 MHz).
(b) ^{13}C NMR data: δ 17.3, 117.6, 121.3, 128.0, 142.5.

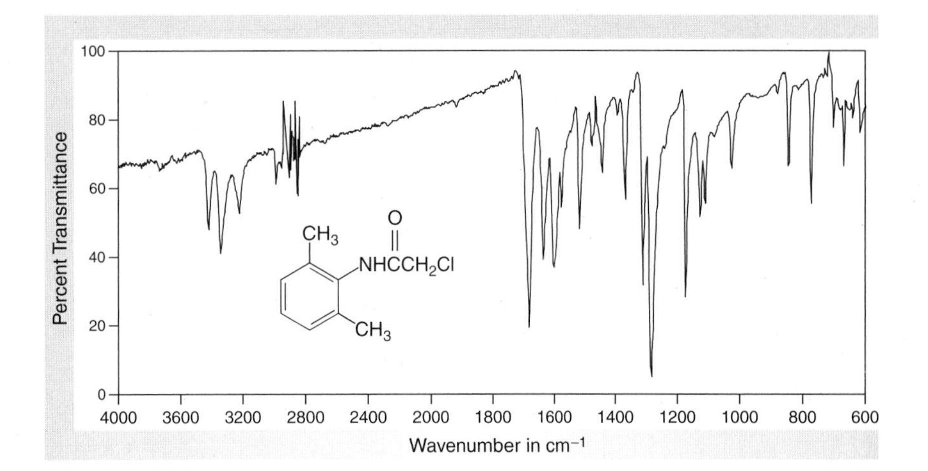

Figure 21.26
IR spectrum of α-chloro-2, 6-dimethylacetanilide (IR card).

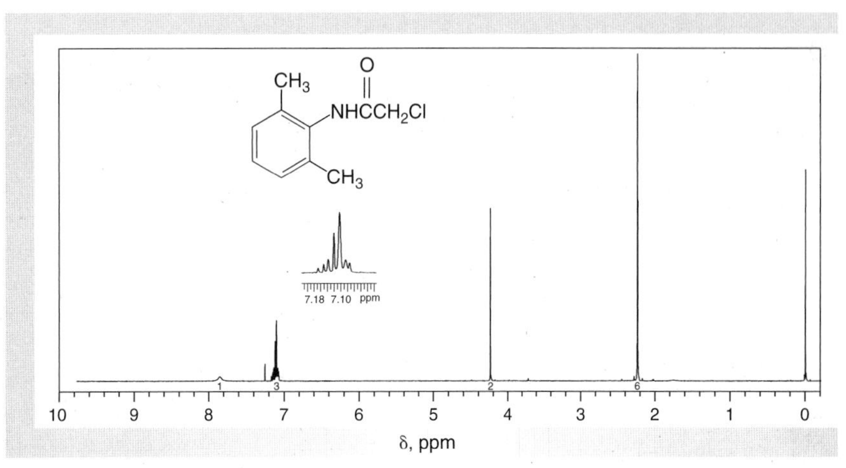

Figure 21.27
NMR data for α-chloro-2, 6-dimethylacetanilide ($CDCl_3$).

(a) ^{1}H NMR spectrum (300 MHz).
(b) ^{13}C NMR data: δ 18.3, 42.8, 127.9, 128.3, 132.6, 135.3, 164.3.

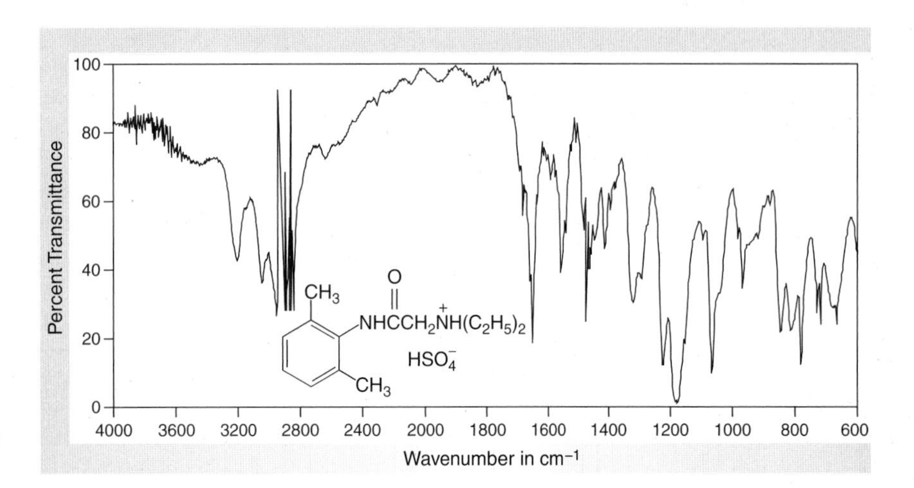

Figure 21.28
IR spectrum of lidocaine bisulfate (IR card).

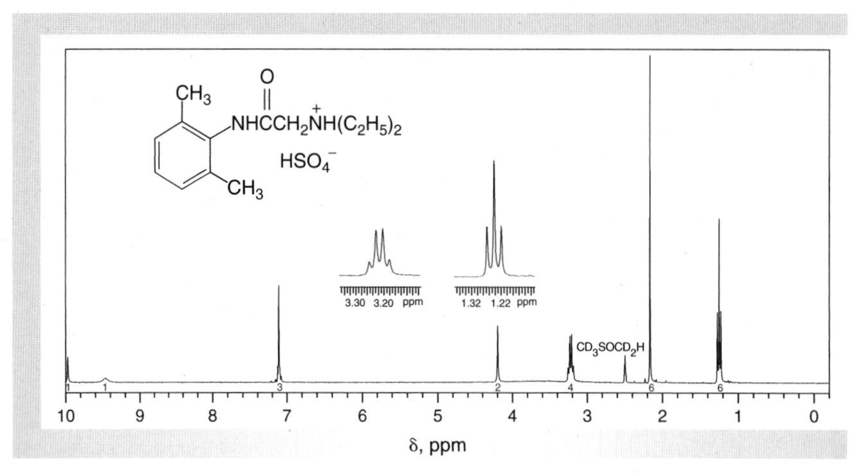

Figure 21.29
NMR data for lidocaine bisulfate (DMSO-d_6).

(a) ^{1}H NMR spectrum (300 MHz).
(b) ^{13}C NMR data: δ 8.8, 18.1, 48.5, 52.7, 127.1, 129.9, 133.6, 135.0, 163.3.

HISTORICAL HIGHLIGHT

Discovery of Sulfa Drugs

The sulfanilamides were the first antibacterial drugs invented by chemists, and a fascinating story underlies the discovery of the medicinal properties of these compounds. Although many scientists played important roles in this discovery, a key individual in the effort was Gerhard Domagk. Born in Lagow, Germany, in 1895, Domagk attended the University of Kiel intending to become a doctor, but World War I interrupted his medical studies. Following the armistice in 1918, he reentered the University of Kiel and earned his medical degree in 1921. After a brief career in academia, he moved to I. G. Farbenindustrie (I.G.F.), the German dye cartel, where his responsibility was testing the pharmacological properties of new dyes being synthesized by chemists.

At the time Domagk joined I.G.F., there were no antibacterial agents. This presented a serious health problem because bacteria were known to be the agents that caused pneumonia, meningitis, gonorrhea, and streptococcic and staphylococcic infections. The I.G.F. team of chemists and pharmacologists set out to find compounds that would kill these microbes without harming their animal or human hosts, and they formulated a plan to determine whether certain dyes might be bactericidal. Their strategy evolved from the observation that particular dyes, specifically those containing a sulfonamido group, seemed to be particularly "fast," or tightly bound, to wool fabrics, indicating their affinity for the protein molecules comprising wool. Because bacteria are proteinaceous in nature, the researchers reasoned that the dyes might fasten to the bacteria in such a way as to inhibit or kill them selectively. As we shall see, this simple hypothesis was partially correct: The sulfonamido group was indeed essential, but the part of the molecule that made it function as a dye was irrelevant to its effectiveness as a bactericide.

One dye Domagk tested on laboratory mice and rabbits infected with streptococci was called Prontosil. This compound was found to be strongly disinfective against these bacteria and could be tolerated by the animals in large doses with no apparent ill effects. This discovery of the bactericidal effect of Prontosil in animals was probably made in early 1932; I.G.F. applied for a patent in December of that year.

Clinical tests on human patients apparently began soon after this, but the record is confused. Some accounts say that before any other tests had been done on humans, Domagk gave a dose of Prontosil in desperation to his deathly ill young daughter, who had developed a serious streptococcal infection following a needle prick; the girl then made a rapid recovery. Others report that the first clinical test was on a ten-month-old boy who was dying of staphylococcal septicemia. His doctor, R. Förster, was a friend of Domagk's

(Continued)

HISTORICAL HIGHLIGHT *Discovery of Sulfa Drugs (Continued)*

superior at I.G.F. and, through him, learned about a red dye (Prontosil) that was miraculously effective in animals against streptococci. Since the baby was close to death anyway and he felt that there was nothing to lose if the dye was not effective against staphylococci, Förster gave the child two doses of the red dye; complete recovery rapidly followed.

Regardless of which of these two stories is correct, or whether both are true, it was widely recognized by the middle 1930s that the discovery of the bactericidal properties of Prontosil was a medical miracle, for which Domagk was awarded the Nobel Prize in Physiology of Medicine in 1939. There were other important developments in the years between 1933 and 1939, however. Domagk did not publish the results of his tests of Prontosil on animal infections until February 1935, more than two years after the work was done. Learning of his results, the Trefouels, a wife-and-husband team at the Pasteur Institute in Paris, were prompted to test the bactericidal properties of several compounds, all of which were "azo" dyes closely related in chemical structure to Prontosil. The feature common to their dyes was the sulfonamido portion, but the other parts of the molecules differed significantly. Remarkably, they found that the antibacterial properties of these dyes were virtually identical to those of Prontosil.

This finding led to an explanation of a puzzling fact about Prontosil: It was ineffective against bacteria *in vitro* but was strongly effective *in vivo*. Apparently, a metabolic process within animals was necessary to make the sulfonamide dyes antibacterial. The Trefouels reasoned that the dye was broken into two parts in animals, and only the sulfonamido portion was effective as an antibacterial. To prove this hypothesis, they synthesized the sulfonamido component of Prontosil, which was the known compound *p*-aminobenzenesulfonamide, or sulfanilamide, and found it to be as effective as Prontosil against bacterial infections. Comparison of the formulas of Prontosil and sulfanilamide makes it clear that cleavage of Prontosil at the azo double bond affords the skeleton of sulfanilamide. This cleavage occurs biochemically when Prontosil is injected or taken orally, and the sulfanilamide so produced is the actual antibacterial agent. The original hypothesis that sulfonamido dyes would be bactericidal was thus partly misconceived, in that only the sulfonamido part of the dye molecule kills microbes; the fact that it was a part of a dye molecule was incidental.

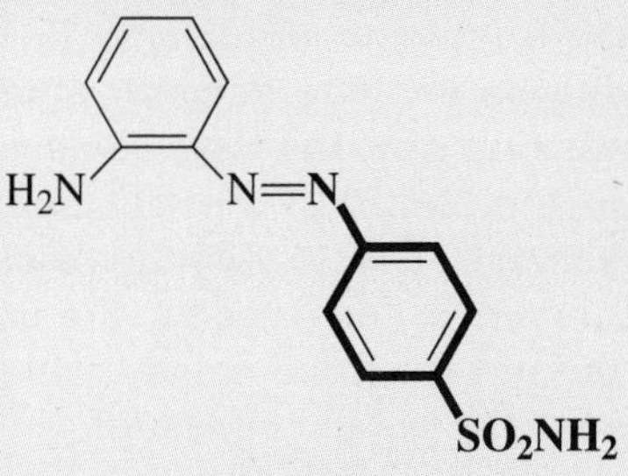

H_2N N=N SO_2NH_2

Prontosil

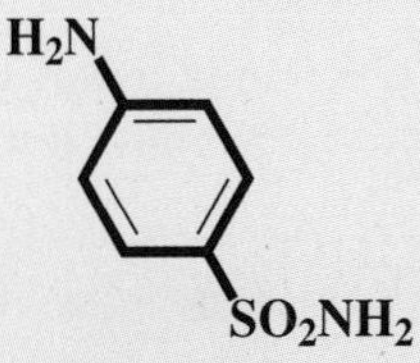

H_2N SO_2NH_2

Sulfanilamide

Interestingly, the Trefouels' observations made the patent on Prontosil filed by I.G.F. useless. Sulfanilamide had been synthesized and patented many years before as a dye intermediate, but the patent had expired by the time the substance was found to be a potent bactericide. Moreover, their findings led to clinical trials of sulfanilamide in France, England, and the United States, all of which were highly successful. One case that gave great publicity to the new drug was the use of Prontosil to save the life of Franklin D. Roosevelt, Jr., son of the president. In 1936, young Roosevelt was dying from a streptococcic infection when his mother convinced a doctor to administer Prontosil, which saved his life.

By 1947, over 5000 sulfonamides related to sulfanilamide had been prepared and tested for their efficacy as antibacterials. Although not all were effective, some were found to be better than sulfanilamide against certain infections. Of the thousands of compounds prepared and tested, the active ones are almost always those in which the only variation in structure is a change in the group of atoms attached to the nitrogen atom of the sulfonamido moiety.

We mentioned that Domagk was awarded the Nobel Prize in 1939. This is not quite correct. He was selected for the prize in that year but did not

(Continued)

HISTORICAL HIGHLIGHT *Discovery of Sulfa Drugs (Continued)*

actually receive it until many years later. When he received notice of the award in October 1939, Domagk sent a letter of acceptance. However, he sent a second letter to Stockholm in November declining the award. The second letter was the result of pressure from the Nazis: Domagk was by this time in the hands of the Gestapo. In 1947, Domagk was finally able to visit Stockholm, deliver the Nobel lecture, and receive the medal and diploma—but not the prize money, which by then had reverted to the Nobel Foundation.

Relationship of Historical Highlight to Experiments

The experiments performed in this chapter illustrate some of the various strategies that are the keys to successful multistep syntheses. In such processes, the product of one reaction serves as the starting material for the next, so it is necessary to develop efficient procedures that afford products of sufficient purity that they may be used directly in the next step with minimal purification. The preparation of the antibiotic sulfanilamide from the readily available starting material nitrobenzene (Sec. 21.2) represents the use of multistep synthesis to produce a biologically important antibacterial agent. That of 1-bromo-3-chloro-5-iodobenzene (Sec. 21.3) shows how a polysubstituted aromatic compound can be selectively formed in a multistep sequence. Finally, the synthesis of lidocaine (Sec. 21.4) illustrates how a common anesthetic agent can be prepared by a short sequence of reactions.

See more on *Sulfa Drugs*
See *Who was Domagk?*

Polymers

When you see this icon, sign in at this book's premium website at **www.cengage.com/login** to access videos, Pre-Lab Exercises, and other online resources.

See more on *Polymers*

See more on *Biopolymers*

Bigger is better? You may think so after performing experiments in this chapter because they will give you the opportunity to explore various aspects of the chemistry of **polymers,** molecules having molar masses in the tens of thousands. In addition to your everyday use of polymers in a variety of forms, including the fibers in the clothing you wear and the adhesive on Post-it™ notes that you stick on everything, you yourself, are constantly manufacturing **biopolymers,** in the form of hair, skin, DNA, and RNA, for example. Although in this chapter we can provide only a small sampling of the wealth of chemistry surrounding polymers, it should give you a sense of this fascinating branch of organic chemistry.

22.1 INTRODUCTION

See more on *Polysaccharides*

See more on *Proteins*

See more on *Nucleic Acids*

See more on *Terpenes*

See more on *Plastics*

See more on *Elastomers*

See more on *Fibers*

Polymers are a class of molecules characterized by their high molar masses, which range from the thousands to the hundreds of thousands, and by the presence of simple repeating structural units called **monomers.** Because of their large size, polymers are often referred to as **macromolecules.** A polymer comprised of a *single* recurring monomer, M, is termed a **homopolymer,** whereas one containing at least two structurally distinct monomeric units, M_1 and M_2, distributed at random in the molecule is called a **heteropolymer** or **copolymer.** These two classes of macromolecules are represented by **1** and **2,** respectively.

$$\underset{\text{Homopolymer}}{\underset{\mathbf{1}}{R—M—(M)_n—M—R}} \qquad \underset{\text{Copolymer}}{\underset{\mathbf{2}}{R—M_1—(M_1—M_2)_n—M_2—R}}$$

Polymers are found in nature and may also be produced by laboratory synthesis. Important examples of naturally occurring macromolecules, or **biopolymers,** are **proteins, polysaccharides, terpenes,** and **nucleic acids.** General representations of these substances are provided by structures **3–6,** respectively, in which the monomeric subunits of an α-amino acid, **3,** a pyranose, **4,** an isoprene, **5,** and a ribonucleotide phosphate, **6,** are seen. Synthetic, or man-made, polymers are represented by the myriad **plastics, elastomers,** and **fibers** that are commonplace in contemporary society.

Two primary methods are commonly used to convert monomers into synthetic polymers. In the older literature, these techniques are referred to as **addition** and **condensation polymerization,** but because of ambiguities in these terms, the preferred names now are **chain-reaction polymerization** and **step-growth**

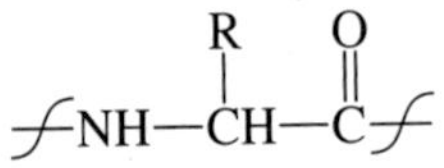

3 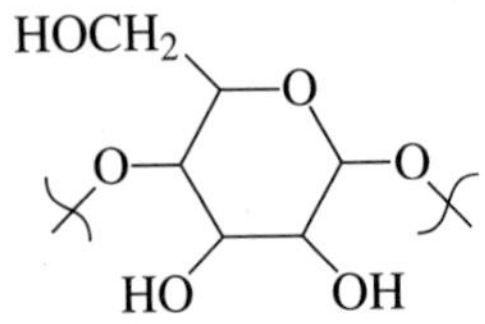 4 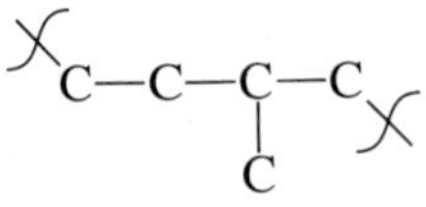 5 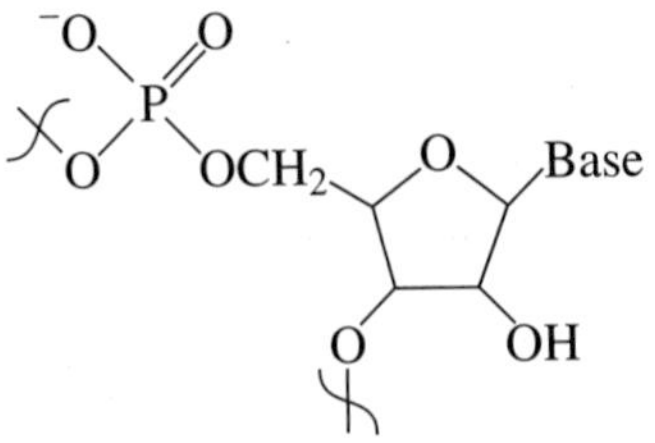 6

(Base = purine or pyrimidine)

See more on *Polymerization*

See more on *Chain-Reaction Polymerization*

See more on *Step-Growth Polymerization*

See more on *Polyethylene*

See more on *Polystyrene*

See more on *Teflon™*

See more on *Plexiglas™*

polymerization, respectively. As discussed in Sections 22.2 and 22.3, the major distinction between these two types of polymerization is the general mechanism by which the polymer forms.

The term *addition polymerization* arose because such polymers are produced by combining a large number of monomer molecules through addition reactions. For example, the self-addition of thousands of ethylene molecules yields polyethylene (**7**; see the Historical Highlight at the end of this chapter), a homopolymer (Eq. 22.1). Another homopolymer is polystyrene (**9**), which is formed by self-addition of styrene (**8**) as shown in Equation 22.2. We note that representing the molecular formula of the polymer as essentially n times that of the monomer, as shown in these two equations, is only a slight oversimplification of the actual formula. Because n is such a large number, inclusion of the elemental compositions of the end-groups, R, that appear at the termini of the polymeric chain makes an insignificant change in the molecular formula in comparison with the rest of the molecule.

$$n\ CH_2{=}CH_2 \xrightarrow[\text{and/or heat}]{\text{catalyst}} \left(CH_2{-}CH_2\right)_n \qquad (22.1)$$

Ethylene — **7**, $n = 10{,}000–30{,}000$, Polyethylene

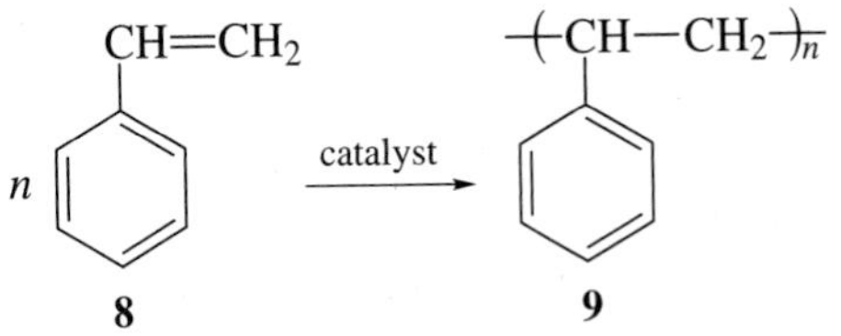

$$n\ C_6H_5CH{=}CH_2 \xrightarrow{\text{catalyst}} \left(CH(C_6H_5){-}CH_2\right)_n \qquad (22.2)$$

8 Styrene — **9** Polystyrene

Some other common addition polymers have trade names that do not indicate their structure. For example, Teflon™ and Plexiglas™ are homopolymers of tetrafluoroethylene (**10**), and methyl methacrylate (**11**), respectively.

$CF_2{=}CF_2$ — $CH_2{=}C(CO_2CH_3)CH_3$

10 Tetrafluoroethylene — **11** Methyl methacrylate

Copolymers are produced from a *mixture* of monomers, as noted earlier. For example, Saran™ (**14**), a widely used plastic film, is made by polymerizing a mixture of vinyl chloride (**12**) and vinylidene chloride (**13**), as depicted in Equation 22.3. You should be aware that the abbreviated formula of **14** is not meant to imply that the two monomeric units appear as a sequence of two distinct blocks, each of which individually represents a homopolymer. Although such **block copolymers** can be produced by special techniques, copolymers usually have the two monomers distributed randomly along the chain, as in **15**.

$$n\,CH_2{=}CH(Cl) + m\,CH_2{=}CCl_2 \longrightarrow \left[\left(CH_2{-}CH(Cl) \right)_n \left(CH_2{-}CCl_2 \right) \right]_m \qquad (22.3)$$

12 Vinyl chloride; **13** Vinylidene chloride; **14** Saran

$$R{-}M_1{-}(M_2{-}M_2{-}M_1{-}M_2{-}M_1{-}M_1{-}M_1)_n{-}M_2{-}R$$

15

The term *condensation polymerization* was originally used because this form of polymerization involves condensation reactions. Such transformations normally involve combining two functionalized molecules through an addition-elimination process, termed nucleophilic acyl substitution (Sec. 20.1), that results in the loss of a small molecule, H–L, such as water (Eq. 22.4). The aldol condensation (Sec. 18.3) and esterification (Sec. 20.2) are examples of such reactions. In the case of condensation or step-growth polymerization, *di*functionalized substrates are required as monomers, as illustrated in Equation 22.5.

$$R{-}C({=}O){-}L + HNu{:} \longrightarrow R{-}C({=}O){-}Nu{:} + HL \qquad (22.4)$$

$$L{-}C({=}O)\sim\sim C({=}O){-}L + H\ddot{N}u\sim\sim \ddot{N}uH \longrightarrow \left(C({=}O)\sim\sim C({=}O){-}\ddot{N}u\sim\sim \ddot{N}u \right)_n \qquad (22.5)$$

The following sections contain discussions of the mechanisms of chain-reaction and step-growth polymerization. The associated experimental procedures illustrate the preparation of polymers derived from each type of polymerization.

22.2 CHAIN-REACTION POLYMERIZATION

As the name implies, *chain-reaction polymerization* is a chain reaction in which the initiator may be a cation, anion, or free radical. An example of **cationic polymerization** is found in the polymerization of isobutylene (2-methylpropene) in the presence of protic or Lewis acid catalysts to give poly(isobutylene) (**16**), as depicted in Equation 22.6. The conversion of acrylonitrile to poly(acrylonitrile) (**17**) using sodium amide, a strong base, represents **anionic polymerization** (Eq. 22.7).

See more on *Poly(isobutylene)*

$$(CH_3)_2C{=}CH_2 \xrightarrow{H^+} (CH_3)_2\overset{+}{C}{-}CH_3 \xrightarrow{H_2C=C(CH_3)_2} (CH_3)_3C{-}CH_2\overset{+}{C}(CH_3)_2$$

Isobutylene

$$\xrightarrow{n\ H_2C=C(CH_3)_2} (CH_3)_3C{-}\left(CH_2C(CH_3)_2\right)_n{-}CH_2C{-}R \qquad (22.6)$$

16

Poly(isobutylene)

See more on *Poly(acrylonitrile)*

$$CH_2{=}CHCN \xrightarrow{H_2N^-} CH_2(NH_2){-}\overset{-}{C}HCN \xrightarrow{CH_2=CHCN} H_2NCH_2CH(CN){-}CH_2C^-(CN){-}H$$

Acrylonitrile

$$\xrightarrow{n\ CH_2=CHCN} H_2N{-}\left(CH_2CH(CN)\right)_n{-}CH_2CH(CN){-}R \qquad (22.7)$$

17

Poly(acrylonitrile)

See more on *Free Radicals*

Free-radical polymerization is a widely used method to induce chain-reaction polymerization, and its mechanistic course is parallel to that of the free-radical halogenation of hydrocarbons (Sec. 9.2). In the experiment in this section, you will perform the free-radical polymerization of styrene to give polystyrene (**24,** Eq. 22.12).

The reaction is started by the thermal decomposition of an *initiator,* which in our experiment is *tert*-butyl peroxybenzoate (**18**), a compound that produces the free radicals **19** and **20** when heated (Eq. 22.8). If In• represents one or both of these free radicals, the course of the polymerization may be illustrated as shown in Equations 22.9–22.12. Equation 22.9 indicates the function of the free radicals in **initiating** the polymerization. Equations 22.10a and 22.10b represent the **propagation** of the growing polymer chain. Equations 22.11 and 22.12 show possible **termination** processes. In Equation 22.11, the free-radical end of one growing polymer chain abstracts a hydrogen atom from the carbon atom next to the end of another polymer radical to produce the unsaturated and saturated polymer molecules **22** and **23,** respectively, in a process termed **disproportionation.** For the termination reaction illustrated by Equation 22.12, Rad• may be one of the initiating radicals, In•, or another growing polymer chain.

$$\underset{\substack{\textbf{18}\\ \textit{tert}\text{-Butyl peroxybenzoate}}}{(CH_3)_3C{-}O{-}O{-}C(=O){-}Ph} \xrightarrow{\text{heat}} \underset{\textbf{19}}{(CH_3)_3C{-}O^{\bullet}} + \underset{\textbf{20}}{{}^{\bullet}O_2C{-}Ph} \quad (22.8)$$

$$In^{\bullet} + \underset{\textbf{8}}{CH_2{=}CHC_6H_5} \longrightarrow In{-}CH_2{-}\dot{C}H(C_6H) \quad (22.9)$$

$$InCH_2{-}\dot{C}H(C_6H_5) + CH_2{=}CHC_6H_5 \longrightarrow \underset{\textbf{21}}{InCH_2CH(C_6H_5){-}CH_2\dot{C}H(C_6H)} \quad (22.10a)$$

$$\underset{\textbf{21}}{InCH_2CH(C_6H_5){-}CH_2\dot{C}H(C_6H_5)} + n\,CH_2{=}CHC_6H_5 \longrightarrow InCH_2CH(C_6H_5){-}(CH_2CH(C_6H_5))_n{-}CH_2\dot{C}H(C_6H_5) \quad (22.10b)$$

$$C_6H_5{-}\dot{C}H{-}CH(H)R + \dot{C}H(C_6H_5){-}CH_2R \longrightarrow \underset{\textbf{22}}{C_6H_5{-}CH{=}CHR} + \underset{\textbf{23}}{C_6H_5{-}CH(H){-}CH_2R} \quad (22.11)$$

$$R = {-}(CH_2CH(C_6H_5))_n{-}$$

$$InCH_2CH(C_6H_5){-}(CH_2CH(C_6H_5))_n{-}CH_2\dot{C}H(C_6H_5) + Rad^{\bullet} \longrightarrow \underset{\textbf{24}}{InCH_2CH(C_6H_5){-}(CH_2CH(C_6H_5))_n{-}CH_2CH(C_6H_5){-}Rad} \quad (22.12)$$

The commercially available styrene used in our experiments contains *tert*-butylcatechol (**25**), a phenol that stabilizes styrene by functioning as a **radical scavenger.** The catechol does this by donating a hydrogen atom to reactive free radicals to convert them into nonradical products as shown in Equation 22.13. The resulting phenoxy radical **26** is relatively unreactive as an initiator of a free-radical chain reaction. The stabilizer is necessary to prevent premature polymerization of styrene during storage or shipment because it is so readily polymerized by traces of substances such as atmospheric oxygen. Thus, the stabilizer must first be removed before polymerizing styrene.

$$(CH_3)_3C\text{-}C_6H_3(OH)\text{-}O\text{-}H + R^{\bullet} \longrightarrow (CH_3)_3C\text{-}C_6H_3(OH)\text{-}O^{\bullet} + RH \quad (22.13)$$

25 *tert*-Butycatechol — **26**

In the discovery experiment that follows, you will explore an important aspect of polymer chemistry by preparing polystyrene (**24**) under different reaction conditions and test whether it, like many polymers, may be produced in a variety of physical forms such as an amorphous solid, a film, and a clear glass.

EXPERIMENTAL PROCEDURES

Preparation of Polystyrene

Purpose To demonstrate the synthesis of polystyrene by free-radical polymerization under different conditions.

SAFETY ALERT

The free-radical initiator *tert*-butyl peroxybenzoate is a safe material to use in this experiment because it decomposes at a moderate rate when heated. Nonetheless, do not heat this catalyst excessively when performing the polymerization.

MINISCALE PROCEDURE

Preparation Sign in at **www.cengage.com/login** to answer Pre-Lab Exercises, access videos, and read the MSDSs for the chemicals used or produced in this procedure. Review Sections 2.9, 2.17, 2.19, 2.21, and 2.22.

Apparatus A separatory funnel, small soft-glass test tube, 25-mL round-bottom flask, microburner, apparatus for magnetic stirring, heating under reflux, and *flameless* heating.

A ■ *Removal of the Inhibitor from Commercial Styrene*

Place about 10 mL of commercial styrene in a small separatory funnel and add 4 mL of 3 *M* sodium hydroxide and 15 mL of water. Shake the mixture thoroughly, allow the layers to separate, and withdraw the aqueous layer. Wash the organic layer sequentially with two 8-mL portions of water, carefully separating the aqueous layers after each wash. Dry the styrene by pouring it into a small Erlenmeyer flask containing a little anhydrous calcium chloride and then swirling the flask. Allow the mixture to stand for 5–10 min, decant the liquid from the drying agent, and use the dried styrene in the following experiments.

Analysis Obtain IR and NMR spectra of styrene and compare them with those of an authentic sample (Figs. 22.1 and 22.2).

WRAPPING IT UP

Allow the volatiles to evaporate from the *calcium chloride* by placing it on a tray in the hood; then discard it in the container for nontoxic solids. Neutralize the *aqueous layers* before flushing them down the drain.

B ■ *Polymerization of Pure Styrene*

Place about 2–3 mL of dry styrene in a small soft-glass test tube, and add 2 or 3 drops of *tert*-butyl peroxybenzoate. Clamp the test tube in a *vertical* position over a wire gauze, insert a thermometer so that its bulb is in the liquid, and heat the styrene and catalyst with a *small* burner flame. When the temperature reaches 140 °C, temporarily remove the flame. If boiling stops, resume heating to maintain gentle boiling. The exothermicity of the polymerization increases the rate of formation of free radicals by thermal decomposition of the initiator, and this in turn increases the rate of polymerization. Thus be watchful for a rapid increase in the rate of boiling and remove the flame if the refluxing liquid nears the top of the test tube.

After the onset of polymerization, the temperature should rise to 180–190 °C, much above the boiling point of styrene. The viscosity of the liquid will increase rapidly during this time. As soon as the temperature begins to decrease, remove the thermometer and pour the polystyrene onto a watchglass. Do not touch the thermometer *before* the temperature decreases, because moving it in the boiling liquid might cause a sudden "bump," which could throw hot liquid out of the tube. Note the formation of fibers as the thermometer is pulled out of the polymer. The rate of solidification of the polystyrene depends on the amount of catalyst used, the temperature, and the length of time the mixture is heated.

C ■ *Solution Polymerization of Styrene*

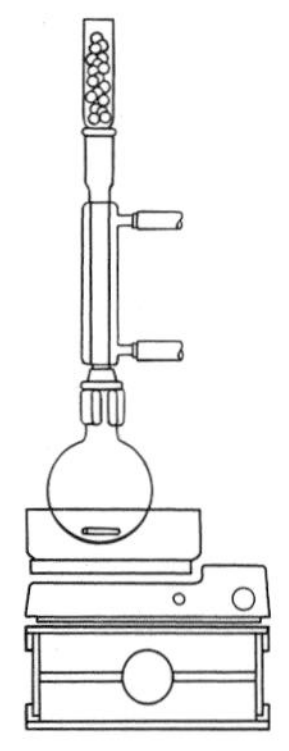

Place about 2 mL of dry styrene and 5 mL of xylene in a 25-mL round-bottom flask and add 7 drops of *tert*-butyl peroxybenzoate from a Pasteur pipet. Assemble the apparatus for heating under reflux and heat the mixture under reflux for 20 min. Cool the solution to room temperature and then pour about *half* of it into 25 mL of methanol. Collect the white precipitate of polystyrene that forms by decantation or by vacuum filtration if decantation is not practical. Resuspend the polystyrene in fresh methanol and stir it vigorously. Collect the polystyrene by filtration and allow it to dry in the hood.

Pour the remaining *half* of the polystyrene solution onto a watchglass or the bottom of a large inverted beaker and allow the solvent to evaporate. A clear film of polystyrene should form.

WRAPPING IT UP

Place the *filtrate* containing a mixture of xylene and methanol in the container for nonhalogenated organic liquids. Flush the *methanolic filtrate* obtained after resuspension of the polystyrene down the drain.

Optional Discovery Experiments

Two of the following experiments allow you to explore whether polystyrene is stable toward different organic solvents and to assess the interaction between an ionic polymer, sodium polyacrylate (**27**), and water. The third lets you investigate the change in the properties of a polymer when it is cross-linked with other strands of itself or other polymers.

Discovery Experiment

Stability of Polystyrene toward Organic Solvents

Styrofoam™ is a "puffed-up" form of **24** that is produced by polymerizing styrene in the presence of a "blowing agent" like pentane. The heat of the polymerization causes the agent to vaporize, and it is temporarily trapped in the polymerizing material, forming a bubble or cell if the viscosity of the material is high enough to prevent the gases from escaping. The final polymer has many of these cells, resulting in a foamlike material. The foam is a good semirigid shock absorber and also serves as an insulator because air is not very effective at transferring heat or cold. As a consequence of these physical properties, Styrofoam may be used in a variety of ways, ranging from packaging materials to ice chests.

Work in groups of at least two on this experiment and carefully observe what happens as you perform it. All of the results should be compiled and separately interpreted by each member of the team.

Working at the fume hood and away from any flames and hot plates, fill a petri dish about half-full with an organic solvent. Place a Styrofoam cup bottom-down in the solvent and record what happens. Use a glass stirring rod to explore what remains of the cup. Some solvents that might be tested are acetone, 95% ethanol, dichloromethane, and toluene. Properly dispose of all liquids and solids remaining at the completion of this experiment.

Discovery Experiment

Polymers and Water

Most polymers are water-insoluble even though they may contain polar, but neutral, substituents along the carbon backbone. Sodium polyacrylate, however, is a cross-linked ionic polymer that is water-soluble or, equivalently, is a solid polymer in which water is soluble. It is used in products as diverse as disposable diapers and as a replacement for plant soil. Just what happens when this polymer and water are combined is the subject of this experiment.

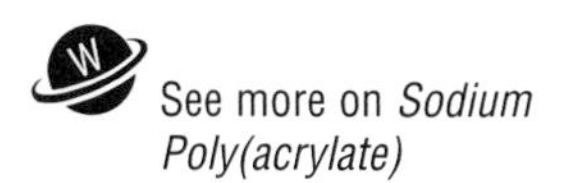

$$-\!\!\left(CH_2-\underset{\displaystyle CO_2^-\ Na^+}{\underset{|}{CH}}\right)_{\!n}\!\!-$$

Sodium poly(acrylate)

Weigh out approximately 0.5 g of sodium polyacrylate and transfer it to a *dry* 250-mL Erlenmeyer flask. Add 100 mL of water and immediately stir the mixture. Record your observations.

Discovery Experiment

See more on *Poly(vinyl alcohol)*

Cross-Linking of Polymers

Poly(vinyl alcohol), like sodium poly(vinyl acrylate), is another water-soluble polymer and is composed of repeating units of vinyl alcohol. In contrast to the acrylate, however, the solubility is not due to a charged polar functionality; rather, extensive hydrogen-bonding involving the hydroxyl groups accounts for the water solubility.

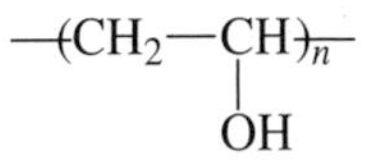

$$-\!\!\left(CH_2-\underset{\displaystyle OH}{\underset{|}{CH}}\right)_{\!n}\!\!-$$

Poly(vinyl alcohol)

When sodium tetraborate is dissolved in water, an equilibrium is established according to the equation below, resulting in the formation of a buffer having a pH of about 9. The protonated borate ion reacts with the hydroxyl groups of one strand of the poly(vinyl alcohol) and subsequently with the hydroxyl groups of another strand, possibly with the elimination of water. The resulting array may be represented in a general way as shown below. By joining strands of the polymer through such cross-links, a measure of rigidity is imparted to the molecular array that has been formed. In addition to the cross-links, the molecules have extensive intra- and intermolecular hydrogen-bonding, which also effects a form of cross-linking, albeit a weak one. Breaking and reforming of the hydrogen bonds presumably accounts for the viscoelastic properties of the material that is produced. This material can be formed into a ball, but you should see what happens when it is left untouched.

$$B_4O_7{}^{2-}{}_{(aq)} + H_2O \rightleftharpoons HB_4O_7{}^{-}{}_{(aq)} + HO^{-}{}_{(l)}$$

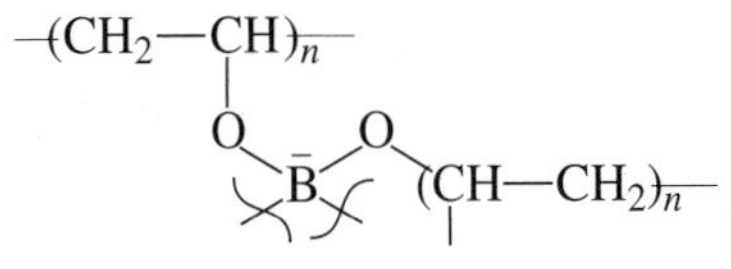

$$-\!(CH_2-CH)_n- \quad | \quad O-\overset{-}{B}-O-(CH-CH_2)_n-$$

Apparatus A 100-mL and a 250-mL beaker, thermometer, glass stirring rod, and apparatus for magnetic stirring and *flameless* heating.

Setting Up Add 50 mL of distilled water into the 250-mL beaker equipped with a stirbar. Heat the water with stirring but do *not* exceed a temperature of 90 °C.

Gel Formation With continued stirring and warming, slowly sprinkle 2 g of poly(vinyl alcohol) having an average molar mass of at least 10^5 g/mol onto the surface of the water; this procedure prevents the formation of a sticky mass of polymer that is difficult to dissolve. Combine the poly(vinyl alcohol) solution and 5 mL of a 4% (by mass) aqueous solution of sodium tetraborate, $Na_2B_4O_7$, in the 100-mL beaker, and stir the mixture vigorously with the glass rod. A material that you may consider to be "slime" should form almost immediately.

This cross-linked polymer is a gel that has interesting physical properties. You may explore some of them by forming your material into a ball and then seeing what happens when you carefully tip it in the palm of your hand. If a long column of polymer forms, jerk it abruptly and record the result. Be creative and test the properties of the gel in other ways!

Wrapping It Up Clean the *beaker* and *stirring rod* used to prepare the gel with soap and water. Mix the *solution of polymer* of polymer with a copious amount of water and flush the *mixture* down the drain. Discard the *gel* itself in the container for non-hazardous waste.

EXERCISES

1. *tert*-Butylcatechol (**25**) is capable of reacting with *two* equivalents of a radical, R•, to produce two moles of RH and a stable non radical oxidation product of **25**. Propose a structure for this product and write a stepwise reaction mechanism for its formation. Use curved "fish-hook" arrows to symbolize the flow of electrons.
2. The use of phenols such as *tert*-butylcatechol as free-radical scavengers is based on the fact that phenolic hydrogens are readily abstracted by radicals, producing relatively stable phenoxyl radicals that interrupt chain processes of oxidation and polymerization. Alcohols such as cyclohexanol, on the other hand, do *not* function as radical scavengers. Explain why the two types of molecules differ in their abilities to donate a hydrogen atom to a radical, R•.
3. Write an equation for the reaction involved in the removal of *tert*-butylcatechol from styrene by extraction with sodium hydroxide.
4. Why is it necessary to remove *tert*-butylcatechol from commercially available styrene prior to preparing polystyrene?
5. Why is *tert*-butyl peroxybenzoate a good radical initiator?
6. Explain why only a catalytic amount of the radical initiator is required in a free-radical-chain polymerization reaction.
7. Why is the polymerization of styrene an exothermic reaction? Explain in terms of a calculation based on the following equation using these bond dissociation energies: PhCH(R)–H, 83 kcal/mol; CH_2=CHPh, 53 kcal/mol (π-bond only); $PhCH_2CH_2$–$CH(CH_3)Ph$, 73 kcal/mol.

$$\underset{\displaystyle CH_3}{PhCH}\text{—H} + CH_2\text{=}\underset{\displaystyle Ph}{CH} \longrightarrow \underset{\displaystyle CH_3}{PhCH}\text{—}CH_2\underset{\displaystyle H}{CHPh}$$

8. Explain why polystyrene is soluble in xylene but insoluble in methanol.
9. What effect would using a smaller proportion of catalyst to styrene have on the average molar mass of polystyrene?
10. In principle, radicals could add to styrene at the carbon atom bearing the phenyl group rather than the other one, yet they do not. Explain the basis of this selectivity for the addition reaction.
11. Specify whether polystyrene is a condensation polymer, a homopolymer, a copolymer, or a block polymer.
12. Some monomers polymerize to produce a polymer having centers of chirality (Chap. 7). If there is no preference for one configuration over another, the

configuration of the centers will be random throughout the polymer. This type of polymer is called *atactic*. When the stereocenters are nonrandom, the polymer may be either *syndiotactic* or *isotactic* (see your lecture textbook for a definition of these terms).

a. Write a portion of polystyrene containing two monomeric units and circle any stereocenters that are present.

b. Would you expect the polystyrene generated in this experiment to be *atactic, syndiotactic,* or *isotactic?*

13. Circle the monomeric unit in the polysaccharide shown below.

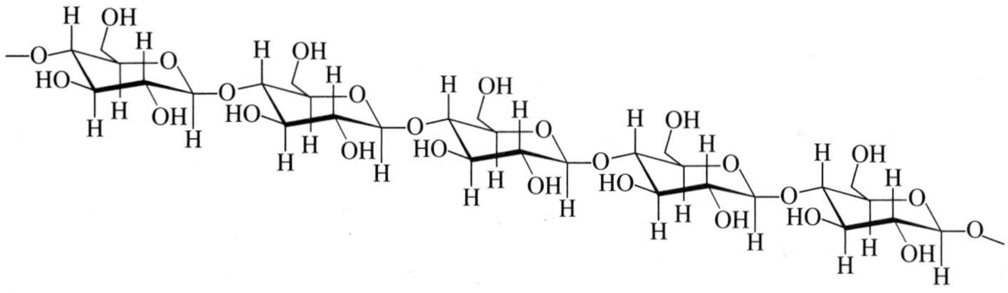

14. Teflon is produced from the polymerization of tetrafluoroethene. Write the structure of Teflon showing the monomeric unit in parentheses.

15. Why does the nucleophilic attack of isobutylene on the $(CH_3)_3C^+$ cation in Equation 22.6 form a new carbon-carbon bond at C(2) of isobutylene instead of C(1)?

16. Super Glue™ is a polymer formed via an anionic polymerization of methyl cyanoacrylate, $CH_2{=}(C{\equiv}N)CO_2CH_3$. Predict the structure of this glue.

Sign in at **www.cengage.com/login** and use the spectra viewer and Tables 8.1–8.8 as needed to answer the blue-numbered questions on spectroscopy.

17. Consider the spectral data for styrene (Figs. 22.1 and 22.2).

a. In the functional group region of the IR spectrum, specify the absorptions associated with the carbon-carbon double bond and the aromatic ring. Also specify the absorptions in the fingerprint region that are characteristic for the terminal vinyl group.

b. In the ^1H NMR spectrum, assign the various resonances to the hydrogen nuclei responsible for them.

c. For the ^{13}C NMR data, assign the various resonances to the carbon nuclei responsible for them.

18. Consider the IR spectrum of polystyrene (Fig. 8.10). In the functional group region of the spectrum, specify the range of absorptions associated with C–H stretching vibrations for the hydrogen atoms attached to the aromatic ring and for those of the methylene groups. Also specify the absorption associated with the carbon-carbon double bond of the aromatic rings.

19. Discuss the differences in the IR spectrum of styrene and polystyrene that are consistent with the loss of the vinyl group during the polymerization of styrene in this experiment. The IR spectrum of polystyrene is presented in Figure 8.10.

SPECTRA

Starting Material and Product

The IR spectrum of polystyrene is provided in Figure 8.10.

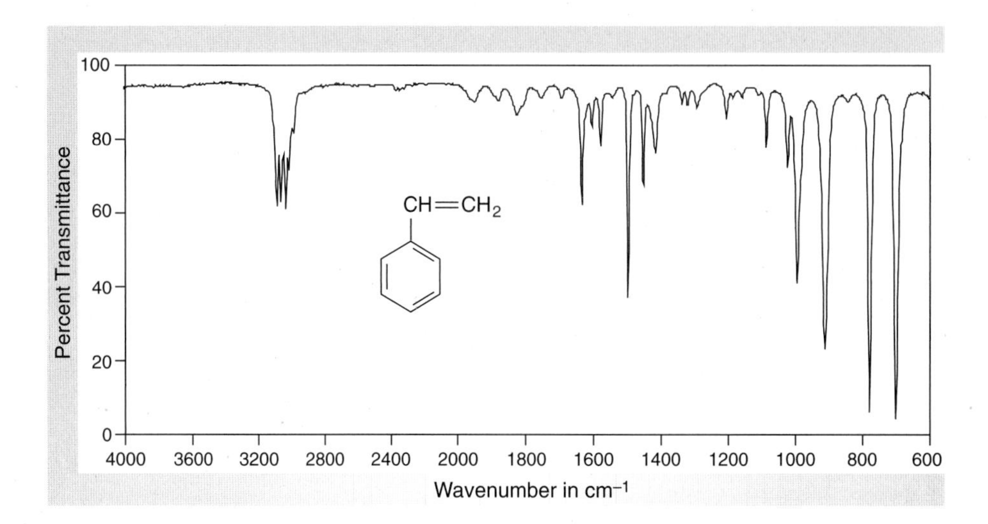

Figure 22.1
IR spectrum of styrene (neat).

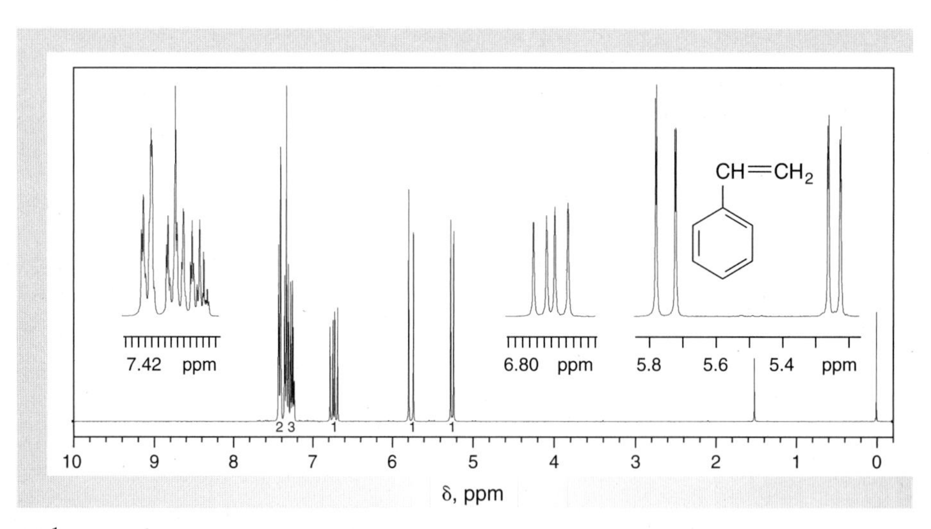

Figure 22.2
NMR data for styrene ($CDCl_3$).

(a) ^{1}H NMR spectrum (300 MHz).
(b) ^{13}C NMR data: δ 113.5, 126.2, 127.8, 128.5, 137.0, 137.7.

22.3 STEP-GROWTH POLYMERIZATION

Step-growth polymerization typically involves the reaction between two different *difunctionalized* monomers. Both functionalities of each monomer react and this leads to the formation of polymers. For example, a diacid such as terephthalic acid (**27**) can react with a diol such as ethylene glycol (**28**) in the presence of an acid catalyst to produce a polyester, as shown in Equation 22.14.

$$n\,HO_2C-C_6H_4-CO_2H \xrightarrow[n\,HOCH_2CH_2OH]{H^+} \tag{22.14}$$

27 Terephthalic acid

28 Ethylene glycol

$$H\!\left(\!O-\overset{O}{\overset{\|}{C}}-C_6H_4-\overset{O}{\overset{\|}{C}}-O-CH_2CH_2\!\right)_{\!n}OH + (2n-1)\,H_2O$$

A polyester

See more on *Step-Growth Polymerization*

See more on *Polyesters*

See more on *Nylon*

See more on *Fibers*

Chain growth is initiated by the acid-catalyzed reaction of a carboxyl group of the diacid with a hydroxy function of the diol to provide an ester and a molecule of water. The free carboxyl group or hydroxyl group of the resulting dimer then reacts with an appropriate functional group in another monomer or dimer, and the process is repeated by such *steps* until all of the monomers are converted into dimers, trimers, tetramers, and, eventually, polymers.

Step-growth polymerization processes are much slower than chain-reaction processes. Because they also typically have higher activation enthalpies, heating is often required to attain satisfactory rates of polymerization. Finally, step-growth polymers generally have lower average molar masses than polymers formed by chain-reaction polymerization.

Polyamides are one type of useful polymer that is produced by a step-growth process, and a variety of such polymers are preparable from various diacids and diamines. Nylon-6,6™ was the first commercially successful polyamide and is derived from the two monomers, hexanedioic acid (**29**) and 1,6-hexanediamine (**30**), as shown in Equation 22.15. Its trade name reflects the presence of six carbon atoms in each of the monomeric units that comprise the polymer. Of interest regarding the commercial importance of the nylons is the fact that the tremendous financial success enjoyed by E. I. Du Pont and Company from these types of polymers stems from the firm's patent on a method to draw the substance into fibers rather than from a patent on the molecular composition of the polymer itself. (See the Historical Highlight at the end of this chapter.)

$$n\,HO-\overset{O}{\overset{\|}{C}}(CH_2)_4\overset{O}{\overset{\|}{C}}-OH + n\,H_2N(CH_2)_6NH_2 \longrightarrow$$

29 Hexanedioic acid

30 1,6-Hexanediamine

$$H\!\left(\!O-\overset{O}{\overset{\|}{C}}(CH_2)_4\overset{O}{\overset{\|}{C}}-HN(CH_2)_6NH\!\right)_{\!n}H + (2n-1)\,H_2O \tag{22.15}$$

Nylon-6,6

In the typical industrial process for preparing polyamides, equimolar amounts of the diacid and diamine are mixed to give a salt that is then heated to high temperature under vacuum to eliminate the water. The resulting polymer has a molar mass of about 10,000 and a melting point of about 250 °C. Fibers may be spun from melted polymer, and if the fibers are stretched to several times their original length, they become very strong. This "cold-drawing" orients the polymer molecules parallel to one another so that hydrogen bonds form between C–O and N–H groups on adjacent polymer chains, as shown in **31**, greatly increasing the strength of the fibers. The strength of the fibers of silk, a well-known biopolymer involving protein molecules, is ascribed to the same stabilizing factor.

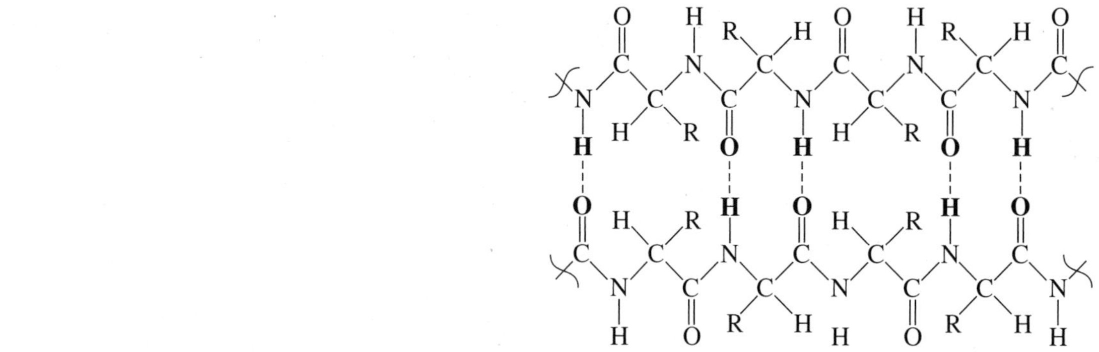

31

The preparation of Nylon-6,10™, rather than Nylon-6,6, has been chosen to illustrate step-growth polymerization for the present experiment. This polyamide is commercially produced from decanedioic acid (**32**) and 1,6-hexanediamine (**30**), as shown in Equation 22.16. To facilitate forming the polyamide under simple laboratory conditions, however, the diacid dichloride of **32** is used because it is more reactive toward diamine **30.** Using the diacid dichloride means that the small molecule eliminated in this step-growth polymerization is hydrogen chloride rather than water (Eq. 22.17). Sodium carbonate is added to neutralize the acid formed to prevent consumption of the expensive diamine via an acid-base reaction. If the base were not added, an excess of diamine would be required for complete polymerization.

$$n\,\mathrm{HO{-}C(=O)(CH_2)_8C(=O){-}OH} + n\,\mathrm{H_2N(CH_2)_6NH_2} \longrightarrow \mathrm{H{-}\!\left(O{-}C(=O)(CH_2)_8C(=O){-}HN(CH_2)_6NH\right)_{\!n}\!{-}H} + (2n-1)\,\mathrm{H_2O} \quad (22.16)$$

32 Decanedioic acid; **30**; Nylon-6,10

$$\mathrm{ClC(=O)(CH_2)_8C(=O)Cl} + \mathrm{H_2NCH_2{-}R} \longrightarrow \mathrm{ClC(=O)(CH_2)_8C(=O){-}NHCH_2{-}R} \quad (22.17)$$

Decanedioyl dichloride (Sebacoyl chloride)

The reactivity of the diacid chloride toward nucleophilic acyl substitution (Eq. 22.17) allows this polymerization to be performed under mild conditions. When a solution of the diacid chloride in a water-immiscible solvent is brought into contact

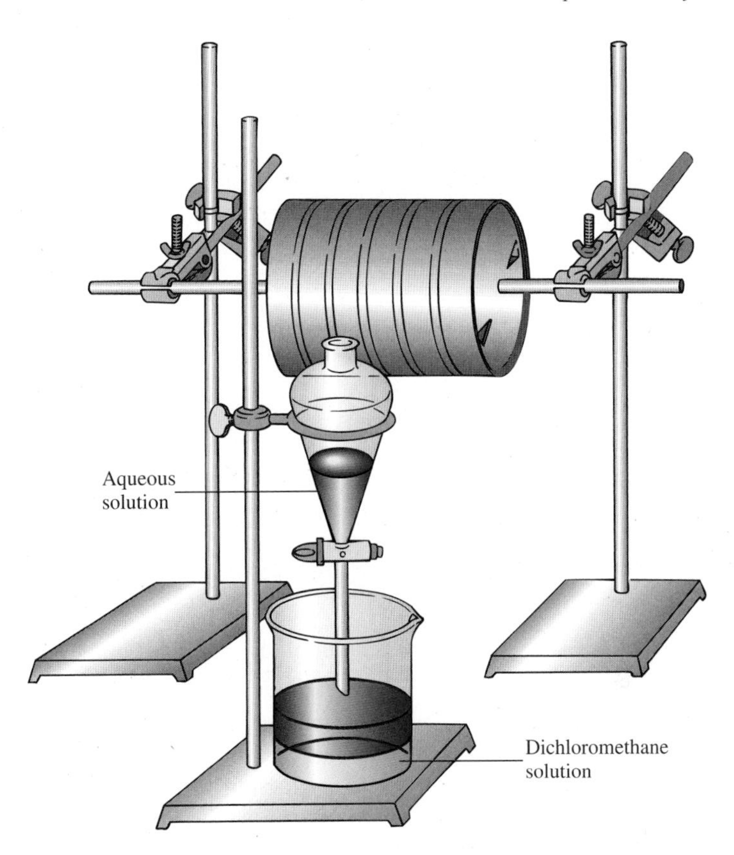

Figure 22.3
Apparatus for the "Nylon Rope Trick."

with an aqueous solution of the aliphatic diamine, a film of polymer of high molar mass instantly forms at the interface between the two solutions. The film is thin but strong, and can be pulled from the interface, where it is immediately and continuously replaced by additional polymer. In this way a long cord or rope of polyamide can be produced, much as a magician pulls a string of silk handkerchiefs out of a top hat. When this experiment was first described by two Du Pont chemists, they characterized it as the "Nylon Rope Trick." It does seem to be almost magic that a polymer can attain an average molar mass in the range 5000–20,000 in a fraction of a second!

To perform this experiment properly, the necessary equipment must be assembled so the polymer rope can be pulled from the reaction zone rapidly. A convenient way to do this is illustrated in Figure 22.3. A can, preferably with a diameter of 10 cm or more, makes a good drum on which to wind the polymer. After puncturing the can in the center of each end, a wooden or metal rod is passed through the center holes to make an axle for the drum. The rod is supported horizontally by clamps attached to ring stands in the usual way.

The circumference of the drum should be measured and a reference point should be marked on the drum so that an estimate of the length of the nylon rope can be made by counting the revolutions made as the rope is produced. A length of 6 m or more can usually be obtained with the procedure described here.

If instructed to do so, a much simpler procedure may be used to draw a polymeric fiber, but it will not allow you to determine the length of the fiber easily. Thus, you may dip a glass stirring rod or copper wire into the polymerizing mixture, lift the rod or wire out, and wind the fiber around it as the polymerization proceeds.

The dry polymer obtained in this experiment does not appear to have the properties expected of a nylon; it is fragile and of low density. However, fibers produced from it are much more dense and have the appearance and strength more characteristic of a typical polyamide. You will make fibers and films from the dry polymer so you can compare the properties of the different forms of the polymer.

EXPERIMENTAL PROCEDURES

Preparation of Nylon-6,10

Purpose To demonstrate step-growth polymerization by the synthesis of Nylon-6,10.

SAFETY ALERT

1. **If Pasteur pipets are used instead of syringes to measure the reactants, use a rubber bulb to draw up the liquid.**
2. **Do not handle the polymer rope with your bare hands more than is necessary until it has been washed free of solvent and reagents. Use latex gloves, tongs, or forceps to manipulate it. If you touch the crude polymer, immediately wash your hands with soap and warm water.**
3. **If you use formic acid to form a film, do not let it get on your skin, because it causes deep skin burns that are not immediately obvious. If the acid does accidentally come in contact with your skin, wash the affected area immediately with 5% sodium bicarbonate solution and then with copious amounts of water.**

MINISCALE PROCEDURE

Preparation Sign in at **www.cengage.com/login** to answer Pre-Lab Exercises, access videos, and read the MSDSs for the chemicals used or produced in this procedure. Discuss the experiment with your lab partner.

Apparatus A 250-mL beaker, a 5-mL syringe, separatory funnel, apparatus for the "Nylon Rope Trick."

Setting Up Measure 2 mL of decanedioyl dichloride into a 250-mL beaker using the syringe. *The size of the beaker is important:* In smaller beakers, the polymer tends to stick to the walls, whereas in larger beakers, poor "ropes" are obtained unless larger amounts of reagents are used. Dissolve the decanedioyl dichloride in 100 mL of dichloromethane. Place 1.1 g of crystalline 1,6-hexanediamine or 1.3 mL of its commercially available 80–95% aqueous solution in a separatory funnel, and add 2.0 g of sodium carbonate and 50 mL of water. Gently shake the mixture to dissolve both reactants. Arrange the drum onto which the polymer is to be wound at a height such that the beaker containing the decanedioyl dichloride solution can be placed on the lab bench about 40 cm beneath and slightly in front of the drum (Fig. 22.3). Support the separatory funnel containing the other reagents so the lower tip of the funnel is centered no more than a centimeter above the surface of the dichloromethane solution of the decanedioyl dichloride.

Reaction Open the stopcock of the separatory funnel slightly so the aqueous solution runs *slowly and gently* onto the surface of the organic solution. A film of polymer will form immediately at the interface of the two solutions. Use long forceps or tongs to grasp the *center* of the polymer film and pull the rope that forms up to the front of the drum, loop it over the drum, and rotate the drum so as to wind the rope onto the drum. For the first turn or two it may be necessary to use your fingers to secure the rope to the drum. Continue to rotate the drum and rapidly wind the nylon rope onto the drum until the reactants are consumed, remembering to count the revolutions of the drum as you wind.

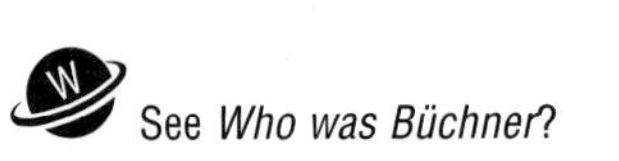

See *Who was Büchner?*

Work-Up and Isolation Replace the beaker with a large dish or pan containing about 200 mL of 50% aqueous ethanol and unwind the nylon rope into the wash solution. After stirring the mixture gently, decant the wash solution, and transfer the polymer to a filter on a Büchner funnel. Press the polymer as dry as possible, and then place it in your desk until the next laboratory period. When the nylon is thoroughly dry, weigh it and calculate the yield. Note how the bulk of polymer is affected upon drying.

Film Formation To produce a film of Nylon-6,10, dissolve the dry polymer in about 10 times its weight of 90–100% formic acid (*Caution:* See item 3 of the Safety Alert) by stirring the mixture at *room temperature;* heating to achieve dissolution degrades the polymer. Spread the viscous solution on a glass plate. Leave the plate *in a hood* until the next laboratory period to allow evaporation of the formic acid.

Alternative Procedure

Apparatus A 100-mL beaker, 10-mL graduated cylinder, Pasteur pipet, glass rod or 15-cm length of copper wire.

Setting Up Pour 10 mL of a 5% aqueous solution of 1,6-hexanediamine into the beaker and add 10 drops of 20% aqueous sodium hydroxide solution. Stir the solution to ensure homogeneity.

Reaction Carefully pour 10 mL of a 5% solution of decanedioyl dichloride into the beaker to produce a biphasic mixture. Touch the tip of the glass rod or end of the copper wire to the interface between the two layers and gently remove the rod or wire from the mixture, pulling the fiber of polymer along with it. Twist the rod or wire to spin the fiber around it.

Work-Up and Isolation Follow the procedure described for the "Nylon Rope Trick."

Fiber Formation Form fibers by carefully melting the dry polymer in a metal spoon or spatula with gentle heating over a very small burner flame or a hot plate, and then drawing fibers from the melt with a small glass rod. If necessary, combine your polymer with that of several students to provide enough material to be melted and drawn successfully. Do not heat the polymer much above the melting temperature because it becomes discolored and charred.

Obtain IR and NMR spectra of your starting materials and compare them to those of authentic specimens (Figs. 22.4–22.7).

WRAPPING IT UP

After the rope has been drawn, stir the remaining reaction mixture thoroughly until no more *polymer* forms. Isolate any additional polymer and, after thoroughly washing it with water, put it in the container for nonhazardous organic solids. Separate the *dichloromethane and aqueous layers* of the reaction mixture. Pour the *dichloromethane* into the container for halogenated organic liquids. Flush the *aqueous layer and all aqueous solutions* down the drain.

EXERCISES

1. Explain why the preparation of Nylon-6,10 occurs under milder conditions when decanedioyl dichloride is used instead of decanedioic acid.
2. Using curved arrows to symbolize the flow of electrons, write the stepwise mechanism for the condensation reaction between decandioyl dichloride and 1,6-hexanediamine.
3. Write an equation for the formation of the salt produced from one molecule of hexanedioic acid and two molecules of 1,6-hexanediamine.
4. Why is sodium carbonate used in the reaction to prepare Nylon-6,10?
5. Explain the large decrease in the bulk of the rope of Nylon-6,10 upon drying.
6. Using full structural formulas, draw a typical portion of a Nylon-6,6 molecule; that is, expand a portion of the formula given in Equation 22.15. Show at least two hexanedioic acid units and two 1,6-hexanediamine units.
7. Draw formulas that illustrate the hydrogen bonding that may exist between two polyamide molecules after fibers have been "cold-drawn."
8. Nylons undergo *de*polymerization when heated in aqueous acid. Propose a reaction mechanism that accounts for this fact, using curved arrows to symbolize the flow of electrons.
9. Nylon-6 is produced from caprolactam by adding a small amount of aqueous base and then heating the mixture to about 270 °C.

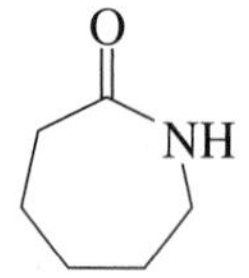

O
NH

Caprolactam

   a. Draw a representative portion of the polyamide molecule.
   b. Suggest a mechanism for the polymerization, using curved arrows to symbolize the flow of electrons, and indicate whether it is of the chain-reaction or step-growth type.

10. Would you expect polyesters to be stabilized by hydrogen bonding? Explain.

11. Proteins are polyamides formed from α-amino acids.

R
H_2N OH
O

An α-amino acid

a. Write a partial structure of a protein by drawing three monomeric units of it.

b. Are proteins *chain-reaction* or *step-growth* polymers?

Sign in at **www.cengage.com/login** and use the spectra viewer and Tables 8.1–8.8 as needed to answer the blue-numbered questions on spectroscopy.

12. Consider the spectral data for decanedioyl dichloride (Figs. 22.4 and 22.5).

a. In the functional group region of the IR spectrum, specify the absorptions associated with the carbonyl component of the acid chloride.

b. In the ^{1}H NMR spectrum, assign the various resonances to the hydrogen nuclei responsible for them.

c. For the ^{13}C NMR data, assign the various resonances to the carbon nuclei responsible for them.

13. Consider the spectral data for 1,6-hexanediamine (Figs. 22.6 and 22.7).

a. In the IR spectrum, specify the absorptions associated with the amino group.

b. In the ^{1}H NMR spectrum, assign the various resonances to the hydrogen nuclei responsible for them.

c. For the ^{13}C NMR data, assign the various resonances to the carbon nuclei responsible for them.

SPECTRA

Starting Materials and Products

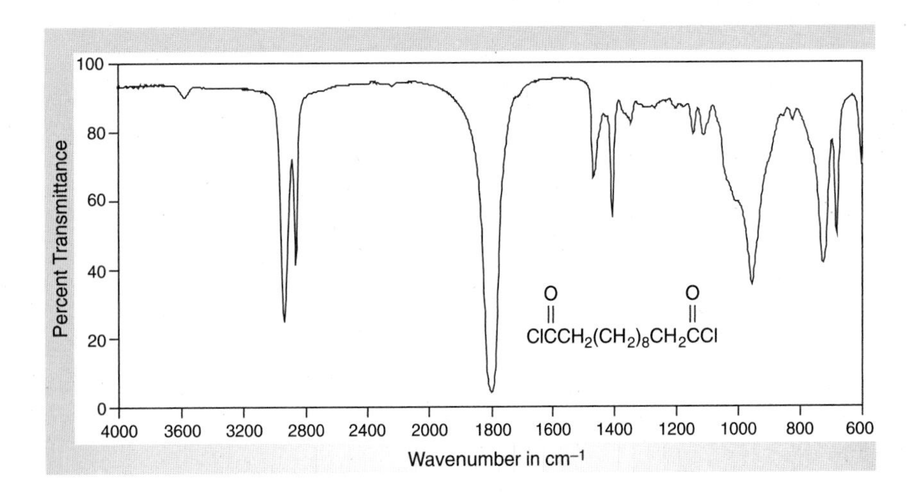

Figure 22.4
IR spectrum of decanedioyl dichloride (neat).

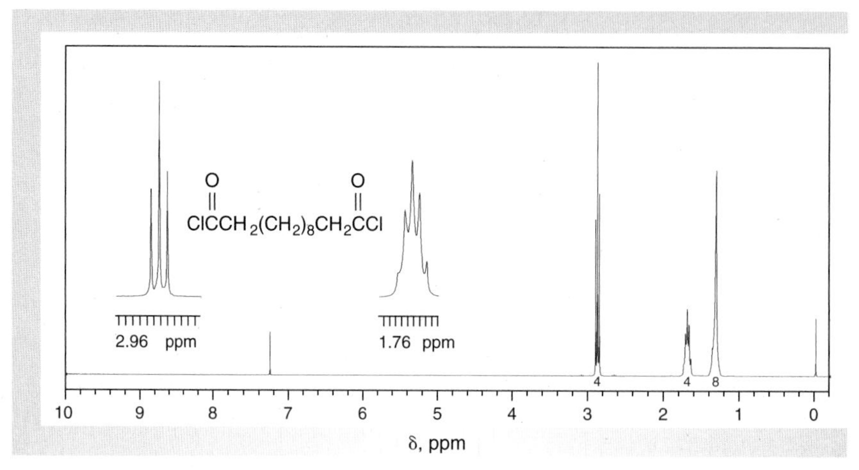

Figure 22.5
NMR data for decanedioyl dichloride ($CDCl_3$).

(a) ^1H NMR spectrum (300 MHz).
(b) ^{13}C NMR data: δ 24.7, 27.9, 28.4, 46.7, 173.2.

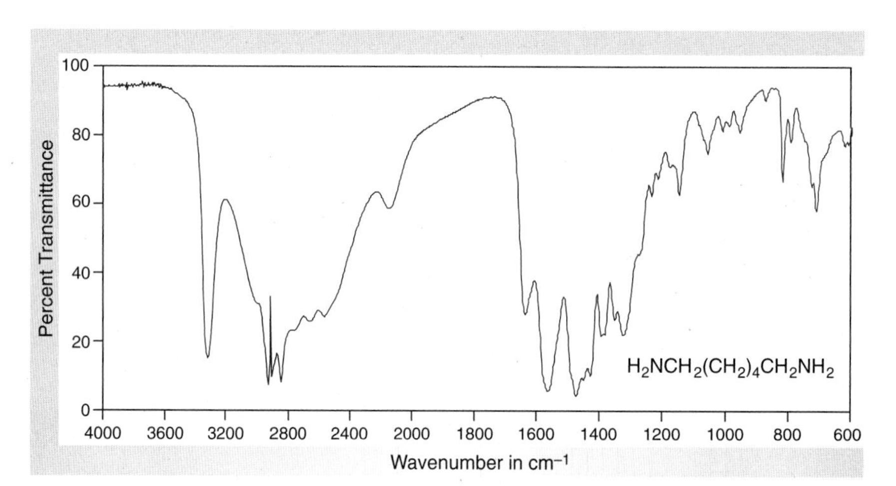

Figure 22.6
IR spectrum of 1,6-hexanediamine (neat).

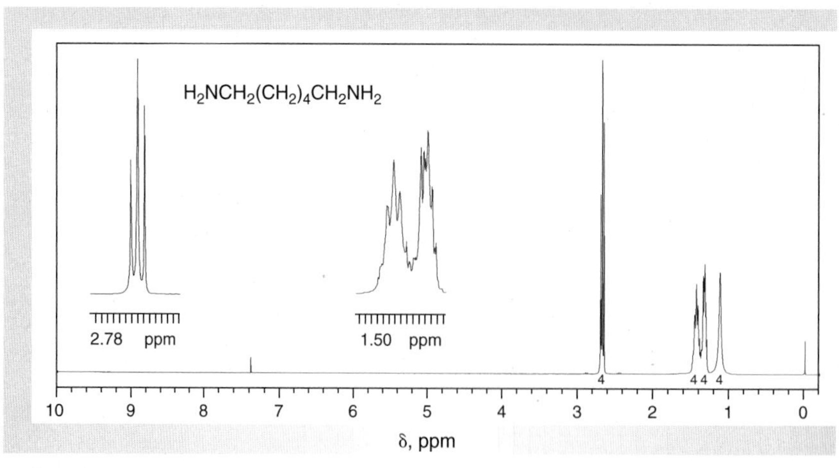

Figure 22.7
NMR data for 1,6-hexanediamine ($CDCl_3$).

(a) ^1H NMR spectrum (300 MHz).
(b) ^{13}C NMR data: δ 26.0, 33.0, 41.4.

HISTORICAL HIGHLIGHT

Discovery of Polyethylene and Nylon

Polyethylene, or "polythene," as its British inventors called it, and the nylons are two types of polymers that have particularly interesting histories of discovery that bear repeating here. Both stories illustrate the role of serendipity in scientific achievements of great importance.

Polyethylene was discovered accidentally by British chemists at Imperial Chemicals Industries (I.C.I.) as an unexpected result of experiments on chemical reactions at very high pressures. In 1933, a reaction of benzaldehyde and ethylene at 170 °C and 1400 atmospheres gave no adducts involving the two reagents and was considered a complete failure. However, an observant chemist noticed a thin layer of "white waxy solid" on the walls of the reaction vessel used for the experiment. This was recognized as a polymer of ethylene, but additional experiments with ethylene alone to produce the same polymer only resulted in violent decompositions that destroyed the equipment.

Two years elapsed before better and stronger equipment was available for further experimentation. When ethylene was heated to 180 °C in this new equipment, the pressure in the apparatus dropped unexpectedly, so more ethylene was pumped in. Then, when the reaction vessel was opened, the I.C.I. chemists found a large amount of white powdery solid, which was the long-sought polyethylene. Because they knew that the polymerization could not account for all of the pressure drop that had been observed, they suspected a leak in one of the joints of the apparatus. This idea led to the proposal that the polymerization had been catalyzed by oxygen in the air that had leaked into the apparatus, and this hypothesis was confirmed by experiments in which air was intentionally included with the ethylene. Oxygen can act as a radical initiator and catalyze the polymerization by a chain-reaction mechanism analogous to Equations 22.7–22.10.

The polyethylene produced by the oxygen-catalyzed, high-pressure, high-temperature process developed by I.C.I. in the mid-1930s was ideal for many applications, including insulation of radar equipment, where it was used to great advantage by the Allies in World War II. Referring to the contribution radar made to naval operations, the British Commander-in-Chief said it enabled the Home Fleet to "find, fix, fight, and finish the *Scharnhorst* (the pride of Hitler's navy)."

The group of polymers called nylons was first produced in 1939 as a material for women's hose and other garments and was the world's first totally synthetic fiber. Nylon hose were an instant hit with the buying public: 800,000 pairs were sold on May 15, 1940, the first day they were on the market! With the onset of World War II and the involvement of the United States, by 1941, nylon was taken off the domestic market because it was found to be the best available material for military parachutes.

The first nylon to be produced industrially was Nylon-6,6. The remarkable fact about the discovery of this polymer is not how it was first prepared from the two monomers, but how it was first prepared in a form suitable for a textile fiber. This depended on invention of the "cold-drawing process," and this technique was discovered almost completely by accident, as we shall see.

Wallace Hume Carothers was brought to Du Pont to direct its new basic chemical research program because his faculty mentors at the University of Illinois, where he earned his Ph.D., and his colleagues at Harvard University, where he served on the faculty, recommended him as the most brilliant chemist they knew. An incentive for his move from academia to industry in 1927 might well have been the fact that his salary was doubled—to all of $6000 per year. Carothers initiated a program aimed at understanding the composition of the high polymers of nature, such as cellulose, silk, and rubber, and producing synthetic materials like them. By 1934 his group had contributed valuable fundamental knowledge in these areas, but Carothers had just about decided that their effort to produce a synthetic fiber like silk was a failure. It was a shrewd observation made during some "horseplay" among Carothers's chemists in the laboratory that turned this failure to compete with nature into the enormous success ultimately advertised at the 1939 New York World's Fair as "Nylon, the Synthetic Silk Made from Coal, Air, and Water!"

The Carothers group had learned how to make Nylon-6,6, but even though this polyamide had a molecular structure similar to that of silk, they had

(Continued)

HISTORICAL HIGHLIGHT *Discovery of Polyethylene and Nylon (Continued)*

"put it on the back shelf" without patent protection because the polymer did not have the tensile strength of silk, a necessary criterion for a good textile fiber. The group continued its research by investigating a series of polyester polymers that were more soluble, easier to handle, and thus simpler to work with in the laboratory. It was while working with one of these softer materials that Julian Hill noted that if he gathered a small ball of such a polymer on the end of a glass stirring rod and drew a thread out of the mass, the thread of polymer so produced became very silky in appearance. This attracted his attention and that of the others working with him, and it is reported that one day while Carothers was in downtown Wilmington, Hill and his cohorts tried to see how far they could stretch one of these samples. One chemist put a little ball of the polymer on a stirring rod, and a second chemist touched a glass rod to the polymer ball and then ran down the hall to see how far he could stretch the thread of polymer. While doing this, they noticed not only the silky appearance of the extended strands but also their increased strength. They soon realized that this additional strength might result from some special orientation of the polymer molecules produced by the stretching procedure.

Because the polyesters they were working with at that time had melting points too low for use in textile products, a deficiency that has since been removed, the researchers returned to the polyamides (nylons) that had earlier been put aside. They soon found that these polymers, too, could be "cold-drawn" to increase their tensile strength so much that they could be made into excellent textiles. Filaments, gears, and other molded objects could also be made from the strong polymer produced by cold-drawing.

The alignment of the long polyamide molecules in a manner that produces extensive intermolecular hydrogen-bonding (**31**, Sec. 22.3) binds the individual polymer molecules together in much the same way that separate strands in a rope, when twisted together, form a cable. This association of linear polymer molecules through hydrogen bonding is responsible for the greatly increased strength of the nylon fibers. We believe that the same principle accounts for the strength of silk fibers; the natural polyamide molecules of silk are oriented in such a way that hydrogen bonds hold the individual molecules together. Interestingly, silkworms accomplish the equivalent of "cold-drawing" as they extrude the viscous silk filaments to produce cocoons!

Relationship of Historical Highlight to Experiments

The experiments in this chapter represent examples of chain-reaction polymerization to produce polystyrene (Sec. 22.2) and step-growth polymerization (Sec. 22.3) to yield a nylon. The procedures of Section 22.2 provide an opportunity to observe formation of the same polymer in three physically different forms, whereas that of Section 22.3 illustrates how strong hydrogen-bonded fibers can result from the "cold-drawing" technique patented by Du Pont.

See more on *Polyethylene*

See more on *Nylon*

See more on *Sharnhorst*

See *Who was Carothers?*

See *Who was Hill?*

See more on *Silk*

Carbohydrates

When you see this icon, sign in at this book's premium website at **www.cengage.com/login** to access videos, Pre-Lab Exercises, and other online resources.

See more on *Carbohydrates*

See more on *Cell Signaling*

See more on *AIDS*

Carbohydrates are one of the important classes of organic biomolecules that are essential to life as we know it and are found in every living organism. For example, carbohydrates comprise an essential part of the backbone of the nucleic acids that carry the genetic information of living cells. The sugar and starch in the food you eat are carbohydrates, as is the cellulose in wood and paper. Modified carbohydrates, especially those linked to proteins, are found in the membranes on the surfaces of many cells, and they play an important role in mediating interactions between cells in a number of signaling processes. Attractions between carbohydrate groups on the surface of white blood cells, or leukocytes, and carbohydrates on the surface of an injured cell are involved in the inflammatory response, and entry of the AIDS virus into a target cell depends upon interactions between carbohydrates on the viral coat and the cell receptor.

In this chapter we'll introduce you to some of the rich chemistry of carbohydrates. For example, you may already know that invert sugar is sweeter than sucrose, or table sugar, but do you know how it's formed? In one of the experiments in this chapter, you will learn how to make invert sugar. In others, you will learn about some of the chemical properties of simple carbohydrates.

23.1 INTRODUCTION

See more on *Monosaccharides*

The term **carbohydrate** originates from the fact that many, but not all, members of this class have the general molecular formula $C_nH_{2n}O_n$ and thus were once considered to be "hydrates of carbon." Examination of the structures of carbohydrates readily reveals that this view is inaccurate, but use of the term persists. The simplest carbohydrates are either polyhydroxy aldehydes, which have the general structure **1** and are referred to as **aldoses,** or polyhydroxy ketones, which have the general structure **2** and are called **ketoses.** Such simple carbohydrates are termed **monosaccharides.**

1 An aldose	**2** A ketose
CHO	CH_2OH
$(CHOH)_n$	$C{=}O$
CH_2OH	$(CHOH)_n$
	CH_2OH

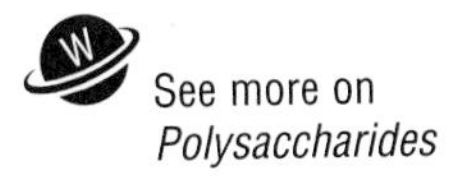

Two monosaccharides may be combined with the loss of one molecule of water to form a **disaccharide. Polysaccharides** and other complex carbohydrates are then produced by the condensation of more monosaccharide subunits with the loss of a molecule of water for each additional monosaccharide. Depending upon its constitution, hydrolysis of a polysaccharide yields either a single monosaccharide or a mixture of monosaccharides. In this chapter, some of the fundamental chemical and physical properties of carbohydrates will be investigated using simple mono- and disaccharides; several classic qualitative tests to characterize and classify carbohydrates will also be performed.

23.2 MONOSACCHARIDES: GENERAL PRINCIPLES

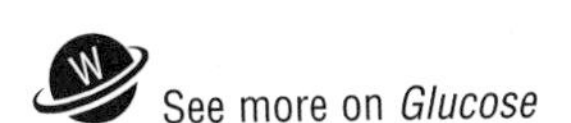

The monosaccharide D-glucose, whose chemistry is representative of all aldoses containing four or more carbon atoms, exists predominantly in the two **pyranose** forms **4** and **5**. These are six-membered **hemiacetals** formed by the reversible cyclization of the acyclic polyhydroxy aldehyde **3** (Eq. 23.1). In the cyclic forms **4** and **5**, the ring carbon that is derived from the carbonyl group is referred to as the **anomeric carbon** atom. The **specific rotation,** $[\alpha]_D^{25}$ (Sec. 7.5), of α-D-(+)-glucose (**4**) is +112° whereas that of the β-anomer **5** is +19°. When crystals of either pure **4** or pure **5** are dissolved in water, the $[\alpha]_D^{25}$ changes to an equilibrium value of +52.7°. This process is termed **mutarotation.** At equilibrium in water, the α- and β-forms are present in the ratio of 36:64; only about 0.03% of D-glucose is in the acyclic form **3**.

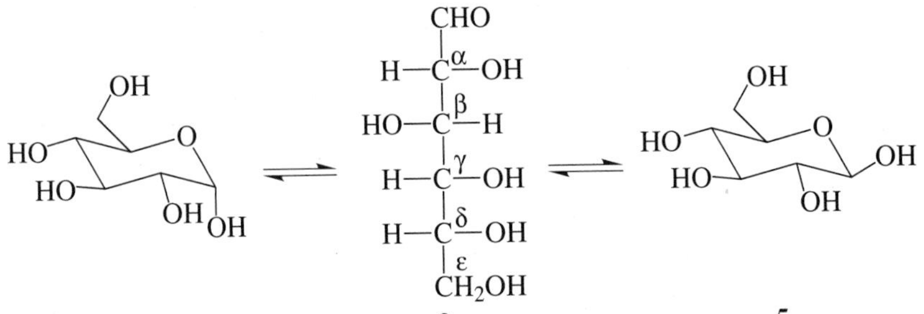

CHO
H—C(α)—OH
HO—C(β)—H
H—C(γ)—OH
H—C(δ)—OH
(ε) CH_2OH

(23.1)

4 α-D-Glucopyranose ⇌ **3** D-Glucose ⇌ **5** β-D-Glucopyranose

Another common monosaccharide is D-fructose (**6**), a ketose. In aqueous solution, D-fructose also undergoes mutarotation to produce a complex equilibrium mixture of the acyclic form **6** (< 1%), the five-membered hemiacetals **7** (31%) and **8** (9%), which are called **furanoses,** and the cyclic pyranoses **9** (57%) and **10** (3%); this mixture exhibits an $[\alpha]_D^{25} = -92°$ (Scheme 23.1).

Scheme 23.1

The ready oxidation of the aldehyde group of an aldose to a carboxylic acid function forms the basis of a number of useful qualitative tests for classifying a carbohydrate as a **reducing sugar.** Ketoses also yield positive tests as reducing sugars even though no aldehyde group is present. This positive test for ketoses arises because the α-hydroxy keto group of the open form of a ketose undergoes base-catalyzed **tautomerization** to give an **enediol** that is protonated to provide a pair of epimeric aldehydes

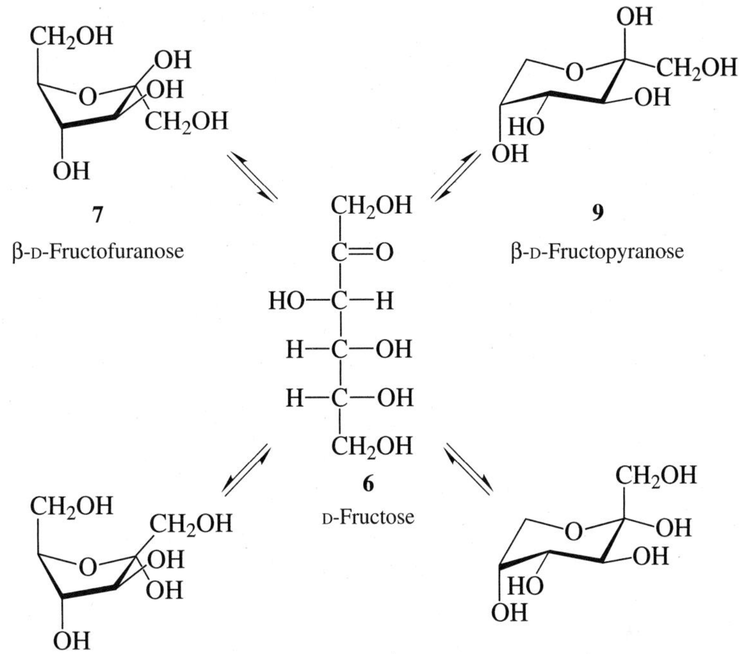

7 β-D-Fructofuranose

9 β-D-Fructopyranose

6 D-Fructose

8 α-D-Fructofuranose

10 α-D-Fructopyranose

differing in configuration at C2. These transformations are outlined in the form of partial structures in Equation 23.2. Since the tests for reducing sugars are performed under basic conditions that allow the equilibria of Equation 23.2 to be established, *all* known *monosaccharides* are reducing sugars.

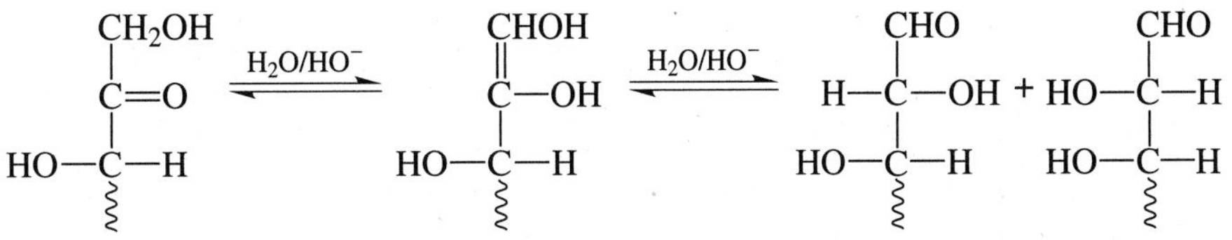

A ketose $\underset{}{\overset{H_2O/HO^-}{\rightleftharpoons}}$ An enediol $\underset{}{\overset{H_2O/HO^-}{\rightleftharpoons}}$ An aldose + An aldose (23.2)

23.3 DISACCHARIDES: HYDROLYSIS OF SUCROSE

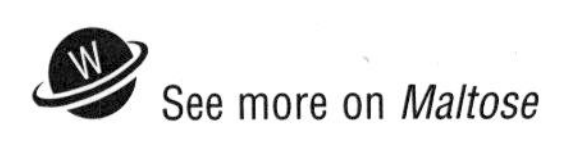
See more on *Maltose*

Disaccharides are ***O*-glycosides** in which the hydroxyl group attached to the anomeric carbon atom of one monosaccharide is replaced with a hydroxyl group of another monosaccharide, thereby forming a cyclic **acetal.** For example, maltose (**11**) is a disaccharide in which the anomeric hydroxyl group of one α-D-glucopyranose has been replaced with the C(4)-hydroxyl group of another D-glucopyranose subunit. Because the other D-glucopyranose ring of **11** is in equilibrium with the open-chain isomer **12** having a free aldehyde function (Eq. 23.3), maltose is a reducing sugar. Indeed, any disaccharide in which *one* of the rings is a hemiacetal or hemiketal is a *reducing sugar* (Sec. 23.2), because the cyclic hemiacetal or hemiketal moiety is in equilibrium with the open-chain isomer in which the aldehyde or the α-hydroxyketone function can be oxidized.

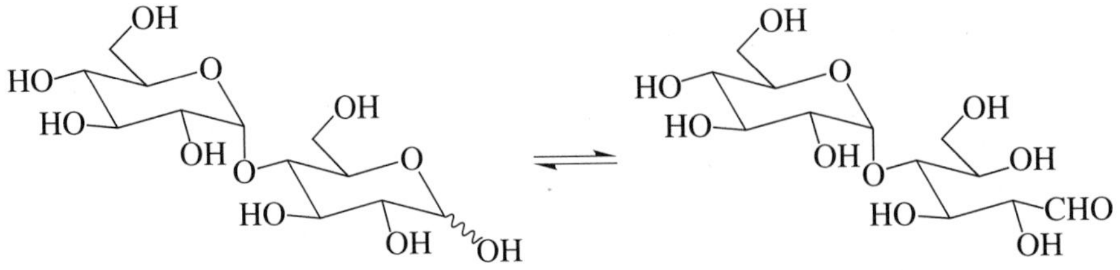

11 ⇌ **12** (23.3)

11
Maltose

12

See more on *Sucrose*

Disaccharides in which *both* rings are in the **acetal** or **ketal** form are *not* reducing sugars because they *cannot* be in equilibrium with an aldehydo or keto form under neutral or basic conditions. For example, sucrose (**13**), a common foodstuff, is a **nonreducing** disaccharide because the glycosidic linkage between the two monosaccharide subunits is formed between the two anomeric carbon atoms, thereby incorporating the potential aldehyde and α-hydroxyketone functions into cyclic acetal and ketal groups. Thus sucrose, $[\alpha]_D^{25} = +66.5°$, does not undergo mutarotation under either neutral or alkaline conditions. However, sucrose hydrolyzes under acidic conditions to give the reducing sugars D-glucose and D-fructose (Eq. 23.4). The 1:1 mixture of D-glucose and D-fructose that is produced upon hydrolysis of sucrose is called **invert sugar** because the specific rotation of the mixture is dominated by the negative rotation of D-fructose.

Fructose is the sweetest common sugar, being about twice as sweet as sucrose; consequently, invert sugar is sweeter than sucrose. The enzyme **invertase,** which bees use in making honey, accomplishes the same chemical result as does the acid-catalyzed hydrolysis of sucrose.

In the experiment in this section, you will examine the acid-catalyzed hydrolysis of sucrose to give a mixture of D-glucose and D-fructose and monitor the consequent change in the specific rotation that occurs using a polarimeter.

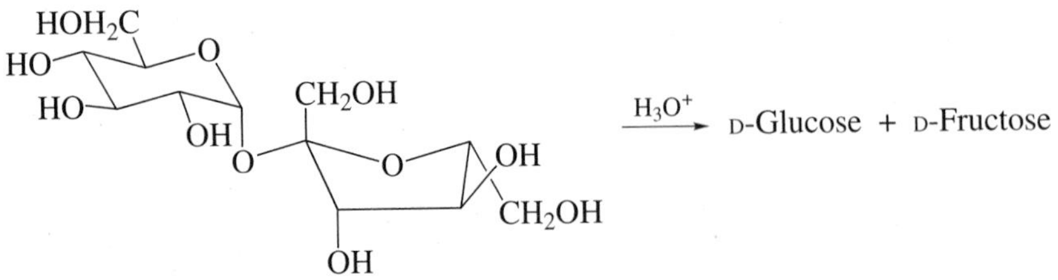

13 $\xrightarrow{H_3O^+}$ D-Glucose + D-Fructose (23.4)

13
Sucrose

EXPERIMENTAL PROCEDURE

Hydrolysis of Sucrose

Purpose To demonstrate the acid-catalyzed hydrolysis of a disaccharide into its component monosaccharides.

SAFETY ALERT

Wear latex gloves when measuring the *concentrated* hydrochloric acid. If any acid spills on your skin, wash it off with large volumes of water and then with dilute sodium bicarbonate solution to neutralize any residual acid.

MINISCALE PROCEDURE

Preparation Sign in at **www.cengage.com/login** to answer Pre-Lab Exercises, access videos, and read the MSDSs for the chemicals used or produced in this procedure. Review Sections 2.9, 2.11, and 2.22.

Apparatus A 100-mL round-bottom flask, separatory funnel, 50-mL volumetric flask, polarimeter, apparatus for magnetic stirring, heating under reflux, and *flameless* heating.

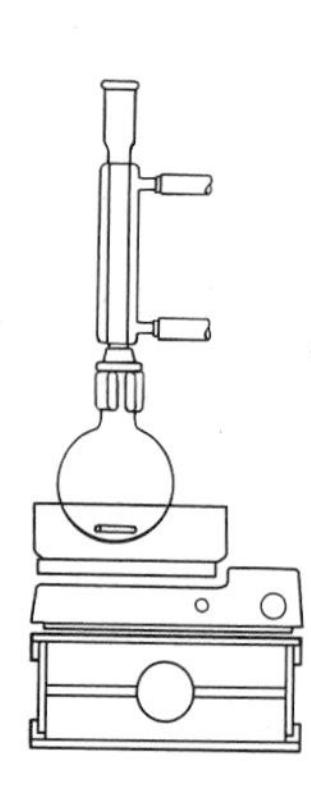

Setting Up and Reaction Accurately weigh about 7.5 g of sucrose and place it in the round-bottom flask containing a stirbar. Add about 40 mL of water, swirl the contents of the flask to effect solution, and add about 0.5 mL of *concentrated* hydrochloric acid. Assemble the apparatus for heating under reflux. Heat the solution under reflux for about 2 h.

Determining the Specific Rotation of Sucrose While the solution of sucrose in aqueous acid is being heated under reflux, determine the specific rotation of sucrose according to the general directions that follow, together with any specific directions of your instructor. Carefully fill the sample tube for the polarimeter with water; *be certain that no air bubbles are trapped in it.* Place the sample tube in the polarimeter and determine the blank reading for the solvent. Record the blank rotation, the temperature, and the length of the tube. Empty and carefully dry the tube. Accurately weigh 5–10 g of sucrose to the nearest 0.05 g, and transfer the sample *quantitatively* to the volumetric flask. Fill the flask to within several milliliters of the volumetric mark, tightly stopper the flask, and shake the flask thoroughly. Carefully fill the flask to the mark and again shake the flask thoroughly. Transfer a portion of this solution to the polarimeter tube, measure the rotation, and record this value.

Analysis At the end of the period of reflux, cool the reaction mixture to room temperature and transfer the solution to a clean volumetric flask. Use small amounts of water to rinse the round-bottom flask and add the rinses to the flask to effect quantitative transfer; fill the flask to the volumetric mark. Using the polarimeter as before, determine the specific rotation of the product mixture from the hydrolysis of sucrose. Compare this value with the specific rotation of pure sucrose determined earlier.

WRAPPING IT UP

Neutralize the *aqueous solutions* with sodium carbonate and flush them down the drain.

EXERCISES

1. Explain the change in sign of the optical rotation that occurs when sucrose undergoes hydrolysis.
2. When determining the rotation, why is it important that no air bubbles be present in the polarimeter tube?
3. Calculate the specific rotation of *invert* sugar from the known rotations of the equilibrium mixtures of the anomers of D-glucose and D-fructose. How does this number compare with that determined experimentally?
4. The specific rotation of invert sugar and of a racemic mixture represent average rotations produced by the molecules in solution. Why is the rotation of a racemic mixture 0°, whereas that of invert sugar is not?
5. Using curved arrows to symbolize the flow of electrons, write the stepwise mechanism for the intramolecular cyclization of D-glucose (**3**) into β-D-glucopyranose (**5**).
6. Using curved arrows to symbolize the flow of electrons, write a stepwise reaction mechanism for the acid-catalyzed hydrolysis of sucrose to D-glucose and D-fructose.
7. Specify which of the structures in Scheme 23.1 are α-anomers and which are β-anomers.
8. Using curved arrows to symbolize the flow of electrons, write a stepwise reaction mechanism for acid-catalyzed isomerization of **9** to **6,** followed by cyclization to produce a five-membered ring as in **7.**
9. Which hydroxyl group (i.e., α-, β-, γ-) of an aldoheptose adds to the aldehyde group to produce a furanose? A pyranose?
10. Would you classify the stereochemical relationship between the α- and β-anomers of a cyclic sugar as being enantiomeric, diastereomeric, *meso,* or none of these? Explain your answer.
11. Explain why D-fructose is considered a D-sugar even though its optical rotation is levorotatory.
12. The Fischer projection of D-galactose is shown below. Write the structure of α-D-galactopyranose and explain how it differs from α-D-glucopyranose.

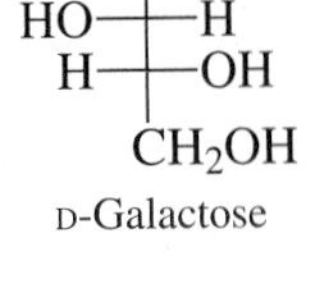

```
       CHO
   H───┼───OH
  HO───┼───H
  HO───┼───H
   H───┼───OH
      CH2OH
```

D-Galactose

13. Explain what is meant by the term *reducing sugar.*

14. What functional group must be present or formed under the reaction conditions for a sugar to be classified as a *reducing sugar*?

15. Using curved arrows to symbolize the flow of electrons, write a stepwise mechanism for the base-catalyzed conversion of a ketose into an aldose.

16. Lactose is a disaccharide in which the anomeric hydroxyl group of β-D-galactopyranose has been replaced with the C(4)-hydroxyl group of one α-D-glucopyranose unit.

- **a.** Write the structure of lactose.
- **b.** Is lactose a reducing sugar? Explain your answer.
- **c.** Would you expect lactose to undergo mutarotation? Explain your answer.

17. Consider the structure raffinose, a trisaccharide, shown.

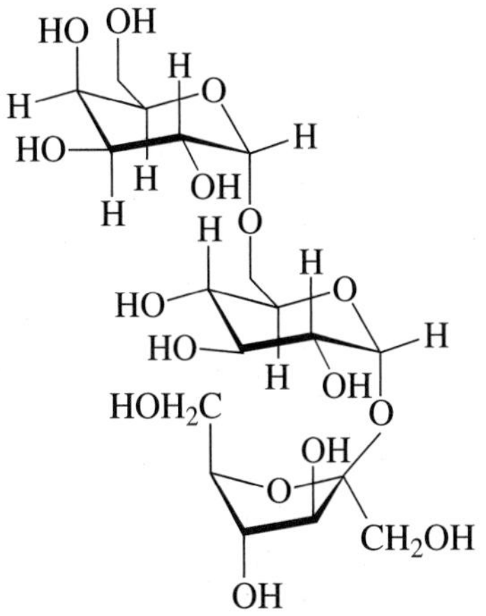

Raffinose

- **a.** Label each of the anomeric carbon atoms with an asterisk.
- **b.** Circle all the α-glycosidic linkages.
- **c.** Why is raffinose considered a *trisaccharide*?
- **d.** Is raffinose a reducing sugar?
- **e.** Write the structure(s) of the monosaccharide(s) that would be produced when raffinose is heated with aqueous acid.
- **f.** Write the structures and names of the monosaccharide and the disaccharide that would be produced when raffinose is incubated with *invertase*, the enzyme used by bees to make honey.
- **g.** Why is the disaccharide in Part **f** not hydrolyzed further into monosaccharides by invertase?

18. The specific rotations for α-D-mannose and β-D-mannose are +29.3° and –17.0°, respectively. When either of these monosaccharides is dissolved in water, the rotation reaches an equilibrium value of +14.2. Calculate the percentage of the α-anomer present at equilibrium. Show your calculations.

23.4 CARBOHYDRATES: THEIR CHARACTERIZATION AND IDENTIFICATION

See more on *Monosaccharides*

Determining the complete structure of an unknown monosaccharide was a formidable challenge to early organic chemists (see the Historical Highlight at the end of this chapter). Many carbohydrates, particularly when impure, have a tendency to form syrups rather than crystallize from solution, and this sometimes makes it difficult to obtain pure compounds for characterization and identification. The number and nature of functional groups and the number of stereocenters present in carbohydrates further exacerbates the problem of assigning their structures.

See more on *Glucose*

Some of these difficulties are illustrated by considering the example of D-(+)-glucose, whose nature in solution is shown in Equation 23.1. The first step toward solving the structure of D-(+)-glucose was to identify D-(+)-glucose as a derivative of a 2,3,4,5,6-pentahydroxyhexanal **14.** Because D-(+)-glucose exists primarily as a cyclic hemiacetal at equilibrium, another issue involved determining the ring size. Namely, did D-(+)-glucose exist in the form of a *furanose,* **15**, or a *pyranose,* **16**? Finally, it was necessary to assign the relative and absolute configuration to each of the centers of chirality (Sec. 7.1), including the anomeric carbon atom (Sec. 23.2). Despite these tremendous experimental challenges, the complete structures of many monosaccharides, some containing up to nine carbon atoms, are now known. It is amazing that much of this structural work was performed *prior* to the availability of the modern spectroscopic methods discussed in Chapter 8.

CHO–CHOH–CHOH–CHOH–CHOH–CH_2OH

14 **15** **16**

See more on *Polysaccharides*

The structural elucidation of polysaccharides represents an even greater challenge. The individual monosaccharide subunits that constitute the unknown polysaccharide must first be identified. Then the ring size and position in the polysaccharide sequence must be elucidated for each monosaccharide. Finally, the nature of the glycosidic linkages that form the polysaccharide backbone must be defined. Toward this end, it is necessary to establish which hydroxyl group on one monosaccharide is involved in the formation of the acetal or ketal that forms the glycosidic bond to the adjoining monosaccharide; the stereochemistry at this anomeric center must also be determined.

Classification Tests

A number of useful qualitative chemical tests have been devised to obtain information for unknown carbohydrates. Some of these tests are used to classify such molecules according to their structural type. In the experiments that follow, you will use such tests to identify certain structural features that are found in mono- and polysaccharides. Data derived from these experiments provide information that may be used to prove the structure of an unknown carbohydrate.

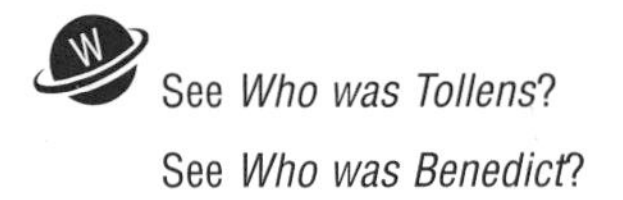

Tollens's Test

Tollens's test is designed to distinguish between aldehydes and ketones. A complete discussion of this test together with an experimental procedure is included with the classification tests for aldehydes in Section 25.7C.

Benedict's Test

This test is based on the fact that cupric ion will oxidize aliphatic aldehydes, including α-hydroxyaldehydes, but not aromatic aldehydes. The reagent used in this test is a solution of cupric sulfate, sodium citrate, and sodium carbonate. The citrate ion forms a complex with the Cu(II) ion so that $Cu(OH)_2$ does not precipitate from the basic solution. A positive test for the presence of the aliphatic aldehyde group in an aldose is evidenced by the formation of a red precipitate of cuprous oxide, Cu_2O (Eq. 23.5). A yellow precipitate is occasionally observed as a positive test. This yellow solid has not been characterized, but its formation seems to depend on the amount of oxidizing agent present.

$$RCHO + 2\,Cu^{2+} + 5\,HO^- \xrightarrow{\text{Citrate}} RCO_2^- + \underset{\text{(brick red)}}{\underline{Cu_2O}} + 3\,H_2O \quad (23.5)$$

Barfoed's Test for Monosaccharides

Like Tollens's and Benedict's tests, this test depends on the *reducing* properties of the saccharides. However, because of the specific conditions employed for the test, it is possible to distinguish between monosaccharides and disaccharides. The test reagent consists of an aqueous solution of cupric acetate and acetic acid. Thus, in contrast to the previous tests, the reaction is carried out under *acidic* conditions. A positive test for monosaccharides is indicated by the formation of the brick-red precipitate of Cu_2O within *two* or *three* minutes. For reasons that are not fully understood, disaccharides require a longer time, providing the precipitate only after about 10 minutes or more. Nonreducing sugars such as sucrose undergo slow hydrolysis under the acidic aqueous conditions of the test, and therefore they give a precipitate after *extended* time.

EXPERIMENTAL PROCEDURES

Classification Tests for Carbohydrates

Purpose To demonstrate several classification tests for carbohydrates.

MICROSCALE PROCEDURE

Preparation Sign in at **www.cengage.com/login** to answer Pre-Lab Exercises, access videos, and read the MSDSs for the chemicals used or produced in this procedure.

Apparatus A number of 13-mm × 100-mm test tubes.

Setting Up Obtain the carbohydrates that you will use in the following classification tests from your instructor. Perform each test on each carbohydrate, and enter your observations and conclusions in your notebook.

Tollens's Test Before performing this test, read the Safety Alert and Wrapping It Up sections in Section 25.7C. Perform Tollens's test according to the procedure outlined in that section using about 0.1 g of the carbohydrate in a *clean* glass test tube and about 1 mL of Tollens's reagent. The formation of a silver mirror or a black precipitate constitutes a positive test.

Benedict's Test Place about 0.1 g of the carbohydrate in a test tube, and add 2 mL of water. Stir the mixture to effect solution, and add 2–3 mL of the stock solution of Benedict's reagent. Heat the solution to boiling. The formation of a yellow to red precipitate is a positive test for aliphatic aldehydes and α-hydroxyaldehydes; the precipitate appears green when viewed in the blue solution of the reagent. For comparison, perform the test simultaneously on an unknown, on glucose, and on sucrose, and record the results.

Barfoed's Test Place 3 mL of Barfoed's reagent and 1 mL of a 1% solution of the carbohydrate in a test tube; place the test tube in a beaker of boiling water for *5 min*. Remove the test tube and cool it under running water. A red precipitate of cuprous oxide is a positive test. To see the precipitate, it may be necessary to view the tube against a dark background in good light. For comparison, run the test simultaneously on the unknown, on glucose, and on lactose, and record the results.

WRAPPING IT UP

Remove any precipitated *cuprous oxide* by vacuum filtration, and place it in the container for nonhazardous solids. Flush the *filtrate* down the drain with an excess of water.

FORMATION OF OSAZONES

Carbohydrates that can exist in solution in an acyclic form with a free aldehyde or ketone function react with three equivalents of phenylhydrazine to form bright yellow crystalline derivatives called **phenylosazones,** together with aniline and ammonia as the other products of the reaction. For example, α- and β-D-glucopyranoses **4** and **5** are in dynamic equilibrium in aqueous solution with the ring-opened form **3** (Sec. 23.2), which, because of its free aldehyde group, will react with phenylhydrazine to form the phenylosazone **17,** as shown in Equation 23.6. Since phenylosazones are readily identified by either their melting points or temperatures of thermal decomposition, these derivatives may be used to characterize carbohydrates.

The accepted mechanism for the formation of phenylosazones is presented in Scheme 23.2. Following initial formation of the phenylhydrazone **18** from the aldose, a double tautomerization leads to the ketone **19.** The newly formed carbonyl group then condenses with a second equivalent of phenylhydrazine to give **20,** which tautomerizes to **21.** After 1,4-elimination of aniline to produce either **22** or **23,** a third equivalent of phenylhydrazine condenses with the imine group of **22** or **23** to yield the phenylosazone **24** and ammonia. Although it may appear that

24 should undergo further reaction with phenylhydrazine via tautomerization involving the secondary alcohol group at C3 and the hydrazone group at C2, the reaction stops at this point so that only two phenylhydrazine units are introduced. The formation of the intramolecular hydrogen bond shown by the dashed line in **24** apparently limits the reaction to the first two carbon atoms of the chain. In support of this hypothesis, when 1-methylphenylhydrazine [$H_2NNMeC_6H_5$] is used in place of phenylhydrazine, the reaction readily proceeds down the chain at least as far as C5.

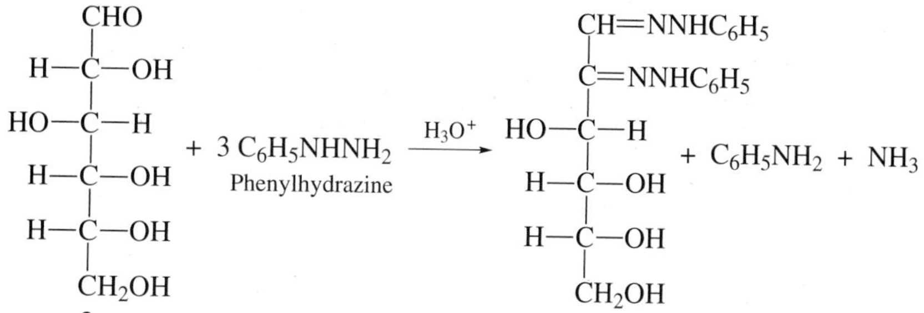

CHO
H—C—OH
HO—C—H
H—C—OH
H—C—OH
CH_2OH
3
D-Glucose

$+ \ 3\ C_6H_5NHNH_2$ (Phenylhydrazine) $\xrightarrow{H_3O^+}$

$CH{=}NNHC_6H_5$
$C{=}NNHC_6H_5$
HO—C—H
H—C—OH
H—C—OH
CH_2OH
17
Phenylosazone of D-glucose

$+ \ C_6H_5NH_2 + NH_3$ (23.6)

Scheme 23.2

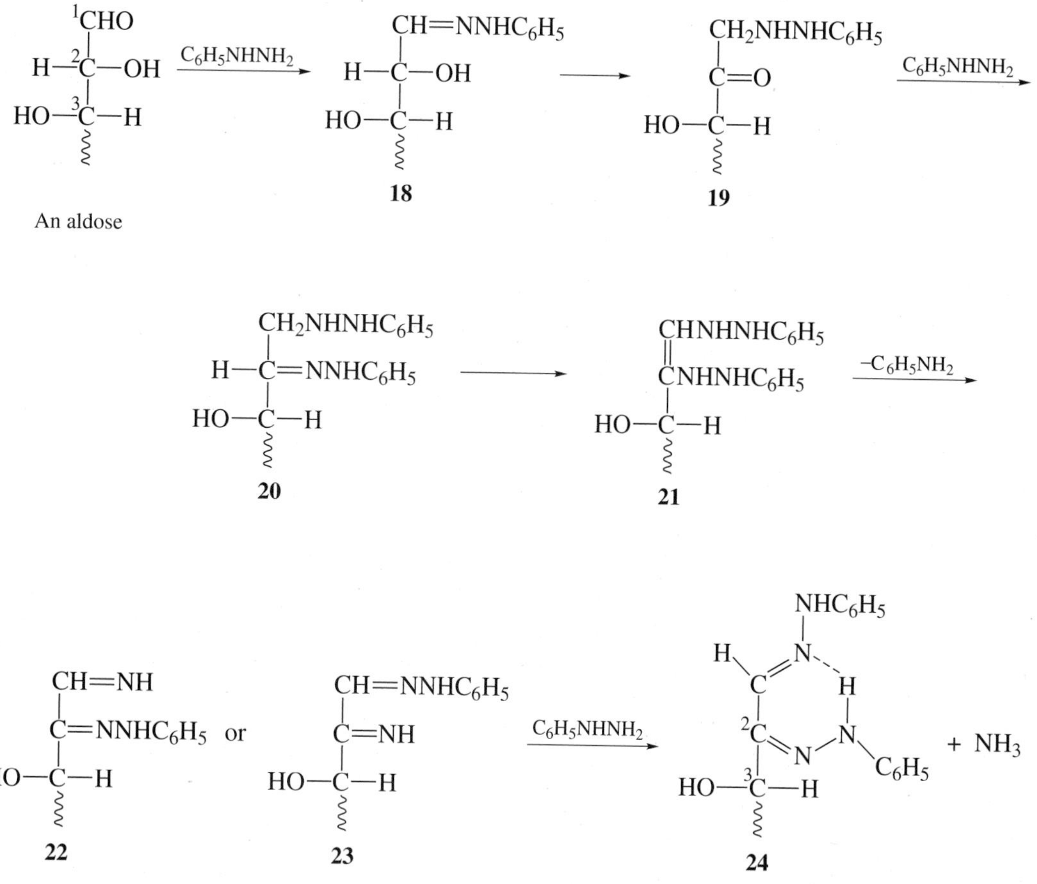

An aldose: ^{1}CHO, H—^{2}C—OH, HO—^{3}C—H $\xrightarrow{C_6H_5NHNH_2}$ **18**: $CH{=}NNHC_6H_5$, H—C—OH, HO—C—H $\longrightarrow$ **19**: $CH_2NHNHC_6H_5$, C=O, HO—C—H $\xrightarrow{C_6H_5NHNH_2}$

20: $CH_2NHNHC_6H_5$, H—$C{=}NNHC_6H_5$, HO—C—H $\longrightarrow$ **21**: $CHNHNHC_6H_5$ ‖ $CNHNHC_6H_5$, HO—C—H $\xrightarrow{-C_6H_5NH_2}$

22: CH=NH, $C{=}NNHC_6H_5$, HO—C—H or **23**: $CH{=}NNHC_6H_5$, C=NH, HO—C—H $\xrightarrow{C_6H_5NHNH_2}$ **24**: H—C(=N—NHC_6H_5)—^{2}C(=N—N(H)—C_6H_5), HO—^{3}C—H $+ \ NH_3$

EXPERIMENTAL PROCEDURE

Formation of Osazones

Purpose To demonstrate the preparation of osazones.

SAFETY ALERT

Phenylhydrazine is toxic; avoid contact with it. If you spill some on your skin, rinse it off thoroughly with household bleach, dilute acetic acid, and then water.

MICROSCALE PROCEDURE

Preparation Sign in at **www.cengage.com/login** to answer Pre-Lab Exercises, access videos, and read the MSDSs for the chemicals used or produced in this procedure. Review Sections 2.9 and 2.17.

Apparatus Test tubes, hot-water bath, apparatus for vacuum filtration.

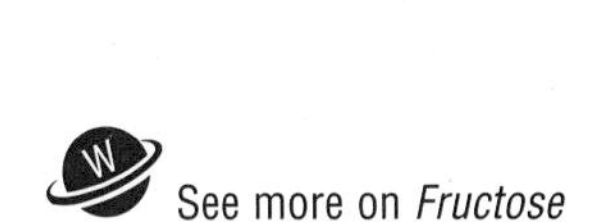

See more on *Fructose*

See more on *Sucrose*

Setting Up Heat a large beaker of water to boiling. In each of three test tubes, *separately* dissolve 0.2-g portions of D-glucose, D-fructose, and sucrose in 4 mL of water. In a fourth test tube, place 4 mL of the solution that was used for rotation experiments for the hydrolysis of sucrose (Sec. 23.3). Add 0.5 mL of saturated sodium bisulfite solution to each test tube to suppress oxidation of the phenylhydrazine during the reaction; this avoids contaminating the phenylosazone with tarry by-products.

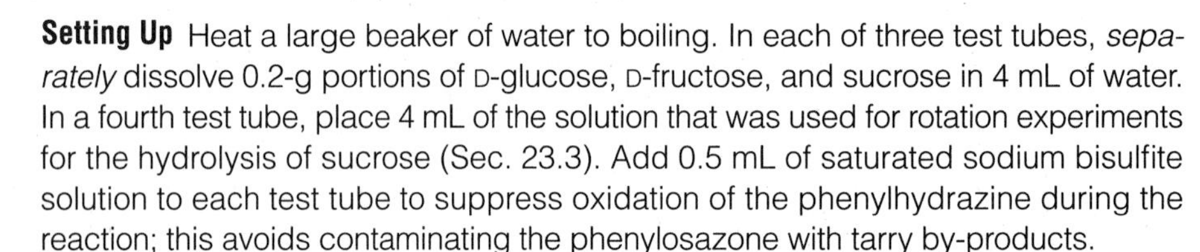

Reactions To each of the test tubes, add *either* 0.6 mL of glacial acetic acid, 0.6 g of sodium acetate, and 0.4 g of phenylhydrazine *or* 0.6 g of sodium acetate and 0.6 g of phenylhydrazine hydrochloride. Stir the solutions thoroughly, place the test tubes in the beaker of boiling water, and discontinue heating. Allow the test tubes to remain in the hot-water bath for 30 min. Remove the test tubes from the hot-water bath and cool the contents to room temperature.★

Isolation and Purification Collect any precipitates by vacuum filtration.★ Using ethanol-water, recrystallize each of the phenylosazones formed.

Analysis Determine the melting points of the purified phenylosazones. Make a mixture containing equal parts of the two phenylosazones obtained from the pure carbohydrates, and determine the melting point of this mixture. Because the melting points of phenylosazones may depend on the rate of heating, perform these determinations simultaneously. Make a mixture containing equal parts of the phenylosazone obtained from the hydrolysis of sucrose and each of the phenylosazones obtained from the pure carbohydrates and determine their melting points as well. Record your observations.

WRAPPING IT UP

Neutralize the *filtrates* obtained from isolating the crude phenylosazones with a small amount of sodium carbonate, and add 2 mL of laundry bleach (5.3% aqueous sodium hypochlorite) for every 1 mL of the filtrate. Heat the mixture at 45–50 °C for 2 h to oxidize and decompose any phenylhydrazine that is present. Cool the mixture and then flush it down the drain with excess water. Flush the *filtrates from the recrystallizations* down the drain.

EXERCISES

1. Consider Tollens's test.
   a. What causes the formation of a silver mirror?
   b. Write the half reaction for silver in the test and indicate whether silver undergoes oxidation or reduction.
   c. Write the half reaction for the sugar in the test and indicate whether the sugar undergoes oxidation or reduction.
2. List the classification tests that may be used to determine whether a sugar is a reducing sugar.
3. What ion is responsible for the blue color of Benedict's reagent?
4. Why does a solution of D-glucose and Benedict's reagent turn brick red when heated?
5. Using curved arrows to symbolize the flow of electrons, write a stepwise mechanism for the acid-catalyzed reaction of an aldose with phenylhydrazine to give a phenylhydrazone like **18.**
6. Which of the pure carbohydrates formed phenylosazones?
7. What conclusions can be drawn from the results of the mixed melting-point determination on the phenylosazones formed from the pure carbohydrates?
8. What conclusions can be drawn from the results of the mixed melting-point determinations on the phenylosazones formed from the pure carbohydrates and that obtained from the hydrolysis product of sucrose?
9. D-Glucose and D-mannose give the same phenylosazone. Explain.
10. Draw the structure of the product that would be expected upon treatment of D-glucose with a large excess of 1-methylphenylhydrazine.
11. In what way does the intramolecular hydrogen bond in the phenylosazone **24** prevent further reaction with phenylhydrazine?

HISTORICAL HIGHLIGHT

Emil Fischer: Carbohydrate Chemist Extraordinaire

One of the great pioneers in the development of carbohydrate chemistry was Emil Fischer, who was born near Bonn, Germany, in 1852. Although he went to the University of Strasbourg to study chemistry under Friedrich Kekulé, he actually obtained his Ph.D. under Adolf von Baeyer in 1874. He moved with von Baeyer to Munich, where he received his *Habilitation* in 1878. He was then appointed as an *Ausserordentlicher,* or assistant professor, in 1879. Afterward he held appointments as Professor of Chemistry at Erlangen (1882) and Würzburg (1885) before succeeding von Hoffmann at the prestigious Chemical Institute of the University of Berlin. His original laboratories and office in Berlin are now a part of the Humboldt University, which has a special exhibit of some of his library, writings, and laboratory equipment.

(Continued)

HISTORICAL HIGHLIGHT *Emil Fischer: Carbohydrate Chemist Extraordinaire (Continued)*

In 1875, Fischer synthesized phenylhydrazine, $C_6H_5NHNH_2$, for the first time. Although the preparation of a totally new chemical entity must itself have been an exhilarating experience, the discovery of its utility in carbohydrate and carbonyl chemistry quickly overshadowed its mere creation. Fischer found that phenylhydrazine reacted with certain sugars to form *osazones.* He not only studied the mechanism of the formation of osazones, but he was able to determine the structures of the osazones of a large number of sugars and to use these derivatives as the basis for developing techniques for the systematic classification of sugars. Fischer was the first to synthesize glucose, an achievement for which he was awarded the Davy medal of the Royal Society in 1890. Shortly thereafter, he prepared unnatural sugars such as the nonoses. Synthetic organic chemistry had thus come a long way since the preparation of urea, H_2NCONH_2, by Wöhler in 1828!

Fischer applied the concept of the tetrahedral carbon atom, which had been proposed in 1874–1875 by van't Hoff and Le Bel, to assigning the relative stereochemistries of *all eight* of the isomeric hexoses. The solution to this difficult problem eluded Fischer for some time, until, on the advice of Victor Meyer, he built molecular models of the different sugars. Having done so, the complex stereochemical relationships became apparent. It was at this time that he developed the protocol for representing stereochemical relationships in sugars via planar structures, known as *Fischer projections,* which are used to this day. The elucidation of the structures of glucose and other known sugars is regarded as Fischer's most important contribution to science and earned him the Nobel Prize in Chemistry in 1902; he was only the second chemist to receive this award.

Fischer is best known for his work in carbohydrate chemistry, but he was also active in the fields of photosynthesis and protein and heterocyclic chemistry. He demonstrated that amino acids were the basic building blocks in proteins. His ingenious contribution of the "lock and key" hypothesis of how enzymes selectively bind substrates of complementary shapes evolved from his work with the proteins known as glucosidases. For example, he found that α-D-methyl glucoside was easily hydrolyzed by a yeast extract termed invertin but not by a preparation of emulsin isolated from almonds. Conversely, β-D-methyl glucoside was hydrolyzed by emulsin, but not by invertin. This important work helped lay the foundation for modern biochemistry. Fischer felt strongly that the great chemical secrets of life would only be unveiled by a cooperative effort in which chemical methods were applied to solving biological problems, a point he made in a seminal lecture to the Faraday Society in 1907.

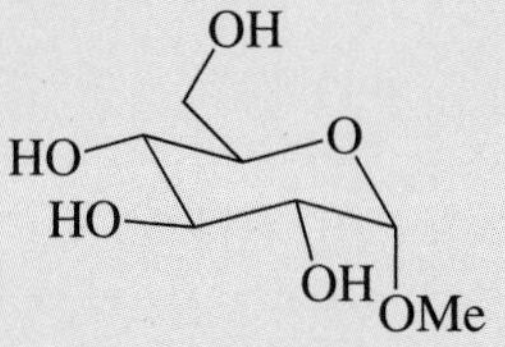

α-D-Methyl glucoside

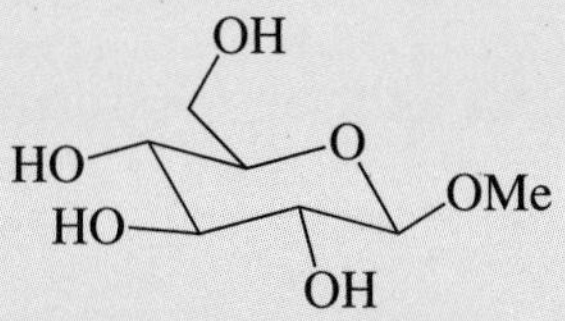

β-D-Methyl glucoside

Fischer's interest in other heterocyclic molecules found in nature led to the first synthesis of caffeine, which is the stimulant in coffee and tea. The purine ring system found in caffeine also comprises the ring systems of the heterocyclic bases adenine and guanine that are found in nucleic acids. He also developed a general synthesis of indole, the heterocyclic ring-containing component in the amino acid tryptophan, and the method is known today as the Fischer indole synthesis.

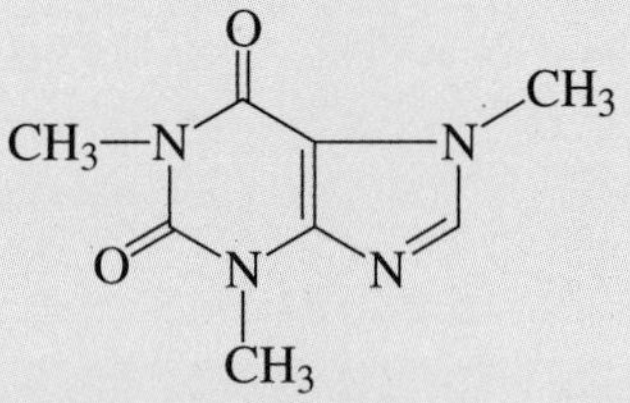

Caffeine

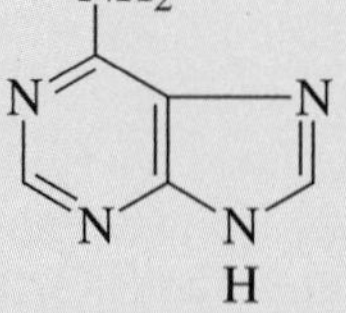

Adenine

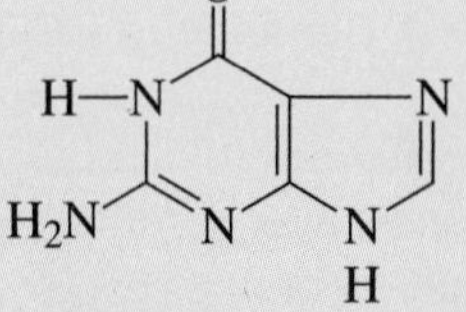

Guanine

(Continued)

HISTORICAL HIGHLIGHT *Emil Fischer: Carbohydrate Chemist Extraordinaire (Continued)*

N H Indole

CO_2H NH_2 N H L-Tryptophan

While Fischer must have gained satisfaction and joy from his scientific accomplishments, his personal life ended in tragedy. Before moving to Berlin, he married Agnes Gerlach, with whom he had three sons. His wife died at the age of 33, and two of his sons were killed in World War I. By the end of the war, Fischer himself suffered from mercury and phenylhydrazine poisoning and advanced intestinal cancer. Depressed by his own poor health, the losses of his wife and two sons, and the socioeconomic conditions of postwar Germany, he committed suicide in 1919. However, his legacy as one of the great bioorganic chemists in the history of the science remains.

Relationship of Historical Highlight to Experiments

Fischer was the first to synthesize and determine the structures of a number of the sugars used in the experiments in this section. He studied the hydrolysis of glucosides and disaccharides in much the same way you may if you perform the hydrolysis of sucrose (Sec. 23.3). He was also the first to synthesize phenylhydrazine, which you may use to form the same osazones he did in his pioneering work in characterizing sugars (Sec. 23.4). You can share his excitement as you observe the formation of the solid osazones from the combination of different monosaccharides with excess phenylhydrazine.

See *Who was Fischer?*
See *Who was van't Hoff?*
See *Who was Davy?*
See *Who was Le Bel?*
See *Who was Meyer?*
See more on *Caffeine*
See more on *Indole*
See more on *Adenine*
See more on *Guanine*
See more on *Tryptophan*

CHAPTER 24

α-Amino Acids and Peptides

When you see this icon, sign in at this book's premium website at **www.cengage.com/login** to access videos, Pre-Lab Exercises, and other online resources.

α-Amino acids are organic molecules that contain both an amine and a carboxylic acid functional group on the same carbon atom. α-Amino acids thus undergo some reactions that are typical of both of these functionalities, but they also exhibit special chemical and physical properties that are a consequence of having these two functional groups attached to one carbon atom. You have an opportunity to learn about some of the chemistry of amines and carboxylic acids in Chapters 17, 20, and 21. In this chapter, you will study several aspects of the chemistry of α-amino acids and how these compounds can be combined to make peptides, substances that comprise two or more amino acids joined by an amide bond between the nitrogen atom of one amino acid and the carboxyl carbon atom of another.

α-Amino acids are the essential building blocks of proteins, which are high-molar-mass biopolymers comprising many α-amino acids linked by peptide bonds. Examples of proteins in living cells include the enzymes that catalyze essential biochemical transformations and the receptors in cell membranes that bind smaller molecules such as neurotransmitters, hormones, and pharmaceuticals. Useful materials like wool or silk are also constituted of proteins. In the experiments in this chapter, you will learn how to combine individual α-amino acids to form peptides in exactly the same way such compounds are prepared in research laboratories.

24.1 INTRODUCTION

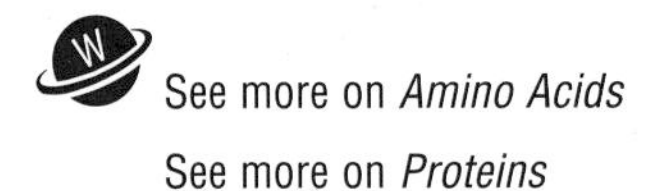

See more on *Amino Acids*

See more on *Proteins*

Amino acids are generally characterized by the presence of an **amine functional group**, $-NH_2$, and a **carboxylic acid functional group**, $-CO_2H$. Although the amino and carboxyl groups may be located anywhere in the molecule, the most important amino acids in biological systems are **α-amino acids**, as these are the monomeric building blocks for forming **proteins** in living organisms. These compounds are commonly represented as **1**, a general structural formula in which R^1 is H or variously substituted alkyl groups. The R^2 substituent on the nitrogen atom is usually an H. Whenever R^1 is an alkyl group, the carbon atom bearing the amino and carboxyl groups is a *center of chirality* or *stereocenter* (see Chap. 7.1), and the vast majority of amino acids found in mammalian proteins have the absolute configuration shown, which may be either *S*- or *R*- depending on the functionality in R^1, and are classified as L-amino acids because of their stereochemical relationship with L-glyceraldehyde.

R^1 CO_2H H NHR^2

1

An α-amino acid

HOH_2C CHO H OH

L-Glyceraldehyde

The R^1 substituent in **1** is commonly referred to as the **side chain,** and the variations in the structure and functionality of these groups play an important role in determining the chemical and physical properties of the individual amino acids. Based upon the nature of the side chain, the 20 common L-amino acids found in proteins may be grouped into four general categories as shown in Table 24.1, in which the structures, names, and three-letter abbreviations are listed. The largest single class consists of those amino acids with nonpolar side chains containing simple alkyl and various aromatic and heteroaromatic groups. A second category comprises polar but neutral or uncharged functional groups such as amides and alcohols. The third and fourth groups possess either acidic or basic side chains that will be at least partially ionized at the physiological pH of about 7.4. Because of the manner in which these side chains interact with water, those amino acids with nonpolar side chains are sometimes referred to as *hydrophobic,* or "water-fearing," whereas those with polar side chains are considered *hydrophilic,* or "water-loving." Although it is beyond the scope of this discussion, those of you who study biochemistry will learn that the amino acid side chains in proteins play important roles in protein folding, in substrate recognition, and in catalyzing biological reactions.

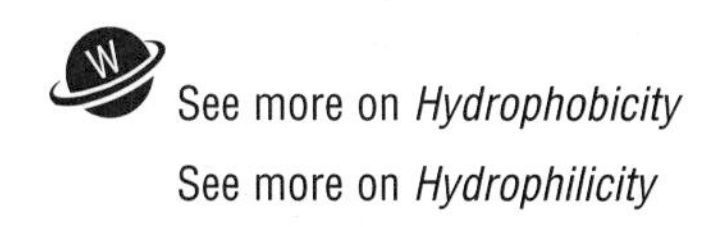

See more on *Hydrophobicity*

See more on *Hydrophilicity*

Because amino acids have both a basic (the amine) and an acidic (the carboxylic acid) functional group, the state of protonation of the molecule varies with pH as shown in Equation 24.1. The presence of a functional group in the side chain may have considerable influence upon the position of this equilibrium at different pHs. The pH at which the vast majority of the molecules are in the **zwitterionic** form, and therefore have a net charge of zero, is referred to as the **isoelectric point, pI.** An amino acid is *least* soluble in water at its isoelectric point, which is different for each amino acid. If the side chain bears an ionizable group such as a carboxylic acid or an amine, the state of protonation of that functional group will also vary as a function of pH. However, the amino acids used in the experiments in this chapter are nonpolar, so we will not concern ourselves with the complexities associated with ionizable side chains.

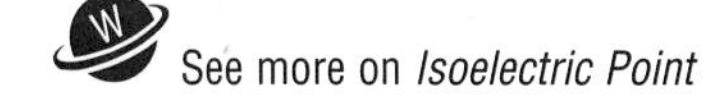

See more on *Isoelectric Point*

$$R^1CH(CO_2^-)NHR^2 \underset{-H^+}{\overset{+H^+}{\rightleftharpoons}} R^1CH(CO_2^-)\overset{+}{N}H_2R^2 \underset{-H^+}{\overset{+H^+}{\rightleftharpoons}} R^1CH(CO_2H)\overset{+}{N}H_2R^2 \quad (24.1)$$

Species present in strong base — Zwitterionic form — Species present in strong acid

An amide bond formed between the nitrogen atom of one amino acid and the acyl carbon atom of another is referred to as a **peptide bond,** a term coined in 1902 by the Nobel laureate Emil Fischer (see the Historical Highlight at the end of Chapter 23 for an account of the life of this famous chemist). For example, **aspartame (2),** the active ingredient in the artificial sweetener NutraSweet®, is the methyl ester of a dipeptide derived from joining a molecule of L-aspartic acid (Asp) and

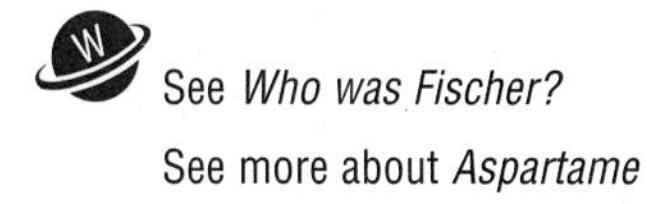

See *Who was Fischer?*

See more about *Aspartame*

Table 24.1 *The 20 Amino Acids Commonly Found in Proteins*

Nonpolar Side Chains

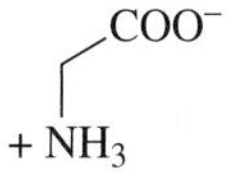

L-Glycine	(Gly)
L-Alanine	(Ala)
L-Valine	(Val)
L-Isoleucine	(Ile)
L-Leucine	(Leu)
L-Methionine	(Met)
L-Proline	(Pro)
L-Phenylalanine	(Phe)
L-Tryptophan	(Trp)

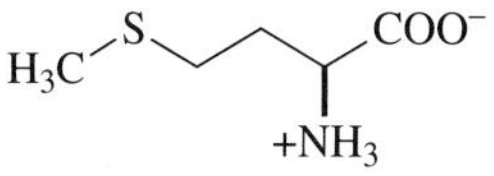

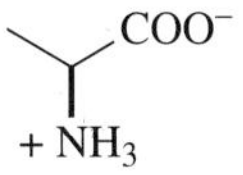

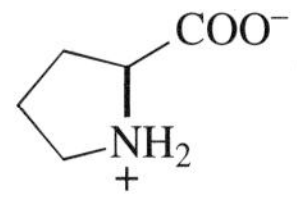

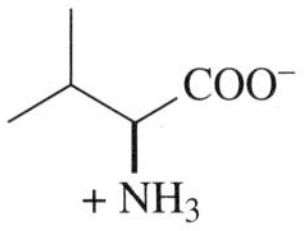

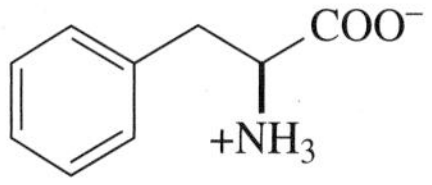

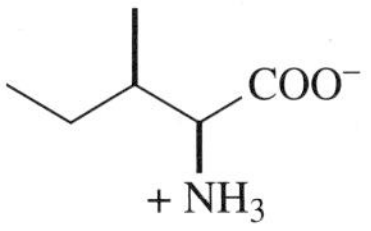

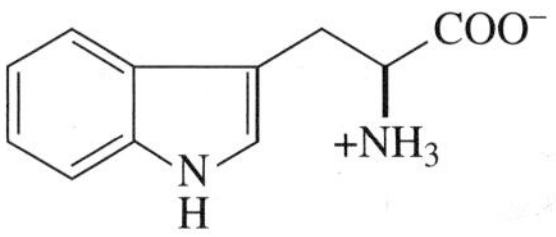

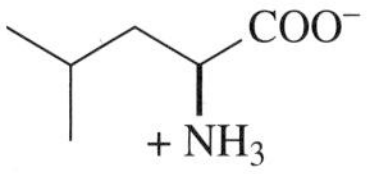

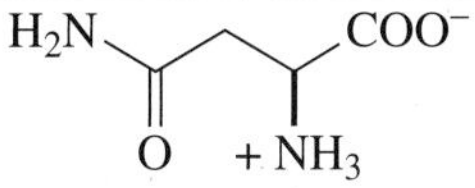

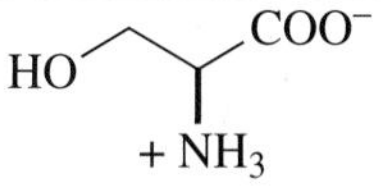

Polar Side Chains

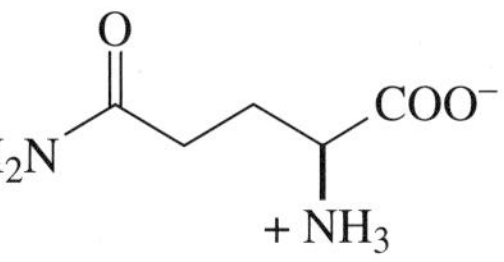

Amino acid	Abbreviation
L-Asparagine	(Asn)
L-Glutamine	(Gln)
L-Serine	(Ser)
L-Threonine	(Thr)

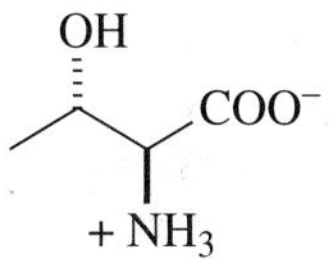

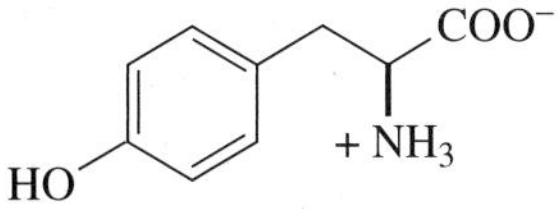

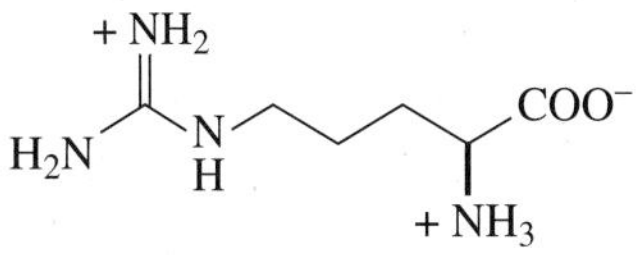

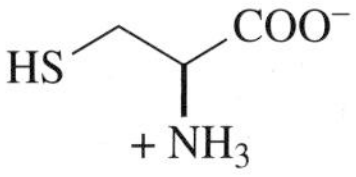

Acidic Side Chains

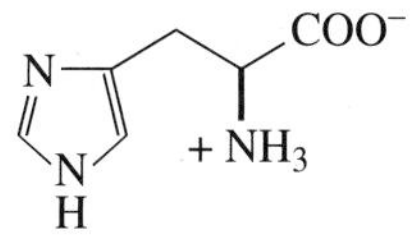

Amino acid	Abbreviation
L-Tyrosine	(Tyr)
L-Cysteine	(Cys)
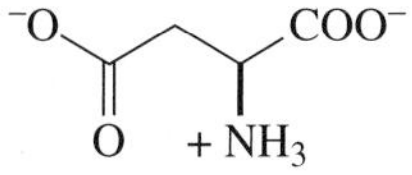 L-Aspartic acid	(Asp)
L-Glutamic acid	(Glu)

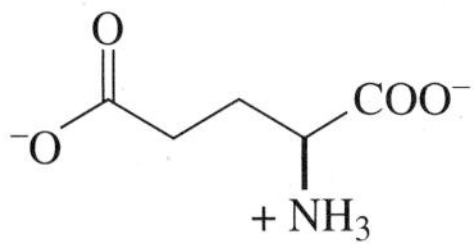

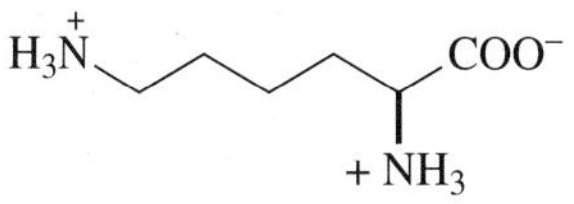

Basic Side Chains

Amino acid	Abbreviation
L-Arginine	(Arg)
L-Histidine	(His)
L-Lysine	(Lys)

L-phenylalanine (Phe) by a peptide bond. By convention, **peptides,** which is the general name given to chains comprising up to approximately 30–40 amino acid residues, are written from the left, beginning with the amino acid having a free $-NH_3^+$ group, the **N-terminus,** and progressing to the amino acid with the free COO^- group, the **C-terminus. Proteins** are biopolymers that are typically formed by joining more than about 100 amino acids in a single polypeptide chain. The definitions of *N*- and *C*-terminus and peptide bonds are illustrated in the structure of **2.**

Asp **Phe**

COO^- H N O $H_3\overset{+}{N}$ O OCH_3

C-Terminal amino acid

N-Terminal amino acid

Peptide bond

Aspartame

24.2 SYNTHESIS OF PEPTIDES AND POLYPEPTIDES

The synthesis of peptides and proteins in nature is a complex process that you may study in detail in a biochemistry course. However, because short peptides and their derivatives may exhibit useful properties and potent biological activities, chemists in academic and pharmaceutical laboratories have developed efficient methods for their synthesis.

Short peptides consisting of four or five amino acids may be easily prepared in solution, but problems associated with the insolubility of larger peptide derivatives cause difficulties with purification and isolation of intermediates as well as of the desired polypeptides. A major advance in polypeptide synthesis was reported by R. Bruce Merrifield in 1962. He discovered that all the steps in the synthesis of polypeptides could be performed on the surface of an insoluble polymer rather than in solution; you can read some of the details of this approach in the Historical Highlight at the end of this chapter. This method for preparing polypeptides revolutionized the field, and it enabled Merrifield to complete the synthesis of ribonuclease, an enzyme containing 124 amino acids, in only six weeks! The significance of Merrifield's work earned him the Nobel Prize in Chemistry in 1984. You will not have enough time to complete the synthesis of such a large protein as ribonuclease, of course, but the experiments in this chapter allow you to learn the essential principles of peptide synthesis by preparing a simple dipeptide derivative.

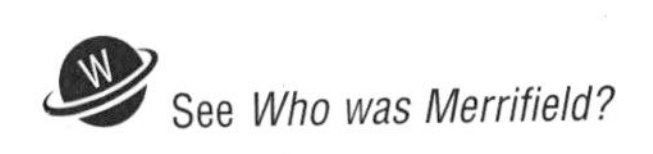

See *Who was Merrifield?*

The fundamental problem in the synthesis of peptides is that amino acids must be connected in a defined sequence by specifically forming a peptide bond between the carboxylic acid group of one amino acid and the amino group of another. The potential difficulties arising in such endeavors may be illustrated by considering the synthesis of the simple dipeptide Ala–Phe from the individual amino acids L-alanine (Ala) (**3**) and L-phenylalanine (Phe) (**4**), as seen in Equation 24.2. Formation of peptide bonds in a random manner could lead to four different dipeptides. This mixture arises because the carboxylic acid of Ala may react with either the amino group of another molecule of Ala or with the amino group of a Phe. Similarly, the carboxyl group of the Phe may react with the amino group of a molecule of Phe or Ala. Obtaining such a mixture is clearly undesirable, as the desired Ala–Phe would be difficult to isolate in pure form and high yield. Hence, the problem facing the early peptide chemist was determining how to make the amino group of one amino acid react selectively with the carboxyl group of another.

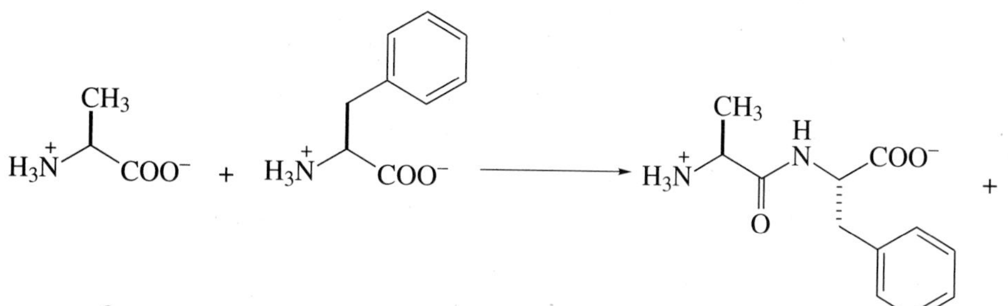

3 L-Alanine (Ala) + 4 L-Phenylalanine (Phe) → Ala–Phe 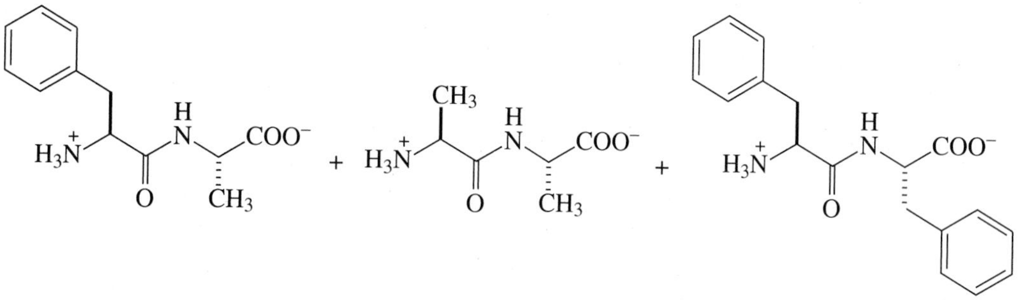 + Phe–Ala + Ala–Ala + Phe–Phe (24.2)

24.3 SYNTHESIS OF THE PROTECTED DIPEPTIDE ALA–PHE–OME

By completing the experimental procedures of this chapter, you will prepare the methyl ester of Ala–Phe (Ala–Phe–OMe). You will be using a general strategy that has been developed to direct the synthesis of polypeptides so that only one specific peptide bond can be formed in the coupling step, thereby avoiding the production of mixtures (Eq. 24.2). The first stage of this process involves protecting (see Sec. 21.2 for more on protecting groups) the amino group of one amino acid so it cannot serve as a *nucleophile*, and also protecting the carboxylic acid group of the other amino acid so it cannot serve as an *electrophile* or an acylating agent. Coupling two such protected α-amino acids furnishes a diprotected dipeptide exclusively, as shown in Equation 24.3.

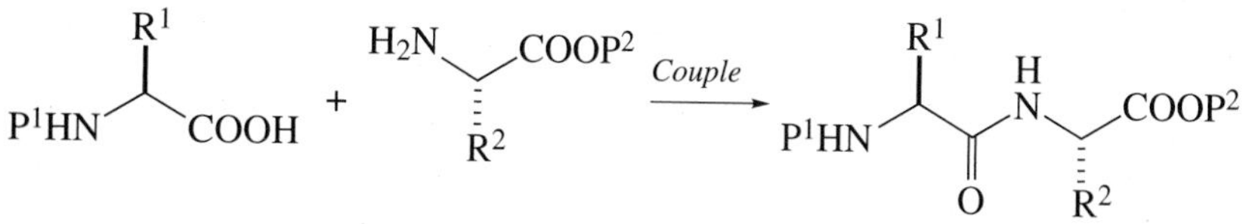

N-Protected α-amino acid + *C*-Protected α-amino acid —Couple→ Diprotected dipeptide (24.3)

In developing a strategy for selecting amine and carboxylic acid protecting groups for the synthesis of peptides, it is essential that each protecting group may be *selectively* removed in the presence of the other; such protecting groups are often referred to as being *orthogonal.* For example, selective removal of the nitrogen-protecting group P^1 from the diprotected dipeptide will give a *C*-protected dipeptide that may then be coupled with another *N*-protected amino acid. The selective deprotection and coupling process may then be repeated to continue chain growth from the *N*-terminal end of the peptide. Alternatively, the protecting group P^2 on the carboxylic acid can be removed first to give a *N*-protected dipeptide that may be coupled with a *C*-protected amino acid to propagate chain growth from the *C*-terminal end. For reasons that are beyond the scope of this discussion, peptide synthesis in the laboratory is normally performed by extending the growing chain from the *N*-terminus. On the other hand, the enzyme-catalyzed synthesis of peptides *in vivo* proceeds by addition of new amino acids to the *C*-terminus of the growing chain.

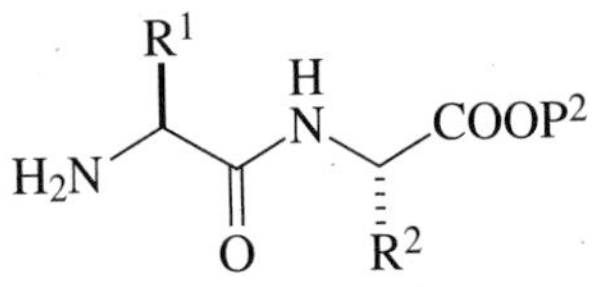

C-Protected dipeptide 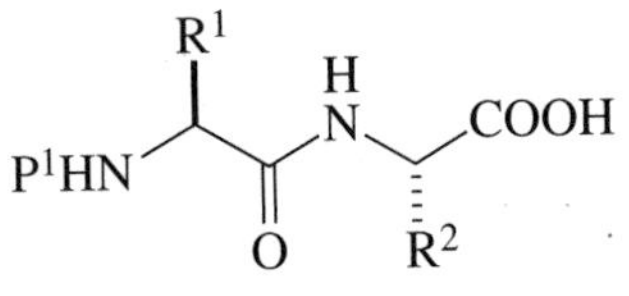   *N*-Protected dipeptide

There are a variety of protecting groups for an amino group that may be used, but you will protect this functionality in L-alanine (**3**) with a *tert*-butoxycarbonyl (Boc) group by treating **3** with di-*tert*-butyl dicarbonate (**5**) in aqueous base to give *N-tert*-butoxycarbonyl-L-alanine (Boc–Ala) (**6**) (Eq. 24.4). The carboxylic acid group of an amino acid is typically protected as an ester, and you will protect L-phenylalanine (**4**) by forming a methyl ester using thionyl chloride, $SOCl_2$, in methanol (Eq. 24.5) to give methyl L-phenylalaninate hydrochloride (**7**). It is left as an exception to provide mechanisms for the reactions depicted in Equations 24.4 and 24.5 (see Exercises 17 and 25 at the end of the chapter). The relative solubilities of **6** and **7** in water or organic solvents depend upon whether or not they are in their ionized or neutral forms.

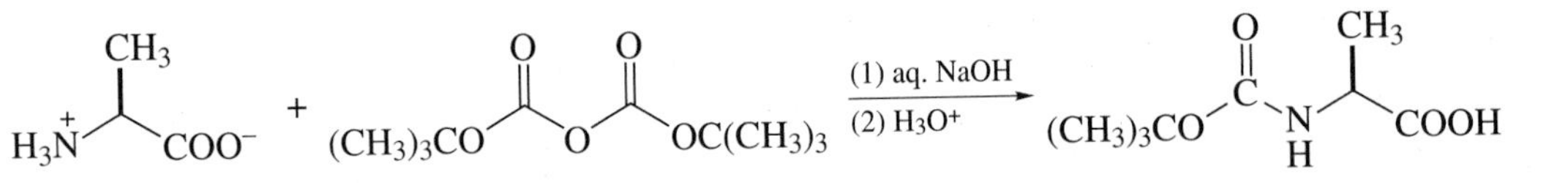

(24.4)

3 + **5** Di-*tert*-butyl dicarbonate → (1) aq. NaOH (2) H_3O^+ → **6** *N-tert*-Butoxycarbonyl-L-alanine (Boc–Ala)

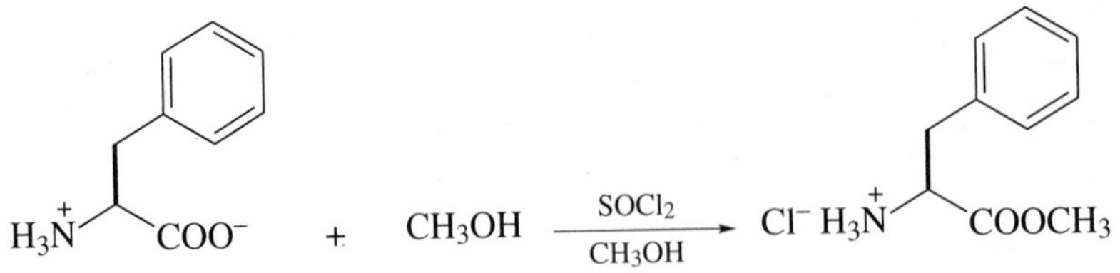

(24.5)

4 + CH_3OH → $SOCl_2$, CH_3OH → **7** Methyl L-phenylalaninate hydrochloride (Phe–OMe • HCl)

See more on *Peptide Coupling Reagents*

The next step in our synthesis of Ala–Phe requires activating the carboxyl group of *N-tert*-butoxy-L-carbonylalanine (**6**) so it can serve as an acylating agent for the amino group of methyl L-phenylalaninate to give the protected dipeptide methyl *N-tert*-butoxycarbonyl-L-alanyl-L-phenylalaninate (**10**). The acid chlorides of carboxylic acids are often used as intermediates for forming amides (Sec. 20.3), but such derivatives are unsuitable for use in procedures for coupling peptides because their high reactivity leads to side reactions and undesirable by-products. Although a variety of specialized methods and reagents for forming peptide bonds have been developed, the simple and effective method used in this experiment involves converting **6** to the *mixed anhydride* **9** by reaction with isobutyl chloroformate in the presence of a tertiary amine such as *N*-methylmorpholine (Eq. 24.6). This mixed anhydride is easily hydrolyzed, so it is not isolated. Rather, a solution of methyl L-phenylalaninate hydrochloride (**7**) and *N*-methylmorpholine is added to a solution of **9** to give **10** (Eq. 24.7). It is left as an exercise to provide mechanisms for the reactions depicted in Equations 24.6 and 24.7 (see Exercises 34 and 35 at the end of the chapter). Unlike free amino acids and peptides, protected peptides are not zwitterionic, so they are more soluble in organic solvents than in water.

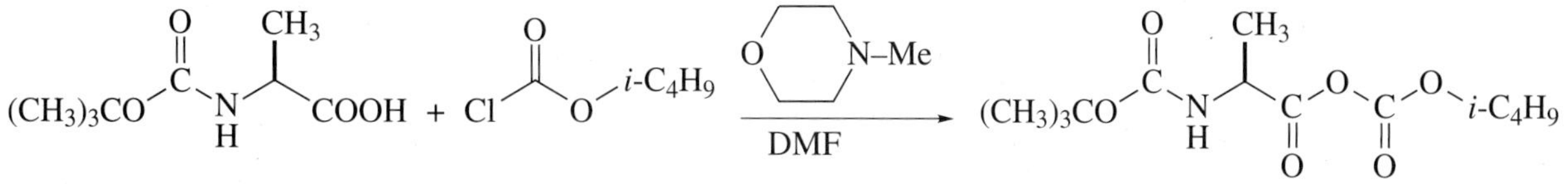

6 + **8** (Isobutyl chloroformate) —(N-methylmorpholine, DMF)→ **9** (24.6)

9 + **7** —(N-methylmorpholine, DMF)→ **10** (24.7)

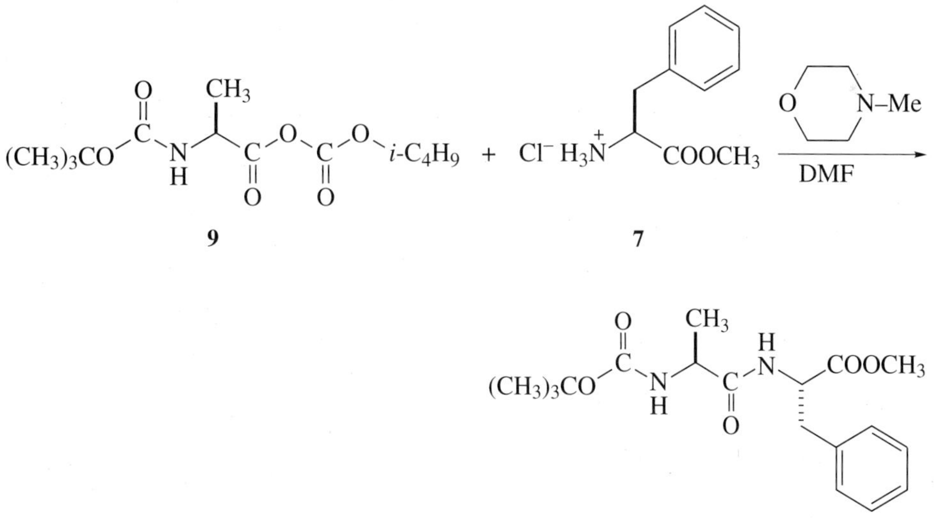

10
Methyl *N-tert*-butoxycarbonyl-L-alanyl-L-phenylalaninate
(Boc–Ala–Phe–OMe)

To convert the diprotected dipeptide Boc–Ala–Phe–OMe (**10**) into Ala–Phe, it is necessary to remove the protecting groups from the nitrogen atom of the *N*-terminal L-alanine and the carboxylic acid of the *C*-terminal L-phenylalanine. This might be accomplished by hydrolysis of both groups under acidic conditions. However, the zwitterionic Ala–Phe is difficult to isolate, so in this experiment you will only selectively remove the Boc group from the L-alanine residue to give methyl L-alanyl-L-phenylalaninate trifluoroacetate (**11**). This involves an acid-catalyzed reaction in which carbon dioxide and isobutylene are also produced (Eq. 24.8). It is left as an exercise to provide a mechanism for this transformation (see Exercise 38 at the end of the chapter).

$$(CH_3)_3CO\text{–}C(=O)\text{–}NH\text{–}CH(CH_3)\text{–}C(=O)\text{–}NH\text{–}CH(CH_2C_6H_5)\text{–}COOCH_3 \xrightarrow[CH_2Cl_2]{CF_3CO_2H}$$

10

$$CF_3COO^- \, H_3\overset{+}{N}\text{–}CH(CH_3)\text{–}C(=O)\text{–}NH\text{–}CH(CH_2C_6H_5)\text{–}COOCH_3 + CO_2 + CH_2{=}C(CH_3)_2 \quad (24.8)$$

Isobutylene

11

Methyl L-alanyl-L-phenylalaninate trifluoroacetate
(Ala–Phe–OMe • CF_3CO_2H)

The conversion of **11** into Ala–Phe (**12**) requires hydrolysis of the methyl ester, a reaction that may be accomplished by hydrolysis of **11** in aqueous acid or base followed by neutralization to provide **12** (Eq. 24.9). You may have an opportunity to develop a procedure to effect this transformation and isolate **12** if you perform the suggested Discovery Experiment.

$$CF_3COO^- \, H_3\overset{+}{N}\text{–}CH(CH_3)\text{–}C(=O)\text{–}NH\text{–}CH(CH_2C_6H_5)\text{–}COOCH_3 \xrightarrow[\text{neutralization}]{\text{hydrolysis}} H_3\overset{+}{N}\text{–}CH(CH_3)\text{–}C(=O)\text{–}NH\text{–}CH(CH_2C_6H_5)\text{–}COO^- \quad (24.9)$$

11

12

L-Alanyl-L-phenylalanine
(Ala–Phe)

Dipeptides are rarely the final products of peptide synthesis, but the principles described for the preparation of **12** may be readily applied to the synthesis of peptides containing more amino acids. In practice, the stepwise extension is performed at the *N*-terminus of a growing peptide chain that is protected with an ester group at the *C*-terminus. *N*-Protected amino acids are sequentially added by a coupling reaction followed by removal of the protecting group on the *N*-terminus until the desired peptide, in its diprotected form, has been prepared. The protecting groups at the *N*- and *C*-termini are then removed to provide the peptide of interest. Indeed, this is precisely the strategy that is used in the **solid-phase synthesis** of large peptides (see the Historical Highlight at the end of this chapter). It is left as an exercise for you to apply these principles to preparing the tripeptide Val–Ala–Phe from the monoprotected dipeptide **11** (see Exercise 10 at the end of the chapter).

EXPERIMENTAL PROCEDURES

A ■ *Preparation of* N-tert-*Butoxycarbonyl* L-*Alanine*

Purpose To demonstrate the protection of the amine group of an α-amino acid by forming a *tert*-butyl carbamate.

SAFETY ALERT

1. **Di-*tert*-butyl dicarbonate is a low-melting solid that is a liquid at or near room temperature and hydrolyzes readily to give *tert*-butyl alcohol and carbon dioxide. Wear latex gloves when handling this chemical and do not breathe it or allow it to come in contact with your skin. If contact occurs, immediately flood the affected area with cold water and then rinse with 5% sodium bicarbonate solution. Dispense di-*tert*-butyl dicarbonate only at a hood and minimize its exposure to atmospheric moisture by keeping the bottle tightly closed when not in use.**
2. **The 3 *M* aqueous sodium hydroxide solution is caustic. *Do not allow it to come in contact with your skin.* If this should happen, wash the affected area with dilute acetic acid and then copious amounts of water.**
3. **Diethyl ether is *extremely* flammable and volatile, and its vapors can easily travel several feet along the bench top or the floor and then be ignited. Consequently, *be certain there are no open flames anywhere in the laboratory* whenever you are working with it. Use a *flameless* heating source *whenever* heating is required.**

MINISCALE PROCEDURE

Preparation Sign in at **www.cengage.com/login** to answer Pre-Lab Exercises, access videos, and read the MSDSs for the chemicals used or produced in this procedure. Review Sections 2.5, 2.10, 2.11, 2.13, 2.17, 2.21, and 2.29.

Apparatus A 50-mL round-bottom flask, 5-mL glass syringe with a needle, separatory funnel, apparatus for magnetic stirring, simple distillation, vacuum filtration, and *flameless* heating.

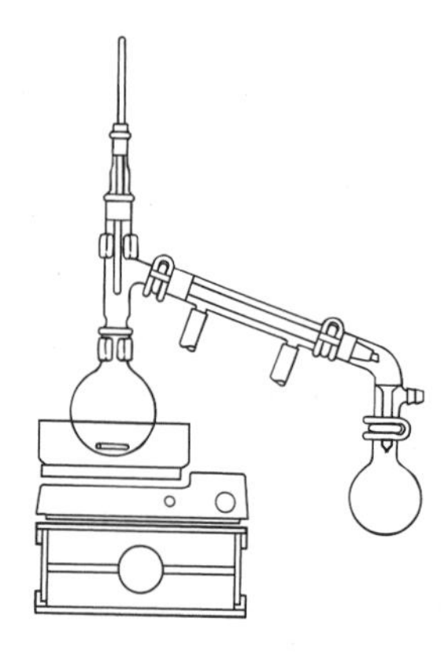

Set-Up and Reaction Place 0.90 g of L-alanine in a 50-mL round-bottom flask containing a stirbar. Add 5 mL of a 3 *M* solution of aqueous sodium hydroxide and 5 mL of *tert*-butyl alcohol to the round-bottom flask and stir the mixture until all of the solids have dissolved. With continued stirring, add 2.5 mL of di-*tert*-butyl dicarbonate and then stir the reaction mixture for 45 min at room temperature.

Work-Up Add 10 mL of water to the reaction mixture and transfer the solution to a separatory funnel. Rinse the round-bottom flask with 1–2 mL of water and transfer the rinse to the separatory funnel. Add 25 mL of diethyl ether to the separatory funnel and shake the funnel gently, *venting as necessary*. Separate the layers and acidify the aqueous layer to pH 2 with 3 *M* HCl. Extract the aqueous layer with two 10-mL portions of diethyl ether and then wash the combined organic layers with

10 mL of brine. Dry the organic layer over several spatula-tips full of *anhydrous* sodium sulfate. Occasionally swirl the mixture over a period of 10–15 min to facilitate drying; add further small portions of anhydrous sodium sulfate if the solution remains cloudy.

Isolation and Purification Filter or carefully decant the dried organic solution into a round-bottom flask equipped for simple distillation and concentrate the solution to a volume of approximately 2 mL. Alternatively, use rotary evaporation or other techniques to concentrate the solution. Allow the solution to cool to room temperature and transfer it to a 50-mL Erlenmeyer flask. Rinse the round-bottom flask with 0.5–1 mL of diethyl ether and transfer this rinse to the Erlenmeyer flask. Add approximately 25 mL of hexanes to the ethereal solution, stir the solution with a glass stirring rod, and place the flask in an ice-water bath for 10 min. If no solid precipitates, use one of the methods described in Chapter 3 to induce crystallization. Isolate the white solid by vacuum filtration and air-dry it. Recrystallize the crude product by a mixed solvent recrystallization according to the second option for dissolution (Sec. 3.2) by suspending the solid in about 20 mL of boiling hexanes and adding ethyl acetate dropwise until all the solids have dissolved.

Analysis Weigh the recrystallized product and calculate the percent yield; determine its melting point. Obtain IR and ^{1}H NMR spectra of your starting materials and product, and compare them with those of authentic samples (Figs. 24.1–24.6).

MICROSCALE PROCEDURE

Preparation Sign in at **www.cengage.com/login** to answer Pre-Lab Exercises, access videos, and read the MSDSs for the chemicals used or produced in this procedure. Review Sections 2.5, 2.10, 2.11, 2.17, 2.21, and 2.29.

Apparatus A 5-mL conical vial, 1-mL glass syringe with a needle, a screw-cap centrifuge tube, apparatus for magnetic stirring, simple distillation, vaccum filtration, and *flameless* heating.

Set-Up and Reaction Place 0.20 g of L-alanine in a 5-mL conical vial containing a spinvane. Add 1 mL of a 3 *M* solution of aqueous sodium hydroxide and 1 mL of *tert*-butyl alcohol to the conical vial, and stir the mixture until all of the solids have dissolved. With continued stirring, add 0.5 mL of di-*tert*-butyl dicarbonate and then stir the reaction mixture for 45 min at room temperature.

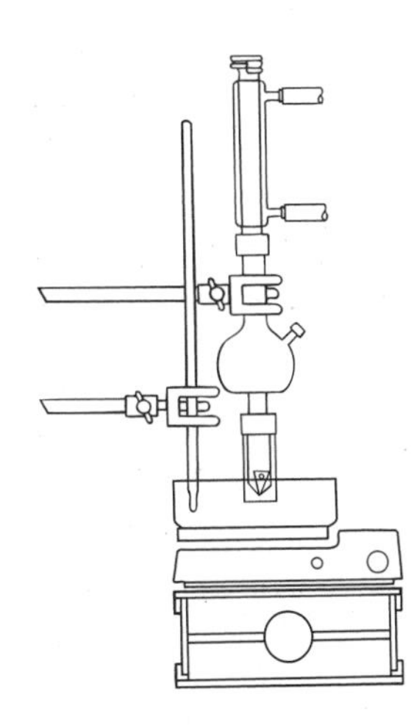

Work-Up Add 2 mL of water to the reaction mixture and transfer the solution to a screw-cap centrifuge tube. Rinse the conical vial with about 5 drops of water and transfer the rinse to the screw-cap centrifuge tube. Add 3 mL of diethyl ether to the screw-cap centrifuge tube, screw the cap onto the centrifuge tube, and shake the tube gently, *venting as necessary*. Separate the layers and acidify the aqueous layer to pH 2 with 3 *M* HCl. Extract the aqueous layer with two 1.5-mL portions of diethyl ether and then wash the combined organic layers with 2 mL of brine. Transfer the organic layer to a dry test tube and add several microspatula-tips full of *anhydrous* sodium sulfate. Occasionally swirl the mixture over a period of 10–15 min to facilitate drying; add further small portions of anhydrous sodium sulfate if the solution remains cloudy.

Isolation and Purification Transfer the organic solution to a conical vial equipped for simple distillation, and remove most of the solvent. Alternatively, use rotary evaporation or other techniques to concentrate the solution. Allow the vial to cool to room temperature. Add about 2 mL of hexanes to the residue in the vial, stir the solution with a glass stirring rod, and place the flask in an ice-water bath for 10 min. If no solid precipitates, use one of the methods described in Chapter 3 to induce crystallization. Isolate the white solid by vacuum filtration and air-dry it. Recrystallize the crude product by a mixed solvent recrystallization according to the second option for dissolution (Sec. 3.2) by suspending the solid in about 4 mL of boiling hexanes and adding ethyl acetate dropwise until all the solids have dissolved.

Analysis Weigh the recrystallized product and calculate the percent yield. Determine its melting point. Obtain IR and ^1H NMR spectra of your starting materials and product and compare them with those of authentic samples (Figs. 24.1–24.6).

WRAPPING IT UP

Combine all of the *aqueous layers*, adjust the pH of the solution to approximately 7 with solid sodium carbonate, and flush it down the drain with excess water. Pour the *organic layer from the first extraction* and the recovered *diethyl ether* and *recrystallization solvent* into a container for nonhalogenated organic liquids. Spread the *sodium sulfate* on a tray in the hood to evaporate residual solvent, and then put it in the container for nonhazardous solids.

B ■ *Preparation of Methyl L-Phenylalaninate Hydrochloride*

Purpose To demonstrate the protection of the carboxylic acid group of an α-amino acid by forming a methyl ester.

SAFETY ALERT

1. **Thionyl chloride is a volatile and *corrosive* chemical that undergoes rapid reaction with water and atmospheric moisture to produce sulfur dioxide and hydrogen chloride, both of which are also corrosive. Wear latex gloves when handling this chemical and do not breathe it or allow it to come in contact with your skin. If contact occurs, immediately flood the affected area with cold water and then rinse with 5% sodium bicarbonate solution. Dispense thionyl chloride only at a hood and minimize its exposure to atmospheric moisture by keeping the bottle tightly closed when not in use.**
2. **This experiment should be performed in a hood if possible.**
3. **Diethyl ether is *extremely* flammable and volatile, and its vapors can easily travel several feet along the bench top or the floor and then be ignited. Consequently, *be certain there are no open flames anywhere in the laboratory* whenever you are working with it. Use a flameless heating source *whenever* heating is required.**

MINISCALE PROCEDURE

Preparation Sign in at **www.cengage.com/login** to answer Pre-Lab Exercises, access videos, and read the MSDSs for the chemicals used or produced in this procedure. Review Sections 2.10, 2.11, 2.17, and 2.22.

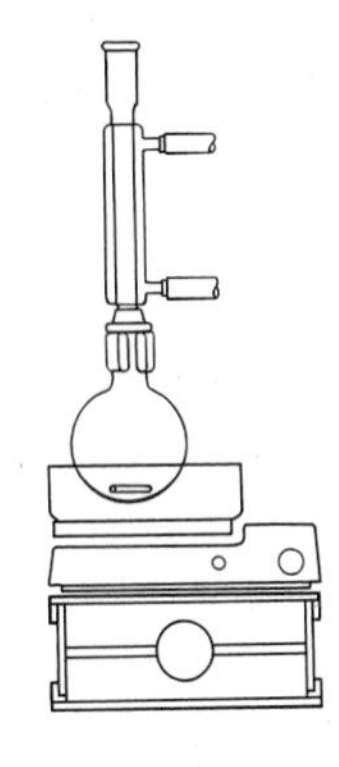

Apparatus A 25-mL round-bottom flask, 1-mL glass syringe with needle, apparatus for heating under reflux, magnetic stirring, vacuum filtration, and *flameless* heating.

Set-Up Place 1.0 g of L-phenylalanine in a 25-mL round-bottom flask containing a stirbar, add 5 mL of MeOH, and begin stirring the mixture.

Reaction Add 0.5 mL of thionyl chloride dropwise using a syringe. After the addition of thionyl chloride is completed, the solution should be homogeneous. Equip the round-bottom flask with a reflux condenser and heat the mixture under gentle reflux for 45 min.

Isolation and Purification Allow the reaction mixture to cool to room temperature and then transfer the solution to an Erlenmeyer flask. Rinse the round-bottom flask with a 0.5–1-mL portion of methanol and transfer this rinse to the Erlenmeyer flask. Place the Erlenmeyer flask in an ice-water bath and add 25 mL of diethyl ether. Using a glass stirring rod, scratch the inside of the flask at the air-liquid interface to induce crystallization. Isolate the white solid by vacuum filtration and air-dry it. Recrystallize the crude product by a mixed solvent recrystallization according to the second option for dissolution (Sec. 3.2) by suspending the solid in about 20 mL of boiling diethyl ether and adding methanol dropwise until all the solids have dissolved.

Analysis Weigh the recrystallized product and calculate the percent yield. Determine its melting point. Obtain IR and ^{1}H NMR spectra of your starting material and product, and compare them with those of authentic samples (Figs. 24.7–24.10).

MICROSCALE PROCEDURE

Preparation Sign in at **www.cengage.com/login** to answer Pre-Lab Exercises, access videos, and read the MSDSs for the chemicals used or produced in this procedure. Review Sections 2.10, 2.11, 2.17, and 2.22.

Apparatus A 3-mL conical vial, 1-mL glass syringe with needle, apparatus for heating under reflux, magnetic stirring, vacuum filtration, and *flameless* heating.

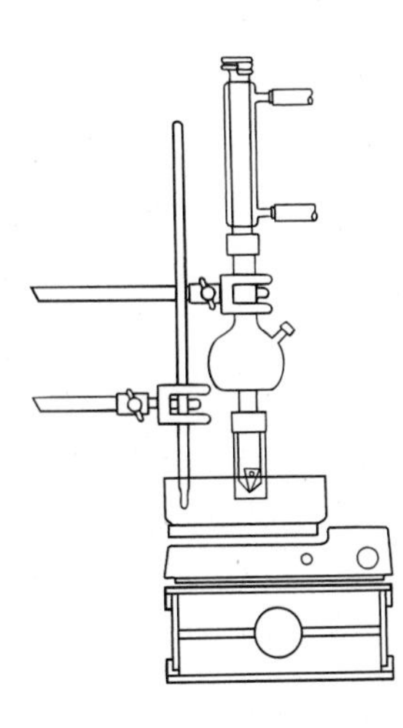

Set-Up Place 0.2 g of L-phenylalanine in a 3-mL conical vial containing a spinvane, add 1 mL of MeOH, and begin stirring the mixture.

Reaction Add 0.1 mL of thionyl chloride dropwise using a syringe. After the addition of thionyl chloride is completed, the solution should be homogeneous. Equip the conical vial with a reflux condenser, and heat the mixture under gentle reflux for 45 min.

Isolation and Purification Allow the reaction mixture to cool to room temperature and then transfer the solution to a small Erlenmeyer flask. Rinse the round-bottom flask with a few drops of methanol and transfer this rinse to the Erlenmeyer flask. Place the Erlenmeyer flask into an ice-water bath and add 5 mL of diethyl ether. Use a glass stirring rod to scratch the inside of the flask at the air-liquid interface to induce crystallization. Isolate the white solid by vacuum filtration and air-dry it. Recrystallize the crude product by a mixed solvent recrystallization according to the second option for dissolution (Sec. 3.2) by suspending the solid in about 4 mL of boiling diethyl ether and adding methanol dropwise until all the solids have dissolved.

Analysis Weigh the recrystallized product and calculate the percent yield. Determine its melting point. Obtain IR and ^1H NMR spectra of your material and product and compare them with those of authentic samples (Figs. 24.7–24.10).

WRAPPING IT UP

Combine all the *organic filtrates* and pour them into a container for nonhalogenated organic liquids.

C ■ *Preparation of Methyl* N-tert-*Butoxycarbonyl* L-*Alanyl*-L-*Phenylalaninate*

Purpose To demonstrate the formation of a diprotected dipeptide by the reaction of the mixed anhydride of an *N*-protected amino acid with the ester of a second amino acid.

SAFETY ALERT

1. **Isobutyl chloroformate is a lachrymator and a volatile, *corrosive* chemical that undergoes rapid reaction with water and atmospheric moisture to produce carbon dioxide, isobutyl alcohol, and hydrogen chloride, which is also corrosive. Wear latex gloves when handling this chemical and do not breathe it or allow it to come in contact with your skin. If contact occurs, immediately flood the affected area with cold water and then rinse with 5% sodium bicarbonate solution. Dispense isobutyl chloroformate only at a hood and minimize its exposure to atmospheric moisture by keeping the bottle tightly closed when not in use.**
2. **This experiment should be performed in a hood if possible.**
3. **Diethyl ether is *extremely* flammable and volatile, and its vapors can easily travel several feet along the bench top or the floor and then be ignited. Consequently, *be certain there are no open flames anywhere in the laboratory* whenever you are working with it. Use a *flameless* heating source *whenever* heating is required.**

MINISCALE PROCEDURE

Preparation Sign in at **www.cengage.com/login** to answer Pre-Lab Exercises, access videos, and read the MSDSs for the chemicals used or produced in this procedure. Review Sections 2.5, 2.10, 2.11, 2.13, 2.17, 2.21, and 2.29.

Apparatus A 100-mL round-bottom flask, 50-mL Erlenmeyer flask, two 1-mL glass syringes with needles, ice-water bath, separatory funnel, apparatus for magnetic stirring, simple distillation, vacuum filtration, and *flameless* heating.

Set-Up and Reaction Place 0.63 g of methyl L-phenylalaninate hydrochloride in a 50-mL Erlenmeyer flask. Add 10 mL of dimethylformamide and 0.3 mL of *N*-methylmorpholine to the flask, swirl the flask to mix the contents, and place the flask in an ice-water bath. Label the flask as Solution A.

Place 0.50 g of *N*-*tert*-butoxycarbonyl-L-alanine in a 100-mL round-bottom flask containing a stirbar. Add 10 mL of dimethylformamide and 0.3 mL of *N*-methylmorpholine to the flask. Place the flask in an ice-water bath and stir the solution for 5 min. Add 0.4 mL of isobutyl chloroformate dropwise and continue stirring the reaction mixture in the ice-water bath for 5–10 min. Transfer Solution A into the round-bottom flask and continue to stir the reaction with cooling in the ice-water bath for 45 min.

Work-Up Add 20 mL of water to the reaction mixture and transfer it to a separatory funnel. Rinse the round-bottom flask with a 30-mL portion of diethyl ether and transfer the rinse to the separatory funnel. Shake the funnel gently, *venting as necessary.* Separate the layers and wash the organic layer with two 25-mL portions of 1 *M* HCl, a 25-mL portion of saturated sodium bicarbonate, and a 25-mL portion of brine. Transfer the organic layer to a 125-mL Erlenmeyer flask and add several spatula-tips full of *anhydrous* sodium sulfate. Occasionally swirl the mixture over a period of 10–15 min to facilitate drying; add further small portions of anhydrous sodium sulfate if the solution remains cloudy.

Isolation and Purification Filter or carefully decant the dried organic solution into a round-bottom flask equipped for simple distillation, and concentrate to a volume of approximately 5 mL. Alternatively, use rotary evaporation or other techniques to concentrate the solution. Allow the solution to cool to room temperature. Transfer this solution to a 50-mL Erlenmeyer flask. Rinse the round-bottom flask with 0.5–1 mL of diethyl ether and transfer this rinse to the Erlenmeyer flask. Add approximately 15 mL of hexanes to the flask and place the flask in an ice-water bath. If crystals do not form within 5–10 min, use one of the methods described in Chapter 3 to induce crystallization. Isolate the crystals by vacuum filtration and air-dry them. Recrystallize the product from diethyl ether/hexanes (2:1).

Analysis Weigh the recrystallized product and calculate the percent yield. Determine its melting point. Obtain IR and ^{1}H NMR spectra of your product and compare them with those of authentic samples (Figs. 24.11 and 24.12).

MICROSCALE PROCEDURE

Preparation Sign in at **www.cengage.com/login** to answer Pre-Lab Exercises, access videos, and read the MSDSs for the chemicals used or produced in this procedure. Review Sections 2.5, 2.10, 2.11, 2.13, 2.17, 2.21, and 2.29.

Apparatus A 5-mL conical vial, 13-mm × 100-mm test tube, two 1-mL glass syringes with needles, ice-water bath, screw-cap centrifuge tube, apparatus for magnetic stirring simple distillation, vacuum filtration, and *flameless* heating.

Set-Up and Reaction Place 0.12 g of methyl L-phenylalaninate hydrochloride in a 13-mm × 100-mm test tube. Add 1.0 mL dimethylformamide and 0.06 mL of *N*-methylmorpholine to the test tube, swirl the tube to mix the contents, and place it in an ice-water bath. Label the test tube as Solution A.

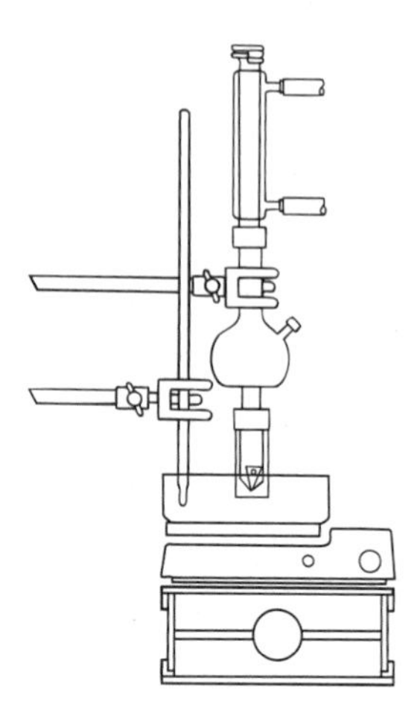

Place 0.10 g of *N-tert*-butoxycarbonyl-L-alanine in a 5-mL conical vial containing a spinvane. Add 1.0 mL of dimethylformamide and 0.06 mL of *N*-methylmorpholine to the vial. Place the vial in an ice-water bath and stir the solution for 5 min. Add 0.08 mL of isobutyl chloroformate dropwise and continue stirring the reaction mixture in the ice-water bath for 5–10 min. Transfer Solution A into the conical vial and continue to stir the reaction with cooling in the ice-water bath for 45 min.

Work-Up Add 20 mL of water to the reaction mixture and transfer it to a screw-cap centrifuge tube. Rinse the conical vial with a 4-mL portion of diethyl ether and transfer the rinse to the screw-cap centrifuge tube, screw the cap onto the centrifuge tube, and shake the tube gently, *venting as necessary.* Separate the layers and wash the organic layer with two 4-mL portions of 1 *M* HCl, a 4-mL portion of saturated sodium bicarbonate, and a 4-mL portion of brine. Transfer the organic layer to a dry test tube and add several spatula-tips full of *anhydrous* sodium sulfate. Occasionally swirl the mixture over a period of 10–15 min to facilitate drying; add further small portions of anhydrous sodium sulfate if the solution remains cloudy.

Isolation and Purification Transfer the organic solution to a conical vial equipped for simple distillation and distill most of the solvent. Alternatively, use rotary evaporation or other techniques to concentrate the solution. Allow the vial and its contents to cool to room temperature, add approximately 3 mL of hexanes to the vial, and place it in an ice-water bath. If crystals do not form within 5–10 min, use one of the methods described in Chapter 3 to induce crystallization. Isolate the crystals by vacuum filtration and air-dry them. Recrystallize the product from diethyl ether/hexanes (2:1).

Analysis Weigh the recrystallized product and calculate the percent yield. Determine its melting point. Obtain IR and ^{1}H NMR spectra of your product and compare them with those of authentic samples (Figs. 24.11 and 24.12).

WRAPPING IT UP

Combine all of the *aqueous layers,* adjust the pH of the solution to approximately 7 with either solid sodium carbonate or 3 *M* HCl, and flush it down the drain with excess water. Pour the recovered *diethyl ether* and *recrystallization solvent* into a container for nonhalogenated organic liquids. Spread the *sodium sulfate* on a tray in the hood to evaporate residual solvent, and then put it in the container for nonhazardous solids.

D ■ *Preparation of Methyl L-Alanylphenyl-L-Alaninate Trifluoroacetate*

Purpose To demonstrate the selective removal of a *tert*-butylcarbamate of a diprotected dipeptide in the presence of a methyl ester.

SAFETY ALERT

1. **Trifluoroacetic acid is a volatile and *corrosive* chemical. Wear latex gloves when handling this chemical and do not breathe it or allow it to come in contact with your skin. If contact occurs, immediately flood the affected area with cold water and then rinse with 5% sodium bicarbonate solution. Dispense trifluoroacetic acid only at a hood.**

2. This experiment should be performed in a hood if possible.

3. Diethyl ether is *extremely* flammable and volatile, and its vapors can easily travel several feet along the bench top or the floor and then be ignited. Consequently, *be certain there are no open flames anywhere in the laboratory* whenever you are working with it. Use a *flameless* heating source *whenever* heating is required.

MINISCALE PROCEDURE

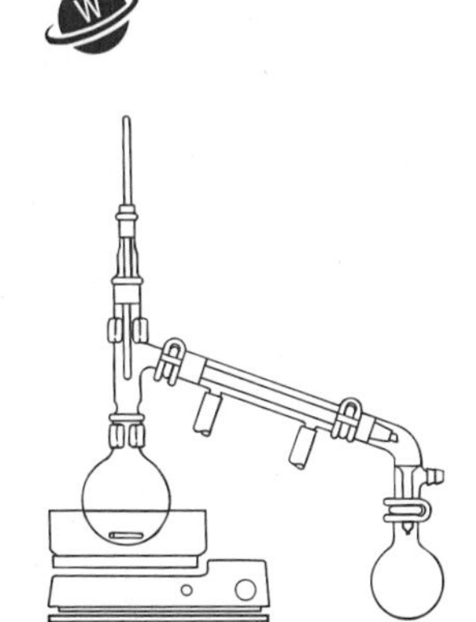

Preparation Sign in at **www.cengage.com/login** to answer Pre-Lab Exercises, access videos, and read the MSDSs for the chemicals used or produced in this procedure. Review Sections 2.5, 2.10, 2.11, 2.13, 2.17, and 2.29.

Apparatus A 10-mL round-bottom flask, 1-mL glass syringe with needle, apparatus for magnetic stirring, simple distillation, vacuum filtration, and *flameless* heating.

Set-Up and Reaction Place 300 mg of methyl *N-tert*-butoxycarbonyl-L-alanyl-L-phenylalaninate in the 10-mL round-bottom flask containing a stirbar, and add 6 mL of dichloromethane. Begin stirring the solution, add 1.5 mL of trifluoroacetic acid to the flask, and continue stirring the mixture for 30 min at room temperature.

Isolation and Purification Equip the flask for simple distillation and distill the solvent. Alternatively, use rotary evaporation or other techniques to concentrate the solution. Allow the flask to cool to room temperature and then add 4 mL of diethyl ether to the flask. Using a spatula, thoroughly mix the solvent with the residue to induce the formation of the solid product. Place the flask in an ice-water bath for 5–10 min to complete the precipitation. Isolate the white solid by vacuum filtration and air-dry it. Recrystallize the crude product from ethyl acetate.

Analysis Weigh the recrystallized product and calculate the percent yield. Determine its melting point. Obtain IR and ^{1}H NMR spectra of your product and compare them with those of authentic samples (Figs. 24.13–24.14).

MICROSCALE PROCEDURE

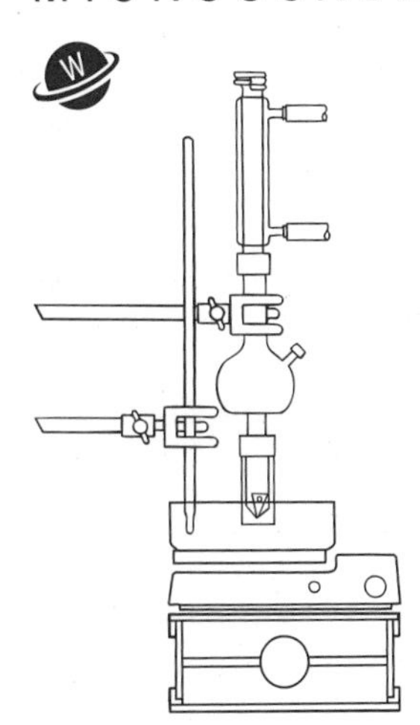

Preparation Sign in at **www.cengage.com/login** to answer Pre-Lab Exercises, access videos, and read the MSDSs for the chemicals used or produced in this procedure. Review Sections 2.5, 2.10, 2.11, 2.13, 2.17, and 2.29.

Apparatus A 3-mL conical vial, 1-mL glass syringe with needle, apparatus for magnetic stirring, simple distillation, vacuum filtration, and *flameless* heating.

Set-Up and Reaction Place 50 mg of methyl *N-tert*-butoxycarbonyl-L-alanyl-L-phenylalaninate in the 3-mL conical vial containing a spinvane and add 1 mL of dichloromethane. Begin stirring the solution, add 0.2 mL of trifluoroacetic acid to the conical vial, and continue stirring the mixture for 30 min at room temperature.

Isolation and Purification Equip the conical vial for simple distillation and distill the solvent. Alternatively, use rotary evaporation or other techniques to concentrate the solution. Allow the conical vial to cool to room temperature and then add 1 mL of diethyl ether to the conical vial. Using a spatula, thoroughly mix the solvent with the residue to induce formation of the solid product. Place the conical vial in an ice-water bath for 5–10 min to complete the precipitation. Isolate the white solid by vacuum filtration and air-dry it. Recrystallize the crude product from ethyl acetate.

Analysis Weigh the recrystallized product and calculate the percent yield. Determine its melting point. Obtain IR and ^1H NMR spectra of your product and compare them with those of authentic samples (Figs. 24.13–24.14).

WRAPPING IT UP

Transfer the distillate to a separatory funnel and wash it *carefully with venting* with a small amount of saturated aqueous sodium bicarbonate. Dilute the *aqueous layer* with water, and flush it down the drain with excess water. Pour the *organic layer* into a container for halogenated organic liquids. Pour the *ethereal filtrate* into a container for nonhalogenated organic liquids.

Discovery Experiment

Synthesis of L*-Alanyl-*L*-phenylalanine*

Consult with your instructor before performing this experiment, in which you will prepare the dipeptide L-alanyl-L-phenylalanine by hydrolysis of the methyl ester of the monoprotected dipeptide Ala–Phe–OMe (**11**). The ester may be hydrolyzed in either aqueous base or acid and you should design procedures for both. You should consider the chemical and physical properties of **11** and develop a reaction and a work-up procedure that will most easily enable you to isolate the product. Obtain the IR, ^1H, and ^{13}C NMR spectra of the purified product for characterization and compare these with the literature data.

EXERCISES

General Questions

1. Of the 19 amino acids with a center of chirality in Table 24.1, identify the one that has the *R*-configuration rather than the *S*-configuration.
2. Amino acids are least soluble in water when the pH of the solution is the same as the isoelectric point for the amino acid. Explain this observation.
3. Write the structures of all the possible products that would be obtained by random peptide bond formation involving L-isoleucine and L-asparagine.
4. Another common protecting group for the amine function of amino acids is the benzyloxycarbonyl (Cbz) group, as exemplified by the protection of L-valine by reaction with benzyloxycarbonyl chloride in the presence of an aqueous base to give *N*-benzyloxycarbonyl-L-valine (Cbz–Val). Using curved arrows to symbolize the flow of electrons, write a stepwise mechanism for this reaction.

$H_3\overset{+}{N}$ ... COO^- + (benzyl)–O–C(=O)–Cl $\xrightarrow[(2)\ H^+]{(1)\ aq.\ NaOH}$ (benzyl)–O–C(=O)–NH– ... COOH

L-Valine | Benzyloxycarbonyl chloride (Cbz–Cl) | *N*-Benzyloxycarbonyl-L-valine (Cbz–Val)

5. One method for removing an *N*-benzyloxycarbonyl group is by catalytic *hydrogenolysis* using hydrogen in the presence of a palladium catalyst. Another method involves use of HBr in glacial acetic acid. Using curved arrows to symbolize the flow of electrons, write a stepwise mechanism for the deprotection of *N*-benzyloxycarbonyl-L-valine by this latter technique.

HBr / CH_3CO_2H → Br^- $H_3\overset{+}{N}$–CH(CH(CH$_3$)$_2$)–COOH + benzyl bromide + CO_2

N-Benzyloxycarbonyl-L-valine → L-Valine

6. Another common ester for protecting amino acids is the *tert*-butyl ester, which may be prepared as exemplified by the reaction of L-valine with isobutylene in the presence of acid to give *tert*-butyl L-valinate. Using curved arrows to symbolize the flow of electrons, write a stepwise mechanism for this reaction.

$H_3\overset{+}{N}$–CH(CH(CH$_3$)$_2$)–COO^- + $(H_3C)_2C{=}CH_2$ → (1) H^+; (2) aq. NaOH → H_2N–CH(CH(CH$_3$)$_2$)–$COOC(CH_3)_3$

L-Valine — Isobutylene — *tert*-Butyl L-valinate (Val–O–*t*-Bu)

7. It is possible to hydrolyze the methyl ester group of **10** selectively using lithium hydroxide in methanol, followed by acidification to give *N*-*tert*-butoxycarbonyl-L-alanyl-L-phenylalanine.

$(CH_3)_3CO$–C(=O)–NH–CH(CH$_3$)–C(=O)–NH–CH(CH$_2$Ph)–$COOCH_3$ (**10**) → (1) LiOH, CH_3OH; (2) H_3O^+ → $(CH_3)_3CO$–C(=O)–NH–CH(CH$_3$)–C(=O)–NH–CH(CH$_2$Ph)–COOH

N-*tert*-Butoxycarbony-L-alanyl-L-phenylalanine (Boc–Ala–Phe)

a. Using curved arrows to symbolize the flow of electrons, write a stepwise mechanism for the hydrolysis.

b. There are three different carbonyl groups in **10** that could undergo hydrolysis, two of which are associated with esters, yet the methyl ester is hydrolyzed preferentially. Provide an explanation for this observation.

8. When exposed to solutions of aqueous base, the *tert*-butyl esters of amino acids and peptides are much more stable toward hydrolysis than the corresponding methyl esters. Provide an explanation for this observation.

9. The *tert*-butyl esters of amino acids and peptides may be readily removed upon treatment with trifluoroacetic acid in dichloromethane. Using curved arrows to symbolize the flow of electrons, write a stepwise mechanism for this reaction.

10. Starting with L-aspartic acid and L-phenylalanine and, using any of the protecting groups presented in this chapter, including those in Exercises 3 and 5, outline a synthesis of the artificial sweetener aspartame (**2**).
11. Starting with the monoprotected dipeptide **11** and the unprotected amino acid L-valine, outline a sequence of reactions for the preparation of the monoprotected tripeptide Val–Ala–Phe.

Questions for Part A

12. Write the balanced equation for the reaction of di-*tert*-butyl dicarbonate with water.
13. Using curved arrows to symbolize the flow of electrons, write a stepwise mechanism for the reaction of di-*tert*-butyl dicarbonate with water in Exercise 12.
14. Write the structure of the charged form of L-alanine in aqueous base.
15. In the work-up of the reaction, the diethyl ether layer and aqueous layers are separated. What compound(s) remain in the diethyl ether layer? The aqueous layer?
16. Why is it necessary to acidify the aqueous layer before extracting it with diethyl ether?
17. What compound(s) might remain in the aqueous layer after this extraction with diethyl ether?
18. Using curved arrows to symbolize the flow of electrons, write a stepwise mechanism for the formation of *N-tert*-butoxycarbonyl-L-alanine from L-alanine and di-*tert*-butyl dicarbonate according to Equation 24.4.

Sign in at **www.cengage.com/login** and use the spectra viewer and Tables 8.1–8.8 as needed to answer the blue-numbered questions on spectroscopy.

19. Consider the spectral data for L-alanine (Figs. 24.1 and 24.2).
    a. In the functional group region of the IR spectrum, specify the absorptions associated with the nitrogen-hydrogen bonds and carbonyl component of the carboxyl group.
    b. In the ^{1}H NMR spectrum, assign as many as possible of the various resonances to the hydrogen nuclei responsible for them.
    c. For the ^{13}C NMR data, assign the various resonances to the carbon nuclei responsible for them.
20. Consider the spectral data for di-*tert*-butyl dicarbonate (Figs. 24.3 and 24.4).
    a. In the functional group region of the IR spectrum, specify the absorption associated with the carbonyl component of the dicarbonate group.
    b. In the ^{1}H NMR spectrum, assign the resonances to the hydrogen nuclei responsible for them.
    c. For the ^{13}C NMR data, assign the various resonances to the carbon nuclei responsible for them.
21. Consider the spectral data for *N-tert*-butoxycarbonyl-L-alanine (Figs. 24.5 and 24.6).
    a. In the functional group region of the IR spectrum, specify the absorptions associated with the nitrogen-hydrogen bond and the carbonyl components of the carboxyl and the carbamate groups.
    b. In the ^{1}H NMR spectrum, assign as many as possible of the various resonances to the hydrogen nuclei responsible for them.
    c. For the ^{13}C NMR data, assign the various resonances to the carbon nuclei responsible for them.
22. Discuss the differences observed in the IR and NMR spectra of L-alanine, di-*tert*-butyl dicarbonate, and *N-tert*-butoxycarbonyl-L-alanine that are consistent with the formation of the latter in this procedure.

Questions for Part B

23. Write the balanced equation for the reaction of thionyl chloride with an excess of methanol.

24. Using curved arrows to symbolize the flow of electrons, write a stepwise mechanism for the reaction of thionyl chloride with methanol.

25. Write the structure of the charged form of L-phenylalanine in a solution of methanolic hydrogen chloride.

26. Assume that the HCl generated by the reaction of thionyl chloride with methanol catalyzes the esterification of L-phenylalanine. Using curved arrows to symbolize the flow of electrons, write a stepwise mechanism for the acid-catalyzed esterification of phenylalanine with methanol according to Equation 24.5.

Sign in at **www.cengage.com/login** and use the spectra viewer and Tables 8.1–8.8 as needed to answer the blue-numbered questions on spectroscopy.

27. Consider the spectral data for L-phenylalanine (Figs. 24.7 and 24.8).

- **a.** In the functional group region of the IR spectrum, specify the absorptions associated with the nitrogen-hydrogen bonds and the carbonyl component of the carboxyl group.
- **b.** In the ^{1}H NMR spectrum, assign as many as possible of the various resonances to the hydrogen nuclei responsible for them.
- **c.** For the ^{13}C NMR data, assign the various resonances to the carbon nuclei responsible for them.

28. Consider the spectral data for methyl L-phenylalaninate hydrochloride (Figs. 24.9 and 24.10).

- **a.** In the functional group region of the IR spectrum, specify the absorptions associated with the nitrogen-hydrogen bonds and the carbonyl component of the ester group.
- **b.** In the ^{1}H NMR spectrum, assign as many as possible of the various resonances to the hydrogen nuclei responsible for them.
- **c.** For the ^{13}C NMR data, assign the various resonances to the carbon nuclei responsible for them.

29. Discuss the differences observed in the IR and NMR spectra of L-phenylalanine and methyl L-phenylalaninate hydrochloride that are consistent with the formation of the latter in this procedure.

Questions for Part C

30. Write the balanced equation for the reaction of isobutyl chloroformate with water.

31. Why is *N*-methylmorpholine required in the conversion of **6** and **8** into **9** (Eq. 24.6) in the first step of the procedure used in this experiment?

32. Why is *N*-methylmorpholine required in the conversion of **7** and **9** into **10** (Eq. 24.7) in the second step of the procedure used in this experiment?

33. In the work-up of the reaction, the organic layer is washed with dilute aqueous acid. What is the purpose of this step?

34. In the work-up of the reaction, the organic layer is washed with dilute aqueous base. What is the purpose of this step?

35. Using curved arrows to symbolize the flow of electrons, write a stepwise mechanism for the formation of **9** from *N*-*tert*-butoxycarbonyl-L-alanine (**6**) according to Equation 24.6.

36. Using curved arrows to symbolize the flow of electrons, write a stepwise mechanism for the formation of methyl *N*-*tert*-butoxycarbonyl-L-alanylphenyl-L-alaninate (**10**) from **7** and **9** according to Equation 24.7.

Sign in at **www.cengage.com/login** and use the spectra viewer and Tables 8.1–8.8 as needed to answer the blue-numbered questions on spectroscopy.

37. Consider the spectral data for methyl *N-tert*-butoxycarbonyl-L-alanyl-L-phenylalaninate (Figs. 24.11 and 24.12).
    a. In the functional group region of the IR spectrum, specify the absorptions associated with the nitrogen-hydrogen bonds and the carbonyl components of the ester, amide, and carbamate groups.
    b. In the ^{1}H NMR spectrum, assign as many as possible of the various resonances to the hydrogen nuclei responsible for them.
    c. For the ^{13}C NMR data, assign the various resonances to the carbon nuclei responsible for them.
38. Discuss the differences observed in the IR and NMR spectra of *N-tert*-butoxycarbonyl-L-alanine, methyl L-phenylalaninate hydrochloride, and methyl *N-tert*-butoxycarbonyl-L-alanyl-L-phenylalaninate that are consistent with the formation of the latter in this procedure.

Questions for Part D

39. Using curved arrows to symbolize the flow of electrons, write a stepwise mechanism for the formation of methyl L-alanyl-L-phenylalaninate (**11**) from **10** according to Equation 24.8.
40. Consider the spectral data for methyl L-alanyl-L-phenylalaninate (Figs. 24.13 and 24.14).
    a. In the functional group region of the IR spectrum, specify the absorptions associated with the nitrogen-hydrogen bonds and the carbonyl components of the ester and amide groups.
    b. In the ^{1}H NMR spectrum, assign as many as possible of the various resonances to the hydrogen nuclei responsible for them.
    c. For the ^{13}C NMR data, assign the various resonances to the carbon nuclei responsible for them.
41. Discuss the differences observed in the IR and NMR spectra of methyl *N-tert*-butoxycarbonyl-L-alanyl-L-phenylalanine and methyl L-alanyl-L-phenylalaninate that are consistent with the formation of the latter in this procedure.

SPECTRA

Starting Materials and Products

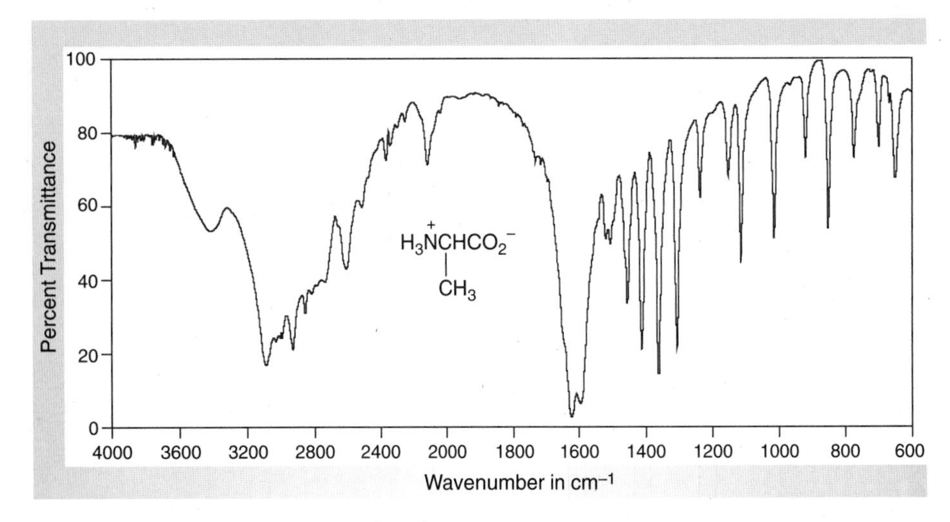

Figure 24.1
IR spectrum of L-alanine (IR card).

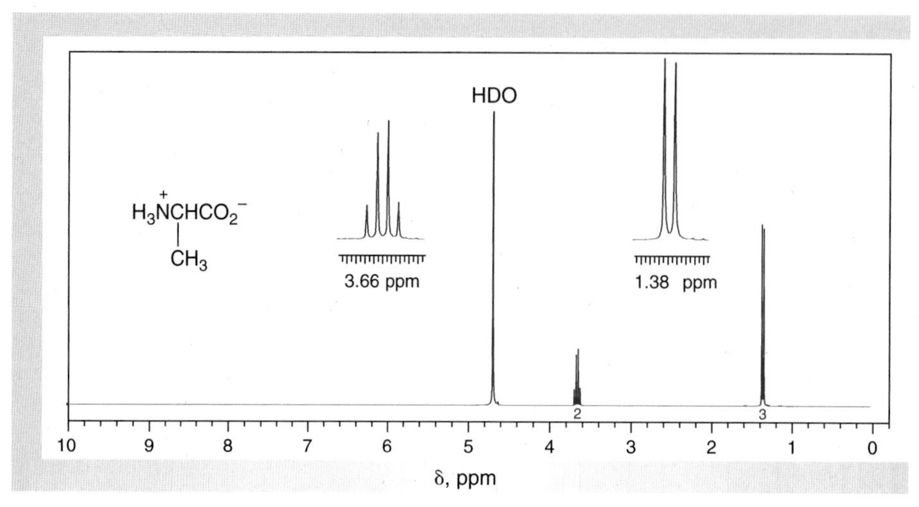

Figure 24.2
NMR data for L-alanine (D_2O).

(a) ^1H NMR spectrum (300 MHz).
(b) ^{13}C NMR data: δ 17.3, 51.7, 176.6.

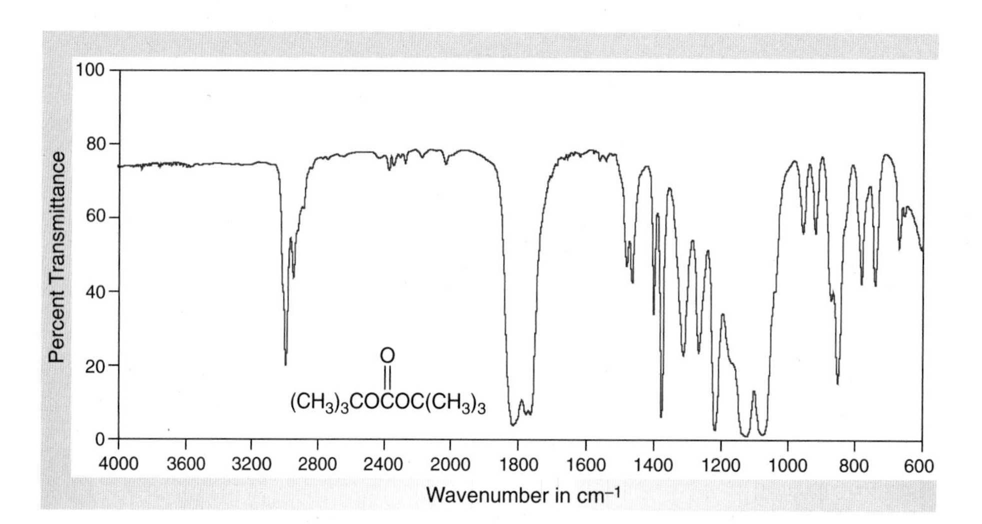

Figure 24.3
*IR spectrum of di-*tert*-butyl dicarbonate (neat).*

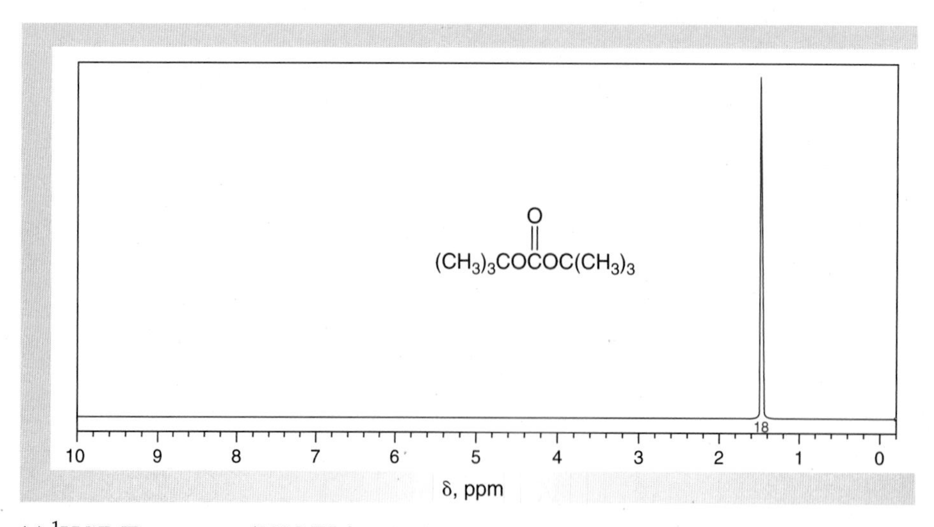

Figure 24.4
*NMR data for di-*tert*-butyl dicarbonate ($CDCl_3$).*

(a) ^1H NMR spectrum (300 MHz).
(b) ^{13}C NMR data: δ 27.5, 85.2, 146.9.

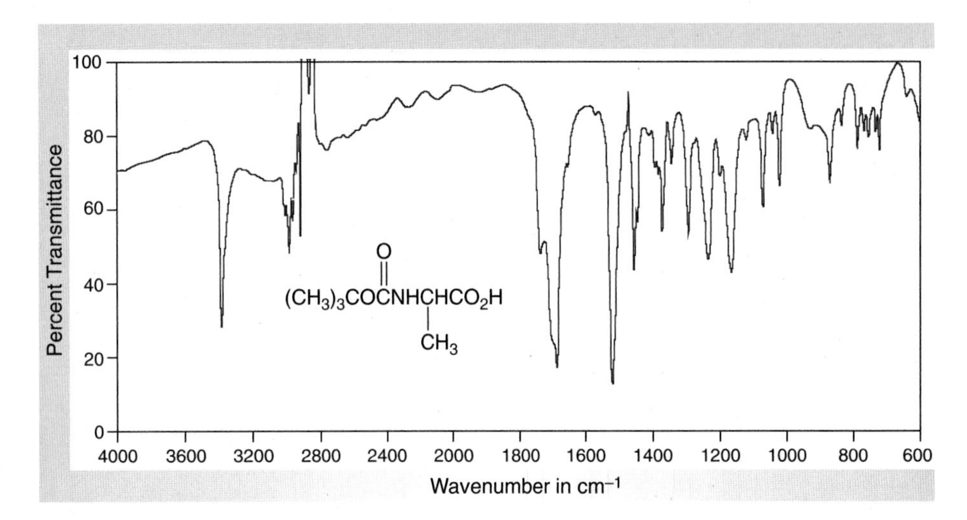

Figure 24.5
IR spectrum of N-tert-*butoxycarbonyl-*L*-alanine (IR card).*

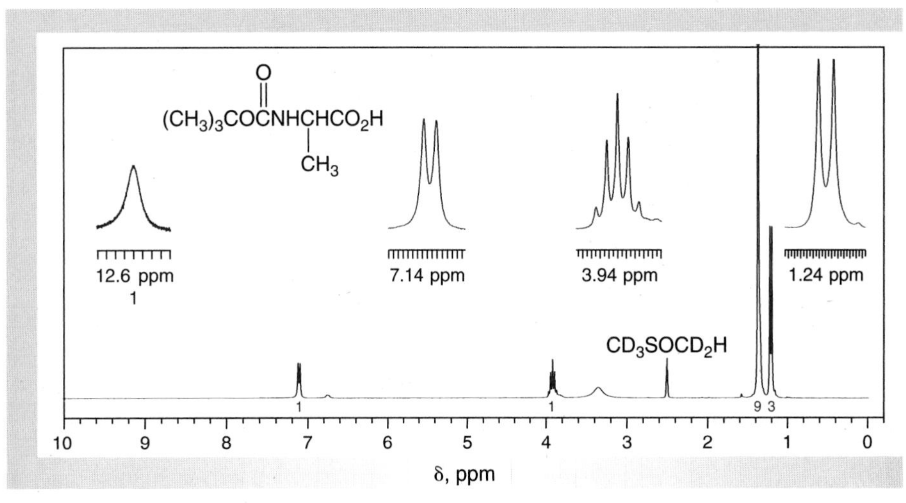

Figure 24.6
NMR data for N-tert-*butoxycarbonyl-*L*-alanine (DMSO-d_6).*

(a) ^{1}H NMR spectrum (300 MHz).
(b) ^{13}C NMR data: δ 17.7, 28.9, 49.5, 78.6, 155.9, 175.4.

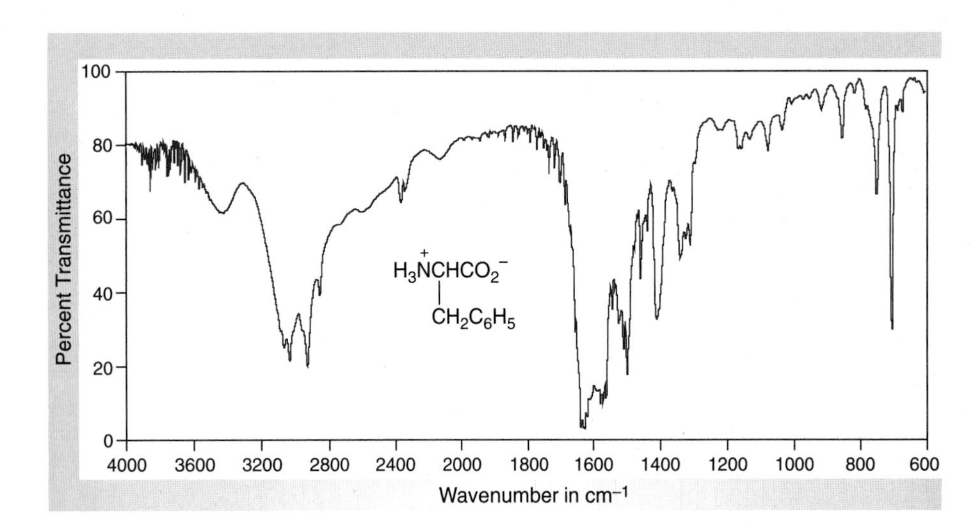

Figure 24.7
IR spectrum of L*-phenylalanine (IR card).*

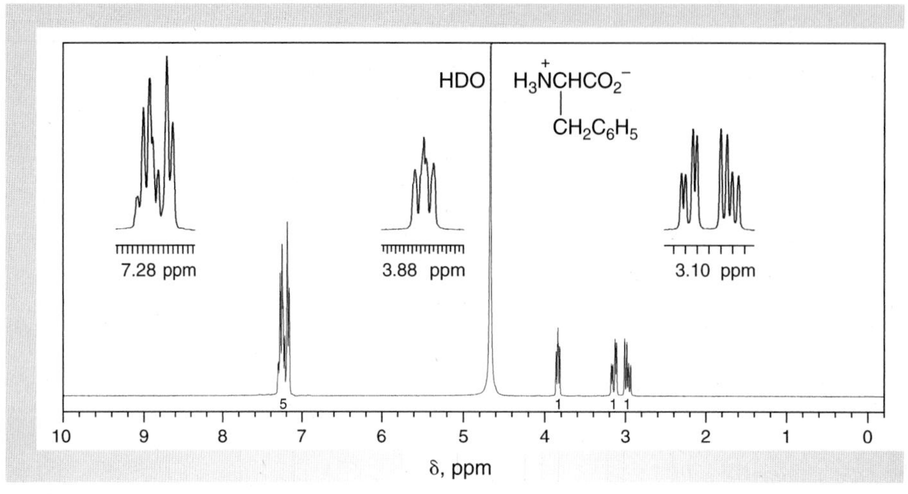

Figure 24.8
NMR data for L-phenylalanine (D_2O).

(a) ^1H NMR spectrum (300 MHz).
(b) ^{13}C NMR data: δ 37.4, 56.9, 128.4, 129.9, 130.2, 136.3, 174.2.

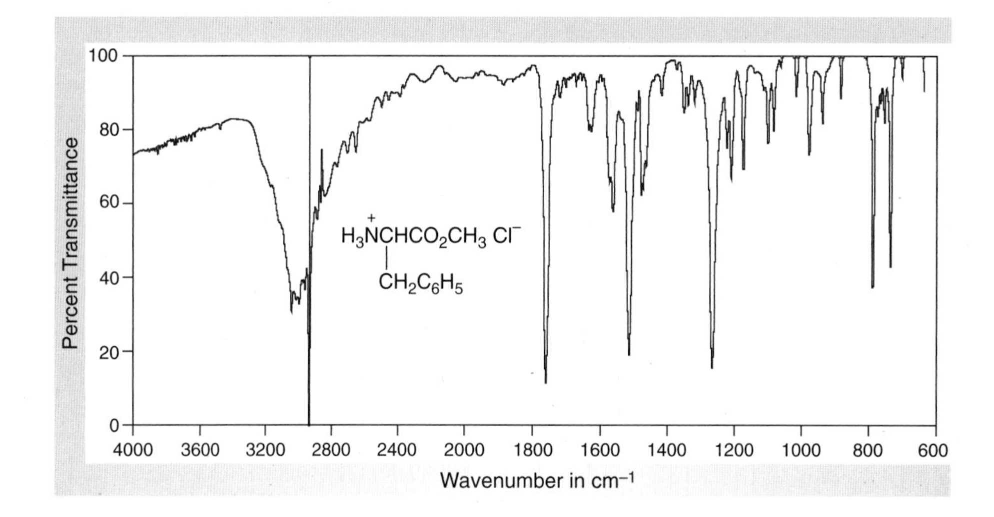

Figure 24.9
IR spectrum of methyl L-phenylalaninate hydrochloride (IR card).

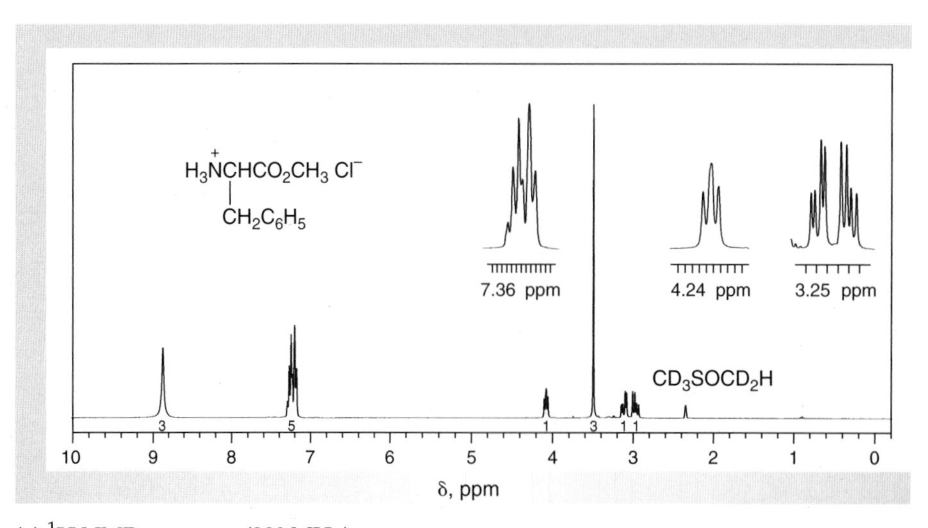

Figure 24.10
NMR data for methyl L-phenylalaninate hydrochloride (DMSO-d_6).

(a) ^1H NMR spectrum (300 MHz).
(b) ^{13}C NMR data: δ 35.7, 52.4, 53.2, 127.2, 128.5, 129.3, 134.7, 169.3.

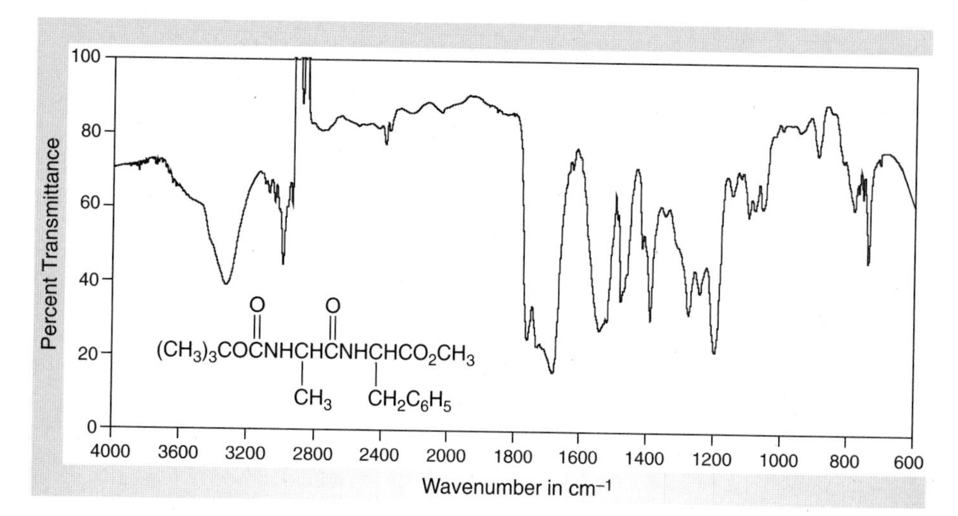

Figure 24.11
IR spectrum of methyl N-tert-*butoxycarbonyl-L-alanyl-L-phenylalaninate (IR card).*

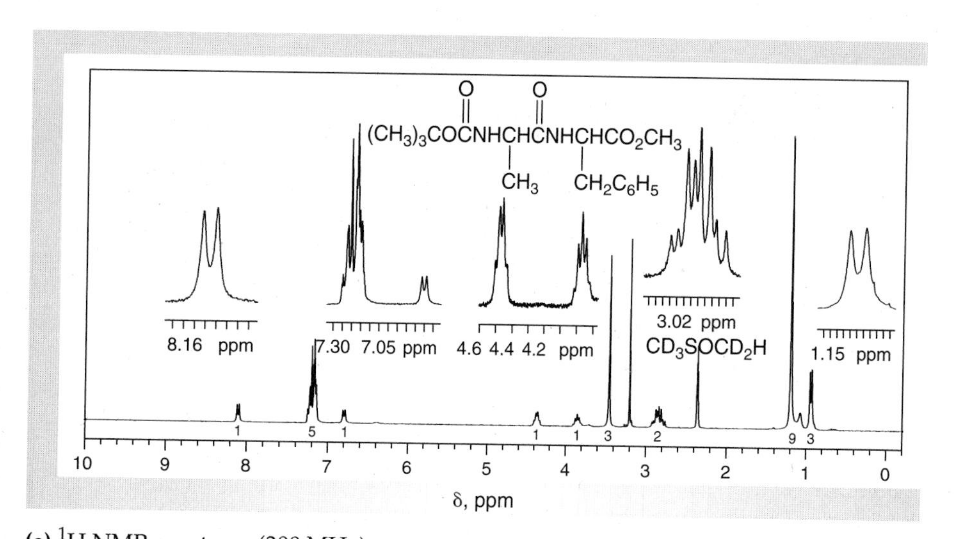

Figure 24.12
NMR data for methyl N-tert-*butoxycarbonyl-L-alanyl-L-phenylalaninate (DMSO-d_6).*

(a) ^{1}H NMR spectrum (300 MHz).
(b) ^{13}C NMR data: δ 18.8, 28.9, 37.5, 50.2, 52.5, 54.1, 78.7, 127.2, 128.9, 129.8, 129.9, 137.7, 155.6, 172.5, 173.4.

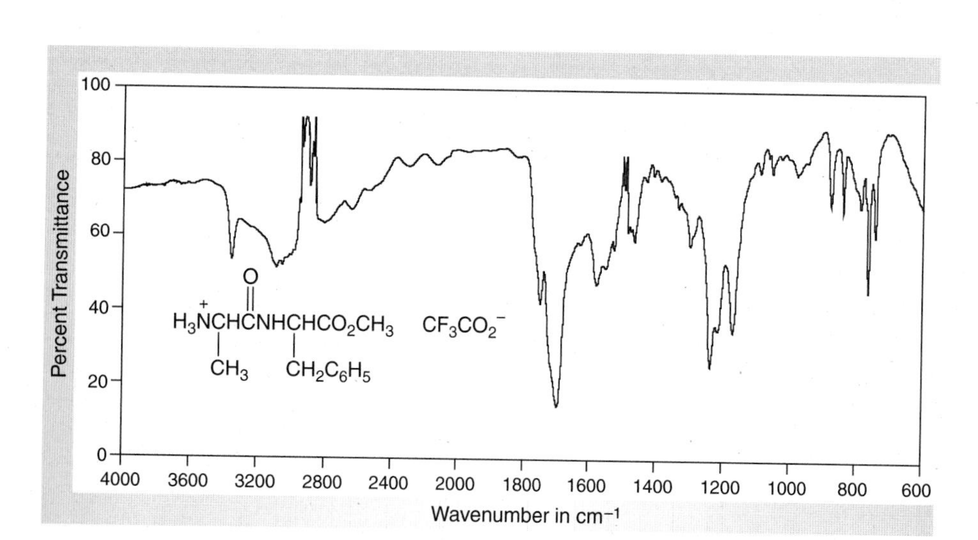

Figure 24.13
IR spectrum of methyl L-alanyl-L-phenylalaninate trifluoroacetate (IR card).

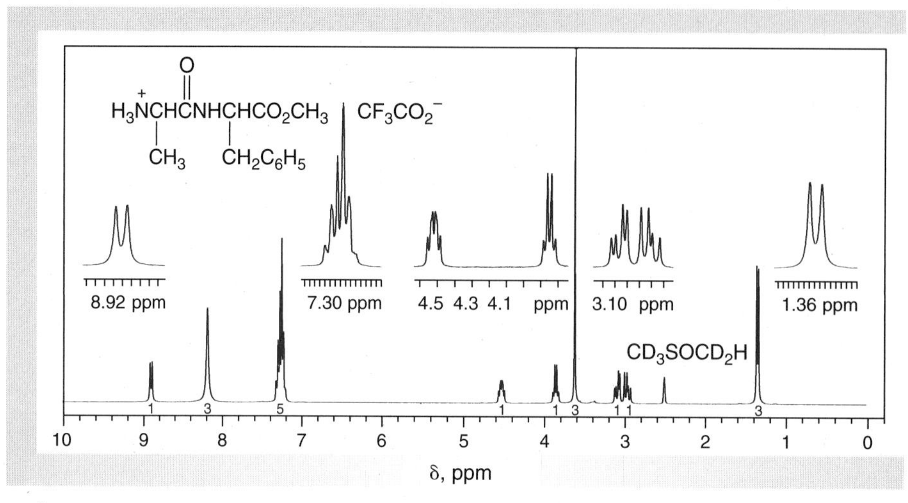

Figure 24.14
NMR data for methyl L-alanyl-L-phenylalaninate trifluoroacetate (DMSO-d_6).

(a) ^{1}H NMR spectrum (300 MHz).
(b) ^{13}C NMR data: δ 17.8, 39.3, 48.6, 52.7, 54.6, 127.4, 129.0, 129.8, 137.6, 170.6, 172.1.

HISTORICAL HIGHLIGHT

Invention of a Method for Solid-Phase Peptide Synthesis

Peptides containing defined sequences of approximately 5–10 amino acids have important roles in a variety of biologically important events. For example, the pentapeptide Leu-enkephalin is a naturally occurring neuropeptide that is found in the brain that, like morphine, serves as a potent analgesic agent or painkiller. Angiotensin II is an octapeptide that is involved in the regulation of several functions, including blood pressure. Bradykinin, which is a major component of bee venom, is a nonapeptide that dilates blood vessels and causes pain. Owing to their biological activities, peptides frequently serve as potential drug candidates or leads in the pharmaceutial industry during the development of new medicines to treat diseases. Proteins are polypeptides or biopolymers having more than 100 amino acids, and these serve a variety of critical functions in living cells. For example, enzymes are proteins that catalyze chemical reactions in biological systems.

It has long been recognized that studying the function of peptides and proteins requires that pure samples of the peptide or protein of interest be available in reasonable quantities. Because isolation of such compounds from natural sources was tedious, early chemists had to invent reliable and efficient methods for their synthesis in the organic laboratory, although today proteins are more readily available using recombinant DNA technology.

The story thus commences at the beginning of the twentieth century with Emil Fischer (see the Historical Highlight at the end of Chapter 23), who prepared the first peptide and coined the term for such compounds. He developed the concepts we commonly use today for synthesizing peptides, although he was never able to develop a suitable protecting group for the amine function in amino acids. That discovery was left to a former student of his, Max Bergmann,

Tyr—Gly—Gly—Phe—Leu
Leu-Enkephalin

Asp—Arg—Val—Tyr—Ile—His—Pro—Phe
Angiotensin II

Arg—Pro—Pro—Gly—Phe—Ser—Pro—Phe—Arg
Bradykinin

(Continued)

HISTORICAL HIGHLIGHT *Invention of a Method for Solid-Phase Peptide Synthesis (Continued)*

who invented the carbobenzyloxy-protecting group. Some 50 years later, the methodology for polypeptide synthesis had advanced to the point that du Vigneaud was able to prepare the nonapeptide hormone oxytocin, an achievement that led to his receiving the Nobel Prize in Chemistry in 1955. However, it soon became clear that there were many problems associated with the technology that had been developed for the synthesis of polypeptides in solution. That peptides became less soluble with increasing size led to problems with purifying intermediates and in executing efficient coupling reactions to produce peptide bonds. A major advance in technology was thus required.

It was left to the ingenuity of R. Bruce Merrifield, an organic chemist at the Rockefeller University, to develop an innovative solution to this problem that totally revolutionized the field of peptide synthesis. His plan was to assemble the peptide chain in a stepwise manner by adding new amino acids at the *N*-terminus while the *C*-terminal end was attached to a solid polymeric support of chloromethylated polystyrene, which is now referred to as the Merrifield. In this fashion, all of the excess reagents, impurities, and by-products could be easily removed by washing the resin after each operation, and the pure polypeptide could be cleaved from the solid support as the last step in the synthesis. Merrifield was awarded the Nobel Prize in Chemistry in 1984 in recognition of his contributions to the invention and development of the solid-phase method for the synthesis of peptides.

The principal elements of the process that was developed by Merrifield for synthesizing peptides on solid supports are outlined in Scheme 24.1. The solid support is a bead composed of a functionalized resin derived from a copolymer (Sec. 22.1) prepared from styrene and divinylbenzene. Individual resins to which *N*-protected derivatives of all 20 amino acids are attached to the polymeric support via an ester linkage are commercially available. Hence, a resin bearing the amino acid corresponding to the *C*-terminal amino acid of the desired polypeptide is selected as the starting material for the solid-phase synthesis. The *N*-terminal protecting group is first removed, and the resin is washed to remove all the reagents and any by-products. An *N*-protected amino acid is then coupled to the amino acid on the resin through formation of a peptide bond, and the polymer is again washed thoroughly to remove all excess reagents and by-products and to give a protected dipeptide that is still linked to the resin via the *C*-terminal amino acid. The *N*-protecting group is again removed, the resin is washed, and then the third *N*-protected amino acid in the sequence is attached to produce a tripeptide. The entire process is repeated over and over until all of the requisite amino acids have been added to the growing chain. The polypeptide is then cleaved from the support and isolated.

Merrifield used this procedure to prepare a number of peptides. For example, he synthesized the nonapeptide bradykinin in 68% yield in only eight days, a remarkable feat at the time. The biological activity of the synthetic peptide was identical with that of the natural peptide. Merrifield was ultimately able to automate all the steps in his technique for solid-phase peptide synthesis and demonstrated its power by using a homemade machine to prepare bovine pancreatic ribonuclease, an enzyme that contains 124 amino acids. This synthesis proceeded in 17% overall yield and required 369 chemical reactions and 11,931 individual operations! The synthetic ribonuclease had a specific activity that was 13–24% that of the native enzyme. The lower activity of the synthetic enzyme can probably be attributed to the fact that each coupling step did not proceed with 100% efficiency, so some polypeptides lacking one or more individual amino acids in the sequence were also produced. Because of their close similarity to ribonuclease, it was not possible to separate these polypeptides from the synthetic enzyme.

Merrifield's revolutionary concept of solid-phase synthesis was not limited to peptides, and similar techniques have been developed for the synthesis of nucleic acids and carbohydrates on solid supports. For each application, specialized instrumentation that is computer-controlled is commercially available. Access to such equipment has enabled researchers in areas of biology, medicine, material science, and biomedical engineering to prepare thousands of peptides and polypeptides for study. In the pharmaceutical industry, for example, solid-phase synthesis has been used to prepare relatively large numbers of related molecules, so-called *compound libraries*, that

(Continued)

HISTORICAL HIGHLIGHT *Invention of a Method for Solid-Phase Peptide Synthesis (Continued)*

Scheme 24.1

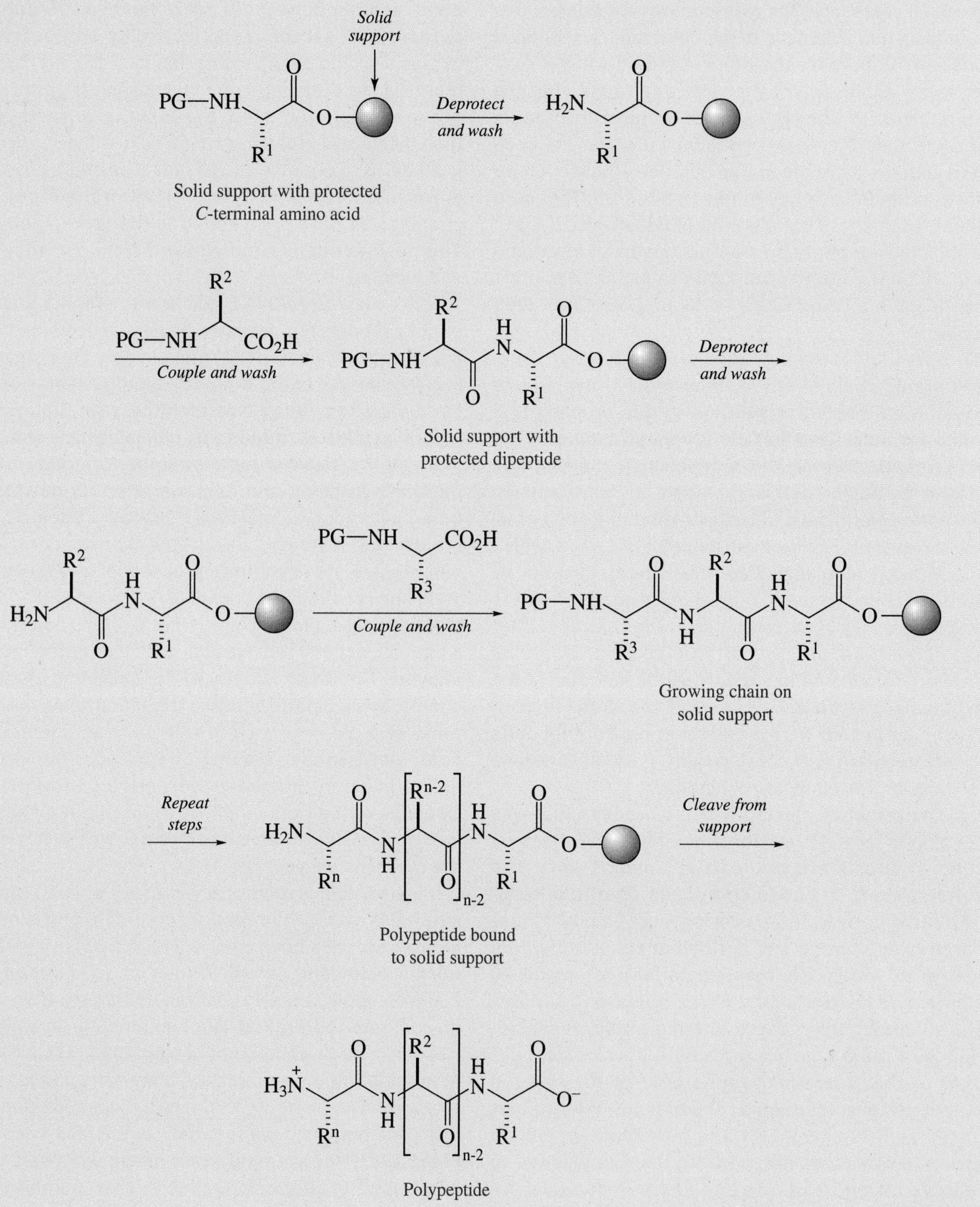
Solid support
PG—NH
O
R1
Deprotect and wash
H2N
Solid support with protected C-terminal amino acid
R2
PG—NH CO2H
Couple and wash
Solid support with protected dipeptide
Deprotect and wash
PG—NH CO2H
R3
Couple and wash
Growing chain on solid support
Repeat steps
Rn-2
n-2
Rn
Polypeptide bound to solid support
Cleave from support
H3N+
R2
Polypeptide

(Continued)

HISTORICAL HIGHLIGHT *Invention of a Method for Solid-Phase Peptide Synthesis (Continued)*

are screened for biological activity. Compounds exhibiting the desired activity then serve as leads for drug development.

Relationship of Historical Highlight to Experiments

The experiments in this chapter illustrate some of the same types of reactions that the Nobel laureates du Vigneaud and Merrifield performed when they prepared oxytocin and ribonuclease, respectively. You will discover how to protect amino acids and link individual amino acids by forming peptide bonds. You will also learn how to deprotect the *N*-terminal amino acid selectively, so that another coupling step with a new amino acid may be performed. You will thus have an opportunity to experience and learn firsthand some of the important techniques that are widely used in research laboratories in academia and industry throughout the world.

See more about *Enkephalins*
See more about *Morphine*
See more about *Bradykinin*
See more about *Angiotensin II*
See *Who was du Vigneaud?*
See more about *Oxytocin*
See *Who was Merrifield?*
See more about *Solid Phase Synthesis*
See more about *Resins*
See *Who was Fischer?*

CHAPTER 25

Identifying Organic Compounds

When you see this icon, sign in at this book's premium website at **www.cengage.com/login** to access videos, Pre-Lab Exercises, and other online resources.

One of the greatest challenges to the practicing organic chemist is identifying the substances that are obtained from chemical reactions or isolated from natural sources. Indeed, this can be one of the more frustrating aspects of organic chemistry that you might experience. Yet there is nothing more exhilarating and fulfilling than being able to tell your classmates, much less your instructor, that you have solved the structural "puzzle" for an unknown compound and are able to assign its structure unambiguously. Elucidating the structure of an unknown may be difficult and time-consuming, but the availability of modern spectroscopic techniques (Chap. 8) in combination with classical chemical methods has greatly facilitated this branch of experimental organic chemistry. Some of the techniques and approaches that are used to accomplish this goal are described in this chapter.

25.1 INTRODUCTION

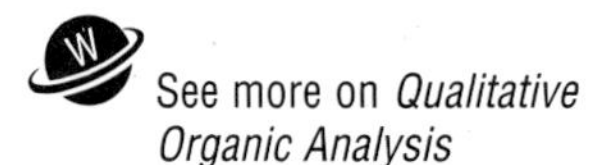
See more on *Qualitative Organic Analysis*

A successful systematic scheme of qualitative analysis for identifying organic compounds was developed early in the twentieth century. This scheme, together with some more recent modifications, is termed **classic qualitative organic analysis,** and it is the basis for most textbook discussions on the subject (Ref. 1 at the end of Section 25.5, for example).

In recent years, the development of chromatographic methods of separation (Chap. 6) and structural analysis by spectroscopic techniques (Chap. 8) have revolutionized the laboratory practice of organic chemistry and have largely supplanted the use of qualitative organic analysis for elucidating structures of unknown compounds. Nevertheless, interest in such qualitative analysis remains high because it is a fun and effective means of applying your knowledge of organic chemistry. This chapter contains an outline of the classic scheme for qualitative organic analysis and identification of functional groups, as well as the necessary experimental procedures so you may appreciate and understand the approach. Because a wealth of information is readily available from spectroscopic techniques, we also provide an introduction to their application to identification and structure determination in Section 25.5. The advantages of the combined use of the classic and instrumental methods are also described there.

25.2 OVERVIEW OF ORGANIC ANALYSIS

An overview of a systematic procedure that can be used to identify an unknown pure organic compound serves to introduce the classic approach for organic analysis. The first step is to ensure that the compound is pure, and this may be done in several ways. In the case of a liquid, gas-liquid chromatography (GLC, Sec. 6.4) may be used to demonstrate that only one component is present; a pure compound produces only a single peak in a gas chromatogram, assuming that no decomposition occurs under the conditions of the analysis. If enough liquid is available, it can be distilled (Chap. 4). A 1–2 °C boiling-point range implies that the compound is pure, although such a narrow range is also consistent with the distillation of an **azeotrope** (Sec. 4.4). However, GLC provides better evidence for purity because it allows detection of low levels of impurities. The purity of a solid substance can be ascertained from its melting point (Sec. 3.3). A 1–2 °C melting-point range usually indicates a pure compound, except in the rare instance of a **eutectic mixture** (Sec. 3.3). Impure liquids can be purified by simple or fractional distillation (Secs. 4.2 and 4.3) and by column or preparative GLC (Secs. 6.3 and 6.4), whereas solids can be purified by recrystallization (Sec. 3.2), by sublimation (Sec. 2.20), or by thin-layer or column chromatography (Secs. 6.2 and 6.3).

Once the purity of the unknown is established, various physical properties are determined. The melting point of a solid or boiling point of a liquid is considered essential. Occasionally, the density and/or refractive index of a liquid may be useful, and for certain compounds, either liquid or solid, the specific rotation can be determined if a substance is optically active.

Establishing what elements other than carbon and hydrogen are present is critical for identifying the compound, and techniques for **elemental analysis** are described later. Molar mass, as determined by cryoscopic techniques or mass spectrometry (Sec. 8.5), or percentage by mass composition of the elements present, also provides important data. The solubility of the unknown compound in water, in dilute acids and bases, or in various organic solvents may signal the presence or absence of various functional groups.

Perhaps the most important step in identifying an unknown substance is determining the functional group(s) that may be present, and IR and NMR spectroscopy (Secs. 8.2 and 8.3) are now commonly used for this purpose. However, before the development of these spectroscopic methods as routine experimental techniques, functional group determination involved performing qualitative chemical tests for each possible group. Although spectral analyses may not provide an *unequivocal* answer about the presence of certain functional groups, they at least permit narrowing the possibilities to a small number, so only one or two chemical tests are then needed to complete the identification of the functionalities present.

Final assignment of a structure to the "unknown" compound is achieved by one of several procedures. The classic method involves the chemical conversion of the substance into a *solid* derivative. The success of this technique depends on the availability of information about the unknown and its various derivatives. Of prime importance is knowledge of the melting or boiling points of possible candidates for the unknown as well as the melting points of solid derivatives. Many tabulations of organic compounds are available for this purpose, and references to two of them are provided at the end of Section 25.5. Abbreviated tables of liquid and solid organic compounds and of their solid derivatives are provided at the website associated with this textbook.

The identification of the compound may also be completed by thoughtful analysis of the spectroscopic data for the compound. Indeed, unequivocal proof of structure based *solely* upon spectral data often is possible; demonstrating that the IR and NMR spectra of an "unknown" and a known compound are *identical* suffices to prove that the substances are identical.

The following sections contain descriptions of the stepwise procedures that may be used to identify an unknown compound using classic methods alone or in combination with spectroscopic methods. As a reminder, these procedures should *not* be performed unless the compound is pure. Since it is possible that you may be given a mixture of unknown compounds to identify, a procedure is provided in Section 25.4 for separating a mixture into its individual components so that each one can be identified.

25.3 CLASSIC QUALITATIVE ANALYSIS PROCEDURE TO IDENTIFY A PURE COMPOUND

The *classic system of qualitative organic analysis* consists of six steps, each of which is discussed in the following subsections. The first four steps, which may be carried out in any order, should be completed *before* performing the qualitative tests for functional groups. The final step must always be the preparation of one or more solid derivatives.

1. Preliminary examination of physical and chemical characteristics.
2. Determining physical constants.
3. Elemental analysis to determine the presence of elements other than carbon, hydrogen, and oxygen.
4. Solubility tests in water, dilute acid, and dilute base.
5. Functional group analysis using classification tests.
6. Derivatization.

It is a tribute to the power of the system that you can identify an unknown organic compound with certainty, even though it may be one of several million known compounds. With the exception of a few general guidelines, there are no rigid directions to be followed. You must rely on good judgment and initiative in selecting a course of attack on the unknown, and it is particularly important to observe and consider each experimental result. Negative results may be as useful as positive ones in your quest to identify an unknown substance.

Preliminary Examination

If it is carried out intelligently, the **preliminary examination** may provide more information with less effort than any other part of the procedure. The simple observation that the unknown is a *crystalline* solid, for example, eliminates a large fraction of all organic compounds from consideration because many are liquids at room temperature. The **color** is also informative: Most pure organic compounds are white or colorless. A brown color is often characteristic of small amounts of impurities; for example, aromatic amines and phenols quickly become discolored by the formation of trace amounts of highly colored air-oxidation products. Color in a pure organic compound is usually attributable to conjugated double bonds (Sec. 8.4).

The **odor** of many organic compounds, particularly those of lower molar mass, is highly distinctive. You should make a conscious effort to learn and recognize the odors that are characteristic of several classes of compounds such as the alcohols, esters, ketones, and aliphatic and aromatic hydrocarbons. The odors of certain compounds demand respect, even when they are encountered in small amounts and at considerable distance; for example, the unpleasant odors of thiols (mercaptans), isonitriles, and higher carboxylic acids and diamines cannot be described definitively, but they are recognizable once encountered. *Be extremely cautious* in smelling unknowns, because some compounds are not only disagreeable but may be toxic as well. Large amounts of organic vapors should *never* be inhaled.

The **ignition test** involves a procedure in which a drop or two of a liquid or about 50 mg of a solid is heated gently on a small spatula or crucible cover with a microburner flame. Whether a solid melts at low temperature or only upon heating more strongly is then noted. The flammability and the nature of any flame from the sample are also recorded. A yellow, sooty flame is indicative of an aromatic or a highly unsaturated aliphatic compound; a yellow but non-sooty flame is characteristic of aliphatic hydrocarbons. The oxygen content of a substance makes its flame more colorless or blue; a high oxygen content lowers or prevents flammability, as does halogen content. The unmistakable and unpleasant odor of sulfur dioxide indicates the presence of sulfur in the compound. If a white, nonvolatile residue is left after ignition, a drop of water is added and the resulting aqueous solution is tested with litmus or pHydrion paper; a metallic salt is indicated if the solution is alkaline.

Physical Constants

If the unknown is a solid, its **melting point** is measured by the capillary tube method (Sec. 2.7). An observed melting-point range of more than 2–3 °C indicates that the sample is impure and should be recrystallized (Sec. 3.1).

For an unknown that is a liquid, the **boiling point** is determined by the micro boiling-point technique (Sec. 2.8). An indefinite or irreproducible boiling point or discoloration or inhomogeneity of the unknown requires that the sample be distilled (Secs. 2.13 and 2.15); the boiling point is obtained during the distillation.

Other physical constants that *may* be of use for liquids are the **refractive index** and the **density**. Consult with your instructor about the advisability of making these measurements and how to perform them.

Elemental Analysis

The technique of **elemental analysis** involves determining *which* elements may be present in a compound. The halogens, sulfur, oxygen, phosphorus, and nitrogen are the elements other than carbon and hydrogen that are most commonly found in organic molecules. Although there is no simple way to test for the presence of oxygen, it is fairly easy to determine the presence of the other heteroatoms, and the appropriate procedures are provided for the halogens, sulfur, and nitrogen.

The chemical basis for the procedures we describe is as follows. Because the bonding found in organic compounds is principally covalent, there rarely are direct methods analogous to those applicable to ionic inorganic compounds for determining the presence of the aforementioned elements. However, the covalent bonds between carbon and these heteroatoms may be broken by heating an organic compound with sodium metal. This process, called **sodium fusion**, results in the formation of inorganic ions involving the heteroatoms if they are present in the original compound; thus, the products are halide ions, X^-, from halogens, sulfide ion, S^{2-}, from sulfur, and cyanide ion, CN^-, from nitrogen. After the organic compound has been heated with sodium metal, the residue is *cautiously* treated with *distilled* water to destroy the excess sodium and to dissolve the inorganic ions that have been formed. The ***f*usion *aq*ueous *s*olution**, designated as **FAQS**, may then be analyzed for the presence of halide, sulfide, and cyanide ions.

The presence of halide is determined by first acidifying a portion of the FAQS with dilute nitric acid and boiling the solution *in the hood* to remove any sulfide or cyanide ions as hydrogen sulfide or hydrogen cyanide, respectively. This is necessary because sulfide and cyanide interfere with the test for halogens. Silver nitrate solution is then added, and the formation of a precipitate of silver halide signals the presence of halide in the FAQS (Eq. 25.1).

$$\mathrm{Ag^+ + X^- \longrightarrow \underset{\text{A silver halide}}{\underline{AgX}}} \tag{25.1}$$

The color of the precipitate provides a *tentative* indication of which halogen is present in the unknown: AgCl is *white* but turns *purple* on exposure to light, AgBr is *light yellow*, and AgI is *dark yellow*. *Definitive* identification is made by standard procedures of inorganic qualitative analysis or by means of thin-layer chromatography (TLC, Sec. 6.2).

Sulfur is detected by taking a second portion of the FAQS and carefully acidifying it. Any sulfide ion that is present will be converted to H_2S gas, which forms a dark precipitate of PbS upon contact with a strip of paper saturated with lead acetate solution (Eq. 25.2).

$$\underset{\text{Lead acetate}}{\mathrm{Pb(OAc)_2}} + \underset{\text{Hydrogen sulfide}}{\mathrm{H_2S}} \longrightarrow \underset{\text{Lead sulfide (black)}}{\underline{\mathrm{PbS\ (solid)}}} + \underset{\text{Acetic acid}}{\mathrm{2\ HOAc}} \tag{25.2}$$

See more on *Prussian Blue*

The presence of nitrogen in the unknown is shown by carefully acidifying a third portion of the FAQS and adding ferrous ion, Fe^{2+}, and ferric ion, Fe^{3+}. This converts the cyanide ion into potassium ferric ferrocyanide, which precipitates as an intensely blue solid called **Prussian blue** (Eq. 25.3).

$$\mathrm{4\,CN^- + Fe^{2+} \longrightarrow Fe(CN)_6^{4-} \xrightarrow[Fe^{3+}]{K^+\ and}} \underset{\text{Potassium ferric ferrocyanide (Prussian blue)}}{\underline{\mathrm{KFeFe(CN)_6}}} \tag{25.3}$$

EXPERIMENTAL PROCEDURES

Elemental Analysis

SAFETY ALERT

1. **Sodium fusion involves heating sodium metal or a sodium-lead alloy to a high temperature and then adding the organic compound. Use extreme care when performing both the heating and addition.**
2. **Perform the sodium fusion in the hood if possible.**

3. **Use a Pyrex™ test tube and check it for cracks or other imperfections before performing the sodium fusion.**
4. **If pure sodium metal is used, hydrolyze the residue very carefully as directed, because any excess metal reacts *vigorously* with alcohol or water.**
5. **Be careful when handling the test tube after the fusion is complete; remember that it may still be hot.**
6. ***Throughout this procedure, point the mouth of the test tube away from yourself and your neighbors;* the organic material may burst into flame when it contacts the hot metal, or it may react so violently that hot materials are splattered from the test tube.**

Preparation Sign in at **www.cengage.com/login** to read the MSDSs for the chemicals used or produced in this procedure.

A ■ *Sodium Fusion*

Sodium fusion may be performed using either sodium metal or a commercially available sodium-lead alloy, which contains nine parts lead and one part sodium. For safety reasons, we recommend using the sodium-lead alloy method because the alloy is easier to handle than is sodium metal and poses less potential danger during hydrolysis.

Sodium-Lead Alloy Method

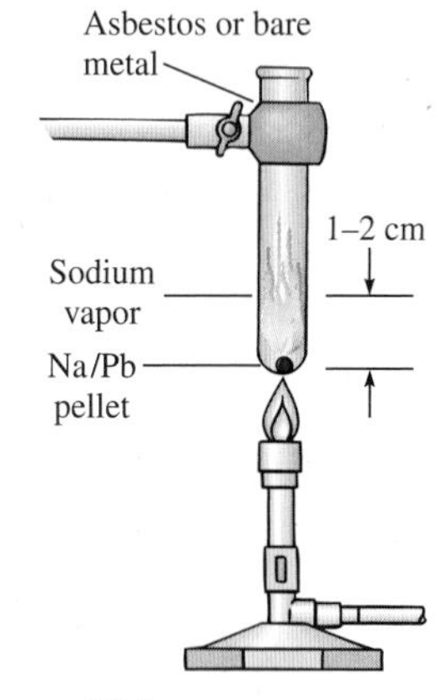

Figure 25.1
Apparatus for sodium fusion.

Support a small Pyrex test tube in a *vertical* position using a clamp whose jaws have either an asbestos liner or none at all; the jaws must *not* be lined with materials such as rubber or plastic (Fig. 25.1). Place a 0.5-g sample of sodium-lead alloy in the test tube and heat the alloy with a flame until it melts and vapors of sodium rise 1–2 cm in the test tube. Do *not* heat the test tube to redness. Then remove the flame and immediately add 2–3 drops of a liquid sample or about 10 mg of a solid sample to the hot alloy, being careful not to allow the sample to touch the sides of the hot test tube during the addition. If there is no visible reaction, heat the fusion mixture gently to initiate the reaction, then discontinue the heating and allow the reaction to subside. Next, heat the test tube to redness for a minute or two, and then *let it cool* to room temperature. Add 3 mL of *distilled water* to the cooled reaction mixture, and heat the mixture gently for a few minutes to complete hydrolysis. Decant or filter the solution. If the latter option is used, wash the filter paper with about 2 mL of water. Otherwise simply dilute the decanted solution with about 2 mL of water. Label the filtrate FAQS and use it in the appropriate tests for detecting sulfur, nitrogen, and the halogens.

WRAPPING IT UP

Put the *lead pellet* that remains after hydrolysis of the reaction mixture in the container for heavy metals.

Sodium Metal Method

Support a small Pyrex test tube in a *vertical* position using a clamp whose jaws have either an asbestos liner or none at all; the jaws must *not* be lined with materials such as rubber or plastic (Fig. 25.1). Place a clean cube of sodium metal about 3–4 mm on an edge in the tube and heat it gently with a microburner flame until the sodium melts and the vapors rise 1–2 cm in the test tube. Do *not* heat the test tube to redness. Then remove the flame and immediately add 2 or 3 drops of a liquid or about 10 mg of a solid sample to the hot test tube, being careful not to allow the sample to touch the sides of the hot test tube during the addition. A brief flash of fire may be observed; this is normal. Reheat the bottom of the tube, remove the flame, and add a second portion of the unknown organic compound in the same amount as for the first addition. Now heat the bottom of the tube until it is a dull red color; then remove the flame and allow the tube to cool to room temperature. Carefully add about 1 mL of 95% ethanol dropwise to the cooled reaction mixture to decompose the excess sodium; stir the contents of the tube with a stirring rod. After a few minutes, add another 1 mL of ethanol with stirring. After the reaction has subsided, apply gentle heat to boil the ethanol; either work at the hood or use an inverted funnel connected to a vacuum source (Fig. 2.72b) over the mouth of the test tube to keep the vapors of ethanol from entering the room. When the ethanol has been removed, allow the tube to cool and add about 10 mL of *distilled water*, stir the mixture and pour it into a small beaker. Rinse the tube with an additional 5 mL of distilled water, and combine the rinse with the main solution. The total amount of water should be 15 to 20 mL. Boil the aqueous mixture briefly, filter it, and label the filtrate FAQS; use this filtrate in the tests for sulfur, nitrogen, and the halogens.

Preparation Sign in at **www.cengage.com/login** to read the MSDSs for the chemicals used or produced in this procedure.

B ■ *Qualitative Analysis for Halogens, Sulfur, and Nitrogen*

Halogens. Acidify about 2 mL of the FAQS by dropwise addition of 6 *M* nitric acid. Working at the hood, boil the solution gently for 2–3 min to expel any hydrogen sulfide or cyanide that may be present. Cool the solution and then add several drops of 0.3 *M* aqueous silver nitrate solution. A *heavy precipitate* of silver halide indicates the presence of chlorine, bromine, or iodine in the original organic compound. A faint turbidity of the solution should *not* be interpreted as a positive test.

Tentative identification of the particular halogen is made on the basis of color. *Definitive* identification must be made by standard inorganic qualitative procedures (Ref. 1 at the end of Section 25.5) or by means of TLC (Sec. 6.2), using the following procedure.

Obtain a 3-cm × 10-cm strip of fluorescent silica gel chromatogram sheet. About 1 cm from one end, place four equivalently spaced spots as follows. Using a capillary, spot the FAQS at the left side of the sheet. Because this solution is likely to be relatively dilute in halide ion, several reapplications of it may be required. Allow the spot to dry following each application, a process that is hastened by blowing on the plate. Do not permit the spot to broaden to a diameter greater than about 2 mm. Next, and in order, spot samples of 1 *M* potassium chloride, 1 *M* potassium bromide, and 1 *M* potassium iodide. Develop the plate in a solvent mixture of acetone, 1-butanol, concentrated ammonium hydroxide, and water in the volume ratio of 13:4:2:1. Following development, allow the plate to air-dry and, working at the hood, spray the plate lightly with an indicator spray prepared by dissolving 1 g of silver nitrate in 2 mL of water and adding this solution to 100 mL of methanol containing 0.1 g of fluorescein and 1 mL of concentrated ammonium hydroxide. Allow the now-yellow strip to dry, and then irradiate it for several minutes with a long-wavelength ultraviolet lamp (366 nm). Compare the spots formed from the test solution with those formed from the solutions of known halides. Be aware that iodide gives two spots according to this procedure.

Sulfur. Acidify a 1–2 mL sample of the FAQS with acetic acid, and add a few drops of 0.15 *M* lead acetate solution. A black precipitate of PbS signifies the presence of sulfur in the original organic compound.

Nitrogen. Determine the pH of a 1-mL sample of the FAQS with pHydrion paper. The pH should be about 13. If it is above 13, add a *small* drop of 3 *M* sulfuric acid to bring the pH down to about 13. If the pH of the fusion solution is below 13, add a *small* drop of 6 *M* NaOH to bring the pH up to about 13. Add 2 drops each of a saturated solution of ferrous ammonium sulfate and of 5 *M* potassium fluoride. Boil the mixture gently for about 30 sec, cool it, and add 2 drops of 5% ferric chloride solution. Then carefully add 3 *M* sulfuric acid dropwise to the mixture until the precipitate of iron hydroxide *just* dissolves. Avoid an excess of acid. At this point, the appearance of the deep-blue color of potassium ferric ferrocyanide (Prussian blue) indicates the presence of nitrogen in the original organic compound. If the solution is green or blue-green, filter it; a blue color remaining on the filter paper is a weak but nonetheless positive test for nitrogen.

WRAPPING IT UP

Transfer precipitated *silver halides* to a container labeled for them so that the silver can be recovered. Put the precipitated *lead sulfide* in a container for heavy metals. Any *FAQS* that remains may be flushed down the drain, as can the *solution* from *the test for nitrogen.*

SOLUBILITY TESTS

The solubility of an organic compound in water, dilute acid, or dilute base can provide useful, but not definitive, information about the presence or absence of certain functional groups. In reality, however, the assignment of an unknown to a formal solubility class may be arbitrary because a large number of compounds exhibit *borderline* behavior. We recommend that the solubility tests be done in the order presented here.

Preparation Sign in at **www.cengage.com/login** to read the MSDSs for the chemicals used or produced in this procedure.

Water

Test the solubility of the unknown in water. For the present purposes, a compound is defined as soluble if it dissolves to the extent of 3 g in 100 mL of water, or more practically, 30 mg in 1 mL of water. As a general rule, few organic compounds exhibit appreciable water solubility as defined here.

Several structural features of the unknown can be deduced if it is water-soluble. It must be of low molar mass and will usually contain no more than four to five carbon atoms, unless it is polyfunctional. It must contain a polar group that will form a hydrogen bond with water, such as the hydroxy group of an alcohol or a carboxylic acid, the amino functionality of an amine, or the carbonyl group of aldehydes or ketones. Esters, amides, and nitriles dissolve to a lesser extent, and acid chlorides or anhydrides react with water rather than simply dissolving in it. On the other hand, alkanes, alkenes, alkynes, and alkyl halides are water-insoluble.

If the unknown is water-soluble as defined above, test its aqueous solution with pHydrion paper. If the solution is acidic, the unknown is likely to be a carboxylic acid of low molar mass such as acetic acid. If the solution is basic, an organic base of low molar mass such as diethylamine is possible. A neutral solution suggests the presence of a neutral polar compound such as an alcohol or a ketone.

The borderline for water solubility of *monofunctional* organic compounds is most commonly at or near the member of the homologous series containing five carbon atoms. Thus, butanoic acid is soluble, pentanoic acid is borderline, and hexanoic acid is insoluble in water; similarly, 1-butanol is soluble, 1-pentanol is borderline, and 1-hexanol is insoluble in this solvent. There are exceptions to this generalization, however. For example, if a molecule is spherically shaped, it can contain a larger number of carbon atoms and still remain soluble in water. A case in point is that 2-methyl-2-butanol (*tert*-pentyl alcohol) is water-soluble to the extent of 12.5 g/100 mL of water even though it contains five carbon atoms. The increased solubility is because the molecular surface areas of spherical molecules are less than those of non-spherical ones, and this decreases the **hydrophobicity** of the molecule.

Solubility in Aqueous Acid and Base

If an unknown is insoluble in water, you should test its solubility first in sodium hydroxide, then in sodium bicarbonate, and finally in hydrochloric acid. Solubility in one or more of these acids and bases is defined in terms of the compound being *more soluble in base or acid than in water* and reflects the presence of an acidic or basic functional group in the water-insoluble unknown compound.

In each of the following solubility tests, *shake the unknown with the test reagent at room temperature.* If it does not dissolve, warm the mixture for several minutes in a hot-water bath and continue shaking it. If the substance still appears not to dissolve, decant or filter the liquid from the undissolved sample and carefully *neutralize* the filtrate; observing a precipitate or turbidity is indicative of greater solubility in the aqueous acid or base than in water itself. The importance of bringing the filtrate to pH 7 arises because an unknown may show enhanced solubility in both acidic and basic solutions if it contains *both* basic and acidic functional groups. Assessing these possibilities necessitates performing *all* the solubility tests.

1. *Sodium hydroxide.* If the compound is water-insoluble, test its solubility in 1.5 *M* NaOH solution. Carboxylic acids, which are strong acids, and phenols, which are weak acids, dissolve in sodium hydroxide because they are converted into their water-soluble sodium salts (Eqs. 25.4a and 25.4b). An unknown that is more soluble in NaOH than in water may be either a phenol or a carboxylic acid, and it must be tested for solubility in the weaker base, 0.6 *M* $NaHCO_3$, which may permit distinction between these two functional groups (Part 2, below). If the unknown does not exhibit solubility in NaOH, its solubility in $NaHCO_3$ need not be tried; rather, it should next be tested for solubility in 1.5 *M* HCl.

Carboxylic acids: $$RCO_2H \xrightarrow{NaOH} RCO_2^-Na^+ \quad (25.4a)$$
A carboxylic acid, *Water-insoluble* → A sodium carboxylate, *Water-soluble*

Phenols: $$ArOH \xrightarrow{NaOH} ArO^- Na^+ \quad (25.4b)$$
A phenol, *Water-insoluble* → A sodium phenoxide, *Water-soluble*

2. *Sodium bicarbonate.* An unknown that is soluble in dilute NaOH solution should also be tested for its solubility in 0.6 *M* $NaHCO_3$. If it is soluble, the *tentative* conclusion is that a carboxylic acid group is present, owing to the formation of the water-soluble sodium salt (Eq. 25.5a); phenols are normally not deprotonated in this medium (Eq. 25.5b). Dissolution should be accompanied by effervescence resulting from decomposition of the carbonic acid, H_2CO_3, formed from reaction of bicarbonate with the carboxylic acid, to carbon dioxide and water (Eq. 25.5a).

Carboxylic acids:

$$\underset{\substack{\text{A carboxylic acid} \\ \textit{Water-insoluble}}}{RCO_2H} \xrightarrow{NaHCO_3} \underset{\substack{\text{A sodium carboxylate} \\ \textit{Water-soluble}}}{RCO_2^-Na^+} + H_2CO_3 \longrightarrow CO_2 + H_2O \qquad (25.5a)$$

Phenols:

$$\underset{\substack{\text{A phenol} \\ \textit{Water-insoluble}}}{ArOH} \xrightarrow{NaHCO_3} \text{No reaction} \qquad (25.5b)$$

Carboxylic acids are *usually* soluble in $NaHCO_3$ *and* in NaOH, whereas phenols usually dissolve *only* in NaOH. However, caution must be used in making *definitive* conclusions about the presence of a carboxylic acid or phenol based upon solubility in $NaHCO_3$. For example, a phenol containing one or more strong electron-withdrawing substituents, such as a nitro group, can be as acidic as a carboxylic acid and thus may form a water-soluble sodium salt by reaction with $NaHCO_3$. Similarly, the salt of a carboxylic acid of high molar mass may *not* be completely soluble in the aqueous medium; nevertheless, evolution of CO_2 should still be observable when the acid comes in contact with the base.

3. *Hydrochloric acid.* You should determine the solubility of the unknown in 1.5 *M* hydrochloric acid. If the unknown is soluble, the presence of an amino group in the compound is indicated because amines are organic bases that react with dilute acids to form ammonium salts that are usually water-soluble (Eq. 25.6). However, this solubility test does *not* permit the distinction between weak and strong organic bases.

$$\underset{\substack{\text{An amine} \\ \textit{Water-insoluble}}}{RNH_2} \xrightarrow{HCl} \underset{\substack{\text{An alkylammonium chloride} \\ \textit{Water-soluble}}}{RNH_3^+\ Cl^-} \qquad (25.6)$$

4. *Concentrated sulfuric acid.* Many compounds that are too weakly basic or acidic to dissolve in dilute aqueous acid or base will dissolve in or react with concentrated H_2SO_4. Such solubility is often accompanied by the observation of a dark solution or the formation of a precipitate; any detectable reaction such as evolution of a gas or formation of precipitate is considered "solubility" in concentrated H_2SO_4. This behavior usually may be attributed to the presence of carbon-carbon π-bonds, oxygen, nitrogen, or sulfur in the unknown. The solubility is normally due to reaction of one of these functional groups with the concentrated acid, which results in the formation of a salt that is soluble in the reagent. For example, an alkene adds the elements of sulfuric acid to form an alkyl hydrogen sulfate (Eq. 25.7) that is soluble in the acid, and an oxygen-containing compound

becomes protonated in concentrated acid to form a soluble oxonium salt (Eq. 25.8). Substances that exhibit this solubility behavior are termed "neutral" compounds.

$$R_2C{=}CR_2 + H_2SO_4 \longrightarrow R_2\underset{\displaystyle H}{\underset{|}{C}}{-}\underset{\displaystyle OSO_3H}{\underset{|}{C}}R_2 \qquad (25.7)$$

An alkene — Sulfuric acid — An alkyl sulfonate

$$R_2C{=}O \text{ or } ROR \xrightarrow{H_2SO_4} R_2C{=}\overset{+}{O}H\ HSO_4^- \text{ or } R_2\overset{+}{O}H\ HSO_4^- \qquad (25.8)$$

An aldehyde or ketone — An ether — Oxonium ions

You do *not* need to test compounds that are soluble in dilute HCl or neutral water-insoluble compounds containing N or S for their solubility in concentrated H_2SO_4 because they will invariably dissolve in or react with it. Note that solubility or insolubility of an unknown in concentrated H_2SO_4 does *not* yield a great deal of evidence for the presence or absence of any *specific* group, whereas the solubility of a compound in dilute HCl or NaOH or $NaHCO_3$ does provide valuable information about the type of functional group present.

Assuming that a compound contains only *one* functional group, the following scheme classifies compounds according to their solubility in acid or base. This picture may be changed dramatically if a compound contains several polar functional groups that cause its solubility properties to be different from those of a monofunctional compound or if an acidic or basic compound contains one or more strongly electron-withdrawing groups.

Acidic Compounds Soluble in NaOH and $NaHCO_3$

Carboxylic acids

Phenols

Acidic Compounds Soluble in NaOH But Not in $NaHCO_3$

Phenols

Basic Compounds Soluble in HCl

Amines

Neutral Compounds Soluble in Concentrated H_2SO_4

Carbonyl compounds (aldehydes and ketones)

Unsaturated compounds (alkenes and alkynes)

Alcohols

Esters

Amides

Nitriles

Nitro compounds

Ethers

Neutral Compounds Insoluble in Cold Concentrated H_2SO_4

Alkyl halides

Aryl halides

Aromatic hydrocarbons

After determining the physical constants, elemental analysis, and solubility properties, provide a **preliminary report** of these findings to your instructor, who may give advice regarding the validity of these observations. This protocol serves to minimize unnecessary loss of time in finding the structure of the unknown.

Classification Tests

The next step in identifying an unknown is to determine which functional groups are present. The classic scheme involves performing a number of chemical tests on a substance, each of which is specific for a type of functional group. You may normally do these tests quickly, and they are designed so that the observation of a color change or the formation of a precipitate indicates the presence of a particular functional group. The results of these tests usually allow the assignment of the unknown to a structural class such as alkene, aldehyde, ketone, or ester, for example. The following factors should be considered when performing qualitative classification tests for functional groups.

1. A compound may contain more than one functional group, so the complete series of tests must be performed unless you have been told that the compound is monofunctional.
2. Careful attention is required when the functional group tests are performed. Record *all* observations, such as the formation and color of any solid produced as a result of a test.
3. Some of the color-forming tests occur for several different functional groups. Although the *expected* color is given in the experimental procedures, the *observed* color may be affected by the presence of other functional groups.
4. It is of utmost importance to perform a qualitative test on *both the unknown and a known compound that contains the group being tested.* Some functional groups may appear to give only a slightly positive test, and you will find it helpful to determine how a compound known to contain a given functional group behaves under the conditions of the test being performed. It is most efficient and reliable to do the tests on standards at the same time as on the unknown. In this manner, inconclusive positive tests may be interpreted correctly. Because aliphatic compounds are sometimes more reactive than aromatic ones, it is wise to perform a test on both of these types of standards along with the unknown.
5. The results obtained from the elemental analysis and the solubility tests can be used in deciding which functional group tests should be performed initially and which should not be done at all. The following examples illustrate the use of the preliminary work in making these decisions:
   a. A classification test for an amine should be applied first if a compound is found to be soluble in dilute hydrochloric acid and to contain nitrogen.
   b. The test for a phenol should be performed on an unknown that is soluble in dilute sodium hydroxide but insoluble in dilute sodium bicarbonate.
   c. The tests for alkyl or aryl halides should be omitted if the elemental analysis indicates the absence of halogen.
   d. The tests for amines, amides, nitriles, and nitro compounds need not be performed if nitrogen is absent, as shown by the elemental analysis.
6. You should use a logical approach for deciding which tests are needed. The result obtained from one test, whether positive or negative, often has a bearing on which additional tests should be done. A random "hit-or-miss" approach is

wasteful of time and often leads to erroneous results. Another error commonly made by beginners in qualitative organic analysis is to omit the tests for functional groups and immediately attempt preparation of a derivative. This tactic has a very low probability for success. For example, trying to make a derivative of a ketone is certain to fail if the unknown is actually an alcohol. You should continue performing the different classification tests until you have defined the nature of the functional group(s) present in the unknown as completely as possible! This will minimize unproductive efforts at derivatization of the compound.

Classification tests for most of the common functional groups are presented in the following sections.

Neutral Compounds	
Alcohols	Section 25.11
Aldehydes	Section 25.7
Alkenes	Section 25.8
Alkyl halides	Section 25.9
Alkynes	Section 25.8
Amides	Section 25.18
Aromatic hydrocarbons	Section 25.10
Aryl halides	Section 25.10
Esters	Section 25.16
Ketones	Section 25.7
Nitriles	Section 25.17
Nitro compounds	Section 25.15
Acidic Compounds	
Carboxylic acids	Section 25.13
Phenols	Section 25.12
Basic Compounds	
Amines	Section 25.14

Derivatization

We mentioned earlier that the classic approach to structure elucidation usually involves converting an "unknown" liquid or solid into a second compound that is a solid, the latter being called a **derivative** of the first compound. It is better to prepare a derivative that is a solid rather than a liquid, because solids can be obtained in pure form by recrystallization and because the melting point of a solid may be determined on a small quantity of material, whereas a larger amount of a liquid derivative would be required in order to determine its boiling point.

You should prepare two solid derivatives of an unknown compound in order to double-check its identity. The melting points of the derivatives, along with the melting point or boiling point of the unknown compound, usually serve to identify the unknown completely. However, the success of this type of identification depends upon the availability of tables listing the melting points and/or boiling points of known compounds and the melting points of suitable derivatives. Extensive listings of compounds are found in the references at the end of Section 25.5 and abbreviated compilations are available at the website for this textbook. These tabulations are by no means comprehensive, and many other compounds that have been identified on the basis of derivatives appear in the scientific literature (Chap. 26).

We repeat the warning given in this section under the heading "Classification Tests": Do *not* proceed directly to the preparation of a derivative merely on the basis of a hunch about the class of compound to which your unknown belongs. Rather, you should make certain of the type of functional group present by obtaining one *or more* positive classification tests *before* attempting the preparation of any derivative.

The following example illustrates how classification tests and preparation of derivatives are used to identify an unknown compound.

Preliminary analysis	Colorless liquid with pleasant odor.
Physical constant	bp 119–120 °C (760 torr).
Elemental analysis	No X, S, or N.
Solubility tests	Slightly soluble in water; no increased solubility in dilute HCl, NaOH, or $NaHCO_3$; soluble in concentrated H_2SO_4.
Preliminary report	Observations confirmed by instructor.
Classification tests	Negative test for aldehyde, alkene, alkyne; positive test for ketone, positive test for methyl ketone, negative tests for all other functional groups.

The experimental error inherent in determining melting or boiling points means it is prudent to consider compounds melting or boiling within about 5 °C of the observed melting or boiling point as possibilities for the unknown. Applying this principle to the entries in Table 25.1 of selected liquid ketones leads to compounds **1–6** as possibilities.

From the experimental facts that were given, some of these compounds may be eliminated as possibilities for the following reasons: **4** because it contains a halogen, **5** because it contains a carbon-carbon double bond, and **6** because it is not a methyl ketone. Now suppose that two derivatives, the 2,4-dinitrophenylhydrazone (Sec. 25.7A) and the semicarbazone (Sec. 25.7F), were prepared from the unknown and found to melt at 93–95 °C and 69–71 °C, respectively. Examining Table 25.1 reveals that the derivatives of only one of the ketones under consideration, namely 3-methyl-2-pentanone (**3**), melt at these temperatures. Hence, the identity of the unknown is deduced. Although it is possible that other ketones with similar boiling points *may* exist and may *not* be listed in Table 25.1, it is highly unlikely that any of these will give two derivatives with the same melting points as those of the derivatives obtained from **3.** This emphasizes the desirability of preparing *two* derivatives.

You can see from the preceding analysis how positive *and* negative information may be utilized in determining the identity of an unknown substance. Of particular note is the importance of the functional group classification tests, which must be done carefully and thoroughly in order to exclude the possible presence of all groups other than the keto group in the unknown.

Table 25.1 *Liquid Ketones with Boiling Points 115–125 °C (760 torr)*

Name and Number	*Structure*	*BP (°C)*	*2,4-DNP (°C)*	*Oxime (°C)*
1-Methoxy-2-propanone (**1**)	$CH_3OCH_2C(=O)CH_3$	115		163
4-Methyl-2-pentanone (**2**)	$(CH_3)_2CHCH_2C(=O)CH_3$	117	135	95
3-Methyl-2-pentanone (**3**)	$(CH_3)CH_2(CH_3)C(=O)CH_3$	118	95	71
Chloroacetone (**4**)	$ClCH_2C(=O)CH_3$	119	150	125
3-Penten-2-one (**5**)	$(CH_3)_2CH=CHC(=O)CH_3$	122	142	155
2,4-Dimethyl-3-pentanone (**6**)	$(CH_3)_2CHC(=O)CH(CH_3)_2$	124	160	188

25.4 SEPARATING MIXTURES OF ORGANIC COMPOUNDS

The preceding section contains information identifying a *pure* organic compound. However, when a chemist is faced with the problem of identifying an organic compound, it is seldom pure; rather, it is often contaminated with by-products or starting materials if it has been synthesized in the laboratory. Modern methods of separation, particularly chromatographic techniques (Chap. 6), make the isolation of a pure compound easier than it once was, but one must not lose sight of the importance of classic techniques of separation and purification, which are treated in detail in Chapters 3–5.

The common basis of the procedures most often used to separate mixtures of organic compounds is the difference in **polarity** that exists or may be induced in the components of the mixture. This difference in polarity is exploited in nearly all the separation techniques, including distillation, recrystallization, extraction, and chromatography. The greatest differences in polarity, which make for the simplest separations, are those that exist between salts and nonpolar organic compounds. Whenever one or more of the components of a mixture is convertible to a salt, it can be separated easily and efficiently from the nonpolar components by distillation (Chap. 4) or extraction (Chap. 5).

When learning the techniques of qualitative organic analysis, you may be given a *mixture* of unknown compounds, each of which is to be identified. Before this can be done, each component of the mixture must be obtained in pure form. The general approach shown in Figure 25.2 illustrates how this may be accomplished using principles based on differences in polarity and functionality. For this scheme to be applicable, each component of the mixture must have a low solubility in water and must not undergo appreciable hydrolysis by reaction with dilute acids or bases at room temperature. The procedure is based primarily on partitioning compounds of significantly different polarities between diethyl ether and water and separating these liquid layers in a separatory funnel. The underlying concept of this process involves extraction, the theory of which is discussed in Section 5.2.

Assuming that a mixture contains a carboxylic acid, a phenol, an amine, and a neutral compound, each of which is water-insoluble, the separation is initiated by dissolving the mixture in a suitable organic solvent such as diethyl ether. The ethereal solution of the mixture is first extracted with sodium bicarbonate solution, which removes the carboxylic acid by converting it to its water-soluble sodium salt. This extraction is followed by one with sodium hydroxide solution, which removes the water-soluble sodium salt of the phenol. Finally, the ethereal solution of the mixture is treated with hydrochloric acid, which reacts with the amine, converting it into a water-soluble ammonium salt. The ethereal solution that remains after removal of the aqueous solution contains the neutral compound.

As you can see, each extraction is performed on the same ethereal solution that originally contained all the components of the mixture, and the *sequence* of extraction, namely, $NaHCO_3$, then NaOH, and finally HCl, is *extremely* important. The bases and acid each remove one type of organic compound from the mixture and leave the neutral compound in the other layer when the extractions have been completed. Each of the basic and acidic extracts is subsequently treated with acid or base to liberate the carboxylic acid, phenol, and amine from its salt; each of these compounds is then removed from the respective aqueous solutions by extraction with ether or by collecting the solid by vacuum filtration.

Many different layers and solutions are obtained in the experimental procedure that follows. This can lead to much confusion unless the flasks containing

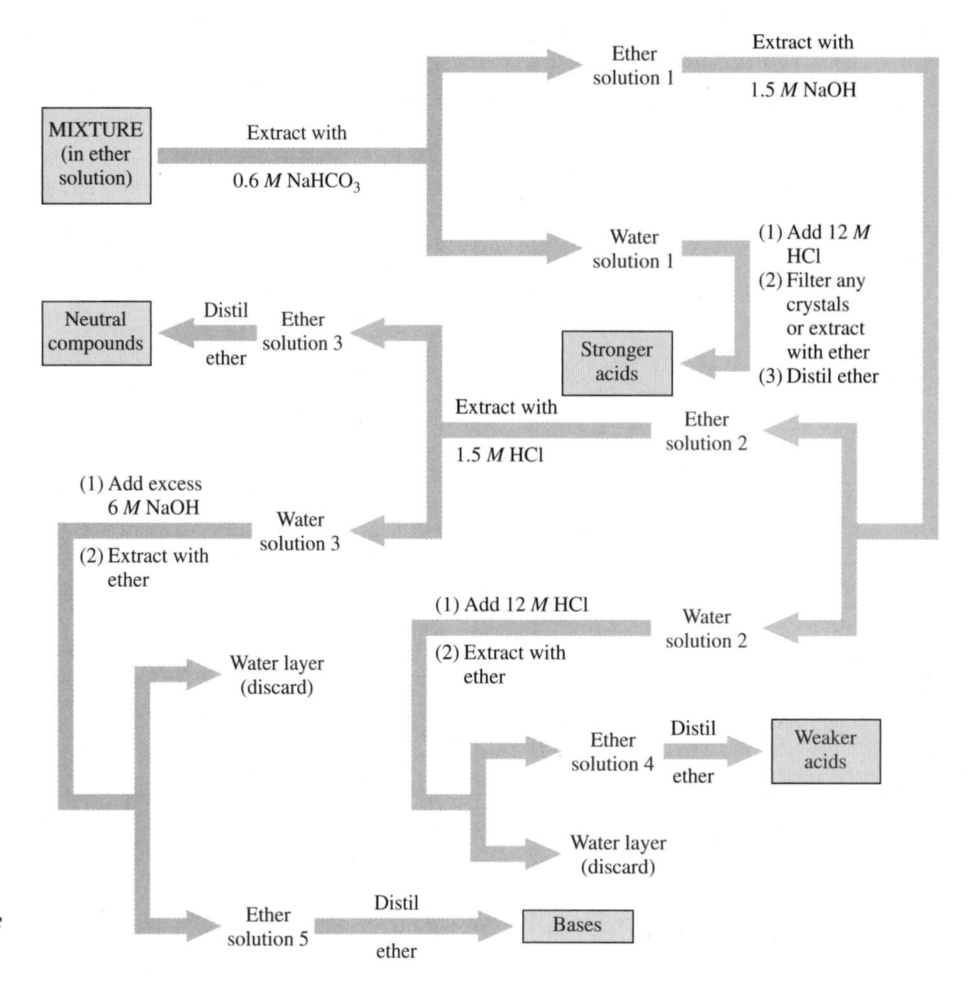

Figure 25.2
General scheme for separating a simple mixture of water-insoluble compounds.

each liquid are clearly labeled as to the identity of their contents. Moreover, *it is prudent to retain the flasks containing all layers and solutions until it is certain that they are no longer needed.*

EXPERIMENTAL PROCEDURE

Separating Mixtures on the Basis of Solubility

SAFETY ALERT

Use *flameless* methods to remove the diethyl ether from solutes in this procedure.

Preparation Sign in at **www.cengage.com/login** to read the MSDSs for the chemicals used or produced in this procedure.

The following procedure, based on the scheme of Figure 25.2, may be used to separate the components of a mixture consisting of three or four compounds, 1–2 g of each, of the types described. It is important that you understand that this generalized procedure may not give complete separation of all compounds of even the limited types for which it is intended. The separated products must be tested for purity by the usual methods of melting- or boiling-point determinations and, if possible, by GLC or TLC. Before attempting identification of the individual compounds by any of the classic or modern instrumental methods described later, the samples should be purified by recrystallization, distillation, or chromatography.

Stir or swirl the mixture (3–8 g) with 25–30 mL of diethyl ether at room temperature. If necessary, collect any solid material that does not dissolve by gravity filtration. Extract the ethereal solution with 5-mL portions of 0.6 *M* $NaHCO_3$ solution until the aqueous extract remains slightly basic. *Caution:* Vent the separatory funnel quickly after the first mixing because of possible buildup of pressure from the formation of carbon dioxide. This solution, **water solution 1** of Figure 25.2, should contain the sodium salt of any carboxylic acid present in the mixture; the ether layer, **ether solution 1**, should contain all other components of the mixture of unknowns. Regenerate the free organic acid by careful acidification of **water solution 1** to pH 2–3 with 12 *M* hydrochloric acid. If a solid acid separates, collect it by vacuum filtration. Otherwise, extract the aqueous acid solution with several 5-mL portions of diethyl ether, dry the ethereal solution over anhydrous sodium sulfate, and remove the ether by one of the techniques described in Section 2.29.

Extract **ether solution 1** with two 5-mL portions of 1.5 *M* NaOH solution to remove a phenol or any other weak acid from the solution and to give **ether solution 2**. Acidify the combined aqueous extracts, **water solution 2**, with 12 *M* hydrochloric acid, and extract the regenerated weak acid with several 5-mL portions of ether to yield **ether solution 4**. If **amphoteric** compounds were included in the unknown mixture, they would be carried through to the water layer separated from **ether solution 4**. Dry this solution and remove the ether from it as already described, and isolate the weak acid that remains as the residue.

Extract **ether solution 2** with one or more 5-mL portions of 1.5 *M* hydrochloric acid until the aqueous extract, **water solution 3**, remains acidic. The ether layer, designated as **ether solution 3** in Figure 25.2, should contain any neutral organic compounds; recover these by drying and removing the ether as described previously.

Add sufficient 5 *M* NaOH solution to **water solution 3** to make the solution strongly basic; then extract it with several 5-mL portions of diethyl ether. The combined ether extracts, **ether solution 5**, should contain any organic base present in the original unknown mixture; isolate the base by drying the solution and removing the ether as described previously.

WRAPPING IT UP

Transfer any recovered *diethyl ether* in the container for nonhalogenated organic liquids. Neutralize all *basic or acidic aqueous solutions* and flush them down the drain. Place the *filter paper* containing any solids that were not soluble in diethyl ether in the container for nonhazardous solids.

25.5 APPLYING SPECTROSCOPIC METHODS TO QUALITATIVE ORGANIC ANALYSIS

A major limitation inherent to the classic system of qualitative organic analysis is that only *known* compounds can be identified. The research chemist is constantly faced with the task of identifying *new* substances. Although information about the type of compound can be derived from this classic approach, the complete identification of a new organic compound traditionally required a combination of degradation and synthesis in order to achieve a correlation with a known substance; this was usually a lengthy and laborious task. Fortunately, the advent of spectroscopic techniques changed this picture dramatically. New and unknown compounds may be identified quickly and with certainty by using a combination of spectroscopic methods such as those described in Chapter 8. A number of examples could be cited of structural elucidations that were completed in a matter of *days* or *weeks* for molecules of greater complexity than those of compounds that were the *lifework* of several of the great nineteenth-century and early twentieth-century organic chemists.

The modern approach to identifying organic compounds typically involves a combination of spectroscopic and classic methods. Ideally, you should be introduced to the application of spectroscopy in organic chemistry by using the instruments that produce the spectra, but this is not feasible in many instances because some of these instruments are expensive. The next best alternative is to have access to the spectra of typical known compounds and then be provided with the spectra of "unknowns" to identify. This textbook contains over 300 IR and NMR spectra of the starting materials and products obtained from various preparative experiments, and the website associated with the textbook has the IR and ^{1}H NMR spectra of these compounds. A careful study of these spectra, aided by the material presented in Chapter 8 and by additional discussion with your instructor, should enable meaningful use of spectroscopic data for identifying unknown organic compounds.

The spectral data serve to complement or supplement the "wet" qualitative classification tests and in many instances substitute for these tests entirely. For example, a strong IR absorption in the 1690–1760 cm^{-1} region is just as indicative of the presence of a carbonyl group as is formation of a 2,4-dinitrophenylhydrazone, and absorptions in the δ 6.0–8.5 region of the ^{1}H NMR spectrum are a more reliable indicator of the presence of an aromatic ring than a color test with $CHCl_3$ and $AlCl_3$.

We emphasize that spectroscopic analysis also requires that you *carefully interpret the data*. Although you may solve some problems quickly and uniquely by modern spectroscopy, you may find that others require the intelligent application of both the modern methods and the classic methods. The remainder of this section illustrates the use of IR and NMR spectroscopy in structure determination.

Because the primary use of IR spectroscopy is for identifying functional groups, the observation of certain IR absorptions provides information about the presence of particular functional groups in a compound. Conversely, the absence of certain peaks may be useful in *excluding* the possibility of specific functional groups. For example, the appearance of a strong IR band in the 1650–1760 cm^{-1} region is strong evidence that a substance contains the carbon-oxygen double bond characteristic of an aldehyde, a ketone, a carboxylic acid, or a derivative of a carboxylic acid. On the other hand, the absence of such an absorption is interpreted to mean that the compound does not contain this functional group. Furthermore, a strong band in the 3500–3650 cm^{-1} region is indicative that a substance is an alcohol, a phenol, a carboxylic acid, or a 1° or 2° amine or amide, whereas the complete absence of such an absorption excludes these functional groups from further consideration.

However, caution must be used in drawing conclusions based on the absence of expected absorption bands. As noted in Section 8.2, absorption of IR energy by a molecule requires a change in dipole moment as a result of the molecular vibration. This means that symmetrically substituted alkynes do *not* have an IR band for the stretching mode of the triple bond, and even disubstituted alkynes in which the two substituents are similar may have only a weak absorption in this region of the spectrum. For similar reasons, the absence of an absorption in the carbon-carbon double-bond stretching region does not necessarily exclude the presence of this functional group, because the nature of substitution at the double bond may render the functionality inactive in the IR spectrum. Finally, because certain structural features in a molecule cause small shifts in the expected absorptions of some functional groups, IR spectroscopy does not always provide a unique answer about the presence of a particular group. However, it is very useful in limiting the possibilities to a small number of functional groups, and you may then resolve the uncertainty by performing just a few qualitative tests in the laboratory.

NMR spectra generally do not permit direct observation of specific functional groups, as noted in Section 8.3, but they do provide indirect evidence regarding the presence or absence of some groups. The three major features of ^{1}H NMR spectra that are useful for identifying a compound are the chemical shift, the splitting pattern, and the relative abundance of each type of hydrogen, as determined by the peak areas. The chemical shifts for carbon atoms in the ^{13}C NMR spectrum are also valuable for defining the nature of functional groups in unknown substances.

The research chemist usually obtains IR and NMR spectra of an unknown substance even before obtaining an elemental analysis or performing solubility tests because these types of spectra are easily and quickly measured on a small amount of sample. Only after analyzing them does the researcher undertake other experimental approaches for determining the structure. This process can save many hours of unnecessary laboratory work. Other useful information about a compound may be obtained from mass spectrometry (Sec. 8.5), which provides the molar mass of the compound and, if high-resolution data are available, its elemental composition.

The following example illustrates applying spectral analysis to identifying an unknown compound.

Example

The unknown **X** has the molecular formula $C_9H_{10}O_2$ and provides the spectra shown in Figure 25.3. Analysis of the molecular formula results in an **index of hydrogen deficiency (IHD)** or degree of unsaturation for the compound of *five*, and thus mandates the presence of rings and/or multiple bonds. Examining the IR spectrum reveals two intense absorptions in the functional group region. Reference to Tables 8.1 and 8.2 shows that the band at about 1700 cm^{-1} is consistent with the presence of a carbonyl group, whereas the broad absorption ranging from 3400 to 2300 cm^{-1} is characteristic of the hydroxyl moiety of a carboxylic acid function: the O–H stretch of an alcohol is not as broad as that of a carboxylic acid. Further evidence for the acid functionality is found in the absorption at 1320 cm^{-1}, which may be assigned to C–O bond stretching. Stretching vibrations associated with the carbon-hydrogen bonds that must be present in the unknown appear to be buried in the hydroxyl absorption, so their absence should be of no concern. Finally, the presence at 1500 cm^{-1} of a sharp

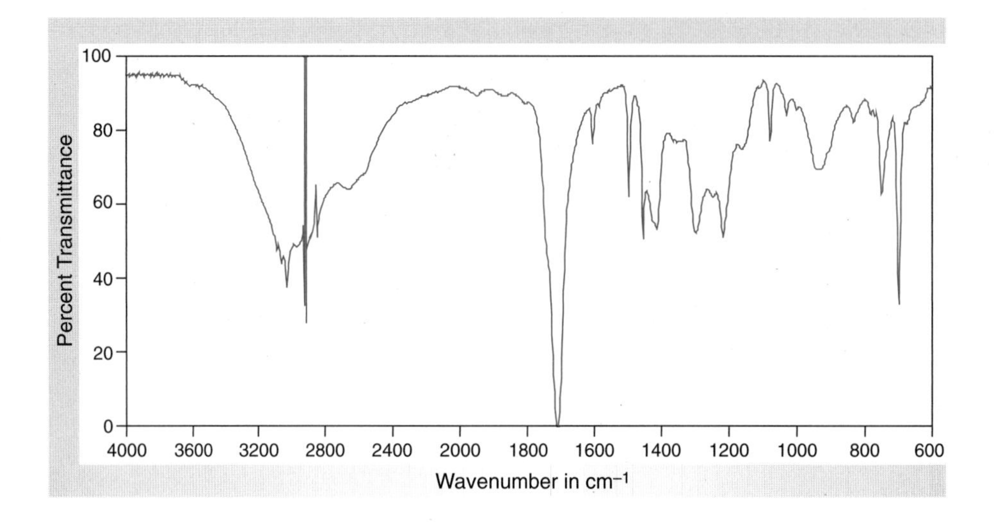

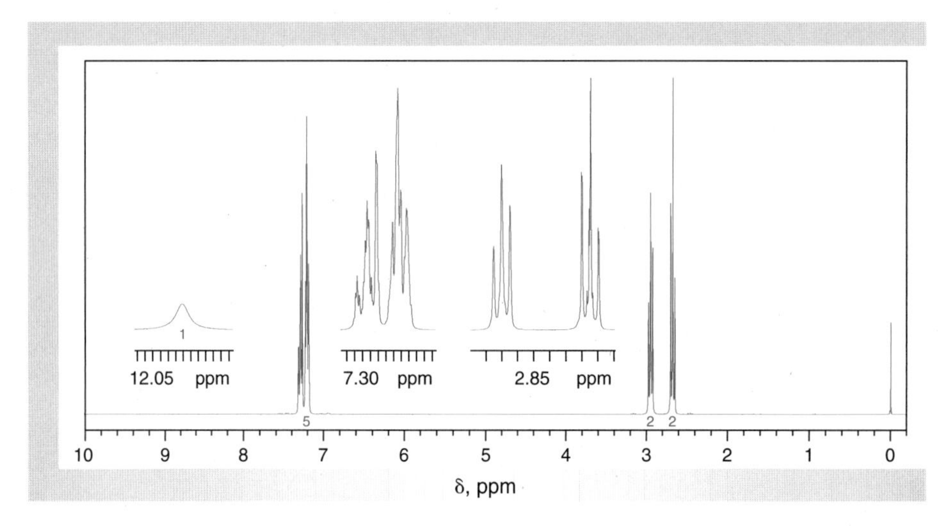

Figure 25.3
IR and ^{1}H NMR spectra of unknown compound ***X****.*

band of intermediate intensity may signify an aromatic ring, a possibility that seems reasonable in light of the value of IHD.

The ^{13}C NMR spectrum shows that there are seven magnetically different carbon atoms in **X**. The resonance at δ 179.5 has a chemical shift appropriate for the carbonyl carbon atom of a carboxylic function (Table 8.5), whereas those at δ 30.5 and 35.5 are clearly associated with sp^3- rather than sp^2- or sp-hybridized carbon atoms. The resonances appearing in the range of δ 126–140 have chemical shifts expected for sp^2-hybridized carbon atoms. Because there are only four such resonances and these must account for six carbon atoms, two of the peaks must each represent two magnetically identical carbon atoms. This requirement is met by proposing that **X** contains either a monosubstituted or a *para*-disubstituted benzene ring, **7** or **8**, respectively. Given the conclusions reached earlier about the nature of the other three carbon atoms present, only structures **9–11** are possible. Which of these is actually the unknown compound could be determined by calculating the expected ^{13}C chemical shifts of each of the possibilities, using the data in Tables 8.6–8.8, and comparing the results with those observed experimentally. However, we shall use the data available from the ^{1}H NMR to differentiate among the three possibilities.

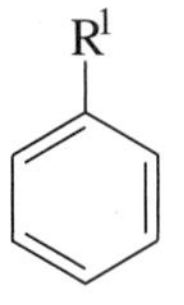

R^1-C$_6$H$_5$ (7); R^1-C$_6$H$_4$-R^2 (8); $C_6H_5CH_2CH_2CO_2H$ (9); $CH_3C_6H_4CH_2CO_2H$ (10); $CH_3CH_2C_6H_4CO_2H$ (11)

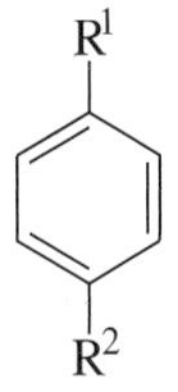

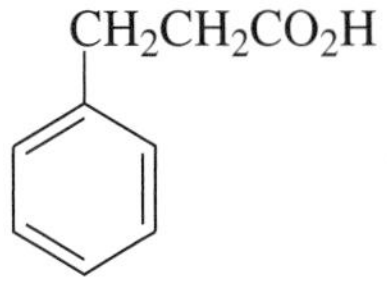

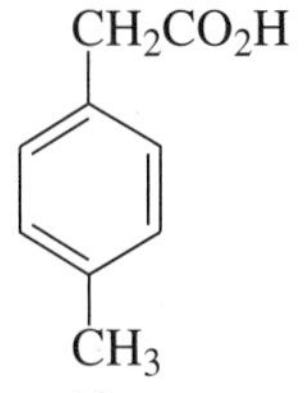

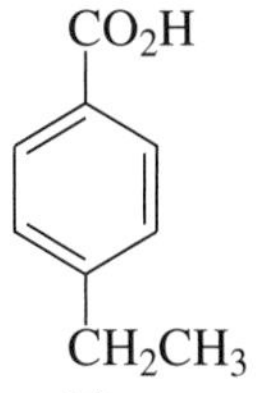

The ^{1}H NMR spectrum reveals that there are *four* sets of chemically nonequivalent protons in the unknown. These are in the *relative ratio* of 1:5:2:2 when going upfield. This is also the *absolute ratio* because there are 10 hydrogen atoms in **X**. The appearance of a resonance at very low field, δ 11.9, confirms the earlier conclusion that a carboxylic acid function is present (Table 8.3). The fact that the broad singlet at δ 7.2, which is in the range of chemical shifts for aromatic protons, integrates for *five* hydrogens means that the ring is *monosubstituted*. Consequently, the unknown compound must have structure **9**, in which the pair of two-proton multiplets centered at δ 2.8 correspond to the two heterotopic sets of methylene protons.

This example provides a logical approach to the use of a molecular formula and spectral data in deducing the structure of an unknown compound. The interplay of the information available from IR, ^{1}H NMR, and ^{13}C NMR spectra is critical to the process of structural elucidation, and you will find that it is commonly necessary to use more than one spectral technique before an assignment is possible. We cannot overemphasize the fact that *all* of the spectral data available must be consistent with the proposed structure of the sample. The existence of even one inconsistency between the data and the proposed structure renders that structure unlikely, unless the inconsistency is produced by an extraneous factor such as an impurity in the sample or a malfunction in the instrumentation used to obtain the spectra.

A second example illustrates the complementarity between spectral analysis and classic qualitative analysis for determining the structure of an unknown compound.

Example

The unknown **Y** is a liquid and the following information is available:

Preliminary analysis	Colorless liquid with pleasant odor.
Physical constant	bp 143–145 °C (760 torr).
Elemental analysis	No X, S, or N.
Solubility tests	Water-soluble to give a neutral solution.
Preliminary report	Observations confirmed by instructor.
Spectra	IR and ^{1}H NMR spectra are shown in Figure 25.4.
Classification tests	Positive test for ester, negative tests for all other functional groups.

Analysis

Owing to its water-solubility, **Y** most likely contains oxygen-bearing polar functional groups; it also probably contains a relatively small number of carbon atoms because compounds of more than five or six carbon atoms are usually water-insoluble. It is immediately evident from the IR spectrum shown in Figure 25.4 that the liquid contains an ester group because of the strong absorption peaks at about 1750 and 1230 cm^{-1}; the former peak is due to C–O stretching and the latter to the C–O–C bond (Table 8.1). Although the 1750 cm^{-1} band could indicate a five-membered cyclic ketone, this and other possibilities are negated by the presence of

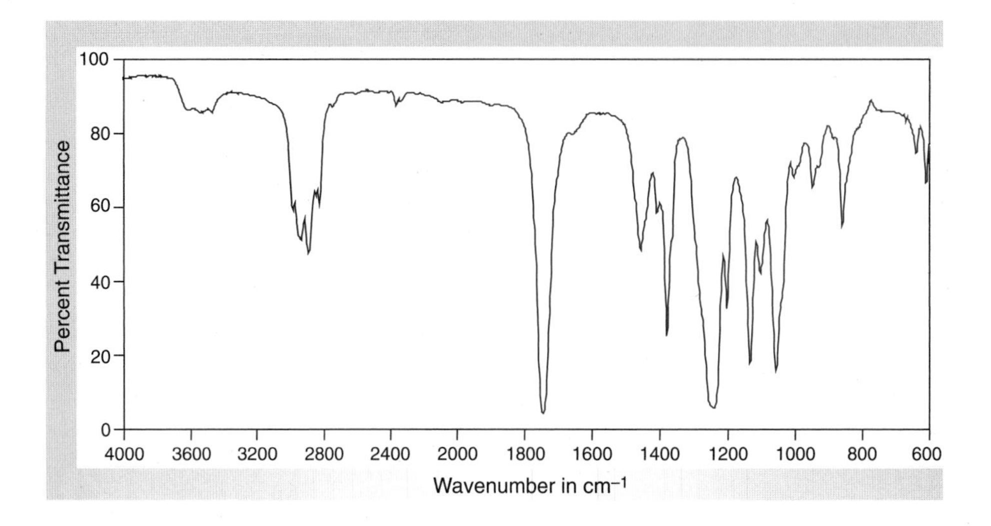

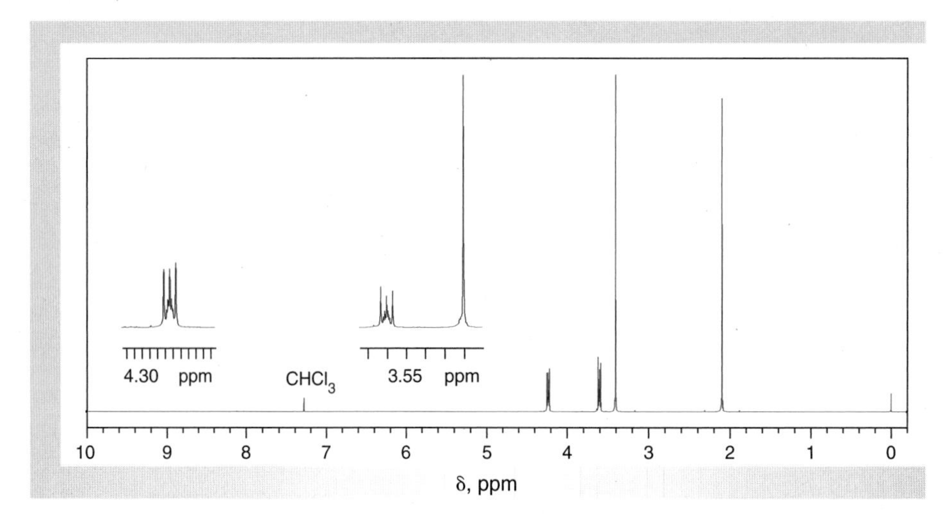

Figure 25.4
IR and ^{1}H NMR spectra of unknown compound **Y**.

numerous peaks in the ^{1}H NMR spectrum. The presence of an ester function was confirmed by obtaining a positive qualitative test for this group (Sec. 25.16).

The ^{1}H NMR spectrum has four different resonances: a multiplet centered at δ 4.2, a second multiplet centered at δ 3.6, a singlet at δ 3.4, and another singlet at δ 2.1, and their relative areas are 2:2:3:3, respectively. This means that *at least* four chemically different carbon atoms must be present in the molecule. In fact, reference to the ^{13}C NMR data shows that *five* chemically distinct carbon atoms are contained in **Y**. The *relative* abundances of different types of protons are likely to correspond to the *absolute* ones, because the water solubility of the unknown signals that **Y** is likely to contain fewer than seven carbon atoms. Consequently, the two singlets at δ 2.1 and 3.4, each with relative integrations of 3, must represent two rather than some larger number of methyl groups, and the two groups have no nearest neighbors. The higher-field peak at δ 2.1 is characteristic of a methyl ketone or acetate (Table 8.3), the latter being consistent with the deduction provided by the IR spectrum and the qualitative test that **Y** is an ester. The second methyl peak at δ 3.4 is shifted downfield, as would be expected if the methyl group were bonded to oxygen. This peak is too far upfield for a methyl ester because methyl esters normally show the methyl absorption at about δ 3.7–4.1. However, it is within the δ 3.3–4.0 range frequently observed for aliphatic α-hydrogens of an alcohol or ether. Because neither spectrum shows any evidence for a hydroxy group, the compound probably contains a methoxy group, and the presence of an aliphatic ether is consistent with the C–O–C absorption observed at 1050 cm^{-1} in the IR spectrum. The multiplets centered at δ 4.2 and 3.6 each integrate for two hydrogens and show mutual spin-spin coupling. This pattern is diagnostic of chemically nonequivalent, adjacent methylene groups such as X–CH_2CH_2–Y. The downfield absorption for the CH_2 groups indicates that each is bonded to oxygen.

Table 25.2 is a listing of some liquid esters, and only a single structure, 2-methoxyethyl acetate (**15**), appears to be consistent with all of the spectral observations for **Y**. You may confirm this structural assignment by applying the classic techniques of derivatization. In this instance, hydrolyzing the suspected ester **15** (Sec. 25.16) should produce acetic acid and 2-methoxyethanol, both of which could be separately characterized through a solid derivative. The alcohol, for example, could be converted into an α-naphthylurethane derivative (Sec. 25.11D). Finding that this derivative has a melting point of 111–113 °C, a range corresponding to that expected for the α-naphthylurethane of 2-methoxyethanol (Table 25.2), verifies that the unknown **Y** contains the 2-methoxyethoxy, $CH_3OCH_2CH_2O$–, fragment, as required in **15**.

This example of proving the structure of an unknown should provide you with an appreciation of the complementary aspects of spectral analysis and the "wet" classification scheme, including the use of physical properties. Although the spectra were not in themselves specifically definitive in the assignment of the structures of the unknown, they provided data regarding the possible presence of certain functional groups, which could be confirmed by performing the appropriate classification tests. Thus, the spectral analyses helped minimize the effort and time spent in performing numerous classification tests, most of which would have been negative. The final decision on the structure of the unknown came from preparing a solid derivative whose melting point could be compared with that of the known compound. Of course, when solving the structure of an unknown, it is often possible to find the IR

Table 25.2 *Liquid Esters with Boiling Points 140–150 °C (760 torr)*

Name and Number	*Structure*	*BP (°C)*	*α-Naphthylurethane (°C)*
3-Methylbutyl acetate (**12**)	$CH_3C(=O)OCH_2CH_2CH(CH_3)_2$	142	82
Methyl lactate (**13**)	$CH_3CH(OH)C(=O)OCH_3$	145	124
Ethyl chloroacetate (**14**)	$ClCH_2C(=O)OCH_2CH_3$	145	79
2-Methoxyethyl acetate (**15**)	$CH_3C(=O)OCH_2CH_2OCH_3$	145	113
Ethyl valerate (**16**)	$CH_3(CH_2)_3C(=O)OCH_2CH_3$	146	79
Ethyl α-chloropropionate (**17**)	$CH_3CH(C1)C(=O)OCH_2CH_3$	146	79
Diisoprophyl carbonate (**18**)	$(CH_3)_2CHOC(=O)OCH(CH_3)_2$	147	106
Pentyl acetate (**19**)	$CH_3C(=O)O(CH_2)_4CH_3$	149	68

and NMR spectra of the suspected compounds in one of the catalogs of spectra given in the references in Chapter 8. If the reported spectra are *identical* to those obtained for the unknown, the identity of the unknown is confirmed.

REFERENCES

1. Shriner, R. L.; Hermann, C. K. F.; Morrill, T. C.; Curtin, D. Y.; Fuson, R. C. *The Systematic Identification of Organic Compounds*, 8th ed., John Wiley & Sons, New York, 2004.
2. *Handbook of Tables for Organic Compound Identification*, 3rd ed., Z. Rappoport, ed., Chemical Rubber Company, Cleveland, OH, 1967. Provides physical properties and derivatives for more than 4,000 compounds, arranged according to functional groups.

25.6 QUALITATIVE CLASSIFICATION TESTS AND PREPARATION OF DERIVATIVES

In this section, we provide discussions of and experimental procedures for the qualitative classification tests and preparation of derivatives for compounds having most of the commonly encountered functional groups. Proper interpretation of the information from the qualitative tests is *crucial* for successfully applying the principles of Section 25.3 to structure elucidation. Consequently, you should exercise great care not only when performing the tests but also in analyzing their results. In the end, you may find solving the structures of unknowns the most intellectually challenging, yet most satisfying, experience you have in the introductory organic laboratory.

25.7 ALDEHYDES AND KETONES

Classification Tests

A. 2,4-Dinitrophenylhydrazine

Arylhydrazines ($ArNHNH_2$), which include phenylhydrazine, 4-nitrophenylhydrazine, and 2,4-dinitrophenylhydrazine, are commonly used to make crystalline derivatives of carbonyl compounds. The formation of a 2,4-dinitrophenylhydrazone **3** is represented by Equation 25.9, and the mechanism of this transformation typifies that followed by a number of compounds, RNH_2, that may be considered derivatives of ammonia. Thus, the overall reaction involves initial acid-catalyzed addition of the elements of N–H across the carbonyl π-bond to afford a tetrahedral intermediate, which subsequently dehydrates to the product (Eq. 25.9).

The arylhydrazines are valuable reagents for both classifying and forming derivatives because the solid products of these test reactions may be used as derivatives of the aldehyde or ketone; 2,4-dinitrophenylhydrazine is particularly useful in this regard.

The arylhydrazines will give a positive test for either an aldehyde or a ketone. Schiff's and Tollens's tests given in Parts B and C, respectively, provide methods for distinguishing between these two types of compounds.

$$R_2C{=}O + H_2NNH{-}Ar \underset{}{\overset{H_3O^+}{\rightleftharpoons}} R_2C(OH)(NH{-}NHAr) \rightleftharpoons R_2C{=}NNHAr + H_2O \quad (25.9)$$

An aldehyde or ketone; 2,4-Dinitrophenylhydrazine; **3** A 2,4-dinitrophenylhydrazone

Ar = 2,4-dinitrophenyl ($C_6H_3(NO_2)_2$)

EXPERIMENTAL PROCEDURE

2,4-Dinitrophenylhydrazine Test for Aldehydes and Ketones

SAFETY ALERT

Concentrated sulfuric acid may produce severe chemical burns, and 2,4-dinitrophenylhydrazine is a possible mutagen and may cause skin sensitization. Wear latex gloves when preparing and handling solutions involving these reagents. If this acid or the ethanolic solution of 2,4-dinitrophenylhydrazine comes in contact with your skin, immediately flood the affected area with water and rinse it with 5% sodium bicarbonate solution.

Preparation Sign in at **www.cengage.com/login** to read the MSDSs for the chemicals used or produced in this procedure.

If the test reagent is not supplied, prepare it by dissolving 0.2 g of 2,4-dinitrophenylhydrazine in 1 mL of concentrated sulfuric acid. Add this solution, with stirring, to 1.5 mL of water and 5 mL of 95% ethanol. Stir the solution vigorously and then filter it to remove any undissolved solids.

Dissolve 1 or 2 drops of a liquid, or about 100 mg of a solid, in 2 mL of 95% ethanol, and add this solution to 2 mL of the 2,4-dinitrophenylhydrazine reagent. Shake the mixture vigorously; if a precipitate does not form immediately, let the solution stand for 15 min.

If more crystals are desired for a melting-point determination, dissolve 200–500 mg of the carbonyl compound in 20 mL of 95% ethanol, and add this solution to 15 mL of the reagent. Recrystallize the product from aqueous ethanol.

WRAPPING IT UP

Neutralize and then filter any excess *2,4-dinitrophenylhydrazine solution.* Put the filter cake in the container for nonhazardous solids. Flush the *filtrate,* as well as *filtrates* obtained from recrystallization, down the drain.

B. Schiff's Test

See *Who Was Schiff?*

A common method for distinguishing between aldehydes and ketones is Schiff's test. Aldehydes give positive tests, whereas ketones do not.

Schiff's reagent is an aqueous solution formed by combining *p*-rosaniline hydrochloride (basic fuchsin), sodium bisulfite, and concentrated hydrochloric acid (Eq. 25.10). The reagent reacts with aldehydes to produce an imine, also called a Schiff base, that is magenta or purple in color (Eq. 25.11). The mechanism for forming the addition product presumably involves nucleophilic attack of sulfur on the aldehydic carbonyl group.

Color development typically occurs in no more than 10 minutes and ketones do not produce a corresponding color change. However, because the reagent may produce a light blue color in the presence of other functional groups, it is important to perform this test with known compounds such as benzaldehyde and acetone.

$H_2N^+ Cl^-$, CH_3, NH_2, H_2N — *p*-Rosaniline hydrochloride *Fuchsia (purple)* $\xrightarrow{(1)\ NaHSO_3\quad (2)\ HCl}$ $H_2\overset{+}{N}SO_2H\ Cl^-$, CH_3, HSO_2NH, $NHSO_2H$, SO_3H — Schiff's reagent *(colorless)* (25.10)

Schiff's reagent + RCH=O ⟶ $H\overset{+}{N}SO_2CH(OH)R\ Cl^-$, CH_3, $RCH(OH)O_2SNH$, $NHSO_2CH(OH)R$ (25.11)

An aldehyde — An addition product *(magenta or purple)*

EXPERIMENTAL PROCEDURE

Schiff's Test for Aldehydes

SAFETY ALERT

p-Rosaniline hydrochloride is a suspected carcinogen and a mutagen. Wear latex gloves to prevent Schiff's reagent from contacting your skin. If you get the reagent on your skin, immediately flood the area with warm water.

Preparation Sign in at **www.cengage.com/login** to read the MSDSs for the chemicals used or produced in this procedure.

If Schiff's reagent is not available as a stock solution, prepare it by dissolving 0.01 g of *p*-rosaniline hydrochloride in 10 mL of water, adding 0.4 mL of saturated aqueous sodium bisulfite, and allowing the resulting solution to stand for 1 h. Then add 0.2 mL of concentrated hydrochloric acid to complete the preparation.

Place 1 mL of Schiff's reagent in a small test tube and add 1 drop of the unknown. Compare any color changes with those observed when a known aldehyde and a known ketone are subjected to the same test.

WRAPPING IT UP

Neutralize the *acidic solutions* by adding solid sodium carbonate and pour the resulting solution into a container for hazardous aqueous waste.

C. Tollens's Test.

Another method for distinguishing between aldehydes and ketones is Tollens's test. A positive test indicates the presence of an aldehyde function, whereas no reaction occurs with ketones. Tollens's reagent consists of silver-ammonia complex, $Ag(NH_3)_2^+$, in an ammonia solution. This reagent oxidizes both aliphatic and aromatic aldehydes to the corresponding carboxylic acids; silver ion is reduced to elemental silver, which is deposited as a silver mirror on the glass wall of a *clean* test tube. Thus, the formation of the silver mirror or of a precipitate is considered a positive test. Equation 25.12 shows the reaction that occurs.

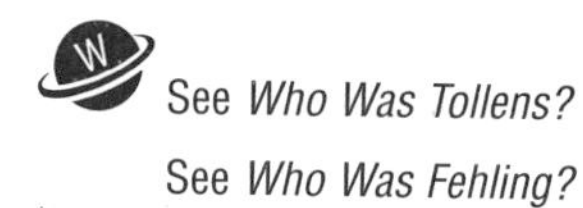

See *Who Was Tollens?*
See *Who Was Fehling?*
See *Who Was Benedict?*

$$\underset{\text{An aldehyde}}{RCHO} + 2\,Ag(NH_3)_2^+ + 2\,HO^- \longrightarrow \underset{\text{(metallic silver)}}{2\,\underline{Ag^o}} + RCO_2^-\,NH_4^+ + H_2O + 3\,NH_3 \quad (25.12)$$

Similar tests for aldehydes make use of Fehling's and Benedict's reagents, which contain complex tartrate and citrate salts, respectively, of cupric ion as the oxidizing agents. With these reagents, a positive test is the formation of a brick-red precipitate of cuprous oxide (Cu_2O), which forms when Cu^{2+} is reduced to Cu^+ by the aldehyde. These two tests are more useful in distinguishing between aliphatic and aromatic aldehydes, because the aliphatic compounds give a faster reaction. They have been used widely to detect reducing sugars (Sec. 23.4), whereas Tollens's test is used to distinguish between aldehydes, both aliphatic and aromatic, and ketones.

EXPERIMENTAL PROCEDURE

Tollens's Test for Aldehydes

SAFETY ALERT

1. **Avoid spilling Tollens's reagent on your skin, because staining may result. If you do come into contact with the solution, immediately flood the affected area with water.**
2. ***Do not store unused Tollens's reagent, because it decomposes on standing and yields silver azide, AgN_3, an explosive precipitate.* Follow the directions under Wrapping It Up for proper disposal of the solution.**

Preparation Sign in at **www.cengage.com/login** to read the MSDSs for the chemicals used or produced in this procedure.

Tollens's reagent must be prepared at the time of use by combining two other solutions, A and B, which should be available as stock solutions. If not, make up solution A by dissolving 0.25 g of silver nitrate in 4.3 mL of distilled water and solution B by dissolving 0.3 g of potassium hydroxide in 4.2 mL of distilled water.

Prepare the test reagent according to the following directions. Add concentrated ammonium hydroxide solution dropwise to 3 mL of solution A until the initial brown precipitate begins to clear. The solution should be grayish and almost clear. Then add 3 mL of solution B to this mixture, and again add concentrated ammonium hydroxide dropwise until the solution is almost clear.

To carry out the test, add 0.5 mL of the reagent to 3 drops or 50–100 mg of the unknown compound; the formation of a silver mirror or black precipitate constitutes a positive test. The silver deposits in the form of a mirror only on a *clean* glass surface. A black precipitate, although not as aesthetically pleasing, still constitutes a positive test. If no reaction occurs at room temperature, warm the solution slightly in a beaker of warm water.

WRAPPING IT UP

Remove the *silver mirrors* from the test tube with nitric acid, and pour the resulting *solution* into a beaker containing unused *Tollens's reagent, ammonium hydroxide,* and *sodium hydroxide.* Acidify this mixture with nitric acid to destroy the unreacted Tollens's reagent. Neutralize the solution with sodium carbonate and add saturated sodium chloride solution to precipitate silver chloride. Collect the silver chloride by vacuum filtration and place the *filter cake* in a container for recovered silver halides; flush the filtrate down the drain.

D. Chromic Acid Test

Another method for distinguishing between aldehydes and ketones is the chromic acid test. Because chromic acid is unstable when stored for extended periods of time, the test reagent is prepared as needed by dissolving chromic anhydride in sulfuric acid (Eq. 25.13) and used in acetone solution. This reagent oxidizes primary and secondary alcohols (Eqs. 25.14 and 25.15, respectively) and aldehydes (Eq. 25.14); it gives no visible reaction with ketones and tertiary alcohols (Eqs. 25.15 and 25.16, respectively) under the test conditions. A distinctive color change from the orange-red of Cr^{6+} to the green of Cr^{3+} occurs as oxidation of the organic substrate proceeds. The stoichiometry of oxidation of an alcohol to an aldehyde or ketone is depicted in Equation 25.17, where you can see that three moles of sulfuric acid are consumed in the process.

$$CrO_3 + H_2O \xrightarrow[\text{(orange-red)}]{H_2CrO_4} H_2CrO_4 \qquad (25.13)$$

$$\underset{1^\circ\text{ alcohol}}{RCH_2OH} \xrightarrow[\text{(orange-red)}]{H_2CrO_4} [RCHO] \xrightarrow[\text{(orange-red)}]{H_2CrO_4} RCO_2H + \textit{Green}\text{ solution or precipitate} \qquad (25.14)$$

$$\underset{\text{A }2^\circ\text{ alcohol}}{R_2CHOH} \xrightarrow[\text{(orange-red)}]{H_2CrO_4} R_2CO + \textit{Green}\text{ solution or precipitate} \xrightarrow{H_2CrO_4}\!\!\!\!\!\!\!\!/\;\; \text{No visible reaction} \qquad (25.15)$$

$$R_3COH \xrightarrow{H_2CrO_4}\!\!\!\!\!\!\!\!/\;\; \text{No visible reaction} \qquad (25.16)$$

$$3\,R^2\text{—}\overset{OH}{\overset{|}{C}H}\text{—}R^1 + 2\,H_2CrO_4 + 3\,H_2SO_4 \longrightarrow 3\,R^2\text{—}\overset{O}{\overset{\|}{C}}\text{—}R^1 + Cr_2(SO_4)_3 + 8\,H_2O \qquad (25.17)$$

Thus, the chromic acid reagent gives a clear-cut distinction between primary and secondary alcohols and aldehydes on the one hand and tertiary alcohols and ketones on the other. Aldehydes may be distinguished from primary and secondary alcohols by means of Schiff's, Tollens's, Benedict's (Sec. 23.4), and Fehling's tests, and primary and secondary alcohols of lower molar mass may be differentiated on the basis of their rates of reaction with concentrated hydrochloric acid containing zinc chloride—the Lucas reagent (Sec. 25.11B).

EXPERIMENTAL PROCEDURE

Chromic Acid Test for Aldehydes and 1° and 2° Alcohols

SAFETY ALERT

1. **When preparing and handling solutions of chromic acid, wear latex gloves to keep the acids from contacting your skin. Chromic acid causes unsightly stains on your hands for several days and may cause severe chemical burns. If the oxidant comes in contact with your skin, immediately flood the affected area with warm water and rinse it with 5% sodium bicarbonate solution.**

2. The preparation of chromic acid requires diluting a paste of chromic anhydride and concentrated sulfuric acid with water. Be certain to add the water *slowly* to the acid and swirl the mixture to ensure continuous mixing. Swirling keeps the denser sulfuric acid from layering at the bottom of the flask and avoids possible splattering because of the heat generated when the two layers are suddenly mixed.

Preparation Sign in at **www.cengage.com/login** to read the MSDSs for the chemicals used or produced in this procedure.

If the chromic acid reagent is not available, prepare it as follows. Add 1 g of chromic anhydride (CrO_3) to 1 mL of concentrated sulfuric acid and stir the mixture until a smooth paste is obtained. Dilute the paste *cautiously* with 3 mL of distilled water, and stir this mixture until a clear orange solution is obtained.

Dissolve 1 drop of a liquid or about 10 mg of a solid alcohol or carbonyl compound in 1 mL of *reagent-grade* acetone. Add 1 drop of the acidic chromic anhydride reagent to the acetone solution and shake the tube to mix the contents. A positive reaction is indicated by disappearance of the orange color of the reagent and the formation of a green or blue-green precipitate or emulsion.

Primary and secondary alcohols and aliphatic aldehydes give a positive test within 5 sec. Aromatic aldehydes require 30–45 sec. Color changes occurring after about 1 min should not be interpreted as positive tests; other functional groups such as ethers and esters may slowly hydrolyze under the conditions of the test, releasing alcohols that in turn provide "false-positive" tests. Tertiary alcohols and ketones produce no visible change in several minutes. Phenols and aromatic amines give dark precipitates, as do aromatic aldehydes and benzylic alcohols having hydroxyl or amino groups on the aromatic ring.

WRAPPING IT UP

Add sodium sulfite to the *aqueous solution* of chromium salts in order to destroy excess Cr^{6+}. Make the solution slightly basic, to form chromium hydroxide, and isolate this salt by vacuum filtration through a bed of a filter-aid. Place the *filter paper* and the *filter cake* in the container for heavy metals; flush the *filtrate* down the drain.

E. Iodoform Test

When an aldehyde or ketone that has α-hydrogen atoms is treated with a halogen in basic medium, halogenation occurs on the α-carbon atom (Eq. 25.18). This reaction involves the formation of an intermediate enolate ion (Eq. 18.3), which subsequently reacts with halogen to produce the substitution product. In the case of acetaldehyde or a methyl ketone, all three of the α-hydrogen atoms of the methyl group are replaced by halogen atoms to give **20** (Eq. 25.19), which then reacts with excess base to give products **21** and **22** (Eq. 25.20).

$$R_2CH{-}C({=}O)R \xrightarrow{\text{Base, } X_2} R_2CX{-}C({=}O)R \qquad (25.18)$$

An aldehyde or ketone

$$\text{H}-\overset{\text{H}}{\underset{\text{H}}{\text{C}}}-\text{C}(=\text{O})\text{R} \xrightarrow{\text{NaOH, X}_2} \text{X}-\overset{\text{X}}{\underset{\text{X}}{\text{C}}}-\text{C}(=\text{O})\text{R} \quad (25.19)$$

An aldehyde or ketone → **20**

$$\text{X}_3\text{C}-\text{C}(=\text{O})-\text{R} \xrightarrow{\text{NaOH}} \text{CHX}_3 + \text{RCO}_2^-\text{Na}^+ \quad (25.20)$$

20 → **21** (A haloform) + **22**

Although chlorine, bromine, and iodine all react in this manner, the qualitative test for a methyl ketone uses iodine because it is safer to handle and because the product is *iodoform*, CHI_3, a highly *insoluble* crystalline *yellow* solid that is readily observed and identified on the basis of its characteristic odor and melting point. We note that a positive iodoform test involves two experimental observations: (1) the disappearance of the characteristic red-brown color of iodine as the test reagent is added to the compound being analyzed, *and* (2) the formation of a yellow precipitate of iodoform. The reason both observations are required is that *all* aldehydes and ketones containing α-hydrogen atoms react with iodine in base to decolorize the test reagent. Only methyl ketones, after being trisubstituted with iodine on the methyl group, react with base to produce iodoform and the salt of a carboxylic acid.

A positive test also occurs for alcohols having a hydroxymethyl ($-CHOHCH_3$) functionality, as in ethanol and isopropyl alcohol, and for acetaldehyde, which is analogous to a methyl ketone in having three α-hydrogen atoms. The reason a hydroxymethyl compound gives a positive iodoform test is that the combination of iodine and sodium hydroxide produces sodium hypoiodite (Eq. 25.21), which is a mild oxidant (Sec. 16.2). Consequently, the test reagent is able to oxidize a hydroxymethyl function to a methyl ketone, and the methyl ketone then reacts with iodine in the presence of base to give iodoform. It is thus always necessary to consider this possibility before making a final decision regarding the functional group responsible for a positive test. The ambiguity in interpreting this test is resolved by determining whether a keto group is present or absent by using another qualitative test, such as that involving 2,4-dinitrophenylhydrazine (Eq. 25.9).

$$I_2 + 2\,NaOH \rightarrow NaI + NaOI + H_2O \quad (25.21)$$

The iodoform test is sometimes called the **hypoiodite test** because sodium hypoiodite is formed according to Equation 25.21. In more general terms, the reactions just described can be called the **haloform test** or the **sodium hypohalite test** when the specific halogen is not specified.

EXPERIMENTAL PROCEDURE

Iodoform Test

SAFETY ALERT

Iodine is a hazardous chemical. Do not breathe its vapors, and wear latex gloves to prevent this reagent or solutions of it from contacting your skin because it may cause serious chemical burns. It you get iodine on your skin, immediately flood the area with warm water and soak it in 0.6 *M* sodium thiosulfate solution for up to 3 h if the burn is particularly serious.

Preparation Sign in at **www.cengage.com/login** to read the MSDSs for the chemicals used or produced in this procedure.

If the iodine reagent is not available as a stock solution, prepare it by dissolving 1 g of iodine in a solution of 2 g of potassium iodide in 8 mL of water. The potassium iodide is added to increase the solubility of iodine in water by formation of potassium triiodide, KI_3 (Eq. 25.22).

$$I_2 + KI \longrightarrow KI_3 \qquad (25.22)$$

If the substance is water-soluble, dissolve 2–3 drops of a liquid or an estimated 50 mg of a solid in 2 mL of water in a small test tube, add 2 mL of 3 *M* sodium hydroxide, and then slowly add 3 mL of iodine solution. In a positive test, the brown color disappears and iodoform separates. If the substance tested is insoluble in water, dissolve it in 2 mL of dioxane, proceed as above, and dilute the mixture with 10 mL of water *after* addition of the iodine solution.

Iodoform is recognizable by its color and, more definitively, by its melting point, 118–119 °C. Isolate this product either by vacuum filtration of the test mixture or by the following sequence: Add 2 mL of dichloromethane to the mixture, shake the stoppered test tube to extract the iodoform into the small lower layer, withdraw the clear part of this layer with a Pasteur pipet, and evaporate it in a small test tube or on a watchglass with a steam bath in the hood. Whichever mode of isolation is used, recrystallize the crude solid from methanol-water.

WRAPPING IT UP

Place the residual *aqueous solutions* in the container for halogenated liquids. Put the *iodoform* in the container for halogenated solids. Flush the *filtrate* down the drain.

Derivatives

Two of the most useful solid derivatives of aldehydes and ketones are the **2,4-dinitrophenylhydrazone** and the **semicarbazone**. The **oxime** is also sometimes useful, but it often forms as an oil rather than a solid.

F. 2,4-Dinitrophenylhydrazone

The qualitative test for aldehydes and ketones is described in Section 25.7A, and the solid that forms may be isolated and purified as indicated in that section.

G. Semicarbazone

Semicarbazide reacts with aldehydes and ketones to produce a derivative that is called a semicarbazone (Eq. 25.23), a reaction that is discussed in Chapter 13. Because semicarbazide is unstable as the free base, it is usually stored in the form of its hydrochloric acid salt. In the procedure that follows, the free base is liberated from the salt by addition of sodium acetate.

$$R^1R^2C{=}O + H_2N{-}NHCNH_2\,(C{=}O) \xrightarrow{H^+} R^1R^2C{=}N{-}NHCNH_2\,(C{=}O) \qquad (25.23)$$

An aldehyde or ketone — Semicarbazide — A semicarbazone

EXPERIMENTAL PROCEDURE

Preparation of Semicarbazones

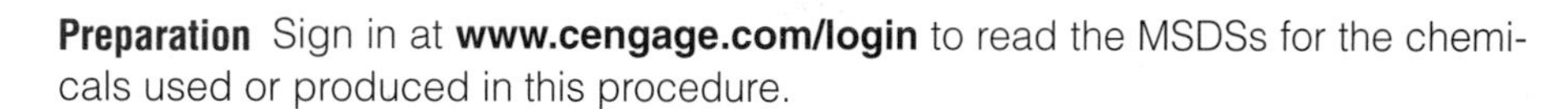

Semicarbazide and its hydrochloride may have carcinogenic, teratogenic, and mutagenic effects. Should either of these substances contact your skin, immediately flood the affected area with water.

Preparation Sign in at **www.cengage.com/login** to read the MSDSs for the chemicals used or produced in this procedure.

Dissolve 0.5 g of semicarbazide hydrochloride and 0.8 g of sodium acetate in 5 mL of water in a test tube and then add about 0.5 mL of the carbonyl compound. Stopper and shake the tube vigorously, remove the stopper, and place the test tube in a beaker of boiling water. Discontinue heating the water, and allow the test tube to cool to room temperature in the beaker of water. Transfer the test tube to an ice-water bath, and if crystals have not formed, scratch the side of the tube with a glass rod at the interface between the liquid and air to induce crystallization. Recrystallize the semicarbazone from water or aqueous ethanol.

If the carbonyl compound is insoluble in water, dissolve it in 5 mL of ethanol. Add water until the solution becomes turbid, then add a little ethanol until the turbidity disappears. Add the semicarbazide hydrochloride and sodium acetate, and continue as above from this point.

Flush *excess reagent* and *filtrates* down the drain.

H. Oximes

Hydroxylamine reacts with aldehydes or ketones to yield an oxime (Eq. 25.24). Hydroxylamine is usually stored as the hydrochloric acid salt because it is not stable as the free base; it is liberated from its salt by addition of base. The use of an oxime as a derivative has the limitation that such compounds frequently are *not* crystalline solids.

$$R^1C(=O)R^2 + H_2N{-}OH \xrightarrow{H^+} R^1C(=N{-}OH)R^2 + H_2O \qquad (25.24)$$

An aldehyde or ketone; Hydroxylamine; An oxime

EXPERIMENTAL PROCEDURE

Preparation of Oximes

SAFETY ALERT

Hydroxylamine and its hydrochloride may have mutagenic effects. Should either of these substances contact your skin, immediately flood the affected area with water.

Preparation Sign in at **www.cengage.com/login** to read the MSDSs for the chemicals used or produced in this procedure.

Dissolve 0.5 g of hydroxylamine hydrochloride in 5 mL of water and 3 mL of 3 *M* sodium hydroxide solution, and add 0.5 g of the aldehyde or ketone. If the carbonyl compound is insoluble in water, add just enough ethanol to give a clear solution. Warm the mixture on a steam bath or boiling-water bath for 10 min; then cool it in an ice-water bath. If crystals do not form immediately, scratch the side of the tube with a glass rod at the air-liquid interface to induce crystallization. Recrystallize the oxime from water or aqueous ethanol.

In some cases, the use of 3 mL of pyridine and 3 mL of *absolute* ethanol in place of the 3 mL of 3 *M* sodium hydroxide solution and 5 mL of water is more effective. A longer heating period is often necessary. After finishing heating, pour the mixture into an evaporating dish and remove the solvent with a current of air in a hood. Grind the solid residue with 3–4 mL of cold water and filter the mixture. Recrystallize the oxime from water or aqueous ethanol.

WRAPPING IT UP

Flush *excess reagent* and *filtrates* down the drain.

25.8 ALKENES AND ALKYNES

Classification Tests

Two common types of unsaturated compounds are alkenes and alkynes, characterized by the carbon-carbon double and triple bonds, respectively, as the functional group. The two common qualitative tests for unsaturation are the reaction of the compounds with *bromine in dichloromethane* and with *potassium permanganate.* In both cases, a positive test is denoted by decoloration of the reagent. There are no simple, direct ways to prepare solid derivatives of unsaturated aliphatic compounds having no other functional groups.

A. Bromine in Dichloromethane

Bromine adds to the carbon-carbon double bond of alkenes to produce dibromoalkanes (Eq. 25.25) and reacts with alkynes to produce tetrabromoalkanes (Eq. 25.26). When this reaction occurs, molecular bromine is rapidly consumed, and its characteristic dark red-orange color disappears if bromine is not added in excess. The nearly *instantaneous* disappearance of the bromine color is a positive test for unsaturation. The test is ambiguous, however, because some alkenes do not react with bromine, and some react very slowly. In the case of a negative test, the Baeyer procedure using potassium permanganate (Sec. 25.8B) should be performed.

$$\underset{\text{An alkene}}{R_2C{=}CR_2} + \underset{\substack{\text{Bromine}\\ \textit{(red-orange)}}}{Br_2} \xrightarrow{CH_2Cl_2} \underset{\substack{\text{A dibromide}\\ \textit{(colorless)}}}{R_2C(Br){-}CR_2(Br)} \tag{25.25}$$

$$\underset{\text{An alkyne}}{RC{\equiv}CR} + Br_2 \xrightarrow{CH_2Cl_2} \underset{\substack{\text{A tetrabromide}\\ \textit{(colorless)}}}{RC(Br)_2{-}C(Br)_2R} \tag{25.26}$$

More details about the reaction of bromine in dichloromethane with alkenes are found in Section 10.4.

EXPERIMENTAL PROCEDURE

Bromine Test for Unsaturation

SAFETY ALERT

1. ***Bromine is a hazardous chemical that may cause serious chemical burns.*** **Do not breathe its vapors or allow it to come into contact with your skin. Perform**

all operations involving the transfer of solutions of bromine at a hood; wear latex gloves. If you get bromine on your skin, immediately flood the area with warm water and soak it in 0.6 *M* sodium thiosulfate solution for up to 3 h if the burn is particularly serious.

2. **Do *not* use acetone to rinse glassware containing residual amounts of bromine. This prevents formation of α-bromoacetone, a severe lachrymator. Rather, follow the procedure described in Wrapping It Up.**

Preparation Sign in at **www.cengage.com/login** to read the MSDSs for the chemicals used or produced in this procedure.

Dissolve 50 mg or 1 or 2 drops of the unknown in dichloromethane, and add to this solution 0.1 *M* bromine in dichloromethane dropwise until a light orange color just persists. *Rapid* disappearance of the bromine color to give a colorless solution is a positive test for unsaturation.

WRAPPING IT UP

Decolorize any *solutions* in which the color of *bromine* is visible by dropwise addition of cyclohexene; then discard the resulting mixtures in the container for halogenated organic liquids.

B. Potassium Permanganate

See *Who Was Baeyer?*

A second qualitative test for unsaturation, the Baeyer test, depends on the ability of potassium permanganate to oxidize the carbon-carbon double bond to give alkanediols (Eq. 25.27) or the carbon-carbon triple bond to yield carboxylic acids (Eq. 25.28). The permanganate is destroyed in the reaction and a brown precipitate of MnO_2 is produced. The disappearance of the characteristic purple color of the permanganate ion is a positive test for unsaturation. However, care is needed in interpreting this test: compounds containing certain other functional groups, such as aldehydes, also decolorize permanganate ion because they undergo oxidation.

$$\underset{\text{An alkene}}{R_2C{=}CR_2} + \underset{\substack{\text{Permanganate}\\(purple)}}{MnO_4^-} \xrightarrow{H_2O} \underset{\substack{\text{A glycol}\\(colorless)}}{R_2C(OH){-}C(OH)R_2} + \underset{\substack{\text{Manganese}\\\text{dioxide}\\(brown)}}{\underline{MnO_2}} \quad (25.27)$$

$$\underset{\text{An alkyne}}{R^1C{\equiv}CR^2} + \underset{\substack{\text{Permanganate}\\(purple)}}{MnO_4^-} \xrightarrow{H_2O} \underset{\substack{\text{Carboxylic acids}\\(colorless)}}{R^1CO_2H + R^2CO_2H} + \underset{\substack{\text{Manganese}\\\text{dioxide}\\(brown)}}{\underline{MnO_2}} \quad (25.28)$$

EXPERIMENTAL PROCEDURE

Baeyer Test for Unsaturation

SAFETY ALERT

Wear latex gloves to prevent solutions of potassium permanganate from contacting your skin. These solutions cause unsightly stains on your hands for several days. If this oxidant comes in contact with the skin, immediately flood the affected area with warm water.

Preparation Sign in at **www.cengage.com/login** to read the MSDSs for the chemicals used or produced in this procedure.

Dissolve 1 or 2 drops of the unknown in 2 mL of 95% ethanol, and then add 0.1 *M* $KMnO_4$ solution dropwise; note the results. Count the number of drops added before the permanganate color persists. For a **blank determination,** count the number of drops of aqueous permanganate that can be added to 2 mL of 95% ethanol before the color persists. A significant difference in the number of drops required in the two cases is a positive test for unsaturation.

WRAPPING IT UP

Filter all *solutions* to remove manganese dioxide. Place the *filter paper* and the *filter cake* containing manganese salts in the container for heavy metals. Flush all *filtrates* down the drain.

25.9 ALKYL HALIDES

Classification Tests

Qualitative tests for alkyl halides are useful in deciding whether the compound in question is a primary, secondary, or tertiary halide. In general, it is quite difficult to prepare solid derivatives of alkyl halides, so this discussion is limited to two qualitative tests: (a) reaction with **alcoholic silver nitrate** solution, and (b) reaction with **sodium iodide in acetone.**

A. Alcoholic Silver Nitrate

If a compound is known to contain bromine, chlorine, or iodine, information concerning the environment of the halogen is obtained from reacting the substance with alcoholic silver nitrate. The overall reaction is shown in Equation 25.29.

$$\underset{\text{An alkyl halide}}{RX} + \underset{\text{Silver nitrate}}{AgNO_3} \xrightarrow{\text{ethanol}} \underset{\text{A silver halide}}{\underline{AgX}} + \underset{\text{An alkyl nitrate}}{RONO_2} \quad (25.29)$$

Such a reaction mechanistically will be of the S_N1 type. As noted in Section 14.2, tertiary halides are more reactive in an S_N1 reaction than are secondary halides, which are in turn more reactive than primary halides. Thus, the rate of precipitation of silver halide in this test is expected to correspond to the ease of ionization of the alkyl halide to a carbocation according to the order primary < secondary < tertiary. From a practical standpoint, these differences are best determined by testing authentic samples of primary, secondary, and tertiary halides with silver nitrate in separate test tubes and comparing the results.

Alkyl bromides and iodides react more rapidly than chlorides, and the latter may even require warming to produce a reaction in a reasonable time period. Aryl halides are *unreactive* toward the test reagent, as are generally vinyl or alkynyl halides. Allylic and benzylic halides, even when primary, show reactivities as great as or greater than tertiary halides because of resonance stabilization of the resulting allyl or benzyl carbocations.

EXPERIMENTAL PROCEDURE

Silver Nitrate Test for Alkyl Halides

SAFETY ALERT

Avoid spilling alcoholic silver nitrate on your skin because staining may result. If you do come into contact with the solution, immediately flood the affected area with water.

Preparation Sign in at **www.cengage.com/login** to read the MSDSs for the chemicals used or produced in this procedure.

Add 1 drop of the alkyl halide to 2 mL of a 0.1 *M* solution of silver nitrate in 95% ethanol. If no reaction is observed within 5 min at room temperature, warm the mixture in a beaker of boiling water and observe any change. Note the color of any precipitates: silver chloride is white but turns purple on exposure to light, silver bromide is pale yellow, and silver iodide is dark yellow. If there is any precipitate, add several drops of 1 *M* nitric acid solution to it, and note any changes; the silver halides are insoluble in acid. To determine expected reactivities, test known primary, secondary, and tertiary halides in this manner. If possible, use alkyl iodides, bromides, and chlorides so that differences in halogen reactivity can also be observed.

WRAPPING IT UP

Neutralize and then filter all test mixtures and place the *filter cake* in a container for recovered silver halides. Flush the *filtrate* down the drain.

B. Sodium Iodide in Acetone (Finkelstein Reaction)

See more on *Finkelstein Reaction*

Another method for distinguishing between primary, secondary, and tertiary halides makes use of sodium iodide dissolved in acetone. This test complements the alcoholic silver nitrate test, and when these two tests are used together, it is possible to determine the gross structure of the attached alkyl group with reasonable accuracy.

The basis of this test is that both sodium chloride and sodium bromide are *not* very soluble in acetone, whereas sodium iodide is. Mechanistically, the reactions occurring are S_N2 substitutions (Secs. 14.2 and 14.4) in which iodide ion is the nucleophile (Eqs. 25.30 and 25.31); the order of reactivity expected is primary > secondary > tertiary.

$$\mathrm{RCl + NAI \xrightarrow{acetone} RI + \underline{NaCl}} \qquad (25.30)$$

$$\mathrm{RBr + NAI \xrightarrow{acetone} RI + \underline{NaBr}} \qquad (25.31)$$

With the test reagent, *primary bromides* give a precipitate of sodium bromide in about 3 min at room temperature, whereas the *primary* and *secondary chlorides* must be heated to about 50 °C before reaction occurs. *Secondary* and *tertiary bromides* react at 50 °C, but the *tertiary chlorides* fail to react in a reasonable time. This test is necessarily limited to bromides and chlorides.

EXPERIMENTAL PROCEDURE

Sodium Iodide Test for Alkyl Chlorides and Bromides

Preparation Sign in at **www.cengage.com/login** to read the MSDSs for the chemicals used or produced in this procedure.

Place 1 mL of the sodium iodide–acetone test solution in a test tube, and add 2 drops of the chloro or bromo compound. If the compound is a solid, dissolve about 50 mg of it in a minimum volume of acetone and add this solution *to* the reagent. Shake the test tube, and allow it to stand for 3 min at room temperature. Note whether a precipitate forms; if no change occurs after 3 min, warm the mixture in a beaker of water at 50 °C. After 6 min of heating, cool the solution to room temperature and note whether any precipitate forms. Occasionally a precipitate forms immediately after combination of the reagents; this represents a positive test *only* if the precipitate remains after the mixture is shaken and allowed to stand for 3 min.

Carry out this reaction with a series of primary, secondary, and tertiary chlorides and bromides. Note in all cases the differences in reactivity as evidenced by the rate of formation of sodium bromide or sodium chloride.

WRAPPING IT UP

Place all *solutions* and *solids* involved in this test in the container for halogenated liquids, unless instructed to do otherwise.

25.10 AROMATIC HYDROCARBONS AND ARYL HALIDES

Classification Test

A. Chloroform and Aluminum Chloride

This test for the presence of an aromatic ring should be performed only on compounds that have been shown to be *insoluble* in concentrated sulfuric acid (Sec. 25.3). The test involves the Friedel-Crafts reaction (Sec. 15.2) between an aromatic compound and chloroform in the presence of anhydrous aluminum chloride as a catalyst. The colors produced in this reaction are often characteristic for certain aromatic compounds, whereas aliphatic compounds give little or no color with this test. Some typical examples are tabulated below. Often these colors change with time and ultimately yield brown-colored solutions.

Type of Compound	*Color*
Benzene and homologs	Orange to red
Aryl halides	Orange to red
Naphthalene	Blue
Biphenyl	Purple

The chemistry of the test involves a series of alkylation reactions (Sec. 15.2); for benzene, the ultimate product is triphenylmethane (Eq. 25.32).

$$\underset{\text{Benzene}}{3\,C_6H_6} + \underset{\text{Chloroform}}{CHCl_3} \xrightarrow{AlCl_3} \underset{\text{Triphenylmethane}}{(C_6H_5)_3CH} + 3\,HCl \qquad (25.32)$$

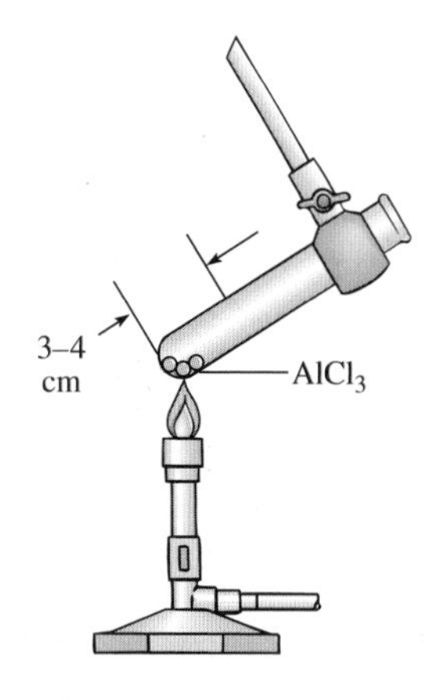

Figure 25.5
Apparatus for subliming aluminum chloride.

The colors arise because species such as triphenylmethyl cations, $(C_6H_5)_3C^+$, form and remain in the solution as their tetrachloroaluminate, $AlCl_4^-$, salts; ions of this sort are highly colored owing to the extensive delocalization of charge that is possible throughout the three aromatic rings.

The test is significant if positive, but a negative test does *not* rule out an aromatic structure; some compounds bearing electron-withdrawing groups are so unreactive that they do not readily undergo Friedel-Crafts alkylation reactions.

Positive tests for aryl halides are difficult to obtain directly, and some of the best evidence for their presence involves indirect methods. Elemental analysis indicates the presence of halogen. If *both* the silver nitrate and sodium iodide–acetone tests are negative, the compound is most likely a vinyl or an aromatic halide, both of which are very unreactive toward silver nitrate and sodium iodide. Distinction between a vinyl and an aromatic halide can be made by means of the aluminum chloride–chloroform test.

EXPERIMENTAL PROCEDURE

Friedel-Crafts Reaction for Detecting Arenes

SAFETY ALERT

Anhydrous aluminum chloride reacts vigorously with water, even the moisture on your hands, producing fumes of hydrogen chloride that are highly corrosive if inhaled. Do not allow aluminum chloride to come in contact with your skin. If it does, immediately flood the affected area with water. Because aluminum chloride is a powdery solid that can easily become airborne, weigh and transfer it into the reaction flask *in the hood.*

Preparation Sign in at **www.cengage.com/login** to Read the MSDSs for the chemicals used or produced in this procedure.

In a *dry* Pyrex test tube held almost horizontally, heat about 100 mg of *anhydrous* aluminum chloride until the material has sublimed to 3 or 4 cm above the bottom of the tube (Fig. 25.5). Allow the tube to cool until it is almost comfortable to touch; then add about 20 mg of a solid, or 1 drop of a liquid, down the side of the tube, followed by 2 or 3 drops of chloroform. The appearance of a bright color ranging from red to blue where the sample and chloroform come in contact with the aluminum chloride is a positive indication of an aromatic ring.

WRAPPING IT UP

Cool the *contents of the test tube* in an ice-water bath and cautiously add a few milliliters of cold water to hydrolyze the aluminum salts. Flush the resulting *aqueous layer* down the drain, and place the *organic layer* in the container for halogenated organic liquids.

Derivatives

Two types of derivatives are used to characterize aromatic hydrocarbons and aryl halides. These are prepared by nitration (Eq. 25.33) and side-chain oxidation (Eq. 25.34).

$$\text{R-C}_6\text{H}_5 \xrightarrow[\text{H}_2\text{SO}_4]{\text{HNO}_3} \text{R-C}_6\text{H}_4\text{-NO}_2 \qquad (25.33)$$

An aromatic hydrocarbon → A nitroaromatic

$$\text{ArC(R(H))(H)—R} \xrightarrow[\text{or H}_2\text{CrO}_4]{\text{KMnO}_4\text{/base}} \text{ArC(=O)—OH} \qquad (25.34)$$

An alkylated or acylated aromatic hydrocarbon → A benzoic acid

B. Nitration

Some of the best solid derivatives of aryl halides are mono- and dinitration products. Two general procedures can be used for nitration.

EXPERIMENTAL PROCEDURES

Preparation of Nitroarenes

SAFETY ALERT

1. **Use care whenever performing a nitration, whether it is with a known or an unknown substance, because many compounds react vigorously under typical nitration conditions.**
2. **Concentrated or fuming sulfuric and nitric acids are used in this procedure. These acids are very corrosive; handle them with care and wear latex gloves. If the acids contact your skin, immediately flood the affected area with water, and rinse it with 5% sodium bicarbonate solution.**

Preparation Sign in at **www.cengage.com/login** to read the MSDSs for the chemicals used or produced in this procedure.

Method A. *Note:* This method yields 1,3-dinitrobenzene from benzene or nitrobenzene and the 4-nitro derivative from chloro- or bromobenzene, benzyl chloride, or toluene; dinitro derivatives are obtained from phenol, acetanilide, naphthalene, and biphenyl.

Perform the reaction in a large test tube or small Erlenmeyer flask. Add about 0.5 g of the compound to 2 mL of concentrated sulfuric acid. Add 2 mL of concentrated nitric acid dropwise to this mixture; stopper the container and shake the mixture after each addition. After the addition is complete, heat the mixture at 45 °C on a water bath for about 5 min. Then pour the reaction mixture onto 15 g of ice, and collect the precipitate by vacuum filtration. Recrystallize the solid from aqueous ethanol.

Method B. *Note:* This method is the better one to use for nitrating halogenated benzenes because dinitration occurs, and these compounds have higher melting points and are easier to purify than are the mononitration products obtained from *Method A.* The xylenes, mesitylene, and pseudocumene yield trinitro compounds using the present method.

Follow the procedure used for *Method A*, except use 2 mL of *fuming* nitric acid in place of the *concentrated* nitric acid and warm the mixture with a steam bath for 10 min. If little or no nitration occurs, substitute *fuming* sulfuric acid for the *concentrated* sulfuric acid. *Perform this reaction in a hood.*

WRAPPING IT UP

Neutralize all *acidic solutions* before flushing them down the drain.

C. Side-Chain Oxidation

A second method for characterizing aromatic hydrocarbons involves oxidation of alkyl side chains to carboxylic acid groups, using either alkaline potassium permanganate or chromic acid solution (Eq. 25.34). Carbon-carbon as well as carbon-hydrogen bonds are cleaved in this process. Moreover, if more than one alkyl side chain is present, each will be oxidized to a carboxyl group, giving a polycarboxylic acid (Eq. 25.35). Identifying the acid provides information regarding the location of the alkyl side chains on the aromatic ring; the acid itself, if it is a solid, serves as a derivative of the unknown.

$$R-C_6H_4-R \xrightarrow[\text{or } H_2CrO_4]{(1)\ KMnO_4/HO^-\ (2)\ \text{Neutralize}} HO_2C-C_6H_4-CO_2H \quad (25.35)$$

(1,2-, 1,3-, or 1,4-) (1,2-, 1,3-, or 1,4-)

The mechanism of this type of oxidation is not well understood, but, may involve benzyl radicals generated by action of the oxidant, followed by conversion to the benzylic alcohol (Eq. 25.36). The alcohol then undergoes further oxidation.

$$C_6H_5CHR_2 \xrightarrow[\text{or } Cr^{6+}]{Mn^{7+}} C_6H_5CR_2OH \xrightarrow{\text{oxidant}} C_6H_5C(=O)R \longrightarrow C_6H_5C(=O)OH \quad (25.36)$$

An aromatic hydrocarbon; A benzylic alcohol; An aromatic aldehyde or ketone; An aromatic carboxylic acid

EXPERIMENTAL PROCEDURES

Side-Chain Oxidation of Arenes

SAFETY ALERT

1. **Wear latex gloves to keep the acids from contacting your skin when you prepare and handle solutions of potassium permanganate or chromic acid. These solutions cause unsightly stains on your skin for several days, and the chromic acid–sulfuric acid solution may cause severe chemical burns. If these oxidants contact your skin, immediately flood the affected area with water. In the case of chromic acid, also rinse the area with 5% sodium bicarbonate solution.**
2. **The preparation of chromic acid requires diluting sulfuric acid with water. Be certain to add the acid *slowly* to the water and swirl the container to ensure continuous mixing. The dissolution of sulfuric acid in water generates heat, and when the acid is added to water, the heat is dispersed through warming of**

the water. Swirling prevents the layering of the denser sulfuric acid at the bottom of the flask and the attendant possibility that hot acid will be splattered by the steam generated when the two layers are suddenly mixed later by agitation.

Preparation Sign in at **www.cengage.com/login** to read the MSDSs for the chemicals used or produced in this procedure.

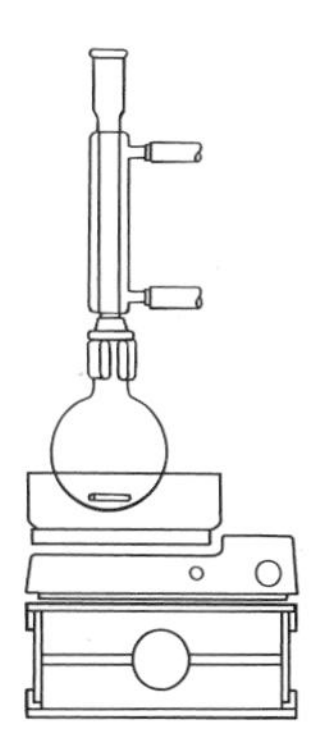

A. Permanganate Method. Add 0.5 g of the compound to a round-bottom flask containing a stirbar and a solution prepared from 80 mL of water and 2 g of potassium permanganate. Add 0.5 mL of 3 *M* sodium hydroxide solution and heat the mixture under reflux until the purple color of permanganate ion has disappeared; this will normally take 0.5–3 h. At the end of the reflux period, cool the mixture and carefully acidify it with 3 *M* sulfuric acid. Now heat the mixture for an additional 30 min and cool it again; remove excess brown manganese dioxide, if any, by addition of sodium bisulfite solution. The bisulfite reduces the manganese dioxide to manganous ion, which is water-soluble. Collect the solid acid that remains by vacuum filtration. Recrystallize the acid from toluene or aqueous ethanol. If little or no solid acid is formed, this may be because the acid is somewhat water-soluble. In this case, extract the aqueous layer with small portions of diethyl ether or dichloromethane, dry the organic extracts over anhydrous sodium sulfate, decant the solution, and remove the organic solvent by simple distillation or rotary evaporation. Recrystallize the acid that remains. In this particular method, the presence of base during the oxidation often means that some silicic acid, derived from the glass of the reaction vessel, forms on acidification. Thus, purify the carboxylic acid by recrystallization prior to determining the melting point.

WRAPPING IT UP

Place the *aqueous filtrate* containing manganese salts in the container for heavy metals. Flush all *ethanolic filtrates* down the drain. Put *filtrates* involving other organic solvents in the container for halogenated or nonhalogenated liquids, as appropriate.

Preparation Sign in at **www.cengage.com/login** to read the MSDSs for the chemicals used or produced in this procedure.

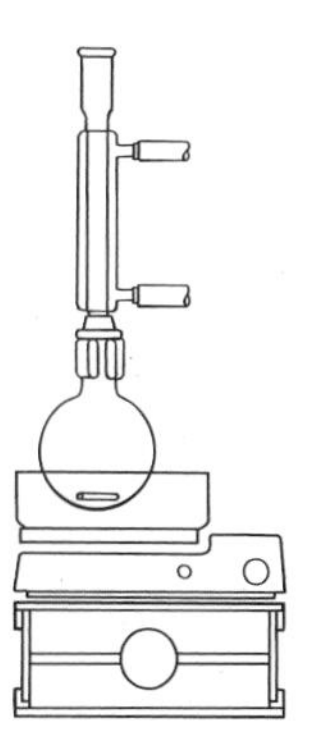

B. Chromic Acid Method. Dissolve 1 g of sodium dichromate in 3 mL of water contained in a round-bottom flask, and add 0.5 g of the compound to be oxidized and a stirbar. Carefully add 1.5 mL of concentrated sulfuric acid to the mixture with mixing and cooling. Attach a reflux condenser to the flask and heat the mixture gently until a reaction ensues. As soon as reaction begins, remove the heating source and cool the mixture if necessary to control the rate of boiling. After spontaneous boiling subsides, heat the mixture under reflux for 2 h. Pour the reaction mixture into 5 mL of water, and collect the precipitate by filtration. Transfer the solid into a flask, add 3 mL of 2 *M* sulfuric acid, and then warm the mixture on a steam cone with stirring. Cool the mixture, collect the precipitate, and wash it with about 3 mL of cold water. Dissolve the residue in 3 mL of 1.5 *M* sodium hydroxide solution and filter the solution. Add the filtrate, with stirring, to 6 mL of 2 *M* sulfuric acid. Collect the new precipitate, wash it with cold water, and recrystallize the carboxylic acid from either toluene or aqueous ethanol.

WRAPPING IT UP

Add sodium sulfite to the *aqueous filtrate* of chromium salts in order to destroy excess Cr^{6+}. Make the solution slightly basic to form chromium hydroxide and isolate this salt by vacuum filtration through a layer of filter-aid. Place the *filter paper* and the *filter cake* in the container for heavy metals; flush the *filtrate* down the drain. Neutralize all other *aqueous filtrates* before flushing them down the drain. Put any *filtrates* containing toluene in the container for nonhalogenated organic liquids.

25.11 ALCOHOLS

Classification Tests

The tests for the presence of a hydroxy group not only detect this functionality but may also indicate whether it is attached to a primary, secondary, or tertiary carbon atom.

A. Chromic Acid

This test may be used to detect the presence of a hydroxy group, provided that it is shown previously that the molecule does *not* contain an aldehyde function. The reactions and experimental procedures for this test are given in Section 25.7C. Chromic acid does *not* distinguish between primary and secondary alcohols because they *both* give a positive test; tertiary alcohols give a negative test.

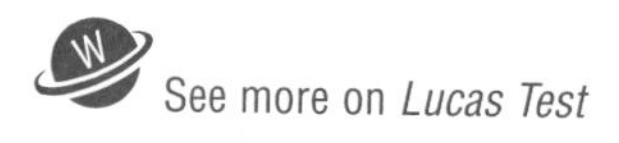

B. Lucas Test

This test is used to distinguish among primary, secondary, and tertiary alcohols. The reagent is a mixture of concentrated hydrochloric acid and zinc chloride, which converts alcohols to the corresponding alkyl chlorides. With this reagent, primary alcohols give no appreciable reaction (Eq. 25.37), secondary alcohols react more rapidly (Eq. 25.38), and tertiary alcohols react very rapidly (Eq. 25.39). A positive test depends on the fact that the alcohol is soluble in the reagent, whereas the alkyl chloride is not; thus the formation of a second layer or an emulsion constitutes a positive test. The *solubility of the alcohol in the reagent places limitations on the scope of the test*, and in general only monofunctional alcohols with six or fewer carbon atoms, as well as polyfunctional alcohols, may be used successfully.

$$\underset{\text{A 1° alcohol}}{RCH_2{-}OH} + HCl \xrightarrow{ZnCl_2} \text{No reaction} \tag{25.37}$$

$$\underset{\text{A 2° alcohol}}{R_2CH{-}OH} + HCl \xrightarrow[\text{slow}]{ZnCl_2} \underset{\text{A 2° alkyl halide}}{R_2CH{-}Cl} + H_2O \tag{25.38}$$

$$\underset{\text{A 3° alcohol}}{R_3C{-}OH} + HCl \xrightarrow[\text{fast}]{ZnCl_2} \underset{\text{A 3° alkyl halide}}{R_3C{-}Cl} + H_2O \tag{25.39}$$

The similarity should be noted between this reaction and the nucleophilic displacement reactions between alcohols and hydrohalic acids as discussed in Sections 14.3 and 14.4. In the Lucas test, the presence of zinc chloride, a Lewis acid, greatly increases the reactivity of alcohols toward hydrochloric acid.

EXPERIMENTAL PROCEDURE

Lucas Test for Alcohols

SAFETY ALERT

The solution of hydrochloric acid and zinc chloride is highly acidic and may produce severe chemical burns. Wear latex gloves when handling solutions of these compounds. If this solution comes in contact with your skin, immediately flood the affected area with water followed by 5% sodium bicarbonate solution.

Preparation Sign in at **www.cengage.com/login** to read the MSDSs for the chemicals used or produced in this procedure.

Add 5 mL of the hydrochloric acid–zinc chloride reagent (Lucas reagent) to about 0.5 mL of the compound in a test tube. Stopper the tube and shake it; allow the mixture to stand at room temperature. Try this test with known primary, secondary, and tertiary alcohols, and note the *time* required for the formation of an alkyl chloride, which appears either as a second layer or as an emulsion. Repeat the test with an unknown, and compare the result with the results from the knowns.

WRAPPING IT UP

Use a Pasteur pipet to separate the layer of alkyl chloride, if formed, from the test reagent. Put the *alkyl chloride* in the container for halogenated liquids. Neutralize the *aqueous soluti.ons* before flushing them down the drain.

C. Ceric Nitrate Test

Although this test is used primarily for detecting phenols, it can also be applied as a qualitative test for alcohols. Discussion about and experimental procedures for this test are given in Section 25.12.

Derivatives

Two common derivatives of alcohols are urethanes and the benzoate esters; the former are best for primary and secondary alcohols, whereas the latter are useful for all types of alcohols.

D. Urethanes

When an alcohol is allowed to react with an aryl substituted isocyanate, ArN=C=O, addition of the alcohol to the carbon-nitrogen double bond occurs to give a urethane (Eq. 25.40). Some commonly used isocyanates are α-naphthyl (1-naphthyl), 4-nitrophenyl, and phenyl.

$$\underset{\text{An aryl isocyanate}}{ArN{=}C{=}O} + \underset{\text{An alcohol}}{ROH} \longrightarrow \underset{\text{A urethane}}{ArNH{-}C({=}O){-}OR} \quad (25.40)$$

A major side reaction involves hydrolysis of the isocyanate to an amine, which then reacts with more isocyanate to give a disubstituted urea (Eqs. 25.41 and 25.42). Because of their symmetry, these ureas are high-melting, and their presence can make purification of the desired urethane difficult. Therefore, take precautions to ensure that the alcohol is *anhydrous* when using this procedure. The method works best for water-insoluble alcohols because they are more easily obtained in anhydrous form.

$$\underset{\text{An aryl isocyanate}}{ArN{=}C{=}O} + H_2O \longrightarrow \underset{\substack{\text{A carbamic acid}\\\text{(unstable)}}}{ArNH{-}C({=}O){-}OH} \xrightarrow{-CO_2} \underset{\text{An arylamine}}{ArNH_2} \quad (25.41)$$

$$\underset{\text{An aryl isocyanate}}{ArN{=}C{=}O} + \underset{\text{An arylamine}}{ArNH_2} \longrightarrow \underset{\text{A disubstituted urea}}{ArNH{-}C({=}O){-}NHAr} \quad (25.42)$$

This type of derivative is also useful for phenols; the procedure given here has been generalized so that it can be used for both alcohols and phenols. Other derivatives of phenols are provided in Section 25.12.

Preparation of Urethanes

SAFETY ALERT

Aryl isocyanates such as those used in the preparation of these derivatives are toxic. Take normal precautions in handling them. If cyanates contact the skin, wash the affected area with soap and warm water.

Preparation Sign in at **www.cengage.com/login** to read the MSDSs for the chemicals used or produced in this procedure.

Place 1 g of the *anhydrous* alcohol or phenol in a round-bottom flask, and add 0.5 mL of phenyl isocyanate or α-naphthyl isocyanate; be sure to recap the bottle of isocyanate tightly to minimize exposure of the reagent to atmospheric moisture. If you are preparing the derivative of a phenol, also add 2 or 3 drops of dry pyridine as a catalyst. Affix a drying tube to the flask and warm the reaction mixture for 5 min. Cool the mixture in an ice-water bath, and scratch the mixture with a stirring rod at the air-liquid interface to induce crystallization. Recrystallize the crude derivative from petroleum ether. *Note:* 1,3-Di(α-naphthyl) urea has a mp of 293 °C, and 1,3-diphenylurea (carbanilide) has a mp of 237 °C; if your product has one of these melting points, repeat the preparation, taking greater care to maintain anhydrous conditions.

WRAPPING IT UP

Flush any *filtrates* that contain the starting isocyanate down the drain. Pour the *filtrates* from the recrystallization of the product into the container for nonhalogenated organic liquids. Put the *dessicant* used in the drying tube in the container for nonhazardous solids.

E. 3,5-Dinitrobenzoates

The reaction between 3,5-dinitrobenzoyl chloride and an alcohol gives the corresponding ester (Eq. 25.43). This method is useful for primary, secondary, and tertiary alcohols, especially those that are water-soluble and thus are likely to contain traces of water.

$$ROH + Cl{-}C(=O){-}C_6H_3(NO_2)_2 \xrightarrow{\text{pyridine}} RO{-}C(=O){-}C_6H_3(NO_2)_2 + HCl \quad (25.43)$$

An alcohol — 3,5-Dinitrobenzoyl chloride — A 3,5-dinitrobenzoate

EXPERIMENTAL PROCEDURES

Preparation of 3,5-Dinitrobenzoates

SAFETY ALERT

1. **3,5-Dinitrobenzoyl chloride releases hydrochloric acid upon hydrolysis, so avoid contact of this reagent with your skin. If contact occurs, immediately flood the area with water and rinse it with 5% sodium bicarbonate solution. Wear latex gloves when handling this reagent.**
2. **Thionyl chloride is a lachrymator. Measure this material in a hood and do not inhale its vapors. Wear latex gloves when handling this reagent.**

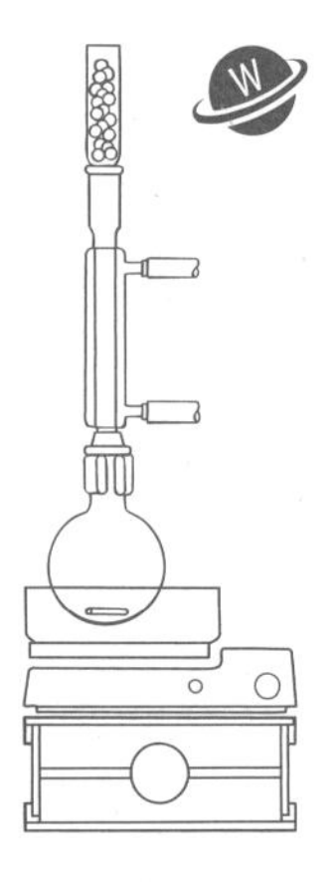

Preparation Sign in at **www.cengage.com/login** to read the MSDSs for the chemicals used or produced in this procedure.

Method A. *Note:* 3,5-Dinitrobenzoyl chloride is reactive toward water; it should be used immediately after weighing. Take care to minimize its exposure to air and to keep the bottle tightly closed.

Mix 0.6 mL of the alcohol with about 0.2 g of 3,5-dinitrobenzoyl chloride and 0.2 mL of pyridine in a small round-bottom flask equipped for magnetic stirring and for heating under reflux with protection from atmospheric moisture. Gently heat the mixture under reflux for 30 min, although 15 min is sufficient for a primary alcohol. Cool the solution and add about 5 mL of 0.6 *M* aqueous sodium bicarbonate solution. Cool this solution in an ice-water bath, and collect the crude crystalline product. Recrystallize the product from aqueous ethanol, using a minimum volume of solvent.

WRAPPING IT UP

Flush all *filtrates* down the drain.

Preparation Sign in at **www.cengage.com/login** to read the MSDSs for the chemicals used or produced in this procedure.

Method B. Perform this reaction *in the hood* if possible; if not, use one of the gas traps shown in Figure 2.68. Add 0.5 g of 3,5-dinitrobenzoic acid, 1.5 mL of thionyl chloride, and 1 drop of pyridine to a small round-bottom flask equipped for magnetic stirring and heating under reflux. Heat the mixture under reflux until the acid has dissolved and then for an additional 10 min. The total reflux time should be about 30 min.

Equip the flask for vacuum distillation. Cool the receiving flask with an ice-salt bath, and attach the vacuum adapter to an aspirator by means of a safety trap such as that shown in Figure 2.42a. Evacuate the system and distill the excess thionyl chloride. When the thionyl chloride has been removed, cautiously release the vacuum. Tightly stopper the flask containing thionyl chloride. To the residue in the stillpot, which is 3,5-dinitrobenzoyl chloride, add 1 mL of the alcohol and 0.3 mL of

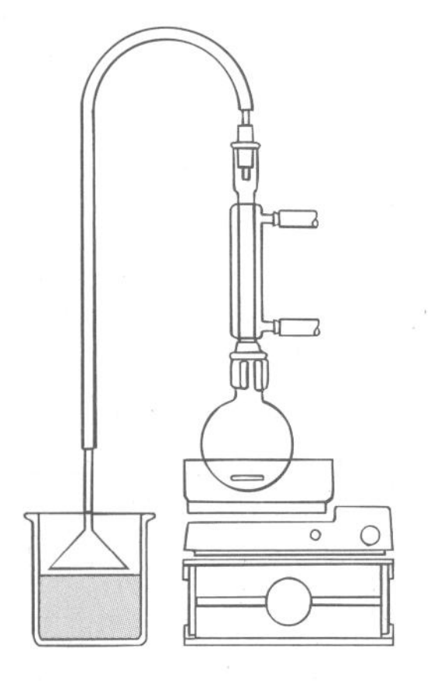

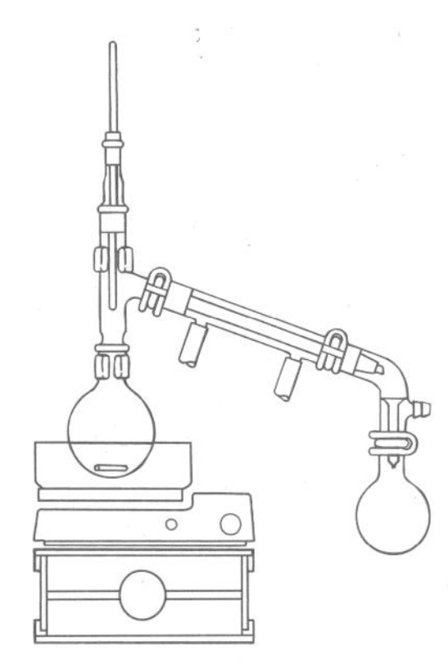

pyridine. Fit the flask with a reflux condenser bearing a calcium chloride drying tube. Proceed with the period of reflux according to the procedure of *Method A*.

WRAPPING IT UP

Flush all *filtrates* down the drain with copious amounts of water. Discard the excess *thionyl chloride* either by pouring it slowly down a drain *in the hood* while running water or by putting it in a container for recovered thionyl chloride. Spread the *dessicant* used in the drying tube on a tray in the hood, and discard it in the container for nonhazardous solids after the volatiles are removed.

25.12 PHENOLS

Classification Tests

Several tests can be used to detect the presence of a phenolic hydroxy group: (a) bromine water, (b) ceric ammonium nitrate reagent, and (c) ferric chloride solution. In addition to these, solubility tests give a preliminary indication of a phenol because phenols are soluble in 1.5 *M* sodium hydroxide solution, but they are generally insoluble in 0.6 *M* sodium bicarbonate solution. You must exercise care in interpreting solubility data, however, because phenols containing highly electronegative groups are stronger acids and *may* be soluble in 0.6 *M* sodium bicarbonate; examples are 2,4,6-tribromophenol and 2,4-dinitrophenol.

A. Bromine Water

Phenols usually are highly reactive toward electrophilic substitution and consequently are brominated readily by bromine water, as illustrated in Equation 25.44. The rate of bromination is much greater in water than in nonpolar organic solvents. Water, being more polar, increases the ionization of bromine and thus enhances the rate of ionic bromination. Although hydrogen bromide is liberated, it is not observed when water is used as the solvent.

$$\text{Phenol} + 3\,Br_2 \xrightarrow{H_2O} \text{2,4,6-Tribromophenol} + 3\,Br_2 \qquad (25.44)$$

Phenols are so reactive toward electrophilic aromatic substitution that all unsubstituted positions *ortho* and *para* to the hydroxy group are brominated. The brominated compounds so formed are often solids and may be used as derivatives (Sec. 25.12E). Aniline and substituted anilines are also very reactive toward bromine and react analogously; however, solubility tests are normally used to distinguish between anilines and phenols.

EXPERIMENTAL PROCEDURE

Bromine Water Test for Phenols

SAFETY ALERT

1. ***Bromine is a hazardous chemical that may cause serious chemical burns.*** **Do not breathe its vapors or allow it to contact your skin. Perform all operations involving the transfer of solutions of bromine in a hood; wear latex gloves when handling this chemical. If you get bromine on your skin, immediately flood the area with warm water and soak the skin in 0.6 *M* sodium thiosulfate solution for up to 3 h if the burn is particularly serious.**
2. **Do *not* use acetone to rinse glassware containing residual bromine. This prevents formation of α-bromoacetone, a severe lachrymator. Rather, follow the procedure described in Wrapping It Up.**

Preparation Sign in at **www.cengage.com/login** to read the MSDSs for the chemicals used or produced in this procedure.

Prepare a 1% aqueous solution of the unknown. If necessary, add dilute sodium hydroxide solution dropwise to effect solution of the phenol. Add a saturated solution of bromine in water dropwise to this solution; continue that addition until the bromine color persists. Note how much bromine water is used and perform this experiment on phenol and aniline for purposes of comparison.

WRAPPING IT UP

Add solid sodium thiosulfate to all *test solutions* to destroy residual bromine, then neutralize and filter them. Flush the *filtrates* down the drain and place the *filter paper* in the container for halogenated organic compounds.

B. Ceric Nitrate Test

Alcohols are capable of replacing nitrate ions in complex cerate anions, resulting in a change in color of the solution from yellow to red (Eq. 25.45). Phenols are oxidized by this reagent to give a brown to greenish-brown precipitate in aqueous solution; a red-to-brown solution is produced in 1,4-dioxane. Because of solubility problems, the alcohol or phenol should have no more than ten carbon atoms for this test. Aromatic amines are also oxidized by the reagent to give a colored solution or precipitate; consequently, the presence of this functionality must be excluded before concluding that a positive test signals an alcohol or phenol moiety.

$$\underset{\substack{\text{Ceric ammonium nitrate}\\ (yellow)}}{(NH_4)_2Ce(NO_3)_6} + \underset{\text{An alcohol}}{ROH} \longrightarrow \underset{\substack{\text{An alkoxy ceric}\\ \text{ammonium nitrate}\\ (red)}}{(NH_4)_2\overset{\overset{\displaystyle OR}{|}}{Ce}(NO_3)_5} + HNO_3 \quad (25.45)$$

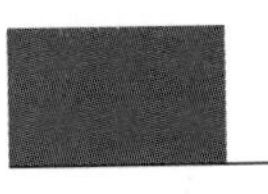

EXPERIMENTAL PROCEDURE

Ceric Nitrate Test for Alcohols and Phenols

Preparation Sign in at **www.cengage.com/login** to read the MSDSs for the chemicals used or produced in this procedure.

Dissolve about 20 mg of a solid or 1 drop of a liquid unknown in 1–2 mL of water, and add 0.5 mL of the ceric ammonium nitrate reagent; shake the mixture and note the color. If the unknown is insoluble in water, dissolve it in 1 mL of 1,4-dioxane, and proceed as before.

WRAPPING IT UP

Flush the *test solutions* down the drain.

C. Ferric Chloride Test

Most phenols and enols react with ferric chloride to give colored complexes. The colors vary, depending on the nature of the phenol or enol and also on the solvent, concentration, and time of observation. Some phenols not giving coloration in aqueous or alcoholic solution do so in chloroform solution, especially after addition of a drop of pyridine. The nature of the colored complexes is still uncertain, but they may be ferric phenoxide salts that absorb visible light. The production of a color is typical of phenols and enols; however, many of them do *not* give colors, so a negative ferric chloride test must *not* be taken as significant without supporting information such as that available from the ceric ammonium nitrate and bromine water tests.

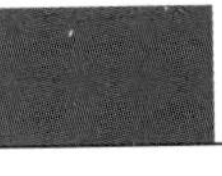

EXPERIMENTAL PROCEDURE

Ferric Chloride Test for Phenols and Enols

Preparation Sign in at **www.cengage.com/login** to read the MSDSs for the chemicals used or produced in this procedure.

Dissolve 30–50 mg of the unknown compound in 1–2 mL of water, or a mixture of water and 95% ethanol if the compound is not water-soluble, and add several drops of a 0.2 *M* aqueous solution of ferric chloride. Most phenols produce red, blue, purple, or green coloration; enols give red, violet, or tan coloration.

WRAPPING IT UP

Flush all *test solutions* down the drain.

Derivatives

Two useful solid derivatives of phenols are α-naphthylurethanes and bromo compounds.

D. Urethanes

The preparation of urethanes is discussed in Section 25.11D. Although either a phenyl- or naphthyl-substituted urethane could be prepared, the majority of the derivatives reported are the α-naphthylurethanes, and they are generally the urethanes of choice.

E. Bromophenols

The high reactivity of phenols toward electrophilic aromatic bromination is discussed in Section 25.12A. This reaction can be used to prepare bromophenols, which may serve as solid derivatives.

EXPERIMENTAL PROCEDURE

Preparation of Bromophenols

SAFETY ALERT

1. ***Bromine is a hazardous chemical that may cause serious chemical burns.*** **Do not breathe its vapors or allow it to contact your skin. Perform all operations involving the transfer of solutions of bromine in a hood; wear latex gloves when handling this chemical. If you get bromine on your skin, immediately flood the area with warm water and soak the skin in 0.6 *M* sodium thiosulfate solution for up to 3 h if the burn is particularly serious.**
2. **Do *not* use acetone to rinse glassware containing residual bromine. This prevents formation of α-bromacetone, a severe lachrymator. Rather, follow the procedure described in Wrapping It Up.**

Preparation Sign in at **www.cengage.com/login** to read the MSDSs for the chemicals used or produced in this procedure.

A stock solution for bromination in this preparation should be supplied; if not, prepare it by dissolving 1 g of potassium bromide in 6 mL of water and adding 1 g of bromine. Dissolve 0.5 g of the phenolic compound in water or 95% ethanol, and add the bromine-containing solution to it *dropwise*. Continue the addition until the reaction mixture begins to develop a yellow color, indicating the presence of excess bromine. Let the mixture stand for about 5 min; if the yellow coloration begins to fade, add another drop or two of the solution of bromine. Add 5 mL of water to the mixture and then a few drops of 0.5 *M* sodium bisulfite solution to destroy the excess bromine. Shake the mixture vigorously, and remove the solid derivative by vacuum filtration. If necessary, neutralize the solution with concentrated hydrochloric acid to promote precipitation of the derivative. Purify the solid by recrystallization from 95% ethanol or aqueous ethanol.

WRAPPING IT UP

Flush all *filtrates* down the drain.

25.13 CARBOXYLIC ACIDS

Classification Test

One of the best qualitative tests for the carboxylic acid group is solubility in basic solutions. Carboxylic acids are soluble both in 1.5 *M* sodium hydroxide solution *and* in 0.6 *M* sodium bicarbonate solution, from which they are regenerated by acidification. Solubility properties are discussed in Section 25.3.

The acidity of carboxylic acids enables ready determination of the **equivalent mass** or **neutralization equivalent** of the acid by titration with standard base. The equivalent weight of an acid is that mass, in grams, of acid that reacts with one equivalent of base. As an example, suppose that 0.1000 g of an unknown acid requires 16.90 mL of 0.1000 *N* sodium hydroxide solution to be titrated to a phenolphthalein endpoint. This means that 0.1000 g of the acid corresponds to (16.90 mL) (0.1000 equivalent/1000 mL) or 0.0016901 equivalent of the acid, or that one equivalent of the acid weighs 0.1000/0.00169 or 59.201 g. Thus the following expression applies:

$$\text{Equivalent mass} = \frac{\text{Grams of acid}}{(\text{Volume of base consumed in liters})(\text{N})}$$

where *N* is the *normality* of the standard base.

Because each carboxylic acid function in a molecule is titrated with base, the equivalent mass corresponds to the molar mass of the acid divided by n, where n is the number of acid functions present in the molecule. For the example given, the molar mass is 59.2 for a single acid function, 118.4 for two, and 177.6 for three. If the molar mass of an unknown compound is known, then the number of acid groups in the molecule is calculated by dividing the molar mass by the equivalent mass. Hence, if the molar mass of the unknown compound is 118 and its equivalent mass is 59.2, the unknown must have *two* titratable acid functions.

EXPERIMENTAL PROCEDURE

Determination of Equivalent Mass of an Acid

Preparation Sign in at **www.cengage.com/login** to read the MSDSs for the chemicals used or produced in this procedure.

Accurately weigh about 0.2 g of the acid and dissolve it in 50–100 mL of water or 95% ethanol or a mixture of the two. It may be necessary to warm the mixture to dissolve the compound completely. Using phenolphthalein as the indicator, titrate the solution with a *standardized* sodium hydroxide solution having a concentration of about 0.1 *M* (Fig. 25.6). From the data obtained, calculate the equivalent mass.

WRAPPING IT UP

Flush the *aqueous solution* down the drain.

Derivatives

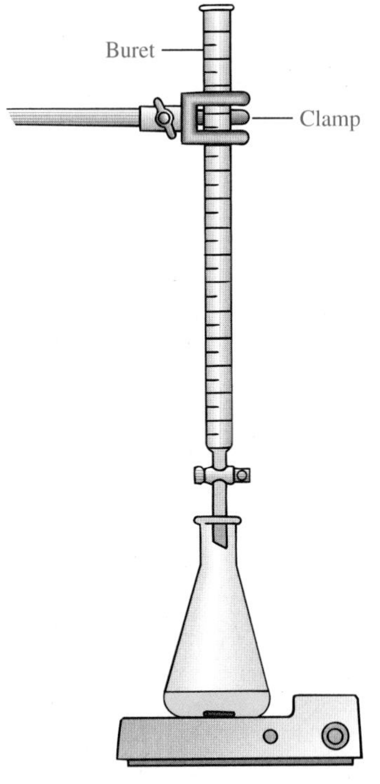

Figure 25.6
Apparatus for quantitative titration.

Three satisfactory solid derivatives of carboxylic acids are amides (Eq. 25.46), anilides (Eq. 25.47), and *p*-toluidides (Eq. 25.48). These derivatives are prepared by treating the corresponding acid chlorides with ammonia, aniline, or *p*-toluidine. The amides are generally less satisfactory derivatives than the other two because they tend to be more soluble in water and thus are harder to isolate. The acid chlorides are most conveniently prepared from the acid, or its salt, and thionyl chloride (Eq. 25.49).

$$\underset{\text{An acid chloride}}{RCOCl} + 2\,NH_3 \xrightarrow{\text{cold}} \underset{\text{An amide}}{RCONH_2} + NH_4Cl \tag{25.46}$$

$$RCOCl + 2\,\underset{\text{Aniline}}{H_2NC_6H_5} \longrightarrow \underset{\text{An anilide}}{RCONHC_6H_5} + C_6H_5NH_3^+\,Cl^- \tag{25.47}$$

$$RCOCl + 2\,\underset{p\text{-Toluidine}}{H_2NAr} \longrightarrow \underset{\text{A }p\text{-toluidide}}{RCONHAr} + ArNH_3\,Cl^- \tag{25.48}$$

$$Ar = -C_6H_4-CH_3$$

$$\underset{\text{A carboxylic acid}}{RCO_2H} + \underset{\text{Thionyl chloride}}{SOCl_2} \longrightarrow \underset{\text{An acid chloride}}{RCOCl} + \underset{\text{Sulfur dioxide}}{SO_2} + HCl \tag{25.49}$$

EXPERIMENTAL PROCEDURES

Preparation of Amides

SAFETY ALERT

1. **Thionyl chloride is a lachrymator. Measure this material in a hood and do not inhale its vapors. Wear latex gloves when handling this chemical.**
2. **Acid chlorides release hydrochloric acid upon hydrolysis, so avoid contact of such reagents with your skin. If contact occurs, immediately flood the area with water and rinse it with 5% sodium bicarbonate solution. Wear latex gloves when handling acid chlorides.**

Preparation Sign in at **www.cengage.com/login** to read the MSDSs for the chemicals used or produced in this procedure.

Perform this reaction *in the hood*: To prepare the acid chloride from a carboxylic acid, place the acid in a small round-bottom flask equipped for magnetic stirring. Add 1 mL of thionyl chloride and 5 drops of *N,N*-dimethylformamide (DMF) and immediately attach a calcium chloride drying tube *directly* to the flask. Warm the flask at 60–65 °C with stirring *in the hood*. Bubbling or fuming usually begins shortly after the addition of the DMF. The reaction is sufficiently complete when the bubbling greatly slows, which typically occurs within 30 min. Use the mixture containing the acid chloride to make the amide, anilides, or *p*-toluidide according to the procedures that follow.

A. Amides. Use 0.5 g of the carboxylic acid to prepare the acid chloride. *At the hood,* pour the mixture containing the acid chloride and unchanged thionyl chloride into 15 mL of *ice-cold*, concentrated ammonium hydroxide solution. Be very careful when performing this addition, because the reaction is quite exothermic. Collect the precipitated amide derivative by vacuum filtration, and recrystallize the crude product from water or aqueous ethanol.

WRAPPING IT UP

Neutralize all *filtrates* and then flush them down the drain.

B. Anilides and p-Toluidides. Use 0.5 g of the carboxylic acid to prepare the acid chloride. After the heating period, cool the reaction mixture to room temperature. Dissolve 0.5 g of either aniline or *p*-toluidine in 15 mL of cyclohexane; slight warming may be necessary to effect complete solution. Pour the acid chloride into the cyclohexane solution of the amine, and heat the resulting mixture for 2–3 min. A heavy white precipitate of the amine hydrochloride forms. Isolate this precipitate by vacuum filtration and set it aside, but do *not* discard it. Transfer the filtrate to a separatory funnel and wash it sequentially with 3-mL portions of water, 1.5 *M* HCl, 1.5 *M* NaOH, and again with water. In some cases, precipitation may occur in the organic layer during one or more of these washings. If this happens, warm the solution gently with a warm-water bath to redissolve the precipitate.

Following the sequential washings, remove the cyclohexane from the organic layer either by distillation or using one of the other techniques described in Section 2.29. Recrystallize the derivative of the carboxylic acid from aqueous ethanol.

If little residue remains after evaporating the cyclohexane, dissolve the precipitate that was removed earlier in about 10 mL of water. Stir the mixture and then remove any undissolved solid by vacuum filtration; any such solid is the desired derivative, which became entrapped in the precipitated amine hydrochloride. Combine this material with the residue obtained from cyclohexane.

WRAPPING IT UP

Neutralize all *filtrates* and then flush them down the drain. Put any recovered *cyclohexane* into the container for nonhalogenated organic liquids.

25.14 AMINES

Classification Tests

Two common qualitative tests for amines are the Hinsberg test and the nitrous acid test. The nitrous acid test is not included here because the *N*-nitroso derivatives of some secondary amines are carcinogenic. The risk of producing an as-yet-unrecognized carcinogenic material in this procedure outweighs any possible benefit of a test that can also be misleading and difficult to interpret. The modified sodium nitroprusside test has been included as an alternative.

A. Hinsberg Test

See more on *Hinsberg Test*

The reaction between primary or secondary amines and benzenesulfonyl chloride (Eqs. 25.50 and 25.52, respectively) yields the corresponding substituted benzenesulfonamide. The reaction is performed in excess aqueous base; if the amine is primary, the sulfonamide, which has an acidic amido hydrogen, is converted by base (Eq. 25.51) to the potassium salt, which is normally soluble in the medium. With few exceptions, which are discussed in the next paragraph, primary amines react with benzenesulfonyl chloride to provide *homogeneous* reaction mixtures. Acidifying this solution regenerates the insoluble primary benzenesulfonamide. On the other hand, the benzenesulfonamides of secondary amines bear no acidic amido hydrogens and thus are normally insoluble in *both* acid and base. Therefore, secondary amines react to yield *heterogeneous* reaction mixtures, with formation of either an oily organic layer or a solid precipitate.

$$RNH_2 + C_6H_5\text{–}SO_2Cl \xrightarrow{KOH} C_6H_5\text{–}SO_2NHR + KCl + H_2O \quad (25.50)$$

A 1° amine; Benzenesulfonyl chloride; A benzenesulfonamide *Insoluble in water*

$$C_6H_5\text{–}SO_2N(H)R \underset{\text{excess HCl}}{\overset{\text{excess KOH}}{\rightleftharpoons}} C_6H_5\text{–}SO_2\bar{N}R\ K^+ + H_2O \quad (25.51)$$

A potassium benzenesulfonamide *Soluble in water*

$$R_2NH + C_6H_5\text{–}SO_2Cl \xrightarrow{KOH} C_6H_5\text{–}SO_2NR_2 + KCl + H_2O \quad (25.52)$$

A 2° amine; Benzenesulfonyl chloride; A benzenesulfonamide *Insoluble in water*

$\downarrow$ excess KOH

No reaction

The distinction between primary and secondary amines thus depends on the different solubility properties of their benzenesulfonamide derivatives. However, the potassium salts of *certain* primary sulfonamides are not completely soluble in basic solution. Examples usually involve primary amines of higher molar mass and those having cyclic alkyl groups. To avoid confusion and possible misassignment of a primary amine as secondary, the basic solution is separated from the oil or solid and acidified. The formation of an oil or a precipitate indicates that the derivative is partially soluble and that the amine is primary. It is important not to overacidify the solution, because this may precipitate certain side-products and result in an ambiguous test. The original oil or solid should be tested for solubility in water and acid to substantiate the test for a primary or a secondary amine.

Tertiary amines behave somewhat differently. Under the conditions of the Hinsberg test, the processes shown in Equation 25.53 typically provide for converting benzenesulfonyl chloride to potassium benzenesulfonate with recovery of the tertiary amine. Tertiary amines are nearly always insoluble in the aqueous potassium hydroxide solution, so the test mixture remains heterogeneous. Proof that the nonaqueous layer is a tertiary amine can generally be obtained simply by noting the relative densities of the oil layer and the aqueous test solution: tertiary amines are less dense than the solution, whereas benzenesulfonamides are generally more dense. Further support for the conclusion that the oil is a tertiary amine is available from testing its solubility in aqueous acid; solubility usually indicates a tertiary amine.

$$R_3N + C_6H_5\text{—}SO_2Cl \longrightarrow C_6H_5\text{—}SO_2\text{—}\overset{+}{N}R_3\ Cl^- \ (\mathbf{23}) \xrightarrow{KOH} C_6H_5\text{—}SO_3^-K^+ + NR_3 + H_2O \quad (25.53)$$

A 3° amine; Benzenesulfonyl chloride; **23**; Potassium benzenesulfonate

$$C_6H_5\text{—}SO_2\overset{+}{N}R_3Cl^- \ (\mathbf{23}) \xrightarrow{R_3N} C_6H_5\text{—}SO_2NR_2 + NR_4^+\ Cl^- \quad (25.54)$$

A benzenesulfonamide; A quaternary ammonium chloride

The procedure for the Hinsberg test must be followed as closely as possible, because it is designed to minimize complications that arise because of side reactions that may occur between tertiary amines and benzenesulfonyl chloride. For example, **23**, the initial adduct between the chloride and the amine, is subject to an S_N2 reaction (Secs. 14.2 and 14.4) with another molecule of amine to produce the benzenesulfonamide of a *secondary* amine and a quaternary ammonium salt (Eq. 25.54). Although this process is normally unimportant, particularly when excess amine is avoided, the observation of *small* amounts of an insoluble product formed by this pathway may erroneously cause designation of a tertiary amine as secondary.

Formation of adduct **23** is usually less of a problem with tertiary *aryl*amines because of their lesser nucleophilicity and lower solubility as compared to *trialkyl* amines. Consequently, the competing hydrolysis of benzenesulfonyl chloride by hydroxide ion (Eq. 25.55) allows recovery of most of the amine. However, this class of amines is often subject to other side reactions that produce a complex mixture of mainly insoluble products (Eq. 25.56). The ambiguity caused by side reactions of tertiary arylamines is minimized by keeping the reaction time short and the temperature low.

$$C_6H_5-SO_2Cl \xrightarrow{KOH} C_6H_5-SO_3^- K^+ + KCl + H_2O \quad (25.55)$$

$$C_6H_5-SO_2Cl \xrightarrow{ArNR_2} \text{Complex mixture including } C_6H_5-SO_2-NRAr \quad (25.56)$$

A further complication with tertiary amines is that they often contain quantities of secondary amines as impurities. If it is not possible to obtain a reliable boiling point or if the amine is not carefully distilled, such contaminants may be present and lead to formation of small quantities of precipitate, making the test results ambiguous.

To summarize the discussion, tertiary amines may produce small amounts of insoluble products if the concentration of the amine in the test solution is too high and if the reaction time is too long. If the directions of the procedure are followed and *care is taken not to interpret small amounts of insoluble product as a positive test for secondary amines,* the Hinsberg test may be used with confidence to designate an amine as primary, secondary, or tertiary.

EXPERIMENTAL PROCEDURE

Hinsberg Test for Amines

SAFETY ALERT

Benzenesulfonyl chloride is a lachrymator and produces strong acids upon hydrolysis. Wear latex gloves when handling this reagent, measure it out in a hood, and avoid inhaling its vapors. Should this reagent come in contact with your skin, immediately flood the area with water and rinse it with 5% sodium bicarbonate solution.

Preparation Sign in at **www.cengage.com/login** to read the MSDSs for the chemicals used or produced in this procedure.

Mix 5 mL of 2 *M* aqueous potassium hydroxide, 0.2 mL (5 drops) or 0.2 g of the amine, and 0.7 mL (15 drops) of benzenesulfonyl chloride in a test tube. Stopper the tube and shake the mixture *vigorously*, with cooling if necessary, until the odor of benzenesulfonyl chloride is gone (*Caution!*). In even the slowest case, this should take no more than about 5 min. Test the solution to see that the mixture is still basic; if it is not, add sufficient 2 *M* potassium hydroxide solution dropwise until it is.

If the mixture forms two layers or a precipitate, note the relative densities, and separate the oil or solid by decantation or filtration. Test any oil for solubility in 0.6 *M* hydrochloric acid. The sulfonamide of a secondary amine is insoluble, whereas an amine is at least partially soluble. A solubility test indicating the presence of an amine could be caused by either a tertiary amine or secondary amines that for steric reasons react very slowly with benzenesulfonyl chloride.

Test any solid for solubility in water and in dilute acid. The potassium salt of a water-insoluble sulfonamide is usually soluble in water; acidifying the salt regenerates the sulfonamide, which is insoluble in aqueous acid. A solid sulfonamide of a secondary amine is insoluble in both water and acid. Acidify the solution from the original reaction mixture to pH 4 as signaled by pHydrion paper or a few drops of Congo red indicator solution; the formation of a precipitate or oil indicates a primary amine.

If the original mixture from the Hinsberg test does not form two layers, a primary amine is indicated. Acidify the solution to pH 4; a sulfonamide of a primary amine will either separate as an oil or precipitate as a solid.

WRAPPING IT UP

Neutralize all *test solutions*. Isolate any solids by filtration and any oils with the aid of a Pasteur pipet. Flush the *filtrates* down the drain and put any *filter cakes* or *oils* in the container for nonhalogenated organic compounds.

B. Sodium Nitroprusside Tests

Two color tests to distinguish primary and secondary *aliphatic* amines have been available for many years, although they have not been widely used. More recently, these tests have been extended to primary, secondary, and tertiary *aromatic* amines. No attempt is made here to explain the complex color-forming reactions that occur. However, they most likely involve the reaction of the amine with either acetone, the Ramini test, or acetaldehyde, the Simon test, and the interaction of the resulting products with sodium nitroprusside to form colored complexes.

To apply these tests to an unknown amine, the *conventional* Ramini or Simon tests are first performed. These will give positive results in the cases of primary and secondary aliphatic amines. If these tests are negative and an aromatic amine is suspected, the *modified* versions of these tests are then performed. Reference to Figure 25.7 is helpful for interpreting the results of these tests.

	1° Aliphatic	2° Aliphatic	1° Aromatic	2° Aromatic	3° Aromatic
Ramini	Deep red	Deep red			
Simon	Pale yellow to red-brown	Deep blue			
Modified Ramini			Orange-red to red-brown	Orange-red to red-brown	Green
Modified Simon			Orange-red to red-brown	Purple	Usually green

Figure 25.7
Colors formed in the Ramini and Simon tests.

EXPERIMENTAL PROCEDURES

Ramini and Simon Tests for Amines

Preparation Sign in at **www.cengage.com/login** to read the MSDSs for the chemicals used or produced in this procedure.

Conventional Tests. Prepare the sodium nitroprusside reagent for use in both the conventional Ramini and Simon tests by dissolving 0.4 g of sodium nitroprusside ($Na_2[Fe(NO)(CN)_5] \cdot 2\ H_2O$ in 10 mL of 50% aqueous methanol.

Ramini Test. To 1 mL of the sodium nitroprusside reagent, add 1 mL of water, 0.2 mL (5 drops) of acetone, and about 30 mg of an amine. In most cases, the characteristic colors given in Figure 25.7 appear in a few seconds, although in some instances up to about 2 min may be necessary.

Simon Test. To 1 mL of the sodium nitroprusside reagent, add 1 mL of water, 0.2 mL (5 drops) of 2.5 *M* aqueous acetaldehyde solution, and about 30 mg of an amine. As in the Ramini test, color formation normally occurs in a few seconds, although times up to 2 min are occasionally necessary.

Modified Tests. Prepare the reagent for use in both the *modified* Ramini and Simon tests by dissolving 0.4 g of sodium nitroprusside in a solution containing 8 mL of dimethyl sulfoxide and 2 mL of water.

Modified Ramini Test. To 1 mL of the *modified* sodium nitroprusside reagent, sequentially add 1 mL of saturated aqueous zinc chloride solution, 0.2 mL (5 drops) of acetone, and about 30 mg of an amine. Primary and secondary aromatic amines provide orange-red to red-brown colors within a period of a few seconds to 5 min. Tertiary aromatic amines give a color that changes from orange-red to green over a period of about 5 min.

Modified Simon Test. To 1 mL of the *modified* sodium nitroprusside reagent, sequentially add 1 mL of saturated aqueous zinc chloride solution, 0.2 mL (5 drops) of 2.5 *M* aqueous acetaldehyde solution, and about 30 mg of an amine. Primary aromatic amines give an orange-red to red-brown color within 5 min; secondary aromatic amines give a color changing from red to purple within 5 min; tertiary aromatic amines give a color that changes from orange-red to green over a period of 5 min.

WRAPPING IT UP

Flush the *test solutions* down the drain.

Derivatives

Suitable derivatives of primary and secondary amines are benzamides and benzenesulfonamides (Eqs. 25.57 and 25.58, respectively).

$$\underset{\text{A 1° amine}}{RNH_2} \text{ or } \underset{\text{A 2° amine}}{R_2NH} + \underset{\text{Benzoyl chloride}}{C_6H_5COCl} \xrightarrow{\text{Pyridine}} \underset{\text{A benzamide}}{C_6H_5\overset{O}{\overset{\|}{C}}—NHR \text{ or } C_6H_5\overset{O}{\overset{\|}{C}}—NR_2} \quad (25.57)$$

$$\underset{\text{A 1° amine}}{RNH_2} \text{ or } \underset{\text{A 2° amine}}{R_2NH} + \underset{\text{Benzenesulfonyl chloride}}{C_6H_5SO_2Cl} \longrightarrow \underset{\text{A benzenesulfonamide}}{C_6H_5SO_2—NHR \text{ or } C_6H_5SO_2—NR_2} \quad (25.58)$$

These transformations are satisfactory for derivatizing most primary and secondary amines, but tertiary amines do not undergo such reactions. However, such amines form salts that constitute solid derivatives. Thus a useful crystalline salt is formed upon reaction with methyl iodide to afford a methiodide (Eq. 25.59).

$$\underset{\text{A 3° amide}}{R_3N} + \underset{\text{Methyl iodine}}{CH_3I} \longrightarrow \underset{\text{A methiodide}}{R_3\overset{+}{N}CH_3\,I^-} \quad (25.59)$$

EXPERIMENTAL PROCEDURES

Preparation of Benzamides, Benzenesulfonamides, and Methiodides

SAFETY ALERT

Benzoyl chloride is a lachrymator and produces hydrochloric and benzoic acids upon hydrolysis. Wear latex gloves when handling this reagent, measure it out in a hood, and avoid inhaling its vapors. Should this reagent come in contact with your skin, immediately flood the area with water and rinse it with 5% sodium bicarbonate solution.

Preparation Sign in at **www.cengage.com/login** to read the MSDSs for the chemicals used or produced in this procedure.

A. Benzamides. In a small round-bottom flask equipped for magnetic stirring, dissolve 0.3 g of the amine in 3 mL of dry pyridine. *Slowly* add 0.3 mL of benzoyl chloride to this solution. Affix a drying tube *directly* to the flask and heat the reaction mixture to 60–70 °C for 30 min; then pour the mixture into 25 mL of water with stirring. If the solid derivative precipitates at this time, isolate it by vacuum filtration, and dissolve it in 10 mL of diethyl ether when it is nearly dry. If no precipitate forms, extract the aqueous mixture twice with 5-mL portions of diethyl ether. Combine the extracts. Wash the ethereal solution sequentially with 5-mL portions of water, 1.5 *M* HCl, and 0.6 *M* sodium bicarbonate solution. Dry the ethereal layer over anhydrous sodium sulfate, filter or decant the dried solution, and remove the diethyl ether by one of the techniques described in Section 2.29. Recrystallize the solid derivative from one of the following solvents: cyclohexane-hexane mixtures, cyclohexane-ethyl acetate mixtures, 95% ethanol, or aqueous ethanol.

WRAPPING IT UP

Transfer any recovered *diethyl ether* into the container for nonhalogenated organic liquids. Flush all *aqueous* and *ethanolic solutions* down the drain, and pour any other *filtrates* containing organic solvents into the container for nonhalogenated liquids. Spread the desiccant used in the drying tube on a tray in the hood and discard it in the container for nonhazardous solids after the volatiles are removed.

Preparation Sign in at **www.cengage.com/login** to read the MSDSs for the chemicals used or produced in this procedure.

B. Benzenesulfonamides. The method for preparing the benzenesulfonamides is discussed with the Hinsberg test (Sec. 25.14A). The derivatives may be prepared using that method, but sufficient amounts of material should be used so that the final product may be purified by recrystallization from 95% ethanol. If the derivative is obtained as an oil, it *may* be induced to crystallize by scratching at the liquid-liquid or air-liquid interface with a stirring rod. If the oil cannot be made to crystallize, separate it, dissolve it in a minimum quantity of hot ethanol, and allow the solution to cool. Note that some amines do not give *solid* benzenesulfonamide derivatives.

SAFETY ALERT

Methyl iodide is a powerful alkylating agent. Avoid inhaling its vapors or contact of this reagent with your skin. Wear latex gloves when handling it.

Preparation Sign in at **www.cengage.com/login** to read the MSDSs for the chemicals used or produced in this procedure.

C. Methiodides. Mix 0.3 g of the amine with 0.3 mL of methyl iodide in a test tube, warm the mixture for several minutes, and then cool it in an ice-water bath. If necessary, scratch at the air-liquid interface with a glass rod to induce crystallization. Isolate the product by vacuum filtration and purify it by recrystallization from absolute ethanol or methanol or from ethyl acetate.

WRAPPING IT UP

Put the *filtrate* from the reaction mixture in the container for halogenated liquids and flush all other *filtrates* down the drain.

25.15 NITRO COMPOUNDS

Classification Test

Ferrous Hydroxide Test

Organic compounds that are oxidizing agents will oxidize ferrous hydroxide (blue) to ferric hydroxide (brown). The most common organic compounds that function in this way are aliphatic and aromatic *nitro* compounds, which are reduced to amines by the reaction (Eq. 25.60). Other less common types of compounds that give the same test are nitroso compounds, hydroxylamines, alkyl nitrates, alkyl nitrites, and quinones.

$$\underset{\text{A nitro compound}}{RNO_2} + \underset{\text{Ferrous hydroxide (blue)}}{Fe(OH)_2} + 4\,H_2O \longrightarrow \underset{\text{A 1° amine}}{RNH_2} + 6\,\underset{\text{Ferric hydroxide (brown)}}{Fe(OH)_3} \quad (25.60)$$

EXPERIMENTAL PROCEDURE

Hydroxide Test for Nitro Compounds

Preparation Sign in at **www.cengage.com/login** to read the MSDSs for the chemicals used or produced in this procedure.

In a small test tube, mix about 20 mg of a solid or 1 drop of a liquid unknown with 1.5 mL of freshly prepared 5% ferrous ammonium sulfate solution. Add 1 drop of 3 *M* sulfuric acid and 1 mL of 2 *M* potassium hydroxide in *methanol*. Stopper the tube immediately and shake it. A positive test is indicated by the blue precipitate turning rust-brown within 1 min. A slight darkening or greenish coloration of the blue precipitate is *not* considered a positive test.

WRAPPING IT UP

Filter the *test mixture* and flush the *filtrates* down the drain. Put the *filter cake* in the container for nonhazardous solids.

Derivatives

Two different types of derivatives of nitro compounds may be prepared. Aromatic nitro compounds can be di- and trinitrated with nitric acid and sulfuric acid. Discussion of and procedures for nitration are given in Section 25.10.

The other method for preparation of a derivative may be utilized for both aliphatic and aromatic nitro compounds. This involves the reduction of the nitro compound to the corresponding primary amine (Eq. 25.61), followed by conversion of the amine to a benzamide or benzenesulfonamide, as described in Section 25.14. The reduction is most often carried out with tin and hydrochloric acid.

$$\underset{\text{A nitro compound}}{RNO_2 \text{ or } ArNO_2} \xrightarrow[\text{(2) NaOH}]{\text{(1) Sn, HCl}} \underset{\text{A 1° amine}}{RNH_2 \text{ or } ArNH_2} \qquad (25.61)$$

EXPERIMENTAL PROCEDURE

Reduction of Nitro Compounds

Preparation Sign in at **www.cengage.com/login** to read the MSDSs for the chemicals used or produced in this procedure.

Combine 1 g of the nitro compound and 2 g of granulated tin in a small round-bottom flask containing a stirbar. Attach a reflux condenser, and add 20 mL of 3 *M* hydrochloric acid in small portions to the stirred mixture through the condenser.

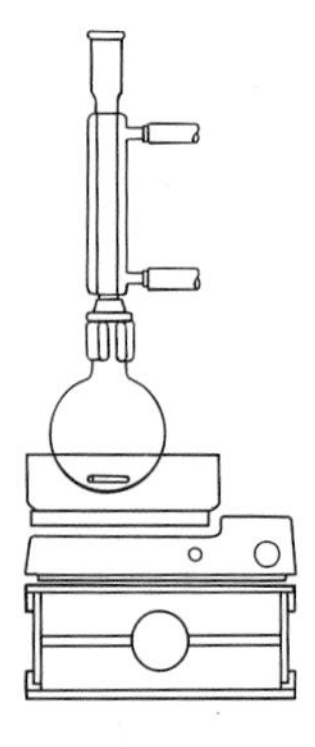

If the nitro compound is insoluble in the aqueous medium, add 5 mL of 95% ethanol to increase its solubility.

After the addition is complete, warm the mixture with stirring for 10 min at 70–80 °C. Decant the warm, homogeneous solution into 10 mL of water, and add sufficient 12 *M* sodium hydroxide solution to dissolve the tin hydroxide completely. Extract the basic solution with several 10-mL portions of diethyl ether. Dry the ethereal solution over potassium hydroxide pellets, decant the liquid, and remove the ether by one of the techniques described in Section 2.29.

The residue contains the primary amine. Convert it to one of the derivatives described in Section 25.14.

WRAPPING IT UP

Transfer any recovered *diethyl ether* into the container for nonhalogenated organic liquids. Neutralize the *aqueous solution* before flushing it down the drain.

25.16 ESTERS

Classification Tests

Two tests for the ester functionality are the hydroxylamine test, a color reaction, and the saponification equivalent. The latter procedure can provide *quantitative* information regarding the number of ester groups present.

A. Hydroxylamine

This test for an ester group involves the use of hydroxylamine and ferric chloride. The former converts the ester to a hydroxamic acid, which then complexes with Fe^{3+} to give a colored species (Eqs. 25.62 and 25.63).

$$\underset{\text{An ester}}{R^1{-}\overset{O}{\overset{\|}{C}}{-}OR^2} + \underset{\text{Hydroxylamine}}{H_2NOH} \longrightarrow \underset{\text{A hydroxamic acid}}{R^1{-}\overset{O}{\overset{\|}{C}}{-}NHOH} + \underset{\text{An alcohol}}{R^2OH} \quad (25.62)$$

$$R^1{-}\overset{O}{\overset{\|}{C}}{-}NHOH + FeCl_3 \longrightarrow \underset{\substack{\text{An iron complex}\\ \textit{(colored)}}}{\left[R^1{-}C{\overset{=O}{\underset{N(H){-}O}{}}}\right]_3 Fe} + 3\,HCl \quad (25.63)$$

All carboxylic acid esters, including polyesters and lactones, give magenta colors that vary in intensity depending on other structural features of the molecule. Acid chlorides and anhydrides also give positive tests. Formic acid produces a red color, but other free acids give negative tests. Primary or secondary aliphatic nitro compounds give a positive test because ferric chloride reacts with the *aci* form, which is equivalent to the enol form of a ketone, that is present in basic solution. Most imides also provide positive tests. Some amides, but not all, produce light magenta coloration, whereas most nitriles give a negative test. A modification of the following procedure provides a positive test for nitriles and amides, and the details of it are provided in Section 25.17.

EXPERIMENTAL PROCEDURE

Hydroxylamine Test for Esters

SAFETY ALERT

Hydroxylamine and its hydrochloride may have mutagenic effects. Wear latex gloves when handling them. Should either of these substances contact your skin, immediately flood the affected area with water.

Preparation Sign in at **www.cengage.com/login** to read the MSDSs for the chemicals used or produced in this procedure.

Before the final test is performed, it is necessary to run a preliminary test as a "blank" or "control" because some compounds produce a color in the absence of hydroxylamine hydrochloride even though they do *not* contain an ester linkage.

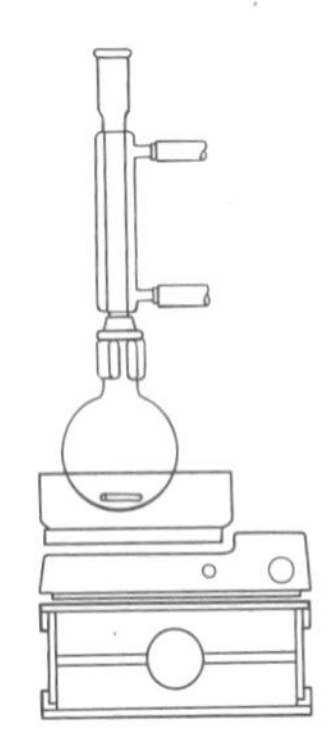

Preliminary Test. Mix 1 mL of 95% ethanol and 50–100 mg of the compound to be tested, and add 1 mL of 1 *M* hydrochloric acid. Note the color that is produced when 1 drop of 0.6 *M* aqueous ferric chloride solution is added. If the color is orange, red, blue, or violet, the following test for the ester group does *not* apply and *cannot* be used.

Final Test. Mix 40–50 mg of the unknown, 1 mL of 0.5 *M* hydroxylamine hydrochloride in 95% ethanol, and 0.2 mL of 6 *M* sodium hydroxide in a test tube. Heat the mixture to boiling and, after cooling it slightly, add 2 mL of hydrochloric acid. If the solution is cloudy, add about 2 mL more of 95% ethanol. Add 1 drop of 0.6 *M* ferric chloride, and observe any color. If needed, add more ferric chloride solution until the color persists. Compare the color obtained here with that from the preliminary test. If the color is burgundy or magenta, as compared to the yellow color in the preliminary experiment, the presence of an ester group is indicated.

WRAPPING IT UP

Neutralize the *aqueous solutions* before flushing them down the drain.

B. Saponification Equivalent

It is possible to carry out the hydrolysis of an ester with alkali in a *quantitative* manner so that the **saponification equivalent, SE,** results. This value is analogous to the equivalent mass of an acid (Sec. 25.13) in that it is the molar mass of the ester divided by the number of ester functions in the molecule. Therefore the SE is the number of grams of ester required to react with one gram-equivalent of alkali.

The SE of an ester is determined by hydrolyzing a weighed amount of the ester with standardized alkali and then titrating the excess alkali to a phenolphthalein endpoint using standardized hydrochloric acid. The SE is then calculated as follows:

$$\textit{Saponification equivalent} = \frac{\textit{Grams of ester}}{\textit{Equivalents of alkali consumed}}$$

$$= \frac{\textit{Grams of ester}}{(\textit{Volume of alkali in liters})(N) - (\textit{Volume of acid in liters})(N')}$$

where N is the normality of the standard base and N' is the normality of the standard acid.

EXPERIMENTAL PROCEDURE

Determination of Saponification Equivalent

Preparation Sign in at **www.cengage.com/login** to read the MSDSs for the chemicals used or produced in this procedure.

Dissolve approximately 3 g of potassium hydroxide in 60 mL of 95% ethanol. Allow the small amount of insoluble material, if any, to settle to the bottom of the container, and, by decantation, fill a 50-mL buret with the clear supernatant. Measure 25.0 mL of the alcoholic solution into each of two round-bottom flasks. Quantitatively transfer an *accurately* weighed 0.3- to 0.4-g sample of pure, dry ester into one of the flasks; the other basic solution is used for a blank determination. Equip each flask with a stirbar and a reflux condenser.

Heat the stirred solutions in both flasks under gentle reflux for 1 h. When the flasks have cooled, rinse each condenser with about 10 mL of distilled water, catching the rinse water in the flask. Add a drop or two of phenolphthalein, and separately titrate the solutions in each flask with *standardized* hydrochloric acid that is approximately 0.5 *M* in concentration.

The difference in the volumes of hydrochloric acid required to neutralize the base in the flask containing the sample and in the flask containing the blank corresponds to the amount of potassium hydroxide that reacted with the ester. The volume difference, in milliliters, multiplied by the molarity of the hydrochloric acid equals the number of millimoles of potassium hydroxide consumed. Using the titration data, calculate the saponification equivalent of the unknown ester.

The ester may not completely saponify in the allotted time, as evidenced by a *non*homogeneous solution. If this is the case, heat the mixture under reflux for a longer period of time (2–4 h). Higher temperatures may be required in some cases. If so, diethylene glycol must be used as a solvent *in place of* the original 60 mL of 95% ethanol.

WRAPPING IT UP

Flush the *neutralized solution* down the drain.

Derivatives

To characterize an ester completely, it is necessary to prepare solid derivatives of both the acid *and* the alcohol components. However, isolating both of these components in pure form so that suitable derivatives can be prepared may present problems. One way to do the isolation is to perform the ester hydrolysis in base (Eq. 25.64) in a high-boiling solvent. If the alcohol is low-boiling, it is distilled from the reaction mixture and characterized. The acid that remains in the mixture as a carboxylate salt is obtained upon acidification of the solution. Derivatives of alcohols and acids are discussed in Sections 25.11 and 25.13, respectively.

$$\underset{\text{An ester}}{R^1CO_2R^2} + HO^- \longrightarrow \underset{\text{A carboxylate salt}}{R^1CO_2^-} + \underset{\text{An alcohol}}{R^2OH} \xrightarrow{H_3O^+} \underset{\text{A carboxylic acid}}{R^1CO_2H} \qquad (25.64)$$

EXPERIMENTAL PROCEDURE

Base-Promoted Hydrolysis of Esters

Preparation Sign in at **www.cengage.com/login** to read the MSDSs for the chemicals used or produced in this procedure.

Mix 3 mL of diethylene glycol, 0.6 g (2 pellets) of potassium hydroxide, and 10 drops of water in a small round-bottom flask equipped with a stirbar. Heat the mixture until the solution is homogeneous and then cool it to room temperature. Add 1 mL of the ester and equip the flask for heating under reflux. Reheat the mixture to reflux with stirring; after the ester layer dissolves (3–5 min), allow the solution to cool. Equip the flask for simple distillation, and heat the solution strongly so that the alcohol distills; all but high-boiling alcohols can be removed by direct distillation. The distillate, which should be an alcohol in a fairly pure and dry state, is used directly for the preparation of a solid derivative (Sec. 25.17).

The residue that remains after distillation contains the salt of the carboxylic acid. Add 10 mL of water to the residue and thoroughly mix the two. Acidify the resulting solution with 6 *M* sulfuric acid. Allow the mixture to stand and collect any crystals by vacuum filtration. If crystals do not form, extract the aqueous acidic solution with small portions of diethyl ether or dichloromethane, dry the organic solution over anhydrous sodium sulfate, decant the liquid, and remove the solvent by distillation or by evaporation (in the hood). Use the residual acid to prepare a derivative (Sec. 25.19).

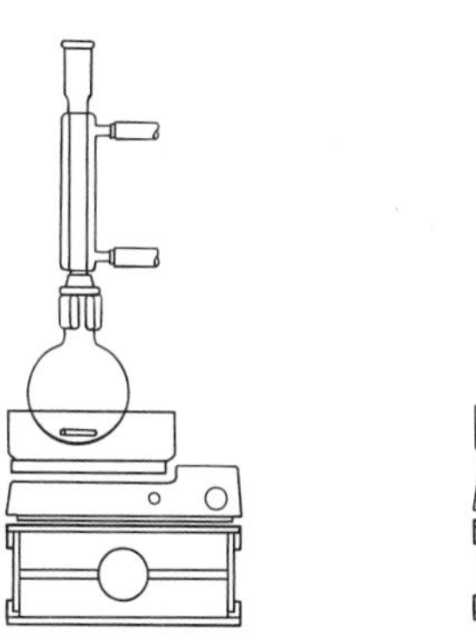

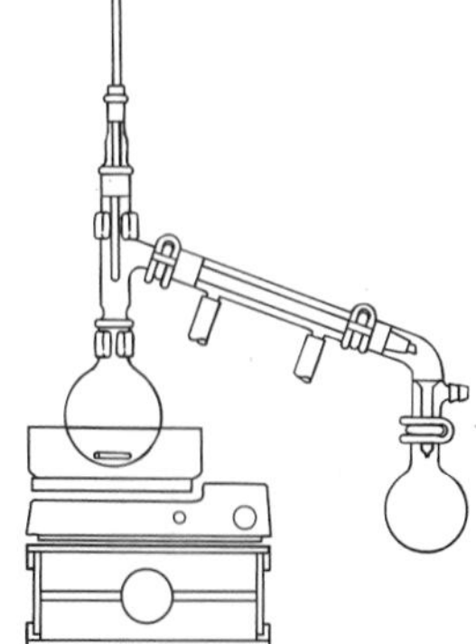

WRAPPING IT UP

Flush the *pot residue* from the distillation of the alcohol down the drain. Neutralize the *aqueous filtrate* and also flush it down the drain. Transfer any *organic solvent* recovered by distillation to the appropriate container, that for nonhalogenated organic liquids in the case of *diethyl ether* and the one for halogenated organic liquids if *dichloromethane* was used.

25.17 NITRILES

Classification Test

A. Hydroxylamine

A qualitative test used for nitriles is similar to the hydroxylamine test for esters (Sec. 25.16A). Nitriles, as well as amides, typically give a colored solution on treatment with hydroxylamine and ferric chloride (Eq. 25.65).

$$\underset{\text{A nitrile}}{R{-}C{\equiv}N} + \underset{\text{Hydroxylamine}}{H_2NOH} \longrightarrow 3\ R{-}\overset{\overset{NH}{\|}}{C}{-}NHOH \xrightarrow{FeCl_3} \underset{\substack{\text{An iron complex}\\(\text{colored})}}{\left[R_1{-}C({=}NH)(N(H){-}O)\right]_3Fe} + 3\ HCl \qquad (25.65)$$

EXPERIMENTAL PROCEDURE

Hydroxylamine Test for Nitriles

SAFETY ALERT

Hydroxylamine and its hydrochloride may have mutagenic effects. Wear latex gloves when handling them. Should either of these substances contact your skin, immediately flood the affected area with water.

Preparation Sign in at **www.cengage.com/login** to read the MSDSs for the chemicals used or produced in this procedure.

In a large test tube, prepare a mixture consisting of 2 mL of 1 *M* hydroxylamine hydrochloride in propylene glycol, 30–50 mg of the compound that has been dissolved in a minimum amount of propylene glycol, and 1 mL of 1 *M* potassium hydroxide. Heat the mixture to boiling for 2 min and then cool it to room temperature. Add 0.5–1.0 mL of a 0.5 *M alcoholic* ferric chloride solution. A red-to-violet color is a positive test. Yellow colors are negative, and brown colors and precipitates are neither positive nor negative.

WRAPPING IT UP

Flush the *solutions* down the drain.

Derivatives

Hydrolysis

On hydrolysis in either acidic or basic solution, nitriles are ultimately converted to the corresponding carboxylic acids (Eqs. 25.66 and 25.67). It is then possible to prepare a derivative of the acid using methods provided in Section 25.13.

$$\textit{Basic hydrolysis: } \underset{\text{A nitrile}}{RC{\equiv}N} + NaOH \xrightarrow{H_2O} \underset{\text{A sodium carboxylate}}{RCO_2^- Na^+} + NH_3 \qquad (25.66)$$

$$\textit{Acidic hydrolysis: } \underset{\text{A nitrile}}{RC{\equiv}N} \xrightarrow[H_2SO_4]{H_2O} \underset{\text{An amide}}{RCONH_2} \xrightarrow[H_2SO_4]{H_2O} \underset{\text{A carboxylic acid}}{RCO_2H} + \overset{+}{N}H_4 \qquad (25.67)$$

EXPERIMENTAL PROCEDURES

Hydrolysis of Nitriles

SAFETY ALERT

The aqueous sodium hydroxide solution is caustic. Wear latex gloves when handling it. Should it contact the skin, immediately flood the area with water.

Preparation Sign in at **www.cengage.com/login** to read the MSDSs for the chemicals used or produced in this procedure.

Basic Hydrolysis. In a small Erlenmeyer flask, mix 10 mL of 3 *M* sodium hydroxide solution and 1 g of the nitrile. Heat the mixture to boiling and note either the odor of ammonia or the color change that occurs when a piece of moist pHydrion paper is held over the flask. After the mixture is homogeneous, cool and then acidify it. If the acid solidifies, collect the crystals by vacuum filtration. If it is a liquid, extract the acidic solution with small portions of diethyl ether, dry and decant the ethereal solution, and then remove the solvent by one of the techniques described in Section 2.29. The residue is the acid. Prepare a suitable derivative of the acid (Sec. 25.13).

WRAPPING IT UP

Neutralize all *aqueous solutions* before flushing them down the drain. Pour any *diethyl ether* isolated into the container for nonhalogenated organic liquids.

SAFETY ALERT

Concentrated sulfuric and nitric acids may produce severe chemical burns. Wear latex gloves when handling these reagents. Should these acids contact your skin,

immediately flood the affected area with water and rinse it with 5% sodium bicarbonate solution.

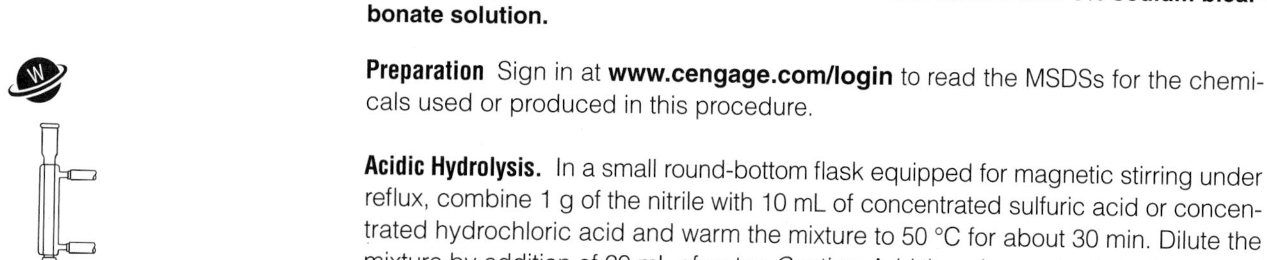

Preparation Sign in at **www.cengage.com/login** to read the MSDSs for the chemicals used or produced in this procedure.

Acidic Hydrolysis. In a small round-bottom flask equipped for magnetic stirring under reflux, combine 1 g of the nitrile with 10 mL of concentrated sulfuric acid or concentrated hydrochloric acid and warm the mixture to 50 °C for about 30 min. Dilute the mixture by addition of 20 mL of water. *Caution:* Add the mixture slowly *to* the water if sulfuric acid has been used. Heat the mixture under gentle reflux for 30 min to 2 h and then allow it to cool. The acid usually forms a separate layer. If the acid solidifies upon cooling, collect it by vacuum filtration. If it is a liquid, extract the acidic mixture with small portions of diethyl ether, dry and decant the ethereal solution, and remove the solvent by one of the techniques described in Section 2.29. Prepare suitable derivatives of the acid (Sec. 25.13).

WRAPPING IT UP

Neutralize the *aqueous solution* before flushing it down the drain. Pour any *diethyl ether* isolated into the container for nonhalogenated organic liquids.

25.18 AMIDES

Classification Test

A. Hydroxylamine

A qualitative test for an amide group (Eq. 25.68) is the same as that for an ester, as given in Section 25.16A. Follow the experimental procedure provided there. The colors observed with amides are the same as those with esters.

$$\underset{\text{An amide}}{R{-}\overset{O}{\overset{\|}{C}}{-}NH_2} + \underset{\text{Hydroxylamine}}{H_2NOH} \longrightarrow \underset{\text{A hydroxamic acid}}{R{-}\overset{O}{\overset{\|}{C}}{-}NHOH} \xrightarrow{FeCl_3} \underset{\text{An iron complex (colored)}}{\left[R{-}C(=O{\rightarrow})(N(H){-}O{-})\right]_3 Fe} + 3\,HCl \qquad (25.68)$$

Derivatives

B. Hydrolysis

Amides may be hydrolyzed under acidic or basic conditions, giving an amine and a carboxylic acid (Eq. 25.69). In the instance of unsubstituted amides, ammonia is liberated. Substituted amides provide a substituted amine as a product. In those cases, it is necessary to classify the amine as being primary or secondary and to prepare derivatives of both the acid (Sec. 25.13) and the amine (Sec. 25.14).

$$\underset{\substack{\text{An amide}\\(R = H,\ \text{alkyl, aryl})}}{R^1CONR^2_2} \xrightarrow[HO^-]{H_3O^+\ \text{or}} \underset{\substack{\text{A carboxylic}\\\text{acid}}}{R^1CO_2H} + \underset{\text{An amine}}{HNR^2_2} \qquad (25.69)$$

EXPERIMENTAL PROCEDURE

Base-Promoted Hydrolysis of Amides

SAFETY ALERT

The aqueous sodium hydroxide solution is caustic. Wear latex gloves when handling it. Should it contact the skin, immediately flood the area with water.

Preparation Sign in at **www.cengage.com/login** to read the MSDSs for the chemicals used or produced in this procedure.

Hydrolysis. Equip a small round-bottom flask containing 10 mL of 3 *M* sodium hydroxide solution for shortpath distillation into a receiver containing a few milliliters of dilute hydrochloric acid. Heat the mixture to boiling and note changes, if any, in the appearance of the contents of the receiver. After the mixture is homogeneous, allow the solution to cool to room temperature.

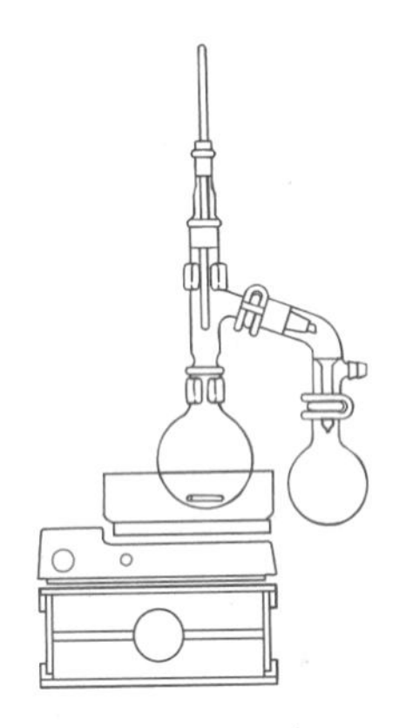

Neutralize the acidic solution in the receiver, perform the Hinsberg test (Sec. 25.14A), and prepare derivatives of the amine (Sec. 25.14). If the amine is *not* volatile enough to distill, and thus is not found in the receiver, use small portions of diethyl ether to extract it from the aqueous layer contained in the stillpot. Dry the combined ethereal extracts over potassium hydroxide pellets, decant the solution, and then remove the solvent by one of the techniques described in Section 2.29. Prepare derivatives of the amine from the residue remaining after solvent removal (Sec. 25.14).

Acidify the alkaline solution from which the amine has been extracted and isolate the carboxylic acid by either vacuum filtration or by extraction with diethyl ether followed by drying over sodium sulfate and removal of solvent as before. Characterize this acid by preparing suitable solid derivatives (Sec. 25.13).

WRAPPING IT UP

Neutralize all *aqueous solutions* before flushing them down the drain. Pour any *diethyl ether* isolated into the container for nonhalogenated organic liquids.

CHAPTER 26

The Literature of Organic Chemistry

Getting a grasp on the chemical literature is a serious challenge, with hundreds of journals already in existence and more being created on a frequent basis. Any one of the publications might contain that kernel of knowledge that will help you solve a problem in the teaching or research laboratory. Before you think you will be buried in paper, however, take heart in knowing that there are both high- and low-technology strategies for efficiently browsing the literature. This chapter is intended to show you how.

26.1 INTRODUCTION

Every serious chemistry student needs to know the library well. In the past this meant knowing one's way around book shelves and printed indexes, but today's library is much more "virtual," and more and more information is now accessed via a computer. This makes it tempting to rely on online information to the exclusion of everything else, however, which can be dangerous. Despite rapid advances in information technology, the bulk of published knowledge in the sciences still exists only in printed form, and the chemist must be careful not to neglect these resources. This chapter is not intended to be a comprehensive guide to the literature of organic chemistry. Rather, it aims to provide a general overview of the important literature sources of organic chemistry that may be used as starting points for research.

26.2 CLASSIFICATION OF THE LITERATURE

Most of the literature of organic chemistry can be divided into eight major categories: (1) primary research journals, (2) review serials, (3) abstracting and indexing sources, (4) handbooks and dictionaries, (5) general multivolume references, (6) reference works on synthetic procedures and techniques, (7) sources of spectral data, and (8) advanced textbooks and monographs. Chemical patents, which form another important segment of the organic literature, are outside the scope of this chapter. In this section, selected examples from each of these classes are given, along with brief explanatory notes. In Section 26.3, examples are provided to illustrate how the literature may be used to find information about a specific organic compound.

Journals have been the lifeblood of scientific communication for centuries. Taken together, they serve as the cumulative memory of scientific research, allowing scientists to build on the work of others and avoid unnecessary repetition of past work. Thus journals are in a sense both a starting point and endpoint for research: Scientists must know the literature of their field well and read continually to keep up to date. This informs their own research from its very beginning. Ultimately, they publish their finished work in the same journals for others to learn from, in what is essentially an unending cycle.

Untold thousands of scientific journals have been published over the centuries, in all languages. Today, Chemical Abstracts Service scans over 9000 journals for new chemical information. But the majority of significant articles are likely to be found in a few hundred of the best and most widely read journals.

From their origins in the mid-seventeenth century, scientific journals remained largely unchanged for over 300 years. In the mid-twentieth century, two major shifts occurred: (1) an explosion in the number of published journals, and (2) the emergence of English as the standard international language of science. Over 85% of new chemistry papers today are written in English, regardless of their country of origin. In the 1990s, the Internet brought profound changes to the way journals are disseminated and read. Many journals in the sciences are now available to subscribers electronically. Most electronic journals, commonly called "e-journals," are digital clones of the printed originals, offering full text of articles and other content in various formats, primarily in HTML (hypertext markup language) and the proprietary PDF (portable document format).

The electronic journal offers a number of obvious advantages over the paper version. An authorized user may access an article any time of day or night, from any location—no trip to the library is necessary. Publishers sometimes make the electronic version of a new issue available days or weeks before the print version appears. Online journals cannot be lost or mutilated like physical volumes can, although they may occasionally be inaccessible due to technical problems. It is easy to search the full text of articles, either within a particular journal or across a number of titles from the same publisher. More significantly, index databases can link directly to the full text of an article, eliminating several steps in obtaining it. The first electronic journals started their coverage in the mid-1990s, and some publishers are now converting their older volumes to this format, in some cases all the way back to the first issue of the journal.

At the present stage of their development, electronic journals mainly save time and reduce inconvenience, but there will be further advantages to this format in the future as technology advances. For example, publishers and scientists are exploring enhancements to the traditional journals that are made possible by the multimedia environment of the Web. Interactive features, manipulatable 3-D chemical structures, spreadsheets, hypertext links, and raw data are appearing in some e-journals. Indeed, the digital copy is now often viewed as the archival copy of record, although predictions that the printed journal itself may eventually disappear altogether have so far not come to pass.

Primary Research Journals

Primary research journals are the ultimate source of most of the information in chemistry. These journals publish original research results in several formats. **Articles** are full papers that provide historical discussions together with a presentation of the important findings, conclusions, and experimental details for preparing new compounds. **Communications** and **letters** are short articles, sometimes with brief experimental details, that are restricted to a single topic or important discovery. Such articles must be especially timely and of general interest to the chemical community. Some journals also publish short papers called **notes** that contain limited discussions and experimental details.

The defining trait of primary scientific journals is the process of **peer review**. Authors submit a paper to the editor of a journal, who in turn sends the manuscript to selected reviewers, or referees, who evaluate the paper for originality, accuracy, relevance, and value to the scientific community. The reviewers may approve the paper for publication as-is, or they may reject it outright, but commonly they request revisions, after which it is accepted. In order to reduce bias, some journals keep the identity of reviewers and authors confidential—this is called double-blind peer review. Others only keep the identity of the reviewers confidential.

The following are some of the most important current journals for the practicing organic chemist, with their CASSI abbreviations in brackets:

1. *Advanced Synthesis & Catalysis* [*Adv. Synth. Catal.*] (2001–present). Formerly *Journal für praktische Chemie*, this German journal, which started in 1834, covers practical synthetic and catalytic methods from both an academic and an industrial perspective.
2. *Angewandte Chemie, International Edition in English* [*Angew. Chem. Int. Ed.*] (1962–present). This is an outstanding high-impact journal that publishes critical reviews of selected topics and communications covering all areas of chemistry.
3. *Biochemistry* [*Biochemistry*] (1962–present). Articles and communications in biochemistry, but many of the contributions tend toward the biological side of organic chemistry. Published by the American Chemical Society.
4. *Bioorganic & Medicinal Chemistry* [*Bioorg. Med. Chem.*] (1993–present). Full articles on bio-organic and medicinal chemistry topics.
5. *Bioorganic & Medicinal Chemistry Letters* [*Bioorg. Med. Chem. Lett.*] (1991–present). As the title implies, this companion to the preceding entry publishes letters and short communications on these topics.
6. *Bulletin of the Chemical Society of Japan* [*Bull. Chem. Soc. Jpn.*] (1926–present). Articles and notes in English covering all areas of chemistry.
7. *Canadian Journal of Chemistry* [*Can. J. Chem.*] (1951–present). Articles in English and French in all areas of chemistry.
8. *Chemical Communications* [*Chem. Commun.* (Cambridge, UK)] (1965–present). This Royal Society of Chemistry title publishes timely brief communications and is one of the most important general chemistry letters journals.
9. *Chemistry: A European Journal* [*Chem.-Eur. J.*] (1995–present). A spin-off from *Angewandte Chemie*, this newer journal has emerged as a leading title.
10. *Chemistry Letters* [*Chem. Lett.*] (1972–present). Communications in English covering all areas of organic chemistry. Published by the Chemical Society of Japan.
11. *European Journal of Organic Chemistry* [*Eur. J. Org. Chem.*] (1998–present). Formed by the merger of the organic sections of a number of established European chemistry journals.
12. *Helvetica Chimica Acta* [*Helv. Chim. Acta*] (1918–present). Articles and notes in English, French, or German covering all areas of organic chemistry. Official journal of the Swiss Chemical Society.

13. *Heterocycles* [*Heterocycles*] (1973–present). Reviews, communications, and articles in all areas of heterocyclic chemistry. Published in Japan by the Sendai Institute of Heterocyclic Chemistry.
14. *Journal of the American Chemical Society* [*J. Am. Chem. Soc.*] (1879–present). Articles and communications covering all areas of chemistry. One of the foremost chemical journals in the world, *JACS* has a broad scope across all of chemistry.
15. *Journal of Medicinal Chemistry* [*J. Med. Chem.*] (1959–present). Articles, communications, and notes in English covering the preparation of new organic compounds having biological activity.
16. *Journal of Organic Chemistry* [*J. Org. Chem.*] (1936–present). Articles, communications, and notes in all areas of organic chemistry; this is arguably the best journal in the world dedicated to publishing work covering organic chemistry. It is published by the American Chemical Society.
17. *Nature* [*Nature* (London, UK)] (1869–present). This prestigious British journal is analogous to the U.S.-based *Science* and publishes papers in all fields of science, including chemistry.
18. *Organic & Biomolecular Chemistry* [*Org. Biomol. Chem.*] (2003–present). Formed by the merger of the two sections of *Journal of the Chemical Society, Perkin Transactions.* Primarily articles and some communications, in all areas of organic and bio-organic chemistry.
19. *Organic Letters* [*Org. Lett.*] (1999–present). This letters journal is published by the American Chemical Society and has rapidly become a leading title.
20. *Organometallics* [*Organometallics*] (1982–present). Published by the American Chemical Society, this is the premier journal covering all aspects of organometallic chemistry.
21. *Proceedings of the National Academy of Sciences of the U.S.A.* [*Proc. Natl. Acad. Sci. U.S.A.*] (1915–present). Major multidisciplinary journal that emphasizes the life sciences.
22. *Science* [*Science* (Washington DC, U.S.)] (1883–present). The official journal of the American Association for the Advancement of Science (AAAS), *Science* is one of the most prestigious, widely read and cited journals in the world. As its title implies, it is multidisciplinary in scope and publishes papers in all fields.
23. *Synlett* [*Synlett*] (1990–present). Articles and communications in the general area of synthetic organic chemistry.
24. *Synthesis* [*Synthesis*] (1969–present). Reviews, articles, and communications in the general area of synthetic organic chemistry.
25. *Tetrahedron* [*Tetrahedron*] (1957–present). An international journal dedicated to publishing articles, reviews, and "symposia in print" in the general areas of organic and bio-organic chemistry.
26. *Tetrahedron*: *Asymmetry* [*Tetrahedron: Asymmetry*] (1990–present). Reviews, articles, and communications in the specialized area of asymmetric synthesis and methods.
27. *Tetrahedron Letters* [*Tetrahedron Lett.*] (1959–present). An international journal dedicated to publishing brief two- or four-page communications in all areas of organic and bio-organic chemistry.

Review Serials

Review serials publish longer articles covering specific topics. They can be journals or periodic book series, such as those published annually. As a rule, review articles do not present new research but rather summarize and synthesize recent original work published previously in primary journals. They can be very useful as a point of entry into the literature on a less familiar topic because they contain substantial bibliographies. Reviews are often commissioned from expert authors and can take a great deal of time and effort to prepare.

Some review serials publish reviews in all areas of chemistry, while others cover only specific areas. In addition to the review serials listed below, a number of the primary research journals listed above also publish excellent reviews. These include *Angewandte Chemie International Edition in English, Heterocycles, Synthesis,* and *Tetrahedron.*

1. *Accounts of Chemical Research* [*Acc. Chem. Res.*] (1968–present). Provides concise reviews of areas of active research in all areas of chemistry. It is published by the American Chemical Society.
2. *Advances in Heterocyclic Chemistry* [*Adv. Heterocycl. Chem.*] (1963–present). Publishes lengthy reviews on this topic.
3. *Advances in Organometallic Chemistry* [*Adv. Organomet. Chem.*] (1964–present). One or two volumes per year with lengthy review articles.
4. *Annual Reports on the Progress of Chemistry* [*Annu. Rep. Prog. Chem.*] (1904–present). A series of review volumes in areas of general interest and importance to organic chemists. (Since 1968, this title has been divided into sections, with Section B devoted to organic chemistry.)
5. *Chemical Reviews* [*Chem. Rev.* (Washington DC, U.S.)] (1924–present). A review journal published by the American Chemical Society that covers all areas of chemistry.
6. *Chemical Society Reviews* [*Chem. Soc. Rev.*] (1972–present). A review journal with broad scope published by the Royal Society of Chemistry.
7. *Natural Product Reports* [*Nat. Prod. Rep.*] (1984–present). Comprehensive reviews of natural products chemistry.
8. *Organic Reaction Mechanisms* [*Org. React. Mech.*] (1965–present). Annual survey of the literature on this topic.
9. *Organic Reactions* [*Org. React.*] (1942–present). Annual. Each volume contains lengthy reviews that deal with an organic reaction of wide applicability. Typical experimental procedures are given in detail, and extensive tables of examples with references are provided. Each volume contains a cumulative author and chapter title index.

Abstracting and Indexing Sources

The vast body of scientific literature would be useless without a way to find what you're looking for within it. Abstracting and indexing sources provide systematic indexing of the literature within a given subject area. Articles, patents, books, conference papers, reports, and other types of documents are regularly scanned by expert indexers, who analyze documents for pertinent content, assign subject terms, and write a concise summary called an **abstract**. This information is then input into a database that can be searched online.

Chemical Abstracts is the world's largest index, providing "the key to the world's chemical literature." It began in 1907 and annually abstracts over a million documents that are drawn from about 9000 technical journals as well as from books, conferences, dissertations, and patents. *Chemical Abstracts* (*CA*), which appeared weekly in print until 2009 provides by far the most complete coverage of the chemical literature after 1940.

Since the entirety of *Chemical Abstracts*, back to 1907, is now available electronically in several different ways, few people use the printed version today. This chapter will therefore focus only on use of the database.

The most important contribution of *Chemical Abstracts* is the systematic registration of chemical compounds in the **Registry** file, a database that is the key piece of the *Chemical Abstracts* online system. Once a new chemical substance has been reported and characterized in the literature, it is assigned a specific **CA Index Name** and a unique **Registry Number**. Index names follow *CA*'s own nomenclature rules, and these may change over time. Registry numbers (CAS RNs), however, do *not* change. They have the recognizable format 1234-56-7, where the first segment contains between two and seven digits. RNs themselves carry no chemical meaning—they merely represent records in the Registry database. Registry numbers are widely used throughout the chemical literature, and by the chemical industry, as a standard way to identify chemical structures without the confusing and contradictory problems of nomenclature. Known synonyms, trade names, and molecular and structural formulas for a given structure are listed in its Registry record, thus providing an effective link to an unambiguous identifier that can be used in indexing the literature. Over 50 million substances, including chemicals, polymers, alloys, and multicomponent mixtures, have been registered by *Chemical Abstracts* since 1965. Nearly a million new substances are registered each year, the majority of them organic. In addition, the database contains records for over 60 million biosequences.

There are several different ways to search *Chemical Abstracts* electronically, and the most common are described below.

1. *SciFinder.* SciFinder is an interface to the Chemical Abstracts family of databases. The **CAPLUS** file is the equivalent to the printed *Chemical Abstracts* from 1907 to the present. The **Registry** file of chemical substances is described above. **CASREACT** is a file of single- and multistep organic reactions drawn from selected organic journals. SciFinder provides a user-friendly, graphical interface that permits searching by topic, author, chemical structure, and substructure. Search results can be refined, analyzed, and displayed in a number of ways. SciFinder also provides links to the electronic full text of many papers.
2. *CAS Online.* This database service is available through STN, the online system operated by Chemical Abstracts Service (CAS) that provides access to several related files with a choice of interfaces. Because there is a charge for using CAS Online, it is advisable to obtain training and assistance from a search specialist before beginning searching. That assistance will improve the quality and the cost-effectiveness of the search.

In addition to *Chemical Abstracts*, many other databases provide access to other specialized types of literature in related fields, including medicine, patents, polymers, toxicology, and pharmacy. These major resources deserve mention here:

1. *Inspec.* The primary index for physics, astronomy, and electrical and computer engineering. Includes strong coverage of chemical physics, physical chemistry, and spectroscopy.
2. *Medline.* Broad coverage of the biomedical journal literature. Available in several different incarnations, including PubMed, which is provided free of charge online by the National Library of Medicine. It is also part of SciFinder.

3. *Science Citation Index*. Produced by Thomson-Reuters, this is a unique reference tool that provides an alternative method to traditional subject-based literature searching. It enables you to search forward from a particular literature reference. For example, to find later applications of a reaction or method that was described in a specific paper, you simply look up the entry for that paper in *Science Citation Index* to identify papers that subsequently cited the original article. A Web version, called **Web of Science**, is now the most commonly used interface to the wealth of citation data gathered from the scientific and technical literature.

Visit your library or its website to learn more about these and other scientific databases that may be available to you.

Handbooks and Dictionaries

Handbooks and dictionaries of organic chemistry provide specific information, such as physical properties and uses of known compounds.

1. *Aldrich Handbook of Fine Chemicals*. Aldrich Chemical Co., Milwaukee, WI, annual editions. The "Handbook" is the sales catalog of the Aldrich Chemical Co., a leading supplier of research chemicals. It is also useful as a source of CAS RNs, formulas, structures, and basic physical data on over 35,000 products, including cross-references to the *Merck Index* and the Aldrich spectra and regulatory reference sets. Most of the data from the Aldrich Catalog can be found on the Sigma-Aldrich website at www.sigmaaldrich.com.
2. *CRC Handbook of Chemistry and Physics*, annual editions. CRC Press, Boca Raton, FL. The *CRC* is probably the best-known single-volume handbook, and it contains a wealth of useful data in all areas of chemistry and physics. The most useful part for organic chemists is the section headed "Physical Constants of Organic Compounds," which provides formulas, structures, molar masses (molecular weights), densities, refractive indexes, solubilities, color, and melting and boiling points for about 11,000 compounds. CAS RNs are also included. The newer editions index this section by synonym, molecular formula, and CAS RN. Many institutions also subscribe to the Web version of the *CRC*.
3. *Dictionary of Organic Compounds*, 6th ed., Buckingham, J., ed. Chapman & Hall, New York, 1996. The *DOC* is an excellent source for physical and chemical data on over 100,000 organic compounds and derivatives. Entries also provide literature references for synthesis, characterization, spectra, and properties, making the *DOC* useful as a starting point in a literature search. The 6th edition is in nine volumes. Volumes 1–6 contain the data for the compounds, Volume 7 is a name index with cross-references, Volume 8 contains a molecular-formula index, and Volume 9 is a CAS RN index. Some institutions subscribe to a Web version of the *DOC* and similar compilations called the Combined Chemical Dictionary (CCD), part of the ChemNetBase system provided by Chapman & Hall/CRC Press.
4. *Handbook of Data on Organic Compounds*, 3rd ed., Lide, D., ed. CRC Press, Boca Raton, FL, 1994. A seven-volume set containing brief factual entries for about 25,000 compounds. A good source for chemical structures and numeric spectral data. The *HODOC* is available as "Properties of Organic Compounds," part of the electronic CRC ChemNetBase system.
5. *Handbook of Tables for Identification of Organic Compounds*, 3rd ed., Rappoport, Z., ed. Chemical Rubber Co., Cleveland, 1967. Gives physical properties and derivatives for more than 4000 compounds, arranged according to functional groups.

6. *Hawley's Condensed Chemical Dictionary*, 15th ed., Lewis, R., ed. Wiley, New York, 2007. One of the better general chemistry dictionaries.
7. *Lange's Handbook of Chemistry*, 16th ed., Dean, J. A., ed. McGraw-Hill, New York, 2005. Gives physical properties for about 4300 organic compounds. Very similar to the *CRC* in scope and arrangement.
8. *Merck Index*, 14th ed. Merck and Co., Rahway, NJ, 2006. Gives a concise summary of the physical, chemical, and pharmacological properties of more than 10,000 compounds, including pharmaceuticals, organic chemicals and reagents, inorganic substances, agricultural chemicals, and naturally occurring substances. Organization is alphabetical by name; it is best to consult the synonym index first. Also includes indexes by CAS RN and therapeutic category.
9. *NIST Chemistry WebBook* (webbook.nist.gov/chemistry). A free and authoritative online database of reliable physical, chemical, and spectral data for several thousand chemical species. Created by the National Institute of Standards and Technology (NIST).
10. *Name Reactions: A Collection of Detailed Reaction Mechanisms*, 3rd ed., Li, J., ed. Springer, Berlin, 2006. A useful compilation of data for name reactions, giving general scheme, mechanism of reaction, examples, and references.

The German *Handbuch* Tradition

Distinct from the familiar American "handbook" publication, which is intended as a handy bench-top reference tool, the German *Handbuch* concept dates from the early nineteenth century, when chemists began to make attempts to catalog, summarize, and organize comprehensively the existing literature on chemical substances. This involved the systematic indexing of compounds, extraction and critical assessment of reported data and methods of preparation, and republishing pertinent data in a single, highly organized reference work. Friedrich Konrad Beilstein (1838–1906) and Leopold Gmelin (1788–1853) were pioneers of the *Handbuch* format, and the works that bear their names are two of the most important resources in the sciences. Today, however, the literature of chemistry is far too vast to be covered completely in handbooks, and many modern chemists are unfamiliar with them, preferring their much faster and more powerful database versions. Gmelin's *Handbook* deals with inorganic and organometallic compounds and is outside the scope of this chapter. *Beilstein's Handbuch der Organischen Chemie* is perhaps the most extensive reference work in any branch of science. It was first published in 1881–1883 in two volumes. Publication of the fourth and last edition began in 1918 and ended in 1998. In contrast to most other collections of physical and chemical data, those included in *Beilstein* have been critically evaluated and checked for internal consistency.

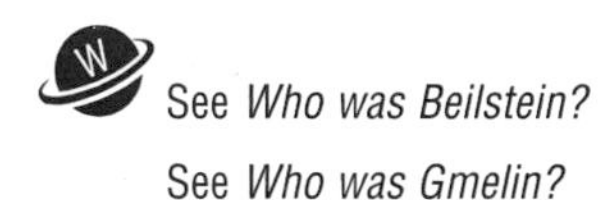
See *Who was Beilstein?*
See *Who was Gmelin?*

Although familiarity with the printed *Beilstein Handbook* is a useful skill for an organic chemist to have, in reality few chemists still use the printed version today. In the mid-1990s, the Beilstein Institute launched an online service that revolutionized the use of this extraordinary tool. The **Crossfire** system provides online access to the database of over eight million compounds, nine million reactions, and 36 million individual data points, through a dedicated client–server system that is accessed via the Internet.

Crossfire is usually searched by drawing a chemical structure or reaction. It is also searchable by identifying elements, such as chemical name, name fragment, CAS RN, and molecular formula, as well as by numeric and keyword searches in various data fields. Crossfire, which will be fully replaced by a web-based successor Reaxys in 2010, is now the tool of choice for many organic and medicinal chemists seeking rapid access to the organic literature. Its primary strength is its thorough retrospective coverage of the organic literature as far back as 1771, up to 1959.

The post-1960 content of Beilstein is more selective and comes from scanning a smaller number of journals in organic and medicinal chemistry, seeking newly reported substances and reactions. Physical and chemical data about new and old compounds are extracted from articles, evaluated critically, and added to the database along with references to the literature source.

General Treatises

Unlike indexes or handbooks, these tools provide encyclopedic review chapters on specific subjects, with plentiful literature references, and act as an overview of the field.

1. *Chemistry of Functional Groups*, Patai, S., ed. Wiley-Interscience, New York. A long-running book series devoted to in-depth reviews of functional groups.
2. *Chemistry of Heterocyclic Compounds* (1950–present). Wiley Interscience, New York. An ongoing series of review volumes that cover specific types of heterocyclic compounds.
3. *Comprehensive Heterocyclic Chemistry*, Katritzky, A. R.; Rees, C. W., eds. Pergamon Press, Oxford, 1984. An eight-volume treatise covering the reactions, structure, synthesis, and uses of heterocyclic compounds. It is supplemented by *Comprehensive Heterocyclic Chemistry II*, Katritzky, A. R.; Rees, C. W.; Scriven, E. F. V., eds. Pergamon Press, Oxford, 1996, an 11-volume treatise updating the literature coverage through 1995; and *Comprehensive Heterocyclic Chemistry III*, Katrizky A.R., ed. Elsevier, Amsterdam, 2008, a 15-volume set.
4. *Comprehensive Organic Chemistry*, Barton, D. H. R.; Ollis, W. D., eds. Pergamon Press, Oxford, 1978. A six-volume treatise on the synthesis and reactions of organic compounds.
5. *Comprehensive Organic Functional Group Transformations*, Katritzky, A. R.; Meth-Cohn, O.; Rees, C. W., eds. Pergamon, Oxford, UK, 1995. A seven-volume set that comprehensively reviews transformations of functional groups. A seven-volume update was published in 2005.
6. *Comprehensive Organic Synthesis*, Trost, B. M.; Fleming, I., eds. Pergamon Press, Oxford, 1991. A nine-volume treatise that covers all aspects of synthetic organic chemistry, including carbon-carbon bond formation, heteroatom manipulation, and oxidation and reduction.
7. *Comprehensive Organometallic Chemistry*, Wilkinson, G., ed. Pergamon Press, Oxford, 1982. A nine-volume treatise covering the synthesis, reactions, and structure of organometallic compounds. It is supplemented by *Comprehensive Organometallic Chemistry II,* Wilkinson, G., ed. Pergamon Press, Oxford, 1996, a 14-volume treatise updating the literature coverage through 1994; and *Comprehensive Organometallic Chemistry III*, Mingos, D. M. P., ed. Elsevier, Amsterdam, 2007, a 13-volume update.
8. *Rodd's Chemistry of Carbon Compounds*, 2nd ed., Coffey, S., ed. Elsevier, Amsterdam, 1964–2001. A survey of all classes of organic compounds, giving properties and syntheses for many individual compounds; consists of four volumes in 30 parts with supplements.

It is often necessary to survey the scope and limitations of a reaction, a synthetic method, a reagent, or a technique to determine whether it may be applied to solving a specific problem. To facilitate access to this information, a number of reference works are available that contain reviews of reactions, reagents, techniques, and methods in organic chemistry; these are often serial in nature.

1. *Comprehensive Organic Transformations: A Guide to Functional Group Preparations*, 2nd ed., Larock, R. C. John Wiley & Sons, New York, 1999. A one-volume compilation of reactions of functional groups.
2. *Fieser's Reagents for Organic Synthesis*. Wiley-Interscience, New York, 1967–present. These volumes describe thousands of reagents and solvents, in terms of methods of preparation or source, purification, and use in typical reactions.
3. *Encyclopedia of Reagents for Organic Synthesis*, 2nd ed. Paquette, L. A. John Wiley & Sons, New York, 2009. A fourteen-volume set, also available online, that covers important reagents used in synthetic organic chemistry.
4. *Organic Syntheses*. John Wiley & Sons, New York, 1921–present. An annual publication collecting detailed synthetic procedures that have been thoroughly checked by independent investigators. The procedures are general and can be applied to the synthesis of related compounds other than those described. The *Collective Volumes* revise, index, and republish the methods originally appearing in the annual volumes. The collective volumes contain indexes of formulas, names, types of reaction, types of compounds, purification of solvents and reagents, and illustrations of special apparatus. This useful series is freely available on the Web at orgsyn.org.
5. *Purification of Laboratory Chemicals*, 5th ed., Armarego, W. L. F. et al. Butterworth- Heinemann, Amsterdam, 2003. Brief procedures for purifying organic solvents and reagents in the laboratory.
6. *Science of Synthesis*. Thieme, Stuttgart, 2000–present. A revised edition of Houben-Weyl *Methoden der Organischen Chemie,* providing systematic coverage of organic synthetic techniques. This resource is also available electronically by subscription.

Sources of Spectral Data

A problem commonly encountered in organic chemistry is identifying a compound that has been obtained from a chemical reaction or isolated from a natural source or the environment. The spectroscopic characteristics of a sample provide important clues to the identity of the substance. Collections of spectral data of known compounds provide a source of such information that may be used either to identify known compounds or to assist in the determination of the structure of an unknown substance.

Two major publishers of chemical reference spectra are the Aldrich Chemical Company and Bio-Rad Sadtler Research Laboratories. Aldrich has published a number of compact compilations of infrared and NMR spectra based largely on organic chemicals sold through their catalog. These sets are commonly found in academic libraries and are easy to use, with indexes by chemical name, molecular formula, CAS RN, and Aldrich catalog number. The Aldrich sets are organized by chemical class, making it relatively easy to locate spectra similar to that of an unknown.

The various Sadtler series, in contrast, are much larger and somewhat more complex to use. The spectra sheets are filed in random order, and separate index volumes must be used to locate desired compounds. The indexes provide access by name, molecular formula, or chemical class.

Some important printed sets of reference spectra found in many libraries are listed below.

1. **^{1}H NMR Spectra**
   - *Aldrich Library of NMR Spectra*, 2nd ed., Pouchert, C. J., ed. Aldrich Chemical Company, Milwaukee, 1983; two volumes. A collection of about 37,000 spectra.
   - *Aldrich Library of ^{13}C and ^{1}H FT-NMR Spectra*, Pouchert, C. J.; Behnke, J., eds. Aldrich Chemical Company, Milwaukee, 1993; three volumes. A collection of about 12,000 spectra.
   - *Nuclear Magnetic Resonance Spectra*. Sadtler Research Laboratories, Philadelphia. Proton NMR spectra of more than 64,000 compounds. Peaks are assigned to the hydrogen nuclei responsible for the absorptions, and integration of the signals is shown on many of the spectra.
2. **^{13}C NMR Spectra**
   - *Atlas of Carbon-13 NMR Data*, Breitmaier, E.; Haas, G.; Voelter, W. Heyden, Philadelphia, 1979; two volumes. Tabular data on 3017 compounds, with chemical shifts for ^{13}C given and ^{1}HY–^{13}C multiplicities indicated.
   - *Sadtler Standard Carbon-13 NMR Spectra*. Sadtler Research Laboratories, Philadelphia, 1976–1996. Compilation of proton-decoupled spectra.
3. **IR Spectra**
   - *Aldrich Library of Infrared Spectra*, 3rd ed., Pouchert, C. J., ed. Aldrich Chemical Company, Milwaukee, 1981. A collection of about 12,000 spectra.
   - *Aldrich Library of FT-IR Spectra*, Pouchert, C. J., ed. Aldrich Chemical Co., Milwaukee, 1985–1989; three volumes.
   - *Sadtler Standard Infrared Prism Spectra*. Sadtler Research Laboratories, Philadelphia. Compilation of 91,000 spectra.
   - *Sadtler Standard Infrared Grating Spectra*. Sadtler Research Laboratories, Philadelphia. Compilation of 91,000 spectra.
   - *Sigma Library of FT-IR Spectra*, Keller, R. J., ed. Sigma Chemical Company, St. Louis, 1986; two volumes. A collection of 10,400 spectra of biological compounds.
4. **Ultraviolet Spectra**
   - *Ultra Violet Spectra*. Sadtler Research Laboratories, Philadelphia, 1968–1996.

The Crossfire system (see Beilstein, above) indexes and provides literature references for spectral data of organic compounds, although the spectra themselves are not included.

While most spectral libraries are fee-based, the Web offers a few free options for locating chemical spectra as well. One is the *NIST Chemistry WebBook* (webbook.nist.gov/chemistry), maintained by the National Institute of Standards and Technology. Another worthwhile free tool is the Integrated Spectral Data Base System for Organic Compounds (SDBS), maintained by the National Institute of Advanced Industrial Science and Technology in Japan (www.aist.go.jp/RIODB/SDBS/menu-e.html).

Advanced Textbooks and Monographs

Advanced textbooks in organic chemistry provide useful information for students who are interested in a more sophisticated or advanced treatment of the information than that found in a typical undergraduate organic textbook.

Monographs are books in which specific topics are examined in depth for a readership of practicing specialists and advanced students. Some monographs are written entirely by one or two authors, but most are collections of chapters written by a variety of experts under the oversight of an editor. Monographs do not present new research results, but rather serve as full-length reviews of current knowledge. Both textbooks and monographs include extensive bibliographies of primary literature references for further reading.

1. Carey, F. A.; Sundberg, R. J. *Advanced Organic Chemistry; Part A: Structure and Mechanisms; Part B: Reactions and Synthesis,* 5th ed. Springer, New York, 2007. An excellent survey of reactions and their applications.
2. Carruthers, W. *Some Modern Methods of Organic Synthesis,* 3rd ed. Cambridge University Press, Cambridge, UK, 1986. A detailed discussion of selected reactions used in synthetic organic chemistry.
3. Collman, J. P.; Hegedus, L. S.; Norton, J. R.; Finke, R. G. *Principles and Applications of Organotransition Metal Chemistry,* 2nd ed. University Science Books, Mill Valley, CA, 1987. An excellent survey of organometallic chemistry.
4. Eliel, E. L.; Wilen, S. H.; Doyle, M. P. *Basic Organic Stereochemistry,* Wiley-Interscience, New York, 2001. An excellent treatise on all aspects of stereochemistry.
5. Kocienski, P. J. *Protecting Groups,* 3rd ed. Thieme, New York, 2005. An extensive compilation of methods for protection and deprotection of various functional groups together with insights on compatibility in polyfunctional molecules.
6. Wuts, P. G. M. *Greene's Protective Groups in Organic Synthesis,* 4th ed. Wiley-Interscience, New York, 2007. An extensive compilation of methods for protection and deprotection of various functional groups.
7. Lowry, T. H.; Richardson, K. S. *Mechanism and Theory in Organic Chemistry,* 3rd ed. Harper and Row, New York, 1987. An excellent treatment of mechanistic and theoretical aspects of organic chemistry.
8. Anslyn, E. V.; Dougherty, D. A. *Modern Physical Organic Chemistry,* University Science Books, Sausalito, CA, 2005. An advanced text that addresses topics in biochemistry and organometallic, materials, and bio-organic chemistry from a mechanistic viewpoint.
9. Smith, M. B.; March, J. *March's Advanced Organic Chemistry: Reactions, Mechanisms, and Structure,* 6th ed. Wiley-Interscience, New York, 2007. An advanced text with broad coverage and many references to the original literature.
10. Smith, M. B. *Organic Synthesis,* 2nd ed. McGraw-Hill, New York, 2002. Chapters on synthetic topics contain extensive bibliographies and homework problems.
11. Crews, P.; Rodríguez, J.; Jaspars, M. *Organic Structure Analysis,* Oxford University Press, Oxford, UK, 1998.

26.3 USING THE LITERATURE OF ORGANIC CHEMISTRY

The literature outline in this chapter may be used in a variety of ways, according to the aims and needs of different courses and the library facilities available. Even in those cases where the pressure of time and/or lack of facilities preclude the use of literature beyond the pages of this textbook itself, this chapter is still valuable if you decide to go further in the study of organic chemistry.

In many organic laboratory courses, instructors make part of the experimental assignments open-ended—encouraging the students to plan and carry out experiments with some independence. This is a desirable objective, but there is an element of risk if the experimental procedure has not been carefully checked; procedures found in the literature are not always easily reproduced. One of the better sources of experiments is *Organic Syntheses;* although the experiments are typically reported on a large scale, they may often be easily scaled down. The experiments that occasionally appear in the *Journal of Chemical Education* also deserve mention.

The synthesis of known and unknown compounds is a task commonly encountered by the research organic chemist. In the course of this work, it is often necessary to determine whether a particular compound has been previously prepared; if it has, then the questions of when, how, and by whom arise. If the compound has not been reported, then the chemist searches for similar compounds, because the same synthetic methods might apply to the substance of interest. The examples that follow will illustrate two standard methods of performing literature searches to solve such problems, and the exercises at the end pose many types of problems that are routinely encountered in research.

When beginning a search for information on a chemical compound, it is always best to start with the readily available, easy-to-use reference tools. If you do not find what you are looking for, you then proceed to larger, more complex tools. Although many students today are inclined to start with free Internet sites or search engines, this strategy will rarely provide the necessary information and may even result in misleading or erroneous data. In the setting of a chemistry laboratory, this can be dangerous. It is always better to stick with reliable, authoritative tools, whether online or in print. A trip to the library can actually be faster and more fruitful than browsing random websites.

What Is Mustard Gas?

See more on *Mustard Gas*

"Mustard gas" is one of the names that has been applied to the compound $ClCH_2CH_2$–S–CH_2CH_2Cl. Let's say that you need to find the answers to the following questions: (1) Has this compound been synthesized or isolated; (2) by whom, and when; (3) where can the most recent information on this compound be found? How do you find this information?

Starting with the *Merck Index* (14th edition), you may look up "mustard gas" in the name index in the back of the book. (Or you can use the electronic version of the *Merck* if it's available to you.) You see the name with the number "6316" next to it. This number is the "monograph" number within the *Merck Index* and indicates where you will find the entry on this substance. Turning to entry 6316, there is a paragraph containing basic information about this compound: its synonyms, molecular formula and molecular weight, and literature references describing its preparation, reactions, and toxicity. (Note the phrase "Deadly vesicant.") Other information includes melting and boiling points, densities, appearance and odor, refractive index, uses, and safety warnings. Note the CAS RN, 505-60-2, as this will come in handy in subsequent searches.

The oldest literature reference in the preparation section is "Meyer, *Ber.* **19**, 3260 (1886)," which refers to the 1886 article by Victor Meyer in *Chemische Berichte* that describes the preparation of this dangerous gas by treating [b, b']-dihydroxyethyl sulfide with HCl gas. The *Merck Index* often cites the original paper or patent that reported the first synthesis of a chemical.

Now that you have the CAS RN, consult the Registry Number Index volume of the *Dictionary of Organic Compounds,* 6th edition. Or you can use the online equivalent of the *DOC,* the Combined Chemical Dictionary, if it's available to you.

Under 505-60-2, you see the chemical name "1,1-Thiobis[2-chloroethane], T0-03220." The *DOC* is organized alphabetically by preferred chemical name, which in this case is not the trivial name "mustard gas." The entry T0-03220 is found in Volume 6. As in the *Merck Index*, *DOC* entries usually provide synonyms, basic chemical and physical data, and literature references to background information on preparation, toxicology, spectra, etc. (In the case of mustard gas, however, note that the *DOC* does not refer to Meyer's original synthesis.)

You may then use the *Handbook of Data on Organic Compounds* and look up the registry number in the appropriate index volume. You get the *HODOC* entry number, 13011. Turn to that entry in Volume 3 to find the mustard gas entry. In addition to basic chemical data, note the chemical structure diagram and the mass spectrometric data. The *HODOC* does not include literature references.

Now, after three quick look-ups in the library, you have obtained a fair amount of basic information about mustard gas. To gather more extensive information and more recent references, some basic database searching is the next step.

Beilstein Crossfire

Beilstein provides a large amount of chemical and physical data, as well as many retrospective references to preparation and reaction articles. Searching in Crossfire is compound-based: it allows searching by chemical name, name segment(s), molecular formula, Registry Number, and (sub)structure.

SciFinder

SciFinder (*Chemical Abstracts*) will provide literature citations from 1907 to last week. Within SciFinder, you may search for information about a compound in different ways. For instance, you can search the phrase "preparation of mustard gas" as a "Research Topic," and this option will automatically locate the Registry Number of mustard gas and return a list of literature references where the preparation of this substance is discussed. Alternately, you can choose the search option "Chemical Substance or Reaction" and enter its Registry Number or its name. The Substance option will locate a group of Registry compound records for you to examine and choose from, after which you can retrieve the relevant literature references. Both approaches work well for this question. It is always a good idea to examine the compound's Registry record in order to confirm the structure and gather synonyms for use elsewhere. In the case of a well-known compound like mustard gas, a search in *Chemical Abstracts* will often retrieve thousands of documents, requiring you to narrow your search with additional terms or parameters.

Medline

If you are also interested in the toxicity and health effects of mustard gas, searching the medical literature is the next step. Medline thoroughly indexes the medical literature, and it contains a great deal of information on the biochemical, toxicological, and clinical aspects of this poison. Medline is included in SciFinder, where it can be searched simultaneously with *Chemical Abstracts*. The free version of Medline is called PubMed.

Web of Science (Science Citation Index)

Another way to find papers that might be relevant to the mustard gas question is to search *Science Citation Index* for articles that have cited Meyer's original synthesis. The *Web of Science* interface allows you to look up the Meyer paper as a cited reference.

Several authors have cited Meyer in the last decade or so, demonstrating that the organic chemistry literature has a long "shelf-life."

NIST Chemistry WebBook

The *Chemistry WebBook* (webbook.nist.gov/chemistry) is an excellent source of physicochemical data on substances. This database contains a wealth of reliable information, especially thermochemical and spectral data. As for most databases, the most straightforward way of searching it is to use the CAS Registry Number of the compound.

Is a Compound of Interest Known? A second example of a search for the preparation of the compound 3-(2-furyl)-1-(3-nitrophenyl) propenone illustrates how you can use Beilstein and Registry to find compounds that have been reported in the literature, especially those that are not likely to be found in standard handbooks such as the *Merck Index* and the *CRC*. It does not really matter which database you consult first, as you should use both. An important point to realize is that the exact name you have for the compound, if you have one at all, is unlikely to be the name used in Beilstein or Registry, which use their own quite different systems of nomenclature. So it is generally unwise to rely solely on a name search in any database or index.

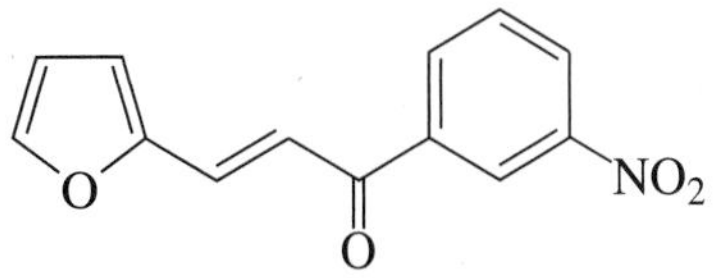

3-(2-Furyl)-1-(3-nitrophenyl)propenone

First, draw the structure of the compound in Crossfire's structure editor and search it against the Beilstein database. This brings up the record for the compound in question. A method of preparation using furfural and 1-(3-nitrophenyl)ethanone is indicated under Reactions. If you follow the link to the Crossfire reaction record, you find two literature references: D. L. Turner, *J. Am. Chem. Soc.* **71**, *1949*, 612, and Li, Synth. *Comm.* **29**, *1999*, 965. These articles describe methods of preparation from 2-furaldehyde and acetophenone.

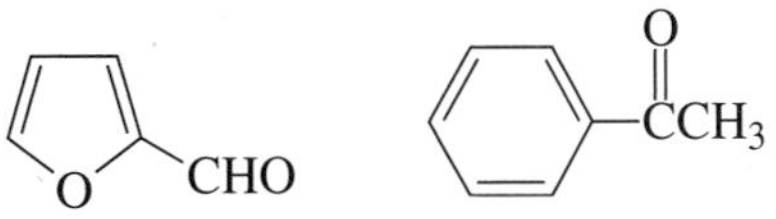

2-Furaldehyde Acetophenone

Next, follow up with a search in *Chemical Abstracts*/Registry. Using SciFinder, you can draw the structure in the structure editor, specify an exact search, and locate Registry records that match. Note that searching the exact name as given above results in no hits, and a molecular formula search on $C_{13}H_9NO_4$ retrieves over 400 hits, neither of which is very helpful. A structure search is thus clearly the best option. From this search we find that the CAS RN for this compound is 15462-51-8, and its CA Index Name is 2-propen-1-one, 3-(2-furanyl)-1-(3-nitrophenyl)-. From this point you can retrieve further literature references, including the two papers cited through Beilstein as noted above.

If thorough searches in both Beilstein and SciFinder turn up no exact matches for the structure in question, it is advisable to search the molecular formula and/or name in the pre-1967 *Chemical Abstracts* printed indexes. If you still are unable to find a match, you can fairly safely assume that the compound has not been fully characterized in the literature. Further searches for similar compounds can then begin.

EXERCISES

1. Find the melting points of the following crystalline derivatives:
   a. 2,4-dinitrophenylhydrazone of trichloroacetaldehyde
   b. semicarbazone of 3-methylcyclohexanone
   c. phenylurethane of 1,3-dichloro-2-propanol
   d. amide of 2-methyl-3-phenylpropanoic acid
   e. benzamide of 4-fluoroaniline
2. Locate an article or a chapter on each of the following types of organic reactions:
   a. the aldol reaction
   b. the Wittig reaction
   c. reactions of diazoacetic esters with unsaturated compound
   d. hydration of alkenes and alkynes through hydroboration
   e. metalation with organolithium compounds
   f. reactions of lithium dialkylcuprates
   g. asymmetric synthesis of amino acids
3. Give a literature reference for a practical synthetic procedure for each of the following compounds and state the yield that may be expected:
   a. 1,2-dibromocyclohexane
   b. a-tetralone
   c. 3-chlorocyclopentene
   d. 2-carboethoxycyclopentanone
   e. norcarane
   f. tropylium fluoborate
   g. 1-methyl-2-tetralone
   h. adamantine
   i. buckminsterfullerene (buckyball)
4. Locate descriptions of procedures for the preparation or purification of the following reagents and solvents used in organic syntheses:
   a. Raney nickel catalysts
   b. sodium borohydride
   c. dimethyl sulfoxide
   d. sodium amide
   e. diazomethane

5. Find IR spectra for the following compounds:

a. *N*-cyclohexylbenzamide

b. 4,5-dihydroxy-2-nitrobenzaldehyde

c. benzyl acetate

d. diisopropyl ether

e. 3,6-diphenyl-2-cyclohexen-1-one

f. 4-amino-1-butanol

6. Find ^{1}H and ^{13}C NMR spectra of the following compounds:

a. benzyl acetate

b. diisopropyl ether

c. 4-amino-1-butanol

d. 1-propanol

e. indan

7. *N*-Mesityl-*N*-phenyl-formamidine was first synthesized between 1950 and 1960. Use Beilstein Crossfire (if available to you) or other indexes to find the primary research article in which this compound is described, and write an equation for the reaction used to prepare it.

8. Determine whether or not each of the following compounds whose names are provided below has ever been synthesized; if it has, give the reference to the first appearance of its synthesis in the literature.

a. Vitamin A

b. Strychnine

c. Testosterone

d. Cephalosporin

e. Reserpine

f. Penicillin V

g. Prostaglandin E_2

h. Lysergic acid

i. Erythromycin B

j. Vinblastine

k. Morphine

l. Vitamin B_{12}

m. Cocaine

n. Taxol

o. Vancomycin

p. Brevitoxin

q. Discodermolide

r. Manzamine A

s. Halichlorine

t. Peloruside

u. Salicylihalamide

v. Spongistatin
w. Didehydrostemofoline
x. Ciguatoxin
y. Epothilone A

9. Determine whether or not each of the following compounds (a)–(l) has ever been synthesized; if it has, give the reference to the first appearance of its synthesis in the literature.

a. $CH_2CH_2NMe_2$, OCH_3, OCH_3

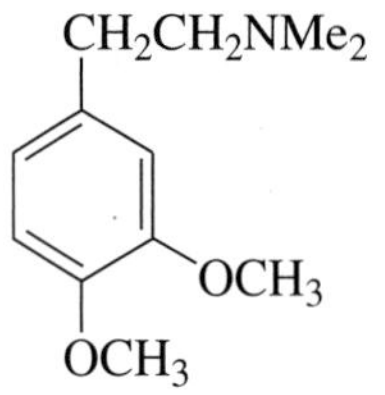

b. H_3C, H_3C, C=C=C=C, CH_3, CH_3

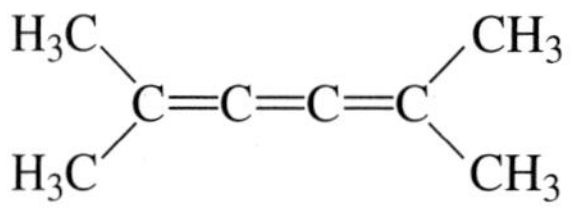

c.

d.

e. O, H_3C, CH_3

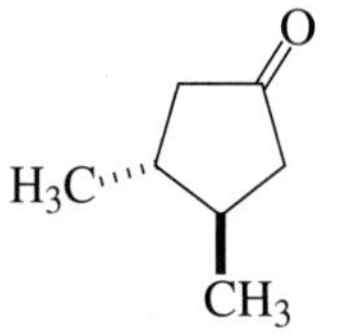

f. H, H

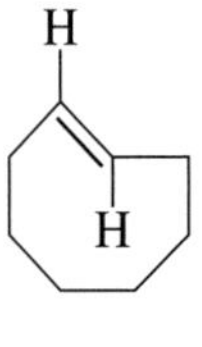

g.

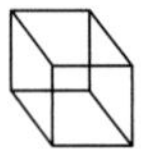

h.

i. H_3CO, H_3CO, C=C, OCH_3, OCH_3

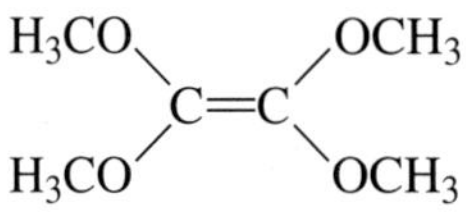

j.

k.

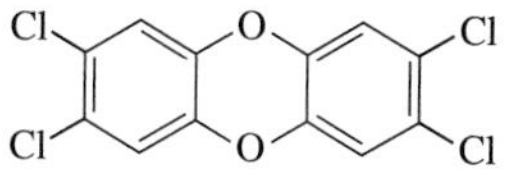

Cl, Cl, O, O, Cl, Cl

l.

CH_2OH

CH_3

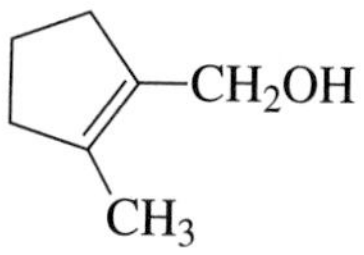

10. Using *Science Citation Index* or its online version (Web of Science), list five research papers by complete title and journal citation that have cited the review by Deiters, A.; Martin, S. F. *Chem. Rev.* **2004**, *104*, 2199.

FURTHER READING

Lowenthal, H. *A Guide for the Perplexed Organic Experimentalist*, 2nd ed., Wiley, Chichester, UK, 1990. Chapter 2.

Maizell, R. E. *How to Find Chemical Information*, 3rd ed., Wiley-Interscience, New York, 1998.

Smith, M. B. *March's Advanced Organic Chemistry*, 6th ed., Wiley-Interscience, New York, 2007. Appendix A: "The Literature of Organic Chemistry."

Schulz, H. *From CA to CAS Online: Databases in Chemistry*, 2nd ed., VCH, New York, 1994.

Index

Note: Boldface numbers indicate a figure.

C

Periodic Table of the Elements

1 1.01 H Hydrogen																	2 4.00 He Helium
3 6.94 Li Lithium	4 9.01 Be Beryllium											5 10.81 B Boron	6 12.01 C Carbon	7 14.01 N Nitrogen	8 16.00 O Oxygen	9 19.00 F Fluorine	10 20.18 Ne Neon
11 22.99 Na Sodium	12 24.31 Mg Magnesium											13 26.98 Al Aluminum	14 28.09 Si Silicon	15 30.97 P Phosphorus	16 32.07 S Sulfur	17 35.45 Cl Chlorine	18 39.95 Ar Argon
19 39.10 K Potassium	20 40.08 Ca Calcium	21 44.96 Sc Scandium	22 47.87 Ti Titanium	23 50.94 V Vanadium	24 52.00 Cr Chromium	25 54.94 Mn Manganese	26 55.85 Fe Iron	27 58.93 Co Cobalt	28 58.69 Ni Nickel	29 63.55 Cu Copper	30 65.39 Zn Zinc	31 69.72 Ga Gallium	32 72.61 Ge Germanium	33 74.92 As Arsenic	34 78.96 Se Selenium	35 79.90 Br Bromine	36 83.80 Kr Krypton
37 85.47 Rb Rubidium	38 87.62 Sr Strontium	39 88.91 Y Yttrium	40 97.22 Zr Zirconium	41 92.91 Nb Niobium	42 95.94 Mo Molybdenum	43 (98) Tc Technicium	44 101.07 Ru Ruthenium	45 102.91 Rh Rhodium	46 106.42 Pd Palladium	47 107.87 Ag Silver	48 112.41 Cd Cadmium	49 114.82 In Indium	50 118.71 Sn Tin	51 121.76 Sb Antimony	52 127.60 Te Tellurium	53 126.90 I Iodine	54 131.29 Xe Xenon
55 132.91 Cs Cesium	56 137.33 Ba Barium	57 138.91 La* Lanthanum	72 178.49 Hf Hafnium	73 180.95 Ta Tantalum	74 183.84 W Tungsten	75 186.21 Re Rhenium	76 190.23 Os Osmium	77 192.22 Ir Iridium	78 195.08 Pt Platinum	79 196.97 Au Gold	80 200.59 Hg Mercury	81 204.38 Tl Thallium	82 207.2 Pb Lead	83 208.98 Bi Bismuth	84 (208.98) Po Polonium	85 (209.99) At Astatine	86 (222.02) Rn Radon
87 (223.02) Fr Francium	88 (226.03) Ra Radium	89 (227) Ac** Actinium	104 (263.11) Rf Rutherfordium	105 (262) Db Dubrium	106 (266.12) Sg Seaborgium	107 (264.12) Bh Bohrium	108 (269.13) Hs Hassium	109 (268.14) Mt Meitnerium	110 (272.15) Ds Darmstadtium	111 (272.15) Rg Roentgenium	112 (277) Uub Ununbium	113 (284) Uut Ununtrium	114 (289) Uuq Ununquadium	115 (288) Uup Ununpentium	116 (292) Uuh Ununhexium	117 (?) Uus Ununseptium	118 (294) Uuo Ununoctium

*Lanthanide Series	58 140.12 Ce Cerium	59 140.91 Pr Praseodymium	60 144.24 Nd Neodymium	61 (145) Pm Promethium	62 150.36 Sm Samarium	63 151.96 Eu Europium	64 157.25 Gd Gadolinium	65 158.93 Tb Terbium	66 162.50 Dy Dysprosium	67 164.93 Ho Holmium	68 167.26 Er Erbium	69 168.93 Tm Thulium	70 173.04 Yb Ytterbium	71 174.97 Lu Lutetium
**Actinide Series	90 232.04 Th Thorium	91 231.04 Pa Protactinium	92 238.03 U Uranium	93 237.05 Np Neptunium	94 (244) Pu Plutonium	95 (243) Am Americium	96 (247) Cm Curium	97 (247) Bk Berkelium	98 (251) Cf Californium	99 (252) Es Einsteinium	100 (257) Fm Fermium	101 (258) Md Mendelevium	102 (259) No Nobelium	103 (260) Lr Lawrencium

http://en.wikipedia.org/wiki/Periodic_table